Biology

Concepts and Applications

Without Physiology

Starr | Evers | Starr

CENGAGE Learning®

Australia • Brazil • Japan • Korea • Mexico • Singapore • Spain • United Kingdom • United States

CENGAGE Learning

Biology: Concepts and Applications Without Physiology

Senior Manager, Student Engagement:

Linda deStefano

Janey Moeller

Manager, Student Engagement:

Julie Dierig

Marketing Manager:

Rachael Kloos

Manager, Production Editorial:

Kim Fry

Manager, Intellectual Property Project Manager:

Brian Methe

Senior Manager, Production and Manufacturing:

Donna M. Brown

Manager, Production:

Terri Daley

Printed in the United States of America

Biology: Concepts and Applications without Physiology, 9th Edition
Cecie Starr

ISBN-13: 978-1-305-28764-8
ISBN-10: 1-305-28764-9

WCN: 01-100-101

Cengage Learning
5191 Natorp Boulevard
Mason, Ohio 45040
USA

Cengage Learning is a leading provider of customized learning solutions with office locations around the globe, including Singapore, the United Kingdom, Australia, Mexico, Brazil, and Japan. Locate your local office at: **international.cengage.com/region.**

Cengage Learning products are represented in Canada by Nelson Education, Ltd.
For your lifelong learning solutions, visit **www.cengage.com/custom.**
Visit our corporate website at **www.cengage.com.**

Biology

Concepts and Applications | 9e

WITHOUT PHYSIOLOGY

Cecie Starr
Christine A. Evers
Lisa Starr

NATIONAL GEOGRAPHIC LEARNING | CENGAGE Learning

Australia • Brazil • Japan • Korea • Mexico • Singapore • Spain • United Kingdom • United States

NATIONAL GEOGRAPHIC LEARNING | CENGAGE Learning

***Biology: Concepts and Applications Without Physiology*, Ninth Edition**
Cecie Starr, Christine A. Evers, Lisa Starr

Senior Product Manager: Peggy Williams

Content Developer: Jake Warde

Product Assistant: Victor Luu

Media Developer: Lauren Oliveira

Executive Brand Manager: Nicole Hamm

Senior Marketing Development Manager: Tom Ziolkowski

Content Project Manager: Hal Humphrey

Senior Art Director: Pamela Galbreath

Manufacturing Planner: Karen Hunt

Senior Rights Acquisitions Specialist: Dean Dauphinais

Production Service: Grace Davidson & Associates, Inc.

Photo Researcher: Christina Ciaramella, PreMedia Global

Text Researcher: Melissa Tomaselli and Sunetra Mukundan, PreMedia Global

Copy Editor: Anita Wagner

Illustrators: Lisa Starr, ScEYEnce Studios, Precision Graphics

Text and Cover Designer: Irene Morris

Cover and Title Page Image: Emperor penguins; PAUL NICKLEN/National Geographic Creative

Compositor: Lachina

Library of Congress Control Number: 2013946060

ISBN-13: 978-1-285-42783-6

ISBN-10: 1-285-42783-1

Cengage Learning
200 First Stamford Place, 4th Floor
Stamford, CT 06902
USA

Printed in the United States of America
1 2 3 4 5 6 7 17 16 15 14 13

Contents in Brief

Detailed Contents

INTRODUCTION

UNIT I PRINCIPLES OF CELLULAR LIFE

2 Life's Chemical Basis

3 Molecules of Life

4 Cell Structure

5 Ground Rules of Metabolism

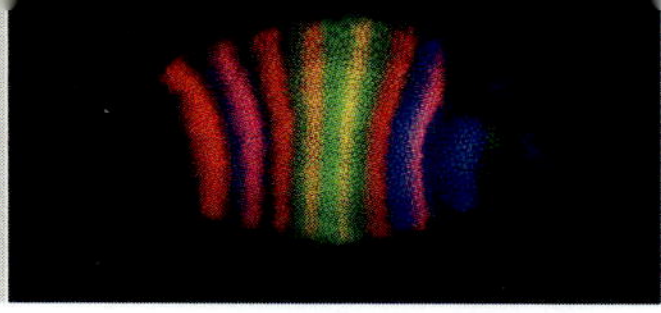
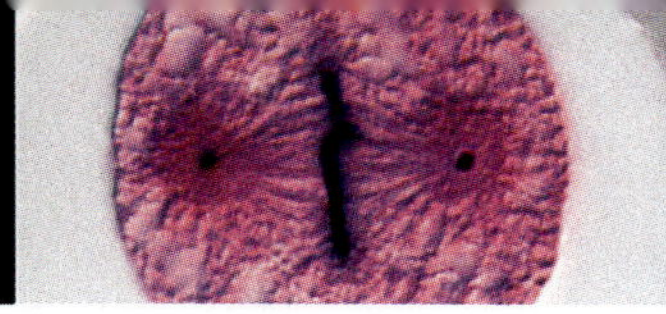
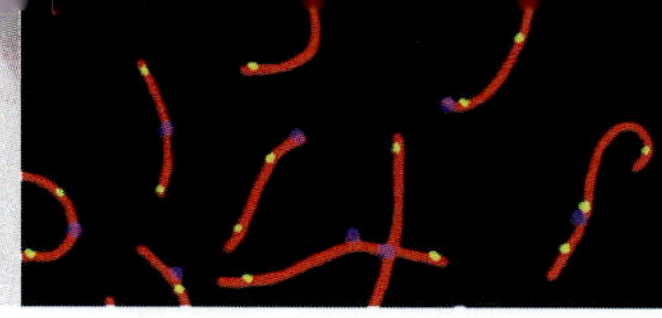
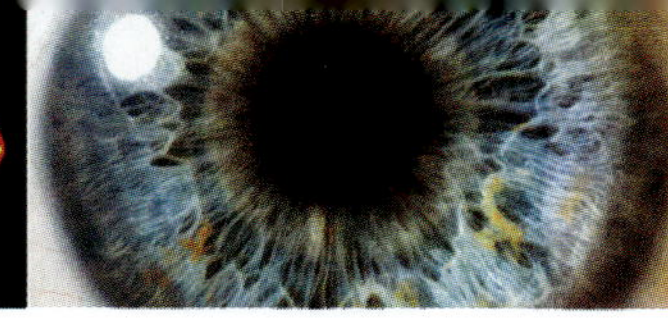

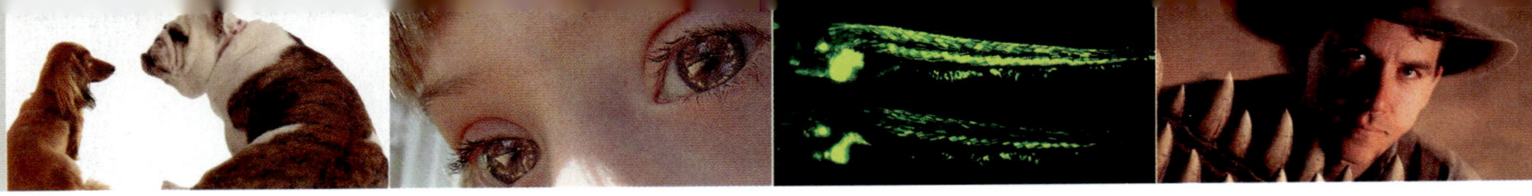

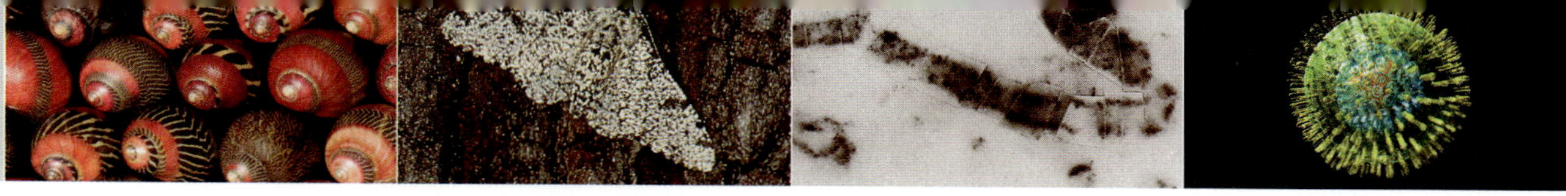

20 The Protists

21 Plant Evolution

22 Fungi

23 Animals I: Major Invertebrate Groups

24 Animals II: The Chordates

UNIT V PRINCIPLES OF ECOLOGY

25 Animal Behavior

Preface

We wrote this book to provide an accessible and appealing introduction to the study of life. Most students who use it will not become biologists, but all can benefit from an enhanced understanding of biological processes. For example, knowing how cells and bodies work helps a person make informed decisions about nutrition, life-style, and medical care. Recognizing the breadth of biodiversity, the mechanisms by which it arises, and the ways in which species interact brings to light the threats posed by human-induced extinctions. Realizing how living and nonliving components of ecosystems interact makes it clear why human activities such as adding greenhouse gases to the atmosphere puts us and other species at risk.

Our quest to educate and edify is shared by the National Geographic Society, with whom we have partnered for this edition. You will see the fruits of this partnership throughout the text—in spectacular new photographs, informative illustrations, and text features that highlight the wide variety of work supported by the society.

FEATURES OF THIS EDITION

SETTING THE STAGE

Each chapter opens with a dramatic two-page photo spread. A brief Links to Earlier Concepts paragraph reminds students of relevant information that has been covered in previous chapters, and concise Key Concept statements summarize the current chapter's content. An eye-catching image that appears in icon form next to each key concept also occurs within a relevant section, as part of a visual message that threads through the chapter.

CONCEPT SPREADS

The content of every chapter is organized as a series of Concepts, each explored in a section that is two pages or less. A section's Concept is reflected in its title, which is posed as a question that the student should be able to answer after reading the text. Bulleted sentences in the Take Home Message summarize and reinforce the Concept and supporting information provided in the section.

PEOPLE MATTER

Our new People Matter feature illustrates the relevance of ongoing research, and also highlights the diversity of the modern scientific community. Individuals whose work is spotlighted in this feature include well-established scientists, young scientists who are just beginning their careers, and a few nonscientists; most are National Geographic Explorers or Grantees.

ON-PAGE GLOSSARY

A new On-Page Glossary comprises boldface key terms introduced in each section. This section-by-section glossary offers definitions in alternate wording, and can also be used as a quick study aid. All glossary terms also appear in boldface in the Chapter Summary.

EMPHASIS ON RELEVANCE

Each chapter ends with an Application section that explains a current topic in light of the chapter content. For example, in Chapter 29, students use what they just learned about neural control to understand how sports-induced concussions permanently injure the brain—which is under intense study by scientific and athletic communities at this writing. Each Application also relates to a core interest of the National Geographic Society: Education, Conservation, Exploration, Get Involved, or Sustainability.

SELF-ASSESSMENT TOOLS

Many figure captions now include a Figure-It-Out Question and answer that allow students to quickly check their understanding of the illustration. At the end of each chapter, Self-Quiz and Critical Thinking Questions provide additional self-assessment material. A new chapter-end Data Analysis Activity sharpens analytical skills by asking the student to interpret data presented in graphic or tabular form. The data is related to the chapter material, and is from a published scientific study in most cases. For example, the Activity in Chapter 13 (Observing Patterns in Inherited Traits) asks the student to interpret data from an experiment revealing the suppressive effect of cystic fibrosis mutations on cellular uptake of *Salmonella* bacteria. This Activity continues the explanation of cystic fibrosis inheritance patterns begun in the chapter's Application section.

CHAPTER-SPECIFIC CHANGES This new edition contains 230 new photographs and almost 190 new or updated illustrations. In addition, the text of every chapter has been updated and revised for clarity. A page-by-page guide to new content and figures is available upon request, but we summarize the highlights here.

Chapter 1, Invitation to Biology
Renewed and updated emphasis on the relevance of new species discovery and the process of science; new features spotlight Smithsonian curator Kris Helgen and marine biologist Tierney Thys.

Chapter 2, Life's Chemical Basis
New graphics illustrate elements and radioactive decay; new feature spotlights volcanologist Ken Sims sampling radioisotopes in lava.

Chapter 3, Molecules of Life
New illustrations of carbon rings and tertiary structure; new feature spotlights discovery of carbohydrates in gas surrounding a sunlike star.

Chapter 4, Cell Structure
New photos illustrate surface-to-volume ratio, prokaryotes, biofilms, food vacuoles, chloroplasts, amyloplasts, basal bodies, and *E. coli* on food. New art illustrates plant cell walls, plasmodesmata, and cell junctions. Comparison of microscopy techniques updated using *Paramecium*. New features highlight electrical engineer Aydogan Ozcan's cell phone microscope, and astrobiologist Kevin Hand's work with NASA.

Chapter 5, Ground Rules of Metabolism
New photos illustrate potential energy, activation energy, energy transfer in redox reactions, turgor, phagocytosis, and alcohol abuse. Temperature-dependent enzyme activity now illustrated with polymerases. New feature highlights pressure-tolerant enzymes of deep sea amphipods. Expanded material on cofactors consolidated with ATP into new section. New activity requires interpretation of pH activity graphs of four enzymes from an extremophile archaean.

Chapter 6, Where It Starts—Photosynthesis
New photos illustrate phycobilins, and adaptations of C4 plants. New feature highlights conservation work of forester Willie Smits.

Chapter 7, How Cells Release Chemical Energy
New photos illustrate alcoholic and lactate fermentation, mitochondria, and mitochondrial disease. Revised art shows aerobic respiration's third stage. New feature highlights Benjamin Rapoport's glucose-driven, implantable fuel cell.

Chapter 8, DNA Structure and Function
Chromosome and DNA artwork has been revised for consistency throughout unit; DNA replication art updated. New photos illustrate DNA, x-ray diffraction, and mutation. New section consolidates material on DNA damage and mutations. New Figure spotlights marine biologist Mariana Fuentes, who studies how global warming is impacting sex ratios in sea turtle populations. Sex determination figure removed (text content remains).

Chapter 9, From DNA to Protein
New feature highlights Jack Horner's discovery of *T. rex* collagen in an ancient fossil. Expanded material on the effects of mutation includes new micrograph of a sickled blood cell. Ricin essay expanded to include other RIPs along with new photos and illustrations. New activity requires interpretation of data on the effects of an engineered RIP on cancer cells.

Chapter 10, Control of Gene Expression
New photos illustrate a polytene chromosome, *antennapedia* gene and mutation, X chromosome inactivation, *Arabidopisis* mutations, and breast cancer survivors; new feature details evolution of lactose tolerance. New section covers epigenetics; new activity requires analysis of retrospective data on an epigenetic effect.

Chapter 11, How Cells Reproduce
New photos illustrate mitosis, the mitotic spindle; new section details telomeres. New feature highlights behavioral ecologist Iain Couzin's hypothesis about collective behavior in metatstatic cells.

Chapter 12, Meiosis and Sexual Reproduction
New features highlight evolutionary biologist Maurine Neiman's work on the selective advantage of asexuality in the New Zealand mud snail, and asexuality in bdelloid rotifers. New photos illustrate crossovers and DNA repair during mitosis and meiosis.

Chapter 13, Observing Patterns in Inherited Traits
New figure illustrates how genotype gives rise to phenotype; new photos illustrate epistasis and continuous variation. Coverage of environmental effects on gene expression expanded and updated with new epigenetics research and new feature highlighting psychologist Gay Bradshaw's work on PTSD in elephants.

Chapter 14, Human Inheritance
New feature spotlights geneticist Nancy Wexler's work on Huntington's disease; new photo illustrates albinism.

Chapter 15, Biotechnology
Coverage of personal genetic testing updated with new medical applications, including photo of Angelina Jolie. New photos show recent examples of genetically modified animals. New "who's the daddy" critical thinking question offers students an opportunity to analyze a paternity test based on SNPs.

Chapter 16, Evidence of Evolution
Photos of analogous plants replaced with classic examples; new photo in morphological convergence section illustrates the difference. Photos of 19th century naturalists added to emphasize the process of science that led to natural selection theory. Expanded coverage of fossils includes how banded iron formations provide evidence of evolution of photosynthesis, and new feature spotlighting paleontologist Paul Sereno. New series of paleogeographic maps from Ron Blakey.

Chapter 17, Processes of Evolution
Added simple graphic to illustrate founder effect, and replaced hypothetical example in text with reduced diversity of ABO alleles in Native Americans. Consolidated and expanded material on antibiotic resistance into new Application section that covers overuse of antibiotics in livestock. New feature highlights evolutionary biologist Julia Day's work on speciation in African cichlids. New photos illustrate behavioral isolation in peacock spiders; new graphics illustrate stasis in coelacanths, and parsimony analysis. Added example of using cladistics to study viral evolution.

Chapter 18, Life's Origin and Early Evolution
Added information about earliest evidence of liquid water on Earth, a new contender for oldest fossil cells, and Robert Ballard's discovery of deep sea hydrothermal vents.

Chapter 19, Viruses, Bacteria, and Archaea
New feature about virologist Nathan Wolfe, increased coverage of viral recombination and of the roles of bacteria in human health and as decomposers.

Chapter 20, The Protists
New graphic illustrating primary and secondary endosymbiosis; added information about diatoms as a source of petroleum; new feature about Ken Banks, who created a text messaging system now used to track malaria outbreaks; coverage of choanoflagellates (the modern protists most closely related to animals) moved to this chapter.

Chapter 21, Plant Evolution
Updated life cycle graphics; improved photos of liverworts and horworts; new feature about Jeff Benca's studies of lycophytes, new coverage of seed banks as stores of plant diversity.

Chapter 22, Fungi
New graphics illustrating fungal phylogeny and a generalized fungal life cycle; new feature about DeeAnn Reeder, who studies white nose syndrome in bats; new coverage of fungi that infect insects and use of fungi in biotechnology and research; new coverage of the chytrid implicated in many amphibian declines.

Chapter 23, Animals I: Major Invertebrate Groups
New graphic comparing body plans in acolomate, pseudocoelomate, and coelomate worms; new feature about David Gruber's studies of biofluorescence in cnidarians; added information about penis fencing in marine flatworms and regeneration in planarians, similarity between larvae of annelids and mollusks as evidence of shared ancestry.

Chapter 24, Animals II: The Chordates
Updated evolutionary tree diagrams for chordates showing monophyly of jawless fishes; dropped discussion of "craniates"; updated and revised discussion of primate subgroups and human evolution with new photos and graphics; new feature about paleontologists Meave and Louise Leakey; new discussion of health problems related to our bipedalism.

Chapter 25, Animal Behavior
Added information about epigenetic effects on behavior. New feature highlights biologist Isabelle Charrier's work studying vocal recognition in animals. Improved discussion of parental care variation among animal groups.

Chapter 26, Population Ecology
New feature highlights the work of biologist Karen DeMatteo, who studies predator populations. Revised coverage of life history strategies.

Chapter 27, Community Ecology
New feature highlights population geneticist Nayuta Yamashita's study of resource partitioning in lemurs.

Chapter 28, Ecosystems
New feature highlights conservationist Jonathan Waterman's journey along the water-deprived Colorado River. Updated information about current level of atmospheric carbon dioxide; expanded discussion of the nitrogen cycle.

Chapter 29, The Biosphere
Improved general description of biomes. Added information about desert crust. New feature highlights biogeochemist Katey Walter Anthony's studies of arctic methane.

Chapter 30, Human Effects on the Biosphere
New feature about Paula Kahumbu's conservation efforts in Africa; sustainable uses of resources. Updated information about acid rain, ozone depletion.

STUDENT AND INSTRUCTOR RESOURCES

INSTRUCTOR COMPANION SITE Everything you need for your course in one place! This collection of book-specific lecture and class tools is available online via *www.cengage.com/login*. Access and download PowerPoint presentations, images, instructor's manuals, videos, and more.

CENGAGE LEARNING TESTING POWERED BY COGNERO
A flexible online system that allows you to:
• author, edit, and manage test bank content from multiple Cengage Learning solutions
• create multiple test versions in an instant
• deliver tests from your LMS, your classroom or wherever you want

STUDENT INTERACTIVE WORKBOOK Labeling exercises, self-quizzes, review questions, and critical thinking exercises help students with retention and better test results.

THE BROOKS/COLE BIOLOGY VIDEO LIBRARY 2009 FEATURING BBC MOTION GALLERY Looking for an engaging way to launch your lectures? The Brooks/Cole series features short high-interest segments: Pesticides: Will More Restrictions Help or Hinder?; A Reduction in Biodiversity; Are Biofuels as Green as They Claim?; Bone Marrow as a New Source for the Creation of Sperm; Repairing Damaged Hearts with Patients' Own Stem Cells; Genetically Modified Virus Used to Fight Cancer; Seed Banks Helping to Save Our Fragile Ecosystem; The Vanishing Honeybee's Impact on Our Food Supply.

MINDTAP A personalized, fully online digital learning platform of authoritative content, assignments, and services that engages your students with interactivity while also offering you choice in the configuration of coursework and enhancement of the curriculum via web-apps known as MindApps. MindApps range from ReadSpeaker (which reads the text out loud to students), to Kaltura (allowing you to insert inline video and audio into your curriculum). MindTap is well beyond an eBook, a homework solution or digital supplement, a resource center website, a course delivery platform, or a Learning Management System. It is the first in a new category—the Personal Learning Experience.

APLIA FOR BIOLOGY The Aplia system helps students learn key concepts via Aplia's focused assignments and active learning opportunities that include randomized, automatically graded questions, exceptional text/art integration, and immediate feedback. Aplia has a full course management system that can be used independently or in conjunction with other course management systems such as MindTap, D2L, or Blackboard. Visit *www.aplia.com/biology*.

Acknowledgments

We are incredibly grateful for the ongoing input of the instructors, listed on the following page, who helped us polish our text and shape our thinking. Key Concepts, Data Analysis Activities, On-Page Glossaries, custom videos—such features are direct responses to their comments and suggestions.

This edition benefits from a collaborative association with the National Geographic Society in Washington D.C. The Society has graciously allowed us to enhance our presentation with its extensive resources, including beautiful maps and images, Society explorer and grantee materials, and online videos.

Thanks to Yolanda Cossio and Peggy Williams at Cengage Learning for continuing to encourage us to improve and innovate, and for suggesting that we collaborate with National Geographic. Jake Warde ensured that this collaboration ran smoothly and Grace Davidson did the same for the project as a whole. Leila Hishmeh, Anna Kistin, Jen Shook, and Wesley Della Volla helped us acquire National Geographic maps and photos; Melissa Tomaselli and Sunetra Mukundan obtained text permissions.

This edition's sleek new look is a product of talented designer Irene Morris. Thanks also to Christina Ciaramella for photoresearch and photo permissions. Copyeditor Anita Wagner and proofreader Diane Miller helped us keep our text clear, concise, and correct; tireless editorial assistant Victor Luu organized meetings, reviews, and paperwork. Lauren Oliveira created a world-class technology package for both students and instructors.

—Lisa Starr, Chris Evers, and Cecie Starr 2013

Influential Class Testers and Reviewers

Brenda Alston-Mills
North Carolina State University

Kevin Anderson
Arkansas State University - Beebe

Norris Armstrong
University of Georgia

Tasneem Ashraf
Coshise College

Dave Bachoon
Georgia College & State University

Neil R. Baker
The Ohio State University

Andrew Baldwin
Mesa Community College

David Bass
University of Central Oklahoma

Lisa Lynn Boggs
Southwestern Oklahoma State University

Gail Breen
University of Texas at Dallas

Marguerite "Peggy" Brickman
University of Georgia

David Brooks
East Central College

David William Bryan
Cincinnati State College

Lisa Bryant
Arkansas State University - Beebe

Katherine Buhrer
Tidewater Community College

Uriel Buitrago-Suarez
Harper College

Sharon King Bullock
Virginia Commonwealth University

John Capehart
University of Houston - Downtown

Daniel Ceccoli
American InterContinental University

Tom Clark
Indiana University South Bend

Heather Collins
Greenville Technical College

Deborah Dardis
Southeastern Louisiana University

Cynthia Lynn Dassler
The Ohio State University

Carole Davis
Kellogg Community College

Lewis E. Deaton
University of Louisiana - Lafayette

Jean Swaim DeSaix
University of North Carolina - Chapel Hill

(Joan) Lee Edwards
Greenville Technical College

Hamid M. Elhag
Clayton State University

Patrick Enderle
East Carolina University

Daniel J. Fairbanks
Brigham Young University

Amy Fenster
Virginia Western Community College

Kathy E. Ferrell
Greenville Technical College

Rosa Gambier
Suffok Community College - Ammerman

Tim D. Gaskin
Cuyahoga Community College - Metropolitan

Stephen J. Gould
Johns Hopkins University

Laine Gurley
Harper College

Marcella Hackney
Baton Rouge Community College

Gale R. Haigh
McNeese State University

John Hamilton
Gainesville State

Richard Hanke
Rose State Community College

Chris Haynes
Shelton St. Community College

Kendra M. Hill
South Dakota State University

Juliana Guillory Hinton
McNeese State University

W. Wyatt Hoback
University of Nebraska, Kearney

Kelly Hogan
University of North Carolina

Norma Hollebeke
Sinclair Community College

Robert Hunter
Trident Technical College

John Ireland
Jackson Community College

Thomas M. Justice
McLennan College

Timothy Owen Koneval
Laredo Community College

Sherry Krayesky
University of Louisiana - Lafayette

Dubear Kroening
University of Wisconsin - Fox Valley

Jerome Krueger
South Dakota State University

Jim Krupa
University of Kentucky

Mary Lynn LaMantia
Golden West College

Dale Lambert
Tarrant County College

Kevin T. Lampe
Bucks County Community College

Susanne W. Lindgren
Sacramento State University

Madeline Love
New River Community College

Dr. Kevin C. McGarry
Kaiser College - Melbourne

Ashley McGee
Alamo College

Jeanne Mitchell
Truman State University

Alice J. Monroe
St. Petersburg College - Clearwater

Brenda Moore
Truman State University

Erin L. G. Morrey
Georgia Perimeter College

Rajkumar "Raj" Nathaniel
Nicholls State University

Francine Natalie Norflus
Clayton State University

Harold Olivey
Indiana University Northwest

Alexander E. Olvido
Virginia State University

John C. Osterman
University of Nebraska, Lincoln

Bob Patterson
North Carolina State University

Shelley Penrod
North Harris College

Carla Perry
Community College of Philadelphia

Mary A. (Molly) Perry
Kaiser College - Corporate

John S. Peters
College of Charleston

Carlie Phipps
SUNY IT

Michael Plotkin
Mt. San Jacinto College

Ron Porter
Penn State University

Karen Raines
Colorado State University

Larry A. Reichard
Metropolitan Community College - Maplewood

Jill D. Reid
Virginia Commonwealth University

Robert Reinswold
University of Northern Colorado

Ashley E. Rhodes
Kansas State University

David Rintoul
Kansas State University

Darryl Ritter
Northwest Florida State College

Amy Wolf Rollins
Clayton State University

Sydha Salihu
West Virginia University

Jon W. Sandridge
University of Nebraska

Robin Searles-Adenegan
Morgan State University

Erica Sharar
IVC; National University

Julie Shepker
Kaiser College - Melbourne

Rainy Shorey
Illinois Central College

Eric Sikorski
University of South Florida

Phoebe Smith
Suffolk County Community College

Robert (Bob) Speed
Wallace Junior College

Tony Stancampiano
Oklahoma City Community College

Jon R. Stoltzfus
Michigan State University

Peter Svensson
West Valley College

Jeffrey L. Travis
University at Albany

Nels H. Troelstrup, Jr.
South Dakota State University

Allen Adair Tubbs
Troy University

Will Unsell
University of Central Oklahoma

Rani Vajravelu
University of Central Florida

Jack Waber
West Chester University of Pennsylvania

Kathy Webb
Bucks County Community College

Amy Stinnett White
Virginia Western Community College

Virginia White
Riverside Community College

Robert S. Whyte
California University of Pennsylvania

Kathleen Lucy Wilsenn
University of Northern Colorado

Penni Jo Wilson
Cleveland State Community College

Robert Wise
University of Wisconsin Oshkosh

Michael L. Womack
Macon State College

Maury Wrightson
Germanna Community College

Mark L. Wygoda
McNeese State University

Lan Xu
South Dakota State University

Poksyn ("Grace") Yoon
Johnson and Wales University

Muriel Zimmermann
Chaffey College

Near a tent serving as a makeshift laboratory, herpetologist Paul Oliver records the call of a frog on an expedition to New Guinea's Foja Mountains cloud forest.

1 INVITATION TO BIOLOGY

Links to Earlier Concepts

Whether or not you have studied biology, you already have an intuitive understanding of life on Earth because you are part of it. Every one of your experiences with the natural world—from the warmth of the sun on your skin to the love of your pet—contributes to that understanding.

KEY CONCEPTS

THE SCIENCE OF NATURE
We can understand life by studying it at many levels, starting with atoms that are components of all matter, and extending to interactions of organisms with their environment.

LIFE'S UNITY
All living things require ongoing inputs of energy and raw materials; all sense and respond to change; and all have DNA that guides their functioning.

LIFE'S DIVERSITY
Observable characteristics vary tremendously among organisms. Various classification systems help us keep track of the differences.

THE NATURE OF SCIENCE
Carefully designing experiments helps researchers unravel cause-and-effect relationships in complex natural systems.

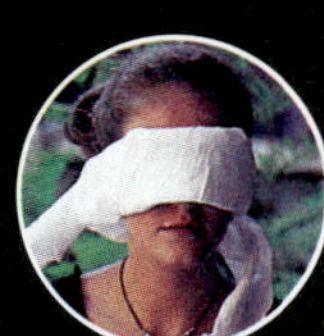

LIMITATIONS OF SCIENCE
Science addresses only testable ideas about observable events and processes. It does not address anything untestable, such as beliefs and opinions.

Photograph by Tim Laman/National Geographic Creative.

1.1 HOW DO LIVING THINGS DIFFER FROM NONLIVING THINGS?

LIFE IS MORE THAN THE SUM OF ITS PARTS

Biology is the study of life, past and present. What, exactly, is the property we call "life"? We may never actually come up with a good definition, because living things are too diverse, and they consist of the same basic components as nonliving things. When we try to define life, we end up only identifying properties that differentiate living from nonliving things.

Complex properties, including life, often emerge from the interactions of much simpler parts. To understand why, take a look at this drawing:

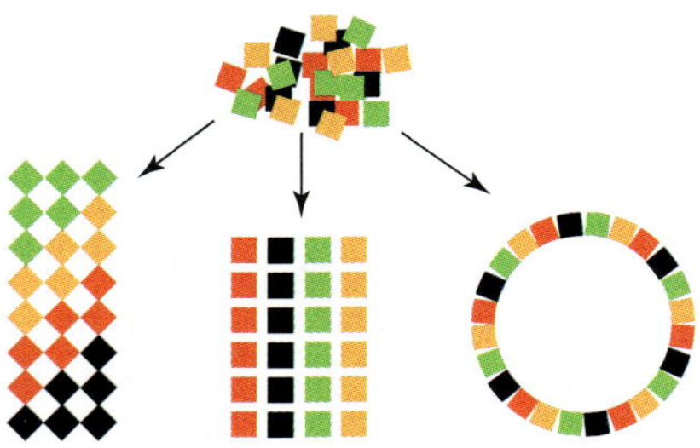

The property of "roundness" emerges when the parts are organized one way, but not other ways. Characteristics of a system that do not appear in any of the system's components are called **emergent properties.** The idea that structures with emergent properties can be assembled from the same basic building blocks is a recurring theme in our world, and also in biology.

LIFE'S ORGANIZATION

Through the work of biologists, we are beginning to understand an overall pattern in the way life is organized. We can look at life in successive levels of organization, with new emergent properties appearing at each level (**FIGURE 1.1**).

Life's organization starts with interactions between atoms. **Atoms** are fundamental building blocks of all substances ❶. Atoms join as **molecules** ❷. There are no atoms unique to living things, but there are unique molecules. In today's world, only living things make the "molecules of life," which are lipids, proteins, DNA, RNA, and complex carbohydrates. The emergent property of "life" appears at the next level, when many molecules of life become organized as a cell ❸. A **cell** is the smallest unit of life. Cells survive and reproduce themselves using energy, raw materials, and information in their DNA.

Some cells live and reproduce independently. Others do so as part of a multicelled organism. An **organism** is an individual that consists of one or more cells. A poppy plant is an example of a multicelled organism ❼.

In most multicelled organisms, cells are organized as tissues ❹. A **tissue** consists of specific types of cells organized in a particular pattern. The arrangement

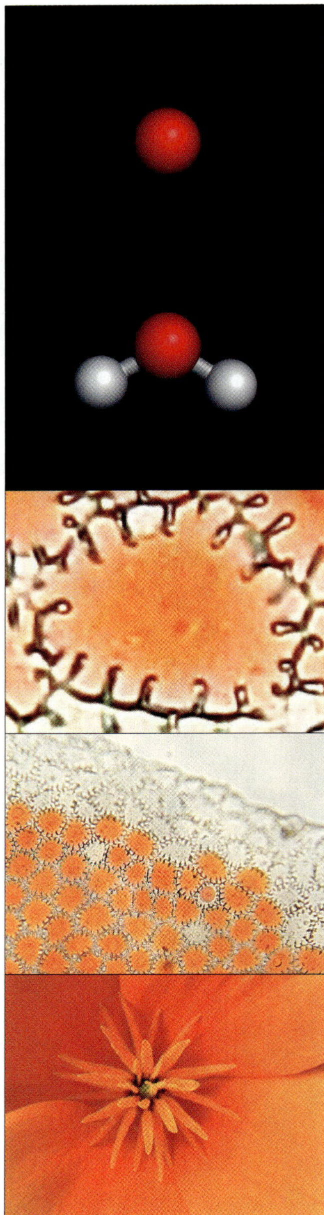

❶ **atom**
Atoms are fundamental units of all substances, living or not. This image shows a model of a single atom.

❷ **molecule**
Atoms join other atoms in molecules. This is a model of a water molecule. The molecules special to life are much larger and more complex than water.

❸ **cell**
The cell is the smallest unit of life. Some, like this plant cell, live and reproduce as part of a multicelled organism; others do so on their own.

❹ **tissue**
Organized array of cells that interact in a collective task. This is epidermal tissue on the outer surface of a flower petal.

❺ **organ**
Structural unit of interacting tissues. Flowers are the reproductive organs of many plants.

FIGURE 1.1 {Animated} An overall pattern in the way life is organized. New emergent properties appear at each successive level.

6 organ system
A set of interacting organs. The shoot system of this poppy plant includes its aboveground parts: leaves, flowers, and stems.

7 multicelled organism
Individual that consists of more than one cell. Cells of this California poppy plant are part of its two organ systems: aboveground shoots and belowground roots.

8 population
Group of single-celled or multicelled individuals of a species in a given area. This population of California poppy plants is in California's Antelope Valley Poppy Reserve.

9 community
All populations of all species in a specified area. These plants are part of a community called the Antelope Valley Poppy Reserve.

10 ecosystem
A community interacting with its physical environment through the transfer of energy and materials. Sunlight and water sustain the community in the Antelope Valley.

11 biosphere
The sum of all ecosystems: every region of Earth's waters, crust, and atmosphere in which organisms live. No ecosystem in the biosphere is truly isolated from any other.

allows the cells to collectively perform a special function such as protection from injury (dermal tissue), movement (muscle tissue), and so on.

An **organ** is an organized array of tissues that collectively carry out a particular task or set of tasks 5. For example, a flower is an organ of reproduction in plants; a heart, an organ that pumps blood in animals. An **organ system** is a set of organs and tissues that interact to keep the individual's body working properly 6. Examples of organ systems include the aboveground parts of a plant (the shoot system), and the heart and blood vessels of an animal (the circulatory system).

A **population** is a group of individuals of the same type, or species, living in a given area 8. An example would be all of the California poppies that are living in California's Antelope Valley Poppy Reserve. At the next level, a **community** consists of all populations of all species in a given area. The Antelope Valley Reserve community includes California poppies and all other organisms—plants, animals, microorganisms, and so on—living in the reserve 9. Communities may be large or small, depending on the area defined.

The next level of organization is the **ecosystem**, which is a community interacting with its environment 10. The most inclusive level, the **biosphere**, encompasses all regions of Earth's crust, waters, and atmosphere in which organisms live 11.

atom Fundamental building block of all matter.
biology The scientific study of life.
biosphere All regions of Earth where organisms live.
cell Smallest unit of life.
community All populations of all species in a given area.
ecosystem A community interacting with its environment.
emergent property A characteristic of a system that does not appear in any of the system's component parts.
molecule An association of two or more atoms.
organ In multicelled organisms, a grouping of tissues engaged in a collective task.
organism Individual that consists of one or more cells.
organ system In multicelled organisms, set of organs engaged in a collective task that keeps the body functioning properly.
population Group of interbreeding individuals of the same species that live in a given area.
tissue In multicelled organisms, specialized cells organized in a pattern that allows them to perform a collective function.

TAKE-HOME MESSAGE 1.1

Biologists study life by thinking about it at different levels of organization, with new emergent properties appearing at each successive level.

All things, living or not, consist of the same building blocks: atoms. Atoms join as molecules.

The unique properties of life emerge as certain kinds of molecules become organized into cells.

Higher levels of life's organization include multicelled organisms, populations, communities, ecosystems, and the biosphere.

CREDITS: (1) 6: Lady Bird Johnson Wildflower Center; 7: Michael Szoenyi/Science Source; 8: Photographers Choice RF/SuperStock; 9: © Sergei Krupnov, www.flickr.com/photos/7969319@N03; 10: © Mark Koberg Photography; 11: NASA.

1.2 HOW ARE ALL LIVING THINGS ALIKE?

FIGURE 1.2 {Animated} The one-way flow of energy and cycling of materials through the world of life.

Even though we cannot precisely define "life," we can intuitively understand what it means because all living things share a set of key features. All require ongoing inputs of energy and raw materials; all sense and respond to change; and all pass DNA to offspring (**TABLE 1.1**).

TABLE 1.1

Three Key Features of Living Things

Requirement for energy and nutrients	Ongoing inputs of energy and nutrients sustain life.
Homeostasis	Each living thing has the capacity to sense and respond to change.
Use of DNA as hereditary material	DNA is passed to offspring during reproduction.

ORGANISMS REQUIRE ENERGY AND NUTRIENTS

Not all living things eat, but all require energy and nutrients on an ongoing basis. Both are essential to maintain the functioning of individual organisms and the organization of life. A **nutrient** is a substance that an organism needs for growth and survival but cannot make for itself.

Organisms spend a lot of time acquiring energy and nutrients (**FIGURE 1.2**). However, the source of energy and the type of nutrients required differ among organisms. These differences allow us to classify all living things into two categories: producers and consumers. **Producers** make their own food using energy and simple raw materials they get from nonbiological sources 1. Plants are producers that use the energy of sunlight to make sugars from water and carbon dioxide (a gas in air), a process called **photosynthesis**. By contrast, **consumers** cannot make their own food. They get energy and nutrients by feeding on other organisms 2. Animals are consumers. So are decomposers, which feed on the wastes or remains of other organisms. The leftovers from consumers' meals end up in the environment, where they serve as nutrients for producers. Said another way, nutrients cycle between producers and consumers.

Unlike nutrients, energy is not cycled. It flows through the world of life in one direction: from the environment 3, through organisms 4, and back to the environment 5. This flow maintains the organization of every living cell and body, and it also influences how individuals interact with one another and their environment. The energy flow is one-way, because with each transfer, some energy escapes as heat, and cells cannot use heat as an energy source. Thus, energy that enters the world of life eventually leaves it (we return to this topic in Chapter 5).

ORGANISMS SENSE AND RESPOND TO CHANGE

An organism cannot survive for very long in a changing environment unless it adapts to the changes. Thus, every living thing has the ability to sense and respond to change both inside and outside of itself (**FIGURE 1.3**). For example, after you eat, the sugars from your meal enter your bloodstream. The added sugars set in motion a series of events that causes cells throughout the body to take up sugar faster, so the sugar level in your blood quickly falls. This response keeps your blood sugar level within a certain range, which in turn helps keep your cells alive and your body functioning.

The fluid portion of your blood is a component of your internal environment, which is all of the body fluids outside of cells. Unless that internal environment is kept within certain ranges of temperature and other conditions, your body cells will die. By sensing and adjusting to change, you and all other organisms keep conditions in the internal environment within a range that favors survival. **Homeostasis** is the name for this process, and it is one of the defining features of life.

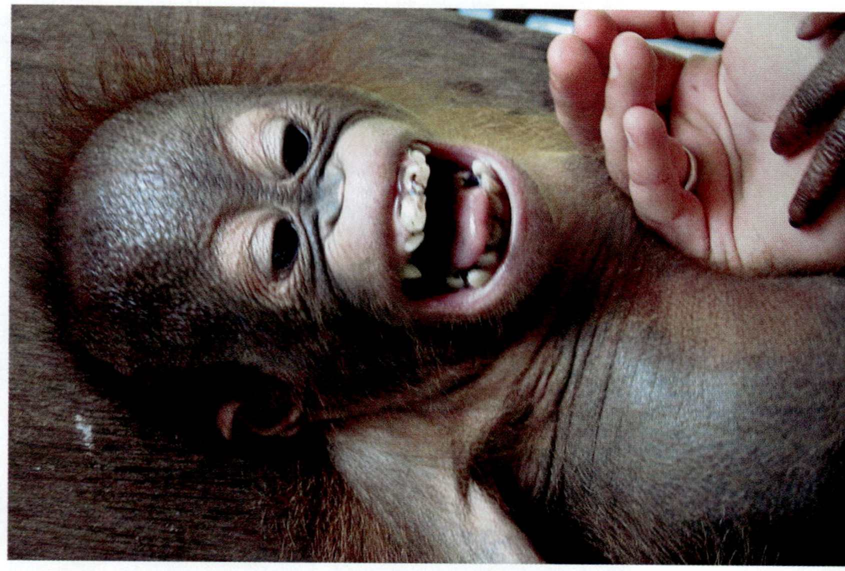

FIGURE 1.3 Living things sense and respond to their environment. This baby orangutan is laughing in response to being tickled. Apes and humans make different sounds when being tickled, but the airflow patterns are so similar that we can say apes really do laugh.

ORGANISMS USE DNA

With little variation, the same types of molecules perform the same basic functions in every organism. For example, information in an organism's **DNA** (deoxyribonucleic acid) guides ongoing functions that sustain the individual through its lifetime. Such functions include **development**: the process by which the first cell of a new individual gives rise to a multicelled adult; **growth**: increases in cell number, size, and volume; and **reproduction**: processes by which individuals produce offspring.

Individuals of every natural population are alike in certain aspects of their body form and behavior because their DNA is very similar: Orangutans look like orangutans and not like caterpillars because they inherited orangutan DNA, which differs from caterpillar DNA in the information it carries. **Inheritance** refers to the transmission of DNA to offspring. All organisms inherit their DNA from one or two parents.

DNA is the basis of similarities in form and function among organisms. However, the details of DNA molecules differ, and herein lies the source of life's diversity. Small variations in the details of DNA's structure give rise to differences among individuals, and also among types of organisms. As you will see in later chapters, these differences are the raw material of evolutionary processes.

TAKE-HOME MESSAGE 1.2

Continual inputs of energy and the cycling of materials maintain life's complex organization.

Organisms sense and respond to change inside and outside themselves. They make adjustments that keep conditions in their internal environment within a range that favors cell survival, a process called homeostasis.

All organisms use information in the DNA they inherited from their parent or parents to develop, grow, and reproduce. DNA is the basis of similarities and differences in form and function among organisms.

consumer Organism that gets energy and nutrients by feeding on tissues, wastes, or remains of other organisms.
development Multistep process by which the first cell of a new multicelled organism gives rise to an adult.
DNA Deoxyribonucleic acid; carries hereditary information that guides development and other activities.
growth In multicelled species, an increase in the number, size, and volume of cells.
homeostasis Process in which an organism keeps its internal conditions within tolerable ranges by sensing and responding to change.
inheritance Transmission of DNA to offspring.
nutrient Substance that an organism needs for growth and survival but cannot make for itself.
photosynthesis Process by which producers use light energy to make sugars from carbon dioxide and water.
producer Organism that makes its own food using energy and nonbiological raw materials from the environment.
reproduction Processes by which parents produce offspring.

1.3 HOW ARE LIVING THINGS DIFFERENT?

A Prokaryotes are single-celled, and have no nucleus. As a group, they are the most diverse organisms.

B Eukaryotes consist of cells that have a nucleus. Eukaryotic cells are typically larger and more complex than prokaryotes.

bacteria are the most numerous organisms on Earth. Top, this bacterium has a row of iron crystals that functions like a tiny compass; bottom, a resident of human intestines.

archaea resemble bacteria, but they are more closely related to eukaryotes. Top: two types from a hydrothermal vent on the seafloor. Bottom, a type that grows in sulfur hot springs.

protists are a group of extremely diverse eukaryotes that range from microscopic single cells (top) to giant multicelled seaweeds (bottom).

fungi are eukaryotic consumers that secrete substances to break down food outside their body. Most are multicelled (top), but some are single-celled (bottom).

FIGURE 1.4 A few representatives of life's diversity: A some prokaryotes; B some eukaryotes.

Living things differ tremendously in their observable characteristics. Various classification schemes help us organize what we understand about the scope of this variation, which we call Earth's **biodiversity**.

For example, organisms can be grouped on the basis of whether they have a **nucleus**, which is a sac with two membranes that encloses and protects a cell's DNA. **Bacteria** (singular, bacterium) and **archaea** (singular, archaeon) are organisms whose DNA is *not* contained within a nucleus. All bacteria and archaea are single-celled, which means each organism consists of one cell (**FIGURE 1.4A**). Collectively, these organisms are the most diverse representatives of life. Different kinds are producers or consumers in nearly all regions of Earth. Some inhabit such extreme environments as frozen desert rocks, boiling sulfurous lakes, and nuclear reactor waste. The first cells on Earth may have faced similarly hostile environments.

Traditionally, organisms without a nucleus have been called **prokaryotes**, but this designation is now used only informally. This is because, despite the similar appearance of bacteria and archaea, the two types of cells are less related to one another than we once thought. Archaea turned out to be more closely related to **eukaryotes**, which are organisms whose DNA is contained within a nucleus. Some eukaryotes live as individual cells; others are multicelled (**FIGURE 1.4B**). Eukaryotic cells are typically larger and more complex than bacteria or archaea.

Structurally, **protists** are the simplest eukaryotes, but as a group they vary dramatically, from single-celled consumers to giant, multicelled producers.

animal Multicelled consumer that develops through a series of stages and moves about during part or all of its life.
archaea Group of single-celled organisms that lack a nucleus but are more closely related to eukaryotes than to bacteria.
bacteria The most diverse and well-known group of single-celled organisms that lack a nucleus.
biodiversity Scope of variation among living organisms.
eukaryote Organism whose cells characteristically have a nucleus.
fungus Single-celled or multicelled eukaryotic consumer that breaks down material outside itself, then absorbs nutrients released from the breakdown.
nucleus Sac that encloses a cell's DNA; has two membranes.
plant A multicelled, typically photosynthetic producer.
prokaryote Single-celled organism without a nucleus.
protist Member of a diverse group of simple eukaryotes.

plants are multicelled eukaryotes. Most are photosynthetic, and have roots, stems, and leaves.

animals are multicelled eukaryotes that ingest tissues or juices of other organisms. All actively move about during at least part of their life.

Fungi (singular, fungus) are eukaryotic consumers that secrete substances to break down food externally, then absorb nutrients released by this process. Many fungi are decomposers. Most fungi, including those that form mushrooms, are multicellular. Fungi that live as single cells are called yeasts.

Plants are multicelled eukaryotes; the majority are photosynthetic producers that live on land. Besides feeding themselves, plants also serve as food for most other land-based organisms.

Animals are multicelled consumers that consume tissues or juices of other organisms. Unlike fungi, animals break down food inside their body. They also develop through a series of stages that lead to the adult form. All kinds actively move about during at least part of their lives.

TAKE-HOME MESSAGE 1.3

Organisms differ in their details; they show tremendous variation in observable characteristics, or traits.

We can divide Earth's biodiversity into broad groups based on traits such as having a nucleus or being multicellular.

PEOPLE MATTER

National Geographic Explorer
KRISTOFER HELGEN

Kristofer Helgen discovers new animals. Deep in a New Guinea rain forest. High on an Andean mountainside. Resting in a museum's specimen drawer. "Conventional wisdom would have it that we know all the mammals of the world," he notes. "In fact, we know so little. Unique species, profoundly different from anything ever discovered, are out there waiting to be found." His own efforts prove this. Helgen himself has discovered approximately 100 new species of mammals previously unknown to science. "Since I was three years old, I've been transfixed by animals," he recalls. "Even then, my excitement revolved around figuring out how many different kinds there were."

Helgen's search plunges him into the wild on almost every continent. Yet about three times as many new finds are made within the walls of museums. "An expert can go into any large natural history museum and identify kinds of animals no one knew existed," he explains. When only a few specimens of a species exist, and reside in museums scattered across the globe, sheer logistics often prevent researchers from connecting the dots and pinpointing a new find. "Collections build up over centuries," he says, "It's virtually impossible to fully interpret that wealth of material. Every day brings surprises." As Curator of Mammals for the Smithsonian Institution's National Museum of Natural History, he oversees not only the collection's use as an invaluable research resource, but also its continued expansion through exploration.

1.4 WHAT IS A SPECIES?

Each time we discover a new **species**, or unique kind of organism, we name it. **Taxonomy**, a system of naming and classifying species, began thousands of years ago, but naming species in a consistent way did not become a priority until the eighteenth century. At the time, European explorers who were just discovering the scope of life's diversity started having more and more trouble communicating with one another because species often had multiple names. For example, the dog rose (a plant native to Europe, Africa, and Asia) was alternately known as briar rose, witch's briar, herb patience, sweet briar, wild briar, dog briar, dog berry, briar hip, eglantine gall, hep tree, hip fruit, hip rose, hip tree, hop fruit, and hogseed—and those are only the English names! Species often had multiple scientific names too, in Latin that was descriptive but often cumbersome. The scientific name of the dog rose was *Rosa sylvestris inodora seu canina* (odorless woodland dog rose), and also *Rosa sylvestris alba cum rubore, folio glabro* (pinkish white woodland rose with smooth leaves).

An eighteenth-century naturalist, Carolus Linnaeus, standardized a naming system that we still use. By the Linnaean system, every species is given a unique two-part scientific name. The first part is the name of the **genus** (plural, genera), a group of species that share a unique set of features. The second part is the **specific epithet**. Together, the genus name and the specific epithet designate one species. Thus, the dog rose now has one official name, *Rosa canina*, that is recognized worldwide.

Genus and species names are always italicized. For example, *Panthera* is a genus of big cats. Lions belong to the species *Panthera leo*. Tigers belong to a different species in the same genus (*Panthera tigris*), and so do leopards (*P. pardus*). Note how the genus name may be abbreviated after it has been spelled out once.

A ROSE BY ANY OTHER NAME . . .

The individuals of a species share a unique set of inherited characteristics, or **traits**. For example, giraffes normally have very long necks, brown spots on white coats, and so on. These are morphological traits (*morpho*– means form). Individuals of a species also share biochemical traits (they make and use the same molecules) and behavioral traits (they respond the same way to certain stimuli, as when hungry giraffes feed on tree leaves).

We can rank species into ever more inclusive categories based on shared sets of traits. Each rank, or **taxon** (plural, taxa), is a group of organisms that share a unique set of traits. Each category above species—genus, family, order, class, phylum (plural, phyla), kingdom, and domain—consists of a group of the next lower taxon (**FIGURE 1.5**). Using this system, we can sort all life into a few categories (**FIGURE 1.6** and **TABLE 1.2**).

domain	Eukarya	Eukarya	Eukarya	Eukarya	Eukarya
kingdom	Plantae	Plantae	Plantae	Plantae	Plantae
phylum	Magnoliophyta	Magnoliophyta	Magnoliophyta	Magnoliophyta	Magnoliophyta
class	Magnoliopsida	Magnoliopsida	Magnoliopsida	Magnoliopsida	Magnoliopsida
order	Apiales	Rosales	Rosales	Rosales	Rosales
family	Apiaceae	Cannabaceae	Rosaceae	Rosaceae	Rosaceae
genus	*Daucus*	*Cannabis*	*Malus*	*Rosa*	*Rosa*
species	*carota*	*sativa*	*domestica*	*acicularis*	*canina*
common name	wild carrot	marijuana	apple	prickly rose	dog rose

FIGURE 1.5 Linnaean classification of five species that are related at different levels. Each species has been assigned to ever more inclusive groups, or taxa: in this case, from genus to domain.

FIGURE IT OUT: Which of the plants shown here are in the same order?

Answer: Marijuana, apple, prickly rose, and dog rose

CREDITS: (5) from left, © xania.g, www.flickr.com/photos/52287712@N00; © kymkemp.com; Nigel Cattlin/Visuals Unlimited, Inc.; Courtesy of Melissa S. Green, www.flickr.com/photos/henkimaa; © Grodana Sarkotic.

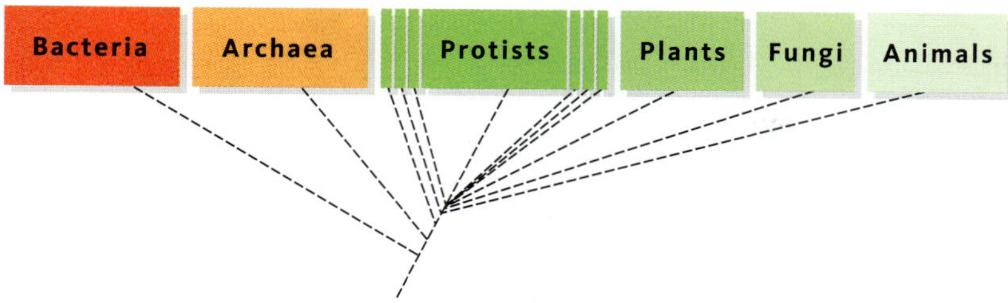

A Six-kingdom classification system. The protist kingdom includes the most ancient multicelled and all single-celled eukaryotes.

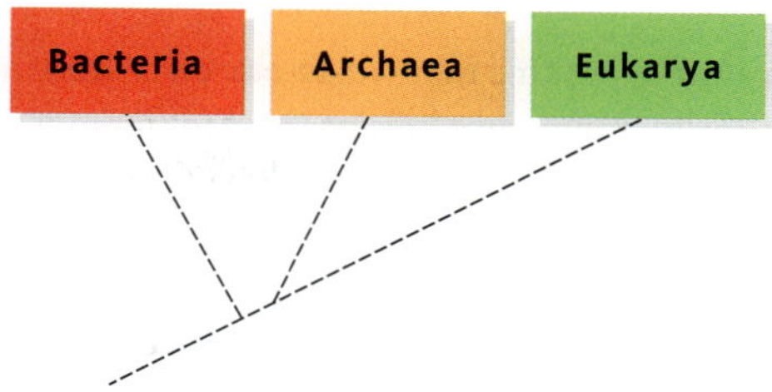

B Three-domain classification system. The Eukarya domain includes protists, plants, fungi, and animals.

FIGURE 1.6 {Animated} Two ways to see the big picture of life. The lines in such diagrams indicate evolutionary connections.

It is easy to tell that orangutans and caterpillars are different species because they appear very different. Distinguishing species that are more closely related may be much more challenging (**FIGURE 1.7**). In addition, traits shared by members of a species often vary a bit among individuals, such as eye color does among people. How do we decide if similar-looking organisms belong to different species or not? The short answer to that question is that we rely on whatever information we have. Early naturalists studied anatomy and distribution—essentially the only methods available at the time—so species were named and classified according to what they looked like and where they lived. Today's biologists are able to compare traits that the early naturalists did not even know about, including biochemical ones.

FIGURE 1.7 Four butterflies, two species: Which are which?

The top row shows two forms of the species *Heliconius melpomene*; the bottom row, two forms of *H. erato*.

H. melpomene and *H. erato* never cross-breed. Their alternate but similar patterns of coloration evolved as a shared warning signal to predatory birds that these butterflies taste terrible.

genus A group of species that share a unique set of traits.
species Unique type of organism.
specific epithet Second part of a species name.
taxon Group of organisms that share a unique set of traits.
taxonomy The science of naming and classifying species.
trait An observable characteristic of an organism or species.

The discovery of new information sometimes changes the way we distinguish a particular species or how we group it with others. For example, Linnaeus grouped plants by the number and arrangement of reproductive parts, a scheme that resulted in odd pairings such as castor-oil plants with pine trees. Having more information today, we place these plants in separate phyla.

Evolutionary biologist Ernst Mayr defined a species as one or more groups of individuals that potentially can interbreed, produce fertile offspring, and do not interbreed with other groups. This "biological species concept" is useful in many cases, but it is not universally applicable. For example, we may never know whether separate populations could interbreed even if they did get together. As another example, populations often continue to interbreed even as they diverge, so the exact moment at which two populations become two species is often impossible to pinpoint. We return to speciation and how it occurs in Chapter 17, but for now it is important to remember that a "species" is a convenient but artificial construct of the human mind.

TABLE 1.2

All of Life in Three Domains

Bacteria	Single cells, no nucleus. Most ancient lineage.
Archaea	Single cells, no nucleus. Evolutionarily closer to eukaryotes than bacteria.
Eukarya	Eukaryotic cells (with a nucleus). Single-celled and multicelled species of protists, plants, fungi, and animals.

TAKE-HOME MESSAGE 1.4

Each type of organism, or species, is given a unique, two-part scientific name.

Classification systems group species on the basis of shared, inherited traits.

1.5 HOW DOES SCIENCE WORK?

Most of us assume that we do our own thinking, but do we, really? You might be surprised to find out how often we let others think for us. Consider how a school's job (which is to impart as much information to students as quickly as possible) meshes perfectly with a student's job (which is to acquire as much knowledge as quickly as possible). In this rapid-fire exchange of information, it is sometimes easy to forget about the quality of what is being exchanged. Anytime you accept information without questioning it, you let someone else think for you.

THINKING ABOUT THINKING

Critical thinking is the deliberate process of judging the quality of information before accepting it. "Critical" comes from the Greek *kriticos* (discerning judgment). When you use critical thinking, you move beyond the content of new information to consider supporting evidence, bias, and alternative interpretations. How does the busy student manage this? Critical thinking does not necessarily require extra time, just a bit of extra awareness. There are many ways to do it. For example, you might ask yourself some of the following questions while you are learning something new:

What message am I being asked to accept?
Is the message based on facts or opinion?
Is there a different way to interpret the facts?
What biases might the presenter have?
How do my own biases affect what I'm learning?

Such questions are a way of being conscious about learning. They can help you decide whether to allow new information to guide your beliefs and actions.

THE SCIENTIFIC METHOD

Critical thinking is a big part of **science**, the systematic study of the observable world and how it works (**FIGURE 1.8**). A scientific line of inquiry usually begins with curiosity about something observable, such as, say, a decrease in the number of birds in a particular area. Typically, a scientist will read about what others have discovered before making a **hypothesis**, a testable explanation for a natural phenomenon. An example of a hypothesis would be, "The number of birds is decreasing because the number of cats is increasing." Making a hypothesis this way is an example of **inductive reasoning**, which means arriving at a conclusion based on one's observations. Inductive reasoning is the way we come up with new ideas about groups of objects or events.

A **prediction**, or statement of some condition that should exist if the hypothesis is correct, comes next. Making predictions is called the if–then process, in which the "if" part is the hypothesis, and the "then" part is the prediction. Using a hypothesis to make a prediction is a form of **deductive reasoning**, the logical process of using a general premise to draw a conclusion about a specific case.

Next, a scientist will devise ways to test a prediction. Tests may be performed on a **model**, or analogous system, if working with an object or event directly is not possible. For example, animal diseases are often used as models of similar human diseases. Careful observations are one way to test predictions that flow from a hypothesis. So are **experiments**: tests designed to support or falsify a prediction. A typical experiment explores a cause-and-effect relationship.

Researchers often investigate causal relationships by changing and observing **variables**, characteristics or events that can differ among individuals or over time.

TABLE 1.3

The Scientific Method

1. **Observe** some aspect of nature.
2. **Think of an explanation** for your observation (in other words, form a hypothesis).
3. **Test the hypothesis.**
   a. Make a prediction based on the hypothesis.
   b. Test the prediction using experiments or surveys.
   c. Analyze the results of the tests (data).
4. **Decide** whether the results of the tests support your hypothesis or not (form a conclusion).
5. **Report** your experiment, data, and conclusion to the scientific community.

control group Group of individuals identical to an experimental group except for the independent variable under investigation.
critical thinking Judging information before accepting it.
data Experimental results.
deductive reasoning Using a general idea to make a conclusion about a specific case.
dependent variable In an experiment, a variable that is presumably affected by an independent variable being tested.
experiment A test designed to support or falsify a prediction.
experimental group In an experiment, a group of individuals who have a certain characteristic or receive a certain treatment.
hypothesis Testable explanation of a natural phenomenon.
independent variable Variable that is controlled by an experimenter in order to explore its relationship to a dependent variable.
inductive reasoning Drawing a conclusion based on observation.
model Analogous system used for testing hypotheses.
prediction Statement, based on a hypothesis, about a condition that should exist if the hypothesis is correct.
science Systematic study of the observable world.
scientific method Making, testing, and evaluating hypotheses.
variable In an experiment, a characteristic or event that differs among individuals or over time.

FIGURE 1.8 Tierney Thys travels the world's oceans to study the giant sunfish (mola). This mola is carrying a satellite tracking device.

An **independent variable** is defined or controlled by the person doing the experiment. A **dependent variable** is an observed result that is supposed to be influenced by the independent variable. For example, an independent variable in an investigation of our observed decrease in the number of birds may be the removal of cats in the area. The dependent variable in this experiment would be the number of birds.

Biological systems are complex, with many interacting variables. It can be difficult to study one variable separately from the rest. Thus, biology researchers often test two groups of individuals simultaneously. An **experimental group** is a set of individuals that have a certain characteristic or receive a certain treatment. This group is tested side by side with a **control group**, which is identical to the experimental group except for one independent variable: the characteristic or the treatment being tested. Any differences in experimental results between the two groups is likely to be an effect of changing the variable.

Test results—**data**—that are consistent with the prediction are evidence in support of the hypothesis. Data inconsistent with the prediction are evidence that the hypothesis is flawed and should be revised.

A necessary part of science is reporting one's results and conclusions in a standard way, such as in a peer-reviewed journal article. The communication gives other scientists an opportunity to evaluate the information for themselves, both by checking the conclusions drawn and by repeating the experiments.

Forming a hypothesis based on observation, and then systematically testing and evaluating the hypothesis, are collectively called the **scientific method** (**TABLE 1.3**).

TAKE-HOME MESSAGE 1.5

- Judging the quality of information before accepting it is called critical thinking.
- The scientific method consists of making, testing, and evaluating hypotheses. It is a way of critical thinking.
- Experiments measure how changing an independent variable affects a dependent variable.

1.6 WHY DO BIOLOGISTS PERFORM EXPERIMENTS?

There are many different ways to do research, particularly in biology. Some biologists make surveys; they observe without making hypotheses. Some make hypotheses and leave experimentation to others. Despite a broad range of approaches, however, researchers typically try to design experiments in a consistent way. They change one independent variable at a time, and carefully measure the effects of the change on a dependent variable.

To give you a sense of how biology experiments work, we summarize two published studies here.

POTATO CHIPS AND STOMACHACHES

In 1996 the U.S. Food and Drug Administration (the FDA) approved Olestra® (a fat replacement manufactured from sugar and vegetable oil) for use as a food additive. Potato chips were the first Olestra-containing food product on the market in the United States. Controversy soon raged. Many people complained of intestinal problems after eating the chips and thought that the Olestra was at fault.

Two years later, researchers at Johns Hopkins University School of Medicine designed an experiment to test the hypothesis that this food additive causes cramps. The researchers predicted *if* Olestra causes cramps, *then* people who eat Olestra will be more likely to get cramps than people who do not. To test their prediction, they used a Chicago theater as a "laboratory," and asked 1,100 people between the ages of thirteen and thirty-eight to eat potato chips while watching a movie. Each person got an unmarked bag that contained 13 ounces of chips. In this experiment, individuals who ate Olestra-containing potato chips constituted the experimental group, and individuals who ate regular chips were the control group. The independent variable was the presence or absence of Olestra in the chips.

A few days after the experiment was finished, the researchers contacted everyone and collected reports of any post-movie cramps (the dependent variable). Of the 563 people in the experimental (Olestra-eating) group, 89 (15.8 percent) complained about cramps. However, so did 93 of the 529 people (17.6 percent) making up the control group—who had eaten the regular chips. In this experiment, people were about as likely to get cramps whether or not they ate chips made with Olestra. These results did not support the prediction, so the researchers concluded that eating Olestra does not cause cramps (**FIGURE 1.9**).

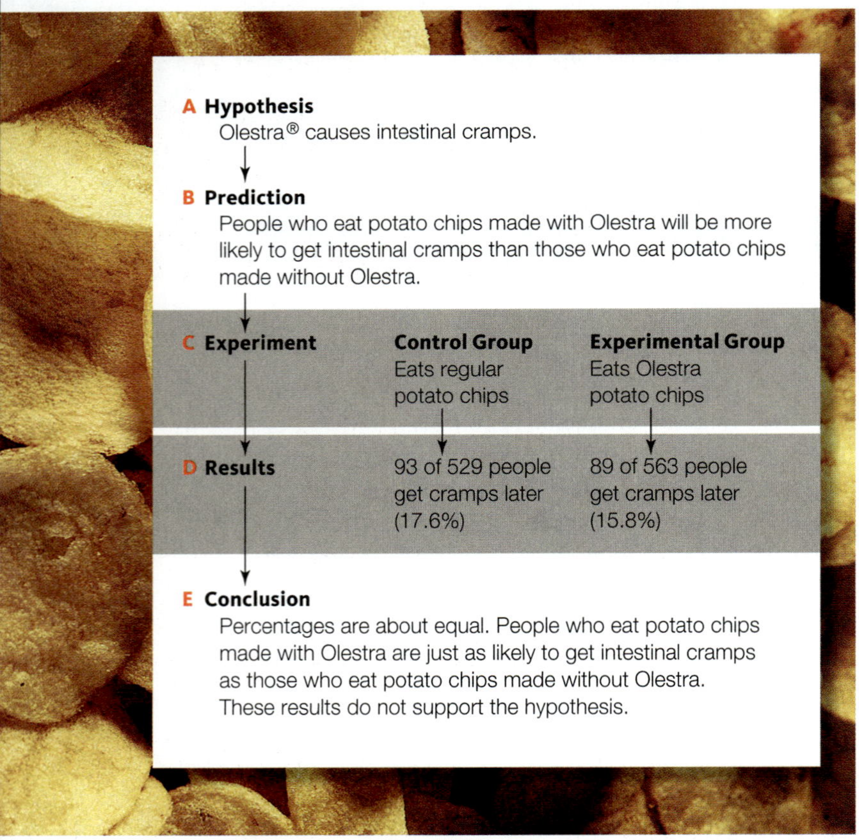

FIGURE 1.9 The steps in a scientific experiment to determine if Olestra causes cramps. A report of this study was published in the *Journal of the American Medical Association* in January 1998.

FIGURE IT OUT: What was the dependent variable in this experiment?

Answer: Whether or not a person got cramps

BUTTERFLIES AND BIRDS

A 2005 experiment investigated whether certain peacock butterfly behaviors defend these insects from predatory birds. The researchers performing this experiment began with two observations. First, when a peacock butterfly rests, it folds its wings, so only the dark underside shows (**FIGURE 1.10A**). Second, when a butterfly sees a predator approaching, it repeatedly flicks its wings open, while also moving them in a way that produces a hissing sound and a series of clicks (**FIGURE 1.10B**).

The researchers were curious about why the peacock butterfly flicks its wings. After they reviewed earlier studies, they came up with two hypotheses that might explain the wing-flicking behavior:

1. Although wing-flicking probably attracts predatory birds, it also exposes brilliant spots that resemble owl eyes. Anything that looks like owl eyes is known to startle small, butterfly-eating birds, so exposing the wing spots might scare off predators.
2. The hissing and clicking sounds produced when the peacock butterfly moves its wings may be an additional defense that deters predatory birds.

A With wings folded, a resting peacock butterfly resembles a dead leaf.

B When a bird approaches, a butterfly repeatedly flicks its wings open. This behavior exposes brilliant spots and also produces hissing and clicking sounds.

C Researchers tested whether peacock butterfly wing flicking and hissing reduce predation by blue tits.

FIGURE 1.10 Testing peacock butterfly defenses. Researchers painted out the spots of some butterflies, cut the sound-making part of the wings on others, and did both to a third group; then exposed each butterfly to a hungry blue tit. Results, listed below in Table 1.4, support the hypotheses that peacock butterfly spots and sounds can deter predatory birds.

FIGURE IT OUT: What was the dependent variable in this series of experiments?

Answer: Being eaten

TABLE 1.4

Results of Peacock Butterfly Experiment*

Wing Spots	Wing Sound	Total Number of Butterflies	Number Eaten	Number Survived
Spots	Sound	9	0	9 (100%)
No spots	Sound	10	5	5 (50%)
Spots	No sound	8	0	8 (100%)
No spots	No sound	10	8	2 (20%)

* *Proceedings of the Royal Society of London, Series B (2005) 272: 1203–1207.*

The researchers then used their hypotheses to make the following predictions:

1. If peacock butterflies startle predatory birds by exposing their brilliant wing spots, then individuals with wing spots will be less likely to get eaten by predatory birds than those without wing spots.
2. If peacock butterfly sounds deter predatory birds, then sound-producing individuals will be less likely to get eaten by predatory birds than silent individuals.

The next step was the experiment. The researchers used a marker to paint the wing spots of some butterflies black, and scissors to cut off the sound-making part of the wings of others. A third group had both treatments: Their wings were painted and cut. The researchers then put each butterfly into a large cage with a hungry blue tit (**FIGURE 1.10C**) and watched the pair for thirty minutes.

TABLE 1.4 lists the results of the experiment. All of the butterflies with unmodified wing spots survived, regardless of whether they made sounds. By contrast, only half of the butterflies that had spots painted out but could make sounds survived. Most of the silenced butterflies with painted-out spots were eaten quickly. The test results confirmed both predictions, so they support the hypotheses. Predatory birds are indeed deterred by peacock butterfly sounds, and even more so by wing spots.

TAKE-HOME MESSAGE 1.6

Natural processes are often influenced by many interacting variables.

Researchers unravel cause-and-effect relationships in complex natural processes by performing experiments in which they change one variable at a time.

1.7 WHAT ARE SOME POTENTIAL PITFALLS IN SCIENTIFIC INQUIRY?

SAMPLING ERROR

When researchers cannot directly observe all individuals of a population, all instances of an event, or some other aspect of nature, they may test or survey a subset. Results from the subset are then used to make generalizations about the whole. For example, a survey team may catalog the number of beetles in a given area of a very large forest. If that given area is one-thousandth of the forest, then an estimate of the number of beetles in the entire forest would be one thousand times their result. However, this type of generalization is risky because the subset may not be representative of the whole. In our beetle survey, for example, if the only nest of beetles in the entire forest happened to be located in the area that was surveyed, then the generalized result would be in error. **Sampling error** is a difference between results obtained from a subset, and results from the whole (**FIGURE 1.11A**).

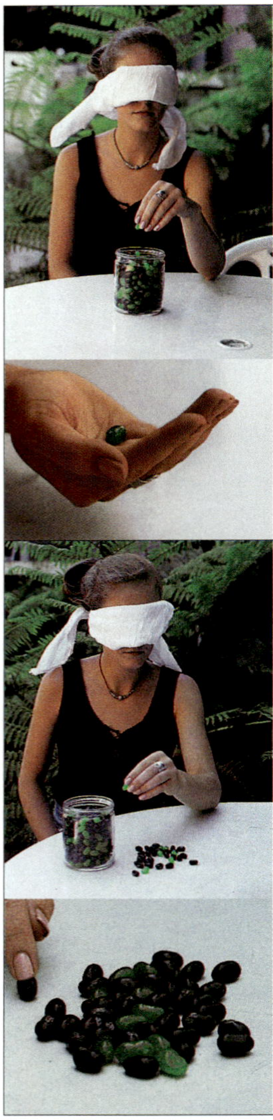

A Natalie chooses a random jelly bean from a jar. She is blindfolded, so she does not know that the jar contains 120 green and 280 black jelly beans.

The jar is hidden from Natalie's view before she removes her blindfold. She sees one green jelly bean in her hand and assumes that the jar must hold only green jelly beans. This assumption is incorrect: 30 percent of the jelly beans in the jar are green, and 70 percent are black. The small sample size has resulted in sampling error.

B Still blindfolded, Natalie randomly picks out 50 jelly beans from the jar. She ends up choosing 10 green and 40 black ones.

The larger sample leads Natalie to assume that one-fifth of the jar's jelly beans are green (20 percent) and four-fifths are black (80 percent). The larger sample more closely approximates the jar's actual green-to-black ratio of 30 percent to 70 percent.

The more times Natalie repeats the sampling, the greater the chance she has of guessing the actual ratio.

FIGURE 1.11 **{Animated}** Demonstration of sampling error, and the effect of sample size on it.

Sampling error may be unavoidable, but knowing how it can occur helps researchers design their experiments to minimize it. For example, sampling error can be a substantial problem with a small subset, so experimenters try to start with a relatively large sample, and they repeat their experiments (**FIGURE 1.11B**). To understand why these practices reduce the risk of sampling error, think about flipping a coin. There are two possible outcomes of each flip: The coin lands heads up, or it lands tails up. Thus, the chance that the coin will land heads up is one in two (1/2), which is a proportion of 50 percent. However, when you flip a coin repeatedly, it often lands heads up, or tails up, several times in a row. With just 3 flips, the proportion of times that heads actually land up may not even be close to 50 percent. With 1,000 flips, however, the overall proportion of times the coin lands heads up is much more likely to approach 50 percent.

In cases such as flipping a coin, it is possible to calculate **probability**, which is the measure, expressed as a percentage, of the chance that a particular outcome will occur. That chance depends on the total number of possible outcomes. For instance, if 10 million people enter a drawing, each has the same probability of winning: 1 in 10 million, or (an extremely improbable) 0.00001 percent.

Analysis of experimental data often includes probability calculations. If a result is very unlikely to have occurred by chance alone, it is said to be **statistically significant**. In this context, the word "significant" does not refer to the result's importance. It means that the result has been subjected to a rigorous statistical analysis that shows it has a very low probability (usually 5 percent or less) of being skewed by sampling error.

Variation in data is often shown as error bars on a graph (**FIGURE 1.12**). Depending on the graph, error bars may indicate variation around an average for one sample set, or the difference between two sample sets.

BIAS IN INTERPRETING RESULTS

Particularly when studying humans, changing a single variable apart from all others is not often possible. For example, remember that the people who participated in the Olestra experiment were chosen randomly. That means the study was not controlled for gender, age, weight, medications taken, and so on. Such variables may have influenced the results.

Human beings are by nature subjective, and scientists are no exception. Experimenters risk interpreting their results in terms of what they want to find out. That is why they often design experiments to yield quantitative results, which are counts or some other data that can be measured or gathered objectively. Such results minimize the potential for bias, and also give other scientists an opportunity to repeat the experiments and check the conclusions drawn from them.

This last point gets us back to the role of critical thinking in science. Scientists expect one another to recognize and put aside bias in order to test their hypotheses in ways that may prove them wrong. If a scientist does not, then others will, because exposing errors is just as useful as applauding insights. The scientific community consists of critically thinking people trying to poke holes in one another's ideas. Their collective efforts make science a self-correcting endeavor.

THE LIMITS OF SCIENCE

Science helps us be objective about our observations in part because of its limitations. For example, science does not address many questions, such as "Why do I exist?" Answers to such questions can only come from within as an integration of all the personal experiences and mental connections that shape our consciousness. This is not to say subjective answers have no value, because no human society can function for long unless its individuals share standards for making judgments, even if they are subjective. Moral, aesthetic, and philosophical standards vary from one society to the next, but all help people decide what is important and good. All give meaning to our lives.

Neither does science address the supernatural, or anything that is "beyond nature." Science neither assumes nor denies that supernatural phenomena occur, but scientists may cause controversy when they discover a natural explanation for something that was thought to have none. Such controversy often arises when a society's moral standards are interwoven with its understanding of nature. For example, Nicolaus Copernicus proposed in 1540 that Earth orbits the sun. Today that idea is generally accepted, but the prevailing belief system had Earth as the immovable center of the universe. In 1610, astronomer Galileo Galilei published evidence for the Copernican model of the solar system, an act that resulted in his imprisonment. He was publicly forced to recant his work, spent the rest of his life under house arrest, and was never allowed to publish again.

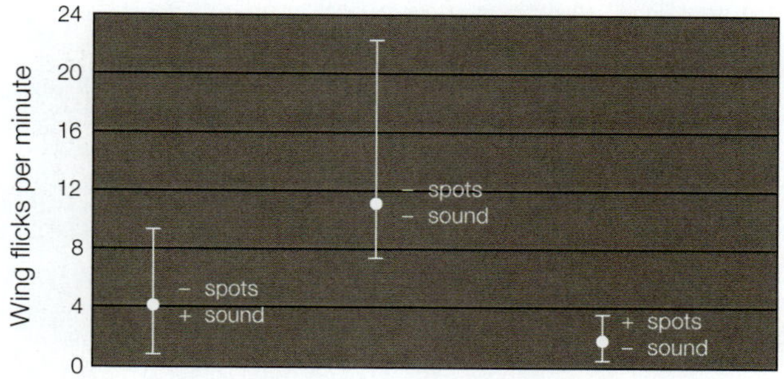

FIGURE 1.12 Example of error bars in a graph. This graph was adapted from the peacock butterfly research described in Section 1.6.

The researchers recorded the number of times each butterfly flicked its wings in response to an attack by a bird.

The dots represent average frequency of wing flicking for each sample set of butterflies. The error bars that extend above and below the dots indicate the range of values—the sampling error.

FIGURE IT OUT: What was the fastest rate at which a butterfly with no spots or sound flicked its wings?

Answer: 22 times per minute

As Galileo's story illustrates, exploring a traditional view of the natural world from a scientific perspective can be misinterpreted as a violation of morality. As a group, scientists are no less moral than anyone else, but they follow a particular set of rules that do not necessarily apply to others: Their work concerns only the natural world, and their ideas must be testable by other scientists.

Science helps us communicate our experiences without bias. As such, it may be as close as we can get to a universal language. We are fairly sure, for example, that the laws of gravity apply everywhere in the universe. Intelligent beings on a distant planet would likely understand the concept of gravity. We might well use gravity or another scientific concept to communicate with them, or anyone, anywhere. The point of science, however, is not to communicate with aliens. It is to find common ground here on Earth.

TAKE-HOME MESSAGE 1.7

- Checks and balances inherent in the scientific process help researchers to be objective about their observations.
- Researchers minimize sampling error by using large sample sizes and by repeating their experiments.
- Probability calculations can show whether a result is likely to have occurred by chance alone.
- Science is a self-correcting process because it is carried out by an aggregate community of people systematically checking one another's ideas.

probability The chance that a particular outcome of an event will occur; depends on the total number of outcomes possible.
sampling error Difference between results derived from testing an entire group of events or individuals, and results derived from testing a subset of the group.
statistically significant Refers to a result that is statistically unlikely to have occurred by chance.

1.8 WHAT IS A "THEORY"?

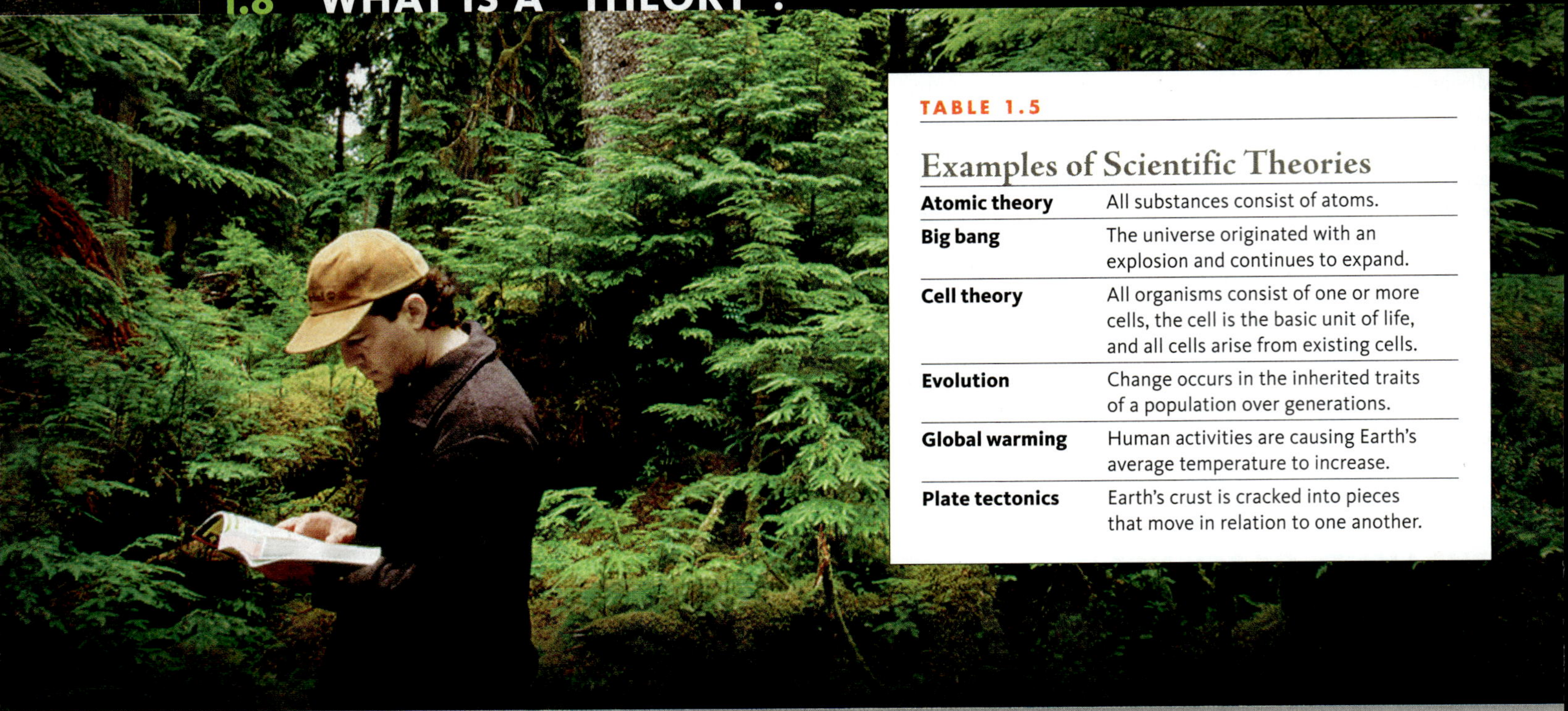

TABLE 1.5

Examples of Scientific Theories

Atomic theory	All substances consist of atoms.
Big bang	The universe originated with an explosion and continues to expand.
Cell theory	All organisms consist of one or more cells, the cell is the basic unit of life, and all cells arise from existing cells.
Evolution	Change occurs in the inherited traits of a population over generations.
Global warming	Human activities are causing Earth's average temperature to increase.
Plate tectonics	Earth's crust is cracked into pieces that move in relation to one another.

Suppose a hypothesis stands even after years of tests. It is consistent with all data ever gathered, and it has helped us make successful predictions about other phenomena. When a hypothesis meets these criteria, it is considered to be a **scientific theory** (**TABLE 1.5**). To give an example, all observations to date have been consistent with the hypothesis that matter consists of atoms. Scientists no longer spend time testing this hypothesis for the compelling reason that, since we started looking 200 years ago, no one has discovered matter that consists of anything else. Thus, scientists use the hypothesis, now called atomic theory, to make other hypotheses about matter.

Scientific theories are our best objective descriptions of the natural world, but they can never be proven absolutely because to do so would necessitate testing under every possible circumstance. For example, in order to prove atomic theory, the composition of all matter in the universe would have to be checked—an impossible task even if someone wanted to try.

Like all hypotheses, a scientific theory can be disproven by one observation or result that is inconsistent with it. For example, if someone discovers a form of matter that does not consist of atoms, atomic theory would be revised until no one could prove it to be incorrect. This potentially falsifiable nature of scientific theories is part of science's built-in system of checks and balances. The theory of evolution, which states that change occurs in a line of descent over time, still holds after a century of observations and testing. As with all other scientific theories, no one can be absolutely sure that it will hold under all possible conditions, but it has a very high probability of not being wrong. Few other theories have withstood as much scrutiny.

You may hear people apply the word "theory" to a speculative idea, as in the phrase "It's just a theory." This everyday usage of the word differs from the way it is used in science. Speculation is an opinion, belief, or personal conviction that is not necessarily supported by evidence. A scientific theory differs because it is supported by a large body of evidence, and it is consistent with all known facts.

A scientific theory also differs from a **law of nature**, which describes a phenomenon that has been observed to occur in every circumstance without fail, but for which we do not have a complete scientific explanation. The laws of thermodynamics, which describe energy, are examples. As you will see in Chapter 5, we understand *how* energy behaves, but not exactly *why* it behaves the way it does.

law of nature Generalization that describes a consistent natural phenomenon for which there is incomplete scientific explanation.
scientific theory Hypothesis that has not been disproven after many years of rigorous testing.

TAKE-HOME MESSAGE 1.8

A scientific theory is a time-tested hypothesis that is consistent with all known facts. It is our most objective way of describing the natural world.

1.9 Application: THE SECRET LIFE OF EARTH

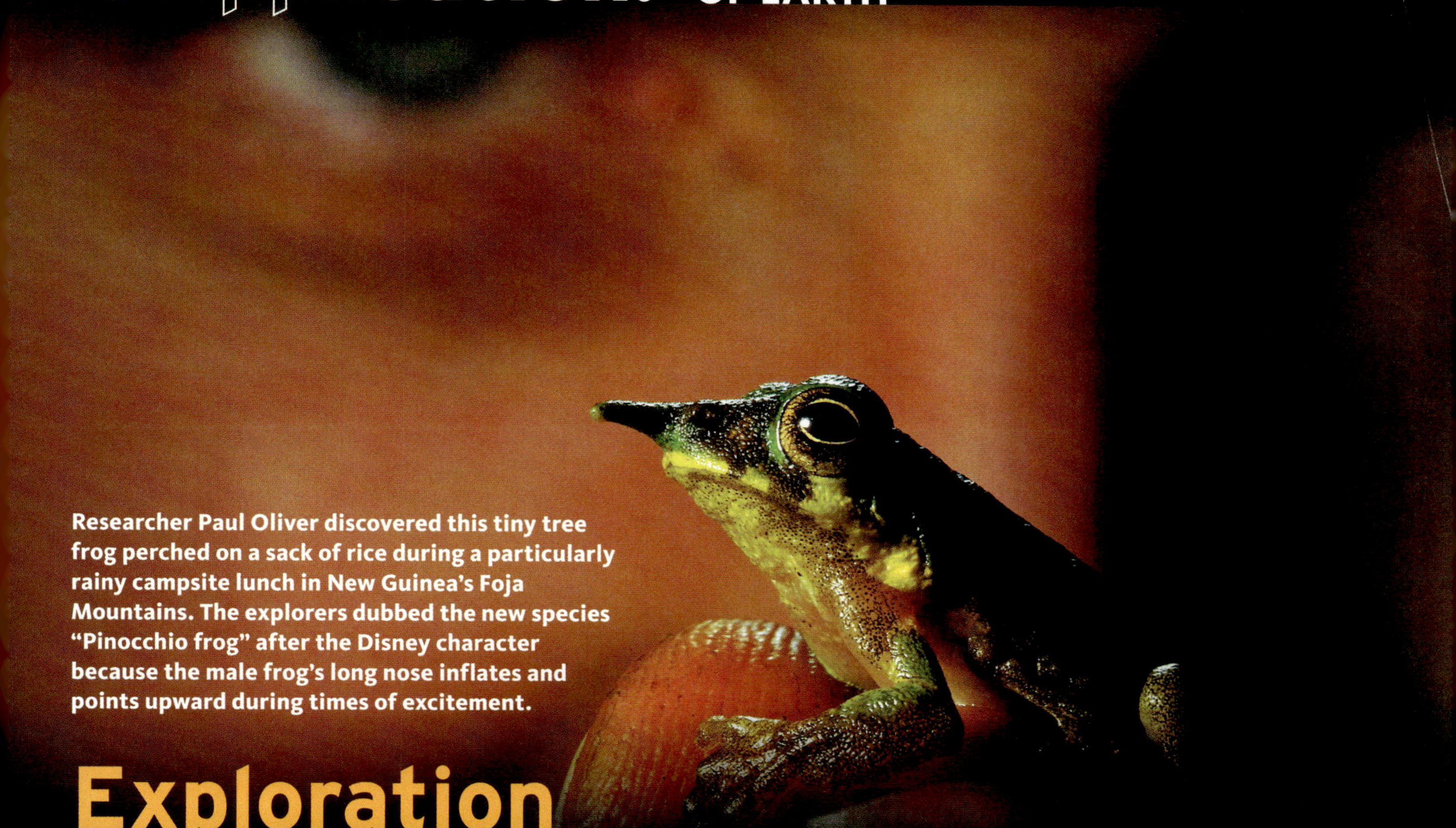

Researcher Paul Oliver discovered this tiny tree frog perched on a sack of rice during a particularly rainy campsite lunch in New Guinea's Foja Mountains. The explorers dubbed the new species "Pinocchio frog" after the Disney character because the male frog's long nose inflates and points upward during times of excitement.

Exploration

IN THIS ERA OF DETAILED CELL PHONE GPS, could there possibly be any places left on Earth that humans have not yet explored? Actually, there are plenty. For example, a 2-million-acre cloud forest in New Guinea was only recently penetrated by explorers. How did the explorers know they had landed in uncharted territory? For one thing, the forest was filled with plants and animals unknown even to native peoples that have long inhabited other parts of the region. Team member Bruce Beehler remarked, "I was shouting. This trip was a once-in-a-lifetime series of shouting experiences." The team members discovered many new species, including a rhododendron plant with flowers the size of plates and a frog the size of a pea. They also came across hundreds of species that are on the brink of extinction in other parts of the world, and some that supposedly had been extinct for decades.

Each new species is a reminder that we do not yet know all of the organisms that share our planet. We don't even know how many to look for. Why does that matter? Understanding the scope of life on Earth gives us perspective on where we fit into it. For example, the current rate of extinctions is about 1,000 times faster than ever recorded, and we now know that human activities are responsible for the acceleration. At this rate, we will never know about most of the species that are alive today. Is that important? Biologists think so. Whether or not we are aware of it, humans are intimately connected with the world around us. Our activities are profoundly changing the entire fabric of life on Earth. The changes are, in turn, affecting us in ways we are only beginning to understand.

Ironically, the more we learn about the natural world, the more we realize we have yet to learn. But don't take our word for it. Find out what biologists know, and what they do not, and you will have a solid foundation upon which to base your own opinions about the human connection—your connection—with all life on Earth.

CREDIT: Tim Laman/National Geographic Creative.

Summary

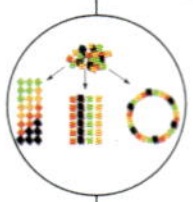

SECTION 1.1 **Biology** is the scientific study of life. Biologists think about life at different levels of organization, with **emergent properties** appearing at successive levels. All matter consists of **atoms**, which combine as **molecules**. **Organisms** are individuals that consist of one or more **cells**, the level at which life emerges. Cells of larger multicelled organisms are organized as **tissues**, **organs**, and **organ systems**. A **population** is a group of interbreeding individuals of a species in a given area; a **community** is all populations of all species in a given area. An **ecosystem** is a community interacting with its environment. The **biosphere** includes all regions of Earth that hold life.

SECTION 1.2 All organisms require energy and **nutrients** to sustain themselves. **Producers** harvest energy from the environment to make their own food by processes such as **photosynthesis**; **consumers** eat other organisms, their wastes, or remains. Organisms keep the conditions in their internal environment within ranges that their cells tolerate—a process called **homeostasis**. **DNA** contains information that guides an organism's **growth**, **development**, and **reproduction**. The passage of DNA from parents to offspring is called **inheritance**.

SECTION 1.3 The many types of organisms that currently exist on Earth differ greatly in details of body form and function. **Biodiversity** is the sum of differences among living things. **Bacteria** and **archaea** are both **prokaryotes**, single-celled organisms whose DNA is not contained within a **nucleus**. The DNA of single-celled or multicelled **eukaryotes** (**protists**, **plants**, **fungi**, and **animals**) is contained within a nucleus.

SECTION 1.4 Each **species** has a two-part name. The first part is the **genus** name. When combined with the **specific epithet**, it designates the particular species. With **taxonomy**, species are ranked into ever more inclusive **taxa** on the basis of shared **traits**.

SECTION 1.5 **Critical thinking**, the self-directed act of judging the quality of information as one learns, is an important part of **science**. Generally, a researcher observes something in nature, uses **inductive reasoning** to form a **hypothesis** (testable explanation) for it, then uses **deductive reasoning** to make a testable **prediction** about what might occur if the hypothesis is correct. **Experiments** with **variables** may be performed on an **experimental group** as compared with a **control group**, and sometimes on **models**. A researcher changes an **independent variable**, then observes the effects of the change on a **dependent variable**. Conclusions are drawn from the resulting **data**. The **scientific method** consists of making, testing, and evaluating hypotheses, and sharing results.

SECTION 1.6 Biological systems are usually influenced by many interacting variables. Research approaches differ, but experiments are typically designed in a consistent way, in order to study a single cause-and-effect relationship in a complex natural system.

SECTION 1.7 Small sample size increases the potential for **sampling error** in experimental results. In such cases, a subset may be tested that is not representative of the whole. Researchers design experiments carefully to minimize sampling error and bias, and they use **probability** rules to check the **statistical significance** of their results. Science is ideally a self-correcting process because scientists check and test one another's ideas. Science helps us be objective about our observations because it is only concerned with testable ideas about observable aspects of nature. Opinion and belief have value in human culture, but they are not addressed by science.

SECTION 1.8 A **scientific theory** is a long-standing hypothesis that is useful for making predictions about other phenomena. It is our best way of describing reality. A **law of nature** describes something that occurs without fail, but has an incomplete scientific explanation.

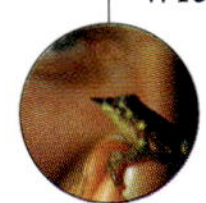

SECTION 1.9 We know about only a fraction of the organisms that live on Earth, in part because we have explored only a fraction of its inhabited regions.

Self-Quiz

Answers in Appendix VII

1. _______ are fundamental building blocks of all matter.
   a. Atoms
   b. Molecules
   c. Cells
   d. Organisms

2. The smallest unit of life is the _______ .
   a. atom
   b. molecule
   c. cell
   d. organism

3. Organisms require _______ and _______ to maintain themselves, grow, and reproduce.

4. By sensing and responding to change, organisms keep conditions in the internal environment within ranges that cells can tolerate. This process is called _______ .

5. DNA _______ .
   a. guides form and function
   b. is the basis of traits
   c. is transmitted from parents to offspring
   d. all of the above

6. A process by which an organism produces offspring is called _______ .

7. _______ is the transmission of DNA to offspring.
   a. Reproduction
   b. Development
   c. Homeostasis
   d. Inheritance

8. A butterfly is a(n) _______ (choose all that apply).
   a. organism
   b. domain
   c. species
   d. eukaryote
   e. consumer
   f. producer
   g. prokaryote
   h. trait

CREDIT: Scientific Paper; Adrian Vallin, Sven Jakobsson, Johan Lind and Christer Wiklund, *Proc. R. Soc. B* (2005 272, 1203, 1207). Used with permission of The Royal Society and the author.

Data Analysis Activities

Peacock Butterfly Predator Defenses The photographs below represent experimental and control groups used in the peacock butterfly experiment discussed in Section 1.6. See if you can identify the experimental groups, and match them up with the relevant control group(s). ***Hint***: Identify which variable is being tested in each group (each variable has a control).

A Wing spots painted out

B Wing spots visible; wings silenced

C Wing spots painted out; wings silenced

D Wings painted but spots visible

E Wings cut but not silenced

F Wings painted, spots visible; wings cut, not silenced

9. _______ move around for at least part of their life.

10. A bacterium is _______ (choose all that apply).
    a. an organism
    b. single-celled
    c. an animal
    d. a eukaryote

11. Bacteria, Archaea, and Eukarya are three _______ .

12. A control group is _______ .
    a. a set of individuals that have a certain characteristic or receive a certain treatment
    b. the standard against which an experimental group is compared
    c. the experiment that gives conclusive results

13. Fifteen randomly selected students are found to be taller than 6 feet. The researchers concluded that the average height of a student is greater than 6 feet. This is an example of _______ .
    a. experimental error
    b. sampling error
    c. a subjective opinion
    d. experimental bias

14. Science only addresses that which is _______ .
    a. alive
    b. observable
    c. variable
    d. indisputable

15. Match the terms with the most suitable description.

___ life	a. if–then statement
___ probability	b. unique type of organism
___ species	c. emerges with cells
___ hypothesis	d. testable explanation
___ prediction	e. measure of chance
___ producer	f. makes its own food

Critical Thinking

1. A person is declared dead upon the irreversible ceasing of spontaneous body functions: brain activity, blood circulation, and respiration. Only about 1% of a body's cells have to die in order for all of these things to happen. How can a person be dead when 99% of his or her cells are alive?

2. Explain the difference between a one-celled organism and a single cell of a multicelled organism.

3. Why would you think twice about ordering from a restaurant menu that lists the specific epithet but not the genus name of its offerings? *Hint:* Look up *Homarus americanus*, *Ursus americanus*, *Ceanothus americanus*, *Bufo americanus*, *Lepus americanus*, and *Nicrophorus americanus*.

4. Once there was a highly intelligent turkey that had nothing to do but reflect on the world's regularities. Morning always started out with the sky turning light, followed by the master's footsteps, which were always followed by the appearance of food. Other things varied, but food always followed footsteps. The sequence of events was so predictable that it eventually became the basis of the turkey's theory about the goodness of the world. One morning, after more than 100 confirmations of this theory, the turkey listened for the master's footsteps, heard them, and had its head chopped off.

Any scientific theory is modified or discarded upon discovery of contradictory evidence. The absence of absolute certainty has led some people to conclude that "theories are irrelevant because they can change." If that is so, should we stop doing scientific research? Why or why not?

5. In 2005, researcher Woo-suk Hwang reported that he had made immortal stem cells from human patients. His research was hailed as a breakthrough for people affected by degenerative diseases, because stem cells may be used to repair a person's own damaged tissues. Hwang published his results in a peer-reviewed journal. In 2006, the journal retracted his paper after other scientists discovered that Hwang's group had faked their data.

Does the incident show that results of scientific studies cannot be trusted? Or does it confirm the usefulness of a scientific approach, because other scientists discovered and exposed the fraud?

A unique set of properties makes water essential to life. These properties arise from interactions among individual water molecules.

2 LIFE'S CHEMICAL BASIS

Links to Earlier Concepts

In this chapter, you will explore the first level of life's organization—atoms—as you encounter the first example of how the same building blocks, arranged different ways, form different products (Section 1.1). You will also see one aspect of homeostasis, the process by which organisms keep themselves in a state that favors cell survival (1.2).

KEY CONCEPTS

ATOMS AND ELEMENTS

Atoms, the building blocks of all matter, differ in their numbers of protons, neutrons, and electrons. Atoms of an element have the same number of protons.

WHY ELECTRONS MATTER

Whether an atom interacts with other atoms depends on the number of electrons it has. An atom with an unequal number of electrons and protons is an ion.

ATOMS BOND

Atoms of many elements interact by acquiring, sharing, and giving up electrons. Interacting atoms may form ionic, covalent, or hydrogen bonds.

WATER

Hydrogen bonding among individual molecules gives water properties that make life possible: temperature stabilization, cohesion, and the ability to dissolve many other substances.

HYDROGEN POWER

Most of the chemistry of life occurs in a narrow range of pH, so most fluids inside organisms are buffered to stay within that range.

Photograph by Paul Nicklen, National Geographic Creative.

2.1 WHAT ARE THE BASIC BUILDING BLOCKS OF ALL MATTER?

Even though atoms are about 20 million times smaller than a grain of sand, they consist of even smaller subatomic particles. Positively charged **protons** (p^+) and uncharged **neutrons** occur in an atom's core, or **nucleus**. Negatively charged **electrons** (e^-) move around the nucleus (**FIGURE 2.1A**). **Charge** is an electrical property: Opposite charges attract, and like charges repel.

A typical atom has about the same number of electrons and protons. The negative charge of an electron is the same magnitude as the positive charge of a proton, so the two charges cancel one another. Thus, an atom with the same number of electrons and protons carries no charge.

All atoms have protons. The number of protons in the nucleus is called the **atomic number**, and it determines the type of atom, or element. **Elements** are pure substances, each consisting only of atoms with the same number of protons in their nucleus (**FIGURE 2.1B**). For example, the atomic number of carbon is 6, so all atoms with six protons in their nucleus are carbon atoms, no matter how many electrons or neutrons they have. Carbon, the substance, consists only of carbon atoms, and all of those atoms have six protons.

Knowing the numbers of electrons, protons, and neutrons in atoms helps us predict how elements will behave. In 1869, chemist Dmitry Mendeleyev arranged the elements known at the time by their chemical properties. The arrangement, which he called the **periodic table** (**FIGURE 2.1C**), turned out to be by atomic number, even though subatomic particles would not be discovered until the early 1900s.

In the periodic table, each element is represented by a symbol that is typically an abbreviation of the element's Latin or Greek name. For instance, Pb (lead) is short for *plumbum*; the word "plumbing" is related—ancient Romans made their water pipes with lead. Carbon's symbol, C, is from *carbo*, the Latin word for coal (which is mostly carbon).

ISOTOPES AND RADIOISOTOPES

All atoms of an element have the same number of protons, but they can differ in the number of other subatomic particles. Those that differ in the number of neutrons are called **isotopes**. We define isotopes by their **mass number**, which is the total number of protons and neutrons in their nucleus. Mass number is written as a superscript to the left of an element's symbol. For example, hydrogen, the simplest atom, has one proton and no neutrons, so it is designated ^{1}H. The most common isotope of carbon has six protons and six neutrons, so it is ^{12}C, or carbon 12. The other naturally occurring isotopes of carbon are ^{13}C (six protons, seven neutrons), and ^{14}C (six protons, eight neutrons).

Carbon 14 is a **radioisotope**, or radioactive isotope. Atoms of a radioisotope have an unstable nucleus that breaks up

FIGURE 2.1 {Animated} Atoms and elements.

A Atoms consist of electrons moving around a nucleus of protons and neutrons. Models such as this one do not show what atoms look like. Electrons move in defined, three-dimensional spaces about 10,000 times bigger than the nucleus.

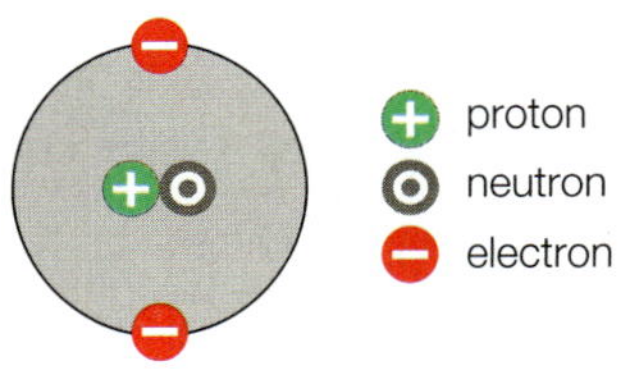

B Example of an element.

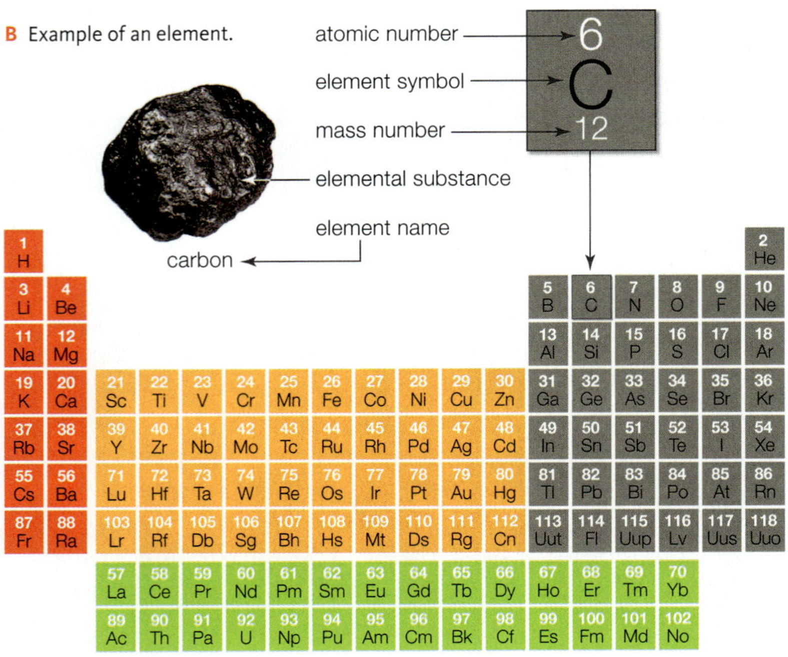

C The periodic table of the elements.

atomic number Number of protons in the atomic nucleus; determines the element.
charge Electrical property. Opposite charges attract, and like charges repel.
electron Negatively charged subatomic particle.
element A pure substance that consists only of atoms with the same number of protons.
isotopes Forms of an element that differ in the number of neutrons their atoms carry.
mass number Of an isotope, the total number of protons and neutrons in the atomic nucleus.
neutron Uncharged subatomic particle in the atomic nucleus.
nucleus Core of an atom; occupied by protons and neutrons.
periodic table Tabular arrangement of all known elements by their atomic number.
proton Positively charged subatomic particle that occurs in the nucleus of all atoms.
radioactive decay Process by which atoms of a radioisotope emit energy and/or subatomic particles when their nucleus spontaneously breaks up.
radioisotope Isotope with an unstable nucleus.
tracer A molecule with a detectable component.

spontaneously. As a nucleus breaks up, it emits radiation—subatomic particles, energy, or both—a process called **radioactive decay**. The atomic nucleus cannot be altered by ordinary means, so radioactive decay is unaffected by external factors such as temperature, pressure, or whether the atoms are part of molecules.

Each radioisotope decays at a predictable rate into predictable products. For example, when carbon 14 decays, one of its neutrons splits into a proton and an electron. The nucleus emits the electron as radiation. Thus, a carbon atom with eight neutrons and six protons (^{14}C) becomes a nitrogen atom, with seven neutrons and seven protons (^{14}N):

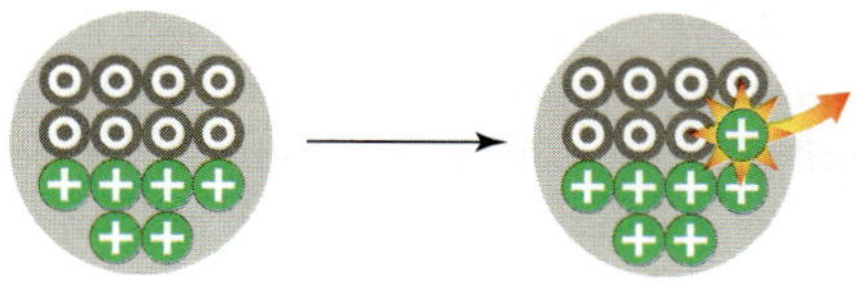

nucleus of ^{14}C, with 6 protons, 8 neutrons — nucleus of ^{14}N, with 7 protons, 7 neutrons

This process is so predictable that we can say with certainty that about half of the atoms in any sample of ^{14}C will be ^{14}N atoms after 5,730 years. The predictable rate of radioactive decay makes it possible for scientists to estimate the age of a rock or fossil by measuring its isotope content (we return to this topic in Section 16.4).

TRACERS

All isotopes of an element generally have the same chemical properties regardless of the number of neutrons in their atoms. This consistent chemical behavior means that organisms use atoms of one isotope the same way that they use atoms of another. Thus, radioisotopes can be used as tracers to study biological processes. A **tracer** is any substance with a detectable component. For example, a molecule in which an atom (such as ^{12}C) has been replaced with a radioisotope (such as ^{14}C) can be used as a radioactive tracer. When delivered into a biological system such as a cell, body, or ecosystem, this tracer may be followed as it moves through the system with instruments that detect radiation.

TAKE-HOME MESSAGE 2.1

All matter consists of atoms, tiny particles that in turn consist of electrons moving around a nucleus of protons and neutrons.

An element is a pure substance that consists only of atoms with the same number of protons. Isotopes are forms of an element that have different numbers of neutrons.

Unstable nuclei of radioisotopes break down spontaneously (decay) at a predictable rate to form predictable products.

PEOPLE MATTER

National Geographic Explorer
KENNETH SIMS

A volcano is a force of nature most of us prefer to observe from a very long and safe distance. Not volcanologist Ken Sims. The National Geographic explorer could never be satisfied with anything less than standing on the edge of an erupting volcano. In fact, even standing on the edge of an erupting volcano wasn't enough for Sims. As part of his research, he rappelled down into the mouth of Nyiragongo, a volcano in the Democratic Republic of the Congo, to gather fresh lava from a molten lake boiling at 1800°F.

Sims says, "While many think of me as a volcanologist, I am actually an isotope geochemist and, as such, I have pondered and written professional papers on a wide range of problems, including chemical oceanography; the Earth's paleo-climate; oceanic and continental crustal growth; continental crustal weathering; ground water transport; and, of course, the genesis and evolution of volcanic systems. The measurement of radioactive isotopes in natural systems allows me to quantify fundamental processes that would otherwise be limited to qualitative observation. This may make me sound like a nerd but to be able to quantify the Earth's processes is truly inspiring."

2.2 WHY DO ATOMS INTERACT?

ELECTRONS MATTER

Electrons are really, really small. How small are they? If they were as big as apples, you would be about 3.5 times taller than our solar system is wide. Simple physics explains the motion of, say, an apple falling from a tree, but electrons are so tiny that such everyday physics cannot explain their behavior. For example, electrons carry energy, but only in incremental amounts. An electron gains energy only by absorbing the exact amount needed to boost it to the next energy level. Likewise, it loses energy only by emitting the exact difference between two energy levels. This concept will be important to remember when you learn how cells harvest and release energy.

A lot of electrons may be zipping around in the same atom. Despite moving really fast (around 3 million meters per second), they never collide. Why not? For one reason, electrons in an atom occupy different orbitals, which are defined volumes of space around the atomic nucleus.

To understand how orbitals work, imagine that an atom is a multilevel apartment building, with the nucleus in the basement. Each "floor" of the building corresponds to a certain energy level, and each has a certain number of "rooms" (orbitals) available for rent. Two electrons can occupy each room. Pairs of electrons populate rooms from the ground floor up; in other words, they fill orbitals from lower to higher energy levels. The farther an electron is from the nucleus in the basement, the greater its energy. An electron can move to a room on a higher floor if an energy input gives it a boost, but it immediately emits the extra energy and moves back down.

A **shell model** helps us visualize how electrons populate atoms (**FIGURE 2.2**). In this model, nested "shells" correspond to successively higher energy levels. Thus, each shell includes all of the rooms (orbitals) on one floor (energy level) of our atomic apartment building.

We draw a shell model of an atom by filling it with electrons (represented as balls or dots), from the innermost shell out, until there are as many electrons as the atom has protons. There is only one room on the first floor, one orbital at the lowest energy level. It fills up first. In hydrogen, the simplest atom, a single electron occupies that room (**FIGURE 2.2A**). Helium, with two protons, has two electrons that fill the room—and the first shell. In larger atoms, more electrons rent the second-floor rooms (**FIGURE 2.2B**). When the second floor fills, more electrons rent third-floor rooms (**FIGURE 2.2C**), and so on.

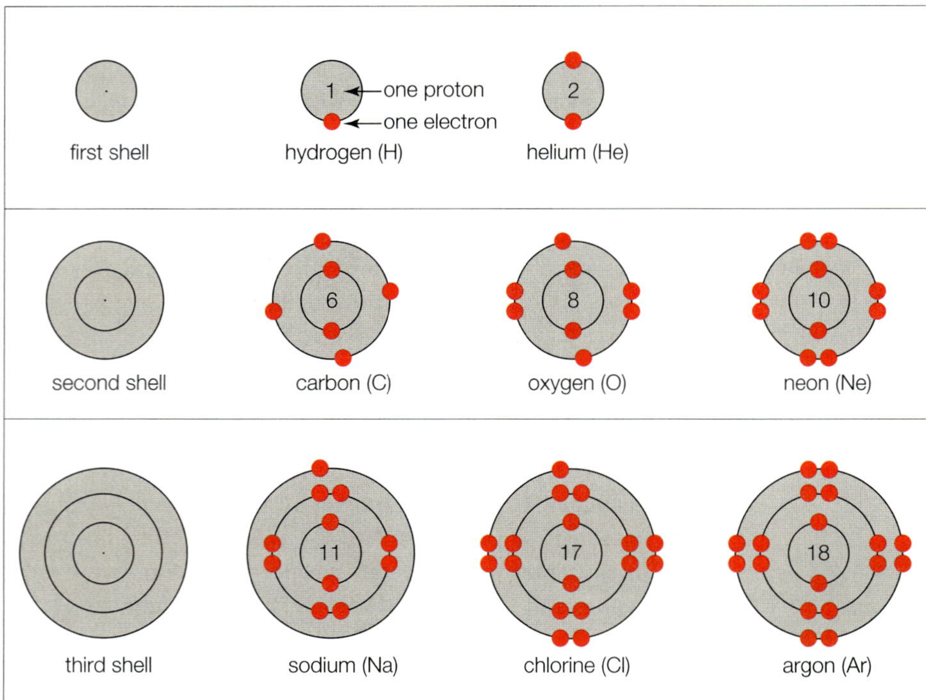

A The first shell corresponds to the first energy level, and it can hold up to 2 electrons. Hydrogen has one proton, so it has 1 electron and 1 vacancy. A helium atom has 2 protons, 2 electrons, and no vacancies.

B The second shell corresponds to the second energy level, and it can hold up to 8 electrons. Carbon has 6 electrons, so its first shell is full. Its second shell has 4 electrons and four vacancies. Oxygen has 8 electrons and two vacancies. Neon has 10 electrons and no vacancies.

C The third shell corresponds to the third energy level, and it can hold up to 8 electrons. A sodium atom has 11 electrons, so its first two shells are full; the third shell has one electron. Thus, sodium has seven vacancies. Chlorine has 17 electrons and one vacancy. Argon has 18 electrons and no vacancies.

FIGURE 2.2 {Animated} Shell models. Each circle (shell) represents one energy level. To make these models, we fill the shells with electrons from the innermost shell out, until there are as many electrons as the atom has protons. The number of protons in each model is indicated.

FIGURE IT OUT: Which of these models have unpaired electrons in their outer shell? Answer: Hydrogen, carbon, oxygen, sodium, and chlorine

ABOUT VACANCIES

When an atom's outermost shell is filled with electrons, we say that it has no vacancies. Any atom is in its most stable state when it has no vacancies. Helium, neon, and argon are examples of elements with no vacancies. Atoms of these elements are chemically stable, which means they have no tendency to interact with other atoms. Thus, these elements occur most frequently in nature as solitary atoms.

By contrast, when an atom's outermost shell has room for another electron, it has a vacancy. Atoms with vacancies tend to get rid of them by interacting with other atoms; in other words, they are chemically active. For example, the sodium atom (Na) depicted in **FIGURE 2.2C** has one electron in its outer (third) shell, which can hold eight. With seven vacancies, we can predict that this atom is chemically active.

In fact, this particular sodium atom is not just active, it is extremely so. Why? The shell model shows that a sodium atom has an unpaired electron, but in the real world, electrons really like to be in pairs when they occupy orbitals. Solitary atoms that have unpaired electrons are called **free radicals**. With some exceptions, free radicals are very unstable, easily forcing electrons upon other atoms or ripping electrons away from them. This property makes free radicals dangerous to life (we return to this topic in Section 5.5).

A free radical sodium atom can easily evict its one unpaired electron, so that its second shell—which is full of electrons—becomes its outermost, and no vacancies remain. This is the atom's most stable state. The vast majority of sodium atoms on Earth are like this one, with 11 protons and 10 electrons.

Atoms with an unequal number of protons and electrons are called **ions**. Ions carry a net (or overall) charge. Sodium ions (Na^+) offer an example of how atoms gain a positive charge by losing an electron (**FIGURE 2.3A**). Other atoms gain a negative charge by accepting an electron. For example, an uncharged chlorine atom has 17 protons and 17 electrons. The outermost shell of this atom can hold eight electrons, but it has only seven. With one vacancy and one unpaired electron, we can predict—correctly—that this atom is chemically very active. An uncharged chlorine atom easily fills its third shell by accepting an electron. When that happens, the atom becomes a chloride ion (Cl^-) with 17 protons, 18 electrons, and a net negative charge (**FIGURE 2.3B**).

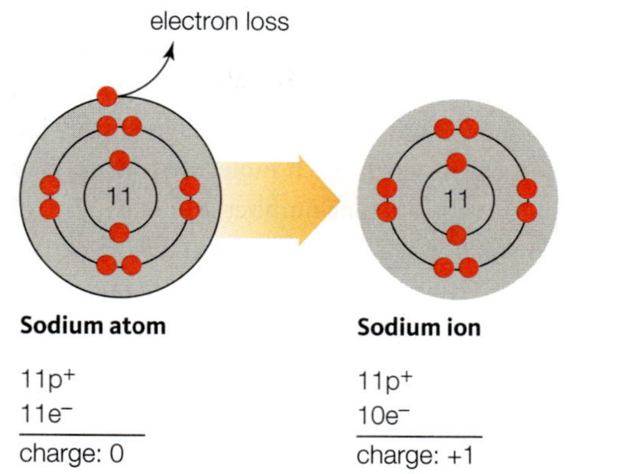

A A sodium atom (Na) becomes a positively charged sodium ion (Na^+) when it loses the single electron in its third shell. The atom's full second shell is now its outermost, so it has no vacancies.

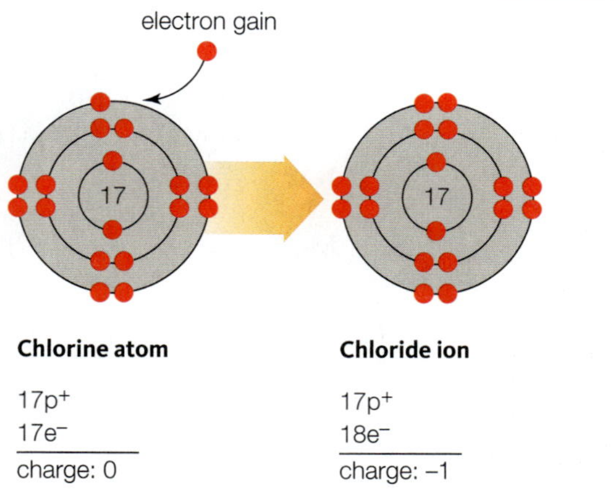

B A chlorine atom (Cl) becomes a negatively charged chloride ion (Cl^-) when it gains an electron and fills the vacancy in its third, outermost shell.

FIGURE 2.3 Ion formation.

TAKE-HOME MESSAGE 2.2

An atom's electrons are the basis of its chemical behavior.

Shells represent all electron orbitals at one energy level in an atom. When the outermost shell is not full of electrons, the atom has a vacancy.

Atoms with vacancies tend to interact with other atoms.

free radical Atom with an unpaired electron.
ion Charged atom.
shell model Model of electron distribution in an atom.

2.3 HOW DO ATOMS INTERACT IN CHEMICAL BONDS?

An atom can get rid of vacancies by participating in a chemical bond with another atom. A **chemical bond** is an attractive force that arises between two atoms when their electrons interact. Chemical bonds link atoms into molecules. In other words, each molecule consists of atoms held together in a particular number and arrangement by chemical bonds. For example, a water molecule consists of three atoms: two hydrogen atoms bonded to the same oxygen atom (**FIGURE 2.4**). A water molecule is also a **compound**, which means it has atoms of two or more elements. Other molecules, including molecular oxygen (a gas in air), have atoms of one element only.

FIGURE 2.4 The water molecule. Each water molecule has two hydrogen atoms bonded to the same oxygen atom.

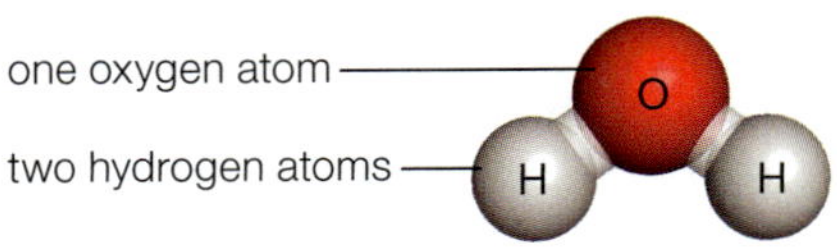

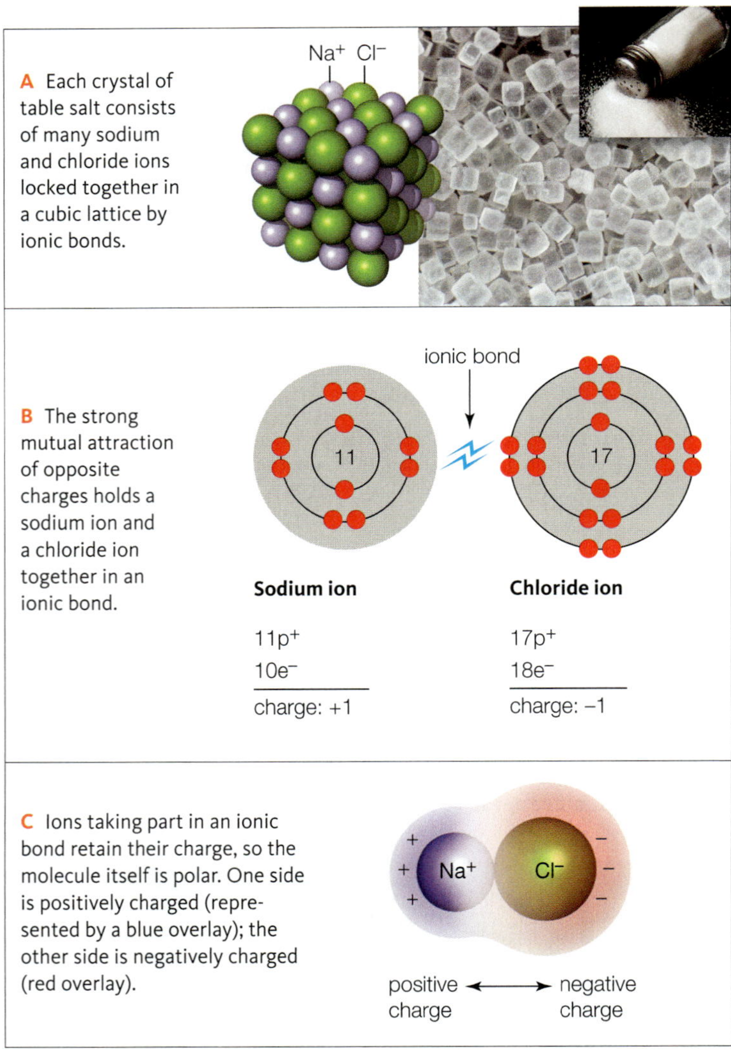

A Each crystal of table salt consists of many sodium and chloride ions locked together in a cubic lattice by ionic bonds.

B The strong mutual attraction of opposite charges holds a sodium ion and a chloride ion together in an ionic bond.

C Ions taking part in an ionic bond retain their charge, so the molecule itself is polar. One side is positively charged (represented by a blue overlay); the other side is negatively charged (red overlay).

FIGURE 2.5 Ionic bonds in table salt, or NaCl.

The term "bond" applies to a continuous range of atomic interactions. However, we can categorize most bonds into distinct types based on their properties. Which type forms depends on the atoms taking part in the molecule.

IONIC BONDS

Two ions may stay together by the mutual attraction of their opposite charges, an association called an **ionic bond**. Ionic bonds can be quite strong. Ionically bonded sodium and chloride ions make up sodium chloride (NaCl), which we know as common table salt. A crystal of this substance consists of a cubic lattice of sodium and chloride ions interacting in ionic bonds (**FIGURE 2.5A**).

Ions retain their respective charges when participating in an ionic bond (**FIGURE 2.5B**). Thus, one "end" of an ionic bond has a positive charge, and the other "end" has a negative charge. Any such separation of charge into distinct positive and negative regions is called **polarity** (**FIGURE 2.5C**). A sodium chloride molecule is polar because the chloride ion keeps a very strong hold on its extra electron. In other words, it is strongly electronegative. **Electronegativity** is a measure of an atom's ability to pull electrons away from another atom. Electronegativity is not the same thing as charge. Rather, an atom's electronegativity depends on its size, how many vacancies it has, and what other atoms it is interacting with.

An ionic bond is very polar because the atoms that are participating in it have a very large difference in electronegativity. When atoms with a lower difference in electronegativity interact, they tend to form chemical bonds that are less polar than ionic bonds.

COVALENT BONDS

Covalent bonds form between atoms with a small difference in electronegativity or none at all. In a **covalent bond**, two atoms share a pair of electrons, so each atom's vacancy is partially filled (**FIGURE 2.6**). Sharing electrons links the two atoms, just as sharing a pair of earphones links two friends (*above*). Covalent bonds can be stronger than ionic bonds, but they are not always so.

TABLE 2.1 shows different ways of representing covalent bonds. In structural formulas, a line between two

TABLE 2.1

Ways of Representing Molecules

Common name:	Water	Familiar term.
Chemical name:	Dihydrogen monoxide	Describes elemental composition.
Chemical formula:	H_2O	Indicates unvarying proportions of elements. Subscripts show number of atoms of an element per molecule. The absence of a subscript means one atom.
Structural formula:	H—O—H	Represents each covalent bond as a single line between atoms.
Structural model:		Shows relative sizes and positions of atoms in three dimensions.
Shell model:	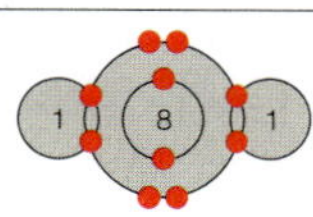	Shows how pairs of electrons are shared in covalent bonds.

MOLECULAR HYDROGEN (H—H)

Two hydrogen atoms, each with one proton, share two electrons in a nonpolar covalent bond.

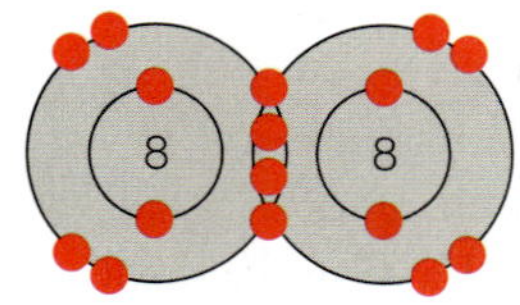

MOLECULAR OXYGEN (O=O)

Two oxygen atoms, each with eight protons, share four electrons in a double covalent bond.

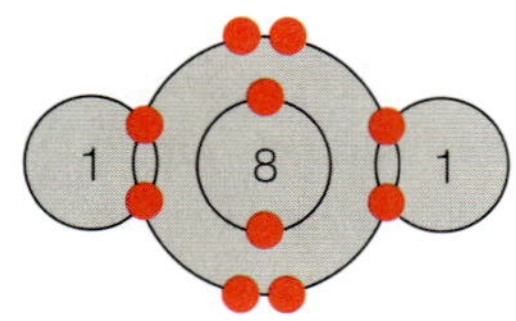

WATER (H—O—H)

Two hydrogen atoms share electrons with an oxygen atom in two covalent bonds. The bonds are polar because the oxygen exerts a greater pull on the shared electrons than the hydrogens do.

FIGURE 2.6 {Animated} Covalent bonds, in which atoms fill vacancies by sharing electrons. Two electrons are shared in each covalent bond. When sharing is equal, the bond is nonpolar. When one atom exerts a greater pull on the electrons, the bond is polar.

atoms represents a single covalent bond, in which two atoms share one pair of electrons. For example, molecular hydrogen (H_2) has one covalent bond between hydrogen atoms (H—H). Two, three, or even four covalent bonds may form between atoms when they share multiple pairs of electrons. For example, two atoms sharing two pairs of electrons are connected by two covalent bonds, which are represented by a double line between the atoms. A double bond links the two oxygen atoms in molecular oxygen (O=O). Three lines indicate a triple bond, in which two atoms share three pairs of electrons. A triple covalent bond links the two nitrogen atoms in molecular nitrogen (N≡N). Comparing bonds between the same two atoms: A triple bond is stronger than a double bond, which is stronger than a single bond.

Double and triple bonds are not distinguished from single bonds in structural models, which show positions and relative sizes of the atoms in three dimensions. The bonds are shown as one stick connecting two balls, which represent atoms. Elements are usually coded by color:

carbon hydrogen oxygen nitrogen phosphorus

Atoms share electrons unequally in a polar covalent bond. A bond between an oxygen atom and a hydrogen atom in a water molecule is an example. One atom (the oxygen, in this case) is a bit more electronegative. It pulls the electrons a little more toward its side of the bond, so that atom bears a slight negative charge. The atom at the other end of the bond (the hydrogen) bears a slight positive charge. Covalent bonds in compounds are usually polar. By contrast, atoms participating in a nonpolar covalent bond share electrons equally. There is no difference in charge between the two ends of such bonds. The bonds in molecular hydrogen (H_2), oxygen (O_2), and nitrogen (N_2) are nonpolar.

chemical bond An attractive force that arises between two atoms when their electrons interact.
compound Molecule that has atoms of more than one element.
covalent bond Chemical bond in which two atoms share a pair of electrons.
electronegativity Measure of the ability of an atom to pull electrons away from other atoms.
ionic bond Type of chemical bond in which a strong mutual attraction links ions of opposite charge.
polarity Separation of charge into positive and negative regions.

TAKE-HOME MESSAGE 2.3

A chemical bond forms between atoms when their electrons interact. A chemical bond may be ionic or covalent depending on the atoms taking part in it.

An ionic bond is a strong mutual attraction between two ions of opposite charge. Ionic bonds are very polar.

Atoms share a pair of electrons in a covalent bond. When the atoms share electrons unequally, the bond is polar.

2.4 WHAT ARE LIFE-SUSTAINING PROPERTIES OF WATER?

HYDROGEN BONDING IN WATER

Water has unique properties that arise from the two polar covalent bonds in each water molecule. Overall, the molecule has no charge, but the oxygen atom carries a slight negative charge; the hydrogen atoms, a slight positive charge. Thus, the molecule itself is polar (**FIGURE 2.7A**).

The polarity of individual water molecules attracts them to one another. The slight positive charge of a hydrogen atom in one water molecule is drawn to the slight negative charge of an oxygen atom in another. This type of interaction is called a hydrogen bond. A **hydrogen bond** is an attraction between a covalently bonded hydrogen atom and another atom taking part in a separate polar covalent bond (**FIGURE 2.7B**). Like ionic bonds, hydrogen bonds form by the mutual attraction of opposite charges. However, unlike ionic bonds, hydrogen bonds do not make molecules out of atoms, so they are not chemical bonds.

Hydrogen bonds are on the weaker end of the spectrum of atomic interactions, and they form and break much more easily than covalent or ionic bonds. Even so, many of them form, and collectively they are quite strong. As you will see, hydrogen bonds stabilize the characteristic structures of biological molecules such as DNA and proteins. They also form in tremendous numbers among water molecules (**FIGURE 2.7C**). Extensive hydrogen bonding among water molecules gives liquid water several special properties that make life possible.

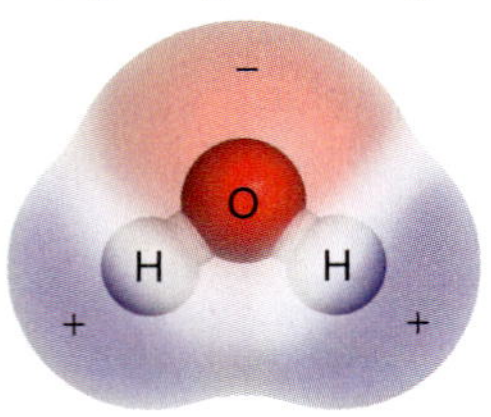

A Polarity of the water molecule. Each of the hydrogen atoms in a water molecule bears a slight positive charge (represented by a blue overlay). The oxygen atom carries a slight negative charge (red overlay).

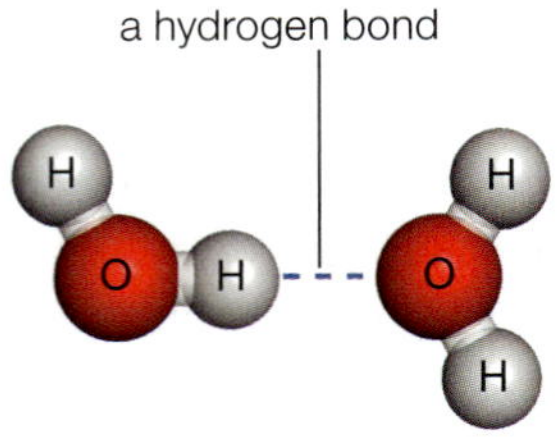

B A hydrogen bond is an attraction between a hydrogen atom and another atom taking part in a separate polar covalent bond.

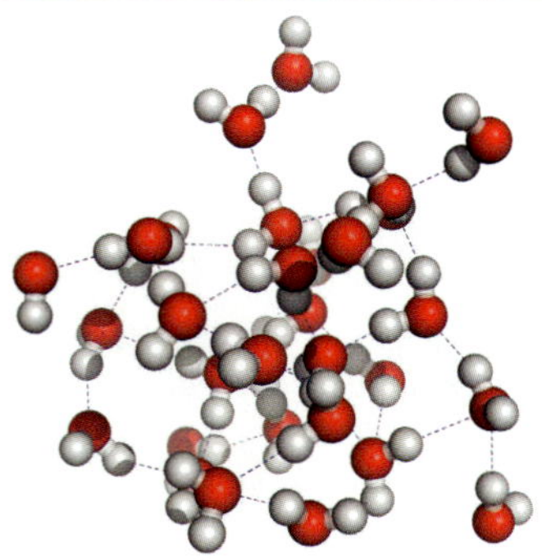

C The many hydrogen bonds that form among water molecules impart special properties to liquid water.

FIGURE 2.7 **{Animated}** **Hydrogen bonds** in water.

WATER'S SPECIAL PROPERTIES

Water Is an Excellent Solvent

The polarity of the water molecule and its ability to form hydrogen bonds make water an excellent **solvent**, which means that many other substances easily dissolve in it. Substances that dissolve easily in water are **hydrophilic** (water-loving). Ionic solids such as sodium chloride (NaCl) dissolve in water because the slight positive charge on each hydrogen atom in a water molecule attracts negatively charged ions (Cl^-), and the slight negative charge on the oxygen atom attracts positively charged ions (Na^+). Hydrogen bonds among many water molecules are collectively stronger than an ionic bond between two ions, so the solid dissolves as water molecules tug the ions apart and surround each one (*right*).

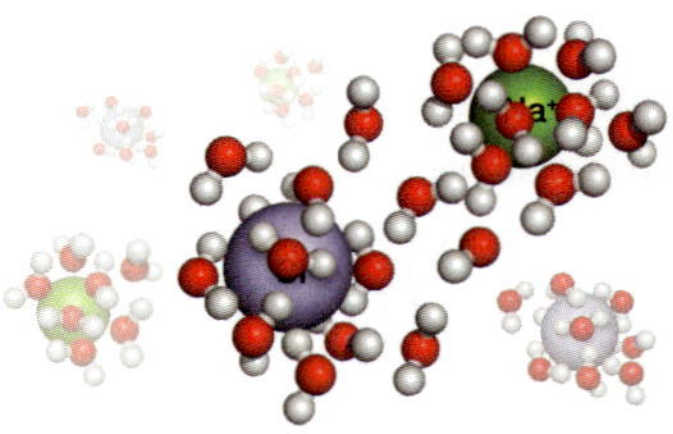

Sodium chloride is called a **salt** because it releases ions other than H^+ and OH^- when it dissolves in water (more about these ions in the next section). When a substance such as NaCl dissolves, its component ions disperse uniformly among the molecules of liquid, and it becomes a **solute**. A uniform mixture such as salt dissolved in water is called a **solution**. Chemical bonds do not form between molecules of solute and solvent, so the proportions of the two substances in a solution can vary.

Nonionic solids such as sugars dissolve easily in water because their molecules can form hydrogen bonds with water molecules. Hydrogen bonding with water does not break the covalent bonds of such molecules; rather, it dissolves the substance by pulling individual molecules away from one another and keeping them apart.

Water does not interact with **hydrophobic** (water-dreading) substances such as oils. Oils consist of nonpolar molecules, and hydrogen bonds do not form between nonpolar molecules and water. When you mix oil and water, the water breaks into small droplets, but quickly begins to cluster into larger drops as new hydrogen bonds form among its molecules. The bonding excludes molecules of oil and pushes them together into drops that rise to the surface of the water. The same interactions occur at the thin, oily membrane that separates the watery fluid inside cells from the watery fluid outside of them. As you will see in Chapter 3, such interactions give rise to the structure of cell membranes.

Water Has Cohesion Molecules of some substances resist separating from one another, and the resistance gives rise to a property called **cohesion**. Water has cohesion because hydrogen bonds collectively exert a continuous pull on its individual molecules. You can see cohesion in water as surface tension, which means that the surface of liquid water behaves a bit like a sheet of elastic (*left*).

Cohesion plays a role in many processes that sustain multicelled bodies. As one example, water molecules constantly escape from the surface of liquid water as vapor, a process called **evaporation**. Evaporation is resisted by hydrogen bonding among water molecules. In other words, overcoming water's cohesion takes energy. Thus, evaporation sucks energy (in the form of heat) from liquid water, and this lowers the water's surface temperature. Evaporative water loss helps you and some other mammals cool off when you sweat in hot, dry weather. Sweat, which is about 99 percent water, cools the skin as it evaporates.

Cohesion works inside organisms, too. Consider how plants absorb water from soil as they grow. Water molecules evaporate from leaves, and replacements are pulled upward from roots. Cohesion makes it possible for columns of liquid water to rise from roots to leaves inside narrow pipelines of vascular tissue. In some trees, these pipelines extend hundreds of feet above the soil.

cohesion Property of a substance that arises from the tendency of its molecules to resist separating from one another.
evaporation Transition of a liquid to a vapor.
hydrogen bond Attraction between a covalently bonded hydrogen atom and another atom taking part in a separate covalent bond.
hydrophilic Describes a substance that dissolves easily in water.
hydrophobic Describes a substance that resists dissolving in water.
salt Compound that releases ions other than H^+ and OH^- when it dissolves in water.
solute A dissolved substance.
solution Uniform mixure of solute completely dissolved in solvent.
solvent Liquid that can dissolve other substances.
temperature Measure of molecular motion.

FIGURE 2.8 {Animated} Hydrogen bonds lock water molecules in a rigid lattice in ice. The molecules in this lattice pack less densely than in liquid water, which is why ice floats on water. A covering of ice can insulate water underneath it, thus keeping aquatic organisms from freezing during harsh winters.

Water Stabilizes Temperature All atoms jiggle nonstop, so the molecules they make up jiggle too. We measure the energy of this motion as degrees of **temperature**. Adding energy (in the form of heat, for example) makes the jiggling faster, so the temperature rises.

Hydrogen bonding keeps water molecules from jiggling as much as they would otherwise, so it takes more heat to raise the temperature of water compared with other liquids. Temperature stability is an important part of homeostasis, because most of the molecules of life function properly only within a certain range of temperature.

Below 0°C (32°F), water molecules do not jiggle enough to break hydrogen bonds, and they become locked in the rigid, lattice-like bonding pattern of ice (**FIGURE 2.8**). Individual water molecules pack less densely in ice than they do in water, which is why ice floats on water. Sheets of ice that form on the surface of ponds, lakes, and streams can insulate the water under them from subfreezing air temperatures. Such "ice blankets" protect aquatic organisms during cold winters.

TAKE-HOME MESSAGE 2.4

Extensive hydrogen bonding among water molecules, which arises from the polarity of the individual molecules, gives water special properties.

Liquid water is an excellent solvent. Hydrophilic substances such as salts and sugars dissolve easily in water to form solutions. Hydrophobic substances do not dissolve in water.

Water also has cohesion, and it stabilizes temperature.

2.5 WHY ARE HYDROGEN IONS IMPORTANT IN BIOLOGICAL SYSTEMS?

FIGURE 2.9 A pH scale. Here, red dots signify hydrogen ions (H^+) and blue dots signify hydroxyl ions (OH^-). Also shown are the approximate pH values for some common solutions.

This pH scale ranges from 0 (most acidic) to 14 (most basic). A change of one unit on the scale corresponds to a tenfold change in the amount of H^+ ions.

FIGURE IT OUT: What is the approximate pH of cola?

Answer: 2.5

When water is liquid, some of its molecules spontaneously separate into hydrogen ions (H^+) and hydroxide ions (OH^-). These ions can combine again to form water:

$$\underset{\text{water}}{H_2O} \longrightarrow \underset{\text{hydrogen ions}}{H^+} + \underset{\text{hydroxide ions}}{OH^-} \longrightarrow \underset{\text{water}}{H_2O}$$

Concentration refers to the amount of a particular solute dissolved in a given volume of fluid. Hydrogen ion (H^+) concentration is a special case. We measure the amount of hydrogen ions in a solution using a value called **pH**. When the number of H^+ ions equals the number of OH^- ions in the liquid, the pH is 7, or neutral. The higher the number of hydrogen ions, the lower the pH. A one-unit decrease in pH corresponds to a tenfold increase in the number of H^+ ions, and a one-unit increase corresponds to a tenfold decrease in the number of H^+ ions (**FIGURE 2.9**).

One way to get a sense of the pH scale is to taste dissolved baking soda (pH 9), pure water (pH 7), and lemon juice (pH 2). Nearly all of life's chemistry occurs near pH 7. Most of your body's internal environment (tissue fluids and blood) stays between pH 7.3 and 7.5.

Substances called **bases** accept hydrogen ions, so they can raise the pH of fluids and make them basic, or alkaline (above pH 7). **Acids** give up hydrogen ions when they dissolve in water, so they lower the pH of fluids and make them acidic (below pH 7).

Strong acids ionize completely in water to give up all of their H^+ ions; weak acids give up only some of them. Hydrochloric acid (HCl) is an example of a strong acid: its H^+ and Cl^- ions stay separated in water. Inside your stomach, the H^+ from HCl makes gastric fluid acidic (pH 1–2). Carbonic acid is an example of a weak acid. It forms when carbon dioxide gas (CO_2) dissolves in the watery, fluid portion of human blood:

$$\underset{\text{carbon dioxide}}{CO_2} + H_2O \longrightarrow \underset{\text{carbonic acid}}{H_2CO_3}$$

A carbonic acid molecule can break apart into a hydrogen ion and a bicarbonate ion, which in turn can recombine to form carbonic acid again:

$$\underset{\text{carbonic acid}}{H_2CO_3} \longrightarrow H^+ + \underset{\text{bicarbonate}}{HCO_3^-} \longrightarrow \underset{\text{carbonic acid}}{H_2CO_3}$$

Together, carbonic acid and bicarbonate constitute a **buffer**, a set of chemicals that can keep the pH of a solution stable by alternately donating and accepting ions that contribute to pH. For example, when a base is added to an unbuffered fluid, the number of OH^- ions increases, so the pH rises. However, if the fluid is buffered, the addition of base causes the buffer to release H^+ ions. These combine with OH^- ions to form water, which has no effect on pH. Excess hydrogen ions combine with the buffer, so they do not contribute to pH. Thus, the pH of a buffered fluid stays the same when base or acid is added.

Under normal circumstances, the fluids inside cells (as well as those inside bodies) stay within a consistent range of pH because they are buffered. For example, excess OH^- in blood combines with the H^+ from carbonic acid to form water, which does not contribute to pH. Excess H^+ in blood combines with bicarbonate, so it does not affect pH. This exchange of ions keeps the blood pH stable, but only up to a certain point. A buffer can neutralize only so many ions; even slightly more than that limit and the pH of the fluid will change dramatically.

Most biological molecules can function properly only within a narrow range of pH. Even a slight deviation from that range can halt cellular processes, so buffer failure can be catastrophic in a biological system. For instance, when breathing is impaired suddenly, carbon dioxide gas accumulates in tissues, so too much carbonic acid forms in blood. The resulting decline in blood pH may cause the person to enter a coma (a dangerous level of unconsciousness). By contrast, hyperventilation (sustained rapid breathing) causes the body to lose too much CO_2. The loss results in a rise in blood pH. If blood pH rises too much, prolonged muscle spasm (tetany) or coma may occur.

Burning fossil fuels such as coal releases sulfur and nitrogen compounds that affect the pH of rain and other forms of precipitation. Rainwater is not buffered, so the addition of acids or bases has a dramatic effect. In places with a lot of fossil fuel emissions, the rain and fog can be more acidic than vinegar. The corrosive effect of this acid rain is visible in urban areas (*left*). Acid rain also drastically changes the pH of water in soil, lakes, and streams. Such changes can overwhelm the buffering capacity of fluids inside organisms, with lethal effects. We return to acid rain in Section 30.4.

acid Substance that releases hydrogen ions in water.
base Substance that accepts hydrogen ions in water.
buffer Set of chemicals that can keep the pH of a solution stable by alternately donating and accepting ions that contribute to pH.
concentration Amount of solute per unit volume of solution.
pH Measure of the number of hydrogen ions in a fluid.

TAKE-HOME MESSAGE 2.5

The number of hydrogen ions in a fluid determines its pH. Most biological systems function properly only within a narrow range of pH.

Acids release hydrogen ions in water; bases accept them. Salts release ions other than H^+ and OH^-.

Buffers help keep pH stable. Inside organisms, they are part of homeostasis.

CREDITS: (in text) W. K. Fletcher/Science Source; (10) Brian J. Skerry/National Geographic Creative.

2.6 MERCURY RISING

FIGURE 2.10 Mercury that falls on Earth's oceans accumulates in the bodies of tuna and other large predatory fish.

MERCURY IS A TOXIC ELEMENT. Most of it is safely locked away in rocks, but volcanic activity and other geologic processes release it into the atmosphere. So do human activities, especially burning coal. Airborne mercury can drift long distances before settling to Earth's surface, where microbes combine it with carbon to form a substance called methylmercury.

Unlike mercury alone, methylmercury easily crosses skin and mucous membranes. In water, it ends up in the tissues of aquatic organisms. All fish and shellfish contain it. Humans contain it too, mainly as a result of eating seafood. When mercury enters the body, it damages the nervous system, brain, kidneys, and other organs. An average-sized adult who ingests as little as 200 micrograms of methylmercury may experience blurred vision, tremors, itching or burning sensations, and loss of coordination. Exposure to larger amounts can result in thought and memory impairment, coma, and death. Methylmercury in a pregnant woman's blood passes to her unborn child, along with a legacy of permanent developmental problems.

It takes months or even years for mercury to be cleared from the body, so the toxin can build up to high levels if even small amounts are ingested on a regular basis. That is why large predatory fish have a lot of mercury in their tissues (FIGURE 2.10). It is also why the U.S. Environmental Protection Agency recommends that adult humans ingest less than 0.1 microgram of mercury per kilogram of body weight per day. For an average-sized person, that limit works out to be about 7 micrograms per day, which is not a big amount if you eat seafood. A typical 6-ounce can of albacore tuna contains about 60 micrograms of mercury, and the occasional can has many times that amount. It does not matter if the fish is canned, grilled, or raw, because methylmercury is unaffected by cooking. Eat a medium-sized tuna steak, and you could be getting more than 700 micrograms of mercury along with it.

Summary

SECTION 2.1 Atoms consist of **electrons**, which carry a negative **charge**, moving about a **nucleus** of positively charged **protons** and uncharged **neutrons** (**TABLE 2.2**). The **periodic table** lists **elements** in order of **atomic number**. **Isotopes** of an element differ in the number of neutrons. The total number of protons and neutrons is the **mass number**. **Tracers** can be made with **radioisotopes**, which, by a process called **radioactive decay**, emit particles and energy when their nucleus spontaneously breaks up.

SECTION 2.2 Which atomic orbital an electron occupies depends on its energy. A **shell model** represents successive energy levels as concentric circles. Atoms tend to get rid of vacancies. Many do so by gaining or losing electrons, thereby becoming **ions**. Unpaired electrons make **free radicals** chemically active.

SECTION 2.3 A **chemical bond** is an attractive force that unites two atoms as a molecule. A **compound** consists of two or more elements. Atoms form different types of bonds depending on their **electronegativity**. The mutual attraction of opposite charges can hold atoms together in an **ionic bond**, which is completely polar (**polarity** is separation of charge). Atoms share a pair of electrons in a **covalent bond**, which is nonpolar if the sharing is equal, and polar if it is not.

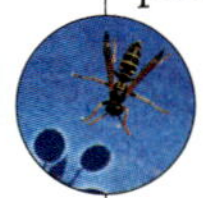

SECTION 2.4 Two polar covalent bonds give each water molecule an overall polarity. **Hydrogen bonds** that form among water molecules in tremendous numbers are the basis of water's unique properties. Water has **cohesion** and a capacity to act as a **solvent** for **salts** and other polar **solutes**; and it resists **temperature** changes. **Hydrophilic** substances dissolve easily in water to form **solutions**; **hydrophobic** substances do not. **Evaporation** is the transition of liquid to vapor.

SECTION 2.5 A solute's **concentration** refers to the amount of solute in a given volume of fluid; **pH** reflects the number of hydrogen ions (H^+). **Acids** release hydrogen ions in water; **bases** accept them. A **buffer** can keep a solution within a consistent range of pH. Most cell and body fluids are buffered because most molecules of life work only within a narrow range of pH.

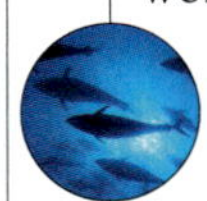

SECTION 2.6 Interactions between atoms make the molecules that sustain life, and also some that destroy it. Mercury in air pollution ends up in the bodies of fish, and in turn, in the bodies of humans.

TABLE 2.2

Players in the Chemisty of Life

Atoms	Particles that are basic building blocks of all matter.
Proton (p^+)	Positively charged particle of an atom's nucleus.
Electron (e^-)	Negatively charged particle that can occupy a defined volume of space (orbital) around an atom's nucleus.
Neutron	Uncharged particle of an atom's nucleus.
Element	Pure substance that consists entirely of atoms with the same, characteristic number of protons.
Isotopes	Atoms of an element that differ in the number of neutrons.
Radioisotope	Unstable isotope that emits particles and energy when its nucleus breaks up.
Tracer	Molecule that has a detectable component such as a radioisotope. Used to track the movement or destination of the molecule in a biological system.
Ion	Atom that carries a charge after it has gained or lost one or more electrons. A single proton without an electron is a hydrogen ion (H^+).
Molecule	Two or more atoms joined in a chemical bond.
Compound	Molecule of two or more different elements in unvarying proportions (for example, water: H_2O).
Solute	Substance dissolved in a solvent.
Hydrophilic	Refers to a substance that dissolves easily in water. Such substances consist of polar molecules.
Hydrophobic	Refers to a substance that resists dissolving in water. Such substances consist of nonpolar molecules.
Acid	Compound that releases H^+ when dissolved in water.
Base	Compound that accepts H^+ when dissolved in water.
Salt	Ionic compound that releases ions other than H^+ or OH^- when dissolved in water.
Solvent	Substance that can dissolve other substances.

Self-Quiz

Answers in Appendix VII

1. What atom has only one proton?
   a. hydrogen
   b. an isotope
   c. a free radical
   d. a radioisotope

2. A molecule into which a radioisotope has been incorporated can be used as a(n) _______ .
   a. compound
   b. tracer
   c. salt
   d. acid

3. Which of the following statements is incorrect?
   a. Isotopes have the same atomic number and different mass numbers.
   b. Atoms have about the same number of electrons as protons.
   c. All ions are atoms.
   d. Free radicals are dangerous because they emit energy.

4. In the periodic table, symbols for the elements are arranged according to _______ .
   a. size
   b. charge
   c. mass number
   d. atomic number

5. An ion is an atom that has _______ .
   a. the same number of electrons and protons
   b. a different number of electrons and protons
   c. electrons, protons, and neutrons

Data Analysis Activities

Radioisotopes in PET Scans Positron-emission tomography (PET) helps us "see" a functional process inside the body. By this procedure, a radioactive sugar or other tracer is injected into a patient, who is then moved into a scanner. Inside the patient's body, cells with differing rates of activity take up the tracer at different rates. The scanner detects radioactive decay wherever the tracer is, then translates that data into an image.

1. What conclusion does **FIGURE 2.11** and its caption suggest about the behavior of a smoker compared with that of a nonsmoker?
2. What is an alternate interpretation of the differences between the results of these two PET scans?
3. This experiment compares two human individuals, who undoubtedly differ in factors other than smoking. What would be an appropriate control for this study?

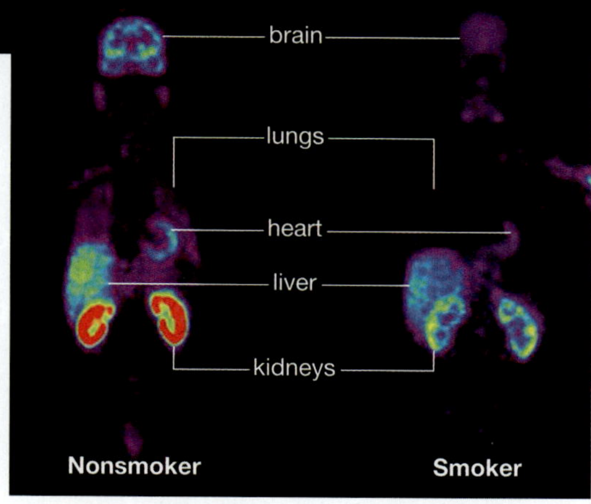

FIGURE 2.11 Two PET scans showing the activity of a molecule called MAO-B in the body of a nonsmoker (*left*) and a smoker (*right*). The activity is color-coded from red (highest activity) to purple (lowest). Low MAO-B activity is associated with violence, impulsiveness, and other behavioral problems.

6. The measure of an atom's ability to pull electrons away from another atom is called _______ .
   a. electronegativity b. charge c. polarity
7. The mutual attraction of opposite charges holds atoms together as molecules in a(n) _______ bond.
   a. ionic c. polar covalent
   b. hydrogen d. nonpolar covalent
8. Atoms share electrons unequally in a(n) _______ bond.
   a. ionic c. polar covalent
   b. hydrogen d. nonpolar covalent
9. A(n) _______ substance repels water.
   a. acidic c. hydrophobic
   b. basic d. polar
10. A salt does not release _______ in water.
   a. ions b. energy c. H^+
11. Hydrogen ions (H^+) are _______ .
   a. indicated by a pH scale c. in blood
   b. protons d. all of the above
12. When dissolved in water, a(n) _______ donates H^+; a(n) _______ accepts H^+.
   a. acid; base c. buffer; solute
   b. base; acid d. base; buffer
13. A(n) _______ can help keep the pH of a solution stable.
   a. covalent bond c. buffer
   b. hydrogen bond d. pH
14. A _______ is dissolved in a solvent.
   a. molecule b. solute c. salt
15. Match the terms with their most suitable description.

___ hydrophilic	a. protons > electrons
___ atomic number	b. number of protons in nucleus
___ hydrogen bonds	c. polar; dissolves easily in water
___ positive charge	d. collectively strong
___ temperature	e. protons < electrons
___ negative charge	f. measure of molecular motion

Critical Thinking

1. Alchemists were medieval scholars and philosophers who were the forerunners of modern-day chemists. Many spent their lives trying to transform lead (atomic number 82) into gold (atomic number 79). Explain why they never did succeed in that endeavor.

2. Draw a shell model of a lithium atom (Li), which has 3 protons, then predict whether the majority of lithium atoms on Earth are uncharged, positively charged, or negatively charged.

3. Polonium is a rare element with 33 radioisotopes. The most common one, ^{210}Po, has 82 protons and 128 neutrons. When ^{210}Po decays, it emits an alpha particle, which is a helium nucleus (2 protons and 2 neutrons). ^{210}Po decay is tricky to detect because alpha particles do not carry very much energy compared to other forms of radiation. They can be stopped by, for example, a sheet of paper or a few inches of air. This property is one reason why authorities failed to discover toxic amounts of ^{210}Po in the body of former KGB agent Alexander Litvinenko until after he died suddenly and mysteriously in 2006. What element does an atom of ^{210}Po change into after it emits an alpha particle?

4. Some undiluted acids are not as corrosive as when they are diluted with water. That is why lab workers are told to wipe off splashes with a towel before washing. Explain.

Rice has been cultivated for thousands of years. Carbohydrate-packed seeds make this grain the most important food source for humans worldwide.

3

MOLECULES OF LIFE

Links to Earlier Concepts

Having learned about atomic interactions (Section 2.3), you are now in a position to understand the structure of the molecules of life. Keep the big picture in mind by reviewing Section 1.1. You will be building on your knowledge of covalent bonding (2.3), acids and bases (2.5), and the effects of hydrogen bonds (2.4).

KEY CONCEPTS

STRUCTURE DICTATES FUNCTION

Complex carbohydrates and lipids, proteins, and nucleic acids are assembled from simpler molecules. Functional groups add chemical character to a backbone of carbon atoms.

CARBOHYDRATES

Cells use carbohydrates as structural materials, for fuel, and to store and transport energy. They can build different complex carbohydrates from the same simple sugars.

LIPIDS

Lipids are the main structural component of all cell membranes. Cells use them to make other compounds, to store energy, and as waterproofing or lubricating substances.

PROTEINS

Proteins are the most diverse molecules of life. They include enzymes and structural materials. A protein's function arises from and depends on its structure.

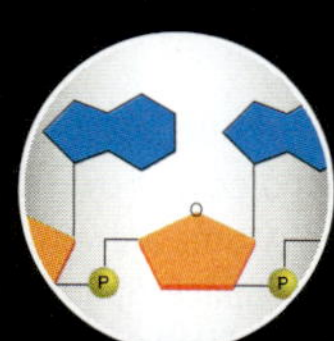

NUCLEIC ACIDS

Nucleotides are building blocks of nucleic acids; some have additional roles in metabolism. DNA stores a cell's heritable information. RNA helps put that information to use.

Photograph by Alex Treadway, National Geographic Creative.

3.1 WHAT ARE THE MOLECULES OF LIFE?

THE STUFF OF LIFE: CARBON

The same elements that make up a living body also occur in nonliving things, but their proportions differ. For example, compared to sand or seawater, a human body contains a much larger proportion of carbon atoms. Why? Unlike sand or seawater, a body consists of a very high proportion of the molecules of life—complex carbohydrates and lipids, proteins, and nucleic acids—which in turn consist of a high proportion of carbon atoms. Molecules that have primarily hydrogen and carbon atoms are said to be **organic**. The term is a holdover from a time when these molecules were thought to be made only by living things, as opposed to the "inorganic" molecules that formed by nonliving processes.

Carbon's importance to life arises from its versatile bonding behavior. Carbon has four vacancies (Section 2.2), so it can form four covalent bonds with other atoms, including other carbon atoms. Many organic molecules have a backbone—a chain of carbon atoms—to which other atoms attach. The ends of a backbone may join to form a carbon ring structure (**FIGURE 3.1**). Carbon's ability to form chains and rings, and also to bond with many other elements, means that atoms of this element can be assembled into a wide variety of organic compounds.

We represent organic molecules in several ways. The structure of many organic molecules is quite complex (**FIGURE 3.2A**). For clarity, we may omit some of the bonds in a structural formula. Hydrogen atoms bonded to a carbon backbone may also be omitted. Carbon rings are often represented as polygons (**FIGURE 3.2B**). If no atom is shown at a corner or at the end of a bond, a carbon is implied there. Ball-and-stick models are useful for representing smaller organic compounds (**FIGURE 3.2C**). Space-filling models show a molecule's overall shape (**FIGURE 3.2D**). Proteins and nucleic acids are often represented as ribbon structures, which, as you will see in Section 3.4, show how the backbone folds and twists.

FIGURE 3.1 Carbon rings.

A Carbon's versatile bonding behavior allows it to form a variety of structures, including rings.

B Carbon rings form the framework of many sugars, starches, and fats (such as those found in doughnuts).

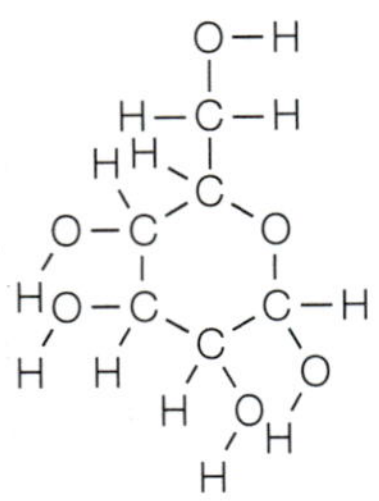

A A structural formula for an organic molecule—even a simple one—can be very complicated. The overall structure is obscured by detail.

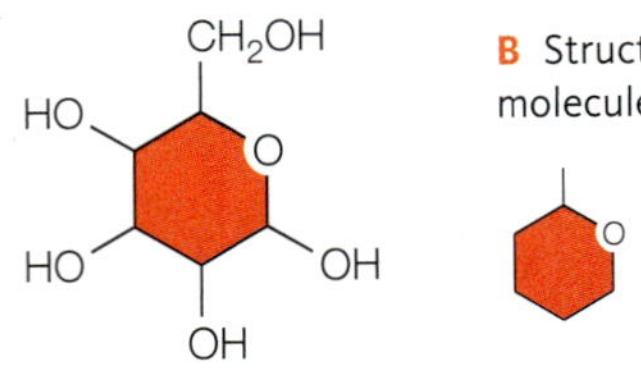

B Structural formulas of organic molecules are typically simplified by using polygons as symbols for rings, omitting some bonds and element labels.

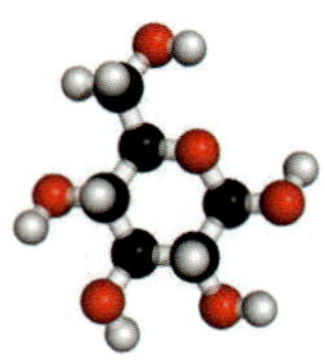

C A ball-and-stick model is often used to show the arrangement of atoms and bonds in three dimensions.

D A space-filling model can be used to show a molecule's overall shape. Individual atoms are visible in this model. Space-filling models of larger molecules often show only the surface contours.

FIGURE 3.2 **Modeling an organic molecule.** All of these models represent the same molecule: glucose.

FROM STRUCTURE TO FUNCTION

An organic molecule that consists only of hydrogen and carbon atoms is called a **hydrocarbon**. Hydrocarbons are generally nonpolar. Methane, the simplest kind, is one carbon atom bonded to four hydrogen atoms. Other organic

condensation Chemical reaction in which an enzyme builds a large molecule from smaller subunits; water also forms.
enzyme Organic molecule that speeds up a reaction without being changed by it.
functional group An atom (other than hydrogen) or a small molecular group bonded to a carbon of an organic compound; imparts a specific chemical property.
hydrocarbon Compound or region of one that consists only of carbon and hydrogen atoms.
hydrolysis Water-requiring chemical reaction in which an enzyme breaks a molecule into smaller subunits.
metabolism All of the enzyme-mediated chemical reactions by which cells acquire and use energy as they build and break down organic molecules.
monomers Molecules that are subunits of polymers.
organic Describes a molecule that consists mainly of carbon and hydrogen atoms.
polymer Molecule that consists of multiple monomers.
reaction Process of molecular change.

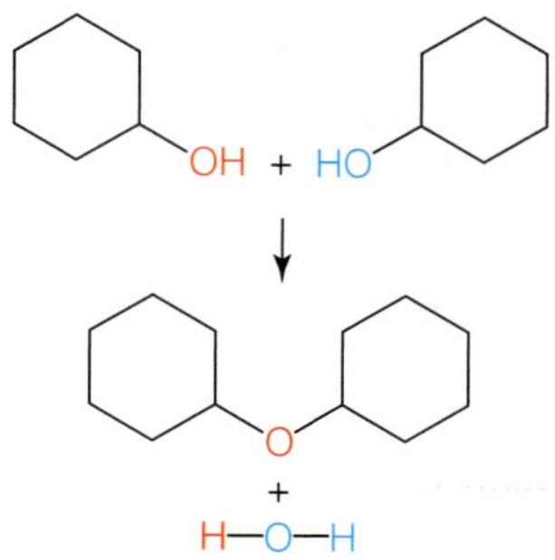

A Condensation. Cells build a large molecule from smaller ones by this reaction. An enzyme removes a hydroxyl group from one molecule and a hydrogen atom from another. A covalent bond forms between the two molecules; water also forms.

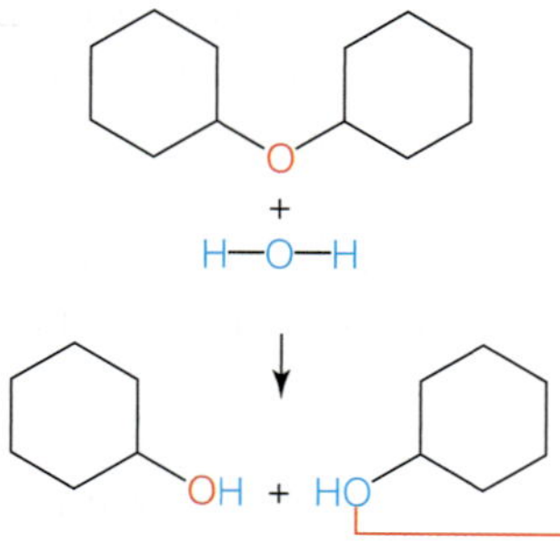

B Hydrolysis. Cells split a large molecule into smaller ones by this water-requiring reaction. An enzyme attaches a hydroxyl group and a hydrogen atom (both from water) at the cleavage site.

FIGURE 3.3 {Animated} Two common metabolic processes by which cells build and break down organic molecules.

TABLE 3.1

Some Functional Groups in Biological Molecules

Group	Structure	Character	Formula	Found in:
acetyl	—C(=O)—CH$_3$	polar, acidic	$-COCH_3$	some proteins, coenzymes
aldehyde	—C—C(=O)—H	polar, reactive	—CHO	simple sugars
amide	—C(=O)—N<	weakly basic, stable, rigid	—C(O)N—	proteins nucleotide bases
amine	—N(H)H	very basic	$-NH_2$	nucleotide bases amino acids
carboxyl	—C—C(=O)—OH	very acidic	—COOH	fatty acids amino acids
hydroxyl	—O—H	polar	—OH	alcohols sugars
ketone	—C—C(=O)—C—	polar, acidic	—CO—	simple sugars nucleotide bases
methyl	—CH$_3$	nonpolar	$-CH_3$	fatty acids some amino acids
sulfhydryl	—S—H	forms rigid disulfide bonds	—SH	cysteine many cofactors
phosphate	—O—P(=O)(OH)—OH	polar, reactive	$-PO_4$	nucleotides DNA, RNA phospholipids proteins

molecules, including the molecules of life, have at least one functional group. A **functional group** is an atom (other than hydrogen) or small molecular group covalently bonded to a carbon atom of an organic compound. These groups impart chemical properties such as acidity or polarity (**TABLE 3.1**). The chemical behavior of the molecules of life arises mainly from the number, kind, and arrangement of their functional groups.

All biological systems are based on the same organic molecules, a similarity that is one of many legacies of life's common origin. However, the details of those molecules differ among organisms. Just as atoms bonded in different numbers and arrangements form different molecules, simple organic building blocks bonded in different numbers and arrangements form different versions of the molecules of life. These small organic molecules—simple sugars, fatty acids, amino acids, and nucleotides—are called **monomers** when they are used as subunits of larger molecules. Molecules that consist of multiple monomers are called **polymers**.

Cells build polymers from monomers, and break down polymers to release monomers. These and any other processes of molecular change are called chemical **reactions**. Cells constantly run reactions as they acquire and use energy to stay alive, grow, and reproduce—activities that are collectively called **metabolism**. Metabolism requires **enzymes**, which are organic molecules (usually proteins) that speed up reactions without being changed by them.

In many metabolic reactions, large organic molecules are assembled from smaller ones. With **condensation**, an enzyme covalently bonds two molecules together. Water (H—O—H) usually forms as a product of condensation when a hydroxyl group (—OH) from one of the molecules combines with a hydrogen atom (—H) from the other molecule (**FIGURE 3.3A**). With **hydrolysis**, the reverse of condensation, an enzyme breaks apart a large organic molecule into smaller ones. During hydrolysis, a bond between two atoms breaks when a hydroxyl group gets attached to one of the atoms, and a hydrogen atom gets attached to the other (**FIGURE 3.3B**). The hydroxyl group and hydrogen atom come from a water molecule, so this reaction requires water.

We will revisit enzymes and metabolic reactions in Chapter 5. The remainder of this chapter introduces the different types of biological molecules and the monomers from which they are built.

TAKE-HOME MESSAGE 3.1

The molecules of life are organic, which means they consist mainly of carbon and hydrogen atoms. Functional groups bonded to their carbon backbone impart chemical characteristics to these molecules.

Cells assemble large polymers from smaller monomer molecules. They also break apart polymers into monomers.

3.2 WHAT IS A CARBOHYDRATE?

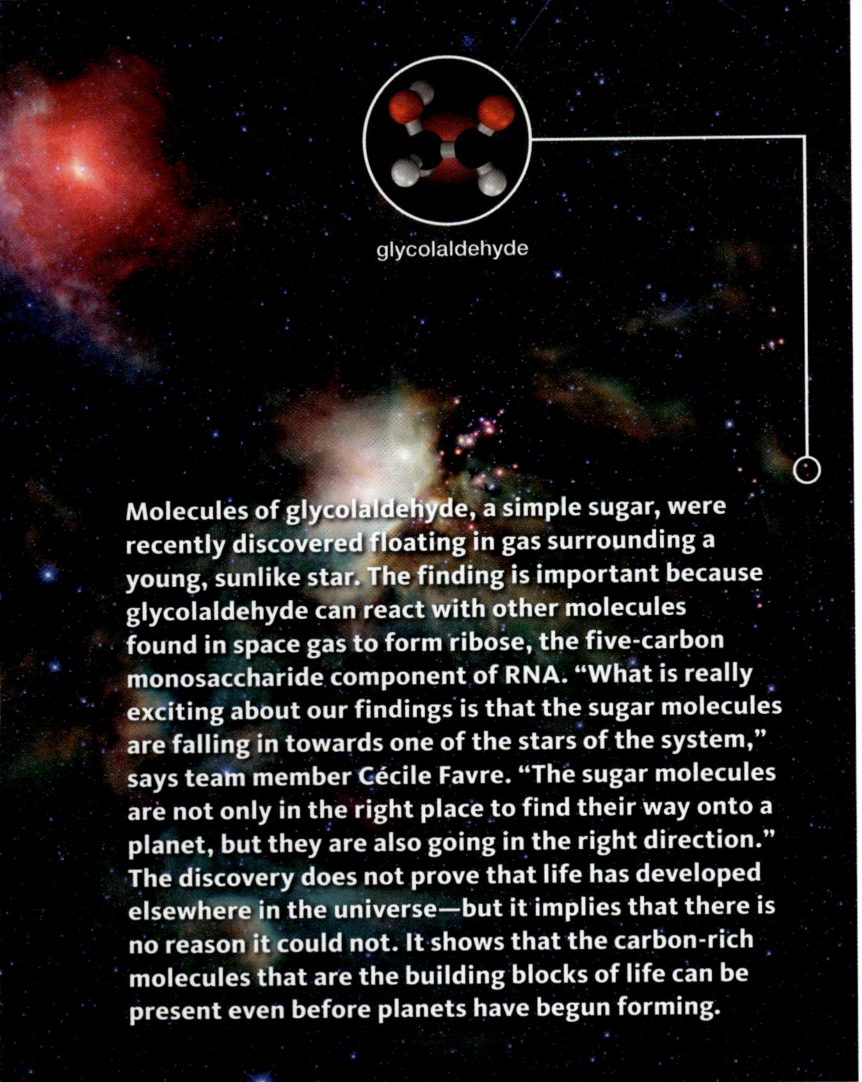

FIGURE 3.4 Astronomers made a sweet discovery in 2012.

Carbohydrates are organic compounds that consist of carbon, hydrogen, and oxygen in a 1:2:1 ratio. Cells use different kinds as structural materials, for fuel, and for storing and transporting energy. The three main types of carbohydrates in living systems are monosaccharides, oligosaccharides, and polysaccharides.

SIMPLE SUGARS

"Saccharide" is from *sacchar*, a Greek word that means sugar. Monosaccharides (one sugar) are the simplest type of carbohydrate. These molecules have extremely important biological roles. Common monosaccharides have a backbone of five or six carbon atoms (carbon atoms of sugars are numbered in a standard way: $1'$, $2'$, $3'$, and so on, as illustrated in the model of glucose on the *right*).

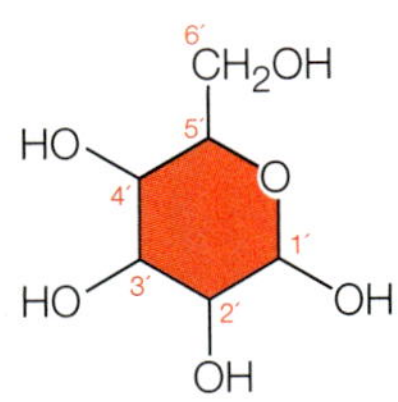

glucose

Glucose has six carbon atoms. Five-carbon monosaccharides are components of the nucleotide monomers of DNA and RNA (**FIGURE 3.4**). Two or more hydroxyl (—OH) groups impart solubility to a sugar molecule, which means that monosaccharides move easily through the water-based internal environments of all organisms.

Cells use monosaccharides for cellular fuel, because breaking the bonds of sugars releases energy that can be harnessed to power other cellular processes (we return to this important metabolic process in Chapter 7). Monosaccharides are also used as precursors, or parent molecules, that are remodeled into other molecules; and as structural materials to build larger molecules.

POLYMERS OF SIMPLE SUGARS

Oligosaccharides are short chains of covalently bonded monosaccharides (*oligo–* means a few). Disaccharides consist of two monosaccharide monomers. The lactose in milk, with one glucose and one galactose, is a disaccharide. Sucrose, the most plentiful sugar in nature, has a glucose and a fructose unit (sucrose extracted from sugarcane or sugar beets is our table sugar). Oligosaccharides attached to lipids or proteins have important functions in immunity.

Foods that we call "complex" carbohydrates consist mainly of polysaccharides: chains of hundreds or thousands of monosaccharide monomers. The chains may be straight or branched, and can consist of one or many types of monosaccharides. The most common polysaccharides are cellulose, starch, and glycogen. All consist only of glucose monomers, but as substances their properties are very different. Why? The answer begins with differences in patterns of covalent bonding that link their monomers.

Cellulose, the major structural material of plants, is the most abundant biological molecule on Earth. Its long, straight chains are locked into tight, sturdy bundles by hydrogen bonds (**FIGURE 3.5A**). The bundles form tough fibers that act like reinforcing rods inside stems and other plant parts, helping these structures resist wind and other forms of mechanical stress. Cellulose does not dissolve in water, and it is not easily broken down. Some bacteria and fungi make enzymes that can break it apart into its component sugars, but humans and other mammals do not. Dietary fiber, or "roughage," usually refers to the cellulose in our vegetable foods. Bacteria that live in the guts of termites and grazers such as cattle and sheep help these animals digest the cellulose in plants.

In starch, a different covalent bonding pattern between glucose monomers makes a chain that coils up into a spiral (**FIGURE 3.5B**). Like

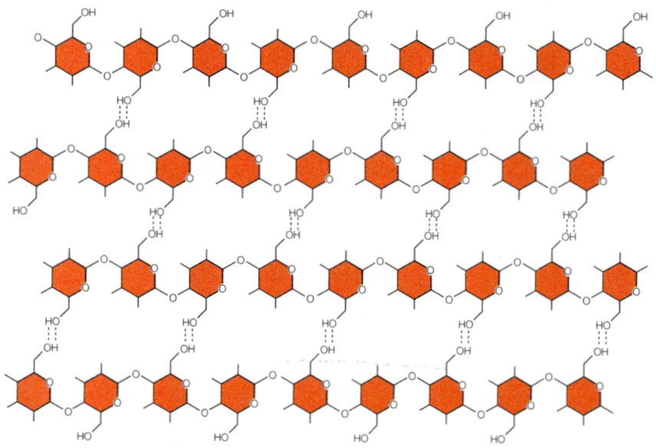

A Cellulose
Cellulose is the main structural component of plants. Above, in cellulose, chains of glucose monomers stretch side by side and hydrogen-bond at many —OH groups. The hydrogen bonds stabilize the chains in tight bundles that form long fibers. Very few types of organisms can digest this tough, insoluble material.

B Starch
Starch is the main energy reserve in plants, which store it in their roots, stems, leaves, seeds, and fruits. Below, in starch, a series of glucose monomers form a chain that coils up.

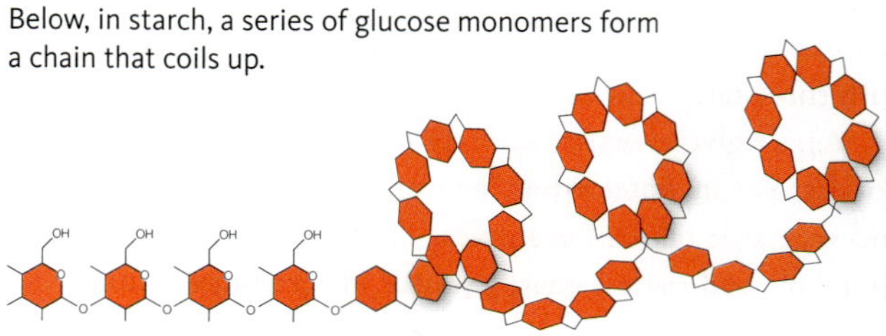

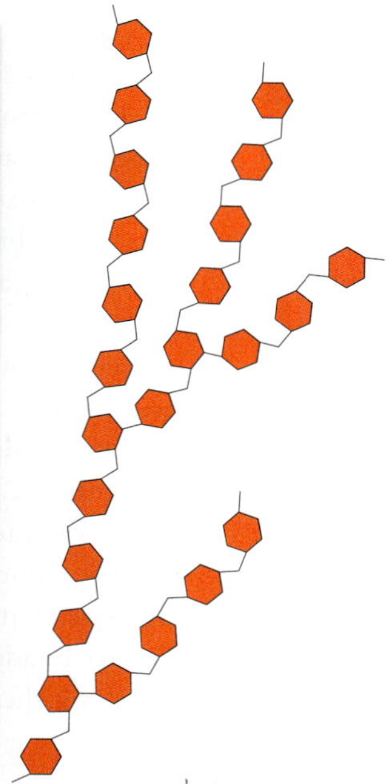

C Glycogen
Glycogen functions as an energy reservoir in animals, including people. It is especially abundant in the liver and muscles. Above, glycogen consists of highly branched chains of glucose monomers.

FIGURE 3.5 {Animated} Three of the most common complex carbohydrates and their locations in a few organisms. Each polysaccharide consists only of glucose subunits, but different bonding patterns result in substances with very different properties.

cellulose, starch does not dissolve easily in water, but it is more easily broken down than cellulose. These properties make starch ideal for storing sugars in the watery, enzyme-filled interior of plant cells. Most plant leaves make glucose during the day, and their cells store it by building starch. At night, hydrolysis enzymes break the bonds between starch's glucose monomers. The released glucose can be broken down immediately for energy, or converted to sucrose that is transported to other parts of the plant. Humans also have hydrolysis enzymes that break down starch, so this carbohydrate is an important component of our food.

Animals store their sugars in the form of glycogen. The covalent bonding pattern between glucose monomers in glycogen forms highly branched chains (**FIGURE 3.5C**). Muscle and liver cells contain most of the body's stored glycogen. When the sugar level in blood falls, liver cells break down stored glycogen, and the released glucose subunits enter the blood.

In chitin, a polysaccharide similar to cellulose, long, unbranching chains of nitrogen-containing monomers are linked by hydrogen bonds. As a structural material, chitin is durable, translucent, and flexible. It strengthens hard parts of many animals, including the outer cuticle of lobsters (*left*), and it reinforces the cell wall of many fungi.

TAKE-HOME MESSAGE 3.2

Cells use simple carbohydrates (sugars) for energy and to build other molecules.

Glucose monomers, bonded in different ways, form complex carbohydrates, including cellulose, starch, and glycogen.

carbohydrate Molecule that consists primarily of carbon, hydrogen, and oxygen atoms in a 1:2:1 ratio.
cellulose Tough, insoluble carbohydrate that is the major structural material in plants.

3.3 WHAT ARE LIPIDS?

Lipids are fatty, oily, or waxy organic compounds. Many lipids incorporate **fatty acids**, which are small organic molecules that consist of a long hydrocarbon "tail" with a carboxyl group "head" (**FIGURE 3.6**). The tail is hydrophobic; the carboxyl group makes the head hydrophilic (and acidic). You are already familiar with the properties of fatty acids because these molecules are the main component of soap. The hydrophobic tails of fatty acids in soap attract oily dirt, and the hydrophilic heads dissolve the dirt in water.

Saturated fatty acids have only single bonds linking the carbons in their tails. In other words, their carbon chains are fully saturated with hydrogen atoms (**FIGURE 3.6A**). Saturated fatty acid tails are flexible and they wiggle freely. Double bonds between carbons of unsaturated fatty acid tails limit their flexibility (**FIGURE 3.6B,C**).

FATS

The carboxyl group head of a fatty acid can easily form a covalent bond with another molecule. When it bonds to a glycerol, a type of alcohol, it loses its hydrophilic character and becomes part of a fat. **Fats** are lipids with one, two, or three fatty acids bonded to the same glycerol. A fat with three fatty acid tails is called a **triglyceride** (**FIGURE 3.7A**). Triglycerides are entirely hydrophobic, so they do not dissolve in water. Most "neutral" fats, such as butter and vegetable oils, are examples. Triglycerides are the most abundant and richest energy source in vertebrate bodies. Gram for gram, fats store more energy than carbohydrates.

Butter, cream, and other high-fat animal products have a high proportion of **saturated fats**, which means they consist mainly of triglycerides with three saturated fatty acid tails. Saturated fats tend to be solid at room temperature because their floppy saturated tails can pack tightly. Most vegetable oils are **unsaturated fats**, which means they consist mainly of triglycerides with one or more unsaturated fatty acid tails. Each double bond in a fatty acid tail makes a rigid kink. Kinky tails do not pack tightly, so unsaturated fats are typically liquid at room temperature.

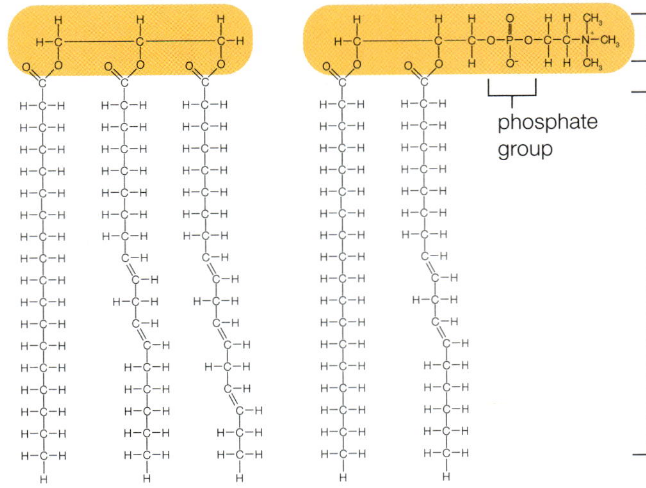

A The three fatty acid tails of a triglyceride are attached to a glycerol head.

B The two fatty acid tails of this phospholipid are attached to a phosphate-containing head.

FIGURE 3.7 {Animated} Lipids with fatty acid tails.

FIGURE 3.6 {Animated} Fatty acids. **A** The tail of stearic acid is fully saturated with hydrogen atoms. **B** Linoleic acid, with two double bonds, is unsaturated. The first double bond occurs at the sixth carbon from the end, so linoleic acid is called an omega-6 fatty acid. Omega-6 and **C** omega-3 fatty acids are "essential fatty acids." Your body does not make them, so they must come from food.

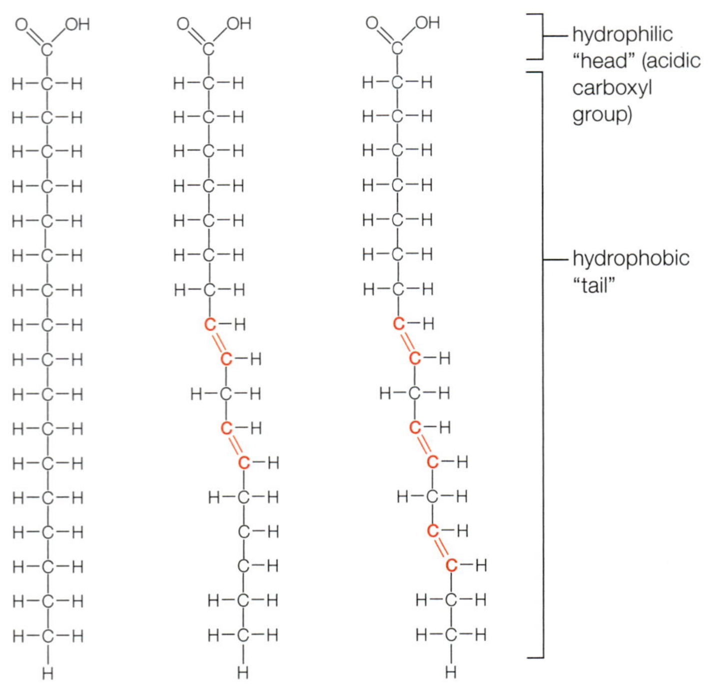

A stearic acid (saturated)

B linoleic acid (omega-6)

C linolenic acid (omega-3)

fat Lipid that consists of a glycerol molecule with one, two, or three fatty acid tails.
fatty acid Organic compound that consists of a chain of carbon atoms with an acidic carboxyl group at one end.
lipid Fatty, oily, or waxy organic compound.
lipid bilayer Double layer of lipids arranged tail-to-tail; structural foundation of cell membranes.
phospholipid A lipid with a phosphate group in its hydrophilic head, and two nonpolar tails typically derived from fatty acids.
saturated fat Triglyceride that has three saturated fatty acid tails.
steroid Type of lipid with four carbon rings and no tails.
triglyceride A fat with three fatty acid tails.
unsaturated fat Triglyceride that has one or more unsaturated fatty acid tails.
wax Water-repellent mixture of lipids with long fatty acid tails bonded to long-chain alcohols or carbon rings.

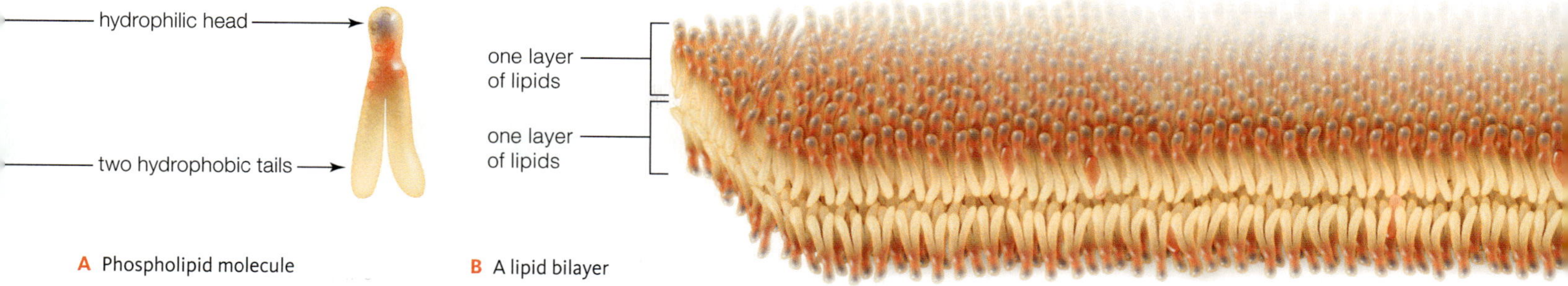

A Phospholipid molecule

B A lipid bilayer

FIGURE 3.8 {Animated} Phospholipids as components of cell membranes. A double layer of phospholipids—the lipid bilayer—is the structural foundation of all cell membranes. You will read more about the structure of cell membranes in Chapter 4.

FIGURE 3.9 **Steroids.** Estrogen and testosterone are steroid hormones that govern reproduction and secondary sexual traits. The two hormones are the source of gender-specific traits in many species, including wood ducks.

PHOSPHOLIPIDS

A **phospholipid** consists of a phosphate-containing head with two long hydrocarbon tails that are typically derived from fatty acids (**FIGURE 3.7B**). The tails are hydrophobic, but the highly polar phosphate group makes the head hydrophilic. These opposing properties give rise to the basic structure of cell membranes, which consist mainly of phospholipids. In a cell membrane, phospholipids are arranged in two layers—a **lipid bilayer** (**FIGURE 3.8**). The heads of one layer are dissolved in the cell's watery interior, and the heads of the other layer are dissolved in the cell's fluid surroundings. All of the hydrophobic tails are sandwiched between the hydrophilic heads.

WAXES

A **wax** is a complex, varying mixture of lipids with long fatty acid tails bonded to alcohols or carbon rings. The molecules pack tightly, so waxes are firm and water-repellent. Plants secrete waxes onto their exposed surfaces to restrict water loss and keep out parasites and other pests. Other types of waxes protect, lubricate, and soften skin and hair. Waxes, together with fats and fatty acids, make feathers waterproof. Bees store honey and raise new generations of bees inside a honeycomb of secreted beeswax.

STEROIDS

Steroids are lipids with no fatty acid tails; they have a rigid backbone that consists of twenty carbon atoms arranged in a characteristic pattern of four rings (**FIGURE 3.9**). Functional groups attached to the rings define the type of steroid. These molecules serve varied and important physiological functions in plants, fungi, and animals. Cells remodel cholesterol, the most common steroid in animal tissue, to produce many other molecules, including bile salts (which help digest fats), vitamin D (required to keep teeth and bones strong), and steroid hormones.

TAKE-HOME MESSAGE 3.3

- Lipids are fatty, waxy, or oily organic compounds.
- Fats have one, two, or three fatty acid tails; triglyceride fats are an important energy reservoir in vertebrate animals.
- Phospholipids arranged in a lipid bilayer are the main component of cell membranes.
- Waxes have complex, varying structures. They are components of water-repelling and lubricating secretions.
- Steroids serve varied and important physiological roles in plants, fungi, and animals.

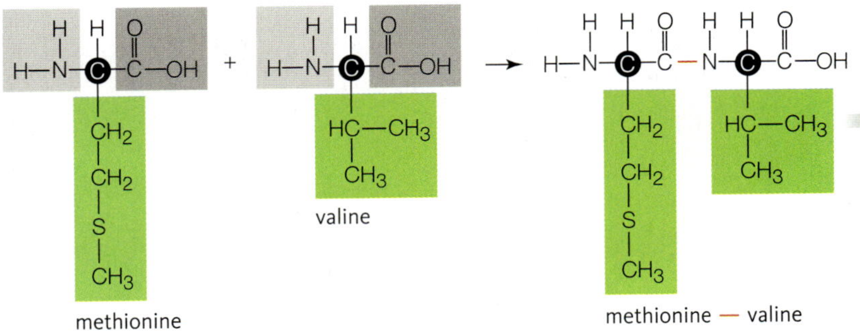

FIGURE 3.11 **{Animated}** Peptide bond formation.
A condensation reaction joins the carboxyl group of one amino acid and the amine group of another to form a peptide bond. In this example, a peptide bond forms between methionine and valine.

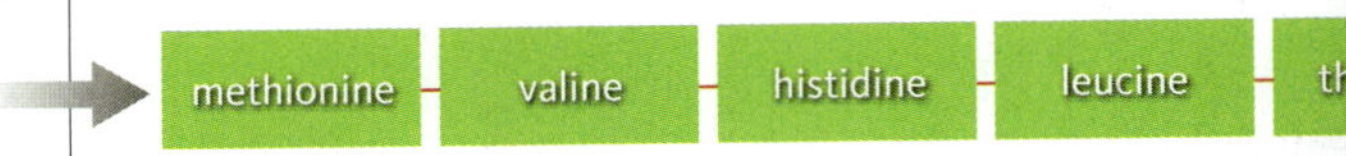

1 A protein's primary structure consists of a linear sequence of amino acids (a polypeptide chain). Each type of protein has a unique primary structure.

FIGURE 3.12 **{Animated}** Protein structure.

AMINO ACID SUBUNITS

With a few exceptions, cells can make all of the thousands of different proteins they need from only twenty kinds of monomers called amino acids. An **amino acid** is a small organic compound with an amine group (—NH_2), a carboxyl group (—COOH, the acid), and a side chain called an "R group" that defines the kind of amino acid. In most amino acids, all three groups are attached to the same carbon atom (**FIGURE 3.10**).

FIGURE 3.10 **Generalized structure of an amino acid.**
See Appendix V for the complete structures of the twenty most common amino acids found in eukaryotic proteins.

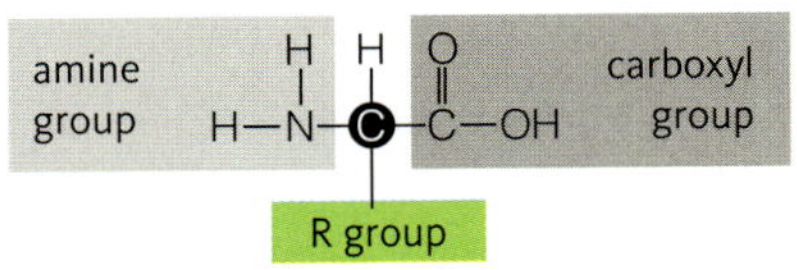

The covalent bond that links amino acids in a protein is called a **peptide bond**. During protein synthesis, a peptide bond forms between the carboxyl group of the first amino acid and the amine group of the second (**FIGURE 3.11**). Another peptide bond links a third amino acid to the second, and so on (you will learn more about the details of protein synthesis in Chapter 9). A short chain of amino acids is called a **peptide**; as the chain lengthens, it becomes a **polypeptide**. **Proteins** consist of polypeptides that are hundreds or even thousands of amino acids long.

STRUCTURE DICTATES FUNCTION

Of all biological molecules, proteins are the most diverse. Structural proteins support cell parts and, as part of tissues, multicelled bodies. Feathers, hooves, and hair, as well as tendons and other body parts, consist mainly of structural proteins. A tremendous number of different proteins, including some structural types, participate in all processes that sustain life. Most enzymes that help cells carry out metabolic reactions are proteins. Proteins also function in movement, defense, and cellular communication.

One of the fundamental ideas in biology is that structure dictates function. This idea is particularly appropriate for proteins, because a protein's biological activity arises from and depends on its structure (**FIGURE 3.12**).

The linear series of amino acids in a polypeptide chain is called primary structure 1, which defines the type of protein. The protein's three-dimensional shape begins to arise during synthesis, when hydrogen bonds that form among amino acids cause the lengthening polypeptide chain to twist and fold. Parts of the polypeptide form loops, helixes (coils), or flat sheets, and these patterns constitute secondary structure 2. The primary structure of each type of protein is unique, but most proteins have similar patterns of secondary structure.

Much as an overly twisted rubber band coils back upon itself, hydrogen bonding between nonadjacent regions of a protein makes its loops, helices, and sheets fold up into even more compact domains (**FIGURE 3.13A**). These domains are called tertiary structure 3. Tertiary structure makes a protein a working molecule. For example, the helices and loops in a globin chain fold up together to form a pocket that can hold a heme, which is a small compound essential to the finished protein's function. Sheets, loops, and helices of other proteins roll up into complex structures that resemble barrels, propellers, sandwiches, and so on. Large proteins typically have several domains, each contributing a particular structural or functional property to the molecule. For example, some barrel domains rotate like motors in

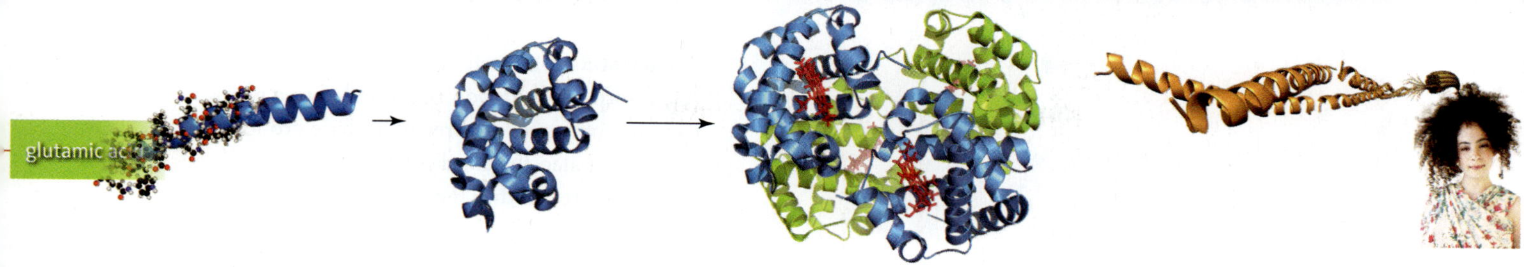

❷ Secondary structure arises as a polypeptide chain twists into a helix (coil), loop, or sheet held in place by hydrogen bonds.

❸ Tertiary structure occurs when loops, helices, and sheets fold up into a domain. In this example, the helices of a globin chain form a pocket.

❹ Many proteins have two or more polypeptide chains (quaternary structure). Hemoglobin, shown here, consists of four globin chains (green and blue). Each globin pocket now holds a heme group (red).

❺ Some types of proteins aggregate into much larger structures. As an example, organized arrays of keratin, a fibrous protein, compose filaments that make up your hair.

small molecular machines (**FIGURE 3.13B**). Other barrels function as tunnels for small molecules, allowing them to pass, for example, through a cell membrane.

Many proteins also have quaternary structure, which means they consist of two or more polypeptide chains that are closely associated or covalently bonded together ❹. Most enzymes are like this, with multiple polypeptide chains that collectively form a roughly spherical shape.

Fibrous proteins aggregate by many thousands into much larger structures, with their polypeptide chains organized into strands or sheets. The keratin in your hair is an example ❺. Some fibrous proteins contribute to the structure and organization of cells and tissues. Others help cells, cell parts, and multicelled bodies move.

Enzymes often attach sugars or lipids to proteins. A glycoprotein forms when oligosaccharides are attached to a polypeptide. The molecules that allow a tissue or a body to recognize its own cells are glycoproteins, as are other molecules that help cells interact in immunity.

Some lipoproteins form when enzymes covalently bond lipids to a protein. Other lipoproteins are aggregate structures that consist of variable amounts and types of proteins and lipids (**FIGURE 3.14**).

A In this protein, loops (green), coils (red), and a sheet (yellow) fold up together into a chemically active pocket. The pocket gives this protein the ability to transfer electrons from one molecule to another. Many other proteins have the same pocket structure.

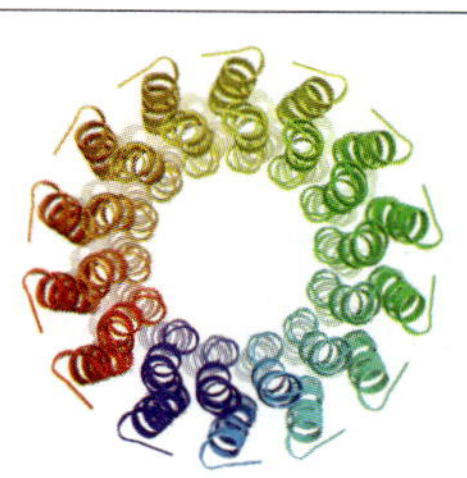

B This barrel domain is part of a rotary mechanism in a larger protein. The protein functions as a molecular motor that pumps hydrogen ions through cell membranes.

FIGURE 3.13 Examples of protein domains.

FIGURE 3.14 A lipoprotein particle. The one depicted here (HDL, often called "good" cholesterol) consists of thousands of lipids lassoed into a clump by two protein chains.

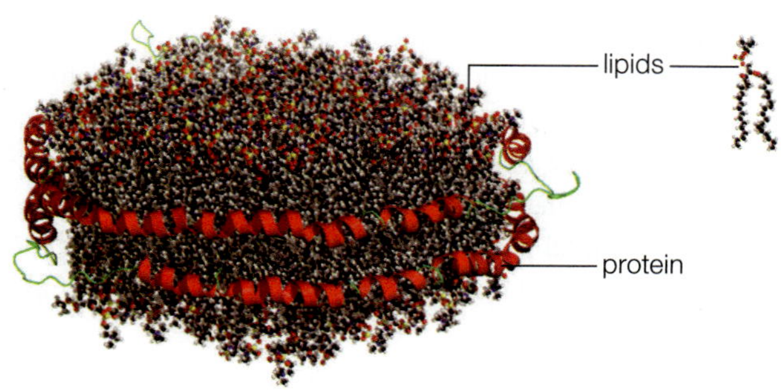

amino acid Small organic compound that is a subunit of proteins. Consists of a carboxyl group, an amine group, and a characteristic side group (R), all typically bonded to the same carbon atom.
peptide bond A bond between the amine group of one amino acid and the carboxyl group of another. Joins amino acids in proteins.
peptide Short chain of amino acids linked by peptide bonds.
polypeptide Long chain of amino acids linked by peptide bonds.
protein Organic molecule that consists of one or more polypeptides.

TAKE-HOME MESSAGE 3.4

Proteins are chains of amino acids. The order of amino acids in a polypeptide chain dictates the type of protein.

Polypeptide chains twist and fold into coils, sheets, and loops, which fold and pack further into functional domains.

A protein's function arises from its three-dimensional shape.

CREDITS: (13A, 14) Castrignanò T, De Meo PD, Cozzetto D, Talamo IG, Tramontano A. (2006). The PMDB Protein Model Database. Nucleic Acids Research, 34: D306-D309. (13B) pdb ID2W5J, Vollmar, M., Schlieper, D., Winn M., Buechner, C., Groth, G. "Structure of the C14 rotor ring of the proton translocating chloroplast ATP synthase." (2009) *J. Biol. Chem.* 284:18228.

3.5 WHY IS PROTEIN STRUCTURE IMPORTANT?

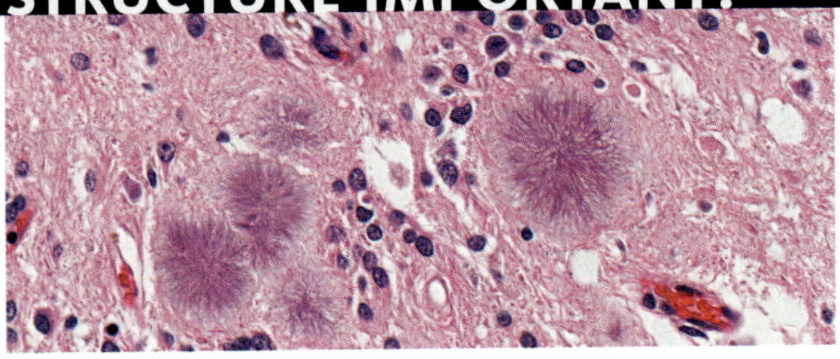

FIGURE 3.15 Variant Creutzfeldt–Jakob disease (vCJD). Characteristic holes and prion protein fibers radiating from several deposits are visible in this slice of brain tissue from a person with vCJD.

Protein shape depends on hydrogen bonding, which can be disrupted by heat, some salts, shifts in pH, or detergents. Such disruption causes proteins to **denature**, which means they lose their three-dimensional shape. Once a protein's shape unravels, so does its function.

Consider three fatal diseases: scrapie in sheep, mad cow disease (BSE, bovine spongiform encephalopathy), and variant Creutzfeldt–Jakob disease (vCJD) in humans. All begin with a glycoprotein called PrPC that occurs normally in cell membranes of the mammalian body. Sometimes, a PrPC protein spontaneously misfolds. A single misfolded protein molecule should not pose much of a threat, but when this particular protein misfolds it becomes a **prion**, or infectious protein. The altered shape of a misfolded PrPC protein causes normally folded PrPC proteins to misfold too. Because each protein that misfolds becomes infectious, the number of prions increases exponentially.

The shape of misfolded PrPC proteins allows them to align tightly into long fibers. In the brain, these fibers accumulate in water-repellent patches that disrupt brain cell function, resulting in relentlessly worsening symptoms of confusion, memory loss, and lack of coordination. Holes form in the brain as its cells die (**FIGURE 3.15**).

In the mid-1980s, an epidemic of mad cow disease in Britain was followed by an outbreak of vCJD in humans. The cattle became infected by the prion after eating feed prepared from the remains of scrapie-infected sheep, and people became infected by eating beef from infected cattle. The use of animal parts in livestock feed is now banned in many countries, and the number of cases of BSE and vCJD has since declined.

denature To unravel the shape of a protein or other large biological molecule.
prion Infectious protein.

TAKE-HOME MESSAGE 3.5

Protein shape can unravel if hydrogen bonds are disrupted.

A protein's function depends on its structure, so conditions that alter a protein's structure also alter its function.

3.6 WHAT ARE NUCLEIC ACIDS?

A **nucleotide** is a small organic molecule that consists of a sugar with a five-carbon ring bonded to a nitrogen-containing base and one, two, or three phosphate groups (**FIGURE 3.16A**). When the third phosphate group of a nucleotide is transferred to another molecule, energy is transferred along with it. The nucleotide **ATP** (adenosine triphosphate) serves an especially important role as an energy carrier in cells.

Nucleic acids are polymers, chains of nucleotides in which the sugar of one nucleotide is bonded to the phosphate group of the next (**FIGURE 3.16B**). An example is ribonucleic acid, or **RNA**, named after the ribose sugar of its component nucleotides. An RNA molecule is a chain of four kinds of nucleotide monomers, one of which is ATP. RNA molecules carry out protein synthesis. Deoxyribonucleic acid, or **DNA**, is a nucleic acid named after the deoxyribose sugar of its component nucleotides. A DNA molecule consists of two chains of nucleotides twisted into a double helix. Hydrogen bonds between the nucleotides hold the chains together. Each cell's DNA holds all information necessary to build a new cell and, in the case of multicelled organisms, a new individual.

FIGURE 3.16 Nucleic acids.

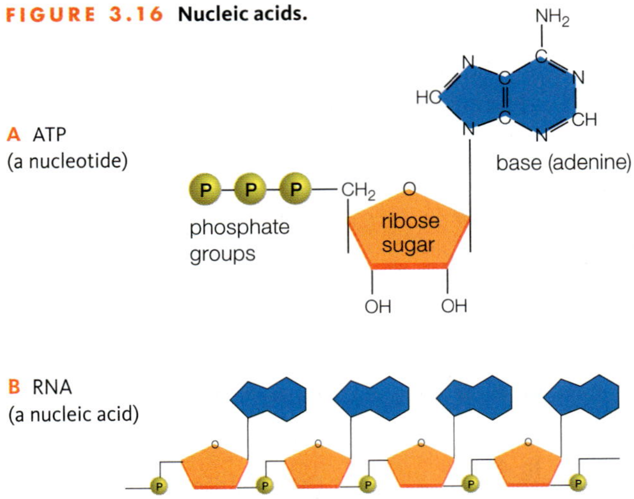

ATP Adenosine triphosphate. Nucleotide that serves an important role as an energy carrier in cells.
DNA Deoxyribonucleic acid. Consists of two chains of nucleotides twisted into a double helix.
nucleic acid Polymer of nucleotides; DNA or RNA.
nucleotide Monomer of nucleic acids; has a five-carbon sugar, a nitrogen-containing base, and one, two, or three phosphate groups.
RNA Ribonucleic acid. Single-stranded chain of nucleotides.

TAKE-HOME MESSAGE 3.6

Nucleotides are monomers of nucleic acids. ATP has an important metabolic role as an energy carrier.

RNA carries out protein synthesis. DNA holds information necessary to build cells and multicelled individuals.

3.7 Application: FEAR OF FRYING

FIGURE 3.17 *Trans* fats, an unhealthy food. Double bonds in the tail of most naturally occurring fatty acids are *cis*, which means that the two hydrogen atoms flanking the bond are on the same side of the carbon backbone. Hydrogenation creates abundant *trans* bonds, with hydrogen atoms on opposite sides of the tail.

FATS ARE NOT INERT MOLECULES THAT SIMPLY ACCUMULATE IN STRATEGIC AREAS OF OUR BODIES. They are major constituents of cell membranes, and as such they have powerful effects on cell function. As you learned in Section 3.3, the long carbon backbone of fatty acid tails can vary a bit in structure. *Trans* fats have unsaturated fatty acid tails with a particular arrangement of hydrogen atoms around the double bonds (FIGURE 3.17). Small amounts of *trans* fats occur naturally, but the main source of these fats in the American diet is an artificial food product called partially hydrogenated vegetable oil. Hydrogenation is a manufacturing process that adds hydrogen atoms to oils in order to change them into solid fats. In 1908, Procter & Gamble Co. developed partially hydrogenated soybean oil as a substitute for the more expensive solid animal fats they had been using to make candles. However, the demand for candles began to wane as more households in the United States became wired for electricity, and P&G looked for another way to sell its proprietary fat. Partially hydrogenated vegetable oil looks like lard, so the company began aggressively marketing it as a revolutionary new food: a solid cooking fat with a long shelf life, mild flavor, and lower cost than lard or butter. By the mid-1950s, hydrogenated vegetable oil had become a major part of the American diet, and it is still found in many manufactured and fast foods. For decades, it was considered healthier than animal fats, but we now know otherwise. *Trans* fats raise the level of cholesterol in our blood more than any other fat, and they directly alter the function of our arteries and veins. The effects of such changes are quite serious. Eating as little as 2 grams a day (about 0.4 teaspoon) of hydrogenated vegetable oil measurably increases a person's risk of atherosclerosis (hardening of the arteries), heart attack, and diabetes. A small serving of french fries made with hydrogenated vegetable oil contains about 5 grams of *trans* fat.

Summary

SECTION 3.1 Complex carbohydrates and lipids, proteins, and nucleic acids are **organic**, which means they consist mainly of carbon and hydrogen atoms. **Hydrocarbons** have only carbon and hydrogen atoms.

Carbon chains or rings form the backbone of the molecules of life. **Functional groups** attached to the backbone influence the chemical character of these compounds, and thus their function.

Metabolism includes chemical **reactions** and all other processes by which cells acquire and use energy as they make and break the bonds of organic compounds. In reactions such as **condensation**, **enzymes** build **polymers** from **monomers** of simple sugars, fatty acids, amino acids, and nucleotides. Reactions such as **hydrolysis** release the monomers by breaking apart the polymers.

SECTION 3.2 Enzymes build complex **carbohydrates** such as **cellulose**, glycogen, and starch from simple carbohydrate (sugar) subunits. Cells use carbohydrates for energy, and as structural materials.

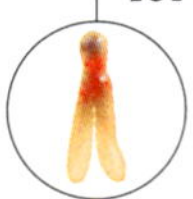

SECTION 3.3 **Lipids** are fatty, oily, or waxy compounds. All are nonpolar. **Fats** have **fatty acid** tails; **triglycerides** have three. **Saturated fats** are mainly triglycerides with three saturated fatty acid tails (only single bonds link their carbons). **Unsaturated fats** are mainly triglycerides with one or more unsaturated fatty acids.

A **lipid bilayer** (that consists primarily of **phospholipids**) is the basic structure of all cell membranes. **Waxes** are part of water-repellent and lubricating secretions. **Steroids** occur in cell membranes, and some are remodeled into other molecules such as hormones.

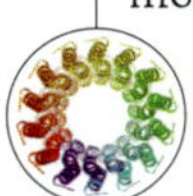

SECTION 3.4 Structurally and functionally, **proteins** are the most diverse molecules of life. The shape of a protein is the source of its function. Protein structure begins as a series of **amino acids** (primary structure) linked by **peptide bonds** into a **peptide**, then a **polypeptide**. Polypeptides twist into helices, sheets, and coils (secondary structure) that can pack further into functional domains (tertiary structure). Many proteins, including most enzymes, consist of two or more polypeptides (quaternary structure). Fibrous proteins aggregate into much larger structures.

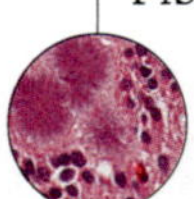

SECTION 3.5 A protein's structure dictates its function, so changes in a protein's structure may also alter its function. A protein's shape may be disrupted by shifts in pH or temperature, or exposure to detergent or some salts. If that happens, the protein unravels, or **denatures**, and so loses its function. **Prion** diseases are a fatal consequence of misfolded proteins.

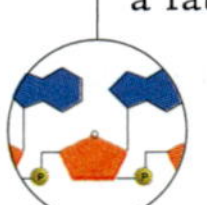

SECTION 3.6 **Nucleotides** are small organic molecules that consist of a five-carbon sugar, a nitrogen-containing base, and one, two, or three phosphate groups. Nucleotides are monomers of **DNA** and **RNA**, which are **nucleic acids**. Some, especially **ATP**, have additional functions such as carrying energy. DNA encodes information necessary to build cells and multicelled individuals. RNA molecules carry out protein synthesis.

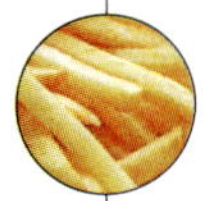

SECTION 3.7 All organisms consist of the same kinds of molecules. Seemingly small differences in the way those molecules are put together can have big effects inside a living organism.

Self-Quiz *Answers in Appendix VII*

1. Organic molecules consist mainly of _______ atoms.
   a. carbon
   b. carbon and oxygen
   c. carbon and hydrogen
   d. carbon and nitrogen

2. Each carbon atom can bond with as many as _______ other atom(s).

3. _______ groups are the "acid" part of amino acids and fatty acids.
   a. Hydroxyl (—OH)
   b. Carboxyl (—COOH)
   c. Methyl ($—CH_3$)
   d. Phosphate ($—PO_4$)

4. _______ is a simple sugar (a monosaccharide).
   a. Glucose
   b. Sucrose
   c. Ribose
   d. Starch
   e. both a and c
   f. a, b, and c

5. Unlike saturated fats, the fatty acid tails of unsaturated fats incorporate one or more _______ .
   a. phosphate groups
   b. glycerols
   c. double bonds
   d. single bonds

6. Is this statement true or false? Unlike saturated fats, all unsaturated fats are beneficial to health because their fatty acid tails kink and do not pack together.

7. Steroids are among the lipids with no _______ .
   a. double bonds
   b. fatty acid tails
   c. hydrogens
   d. carbons

8. Name three kinds of carbohydrates that can be built using only glucose monomers.

9. Which of the following is a class of molecules that encompasses all of the other molecules listed?
   a. triglycerides
   b. fatty acids
   c. waxes
   d. steroids
   e. lipids
   f. phospholipids

10. ______ are to proteins as ______ are to nucleic acids.
    a. Sugars; lipids
    b. Sugars; proteins
    c. Amino acids; hydrogen bonds
    d. Amino acids; nucleotides

11. A denatured protein has lost its _______ .
    a. hydrogen bonds
    b. shape
    c. function
    d. all of the above

12. _______ consist(s) of nucleotides.
    a. Sugars
    b. DNA
    c. RNA
    d. b and c

Data Analysis Activities

Effects of Dietary Fats on Lipoprotein Levels Cholesterol that is made by the liver or that enters the body from food cannot dissolve in blood, so it is carried through the bloodstream by lipoproteins. Low-density lipoprotein (LDL) carries cholesterol to body tissues such as artery walls, where it can form deposits associated with cardiovascular disease. Thus, LDL is often called "bad" cholesterol. High-density lipoprotein (HDL) carries cholesterol away from tissues to the liver for disposal, so HDL is often called "good" cholesterol.

In 1990, Ronald Mensink and Martijn Katan published a study that tested the effects of different dietary fats on blood lipoprotein levels. Their results are shown in **FIGURE 3.18**.

1. In which group was the level of LDL ("bad" cholesterol) highest?
2. In which group was the level of HDL ("good" cholesterol) lowest?
3. An elevated risk of heart disease has been correlated with increasing LDL-to-HDL ratios. Which group had the highest LDL-to-HDL ratio?
4. Rank the three diets from best to worst according to their potential effect on heart disease.

	Main Dietary Fats			
	cis fatty acids	*trans* fatty acids	saturated fats	optimal level
LDL	103	117	121	<100
HDL	55	48	55	>40
ratio	1.87	2.44	2.2	<2

FIGURE 3.18 Effect of diet on lipoprotein levels. Researchers placed 59 men and women on a diet in which 10 percent of their daily energy intake consisted of *cis* fatty acids, *trans* fatty acids, or saturated fats.

Blood LDL and HDL levels were measured after three weeks on the diet; averaged results are shown in mg/dL (milligrams per deciliter of blood). All subjects were tested on each of the diets. The ratio of LDL to HDL is also shown.

13. In the following list, identify the carbohydrate, the fatty acid, the amino acid, and the polypeptide:
    a. $NH_2—CH_2—COOH$ c. $(\text{methionine})_{20}$
    b. $C_6H_{12}O_6$ d. $CH_3(CH_2)_{16}COOH$

14. Match the molecules with the best description.
    ___ wax a. sugar storage in plants
    ___ starch b. richest energy source
    ___ triglyceride c. water-repellent secretions

15. Match each polymer with the appropriate monomer(s).
    ___ protein a. phosphate, fatty acids
    ___ phospholipid b. amino acids, sugars
    ___ glycoprotein c. glycerol, fatty acids
    ___ fat d. nucleotides
    ___ nucleic acid e. glucose only
    ___ wax f. sugar, phosphate, base
    ___ nucleotide g. amino acids
    ___ lipoprotein h. glucose, fructose
    ___ sucrose i. lipids, amino acids
    ___ glycogen j. fatty acids, carbon rings

Critical Thinking

1. Lipoproteins are relatively large, spherical clumps of protein and lipid molecules (see **FIGURE 3.14**) that circulate in the blood of mammals. They are like suitcases that move cholesterol, fatty acid remnants, triglycerides, and phospholipids from one place to another in the body. Given what you know about the insolubility of lipids in water, which of the four kinds of lipids would you predict to be on the outside of a lipoprotein clump, bathed in the water-based fluid portion of blood?

2. In 1976, a team of chemists in the United Kingdom was developing new insecticides by modifying sugars with chlorine (Cl_2), phosgene (Cl_2CO), and other toxic gases. One young member of the team misunderstood his verbal instructions to "test" a new molecule. He thought he had been told to "taste" it. Luckily for him, the molecule was not toxic, but it was very sweet. It became the food additive sucralose.

Sucralose has three chlorine atoms substituted for three hydroxyl groups of sucrose (table sugar). It binds so strongly to the sweet-taste receptors on the tongue that the human brain perceives it as 600 times sweeter than sucrose. Sucralose was originally marketed as an artificial sweetener called Splenda®, but it is now available under several other brand names.

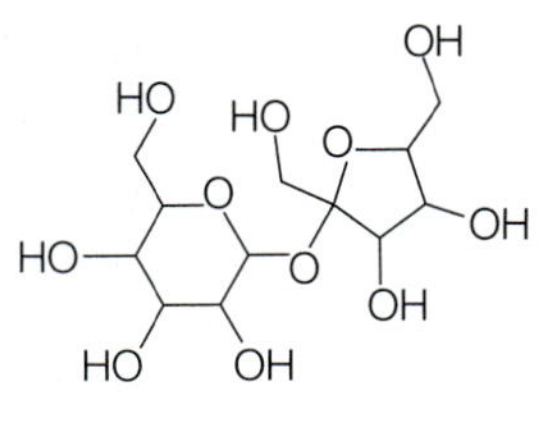

sucrose

Researchers investigated whether the body recognizes sucralose as a carbohydrate by feeding sucralose labeled with ^{14}C to volunteers. Analysis of the radioactive molecules in the volunteers' urine and feces showed that 92.8 percent of the sucralose passed through the body without being altered. Many people are worried that the chlorine atoms impart toxicity to sucralose. How would you respond to that concern?

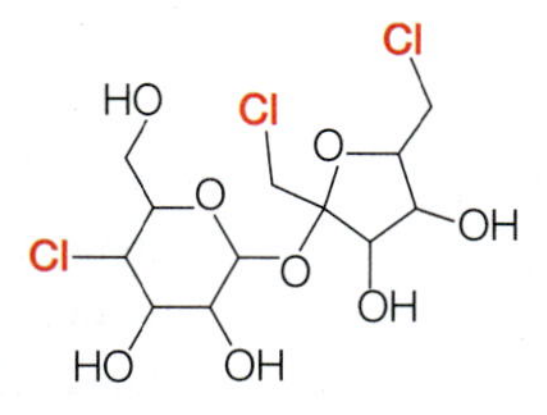

sucralose

CREDITS: (18) Source, Mensink RP, Katan MB., "Effect of dietary trans fatty acids on high-density and low-density lipoprotein cholesterol levels in healthy subjects." NEJM 323(7):439-45, 1990; From Starr/Taggart/Evers/Starr, Biology, 13E. © Cengage Learning; (insets) © Cengage Learning.

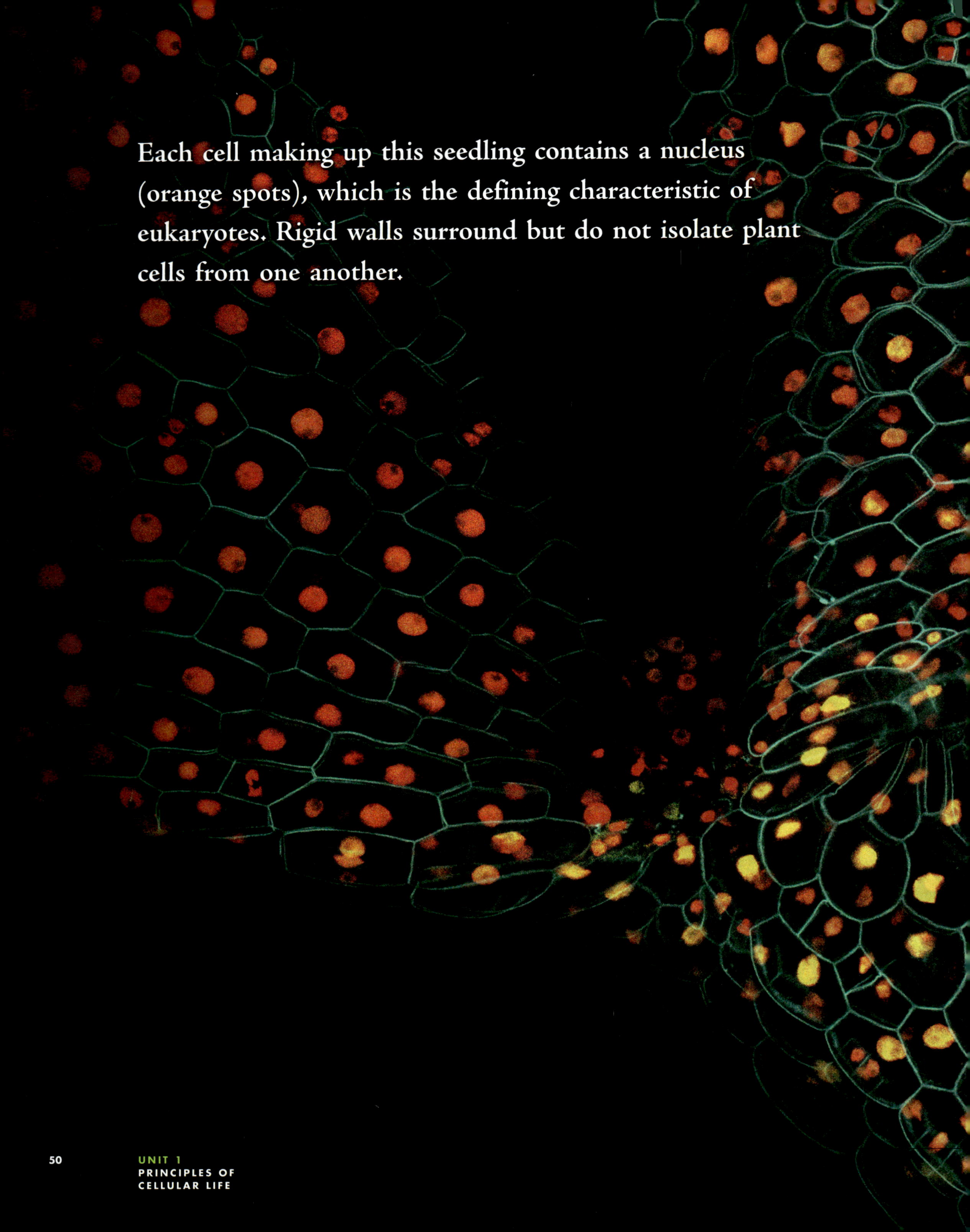

Each cell making up this seedling contains a nucleus (orange spots), which is the defining characteristic of eukaryotes. Rigid walls surround but do not isolate plant cells from one another.

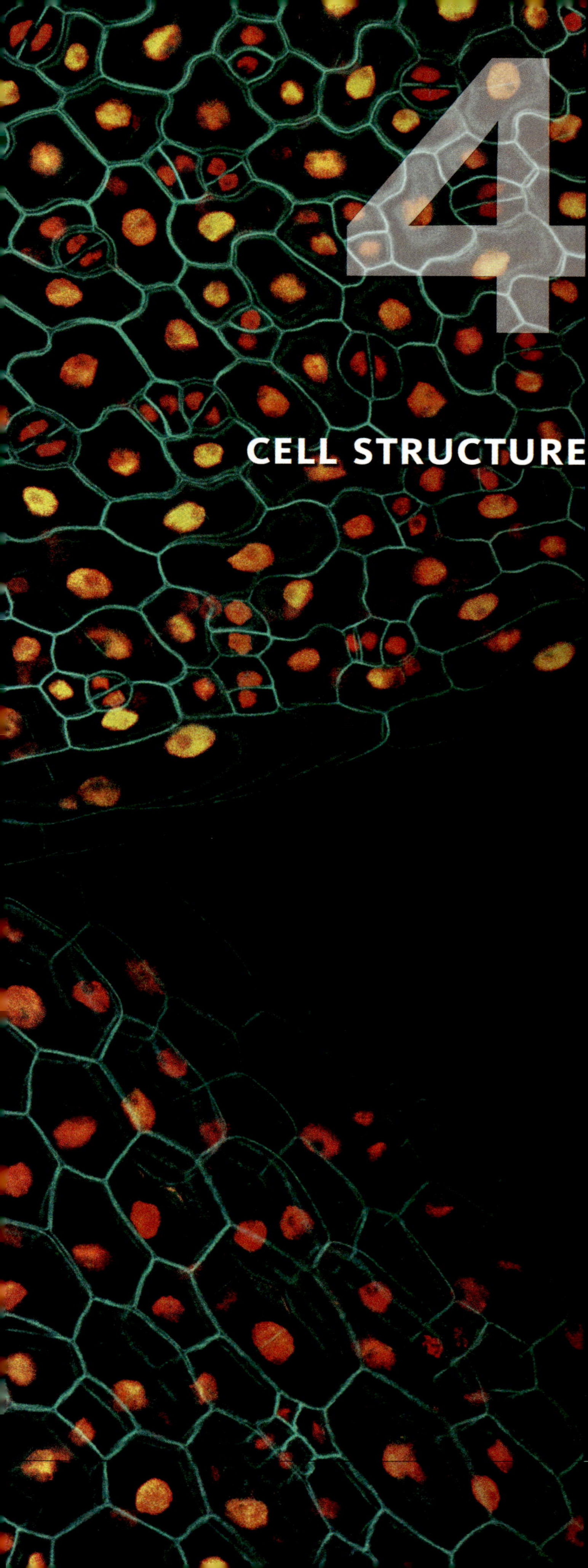

4 CELL STRUCTURE

Links to Earlier Concepts

Reflect on the overview of life's levels of organization in Section 1.1. In this chapter, you will see how the properties of lipids (3.3) give rise to cell membranes; consider the location of DNA (3.6) and the sites where carbohydrates are built and broken apart (3.1, 3.2); and expand your understanding of the vital roles of proteins in cell function (3.4, 3.5). You will also revisit the philosophy of science (1.5, 1.8) and tracers (2.1).

KEY CONCEPTS

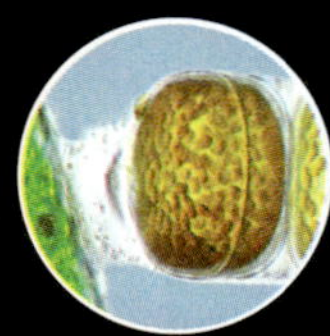

COMPONENTS OF ALL CELLS

Every cell has a plasma membrane separating its interior from the exterior environment. Its interior contains cytoplasm, DNA, and other structures.

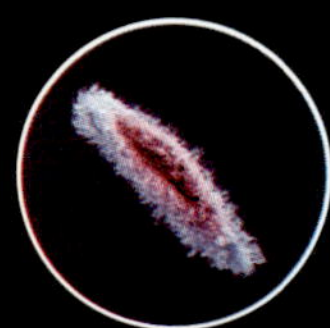

THE MICROSCOPIC WORLD

Most cells are too small to see with the naked eye. We use different types of microscopes to reveal different details of their structure.

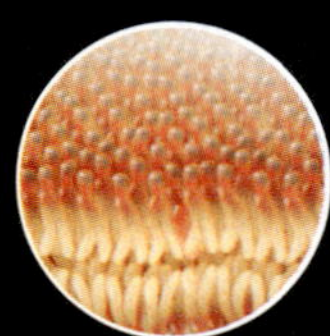

CELL MEMBRANES

All cell membranes consist of a lipid bilayer with various proteins embedded in it and attached to its surfaces. A membrane controls the kinds and amounts of substances that cross it.

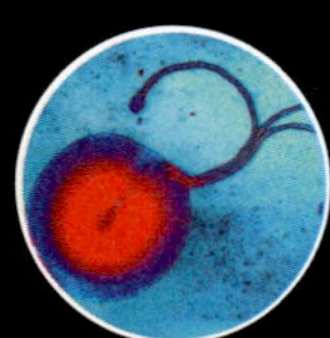

PROKARYOTIC CELLS

Archaea and bacteria have no nucleus. In general, they are smaller and structurally more simple than eukaryotic cells, but they are by far the most numerous and diverse organisms.

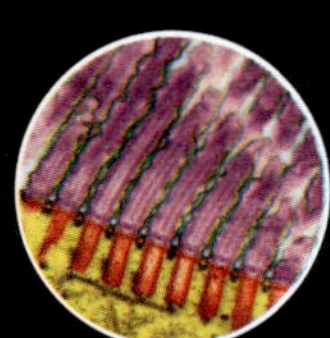

EUKARYOTIC CELLS

Protists, plants, fungi, and animals are eukaryotes. Cells of these organisms differ in internal parts and surface specializations, but all start out life with a nucleus.

4.1 WHAT IS A CELL?

CELL THEORY

No one knew cells existed until after the first microscopes were invented. By the mid-1600s, Antoni van Leeuwenhoek had constructed a crude instrument, and was writing about the tiny moving organisms he spied in rainwater, insects, fabric, sperm, feces, and other samples. In scrapings of tartar from his teeth, Leeuwenhoek saw "many very small animalcules, the motions of which were very pleasing to behold." He (incorrectly) assumed that movement defined life, and (correctly) concluded that the moving "beasties" he saw were alive. Leeuwenhoek might have been less pleased to behold his animalcules if he had grasped the implications of what he saw: Our world, and our bodies, teem with microbial life.

Today we know that a cell carries out metabolism and homeostasis, and reproduces either on its own or as part of a larger organism. By this definition, each cell is alive even if it is part of a multicelled body, and all living organisms consist of one or more cells. We also know that cells reproduce by dividing, so it follows that all existing cells must have arisen by division of other cells (later chapters discuss the processes by which cells divide). As a cell divides, it passes its hereditary material—its DNA—to offspring. Taken together, these generalizations constitute the **cell theory**, which is one of the foundations of modern biology (**TABLE 4.1**).

TABLE 4.1

Cell Theory
1. Every living organism consists of one or more cells.
2. The cell is the structural and functional unit of all organisms. A cell is the smallest unit of life, individually alive even as part of a multicelled organism.
3. All living cells arise by division of preexisting cells.
4. Cells contain hereditary material, which they pass to their offspring when they divide.

COMPONENTS OF ALL CELLS

Cells vary in shape and function, but all have at least three components in common: a plasma membrane, cytoplasm, and DNA (**FIGURE 4.1**). A cell's **plasma membrane** is its outermost, separating the cell's contents from the external environment. Like all other cell membranes, a plasma membrane is selectively permeable, which means that only certain materials can cross it. Thus, a plasma membrane controls exchanges between the cell and its environment.

The plasma membrane encloses a jellylike mixture of water, sugars, ions, and proteins called **cytoplasm**. A major part of a cell's metabolism occurs in the cytoplasm, and the cell's internal components, including organelles, are suspended in it. **Organelles** are structures that carry out special functions inside a cell. Membrane-enclosed organelles allow a cell to compartmentalize activities.

All cells start out life with DNA, though a few types lose it as they mature. In nearly all bacteria and archaea, the DNA is suspended directly in cytoplasm. By contrast, all eukaryotic cells start out life with a **nucleus** (plural, nuclei), an organelle with a double membrane that contains the cell's DNA. All protists, fungi, plants, and animals are eukaryotes. Some of these organisms are independent, free-living cells; others consist of many cells working together as a body.

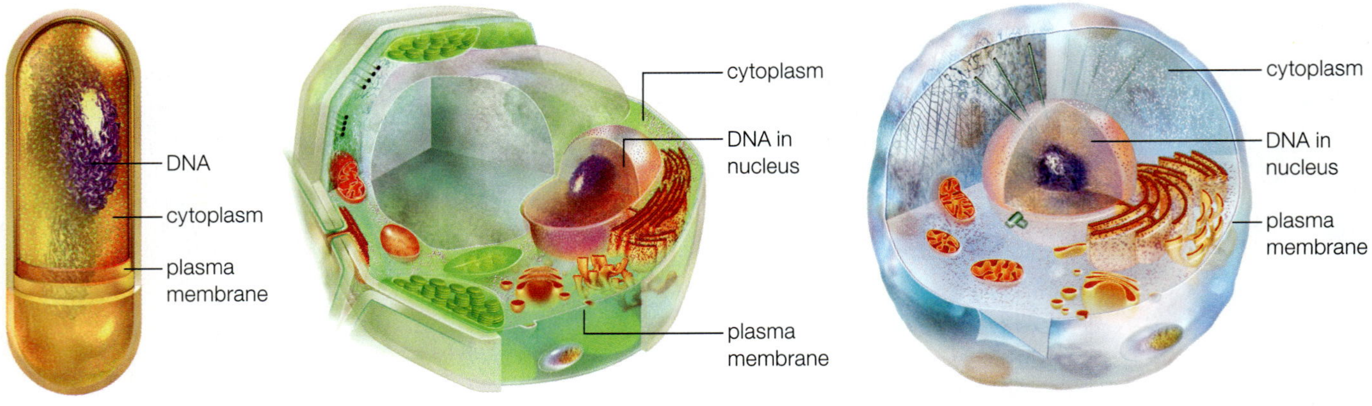

FIGURE 4.1 **{Animated}** All cells start out life with a plasma membrane, cytoplasm, and DNA. Archaea are similar to bacteria in overall structure; both are typically much smaller than eukaryotic cells. If the cells depicted here had been drawn to the same scale, the bacterium would be about this big: 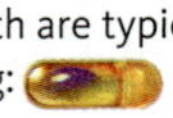

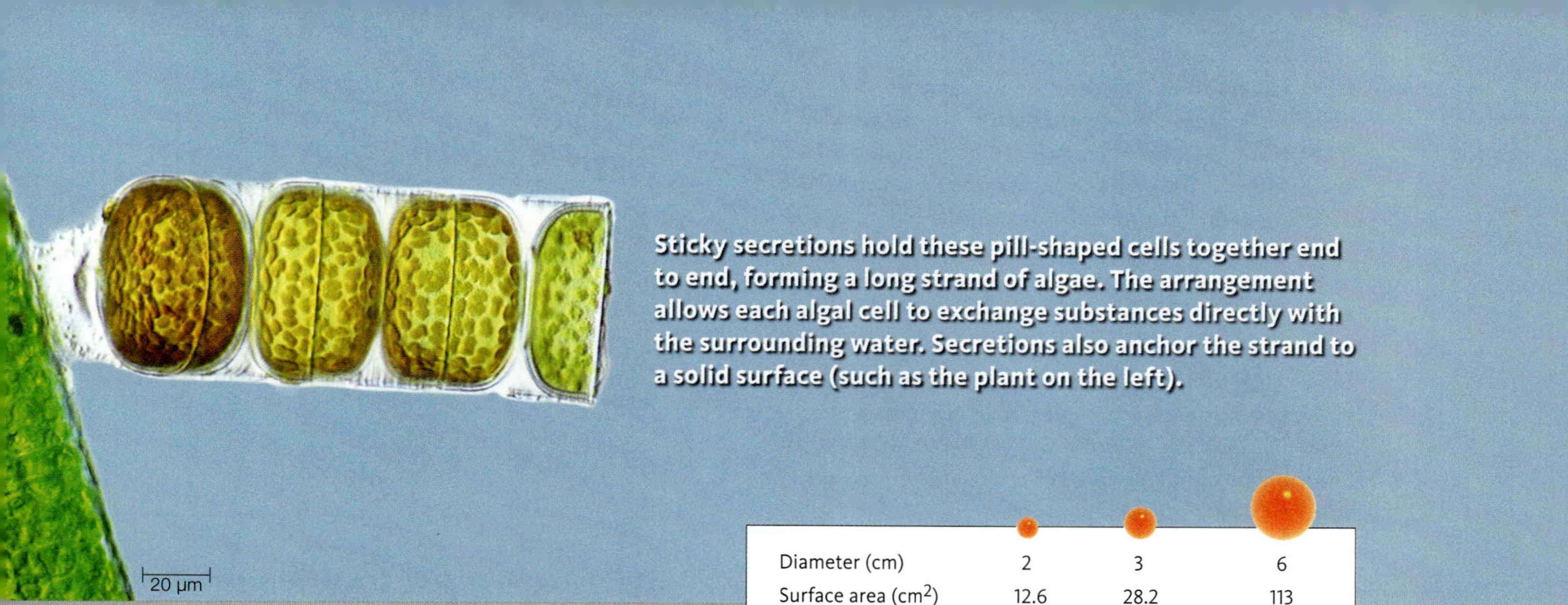

FIGURE 4.3 An example of a colonial algae.

Diameter (cm)	2	3	6
Surface area (cm^2)	12.6	28.2	113
Volume (cm^3)	4.2	14.1	113
Surface-to-volume ratio	3:1	2:1	1:1

FIGURE 4.2 Examples of surface-to-volume ratio. This physical relationship between increases in volume and surface area limits the size and influences the shape of cells.

CONSTRAINTS ON CELL SIZE

Almost all cells are too small to see with the naked eye. Why? The answer begins with the processes that keep a cell alive. A living cell must exchange substances with its environment at a rate that keeps pace with its metabolism. These exchanges occur across the plasma membrane, which can handle only so many exchanges at a time. The rate of exchange across a plasma membrane depends on its surface area: the bigger it is, the more substances can cross it during a given interval. Thus, cell size is limited by a physical relationship called the **surface-to-volume ratio**. By this ratio, an object's volume increases with the cube of its diameter, but its surface area increases only with the square.

Apply the surface-to-volume ratio to a round cell. As **FIGURE 4.2** shows, when a cell expands in diameter, its volume increases faster than its surface area does. Imagine that a round cell expands until it is four times its original diameter. The volume of the cell has increased 64 times (4^3), but its surface area has increased only 16 times (4^2). Each unit of plasma membrane must now handle exchanges with four times as much cytoplasm ($64 \div 16 = 4$). If the cell gets too big, the inward flow of nutrients and the outward flow of wastes across that membrane will not be fast enough to keep the cell alive.

Surface-to-volume limits also affect the form of colonial types and multicelled ones too. For example, small cells attach end to end to form strandlike algae, so each can interact directly with the environment (**FIGURE 4.3**). Muscle cells in your thighs are as long as the muscle in which they occur, but each is thin, so it exchanges substances efficiently with fluids in the surrounding tissue.

cell theory Theory that all organisms consist of one or more cells, which are the basic unit of life; all cells come from division of preexisting cells; and all cells pass hereditary material to offspring.
cytoplasm Semifluid substance enclosed by a cell's plasma membrane.
nucleus Of a eukaryotic cell, organelle with a double membrane that holds the cell's DNA.
organelle Structure that carries out a specialized metabolic function inside a cell.
plasma membrane A cell's outermost membrane.
surface-to-volume ratio A relationship in which the volume of an object increases with the cube of the diameter, and the surface area increases with the square.

TAKE-HOME MESSAGE 4.1

Observations of cells led to the cell theory: All organisms consist of one or more cells; the cell is the smallest unit of life; each new cell arises from another cell; and a cell passes hereditary material to its offspring.

All cells start life with a plasma membrane, cytoplasm, and a region of DNA, which, in eukaryotic cells only, is enclosed by a nucleus.

The surface-to-volume ratio limits cell size and influences cell shape.

4.2 HOW DO WE SEE CELLS?

Most cells are 10–20 micrometers in diameter, about fifty times smaller than the unaided human eye can perceive (**FIGURE 4.4**). One micrometer (μm) is one-thousandth of a millimeter, which is one-thousandth of a meter (**TABLE 4.2**). We use microscopes to observe cells and other objects in the micrometer range of size.

Light microscopes use visible light to illuminate samples. As you will learn in Chapter 6, all light travels in waves, a property that makes it bend when it passes through a curved glass lens. Curved lenses inside a light microscope focus light that passes through a specimen, or bounces off of one, into a magnified image (**FIGURE 4.5A**). Photographs of images enlarged with a microscope are called micrographs. Microscopes that use polarized light can yield images in which the edges of some structures appear in three-dimensional relief (**FIGURE 4.5B**).

Most cells are nearly transparent, so their internal details may not be visible unless they are first stained (exposed to dyes that only some cell parts soak up). Parts that absorb the most dye appear darkest. Staining results in an increase in contrast (the difference between light and dark parts) that allows us to see a greater range of detail.

Researchers often use light-emitting tracers (Section 2.1) to pinpoint the location of a molecule of interest within a cell. When illuminated with laser light, the tracer fluoresces (emits light), and an image of the emitted light can be captured with a fluorescence microscope (**FIGURE 4.5C**).

Other microscopes can reveal even finer details. For example, electron microscopes use magnetic fields to focus a beam of electrons onto a sample; these instruments resolve details thousands of times smaller than light microscopes do. Transmission electron microscopes direct electrons through a thin specimen, and the specimen's internal details appear as shadows in the resulting image—a transmission electron micrograph, or TEM (**FIGURE 4.5D**). Scanning electron microscopes direct a beam of electrons across the surface of a specimen that has been coated with a thin layer of metal. The irradiated metal emits electrons and x-rays, which can be converted into an image (a scanning electron micrograph, or SEM) of the surface (**FIGURE 4.5E**). SEMs and TEMs are always black and white; colored versions have been digitally altered to highlight specific details.

TABLE 4.2

Equivalent Units of Length

Unit	Equivalent: Meter	Equivalent: Inch
centimeter (cm)	1/100	0.4
millimeter (mm)	1/1000	0.04
micrometer (μm)	1/1,000,000	0.00004
nanometer (nm)	1/1,000,000,000	0.00000004
meter (m)	100 cm 1,000 mm 1,000,000 μm 1,000,000,000 nm	

TAKE-HOME MESSAGE 4.2

Most cells are visible only with the help of microscopes.

We use different microscopes and techniques to reveal different aspects of cell structure.

FIGURE 4.4 Relative sizes. Most cells are between 1 and 100 micrometers in diameter. See also Units of Measure, Table 4.2 and Appendix VI.

FIGURE IT OUT: Which one is smallest: a protein, a virus, or a bacterium?

Answer: A protein

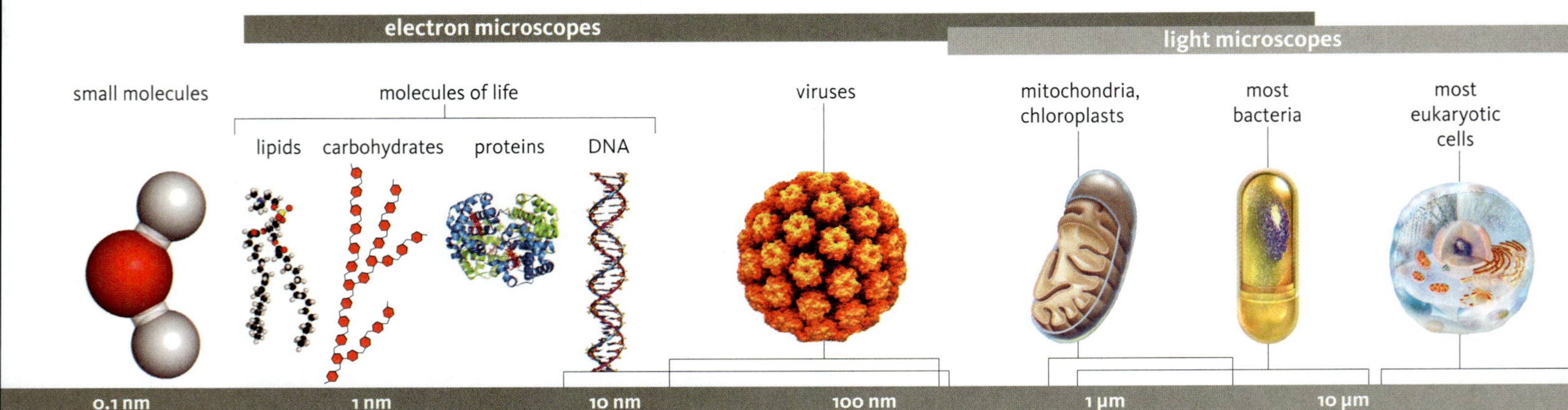

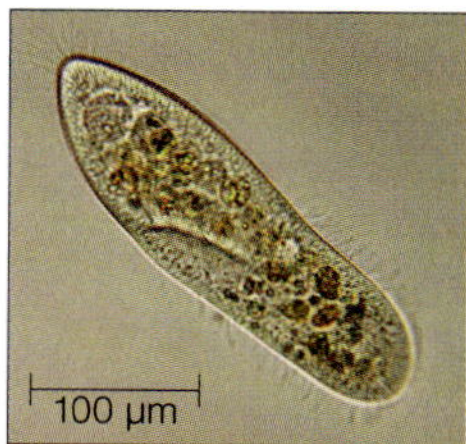

A The green blobs visible in this light micrograph of a living cell are ingested algal cells (also visible in **B** and **D**). Fine, hairlike structures on the cell's surface (also visible in **E**) are waving cilia that propel this motile organism through its fluid surroundings.

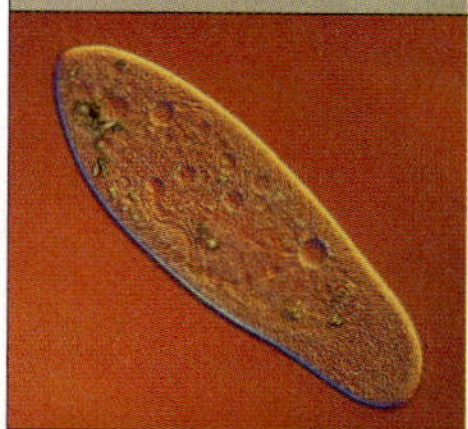

B A light micrograph taken with polarized light shows edges in relief. This technique reveals some internal structures not visible in **A**.

C In this fluorescence micrograph, yellow pinpoints the location of a particular type of protein in the membrane of organelles called contractile vacuoles. These organelles are also visible in **B**.

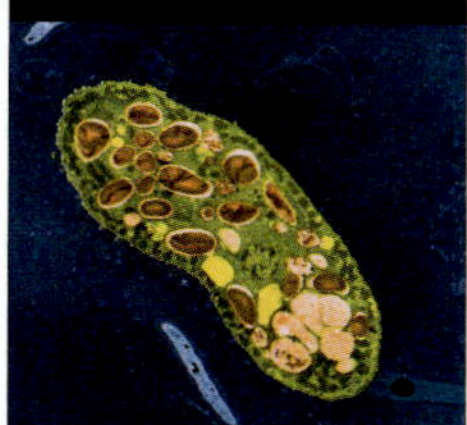

D A colorized transmission electron micrograph (TEM) reveals several types of internal structures in a plane (slice) through the sample. Ingested algal cells are clearly visible.

E A colorized scanning electron micrograph (SEM) shows details of the cell's surface. The cell ingests its food via the indentation (also visible in **A**).

FIGURE 4.5 Different microscopy techniques reveal different characteristics of the same organism, a protist (*Paramecium*).

FIGURE IT OUT: About how big are these cells?

Answer: About 250 μm long

PEOPLE MATTER

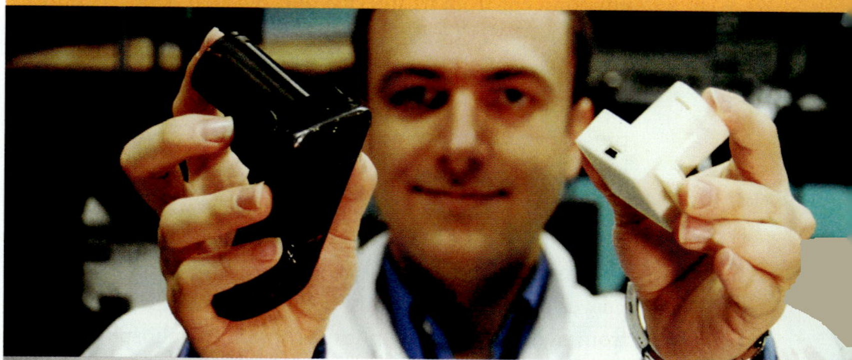

National Geographic Explorer
DR. AYDOGAN OZCAN

Aydogan Ozcan is developing a revolutionary global health solution using one of the most common forms of technology available—the smart phone. Using readily available parts that cost less than $50, Ozcan builds adapters that transform a smart phone into a mobile medical lab with the capability to test and diagnose diseases like HIV, malaria, and tuberculosis in remote communities.

Conventional microscopes, the mainstay of diagnosis for centuries, are impractical on a global level. "They are too heavy and powerful to be cost-effectively miniaturized. They also can't quickly capture and screen the large number of cells needed for statistically viable diagnoses," he says. What's more, because technicians in remote areas may be poorly trained, they often interpret images inaccurately. "In some parts of Africa, 70 percent of malaria diagnoses are incorrect false-positives." Ozcan's invention solves these problems by arming cell phones with sophisticated algorithms that do the interpreting. To tackle the most expensive part of microscopes—lenses—his team simply eliminated them. Ozcan's modified phone uses a special light source and the phone's camera to capture an image of a blood sample, essentially turning the phone into a lens-free microscope able to resolve structures smaller than one micron.

4.3 WHAT IS A MEMBRANE?

A plasma membrane physically separates a cell's external environment from its internal one, but that is not its only function. For example, you learned in Section 4.1 that a cell's plasma membrane allows some substances, but not others, to cross it. Membranes around organelles do this too. We return to membrane functions in Chapter 5; here, we explore the structure that gives rise to these functions.

THE FLUID MOSAIC MODEL

The foundation of almost all cell membranes is a lipid bilayer that consists mainly of phospholipids. Remember from Section 3.3 that a phospholipid has a phosphate-containing head and two fatty acid tails. The polar head is hydrophilic, which means that it interacts with water molecules. The nonpolar tails are hydrophobic, so they do not interact with water molecules. As a result of these opposing properties, phospholipids swirled into water will spontaneously organize themselves into lipid bilayer sheets or bubbles (*left*), with hydrophobic tails together, hydrophilic heads facing the watery surroundings (**FIGURE 4.6A**).

Other molecules, including cholesterol, proteins, glycoproteins, and glycolipids, are embedded in or attached to the lipid bilayer of a cell membrane. Many of these molecules move around the membrane more or less freely. We describe a eukaryotic or bacterial cell membrane as a **fluid mosaic** because it behaves like a two-dimensional liquid of mixed composition. The "mosaic" part of the name comes from the many different types of molecules in the membrane. A cell membrane is fluid because its phospholipids are not chemically bonded to one another; they stay organized in a bilayer as a result of collective hydrophobic and hydrophilic attractions. These interactions are, on an individual basis, relatively weak. Thus, individual phospholipids in the bilayer drift sideways and spin around their long axis, and their tails wiggle.

A cell membrane's properties vary depending on the types and proportions of molecules composing it. For example, membrane fluidity decreases with increasing cholesterol content. A membrane's fluidity also depends on the length and saturation of its phospholipids' fatty acid tails (Section 3.3). Archaea do not even use fatty acids to build their phospholipids. Instead, they use molecules with reactive side chains, so the tails of archaeal phospholipids form covalent bonds with one another. As a result of this rigid crosslinking, archaeal phospholipids do not drift, spin, or wiggle in a bilayer. Thus, membranes of archaea are stiffer than those of bacteria or eukaryotes, a characteristic that may help these cells survive in extreme habitats.

A In a watery fluid, phospholipids spontaneously line up into two layers: the hydrophobic tails cluster together, and the hydrophilic heads face outward, toward the fluid. This lipid bilayer forms the framework of all cell membranes. Many types of proteins intermingle among the lipids; a few that are typical of plasma membranes are shown *opposite*.

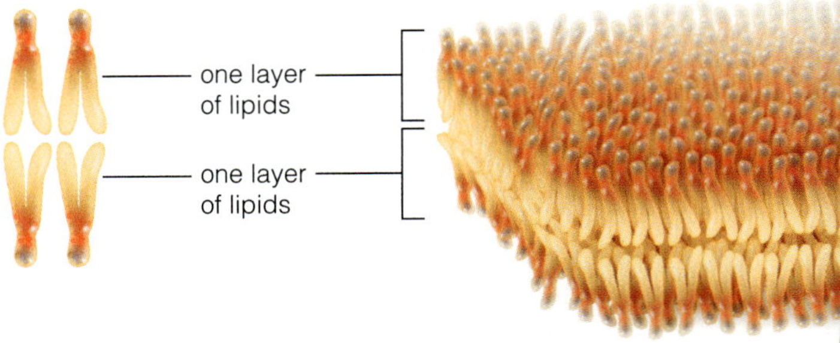

FIGURE 4.6 {Animated} Cell membrane structure.
A Organization of phospholipids in cell membranes.
B–E Examples of common membrane proteins.

TABLE 4.3

Common Membrane Proteins

Category	Function	Examples
Passive transport protein	Allows ions or small molecules to cross a membrane to the side where they are less concentrated.	Porin; glucose transporter
Active transport protein	Pumps ions or molecules through membranes to the side where they are more concentrated. Requires energy input, as from ATP.	Calcium pump; serotonin transporter
Receptor	Initiates change in a cell activity by responding to an outside signal (e.g., by binding a signaling molecule or absorbing light energy).	Insulin receptor; B cell receptor
Adhesion protein	Helps cells stick to one another, to cell junctions, and to extracellular matrix.	Integrins; cadherins
Recognition protein	Identifies a cell as self (belonging to one's own body or tissue) or nonself (foreign to the body).	MHC molecule
Enzyme	Speeds a specific reaction. Membranes provide a relatively stable reaction site for enzymes that work in series with other molecules.	Cytochrome *c* oxidase

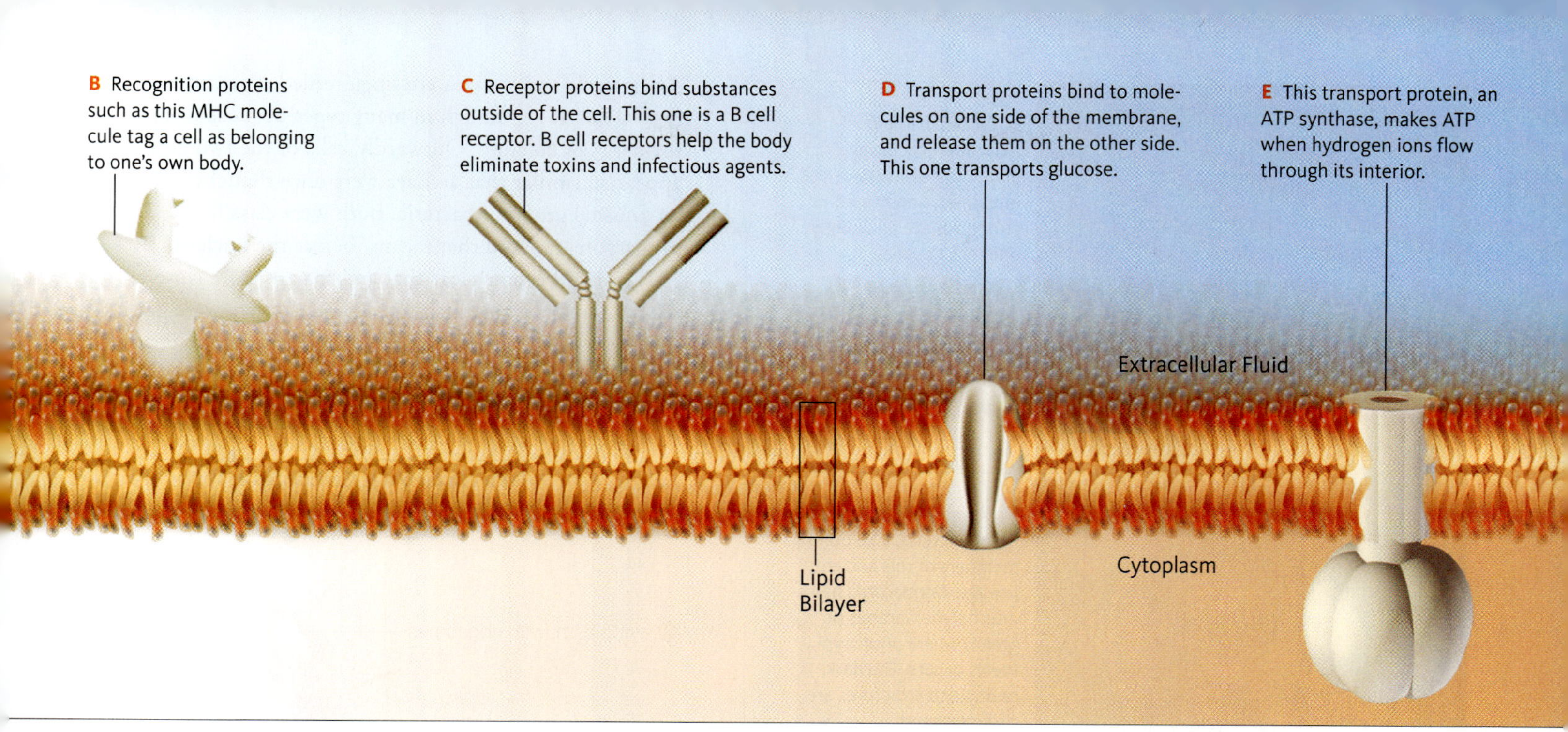

PROTEINS ADD FUNCTION

Many types of proteins are associated with a cell membrane (**TABLE 4.3**). Some are temporarily or permanently attached to one of the lipid bilayer's surfaces. Others have a hydrophobic domain that anchors the protein in the bilayer. Filaments inside the cell fasten some membrane proteins in place, including those that cluster as rigid pores.

Each type of protein in a membrane imparts a specific function to it. Thus, different cell membranes can have different functions depending on which proteins are associated with them. A plasma membrane has certain proteins that no internal cell membrane has. For example, cells in some animal tissues are fastened together by **adhesion proteins** in their plasma membranes, an arrangement that strengthens these tissues. **Recognition proteins** in the plasma membrane function as unique identity tags for an individual or a species (**FIGURE 4.6B**). Being able to recognize "self" imparts the potential ability to distinguish nonself (foreign) cells or particles.

Plasma membranes and some internal membranes incorporate **receptor proteins**, which trigger a change in the cell's activities upon binding a particular substance (**FIGURE 4.6C**). Different receptors bind to hormones or other signaling molecules, toxins, or molecules on another cell. The response triggered may involve metabolism, movement, division, or even cell death.

All cell membranes have some types of proteins, including enzymes. **Transport proteins** move specific substances across a membrane, typically by forming a channel through it (**FIGURE 4.6D,E**). These proteins are important because lipid bilayers are impermeable to most substances, including ions and polar molecules. Some transport proteins are open channels through which a substance moves on its own across a membrane. Others use energy to actively pump a substance across.

adhesion protein Protein that helps cells stick together in animal tissues.
fluid mosaic Model of a cell membrane as a two-dimensional fluid of mixed composition.
receptor protein Membrane protein that triggers a change in cell activity after binding to a particular substance.
recognition protein Plasma membrane protein that identifies a cell as belonging to self (one's own body or species).
transport protein Protein that passively or actively assists specific ions or molecules across a membrane.

TAKE-HOME MESSAGE 4.3

A cell membrane selectively controls exchanges between the cell and its surroundings.

The foundation of almost all cell membranes is the lipid bilayer—two layers of lipids (mainly phospholipids), with tails sandwiched between heads.

Proteins embedded in or attached to a lipid bilayer add specific functions to each type of cell membrane.

4.4 HOW ARE BACTERIA AND ARCHAEA ALIKE?

All bacteria and archaea are single-celled organisms (**FIGURE 4.7**), though in many types the cells form filaments or colonies. Outwardly, cells of the two groups appear so similar that archaea were once thought to be an unusual group of bacteria. Both were classified as prokaryotes, a word that means "before the nucleus." By 1977, it had become clear that archaea are more closely related to eukaryotes than to bacteria, so they were given their own separate domain. The term "prokaryote" is now an informal designation.

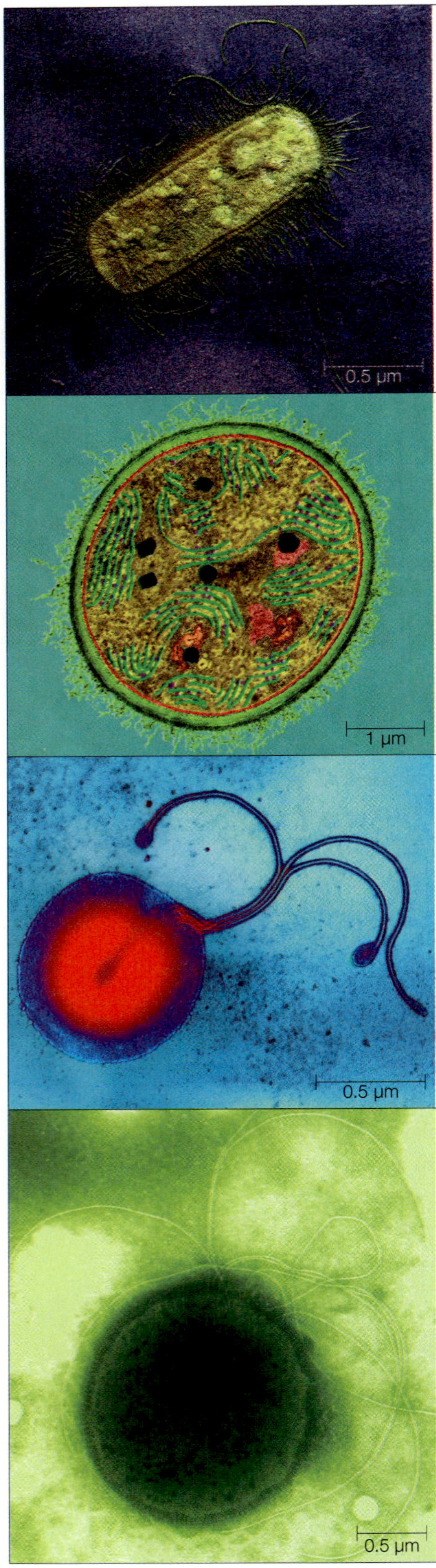

A *Escherichia coli*, a common bacterial inhabitant of human intestines. Short, hairlike structures are pili; longer ones are flagella.

B *Dermocarpa*, a type of cyanobacteria. Like other members of this ancient lineage, *Dermocarpa* has internal membranes (in green) where photosynthesis occurs. The dark, multisided structures are carboxysomes, protein-enclosed organelles that assist photosynthesis.

C *Helicobacter pylori*, a bacterium that can cause stomach ulcers when it infects the lining of the stomach. In unfavorable conditions, this species takes on a ball-shaped form (shown) that may offer the cells protection from environmental challenges such as antibiotic treatment.

D *Thermococcus gammatolerans*, an archaeon discovered at a deep-sea hydrothermal vent, where it lives under extreme conditions of salt, temperature, and pressure. It is by far the most radiation-resistant organism ever discovered, capable of withstanding thousands of times more radiation than humans can.

FIGURE 4.7 Some representatives of bacteria (**A–C**) and an archaeon (**D**).

FIGURE 4.8 {Animated} Generalized body plan of a prokaryote (a bacterium or archaeon).

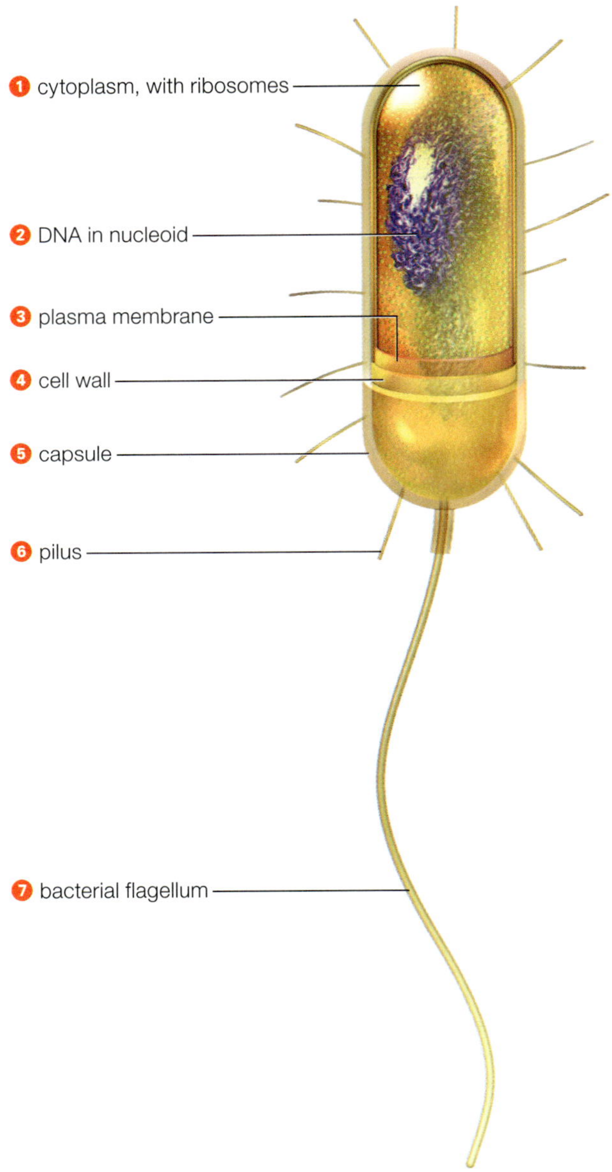

Bacteria and archaea are the smallest and most metabolically diverse forms of life that we know about. They inhabit nearly all of Earth's environments, including some extremely hostile places. The two kinds of cells differ in structure and metabolism. Chapter 19 revisits them in more detail; here we present an overview of structures shared by both groups (**FIGURE 4.8**).

Compared with eukaryotic cells, prokaryotes have little in the way of internal framework, but they do have protein filaments under the plasma membrane that reinforce the cell's shape and act as scaffolding for internal structures. The cytoplasm of these cells ❶ contains many **ribosomes** (organelles upon which polypeptides are assembled), and in some species, additional organelles. Cytoplasm also contains **plasmids**, small circles of DNA that carry a few genes (units of inheritance) that can provide advantages, such as resistance to antibiotics. The cell's remaining genes typically occur on one large circular molecule of DNA located in an irregularly shaped region of cytoplasm called the **nucleoid** ❷. In a few species, the nucleoid is enclosed by a membrane. Other internal membranes carry out special metabolic processes such as photosynthesis in some prokaryotes (**FIGURE 4.7B**).

Like all cells, bacteria and archaea have a plasma membrane ❸. In nearly all prokaryotes, a rigid **cell wall** ❹ surrounding the plasma membrane protects the cell and supports its shape. Most archaeal cell walls consist of proteins; most bacterial cell walls consist of a polymer of peptides and polysaccharides. Both types are permeable to water, so dissolved substances easily cross.

Polysaccharides form a slime layer or capsule ❺ around the wall of many types of bacteria. These sticky structures help the cells adhere to many types of surfaces, and they also offer protection against some predators and toxins.

Protein filaments called **pili** (singular, pilus) ❻ project from the surface of some prokaryotes. Pili help these cells move across or cling to surfaces. Many prokaryotes also have one or more **flagella** (singular, flagellum) ❼, which are long, slender cellular structures used for motion. A bacterial flagellum rotates like a propeller that drives the cell through fluid habitats.

BIOFILMS

Bacterial cells often live so close together that an entire community shares a layer of secreted polysaccharides and proteins. A communal living arrangement in which single-celled organisms live in a shared mass of slime is called a **biofilm**. A biofilm is often attached to a solid surface, and may include bacteria, algae, fungi, protists, and/or archaea. Participating in a biofilm allows the cells to linger in a favorable spot rather than be swept away by fluid currents, and to reap the benefits of living communally. For example, rigid or netlike secretions of some species serve as permanent scaffolding for others; species that break down toxic chemicals allow more sensitive ones to thrive in habitats that they could not withstand on their own; and waste products of some serve as raw materials for others. Later chapters discuss medical implications of biofilms, including the dental plaque that forms on teeth (**FIGURE 4.9**).

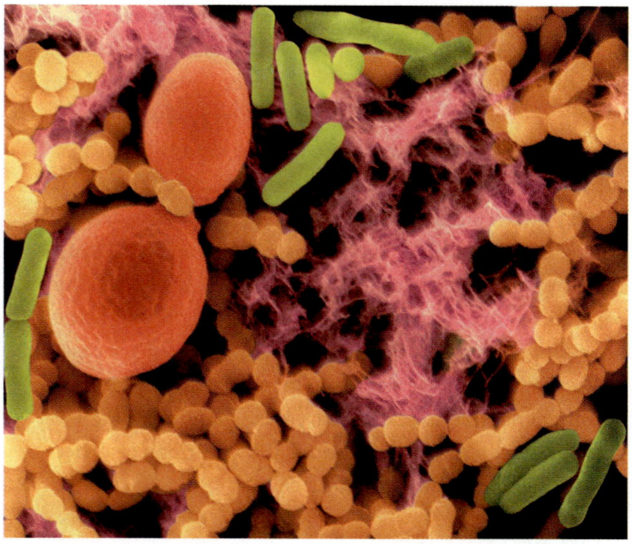

FIGURE 4.9 Oral bacteria in dental plaque, a biofilm. This micrograph shows two species of bacteria (tan, green) and a yeast (red) sticking to one another and to teeth via a gluelike mass of shared, secreted polysaccharides (pink). Other secretions of these organisms cause cavities and periodontal disease.

biofilm Community of microorganisms living within a shared mass of secreted slime.
cell wall Rigid but permeable structure that surrounds the plasma membrane of some cells.
flagellum Long, slender cellular structure used for motility.
nucleoid Of a bacterium or archaeon, region of cytoplasm where the DNA is concentrated.
pilus A protein filament that projects from the surface of some prokaryotic cells.
plasmid Small circle of DNA in some bacteria and archaea.
ribosome Organelle of protein synthesis.

TAKE-HOME MESSAGE 4.4

Bacteria and archaea do not have a nucleus. Most kinds have a cell wall around their plasma membrane. The permeable wall reinforces and imparts shape to the cell body.

The structure of bacteria and archaea is relatively simple, but as a group these organisms are the most diverse forms of life.

In addition to the nucleus, a typical eukaryotic cell has many other organelles, including endoplasmic reticulum, Golgi bodies, ribosomes, and at least one mitochondrion (**TABLE 4.4** and **FIGURE 4.10**). Organelles with membranes can regulate the types and amounts of substances that enter and exit. Such control maintains a special internal environment that allows the organelle to carry out its particular function—for example, isolating toxic or sensitive substances from the rest of the cell, moving substances through cytoplasm, maintaining fluid balance, or providing a favorable environment for a special process.

In this section, we detail the nucleus, which is the defining characteristic of eukaryotes. The remaining sections of the chapter introduce the functions of other organelles typical of eukaryotic cells.

TABLE 4.4

Some Organelles in Eukaryotic Cells

Organelles with membranes	
Nucleus	Protecting and controlling access to DNA
Endoplasmic reticulum (ER)	Making, modifying new polypeptides and lipids; other tasks
Golgi body	Modifying and sorting new polypeptides and lipids
Vesicle	Transporting, storing, or breaking down substances
Mitochondrion	Making ATP by glucose breakdown
Chloroplast	Making sugars in plants, some protists
Lysosome	Intracellular digestion
Peroxisome	Breaking down fatty acids, amino acids, toxins
Vacuole	Storage, breaking down food or waste
Organelles without membranes	
Ribosome	Assembling polypeptides
Centriole	Anchor for cytoskeleton
Other components	
Cytoskeleton	Contributes to cell shape, internal organization, movement

THE NUCLEUS

A cell nucleus (**FIGURE 4.11**) serves multiple functions. First, it keeps the cell's genetic material—its one and only copy of DNA—safe from metabolic processes that might damage it. Isolated in its own compartment, the DNA stays separated from the bustling activity of the cytoplasm. The nucleus also allows some molecules, but not others, to access the DNA. The nuclear membrane, which is called

FIGURE 4.10 Some components of eukaryotic cells.

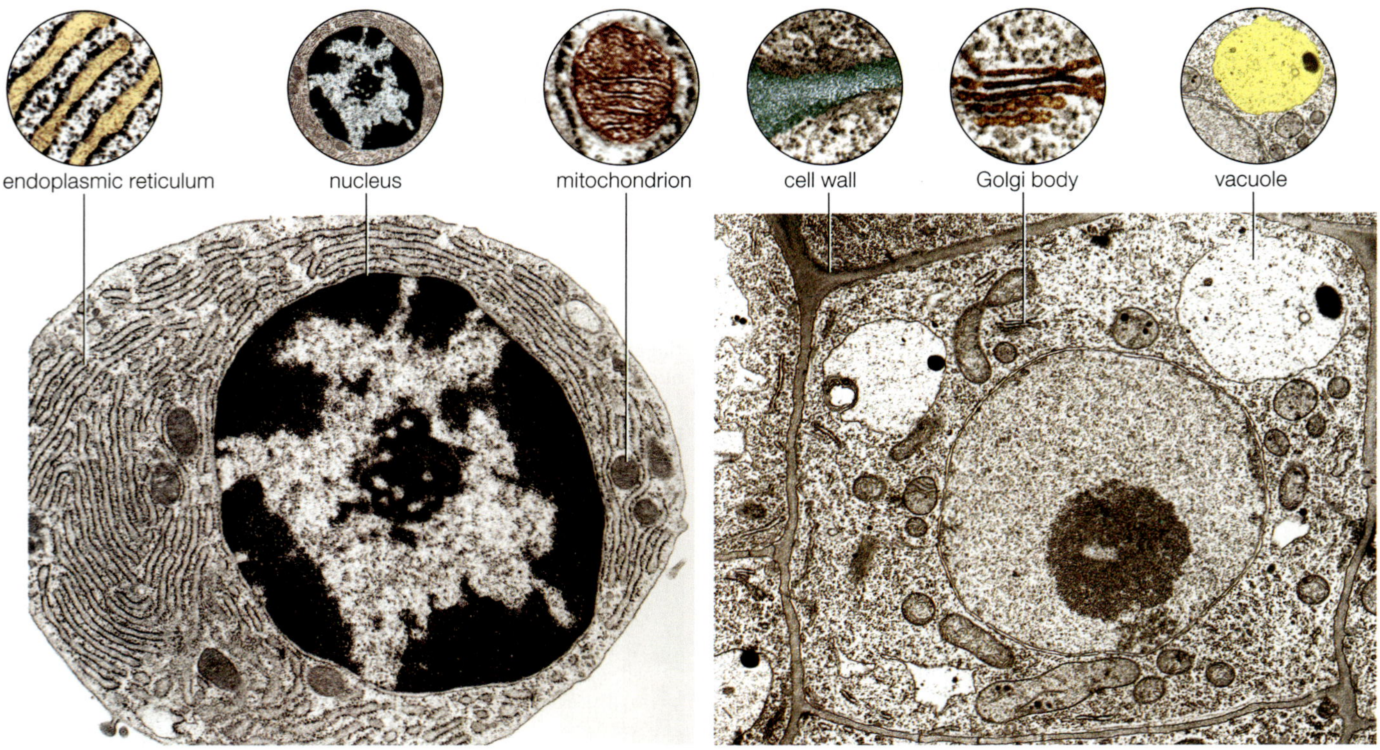

An animal cell (a white blood cell of a guinea pig)

A plant cell (from a root of thale cress)

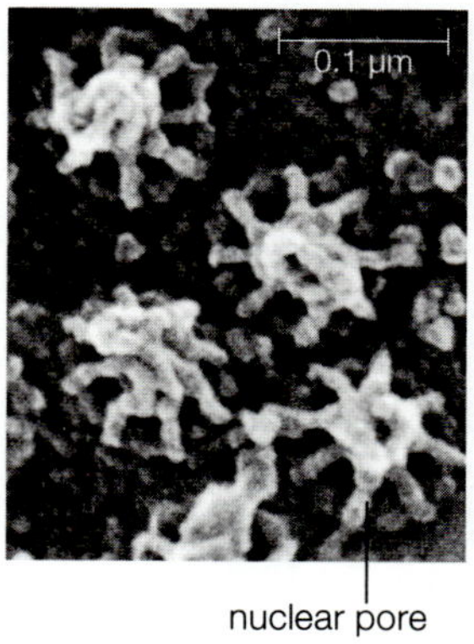

the **nuclear envelope**, carries out this function. A nuclear envelope consists of two lipid bilayers folded together as a single membrane. Membrane proteins aggregate into thousands of tiny pores (*left*) that span the nuclear envelope. The pores are anchored by the nuclear lamina, a dense mesh of fibrous proteins that supports the inner surface of the membrane. Some bacteria have membranes around their DNA, but we do not consider the bacteria to have nuclei because there are no pores in these membranes.

As you will see in Chapter 5, large molecules, including RNA and proteins, cannot cross a lipid bilayer on their own. Nuclear pores function as gateways for these molecules to enter and exit a nucleus. Protein synthesis offers an example of why this movement is important. Protein synthesis occurs in cytoplasm, and it requires the participation of many molecules of RNA. RNA is produced in the nucleus. Thus, RNA molecules must move from nucleus to cytoplasm, and they do so through nuclear pores. Proteins that carry out RNA synthesis must move in the opposite direction, because this process occurs in the nucleus. A cell can regulate the amounts and types of proteins it makes at a given time by selectively restricting the passage of certain molecules through nuclear pores. (Later chapters return to details of protein synthesis and controls over it.)

The nuclear envelope encloses **nucleoplasm**, a viscous fluid similar to cytoplasm, in which the cell's DNA is suspended. The nucleus contains at least one **nucleolus** (plural, nucleoli), a dense, irregularly shaped region of proteins and nucleic acid where subunits of ribosomes are produced.

---

nuclear envelope A double membrane that constitutes the outer boundary of the nucleus. Pores in the membrane control which substances can cross.
nucleolus In a cell nucleus, a dense, irregularly shaped region where ribosomal subunits are assembled.
nucleoplasm Viscous fluid enclosed by the nuclear envelope.

TAKE-HOME MESSAGE 4.5

All eukaryotic cells start life with a nucleus and other membrane-enclosed organelles.

A nucleus protects and controls access to a eukaryotic cell's DNA.

The nuclear envelope is a double lipid bilayer. Proteins embedded in the bilayer form pores that control the passage of molecules between the nucleus and cytoplasm.

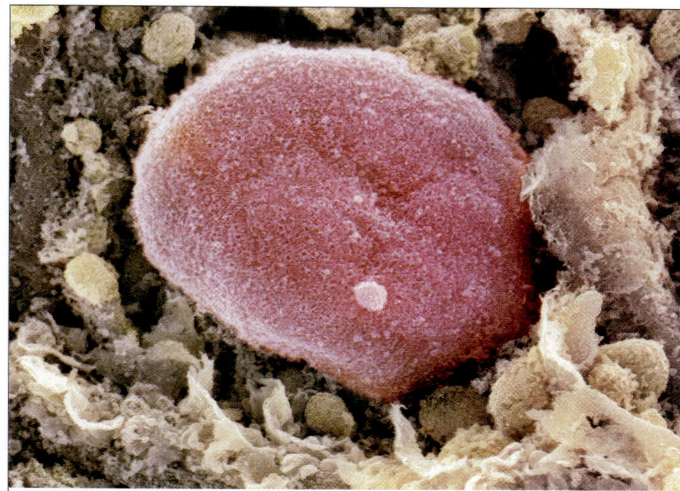

A Nucleus of a liver cell (in pink).

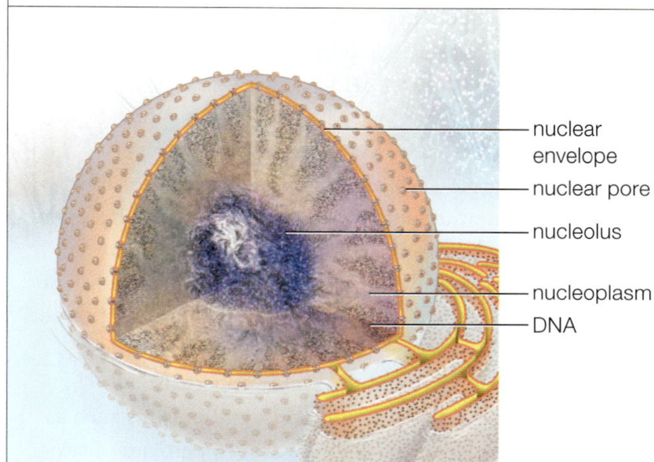

B Components of the cell nucleus.

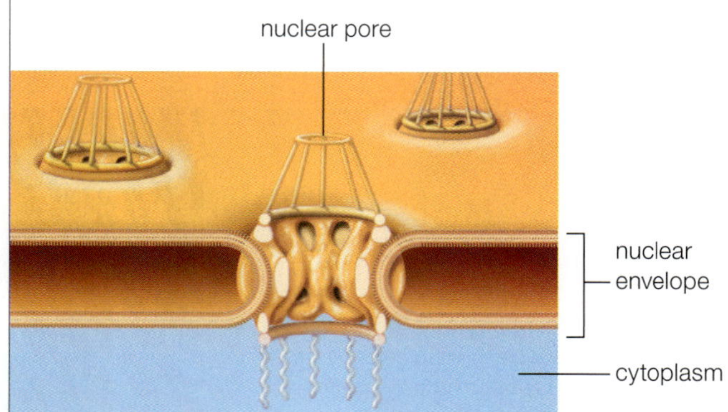

C The nuclear envelope consists of two lipid bilayers folded together as a single membrane and studded with thousands of nuclear pores. Each nuclear pore is an organized cluster of membrane proteins that selectively allows certain substances to cross it on their way into and out of the nucleus.

FIGURE 4.11 {Animated} The cell nucleus.

4.6 WHAT IS THE ENDOMEMBRANE SYSTEM?

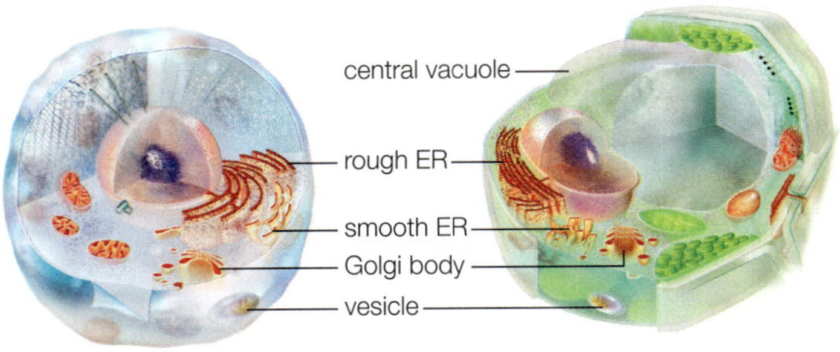

The **endomembrane system** is a series of interacting organelles between the nucleus and the plasma membrane (*above*). Its main function is to make lipids, enzymes, and proteins for insertion into the cell's membranes or secretion to the external environment. The endomembrane system also destroys toxins, recycles wastes, and has other special functions. Components of the system vary among different types of cells, but here we present an overview of the most common ones (**FIGURE 4.12**).

A VARIETY OF VESICLES

Small, membrane-enclosed sacs called **vesicles** form by budding from other organelles or when a patch of plasma membrane sinks into the cytoplasm ❶. Vesicles have a variety of functions. Many transport substances from one organelle to another, or to and from the plasma membrane. Some are a bit like trash cans that collect and dispose of waste, debris, or toxins. Enzymes in **peroxisomes** break down fatty acids, amino acids, and poisons such as alcohol. They also break down hydrogen peroxide, a toxic by-product of fatty acid metabolism. **Lysosomes** take part in intracellular digestion. They contain powerful enzymes that can break down cellular debris and wastes (carbohydrates, proteins, nucleic acids, and lipids). Vesicles in cells such as amoebas or white blood cells deliver ingested bacteria, cell parts, and other debris to lysosomes for breakdown.

Vacuoles form by the fusion of multiple vesicles. They have different functions in different kinds of cells. Many isolate or break down waste, debris, toxins, or food (**FIGURE 4.13**). Amino acids, sugars, ions, wastes, and toxins accumulate in the water-filled interior of a plant cell's large **central vacuole**. Fluid pressure in a central vacuole keeps plant cells plump, so stems, leaves, and other plant parts stay firm.

central vacuole Fluid-filled vesicle in many plant cells.
endomembrane system Series of interacting organelles (endoplasmic reticulum, Golgi bodies, vesicles) between nucleus and plasma membrane; produces lipids, proteins.
endoplasmic reticulum (ER) Organelle that is a continuous system of sacs and tubes extending from the nuclear envelope. Smooth ER makes lipids and breaks down carbohydrates and fatty acids; ribosomes on the surface of rough ER synthesize proteins.
Golgi body Organelle that modifies proteins and lipids, then packages the finished products into vesicles.
lysosome Enzyme-filled vesicle that breaks down cellular wastes and debris.
peroxisome Enzyme-filled vesicle that breaks down amino acids, fatty acids, and toxic substances.
vacuole A fluid-filled organelle that isolates or disposes of waste, debris, or toxic materials.
vesicle Small, membrane-enclosed organelle; different kinds store, transport, or break down their contents.

FIGURE 4.12 Some interactions among components of the endomembrane system.

❶ **Vesicles**
Vesicles are membrane-enclosed sacs that often form by budding from other components of the endomembrane system. Many transport substances among organelles of the ER, and to and from the plasma membrane. Other vesicles store or break down substances.

❷ **Rough ER**
Ribosomes attached to the rough ER use RNA in the cytoplasm to make polypeptides. The newly synthesized polypeptide chains enter rough ER, where they take on tertiary structure and assemble with other polypeptide chains.

ENDOPLASMIC RETICULUM

The membrane of the **endoplasmic reticulum** (**ER**) is an extension of the nuclear envelope. Its interconnected tubes and flattened sacs form a single compartment that houses many enzymes. Two kinds of ER, rough and smooth, are named for their appearance. Thousands of ribosomes that attach to the outer surface of rough ER give this organelle its "rough" appearance. These ribosomes make polypeptides that thread into the interior of the ER as they are assembled ❷. Inside the ER, the polypeptide chains fold and take on their tertiary structure, and many assemble with other polypeptide chains (Section 3.4). Cells that make, store, and secrete proteins have a lot of rough ER. For example, ER-rich cells in the pancreas make digestive enzymes that they secrete into the small intestine.

Some proteins made in rough ER become part of its membrane. Others migrate through the ER compartment to smooth ER. Smooth ER has no ribosomes, so it does not make its own proteins ❸. Some proteins that arrive in smooth ER are immediately packaged into vesicles for delivery elsewhere. Others are enzymes that stay and become part of the smooth ER. Some of these enzymes break down carbohydrates, fatty acids, and some drugs and poisons. Others make lipids for the cell's membranes.

GOLGI BODIES

A **Golgi body** has a folded membrane that often looks like a stack of pancakes ❹. Enzymes inside of it put finishing touches on proteins and lipids that have been delivered from ER. These enzymes attach phosphate groups or carbohydrates, and cleave certain proteins. The finished products (such as membrane proteins, proteins for secretion, and enzymes) are sorted and packaged in new vesicles. Some of the vesicles deliver their cargo to the plasma membrane; others become lysosomes.

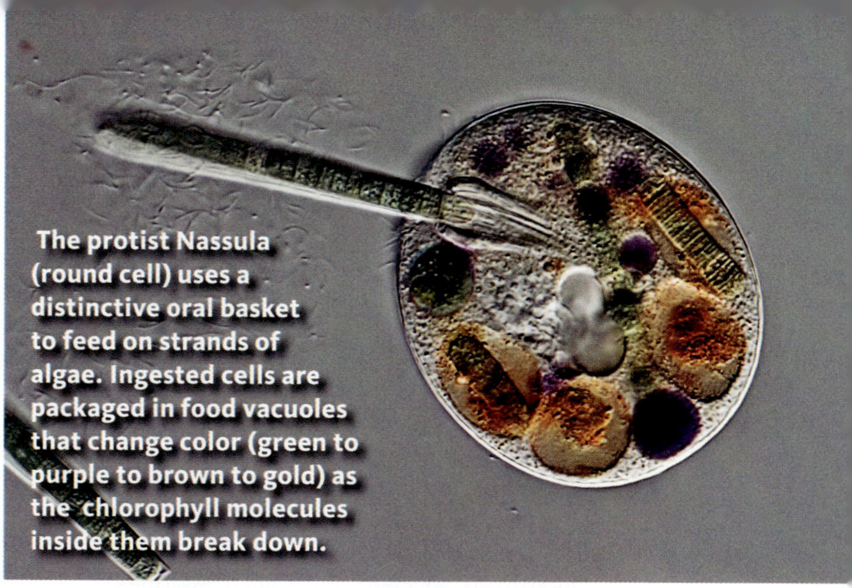

FIGURE 4.13 An example of vacuole function.

TAKE-HOME MESSAGE 4.6

Rough ER produces enzymes, membrane proteins, and secreted proteins. Smooth ER produces lipids and breaks down carbohydrates, fatty acids, and toxins.

Golgi bodies modify proteins and lipids.

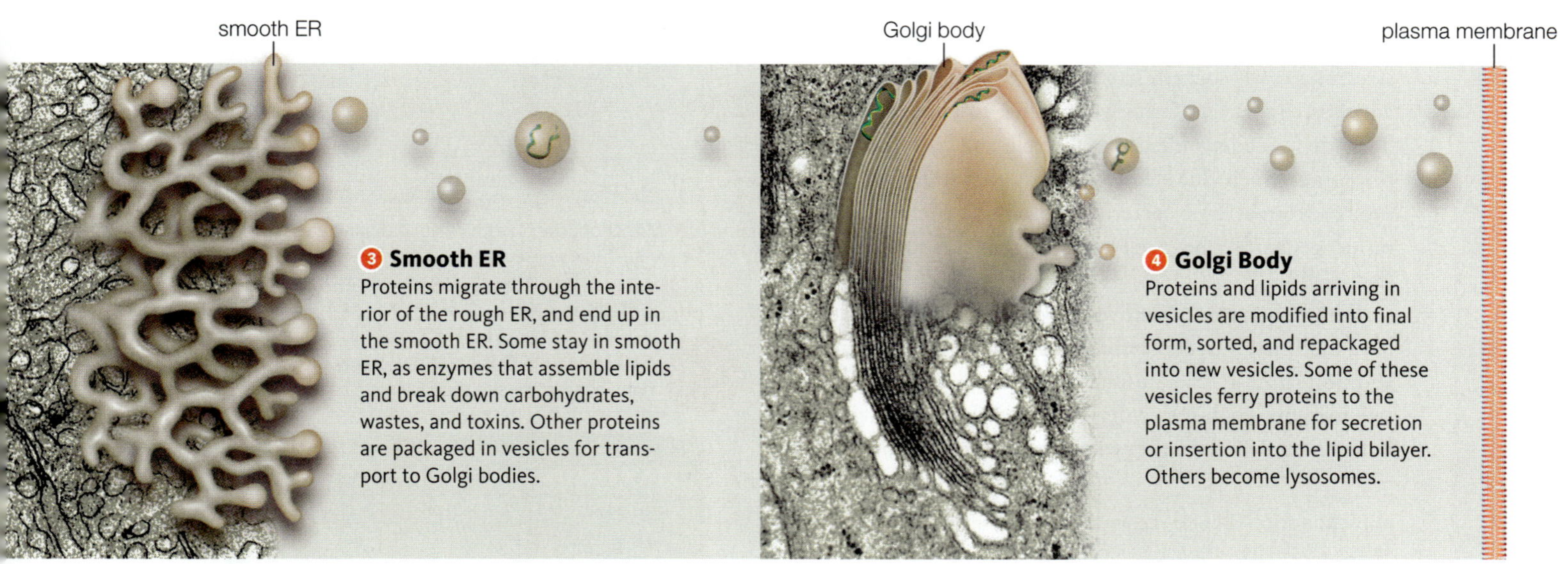

❸ **Smooth ER**
Proteins migrate through the interior of the rough ER, and end up in the smooth ER. Some stay in smooth ER, as enzymes that assemble lipids and break down carbohydrates, wastes, and toxins. Other proteins are packaged in vesicles for transport to Golgi bodies.

❹ **Golgi Body**
Proteins and lipids arriving in vesicles are modified into final form, sorted, and repackaged into new vesicles. Some of these vesicles ferry proteins to the plasma membrane for secretion or insertion into the lipid bilayer. Others become lysosomes.

CREDITS: (12) art, © Cengage Learning; (13) © Michael Plewka.

4.7 WHAT DO MITOCHONDRIA DO?

As you will see in Chapter 5, biologists think of the nucleotide ATP as a type of cellular currency because it carries energy between reactions. Cells require a lot of ATP. The most efficient way they can produce it is by aerobic respiration, a series of oxygen-requiring reactions that harvests the energy in sugars by breaking their bonds. In eukaryotes, aerobic respiration occurs inside organelles called **mitochondria** (singular, mitochondrion). With each breath, you are taking in oxygen mainly for the mitochondria in your trillions of aerobically respiring cells.

The structure of a mitochondrion is specialized for carrying out reactions of aerobic respiration. Each mitochondrion has two membranes, one highly folded inside the other (**FIGURE 4.14**). This arrangement creates two compartments: an outer one (between the two membranes), and an inner one (inside the inner membrane). Hydrogen ions accumulate in the outer compartment. The buildup pushes the ions across the inner membrane, into the inner compartment, and this flow drives ATP formation. Chapter 7 returns to the details of aerobic respiration.

Nearly all eukaryotic cells (including plant cells) have mitochondria, but the number varies by the type of cell and by the organism. For example, single-celled organisms such as yeast often have only one mitochondrion, but human skeletal muscle cells have a thousand or more. In general, cells that have the highest demand for energy tend to have the most mitochondria.

Typical mitochondria are between 1 and 4 micrometers in length. These organelles can change shape, split in two, branch, or fuse together. They resemble bacteria in size, form, and biochemistry. They have their own DNA, which is circular and otherwise similar to bacterial DNA. They divide independently of the cell, and have their own ribosomes. Such clues led to a theory that mitochondria evolved from aerobic bacteria that took up permanent residence inside a host cell (we return to this topic in Section 18.5).

Some eukaryotes that live in oxygen-free environments have modified mitochondria that produce hydrogen in addition to ATP. Like mitochondria, these organelles have two membranes, but they have lost the ability to divide independently because they lack their own DNA.

mitochondrion Double-membraned organelle that produces ATP by aerobic respiration in eukaryotes.

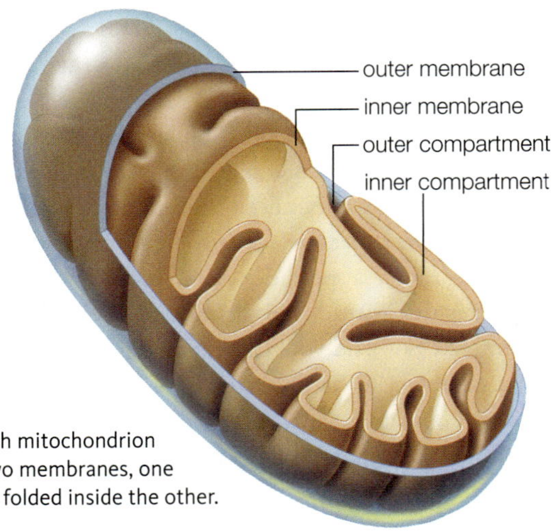

A Each mitochondrion has two membranes, one highly folded inside the other.

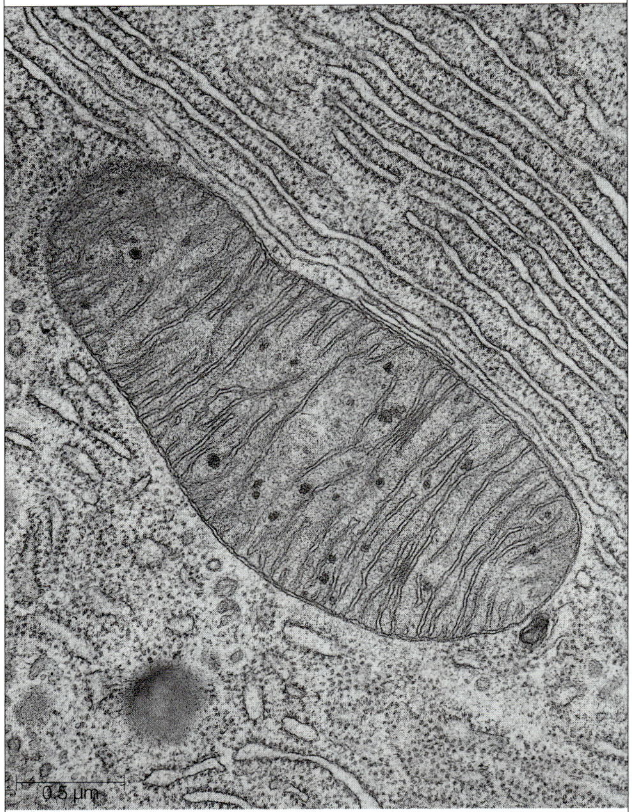

B Mitochondrion in a cell from bat pancreas.

FIGURE 4.14 {Animated} The mitochondrion, a eukaryotic organelle that specializes in producing ATP.

FIGURE IT OUT: What organelle is visible in the upper right-hand corner of the TEM?

Answer: Rough ER

TAKE-HOME MESSAGE 4.7

Mitochondria are eukaryotic organelles specialized to produce ATP by aerobic respiration.

4.8 WHAT ARE PLASTIDS?

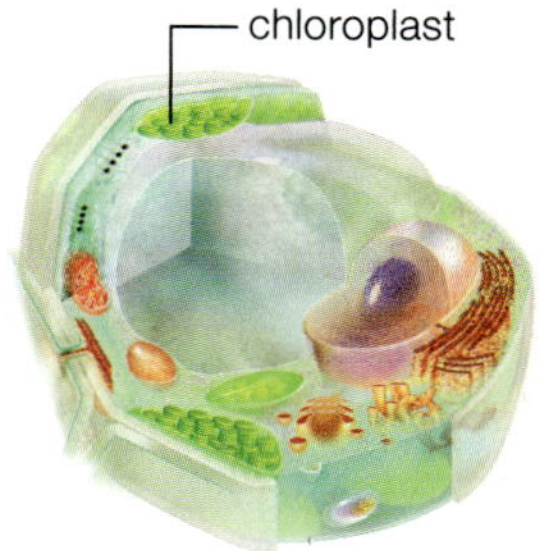

Plastids are double-membraned organelles that function in photosynthesis, storage, or pigmentation in plant and algal cells. Photosynthetic cells of plants and many protists contain **chloroplasts**, which are plastids specialized for photosynthesis (FIGURE 4.15). Most chloroplasts are oval or disk-shaped. Each has two outer membranes enclosing a semifluid interior, the stroma, that contains enzymes and the chloroplast's own DNA. In the stroma, a third, highly folded membrane forms a single, continuous compartment. Photosynthesis occurs at this inner membrane.

The innermost membrane of a chloroplast incorporates many pigments, including a green one called chlorophyll (the abundance of chlorophyll in plant cell chloroplasts is the reason most plants are green). During photosynthesis, these pigments capture energy from sunlight, and pass it to other molecules that require energy to make ATP. The resulting ATP is used inside the stroma to build sugars from carbon dioxide and water. (Chapter 6 returns to details of these processes.) In many ways, chloroplasts resemble the photosynthetic bacteria that they evolved from.

Chromoplasts are plastids that make and store pigments other than chlorophylls. They often contain red or orange carotenoids that color flowers, leaves, roots, and fruits (FIGURE 4.16). Chromoplasts are related to chloroplasts, and the two types of plastids are interconvertible. For example, as fruits such as tomatoes ripen, green chloroplasts in their cells are converted to red chromoplasts, so the color of the fruit changes.

Amyloplasts are unpigmented plastids that make and store starch grains. They are notably abundant in cells of stems, tubers (underground stems), fruits, and seeds. Like chromoplasts, amyloplasts are related to chloroplasts, and one type can change into the other. Starch-packed amyloplasts are dense and heavy compared to cytoplasm; in some plant cells, they function as gravity-sensing organelles.

---

chloroplast Organelle of photosynthesis in the cells of plants and photosynthetic protists.
plastid One of several types of double-membraned organelles in plants and algal cells; for example, a chloroplast or amyloplast.

TAKE-HOME MESSAGE 4.8

Plastids occur in plants and some protists; they function in photosynthesis, storage, and pigmentation.

Chloroplasts are plastids that carry out photosynthesis.

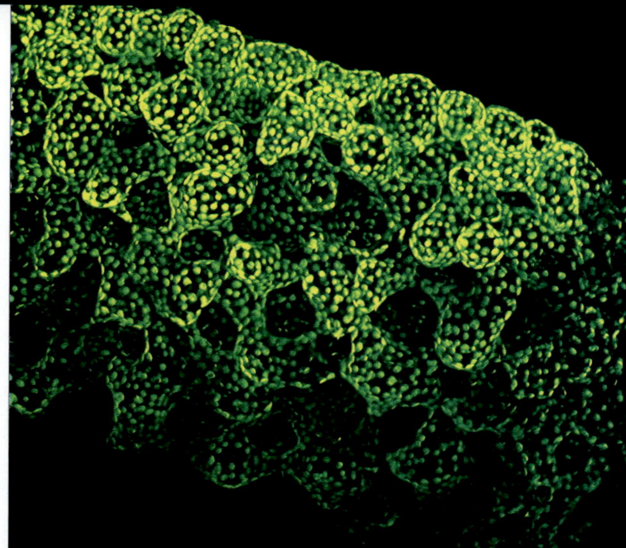

A Chloroplast-packed cells make up a leaf of a flowering plant.

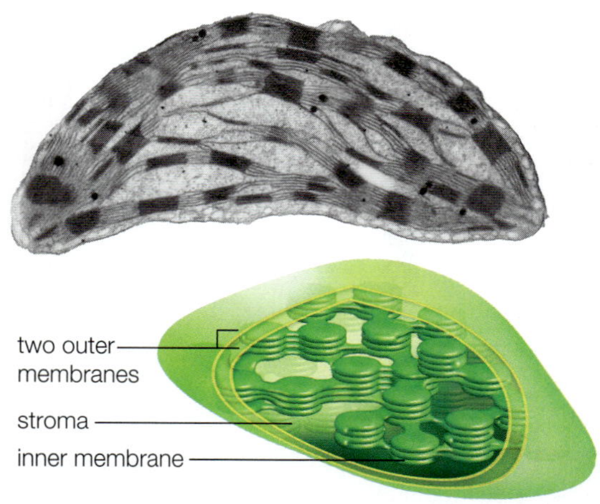

B Each chloroplast has two outer membranes. Photosynthesis occurs at a much-folded inner membrane. The electron micrograph shows a chloroplast from a leaf of corn.

FIGURE 4.15 **{Animated}** The chloroplast.

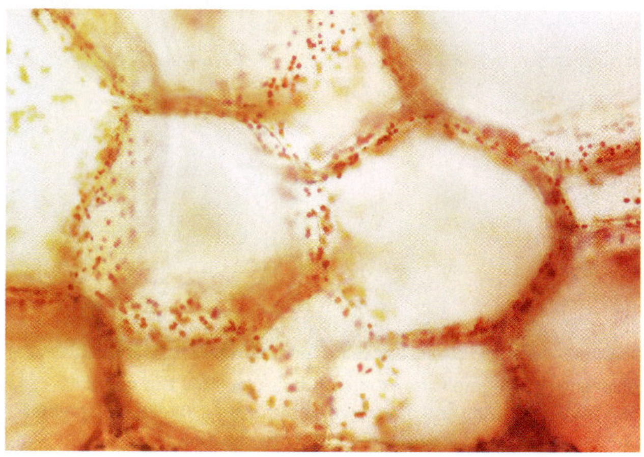

FIGURE 4.16 **Chromoplasts.** The color of a red bell pepper arises from chromoplasts in its cells.

4.9 WHAT IS A CYTOSKELETON?

Between the nucleus and plasma membrane of all eukaryotic cells is a system of interconnected protein filaments collectively called the **cytoskeleton**. Elements of the cytoskeleton reinforce, organize, and move cell structures, and often the whole cell. Some are permanent; others form only at certain times.

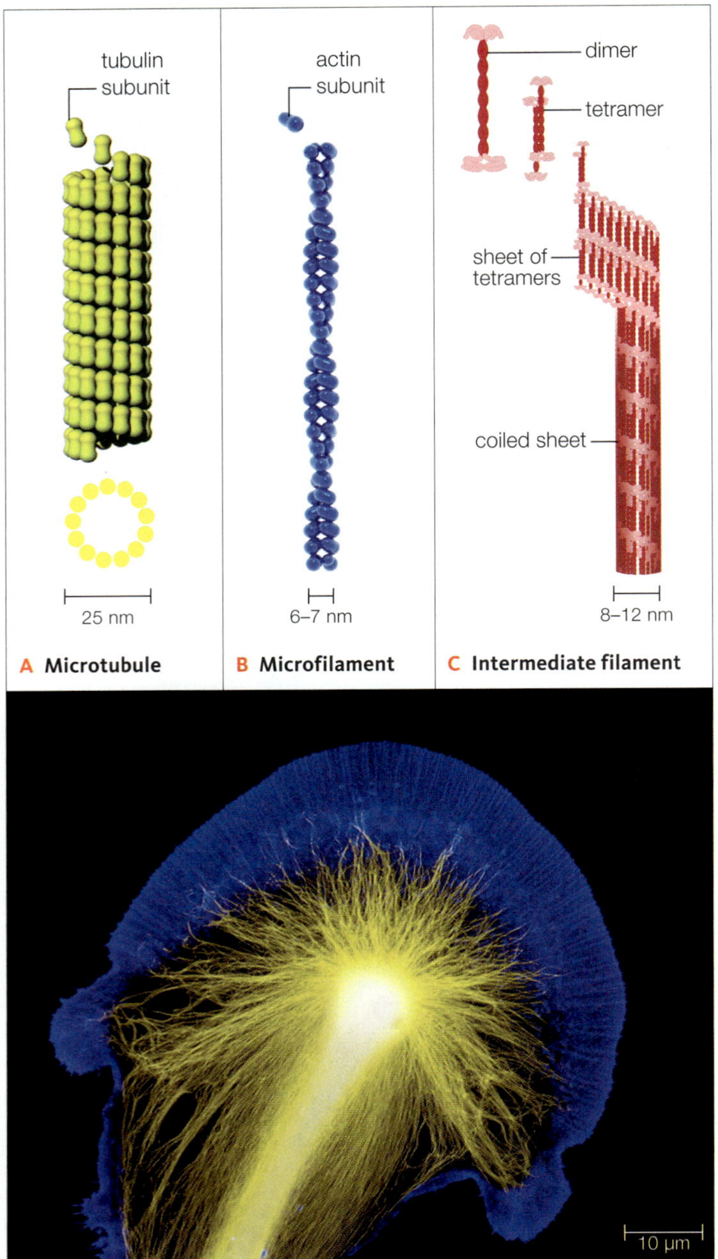

D A fluorescence micrograph shows microtubules (yellow) and microfilaments (blue) in the growing end of a nerve cell. These cytoskeletal elements support and guide the cell's lengthening in a particular direction.

FIGURE 4.17 {Animated} Cytoskeletal elements.

Microtubules are long, hollow cylinders that consist of subunits of the protein tubulin (**FIGURE 4.17A**). They form a dynamic scaffolding for many cellular processes, rapidly assembling when they are needed, disassembling when they are not. For example, before a eukaryotic cell divides, microtubules assemble, separate the cell's duplicated DNA molecules, then disassemble. As another example, microtubules that form in the growing end of a young nerve cell support its lengthening in a particular direction (**FIGURE 4.17D**).

Microfilaments are fibers that consist primarily of subunits of the globular protein actin (**FIGURE 4.17B**). These fine fibers strengthen or change the shape of eukaryotic cells, and have a critical function in cell migration, movement, and contraction. Crosslinked, bundled, or gel-like arrays of them make up the **cell cortex**, a reinforcing mesh under the plasma membrane. Microfilaments also connect plasma membrane proteins to other proteins inside the cell.

Intermediate filaments are the most stable elements of the cytoskeleton, forming a framework that lends structure and resilience to cells and tissues in multicelled organisms. Several types of intermediate filaments are assembled from different proteins (**FIGURE 4.17C**). For example, intermediate filaments that make up your hair consist of keratin, a fibrous protein (Section 3.4). Intermediate filaments that form the nuclear lamina consist of lamins, another type of fibrous protein.

Motor proteins that associate with cytoskeletal elements move cell parts when energized by a phosphate-group transfer from ATP (Section 3.6). A cell is like a bustling train station, with molecules and structures being moved continuously throughout its interior. Motor proteins are like freight trains, dragging cellular cargo along tracks of microtubules and microfilaments (**FIGURE 4.18**). The motor protein myosin interacts with microfilaments to bring about muscle cell contraction. Another motor protein, dynein, interacts with microtubules to bring about movement of flagella and cilia in eukaryotes. Eukaryotic flagella whip back and forth to propel cells such as sperm (*right*) through fluid. **Cilia** (singular, cilium) are short, hairlike structures that project from the surface of some cells. The coordinated

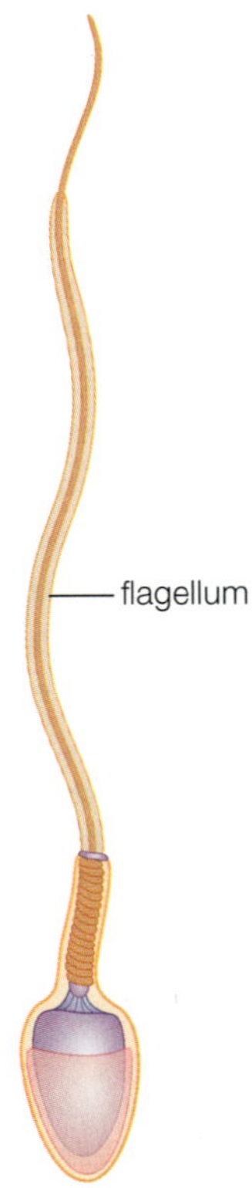

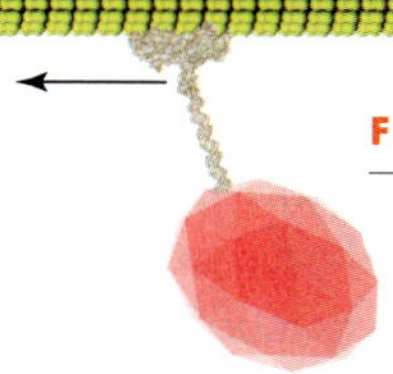

FIGURE 4.18 {Animated} Motor proteins. Here, kinesin (tan) drags a pink vesicle as it inches along a microtubule.

waving of many cilia propels some cells through fluid, and stirs fluid around other cells that are stationary. The waving movement of eukaryotic flagella and cilia, which differs from the propeller-like rotation of prokaryotic flagella, arises from their internal architecture. Microtubules extend lengthwise through them, in what is called a 9+2 array (**FIGURE 4.19**). The array consists of nine pairs of microtubules ringing another pair in the center. The microtubules grow from a barrel-shaped organelle called the **centriole**, which remains below the finished array as a **basal body**.

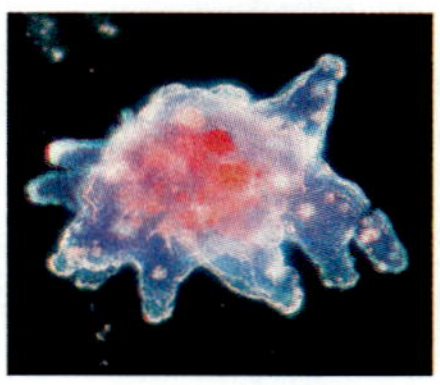

Amoebas (*left*) and other types of eukaryotic cells form **pseudopods**, or "false feet." As these temporary, irregular lobes bulge outward, they move the cell and engulf a target such as prey. Elongating microfilaments force the lobe to advance in a steady direction. Motor proteins that are attached to the microfilaments drag the plasma membrane along with them.

basal body Organelle that develops from a centriole.
cell cortex Mesh of cytoskeletal elements under a plasma membrane.
centriole Barrel-shaped organelle from which microtubules grow.
cilium Short, movable structure that projects from the plasma membrane of some eukaryotic cells.
cytoskeleton Network of interconnected protein filaments that support, organize, and move eukaryotic cells and their parts.
intermediate filament Stable cytoskeletal element that structurally supports cell membranes and tissues.
microfilament Cytoskeletal element that is a fiber of actin subunits. Reinforces cell membranes; functions in muscle contractions.
microtubule Cytoskeletal element involved in movement; hollow filament of tubulin subunits.
motor protein Type of energy-using protein that interacts with cytoskeletal elements to move the cell's parts or the whole cell.
pseudopod A temporary protrusion that helps some eukaryotic cells move and engulf prey.

TAKE-HOME MESSAGE 4.9

A cytoskeleton of protein filaments is the basis of eukaryotic cell shape, internal structure, and movement.

Microtubules organize eukaryotic cells and help move their parts. Networks of microfilaments reinforce cell shape and function in movement. Intermediate filaments strengthen and maintain the shape of cell membranes and tissues, and form external structures such as hair.

When energized by ATP, motor proteins move along tracks of microtubules and microfilaments. As part of cilia, flagella, and pseudopods, they can move the whole cell.

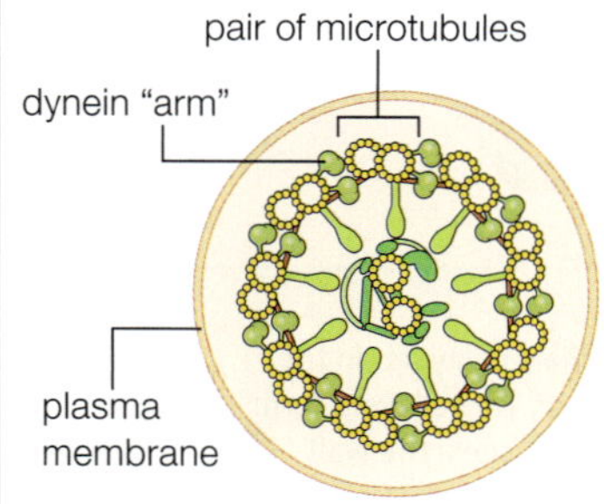

A A 9+2 array, which consists of a ring of nine pairs of microtubules plus one pair at their core, runs lengthwise through a eukaryotic flagellum or cilium. Stabilizing spokes and linking elements connect the microtubules and keep them aligned in this pattern. Projecting from each pair of microtubules in the outer ring are "arms" of the motor protein dynein.

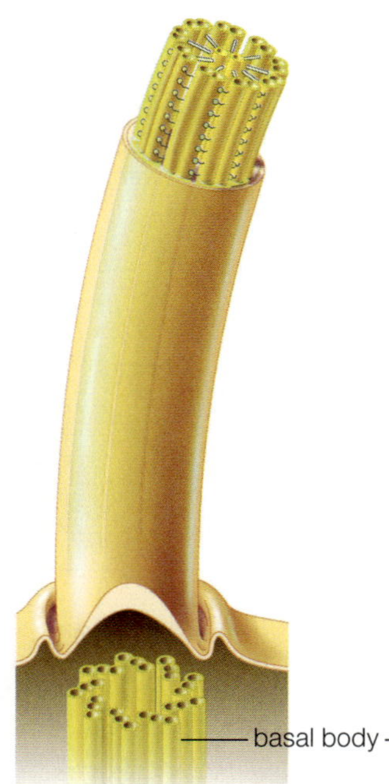

B Microtubules of a developing 9+2 array grow from a centriole, which remains below the finished array as a basal body. The micrograph *below* shows basal bodies underlying cilia of the protist pictured in **FIGURE 4.5**.

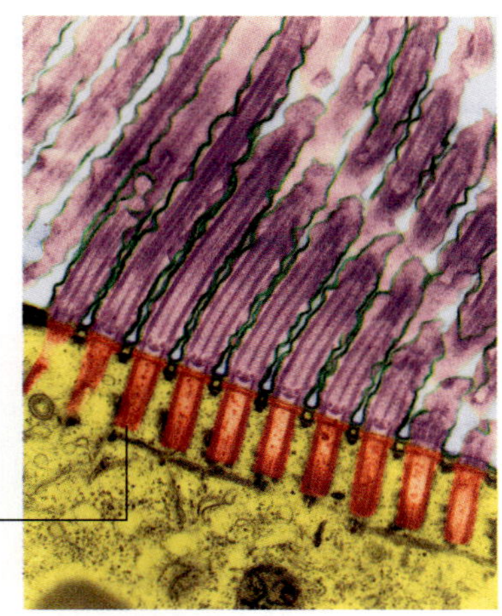

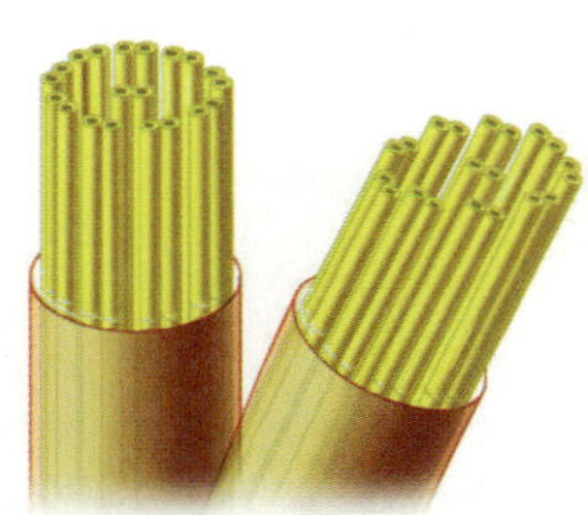

C Phosphate-group transfers from ATP cause the dynein arms in a 9+2 array to repeatedly bind the adjacent pair of microtubules, bend, and then disengage. The dynein arms "walk" along the microtubules, so adjacent microtubule pairs slide past one another. The short, sliding strokes of the dynein arms occur in a coordinated sequence around the ring, down the length of the microtubules. The movement causes the entire structure to bend.

FIGURE 4.19 {Animated} How eukaryotic flagella and cilia move.

4.10 WHAT STRUCTURES FORM ON THE OUTSIDE OF EUKARYOTIC CELLS?

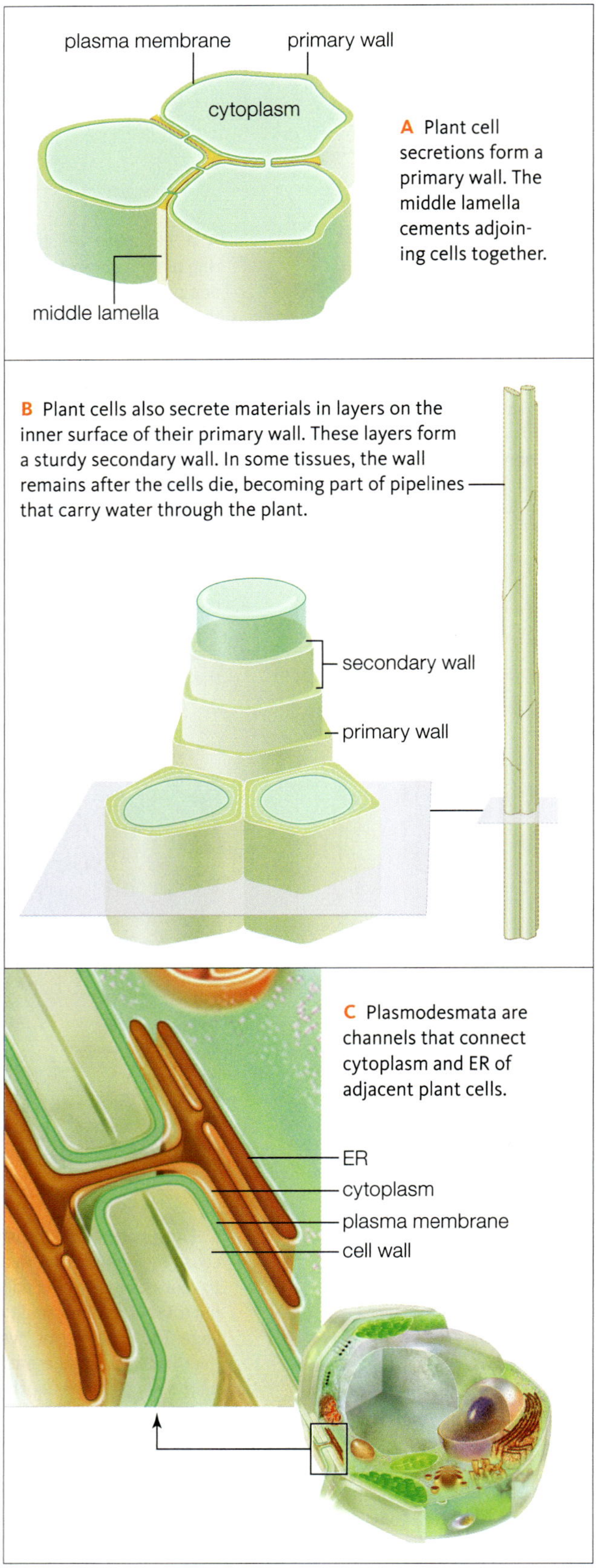

FIGURE 4.20 {Animated} Plant cell walls.

CELL MATRIXES

Many cells secrete an **extracellular matrix (ECM)**, a complex mixture of molecules that often includes polysaccharides and fibrous proteins. The composition and function of ECM vary by the type of cell that secretes it.

A cell wall is an example of ECM. Among eukaryotes, fungi and some protists have walls, as do plant cells (as shown in this chapter's opening photo). The composition of the wall differs among these groups. Like a prokaryotic cell wall, a eukaryotic cell wall is porous: Water and solutes easily cross it on the way to and from the plasma membrane.

In plants, the cell wall forms as a young cell secretes pectin and other polysaccharides onto the outer surface of its plasma membrane. The sticky coating is shared between adjacent cells, and it cements them together. Each cell then forms a **primary wall** by secreting strands of cellulose into the coating. Some of the pectin coating remains as the middle lamella, a sticky layer in between the primary walls of abutting plant cells (**FIGURE 4.20A**).

Being thin and pliable, a primary wall allows a growing plant cell to enlarge and change shape. In some plants, mature cells secrete material onto the primary wall's inner surface. These deposits form a firm **secondary wall** (**FIGURE 4.20B**). One of the materials deposited is **lignin**, an organic compound that makes up as much as 25 percent of the secondary wall of cells in older stems and roots. Lignified plant parts are stronger, more waterproof, and less susceptible to plant-attacking organisms than younger tissues.

Animal cells have no walls, but some types secrete an extracellular matrix called basement membrane. Despite the name, basement membrane is not a cell membrane because it does not consist of a lipid bilayer. Rather, it is a sheet of fibrous material that structurally supports and organizes

FIGURE 4.21 A plant ECM. Section through a plant leaf showing cuticle, a protective covering secreted by living cells.

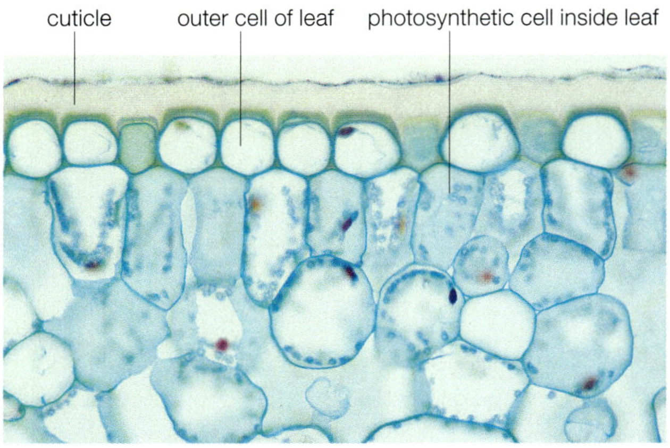

tissues, and it has roles in cell signaling. Bone is an ECM composed mostly of the fibrous protein collagen, and hardened by deposits of calcium and phosphorus.

A **cuticle** is a type of ECM secreted by cells at a body surface. In plants, a cuticle of waxes and proteins helps stems and leaves fend off insects and retain water (**FIGURE 4.21**). Crabs, spiders, and other arthropods have a cuticle that consists mainly of chitin (Section 3.2).

CELL JUNCTIONS

In multicelled species, cells can interact with one another and their surroundings by way of cell junctions. **Cell junctions** are structures that connect a cell directly to other cells and to its environment. Cells send and receive substances and signals through some junctions. Other junctions help cells recognize and stick to each other and to ECM.

Three types of cell junctions are common in animal tissues (**FIGURE 4.22A**). In tissues that line body surfaces and internal cavities, rows of adhesion proteins form **tight junctions** between the plasma membranes of adjacent cells. These junctions prevent body fluids from seeping between the cells (**FIGURE 4.22B**). For example, the lining of the stomach is leak-proof because tight junctions seal its cells together. These junctions keep gastric fluid, which contains acid and destructive enzymes, safely inside the stomach. If a bacterial infection damages the stomach lining, gastric fluid leaks into and damages the underlying layers. A painful peptic ulcer is the result.

Adhering junctions, which fasten cells to one another and to basement membrane, also consist of adhesion proteins. These junctions make a tissue quite strong because they connect to cytoskeletal elements inside the cells. Contractile tissues (such as heart muscle) have a lot of adhering junctions, as do tissues subject to abrasion or stretching (such as skin).

A Three types of cell junctions in animal tissues.

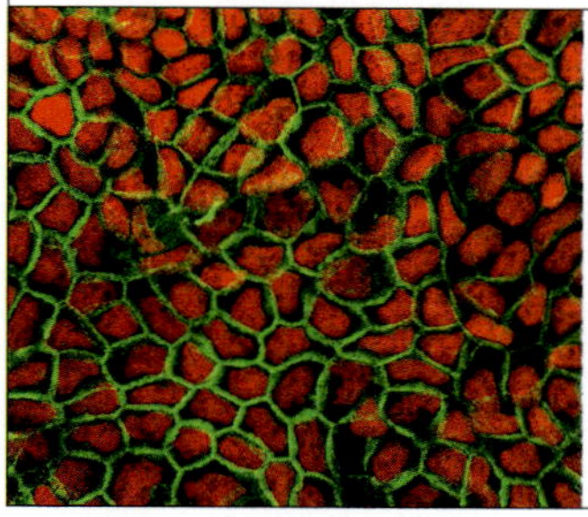

B In this fluorescence micrograph, a profusion of tight junctions (green) seals abutting surfaces of kidney cell membranes and forms a leak-proof tissue. The DNA in each cell nucleus appears red.

FIGURE 4.22 {Animated} Cell junctions.

Gap junctions are closable channels that connect the cytoplasm of adjoining animal cells. When open, they permit water, ions, and small molecules to pass directly from the cytoplasm of one cell to another. These channels allow entire regions of cells to respond to a single stimulus. Heart muscle and other tissues in which the cells perform a coordinated action have many gap junctions.

In plants, open channels called **plasmodesmata** (singular, plasmodesma) extend across plant cell walls to connect the cytoplasm of adjacent cells. Like gap junctions, plasmodesmata also allow substances to flow quickly from cell to cell.

adhering junction Cell junction composed of adhesion proteins that connect to cytoskeletal elements. Fastens cells to each other and basement membrane.
cell junction Structure that connects a cell to another cell or to extracellular matrix.
cuticle Secreted covering at a body surface.
extracellular matrix (ECM) Complex mixture of cell secretions; its composition and function vary by cell type.
gap junction Cell junction that forms a closable channel across the plasma membranes of adjoining animal cells.
lignin Material that strengthens cell walls of vascular plants.
plasmodesmata Cell junctions that form an open channel between the cytoplasm of adjacent plant cells.
primary wall The first cell wall of young plant cells.
secondary wall Lignin-reinforced wall that forms inside the primary wall of a plant cell.
tight junctions Arrays of adhesion proteins that join epithelial cells and collectively prevent fluids from leaking between them.

TAKE-HOME MESSAGE 4.10

Many cells secrete an extracellular matrix (ECM). ECM varies in composition and function depending on the cell type.

Plant cells, fungi, and some protists have a porous wall around their plasma membrane. Animal cells do not have walls.

Cell junctions structurally and functionally connect cells in tissues. In animal tissues, cell junctions also connect cells with basement membrane.

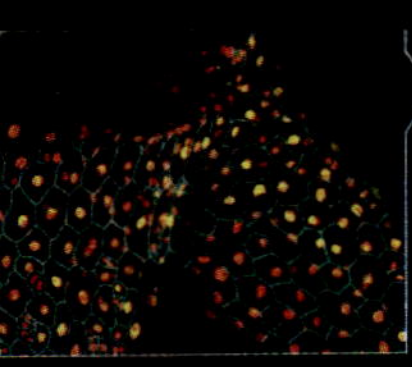

4.11 WHAT IS LIFE?

You learned in Section 1.1 that the cell is the smallest unit with the properties of life. In this chapter, you learned that a living cell has at minimum a plasma membrane, cytoplasm, and a region of DNA; most cells have many other components in addition to these things. So what is it, exactly, that makes it alive? A cell does not spring to life from cellular components mixed in the right amounts and proportions. According to evolutionary biologist Gerald Joyce, the simplest definition of life might well be "that which is squishy." He says, "Life, after all, is protoplasmic and cellular. It is made up of cells and organic stuff and is undeniably squishy."

However, defining life more unambiguously than "squishy" is challenging, if not impossible. We can more easily describe what sets the living apart from the nonliving, but even that can be tricky. For example, living things have a high proportion of the organic molecules of life, but so do the remains of dead organisms in seams of coal. Living things use energy to reproduce themselves, but computer viruses, which are arguably not alive, can do that too.

So how do biologists, who study life as a profession, define it? The short answer is that their best definition is a long list of properties that collectively describe living things. You already know about two of these properties:

1. They make and use the organic molecules of life.
2. They consist of one or more cells.

The remainder of this book details the others:

3. They engage in self-sustaining biological processes such as metabolism and homeostasis.
4. They change over their lifetime, for example by growing, maturing, and aging.
5. They use DNA as their hereditary material when they reproduce.
6. They have the collective capacity to change over successive generations, for example by adapting to environmental pressures.

Collectively, these properties characterize living things as different from nonliving things.

TAKE-HOME MESSAGE 4.11

- We describe the characteristic of "life" in terms of a set of properties. The set is unique to living things.
- In living things, the molecules of life are organized as one or more cells that engage in self-sustaining biological processes.
- Organisms make and use the organic molecules of life.
- Living things change over lifetimes, and over generations.

PEOPLE MATTER

National Geographic Explorer
DR. KEVIN PETER HAND

Today's weather forecast for Europa, Jupiter's fourth-largest moon, is –280°F. A layer of ice several miles thick coats its fractured surface, with 1,000-foot ice cliffs piercing a pitch-black sky. It is devoid of atmosphere, bombarded by fierce radiation—and National Geographic Explorer Kevin Hand can hardly wait to get there.

Hand works at the Jet Propulsion Laboratory (JPL), where he is helping NASA plan a mission to Jupiter's moons—an orbiting probe that will give Earthlings a closer look at Europa. Beneath Europa's icy shell lies a vast global liquid water ocean that Hand thinks could be a great place for life. "I want to know if DNA is the only game in town. Are there different biochemical pathways that could lead to other kinds of life? That's at the heart of why I want to go to Europa—to find something living in that ocean we can poke at and use to understand and define life in a much more comprehensive way."

For the first time, we have the technological capability of taking our search for life to distant worlds. "Nevertheless, our understanding of life as a phenomenon remains largely qualitative and poorly constrained," Hand says. In other words, without an exact definition of "life," how do we determine whether it exists on Europa? "Biology preferentially uses specific organic subunits to build larger compounds while abiotic organic chemistry proceeds randomly," says Hand. "The structures of life arise from a relatively small set of universal building blocks; thus, when we search for life we look for patterns indicative of life's structural biases."

4.12 Application: FOOD FOR THOUGHT

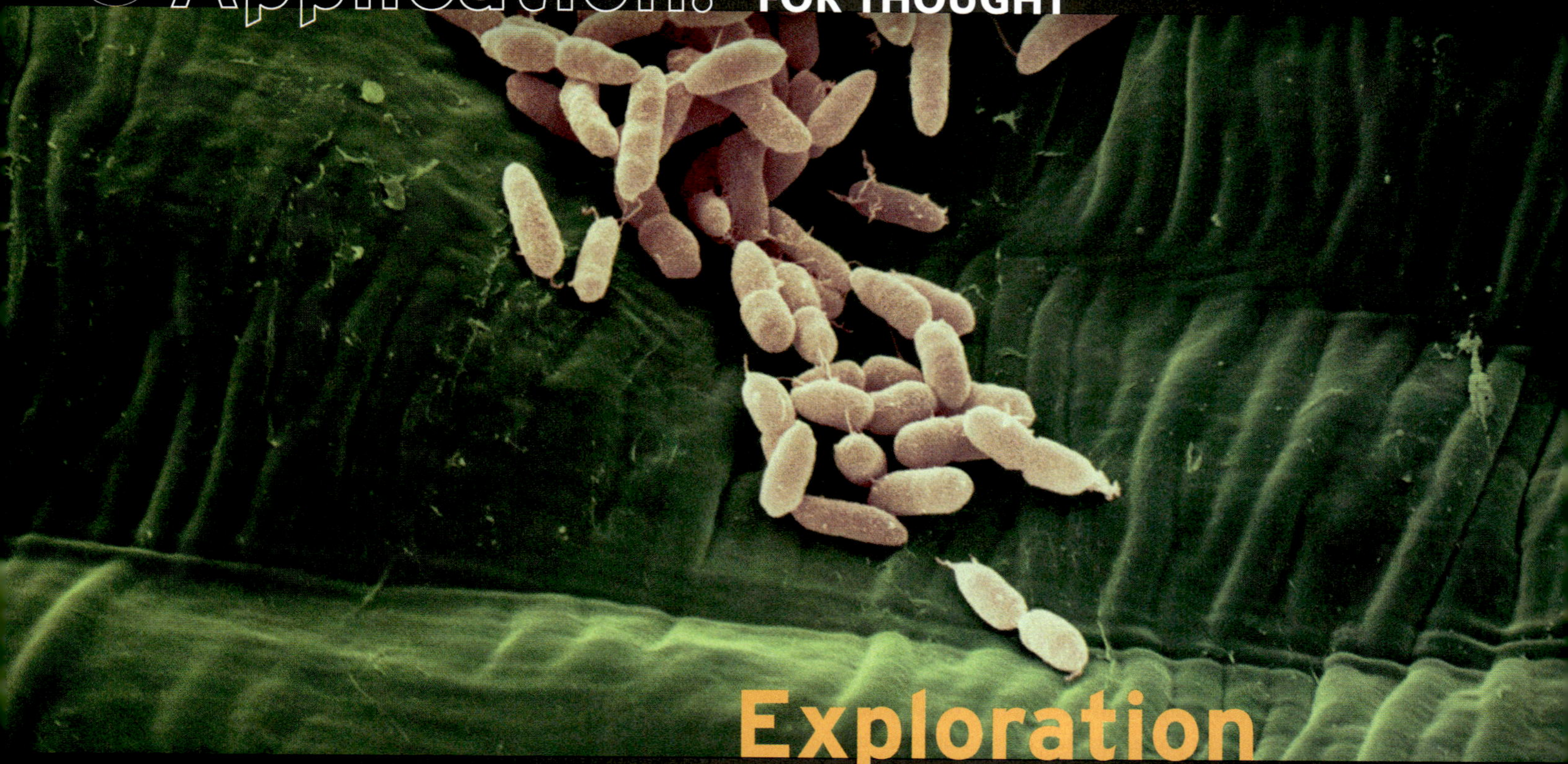

FIGURE 4.23 *Escherichia coli* cells sticking to the surface of a lettuce leaf. Some strains of this bacteria can cause a serious intestinal illness when they contaminate human food.

CELL FOR CELL, BACTERIA THAT LIVE IN AND ON A HUMAN BODY OUTNUMBER THE PERSON'S OWN CELLS BY ABOUT TEN TO ONE. One of the most common intestinal bacteria of warm-blooded animals (including humans) is *Escherichia coli*. Most of the hundreds of types, or strains, of *E. coli* are harmless, but a few strains make a toxic protein that can severely damage the lining of the intestine. After ingesting as few as ten cells of a toxic strain, a person may become ill with severe cramps and bloody diarrhea that lasts up to ten days. In some people, complications of infection result in kidney failure, blindness, paralysis, and death. Each year, about 265,000 people in the United States become infected with toxin-producing *E. coli*.

Strains of *E. coli* that are toxic to people live in the intestines of other animals—mainly cattle, deer, goats, and sheep—apparently without sickening them. Humans are exposed to the bacteria when they come into contact with feces of animals that harbor it, for example, by eating contaminated ground beef. During slaughter, meat can come into contact with feces. Bacteria in the feces stick to the meat, then get thoroughly mixed into it during the grinding process. Unless contaminated meat is cooked to at least 71°C (160°F), live bacteria will enter the digestive tract of whoever eats it.

People also become infected with toxic *E. coli* by eating fresh fruits and vegetables that have come into contact with animal feces. Washing produce with water does not remove all of the bacteria because they are sticky (FIGURE 4.23). In June 2011, more than 4,000 people in Germany and France were sickened after eating sprouts, and 49 of them died. The outbreak was traced to a single shipment of contaminated sprout seeds from Egypt.

The impact of such outbreaks, which occur with unfortunate regularity, extends beyond casualties. The contaminated sprouts cost growers in the European Union at least $600 million in lost sales. In 2011 alone, the United States Department of Agriculture (USDA) recalled 36.7 million pounds of ground meat products contaminated with toxic bacteria, at a cost in the billions of dollars. Such costs are eventually passed to taxpayers and consumers.

Food growers and processors are implementing new procedures intended to reduce the number and scope of these outbreaks. Meat and produce are being tested for some bacteria before sale, and improved documentation should allow a source of contamination to be pinpointed more quickly.

Summary

SECTION 4.1 **Cell theory** is the foundation of modern biology. By this theory, all organisms consist of one or more cells; the cell is the smallest unit of life; each new cell arises from another, preexisting cell; and a cell passes hereditary material to its offspring.

All cells start out life with **cytoplasm**, DNA, and a **plasma membrane** that controls the types and kinds of substances that cross it. Most cells have many additional components (**TABLE 4.5** and **FIGURE 4.24**). In eukaryotes, a cell's DNA is contained within a **nucleus**, which is a membrane-enclosed **organelle**.

A cell's surface area increases with the square of its diameter, while its volume increases with the cube. This **surface-to-volume ratio** limits cell size and influences cell (and body) shape.

SECTION 4.2 Most cells are far too small to see with the naked eye, so we use microscopes to observe them. Different types of microscopes and techniques reveal different internal and external details of cells.

SECTION 4.3 A cell membrane is a mosaic of proteins and lipids (mainly phospholipids) organized as a lipid bilayer. The membranes of bacteria and eukaryotic cells can be described as a **fluid mosaic**; those of archaea are not fluid. Proteins contribute to membrane function. All cell membranes have enzymes, and all have **transport proteins** that help substances move across the membrane. Plasma membranes also incorporate **receptor proteins** that bind specific substances, **adhesion proteins** that lock cells together in tissues, and **recognition proteins** that identify a cell as belonging to a tissue or body.

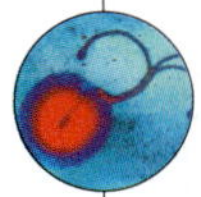

SECTION 4.4 Bacteria and archaea, informally grouped as prokaryotes, are the most diverse forms of life that we know about. These single-celled organisms have no nucleus, but they do have **nucleoids** and **ribosomes**. Many also have a protective, rigid **cell wall** and a sticky capsule, and some have motile structures (**flagella**) and other projections (**pili**). There are often **plasmids** in addition to the single circular molecule of DNA. Bacteria and other microbial organisms may live together in a shared mass of slime as **biofilms**.

SECTION 4.5 All eukaryotic cells start out life with a nucleus and other membrane-enclosed organelles. Membranes allow organelles to compartmentalize tasks and substances that may be sensitive or dangerous to the rest of the cell. A nucleus protects and controls access to a eukaryotic cell's DNA. A double membrane studded with pores constitutes the **nuclear envelope**. The pores serve as gateways for molecules passing into and out of the nucleus.

TABLE 4.5

Summary of Typical Components of Cells

Cell Component	Main Function(s)	Bacteria, Archaea	Eukaryotes: Protists	Fungi	Plants	Animals
Cell wall	Protection, structural support	✔	✔	✔	✔	None
Plasma membrane	Control of substances moving into and out of cell	✔	✔	✔	✔	✔
Nucleus	Protecting and controlling access to DNA	None	✔	✔	✔	✔
DNA	Encoding of hereditary information	✔	✔	✔	✔	✔
RNA	Protein synthesis	✔	✔	✔	✔	✔
Ribosome	Protein synthesis	✔	✔	✔	✔	✔
Endoplasmic reticulum (ER)	Protein, lipid synthesis; carbohydrate and fatty acid breakdown	None	✔	✔	✔	✔
Golgi body	Final modification of proteins; lipid assembly	None	✔	✔	✔	✔
Lysosome	Intracellular digestion	None	✔	✔	✔	✔
Peroxisome	Breakdown of fatty acids, amino acids, and toxins	None	✔	✔	✔	✔
Mitochondrion	Production of ATP by aerobic respiration	None	✔	✔	✔	✔
Photosynthetic pigments	Capturing light for photosynthesis	✔	✔	None	✔	None
Chloroplast	Photosynthesis; starch storage	None	✔	None	✔	None
Vacuole	Isolation and breakdown of food, wastes, toxins	None	✔	✔	✔	None
Vesicle	Storage, transport, or breakdown of contents	✔	✔	✔	✔	✔
Flagellum	Locomotion through fluid surroundings	✔	✔	✔	✔	✔
Cilium	Movement through (and of) fluid	✔	✔	None	✔	✔
Cytoskeleton	Physical reinforcement; internal organization; movement of the cell and its parts	✔	✔	✔	✔	✔

A Typical plant cell components.

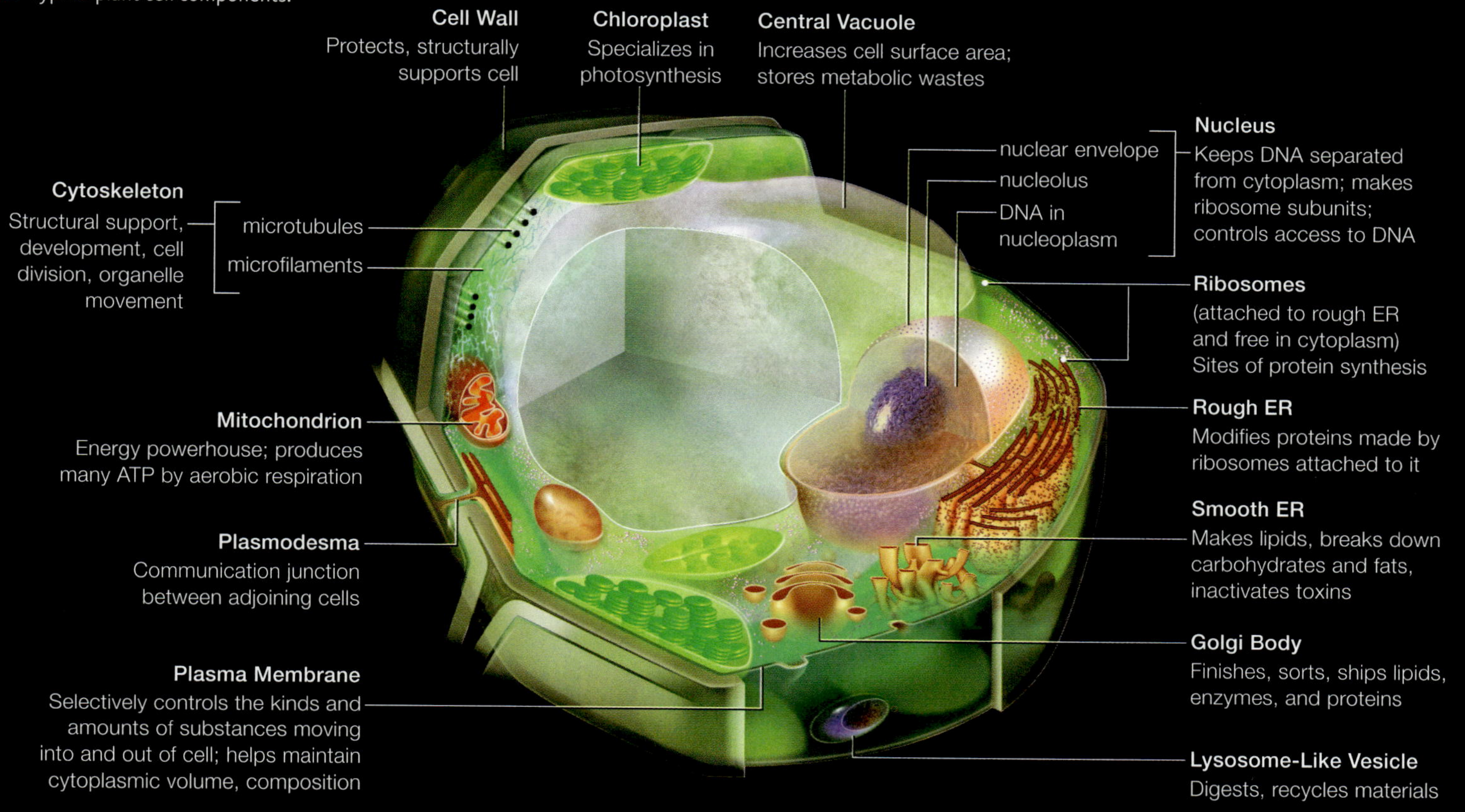

B Typical animal cell components.

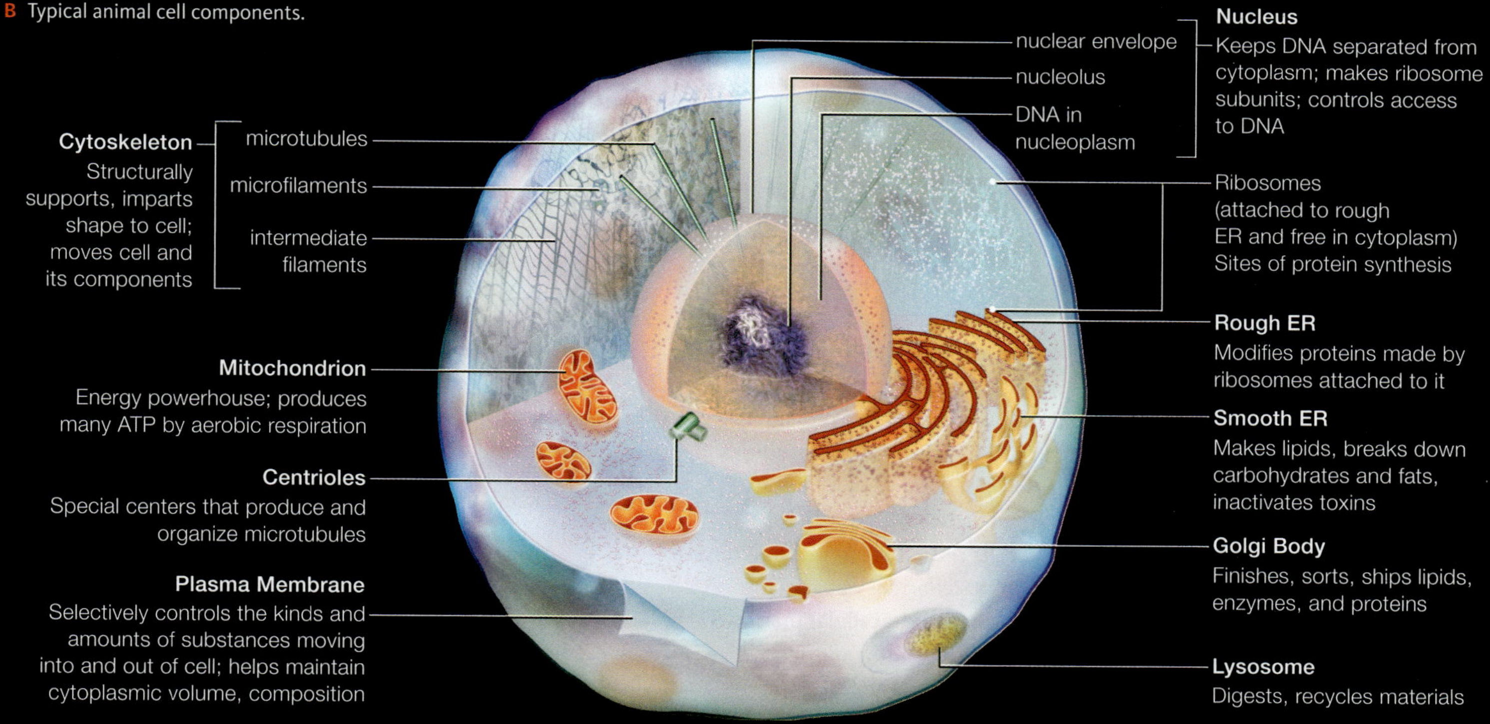

FIGURE 4.24 {Animated} Organelles and structures typical of A plant cells and B animal cells.

Summary continued

Inside the nuclear envelope, the cell's DNA is suspended in viscous **nucleoplasm**. Also inside the nucleus, ribosome subunits are assembled in dense, irregularly shaped areas called **nucleoli**.

SECTION 4.6 The **endomembrane system** is a series of organelles (endoplasmic reticulum, Golgi bodies, vesicles) that interact mainly to make lipids, enzymes, and proteins for insertion into membranes or secretion. **Endoplasmic reticulum (ER)** is a continuous system of sacs and tubes extending from the nuclear envelope. Ribosome-studded rough ER makes proteins; smooth ER makes lipids and breaks down carbohydrates and fatty acids. **Golgi bodies** modify proteins and lipids before sorting them into vesicles. Different types of **vesicles** store, break down, or transport substances through the cell. Enzymes in **peroxisomes** break down substances such as amino acids, fatty acids, and toxins. **Lysosomes** contain enzymes that break down cellular wastes and debris. Fluid-filled **vacuoles** store or break down waste, food, and toxins. Fluid pressure inside a **central vacuole** keeps plant cells plump, thus keeping plant parts firm.

SECTION 4.7 Double-membraned **mitochondria** specialize in making ATP by breaking down organic compounds in the oxygen-requiring metabolic pathway of aerobic respiration.

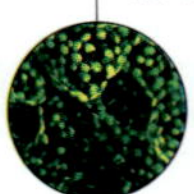

SECTION 4.8 Different types of **plastids** are specialized for photosynthesis or storage in plants and algal cells. In eukaryotes, photosynthesis takes place inside **chloroplasts**. Pigment-filled chromoplasts and starch-filled amyloplasts are used for storage; many of these plastids serve additional roles.

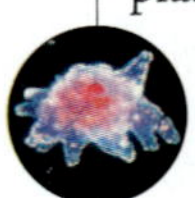

SECTION 4.9 Elements of a **cytoskeleton** reinforce, organize, and move cell structures, and often the whole cell. Cytoskeletal elements include **microtubules**, **microfilaments**, and **intermediate filaments**.

Interactions between ATP-driven **motor proteins** and hollow, dynamically assembled microtubules bring about the movement of cell parts. A microfilament mesh called the **cell cortex** reinforces plasma membranes. Elongating microfilaments bring about movement of **pseudopods**. Intermediate filaments lend structural support to cells and tissues, and they help support the nuclear membrane. **Centrioles** give rise to a special 9+2 array of microtubules inside **cilia** and eukaryotic flagella, then remain beneath these motile structures as **basal bodies**.

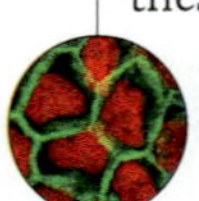

SECTION 4.10 Many cells secrete a complex mixture of fibrous proteins and polysaccharides onto their surfaces. The secretions form an **extracellular matrix (ECM)** that has different functions depending on the cell type. In animals, a secreted basement membrane supports and organizes cells in tissues. Among the eukaryotes, plant cells, fungi, and many protists secrete a cell wall around their plasma membrane. Older plant cells secrete a rigid, **lignin-containing secondary wall** inside their pliable **primary wall.** Many eukaryotic cell types also secrete a protective **cuticle.**

Plasmodesmata are open **cell junctions** that connect the cytoplasm of adjacent plant cells. In animals, **gap junctions** are closable channels between adjacent cells. **Adhering junctions** that connect to cytoskeletal elements fasten cells to one another and to basement membrane. **Tight junctions** form a waterproof seal between cells.

SECTION 4.11 We describe the quality of "life" as a set of properties that are collectively unique to living things. Living things consist of cells that engage in self-sustaining biological processes, pass their hereditary material (DNA) to offspring by mechanisms of reproduction, and have the capacity to change over successive generations.

SECTION 4.12 Bacteria are found in all parts of the biosphere, including the human body. Huge numbers inhabit our intestines, but most of these are beneficial. A few can cause disease. Contamination of food with disease-causing bacteria can result in food poisoning that is sometimes fatal.

Self-Quiz

Answers in Appendix VII

1. Despite the diversity of cell type and function, all cells have these three things in common:
   a. cytoplasm, DNA, and organelles with membranes.
   b. a plasma membrane, DNA, and a nuclear envelope.
   c. cytoplasm, DNA, and a plasma membrane.
   d. a cell wall, cytoplasm, and DNA.

2. Every cell is descended from another cell. This idea is part of _______ .
   a. evolution
   b. the theory of heredity
   c. the cell theory
   d. cell biology

3. Unlike eukaryotic cells, prokaryotic cells _______ .
   a. have no plasma membrane
   b. have RNA but not DNA
   c. have no nucleus
   d. a and c

4. The surface-to-volume ratio _______ .
   a. does not apply to prokaryotic cells
   b. constrains cell size
   c. is part of the cell theory
   d. b and c

5. Cell membranes consist mainly of _______ and _______ .
   a. lipids; carbohydrates
   b. phospholipids; protein
   c. lipids; carbohydrates
   d. phospholipids; ECM

6. In a lipid bilayer, the _______ of all the lipid molecules are sandwiched between all of the _______ .
   a. hydrophilic tails; hydrophobic heads
   b. hydrophilic heads; hydrophilic tails
   c. hydrophobic tails; hydrophilic heads
   d. hydrophobic heads; hydrophilic tails

Data Analysis Activities

Abnormal Motor Proteins Cause Kartagener Syndrome An abnormal form of a motor protein called dynein causes Kartagener syndrome, a genetic disorder characterized by chronic sinus and lung infections. Biofilms form in the thick mucus that collects in the airways, and the resulting bacterial activities and inflammation damage tissues.

Affected men can produce sperm but are infertile (**FIGURE 4.25**). They can become fathers after a doctor injects their sperm cells directly into eggs. Review **FIGURE 4.20**, then explain how abnormal dynein could cause these observed effects.

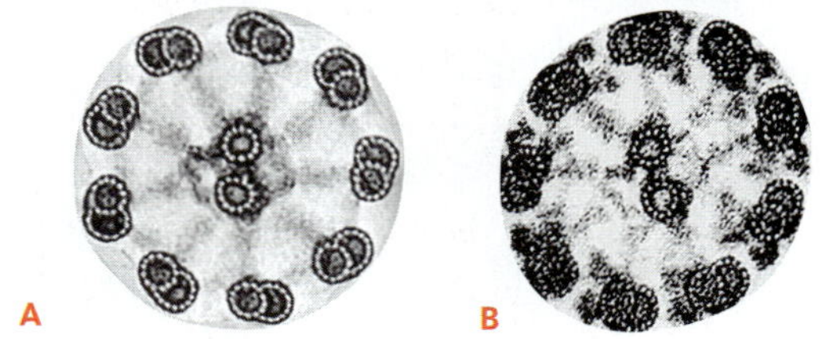

FIGURE 4.25 Cross-section of the flagellum of a sperm cell from **A** a man affected by Kartagener syndrome and **B** an unaffected man.

7. Most of a membrane's diverse functions are carried out by _______ .
   a. proteins c. nucleic acids
   b. phospholipids d. hormones

8. What controls the passage of molecules into and out of the nucleus?
   a. endoplasmic reticulum, an extension of the nucleus
   b. nuclear pores, which consist of membrane proteins
   c. nucleoli, in which ribosome subunits are made

9. The main function of the endomembrane system is _______ .
   a. building and modifying proteins and lipids
   b. isolating DNA from toxic substances
   c. secreting extracellular matrix onto the cell surface
   d. producing ATP by aerobic respiration

10. Which of the following statements is correct?
    a. Ribosomes are only found in bacteria and archaea.
    b. Some animal cells are prokaryotic.
    c. Only eukaryotic cells have mitochondria.
    d. The plasma membrane is the outermost boundary of all cells.

11. Enzymes contained in _______ break down worn-out organelles, bacteria, and other particles.
    a. lysosomes c. endoplasmic reticulum
    b. mitochondria d. peroxisomes

12. Put the following structures in order according to the pathway of a secreted protein:
    a. plasma membrane c. endoplasmic reticulum
    b. Golgi bodies d. post-Golgi vesicles

13. No animal cell has a _______ .
    a. plasma membrane c. lysosome
    b. flagellum d. cell wall

14. _______ connect the cytoplasm of plant cells.
    a. Plasmodesmata c. Tight junctions
    b. Adhering junctions d. Adhesion proteins

15. Match each cell component with its function.

___ mitochondrion	a. connects cells
___ chloroplast	b. movement
___ ribosome	c. ATP production
___ nucleus	d. protects DNA
___ cell junction	e. protein synthesis
___ flagellum	f. maintains internal environment
___ cell membrane	g. photosynthesis

Critical Thinking

1. In a classic episode of *Star Trek*, a gigantic amoeba engulfs an entire starship. Spock blows the cell to bits before it can reproduce. Think of at least one inaccuracy that a biologist would identify in this scenario.

2. In plants, the cell wall forms as a young plant cell secretes polysaccharides onto the outer surface of its plasma membrane. Being thin and pliable, this primary wall allows the cell to enlarge and change shape. At maturity, cells in some plant tissues deposit material onto the primary wall's inner surface. Why doesn't this secondary wall form on the outer surface of the primary wall?

3. Which structures can you identify in the organism *below*? Is it prokaryotic or eukaryotic? How can you tell?

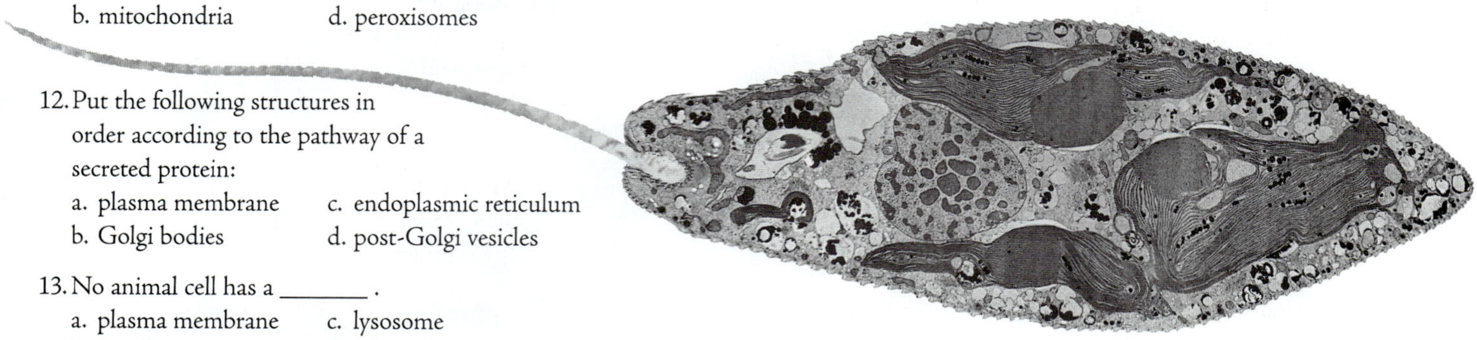

CREDITS: (25) From "Tissue & Cell", Vol. 27, pp.421–427, Courtesy of Bjorn Afzelius, Stockholm University; (in text) P.L. Walne and J. H. Arnott, *Planta*, 77:325–354, 1967.

A single-celled protist with the common name Sea Sparkle glows blue when agitated—for example, by breaking waves. The light is energy released from chemical reactions that run inside these cells. Light emitted by a living organism is called bioluminescence.

5 GROUND RULES OF METABOLISM

Links to Earlier Concepts

In this chapter, you will gain insight into the one-way flow of energy through the world of life (Sections 1.1, 1.2) as you learn more about specific types of energy (2.4) and the laws of nature (1.8) that describe it. The chapter also revisits the structure and function of atoms (2.2), molecules (2.3, 3.1–3.6), and cells (4.3, 4.6, 4.7, 4.9).

KEY CONCEPTS

ENERGY FLOW

Each time energy is transferred, some of it disperses. An organism can sustain its life only as long as it continues to harvest energy from the environment.

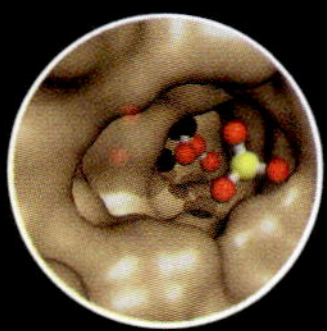

HOW ENZYMES WORK

Enzymes increase the rate of chemical reactions. They are assisted by cofactors, and affected by temperature, salinity, pH, and other environmental factors.

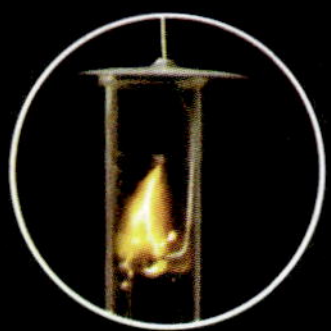

THE NATURE OF METABOLISM

Sequences of enzyme-mediated reactions build, remodel, and break down organic molecules. Controls that govern steps in these pathways quickly shift cell activities.

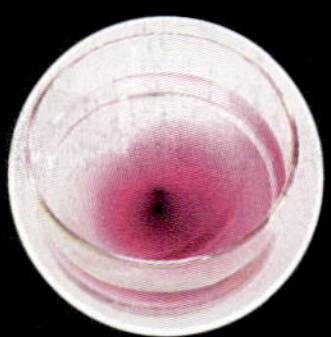

MOVEMENT OF FLUIDS

Gradients drive the directional movements of solutes. Water moves across cell membranes to regions where solute concentration is higher.

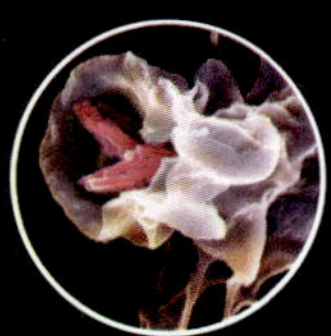

MEMBRANE TRANSPORT

Transport proteins control solute concentrations in cells and organelles by helping substances move across membranes. Substances also move across cell membranes inside vesicles.

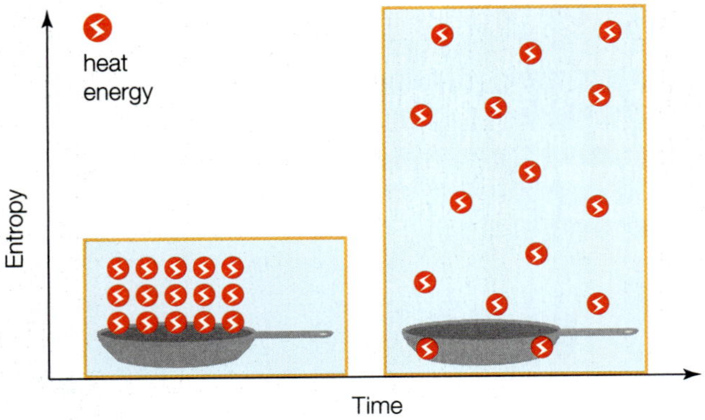

FIGURE 5.1 **Entropy**. Entropy tends to increase, which means that energy tends to spread out spontaneously.

FIGURE 5.2 It takes more than 10,000 pounds of soybeans and corn to raise a 1,000-pound steer. Where do the other 9,000 pounds go? About half of the steer's food is indigestible and passes right through it. The animal's body breaks down molecules in the remaining half to access energy stored in chemical bonds. Only about 15% of that energy goes toward building body mass. The rest is lost during energy conversions, as metabolic heat.

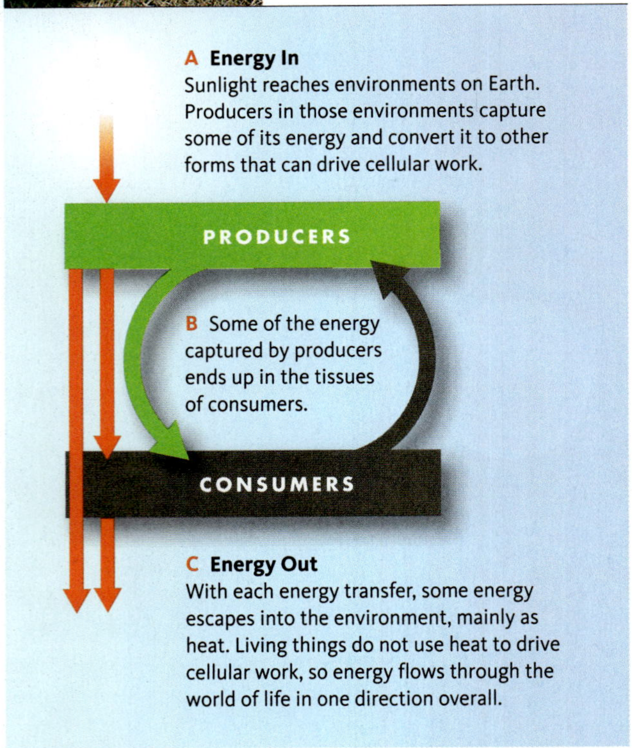

FIGURE 5.3 **{Animated}** Energy flows from the environment into living organisms, and then back to the environment. The flow drives a cycling of materials among producers and consumers.

ENERGY DISPERSES

Energy is formally defined as the capacity to do work, but this definition is not very satisfying. Even brilliant physicists who study energy cannot say exactly what it is. However, we do have an intuitive understanding of energy just by thinking about familiar forms of it, such as light, heat, electricity, and motion. We also understand intuitively that one form of energy can be converted to another. Think about how a lightbulb changes electricity into light, or how an automobile changes gasoline into the energy of motion, which is also called **kinetic energy**.

The formal study of heat and other forms of energy is thermodynamics (*therm* is a Greek word for heat; *dynam* means energy). By making careful measurements, thermodynamics researchers discovered that the total amount of energy before and after every conversion is always the same. In other words, energy cannot be created or destroyed—a phenomenon that is the **first law of thermodynamics**. Remember, a law of nature describes something that occurs without fail, but our explanation of why it occurs is incomplete (Section 1.8).

Energy also tends to spread out, or disperse, until no part of a system holds more than another part. In a kitchen, for example, heat always flows from a hot pan to cool air until the temperature of both is the same. We never see cool air raising the temperature of a hot pan. **Entropy** is a measure of how much the energy of a particular system has become dispersed. We can use the hot pan in a cool kitchen as an example of a system. As heat flows from the pan into the air, the entropy of the system increases (**FIGURE 5.1**). Entropy continues to increase until the heat is evenly distributed throughout the kitchen, and there is no longer a net (or overall) flow of heat from one area to another. Our system has now reached its maximum entropy with respect to heat. The tendency of entropy to increase is the **second law of thermodynamics**. This is the formal way of saying that energy tends to spread out spontaneously.

Biologists use the concept of entropy as it applies to chemical bonding, because energy flow in living things occurs mainly by the making and breaking of chemical bonds. How is entropy related to chemical bonding? Think about it just in terms of motion. Two unbound atoms can vibrate, spin, and rotate in every direction, so they are at high entropy with respect to motion. A covalent bond between the atoms restricts their movement, so they are able to move in fewer ways than they did before bonding. Thus, the entropy of two atoms decreases when a bond forms between them. Such entropy changes are part of the reason why some reactions occur spontaneously and others require an energy input, as you will see in the next section.

ENERGY'S ONE-WAY FLOW

Work occurs as a result of energy transfers. Consider how it takes work to push a box across a floor. In this case, a body (you) transfers energy to another body (the box) to make it move. Similarly, a plant cell works to make sugars. Inside the cell, one set of molecules harvests energy from light, then transfers it to another set of molecules. The second set of molecules uses the energy to build the sugars from carbon dioxide and water. This particular energy transfer involves the conversion of light energy to chemical energy. Most other types of cellular work occur by the transfer of chemical energy from one molecule to another.

As you learn about such processes, remember that every time energy is transferred, a bit of it disperses. Energy lost from a transfer is usually in the form of heat. As a simple example, a typical incandescent lightbulb converts only about 5 percent of the energy of electricity into light. The remaining 95 percent of the energy ends up as heat that disperses from the bulb.

Dispersed heat is not very useful for doing work, and it is not easily converted to a more useful form of energy (such as electricity). Because some of the energy in every transfer disperses as heat, and heat is not useful for doing work, we can say that the total amount of energy available for doing work in the universe is always decreasing.

Is life an exception to this inevitable flow? An organized body is hardly dispersed. Energy becomes concentrated in each new organism as the molecules of life organize into cells. Even so, living things constantly use energy—to grow, to move, to acquire nutrients, to reproduce, and so on—and some energy is lost in every one of these processes (**FIGURE 5.2**). Unless those losses are replenished with energy from another source, the complex organization of life will end.

The energy that fuels most life on Earth comes from the sun. That energy flows through producers such as plants, then consumers such as animals (**FIGURE 5.3**). During this journey, the energy is transferred many times. With each transfer, some energy escapes as heat until, eventually, all of it is permanently dispersed. However, the second law of thermodynamics does not say how quickly the dispersal has to happen. Energy's spontaneous dispersal is resisted by chemical bonds. The energy in chemical bonds is a type of **potential energy**, which is energy stored in the position or arrangement of objects in a system (**FIGURE 5.4**). Think of all the bonds in the countless molecules that make up your skin, heart, liver, fluids, and other body parts. Those bonds hold the molecules, and you, together—at least for the time being.

FIGURE 5.4 Illustration of potential energy. By opposing the downward pull of gravity, the rope attached to the rock prevents the man from falling. Similarly, a chemical bond keeps two atoms from moving apart.

---

energy The capacity to do work.
entropy Measure of how much the energy of a system is dispersed.
first law of thermodynamics Energy cannot be created or destroyed.
kinetic energy The energy of motion.
potential energy Stored energy.
second law of thermodynamics Energy disperses spontaneously.

TAKE-HOME MESSAGE 5.1

- Energy, which is the capacity to do work, cannot be created or destroyed.
- Energy disperses spontaneously.
- Energy can be transferred between systems or converted from one form to another, but some is lost (as heat, typically) during every such exchange.
- Sustaining life's organization requires ongoing energy inputs to counter energy loss. Organisms stay alive by replenishing themselves with energy they harvest from someplace else.

5.2 HOW DO CELLS USE ENERGY?

Remember from Section 3.1 that chemical reactions change molecules into other molecules. During a reaction, one or more **reactants** (molecules that enter a reaction and become changed by it) become one or more **products** (molecules that are produced by the reaction). Intermediate molecules may form between reactants and products.

We show a chemical reaction as an equation in which an arrow points from reactants to products:

$$\underset{\text{(hydrogen)}}{2H_2} + \underset{\text{(oxygen)}}{O_2} \longrightarrow \underset{\text{(water)}}{2H_2O}$$

A number before a chemical formula in such equations indicates the number of molecules; a subscript indicates the number of atoms of that element per molecule. Note that atoms shuffle around in a reaction, but they never disappear: The same number of atoms that enter a reaction remain at the reaction's end (**FIGURE 5.5**).

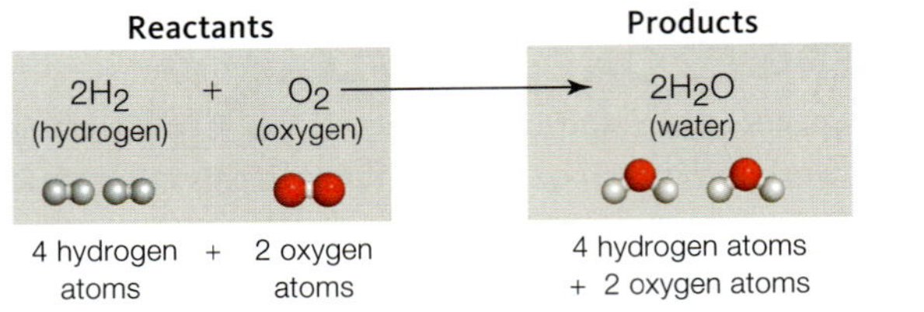

FIGURE 5.5 Chemical bookkeeping. In equations that represent chemical reactions, reactants are written to the left of an arrow that points to the products. A number before a formula indicates the number of molecules. Atoms may shuffle around in a reaction, but the same number of atoms that enter the reaction remain at the reaction's end.

CHEMICAL BOND ENERGY

Every chemical bond holds a certain amount of energy. That is the amount of energy required to break the bond, and it is also the amount of energy released when the bond forms. The particular amount of energy held by a bond depends on which elements are taking part in it. For example, two covalent bonds—one between an oxygen and a hydrogen atom in a water molecule, the other between two oxygen atoms in molecular oxygen (O_2)—both hold energy, but different amounts of it.

Bond energy and entropy both contribute to a molecule's free energy, which is the amount of energy that is available ("free") to do work. In most reactions, the free energy of reactants differs from the free energy of products. If the reactants have less free energy than the products, the reaction will not proceed without a net energy input. Such reactions are **endergonic**, which means "energy in" (**FIGURE 5.6A**). If the reactants have more free energy than the products, the reaction will end with a net release of energy. Such reactions are **exergonic**, which means "energy out" (**FIGURE 5.6B**).

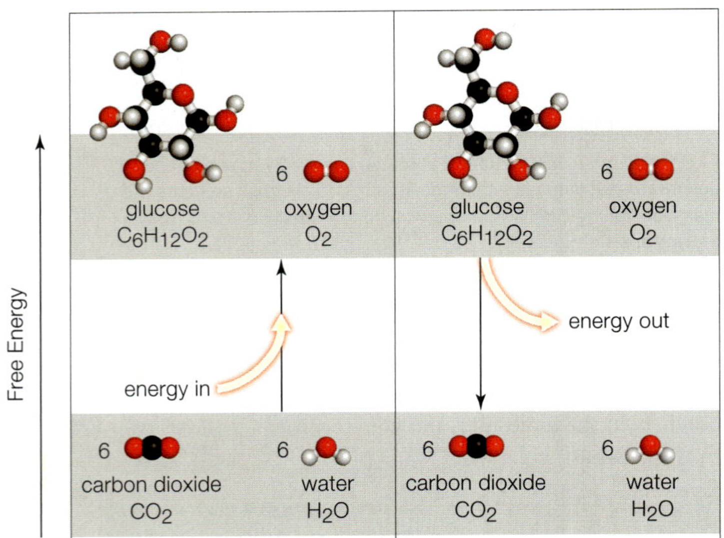

A Endergonic reactions convert molecules with lower free energy to molecules with higher free energy, so they require a net energy input in order to proceed.

B Exergonic reactions convert molecules with higher free energy to molecules with lower free energy, so they end with an energy release.

FIGURE 5.6 Energy inputs and outputs in chemical reactions.

FIGURE IT OUT: Which thermodynamics law explains energy inputs and outputs in chemical reactions?

Answer: The first law

WHY EARTH DOES NOT GO UP IN FLAMES

The molecules of life release energy when they combine with oxygen. For example, think of how a spark ignites wood. Wood is mostly cellulose, which consists of long chains of repeating glucose monomers (Section 3.2). A spark starts a reaction that converts cellulose (in wood) and oxygen (in air) to water and carbon dioxide. The reaction is highly exergonic, which means it releases a lot of energy—enough to initiate the same reaction with other cellulose and oxygen molecules. That is why wood keeps burning after it has been lit.

Earth is rich in oxygen—and in potential exergonic reactions. Why doesn't it burst into flames? Luckily, chemical bonds do not break without at least a small input of energy, even in an energy-releasing reaction. We call this input activation energy. **Activation energy**, the minimum amount of energy required to get a chemical reaction started, is a bit like a hill that reactants must climb before they can coast down the other side to become products (**FIGURE 5.7**).

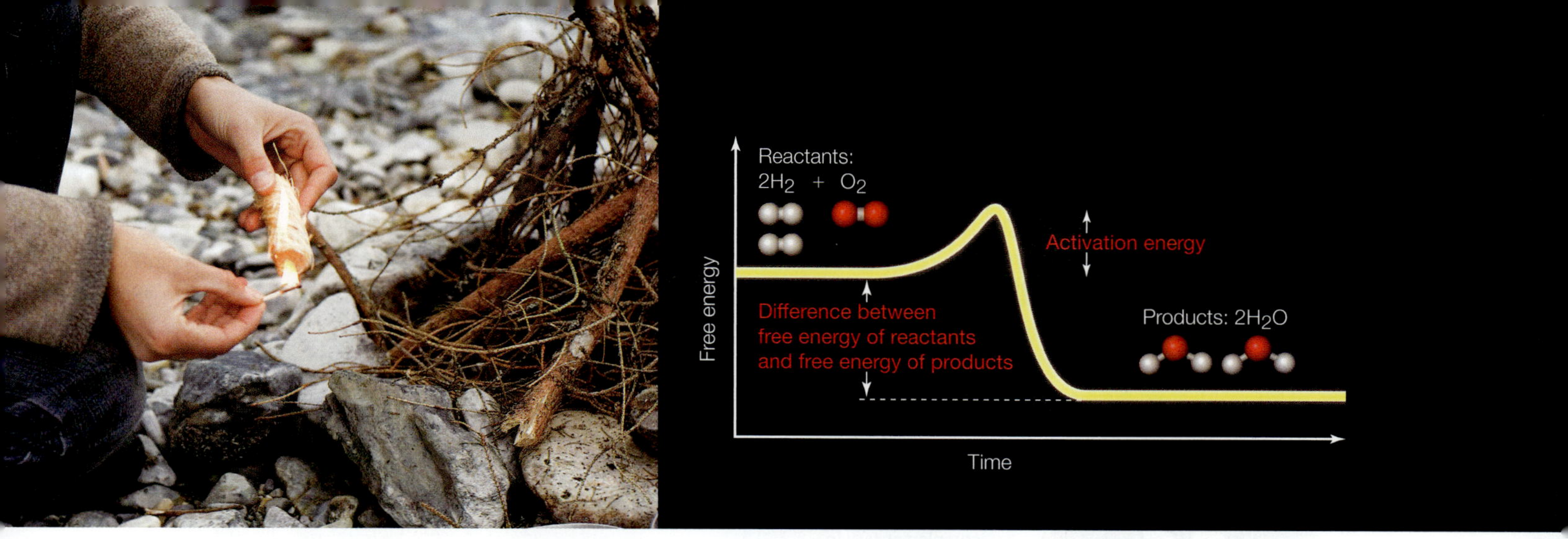

FIGURE 5.7 {Animated} Activation energy. Most reactions will not begin without an input of activation energy, which is shown in the graph as a bump in a free energy hill. Reactants in this example have more energy than the products. Activation energy keeps this and other reactions, including exergonic ones such as burning wood cellulose, from starting spontaneously.

Both endergonic and exergonic reactions have activation energy, but the amount varies with the reaction. Consider guncotton (nitrocellulose), a highly explosive derivative of cellulose. Christian Schönbein accidentally discovered a way to manufacture it when he used his wife's cotton apron to wipe up a nitric acid spill on his kitchen table, then hung it up to dry next to the oven. The apron exploded. Being a chemist in the 1800s, Schönbein immediately thought of marketing guncotton as a firearm explosive, but it proved to be too unstable to manufacture. So little activation energy is needed to make guncotton react with oxygen that it tends to explode unexpectedly. Several manufacturing plants burned to the ground before guncotton was abandoned for use as a firearm explosive. The substitute? Gunpowder, which has a higher activation energy for a reaction with oxygen.

ENERGY IN, ENERGY OUT

Cells store energy by running endergonic reactions that build organic compounds (**FIGURE 5.8A**). For example, light energy drives the overall reactions of photosynthesis, which produce sugars such as glucose from carbon dioxide and water. Unlike light, glucose can be stored in a cell. Cells harvest energy by running exergonic reactions that break the bonds of organic compounds (**FIGURE 5.8B**). Most cells do this when they carry out the overall reactions of aerobic respiration, which releases the energy of glucose by breaking the bonds between its carbon atoms. You will see in the next few sections how cells use energy released from some reactions to drive others.

FIGURE 5.8 {Animated} Cells store and retrieve energy in the chemical bonds of organic molecules.

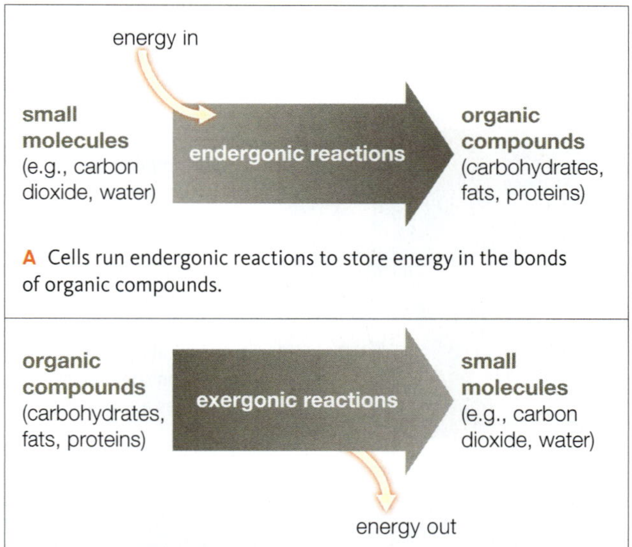

A Cells run endergonic reactions to store energy in the bonds of organic compounds.

B Cells run exergonic reactions to retrieve energy stored in the bonds of organic compounds.

activation energy Minimum amount of energy required to start a reaction.
endergonic Describes a reaction that requires a net input of free energy to proceed.
exergonic Describes a reaction that ends with a net release of free energy.
product A molecule that is produced by a reaction.
reactant A molecule that enters a reaction and is changed by participating in it.

TAKE-HOME MESSAGE 5.2

Endergonic reactions will not run without a net input of energy. Exergonic reactions end with a net release of energy.

Both endergonic and exergonic reactions require an input of activation energy to begin.

Cells store energy in chemical bonds by running endergonic reactions that build organic compounds. To release this stored energy, they run exergonic reactions that break the bonds.

THE NEED FOR SPEED

Metabolism requires enzymes. Why? Consider that sugar can break down to carbon dioxide and water on its own, but it might take decades. That same conversion takes just seconds inside your cells. Enzymes make the difference.

FIGURE 5.9 How an active site works.

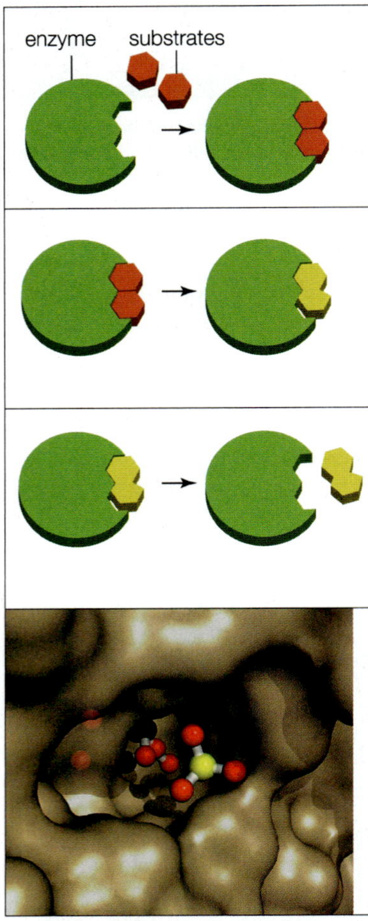

A An active site binds substrates that are complementary in shape, size, polarity, and charge.

B The binding squeezes substrates together, influences their charge, or causes some change that lowers activation energy, so the reaction proceeds.

C The product leaves the active site after the reaction is finished. The enzyme is unchanged, so it can work again.

D For simplicity, enzymes are often depicted as blobs or geometric shapes. This model shows the actual contours of an active site in an enzyme (hexokinase) that adds a phosphate group to a six-carbon sugar. Both substrates are shown.

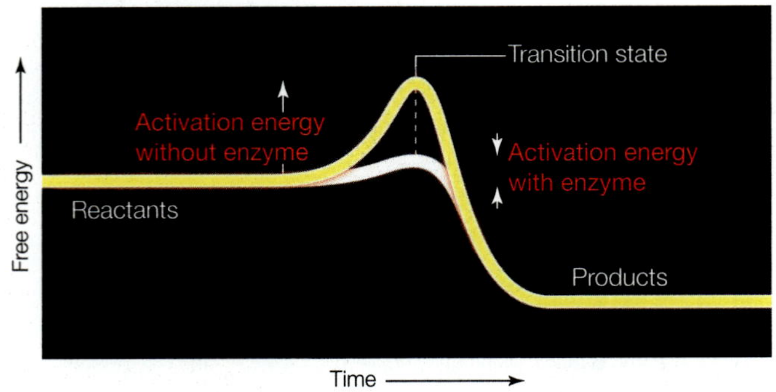

FIGURE 5.10 {Animated} The transition state. An enzyme enhances the rate of a reaction by lowering activation energy.

FIGURE IT OUT: Is the reaction shown in this graph endergonic or exergonic?

Answer: Exergonic

In a process called **catalysis**, an enzyme makes a reaction run much faster than it would on its own. The enzyme is unchanged by participating in the reaction, so it can work again and again.

Some enzymes are RNAs, but most are proteins. Each kind of enzyme recognizes specific reactants, or **substrates**, and alters them in a specific way. For instance, the enzyme hexokinase adds a phosphate group to the hydroxyl group on the sixth carbon of glucose. Such specificity occurs because an enzyme's polypeptide chains fold up into one or more **active sites**, which are pockets where substrates bind and where reactions proceed (**FIGURE 5.9**). An active site is complementary in shape, size, polarity, and charge to the enzyme's substrate. This fit is the reason why each enzyme acts in a specific way on a specific substrate.

When we talk about activation energy, we are really talking about the energy required to bring reactant bonds to their breaking point. At that point, which is called the transition state, the reaction can run without any additional energy input. Enzymes help bring on the transition state by lowering activation energy (**FIGURE 5.10**). They do so by the following four mechanisms.

Forcing Substrates Together Binding at an active site brings substrates together. The closer the substrates are to one another, the more likely they are to react.

Orienting Substrates Substrate molecules in a solution collide from random directions. By contrast, binding at an active site positions substrates optimally for reaction.

Inducing Fit By the **induced-fit model**, an enzyme's active site is not quite complementary to its substrate. Interacting with a substrate molecule causes the enzyme to change shape so that the fit between them improves. The improved fit may result in a stronger bond between enzyme and substrate, or it may better bring on the transition state.

Shutting Out Water Metabolism occurs in water-based fluids, but water molecules can interfere with certain reactions. The active sites of some enzymes repel water, and keep it away from the reactions.

active site Pocket in an enzyme where substrates bind and a reaction occurs.
catalysis The acceleration of a reaction rate by a molecule that is unchanged by participating in the reaction.
induced-fit model Substrate binding to an active site improves the fit between the two.
substrate Of an enzyme, a reactant that is specifically acted upon by the enzyme.

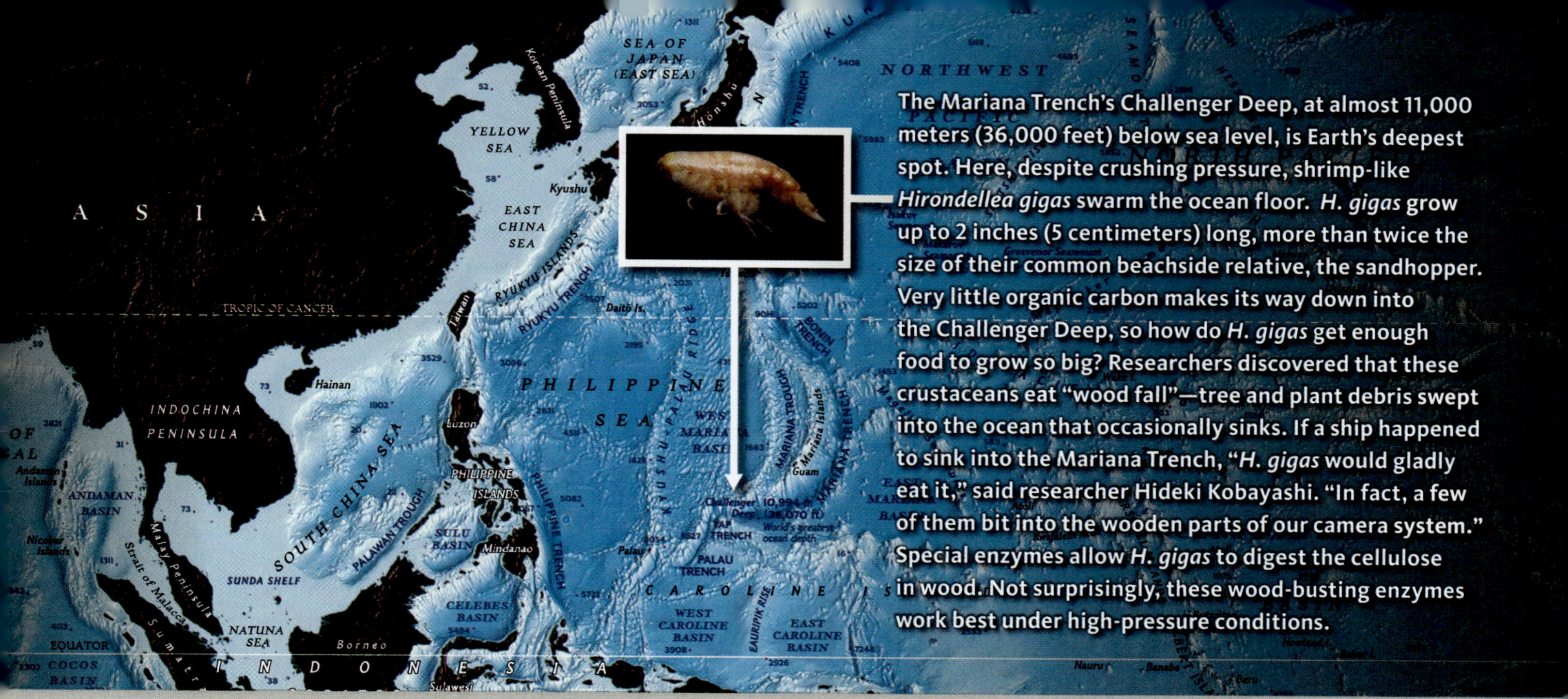

FIGURE 5.11 An organism (and its enzymes) adapted to life in a particular environment.

ENZYME ACTIVITY

Environmental factors such as pH, temperature, and salt influence an enzyme's shape, which in turn influences its function (Sections 3.4 and 3.5). Each enzyme functions best in a particular range of conditions that reflect the environment in which it evolved (**FIGURE 5.11**).

Consider pepsin, a digestive enzyme that works best at low pH (**FIGURE 5.12A**). Pepsin begins the process of protein digestion in the very acidic environment of the stomach (pH 2). During digestion, the stomach's contents pass into the small intestine, where the pH rises to about 9. Pepsin denatures (unfolds) above pH 5.5, so this enzyme becomes inactivated in the small intestine. Here, protein digestion continues with the assistance of trypsin, an enzyme that functions well at the higher pH.

Adding heat boosts free energy, which is why molecular motion increases with temperature. The greater the free energy of reactants, the closer they are to activation energy. Thus, the rate of an enzymatic reaction typically increases with temperature—but only up to a point. An enzyme denatures above a characteristic temperature. Then, the reaction rate falls sharply as the shape of the enzyme changes and it stops working (**FIGURE 5.12B**). Body temperatures above 42°C (107.6°F) adversely affect the function of many of your enzymes, which is why severe fevers are dangerous.

The activity of many enzymes is also influenced by the amount of salt in the surrounding fluid. Too little salt, and polar parts of the enzyme attract one another so strongly that the enzyme's shape changes. Too much salt interferes with the hydrogen bonds that hold the enzyme in its characteristic shape, and the enzyme denatures.

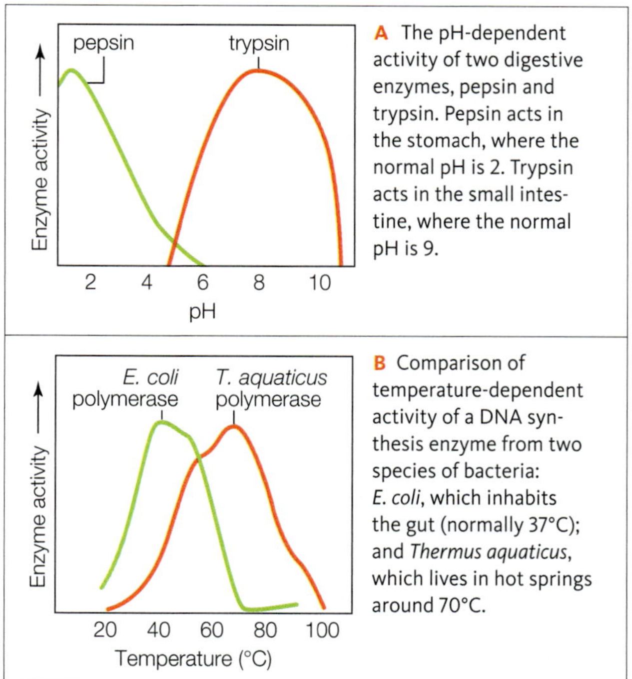

A The pH-dependent activity of two digestive enzymes, pepsin and trypsin. Pepsin acts in the stomach, where the normal pH is 2. Trypsin acts in the small intestine, where the normal pH is 9.

B Comparison of temperature-dependent activity of a DNA synthesis enzyme from two species of bacteria: *E. coli*, which inhabits the gut (normally 37°C); and *Thermus aquaticus*, which lives in hot springs around 70°C.

FIGURE 5.12 Each enzyme functions best within a characteristic range of conditions—generally, the same conditions that occur in the environment in which the enzyme evolved.

TAKE-HOME MESSAGE 5.3

- Enzymes greatly enhance the rate of specific reactions.
- Binding at an enzyme's active site causes a substrate to reach its transition state. In this state, the substrate's bonds are at the breaking point, and the reaction can run spontaneously.
- Each enzyme works best at certain environmental conditions that include temperature, pH, and salt concentration.

5.4 WHAT IS A METABOLIC PATHWAY?

Metabolism, remember, refers to the activities by which cells acquire and use energy as they build, break down, or remodel organic molecules. These activities often occur stepwise, in a series of enzymatic reactions called a **metabolic pathway**. Some metabolic pathways are linear, meaning that the reactions run straight from reactant to product (**FIGURE 5.13A**), and others are cyclic. In a cyclic pathway, the last step regenerates a reactant for the first step (**FIGURE 5.13B**). Both linear and cyclic pathways are common in cells; both can involve thousands of molecules and be quite complex. Later chapters detail the steps in some important pathways.

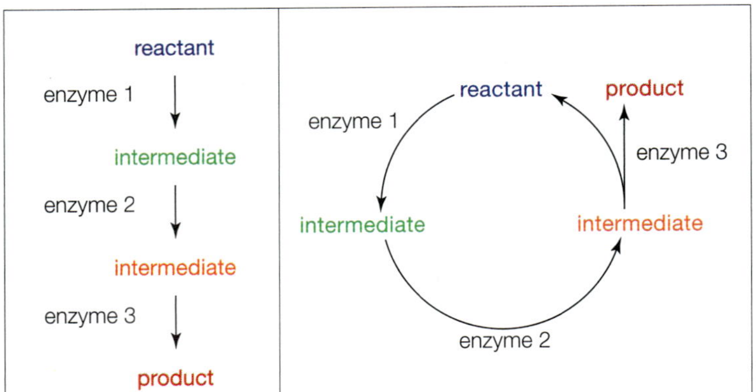

A A linear pathway runs straight from reactant to product.

B The last step of a cyclic pathway regenerates a reactant for the first step.

FIGURE 5.13 Metabolic pathways.

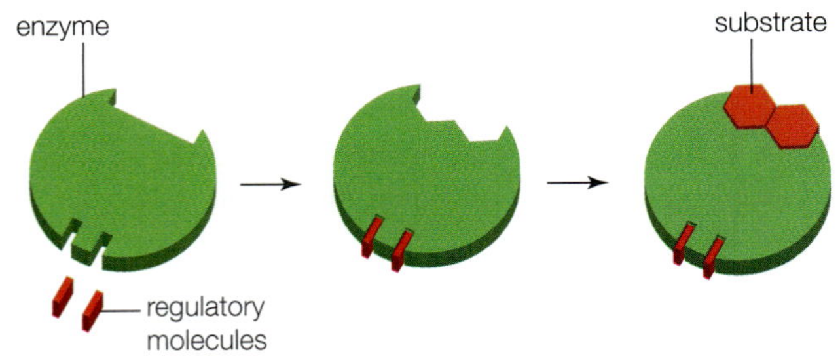

FIGURE 5.14 {Animated} Allosteric regulation, in which regulatory molecules bind to a region of an enzyme that is not the active site. The binding changes the shape of the enzyme, and thus alters its activity.

FIGURE IT OUT: Does the binding of regulatory molecules help or hinder this enzyme's function?

Answer: It helps

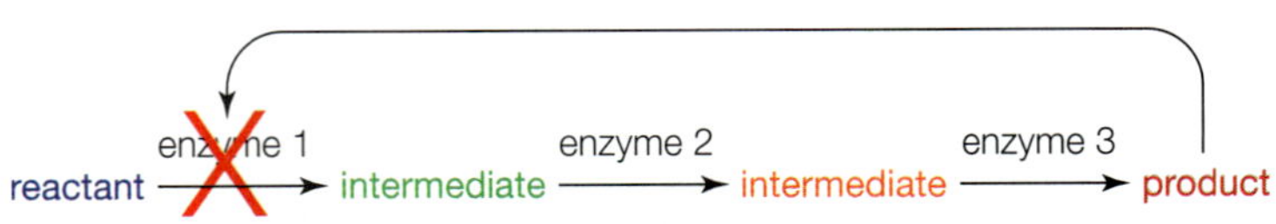

FIGURE 5.15 {Animated} Feedback inhibition. In this example, three different enzymes act in sequence to convert a substrate to a product. The product inhibits the activity of the first enzyme.

FIGURE IT OUT: Is this metabolic pathway cyclic or linear?

Answer: Linear

CONTROLS OVER METABOLISM

Cells conserve energy and resources by making only what they need at any given moment—no more, no less. Several mechanisms help a cell maintain, raise, or lower its production of thousands of different substances. Consider that reactions do not only run from reactants to products. Many also run in reverse at the same time, with some of the products being converted back to reactants. The rates of the forward and reverse reactions often depend on the concentrations of reactants and products: A high concentration of reactants pushes the reaction in the forward direction, and a high concentration of products pushes it in the reverse direction.

Other mechanisms more actively regulate enzymatic reactions. Certain substances—regulatory molecules or ions—can influence enzyme activity. In some cases, a regulatory substance activates or inhibits an enzyme by binding directly to the active site. In other cases, the regulatory substance binds outside of the active site, a mechanism called **allosteric regulation** (*allo*– means other; steric means structure). Binding of an allosteric regulator alters the shape of the enzyme in a way that enhances or inhibits its function (**FIGURE 5.14**).

Regulation of a single enzyme can affect an entire metabolic pathway. For example, the end product of a series of enzymatic reactions often inhibits the activity of one of the enzymes in the series (**FIGURE 5.15**). This regulatory mechanism is an example of **feedback inhibition**, in which a change that results from an activity decreases or stops the activity.

allosteric regulation Control of enzyme activity by a regulatory molecule or ion that binds to a region outside the enzyme's active site.
electron transfer chain Array of enzymes and other molecules that accept and give up electrons in sequence, thus releasing the energy of the electrons in steps.
feedback inhibition Regulatory mechanism in which a change that results from some activity decreases or stops the activity.
metabolic pathway Series of enzyme-mediated reactions by which cells build, remodel, or break down an organic molecule.
redox reaction Oxidation–reduction reaction, in which one molecule accepts electrons (it becomes reduced) from another molecule (which becomes oxidized). Also called electron transfer.

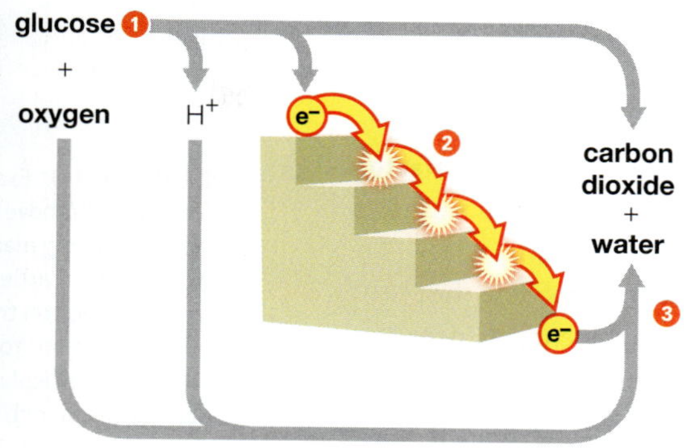

A Left, glucose in a metal spoon reacts (burns) with oxygen inside a glass jar. Energy in the form of light and heat is released all at once as CO_2 and water form.

B In cells, the same overall reaction occurs in a stepwise fashion that involves an electron transfer chain, represented here by a staircase. Energy is released in amounts that cells are able to use.

1 An input of activation energy splits glucose into carbon dioxide, electrons, and hydrogen ions (H^+).
2 Electrons lose energy as they move through an electron transfer chain. Energy released by electrons — is harnessed for cellular work.
3 Electrons, hydrogen ions, and oxygen combine to form water.

FIGURE 5.16 {Animated} Comparing uncontrolled and controlled energy release.

ELECTRON TRANSFERS

The bonds of organic molecules hold a lot of energy that can be released in a reaction with oxygen. Burning is one type of reaction with oxygen, and it releases the energy of organic molecules all at once—explosively (**FIGURE 5.16A**). Cells use oxygen to break the bonds of organic molecules, but they have no way to harvest the explosive burst of energy that occurs during burning. Instead, they break the molecules apart in pathways that release the energy in small, manageable steps. Most of these steps are oxidation–reduction reactions, or redox reactions for short. In a typical **redox reaction**, one molecule accepts electrons (it becomes reduced) from another molecule (which becomes oxidized). To remember what reduced means, think of how the negative charge of an electron "reduces" the charge of a recipient molecule.

An oxidation always occurs together with a reduction (**FIGURE 5.17**). In the next two chapters, you will learn about the importance of this concept in electron transfer chains. An **electron transfer chain** is a series of membrane-bound enzymes and other molecules that give up and accept electrons in turn. Electrons are at a higher energy level (Section 2.2) when they enter a chain than when they leave. Energy given off by an electron as it drops to a lower energy level is harvested by molecules of the electron transfer chain (**FIGURE 5.16B**).

Many electrons are delivered to electron transfer chains in photosynthesis and aerobic respiration. Energy released at certain steps in those chains helps drive the synthesis of ATP. These pathways will occupy our attention in chapters to come.

FIGURE 5.17 Visible evidence of oxidation–reduction: a glowing protist, *Noctiluca scintillans* (*below*). The pathway that produces the glow involves an enzyme, luciferase, and its substrate, luciferin. It occurs when the cells are mechanically stimulated, as by waves (shown in the chapter opening photo) or an attack by a protist-eating predator.

$$\text{luciferin—H}_2 + \text{O}_2 \xrightarrow{\text{luciferase}} \text{luciferin=O} + \text{H}_2\text{O} + \textbf{light}$$

Luciferin is oxidized during the reactions, which are summarized above. At the same time, an oxygen atom is reduced when it combines it with electrons and hydrogen, so water forms.

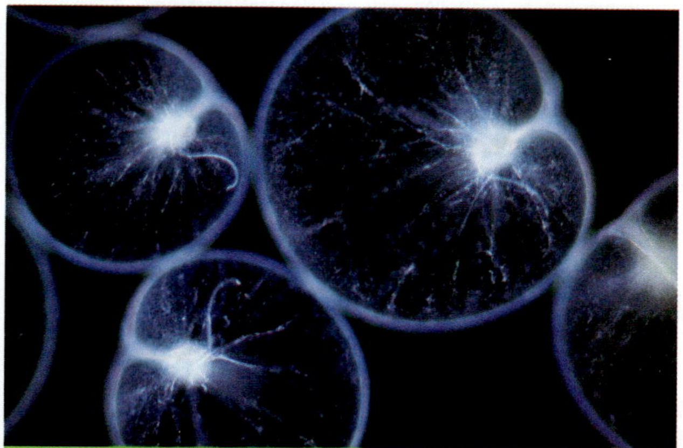

TAKE-HOME MESSAGE 5.4

A metabolic pathway is a series of enzyme-mediated reactions that builds, breaks down, or remodels an organic molecule.

Cells conserve energy and resources by producing only what they require at a given time. This metabolic control arises from regulatory molecules and other mechanisms that influence metabolic pathways and individual reactions.

Many metabolic pathways involve electron transfers. Electron transfer chains are important sites of energy exchange.

5.5 HOW DO COFACTORS WORK?

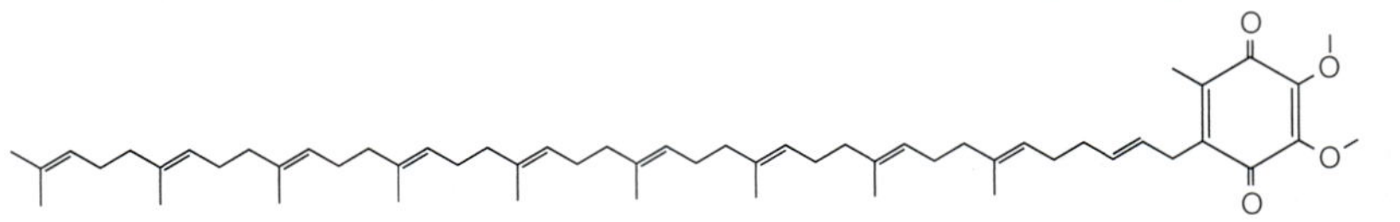

Most enzymes cannot function properly without assistance from metal ions or small organic molecules. Such enzyme helpers are called **cofactors.** Many dietary vitamins and minerals are essential because they are cofactors or are precursors for them.

Some metal ions that act as cofactors stabilize the structure of an enzyme, in which case the enzyme denatures if the ions are removed. In other cases, metal cofactors play a functional role in a reaction by interacting with electrons in nearby atoms. Atoms of metal elements readily lose or gain electrons, so a metal cofactor can help bring on the transition state by donating electrons, accepting them, or simply tugging on them.

FIGURE 5.18 Example of a coenzyme. Coenzyme Q_{10} (above) is an essential part of the ATP-making machinery in your mitochondria. It carries electrons between enzymes of electron transfer chains during aerobic respiration. Your body makes it, but some foods—particularly red meats, soy oil, and peanuts—are rich dietary sources.

TABLE 5.1

Some Common Coenzymes

Coenzyme	Example of Function
ATP	Transfers energy with a phosphate group
NAD, NAD^+	Carries electrons during glycolysis
NADP, NADPH	Carries electrons, hydrogen atoms during photosynthesis
FAD, FADH, $FADH_2$	Carries electrons during aerobic respiration
CoA	Carries acetyl group ($COCH_3$) during glycolysis
Coenzyme Q_{10}	Carries electrons in electron transfer chains of aerobic respiration
Heme	Accepts and donates electrons
Ascorbic acid	Carries electrons during peroxide breakdown (in lysosomes)
Biotin (vitamin B_7)	Carries CO_2 during fatty acid synthesis

Organic cofactors are called **coenzymes** (**TABLE 5.1** and **FIGURE 5.18**). Coenzymes carry chemical groups, atoms, or electrons from one reaction to another, and often into or out of organelles. Unlike enzymes, many coenzymes are modified by taking part in a reaction. They are regenerated in separate reactions.

Consider NAD^+ (nicotinamide adenine dinucleotide), a coenzyme derived from niacin (vitamin B_3). NAD^+ can accept electrons and hydrogen atoms, thereby becoming reduced to NADH. When electrons and hydrogen atoms are removed from NADH (an oxidation reaction), NAD^+ forms again:

$$NAD^+ + \text{electrons} + H^+ \longrightarrow \textbf{NADH} \longrightarrow NAD^+ + \text{electrons} + H^+$$

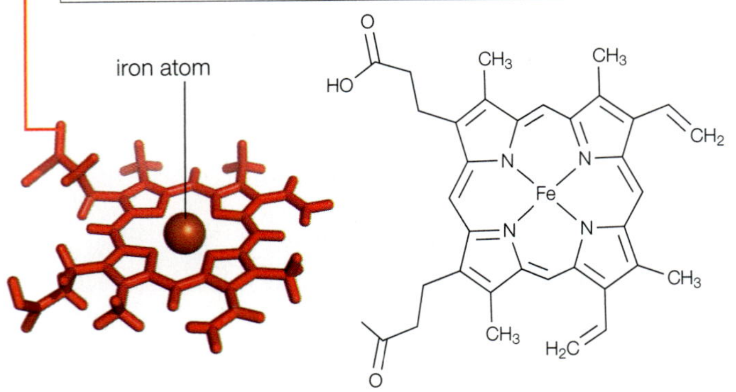

FIGURE 5.19 Heme. This organic molecule is part of the active site in many enzymes (such as catalase). In other contexts, it carries oxygen (e.g., in hemoglobin), or electrons (e.g., in molecules of electron transfer chains).

FIGURE IT OUT: Is heme a cofactor or a coenzyme?

Answer: It is both.

In some reactions, cofactors participate as separate molecules. In others, they stay tightly bound to the enzyme. Catalase, an enzyme of peroxisomes, has four tightly bound cofactors called hemes. A heme is a small organic compound with an iron atom at its center (**FIGURE 5.19**). Catalase's substrate is hydrogen peroxide (H_2O_2), a highly reactive molecule that forms during some normal metabolic reactions. Hydrogen peroxide is dangerous because it can easily oxidize and destroy the organic molecules of life, or form free radicals that do. Catalase neutralizes this threat. When the enzyme binds to hydrogen peroxide, it holds the molecule close to a heme. Interacting with the iron atom in the heme causes peroxide molecules to break down to water.

Substances such as catalase that interfere with the oxidation of other molecules are called **antioxidants**. Antioxidants are essential to health because they reduce the amount of damage that cells sustain as a result of oxidation by free radicals or other molecules. Oxidative damage is associated with many diseases, including cancer, diabetes, atherosclerosis, stroke, and neurodegenerative problems such as Alzheimer's disease.

ATP—A SPECIAL COENZYME

In cells, the nucleotide ATP (adenosine triphosphate, Section 3.6) functions as a cofactor in many reactions. Bonds between phosphate groups hold a lot of energy compared to other bonds. ATP has two of of these bonds holding its three phosphate groups together (**FIGURE 5.20A**). When a phosphate group is transferred to or from a nucleotide, energy is transferred along with it. Thus, the nucleotide can receive energy from an exergonic reaction, and it can contribute energy to an endergonic one. ATP is such an important currency in a cell's energy economy that we use a cartoon coin to symbolize it.

A reaction in which a phosphate group is transferred from one molecule to another is called a **phosphorylation**. ADP (adenosine diphosphate) forms when an enzyme transfers a phosphate group from ATP to another molecule (**FIGURE 5.20B**). Cells constantly run this reaction in order to drive a variety of endergonic reactions. Thus, they must constantly replenish their stockpile of ATP—by running exergonic reactions that phosphorylate ADP. The cycle of using and replenishing ATP is called the **ATP/ADP cycle** (**FIGURE 5.20C**).

The ATP/ADP cycle couples endergonic reactions with exergonic ones (**FIGURE 5.21**). As you will see in Chapter 7, cells harvest energy from organic compounds by running metabolic pathways that break them down. Energy that cells harvest in these pathways is not released to the environment, but rather stored in the high-energy phosphate bonds of ATP molecules and in electrons carried by reduced coenzymes. Both the ATP and the reduced cofactors that form in these pathways can be used to drive many of the different kinds of endergonic reactions that a cell runs.

antioxidant Substance that prevents oxidation of other molecules.
ATP/ADP cycle Process by which cells regenerate ATP. ADP forms when a phosphate group is removed from ATP, then ATP forms again as ADP gains a phosphate group.
coenzyme An organic cofactor.
cofactor A metal ion or organic compound that associates with an enzyme and is necessary for its function.
phosphorylation A phosphate-group transfer.

TAKE-HOME MESSAGE 5.5

- Cofactors associate with enzymes and assist their function.
- Many coenzymes carry chemical groups, atoms, or electrons from one reaction to another.
- When a phosphate group is transferred from ATP to another molecule, energy is transferred along with it. This energy drives cellular work.

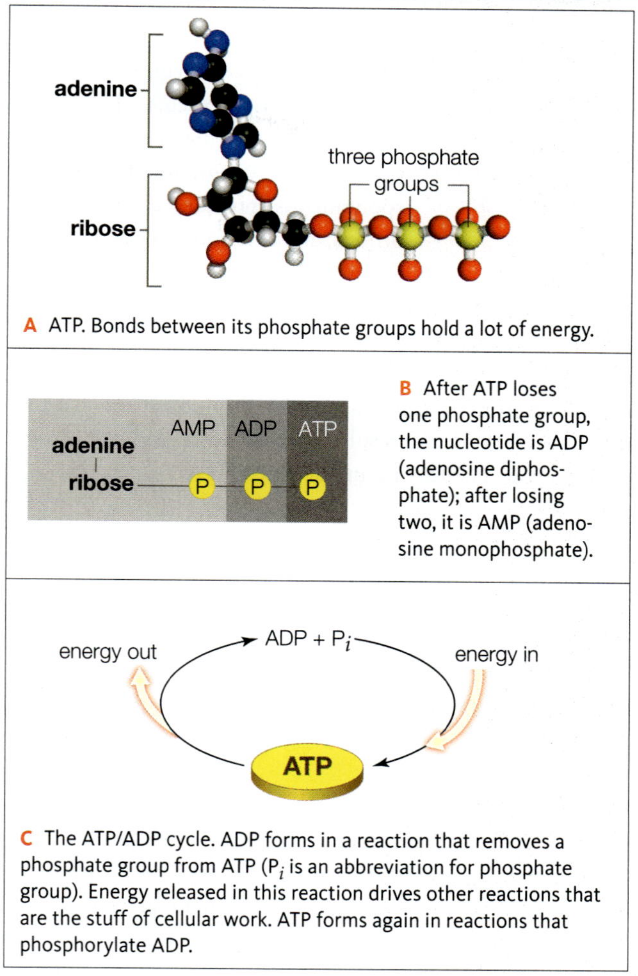

A ATP. Bonds between its phosphate groups hold a lot of energy.

B After ATP loses one phosphate group, the nucleotide is ADP (adenosine diphosphate); after losing two, it is AMP (adenosine monophosphate).

C The ATP/ADP cycle. ADP forms in a reaction that removes a phosphate group from ATP (P_i is an abbreviation for phosphate group). Energy released in this reaction drives other reactions that are the stuff of cellular work. ATP forms again in reactions that phosphorylate ADP.

FIGURE 5.20 ATP, an important energy currency in metabolism.

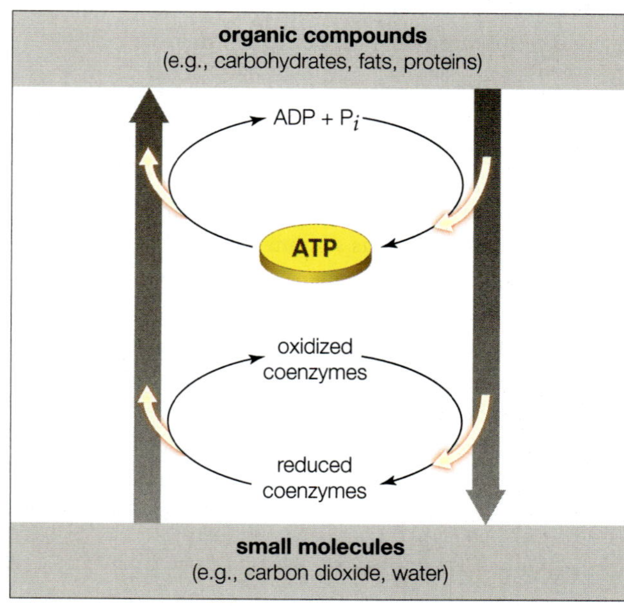

FIGURE 5.21 How ATP and coenzymes couple endergonic reactions with exergonic reactions. Yellow arrows indicate energy flow. Compare **FIGURES 5.8** and **5.20C**.

5.6 WHAT INFLUENCES THE MOVEMENT OF IONS AND MOLECULES?

Metabolic pathways require the participation of molecules that must move across membranes and through cells. **Diffusion** (*left*) is the spontaneous spreading of molecules or ions, and it is an essential way in which substances move into, through, and out of cells. An atom or molecule is always jiggling, and this internal movement causes it to randomly bounce off of nearby objects, including other atoms or molecules. Rebounds from such collisions propel solutes through a liquid or gas, with the result being a gradual and complete mixing. How fast this occurs depends on five factors:

Size It takes more energy to move a large object than it does to move a small one. Thus, smaller molecules diffuse more quickly than larger ones.

Temperature Atoms and molecules jiggle faster at higher temperature, so they collide more often. Thus, the higher the temperature, the faster the rate of diffusion.

Concentration A difference in solute concentration (Section 2.4) between adjacent regions of solution is called a concentration gradient. Solutes tend to diffuse "down" their concentration gradient, from a region of higher concentration to one of lower concentration. Why? Consider that moving objects (such as molecules) collide more often as they get more crowded. Thus, during a given interval, more molecules get bumped out of a region of higher concentration than get bumped into it.

Charge Each ion or charged molecule in a fluid contributes to the fluid's overall electric charge. A difference in charge between two regions of the fluid can affect the rate and direction of diffusion between them. For example, positively charged substances (such as sodium ions) will tend to diffuse toward a region with an overall negative charge.

Pressure Diffusion may be affected by a difference in pressure between two adjoining regions. Pressure squeezes objects—including atoms and molecules—closer together. Atoms and molecules that are more crowded collide and rebound more frequently. Thus, diffusion occurs faster at higher pressures.

SEMIPERMEABLE MEMBRANES

Remember from Section 4.3 that lipid bilayers are selectively permeable: Water can cross them, but ions and most polar molecules cannot (**FIGURE 5.22**). When two

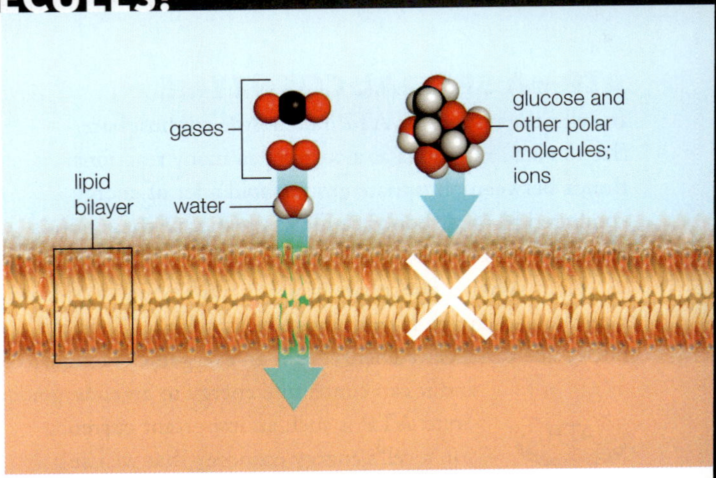

FIGURE 5.22 {Animated} Selective permeability of lipid bilayers. Hydrophobic molecules, gases, and water molecules can cross a lipid bilayer on their own. Ions in particular and most polar molecules such as glucose cannot.

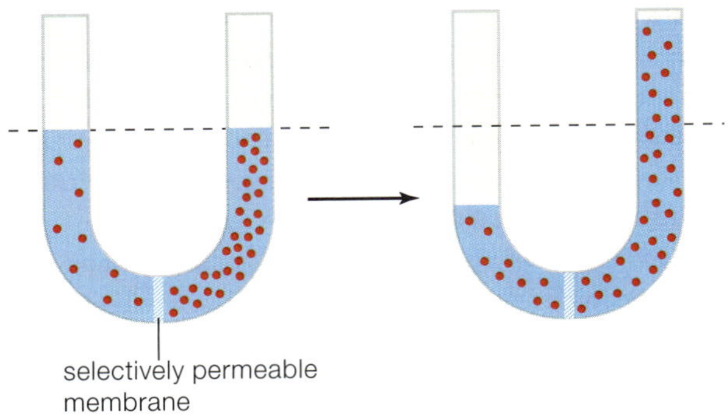

FIGURE 5.23 Osmosis. Water moves across a selectively permeable membrane that separates two fluids of differing solute concentration. The fluid volume changes in the two compartments as water diffuses across the membrane from the hypotonic solution to the hypertonic one.

fluids with different solute concentrations are separated by a selectively permeable membrane, water will diffuse across the membrane. The direction of water movement depends on the relative solute concentration of the two fluids. Tonicity refers to the solute concentration of one fluid relative to another that is separated by a selectively permeable membrane. Fluids that are **isotonic** have the same overall solute concentration. If the overall solute concentrations of the two fluids differ, the fluid with the lower concentration of solutes is said to be **hypotonic** (*hypo*–, under). The other one, with the higher solute concentration, is **hypertonic** (*hyper*–, over).

When a selectively permeable membrane separates two fluids that are not isotonic, water will move across the membrane from the hypotonic fluid into the hypertonic one (**FIGURE 5.23**). The diffusion will continue until the two fluids are isotonic, or until pressure against the

hypertonic fluid counters it. The movement of water across membranes is so important in biology that it is given a special name: **osmosis**.

If a cell's cytoplasm is hypertonic with respect to the fluid outside of its plasma membrane, water diffuses into it. If the cytoplasm is hypotonic with respect to the fluid on the outside, water diffuses out. In either case, the solute concentration of the cytoplasm may change. If it changes enough, the cell's enzymes will stop working, with potentially lethal results. Many cells have built-in mechanisms that compensate for differences in solute concentration between cytoplasm and extracellular (external) fluid. In cells with no such mechanism, the volume—and solute concentration—of cytoplasm will change as water diffuses into or out of the cell (**FIGURE 5.24**).

TURGOR

The rigid cell walls of plants and many protists, fungi, and bacteria can resist an increase in the volume of cytoplasm even in hypotonic environments. In the case of plant cells, cytoplasm usually contains more solutes than soil water does. Thus, water usually diffuses from soil into a plant—but only up to a point. Stiff walls keep plant cells from expanding very much, so an inflow of water causes pressure to build up inside them. Pressure that a fluid exerts against a structure that contains it is called **turgor**. When enough pressure builds up inside a plant cell, water stops diffusing into its cytoplasm. The amount of turgor that is enough to stop osmosis is called **osmotic pressure**.

Osmotic pressure keeps walled cells plump, just as high air pressure inside a tire keeps it inflated. A young land plant can resist gravity to stay erect because its cells are plump with cytoplasm (**FIGURE 5.25A**). When soil dries out, it loses water but not solutes, so the concentration of solutes increases in soil water. If soil water becomes hypertonic with respect to cytoplasm, water will start diffusing out of the plant's cells, so their cytoplasm shrinks (**FIGURE 5.25B**). As turgor inside the cells decreases, the plant wilts.

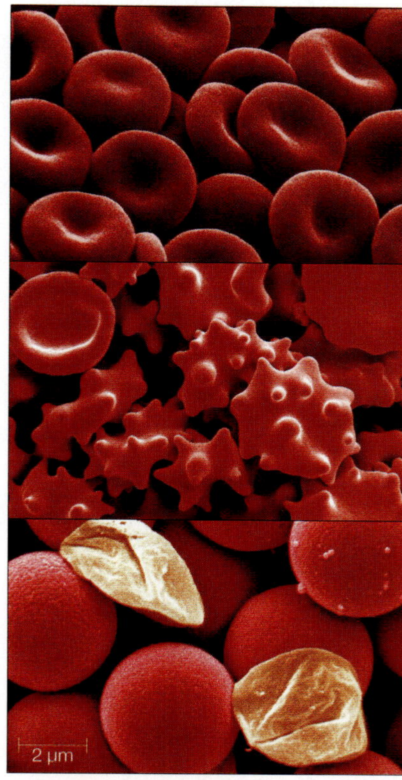

A Red blood cells in an isotonic solution (such as the fluid portion of blood) have a normal, indented disk shape.

B Water diffuses out of red blood cells immersed in a hypertonic solution, so they shrivel up.

C Water diffuses into red blood cells immersed in a hypotonic solution, so they swell up. Some of these have burst.

FIGURE 5.24 {Animated} Effects of tonicity in human red blood cells. These cells have no mechanism to compensate for differences in solute concentration between cytoplasm and extracellular fluid.

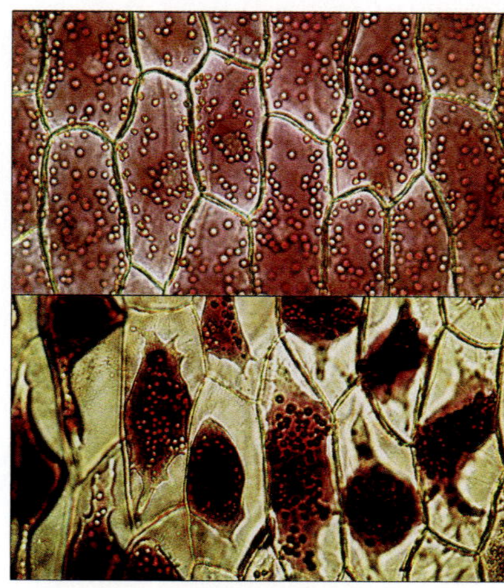

A Osmotic pressure keeps plant parts erect. These cells in an iris petal are plump with cytoplasm.

B Cells from a wilted iris petal. The cytoplasm shrank, and the plasma membrane moved away from the wall.

FIGURE 5.25 Turgor, as illustrated in cells of iris petals.

TAKE-HOME MESSAGE 5.6

Molecules or ions tend to diffuse into an adjoining region of fluid in which they are not as concentrated. The steepness of a concentration gradient as well as temperature, molecular size, charge, and pressure affect the rate of diffusion.

When two fluids of different solute concentration are separated by a selectively permeable membrane, water diffuses from the hypotonic to the hypertonic fluid. This movement, osmosis, is opposed by turgor.

---

diffusion Spontaneous spreading of molecules or ions.
hypertonic Describes a fluid that has a high solute concentration relative to another fluid separated by a semipermeable membrane.
hypotonic Describes a fluid that has a low solute concentration relative to another fluid separated by a semipermeable membrane.
isotonic Describes two fluids with identical solute concentrations and separated by a semipermeable membrane.
osmosis Diffusion of water across a selectively permeable membrane; occurs in response to a difference in solute concentration between the fluids on either side of the membrane.
osmotic pressure Amount of turgor that prevents osmosis into cytoplasm or other hypertonic fluid.
turgor Pressure that a fluid exerts against a structure that contains it.

CREDITS: (24A) Annie Cavanagh/Wellcome Images; (24B,C) CMSP/Getty Images; (25A,B) Claude Nuridsany & Marie Perennou/Science Source; (25 inset) © Evgenyi/Shutterstock.com.

5.7 HOW DO IONS AND CHARGED MOLECULES CROSS CELL MEMBRANES?

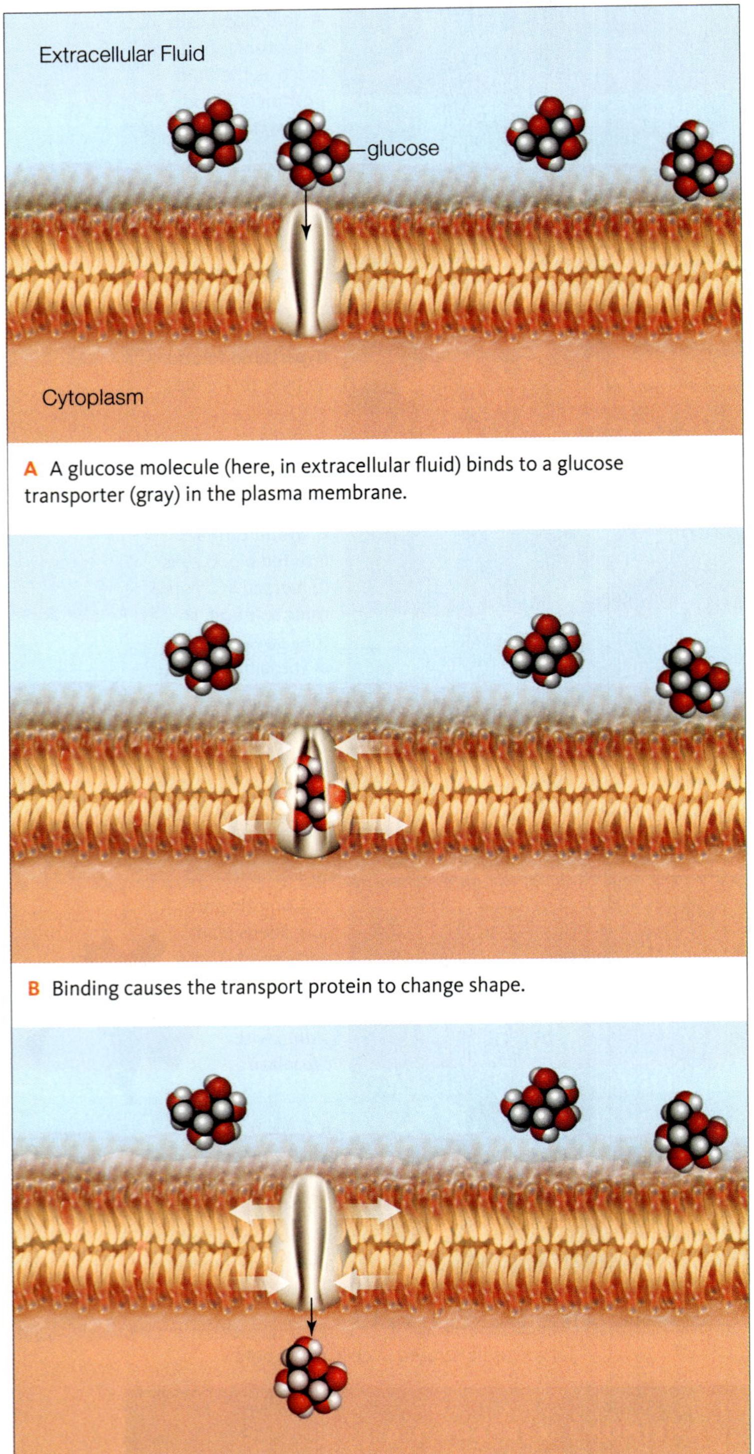

A A glucose molecule (here, in extracellular fluid) binds to a glucose transporter (gray) in the plasma membrane.

B Binding causes the transport protein to change shape.

C The transport protein releases the glucose on the other side of the membrane (here, in cytoplasm) and resumes its original shape.

FIGURE 5.26 {Animated} Facilitated diffusion.

FIGURE IT OUT: In this example, which fluid is hypotonic: the extracellular fluid or cytoplasm?
Answer: Cytoplasm

TRANSPORT PROTEIN SPECIFICITY

Substances that cannot diffuse directly through lipid bilayers—ions in particular—cross cell membranes only with the help of transport proteins (Section 4.3). Each transport protein allows a specific substance to cross: Calcium pumps pump only calcium ions; glucose transporters transport only glucose; and so on. This specificity is an important part of homeostasis. For example, the composition of cytoplasm depends on the movement of particular solutes across the plasma membrane, which in turn depends on the transporters embedded in it. Glucose is an important source of energy for most cells, so they normally take up as much as they can from extracellular fluid. They do so with the help of glucose transporters in the plasma membrane. As soon as a molecule of glucose enters cytoplasm, an enzyme (hexokinase) phosphorylates it. Phosphorylation traps the molecule inside the cell because the transporters are specific for glucose, not phosphorylated glucose. Thus, phosphorylation prevents the molecule from moving back through the transporter and leaving the cell.

FACILITATED DIFFUSION

Osmosis is an example of **passive transport**, which is a membrane-crossing mechanism that requires no energy input. Diffusion of solutes through transport proteins is another example. In this case, the movement of the solute (and the direction of its movement) is driven entirely by the solute's concentration gradient. Some transport proteins form permanently open channels through a membrane. Others are gated, which means they open and close in response to a stimulus such as a shift in electric charge or binding to a particular signaling molecule.

With a passive transport mechanism called **facilitated diffusion**, a solute binds to a transport protein, which then changes shape so the solute is released to the other side of the membrane. A glucose transporter is an example of a transport protein that works in facilitated diffusion (**FIGURE 5.26**). This protein changes shape when it binds to a molecule of glucose. The shape change moves the solute to the opposite side of the membrane, where it detaches from the transport protein. Then, the transporter reverts to its original shape.

ACTIVE TRANSPORT

Maintaining a particular solute's concentration at a certain level often means transporting the solute against its gradient, to the side of the membrane where it is more concentrated. Pumping a solute against its gradient takes energy. In **active transport**, a transport protein uses energy to pump a solute against its gradient across a cell membrane.

After a solute binds to an active transport protein, an energy input (for example, in the form of a phosphate-group transfer from ATP) changes the shape of the protein. The change causes the transporter to release the solute to the other side of the membrane.

A calcium pump moves calcium ions across cell membranes by active transport (**FIGURE 5.27**). Calcium ions act as potent messengers inside cells, and they affect the activity of many enzymes, so their concentration in cytoplasm is very tightly regulated. Calcium pumps in the plasma membrane of all eukaryotic cells can keep the concentration of calcium ions in cytoplasm thousands of times lower than it is in extracellular fluid.

Another example of active transport involves sodium–potassium pumps (**FIGURE 5.28**). Nearly all of the cells in your body have these transport proteins. Sodium ions in cytoplasm diffuse into the pump's open channel and bind to its interior. A phosphate-group transfer from ATP causes the pump to change shape so that its channel opens to extracellular fluid, where it releases the sodium ions. Then, potassium ions from extracellular fluid diffuse into the channel and bind to its interior. The transporter releases the phosphate group and reverts to its original shape. The channel opens to the cytoplasm, where it releases the potassium ions.

Bear in mind that the membranes of all cells, not just those of animals, have active transport proteins. In plants, for example, active transport proteins in the plasma membranes of leaf cells pump sucrose into tubes that thread throughout the plant body.

active transport Energy-requiring mechanism in which a transport protein pumps a solute across a cell membrane against its concentration gradient.
facilitated diffusion Passive transport mechanism in which a solute follows its concentration gradient across a membrane by moving through a transport protein.
passive transport Membrane-crossing mechanism that requires no energy input.

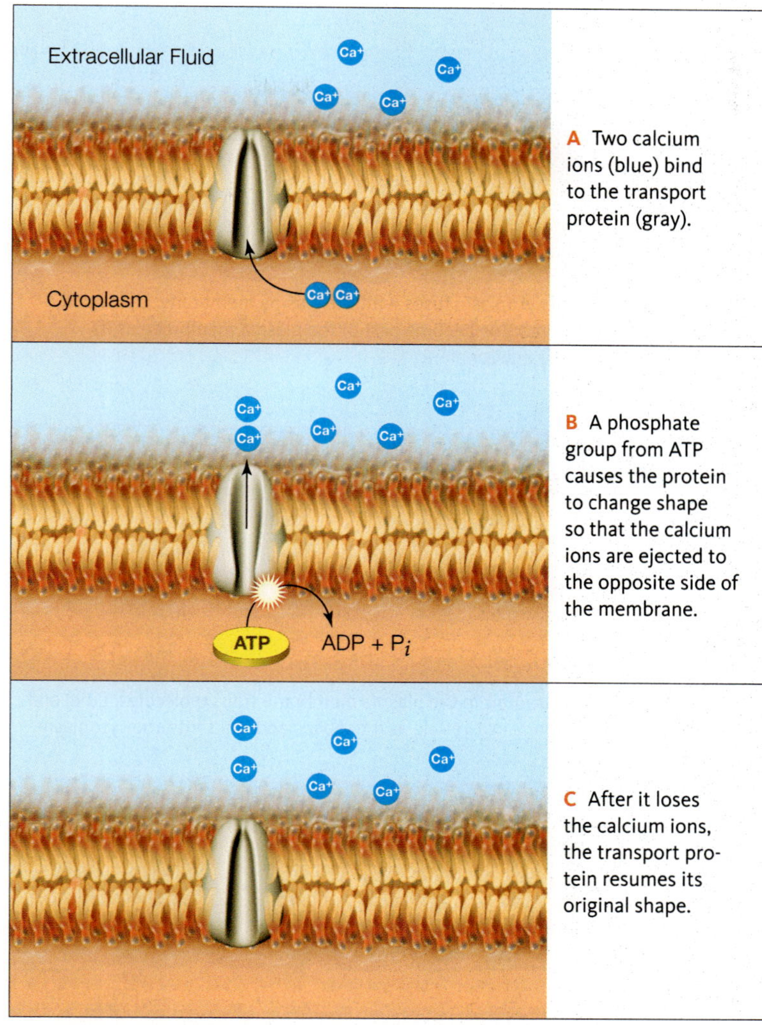

FIGURE 5.27 {Animated} Active transport of calcium ions.

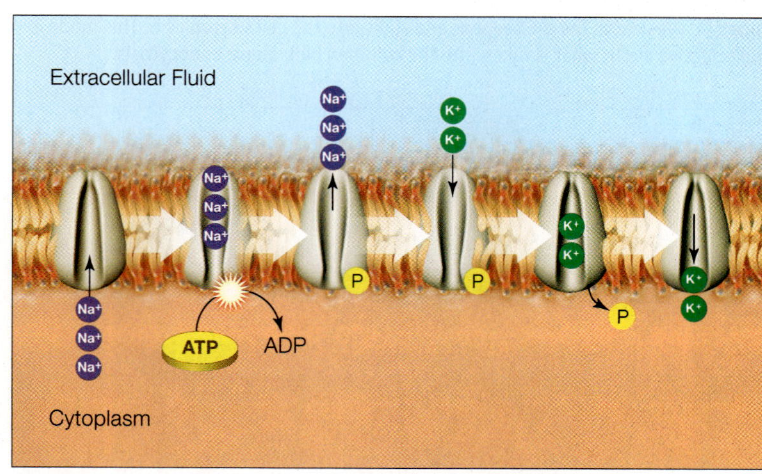

FIGURE 5.28 The sodium–potassium pump. This transport protein (gray) actively transports sodium ions (Na^+) from cytoplasm to extracellular fluid, and potassium ions (K^+) in the other direction. The transfer of a phosphate group (P) from ATP provides energy required for transporting the ions against their concentration gradient.

TAKE-HOME MESSAGE 5.7

Transport proteins move specific ions or molecules across a cell membrane. The amounts and types of these substances that cross a membrane depend on the transport proteins embedded in it.

In a type of passive transport called facilitated diffusion, a solute binds to a transport protein that releases it on the opposite side of the membrane. The movement is driven by the solute's concentration gradient.

In active transport, a transport protein pumps a solute across a membrane against its concentration gradient. The movement is driven by an energy input, as from ATP.

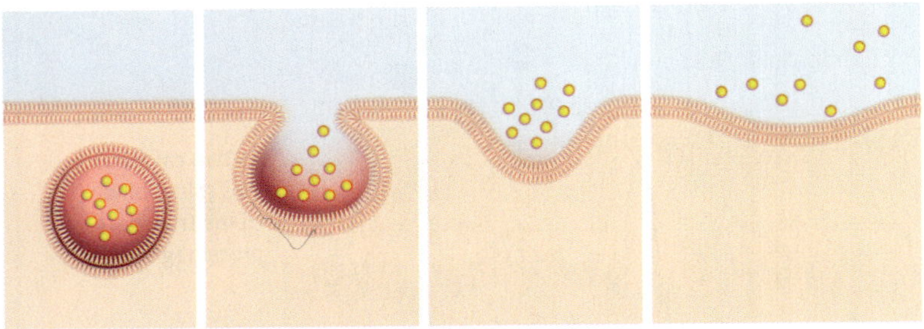

A **Exocytosis**. A vesicle in cytoplasm fuses with the plasma membrane. Lipids and proteins of the vesicle's membrane become part of the plasma membrane as its contents are expelled to the environment.

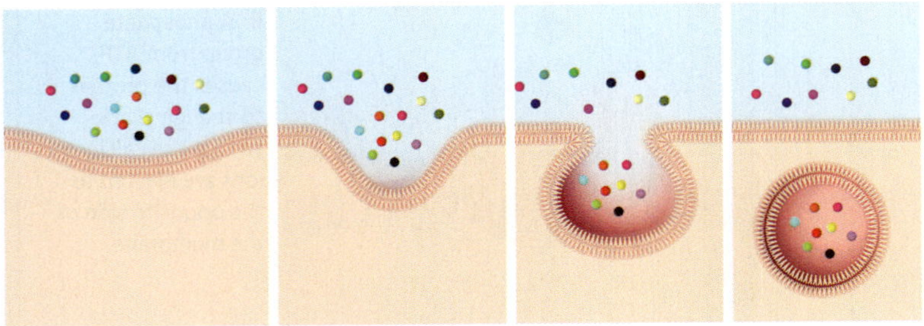

B **Bulk-phase endocytosis**. A pit in the plasma membrane traps molecules, fluid, and particles near the cell's surface in a vesicle as it deepens and sinks into the cytoplasm.

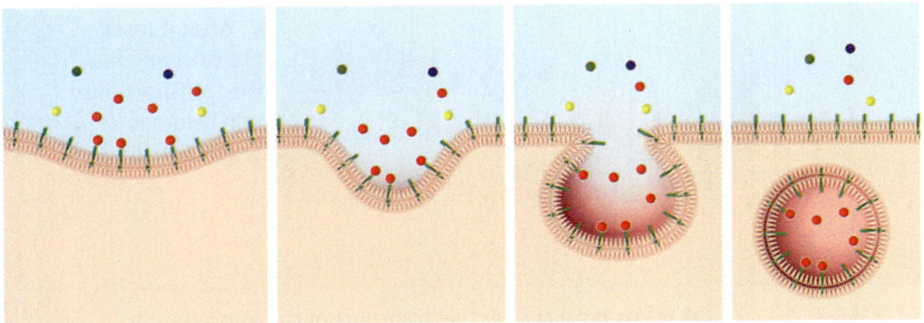

C **Receptor-mediated endocytosis**. Receptors on the cell surface bind a target molecule and trigger a pit to form in the plasma membrane. The target molecules are trapped in a vesicle as the pit deepens and sinks into the cell's cytoplasm. This mode is more selective about what is taken into the cell than bulk-phase endocytosis.

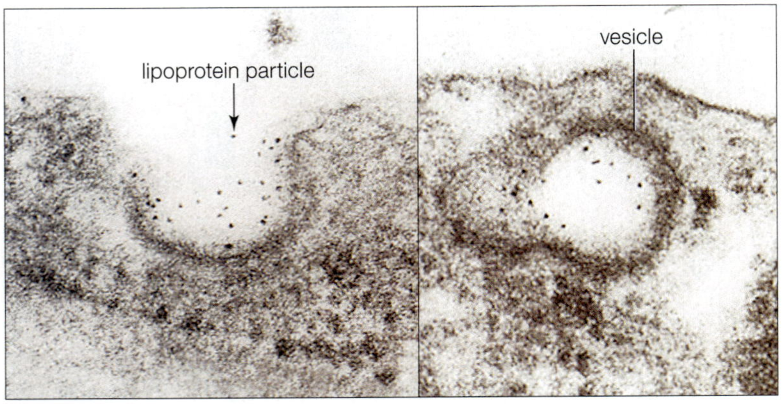

D Receptor-mediated endocytosis of lipoprotein particles.

FIGURE 5.29 {Animated} Exocytosis and endocytosis.

VESICLE MOVEMENT

Think back on the structure of a lipid bilayer (Section 4.3). When a bilayer is disrupted, it seals itself. Why? The disruption exposes the fatty acid tails of the phospholipids to their watery surroundings. Remember, in water, phospholipids spontaneously rearrange themselves so that their nonpolar tails stay together. A vesicle forms when a patch of membrane bulges into the cytoplasm because the hydrophobic tails of the lipids in the bilayer are repelled by the watery fluid on both sides. The fluid "pushes" the phospholipid tails together, which helps round off the bud as a vesicle, and also seals the rupture in the membrane.

Vesicles are constantly carrying materials to and from a cell's plasma membrane. This movement typically requires ATP because it involves motor proteins that drag the vesicles along cytoskeletal elements. We describe the movement based on where and how the vesicle originates, and where it goes.

By **exocytosis**, a vesicle in the cytoplasm moves to the cell's surface and fuses with the plasma membrane. As the exocytic vesicle loses its identity, its contents are released to the surroundings (FIGURE 5.29A).

There are several pathways of **endocytosis**, but all take up substances in bulk near the cell's surface (as opposed to one molecule or ion at a time via transport proteins). In bulk-phase endocytosis, a small patch of plasma membrane balloons inward, and then it pinches off after sinking farther into the cytoplasm. The membrane patch becomes the outer boundary of a vesicle (FIGURE 5.29B).

With receptor-mediated endocytosis, molecules of a hormone, vitamin, mineral, or another substance bind to receptors on the plasma membrane. The binding triggers a shallow pit to form in the membrane patch under the receptors. The pit sinks into the cytoplasm and traps the target substance in a vesicle as it closes back on itself (FIGURE 5.29C,D). LDL and other lipoproteins (Section 3.4) enter cells this way.

Phagocytosis (which literally means "cell eating") is a type of receptor-mediated endocytosis in which motile cells engulf microorganisms, cellular debris, or other large particles (FIGURE 5.30). Many single-celled protists such as amoebas feed by phagocytosis. Some of your white blood cells use phagocytosis to engulf viruses and bacteria, cancerous body cells, and other threats.

Phagocytosis begins when receptor proteins bind to a particular target. The binding causes microfilaments to assemble in a mesh under the plasma membrane. The microfilaments contract, forcing a lobe of membrane-enclosed cytoplasm to bulge outward as a pseudopod (Section 4.9). Pseudopods that merge around a target

FIGURE 5.30 {Animated} Phagocytosis. A phagocytic white blood cell's pseudopods surround bacteria. A vesicle will form around the bacteria as the pseudopod membranes fuse around them. The vesicle will sink into the cytoplasm, where it will fuse with a lysosome that digests its contents.

trap it inside a vesicle that sinks into the cytoplasm. Material taken in by phagocytosis is typically digested by lysosomes, and the resulting molecular bits may be recycled by the cell, or expelled by exocytosis.

MEMBRANE TRAFFICKING

The composition of a plasma membrane begins in the ER. There, membrane proteins and lipids are made and modified, and both become part of vesicles that transport them to Golgi bodies for final modification. New plasma membrane forms when the finished proteins and lipids are repackaged as vesicles that travel to the plasma membrane and fuse with it.

As long as a cell is alive, exocytosis and endocytosis continually replace and withdraw patches of its plasma membrane. If the cell is not enlarging, the total area of the plasma membrane remains more or less constant. Membrane lost as a result of endocytosis is replaced by membrane arriving as exocytic vesicles.

endocytosis Process by which a cell takes in a small amount of extracellular fluid (and its contents) by the ballooning inward of the plasma membrane.
exocytosis Process by which a cell expels a vesicle's contents to extracellular fluid.
phagocytosis "Cell eating"; an endocytic pathway by which a cell engulfs particles such as microbes or cellular debris.

TAKE-HOME MESSAGE 5.8

- Exocytosis and endocytosis move materials in bulk across plasma membranes.
- In exocytosis, a cytoplasmic vesicle fuses with the plasma membrane and releases its contents to the outside of the cell.
- In endocytosis, a patch of plasma membrane sinks inward and forms a vesicle in the cytoplasm.
- Some cells can engulf large particles by phagocytosis.

5.9 Application: A TOAST TO ALCOHOL DEHYDROGENASE

FIGURE 5.31 A tailgate party at a Notre Dame–Alabama football game. During 2012 alone, Indiana State police arrested 138 Notre Dame students for underage drinking at tailgate parties.

MOST COLLEGE STUDENTS ARE UNDER THE LEGAL DRINKING AGE, but alcohol abuse continues to be the most serious drug problem on college campuses throughout the United States (FIGURE 5.31). Every year, drinking kills more than 1,700 students and injures about 500,000 more; it is also a factor in 600,000 assaults and 100,000 rapes on college campuses.

Tens of thousands of undergraduate students have been polled about their drinking habits in recent surveys. More than half of them reported that they regularly drink five or more alcoholic beverages within a two-hour period.

Each alcoholic drink—a bottle of beer, glass of wine, shot of vodka, and so on—contains the same amount of alcohol or, more precisely, ethanol. Ethanol molecules move quickly from the stomach and small intestine into the bloodstream. Almost all of the ethanol ends up in the liver, a large organ in the abdomen. Liver cells have impressive numbers of enzymes. One of them, ADH (alcohol dehydrogenase), helps break down ethanol and other toxic compounds.

ADH converts ethanol to acetaldehyde, an organic molecule even more toxic than ethanol and the most likely source of various hangover symptoms. A different enzyme, ALDH, very quickly converts acetaldehyde to nontoxic acetate. Both ADH and ALDH use the coenzyme NAD^+ to accept electrons and hydrogen atoms. Thus, the overall pathway of ethanol metabolism in humans is:

$$\text{ethanol} \xrightarrow[\text{NAD}^+ \;\to\; \text{NADH}]{\text{ADH}} \text{acetaldehyde} \xrightarrow[\text{NAD}^+ \;\to\; \text{NADH}]{\text{ALDH}} \text{acetate}$$

In the average healthy adult, this metabolic pathway can detoxify between 7 and 14 grams of ethanol per hour. The typical alcoholic beverage contains between 10 and 20 grams of ethanol.

If you put more ethanol into your body than your enzymes can deal with, then you will damage it. Ethanol and acetaldehyde kill liver cells, so the more

a person drinks, the fewer liver cells are left for detoxification . Ethanol also interferes with normal metabolic processes. For example, in the presence of ethanol, oxygen that would ordinarily take part in breaking down fatty acids is diverted to breaking down the ethanol. This is why fats accumulate as large globules in the tissues of heavy drinkers.

Long-term heavy drinking causes alcoholic hepatitis, a disease characterized by inflammation and destruction of liver tissue; and cirrhosis, a condition in which the liver becomes so scarred, hardened, and filled with fat that it loses its function. (The term cirrhosis is from the Greek *kirros*, meaning orange-colored, after the abnormal skin color of people with the disease.) The liver is the largest gland in the human body, and it has many important functions. A cirrhotic liver stops making the protein albumin, so the solute balance of body fluids is disrupted, and the legs and abdomen swell with watery fluid. It can no longer remove drugs and other toxins from the blood, so they accumulate in the brain—which impairs mental functioning and alters personality. Restricted blood flow through the liver causes veins to enlarge and rupture, so internal bleeding is a risk. The damage to the body results in a heightened susceptibility to diabetes and liver cancer. Once cirrhosis has been diagnosed, a person has about a 50 percent chance of dying within 10 years (FIGURE 5.32).

FIGURE 5.32 Gary Reinbach, who died at the age of 22 from alcoholic liver disease shortly after this photograph was taken, in 2009. The odd color of his skin is a symptom of cirrhosis.

Transplantation is a last-resort treatment for a failed liver, but there are not nearly enough liver donors for everyone who needs a transplant. Reinbach was refused a transplant that would have saved his life because he had not abstained from drinking for the prior 6 months.

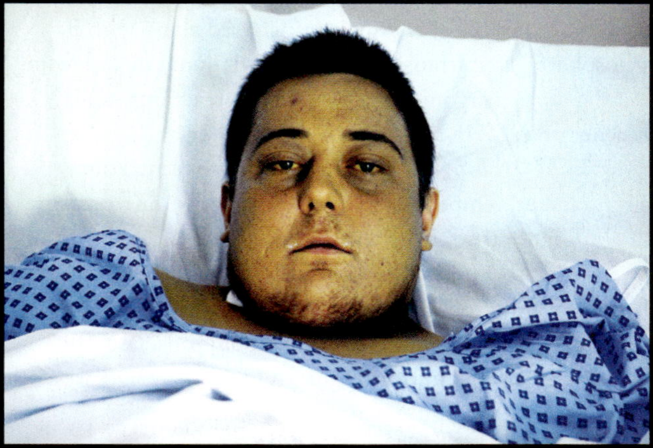

CREDITS: (31) left inset, Southern Illinois University/Science Source; right inset, Martin M. Rotker/Science Source; (32) Stuart Clark/The Sunday Times/nisyndication.

Summary

SECTION 5.1 **Kinetic energy** and **potential energy** are different forms of **energy**, the capacity to do work. Energy, which cannot be created or destroyed (**first law of thermodynamics**), tends to disperse spontaneously (**second law of thermodynamics**). **Entropy** is a measure of how much the energy of a system is dispersed. A bit disperses at each energy transfer, usually in the form of heat.

SECTION 5.2 Cells store and retrieve energy by making and breaking chemical bonds in reactions that convert **reactants** to **products**. **Endergonic** reactions require a net input of energy to proceed. **Exergonic** reactions end with a net release of energy. **Activation energy** is the minimum energy required to start a reaction.

SECTION 5.3 Enzymes greatly enhance the rate of reactions without being changed by them, a process called **catalysis**. They lower a reaction's activation energy, for example by boosting local concentrations of **substrates** or improving the fit between a substrate and the enzyme's **active site** (**induced-fit model**). Each enzyme works best within a characteristic range of conditions that reflect its evolutionary context.

SECTION 5.4 Cells build, convert, and dispose of substances in enzyme-mediated reaction sequences called **metabolic pathways**. Regulating these pathways allows a cell to conserve energy and resources. With **allosteric regulation**, a regulatory molecule or ion alters the activity of an enzyme by binding to it in a region other than the active site. Some products of metabolic pathways inhibit their own production, a regulatory mechanism called **feedback inhibition**. **Redox** (oxidation–reduction) **reactions** in **electron transfer chains** allow cells to harvest energy in small, manageable steps.

SECTION 5.5 Most enzymes require **cofactors**, which are metal ions or organic **coenzymes**. Some enzymes that act as **antioxidants** have cofactors that help them prevent dangerous oxidation reactions.

When a phosphate group is transferred from ATP to another molecule, energy is transferred along with it. Phosphate-group transfers (**phosphorylations**) to and from ATP couple exergonic with endergonic reactions. Cells regenerate ATP in the **ATP/ADP cycle**.

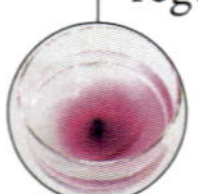

SECTION 5.6 The steepness of a concentration gradient, temperature, solute size, charge, and pressure influence the rate of **diffusion**. **Osmosis** is the diffusion of water across a selectively permeable membrane, from a **hypotonic** fluid toward a **hypertonic** fluid. There is no net movement of water between **isotonic** solutions. **Osmotic pressure** is the amount of **turgor** (fluid pressure against a cell membrane or wall) sufficient to halt osmosis.

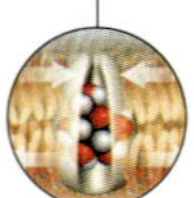

SECTION 5.7 The types of transport proteins in a membrane determine which substances cross it. Ions and most polar molecules cross membranes with the help of a transport protein. With **facilitated diffusion**, a solute follows its concentration gradient across a membrane through a transport protein. Facilitated diffusion is a type of **passive transport** (no energy input is required). With **active transport**, a transport protein uses energy to pump a solute across a membrane against its concentration gradient. A phosphate-group transfer from ATP often supplies the necessary energy for active transport.

SECTION 5.8 Substances can be moved across plasma membranes in bulk by two ATP-requiring processes. With **exocytosis**, a cytoplasmic vesicle fuses with the plasma membrane, and its contents are released to the outside of the cell. With **endocytosis**, a patch of plasma membrane balloons into the cell, and forms a vesicle that sinks into the cytoplasm. Some cells engulf large particles such as prey or cell debris by the endocytic pathway of **phagocytosis**.

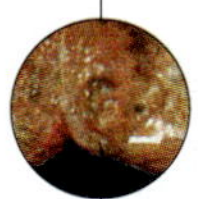

SECTION 5.9 Currently the most serious drug problem on college campuses is binge drinking, which is often a symptom of alcoholism. Drinking more alcohol than the body's enzymes can detoxify can be lethal in both the short term and the long term.

Self-Quiz Answers in Appendix VII

1. _______ is life's primary source of energy.
   a. Food   b. Water   c. Sunlight   d. ATP

2. Which of the following statements is not correct?
   a. Energy cannot be created or destroyed.
   b. Energy cannot change from one form to another.
   c. Energy tends to disperse spontaneously.

3. Entropy _______ .
   a. disperses   c. always increases, overall
   b. is a measure of disorder   d. b and c

4. If we liken a chemical reaction to an energy hill, then a(n) _______ reaction is an uphill run.
   a. endergonic   c. catalytic
   b. exergonic   d. both a and c

5. If we liken a chemical reaction to an energy hill, then activation energy is like _______ .
   a. a burst of speed
   b. coasting downhill
   c. a bump at the top of the hill

6. _______ are always changed by participating in a reaction. (Choose all that are correct.)
   a. Enzymes   b. Cofactors   c. Reactants   d. Coenzymes

7. Name one environmental factor that typically influences enzyme function.

8. A metabolic pathway may _______ .
   a. build or break down molecules   c. generate heat
   b. include an electron transfer chain   d. all of the above

Data Analysis Activities

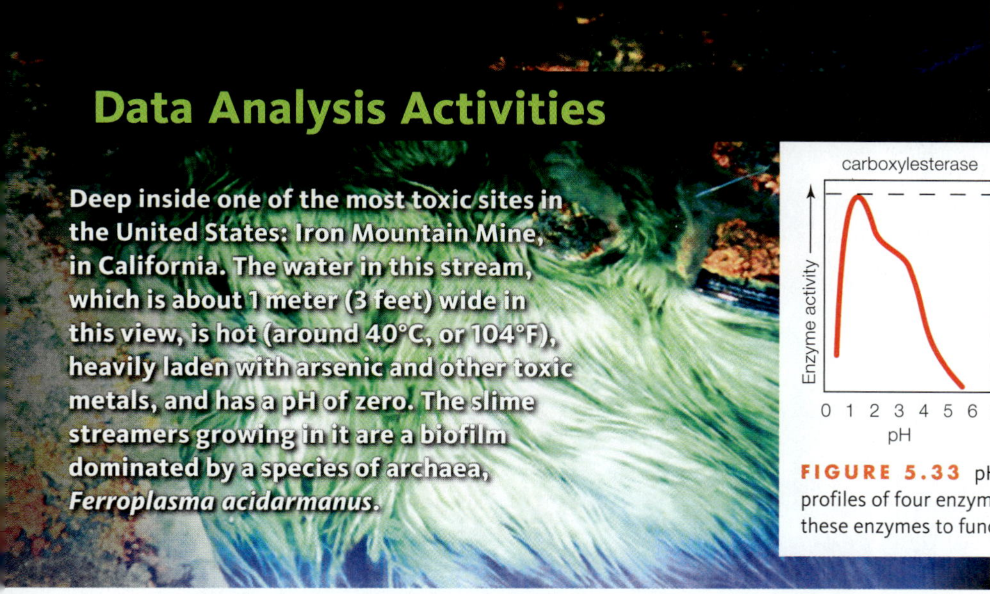

Deep inside one of the most toxic sites in the United States: Iron Mountain Mine, in California. The water in this stream, which is about 1 meter (3 feet) wide in this view, is hot (around 40°C, or 104°F), heavily laden with arsenic and other toxic metals, and has a pH of zero. The slime streamers growing in it are a biofilm dominated by a species of archaea, *Ferroplasma acidarmanus*.

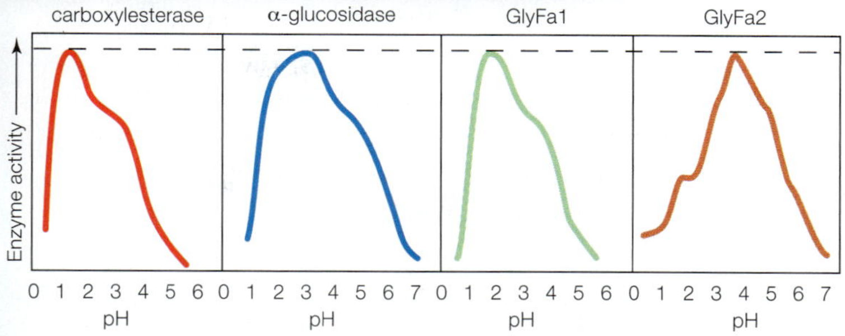

FIGURE 5.33 pH anomaly of *Ferroplasma* enzymes. Above are pH activity profiles of four enzymes isolated from *Ferroplasma*. Researchers had expected these enzymes to function best at the cells' cytoplasmic pH (5.0).

One Tough Bug *Ferroplasma acidarmanus* is a species of archaea discovered in an abandoned California copper mine. These cells use an energy-harvesting pathway that combines oxygen with iron–sulfur compounds in minerals such as pyrite. The reaction dissolves the minerals, so groundwater that seeps into the mine ends up with high concentrations of metal ions such as copper, zinc, cadmium, and arsenic. The reaction also produces sulfuric acid, which lowers the pH of the water around the cells to zero. *F. acidarmanus* cells maintain their internal pH at a cozy 5.0 despite living in an environment similar to hot battery acid. Thus, researchers investigating *Ferroplasma* were surprised to discover that most of the cells' enzymes function best at very low pH (**FIGURE 5.33**).

1. What does the dashed line in the graph signify?
2. Of the four enzymes, how many function optimally at a pH lower than 5?
3. What is the optimal pH for the carboxylesterase?

---

9. All antioxidants _______ .
   a. prevent other molecules from being oxidized
   b. are necessary in the human diet
   c. balance charge
   d. deoxidize free radicals

10. Solutes tend to diffuse from a region where they are _______ (more/less) concentrated to an adjacent region where they are _______ (more/less) concentrated.

11. _______ cannot easily diffuse across a lipid bilayer.
    a. Water c. Ions
    b. Gases d. all of the above

12. A transport protein requires ATP to pump sodium ions across a membrane. This is a case of _______ .
    a. passive transport c. facilitated diffusion
    b. active transport d. a and c

13. Immerse a human blood cell in a hypotonic solution, and water _______ .
    a. diffuses into the cell c. shows no net movement
    b. diffuses out of the cell d. moves in by endocytosis

14. Vesicles form during _______ .
    a. endocytosis c. phagocytosis
    b. exocytosis d. a and c

15. Match each term with its most suitable description.

___ reactant	a. assists enzymes
___ phagocytosis	b. forms at reaction's end
___ first law of thermodynamics	c. enters a reaction
	d. requires energy input
___ product	e. one cell engulfs another
___ cofactor	f. energy cannot be created or destroyed
___ concentration gradient	g. basis of diffusion
___ passive transport	h. no energy input required
___ active transport	

Critical Thinking

1. Often, beginning physics students are taught the basic concepts of thermodynamics with two phrases: First, you can never win. Second, you can never break even. Explain.

2. How do you think a cell regulates the amount of glucose it brings into its cytoplasm from the extracellular environment?

3. The enzyme trypsin is sold as a dietary enzyme supplement. Explain what happens to trypsin that is taken with food.

4. Catalase combines two hydrogen peroxide molecules ($H_2O_2 + H_2O_2$) to make two molecules of water. A gas also forms. What is the gas?

Raindrops can separate light from the sun into its different component wavelengths, which we see as different colors in a rainbow.

6 WHERE IT STARTS— PHOTOSYNTHESIS

Links to Earlier Concepts

This chapter explores the main metabolic pathways (Section 5.4) by which organisms harvest energy from the sun (5.1). We revisit experimental design (1.6), electrons and energy levels (2.2), bonds (2.3), carbohydrates (3.2), membrane proteins (4.3), plastids (4.8), antioxidants (5.5), and concentration gradients (5.6).

KEY CONCEPTS

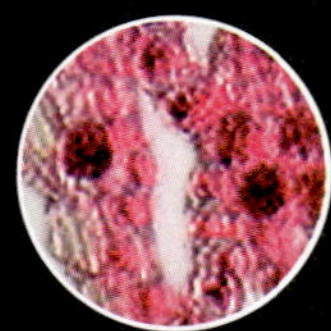

THE RAINBOW CATCHERS

The main flow of energy through the biosphere starts when photosynthetic pigments absorb light. In plants and other eukaryotes, these pigments occur in chloroplasts.

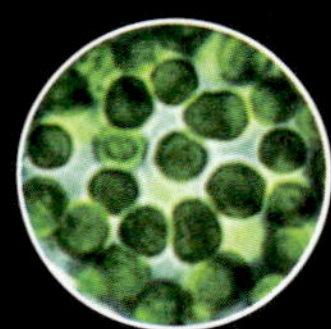

WHAT IS PHOTOSYNTHESIS?

Photosynthesis is a metabolic pathway that occurs in two stages. Light energy harvested in the first stage is used to make molecules that power sugar formation in the second.

MAKING ATP AND NADPH

In the first stage of photosynthesis, light drives ATP synthesis in one of two pathways. A cyclic pathway makes ATP alone. A noncyclic pathway makes ATP and NADPH, and it releases oxygen.

MAKING SUGARS

Sugars are assembled from carbon dioxide (CO_2) during the second stage of photosynthesis. In plants, the reactions run on ATP and NADPH—molecules that formed in the first stage.

ALTERNATE PATHWAYS

Metabolic pathways are shaped by evolution. Variations in photosynthetic pathways are evolutionary adaptations that allow plants to thrive in a variety of environments.

Photograph by Robbie George, National Geographic Creative.

6.1 HOW DO PHOTOSYNTHESIZERS ABSORB LIGHT?

Energy flow through nearly all ecosystems on Earth begins when plants and other photosynthesizers intercept sunlight. These organisms are called producers because they produce the food that sustains ecosystems. Producers are **autotrophs**, organisms that make their own food—sugars—by harvesting energy directly from the environment (*auto–* means self; *–troph* refers to nourishment). All organisms need carbon; autotrophs obtain it from inorganic molecules such as carbon dioxide (CO_2). By contrast, **heterotrophs** get their carbon by breaking down organic molecules assembled by other organisms (*hetero–* means other). Heterotrophs are an ecosystem's consumers.

PROPERTIES OF LIGHT

Photosynthesizers make their own food by converting light energy to chemical energy. In order to understand how that happens, you have to know a little about the nature of light. Light is electromagnetic radiation, a type of energy that moves through space in waves, a bit like waves moving across an ocean. The distance between the crests of two successive waves is a **wavelength**, measured in nanometers (nm). Light that humans can see is a small part of the spectrum of electromagnetic radiation emitted by the sun (**FIGURE 6.1A**). Visible light travels in wavelengths between 380 and 750 nm, and this is the main form of energy that drives photosynthesis. Our eyes perceive all of these wavelengths combined as white light, and particular wavelengths in this range as different colors. White light separates into its component colors when it passes through a prism, or raindrops that act as tiny prisms. A prism bends longer wavelengths more than it bends shorter ones, so a rainbow of colors forms.

Light travels in waves, but it is also organized in packets of energy called photons. A photon's energy and its wavelength are related, so all photons traveling at the same wavelength carry the same amount of energy. Photons that carry the least amount of energy travel in longer wavelengths; those that carry the most energy travel in shorter wavelengths (**FIGURE 6.1B**).

CAPTURING A RAINBOW

Photosynthesizers use pigments to capture light. A **pigment** is an organic molecule that selectively absorbs light of specific wavelengths. Wavelengths of light that are not absorbed are reflected, and that reflected light gives each pigment its characteristic color.

Chlorophyll *a* is the most common photosynthetic pigment in plants and photosynthetic protists. It also occurs in some bacteria. Chlorophyll *a* absorbs violet, red, and orange light, and it reflects green light, so it appears green to us. Accessory pigments, including other chlorophylls, collectively harvest a wide range of additional light wavelengths for photosynthesis (**FIGURE 6.2**).

A pigment molecule is a bit like an antenna specialized for receiving light. Each has a light-trapping part in which single bonds alternate with double bonds; electrons

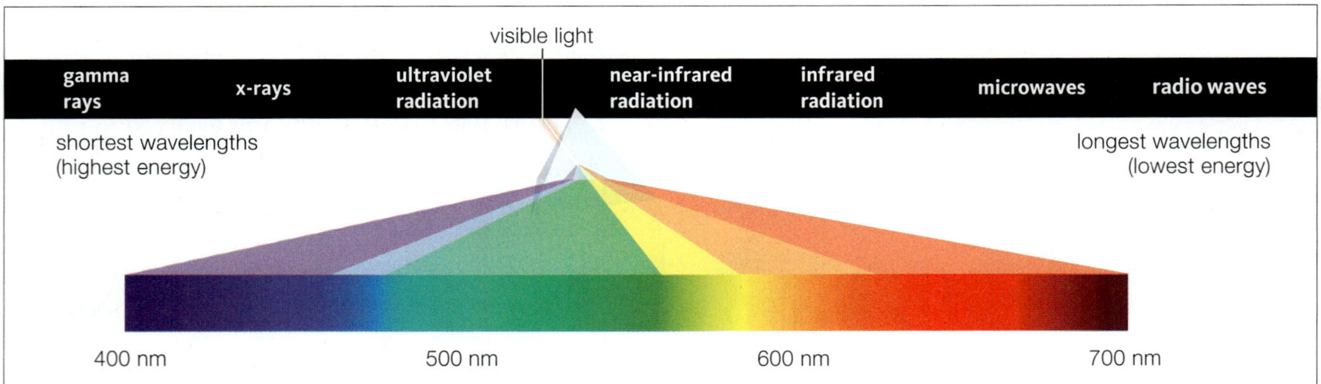

A Electromagnetic radiation moves through space in waves that we measure in nanometers (nm). Visible light makes up a very small part of this energy. Raindrops or a prism can separate visible light's different wavelengths, which we see as different colors. About 25 million nanometers are equal to 1 inch.

B Light is organized as packets of energy called photons. The shorter a photon's wavelength, the greater its energy.

FIGURE 6.1 Properties of light.

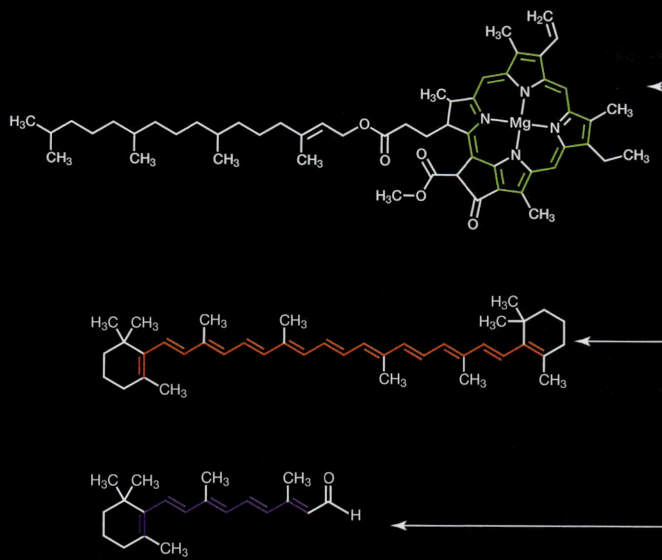

Pigment	Color	Plants	Protists	Bacteria	Archaea
		Produced by			
Chlorophyll *a*	green	●	●	●	
Other chlorophylls	green	●	●	●	
Phycobilins					
phycocyanobilin	blue		●	●	
phycoerythrobilin	red		●	●	
phycoviolobilin	violet		●	●	
Carotenoids					
beta-carotene	orange	●	●	●	
lycopene	red	●	●		
lutein	yellow	●	●	●	
zeaxanthin	yellow	●	●	●	
fucoxanthin	brown	●	●		
Anthocyanins	red, blue	●●	●●	●●	
Retinal	violet				●

FIGURE 6.2 Examples of photosynthetic pigments. Photosynthetic pigments can collectively absorb almost all visible light wavelengths. *Left*, the light-catching part of a pigment (shown in color) is the region in which single bonds alternate with double bonds. These and many other pigments (including heme, Section 5.5) are derived from evolutionary remodeling of the same compound. Animals convert dietary beta-carotene into a similar pigment (retinal) that is the basis of vision.

populating the atoms in such arrays easily absorb a photon—but not just any photon. Only a photon with exactly enough energy to boost an electron to a higher energy level is absorbed (Section 2.2). This is why a pigment absorbs light of only certain wavelengths.

An excited electron (one that has been boosted to a higher energy level) quickly emits its extra energy and returns to a lower energy level. As you will see in Section 6.4, photosynthetic cells capture energy emitted from an electron returning to a lower energy level.

Most photosynthetic organisms use a combination of pigments to capture light for photosynthesis—and often for additional purposes. Many accessory pigments are antioxidants that protect cells from the damaging effects of UV light in the sun's rays (Section 5.5). Appealing colors attract animals to ripening fruit or pollinators to flowers. You may already be familiar with some of these molecules. Carrots, for example, are orange because they contain beta-carotene (β-carotene); roses are red and violets are blue because of their anthocyanin content.

In green plants, chlorophylls are usually so abundant that they mask the colors of the other pigments. Plants that change color during autumn are preparing for a period of dormancy; they conserve resources by moving nutrients from tender parts that would be damaged by winter cold (such as leaves) to protected parts (such as roots). Chlorophylls are not needed during dormancy, so they are disassembled and their components recycled. Yellow and orange accessory pigments are also recycled, but not as quickly as chlorophylls. Their colors begin to show as the chlorophyll content declines in leaves. Anthocyanin synthesis also increases in some plants, adding red and purple tones to turning leaf colors.

autotroph Organism that makes its own food using energy from the environment and carbon from inorganic molecules such as CO_2.
chlorophyll *a* Main photosynthetic pigment in plants.
heterotroph Organism that obtains carbon from organic compounds assembled by other organisms.
pigment An organic molecule that can absorb light of certain wavelengths.
wavelength Distance between the crests of two successive waves.

TAKE-HOME MESSAGE 6.1

The sun emits electromagnetic radiation (light). Visible light is the main form of energy that drives photosynthesis.

Light travels in waves and is organized as photons. We see different wavelengths of visible light as different colors.

Pigments absorb light at specific wavelengths. Photosynthetic species use pigments such as chlorophyll *a* to harvest the energy of light for photosynthesis.

6.2 WHY DO CELLS USE MORE THAN ONE PHOTOSYNTHETIC PIGMENT?

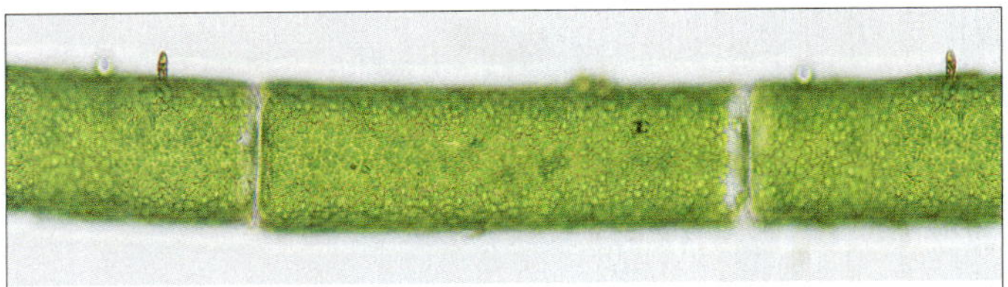

A Light micrograph of a filament of green algae. Each strand is a stack of individual photosynthetic cells. Theodor Engelmann used several species of algae in a series of experiments to determine whether some colors of light are better for photosynthesis than others.

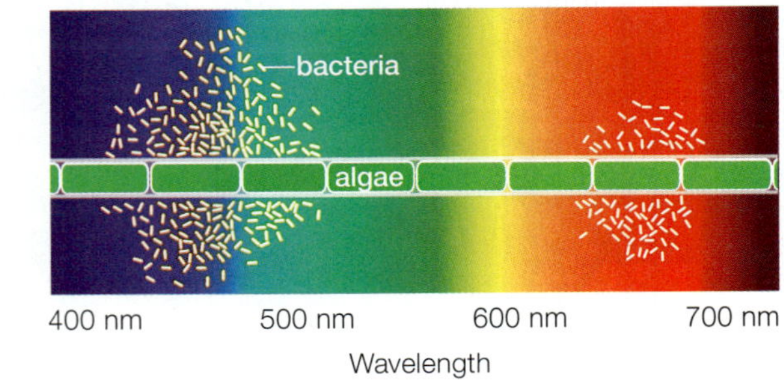

B Engelmann directed light through a prism so that bands of colors crossed a water droplet on a microscope slide. The water held a strand of photosynthetic algae, and also oxygen-requiring bacteria. The bacteria clustered around the algal cells that were releasing the most oxygen—the ones most actively engaged in photosynthesis. Those cells were under blue and red light.

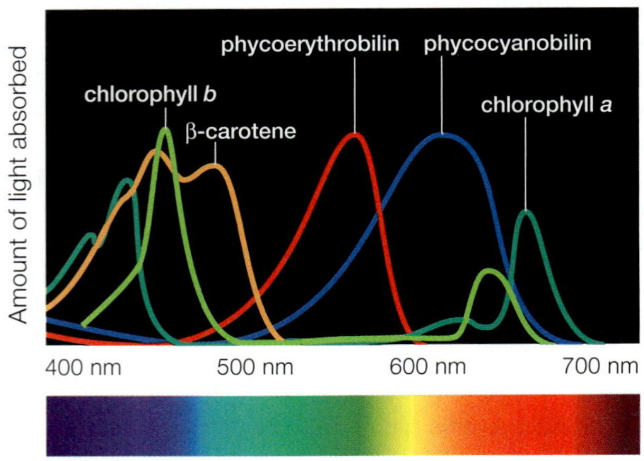

C Absorption spectra of chlorophylls *a* and *b*, β-carotene, and two phycobilins reveal the efficiency with which these pigments absorb different wavelengths of visible light. Line color indicates the characteristic color of each pigment.

FIGURE 6.3 {Animated} Discovery that photosynthesis is driven best by particular wavelengths of visible light.

FIGURE IT OUT: Which three pigments in **C** would you conclude are the main ones in the green algae tested in **B**?

Answer: Chlorophyll *a*, chlorophyll *b*, and β-carotene

In 1882, botanist Theodor Engelmann designed a set of experiments to test his hypothesis that the color of light affects the rate of photosynthesis. It had long been known that photosynthesis releases oxygen, so Engelmann used oxygen emission as an indirect measurement of photosynthetic activity. He directed a spectrum of light across individual strands of green algae suspended in water (**FIGURE 6.3A**). Oxygen-sensing equipment had not yet been invented, so Engelmann used motile, oxygen-requiring bacteria to show him where the oxygen concentration in the water was highest. The bacteria moved through the water and gathered mainly where blue and red light fell across the algal cells (**FIGURE 6.3B**). Engelmann concluded that photosynthetic cells illuminated by light of these colors were releasing the most oxygen—a sign that blue and red light are the best for driving photosynthesis in these algal cells.

Today we have equipment that can directly measure how efficiently a photosynthetic pigment absorbs different wavelengths of light. A graph that shows this efficiency is called an absorption spectrum. Peaks in the graph indicate wavelengths absorbed best (**FIGURE 6.3C**). Engelmann's results—where the bacteria clustered around the algal cells—represent the combined spectra of all the photosynthetic pigments present in the tested algae.

The combination of pigments used for photosynthesis differs among species. Why? Photosynthetic species are adapted to the environment in which they evolved, and light that reaches different environments varies in its proportions of wavelengths. Consider that seawater absorbs green and blue-green light less efficiently than other colors. Thus, more green and blue-green light penetrates deep ocean water. Algae that live in in this environment tend to have pigments—mainly phycobilins—that absorb green and blue-green light (*below*).

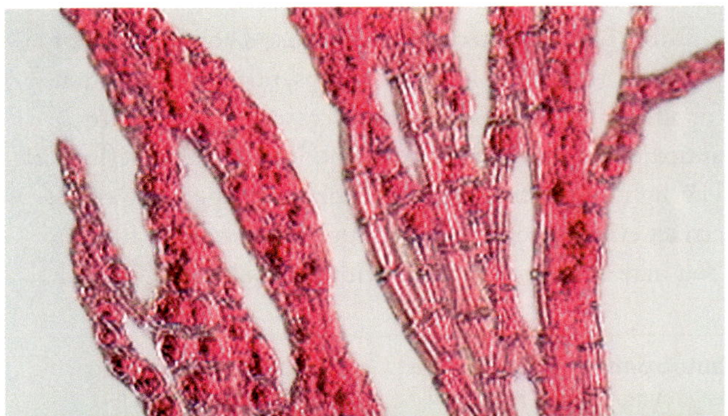

TAKE-HOME MESSAGE 6.2

A combination of pigments allows a photosynthetic organism to most efficiently capture the particular range of light wavelengths that reaches the habitat in which it evolved.

CREDITS: (3A) Jason Sonneman; (3B,C) © Cengage Learning; (in text) © Michael Davidson/The Florida State University.

6.3 WHAT HAPPENS DURING PHOTOSYNTHESIS?

All life is sustained by inputs of energy, but not all forms of energy can sustain life. Sunlight, for example, is abundant here on Earth, but it cannot be used to directly power protein synthesis or other energy-requiring reactions that keep organisms alive. Photosynthesis converts the energy of light into the energy of chemical bonds. Unlike light, chemical energy can power the reactions of life, and it can be stored for use at a later time.

In eukaryotes, photosynthesis takes place in chloroplasts (Section 4.8). Plant chloroplasts have two outer membranes, and they are filled with a thick, cytoplasm-like fluid called **stroma** (**FIGURE 6.4**). Suspended in the stroma are the chloroplast's own DNA, some ribosomes, and an inner, much-folded **thylakoid membrane**. The folds of a thylakoid membrane typically form stacks of interconnected disks called thylakoids. The space enclosed by the thylakoid membrane is a single, continuous compartment.

Photosynthesis is often summarized by this equation:

$$CO_2 + \text{water} \xrightarrow{\textit{light energy}} \text{sugars} + O_2$$

CO_2 is carbon dioxide, and O_2 is oxygen; both are gases abundant in the atmosphere. Keep in mind that photosynthesis is not a single reaction. Rather, it is a metabolic pathway (Section 5.4), a series of many reactions that occur in two stages. Molecules embedded in the thylakoid membrane carry out the reactions of the first stage, which are driven by light and thus called the **light-dependent reactions**. The "photo" in photosynthesis means light, and it refers to the conversion of light energy to the chemical bond energy of ATP during this stage. In addition to making ATP, the main light-dependent pathway in chloroplasts splits water molecules and releases O_2. Hydrogen ions and electrons from the water molecules end up in the coenzyme NADPH:

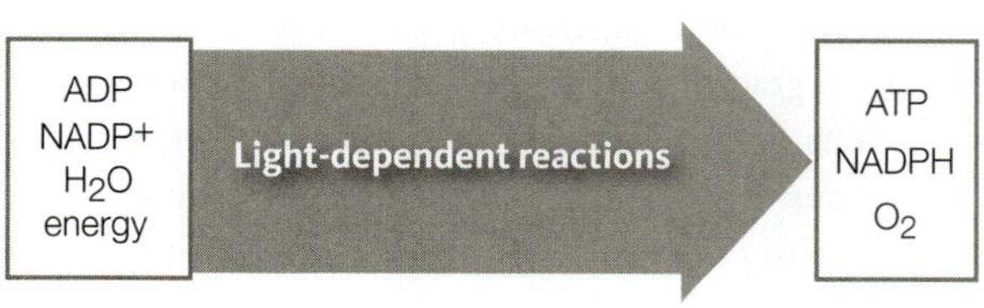

The "synthesis" part of photosynthesis refers to the reactions of the second stage, which build sugars from CO_2 and water. These sugar-building reactions run in the stroma. They are collectively called the **light-independent reactions** because light energy does not power them. Instead, they run on energy delivered by NADPH and ATP that formed during the first stage:

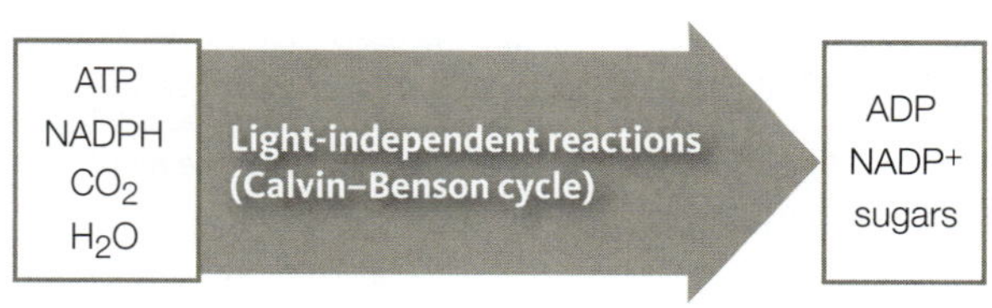

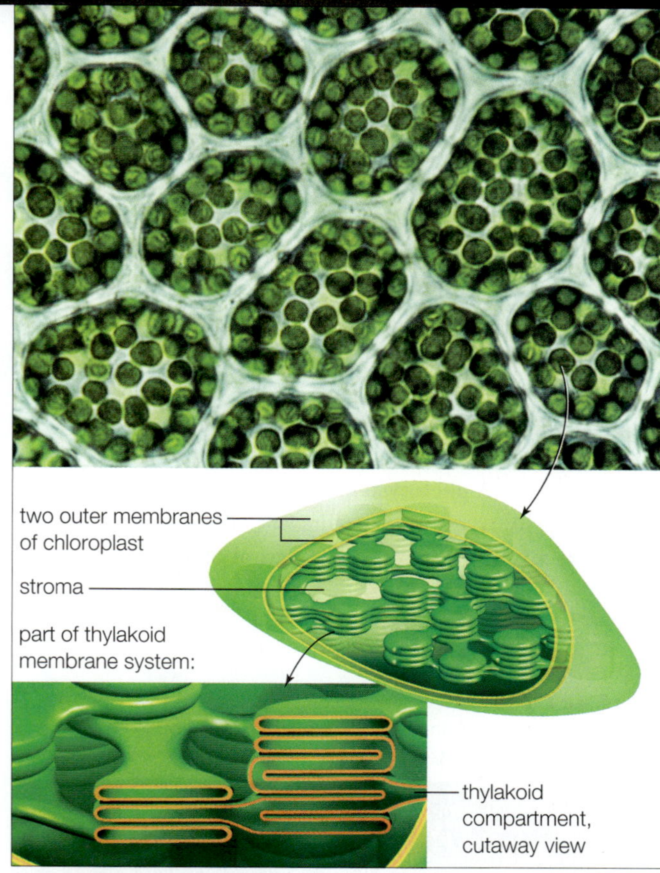

FIGURE 6.4 **{Animated}** Zooming in on the site of photosynthesis in a plant cell. The micrograph shows chloroplasts in cells of a moss leaf.

TAKE-HOME MESSAGE 6.3

In eukaryotic cells, the first stage of photosynthesis occurs at the thylakoid membrane of chloroplasts. During these light-dependent reactions, light energy drives the formation of ATP and NADPH.

In eukaryotic cells, the second stage of photosynthesis occurs in the stroma of chloroplasts. During these light-independent reactions, ATP and NADPH drive the synthesis of sugars from water and carbon dioxide.

---

light-dependent reactions First stage of photosynthesis; convert light energy to chemical energy.
light-independent reactions Second stage of photosynthesis; use ATP and NADPH to assemble sugars from water and CO_2.
stroma The cytoplasm-like fluid between the thylakoid membrane and the two outer membranes of a chloroplast.
thylakoid membrane A chloroplast's highly folded inner membrane system; forms a continuous compartment in the stroma.

6.4 HOW DO THE LIGHT-DEPENDENT REACTIONS WORK?

A chloroplast's thylakoid membrane contains millions of light-harvesting complexes, which are circular arrays of chlorophylls, various accessory pigments, and proteins (**FIGURE 6.5**). When a chlorophyll or accessory pigment in a light-harvesting complex absorbs light, one of its electrons jumps to a higher energy level, or shell (Section 2.2). The electron quickly drops back down to a lower shell by emitting its extra energy. Light-harvesting complexes hold on to that emitted energy by passing it back and forth, a bit like volleyball players pass a ball among team members. The reactions of photosynthesis begin when energy being passed around the thylakoid membrane reaches a photosystem. A **photosystem** is a group of hundreds of chlorophylls, accessory pigments, and other molecules that work as a unit to begin the reactions of photosynthesis.

FIGURE 6.5 A view of some components of the thylakoid membrane as seen from the stroma. Molecules of electron transfer chains and ATP synthases are also present, but not shown for clarity.

FIGURE 6.6 Summary of the inputs and outputs of the two pathways of light-dependent reactions.

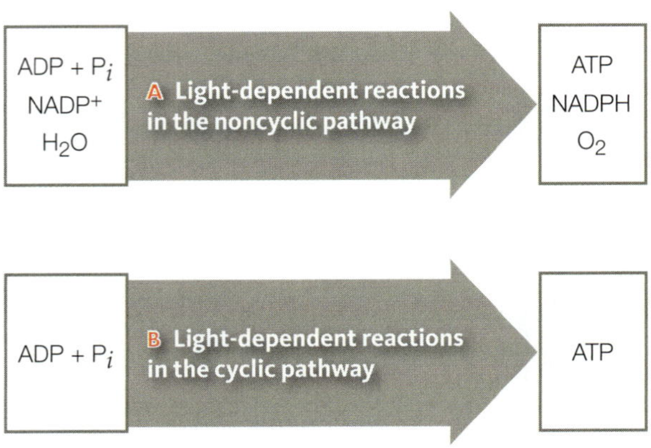

THE NONCYCLIC PATHWAY

Thylakoid membranes contain two kinds of photosystems, type I and type II, that were named in the order of their discovery. In cyanobacteria, plants, and all photosynthetic protists, both photosystem types work together in the noncyclic pathway of photosynthesis (**FIGURE 6.6A**). The pathway begins when energy being passed among light-harvesting complexes reaches a photosystem II (**FIGURE 6.7**). At the center of each photosystem are two very closely associated chlorophyll *a* molecules (a "special pair"). When a photosystem absorbs energy, electrons are ejected from its special pair ①.

A photosystem that loses electrons must be restocked with more. Photosystem II restocks itself with electrons by pulling them off of water molecules in the thylakoid compartment. Pulling electrons off of water molecules causes them to split into hydrogen ions and oxygen atoms ②. The oxygen atoms combine and diffuse out of the cell as oxygen gas (O_2). This and any other process by which a molecule is broken apart by light energy is called **photolysis**.

The actual conversion of light energy to chemical energy occurs when electrons ejected from photosystem II enter an electron transfer chain in the thylakoid membrane ③. Remember that electron transfer chains can harvest the energy of electrons in a series of redox reactions, releasing a bit of their extra energy with each step (Section 5.4). In this case, molecules of the electron transfer chain use the released energy to actively transport hydrogen ions (H^+) across the membrane, from the stroma to the thylakoid compartment ④. Thus, the flow of electrons through electron transfer chains sets up and maintains a hydrogen ion gradient across the thylakoid membrane.

The hydrogen ion gradient is a type of potential energy (Section 5.1) that can be tapped to make ATP. Hydrogen ions in the thylakoid compartment want to follow their concentration gradient by moving back into the stroma. However, ions cannot diffuse through lipid bilayers (Section 5.6). H^+ leaves the thylakoid compartment only by flowing through proteins called ATP synthases embedded in the thylakoid membrane ⑦. An ATP synthase is both a transport protein and an enzyme. When hydrogen ions flow through its interior, the protein phosphorylates ADP, so ATP forms in the stroma ⑧. The process by which the flow of electrons through electron transfer chains drives ATP formation is called **electron transfer phosphorylation**.

After the electrons have moved through the first electron transfer chain, they are accepted by a photosystem I. When this photosystem absorbs light energy, its special pair of chlorophylls emits electrons ⑤. These electrons enter a second, different electron transfer chain. At the end of this

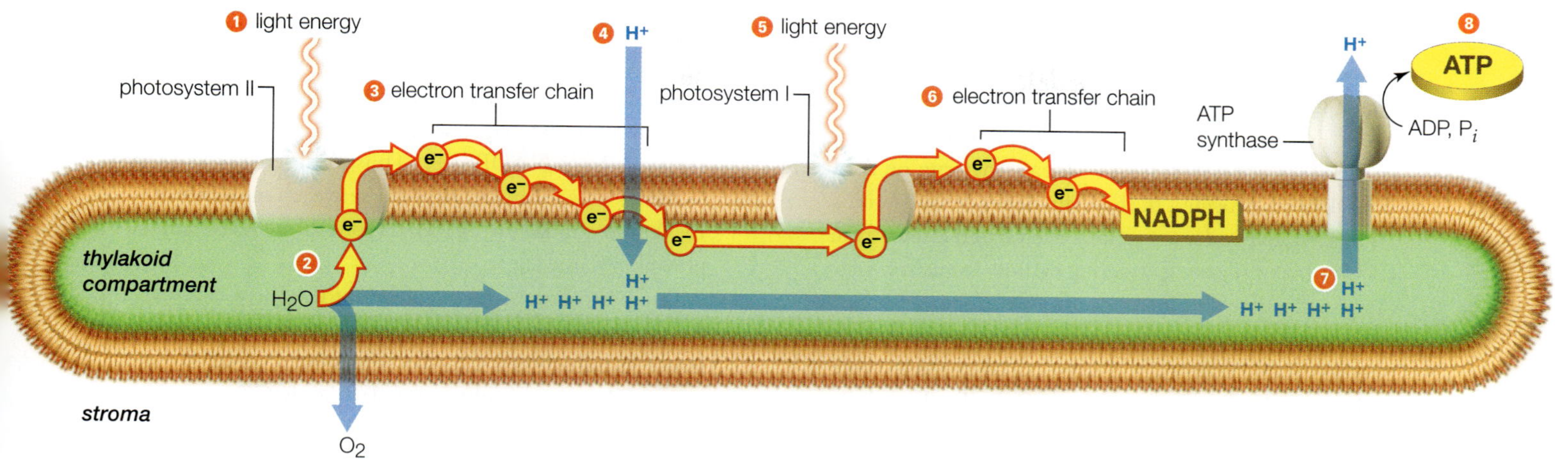

1 Light energy ejects electrons from a photosystem II.

2 The photosystem pulls replacement electrons from water molecules, which then break apart into oxygen and hydrogen ions. The oxygen leaves the cell as O_2.

3 The electrons enter an electron transfer chain in the thylakoid membrane.

4 Energy lost by the electrons as they move through the chain is used to actively transport hydrogen ions from the stroma into the thylakoid compartment. A hydrogen ion gradient forms across the thylakoid membrane.

5 Light energy ejects electrons from a photosystem I. Replacement electrons come from an electron transfer chain.

6 The ejected electrons move through a second electron transfer chain, then combine with $NADP^+$ and H^+, so NADPH forms.

7 Hydrogen ions in the thylakoid compartment are propelled through the interior of ATP synthases by their gradient across the thylakoid membrane.

8 Hydrogen ion flow causes ATP synthases to phosphorylate ADP, so ATP forms in the stroma.

FIGURE 6.7 {Animated} Light-dependent reactions, noncyclic pathway. ATP and oxygen gas are produced in this pathway. Electrons that travel through two different electron transfer chains end up in NADPH.

chain, the coenzyme $NADP^+$ accepts the electrons along with H^+, so NADPH forms 6:

$$NADP^+ + 2e^- + H^+ \longrightarrow \text{NADPH}$$

NADPH is a powerful reducing agent (electron donor).

THE CYCLIC PATHWAY

As you will see in the next section, ATP and NADPH produced in the light-dependent reactions are used to make sugars. On its own, the noncyclic pathway does not yield enough ATP to balance NADPH use in sugar production pathways. The cyclic pathway produces additional ATP for this purpose (**FIGURE 6.6B**).

In the cyclic pathway, electrons that are ejected from photosystem I enter an electron transfer chain, and then return to photosystem I. As in the noncyclic pathway, the electron transfer chain uses electron energy to move hydrogen ions into the thylakoid compartment, and the resulting hydrogen ion gradient drives ATP formation. However, the cyclic pathway does not produce NADPH or oxygen gas.

The cyclic pathway also allows light-dependent reactions to continue when the noncyclic pathway stops, for example under intense illumination. Light energy in excess of what can be used for photosynthesis can result in the formation of dangerous free radicals (Section 2.2). A light-induced structural change in photosystem II prevents this from happening. The photosystem stops initiating the noncyclic pathway, and traps excess energy instead. At such times, the cyclic pathway predominates.

electron transfer phosphorylation Process in which electron flow through electron transfer chains sets up a hydrogen ion gradient that drives ATP formation.
photolysis Process by which light energy breaks down a molecule.
photosystem Cluster of pigments and proteins that converts light energy to chemical energy in photosynthesis.

TAKE-HOME MESSAGE 6.4

Photosynthetic pigments in the thylakoid membrane transfer the energy of light to photosystems, which eject electrons that enter electron transfer chains.

In both noncyclic and cyclic pathways, the flow of electrons through the transfer chains sets up hydrogen ion gradients that drive ATP formation.

In the noncyclic pathway, water molecules are split, oxygen is released, and electrons end up in NADPH.

In the cyclic pathway, no NADPH forms, and no oxygen is released.

6.5 HOW DO THE LIGHT-INDEPENDENT REACTIONS WORK?

THE CALVIN–BENSON CYCLE

The enzyme-mediated reactions of the **Calvin–Benson cycle** build sugars in the stroma of chloroplasts (FIGURE 6.8). These reactions are light-independent because light energy does not power them. Instead, they run on ATP and NADPH that formed in the light-dependent reactions.

Light-independent reactions use carbon atoms from CO_2 to make sugars. Extracting carbon atoms from an inorganic source and incorporating them into an organic molecule is a process called **carbon fixation**. In most plants, photosynthetic protists, and some bacteria, the enzyme **rubisco** fixes carbon by attaching CO_2 to RuBP (ribulose bisphosphate), a five-carbon molecule ❶.

FIGURE 6.8 {Animated} The Calvin–Benson cycle. This sketch shows a cross-section of a chloroplast with the reactions cycling in the stroma. The steps shown are a summary of six cycles of reactions (see Appendix III for details). Black balls are carbon atoms.

❶ Six CO_2 diffuse into a photosynthetic cell, and then into a chloroplast. Rubisco attaches each to a RuBP molecule. The resulting intermediates split, so twelve molecules of PGA form.

❷ Each PGA molecule gets a phosphate group from ATP, plus hydrogen and electrons from NADPH. Twelve PGAL form.

❸ Two PGAL may combine to form one six-carbon sugar (such as glucose).

❹ The remaining ten PGAL receive phosphate groups from ATP. The transfer primes them for endergonic reactions that regenerate the 6 RuBP.

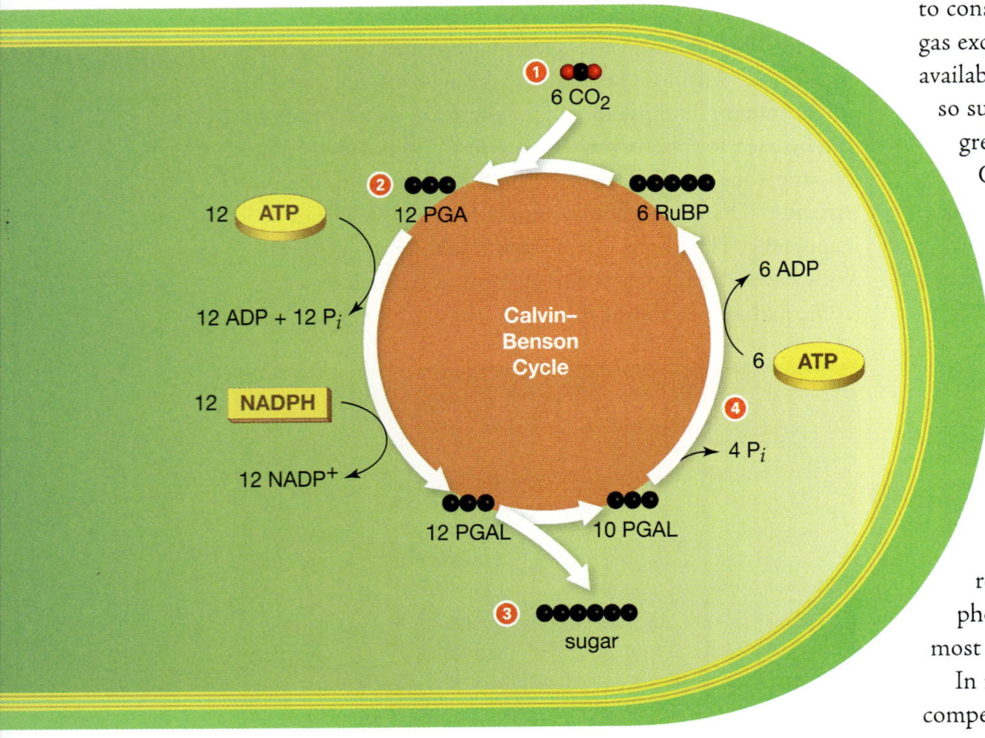

The six-carbon intermediate that forms by this reaction is unstable, so it splits right away into two three-carbon molecules of PGA (phosphoglycerate). Each PGA receives a phosphate group from ATP, and hydrogen and electrons from NADPH ❷. Thus, ATP energy and the reducing power of NADPH convert each molecule of PGA into a molecule of PGAL (phosphoglyceraldehyde), a phosphorylated sugar.

In later reactions, two or more of the three-carbon PGAL molecules can be combined and rearranged to form larger carbohydrates. Glucose, remember, has six carbon atoms. To make one glucose molecule, six CO_2 must be attached to six RuBP molecules, so twelve PGAL form. Two PGAL may combine to form one six-carbon glucose ❸. The ten remaining PGAL regenerate the starting compound of the cycle, RuBP ❹. Most of the glucose that a plant makes is converted at once to sucrose or starch by other pathways.

ADAPTATIONS TO CLIMATE

Mechanisms that help a plant prevent water loss also limit the gas exchange needed for photosynthesis. Most plants have a thin, waterproof cuticle that limits evaporation of water from their aboveground parts. Gases cannot diffuse across the cuticle, so the surfaces of leaves and stems are studded with tiny, closable gaps called stomata. Stomata are tiny gateways for gases. When they are open, CO_2 can diffuse from the air into photosynthetic tissues, and O_2 can diffuse out of the tissues into the air. Stomata close to conserve water on hot, dry days. When that happens, gas exchange comes to a halt. Closed stomata limit the availability of CO_2 for the light-independent reactions, so sugar synthesis slows. This detrimental effect is greatest in **C3 plants**, which fix carbon only by the Calvin–Benson cycle. They are called C3 plants because a three-carbon molecule (PGA) is the first stable intermediate in their light-independent reactions. When the CO_2 concentration inside a C3 plant declines, rubisco uses oxygen as a substrate in **photorespiration**, an alternate pathway that produces carbon dioxide. Thus, when a C3 plant closes stomata during the day, it loses carbon instead of fixing it (FIGURE 6.9A). In addition, ATP and NADPH are used to convert the pathway's intermediates to a molecule that can enter the Calvin–Benson cycle, so extra energy is required to make sugars. C3 plants compensate for photorespiration by making a lot of rubisco: It is the most abundant protein on Earth.

In many plant lineages, an additional set of reactions compensates for rubisco's inefficiency. Plants that use

A When the CO_2 level declines in leaves of a C3 plant, rubisco uses oxygen as a substrate in photorespiration, and sugar production becomes inefficient.

B In a C3 plant (barley, *left*), chloroplasts—the sites of carbon fixation—occur mainly in mesophyll cells. In a C4 plant (millet, *right*), carbon fixation occurs first in mesophyll cells, then in bundle-sheath cells. C4 adaptations maintain a high CO_2/O_2 ratio near rubisco.

FIGURE 6.9 {Animated} Anatomical and biochemical specializations minimize photorespiration in C4 plants. Micrographs show leaf cross sections.

these reactions also close stomata on dry days, but their sugar production does not decline. Examples include corn, switchgrass, and bamboo. We call these **C4 plants** because the first stable intermediate to form in their light-independent reactions is a four-carbon compound. C4 plants fix carbon twice, in two kinds of cells (**FIGURE 6.9B**). The first set of reactions occurs in mesophyll cells, where carbon is fixed by an enzyme that does not use oxygen even when the carbon dioxide level is low. The resulting intermediate is transported to bundle-sheath cells, where it is converted to CO_2. Rubisco fixes carbon for the second time as the CO_2 enters the Calvin–Benson cycle.

Bundle-sheath cells of C4 plants have chloroplasts that carry out light-dependent reactions, but only in the cyclic pathway. No oxygen is released, so the O_2 level near rubisco stays low. This, along with the high CO_2 level provided by the C4 reactions, minimizes photorespiration, so sugar production stays efficient in these plants even in hot, dry weather (**FIGURE 6.10**).

Succulents, cacti, and other **CAM plants** use a carbon-fixing pathway that allows them to conserve water even in desert regions with extremely high daytime temperatures. CAM stands for crassulacean acid metabolism, after the Crassulaceae family of plants in which this pathway was first studied. Like C4 plants, CAM plants fix carbon twice, but the reactions occur at different times rather than in different cells. Stomata on a CAM plant open at night, when typically lower temperatures minimize evaporative water loss. The plants fix carbon from CO_2 in the air at this time. The product of the cycle, a four-carbon acid, is stored in the cell's central vacuole. When the stomata close the next day, the acid moves out of the vacuole and becomes broken down to CO_2, which is fixed for the second time when it enters the Calvin–Benson cycle.

FIGURE 6.10 Crabgrass "weeds" overgrowing a lawn. Crabgrasses, which are C4 plants, thrive in hot, dry summers, when they easily outcompete Kentucky bluegrass and other fine-leaved C3 grasses commonly planted in residential lawns.

TAKE-HOME MESSAGE 6.5

The light-independent reactions build sugars using carbon fixed from CO_2. They run on ATP and electrons from NADPH.

C3 plants use only the Calvin–Benson cycle. In these plants, photorespiration on hot, dry days reduces the efficiency of sugar production, so it limits growth.

Plants adapted to hot, dry conditions minimize photorespiration by fixing carbon twice. C4 plants separate the two sets of reactions in space; CAM plants separate them in time.

C3 plant Type of plant that uses only the Calvin–Benson cycle to fix carbon.
C4 plant Type of plant that minimizes photorespiration by fixing carbon twice, in two cell types.
Calvin–Benson cycle Cyclic carbon-fixing pathway that builds sugars from CO_2; the light-independent reactions of photosynthesis.
CAM plant Type of plant that conserves water by fixing carbon twice, at different times of day.
carbon fixation Process by which carbon from an inorganic source such as carbon dioxide gets incorporated into an organic molecule.
photorespiration Reaction in which rubisco attaches oxygen instead of carbon dioxide to ribulose bisphosphate.
rubisco Ribulose bisphosphate carboxylase. Carbon-fixing enzyme of the Calvin–Benson cycle.

CREDITS: (9A) © Cengage Learning; (9B left) Masahiro Yamada, Michio Kawasaki, Tatsuo Sugiyama, Hiroshi Miyake, Mitsutaka Taniguchi; Differential Positioning of C4 Mesophyll and Bundle Sheath Chloroplasts: Aggregative Movement of C4 Mesophyll Chloroplasts in Response to Environmental Stresses: Plant and Cell Physiology; (2009) 50(10): 1736–1749; (9B right) Eri Maai, Shouu Shimada, Masahiro Yamada, Tatsuo Sugiyama, Hiroshi Miyake, Mitsutaka Taniguchi; The avoidance and aggregative movements of mesophyll chloroplasts in C4 monocots in response to blue light and abscisic acid; *Journal of Experimental Botany*, doi:10.1093/jxb/err008, by permission of Oxford University Press; (10) Image courtesy msuturfweeds.net.

6.6 Application: GREEN ENERGY

Sustainability

FIGURE 6.11 Switchgrass growing wild in a North American prairie.

TODAY, THE EXPRESSION "FOOD IS FUEL" IS NOT JUST ABOUT EATING. With fossil fuel prices soaring, there is an increasing demand for biofuels, which are oils, gases, or alcohols made from organic matter that is not fossilized. Most materials we use for biofuel production today consist of food crops—mainly corn, soybeans, and sugarcane. Growing these crops in large quantities is typically damaging to the environment, and using them to make biofuel competes with our food supply.

How did we end up competing with our vehicles for food? We both run on the same fuel: energy that plants have stored in chemical bonds. Fossil fuels such as petroleum, coal, and natural gas formed from the remains of ancient swamp forests that decayed and compacted over millions of years. These fuels consist of molecules originally assembled by ancient plants. Biofuels—and foods—consist mainly of molecules originally assembled by modern plants.

A lot of energy is locked up in the chemical bonds of molecules made by plants. That energy can fuel heterotrophs, as when an animal cell powers ATP synthesis by breaking the bonds of sugars (a topic detailed in the next chapter). It can also fuel our cars, which run on energy released by burning biofuels or fossil fuels. Both processes are fundamentally the same: They release energy by breaking the bonds of organic molecules. Both use oxygen to break those bonds, and both produce carbon dioxide.

Photosynthesis removes carbon dioxide from the atmosphere, and fixes its carbon atoms in organic compounds. When we burn fossil fuels, the carbon that has been sequestered in them for hundreds of millions of years is released back into the atmosphere, mainly as CO_2 that reenters the atmosphere. Our extensive use of fossil fuels has put Earth's atmospheric cycle of carbon dioxide out of balance: We are adding far more CO_2 to the atmosphere than photosynthetic organisms are removing from it. Atmospheric carbon dioxide affects Earth's climate, so this increase in CO_2 is contributing to global climate change.

Ratna Sharma and Mari Chinn researching ways to reduce the cost of producing biofuel from renewable sources such as wild grasses and agricultural waste.

Unlike fossil fuels, biofuels are a renewable source of energy: We can always make more of them simply by growing more plants. Also unlike fossil fuels, biofuels do not contribute to global climate change, because growing plant matter for fuel recycles carbon that is already in the atmosphere.

Corn and other food crops are rich in oils, starches, and sugars that can be easily converted to biofuels. The starch in corn kernels, for example, can be enzymatically broken down to glucose, which is converted to ethanol by heterotrophic bacteria or yeast. Making biofuels from other types of plant matter requires additional steps, because these materials contain a higher proportion of cellulose. Breaking down this tough, insoluble carbohydrate to its glucose monomers adds a lot of cost to the biofuel product. Researchers are currently working on cost-effective ways to break down the abundant cellulose in fast-growing weeds such as switchgrass (FIGURE 6.11), and agricultural wastes such as wood chips, wheat straw, cotton stalks, and rice hulls.

PEOPLE MATTER

National Geographic Grantee
WILLIE SMITS

One of Indonesia's most ardent rain forest protection activists is in what may seem an unlikely position: spearheading a project to produce biofuel from trees. But tropical forest scientist Willie Smits, after 30 years of studying fragile ecosystems in these Southeast Asian islands, wants to draw world attention to a powerhouse of a tree—the Arenga sugar palm, *Arenga pinnata*. Smits says that this deep-rooted palm could serve as the core of a waste-free system that produces a premium organic sugar as well as ethanol for fuel, providing food products and jobs to villagers while it helps preserve the existing native rain forest. Scientists who have studied the unique harvesting and production process developed by Smits agree the system would protect the atmosphere rather than add to Earth's growing carbon dioxide burden. "The palm juice chiefly consists of water and sugar—made from rain, sunshine, carbon dioxide, and nothing else," says Smits. "You are basically only harvesting sunshine." The project, being funded in part by a grant from National Geographic's Great Energy Challenge initiative, has the potential to disrupt a cycle of poverty and environmental devastation that has gripped one of the most vulnerable and remote areas of the planet, while providing a new source of sustainable fuel.

CREDITS: (11 inset) © Roger W. Winstead/NC State University; (in text right) Photograph courtesy Eric Rasmussen.

Summary

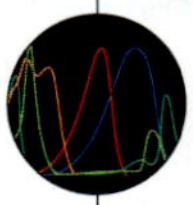

SECTIONS 6.1, 6.2 Plants and other **autotrophs** make their own food using energy from the environment and carbon from inorganic sources such as CO_2. **Heterotrophs** get carbon from molecules that other organisms have already assembled. Visible light drives photosynthesis, which begins when light is absorbed by photosynthetic pigments. **Pigments** absorb light of particular **wavelengths** only; wavelengths not captured are reflected as its characteristic color. The main photosynthetic pigment, **chlorophyll *a***, absorbs violet and red light, so it appears green. Accessory pigments absorb additional wavelengths.

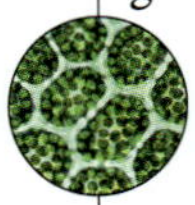

SECTION 6.3 In chloroplasts, the **light-dependent reactions** of photosynthesis occur at a much-folded **thylakoid membrane**. The membrane forms a compartment in the chloroplast's interior (**stroma**), in which the **light-independent reactions** occur:

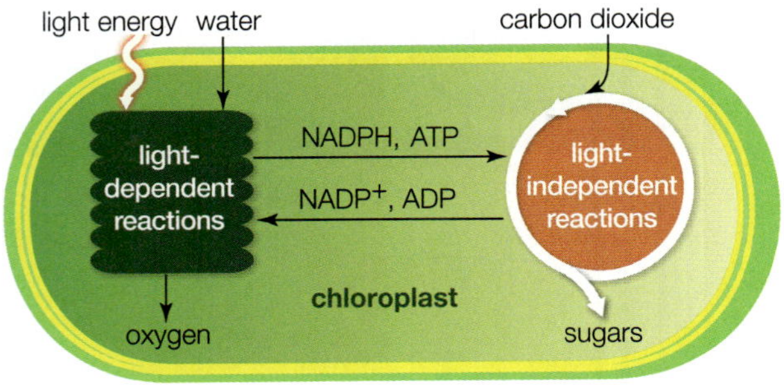

SECTION 6.4 In the light-dependent reactions, light-harvesting complexes in the thylakoid membrane absorb photons and pass the energy to **photosystems**. Receiving energy causes photosystems to release electrons.

In the noncyclic pathway, electrons released from photosystem II flow through an electron transfer chain, then to photosystem I. Photosystem II replaces lost electrons by pulling them from water, which then splits into H^+ and O_2 (an example of **photolysis**). Electrons released from photosystem I end up in NADPH.

In the cyclic pathway, electrons released from photosystem I enter an electron transfer chain, then cycle back to photosystem I. NADPH does not form and O_2 is not released.

ATP forms by **electron transfer phosphorylation** in both pathways. Electrons flowing through electron transfer chains cause H^+ to accumulate in the thylakoid compartment. The H^+ follows its gradient back across the membrane through ATP synthases, driving ATP synthesis.

SECTION 6.5 **Carbon fixation** occurs in light-independent reactions. Inside the stroma, the enzyme **rubisco** attaches a carbon from CO_2 to RuBP to start the **Calvin–Benson cycle**. This cyclic pathway uses energy from ATP, carbon and oxygen from CO_2, and hydrogen and electrons from NADPH to make sugars.

On hot, dry days, plants conserve water by closing stomata, so CO_2 for light-independent reactions cannot enter their tissues. In **C3 plants**, the resulting low CO_2 level causes **photorespiration**, which reduces the efficiency of sugar production. Other types of plants minimize photorespiration by fixing carbon twice, thus keeping the CO_2 level high and the O_2 level low near rubisco. **C4 plants** carry out the two sets of reactions in different cell types; **CAM plants** carry them out at different times.

SECTION 6.6 Autotrophs remove CO_2 from the atmosphere; the metabolic activity of most organisms puts it back. Humans disrupt this cycle by burning fossil fuels, which adds extra CO_2 to the atmosphere. The resulting imbalance is contributing to global warming.

Self-Quiz

Answers in Appendix VII

1. A cat eats a bird, which ate a caterpillar that chewed on a weed. Which of these organisms are autotrophs? Which ones are heterotrophs?

2. Which of the following statements is incorrect?
   a. Pigments absorb light of certain wavelengths only.
   b. Many accessory pigments are multipurpose molecules.
   c. Chlorophyll is green because it absorbs green light.

3. In plants, the light-dependent reactions proceed at/in the ________ .
   a. thylakoid membrane
   b. plasma membrane
   c. stroma
   d. cytoplasm

4. When a photosystem absorbs light, ________ .
   a. sugar phosphates are produced
   b. electrons are transferred to ATP
   c. RuBP accepts electrons
   d. electrons are ejected from its special pair

5. In the light-dependent reactions, ________ .
   a. carbon dioxide is fixed
   b. ATP forms
   c. sugars form
   d. CO_2 accepts electrons
   e. b and c
   f. a and c

6. What accumulates inside the thylakoid compartment during the light-dependent reactions?
   a. sugars
   b. hydrogen ions
   c. O_2
   d. CO_2

7. The atoms in the molecular oxygen released during photosynthesis come from split ________ molecules.
   a. sugar
   b. CO_2
   c. water
   d. O_2

8. Light-independent reactions in plants proceed at/in the ________ of chloroplasts.
   a. thylakoid membrane
   b. plasma membrane
   c. stroma
   d. cytoplasm

Data Analysis Activities

Energy Efficiency of Biofuel Production Most material currently used for biofuel production in the United States consists of food crops—mainly corn, soybeans, and sugarcane. In 2006, David Tilman and his colleagues published the results of a 10-year study comparing the net energy output of various biofuels. The researchers grew a mixture of native perennial grasses without irrigation, fertilizer, pesticides, or herbicides, in sandy soil that was so depleted by intensive agriculture that it had been abandoned. They measured the usable energy in biofuels made from the grasses, from corn, and from soy. They also measured the energy it took to grow and produce each kind of biofuel (**FIGURE 6.12**).

1. About how much energy did ethanol produced from one hectare of corn yield? How much energy did it take to grow the corn to make that ethanol?
2. Which of the biofuels tested had the highest ratio of energy output to energy input?
3. Which of the three crops would require the least amount of land to produce a given amount of biofuel energy?

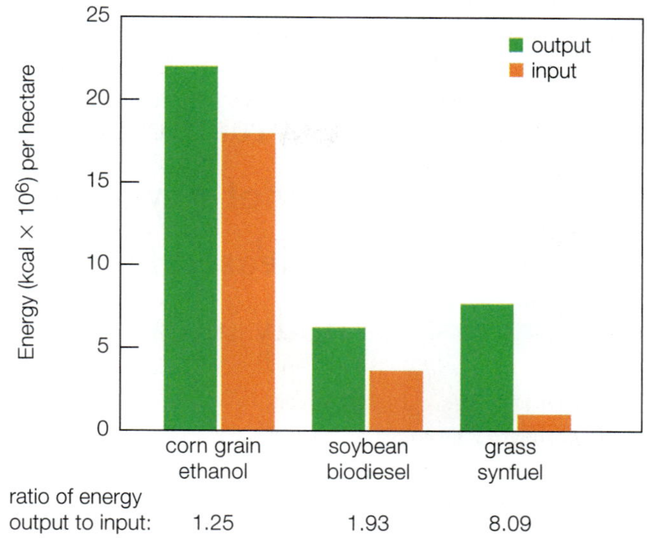

FIGURE 6.12 Energy inputs and outputs of biofuels from different sources. One hectare is about 2.5 acres.

9. The Calvin–Benson cycle starts when ______ .
   a. light is available
   b. carbon dioxide is attached to RuBP
   c. electrons leave photosystem II

10. Which of the following substances does *not* participate in the Calvin–Benson cycle?
    a. ATP c. NADPH e. PGAL
    b. O_2 d. RuBP f. CO_2

11. Plants use energy in ______ to drive photosynthesis.
    a. light b. hydrogen ions c. O_2 d. CO_2

12. Most of the carbon dioxide used in photosynthesis comes from ______ .
    a. glucose c. rainwater
    b. the atmosphere d. photolysis

13. Match each term with its most suitable description.

___ PGAL formation	a. absorbs light
___ CO_2 fixation	b. converts light to chemical energy
___ photolysis	c. self-feeder
___ ATP forms; NADPH does not	d. electrons cycle back to photosystem I
___ photorespiration	e. problem in C3 plants
___ photosynthesis	f. ATP, NADPH needed
___ pigment	g. water molecules split
___ autotroph	h. rubisco function

Critical Thinking

1. About 200 years ago, Jan Baptista van Helmont wanted to know where growing plants get the materials necessary for increases in size. He planted a tree seedling weighing 5 pounds in a barrel filled with 200 pounds of soil and then watered the tree regularly. After five years, the tree weighed 169 pounds, 3 ounces, and the soil weighed 199 pounds, 14 ounces. Because the tree had gained so much weight and the soil had lost so little, he concluded that the tree had gained all of its additional weight by absorbing the water he had added to the barrel, but of course he was incorrect. What really happened?

2. While gazing into an aquarium, you observe bubbles coming from an aquatic plant (*left*). What are the bubbles and where do they come from?

3. A C3 plant absorbs a carbon radioisotope (as part of $^{14}CO_2$). In which stable, organic compound does the labeled carbon appear first?

4. As you learned in this chapter, cell membranes are required for electron transfer phosphorylation. Thylakoid membranes in chloroplasts serve this purpose in photosynthetic eukaryotes. Prokaryotic cells do not have this organelle, but many are photosynthesizers. How do you think they carry out the light-dependent reactions, given that they have no chloroplasts?

Like you, a whale breathes air to provide its cells with a fresh supply of oxygen for aerobic respiration. Carbon dioxide released from aerobically respiring cells leaves the body in each exhalation.

7 HOW CELLS RELEASE CHEMICAL ENERGY

Links to Earlier Concepts

This chapter focuses on metabolic pathways (Section 5.4) that harvest energy (5.1) stored in the chemical bonds of sugars (3.2). Some reactions (3.1) of these pathways occur in mitochondria (4.7). You will revisit free radicals (2.2), lipids (3.3), proteins (3.4), electron transfer chains (5.4), coenzymes (5.5), membrane transport (5.6, 5.7), and photosynthesis (6.4).

KEY CONCEPTS

ENERGY FROM SUGARS

Most cells can make ATP by breaking down sugars in either aerobic respiration or anaerobic fermentation pathways. Aerobic respiration yields the most ATP.

GLYCOLYSIS

Aerobic respiration and fermentation start in the cytoplasm with glycolysis, a pathway that splits glucose into two pyruvate molecules and yields two ATP.

AEROBIC RESPIRATION

Eukaryotes break down pyruvate to CO_2 in mitochondria. Many coenzymes are reduced; these deliver electrons and hydrogen ions to electron transfer chains that drive ATP formation.

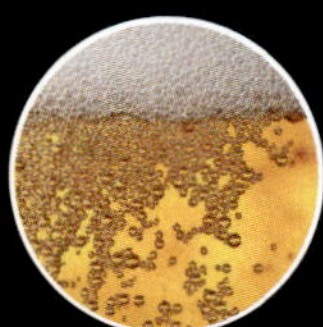

FERMENTATION

Fermentation ends in the cytoplasm, where organic molecules accept electrons from pyruvate. The net yield of ATP is small compared with that from aerobic respiration.

OTHER METABOLIC PATHWAYS

Cells also make ATP by breaking down molecules other than sugars. Dietary lipids and proteins are converted to molecules that enter glycolysis or another step in the aerobic respiration pathway.

Photograph by Ralph Lee Hopkins, National Geographic Creative.

7.1 HOW DO CELLS ACCESS THE CHEMICAL ENERGY IN SUGARS?

Photosynthetic organisms capture energy from the sun and store it in the form of sugars. They and most other organisms use energy stored in sugars to run the diverse reactions that sustain life. However, sugars rarely participate in such reactions, so how do cells harness their energy? In order to use the energy stored in sugars, cells must first transfer it to molecules—ATP in particular—that do participate in energy-requiring reactions. The energy transfer occurs when cells break the bonds of a sugar's carbon backbone. Energy released as those bonds are broken drives ATP synthesis. The two main mechanisms by which organisms break down sugars to make ATP are aerobic respiration and fermentation.

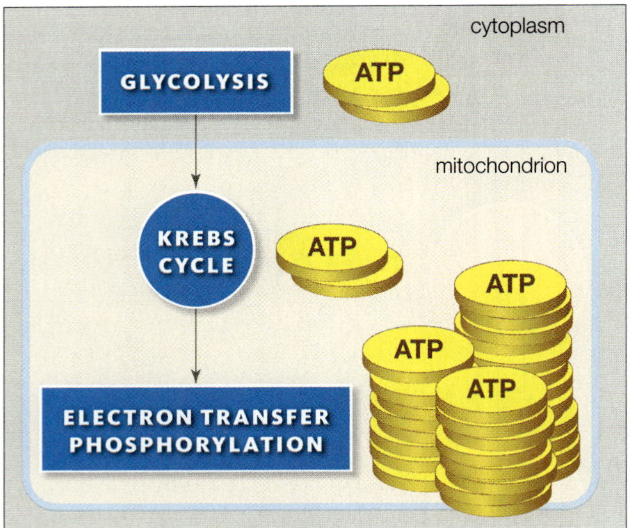

A Aerobic respiration.

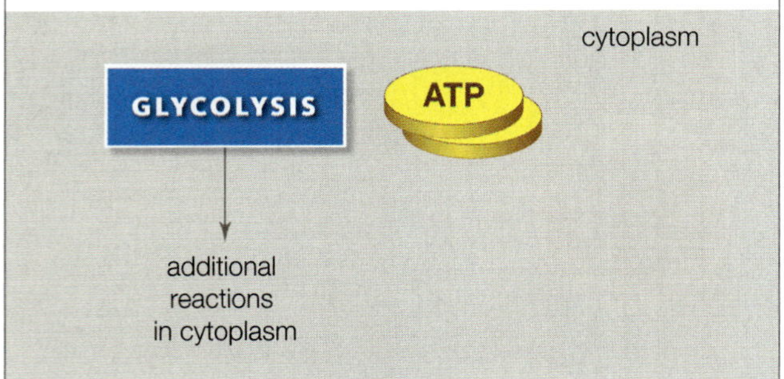

B Fermentation.

FIGURE 7.1 {Animated} Comparison of aerobic respiration and fermentation.

FIGURE IT OUT: Which pathway produces more ATP?

Answer: Aerobic respiration

aerobic respiration Oxygen-requiring metabolic pathway that breaks down sugars to produce ATP.

fermentation A metabolic pathway that breaks down sugars to produce ATP and does not require oxygen.

AEROBIC RESPIRATION AND FERMENTATION COMPARED

Aerobic respiration is an oxygen-requiring metabolic pathway that breaks down sugars to make ATP. It is the main energy-releasing pathway in nearly all eukaryotes and some bacteria. Aerobic respiration occurs in three stages (**FIGURE 7.1A**). The first stage, glycolysis, is a linear pathway that occurs in cytoplasm. Glycolysis begins the breakdown of one sugar molecule for a net yield of 2 ATP. In eukaryotes, the next two stages occur in mitochondria. The second stage, the Krebs cycle, completes the breakdown of the sugar molecule to CO_2. This cyclic pathway produces 2 ATP and reduces many coenzymes. In the third stage, electron transfer phosphorylation (Section 6.4), coenzymes reduced during glycolysis and the Krebs cycle deliver electrons and hydrogen ions to electron transfer chains. Energy released by electrons as they move through the chains drives the formation of as many as 32 ATP. Water forms when oxygen accepts hydrogen ions and electrons at the end of the electron transfer chains. Aerobic respiration, which means "taking a breath of air," refers to this pathway's requirement for oxygen as the final acceptor of electrons.

Fermentation refers to sugar breakdown pathways that do not require oxygen to make ATP (**FIGURE 7.1B**). Like aerobic respiration, fermentation begins with glycolysis in cytoplasm. Unlike aerobic respiration, fermentation occurs entirely in cytoplasm, and does not include electron transfer chains. The reactions that conclude fermentation produce no additional ATP, and the final acceptor of electrons is an organic molecule (not oxygen).

Aerobic respiration produces about 36 ATP per sugar molecule; fermentation produces only 2. Fermentation provides enough ATP to sustain many single-celled species. It also helps cells of multicelled species produce ATP under anaerobic conditions, but aerobic respiration is a much more efficient way of harvesting energy from sugars. You and other large, multicelled organisms could not live without its higher yield.

TAKE-HOME MESSAGE 7.1

Most cells can make ATP by breaking down sugars, either in aerobic respiration or fermentation.

Aerobic respiration and fermentation begin in cytoplasm.

Fermentation does not require oxygen and ends in cytoplasm.

Aerobic respiration requires oxygen and, in eukaryotes, it ends in mitochondria. This pathway yields much more ATP than fermentation.

7.2 HOW DID ENERGY-RELEASING PATHWAYS EVOLVE?

PEOPLE MATTER

National Geographic Grantee
DR. J. WILLIAM SCHOPF

For the first 85 percent of its history, Earth was populated by what was essentially pond scum. In 1987, National Geographic grantee J. William Schopf discovered what may be evidence of cells 3.465 billion years old, opening the floodgates to decades of controversy—and research that is filling in the holes in our understanding of how and when life evolved.

The first cells we know of appeared on Earth about 3.4 billion years ago. Like some modern prokaryotes, these ancient organisms did not tap into sunlight: They extracted the energy they needed from simple molecules such as methane and hydrogen sulfide. When the cyclic pathway of photosynthesis first evolved (Section 6.4), sunlight offered cells that used it an essentially unlimited supply of energy. Shortly afterward, this pathway became modified. The new noncyclic pathway split water molecules into hydrogen and oxygen. Cells that used the pathway were very successful. Oxygen gas (O_2) released from uncountable numbers of water molecules began seeping out of photosynthetic prokaryotes. The gas reacts easily with metals, so at first, most of it combined with metal atoms in exposed rocks. After the exposed minerals became saturated with oxygen, the gas began to accumulate in the ocean and the atmosphere. From that time on, the world of life would never be the same.

Before photosynthesis evolved, molecular oxygen had been a very small component of Earth's early atmosphere. In what may have been the earliest case of catastrophic pollution, the new abundance of this gas exerted tremendous pressure on all life at the time. Why? Then, like now, enzymes that require metal cofactors were a critical part of metabolism. Oxygen reacts with metal cofactors, and free radicals (Section 2.2) form during those reactions. Free radicals damage DNA and other biological molecules, so they are dangerous to life. Cells with no way to cope with them quickly died out. Only a few lineages persisted in deep water, muddy sediments, and other **anaerobic** (oxygen-free) habitats.

Antioxidants evolved in the survivors. Cells that made these molecules were the first **aerobic** organisms—they could live in the presence of oxygen. As they evolved, their antioxidant molecules became incorporated into new metabolic pathways that put oxygen's reactive properties to use. One of the new pathways was aerobic respiration. This pathway requires oxygen, and it produces carbon dioxide—the raw materials from which photosynthetic organisms make sugars. It also combines molecular oxygen with hydrogen ions and electrons—exactly the reverse of the reaction that splits water during photosynthesis:

$$O_2 + 4e^- + 4H^+ \longrightarrow 2H_2O$$

With this connection, the cycling of carbon, hydrogen, and oxygen through living things came full circle (*left*).

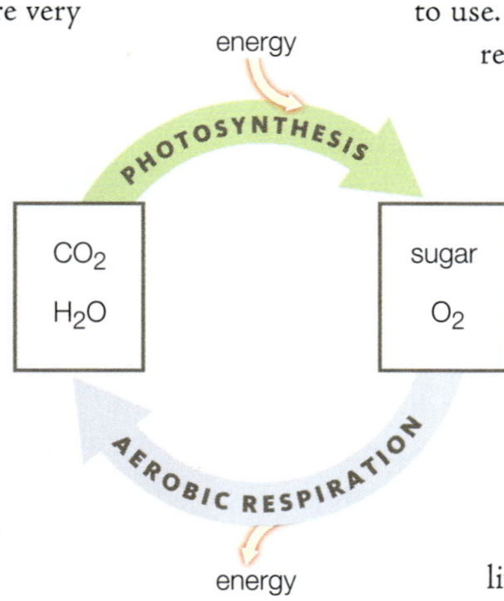

---

aerobic Involving or occurring in the presence of oxygen.
anaerobic Occurring in the absence of oxygen.

TAKE-HOME MESSAGE 7.2

- Molecular oxygen produced by early photosynthetic prokaryotes put severe pressure on all life.
- The evolution of antioxidants allowed the reactive properties of oxygen to be put to use in new metabolic pathways.
- Today, carbon dioxide, water, sugar, and oxygen cycle through the world of life via photosynthesis (energy capture) and aerobic respiration (energy release).

7.3 WHAT IS GLYCOLYSIS?

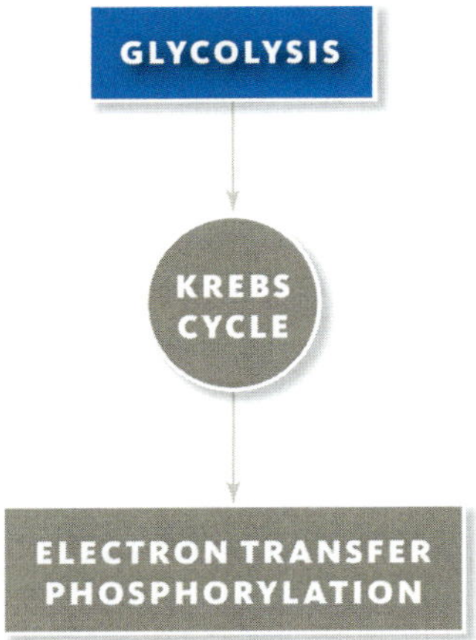

Glycolysis is a series of reactions that begin the sugar breakdown pathways of aerobic respiration (*above*) and fermentation. The reactions of glycolysis, which occur with some variation in the cytoplasm of almost all cells, convert one six-carbon molecule of sugar (such as glucose) into two molecules of **pyruvate**, an organic compound with a three-carbon backbone:

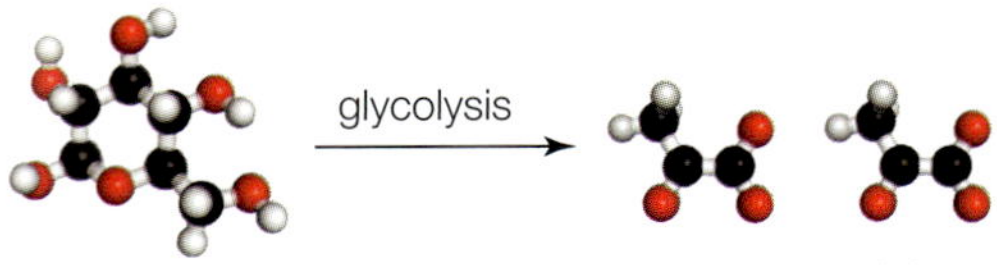

The word glycolysis comes from two Greek words: *glyk–*, sweet, and *–lysis*, loosening; it refers to the release of chemical energy from sugars. Other six-carbon sugars such as fructose and galactose can enter glycolysis, but for clarity we focus here on glucose.

Glycolysis begins when a molecule of glucose enters a cell through a glucose transporter, a passive transport protein you encountered in Section 5.7. The cell invests two ATP in the endergonic reactions that begin the pathway (**FIGURE 7.2**). In the first reaction, a phosphate group is transferred from ATP to the glucose, thus forming glucose-6-phosphate (1). A model of hexokinase, the enzyme that catalyzes this reaction, is pictured in Section 5.3.

Glycolysis continues as the glucose-6-phosphate accepts a phosphate group from another ATP, then splits (2) to form two PGAL (phosphoglyceraldehyde). Remember from Section 6.5 that this phosphorylated sugar also forms during the Calvin–Benson cycle.

glycolysis Set of reactions in which a six-carbon sugar (such as glucose) is converted to two pyruvate for a net yield of two ATP.
pyruvate Three-carbon end product of glycolysis.
substrate-level phosphorylation The formation of ATP by the direct transfer of a phosphate group from a substrate to ADP.

FIGURE 7.2 {Animated} Glycolysis (*opposite*).

For clarity, we track only the six carbon atoms (black balls) that enter the reactions as part of glucose. Cells invest two ATP to start glycolysis, so the net yield from one glucose molecule is two ATP. Two NADH also form, and two pyruvate molecules are the end products. Appendix III has more details for interested students.

In the next reaction, each PGAL receives a second phosphate group, and each gives up two electrons and a hydrogen ion. Two molecules of PGA (phosphoglycerate) form as products of this reaction (3). The electrons and hydrogen ions are accepted by two NAD^+, which thereby become reduced to NADH. Aerobic respiration's third stage requires this NADH, as does fermentation.

Next, a phosphate group is transferred from each PGA to ADP, so two ATP form (4). The direct transfer of a phosphate group from a substrate to ADP is called **substrate-level phosphorylation**. Substrate-level phosphorylation is a completely different process from the way ATP forms during electron transfer phosphorylation (Section 6.4).

Glycolysis ends with the formation of two more ATP by substrate-level phosphorylation (5). Remember, two ATP were invested to begin the reactions of glycolysis. A total of four ATP form, so the net yield is two ATP per molecule of glucose (6). The pathway also produces two three-carbon pyruvate molecules. Pyruvate is a substrate for the second-stage reactions of aerobic respiration, and also for fermentation reactions.

TAKE-HOME MESSAGE 7.3

Glycolysis is the first stage of sugar breakdown in both aerobic respiration and fermentation.

The reactions of glycolysis occur in the cytoplasm.

Glycolysis converts one molecule of glucose to two molecules of pyruvate, with a net yield of two ATP. Two NADH also form.

GLYCOLYSIS

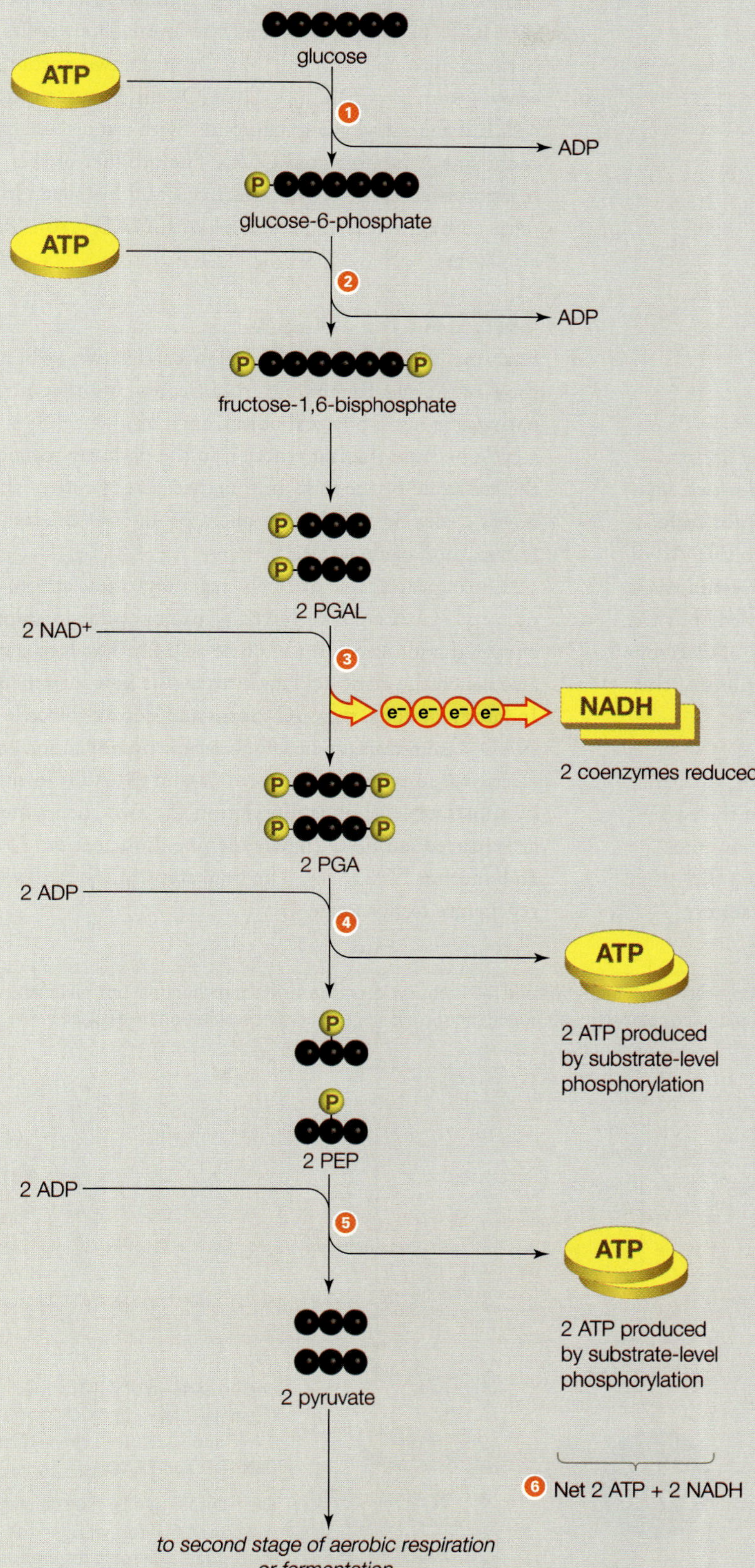

ATP-Requiring Steps

1 A phosphate group is transferred from ATP to glucose, forming glucose-6-phosphate. (You learned about the enzyme that catalyzes this reaction, hexokinase, in **FIGURE 5.10** and Section 5.7).

2 A phosphate group from a second ATP is transferred to the glucose-6-phosphate. The resulting molecule is unstable, and it splits into two three-carbon molecules. The molecules are interconvertible, so we will call them both PGAL (phosphoglyceraldehyde).

Two ATP have now been invested in the reactions.

ATP-Generating Steps

3 An enzyme attaches a phosphate to the two PGAL, so two PGA (phosphoglycerate) form. Two electrons and a hydrogen ion (not shown) from each PGAL are accepted by NAD^+, so two NADH form.

4 An enzyme transfers a phosphate group from each PGA to ADP, forming two ATP and two intermediate molecules (PEP).

The original energy investment of two ATP has now been recovered.

5 An enzyme transfers a phosphate group from each PEP to ADP, forming two more ATP and two molecules of pyruvate.

6 Summing up, glycolysis yields two NADH, two ATP (net), and two pyruvate for each glucose molecule.

Depending on the type of cell and environmental conditions, the pyruvate may enter the second stage of aerobic respiration or it may be used in other ways, such as in fermentation.

7.4 WHAT HAPPENS DURING THE SECOND STAGE OF AEROBIC RESPIRATION?

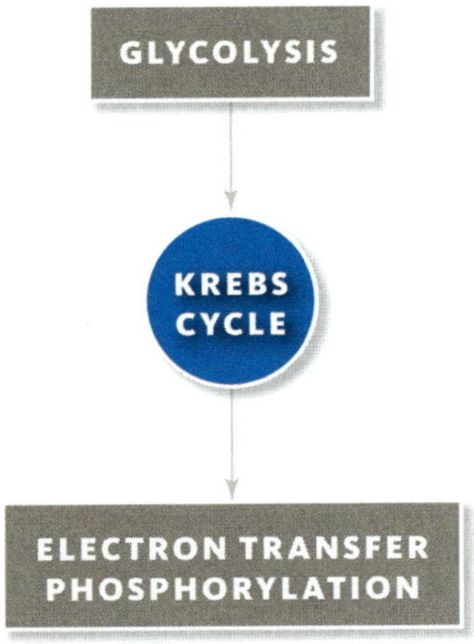

The second stage of aerobic respiration (*above*) occurs inside mitochondria (**FIGURE 7.3**). It includes two sets of reactions, acetyl–CoA formation and the **Krebs cycle**, that break down pyruvate, the product of glycolysis. All of the carbon atoms that were once part of glucose end up in CO_2, which departs the cell. Only two ATP form, but the reactions reduce many coenzymes. The energy of electrons carried by these coenzymes will drive the reactions of the third stage of aerobic respiration.

ACETYL–CoA FORMATION

Aerobic respiration's second stage begins when the two pyruvate molecules that formed during glycolysis enter a mitochondrion. Pyruvate is transported across the mitochondrion's two membranes and into the inner compartment, which is called the mitochondrial matrix (**FIGURE 7.4**). There, an enzyme immediately splits each pyruvate into one molecule of CO_2 and a two-carbon acetyl group ($—COCH_3$). The CO_2 diffuses out of the cell, and the acetyl group combines with a molecule called coenzyme A (abbreviated CoA). The product of this reaction is acetyl–CoA ❶. Electrons and hydrogen ions released by the reaction combine with NAD^+, so NADH also forms.

THE KREBS CYCLE

Each molecule of acetyl–CoA now carries two carbons into the Krebs cycle. Remember from Section 5.4 that a cyclic pathway is not a physical object, such as a wheel. It is called a cycle because the last reaction in the pathway regenerates the substrate of the first. In this case, a substrate of the Krebs cycle's first reaction—and a product of the last—is four-carbon oxaloacetate.

During each cycle of Krebs reactions, two carbon atoms of acetyl–CoA are transferred to oxaloacetate, forming citrate, the ionized form of citric acid ❷. The Krebs cycle is also called the citric acid cycle after this first intermediate. In later reactions, two CO_2 form and depart the cell. Two NAD^+ are reduced when they accept hydrogen ions and electrons, so two NADH form ❸ and ❹. ATP forms by substrate-level phosphorylation ❺. Two coenzymes are reduced: an FAD (flavin adenine dinucleotide) ❻, and another NAD^+ ❼. The final steps of the pathway regenerate oxaloacetate ❽.

FIGURE 7.3 The second stage of aerobic respiration, acetyl–CoA formation and the Krebs cycle, occurs inside mitochondria. *Left*, an inner membrane divides a mitochondrion's interior into two fluid-filled compartments. *Right*, the second stage of aerobic respiration takes place in the mitochondrion's innermost compartment, or matrix.

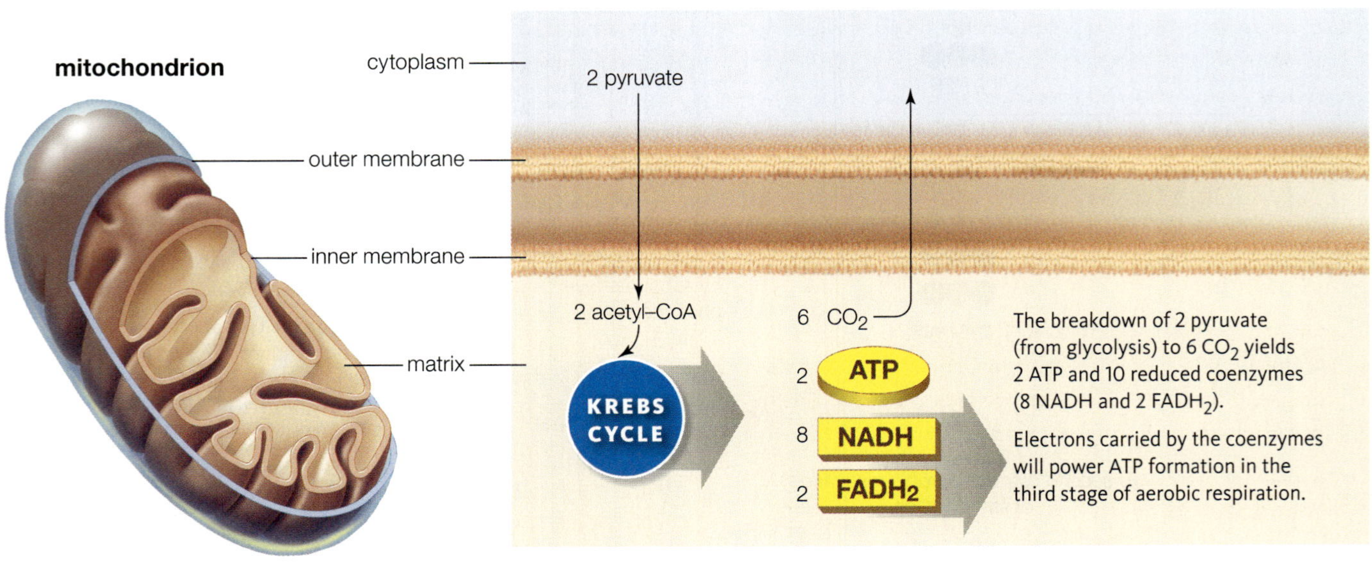

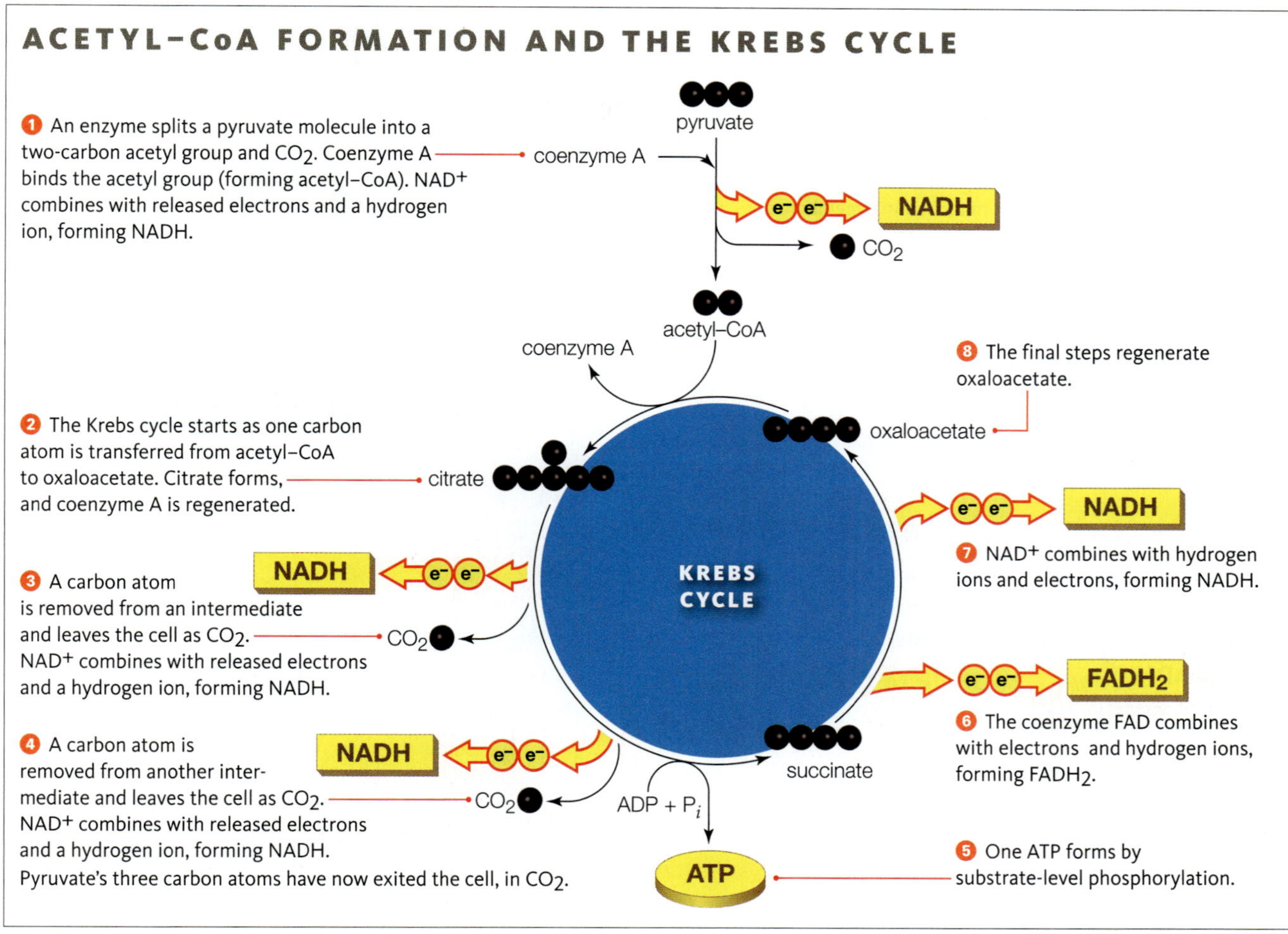

FIGURE 7.4 {Animated} Acetyl–CoA formation and the Krebs cycle. It takes two cycles of Krebs reactions to break down two pyruvate molecules that formed during glycolysis of one glucose molecule. After two cycles, all six carbons that entered glycolysis in one glucose molecule have left the cell, in six CO_2. Electrons and hydrogen ions are released as each carbon is removed from the backbone of intermediate molecules; ten coenzymes will carry them to the third and final stage of aerobic respiration. Not all reactions are shown; see Appendix III for details.

After two cycles of Krebs reactions, the two carbon atoms carried by each acetyl–CoA end up in CO_2. Thus, the combined second-stage reactions of aerobic respiration break down two pyruvate to six CO_2:

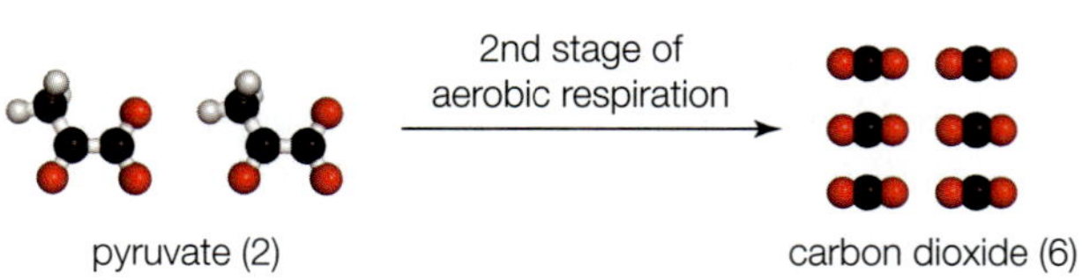

Remember, the two pyruvate were a product of glycolysis. So, at this point in aerobic respiration, the carbon backbone of one glucose molecule has been broken down completely, its six carbon atoms having exited the cell in CO_2.

Krebs cycle Cyclic pathway that, along with acetyl–CoA formation, breaks down pyruvate to carbon dioxide during aerobic respiration.

Two ATP that form during the second stage add to the small net yield of 2 ATP from glycolysis. However, ten coenzymes (eight NAD^+ and two FAD) are reduced during this stage. Add in the two NAD^+ that were reduced in glycolysis, and the full breakdown of each glucose molecule has a big potential payoff. Twelve reduced coenzymes will deliver electrons—and the energy they carry—to the third stage of aerobic respiration.

TAKE-HOME MESSAGE 7.4

The second stage of aerobic respiration, acetyl–CoA formation and the Krebs cycle, occurs in the inner compartment (matrix) of mitochondria.

The second-stage reactions convert the two pyruvate that formed in glycolysis to six CO_2. Two ATP form, and ten coenzymes (eight NAD^+ and two FAD) are reduced.

7.5 WHAT HAPPENS DURING THE THIRD STAGE OF AEROBIC RESPIRATION?

FIGURE 7.5 **{Animated}** The third and final stage of aerobic respiration, electron transfer phosphorylation.

❶ NADH and $FADH_2$ deliver their cargo of electrons and hydrogen ions to electron transfer chains in the inner mitochondrial membrane.

❷ Electron flow through the chains causes the hydrogen ions (H^+) to be pumped from the matrix to the intermembrane space. A hydrogen ion gradient forms across the inner mitochondrial membrane.

❸ Hydrogen ion flow back to the matrix through ATP synthases drives the formation of ATP from ADP and phosphate (P_i).

❹ Oxygen combines with electrons and hydrogen ions at the end of the electron transfer chains, so water forms.

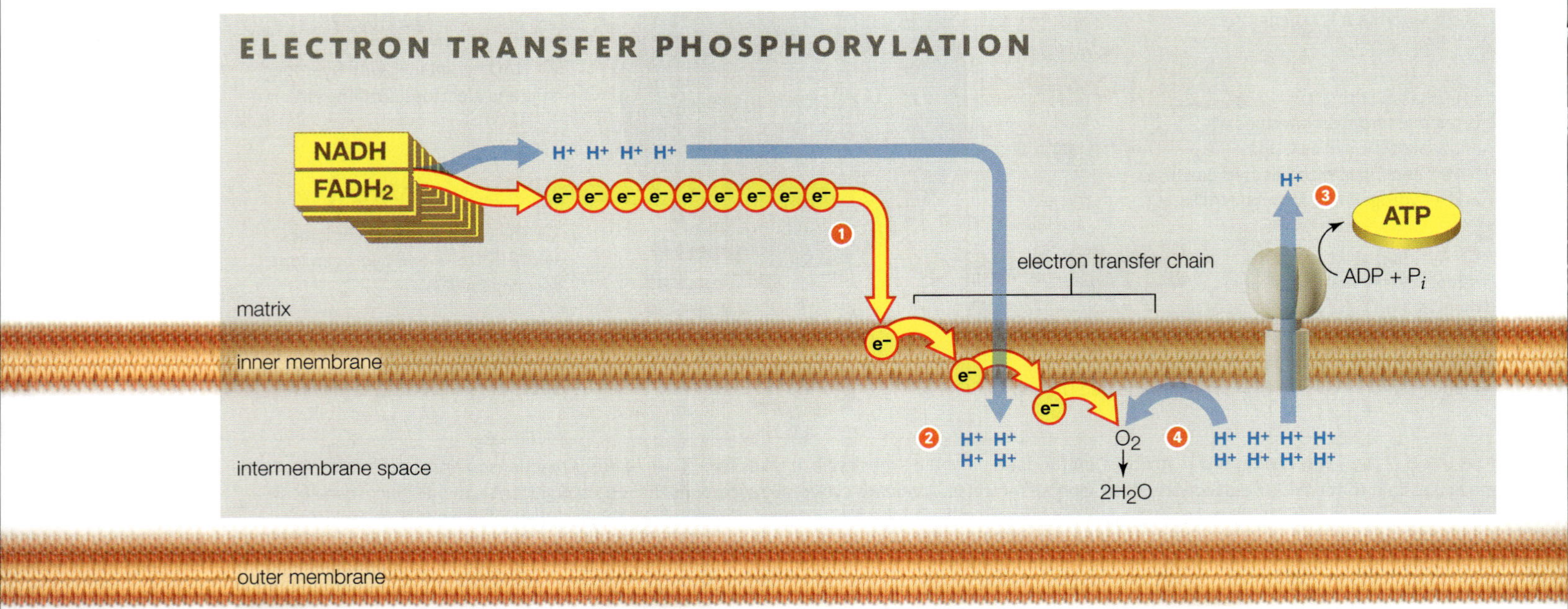

The third stage of aerobic respiration occurs at the inner mitochondrial membrane (**FIGURE 7.5**). Electron transfer phosphorylation reactions begin with the coenzymes NADH and $FADH_2$, which became reduced during the first two stages of aerobic respiration. These coenzymes now deliver their cargo of electrons and hydrogen ions to electron transfer chains embedded in the inner mitochondrial membrane ❶.

As the electrons move through the chains, they give up energy little by little (Section 5.4). Some molecules of the transfer chains harness that energy to actively transport the hydrogen ions across the inner membrane, from the matrix to the intermembrane space ❷. The accumulating ions form a hydrogen ion gradient across the inner mitochondrial membrane. This gradient attracts the ions back toward the matrix. However, ions cannot diffuse through a lipid bilayer on their own (Section 5.7). Hydrogen ions cross the inner mitochondrial membrane only by flowing through ATP synthases embedded in the membrane. The flow of hydrogen ions through ATP synthases causes these proteins to attach phosphate groups to ADP, so ATP forms ❸.

Oxygen accepts electrons at the end of mitochondrial electron transfer chains ❹. When oxygen accepts electrons, it combines with hydrogen ions to form water, which is a product of the third-stage reactions.

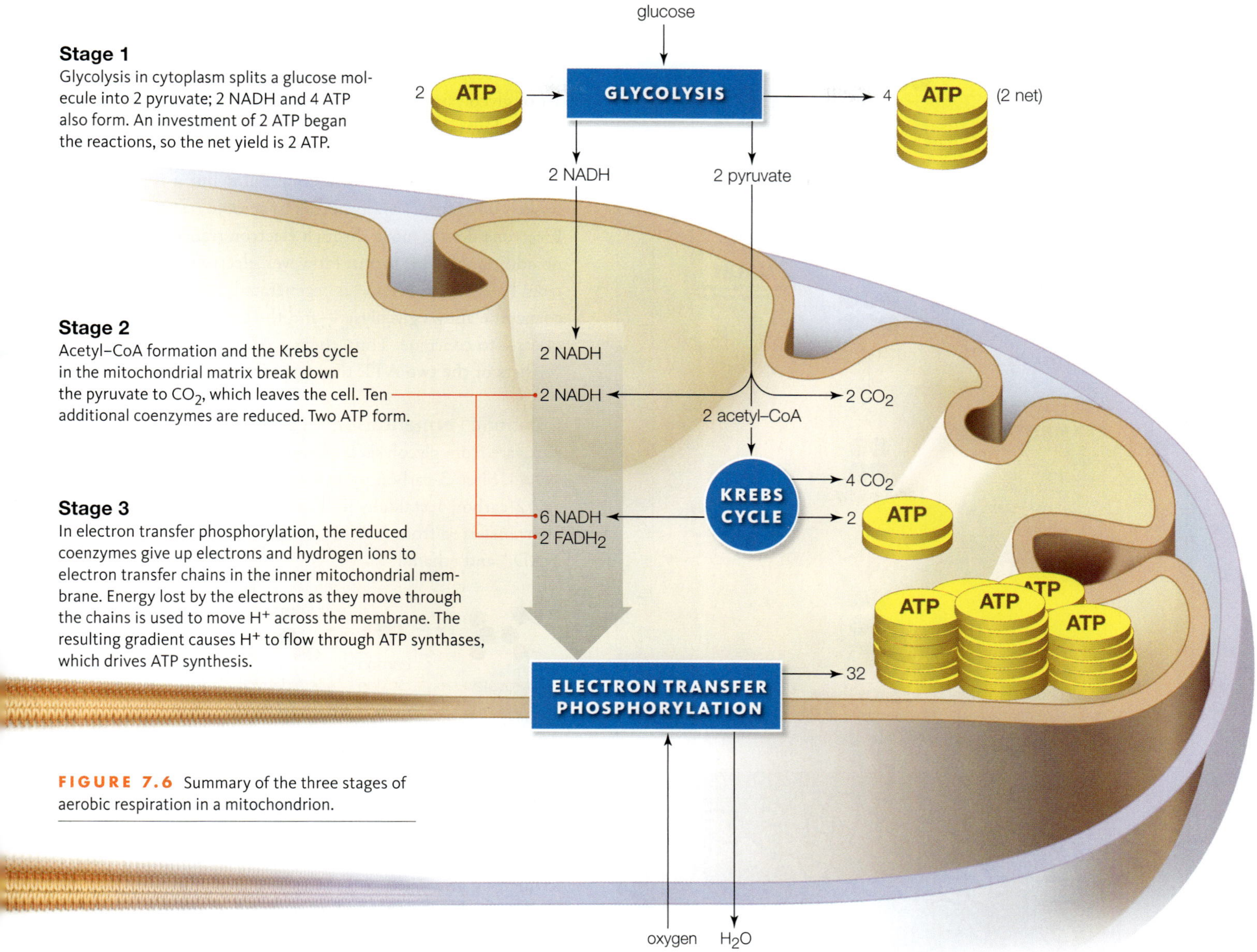

FIGURE 7.6 Summary of the three stages of aerobic respiration in a mitochondrion.

For each glucose molecule that enters aerobic respiration, four ATP form in the first- and second-stage reactions. The twelve coenzymes reduced in these two stages deliver enough electrons to fuel synthesis of about thirty-two additional ATP during the third stage. Thus, the breakdown of one glucose molecule yields about thirty-six ATP (**FIGURE 7.6**). The ATP yield varies depending on cell type. For example, the typical yield of aerobic respiration in brain and skeletal muscle cells is thirty-eight ATP, not thirty-six.

Remember that some energy dissipates with every transfer (Section 5.2). Even though aerobic respiration is a very efficient way of retrieving energy from sugars, about 60 percent of the energy harvested in this pathway disperses as metabolic heat.

TAKE-HOME MESSAGE 7.5

In aerobic respiration's third stage, electron transfer phosphorylation, energy released by electrons moving through electron transfer chains is ultimately captured in the attachment of phosphate to ADP.

The third-stage reactions begin when coenzymes that were reduced in the first and second stages deliver electrons and hydrogen ions to electron transfer chains in the inner mitochondrial membrane.

Energy released by electrons as they pass through electron transfer chains is used to pump H^+ from the mitochondrial matrix to the intermembrane space. The H^+ gradient that forms across the inner mitochondrial membrane drives the flow of hydrogen ions through ATP synthases, which results in ATP formation.

About thirty-two ATP form during the third-stage reactions, so a typical net yield of all three stages of aerobic respiration is thirty-six ATP per glucose.

7.6 WHAT IS FERMENTATION?

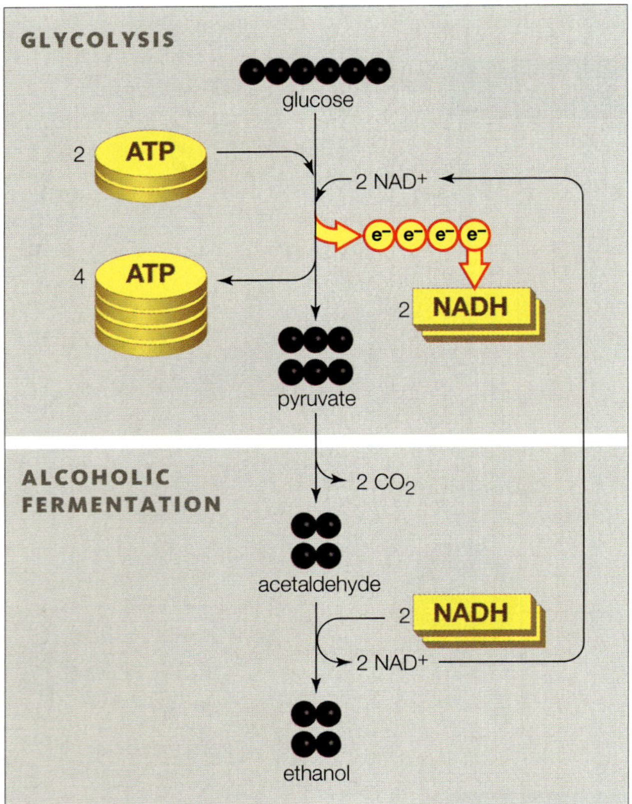

A Alcoholic fermentation begins with glycolysis, and the final steps regenerate NAD^+. The net yield of these reactions is two ATP per molecule of glucose (from glycolysis).

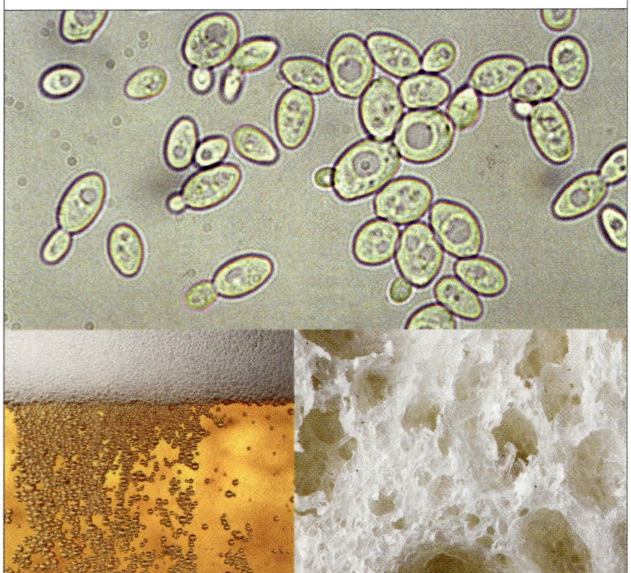

B *Saccharomyces* cells (*top*). One product of alcoholic fermentation in these cells (ethanol) makes beer alcoholic; another (CO_2) makes it bubbly. Holes in bread are pockets where CO_2 released by fermenting *Saccharomyces* cells accumulated in the dough.

FIGURE 7.7 {Animated} Alcoholic fermentation.

TWO FERMENTATION PATHWAYS

Aerobic respiration and fermentation begin with the same set of glycolysis reactions in cytoplasm. After glycolysis, the two pathways differ. The final steps of fermentation occur in the cytoplasm and do not require oxygen. In these reactions, pyruvate is converted to other molecules, but it is not fully broken down to CO_2 (as occurs in aerobic respiration). Electrons do not move through electron transfer chains, so no additional ATP forms. However, electrons are removed from NADH, so NAD^+ is regenerated. Regenerating this coenzyme allows glycolysis—and the small ATP yield it offers—to continue. Thus, the net ATP yield of fermentation consists of the two ATP that form in glycolysis.

Alcoholic Fermentation

In **alcoholic fermentation**, the pyruvate from glycolysis is converted to ethanol (**FIGURE 7.7A**). First, 3-carbon pyruvate is split into carbon dioxide and 2-carbon acetaldehyde. Then, electrons and hydrogen are transferred from NADH to the acetaldehyde, forming NAD^+ and ethanol:

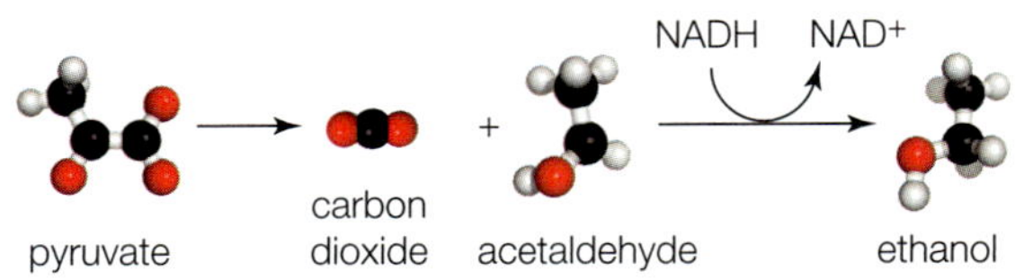

Alcoholic fermentation in a fungus, *Saccharomyces cerevisiae*, sustains these yeast cells as they grow and reproduce. It also helps us produce beer, wine, and bread (**FIGURE 7.7B**). Beer brewers typically use germinated, roasted, and crushed barley as a sugar source for *Saccharomyces* fermentation. Ethanol produced by the fermenting yeast cells makes the beer alcoholic, and CO_2 makes it bubbly. Flowers of the hop plant add flavor and help preserve the finished product. Winemakers start with crushed grapes for *Saccharomyces* fermentation. The yeast cells convert sugars in the grape juice to ethanol.

Bakers take advantage of alcoholic fermentation by *Saccharomyces* cells to make bread from flour, which contains starches and a protein called gluten. When flour is kneaded with water, the gluten forms polymers in long, interconnected strands that make the resulting dough stretchy and resilient. Yeast cells in the dough produce CO_2 as they ferment the starches. The gas accumulates in bubbles that are trapped by the mesh of gluten strands. As the bubbles expand, they cause the dough to rise. Ethanol produced by the fermentation reactions evaporates during baking.

Lactate Fermentation

In **lactate fermentation**, the electrons and hydrogen ions carried by NADH are transferred directly to pyruvate (**FIGURE 7.8A**). This reaction converts pyruvate to 3-carbon lactate (the ionized form of lactic acid), and also converts NADH to NAD^+:

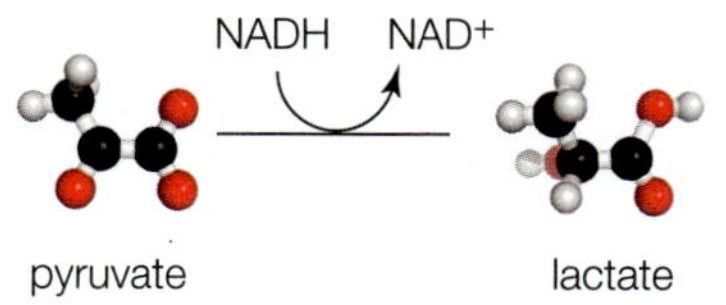

Some lactate fermenters spoil food, but we use others to preserve it. For instance, *Lactobacillus* bacteria break down lactose in milk by fermentation. We use this bacteria to produce dairy products such as buttermilk, cheese, and yogurt, and also to pickle vegetables and other foods.

Cells in animal skeletal muscles are fused as long fibers that carry out aerobic respiration, lactate fermentation, or both. Red fibers have many mitochondria and produce ATP mainly by aerobic respiration. These fibers sustain prolonged activity such as marathon runs. They are red because they have an abundance of myoglobin, a protein that stores oxygen for aerobic respiration (**FIGURE 7.8B**). White muscle fibers contain few mitochondria and no myoglobin, so they do not carry out a lot of aerobic respiration. Instead, they make most of their ATP by lactate fermentation. This pathway makes ATP quickly but not for long, so it is useful for quick, strenuous activities such as weight lifting or sprinting (**FIGURE 7.8C**). The low ATP yield does not support prolonged activity.

Most human muscles are a mixture of white and red fibers, but the proportions vary among muscles and among individuals. Great sprinters tend to have more white fibers. Great marathon runners tend to have more red fibers. Chickens cannot fly very far because their flight muscles consist mostly of white fibers (thus, the "white" breast meat). A chicken most often walks or runs. Its leg muscles consist mostly of red muscle fibers, the "dark meat."

alcoholic fermentation Anaerobic sugar breakdown pathway that produces ATP, CO_2, and ethanol.
lactate fermentation Anaerobic sugar breakdown pathway that produces ATP and lactate.

TAKE-HOME MESSAGE 7.6

- ATP can form by sugar breakdown in fermentation pathways, which are anaerobic.
- The end product of lactate fermentation is lactate. The end product of alcoholic fermentation is ethanol.
- Both pathways have a net yield of two ATP per glucose molecule. The ATP forms during glycolysis.
- Fermentation reactions regenerate the coenzyme NAD^+, without which glycolysis (and ATP production) would stop.

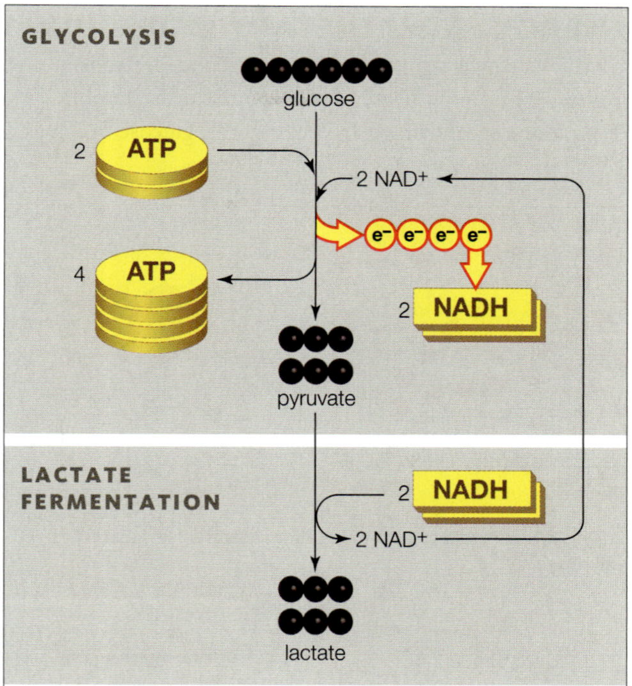

A Lactate fermentation begins with glycolysis, and the final steps regenerate NAD^+. The net yield of these reactions is two ATP per molecule of glucose (from glycolysis).

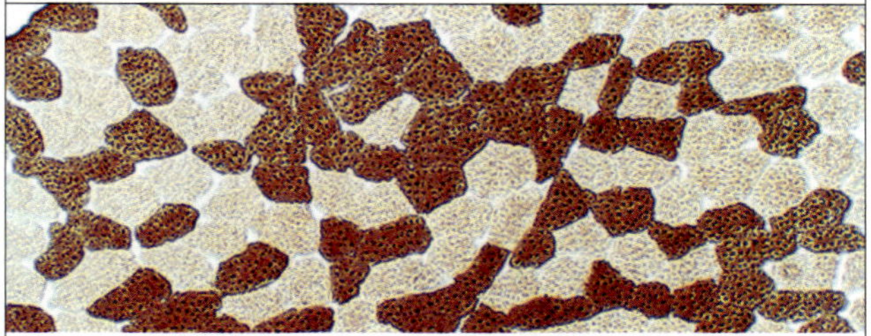

B Lactate fermentation occurs in white muscle fibers, visible in this cross-section of human thigh muscle. The red fibers, which make ATP by aerobic respiration, sustain endurance activities.

C Intense activity such as sprinting quickly depletes oxygen in muscles. Under anaerobic conditions, ATP is produced mainly by lactate fermentation in white muscle fibers. Fermentation does not make enough ATP to sustain this type of activity for long.

FIGURE 7.8 Lactate fermentation.

7.7 CAN THE BODY USE ANY ORGANIC MOLECULE FOR ENERGY?

ENERGY FROM DIETARY MOLECULES

Glycolysis converts glucose to pyruvate, and electrons are transferred from pyruvate to coenzymes during aerobic respiration's second stage. In other words, glucose becomes oxidized (it gives up electrons) and coenzymes become reduced (they accept electrons). Oxidizing an organic molecule can break the covalent bonds of its carbon backbone. Aerobic respiration generates a lot of ATP by fully oxidizing glucose, completely dismantling it carbon by carbon.

Cells also dismantle other organic molecules by oxidizing them. Complex carbohydrates, fats, and proteins in food can be converted to molecules that enter glycolysis or the Krebs cycle (**FIGURE 7.9**). As in glucose metabolism, many coenzymes are reduced, and the energy of the electrons they carry ultimately drives the synthesis of ATP in electron transfer phosphorylation.

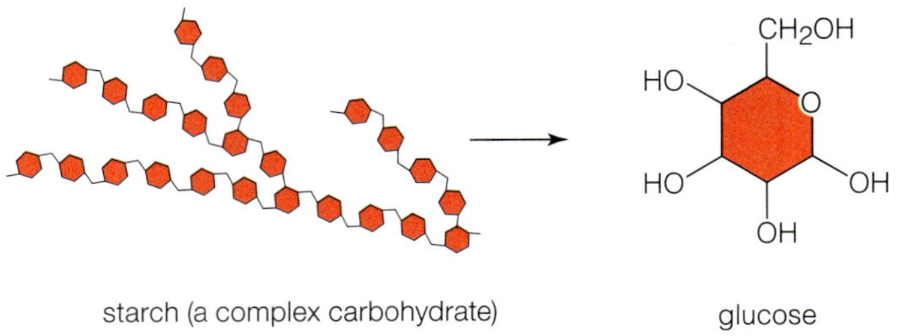

A Complex carbohydrates are broken down to their monosaccharide subunits, which can enter glycolysis ①.

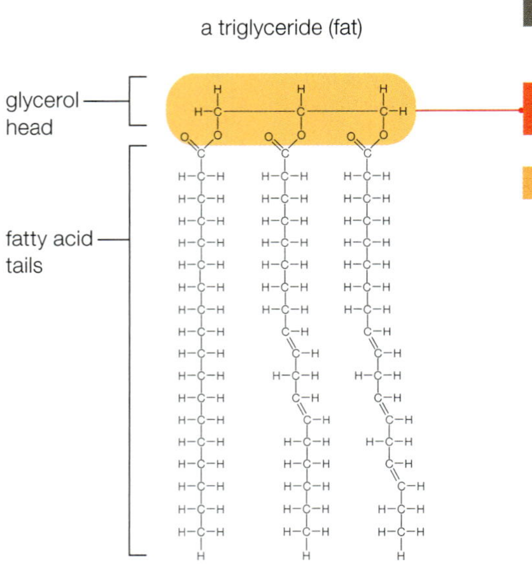

B Fats are broken down by separating the glycerol head from the fatty acid tails. The fatty acids are converted to acetyl–CoA ②, and the glycerol is converted to PGAL ③.

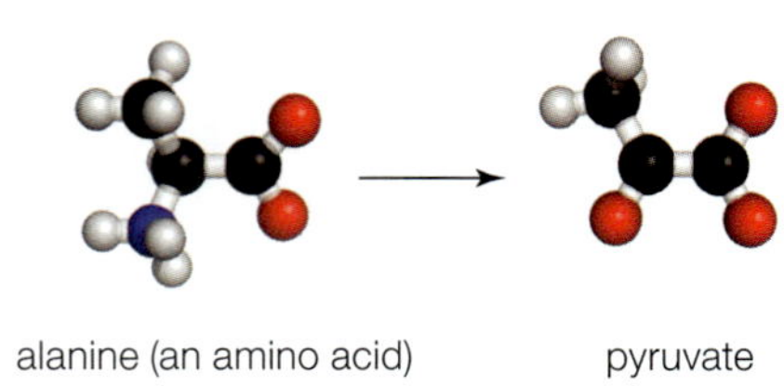

C Amino acids are converted to acetyl–CoA, pyruvate, or an intermediate of the Krebs cycle ④.

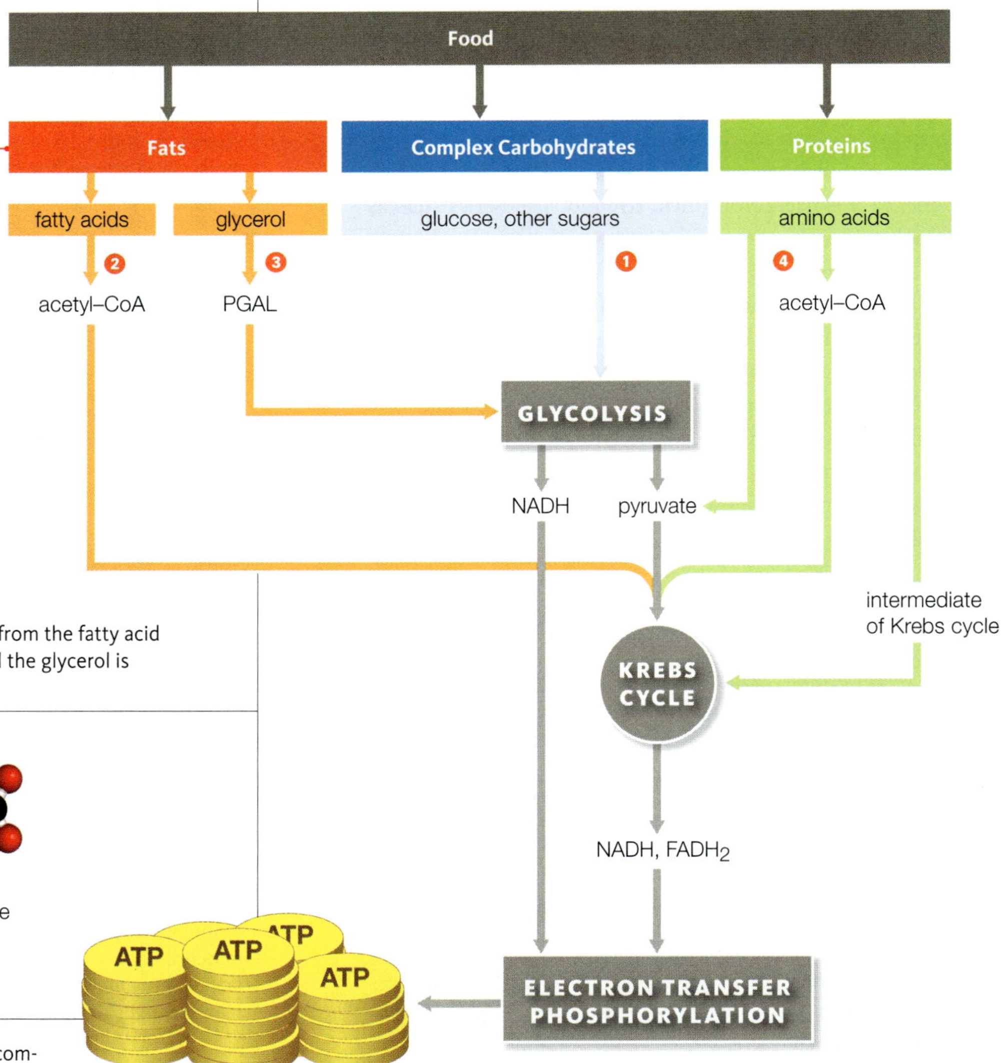

FIGURE 7.9 **{Animated}** A variety of organic compounds from food can enter the reactions of aerobic respiration.

Complex Carbohydrates In humans and other mammals, the digestive system breaks down starch and other complex carbohydrates to monosaccharides (**FIGURE 7.9A**). These sugars are quickly taken up by cells and converted to glucose-6-phosphate for glycolysis ❶. When a cell produces more ATP than it uses, the concentration of ATP rises in the cytoplasm. A high concentration of ATP causes glucose-6-phosphate to be diverted away from glycolysis and into a pathway that forms glycogen. Liver and muscle cells especially favor the conversion of glucose to glycogen, and these cells contain the body's largest stores of it. Between meals, the liver maintains the glucose level in blood by converting the stored glycogen to glucose.

Fats A fat molecule has a glycerol head and one, two, or three fatty acid tails (Section 3.3). Cells dismantle these molecules by first breaking the bonds that connect the fatty acid tails to the glycerol head (**FIGURE 7.9B**). Nearly all cells in the body can oxidize free fatty acids by splitting their long backbones into two-carbon fragments. These fragments are converted to acetyl–CoA, which can enter the Krebs cycle ❷. Enzymes in liver cells convert the glycerol to PGAL, an intermediate of glycolysis ❸.

On a per carbon basis, fats are a richer source of energy than carbohydrates. Carbohydrate backbones have many oxygen atoms, so they are partially oxidized. A fat's long fatty acid tails are hydrocarbon chains that typically have no oxygen atoms bonded to them, so they have a longer way to go to become oxidized—more reactions are required to fully break them down. Coenzymes accept electrons in these oxidation reactions. The more reduced coenzymes that form, the more electrons can be delivered to the ATP-forming machinery of electron transfer phosphorylation.

What happens if you eat too many carbohydrates? When the blood level of glucose gets too high, acetyl–CoA is diverted away from the Krebs cycle and into a pathway that makes fatty acids. That is why excess dietary carbohydrate ends up as fat.

Proteins Enzymes in the digestive system split dietary proteins into their amino acid subunits, which are absorbed into the bloodstream. Cells use the amino acids to build proteins or other molecules. When you eat more protein than your body needs for this purpose, the amino acids are broken down. The amino (NH_3^+) group is removed, and it becomes ammonia (NH_3), a waste product that is eliminated in urine. The carbon backbone is split, and acetyl–CoA, pyruvate, or an intermediate of the Krebs cycle forms, depending on the amino acid (**FIGURE 7.9C**). These molecules enter aerobic respiration's second stage ❹.

PEOPLE MATTER

DR. BENJAMIN RAPOPORT

MIT engineers Benjamin Rapoport, Jakub Kedzierski, and Rahul Sarpeshkar have developed a tiny fuel cell that runs on the same sugar that powers human cells: glucose.

The fuel cell strips electrons from glucose molecules to create a small electric current. In this way, it mimics the activity of cellular enzymes that break down glucose to generate ATP.

The researchers fabricated the fuel cell on a silicon chip, so it can be integrated with other implantable circuits that could, for example, be implanted in the spinal cord to help paralyzed patients move their arms and legs again. Current devices can do this too, but they require an external power source. The new fuel cell can use glucose in the fluid that bathes the brain. Glucose is normally the brain's only fuel, and this fluid contains a lot of it. The fuel cell uses only a tiny amount of glucose, so it would have a minimal impact on brain function.

"It will be a few more years into the future before you see people with spinal-cord injuries receive such implantable systems in the context of standard medical care, but those are the sorts of devices you could envision powering from a glucose-based fuel cell," says Rapoport.

TAKE-HOME MESSAGE 7.7

Oxidizing organic molecules can break their carbon backbones, releasing electrons whose energy can be harnessed to drive ATP formation in aerobic respiration.

Fats, complex carbohydrates, and proteins can be oxidized in aerobic respiration to yield ATP. First the digestive system and then individual cells convert molecules in food into substrates of glycolysis or aerobic respiration's second-stage reactions.

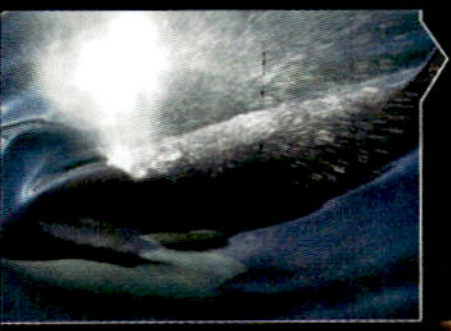

7.8 Application: MITOCHONDRIAL MALFUNCTION

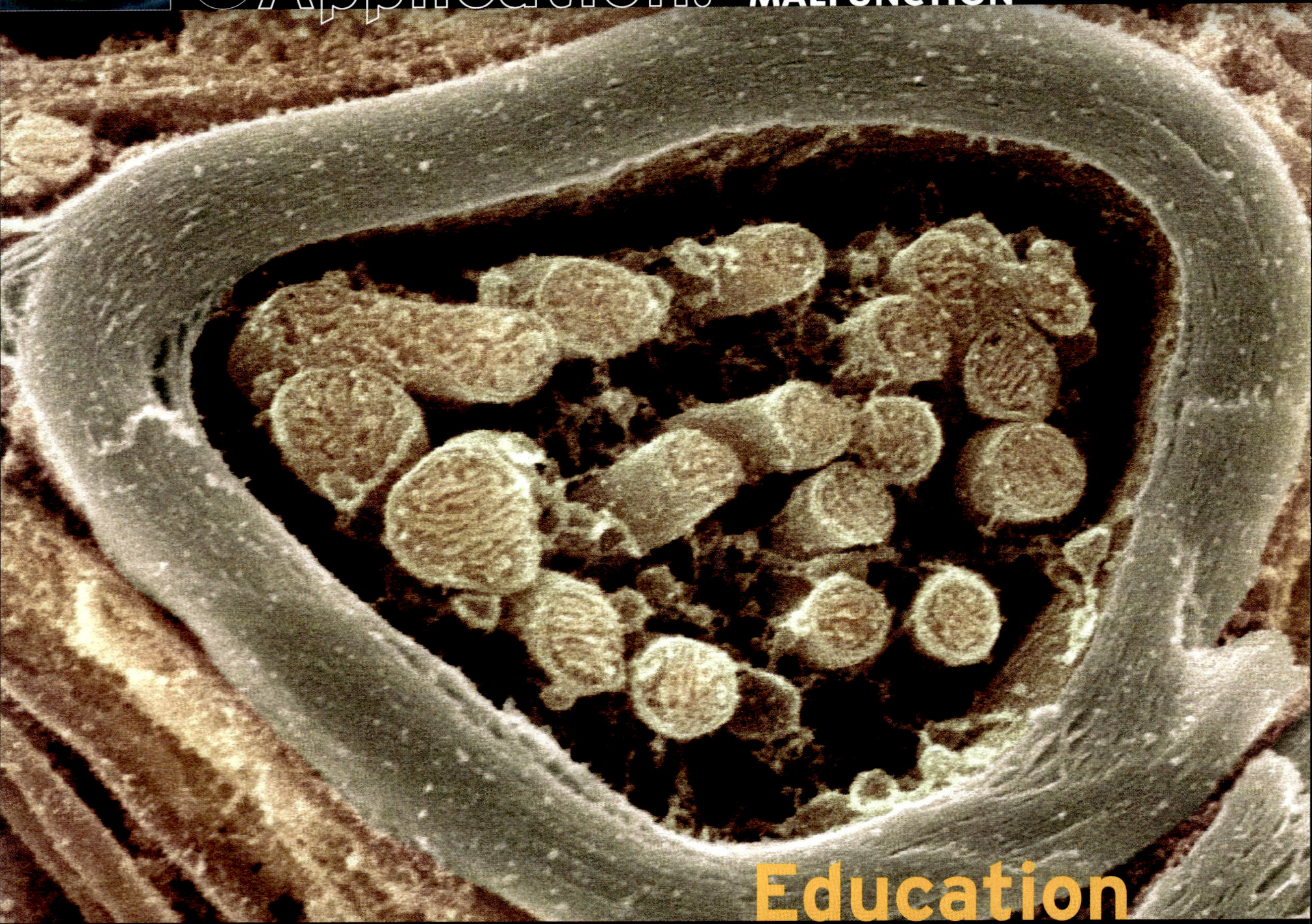

FIGURE 7.10 This cross-section of a nerve shows how these cells are packed with mitochondria. Mitochondria are powerhouses of all eukaryotic cells. When they malfunction, the lights go off in cellular businesses.

AEROBIC RESPIRATION IS A DANGEROUS OCCUPATION. When an oxygen molecule accepts electrons from an electron transfer chain in a mitochondrion, it dissociates into oxygen atoms. These atoms immediately combine with hydrogen ions and end up in water molecules. Occasionally, however, an oxygen atom escapes this final reaction. The atom has an unpaired electron, so it is a free radical. Free radicals can easily strip electrons from (oxidize) biological molecules and break their carbon backbones.

Mitochondria cannot detoxify free radicals, so they rely on antioxidant enzymes and vitamins in the cell's cytoplasm to do it for them. The system works well, at least most of the time. However, a genetic disorder or an unfortunate encounter with a toxin or pathogen can result in a missing antioxidant, or a defective component of the mitochondrial electron transfer chain. In either case, the normal cellular balance of aerobic respiration and free radical formation is tipped. Free radicals accumulate and destroy first the function of mitochondria, then the cell. The resulting tissue damage is called oxidative stress.

At least 83 proteins are directly involved in mitochondrial electron transfer chains. A defect in any one of them—or in any of the thousands of other

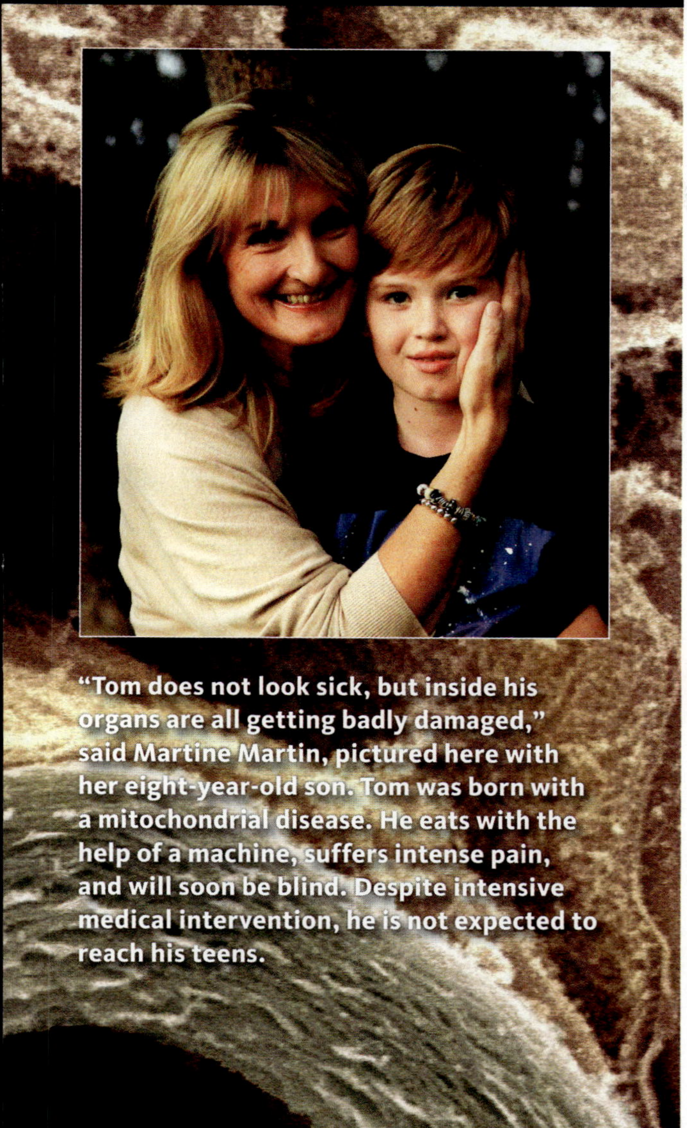

"Tom does not look sick, but inside his organs are all getting badly damaged," said Martine Martin, pictured here with her eight-year-old son. Tom was born with a mitochondrial disease. He eats with the help of a machine, suffers intense pain, and will soon be blind. Despite intensive medical intervention, he is not expected to reach his teens.

proteins used by mitochondria—can wreak havoc in the body. Hundreds of incurable disorders are associated with such defects (FIGURE 7.10), and more are being discovered all the time. Nerve and brain cells, which require a lot of ATP, are particularly affected. Symptoms range from mild to major progressive neurological deficits, blindness, deafness, diabetes, strokes, seizures, gastrointestinal malfunction, and disabling muscle weakness. New research is showing that mitochondrial malfunction is also involved in many other illnesses, including cancer, hypertension, and Alzheimer's and Parkinson's diseases.

Summary

SECTIONS 7.1, 7.2 Most organisms can make ATP by breaking down sugars in fermentation or aerobic respiration. Both pathways begin in cytoplasm. **Aerobic respiration** requires oxygen and, in eukaryotes, ends in mitochondria. It includes electron transfer chains, and ATP forms by electron transfer phosphorylation. **Fermentation** pathways end in cytoplasm and do not require oxygen. Aerobic respiration yields much more ATP per glucose molecule than fermentation.

Photosynthesis by early prokaryotes changed the composition of Earth's atmosphere, with profound effects on life's evolution. Organisms that could not tolerate the increased atmospheric oxygen persisted only in **anaerobic** habitats. The evolution of antioxidants allowed organisms to tolerate the increase in the atmospheric content of oxygen, and to thrive under **aerobic** conditions. Over time, the antioxidants became incorporated into aerobic respiration and other pathways that harnessed the reactive properties of oxygen.

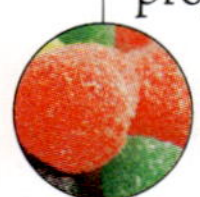

SECTION 7.3 **Glycolysis**, the first stage of aerobic respiration and fermentation, occurs in cytoplasm. In the reactions, enzymes use two ATP to convert one molecule of glucose or another six-carbon sugar to two molecules of 3-carbon **pyruvate**. Electrons and hydrogen ions are transferred to two NAD^+, which are thereby reduced to NADH. Four ATP also form by **substrate-level phosphorylation**.

SECTION 7.4 In eukaryotes, aerobic respiration continues in mitochondria. The second stage of aerobic respiration, acetyl–CoA formation and the **Krebs cycle**, takes place in the inner compartment (matrix) of the mitochondrion. The first steps convert the two pyruvate from glycolysis to two acetyl–CoA and two CO_2. The acetyl–CoA delivers carbon atoms to the Krebs cycle. Electrons and hydrogen ions are transferred to NAD^+ and FAD, which are thereby reduced to NADH and $FADH_2$. ATP forms by substrate-level phosphorylation. Two cycles of Krebs reactions break down the two pyruvate from glycolysis. At this stage of aerobic respiration, the glucose molecule that entered glycolysis has been dismantled completely: All of its carbon atoms have exited the cell in CO_2.

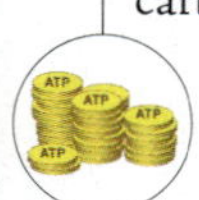

SECTION 7.5 In the third and final stage of aerobic respiration, electron transfer phosphorylation, the many coenzymes that were reduced in the first two stages now deliver their cargo of electrons and hydrogen ions to electron transfer chains in the inner mitochondrial membrane. The electrons move through the chains, releasing energy bit by bit; molecules of the chain use that energy to move H^+ from the matrix to the intermembrane space. Hydrogen ions accumulate in the intermembrane space, forming a gradient across the inner

Summary continued

membrane. The ions follow the gradient back to the matrix through ATP synthases. H^+ flow through these transport proteins drives ATP synthesis.

Oxygen accepts electrons at the end of the chains and combines with hydrogen ions, so water forms.

The ATP yield of aerobic respiration varies, but typically it is about thirty-six ATP for each glucose molecule that enters glycolysis.

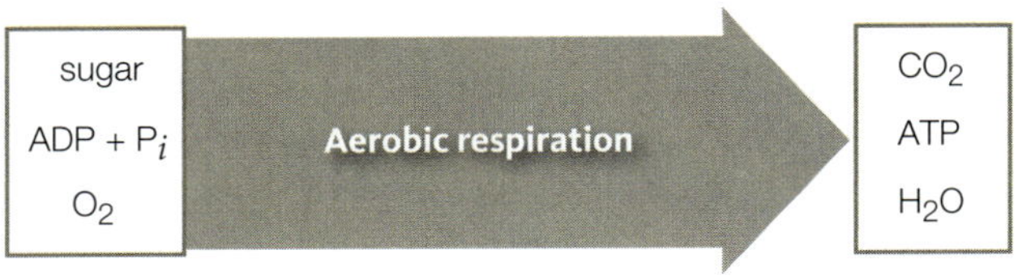

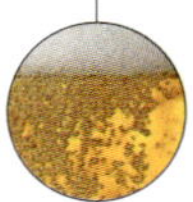

SECTION 7.6 Anaerobic fermentation pathways begin with glycolysis, and they run entirely in the cytoplasm. An organic molecule, rather than oxygen, accepts electrons at the end of these reactions. The end product of **alcoholic fermentation** is ethyl alcohol, or ethanol. The end product of **lactate fermentation** is lactate.

The final steps of fermentation regenerate NAD^+, which is required for glycolysis to continue, but they produce no ATP. Thus, the breakdown of one glucose molecule in either alcoholic or lactate fermentation yields only the two ATP from glycolysis.

Skeletal muscle consists of two types of fiber. ATP produced primarily by aerobic respiration in red fibers sustains activities that require endurance. Lactate fermentation in white fibers supports activities that occur in short, intense bursts.

SECTION 7.7 Oxidizing an organic molecule can break its carbon backbone. Aerobic respiration fully oxidizes glucose, dismantling its backbone carbon by carbon. Each carbon removed releases electrons that drive ATP formation in electron transfer phosphorylation. Organic molecules other than sugars are also broken down (oxidized) for energy. In humans and other mammals, first the digestive system and then individual cells convert fats, proteins, and complex carbohydrates in food to molecules that are substrates of glycolysis or the second-stage reactions of aerobic respiration.

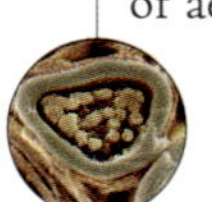

SECTION 7.8 Free radicals that form during aerobic respiration are detoxified by antioxidant molecules in the cell's cytoplasm. Missing antioxidant molecules or heritable defects in mitochondrial electron transfer chain components can cause a buildup of free radicals that damage the cell—and ultimately, the individual. Symptoms can be lethal. Oxidative stress due to mitochondrial malfunction plays a role in many illnesses such as cancer, and Alzheimer's and Parkinson's diseases.

Self-Quiz

Answers in Appendix VII

1. Is the following statement true or false? Unlike animals, which make many ATP by aerobic respiration, plants make all of their ATP by photosynthesis.

2. Glycolysis starts and ends in the ________ .
   a. nucleus   c. plasma membrane
   b. mitochondrion   d. cytoplasm

3. Which of the following metabolic pathways require(s) molecular oxygen (O_2)?
   a. aerobic respiration
   b. lactate fermentation
   c. alcoholic fermentation
   d. all of the above

4. Which molecule does not form during glycolysis?
   a. NADH   c. oxygen (O_2)
   b. pyruvate   d. ATP

5. In eukaryotes, aerobic respiration is completed in the ________ .
   a. nucleus   c. plasma membrane
   b. mitochondrion   d. cytoplasm

6. In eukaryotes, fermentation is completed in the ________ .
   a. nucleus   c. plasma membrane
   b. mitochondrion   d. cytoplasm

7. Which of the following reaction pathways is not part of the second stage of aerobic respiration?
   a. electron transfer phosphorylation   c. Krebs cycle
   b. acetyl–CoA formation   d. glycolysis
   e. a and d

8. After the Krebs reactions run through ________ cycle(s), one glucose molecule has been completely broken down to CO_2.
   a. one   b. two   c. three   d. six

9. In the third stage of aerobic respiration, ________ is the final acceptor of electrons.
   a. water   c. oxygen (O_2)
   b. hydrogen   d. NADH

10. Most of the energy that is released by the full breakdown of glucose to CO_2 and water ends up in ________ .
   a. NADH   c. heat
   b. ATP   d. electrons

11. ________ accepts electrons in alcoholic fermentation.
   a. Oxygen   c. Acetaldehyde
   b. Pyruvate   d. Ethanol

12. Your body cells can break down ________ as a source of energy to fuel ATP production.
   a. fatty acids   c. amino acids
   b. glycerol   d. all of the above

13. Which of the following is *not* produced by an animal muscle cell operating under anaerobic conditions?

a. heat c. ATP e. pyruvate
b. lactate d. NAD^+ f. all are produced

14. Hydrogen ion flow drives ATP synthesis during ______ .

a. glycolysis
b. the Krebs cycle
c. aerobic respiration
d. fermentation
e. a and c

15. Match the term with the best description.

___ mitochondrial matrix a. needed for glycolysis
___ pyruvate b. inner space
___ NAD^+ c. makes many ATP
___ mitochondrion d. product of glycolysis
___ NADH e. reduced coenzyme
___ anaerobic f. no oxygen required

Critical Thinking

1. The higher the altitude, the lower the oxygen level in air. Climbers of very tall mountains risk altitude sickness, which is characterized by shortness of breath, weakness, dizziness, and confusion.

The early symptoms of cyanide poisoning are the same as those for altitude sickness. Cyanide binds tightly to cytochrome *c* oxidase, the protein that reduces oxygen molecules in the final step of mitochondrial electron transfer chains. Cytochrome *c* oxidase with bound cyanide can no longer transfer electrons. Explain why cyanide poisoning starts with the same symptoms as altitude sickness.

2. As you learned, membranes impermeable to hydrogen ions are required for electron transfer phosphorylation. Membranes in mitochondria serve this purpose in eukaryotes. Bacteria do not have this organelle, but they do make ATP by electron transfer phosphorylation. How do you think they do it, given that they have no mitochondria?

3. The bar-tailed godwit is a type of shorebird that makes an annual migration from Alaska to New Zealand and back. The birds make each 11,500-kilometer (7,145-mile) trip by flying over the Pacific Ocean in about nine days, depending on weather, wind speed, and direction of travel. One bird was observed to make the entire journey uninterrupted, a feat that is comparable to a human running a nonstop seven-day marathon at 70 kilometers per hour (43.5 miles per hour). Would you expect the flight (breast) muscles of bar-tailed godwits to be light or dark colored? Explain your answer.

Data Analysis Activities

Mitochondrial Abnormalities in Tetralogy of Fallot

Tetralogy of Fallot (TF) is a genetic disorder characterized by four major malformations of the heart. The circulation of blood is abnormal, so TF patients have too little oxygen in their blood. Inadequate oxygen levels result in damaged mitochondrial membranes, which in turn cause cells to self-destruct.

In 2004, Sarah Kuruvilla and her colleagues looked at abnormalities in the mitochondria of heart muscle in TF patients. Some of their results are shown in **FIGURE 7.11**.

1. Which abnormality was most strongly associated with tetralogy of Fallot?
2. Can you make any correlations between blood oxygen content and mitochondrial abnormalities in these TF patients?

FIGURE 7.11 Mitochondrial changes in tetralogy of Fallot (TF).

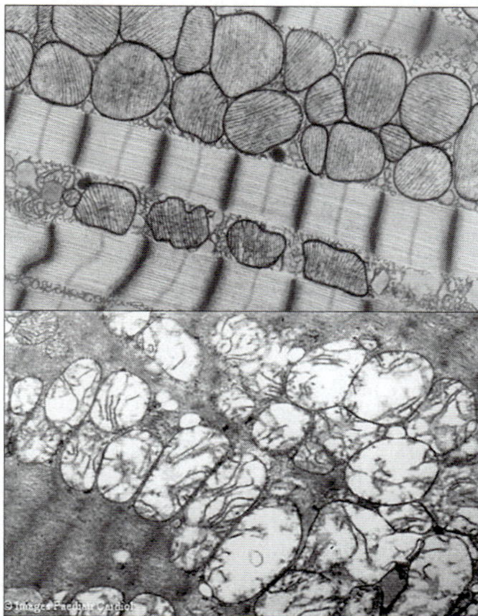

A Normal heart muscle. Many mitochondria between the fibers provide muscle cells with ATP for contraction.

B Heart muscle from a person with TF has swollen, broken mitochondria.

Patient (age)	SPO_2 (%)	Mitochondrial Abnormalities in TF			
		Number	Shape	Size	Broken
1 (5)	55	+	+	–	–
2 (3)	69	+	+	–	–
3 (22)	72	+	+	–	–
4 (2)	74	+	+	–	–
5 (3)	76	+	+	–	+
6 (2.5)	78	+	+	–	+
7 (1)	79	+	+	–	–
8 (12)	80	+	–	+	–
9 (4)	80	+	+	–	–
10 (8)	83	+	–	+	–
11 (20)	85	+	+	–	–
12 (2.5)	89	+	–	+	–

C Types of mitochondrial abnormalities in TF patients. SPO_2 is oxygen saturation of the blood. A normal value of SPO_2 is 96%. Abnormalities are marked +.

Information encoded in DNA is the basis of visible traits that define species and distinguish individuals. Identical twins are identical because they inherited copies of the same DNA.

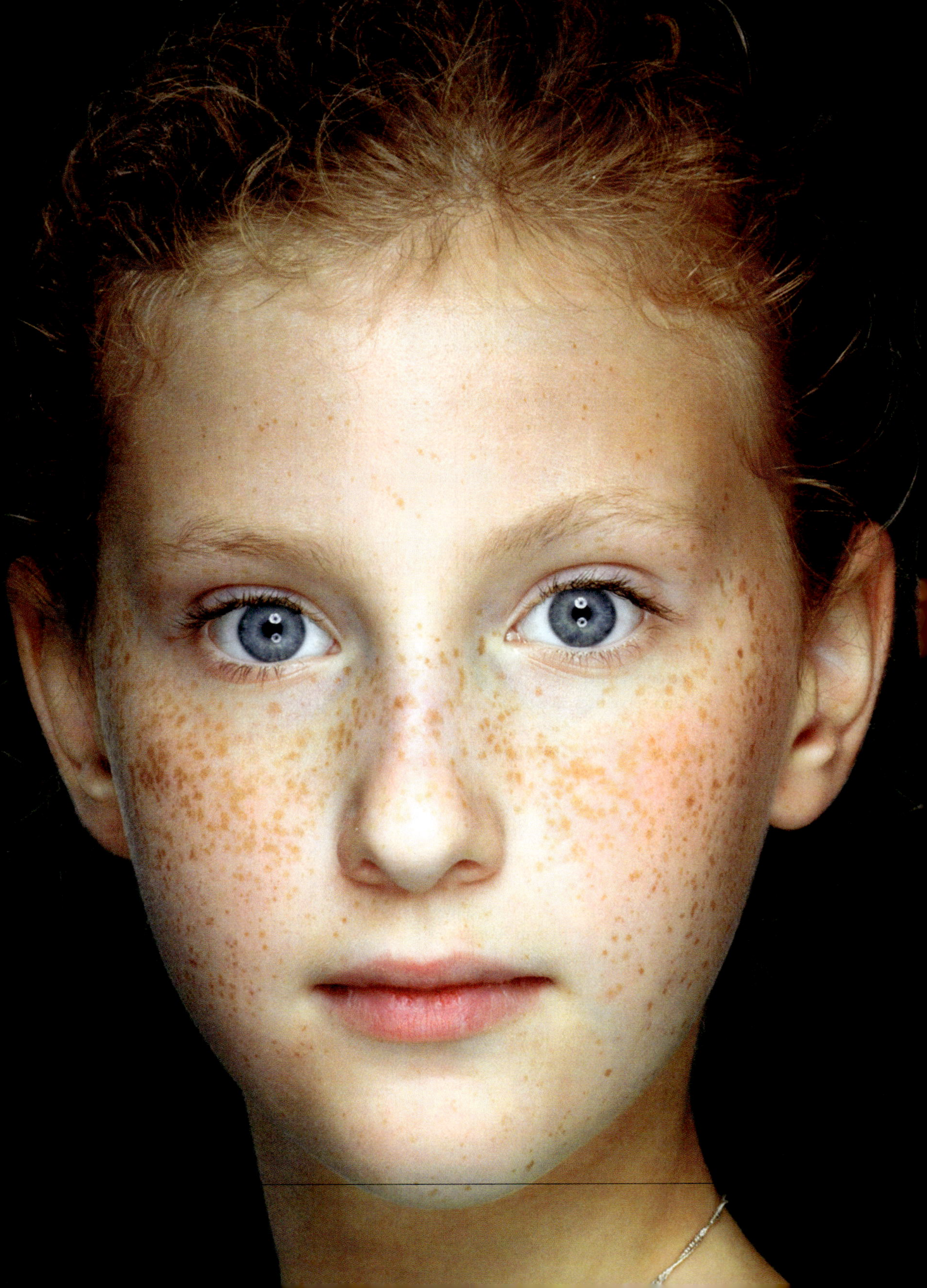

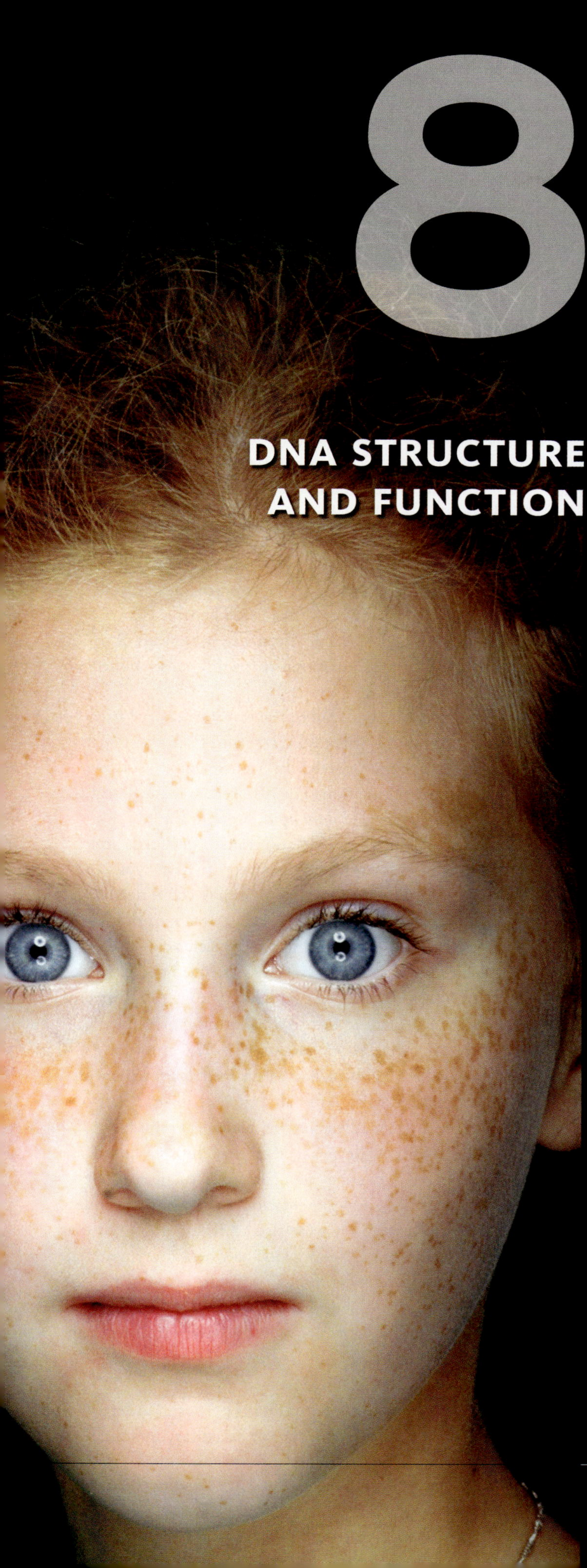

8 DNA STRUCTURE AND FUNCTION

Links to Earlier Concepts

Radioisotope tracers (Section 2.1) were used in research that led to the discovery that DNA (3.6), not protein (3.4, 3.5), is the hereditary material (1.2) of all organisms. This chapter revisits free radicals (2.2), the cell nucleus (4.5), and metabolism (5.3–5.5).

KEY CONCEPTS

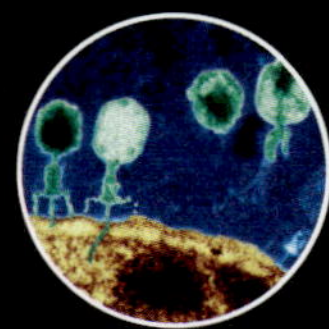

DISCOVERY OF DNA'S FUNCTION

The work of many scientists over nearly a century led to the discovery that DNA, not protein, stores hereditary information in all living things.

STRUCTURE OF DNA MOLECULES

A DNA molecule consists of two long chains of nucleotides coiled into a double helix. The order of the four types of nucleotides in a chain differs among individuals and species.

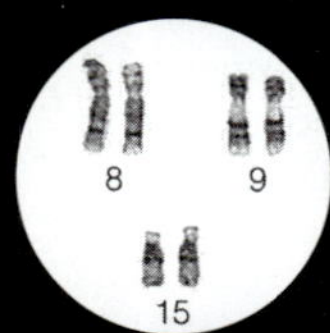

CHROMOSOMES

The DNA of eukaryotes is divided among a characteristic number of chromosomes. A living cell's chromosomes contain all of the information necessary to build a new individual.

DNA REPLICATION

Before a cell divides, it copies its DNA so both descendant cell's will inherit a full complement of chromosomes. Replication of each DNA molecule produces two duplicates.

MUTATIONS

DNA damage by environmental agents can cause replication errors. Newly forming DNA is monitored for errors, most of which are corrected. Uncorrected errors become mutations.

Photograph by Martin Schoeller/AUGUST.

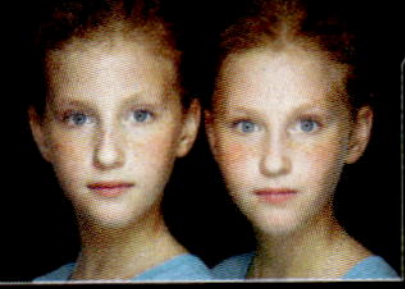

8.1 HOW WAS DNA'S FUNCTION DISCOVERED?

FIGURE 8.1 DNA extracted from human cells.

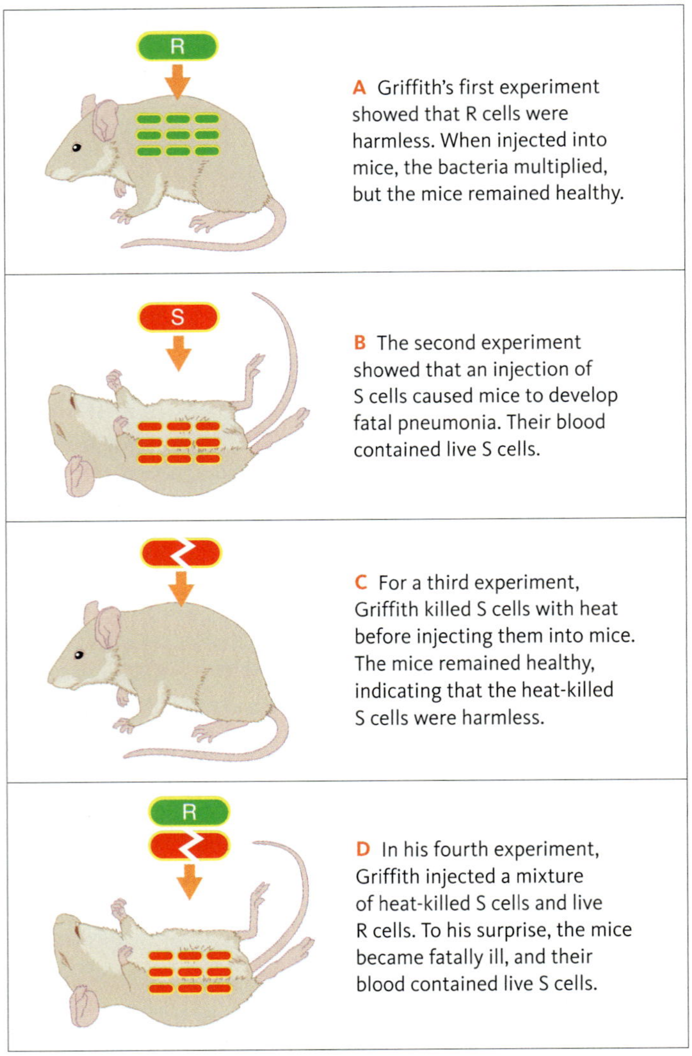

FIGURE 8.2 **{Animated}** Fred Griffith's experiments with two strains (R and S) of *Streptococcus pneumoniae* bacteria.

The substance we now call DNA (**FIGURE 8.1**) was first described in 1869 by Johannes Miescher, a chemist who extracted it from cell nuclei. Miescher determined that DNA is not a protein, and that it is rich in nitrogen and phosphorus, but he never learned its function. That would take many more years and experiments by many scientists.

Sixty years after Miescher's work, Frederick Griffith unexpectedly uncovered a clue about DNA's function. Griffith was studying pneumonia-causing bacteria in the hope of creating a vaccine. He isolated two strains (types) of the bacteria, one harmless (R), the other lethal (S). Griffith used R and S cells in a series of experiments testing their ability to cause pneumonia in mice (**FIGURE 8.2**). He discovered that heat destroyed the ability of lethal S bacteria to cause pneumonia, but it did not destroy their hereditary material, including whatever specified "kill mice." That material could be transferred from the dead S cells to the live R cells, which put it to use. The transformation was permanent and heritable: Even after hundreds of generations, descendants of transformed R cells retained the ability to kill mice.

What substance had caused the transformation? In 1940, Oswald Avery and Maclyn McCarty set out to identify that substance, which they termed the "transforming principle," by a process of elimination. The researchers made an extract of S cells that contained only lipid, protein, and nucleic acids. The S cell extract could still transform R cells after it had been treated with lipid- and protein-destroying enzymes. Thus, the transforming principle could not be lipid or protein, and Avery and McCarty realized that the substance they were seeking must be nucleic acid—DNA or RNA. DNA-degrading enzymes destroyed the extract's ability to transform cells, but RNA-degrading enzymes did not. Thus, DNA had to be the transforming principle.

The result surprised Avery and McCarty, who, along with most other scientists, had assumed that proteins were the material of heredity. After all, traits are diverse, and proteins are the most diverse of all biological molecules. The two scientists were so skeptical that they published their results only after they had convinced themselves, by years of painstaking experimentation, that DNA was indeed hereditary material. They were also careful to point out that they had not proven DNA was the *only* hereditary material.

Avery and McCarty's tantalizing results prompted a stampede of other scientists into the field of DNA research. The resulting explosion of discovery confirmed the molecule's role as carrier of hereditary information. Key in this advance was the realization that any molecule—DNA or otherwise—had to have certain properties in order to function as the sole repository of hereditary material.

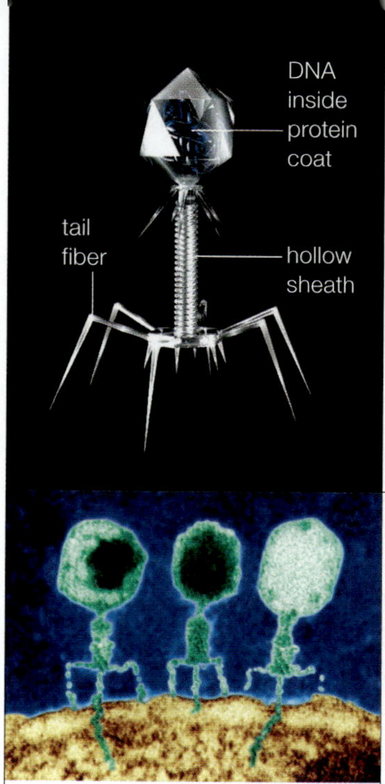

A Top, a model of a bacteriophage. Bottom, micrograph of three viruses injecting DNA into an *E. coli* cell.

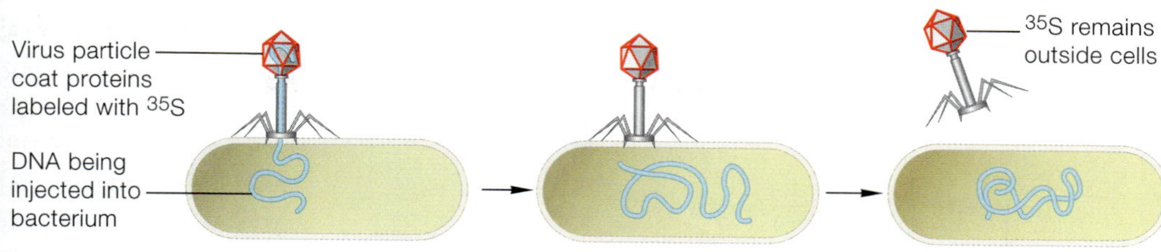

B In one experiment, bacteria were infected with virus particles that had been labeled with a radioisotope of sulfur (^{35}S). The sulfur had labeled mainly viral proteins. The viruses were dislodged from the bacteria by whirling the mixture in a kitchen blender. Most of the radioactive sulfur was detected in the viruses, not in the bacterial cells. The viruses had not injected protein into the bacteria.

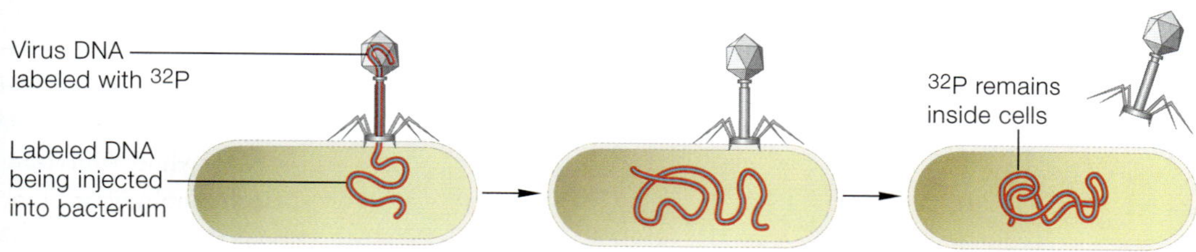

C In another experiment, bacteria were infected with virus particles that had been labeled with a radioisotope of phosphorus (^{32}P). The phosphorus had labeled mainly viral DNA. When the viruses were dislodged from the bacteria, the radioactive phosphorus was detected mainly inside the bacterial cells. The viruses had injected DNA into the cells—evidence that DNA is the genetic material of this virus.

FIGURE 8.3 {Animated} The Hershey–Chase experiments. Alfred Hershey and Martha Chase carried out experiments to determine the composition of the hereditary material that bacteriophage inject into bacteria. The experiments were based on the knowledge that proteins contain more sulfur (S) than phosphorus (P), and DNA contains more phosphorus than sulfur.

First, a full complement of hereditary information must be transmitted along with the molecule; second, each cell of a given species should contain the same amount of it; third, because the molecule functions as a genetic bridge between generations, it has to be exempt from change; and fourth, it must be capable of encoding the almost unimaginably huge amount of information required to build a new individual.

In the late 1940s, Alfred Hershey and Martha Chase proved that DNA, and not protein, satisfies the first property of a hereditary molecule: It transmits a full complement of hereditary information. Hershey and Chase specialized in working with **bacteriophage**, a type of virus that infects bacteria (**FIGURE 8.3**). Like all viruses, these infectious particles carry information about how to make new viruses in their hereditary material. After one injects a cell with this material, the cell starts making new virus particles. Hershey and Chase carried out an elegant series of experiments proving that the material bacteriophage injects into bacteria is DNA, not protein (**FIGURE 8.3B,C**).

The second property expected of a hereditary molecule was pinned on DNA by André Boivin and Roger Vendrely, who meticulously measured the amount of DNA in cell nuclei from a number of species. In 1948, they proved that body cells of any individual of a species contain precisely the same amount of DNA. Daniel Mazia's laboratory discovered that the protein and RNA content of cells varies over time, but not the DNA content, demonstrating that DNA is not involved in metabolism (and proving DNA has the third property expected of a hereditary molecule). The fourth property—that a hereditary molecule must somehow encode a huge amount of information—would be proven along with the elucidation of DNA's structure, a topic we continue in the next section.

TAKE-HOME MESSAGE 8.1

DNA, the molecule of inheritance, was first discovered in the late 1800s. Its role as the carrier of hereditary information was uncovered over many years, as scientists built upon one another's discoveries.

bacteriophage Virus that infects bacteria.

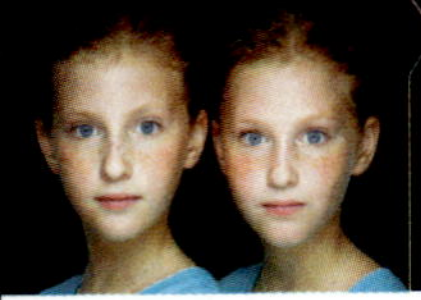

8.2 HOW WAS DNA'S STRUCTURE DISCOVERED?

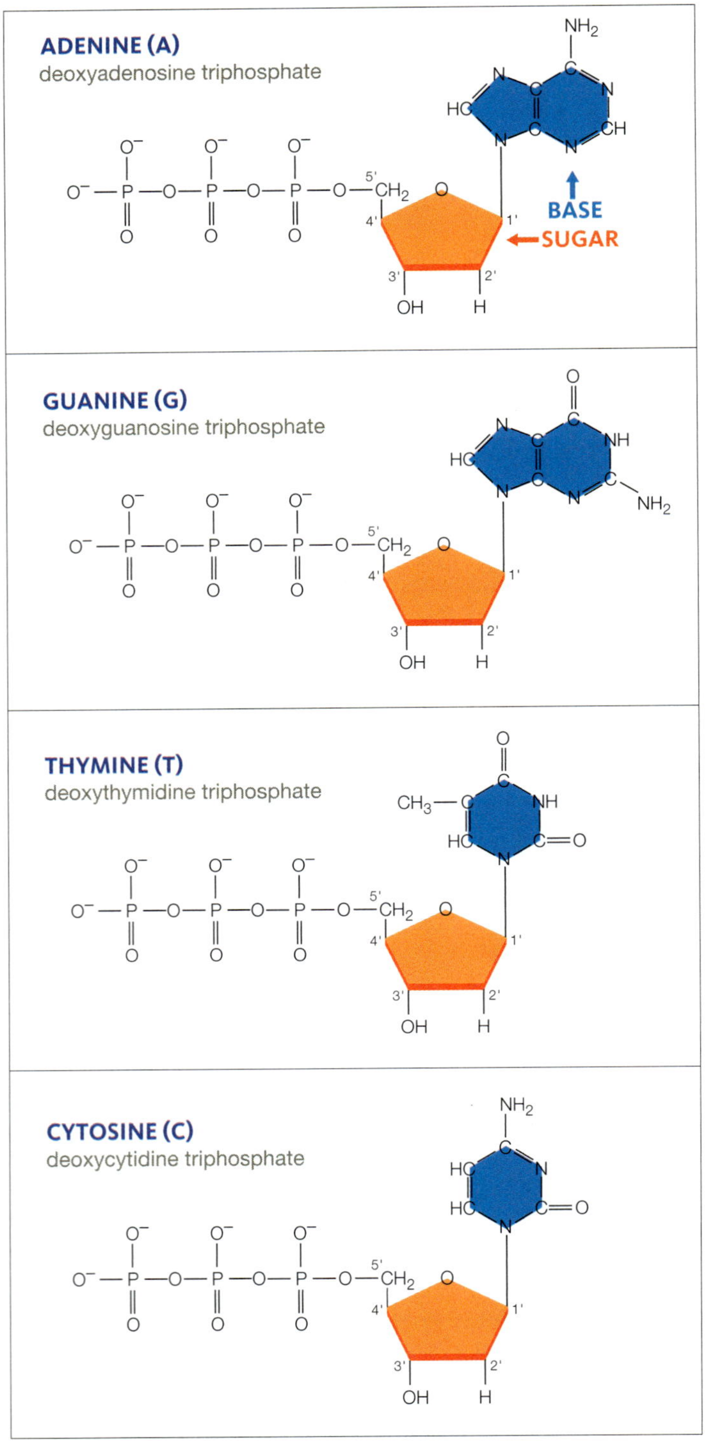

FIGURE 8.4 {Animated} The four nucleotides in DNA. All four have three phosphate groups, a deoxyribose sugar (orange), and a nitrogen-containing base (blue) after which it is named. Biochemist Phoebus Levene identified the structure of these bases and how they are connected in nucleotides in the early 1900s. Levene worked with DNA for almost 40 years.

Adenine and guanine bases are purines; thymine and cytosine, pyrimidines. Numbering the carbons in the sugars allows us to keep track of the orientation of nucleotide chains (compare **FIGURE 8.6**).

BUILDING BLOCKS OF DNA

DNA is a polymer of nucleotides, each with a five-carbon sugar, three phosphate groups, and one of four nitrogen-containing bases (**FIGURE 8.4**). Just how those four nucleotides—adenine (A), guanine (G), thymine (T), and cytosine (C)—are arranged in a DNA molecule was a puzzle that took over 50 years to solve.

Clues about DNA's structure started coming together around 1950, when Erwin Chargaff, one of many researchers investigating DNA's function, made two important discoveries about the molecule. First, the amounts of thymine and adenine are identical, as are the amounts of cytosine and guanine (A = T and G = C). We call this discovery Chargaff's first rule. Chargaff's second discovery, or rule, is that the DNA of different species differs in its proportions of adenine and guanine.

Meanwhile, American biologist James Watson and British biophysicist Francis Crick had been sharing ideas about the structure of DNA. The helical (coiled) pattern of secondary structure that occurs in many proteins (Section 3.4) had just been discovered, and Watson and Crick suspected that the DNA molecule was also a helix. The two spent many hours arguing about the size, shape, and bonding requirements of the four kinds of nucleotides that make up DNA. They pestered chemists to help them identify bonds they might have overlooked, fiddled with cardboard cutouts, and made models from scraps of metal connected by suitably angled "bonds" of wire.

Biochemist Rosalind Franklin had also been working on the structure of DNA. Like Crick, Franklin specialized in x-ray crystallography, a technique in which x-rays are directed through a purified and crystallized substance. Atoms in the substance's molecules scatter the x-rays in a pattern that can be captured as an image. Researchers can use the pattern to calculate the size, shape, and spacing between any repeating elements of the molecules—all of which are details of molecular structure.

As molecules go, DNA is gigantic, and it was difficult to crystallize given the techniques of the time. Franklin made the first clear x-ray diffraction image (*left*) of DNA as it occurs in cells. From the information in this image, she calculated that DNA is very long compared to its 2-nanometer diameter. She also identified a repeating pattern every 0.34 nanometer along its length, and another every 3.4 nanometers.

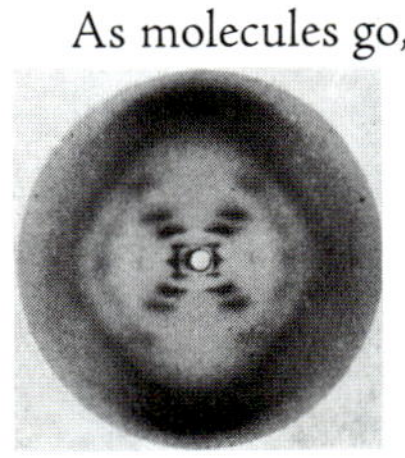

Franklin's image and data came to the attention of Watson and Crick, who now had all the information they needed to build a model of the DNA helix (**FIGURE 8.5**), one with two sugar–phosphate chains running in opposite directions, and paired bases inside (**FIGURE 8.6**). Bonds between the sugar of one nucleotide and the phosphate of the next form the backbone of each chain (or strand). Hydrogen bonds between the internally positioned bases hold the two strands together. Only two kinds of base pairings form: A to T, and G to C, which explains the first of Chargaff's rules. Most scientists had assumed (incorrectly) that the bases had to be on the outside of the helix, because they would be more accessible to DNA-copying enzymes that way. You will see in Section 8.4 how DNA replication enzymes access the bases on the inside of the double helix.

DNA'S BASE SEQUENCE

A small piece of DNA from a tulip, a human, or any other organism might be:

Notice how the two strands of DNA match. They are complementary—the base of each nucleotide on one strand pairs with a suitable partner base on the other. This base-pairing pattern (A to T, G to C) is the same in all molecules of DNA. How can just two kinds of base pairings give rise to the incredible diversity of traits we see among living things? Even though DNA is composed of only four nucleotides, the *order* in which one nucleotide follows the next along a strand—the **DNA sequence**—varies tremendously among species (which explains Chargaff's second rule). DNA molecules can be hundreds of millions of nucleotides long, so their sequence can encode a massive amount of information (we return to the nature of that information in the next chapter). DNA sequence variation is the basis of traits that define species and distinguish individuals. Thus DNA, the molecule of inheritance in every cell, is the basis of life's unity. Variations in its nucleotide sequence are the foundation of life's diversity.

DNA sequence Order of nucleotides in a strand of DNA.

TAKE-HOME MESSAGE 8.2

A DNA molecule consists of two nucleotide chains (strands) running in opposite directions and coiled into a double helix. Internally positioned nucleotide bases hydrogen-bond between the two strands. A pairs with T, and G with C.

The sequence of bases along a DNA strand varies among species and among individuals. This variation is the basis of life's diversity.

FIGURE 8.5 Watson and Crick with their model of DNA.

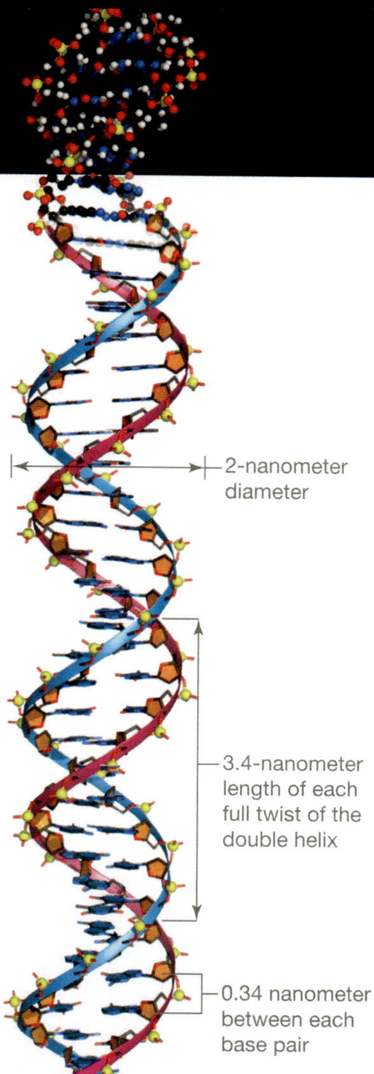

FIGURE 8.6 {Animated} Structure of DNA, as illustrated by a composite of different models. Numbering the carbons in the nucleotide sugars (see **FIGURE 8.4**) allows us to keep track of the orientation of each DNA strand. This orientation is important in DNA replication.

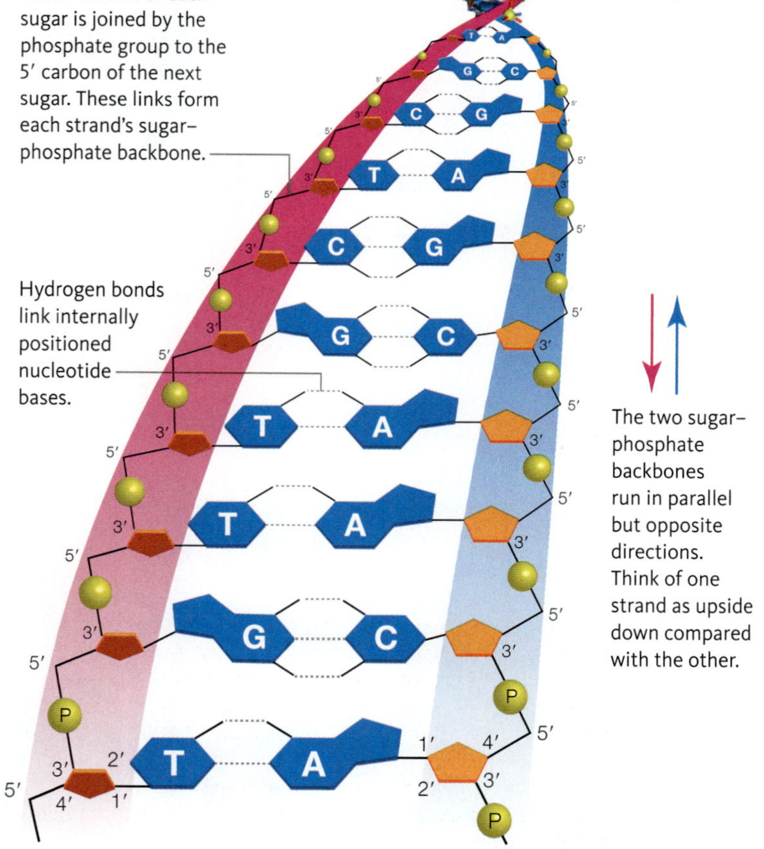

CREDITS: (in text, 6) © Cengage Learning; (5) A.C. Barrington Brown, © 1968 J. D. Watson.

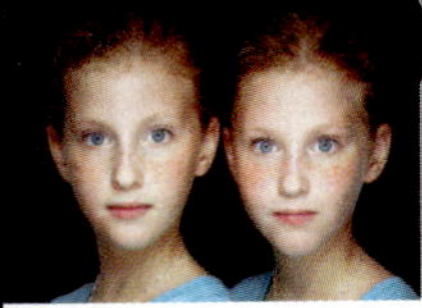

8.3 WHAT IS A CHROMOSOME?

Stretched out end to end, the DNA in a single human cell would be about 2 meters (6.5 feet) long. How can that much DNA pack into a nucleus that is less than 10 micrometers in diameter? Such tight packing is possible because proteins associate with the DNA and help keep it organized. In cells, DNA molecules and their associated proteins form structures

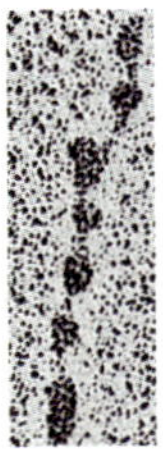

called **chromosomes** (**FIGURE 8.7**). In a chromosome of a eukaryotic cell, the DNA double helix ❶ wraps twice at regular intervals around "spools" of proteins called **histones** ❷. These DNA–histone spools, which are called **nucleosomes**, look like beads on a string in micrographs (*left*). Interactions among histones and other proteins twist the spooled DNA into a tight fiber ❸. This fiber coils, and then it coils again into a hollow cylinder, a bit like an old-style telephone cord ❹.

During most of the cell's life, each chromosome consists of one DNA molecule. When the cell prepares to divide, it duplicates its chromosomes by DNA replication (more about this process in the next section). After replication, each chromosome consists of two DNA molecules, or **sister chromatids**, attached to one another at a constricted region called the **centromere**:

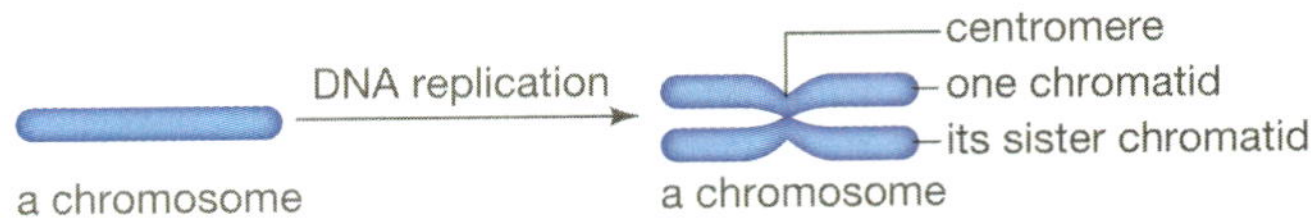

The chromosomes condense into their familiar "X" shapes ❺ just before the cell divides.

CHROMOSOME NUMBER AND TYPE

The DNA of a eukaryotic cell is divided among several chromosomes that differ in length and shape ❻. Each species has a characteristic **chromosome number**—the number of chromosomes in its cells. For example, the chromosome

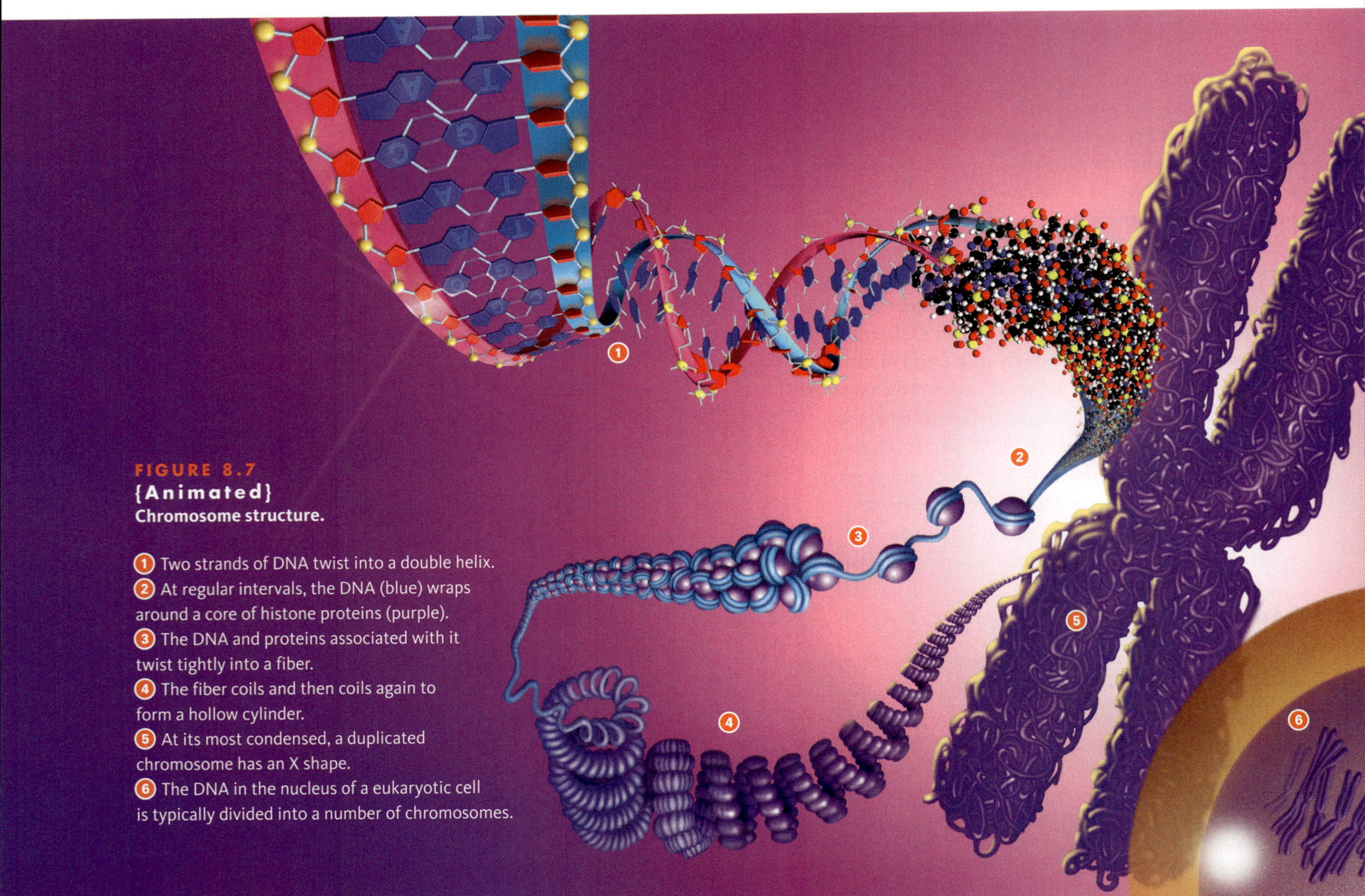

FIGURE 8.7
{Animated}
Chromosome structure.

❶ Two strands of DNA twist into a double helix.
❷ At regular intervals, the DNA (blue) wraps around a core of histone proteins (purple).
❸ The DNA and proteins associated with it twist tightly into a fiber.
❹ The fiber coils and then coils again to form a hollow cylinder.
❺ At its most condensed, a duplicated chromosome has an X shape.
❻ The DNA in the nucleus of a eukaryotic cell is typically divided into a number of chromosomes.

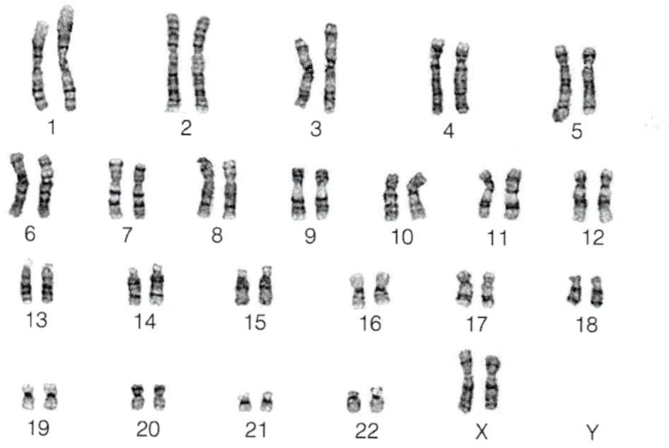

FIGURE 8.8 {Animated} A karyotype of a human female, showing 22 pairs of autosomes and a pair of X chromosomes (XX).

number of oak trees is 12, so the nucleus of a cell from an oak tree contains 12 chromosomes. The chromosome number of humans is 46, so our cells have 46 chromosomes.

Actually, human body cells have two sets of 23 chromosomes—two of each type. Having two sets of chromosomes means these cells are **diploid**, or $2n$. A **karyotype** is an image of an individual's diploid set of chromosomes (**FIGURE 8.8**). To create a karyotype, cells taken from an individual are treated to make the chromosomes condense, and then stained so the chromosomes can be distinguished under a microscope. A micrograph of a single cell is digitally rearranged so the images of the chromosomes are lined up by centromere location, and arranged according to size, shape, and length.

In a human body cell, all but one pair of chromosomes are **autosomes**, which are the same in both females and males. The two autosomes of a pair have the same length, shape, and centromere location. They also hold information about the same traits. Think of them as two sets of books on how to build a house. Your father gave you one set. Your mother had her own ideas about wiring, plumbing, and so on. She gave you an alternate set that says slightly different things about many of those tasks.

Members of a pair of **sex chromosomes** differ between females and males, and the differences determine an individual's sex. The sex chromosomes of humans are called X and Y. The body cells of typical human females have two X chromosomes (XX); those of typical human males have one X and one Y chromosome (XY). XX females and XY males are the rule among fruit flies, mammals, and many other animals, but there are other patterns. In butterflies, moths, birds, and certain fishes, males are the ones with identical sex chromosomes. Environmental factors (not sex chromosomes) determine sex in some invertebrates and reptiles (**FIGURE 8.9**).

FIGURE 8.9 National Geographic Explorer Mariana Fuentes studies sea turtle populations. Climate change is an immediate, serious threat to these reptiles, in part because the temperature of the sand in which their eggs are buried—not sex chromosomes—determines the sex of the hatchlings. Fuentes predicts that rising global temperatures will soon skew the gender ratio of hatchlings toward all female, with disastrous results for sea turtle populations.

autosome A chromosome that is the same in males and females.
centromere Of a duplicated eukaryotic chromosome, constricted region where sister chromatids attach to each other.
chromosome A structure that consists of DNA and associated proteins; carries part or all of a cell's genetic information.
chromosome number The total number of chromosomes in a cell of a given species.
diploid Having two of each type of chromosome characteristic of the species ($2n$).
histone Type of protein that structurally organizes eukaryotic chromosomes.
karyotype Image of an individual's set of chromosomes arranged by size, length, shape, and centromere location.
nucleosome A length of DNA wound twice around a spool of histone proteins.
sex chromosome Member of a pair of chromosomes that differs between males and females.
sister chromatids The two attached DNA molecules of a duplicated eukaryotic chromosome.

TAKE-HOME MESSAGE 8.3

In cells, DNA and associated proteins are organized as chromosomes.

A eukaryotic cell's DNA is divided among some characteristic number of chromosomes, which differ in length and shape.

Members of a pair of sex chromosomes differ between males and females. Chromosomes that are the same in males and females are called autosomes.

CREDITS: (8) © University of Washington Department of Pathology; (9) © Mariana Fuentes.

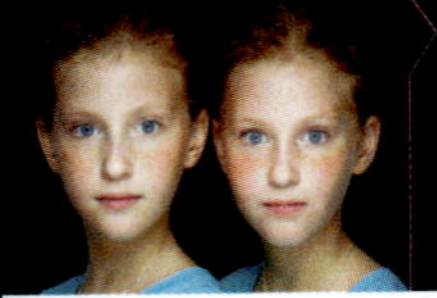

8.4 HOW DOES A CELL COPY ITS DNA?

During most of its life, a typical cell contains one set of chromosomes. When the cell reproduces, it divides. The two descendant cells must inherit a full complement of chromosomes—a complete copy of genetic information—or they will not function properly. Thus, in preparation for division, the cell copies its chromosomes so that it contains two sets: one for each of its future offspring.

The process by which a cell copies its DNA is called **DNA replication**. During this energy-intensive metabolic pathway, enzymes and other molecules open the double helix of a DNA molecule to expose the internally positioned bases, then link nucleotides into new strands of DNA according to the sequence of those bases.

Each chromosome is replicated in its entirety. Two identical molecules of DNA are the result. In eukaryotes, these molecules are sister chromatids that remain attached at the centromere until cell division occurs.

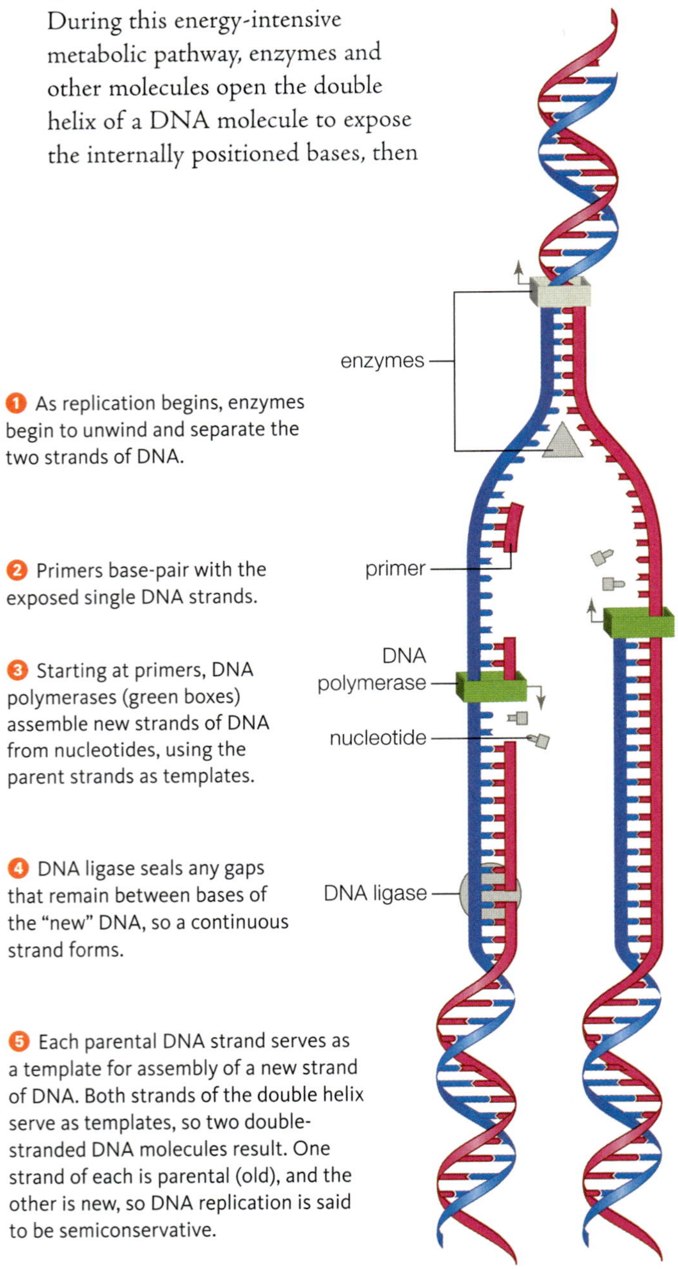

FIGURE 8.10 {Animated} DNA replication, in which a double-stranded molecule of DNA is copied in entirety. The Y-shaped structure of a DNA molecule undergoing replication is called a replication fork.

SEMICONSERVATIVE REPLICATION

Before DNA replication, a chromosome consists of one molecule of DNA—one double helix (**FIGURE 8.10**). As replication begins, enzymes break the hydrogen bonds that hold the double helix together, so the two DNA strands unwind and separate ❶. Another enzyme constructs **primers**—short, single strands of nucleotides that serve as attachment points for **DNA polymerase**, the enzyme that assembles new strands of DNA. The nucleotide bases of a primer can form hydrogen bonds with exposed bases of a single strand of DNA ❷. Thus, a primer can base-pair with a complementary strand of DNA:

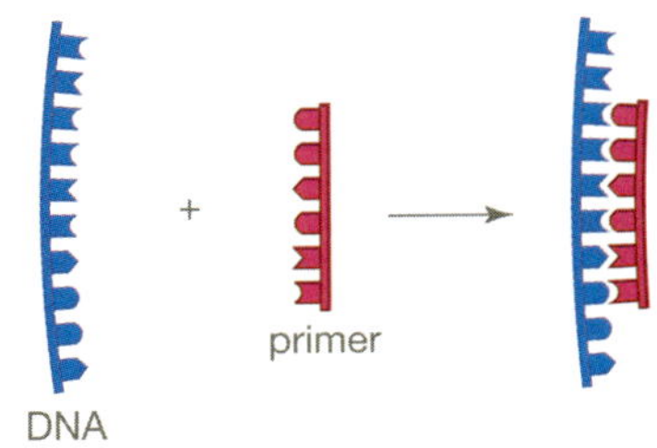

The establishment of base-pairing between two strands of DNA (or DNA and RNA) is called **nucleic acid hybridization**. Hybridization is spontaneous, driven by hydrogen bonding between bases of complementary strands.

DNA polymerases attach to the hybridized primers and begin DNA synthesis. As a DNA polymerase moves along a strand, it uses the sequence of exposed nucleotide bases as a template, or guide, to assemble a new strand of DNA from free nucleotides ❸.

Each nucleotide provides energy for its own attachment to the end of a growing strand of DNA. Remember from Section 5.5 that the bonds between a nucleotide's phosphate groups hold a lot of free energy. Two of the three phosphate groups are removed when the nucleotide is added to a DNA strand. Breaking those bonds releases enough free energy to drive the attachment.

A DNA polymerase follows base-pairing rules: It adds a T to the end of the new DNA strand when it reaches an A in the template strand; it adds a G when it reaches a C; and so on. Thus, the nucleotide sequence of each new strand of DNA is complementary to its template (parental) strand. The enzyme **DNA ligase** seals any gaps, so the new DNA strands are continuous ❹.

Both of the two strands of the parent molecule are copied at the same time. As each new DNA strand lengthens, it winds up with its template strand into a double helix. So, after replication, two double-stranded molecules of DNA have formed ❺. One strand of each molecule is parental (old), and the other is new; hence the name of the process, **semiconservative replication.** Each new strand of DNA is complementary in sequence to one of the two parent strands, so both double-stranded molecules produced by DNA replication are duplicates of the parent molecule.

DIRECTIONAL SYNTHESIS

Numbering the carbons of the sugars in nucleotides allows us to keep track of the orientation of DNA strands in a double helix (see **FIGURES 8.4** and **8.6**). Each strand has two ends. The last carbon atom on one end of the strand is a 5′ (5 prime) carbon of a sugar; the last carbon atom on the other end is a 3′ (three prime) carbon of a sugar:

DNA polymerase can attach a nucleotide only to a 3′ end. Thus, during DNA replication, only one of two new strands of DNA can be constructed in a single piece (**FIGURE 8.11**). Synthesis of the other strand occurs in segments that must be joined by DNA ligase where they meet up. This is why we say that DNA synthesis proceeds only in the 5′ to 3′ direction.

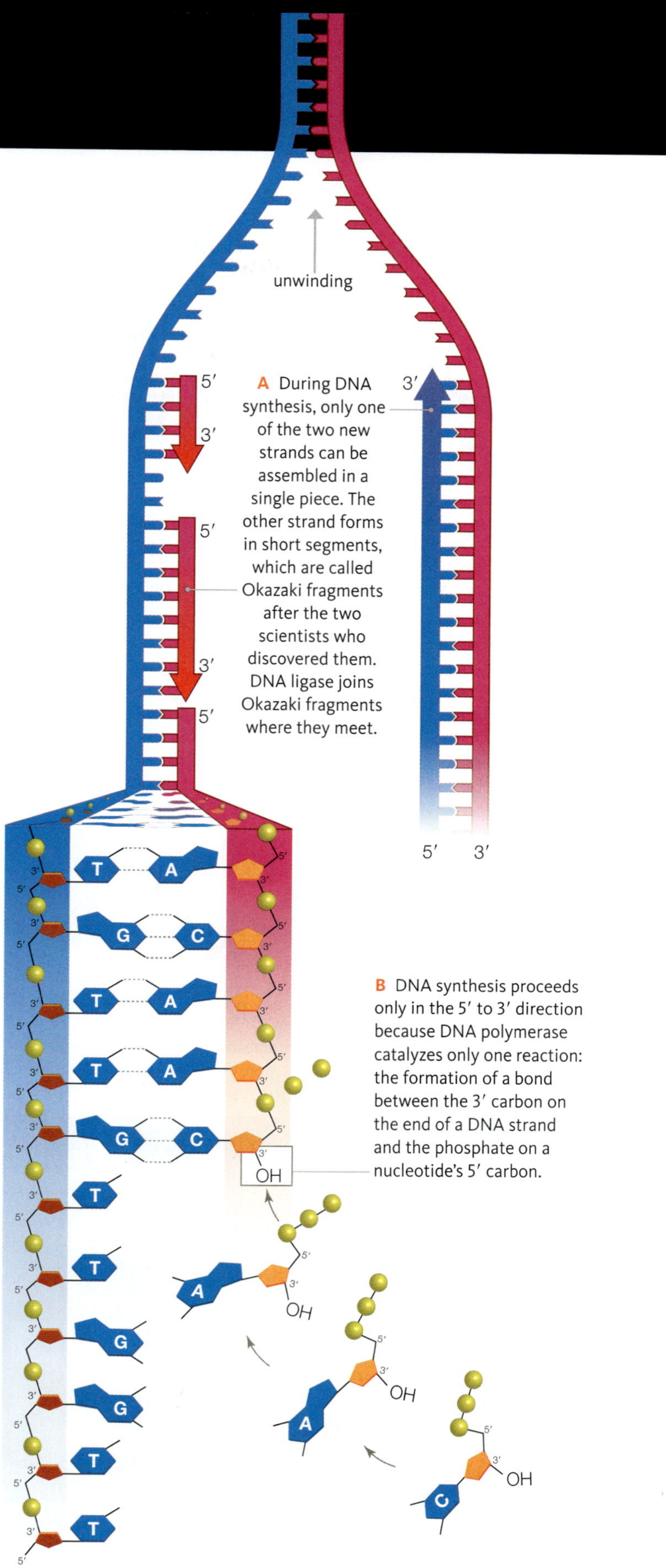

FIGURE 8.11 Discontinuous synthesis of DNA. This close-up of a replication fork shows that only one of the two new DNA strands is assembled in one piece.

DNA ligase Enzyme that seals gaps in double-stranded DNA.
DNA polymerase DNA replication enzyme. Uses one strand of DNA as a template to assemble a complementary strand of DNA from nucleotides.
DNA replication Process by which a cell duplicates its DNA before it divides.
nucleic acid hybridization Convergence of complementary nucleic acid strands. Arises because of base-pairing interactions.
primer Short, single strand of DNA that base-pairs with a targeted DNA sequence.
semiconservative replication Describes the process of DNA replication, which produces two copies of a DNA molecule: one strand of each copy is new, and the other is parental.

TAKE-HOME MESSAGE 8.4

- DNA replication is an energy-intensive metabolic pathway by which a cell copies its chromosomes.
- During replication of a molecule of DNA, each strand of its double helix serves as a template for synthesis of a new, complementary strand of DNA.
- Replication of a molecule of DNA produces two double helices that are duplicates of the parent molecule. One strand of each is parental; the other is new.

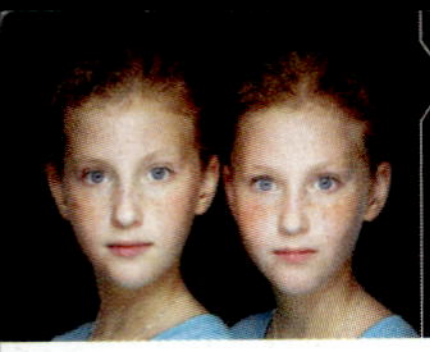

8.5 WHAT CAUSES MUTATIONS?

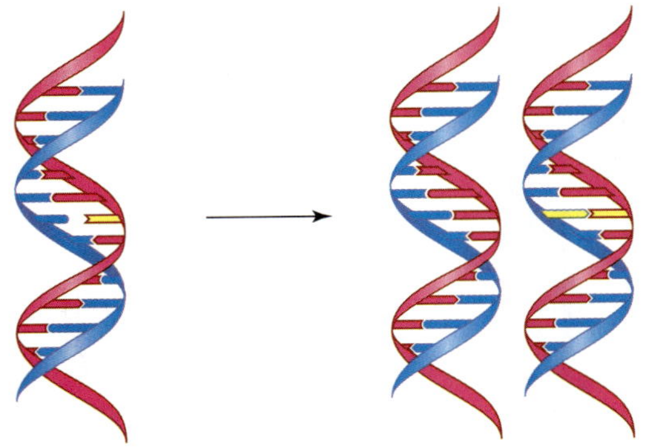

A Repair enzymes can recognize a mismatched base (yellow), but sometimes fail to correct it before DNA replication.

B After replication, both strands base-pair properly. Repair enzymes can no longer recognize the error, which has now become a mutation that will be passed on to the cell's descendants.

FIGURE 8.12 How a replication error can become a mutation.

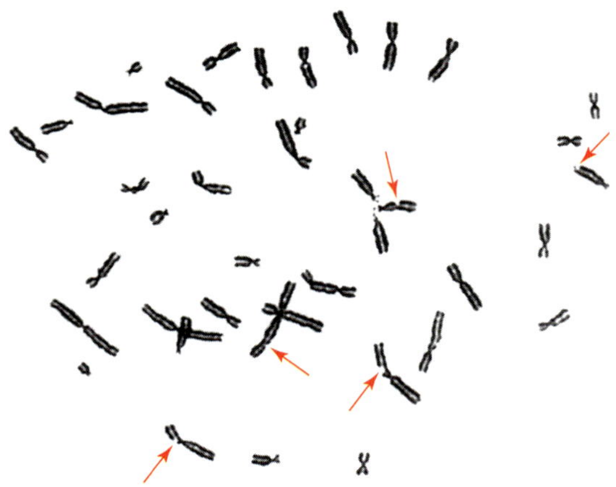

A Major breaks (red arrows) in chromosomes of a human white blood cell after exposure to ionizing radiation. Pieces of broken chromosomes often become lost during DNA replication.

B These *Ranunculus* flowers were grown from plants harvested around Chernobyl, Ukraine, where in 1986 an accident at a nuclear power plant released huge amounts of radiation. A normal flower is shown for comparison, in the inset.

FIGURE 8.13 Exposure to ionizing radiation causes mutations.

REPLICATION ERRORS

Mistakes can and do occur during DNA replication. Sometimes, the wrong base is added to a growing DNA strand; at other times, a nucleotide gets lost, or an extra one slips in. Either way, the newly synthesized DNA strand will no longer be complementary to its parent strand. Most of these replication errors occur simply because DNA polymerases work very fast, copying about 50 nucleotides per second in eukaryotes, and up to 1,000 per second in bacteria. Mistakes are inevitable, and some types of DNA polymerases make a lot of them. Luckily, most DNA polymerases also proofread their work. They can correct a mismatch by reversing the synthesis reaction to remove the mispaired nucleotide, then resuming synthesis in the forward direction.

Replication errors also occur after the cell's DNA gets broken or otherwise damaged, because DNA polymerases do not copy damaged DNA very well. In most cases, repair enzymes and other proteins remove and replace damaged or mismatched bases in DNA before replication begins.

When proofreading and repair mechanisms fail, an error becomes a **mutation**, a permanent change in the DNA sequence of a cell's chromosome(s). Repair enzymes cannot fix a mutation after DNA replication has occurred, because they do not recognize correctly paired bases (**FIGURE 8.12**). Thus, a mutation is passed to the cell's descendants, their descendants, and so on.

Mutations can form in any type of cell. Those that occur during egg or sperm formation can be passed to offspring, and in fact each human child is born with an average of 36 new ones. Mutations that alter DNA's instructions may have a harmful or lethal outcome; most cancers begin with them (we return to this topic in Section 11.5). However, not all mutations are dangerous: As you will see in Chapter 17, they give rise to the variation in traits that is the raw material of evolution.

AGENTS OF DNA DAMAGE

Electromagnetic energy with a wavelength shorter than 320 nanometers, including x-rays, most ultraviolet (UV) light, and gamma rays, can knock electrons out of atoms. Such ionizing radiation damages DNA, breaking it into pieces that get lost during replication (**FIGURE 8.13A**). Ionizing radiation can also cause covalent bonds to form between bases on opposite strands of the double helix, an outcome that permanently blocks replication. (We consider cancer-causing effects of such cell cycle interruptions in Chapter 11.) High-energy radiation also fatally alters nucleotide bases. Repair enzymes can remove bases damaged in this way, but they leave an empty space in the double helix or even a strand break. Sometimes the enzymes cut out the entire nucleotide from the strand, leaving an unpaired nucleotide on the opposite strand. Any of these events can result in mutations (**FIGURE 8.13B**).

UV light in the range of 320–380 nanometers can boost electrons to a higher energy level, but not enough to knock them out of atoms. UV light in this range is still dangerous, because it has enough energy to open up the double bond in the ring of a cytidine or thymine base. The open ring can form a covalent bond with the ring of an adjacent cytidine or thymine (*left*). The resulting dimer kinks the DNA strand. DNA polymerase tends to copy the kinked part incorrectly during replication, and mutations are the outcome. Mutations that arise as a result of nucleotide dimers are the primary cause of skin cancer. Exposing unprotected skin to sunlight increases the risk of cancer because its UV wavelengths cause dimers to form. For every second a skin cell spends in the sun, 50–100 of these dimers form in its DNA.

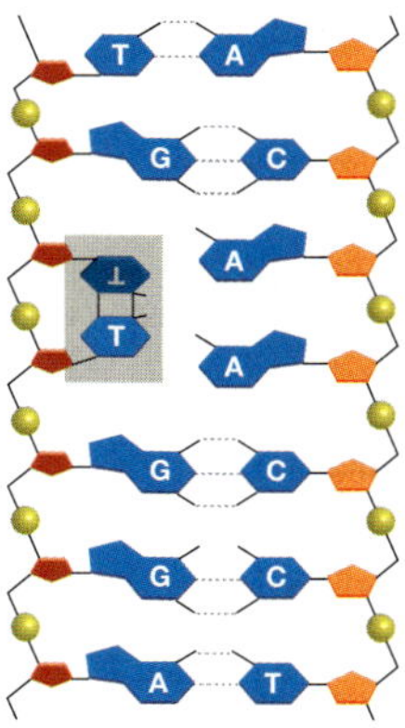

a thymine dimer

Exposure to some natural or synthetic chemicals also causes mutations. For instance, several of the fifty-five or more cancer-causing chemicals in tobacco smoke transfer methyl groups ($—CH_3$) to the nucleotide bases in DNA. Nucleotides altered in this way do not base-pair correctly. Other chemicals in the smoke are converted by the body to compounds that are easier to excrete, and the breakdown products bind irreversibly to DNA. Replication errors that can lead to mutation may be the outcome in both cases. Cigarette smoke also contains free radicals, which inflict the same damage on DNA as ionizing radiation.

mutation Permanent change in the nucleotide sequence of DNA.

TAKE-HOME MESSAGE 8.5

Proofreading and repair mechanisms usually maintain the integrity of a cell's genetic information by correcting mispaired bases and fixing damaged DNA before replication.

Mismatched or damaged nucleotides that are not repaired can become mutations—permanent changes in the DNA sequence of a chromosome.

DNA damage by environmental agents such as UV light and chemicals can result in mutations, because damaged DNA is not replicated very well.

PEOPLE MATTER

DR. ROSALIND FRANKLIN

Rosalind Franklin had been told she would be the only one in her department working on the structure of DNA, so she did not know that Maurice Wilkins was already doing the same thing just down the hall. Franklin's meticulous work yielded the first clear x-ray diffraction image of DNA as it occurs inside cells, and she gave a presentation on this work in 1952. DNA, she said, had two chains twisted into a double helix, with a backbone of phosphate groups on the outside, and bases arranged in an as-yet unknown way on the inside. She had calculated DNA's diameter, the distance between its chains and between its bases, the angle of the helix, and the number of bases in each coil. Francis Crick, with his crystallography background, would have recognized the significance of the work—if he had been there. James Watson was in the audience but he did not fully understand the implications of Franklin's x-ray diffraction image or her calculations.

Franklin started to write a research paper on her findings. Meanwhile, and perhaps without her knowledge, Watson reviewed Franklin's x-ray diffraction image with Wilkins, and Watson and Crick read Franklin's unpublished data. That data provided Watson and Crick with the last piece of the DNA puzzle. In 1953, they put together all of the clues that had been accumulating for fifty years and built the first accurate model of DNA structure. On April 25, 1953, Rosalind Franklin's work appeared third in a series of articles about the structure of DNA in the journal *Nature*. Wilkins's research paper was the second article in the series. The work of Franklin and Wilkins supported with experimental evidence Watson and Crick's theoretical model, which was presented in the first article.

Rosalind Franklin died in 1958 at the age of 37, of ovarian cancer probably caused by extensive exposure to x-rays during her work. At the time, the link between x-rays, mutations, and cancer was not understood. Because the Nobel Prize is not given posthumously, Franklin did not share in the 1962 honor that went to Watson, Crick, and Wilkins for the discovery of the structure of DNA.

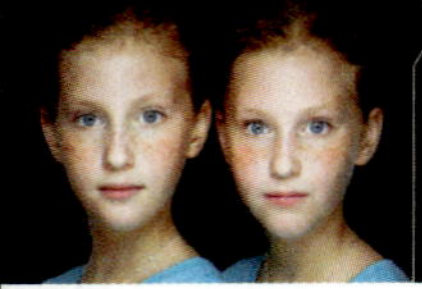

8.6 HOW DOES CLONING WORK?

The word "cloning" means making an identical copy of something, and it can refer to deliberate interventions in reproduction that produce an exact genetic copy of an organism. Genetically identical organisms occur all the time in nature, arising most often by the process of asexual reproduction (which we discuss in Chapter 11). Embryo splitting, another natural process, results in identical twins. The first few divisions of a fertilized egg form a ball of cells that sometimes splits spontaneously. If both halves of the ball continue to develop independently, identical twins result.

Artificial embryo splitting has been used in research and animal husbandry for decades. With this technique, a ball of cells is grown from a fertilized egg in a laboratory. The tiny ball is teased apart into two halves, each of which goes on to develop as a separate embryo. The embryos are implanted in surrogate mothers, who give birth to identical twins. Artificial twinning and any other technology that yields genetically identical individuals is called **reproductive cloning**.

Twins get their DNA from two parents that typically differ in their DNA sequence. Thus, although twins produced by embryo splitting are identical to one another, they are not identical to either parent. Animal breeders who want an exact copy of a specific individual may turn to a cloning method that starts with a somatic cell taken from an adult organism (a somatic cell is a body cell, as opposed to a reproductive cell; *soma* is a Greek word for body). All cells descended from a fertilized egg inherit the same DNA. Thus, the DNA in each living cell of an individual is like a master blueprint that contains enough information to build an entirely new individual. However, a somatic cell taken from an adult will not automatically start dividing to produce an embryo. It must first be tricked into rewinding its developmental clock. During development, cells in an embryo start using different subsets of their DNA. As they do, the cells become different in form and function, a process called **differentiation**. Differentiation is usually a one-way path in animal cells. Once a cell has become specialized, all of its descendant cells will be specialized the same way. By the time a liver cell, muscle cell, or other differentiated cell forms, most of its DNA has been turned off, and is no longer used. To clone an adult, scientists transform one of its differentiated cells into an undifferentiated cell by turning its unused DNA back on. One way to do this is **somatic cell nuclear transfer (SCNT)**, a laboratory procedure in which an unfertilized egg's nucleus is replaced with the nucleus of a donor's somatic cell (**FIGURE 8.14**). If all goes well, the egg's cytoplasm reprograms the transplanted DNA to direct the development of an embryo, which is then implanted into a surrogate mother. The animal that is born to the surrogate is genetically identical with the donor of the nucleus—a clone.

SCNT is now a common practice among people who breed prized livestock. Among other benefits, many more offspring can be produced in a given time frame by cloning than by traditional breeding methods. Cloned animals have the same championship features as their DNA donors

A A cow's egg is held in place by suction through a hollow glass tube called a micropipette. DNA is identified by a purple stain.

B Another micropipette punctures the egg and sucks out the DNA. All that remains inside the egg's plasma membrane is cytoplasm.

C A new micropipette prepares to enter the egg at the puncture site. The pipette contains a cell grown from the skin of a donor animal.

D The micropipette enters the egg and delivers the skin cell to a region between the cytoplasm and the plasma membrane.

E After the pipette is withdrawn, the donor's skin cell is visible next to the cytoplasm of the egg. The transfer is now complete.

F An electric current causes the foreign cell to fuse with and deposit its nucleus into the cytoplasm of the egg. The egg begins to divide, and an embryo forms.

FIGURE 8.14 {Animated} An example of somatic cell nuclear transfer, using cattle cells. This series of micrographs was taken at a company that specializes in cloning livestock.

CREDIT: (14) Courtesy of Cyagra, Inc., www.cyagra.com.

FIGURE 8.15 Champion Holstein dairy cow (*right*) and her clone (*left*), who was produced by somatic cell nuclear transfer in 2003.

(**FIGURE 8.15**). Offspring can also be produced from a donor animal that is castrated or even dead.

As the techniques become routine, cloning humans is no longer only within the realm of science fiction. SCNT is already being used to produce human embryos for medical purposes, a practice called **therapeutic cloning**. Undifferentiated (stem) cells taken from the cloned human embryos are used to treat human patients and to study human diseases. For example, embryos created using cells from people with genetic heart defects are allowing researchers to study how the defect causes developing heart cells to malfunction. Such research may ultimately lead to treatments for people who suffer from fatal diseases. Human cloning is not the intent of such research, but if it were, SCNT would indeed be the first step toward that end.

differentiation Process by which cells become specialized during development; occurs as different cells in an embryo begin to use different subsets of their DNA.
reproductive cloning Technology that produces genetically identical individuals.
somatic cell nuclear transfer (SCNT) Reproductive cloning method in which the DNA of an adult donor's body cell is transferred into an unfertilized egg.
therapeutic cloning The use of SCNT to produce human embryos for research purposes.

TAKE-HOME MESSAGE 8.6

Reproductive cloning technologies produce genetically identical individuals.

The DNA inside a living cell contains all the information necessary to build a new individual.

In somatic cell nuclear transfer (SCNT), the DNA of an adult donor's body cell is transferred to an egg with no nucleus. The hybrid cell may develop into an embryo that is genetically identical to the donor's.

8.7 A HERO DOG'S GOLDEN CLONES

FIGURE 8.16 James Symington and his dog Trakr at Ground Zero, 9/11/2001.

WHY CLONE ANIMALS? Consider the story of Canadian police officer James Symington and his search dog Trakr. On September 11, 2001, Symington drove Trakr from Nova Scotia to Manhattan. Within hours of arriving, the dog led rescuers to the area where the final survivor of the World Trade Center attacks was buried. She had been clinging to life, pinned under rubble from the building where she had worked. Symington and Trakr helped with the search and rescue efforts for three days nonstop, until Trakr collapsed from smoke and chemical inhalation, burns, and exhaustion (**FIGURE 8.16**).

Trakr survived the ordeal, but later lost the use of his limbs, probably because of toxic smoke exposure at Ground Zero. The hero dog died in April 2009, but his DNA lives on—in his clones. Symington's essay about Trakr's superior nature and abilities as a search and rescue dog won the Golden Clone Giveaway, a contest to find the world's most clone-worthy dog. Trakr's DNA was inserted into donor dog eggs, which were then implanted into surrogate mother dogs. Five puppies, all clones of Trakr, were delivered to Symington in July 2009. Today, the clones are search and rescue dogs for Team Trakr Foundation, Symington's international humanitarian organization.

Cloning animals raises uncomfortable ethical questions about cloning humans. For example, if cloning a lost animal for a grieving owner is acceptable, why would it not be acceptable to clone a lost child for a grieving parent? Different people have very different answers to such questions, so controversy over cloning continues to rage even as techniques improve.

Summary

SECTION 8.1 Eighty years of experimentation with cells and **bacteriophage** offered solid evidence that deoxyribonucleic acid (DNA), not protein, is the hereditary material of all life.

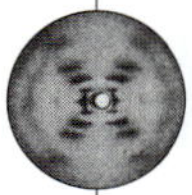

SECTION 8.2 A DNA nucleotide has a five-carbon sugar (deoxyribose), three phosphate groups, and one of four nitrogen-containing bases after which the nucleotide is named: adenine, thymine, guanine, or cytosine. DNA is a polymer that consists of two strands of these nucleotides coiled into a double helix. Hydrogen bonding between the internally positioned bases holds the strands together. The bases pair in a consistent way: adenine with thymine (A–T), and guanine with cytosine (G–C). The order of bases along a strand of DNA—the **DNA sequence**—varies among species and among individuals, and this variation is the basis of life's diversity.

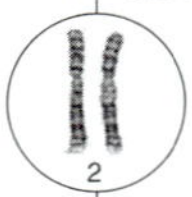

SECTION 8.3 The DNA of eukaryotes is typically divided among a number of **chromosomes** that differ in length and shape. In eukaryotic chromosomes, the DNA wraps around **histone** proteins to form **nucleosomes.** When duplicated, a eukaryotic chromosome consists of two **sister chromatids** attached at a **centromere. Diploid** cells have two of each type of chromosome. **Chromosome number** is the total number of chromosomes in a cell of a given species. A human body cell has twenty-three pairs of chromosomes. Members of a pair of **sex chromosomes** differ among males and females. Chromosomes that are the same in males and females are **autosomes.** Autosomes of a pair have the same length, shape, and centromere location. A **karyotype** is an individual's complete set of chromosomes.

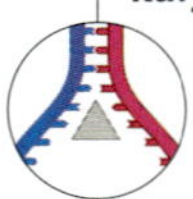

SECTION 8.4 A cell copies its chromosomes before it divides so each of its offspring will inherit a complete set of genetic information. **DNA replication** is the energy-intensive metabolic pathway in which a cell copies its chromosomes. For each double-stranded molecule of DNA that is copied, two double-stranded DNA molecules that are duplicates of the parent are produced. One strand of each molecule is new, and the other is parental; thus the name **semiconservative replication.** During DNA replication, enzymes unwind the double helix. **Primers** base-pair with the exposed single strands of DNA, a process called **nucleic acid hybridization.** Starting at the primers, **DNA polymerase** enzymes use each strand as a template to assemble new, complementary strands of DNA from free nucleotides. Synthesis of one strand necessarily occurs discontinuously. **DNA ligase** seals any gaps to form continuous strands.

SECTION 8.5 Proofreading by DNA polymerases corrects most DNA replication errors as they occur. DNA damage by environmental agents, including ionizing and nonionizing radiation, free radicals, and some other natural and synthetic chemicals, can lead to replication errors because DNA polymerase does not copy damaged DNA very well. Most types of DNA damage can be repaired before replication begins. Uncorrected replication errors become **mutations,** which are permanent changes in the nucleotide sequence of a cell's DNA. Cancer begins with mutations, but not all mutations are harmful.

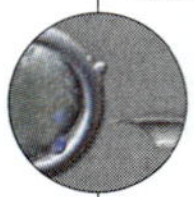

SECTIONS 8.6, 8.7 **Somatic cell nuclear transfer (SCNT)** and other types of **reproductive cloning** technologies produce genetically identical individuals (clones). SCNT using human cells is called **therapeutic cloning.** The DNA in each living cell contains all the information necessary to build a new individual. During development, cells of an embryo become specialized as they begin to use different subsets of their DNA (a process called **differentiation**).

Self-Quiz Answers in Appendix VII

1. Which is not a nucleotide base in DNA?
   a. adenine c. glutamine e. cytosine
   b. guanine d. thymine f. All are in DNA.

2. What are the base-pairing rules for DNA?
   a. A–G, T–C b. A–C, T–G c. A–T, G–C

3. Variation in _______ is the basis of variation in traits.
   a. karyotype c. the double helix
   b. the DNA sequence d. chromosome number

4. One species' DNA differs from others in its _______ .
   a. nucleotides c. sugar–phosphate backbone
   b. DNA sequence d. all of the above

5. In eukaryotic chromosomes, DNA wraps around _______ .
   a. histone proteins c. centromeres
   b. nucleosomes d. none of the above

6. Chromosome number _______ .
   a. refers to a particular chromosome in a cell
   b. is an identifiable feature of a species
   c. is the number of autosomes in cells of a given type

7. Human body cells are diploid, which means _______ .
   a. they are complete
   b. they have two sets of chromosomes
   c. they contain sex chromosomes

8. When DNA replication begins, _______ .
   a. the two DNA strands unwind from each other
   b. the two DNA strands condense for base transfers
   c. old strands move to find new strands

9. DNA replication requires _______ .
   a. DNA polymerase c. primers
   b. nucleotides d. all are required

10. Energy that drives DNA synthesis comes from _______ .
   a. ATP only c. DNA nucleotides
   b. DNA polymerase d. a and c

Data Analysis Activities

Hershey-Chase Experiments The graph in **FIGURE 8.17** is reproduced from Hershey and Chase's original publication. The data are from the two experiments described in Section 8.1, in which bacteriophage DNA and protein were labeled with radioactive tracers and allowed to infect bacteria. The virus–bacteria mixtures were whirled in a blender to dislodge the viruses, and the tracers were tracked inside and outside of the bacteria.

1. Before blending, what percentage of each isotope, ^{35}S and ^{32}P, was outside the bacteria?
2. After 4 minutes in the blender, what percentage of each isotope was outside the bacteria?
3. How did the researchers know that the radioisotopes in the fluid came from outside of the bacterial cells (extracellular) and not from bacteria that had been broken apart by whirling in the blender?
4. The extracellular concentration of which isotope increased the most with blending?
5. Do these results imply that viruses inject DNA or protein into bacteria? Why or why not?

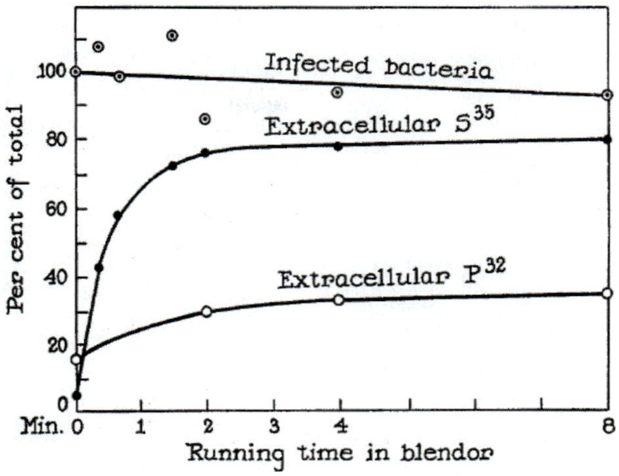

FIGURE 8.17 Detail of Alfred Hershey and Martha Chase's 1952 publication describing their experiments with bacteriophage. "Infected bacteria" refers to the percentage of bacteria that survived the blender.

11. The phrase "5′ to 3′" refers to the _______ .
    a. timing of DNA replication
    b. directionality of DNA synthesis
    c. number of phosphate groups

12. After DNA replication, a eukaryotic chromosome ______ .
    a. consists of two sister chromatids
    b has a characteristic X shape
    c. is constricted at the centromere
    d. all of the above

13. All mutations _______ .
    a. cause cancer
    b. lead to evolution
    c. are caused by radiation
    d. change the DNA sequence

14. _______ is an example of reproductive cloning.
    a. Somatic cell nuclear transfer (SCNT)
    b. Multiple offspring from the same pregnancy
    c. Artificial embryo splitting
    d. a and c

15. Match the terms appropriately.

___ bacteriophage	a. nitrogen-containing base, sugar, phosphate group(s)
___ clone	b. copy of an organism
___ nucleotide	c. does not determine sex
___ diploid	d. only DNA and protein
___ DNA ligase	e. seals breaks in a DNA strand
___ DNA polymerase	f. can cause cancer
___ autosome	g. two chromosomes of each type
___ mutation	h. adds nucleotides to a growing DNA strand

Critical Thinking

1. Show the complementary strand of DNA that forms on this template DNA fragment during replication:

5′—GGTTTCTTCAAGAGA—3′

2. Woolly mammoths have been extinct for about 10,000 years, but we often find their well-preserved remains in Siberian permafrost. Research groups are now planning to use SCNT to resurrect these huge elephant-like mammals. No mammoth eggs have been recovered so far, so elephant eggs would be used instead. An elephant would also be the surrogate mother for the resulting embryo. The researchers may try a modified SCNT technique used to clone a mouse that had been dead and frozen for sixteen years. Ice crystals that form during freezing break up cell membranes, so cells from the frozen mouse were in bad shape. Their DNA was transferred into donor mouse eggs, and cells from the resulting embryos were fused with mouse stem cells. Four healthy clones were born from the hybrid embryos. What are some of the pros and cons of cloning an extinct animal?

3. Xeroderma pigmentosum is an inherited disorder characterized by rapid formation of skin sores that develop into cancers. All forms of radiation trigger these symptoms, including fluorescent light, which contains UV light in the range of 320–400 nm. What normal function has been compromised in affected individuals?

CREDIT: (17) *Journal of General Physiology*, 36(1), Sept. 20, 1952.

The hairless appearance of a sphynx cat arises from a single base-pair mutation in its DNA. The change results in an altered form of the keratin protein that makes up cat fur.

9

FROM DNA TO PROTEIN

Links to Earlier Concepts

Your knowledge of base pairing (Section 8.2) and chromosomes (8.3) will help you understand how cells use nucleic acids (3.6) to build proteins (3.4). You will revisit cell structure, including membrane proteins (4.3), the nucleus (4.5) and endomembrane system (4.6); as well as concepts of hydrophobicity (2.4), pathogenic bacteria (4.12), cofactors (5.5), enzyme function (5.3), DNA replication (8.4), and mutation (8.5).

KEY CONCEPTS

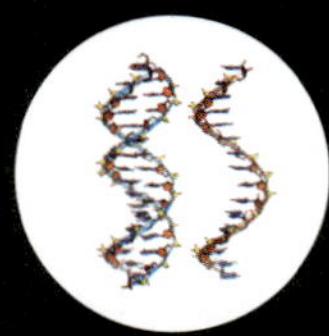

GENE EXPRESSION

The information encoded in DNA occurs in subsets called genes. The conversion of genetic information to a protein product occurs in two steps: transcription and translation.

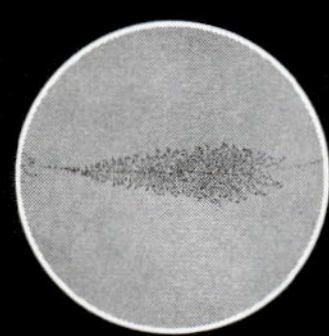

DNA TO RNA: TRANSCRIPTION

During transcription, a gene region in one strand of DNA serves as a template for assembling a strand of RNA. In eukaryotes, a new RNA is modified before leaving the nucleus.

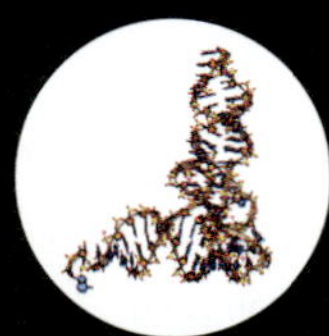

RNA

A messenger RNA carries a gene's protein-building instructions as a string of three-nucleotide codons. Transfer RNA and ribosomal RNA translate those instructions into a protein.

RNA TO PROTEIN: TRANSLATION

During translation, amino acids are assembled into a polypeptide in the order determined by the sequence of codons in an mRNA.

ALTERED PROTEINS

Mutations that change a gene's DNA sequence alter the instructions it encodes. A protein built using altered instructions may function improperly or not at all.

Photograph by Glennis Siverson, National Geographic Creative.

9.1 WHAT IS THE INFORMATION CARRIED IN DNA?

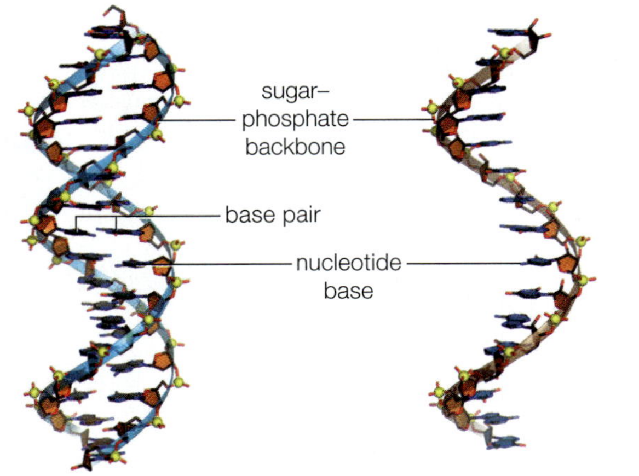

A DNA has one function: It permanently stores a cell's genetic information, which guides cell function and is passed to offspring.

B RNAs have various functions. Some serve as disposable copies of DNA's genetic message; others are catalytic. Still others have roles in gene control.

FIGURE 9.1 Comparing structure and function of DNA and RNA.

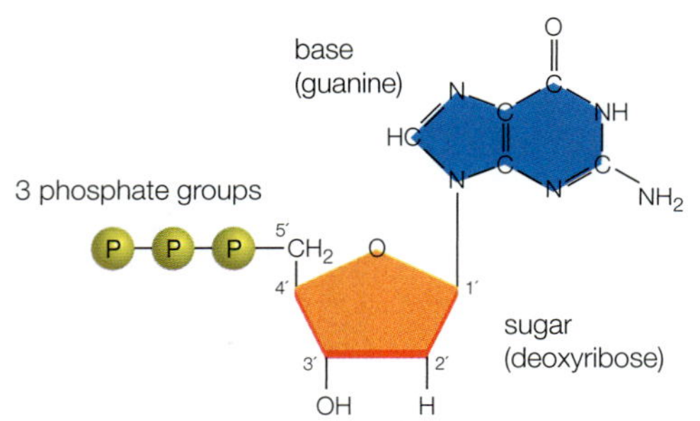

A **The DNA nucleotide guanine (G)**, or deoxyguanosine triphosphate, one of the four nucleotides in DNA. The other nucleotides—adenine, uracil, and cytosine—differ only in their component bases (blue). Three of the four bases in RNA nucleotides are identical to the bases in DNA nucleotides.

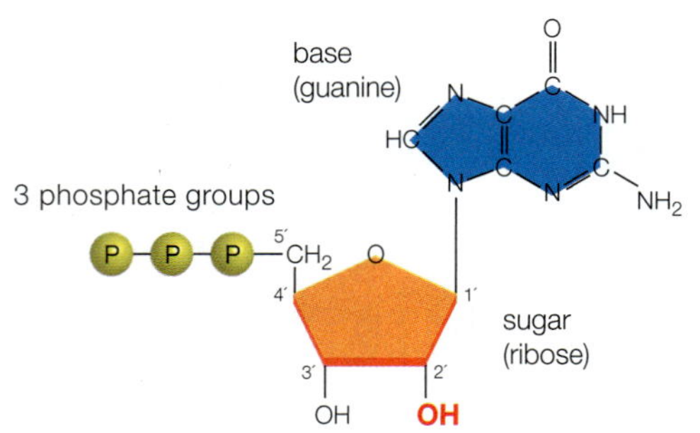

B **The RNA nucleotide guanine (G)**, or guanosine triphosphate. The only difference between the DNA and RNA versions of guanine (or adenine, or cytosine) is that RNA has a hydroxyl group (shown in red) at the 2′ carbon of the sugar.

FIGURE 9.2 Comparing nucleotides of DNA and RNA.

You learned in Chapter 8 that an individual's chromosomes are like a set of books that provide building and operating instructions. You already know the alphabet used to write that book: the four letters A, T, G, and C, for the four nucleotides in DNA: adenine, thymine, guanine, and cytosine. In this chapter, we investigate the nature of information represented by the sequence of nucleotides in DNA, and how a cell uses that information.

DNA TO RNA

Information encoded within a chromosome's DNA sequence occurs in hundreds or thousands of units called genes. The DNA sequence of a **gene** encodes (contains instructions for building) an RNA or protein product. Converting the information encoded by a gene into a product starts with RNA synthesis, or transcription. During **transcription**, enzymes use the gene's DNA sequence as a template to assemble a strand of RNA:

$$\text{DNA} \xrightarrow{\text{transcription}} \text{RNA}$$

Most of the RNA inside cells occurs as a single strand that is similar in structure to a single strand of DNA (**FIGURE 9.1**). Both RNA and DNA are chains of nucleotides. Like a DNA nucleotide, an RNA nucleotide has three phosphate groups, a sugar, and one of four bases. However, the sugar in an RNA nucleotide is a ribose, which differs just a bit from deoxyribose, the sugar in a DNA nucleotide (**FIGURE 9.2**). Three bases (adenine, cytosine, and guanine) occur in both RNA and DNA nucleotides, but the fourth base differs. In DNA, the fourth base is thymine (T); in RNA, it is uracil (U).

DNA's important but only role is to store a cell's genetic information. By contrast, a cell makes several kinds of RNAs, each with a different function. Three types of RNA have roles in protein synthesis. **Ribosomal RNA (rRNA)** is the main component of ribosomes (Section 4.4), which assemble amino acids into polypeptide chains (Section 3.4). **Transfer RNA (tRNA)** delivers the amino acids to ribosomes, one by one, in the order specified by a **messenger RNA (mRNA)**.

RNA TO PROTEIN

Messenger RNA was named for its function as the "messenger" between DNA and protein. An mRNA's protein-building message is encoded by sets of three nucleotides, "genetic words" that occur one after another along its length. Like the words of a sentence, a series of these genetic words can form a meaningful parcel of information—in this case, the sequence of amino acids of a protein.

National Geographic Grantee
DR. JOHN "JACK" HORNER

Excavating a *Tyrannosaurus rex* proved even more exciting than legendary paleontologist Jack Horner and his colleagues had anticipated when they discovered branching blood vessels and bone matrix inside its thigh bone. The team had never expected to find unfossilized tissues in the 68-million-year-old remains, because the molecules of life tend to break down relatively quickly. The tightly wound, durable structure of collagen (the main protein component of bone) may hold the key to the seemingly inexplicable preservation of the ancient tissue.

Fragments of collagen protein isolated from the tissue have a primary structure very similar to that of chicken bone collagen, providing the first molecular support for the hypothesis that modern birds are descended from dinosaurs. Until this discovery, the dinosaur–bird connection had been entirely based on physical similarities in fossils' body structures.

Researchers often compare DNA sequences to investigate evolutionary relationships, but no one has found DNA in such an ancient fossil. The sequence of amino acids in a protein is encoded by a gene, so protein similarities can also be used as evidence of hereditary connection. "If we spend time getting as deep into the sediment as we can, I think we're going to find that many specimens are like this," Horner said.

By the process of **translation**, the protein-building information in an mRNA is decoded (translated) into a sequence of amino acids. The result is a polypeptide chain that twists and folds into a protein:

mRNA —*translation*→ **protein**

Transcription and translation are part of **gene expression**, the multistep process by which information encoded in a gene guides the assembly of an RNA or protein product. During gene expression, this information flows from DNA to RNA to protein:

DNA —*transcription*→ **mRNA** —*translation*→ **protein**

A cell's DNA sequence contains all the information it needs to make the molecules of life. Each gene encodes an RNA, and RNAs interact to assemble proteins from amino acids (Section 3.4). Proteins (enzymes, in particular) assemble lipids and carbohydrates, replicate DNA, make RNA, and perform many other functions that keep the cell alive.

gene A part of a chromosome that encodes an RNA or protein product in its DNA sequence.
gene expression Process by which the information in a gene guides assembly of an RNA or protein product.
messenger RNA (mRNA) RNA that has a protein-building message.
ribosomal RNA (rRNA) RNA that becomes part of ribosomes.
transcription Process by which enzymes assemble an RNA using the nucleotide sequence of a gene as a template.
transfer RNA (tRNA) RNA that delivers amino acids to a ribosome during translation.
translation Process by which a polypeptide chain is assembled from amino acids in the order specified by an mRNA.

TAKE-HOME MESSAGE 9.1

Information in a DNA sequence occurs in units called genes. A cell uses the information encoded in a gene to make an RNA or protein product, a process called gene expression.

The DNA sequence of a gene is transcribed into RNA.

Information carried by a messenger RNA (mRNA) is translated into a protein.

9.2 HOW IS RNA ASSEMBLED?

Remember that DNA replication begins with one DNA double helix and ends with two DNA double helices (Section 8.4). The two double helices are identical to the parent molecule because base-pairing rules are followed during DNA replication. A nucleotide can be added to a growing strand of DNA only if it base-pairs with the corresponding nucleotide of the parent strand: G pairs with C, and A pairs with T (Section 8.2):

The same base-pairing rules also govern RNA synthesis in transcription. An RNA strand is structurally so similar to a DNA strand that the two can base-pair if their nucleotide sequences are complementary. In such hybrid molecules, G pairs with C, and A pairs with U (uracil):

During transcription, a strand of DNA acts as a template upon which a strand of RNA is assembled from nucleotides. A nucleotide can be added to a growing RNA only if it is complementary to the corresponding nucleotide of the parent strand of DNA. Thus, a new RNA is complementary in sequence to the DNA strand that served as its template. As in DNA replication, each nucleotide provides the energy for its own attachment to the end of a growing strand.

Transcription is similar to DNA replication in that one strand of a nucleic acid serves as a template for synthesis of another. However, in contrast with DNA replication, only part of one DNA strand, not the whole molecule, is used as a template for transcription. The enzyme **RNA polymerase**, not DNA polymerase, adds nucleotides to the end of a growing RNA. Also, transcription produces a single strand of RNA, not two DNA double helices.

In eukaryotic cells, transcription occurs in the nucleus; in prokaryotes, it occurs in cytoplasm. The process begins when an RNA polymerase and regulatory proteins attach to the DNA at a site called a **promoter** (**FIGURE 9.3** ❶). Binding positions the polymerase close to the gene that will be transcribed. The strand that is complementary to the gene sequence (the noncoding strand) is the one that serves as the template for transcription.

Like DNA polymerase, RNA polymerase moves along DNA (Section 8.4). As the RNA polymerase moves over a gene region, it unwinds the double helix just a bit so it can "read" the base sequence of the DNA strand ❷. The polymerase joins free RNA nucleotides into a chain, in the order dictated by that DNA sequence. As in

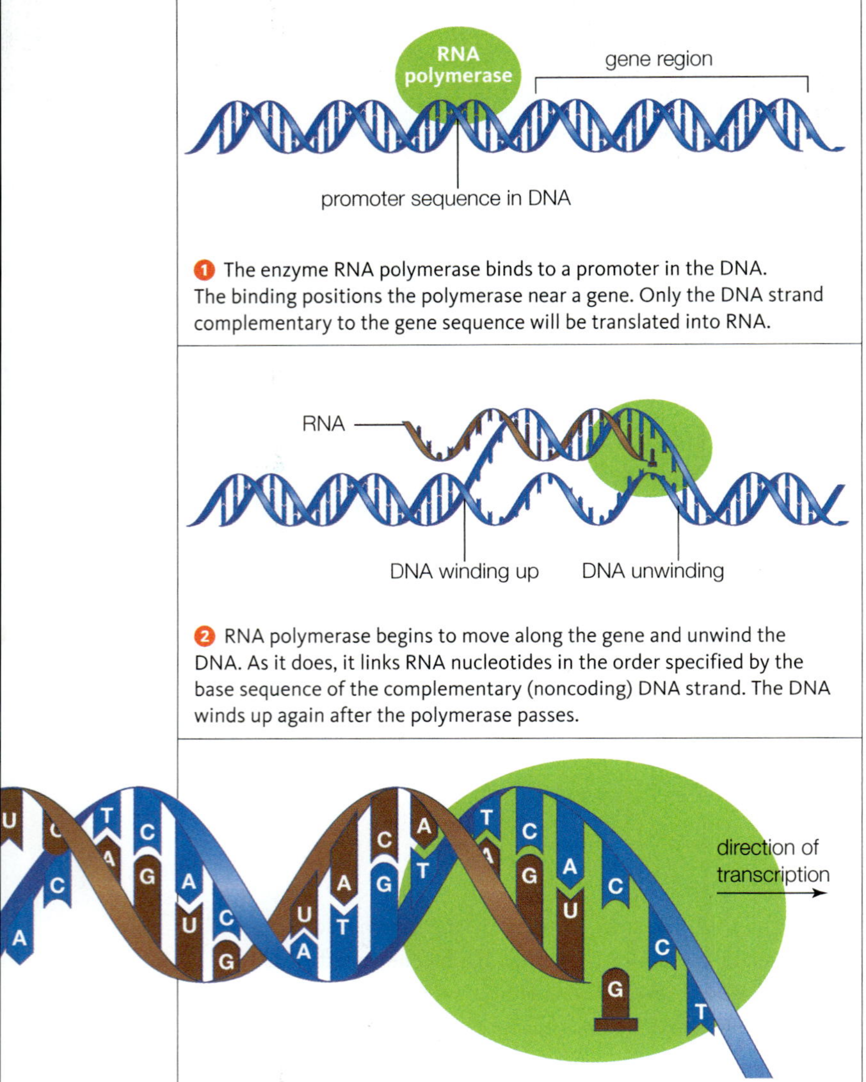

❶ The enzyme RNA polymerase binds to a promoter in the DNA. The binding positions the polymerase near a gene. Only the DNA strand complementary to the gene sequence will be translated into RNA.

❷ RNA polymerase begins to move along the gene and unwind the DNA. As it does, it links RNA nucleotides in the order specified by the base sequence of the complementary (noncoding) DNA strand. The DNA winds up again after the polymerase passes.

❸ Zooming in on the site of transcription, we can see that RNA polymerase covalently bonds successive nucleotides into an RNA strand. The base sequence of the new RNA strand is complementary to the base sequence of its DNA template strand, so it is an RNA copy of the gene.

FIGURE 9.3 **{Animated}** Transcription. By this process, a strand of RNA is assembled from nucleotides. A gene region in the DNA serves as the template for RNA synthesis.

FIGURE IT OUT: After the guanine (G), what nucleotide will be added to this growing strand of RNA?

Answer: Another guanine

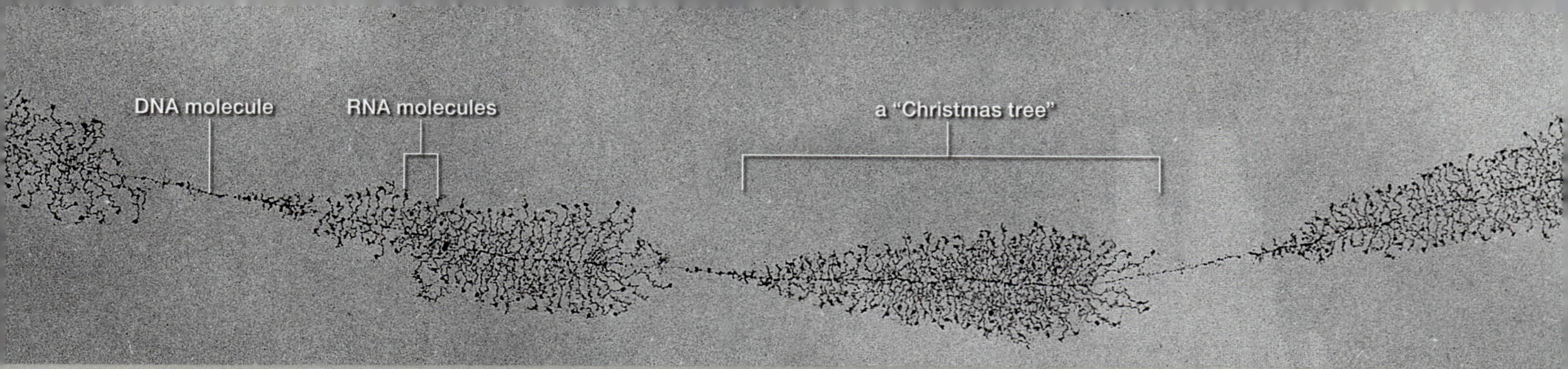

FIGURE 9.4 Typically, many RNA polymerases simultaneously transcribe the same gene, producing a structure often called a "Christmas tree" after its shape. Here, four genes next to one another on the same chromosome are being transcribed.

FIGURE IT OUT: Are the polymerases transcribing this DNA molecule moving from left to right or from right to left?

Answer: Left to right (the RNAs get longer as the polymerases move along the DNA)

DNA replication, the synthesis is directional: An RNA polymerase adds nucleotides only to the 3′ end of the growing strand of RNA.

When the polymerase reaches the end of the gene region, it releases the DNA and the new RNA. RNA polymerase follows base-pairing rules, so the new RNA strand is complementary in base sequence to the DNA strand from which it was transcribed ❸. It is an RNA copy of a gene, the same way that a paper transcript of a conversation carries the same information in a different format. Typically, many polymerases transcribe a particular gene region at the same time, so many new RNA strands can be produced very quickly (**FIGURE 9.4**).

FIGURE 9.5 {Animated} Post-transcriptional modification of RNA. Introns are removed and exons spliced together. Messenger RNAs also get a poly-A tail and modified guanine "cap."

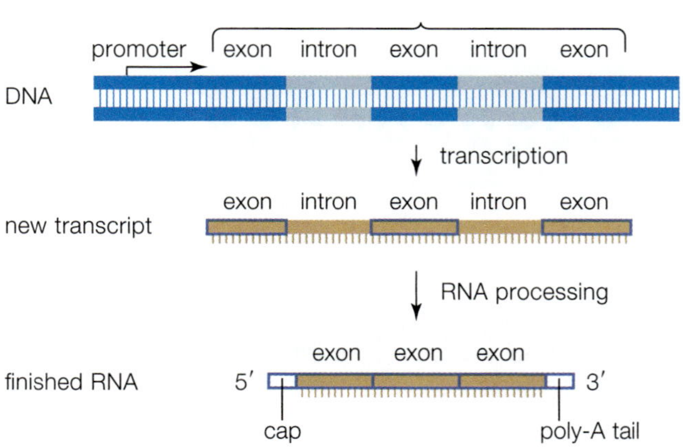

alternative splicing Post-translational RNA modification process in which some exons are removed or joined in various combinations.
exon Nucleotide sequence that remains in an RNA after post-transcriptional modification.
intron Nucleotide sequence that intervenes between exons and is removed during post-transcriptional modification.
promoter In DNA, a sequence to which RNA polymerase binds.
RNA polymerase Enzyme that carries out transcription.

POST-TRANSCRIPTIONAL MODIFICATIONS

Just as a dressmaker may snip off loose threads or add bows to a dress before it leaves the shop, so do eukaryotic cells tailor their RNA before it leaves the nucleus. Consider that most eukaryotic genes contain intervening sequences called **introns**. Introns are removed in chunks from a newly transcribed RNA before it leaves the nucleus. Sequences that stay in the RNA are called **exons** (**FIGURE 9.5**). Exons can be rearranged and spliced together in different combinations—a process called **alternative splicing**—so one gene may encode different proteins.

A newly transcribed RNA that will become an mRNA is further tailored after splicing. Enzymes attach a modified guanine "cap" to the 5′ end; later, this cap will help the finished mRNA bind to a ribosome. Between 50 and 300 adenines are also added to the 3′ end of a new mRNA. This poly-A tail is a signal that allows an mRNA to be exported from the nucleus, and as you will see in Chapter 10, it helps regulate the timing and duration of the mRNA's translation.

TAKE-HOME MESSAGE 9.2

Transcription is an energy-requiring process that uses the information in a gene to produce an RNA.

RNA polymerase uses a gene region in a chromosome as a template to assemble a strand of RNA. The new strand is an RNA copy of the gene from which it was transcribed.

Post-transcriptional modification of RNA occurs in the nucleus of eukaryotes.

9.3 WHAT ROLES DO mRNA, rRNA, AND tRNA PLAY DURING TRANSLATION?

first base	second base U	second base C	second base A	second base G	third base
U	UUU phe	UCU ser	UAU tyr	UGU cys	U
U	UUC phe	UCC ser	UAC tyr	UGC cys	C
U	UUA leu	UCA ser	UAA stop	UGA stop	A
U	UUG leu	UCG ser	UAG stop	UGG trp	G
C	CUU leu	CCU pro	CAU his	CGU arg	U
C	CUC leu	CCC pro	CAC his	CGC arg	C
C	CUA leu	CCA pro	CAA gln	CGA arg	A
C	CUG leu	CCG pro	CAG gln	CGG arg	G
A	AUU ile	ACU thr	AAU asn	AGU ser	U
A	AUC ile	ACC thr	AAC asn	AGC ser	C
A	AUA ile	ACA thr	AAA lys	AGA arg	A
A	AUG met	ACG thr	AAG lys	AGG arg	G
G	GUU val	GCU ala	GAU asp	GGU gly	U
G	GUC val	GCC ala	GAC asp	GGC gly	C
G	GUA val	GCA ala	GAA glu	GGA gly	A
G	GUG val	GCG ala	GAG glu	GGG gly	G

A codon table. Each codon in mRNA is a set of three nucleotide bases. Left column lists a codon's first base, the top row lists the second, and the right column lists the third. Sixty-one of the triplets encode amino acids; one of those, AUG, both codes for methionine and serves as a signal to start translation. Three codons are signals that stop translation.

ala alanine (A)	**gly** glycine (G)	**pro** proline (P)
arg arginine (R)	**his** histidine (H)	**ser** serine (S)
asn asparagine (N)	**ile** isoleucine (I)	**thr** threonine (T)
asp aspartic acid (D)	**leu** leucine (L)	**trp** tryptophan (W)
cys cysteine (C)	**lys** lysine (K)	**tyr** tyrosine (Y)
glu glutamic acid (E)	**met** methionine (M)	**val** valine (V)
gln glutamine (Q)	**phe** phenylalanine (F)	

Amino acid names and abbreviations.

FIGURE 9.6 The genetic code.

FIGURE IT OUT: Which codons specify the amino acid lysine (lys)?

Answer: AAA and AAG

DNA stores heritable information about proteins, but making those proteins requires messenger RNA (mRNA), transfer RNA (tRNA), and ribosomal RNA (rRNA). The three types of RNA interact to translate DNA's information into a protein.

THE MESSENGER: mRNA

An mRNA is essentially a disposable copy of a gene. Its job is to carry the gene's protein-building information to the other two types of RNA during translation. That protein-building information consists of a linear sequence of genetic "words" spelled with an alphabet of the four nucleotide bases A, C, G, and U. Each of the genetic "words" carried by an mRNA is three bases long, and each is a code—a **codon**—for a particular amino acid. With four possible nucleotides in each of the three positions of a codon, there are a total of sixty-four (or 4^3) mRNA codons. Collectively, the sixty-four codons constitute the **genetic code** (FIGURE 9.6). The sequence of bases in a triplet determines which amino acid the codon specifies. For instance, the codon UUU codes for the amino acid phenylalanine (phe), and UUA codes for leucine (leu).

Codons occur one after another along the length of an mRNA. When an mRNA is translated, the order of its codons determines the order of amino acids in the resulting polypeptide. Thus, the base sequence of a gene is transcribed into the base sequence of an mRNA, which is in turn translated into an amino acid sequence (FIGURE 9.7).

With a few exceptions, twenty naturally occurring amino acids are encoded by the sixty-four codons in the genetic

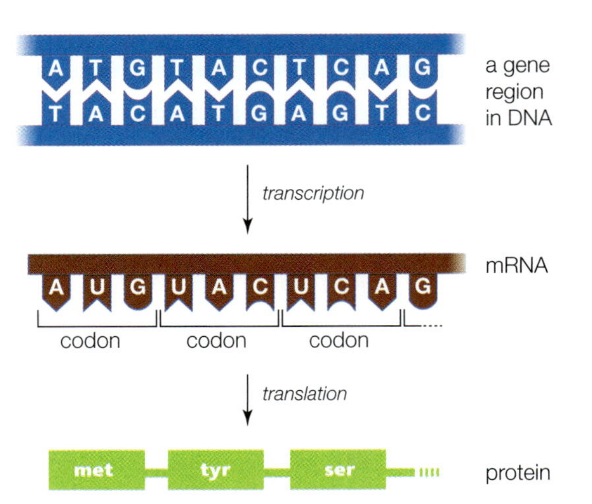

FIGURE 9.7 Example of the correspondence between DNA, RNA, and protein. A gene region in a strand of chromosomal DNA is transcribed into an mRNA, and the codons of the mRNA specify a chain of amino acids—a protein.

code. Sixty-four codons are more than are needed to specify twenty amino acids, so some amino acids are specified by more than one codon. For instance, the amino acid tyrosine (tyr) is specified by two codons: UAA and UAC.

Other codons signal the beginning and end of a protein-coding sequence. In most species, the first AUG in an mRNA serves as the signal to start translation. AUG is the codon for methionine, so methionine is always the first amino acid in new polypeptides of such organisms. The codons UAA, UAG, and UGA do not specify an amino acid. These are signals that stop translation, so they are called stop codons. A stop codon marks the end of the protein-coding sequence in an mRNA.

The genetic code is highly conserved, which means that most organisms use the same code and probably always have. Bacteria, archaea, and some protists have a few codons that differ from the eukaryotic code, as do mitochondria and chloroplasts—a clue that led to a theory of how these two organelles evolved (we return to this topic in Section 18.5).

THE TRANSLATORS: rRNA AND tRNA

Ribosomes interact with transfer RNAs (tRNAs) to translate the sequence of codons in an mRNA into a polypeptide. A ribosome has two subunits, one large and one small (**FIGURE 9.8**). Both subunits consist mainly of rRNA, with some associated structural proteins. During translation, a large and a small ribosomal subunit converge as an intact ribosome on an mRNA. Ribosomal RNA is one example of RNA with enzymatic activity: rRNA catalyzes formation of a peptide bond between amino acids as they are delivered to the ribosome.

Each tRNA has two attachment sites. The first is an **anticodon**, which is a triplet of nucleotides that base-pairs with an mRNA codon (**FIGURE 9.9A**). The other attachment site binds to an amino acid—the one specified by the codon. Transfer RNAs with different anticodons carry different amino acids.

During translation, tRNAs deliver amino acids to a ribosome, one after the next in the order specified by the codons in an mRNA (**FIGURE 9.9B**). As the amino acids are delivered, the ribosome joins them via peptide bonds into a new polypeptide (Section 3.4). Thus, the order of codons in an mRNA—DNA's protein-building message—becomes translated into a new protein.

anticodon In a tRNA, set of three nucleotides that base-pairs with an mRNA codon.
codon In an mRNA, a nucleotide base triplet that codes for an amino acid or stop signal during translation.
genetic code Complete set of sixty-four mRNA codons.

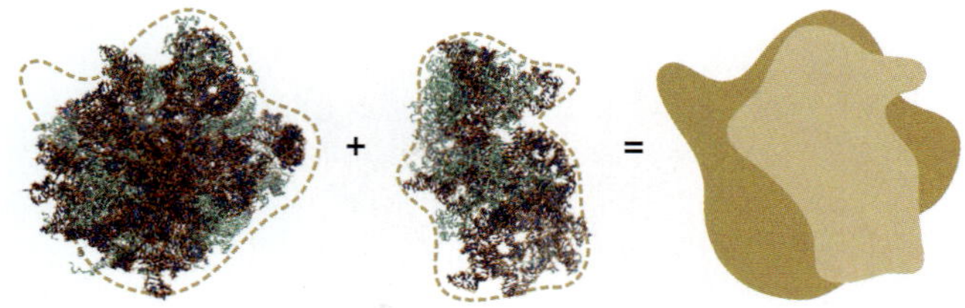

FIGURE 9.8 {Animated} Ribosome structure. Each intact ribosome consists of a large and a small subunit. The structural protein components of the two subunits are shown in green; the catalytic rRNA components, in brown.

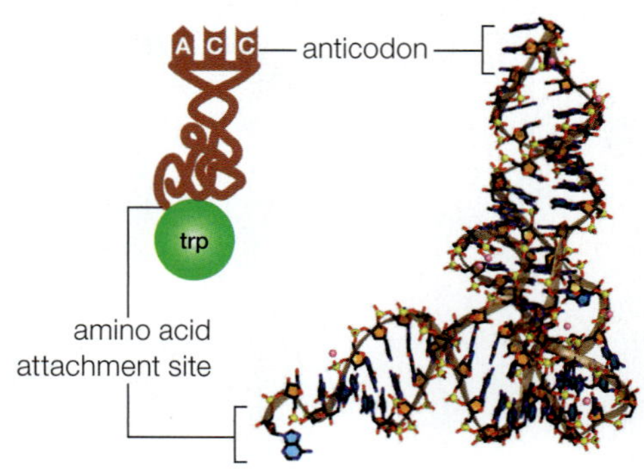

A Icon and model of the tRNA that carries the amino acid tryptophan. Each tRNA's anticodon is complementary to an mRNA codon. Each also carries the amino acid specified by that codon.

B During translation, tRNAs dock at an intact ribosome (for clarity, only the small subunit is shown, in tan). Here, the anticodons of two tRNAs have base-paired with complementary codons on an mRNA (red). The amino acids they carry are not shown, for clarity.

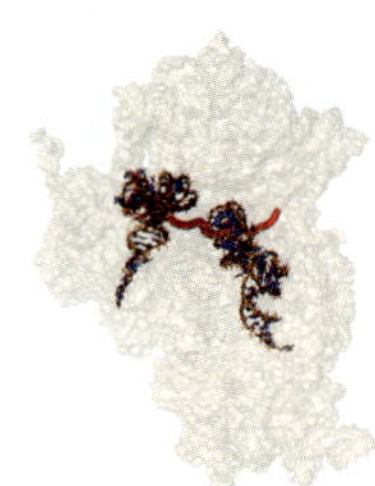

FIGURE 9.9 **tRNA structure.**

TAKE-HOME MESSAGE 9.3

The sequence of nucleotide triplets (codons) in an mRNA encode a gene's protein-building message.

The genetic code consists of sixty-four codons. Three are signals that stop translation; the remaining codons specify an amino acid. In most mRNAs, the first occurrence of the codon that specifies methionine is a signal to begin translation.

Ribosomes, which consist of two subunits of rRNA and proteins, assemble amino acids into polypeptide chains.

A tRNA has an anticodon complementary to an mRNA codon, and a binding site for the amino acid specified by that codon. During translation, tRNAs deliver amino acids to ribosomes.

9.4 HOW IS mRNA TRANSLATED INTO PROTEIN?

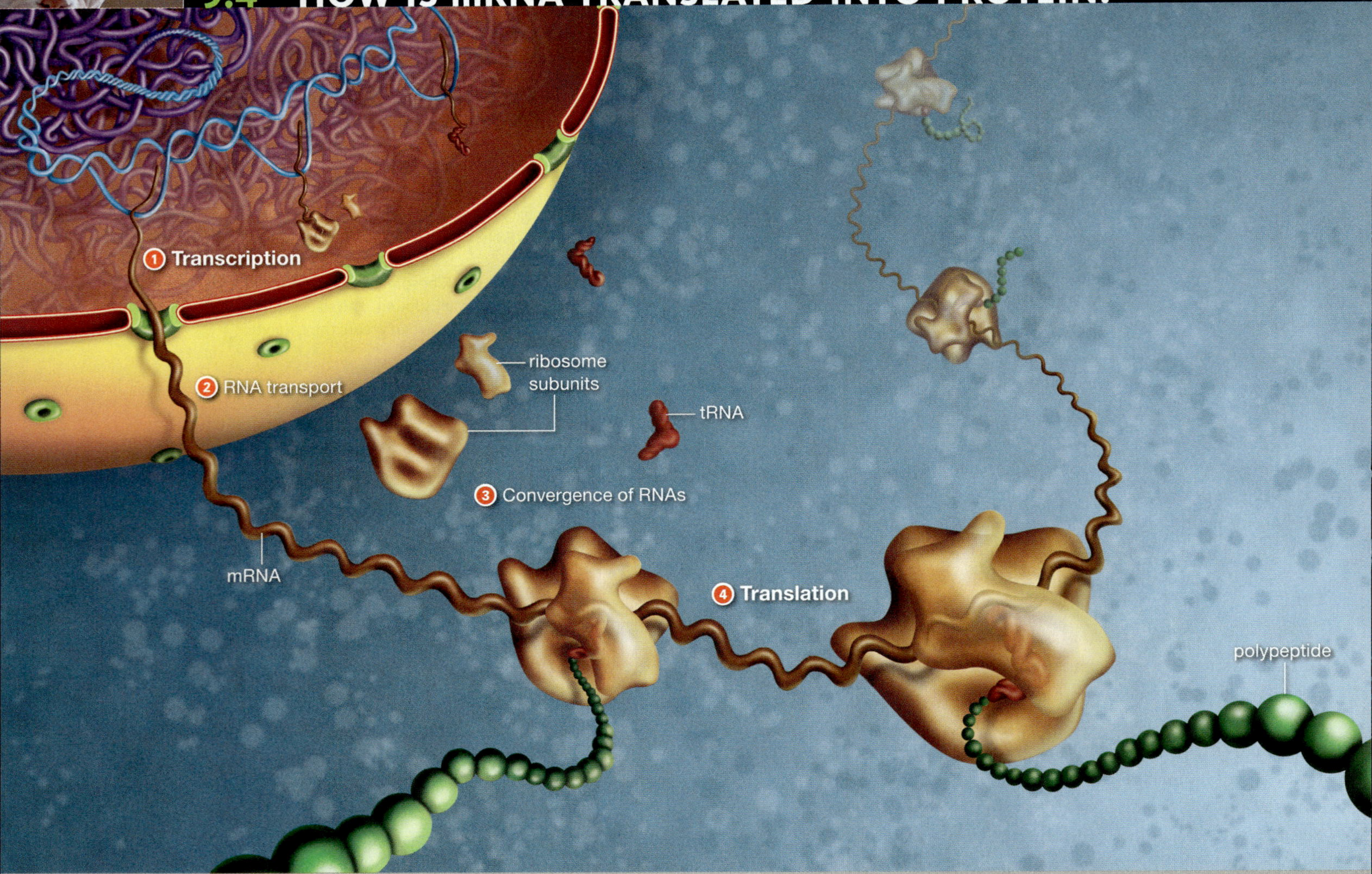

FIGURE 9.10 Overview of translation. In eukaryotes, RNA transcribed in the nucleus moves into the cytoplasm through nuclear pores. Translation occurs in the cytoplasm. Ribosomes simultaneously translating the same mRNA are called polysomes.

Translation, the second part of protein synthesis, occurs in the cytoplasm of all cells. Cytoplasm has many free amino acids, tRNAs, and ribosomal subunits available to participate in the process.

FIGURE 9.10 shows an overview of translation as it occurs in eukaryotes. An mRNA is transcribed in the nucleus ❶, and then transported through nuclear pores into the cytoplasm ❷. Translation begins when a small ribosomal subunit binds to the mRNA. Next, the anticodon of a special tRNA called an initiator base-pairs with the first AUG codon of the mRNA. Then, a large ribosomal subunit joins the small subunit ❸, and the intact ribosome begins to assemble a polypeptide chain as it moves along the mRNA ❹.

FIGURE 9.11 shows details of translation. Initiator tRNAs carry methionine, so the first amino acid of the new polypeptide chain is methionine. Another tRNA joins the complex when its anticodon base-pairs with the second codon in the mRNA ❺. This tRNA brings with it the second amino acid. The ribosome then catalyzes formation of a peptide bond between the first two amino acids ❻.

As the ribosome moves to the next codon, it releases the first tRNA. Another tRNA brings the third amino acid to the complex as its anticodon base-pairs with the third codon of the mRNA ❼. A peptide bond forms between the second and third amino acids ❽.

The second tRNA is released and the ribosome moves to the next codon. Another tRNA brings the fourth amino acid to the complex as its anticodon base-pairs with the fourth codon of the mRNA ❾. A peptide bond forms between the third and fourth amino acids. Elongation of the new polypeptide chain continues as the ribosome catalyzes peptide bonds between amino acids delivered by successive tRNAs.

Termination occurs when the ribosome reaches a stop codon in the mRNA ⑩. The mRNA and the polypeptide detach from the ribosome, and the ribosomal subunits separate from each other. Translation is now complete. The new polypeptide either joins the pool of proteins in the cytoplasm, or it enters rough ER of the endomembrane system (Section 4.6).

In cells that are making a lot of protein, many ribosomes may simultaneously translate the same mRNA, in which case they are called polysomes:

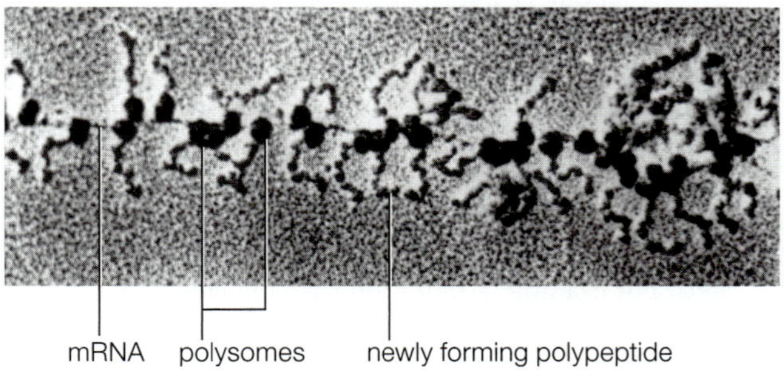

In bacteria and archaea, transcription and translation both occur in the cytoplasm, and these processes are closely linked in time and space. Translation begins before transcription ends, so a transcription "Christmas tree" is often decorated with polysome "balls."

Given that many polypeptides can be translated from one mRNA, why would any cell also make many copies of an mRNA? Compared with DNA, RNA is not very stable. An mRNA may last only a few minutes in cytoplasm before enzymes disassemble it. The fast turnover allows cells to adjust their protein synthesis quickly in response to changing needs.

Translation is energy intensive. That energy is provided mainly in the form of phosphate-group transfers from the RNA nucleotide GTP (shown in **FIGURE 9.2B**) to molecules involved in the process.

TAKE-HOME MESSAGE 9.4

Translation is an energy-requiring process that converts the protein-building information carried by an mRNA into a polypeptide.

Translation begins when an mRNA joins with an initiator tRNA and two ribosomal subunits.

Amino acids are delivered to the complex by tRNAs in the order dictated by successive mRNA codons. As amino acids arrive, the ribosome joins each to the end of the polypeptide.

Translation ends when the ribosome encounters a stop codon in the mRNA. The mRNA and the polypeptide are released, and the ribosome disassembles.

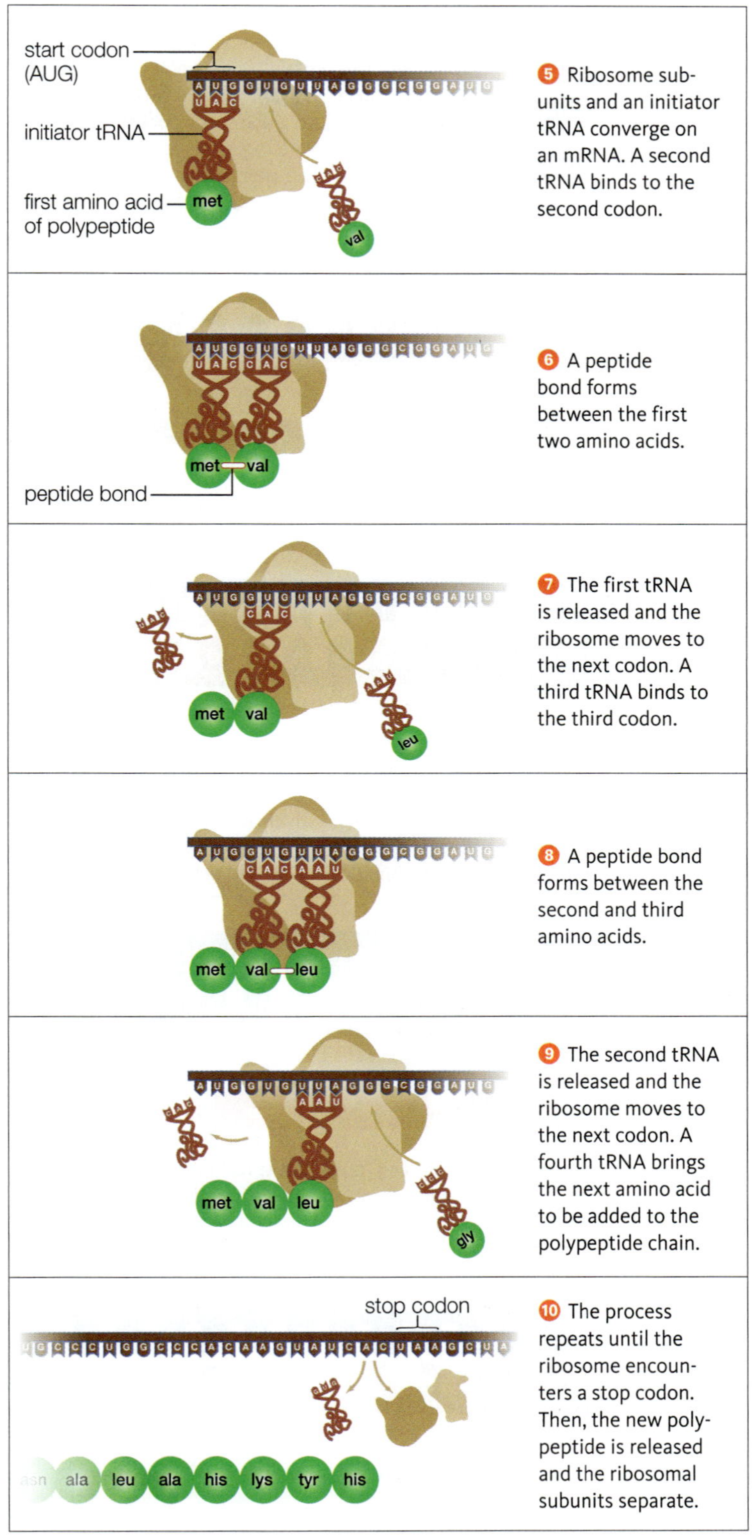

FIGURE 9.11 {Animated} Translation. Translation begins when ribosomal subunits and an initiator tRNA converge on an mRNA. tRNAs deliver amino acids in the order dictated by successive codons in the mRNA. The ribosome links the amino acids together as it moves along the mRNA, so a polypeptide forms and elongates. Translation ends when the ribosome reaches a stop codon.

9.5 WHAT HAPPENS AFTER A GENE BECOMES MUTATED?

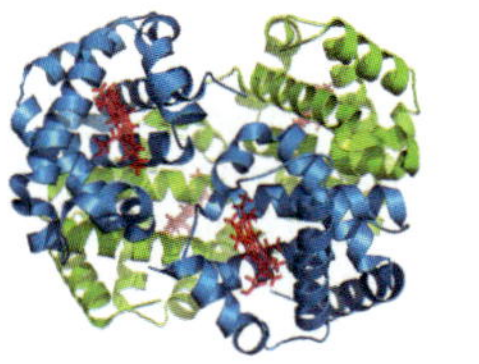

A Hemoglobin, an oxygen-binding protein in red blood cells. This protein consists of four polypeptides: two alpha globins (blue) and two beta globins (green). Each globin has a pocket that cradles a heme (red). Oxygen molecules bind to the iron atom at the center of each heme.

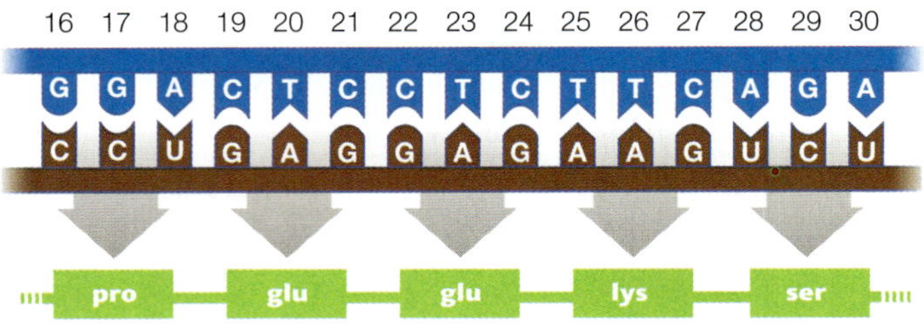

B Part of the DNA (blue), mRNA (brown), and amino acid sequence of human beta globin. Numbers indicate nucleotide position in the mRNA.

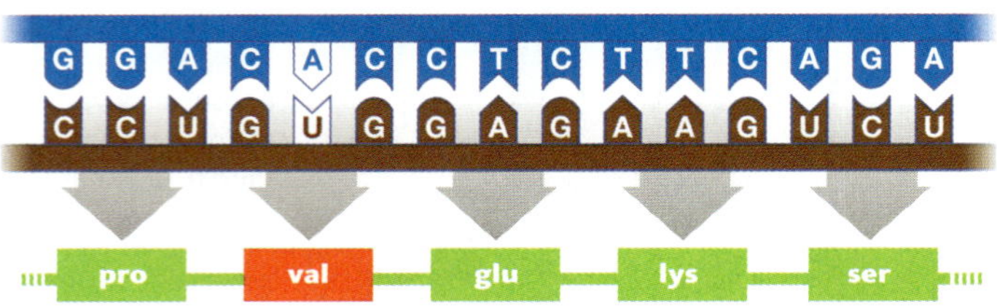

C A base-pair substitution replaces a thymine with an adenine. When the altered mRNA is translated, valine replaces glutamic acid as the sixth amino acid. Hemoglobin with this form of beta globin is called HbS, or sickle hemoglobin.

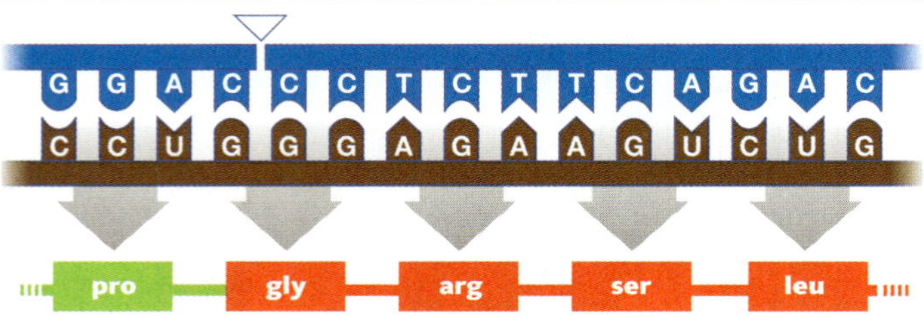

D A base-pair deletion shifts the reading frame for the rest of the mRNA, so a completely different protein product forms. The mutation shown results in a defective beta globin. The outcome is beta thalassemia, a genetic disorder in which a person has an abnormally low amount of hemoglobin.

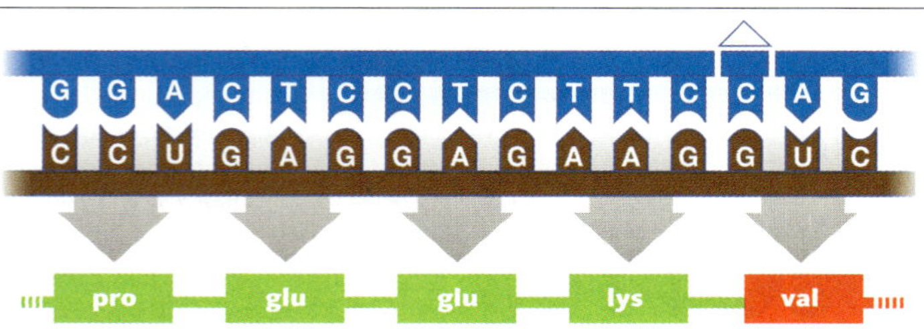

E An insertion of one nucleotide causes the reading frame for the rest of the mRNA to shift. The protein translated from this mRNA is too short and does not assemble correctly into hemoglobin molecules. As in D, the outcome is beta thalassemia.

FIGURE 9.12 {Animated} Examples of mutations.

Mutations, remember, are permanent changes in a DNA sequence (Section 8.5). A mutation in which one base pair is replaced by a different base pair is a **base-pair substitution**. Other mutations may involve the loss of one or more nucleotides (a **deletion**) or the addition of one or more extra nucleotides (an **insertion**).

Mutations are relatively uncommon events in a normal cell. Consider that the chromosomes in a diploid human cell collectively consist of about 6.5 billion nucleotides, any of which may become mutated each time that cell divides. On average, about 175 nucleotides do change during DNA replication. However, only about 3 percent of the cell's DNA encodes protein products, so there is a low probability that any of those mutations will be in a protein-coding region.

When a mutation does occur in a protein-coding region, the redundancy of the genetic code offers the cell a margin of safety. For example, a mutation that changes a CCC codon to CCG may not have further effects, because both of these codons specify serine. Other mutations may change an amino acid in a protein, or result in a premature stop codon that shortens it.

Mutations that alter a protein can have drastic effects on an organism. Consider the effects of mutations on hemoglobin, an oxygen-transporting protein in your red blood cells. Hemoglobin's structure allows it to bind and release oxygen. In adult humans, a hemoglobin molecule consists of four polypeptides called globins: two alpha globins and two beta globins (**FIGURE 9.12A**). Each globin folds around a heme, a cofactor with an iron atom at its center (Section 5.5). Oxygen molecules bind to hemoglobin at those iron atoms.

Mutations in the genes for alpha or beta globin cause a condition called anemia, in which a person's blood is deficient in red blood cells or in hemoglobin. Both outcomes limit the blood's ability to carry oxygen, and the resulting symptoms range from mild to life-threatening.

Sickle-cell anemia, a type of anemia that is most common in people of African ancestry, arises because of a base-pair substitution in the beta globin gene. The substitution causes the body to produce a version of beta globin in which the sixth amino acid is valine instead of glutamic acid (**FIGURE 9.12B,C**). Hemoglobin assembled with this altered beta globin chain is called sickle hemoglobin, or HbS.

Unlike glutamic acid, which carries a negative charge, valine carries no charge. As a result of that one base-pair substitution, a tiny patch of the beta globin polypeptide that is normally hydrophilic becomes hydrophobic. This change slightly alters the hemoglobin's behavior. Under certain

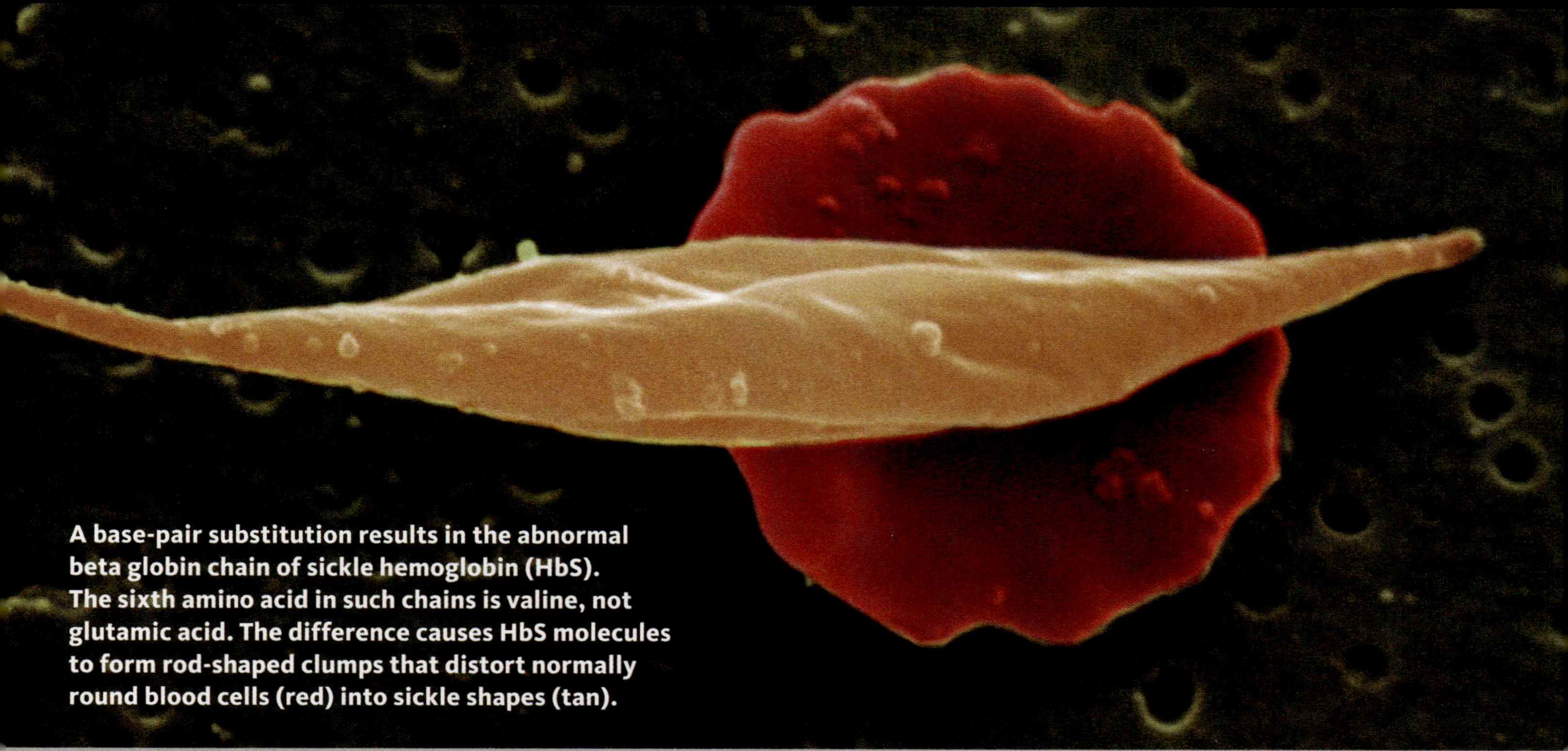

FIGURE 9.13 An amino acid substitution results in abnormally shaped red blood cells characteristic of sickle-cell anemia.

conditions, HbS molecules stick together and form large, rodlike clumps. Red blood cells that contain the clumps become distorted into a crescent (sickle) shape (**FIGURE 9.13**). Sickled cells clog tiny blood vessels, thus disrupting blood circulation throughout the body. Over time, repeated episodes of sickling can damage organs and cause death.

A different type of anemia, beta thalassemia, is caused by the deletion of the twentieth nucleotide in the coding region of the beta globin gene (**FIGURE 9.12D**). Like many other deletions, this one causes the reading frame of the mRNA codons to shift. A frameshift usually has drastic consequences because it garbles the genetic message, just as incorrectly grouping a series of letters garbles the meaning of a sentence:

The fat cat ate the sad rat
T hef atc ata tet hes adr at

The frameshift caused by the beta globin deletion results in a polypeptide that differs drastically from normal beta globin in amino acid sequence and length. This outcome is the source of the anemia. Beta thalassemia can also be caused by insertion mutations, which, like deletions, often result in frameshifts (**FIGURE 9.12E**).

Not all mutations that affect protein structure disrupt codons for amino acids. DNA also contains special nucleotide sequences that influence the expression of nearby genes (we return to this topic in the next chapter). A promoter is one example; an intron–exon splice site is another. Consider the mutation that causes hairlessness in cats (as shown in the chapter opening photo). In this case, a base-pair substitution disrupts an intron–exon splice site in the gene for keratin, a fibrous protein (Section 3.4). The intron, which is not correctly removed from the RNA, becomes an insertion in the mRNA. The altered protein translated from this mRNA cannot properly assemble into filaments that make up cat fur.

TAKE-HOME MESSAGE 9.5

Mutations that result in an altered protein can have drastic consequences.

A base-pair substitution may change an amino acid in a protein, or it may introduce a premature stop codon.

Frameshifts that occur after an insertion or deletion can change an mRNA's codon reading frame, thus garbling its protein-building instructions.

base-pair substitution Type of mutation in which a single base pair changes.
deletion Mutation in which one or more nucleotides are lost.
insertion Mutation in which one or more nucleotides become inserted into DNA.

CREDIT: (13) EM Unit, UCL Medical School, Royal Free Campus/Wellcome Images.

9.6 Application: THE APTLY ACRONYMED RIPs

FIGURE 9.14 Bulgarian spy's weapon: an umbrella modified to fire a tiny pellet of ricin into a victim. An umbrella like this one was used to assassinate Georgi Markov on the streets of London in 1978.

A DOSE OF RICIN AS SMALL AS A FEW GRAINS OF SALT CAN KILL AN ADULT HUMAN, and there is no antidote. Ricin is a protein that deters beetles, birds, mammals, and other animals from eating seeds of the castor-oil plant (*Ricinus communis*), which grows wild in tropical regions worldwide and is widely cultivated. Castor-oil seeds are the source of castor oil, an ingredient in plastics, cosmetics, paints, soaps, polishes, and many other items. After the oil is extracted from the seeds, the ricin is typically discarded along with the leftover seed pulp.

Lethal effects of ricin were being exploited as long ago as 1888, but using ricin as a weapon is now banned by most countries under the Geneva Protocol. However, controlling its production is impossible, because it takes no special skills or equipment to manufacture the toxin from easily obtained raw materials. Thus, ricin appears periodically in the news as a tool of terrorists. For example, in June, 2013, a Texas actress was arrested for sending ricin-laced letters to President Obama and the mayor of New York City. Perhaps the most famous example occurred in 1978, at the height of the Cold War when defectors from countries under Russian control were targets for assassination. Bulgarian journalist Georgi Markov had defected to England and was working for the BBC. As he made his way to a bus stop on a London street, an assassin used a modified umbrella (**FIGURE 9.14**) to fire a tiny, ricin-laced ball into Markov's leg. Markov died in agony three days later.

Ricin is called a ribosome-inactivating protein (RIP) because it inactivates ribosomes. RIPs are enzymes that remove a particular adenine base from one of the rRNAs in the ribosome's heavy subunit. The adenine is part of a binding site for proteins involved in GTP-requiring steps of elongation. After the base has been removed, the ribosome can no longer bind to these proteins, and elongation stops. If enough ribosomes are affected, protein synthesis grinds to a halt. Proteins are critical to all life processes, so cells that cannot make them die very quickly. Someone who inhales ricin can die from low

CREDIT: (14) Cary Wolinsky/National Geographic Creative.

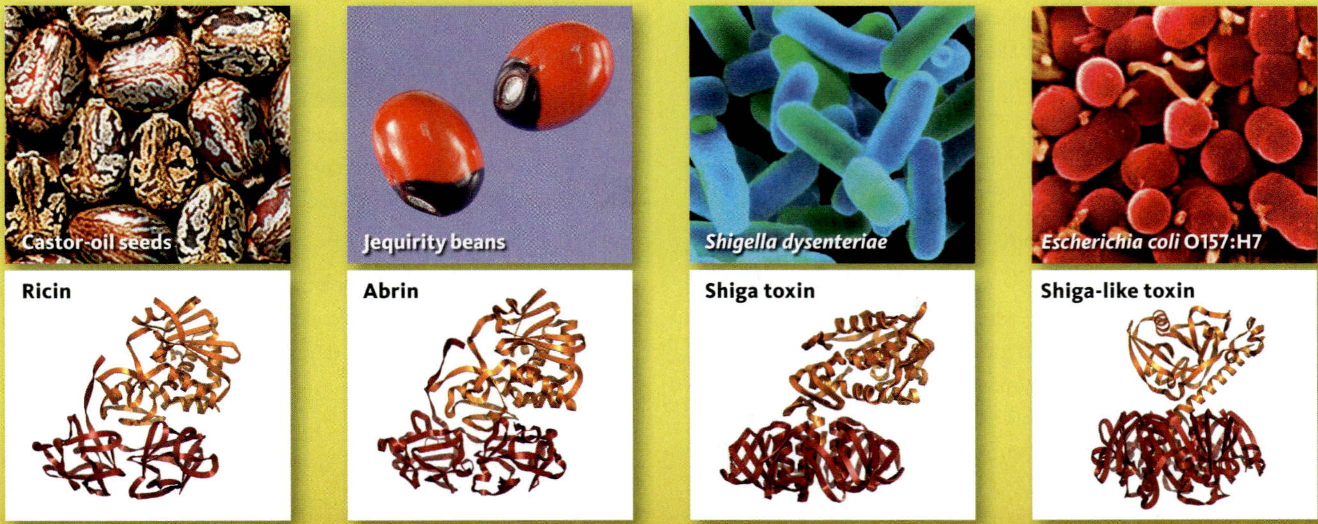

FIGURE 9.15 Lethal lineup: a few toxic ribosome-inactivating proteins (RIPs) and their sources. One of the two chains of a toxic RIP helps the molecule cross a cell's plasma membrane; the other is an enzyme that inactivates ribosomes.

blood pressure and respiratory failure within a few days of exposure.

Other RIPs are made by some bacteria, mushrooms, algae, and many plants (including food crops such as tomatoes, barley, and spinach). Most of these proteins are not particularly toxic in humans because they do not cross intact cell membranes very well. Those that do, including ricin, have a domain that binds tightly to glycolipids attached to proteins on our plasma membranes (**FIGURE 9.15**). Binding causes the cell to take up the RIP by endocytosis (Section 5.8). Once inside cytoplasm, the enzyme part of the molecule quickly goes to work: One molecule of ricin can inactivate more than 1,000 ribosomes per minute.

Fortunately, few people actually encounter ricin. Other RIPs are more prevalent. Bracelets made from beautiful seeds were recalled from stores in 2011 after a botanist recognized the seeds as jequirity beans. These beans contain abrin, an RIP even more toxic than ricin. Shiga toxin, an RIP made by *Shigella dysenteriae* bacteria, causes a severe bloody diarrhea (dysentery) that can be lethal. Some strains of *E. coli* bacteria make Shiga-like toxin, an RIP that is the source of intestinal illness (Section 4.12).

Despite their toxicity, the main function of RIPs may not be destroying ribosomes. Many are part of the immune system in plants, but it is their antiviral and anticancer activity that has researchers abuzz. Plants that make RIPs have been used as traditional medicines for many centuries. Western scientists are now investigating RIPs as potential weapons to combat HIV and cancer.

The unique properties of RIPs are proving particularly useful in drug design. For example, researchers who design drugs for cancer therapy have modified ricin's glycolipid-binding domain to recognize plasma membrane proteins (Section 4.3) especially abundant in cancer cells. The modified ricin preferentially enters—and kills—cancer cells. Ricin's toxic enzyme has also been attached to an antibody that can find cancer cells in a person's body. The intent of both strategies: to assassinate the cancer cells without harming normal ones.

Summary

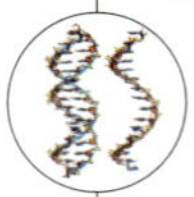

SECTION 9.1 Information encoded within the nucleotide sequence of DNA occurs in subsets called **genes**. Converting the information in a gene to an RNA or protein product is called **gene expression**. RNA is produced during **transcription**. **Ribosomal RNA** (**rRNA**) and **transfer RNA** (**tRNA**) interact during **translation** of a **messenger RNA** (**mRNA**) into a protein product:

DNA —*transcription*→ **mRNA** —*translation*→ **protein**

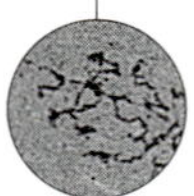

SECTION 9.2 During transcription, the enzyme **RNA polymerase** binds to a **promoter** near a gene on a chromosome. The polymerase moves over the gene region, linking RNA nucleotides in the order dictated by the nucleotide sequence of the DNA. The new RNA strand is an RNA copy of the gene.

The RNA of eukaryotes is modified before it leaves the nucleus. **Introns** are removed, and the remaining **exons** may be rearranged and spliced in different combinations, a process called **alternative splicing**. A cap and poly-A tail are also added to a new mRNA.

SECTION 9.3 An mRNA carries DNA's protein-building information. The information consists of a series of **codons**, which are sets of three nucleotides. Sixty-four codons constitute the **genetic code**. Three codons function as signals that terminate translation. The remaining codons specify a particular amino acid. Some amino acids are specified by multiple codons.

Each tRNA has an **anticodon** that can base-pair with a codon, and it binds to the amino acid specified by that codon. Enzymatic rRNA and proteins make up the two subunits of ribosomes.

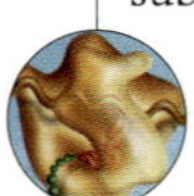

SECTION 9.4 During translation, protein-building information that is carried by an mRNA directs the synthesis of a polypeptide. First, an mRNA, an initiator tRNA, and two ribosomal subunits converge. Next, amino acids are delivered by tRNAs in the order specified by the codons in the mRNA. The intact ribosome catalyzes formation of a peptide bond between the successive amino acids, so a polypeptide forms. Translation ends when the ribosome encounters a stop codon in the mRNA.

SECTION 9.5 **Insertions**, **deletions**, and **base-pair substitutions** are mutations. A mutation that changes a gene's product may have harmful effects. Sickle-cell anemia, which is caused by a base-pair substitution in the gene for the beta globin chain of hemoglobin, is one example. Beta thalassemia is an outcome of frameshift mutations in the beta globin gene.

SECTION 9.6 The ability to make proteins is critical to all life processes. Ribosome-inactivating proteins (RIPs) have an enzyme domain that permanently disables ribosomes. Ricin and other toxic RIPs have an additional protein domain that triggers endocytosis. Once inside cytoplasm, the molecule's enzyme domain destroys the cell's ability to make proteins.

Self-Quiz Answers in Appendix VII

1. A chromosome contains many different gene regions that are transcribed into different ______ .
   a. proteins
   b. polypeptides
   c. RNAs
   d. a and b

2. A binding site for RNA polymerase is called a ______ .
   a. gene
   b. promoter
   c. codon
   d. protein

3. An RNA molecule is typically ______ ; a DNA molecule is typically ______ .
   a. single-stranded; double-stranded
   b. double-stranded; single-stranded
   c. both are single-stranded
   d. both are double-stranded

4. RNAs form by ______ ; proteins form by ______ .
   a. replication; translation
   b. translation; transcription
   c. transcription; translation
   d. replication; transcription

5. The main function of a DNA molecule is to ______ .
   a. store heritable information
   b. carry RNA's message for translation
   c. form peptide bonds between amino acids
   d. carry amino acids to ribosomes

6. The main function of an mRNA molecule is to ______ .
   a. store heritable information
   b. carry DNA's genetic message for translation
   c. form peptide bonds between amino acids
   d. carry amino acids to ribosomes

7. Energy that drives transcription is provided mainly by ______ .
   a. ATP
   b. RNA nucleotides
   c. GTP
   d. all are correct

8. Most codons specify a(n) ______ .
   a. protein
   b. polypeptide
   c. amino acid
   d. mRNA

9. Anticodons pair with ______ .
   a. mRNA codons
   b. DNA codons
   c. RNA anticodons
   d. amino acids

10. Up to ______ amino acids can be encoded by an mRNA that consists of 45 nucleotides plus a stop codon.
    a. 15
    b. 45
    c. 90
    d. 135

Data Analysis Activities

RIPs as Cancer Drugs Researchers are taking a page from the structure–function relationship of RIPs in their quest for cancer treatments. The most toxic RIPs, remember, have one domain that interferes with ribosomes, and another that carries them into cells. Melissa Cheung and her colleagues incorporated a peptide that binds to skin cancer cells into the enzymatic part of an RIP, the *E. coli* Shiga-like toxin. The researchers created a new RIP that specifically kills skin cancer cells, which are notoriously resistant to established therapies. Some of their results are shown in **FIGURE 9.16**.

1. Which cells had the greatest response to an increase in concentration of the engineered RIP?
2. At what concentration of RIP did all of the different kinds of cells survive?
3. Which cells survived best at 10^{-6} grams per liter RIP?
4. On which type of cancer cells did the RIP have the least effect?

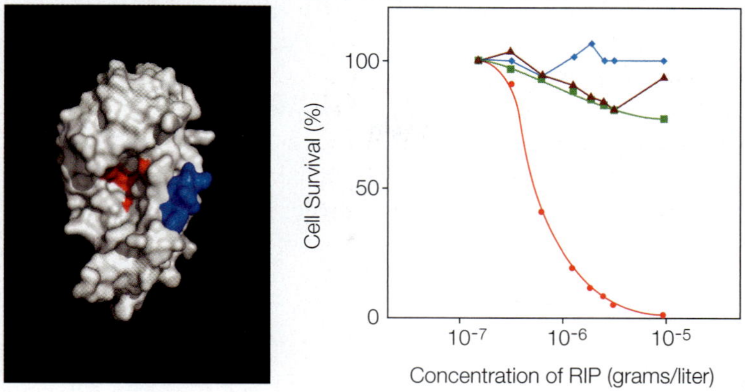

FIGURE 9.16 Effect of an engineered RIP on cancer cells.

The model on the *left* shows the enzyme portion of *E. coli* Shiga-like toxin that has been engineered to carry a small sequence of amino acids (in blue) that targets skin cancer cells. (Red indicates the active site.)

The graph on the *right* shows the effect of this engineered RIP on human cancer cells of the skin (●); breast (◆); liver (▲); and prostate (■).

11. _______ are removed from new mRNAs.
    a. Introns c. Telomeres
    b. Exons d. Amino acids

12. Where does transcription take place in a typical eukaryotic cell?
    a. the nucleus c. the cytoplasm
    b. ribosomes d. b and c are correct

13. Where does translation take place in a typical eukaryotic cell?
    a. the nucleus c. a and b
    b. the cytoplasm d. neither a nor b

14. Energy that drives translation is provided mainly by _______ .
    a. ATP c. GTP
    b. amino acids d. all are correct

15. Match the terms with the best description.

___ genetic message	a. protein-coding segment
___ promoter	b. RNA polymerase binding site
___ polysome	c. read as base triplets
___ exon	d. removed before translation
___ genetic code	e. occurs only in groups
___ intron	f. complete set of 64 codons

Critical Thinking

1. Researchers are designing and testing antisense drugs as therapies for a variety of diseases, including cancer, AIDS, diabetes, and muscular dystrophy. The drugs are also being tested to fight infection by deadly viruses such as Ebola. Antisense drugs consist of short mRNA strands that are complementary in base sequence to mRNAs linked to the diseases. Speculate on how these drugs work.

2. An anticodon has the sequence GCG. What amino acid does this tRNA carry? What would be the effect of a mutation that changed the C of the anticodon to a G?

3. Each position of a codon can be occupied by one of four (4) nucleotides. What is the minimum number of nucleotides per codon necessary to specify all 20 of the amino acids that are found in proteins?

4. Refer to **FIGURE 9.6**, then translate the following mRNA nucleotide sequence into an amino acid sequence, starting at the first base:

(5′) UGUCAUGCUCGUCUUGAAUCUUGU
GAUGCUCGUUGGAUUAAUUGU (3′)

5. Translate the sequence of bases in the previous question, starting at the second base.

6. Can you spell your name using the one-letter amino acid abbreviations shown in **FIGURE 9.6**? If so, construct an mRNA sequence that encodes your "protein" name.

CREDIT: Source: Cheung et al., *Molecular Cancer*, 9:28, 2010.

National Geographic Explorer Jane Goodall observes the behavior of two chimpanzees. About 97 percent of chimpanzee DNA is identical with human DNA, but how that DNA is used differs between the two species. The differences in gene expression patterns give rise to major differences in biology and behavior.

10 CONTROL OF GENE EXPRESSION

Links to Earlier Concepts

This chapter explores metabolism (Sections 5.3, 5.4, 5.5) in the context of gene expression (9.1). You will be applying what you know about DNA: structure (8.2), chromosomes (8.3), replication and hybridization (8.4), mutations (8.5, 9.5), and differentiation (8.6), as well as transcription (9.2) and translation (9.4). You will also revisit functional groups (3.1), carbohydrates (3.2), active transport (5.7), glycolysis (7.1), and fermentation (7.6).

KEY CONCEPTS

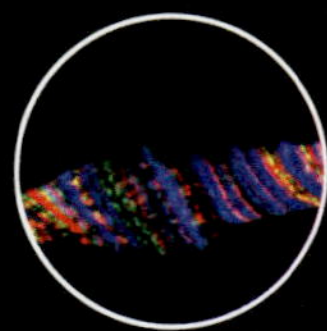

MECHANISMS OF GENE CONTROL

Every step of gene expression is regulated. This control is critical for development, and it allows individual cells to respond to changes in extracellular conditions.

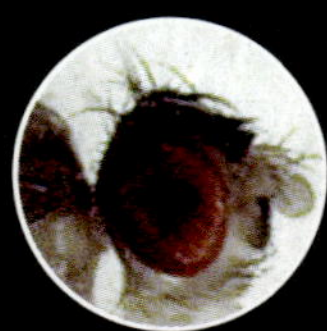

MASTER GENES

During development, the orderly, localized expression of master genes gives rise to the body plan of complex multicelled animals.

GENE CONTROL IN EUKARYOTES

Genes that govern X chromosome inactivation and male sex determination in mammals, and flower formation in plants, offer examples of gene control in eukaryotes.

GENE CONTROL IN PROKARYOTES

Most gene control in prokaryotes occurs at the level of transcription. Fast adjustment of transcription allows these cells to respond quickly to changes in external conditions.

EPIGENETICS

New research is revealing how gene expression patterns that arise during an individual's lifetime in response to environmental pressures can be passed to descendants.

Photograph by Michael Nichols, National Geographic Creative.

10.1 WHAT IS GENE CONTROL?

A typical cell in your body uses only about 10 percent of its genes at one time. Some of the active genes affect structural features and metabolic pathways common to all of your cells; others are expressed only by certain subsets of cells. For example, most body cells express genes that encode the enzymes of glycolysis, but only immature red blood cells express genes that encode globin.

Control over which genes are expressed at a particular time is necessary for cell differentiation (Section 8.6), and for proper development of complex, multicelled bodies. Such control also allows individual cells—prokaryotic and eukaryotic types—to respond appropriately to changes in their external environment.

GENE EXPRESSION CONTROL

The "switches" that turn a gene on or off are molecules or processes that trigger or inhibit the individual steps of its expression (**FIGURE 10.1**).

1 Transcription In prokaryotes, most control over gene expression occurs at the level of transcription, but eukaryotes regulate this step too. Proteins called **transcription factors** affect whether and how fast a gene is transcribed by binding directly to the DNA. Transcription of a eukaryotic gene is typically governed by many interacting transcription factors, giving eukaryotic cells a nuanced level of control over RNA production. A simpler level of control allows prokaryotes to adapt very quickly to changes in their environment.

FIGURE 10.1 {Animated} Points of control over gene expression.

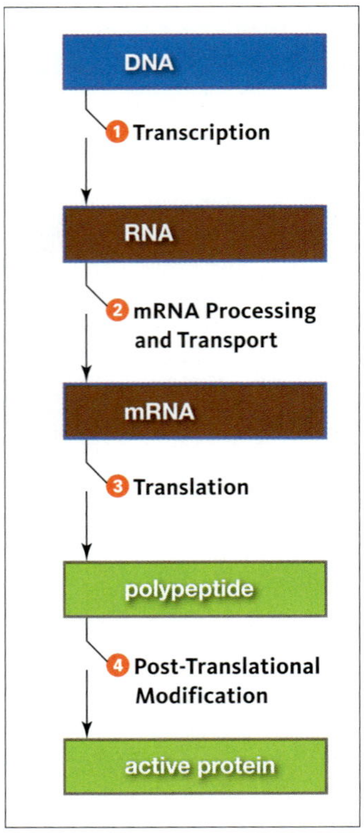

There are many types of transcription factors. Those called **repressors** shut off transcription or slow it down, either by preventing RNA polymerase from accessing the promoter or by impeding its progress along the DNA strand. Eukaryotic repressors work by binding directly to a promotor, or to a silencer—a site in the DNA that may be thousands of base pairs away from the gene. Prokaryotic silencers are called **operators** (we discuss operators in Section 10.4). **Activators** are transcription factors that recruit RNA polymerase to a promoter or help it bind, so they speed up transcription. Some eukaryotic activators work by binding to DNA sequences called **enhancers**, which, like silencers, may be far away from the gene they affect. Transcription factors that bind to regions of DNA called insulators prevent nearby genes from being affected by enhancers or silencers (**FIGURE 10.2**).

Chromatin structure also affects transcription. Almost all eukaryotic cells and some archaea have histones (Section 8.3). In these cells, only DNA regions that have been unwound from histones are accessible to RNA polymerase. Modifications to histone proteins change the way they interact with the DNA that wraps around them. Some modifications make histones release their grip on the DNA; others make them tighten it. For example, adding acetyl groups ($—COCH_3$) to a histone loosens the DNA, so enzymes that acetylate histones allow transcription to proceed in that region of DNA. Conversely, adding methyl groups ($—CH_3$) to a histone tightens the DNA, so enzymes that methylate histones shut down transcription.

In some specialized cells of eukaryotes, a very high level of gene expression can be achieved when DNA is copied repeatedly without cell division. The result is a set of polytene chromosomes, each consisting of hundreds or thousands of side-by-side copies of the same DNA molecule (the chromosome in **FIGURE 10.2** is polytene). Transcription of one gene occurs simultaneously on all of the DNA strands, quickly producing a lot of mRNA.

2 mRNA Processing and Transport As you know, transcription in eukaryotes occurs in the nucleus, and translation occurs in cytoplasm (Section 9.2). An mRNA can pass through pores of a nuclear envelope only after it has been processed appropriately—spliced, capped, and finished with a poly-A tail. Mechanisms that delay these post-transcriptional modifications also delay the mRNA's appearance in cytoplasm for translation.

Control over post-transcriptional modification can also affect the form of a protein. Consider RNA transcribed from the gene for fibronectin, a protein produced by cells of vertebrate animals. Two cell types splice this RNA alternatively, so they produce different mRNAs—and different forms of fibronectin. Liver cells produce a soluble form that circulates in blood plasma. Fibroblasts produce an insoluble form that is a major protein component of extracellular matrix (Section 4.10).

The majority of eukaryotic mRNAs are delivered to organelles or specific regions of cytoplasm. This localization allows translation of an mRNA to occur close to where its protein product is being used. In an egg, mRNA localization is crucial for proper development of the future

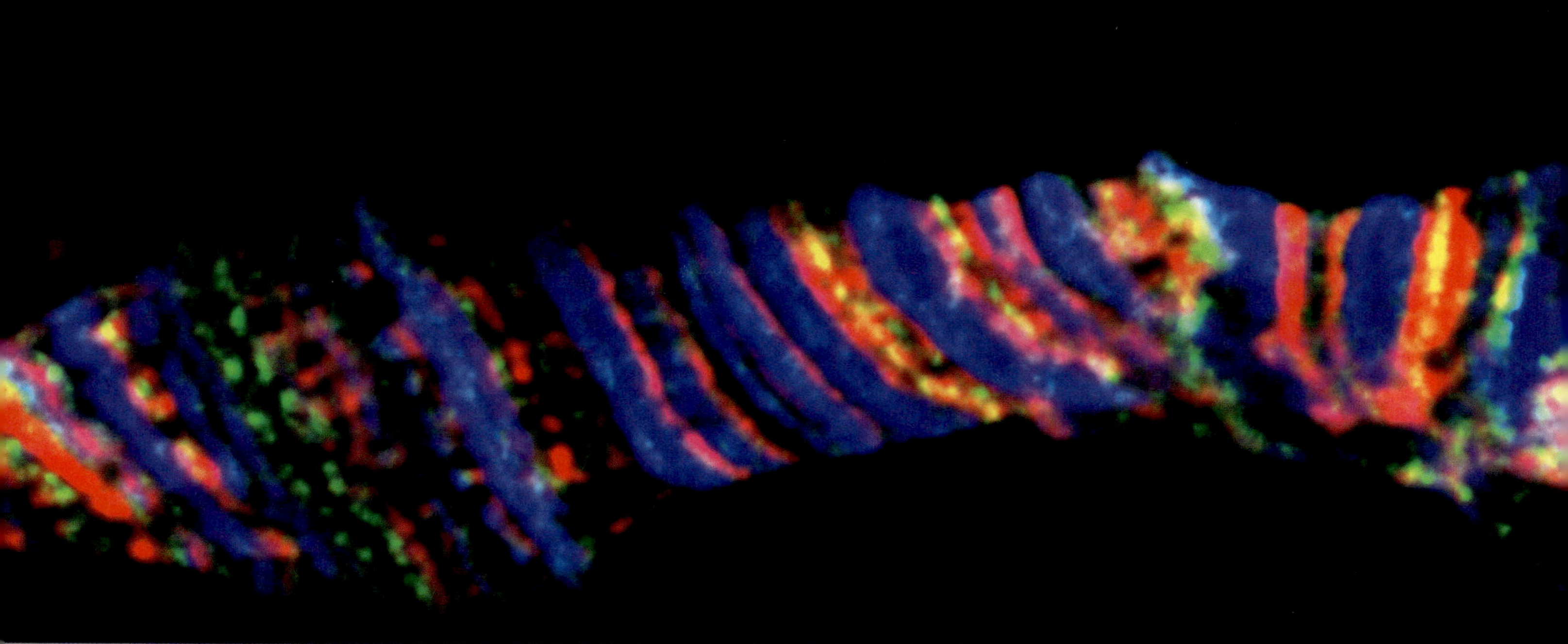

FIGURE 10.2 Part of a chromosome in a salivary gland cell of a fruit fly. DNA appears blue. Red and green show the locations of two transcription factors bound to insulator sequences (yellow shows where these two colors overlap). These proteins are restricting enhancer access to the DNA.

embryo. A short base sequence near an mRNA's poly-A tail is like a zip code that specifies a particular destination. Proteins that attach to the zip code drag the mRNA along cytoskeletal elements to that destination. Other proteins influence localization by interacting with mRNA-binding proteins. Researchers recently discovered that mRNA localization also occurs in prokaryotes, but the mechanism is not yet understood.

❸ Translation In eukaryotes, most control over gene expression occurs at the level of translation. Production of the many molecules that participate in translation is a major point of control in these cells. An mRNA's sequence also affects translation. For example, proteins that bind to a zip code region prevent transcription from occurring before the mRNA is delivered to its final destination. As another example, consider that the longer an mRNA lasts, the more protein can be made from it. Enzymes begin to disassemble a new mRNA as soon as it arrives in cytoplasm. The fast turnover allows a cell to adjust its protein synthesis quickly in response to changing needs. How long an mRNA persists depends on its base sequence, the length of its poly-A tail, and which proteins are attached to it.

In eukaryotes, translation of a particular mRNA can be shut down by tiny bits of noncoding RNA called microRNAs. A microRNA is complementary in sequence to part of an mRNA, and when the two show up together in cytoplasm they base-pair to form a small double-stranded region of RNA. By a process called RNA interference, any double-stranded RNA is cut up into small bits that are taken up by special enzyme complexes. These complexes then destroy every RNA in a cell that can base-pair with the bits. Thus, expression of a microRNA results in the destruction of all mRNA complementary to it.

Double-stranded RNA is also a factor in control over translation in prokaryotes. For example, bacteria can shut off translation of a particular mRNA by expressing an antisense RNA (one that is complementary in sequence to the mRNA). When the two molecules base-pair, ribosomes cannot initiate translation on the resulting double-stranded RNA. As another example, some bacterial mRNAs can loop back on themselves to form a small double-stranded region. Translation only occurs when this structure is unraveled, for example by heat.

❹ Post-Translational Modification Many newly synthesized polypeptide chains must be modified before they become functional. For example, some enzymes become active only after they have been phosphorylated (Section 5.5). Such post-translational modifications inhibit, activate, or stabilize many molecules, including enzymes that participate in transcription and translation.

TAKE-HOME MESSAGE 10.1

Gene control is necessary for individual cells to respond to changes in their extracellular environment. It is also crucial for proper development of complex, multicelled eukaryotes.

Gene expression can be switched on or off, or speeded up or slowed down, by molecules and processes that operate at each step.

activator Transcription factor that increases the rate of transcription.
enhancer In eukaryotic cells, a binding site in DNA for an activator.
operator In prokaryotes, a binding site in DNA for a repressor.
repressor Transcription factor that reduces the rate of transcription.
transcription factor Protein that influences transcription by binding directly to DNA; for example, an activator or repressor.

CREDIT: (2) *Journal of Biosciences*, Volume 36, Number 3, August 2011, Indian Academy of Sciences, Springer.

10.2 HOW DO GENES CONTROL DEVELOPMENT IN ANIMALS?

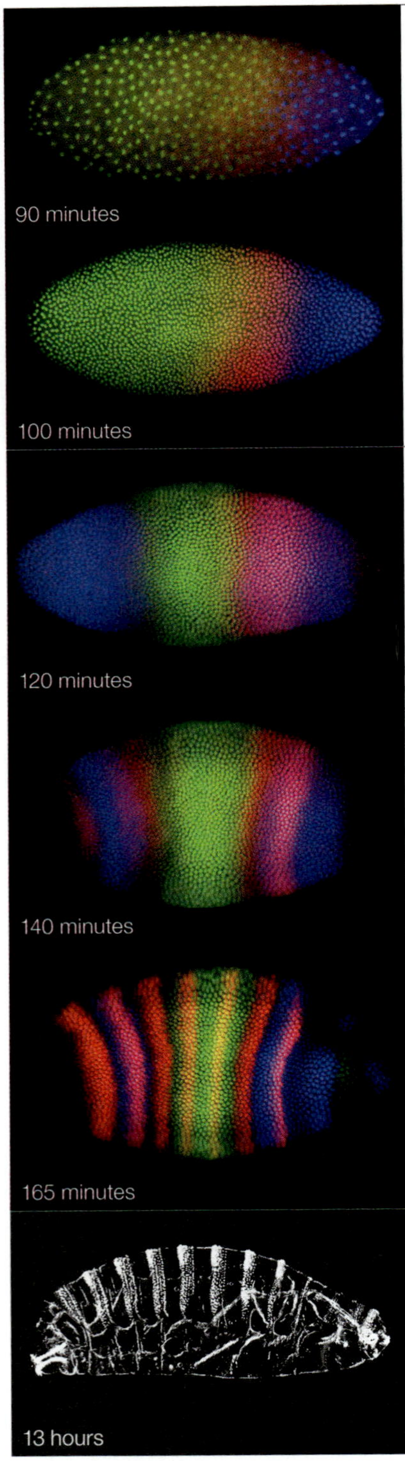

A The master gene *even-skipped* is expressed (in red) only where two maternal gene products (blue and green) overlap.

B By 165 minutes after fertilization, the products of several master genes, including the two shown here in green and blue, have confined the expression of *even-skipped* (red) to seven stripes. (Pink and yellow areas are regions in which red fluorescence has overlapped with blue or green.)

C One day later, seven segments have developed. The position of the segments corresponds to the position of the *even-skipped* stripes.

FIGURE 10.3 How gene expression control makes a fly, as illuminated by the formation of segments in a *Drosophila* embryo.

Expression of different master genes is shown by different colors in fluorescence microscopy images of whole embryos at successive stages of development (time after fertilization is indicated). Bright dots are individual nuclei.

MASTER GENES

As an animal embryo develops, its cells differentiate and form tissues, organs, and body parts. The entire process is driven by cascades of master gene expression. The products of **master genes** affect the expression of many other genes. Expression of a master gene causes other genes to be expressed, which in turn cause other genes to be expressed, and so on. The final outcome is the completion of an intricate task such as the formation of an eye.

The orchestration of gene expression during development begins when maternal mRNAs are delivered to opposite ends of an unfertilized egg as it forms. These mRNAs are translated only after the egg is fertilized. Then, their protein products diffuse away, forming gradients that span the entire developing embryo. The position of a nucleus within the embryo determines how much of these proteins it is exposed to. This in turn determines which master genes it turns on. The products of those master genes also form in gradients that span the embryo. Still other master genes are transcribed depending on where a nucleus falls within these gradients, and so on. Eventually, the products of master genes cause undifferentiated cells to differentiate, and specialized structures form in specific regions of the embryo (**FIGURE 10.3**).

HOMEOTIC GENES

A **homeotic gene** is a type of master gene that governs the formation of a body part such as an eye, leg, or wing. Animal homeotic genes encode transcription factors with a homeodomain, which is a region of about sixty amino acids that can bind directly to a promoter or some other sequence of nucleotides in a chromosome (**FIGURE 10.4A**).

Homeotic genes are often named for what happens when a mutation alters their function. For example, fruit flies with a mutation that affects their *antennapedia* gene (*ped* means foot) have legs in place of antennae (**FIGURE 10.4B**). The *dunce* gene is required for learning and memory. *Wingless, wrinkled,* and *minibrain* are self-explanatory. *Tinman* is necessary for development of a heart. Flies with a mutated *groucho* gene have extra bristles above their eyes (**FIGURE 10.4C**). Flies with a mutated *eyeless* gene develop with no eyes (**FIGURE 10.5A,B**). One gene was named *toll*, after what its German discoverer exclaimed upon seeing the disastrous effects of the mutation (*toll* is German slang that means "cool!").

The function of many homeotic genes has been discovered by deliberately manipulating their expression. Researchers can inactivate a gene by introducing a mutation that prevents its expression, or by deleting it entirely, an experiment called a **knockout**. A knockout organism (one

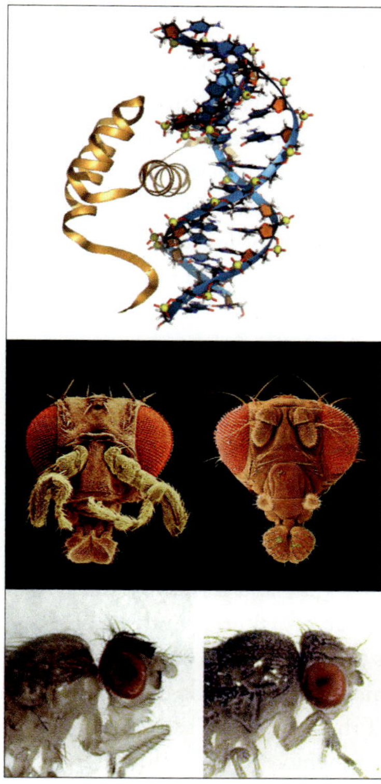

A The protein product (in gold) of the homeotic gene *antennapedia* attached to a promoter. The homeodomain is the region that binds to the DNA. Expression of *antennapedia* in embryonic tissues of the insect thorax causes legs to form.

B A mutation that triggers expression of the *antennapedia* gene in embryonic tissues of the head causes legs to form there too. Compare a normal fly, *right*.

C *Groucho* mutation *left*. Normal fly, *right*.

FIGURE 10.4 Effects of mutations in homeotic genes.

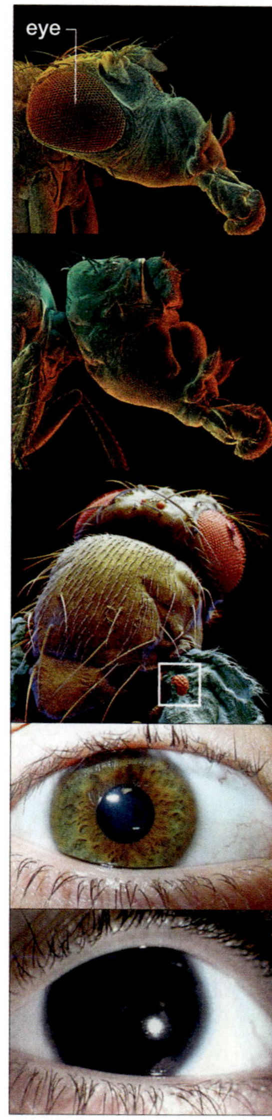

A Normal fruit fly with large, round eyes.

B A fruit fly with a mutation in its *eyeless* gene develops with no eyes.

C Eyes form wherever the *eyeless* gene is expressed in fly embryos—here, on the head and wing.

The *PAX6* gene of humans, mice, squids, and some other animals is so similar to *eyeless* that it also triggers eye development in flies.

D Normal human eye with a colored iris surrounding the pupil (dark area where light enters).

E Eye that developed without an iris, a result of a mutation in *PAX6*. This condition is called aniridia.

FIGURE 10.5 Eyes and *eyeless*.

that has had a gene knocked out) may differ from normal individuals, and the differences are clues to the function of the missing gene product.

Homeotic genes affect development by the same mechanisms in all multicelled eukaryotes, and many are interchangeable among different species. Thus, we can infer that they evolved in the most ancient eukaryotic cells. Homeodomains often differ among species only in conservative substitutions (one amino acid has replaced another with similar chemical properties). Consider the *eyeless* gene. Eyes form in embryonic fruit flies wherever this gene is expressed, which, in normal flies, is only in tissues of the head. If the *eyeless* gene is expressed in another part of the developing embryo, eyes form there too (**FIGURE 10.5C**). Humans, squids, mice, fish, and many other animals have a gene called *PAX6*, which is very similar in DNA sequence to the *eyeless* gene in flies. In humans, mutations in *PAX6* cause eye disorders such as aniridia, in which a person's irises are underdeveloped or missing (**FIGURE 10.5D,E**). If the *PAX6* gene from a human or mouse is inserted into a fly, it has the same effect as the *eyeless* gene: An eye forms wherever it is expressed. Such studies are evidence of shared ancestry among these evolutionarily distant animals.

homeotic gene Type of master gene; its expression controls formation of specific body parts during development.
knockout An experiment in which a gene is deliberately inactivated in a living organism; also, an organism that carries a knocked-out gene.
master gene Gene encoding a product that affects the expression of many other genes.

TAKE-HOME MESSAGE 10.2

- Animal development is orchestrated by cascades of master gene expression in embryos.
- The expression of homeotic genes during development governs the formation of specific body parts.
- Homeotic genes that function in similar ways in evolutionarily distant animals are evidence of shared ancestry.

CREDITS: (4A) From Starr/Taggart/Evers/Starr, Biology, 13E. © 2013 Cengage Learning; (4B) left, © Visuals Unlimited; right, © Jürgen Berger, Max-Planck-Institut for Developmental Biology, Tübingen; (4C) Courtesy of Dr. Barbara Jennings, UCL Cancer Institute, www.ucl.ac.uk; (5A,B) David Scharf/Science Source; (5C) Eye of Science/Science Source; (5D) M. Bloch; (5E) Courtesy of the Aniridia Foundation International, www.aniridia.net.

10.3 WHAT ARE SOME OUTCOMES OF GENE CONTROL IN EUKARYOTES?

Gene control influences many traits that are characteristic of humans and other eukaryotic organisms, as the following examples illustrate.

X MARKS THE SPOT

In humans and other mammals, a female's cells contain two X chromosomes, one inherited from her mother, the other one from her father. In each cell, one X chromosome is always tightly condensed (FIGURE 10.6A). We call the condensed X chromosomes **Barr bodies**, after Murray Barr, who discovered them. Condensation prevents transcription, so most of the genes on a Barr body are not expressed. This **X chromosome inactivation** ensures that only one of the two X chromosomes in a female's cells is active. According to a mechanism called **dosage compensation**, the inactivation equalizes expression of X chromosome genes between the sexes. The body cells of male mammals (XY) have one set of X chromosome genes. Body cells of female mammals (XX) have two sets, but female embryos do not develop properly when both sets are expressed.

X chromosome inactivation occurs when an embryo is a ball of about 200 cells. In humans and most other mammals, it occurs independently in every cell of a female embryo. The maternal X chromosome may get inactivated in one cell, and the paternal or maternal X chromosome may get inactivated in a cell next to it. Once the selection is made in a cell, all of that cell's descendants make the same selection as they continue dividing and forming tissues. As a result of random inactivation of maternal and paternal X chromosomes, an adult female mammal is a "mosaic" for the expression of X chromosome genes. She has patches of tissue in which genes of the maternal X chromosome are expressed, and patches in which genes of the paternal X chromosome are expressed (FIGURE 10.6B).

How does just one of two X chromosomes get inactivated? An X chromosome gene called *XIST* is transcribed on only one of the two X chromosomes. The gene's product, a long noncoding RNA, sticks to the chromosome that expresses the gene. The RNA coats the chromosome, and by an unknown mechanism causes it to condense into a Barr body. Thus, transcription of the *XIST* gene keeps the chromosome from transcribing other genes. The other chromosome does not express *XIST*, so it does not get coated with RNA; its genes remain available for transcription. How a cell "chooses" which X chromosome will express *XIST* is still unknown.

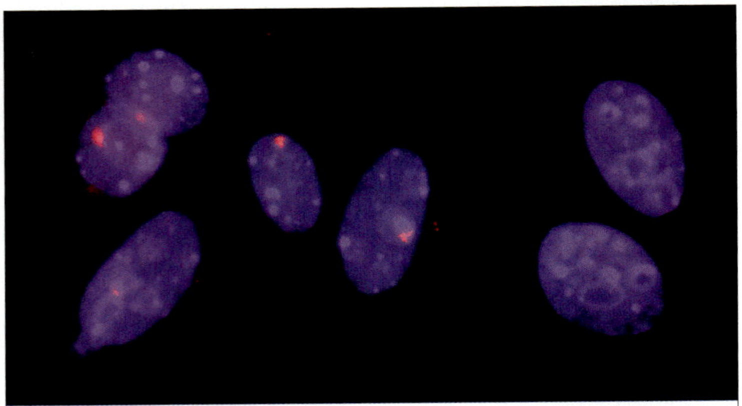

A Barr bodies are visible as red spots in the nucleus of the four XX cells on the *left*. Compare the nucleus of two XY cells to the *right*.

B When this cat was an embryo, one of the two X chromosomes was inactivated in each of her cells. The descendants of the cells formed her adult body, which is a mosaic for expression of X chromosome genes. Black fur arises in patches where genes on the X chromosome inherited from one parent are expressed; orange fur arises in patches where genes on the X chromosome inherited from the other parent are expressed.

FIGURE 10.6 {Animated} X chromosome inactivation.

MALE SEX DETERMINATION IN HUMANS

The human X chromosome carries 1,336 genes. Some of those genes are associated with sexual traits, such as the distribution of body fat and hair. However, most of the genes on the X chromosome govern nonsexual traits such as blood clotting and color perception. Such genes are expressed in both males and females. Males, remember, also inherit one X chromosome.

The human Y chromosome carries only 307 genes, but one of them is *SRY*—the master gene for male sex determination in mammals. Its expression in XY embryos triggers the formation of testes, which are male gonads. Some of the cells in these primary male reproductive

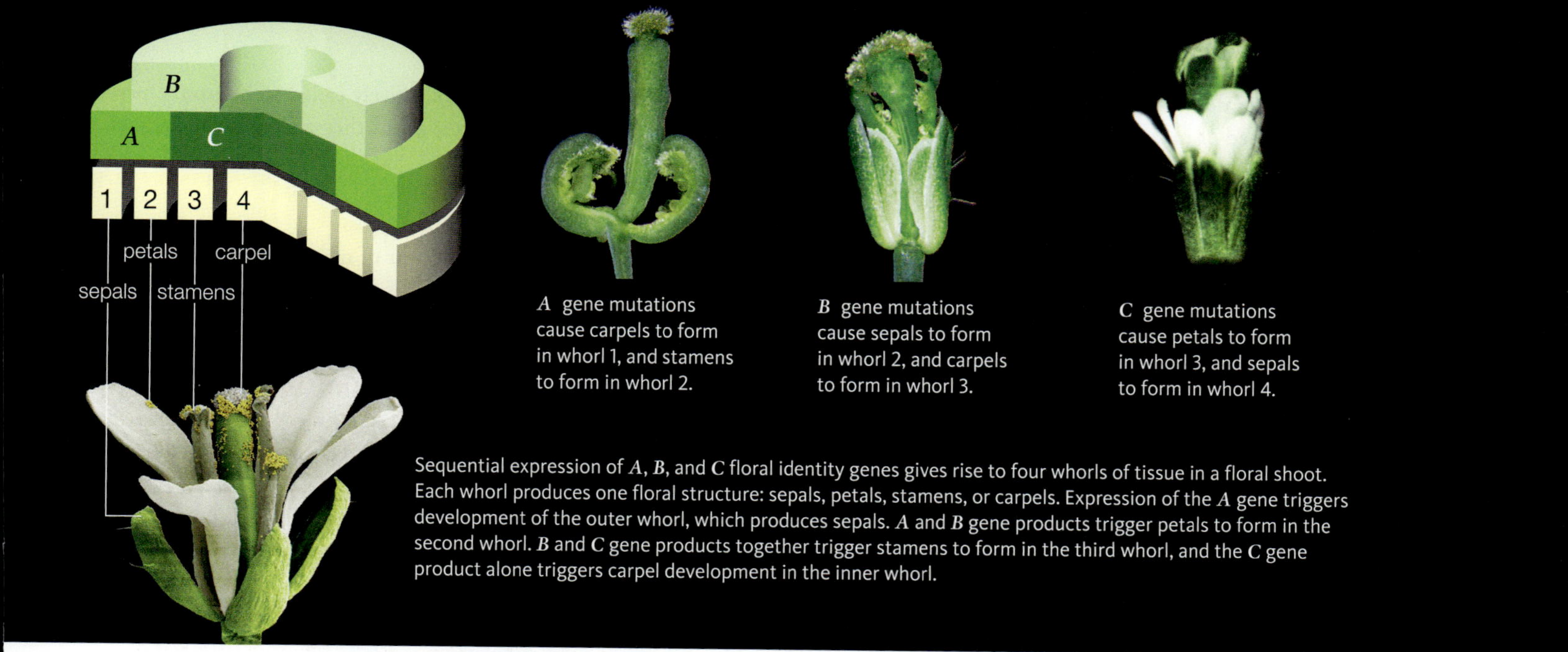

Sequential expression of *A*, *B*, and *C* floral identity genes gives rise to four whorls of tissue in a floral shoot. Each whorl produces one floral structure: sepals, petals, stamens, or carpels. Expression of the *A* gene triggers development of the outer whorl, which produces sepals. *A* and *B* gene products trigger petals to form in the second whorl. *B* and *C* gene products together trigger stamens to form in the third whorl, and the *C* gene product alone triggers carpel development in the inner whorl.

FIGURE 10.7 {Animated} Control of flower formation, as revealed by mutations in *Arabidopsis thaliana*.

FIGURE IT OUT: What type of genes are the floral identity genes?

Answer: They are homeotic genes.

organs make testosterone, a sex hormone that controls the emergence of male secondary sexual traits such as facial hair, increased musculature, and a deep voice. We know that *SRY* is the master gene that controls emergence of male sexual traits because mutations in this gene cause XY individuals to develop external genitalia that appear female. An XX embryo has no Y chromosome, no *SRY* gene, and much less testosterone, so primary female reproductive organs (ovaries) form instead of testes. Ovaries make estrogens and other sex hormones that will govern the development of female secondary sexual traits, such as enlarged, functional breasts, and fat deposits around the hips and thighs.

FLOWER FORMATION

In flowering plants, populations of cells in a shoot tip may give rise to a flower instead of leaves. Studies of mutations in thale cress, *Arabidopsis thaliana*, revealed the gene control behind this switch in development. Transcription factors produced by three sets of floral identity genes (called *A*, *B*, and *C*) guide the process. These genes are switched on by environmental cues such as seasonal changes in the length of night.

When a flower forms at the tip of a shoot, differentiating cells form whorls of tissue, one over the other like layers of an onion. Each whorl produces one type of floral structure—sepals, petals, stamens, or carpels. This pattern is dictated by sequential, overlapping expression of the *ABC* genes (**FIGURE 10.7**). The *A* genes are switched on first in a shoot tip, and their products trigger events that cause the outer whorl to form and give rise to sepals. *B* genes switch on before *A* genes turn off. Together, *A* and *B* gene products cause the second whorl to form and produce petals. Next, *A* genes turn off and the *C* gene switches on (there is only one *C* gene). Together, the products of the *B* and *C* genes trigger formation of the third whorl, which makes stamens. Finally, the *B* gene turns off, and the *C* gene product on its own gives rise to the fourth, inner whorl, which produces the carpel.

Barr body Inactivated X chromosome in a cell of a female mammal. The other X chromosome is active.

dosage compensation Mechanism in which X chromosome inactivation equalizes gene expression between males and females.

X chromosome inactivation Developmental shutdown of one of the two X chromosomes in the cells of female mammals.

TAKE-HOME MESSAGE 10.3

- X chromosome inactivation balances expression of X chromosome genes between female (XX) and male (XY) mammals.
- *SRY* gene expression triggers the development of male traits in mammals.
- In plants, expression of *ABC* floral identity genes governs development of the specialized parts of a flower.

10.4 WHAT ARE SOME OUTCOMES OF GENE CONTROL IN PROKARYOTES?

Prokaryotes do not undergo development, so these cells have no need of master genes. However, they do respond to environmental fluctuations by adjusting gene expression. For example, when a preferred nutrient becomes available, a bacterium begins transcribing genes whose products allow the cell to use that nutrient. When the nutrient is no longer available, transcription of those genes stops. Thus, the cell does not waste energy and resources producing gene products that are not needed at a particular moment.

Bacteria control gene expression mainly by adjusting the rate of transcription. Genes that are used together often occur together on the chromosome, one after the other. A single promoter precedes the genes, so all are transcribed together into a single RNA strand. Thus, their transcription is controllable in a single step that typically involves repressor binding to an operator. A group of genes together with a promoter and one or more operators that control their transcription are collectively called an **operon**. Operons were discovered in bacteria, but they also occur in archaea and eukaryotes.

THE *lac* OPERON

Escherichia coli lives in the gut of mammals, where it dines on nutrients traveling past. Its carbohydrate of choice is glucose, but it can make use of other sugars such as the lactose in milk. An operon called *lac* allows *E. coli* cells to metabolize lactose (**FIGURE 10.8**). The *lac* operon includes three genes and a promoter flanked by two operators ❶. One gene encodes an active transport protein (Section 5.7) that brings lactose across the plasma membrane into the cell. Another encodes an enzyme that breaks the bond between lactose's two monosaccharide monomers, glucose and galactose. (The third gene encodes an enzyme whose function is still being investigated.)

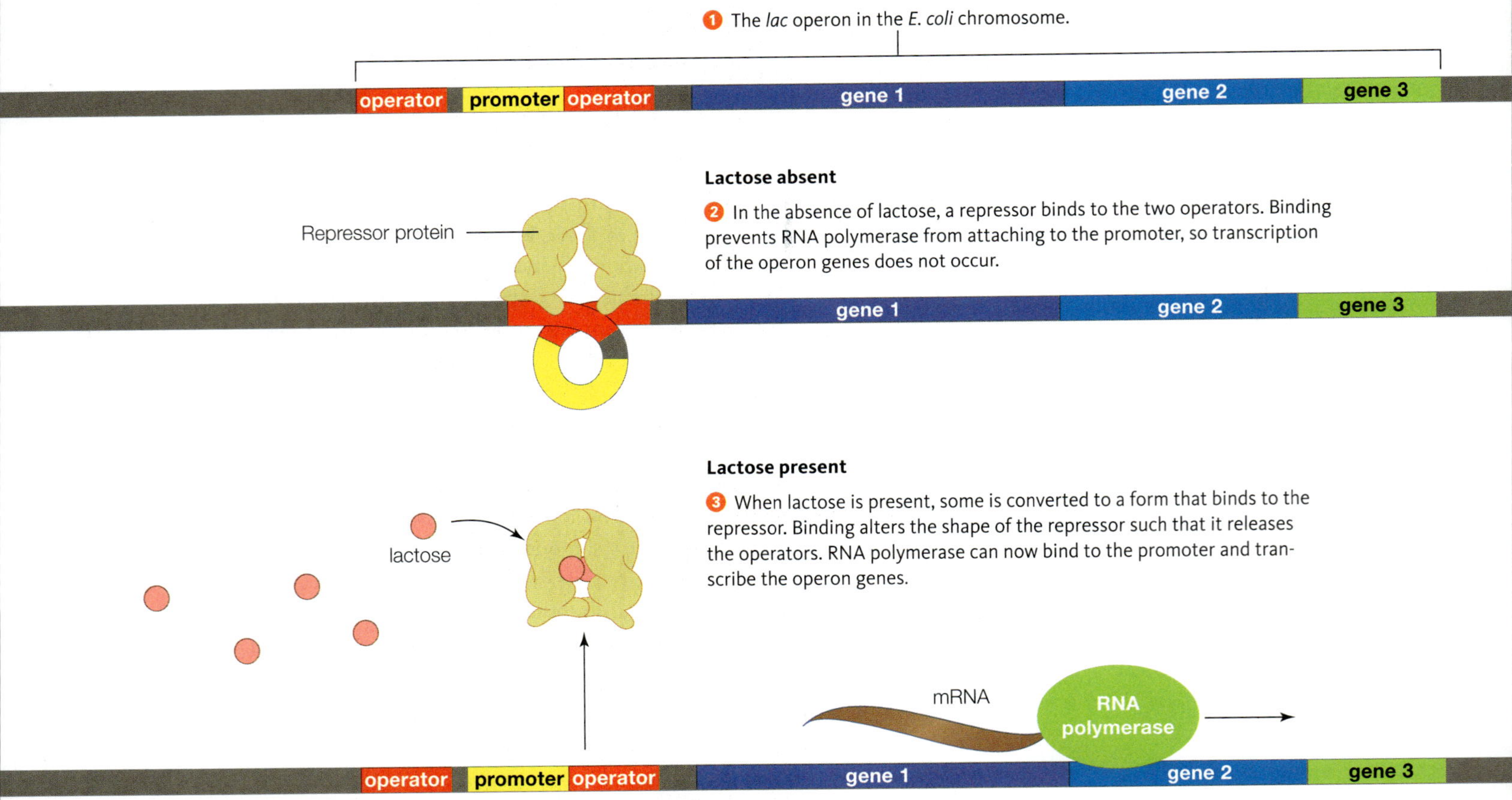

FIGURE 10.8 {Animated} Example of gene control in bacteria: the lactose (*lac*) operon on a bacterial chromosome. The operon consists of a promoter flanked by two operators, and three genes that allow lactose to be metabolized.

FIGURE IT OUT: What portion of the operon binds RNA polymerase when lactose is present?

Answer: The promoter

Recent analyses of DNA from well-preserved skeletons shows that the vast majority of adult humans living about 8,000 years ago in Europe were lactose intolerant. Around that time, a mutation appeared in the DNA of prehistoric people inhabiting a region between what is now central Europe and the Balkans. This mutation allowed its bearers to continue digesting milk as adults, and it spread rapidly to the rest of the continent along with the practice of dairy farming.

Today, most adults of northern and central European ancestry are able to digest milk because they carry this mutation, a single base-pair substitution in an enhancer that controls the lactase gene promoter. Other mutations in the same enhancer arose independently in North Africa, southern Asia, and the Middle East. Some people descended from these populations can continue to digest milk as adults.

FIGURE 10.9 Milk wasn't on the Stone Age menu.

Bacteria conserve resources by making these proteins only when lactose is present. When lactose is not present, a repressor binds to the two operators and twists the region of DNA with the promoter into a loop ❷. RNA polymerase cannot bind to the twisted-up promoter, so the *lac* operon's genes cannot be transcribed. When lactose is present, some of it is converted to another sugar that binds to the repressor and changes its shape. The altered repressor releases the operators and the looped DNA unwinds. The promoter is now accessible to RNA polymerase, and transcription of lactose-metabolizing genes begins ❸.

LACTOSE INTOLERANCE

Like *E. coli*, humans and other mammals break down lactose into monosaccharide subunits, but most do so only when they are young. An individual's ability to digest lactose declines at a species-specific age. In the majority of humans worldwide, the switch occurs at about age five, when expression of the gene for lactase shuts off. The resulting decline in production of this lactose-metabolizing enzyme results in a common condition known as lactose intolerance.

Cells in the intestinal lining secrete lactase into the small intestine, where the enzyme cleaves lactose into its glucose and galactose monomers. The monosaccharides are absorbed directly by the small intestine, but lactose and other disaccharides are not. Thus, when lactase production slows, lactose passes undigested through the small intestine. The lactose ends up in the large intestine, which hosts huge numbers of *E. coli* and a variety of other bacteria. These resident organisms respond to the abundant supply of this sugar by switching on their *lac* operons. Carbon dioxide, methane, hydrogen, and other gaseous products of their various fermentation reactions accumulate quickly in the large intestine, distending its wall and causing pain. Other products of their metabolism disrupt the solute–water balance inside the large intestine, and diarrhea results.

Not everybody is lactose intolerant. About one-third of human adults carry a mutation that allows them to digest milk; this mutation is more common in some populations than in others (**FIGURE 10.9**).

operon Group of genes together with a promoter–operator DNA sequence that controls their transcription.

TAKE-HOME MESSAGE 10.4

The bacterial *lac* operon allows expression of lactose-metabolizing proteins only when lactose is present.

Most adult humans do not produce the enzyme that breaks down lactose. When undigested lactose enters the large intestine, resident bacteria give rise to symptoms of lactose intolerance.

10.5 CAN GENE EXPRESSION PATTERNS BE INHERITED?

You learned in Section 10.1 that methylation of histone proteins silences DNA transcription. Direct methylation of DNA also suppresses gene expression, but in a way that is often more permanent than histone modification. Once a particular nucleotide has become methylated in a cell's DNA, it will usually stay methylated in all of the cell's descendants. Methylation and other heritable modifications to DNA that affect its function but do not involve changes in the nucleotide sequence are said to be **epigenetic**.

DNA methylation is a necessary part of differentiation, so it begins early in embryonic development: Genes actively expressed in cells of a zygote become silenced as their promoters get methylated. An individual's DNA continues to acquire methylations during its lifetime.

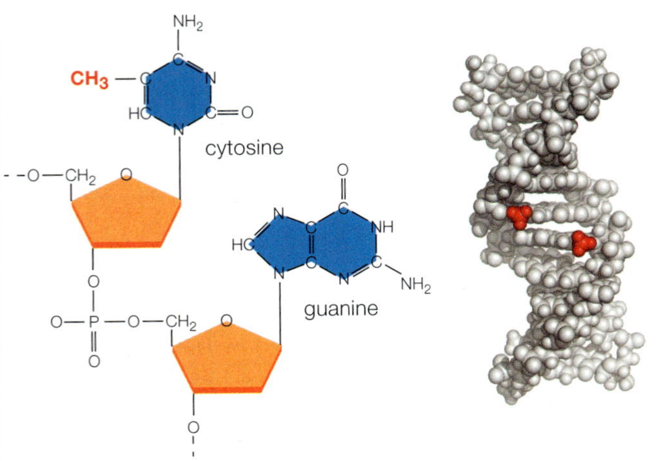

A With DNA methylation, a methyl group (red) is most often attached to a cytosine that is followed by a guanine. The model on the *right* shows methyl groups attached to a cytosine–guanine pair on complementary DNA strands. When the cytosine on one strand is methylated, enzymes methylate the cytosine on the other strand. This is why a methylation tends to persist in a cell's descendants.

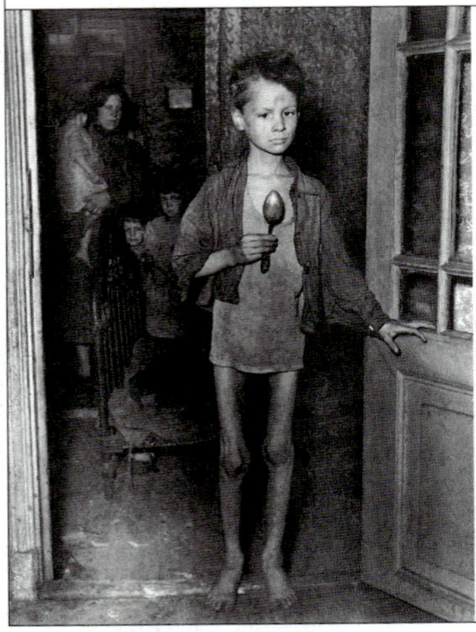

B DNA methylation and other epigenetic changes can be heritable. In 1944, a supply blockade followed by a very harsh winter caused a severe famine in the Nazi-occupied Netherlands. Grandsons of boys who endured the famine (such as the one pictured here) lived about 32 years longer than grandsons of boys who ate well during the same winter. These results were corrected for other factors so the effect is thought to be epigenetic.

FIGURE 10.10 DNA methylation.

In eukaryotes, methyl groups are usually added to a cytosine that is followed by a guanine (**FIGURE 10.10A**), but which of these cytosines are methylated varies by the individual. This is because methylation is influenced by environmental factors. For instance, humans conceived during a famine end up with an unusually low number of methyl groups attached to the nucleotides of certain genes. The product of one of those genes is a hormone that fosters prenatal growth and development. The resulting increase in expression of this gene may offer a survival advantage in a poor nutritional environment.

Methyl groups are added to nucleotides by chance during DNA replication, so cells that divide a lot tend to have more methyl groups in their DNA than inactive cells. Free radicals and toxic chemicals add more methyl groups. These and other factors that influence DNA methylation can have intergenerational effects. When an organism reproduces, it passes its DNA to offspring. Methylation of parental DNA is normally "reset" in the first cell of the new individual, with new methyl groups being added and old ones being removed. This reprogramming does not remove all of the parental methyl groups, however, so methylations acquired during an individual's lifetime can be passed to future offspring.

Inheritance of epigenetic modifications can adapt offspring to an environmental challenge much more quickly than evolution (we return to evolutionary processes in Chapter 17). Such modifications are not considered to be evolutionary because the underlying DNA sequence does not change. Even so, they may persist for generations after an environmental challenge has faded. For example, a recent study showed that grandsons of boys who endured a winter of famine tend to outlive—by far—grandsons of boys who overate at the same age (**FIGURE 10.10B**). The effect is presumed to be due to epigenetic modification because these results were corrected for socioeconomic and genetic factors. In a similar study, nine-year-old boys whose fathers smoked cigarettes before age 11 were overweight compared with boys whose fathers did not smoke in childhood. In these and other studies, the effect was sex-limited: Boys were affected by lifestyle of individuals in the paternal line; girls, by individuals in the maternal line.

epigenetic Refers to heritable changes in gene expression that are not the result of changes in DNA sequence.

TAKE-HOME MESSAGE 10.5

Epigenetic modifications of chromosomal DNA, including DNA methylations acquired during an individual's lifetime, can be passed to offspring.

10.6 Application: BETWEEN YOU AND ETERNITY

FIGURE 10.11 Pink Steel, a dragon boat team composed of breast cancer survivors.

"THERE IS A MOMENT WHEN EVERYTHING CHANGES—WHEN THE WIDTH OF TWO FINGERS CAN SUDDENLY BE THE TOTAL DISTANCE BETWEEN YOU AND ETERNITY." Seventeen-year-old Robin Shoulla wrote those words after being diagnosed with breast cancer. At an age when most young women are thinking about school, friends, parties, and potential careers, Robin was dealing with radical mastectomy: the removal of a breast, all lymph nodes under the arm, and skeletal muscles in the chest wall under the breast. She was pleading with her oncologist not to use her jugular vein for chemotherapy and wondering if she would survive to see the next year.

Cancer typically begins with a mutation in a gene whose product is part of a system of stringent control over cell growth and division. Two examples are *BRCA1* and *BRCA2*: A mutated version of one or both of these genes is often found in breast and ovarian cancer cells. Because such mutations can be inherited, cancer is not only a disease of the elderly (FIGURE 10.11).

BRCA1 and *BRCA2* are master genes whose protein products help maintain the structure and number of chromosomes in a dividing cell. They also participate directly in DNA repair, so any mutations that alter this function also alter the cell's capacity to repair damaged DNA. Other mutations are likely to accumulate, and that sets the stage for cancer.

Researchers recently found that the RNA product of the *XIST* gene does not properly coat one of the two X chromosomes in breast cancer cells. In these cells, both X chromosomes are active. It makes sense that two active X chromosomes would have something to do with abnormal gene expression, but why the RNA product of an unmutated *XIST* gene does not properly coat an X chromosome in cancer cells remains a mystery. Mutations in the *BRCA1* gene may be part of the answer. The protein product of this gene physically associates with the RNA product of the *XIST* gene. Researchers were able to restore proper *XIST* RNA coating—and proper X chromosome inactivation—by restoring the function of the *BRCA1* gene product in breast cancer cells.

Robin Shoulla is one of the unlucky people who carry mutations in both *BRCA1* and *BRCA2*, but she survived. Now, she has what she calls a normal life: career, husband, children. Her goal as a cancer survivor: "To grow very old with gray hair and spreading hips, smiling."

Summary

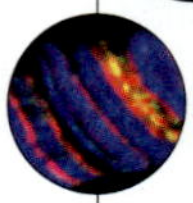

SECTION 10.1 Which genes a cell uses depends on the type of organism, the type of cell, conditions inside and outside the cell, and, in complex multicelled species, the organism's stage of development.

All organisms control gene expression. The control is necessary for cell differentiation during development in multicelled eukaryotes, and it also allows individual cells to respond appropriately to changes in their environment.

Different molecules and processes govern every step between transcription of a gene and delivery of the gene's product to its final destination. **Transcription factors** such as **activators** and **repressors** influence transcription by binding to chromosomal DNA at sites such as promoters, **enhancers**, **operators**, and silencers.

SECTION 10.2 Control over gene expression governs embryonic development of complex, multicelled animal bodies. Various **master genes** are expressed locally in different parts of an embryo as it develops. Their products, which diffuse through the embryo in gradients, affect expression of other master genes, which in turn affect the expression of others, and so on. Cells differentiate according to their exposure to these gradients. Eventually, master gene expression induces the expression of **homeotic genes**, the products of which govern the development of body parts. The function of many homeotic genes was revealed by **knockouts** in fruit flies.

SECTION 10.3 In cells of female mammals, one of the two X chromosomes is condensed as a **Barr body**, rendering most of its genes permanently inaccessible. By the theory of **dosage compensation**, this **X chromosome inactivation** balances gene expression between the sexes. The inactivation occurs because the *XIST* gene's RNA product shuts down the one X chromosome that transcribes it. The *SRY* gene determines male sex in humans.

Overlapping expression of master genes guides expression of flower formation in plants. Cells differentiate and form sepals, petals, stamens, or carpels depending on which floral identity gene products they are exposed to.

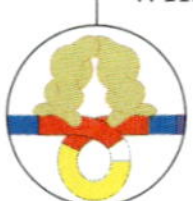

SECTION 10.4 Prokaryotes do not undergo development. Most of their gene control reversibly adjusts transcription rates in response to environmental conditions, especially nutrient availability. The *lac* **operon** governs expression of three genes, the three products of which allow a bacterial cell to metabolize lactose. Two operators that flank the promoter are binding sites for a repressor that blocks transcription. In most humans, the ability to break down lactose ends in early childhood.

SECTION 10.5 **Epigenetic** refers to heritable modifications of DNA that affect gene expression but do not involve changes to the DNA sequence. DNA methylations and other epigenetic modifications acquired during an individual's lifetime can persist for generations.

SECTION 10.6 A complex interplay of gene expression is a critical part of normal functioning of a multicelled body. Cancer typically begins with mutations in master genes that govern cell division.

Self-Quiz

Answers in Appendix VII

1. The expression of a gene may depend on ______ .
   a. the type of organism
   b. environmental conditions
   c. the type of cell
   d. all of the above

2. Gene expression in multicelled eukaryotic organisms changes in response to ______ .
   a. extracellular conditions
   b. master gene products
   c. operons
   d. a and b

3. Binding of ______ to ______ in DNA can increase the rate of transcription of specific genes.
   a. activators; repressors
   b. activators; enhancers
   c. repressors; operators
   d. both a and b

4. Proteins that influence RNA synthesis by binding directly to DNA are called ______ .
   a. promoters
   b. transcription factors
   c. operators
   d. enhancers

5. Eukaryotic gene control governs ______ .
   a. transcription
   b. RNA processing
   c. RNA transport
   d. mRNA degradation
   e. translation
   f. protein modification
   g. a through e
   h. all of the above

6. Muscle cells differ from bone cells because ______ .
   a. they carry different genes
   b. they use different genes
   c. both a and b

7. Control over eukaryotic gene expression guides ______ .
   a. natural selection
   b. nutrient availability
   c. development
   d. all of the above

8. Homeotic gene products ______ .
   a. flank a bacterial operon
   b. map out the overall body plan in embryos
   c. control the formation of specific body parts

9. A gene that is knocked out is ______ .
   a. deleted
   b. inactivated
   c. expressed
   d. either a or b

10. Which of the following includes all of the others?
    a. homeotic genes
    b. master genes
    c. *SRY* gene
    d. *PAX6*

11. The expression of *ABC* genes ______ .
    a. occurs in layers of an onion
    b. controls flower formation
    c. causes mutations in flowers

Data Analysis Activities

Effect of Paternal Grandmother's Food Supply on Infant Mortality Researchers are investigating long-reaching epigenetic effects of starvation, in part because historical data on periods of famine are widely available. Before the industrial revolution, a failed harvest in one autumn typically led to severe food shortages the following winter. A retrospective study has correlated female infant mortality at certain ages with the abundance of food during the paternal grandmother's childhood. **FIGURE 10.12** shows some of the results of this study.

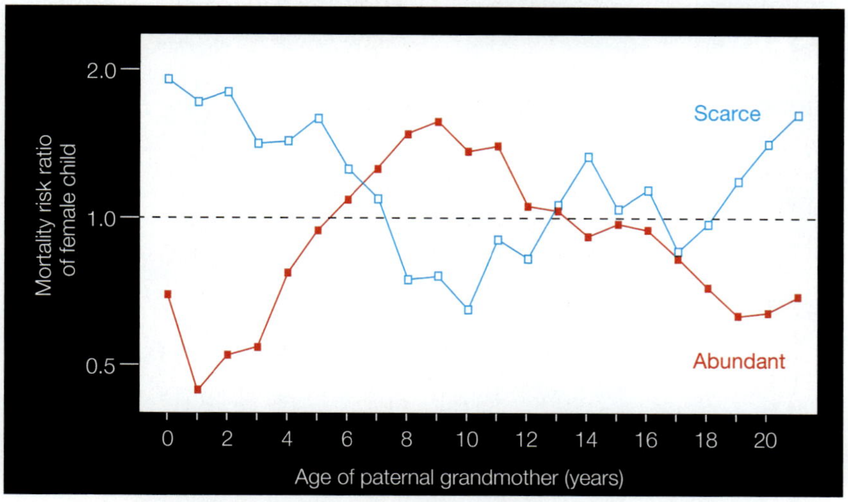

FIGURE 10.12 Graph showing the relative risk of early death of a female child, correlated with the age at which her paternal grandmother experienced a winter with a food supply that was scarce (blue) or abundant (red) during childhood. The dotted line represents no difference in risk of mortality. A value above the line means an increased risk; one below the line indicates a reduced risk.

1. Compare the mortality of girls whose paternal grandmothers ate well at age 2 with that of those who experienced famine at the same age. Which girl was more likely to die early? How much more likely was she to die?
2. Children have a period of slow growth around age 9. What trend in this data can you see around that age?
3. There was no correlation between early death of a male child and eating habits of his paternal grandmother, but there was a strong correlation with the eating habits of his paternal grandfather. What does this tell you about the probable location of epigenetic changes that gave rise to these data?

12. During X chromosome inactivation, _______ .
    a. female cells shut down
    b. RNA coats a chromosome
    c. pigments form
    d. both a and b

13. A cell with a Barr body is _______ .
    a. a bacterium
    b. a sex cell
    c. from a female mammal
    d. infected by the Barr virus

14. Operons _______ .
    a. only occur in bacteria
    b. have multiple genes
    c. involve selective gene expression

15. Match the terms with the most suitable description.

___ knockout	a. makes a man out of you
___ *SRY* gene	b. binding site for repressor
___ operator	c. requires gene expression control
___ Barr body	d. may have epigenetic effects
___ differentiation	e. inactivated X chromosome
___ methylation	f. gene deleted

Critical Thinking

1. Why are some genes expressed and some not?
2. Do the same types of gene control operate in bacterial cells and eukaryotic cells?
3. Suppose a breeder decides to separate baby guinea pigs from their mothers three weeks after they were born. However, he has trouble identifying the sex of young guinea pigs. Suggest how a quick look through a microscope can help him identify the females.

4. Almost all calico cats are female. Why?
5. The photos on the *left* show flowers from *Arabidopsis* plants. One plant is wild-type (unmutated); the other carries a mutation in one of its *ABC* floral identity genes. This mutation causes sepals and petals to form instead of stamens and carpels. Refer to **FIGURE 10.7** to decide which gene (*A*, *B*, or *C*) has been inactivated by the mutation.

CREDITS: (12) Pembrey et al., *European Journal of Human Genetics* (2006) 14, 159–166; (in text) top, Science Source; bottom, © Jose Luis Riechmann.

A multicelled eukaryote develops by repeated cell divisions. These are early frog embryos, each a product of three divisions of one fertilized egg.

11 HOW CELLS REPRODUCE

Links to Earlier Concepts

Before beginning this chapter, be sure you understand cell structure (Sections 4.1, 4.5, 4.9, 4.10); chromosomes (8.3); and DNA replication (8.4) and repair (8.5). What you know about receptors and recognition proteins (4.3), free radicals (5.5) and mutations (9.5), fermentation (7.6), and eukaryotic gene control (10.1) will help you understand how cancer develops.

KEY CONCEPTS

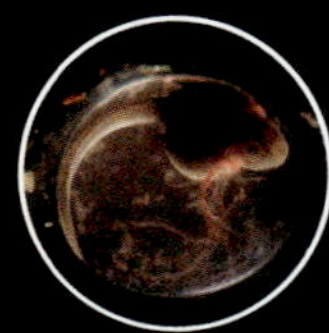

THE CELL CYCLE

A cell cycle starts when a new cell forms by division of a parent cell, and ends when the cell completes its own division. Built-in checkpoints control the timing and rate of the cycle.

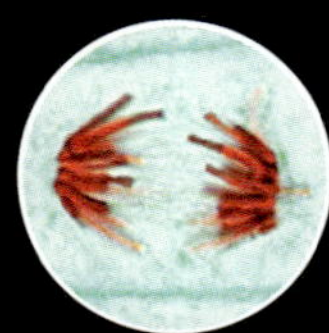

MITOSIS

Mitosis, a mechanism by which a cell's nucleus divides, maintains the chromosome number. Four sequential stages parcel the cell's duplicated chromosomes into two new nuclei.

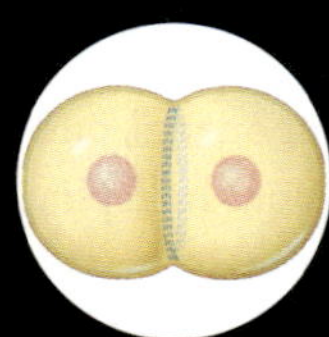

CYTOPLASMIC DIVISION

After nuclear division, the cytoplasm may divide, so one nucleus ends up in each of two new cells. The division proceeds by different mechanisms in animal and plant cells.

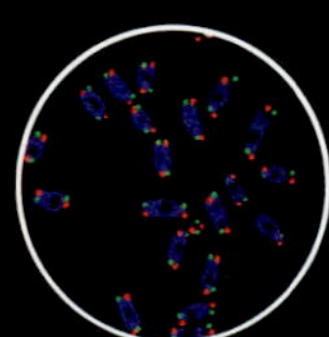

MITOTIC CLOCKS

Built into eukaryotic chromosomes are DNA sequences that protect the cell's genetic information. Degradation of these sequences is associated with cell death and aging.

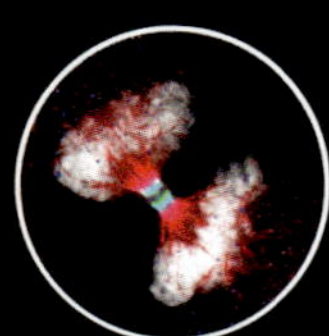

THE CELL CYCLE GONE AWRY

On rare occasions, checkpoint mechanisms fail, and cell division becomes uncontrollable. Tumor formation and cancer are outcomes.

Photograph by Carolina Biological Supply Company/Phototake.

11.1 HOW DO CELLS REPRODUCE?

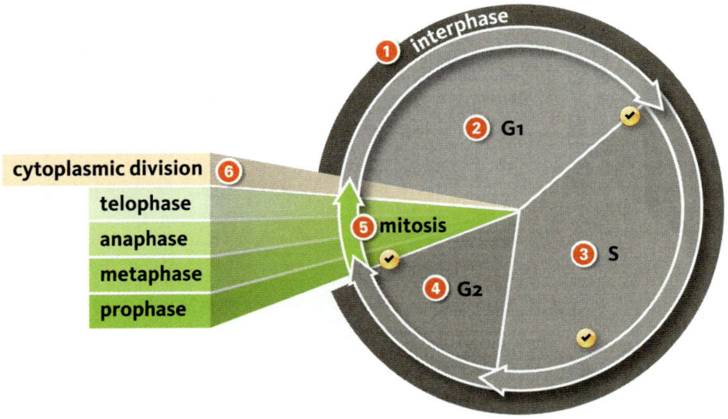

1 A cell spends most of its life in interphase, which includes three stages: G1, S, and G2.

2 G1 is the interval of growth before DNA replication. The cell's chromosomes are unduplicated.

3 S is the time of synthesis, during which the cell copies its DNA (duplicates its chromosomes).

4 G2 is the interval after DNA replication and before mitosis. The cell prepares to divide during this stage.

5 The nucleus divides during mitosis, the four stages of which are detailed in the next section. After mitosis, the cytoplasm may divide. Each descendant cell begins the cycle anew, in interphase.

✔ Built-in checkpoints stop the cycle from proceeding until certain conditions are met.

FIGURE 11.1 {Animated} The eukaryotic cell cycle. The length of the intervals differs among cells. G1, S, and G2 are part of interphase.

MULTIPLICATION BY DIVISION

A life cycle is the collective series of events that an organism passes through during its lifetime. Multicelled organisms and free-living cells have life cycles, but what about cells that make up a multicelled body? Biologists consider such cells to be individually alive, each with its own life that passes through a series of recognizable stages. The events that occur from the time a cell forms until the time it divides are collectively called the **cell cycle** (**FIGURE 11.1**).

A typical cell spends most of its life in **interphase** 1. During this phase, the cell increases its mass, roughly doubles the number of its cytoplasmic components, and replicates its DNA in preparation for division. Interphase is typically the longest part of the cycle, and it consists of three stages: G1, S, and G2. G1 and G2 were named "Gap" intervals because outwardly they seem to be periods of inactivity, but they are not.

Most cells going about their metabolic business are in G1 2. Cells preparing to divide enter S 3, the time of DNA synthesis, when they duplicate their chromosomes (Section 8.4). During G2 4, the cell prepares to divide by making the proteins that will drive the process of division. Once S begins, DNA replication usually proceeds at a predictable rate until division begins.

The remainder of the cycle consists of the division process itself. When a cell divides, both of its cellular offspring end up with DNA and a blob of cytoplasm. Each of the offspring of a eukaryotic cell inherits its DNA packaged inside a nucleus. Thus, a eukaryotic cell's nucleus has to divide before its cytoplasm does.

Mitosis is a nuclear division mechanism that maintains the chromosome number 5. In multicelled organisms, mitosis and cytoplasmic division 6 are the basis of increases in body size and tissue remodeling during development (**FIGURE 11.2**), as well as ongoing replacements of damaged or dead cells. Mitosis and cytoplasmic division are also part of **asexual reproduction**, a reproductive mode by which offspring are produced by one parent only. This mode of reproduction is used by some multicelled eukaryotes and many single-celled ones. (Prokaryotes do not have a nucleus and do not undergo mitosis. We discuss their reproduction in Section 19.4.)

When a cell divides by mitosis, it produces two descendant cells, each with the chromosome number of the parent. However, if only the total number of chromosomes mattered, then one of the descendant cells might get, say, two pairs of chromosome 22 and no chromosome 9. A cell cannot function properly without a full complement of DNA, which means it needs to have *a copy of each*

FIGURE 11.2 A tadpole develops from repeated mitotic divisions of an egg. After it hatches, the individual will grow and undergo metamorphosis to develop into a frog.

chromosome. Thus, the two cells produced by mitosis have the same number and types of chromosomes as the parent.

Remember from Section 8.3 that your body's cells are diploid, which means their nuclei contain pairs of chromosomes—two of each type. One chromosome of a pair was inherited from your father; the other, from your mother. Except for a pairing of nonidentical sex chromosomes (XY) in males, the chromosomes of each pair are homologous. **Homologous chromosomes** have the same length, shape, and genes (*hom*– means alike).

FIGURE 11.3 shows how homologous chromosomes are distributed to descendant cells when a diploid cell divides by mitosis. When a cell is in G1, each of its chromosomes consists of one double-stranded DNA molecule. The cell replicates its DNA in S, so by G2, each of its chromosomes consists of two double-stranded DNA molecules. These molecules stay attached to one another at the centromere as sister chromatids until mitosis is almost over, and then they are pulled apart and packaged into two separate nuclei. The next section details this process.

When sister chromatids are pulled apart, each becomes an individual chromosome that consists of one double-stranded DNA molecule. Thus, each of the two new nuclei that form in mitosis contains a full complement of (unduplicated) chromosomes. When the cytoplasm divides, these nuclei are packaged into separate cells. Each new cell starts the cell cycle over again in G1 of interphase.

CONTROL OVER THE CELL CYCLE

When a cell divides—and when it does not—is determined by mechanisms of gene expression control (Section 10.1). Like the accelerator of a car, some of these mechanisms cause the cell cycle to advance. Others are like brakes, preventing the cycle from proceeding. In the adult body, brakes on the cell cycle normally keep the vast majority of cells in G1. Most of your nerve cells, skeletal muscle cells, heart muscle cells, and fat-storing cells have been in G1 since you were born, for example.

Control over the cell cycle also ensures that a dividing cell's descendants receive intact copies of its chromosomes.

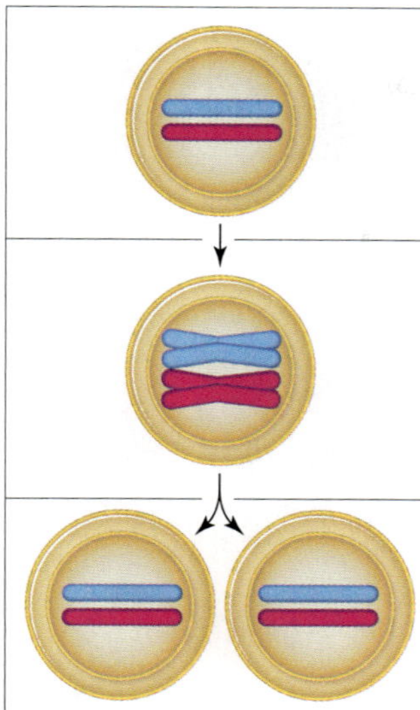

FIGURE 11.3 How mitosis maintains chromosome number in a diploid cell. For clarity, only one homologous pair is shown. The maternal chromosome is shown in pink, the paternal one in blue.

Built-in checkpoints monitor whether the cell's DNA has been copied completely, whether it is damaged, or even whether enough nutrients to support division are available. Protein products of "checkpoint genes" interact to carry out this type of control. For example, a checkpoint that operates in S puts the brakes on the cycle if the cell's chromosomes are damaged during DNA replication (Section 8.5). Checkpoint proteins that function as sensors recognize damaged DNA and bind to it. Upon binding, they trigger other events that stall the cell cycle, and also enhance expression of genes involved in DNA repair. After the problem has been corrected, the brakes are lifted and the cell cycle proceeds. If the problem remains uncorrected, other checkpoint proteins may initiate a series of events that eventually cause the cell to self-destruct.

TAKE-HOME MESSAGE 11.1

A cell cycle is the sequence of stages through which a cell passes during its lifetime (interphase, mitosis, and cytoplasmic division).

A eukaryotic cell reproduces by division: nucleus first, then cytoplasm. Each descendant cell receives a set of chromosomes and some cytoplasm.

When a nucleus divides by mitosis, each new nucleus has the same chromosome number as the parent cell.

Mechanisms of gene expression control can advance, delay, or block the cell cycle in response to internal and external conditions. Checkpoints built into the cycle allow problems to be corrected before the cycle proceeds.

asexual reproduction Reproductive mode of eukaryotes by which offspring arise from a single parent only.
cell cycle A series of events from the time a cell forms until its cytoplasm divides.
homologous chromosomes Chromosomes with the same length, shape, and genes.
interphase In a eukaryotic cell cycle, the interval between mitotic divisions when a cell grows, roughly doubles the number of its cytoplasmic components, and replicates its DNA.
mitosis Nuclear division mechanism that maintains the chromosome number.

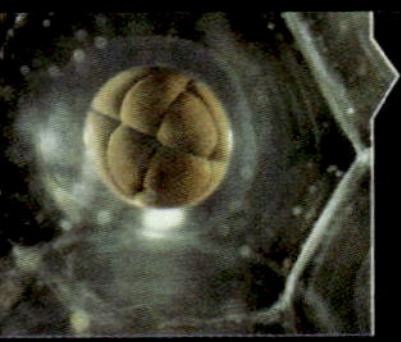

11.2 WHAT IS THE SEQUENCE OF EVENTS DURING MITOSIS?

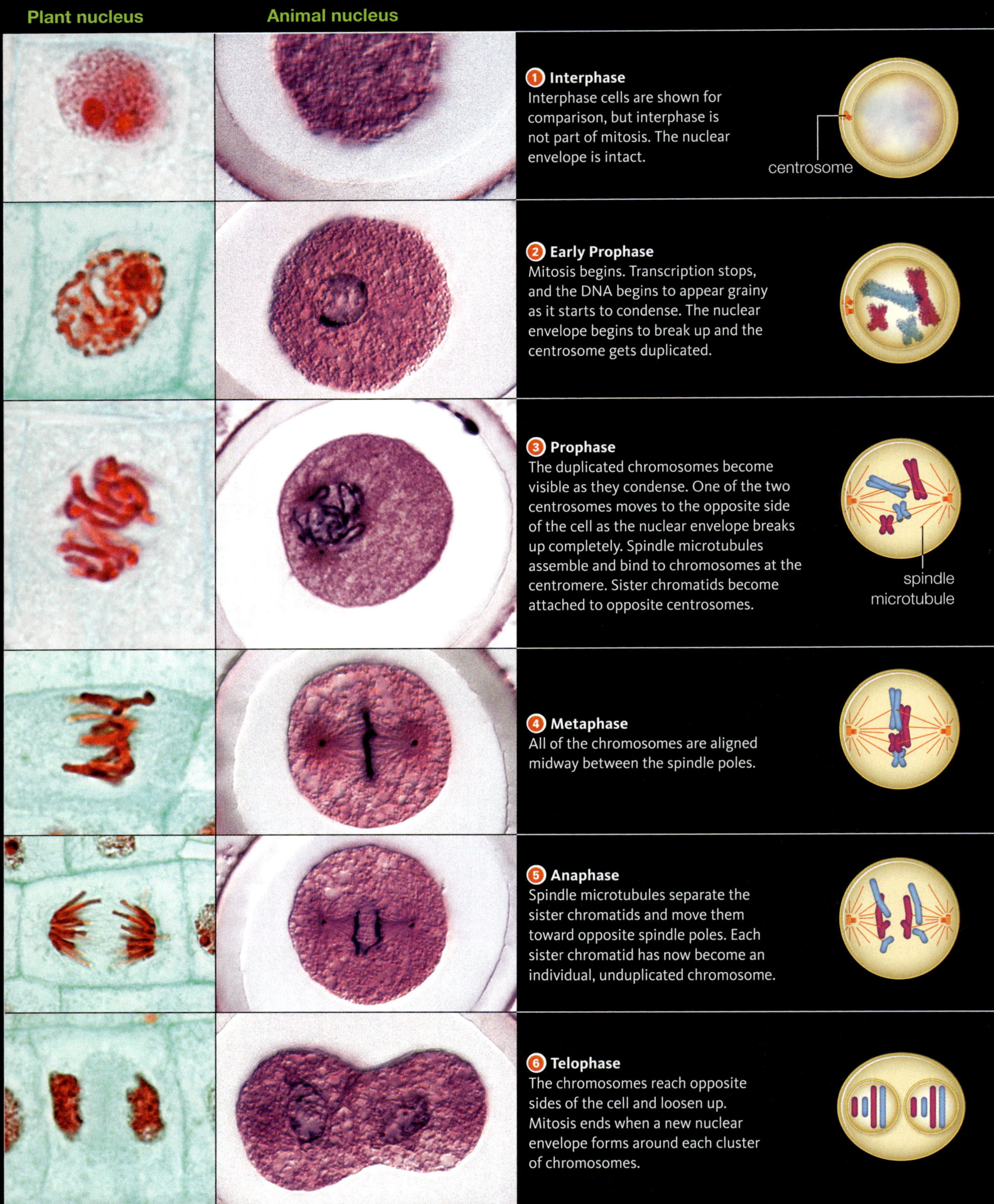

FIGURE 11.4 {Animated} Mitosis. Micrographs show nuclei of plant cells (onion root, *left*), and animal cells (fertilized eggs of a roundworm, *right*). A diploid (2*n*) animal cell with two chromosome pairs is illustrated.

CREDITS: (4) Art, © Cengage Learning; plant cell photos, Michael Clayton/ University of Wisconsin, Department of Botany; animal cell photos, ISM/Phototake.

During interphase, a cell's chromosomes are loosened to allow transcription and DNA replication. Loosened chromosomes are spread out, so they are not easily visible under a light microscope (**FIGURE 11.4** ❶). In preparation for nuclear division, the chromosomes begin to pack tightly ❷. Transcription and DNA replication stop as the chromosomes condense into their most compact "X" forms (Section 8.3). Tight condensation keeps the chromosomes from getting tangled and breaking during nuclear division.

Just before prophase, the centrosome becomes duplicated. Most animal cells have these structures, which typically consist of a pair of centrioles surrounded by a region of dense cytoplasm. (Remember from Section 4.9 that barrel-shaped centrioles help microtubules assemble.)

A cell reaches **prophase**, the first stage of mitosis, when its chromosomes have condensed so much that they are visible under a light microscope ❸. "Mitosis" is from *mitos*, the Greek word for thread, after the threadlike appearance of the chromosomes during nuclear division. If the cell has centrosomes, one of them now moves to the opposite side of the cell. Microtubules begin to assemble and lengthen from the centrosomes (or from other structures in cells with no centrosomes). The lengthening microtubules form a **spindle**, which is a temporary structure for moving chromosomes during nuclear division (**FIGURE 11.5**). The general area from which the spindle forms on each side of the cell is now called a spindle pole.

Spindle microtubules penetrate the nuclear region as the nuclear envelope breaks up. Some of the microtubules stop lengthening when they reach the middle of the cell. Others lengthen until they reach a chromosome and attach to it at the centromere. By the end of prophase, one sister chromatid of each chromosome has become attached to microtubules extending from one spindle pole, and the other sister has become attached to microtubules extending from the other spindle pole.

The opposing sets of microtubules then begin a tug-of-war by adding and losing tubulin subunits. As the microtubules lengthen and shorten, they push and pull the chromosomes. When all the microtubules are the same length, the chromosomes are aligned midway between spindle poles ❹. The alignment marks **metaphase** (from *meta*, the ancient Greek word for between).

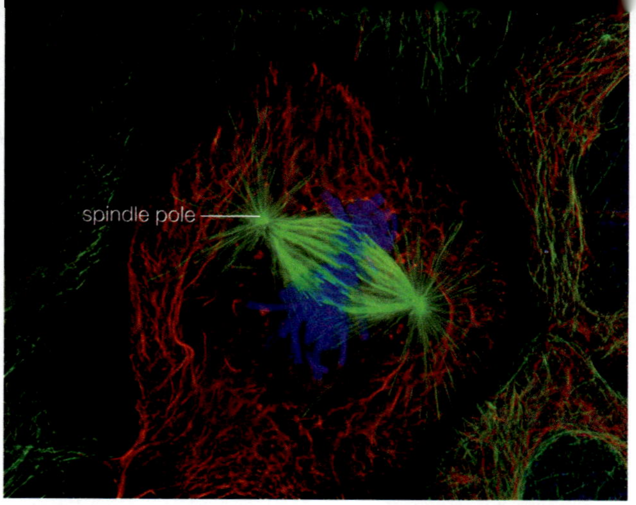

FIGURE 11.5 The spindle in a dividing cell of an amphibian. Microtubules (green) have extended from two centrosomes to form the spindle, which has attached to and aligned the chromosomes (blue) midway between its two poles. Red shows actin microfilaments.

During **anaphase**, the spindle pulls the sister chromatids of each duplicated chromosome apart and moves them toward opposite spindle poles ❺. Each DNA molecule has now become a separate chromosome.

Telophase begins when two clusters of chromosomes reach the spindle poles ❻. Each cluster has the same number and kinds of chromosomes as the parent cell nucleus had: two of each type of chromosome, if the parent cell was diploid. A new nuclear envelope forms around each set of chromosomes as they loosen up again. At this point, telophase—and mitosis—are finished.

anaphase Stage of mitosis during which sister chromatids separate and move toward opposite spindle poles.
metaphase Stage of mitosis at which all chromosomes are aligned midway between spindle poles.
prophase Stage of mitosis during which chromosomes condense and become attached to a newly forming spindle.
spindle Temporary structure that moves chromosomes during nuclear division; consists of microtubules.
telophase Stage of mitosis during which chromosomes arrive at opposite spindle poles and decondense, and two new nuclei form.

TAKE-HOME MESSAGE 11.2

Chromosomes are duplicated before mitosis begins. Each now consists of two DNA molecules attached as sister chromatids.

In prophase, the chromosomes condense and a spindle forms. Spindle microtubules attach to the chromosomes as the nuclear envelope breaks up.

At metaphase, the spindle has aligned all of the (still duplicated) chromosomes in the middle of the cell.

In anaphase, sister chromatids separate and move toward opposite spindle poles. Each DNA molecule is now an individual chromosome.

In telophase, two clusters of chromosomes reach opposite spindle poles. A new nuclear envelope forms around each cluster, so two new nuclei form.

At the end of mitosis, each new nucleus has the same number and types of chromosomes as the parent cell's nucleus.

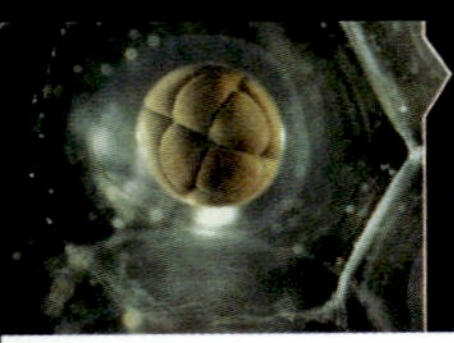

11.3 HOW DOES A EUKARYOTIC CELL DIVIDE?

In most eukaryotes, the cell cytoplasm divides between late anaphase and the end of telophase, so two cells form, each with their own nucleus. The mechanism of cytoplasmic division, which is called **cytokinesis**, differs between plants and animals.

Typical animal cells pinch themselves in two after nuclear division ends (**FIGURE 11.6**). How? The spindle begins to disassemble during telophase (1). The cell cortex, which is the mesh of cytoskeletal elements just under the plasma membrane (Section 4.9), includes a band of actin and myosin filaments that wraps around the cell's midsection. The band is called a contractile ring because it contracts when its component proteins are energized by phosphate-group transfers from ATP. When the ring contracts, it drags the attached plasma membrane inward (2). The sinking plasma membrane becomes visible on the outside of the cell as an indentation between the former spindle poles (3). The indentation, which is called a **cleavage furrow**, advances around the cell and deepens until the cytoplasm (and the cell) is pinched in two (4). Each of the two cells formed by this division has its own nucleus and some of the parent cell's cytoplasm, and each is enclosed by a plasma membrane.

Dividing plant cells face a particular challenge because a stiff cell wall surrounds their plasma membrane (Section 4.10). Accordingly, plant cells have their own mechanism of cytokinesis. By the end of anaphase, a set of short microtubules has formed on either side of the future plane of division. These microtubules guide vesicles from Golgi bodies and the cell surface to the division plane (5). After mitosis, the vesicles start to fuse into a disk-shaped **cell plate** (6). The plate expands at its edges until it reaches the plasma membrane and attaches to it, thus partitioning the cytoplasm (7). In time, the cell plate will develop into two new cell walls, so each of the descendant cells will be enclosed by its own plasma membrane and wall (8).

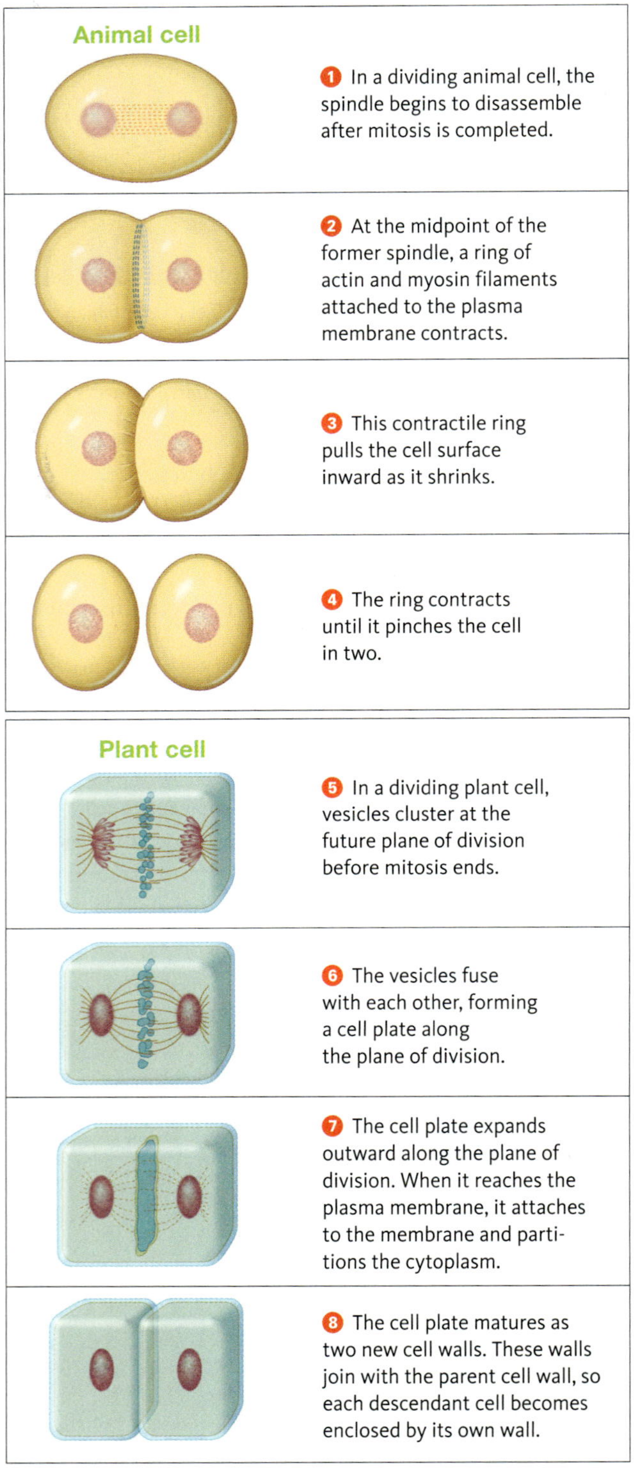

FIGURE 11.6 {Animated} Cytoplasmic division of animal cells (*top*) and plant cells (*bottom*).

cell plate A disk-shaped structure that forms during cytokinesis in a plant cell; matures as a cross-wall between the two new nuclei.
cleavage furrow In a dividing animal cell, the indentation where cytoplasmic division will occur.
cytokinesis Cytoplasmic division.

TAKE-HOME MESSAGE 11.3

In most eukaryotes, the cell cytoplasm divides between late anaphase and the end of telophase. Two descendant cells form, each with its own nucleus.

The mechanism of cell division differs between plants and animals.

In animal cells, a contractile ring pinches the cytoplasm in two. In plant cells, a cell plate that forms in the middle of the cell partitions the cytoplasm when it reaches and connects to the parent cell wall.

11.4 WHAT IS THE FUNCTION OF TELOMERES?

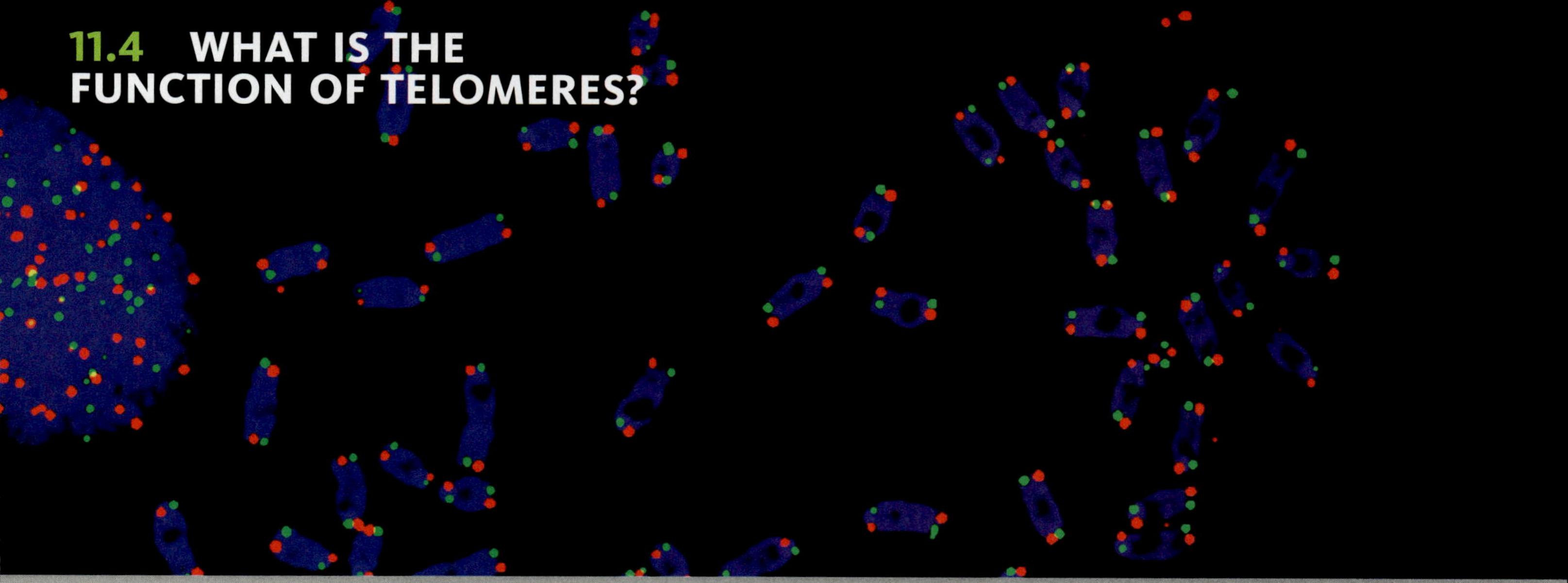

FIGURE 11.7 Telomeres. The bright dots at the end of each DNA strand in these duplicated chromosomes show telomere sequences.

In 1997, Scottish geneticist Ian Wilmut made worldwide headlines after his team cloned the first mammal from an adult somatic cell (SCNT, Section 8.6). The animal, a lamb named Dolly, was genetically identical to the sheep that had donated an udder cell. At first, Dolly looked and acted like a normal sheep, but she died early. By the time Dolly was five, she was as fat and arthritic as a twelve-year-old sheep. The following year, she contracted a lung disease that is typical of geriatric sheep, and had to be euthanized.

Dolly's early demise may have been the result of abnormally short telomeres. **Telomeres** are noncoding DNA sequences that occur at the ends of eukaryotic chromosomes (**FIGURE 11.7**). Vertebrate telomeres consist of a short DNA sequence, 5′-TTAGGG-3′, repeated perhaps thousands of times. These "junk" repeats provide a buffer against the loss of more valuable DNA internal to the chromosomes.

A telomere buffer is particularly important because, under normal circumstances, a eukaryotic chromosome shortens by about 100 nucleotides with each DNA replication. When a cell's offspring receive chromosomes with too-short telomeres, checkpoint gene products halt the cell cycle, and the descendant cells die shortly thereafter. Most body cells can divide only a certain number of times before this happens. This cell division limit may be a fail-safe mechanism in case a cell loses control over the cell cycle and begins to divide again and again. A limit on the number of divisions keeps such cells from overrunning the body (an outcome that, as you will see in the next section, has dangerous consequences to health). The cell division limit varies by species, and it may be part of the mechanism that sets an organism's life span. When Dolly was only two years old, her telomeres were as short as those of a six-year-old sheep—the exact age of the adult animal that had been her genetic donor. Scientists are careful to point out that shortening telomeres could be an effect of aging rather than a cause.

A few normal cells in an adult retain the ability to divide indefinitely. Their descendants replace cell lineages that eventually die out when they reach their division limit. These cells, which are called stem cells, are immortal because they continue to make an enzyme called telomerase. Telomerase reverses the telomere shortening that normally occurs after DNA replication.

Mice that have had their telomerase enzyme knocked out age prematurely. Their tissues degenerate much more quickly than those of normal mice, and their life expectancy declines to about half that of a normal mouse. When one of these knockout mice is close to the end of its shortened life span, rescuing the function of its telomerase enzyme results in lengthened telomeres. The rescued mouse also regains vitality: Decrepit tissue in its brain and other organs repairs itself and begins to function normally, and the once-geriatric individual even begins to reproduce again. While telomerase holds therapeutic promise for rejuvenation of aged tissues, it can also be dangerous: Cancer cells characteristically express high levels of the molecule.

TAKE-HOME MESSAGE 11.4

Telomeres protect eukaryotic chromosomes from losing genetic information at their ends.

Telomeres shorten with every cell division in normal body cells. When they are too short, the cell stops dividing and dies. Thus, telomeres are associated with aging.

telomere Noncoding, repetitive DNA sequence at the end of chromosomes; protects the coding sequences from degradation.

11.5 WHAT HAPPENS WHEN CONTROL OVER THE CELL CYCLE IS LOST?

THE ROLE OF MUTATIONS

Sometimes a checkpoint gene mutates so that its protein product no longer works properly. In other cases, the controls that regulate its expression fail, and a cell makes too much or too little of its product. When enough checkpoint mechanisms fail, a cell loses control over its cell cycle. Interphase may be skipped, so division occurs over and over with no resting period. Signaling mechanisms that cause abnormal cells to die may stop working. The problem is compounded because checkpoint malfunctions are passed along to the cell's descendants, which form a **neoplasm**, an accumulation of cells that lost control over how they grow and divide.

A neoplasm that forms a lump in the body is called a **tumor**, but the two terms are sometimes used interchangeably. Once a tumor-causing mutation has occurred, the gene it affects is called an oncogene. An **oncogene** is any gene that can transform a normal cell into a tumor cell (Greek *onkos*, or bulging mass). Oncogene mutations in reproductive cells can be passed to offspring, which is a reason that some types of tumors run in families.

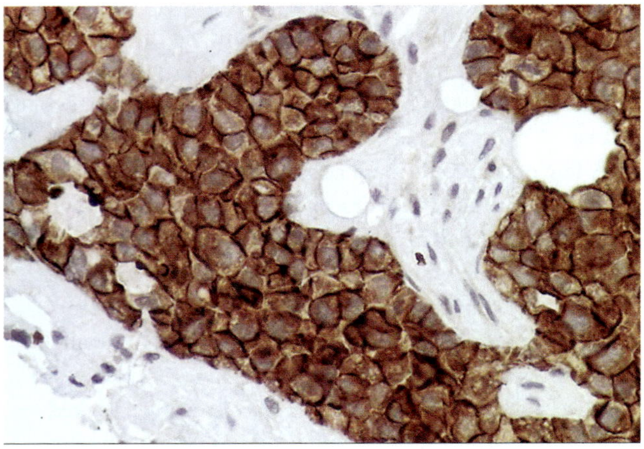

FIGURE 11.8 Effects of an oncogene. In this section of human breast tissue, a brown-colored tracer shows the active form of the EGF receptor. Normal cells are lighter in color. The dark cells have an overactive EGF receptor that is constantly stimulating mitosis; these cells have formed a neoplasm.

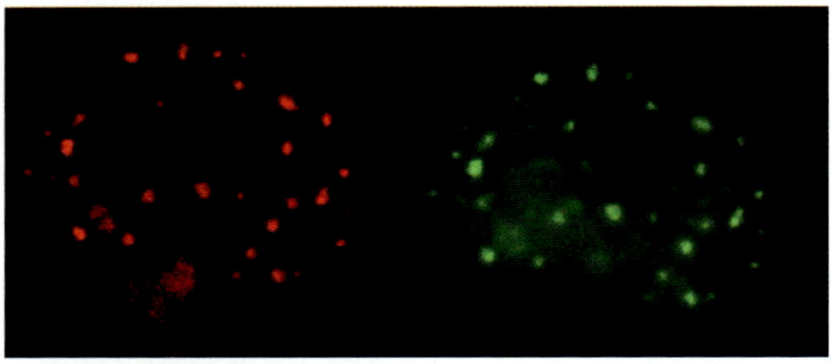

A Red dots show the location of the *BRCA1* gene product.

B Green dots pinpoint the location of another checkpoint gene product.

FIGURE 11.9 Checkpoint genes in action. Radiation damaged the DNA inside this nucleus. Two proteins have clustered around the same chromosome breaks in the same nucleus; both function to recruit DNA repair enzymes. The integrated action of these and other checkpoint gene products blocks mitosis until the DNA breaks are fixed.

Genes encoding proteins that promote mitosis are called **proto-oncogenes** because mutations can turn them into oncogenes. A gene that encodes the epidermal growth factor (EGF) receptor is an example of a proto-oncogene. **Growth factors** are molecules that stimulate a cell to divide and differentiate. The EGF receptor is a plasma membrane protein; when it binds to EGF, it becomes activated and triggers the cell to begin mitosis. Mutations can result in an EGF receptor that stimulates mitosis even when EGF is not present. Most neoplasms carry mutations resulting in an overactivity or overabundance of this particular receptor (FIGURE 11.8).

Checkpoint gene products that inhibit mitosis are called tumor suppressors because tumors form when they are missing. The products of the *BRCA1* and *BRCA2* genes (Section 10.6) are examples of tumor suppressors. These proteins regulate, among other things, the expression of DNA repair enzymes (FIGURE 11.9). Tumor cells often have mutations in their *BRCA* genes.

Viruses such as HPV (human papillomavirus) cause a cell to make proteins that interfere with its own tumor suppressors. Infection with HPV causes skin growths called warts, and some kinds are associated with neoplasms that form on the cervix.

CANCER

Benign neoplasms such as warts are not usually dangerous (FIGURE 11.10). They grow very slowly, and their cells retain the plasma membrane adhesion proteins that keep them properly anchored to the other cells in their home tissue ①.

A malignant neoplasm is one that gets progressively worse, and is dangerous to health. Malignant cells typically display the following three characteristics:

First, like cells of all neoplasms, malignant cells grow and divide abnormally. Controls that usually keep cells from

cancer Disease that occurs when a malignant neoplasm physically and metabolically disrupts body tissues.
growth factor Molecule that stimulates mitosis and differentiation.
metastasis The process in which malignant cells spread from one part of the body to another.
neoplasm An accumulation of abnormally dividing cells.
oncogene Gene that helps transform a normal cell into a tumor cell.
proto-oncogene Gene that, by mutation, can become an oncogene.
tumor A neoplasm that forms a lump.

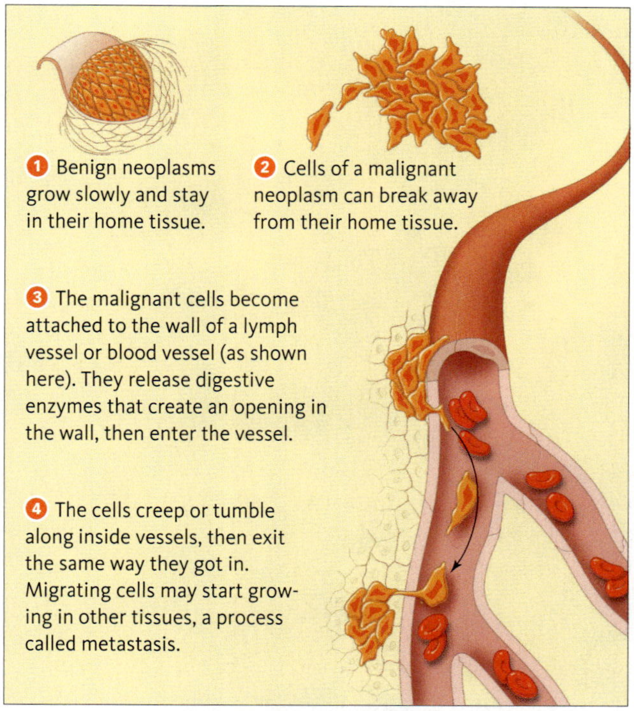

FIGURE 11.10 {Animated} Neoplasms and malignancy.

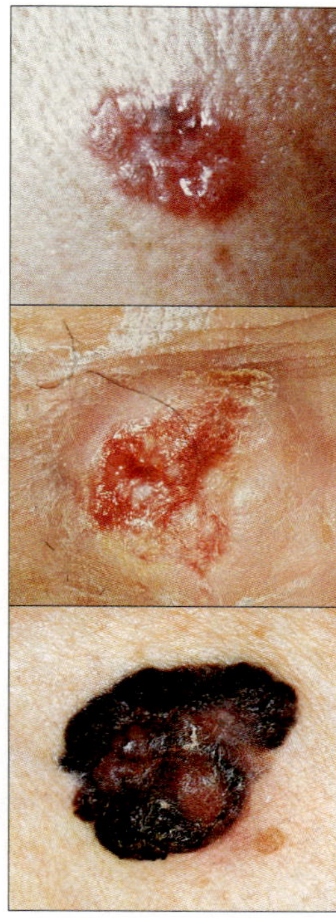

A Basal cell carcinoma is the most common type of skin cancer. This slow-growing, raised lump may be uncolored, reddish-brown, or black.

B Squamous cell carcinoma is the second most common form of skin cancer. This pink growth, firm to the touch, grows under the surface of skin.

C Melanoma spreads fastest. Cells form dark, encrusted lumps that may itch or bleed easily.

FIGURE 11.11 Skin cancer can be detected with early screening.

getting overcrowded in tissues are lost in malignant cells, so their populations may reach extremely high densities with cell division occurring very rapidly. The number of small blood vessels that transport blood to the growing cell mass also increases abnormally.

Second, the cytoplasm and plasma membrane of malignant cells are altered. The cytoskeleton may be shrunken, disorganized, or both. Malignant cells typically have an abnormal chromosome number, with some chromosomes present in multiple copies, and others missing or damaged. The balance of metabolism is often shifted, as in an amplified reliance on ATP formation by fermentation rather than aerobic respiration.

Altered or missing proteins impair the function of the plasma membrane of malignant cells. For example, these cells do not stay anchored properly in tissues because their plasma membrane adhesion proteins are defective or missing 2. Malignant cells can slip easily into and out of vessels of the circulatory and lymphatic systems 3. By migrating through these vessels, the cells can establish neoplasms elsewhere in the body 4. The process in which malignant cells break loose from their home tissue and invade other parts of the body is called **metastasis**. Metastasis is the third hallmark of malignant cells.

The disease called **cancer** occurs when the abnormally dividing cells of a malignant neoplasm disrupt body tissues, both physically and metabolically. Unless chemotherapy, surgery, or another procedure eliminates malignant cells from the body, they can put an individual on a painful road to death. Each year, cancer causes 15 to 20 percent of all human deaths in developed countries. The good news is that mutations in multiple checkpoint genes are required to transform a normal cell into a malignant one, and such mutations may take a lifetime to accumulate. Lifestyle choices such as not smoking and avoiding exposure of unprotected skin to sunlight can reduce one's risk of acquiring mutations in the first place. Some neoplasms can be detected with periodic screening such as gynecology or dermatology exams (**FIGURE 11.11**). If detected early enough, many types of malignant neoplasms can be removed before metastasis occurs.

TAKE-HOME MESSAGE 11.5

- Neoplasms form when cells lose control over their cell cycle and begin dividing abnormally.
- Mutations in multiple checkpoint genes can give rise to a malignant neoplasm that gets progressively worse.
- Cancer is a disease that occurs when the abnormally dividing cells of a malignant neoplasm physically and metabolically disrupt body tissues.
- Although some mutations are inherited, lifestyle choices and early intervention can reduce one's risk of cancer.

CREDITS: (10) From Starr/Taggart/Evers/Starr, Biology, 13E. © 2013 Cengage Learning; (11A) © Ken Greer/Visuals Unlimited; (11B) Biophoto Associates/Photo Researchers, Inc.; (11C) James Stevenson/Photo Researchers, Inc.

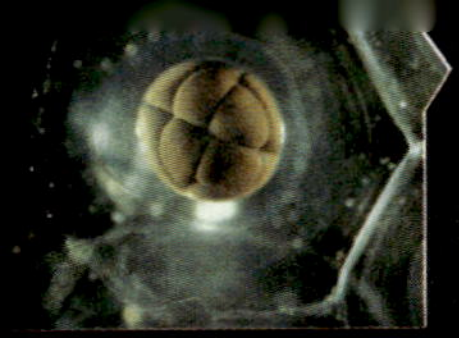

11.6 Application: HENRIETTA'S IMMORTAL CELLS

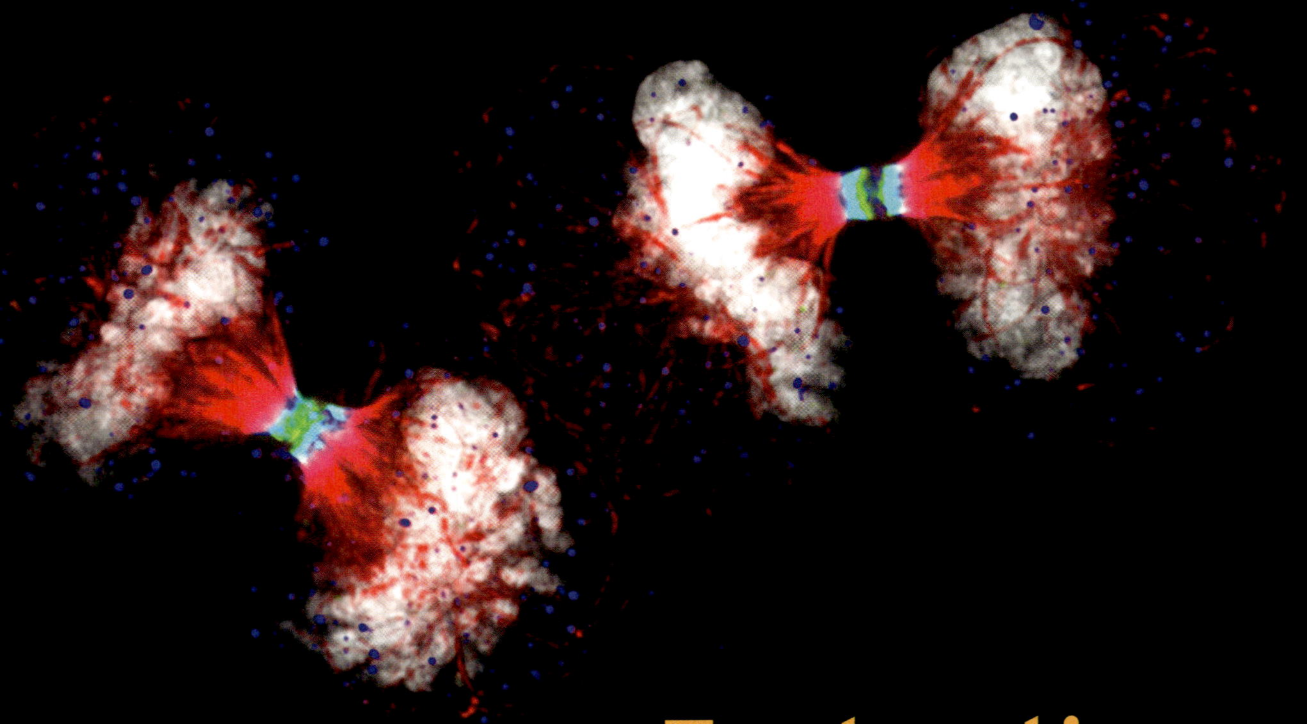

Exploration

FIGURE 11.12 HeLa cells, a legacy of cancer victim Henrietta Lacks. The cells in this fluorescent micrograph are undergoing mitosis.

FINDING HUMAN CELLS THAT GROW IN A LABORATORY TOOK GEORGE AND MARGARET GEY NEARLY 30 YEARS. In 1951, their assistant Mary Kubicek prepared yet another sample of human cancer cells. Mary named the cells HeLa, after the first and last names of the patient from whom the cells had been taken. The HeLa cells began to divide, again and again. The cells were astonishingly vigorous, quickly coating the inside of their test tube and consuming their nutrient broth. Four days later, there were so many cells that the researchers had to transfer them to more tubes. The cell populations increased at a phenomenal rate. The cells were dividing every twenty-four hours and coating the inside of the tubes within days.

Sadly, cancer cells in the patient were dividing just as fast. Only six months after she had been diagnosed with cervical cancer, malignant cells had invaded tissues throughout her body. Two months after that, Henrietta Lacks, a young African American woman from Baltimore, was dead.

Although Henrietta passed away, her cells lived on in the Geys' laboratory. The Geys were able to grow poliovirus in HeLa cells, a practice that enabled them to determine which strains of the virus cause polio. That work was a critical step in the development of polio vaccines, which have since saved millions of lives.

Henrietta Lacks was just thirty-one, a wife and mother of five, when runaway cell divisions of cancer killed her. Her cells, however, are still dividing, again and again, more than fifty years after she died. Frozen away in tiny tubes and packed in Styrofoam boxes, HeLa cells continue to be shipped among laboratories all over the world. They are still widely used to investigate cancer (FIGURE 11.12), viral growth, protein synthesis, the effects of radiation, and countless other processes important in medicine and research. HeLa cells helped several researchers win Nobel Prizes, and some even traveled into space for experiments on satellites.

CREDIT: (12) main, Dr. Paul D. Andrews/ University of Dundee.

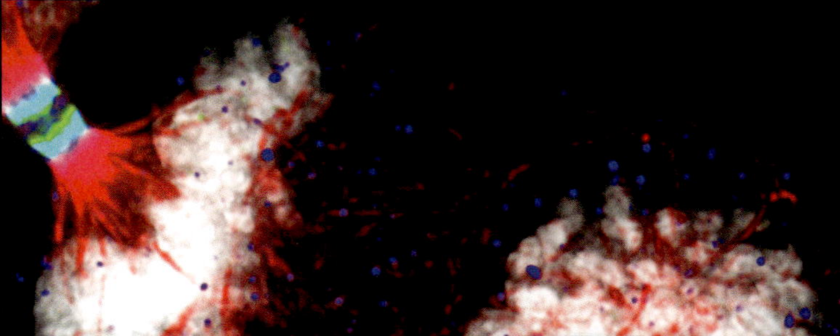

Henrietta Lacks

In these mitotic HeLa cells, chromosomes appear white and the spindle is red. Green identifies an enzyme that helps attach spindle microtubules to centromeres. Blue pinpoints a protein that helps sister chromatids stay attached to one another at the centromere. At this stage of telophase, the blue and green proteins should be closely associated midway between the two clusters of chromosomes. The abnormal distribution means that the spindle microtubules are not properly attached to the centromeres.

These days, physicians and researchers are required to obtain a signed consent form before they take tissue samples from a patient. No such requirement existed in the 1950s. It was common at that time for doctors to experiment on patients without their knowledge or consent. Thus, the young resident who was treating Henrietta Lacks's cancerous cervix probably never even thought about asking permission before he took a sample of it. That sample was the one that the Geys used to establish the HeLa cell line. No one in Henrietta's family knew about the cells until 25 years after her death. HeLa cells are still being sold worldwide, but her family has not received any share of profits.

Ongoing research with HeLa cells may one day allow researchers to identify drugs that target and destroy malignant cells or stop them from dividing. The research is far too late to have saved Henrietta Lacks, but it may one day yield drugs that put the brakes on cancer.

PEOPLE MATTER

National Geographic Explorer
DR. IAIN COUZIN

Can a million migrating wildebeests help explain why cancer spreads? Ask Iain Couzin. He is exploring how collective behavior in animals can be quantified and analyzed to give us new insights into the patterns of nature—and ourselves.

"We're realizing that animals have highly coordinated social systems and make decisions together," Couzin says. "They can do things collectively that no individual could do alone. It's still a very unexplored area of animal behavior." Couzin blends fieldwork, lab experiments, computer simulations, and complex mathematical models to test theories about how and why cells, animals, and humans organize and work together.

Couzin recently coauthored a paper hypothesizing that cancer cells migrating during metastasis—when the cells leave the primary tumor and journey elsewhere in the body—may have some parallels to animal swarms. As animals do in swarms, metastatic cells collectively can sense the environment and make decisions in response to it. It is very early theoretical work, but it does point to a new angle for cancer research: trying to knock out the mechanisms behind such collective migration.

CREDITS: (12 inset) Courtesy of the family of Henrietta Lacks; (right inset) Courtesy © Iain Couzin.

Summary

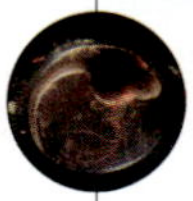

SECTION 11.1 A **cell cycle** includes all the stages through which a eukaryotic cell passes during its lifetime; it starts when a new cell forms, and ends when the cell reproduces. Most of a cell's activities, including replication of the cell's **homologous chromosomes**, occur during **interphase**.

A eukaryotic cell reproduces by dividing: nucleus first, then cytoplasm. **Mitosis** is a mechanism of nuclear division that maintains the chromosome number. It is the basis of growth, cell replacements, and tissue repair in multicelled species, and **asexual reproduction** in many species.

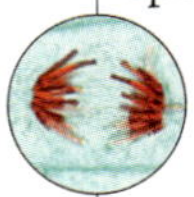

SECTION 11.2 Mitosis proceeds in four stages.

In **prophase**, the duplicated chromosomes start to condense. Microtubules assemble and form a **spindle**, and the nuclear envelope breaks up. Some microtubules that extend from one spindle pole attach to one chromatid of each chromosome; some that extend from the opposite spindle pole attach to its sister chromatid. These microtubules drag each chromosome toward the center of the cell.

At **metaphase**, all chromosomes are aligned at the spindle's midpoint.

During **anaphase**, the sister chromatids of each chromosome detach from each other, and the spindle microtubules move them toward opposite spindle poles.

During **telophase**, a complete set of chromosomes reaches each spindle pole. A nuclear envelope forms around each cluster. Two new nuclei, each with the parental chromosome number, are the result.

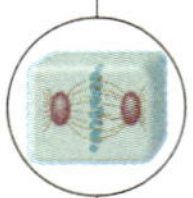

SECTION 11.3 **Cytokinesis** typically follows nuclear division. In animal cells, a contractile ring of microfilaments pulls the plasma membrane inward, forming a **cleavage furrow** that pinches the cytoplasm in two. In plant cells, vesicles guided by microtubules to the future plane of division merge as a **cell plate**. The plate expands until it fuses with the parent cell wall, thus becoming a cross-wall that partitions the cytoplasm.

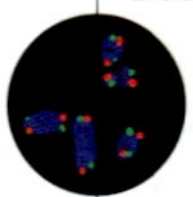

SECTION 11.4 **Telomeres** that protect the ends of eukaryotic chromosomes shorten with every DNA replication. Cells that inherit too-short telomeres die, and in most cells this limits the number of divisions that can occur.

SECTION 11.5 The products of checkpoint genes, including receptors for **growth factors**, work together to control the cell cycle. These molecules monitor the integrity of the cell's DNA, and can pause the cycle until breaks or other problems are fixed. When checkpoint mechanisms fail, a cell loses control over its cell cycle, and the cell's descendants form a **neoplasm**. Neoplasms may form lumps called **tumors**.

Checkpoint genes are examples of **proto-oncogenes**, which means mutations can turn them into tumor-causing **oncogenes**. Mutations in multiple checkpoint genes can transform benign neoplasms into malignant ones. Cells of malignant neoplasms can break loose from their home tissues and colonize other parts of the body, a process called **metastasis**. **Cancer** occurs when malignant neoplasms physically and metabolically disrupt normal body tissues.

SECTION 11.6 An immortal line of human cells (HeLa) is a legacy of cancer victim Henrietta Lacks. Researchers all over the world continue to work with these cells as they try to unravel the mechanisms of cancer.

Self-Quiz

Answers in Appendix VII

1. Mitosis and cytoplasmic division function in _______ .
   a. asexual reproduction of single-celled eukaryotes
   b. growth and tissue repair in multicelled species
   c. gamete formation in bacteria and archaea
   d. sexual reproduction in plants and animals
   e. both a and b

2. A duplicated chromosome has _______ chromatid(s).
   a. one c. three
   b. two d. four

3. Homologous chromosomes _______ .
   a. carry the same genes c. are the same length
   b. are the same shape d. all of the above

4. Most cells spend the majority of their lives in _______ .
   a. prophase d. telophase
   b. metaphase e. interphase
   c. anaphase f. d and e

5. The spindle attaches to chromosomes at the _______ .
   a. centriole c. centromere
   b. contractile ring d. centrosome

6. Only _______ is not a stage of mitosis.
   a. prophase c. interphase
   b. metaphase d. anaphase

7. In intervals of interphase, G stands for _______ .
   a. gap b. growth c. Gey d. gene

8. Interphase is the part of the cell cycle when _______ .
   a. a cell ceases to function
   b. a cell forms its spindle apparatus
   c. a cell grows and duplicates its DNA
   d. mitosis proceeds

9. After mitosis, the chromosome number of a descendant cell is _______ the parent cell's.
   a. the same as c. rearranged compared to
   b. one-half of d. doubled compared to

10. A plant cell divides by the process of _______ .
    a. telekinesis c. fission
    b. nuclear division d. cytokinesis

Data Analysis Activities

HeLa Cells Are a Genetic Mess HeLa cells continue to be an extremely useful tool in cancer research. One early finding was that HeLa cells can vary in chromosome number. Defects in proteins that orchestrate cell division result in descendant cells with too many or too few chromosomes, an outcome that is one of the hallmarks of cancer cells.

The panel of chromosomes in **FIGURE 11.13**, originally published in 1989 by Nicholas Popescu and Joseph DiPaolo, shows all of the chromosomes in a single metaphase HeLa cell.

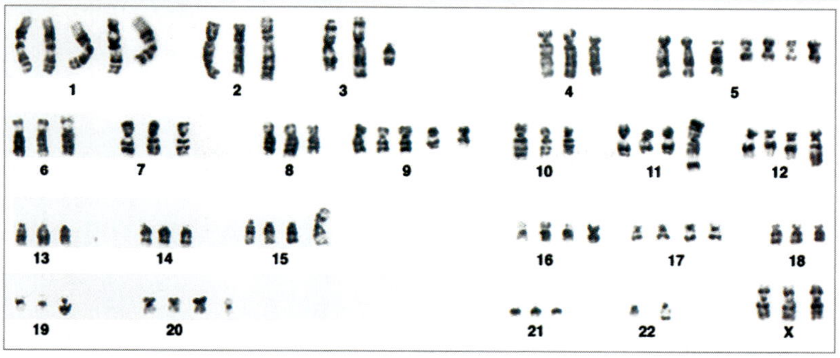

FIGURE 11.13 Chromosomes in a HeLa cell.

1. What is the chromosome number of this HeLa cell?
2. How many extra chromosomes does this cell have, compared to a normal human body cell?
3. Can you tell that this cell came from a female? How?

11. In the diagram of the nucleus *below*, fill in the blanks with the name of each interval.

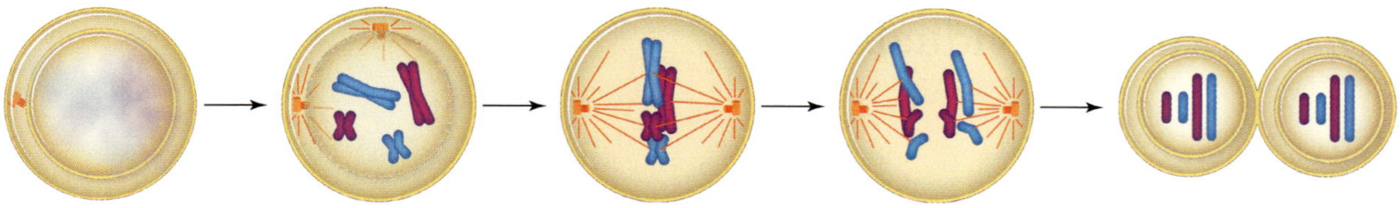

__________ __________ __________ __________ __________

12. *BRCA1* and *BRCA2* _______ .
    a. are checkpoint genes c. encode tumor suppressors
    b. are proto-oncogenes d. all of the above

13. _______ are characteristic of cancer.
    a. Malignant cells b. Neoplasms c. Tumors

14. Match each term with its best description.

___ cell plate	a. lump of cells
___ spindle	b. made of microfilaments
___ tumor	c. divides plant cells
___ cleavage furrow	d. organize(s) the spindle
___ contractile ring	e. caused by metastatic cells
___ cancer	f. made of microtubules
___ centrosomes	g. indentation
___ telomere	h. shortens with age

15. Match each stage with the events listed.

___ metaphase	a. sister chromatids move apart
___ prophase	b. chromosomes start to condense
___ telophase	c. new nuclei form
___ interphase	d. all duplicated chromosomes are aligned at the spindle equator
___ anaphase	e. DNA replication
___ cytokinesis	f. cytoplasmic division

Critical Thinking

1. When a cell reproduces by mitosis and cytoplasmic division, does its life end?

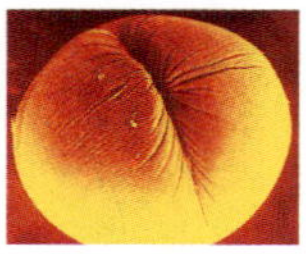

2. The eukaryotic cell in the photo on the *left* is in the process of cytoplasmic division. Is this cell from a plant or an animal? How do you know?

3. Exposure to radioisotopes or other sources of radiation can damage DNA. Humans exposed to high levels of radiation face a condition called radiation poisoning. Why do you think that hair loss and damage to the lining of the gut are early symptoms of radiation poisoning? Speculate about why exposure to radiation is used as a therapy to treat some kinds of cancers.

4. Suppose you have a way to measure the amount of DNA in one cell during the cell cycle. You first measure the amount at the G1 phase. At what points in the rest of the cycle will you see a change in the amount of DNA per cell?

New Zealand mud snails can reproduce on their own (asexually) or with a partner (sexually). Natural populations of this species vary in the frequency of sexual and asexual individuals.

12

MEIOSIS AND SEXUAL REPRODUCTION

Links to Earlier Concepts

Before you begin this chapter, be sure you understand how eukaryotic chromosomes are organized (Section 8.3) and how genes work (9.1). You will draw on your knowledge of DNA replication (8.4), cytoplasmic division (11.3), and cell cycle controls (11.5) as we compare meiosis with mitosis (11.2). This chapter also revisits clones (8.6), the effects of mutation (9.5), and the function of telomeres (11.4).

KEY CONCEPTS

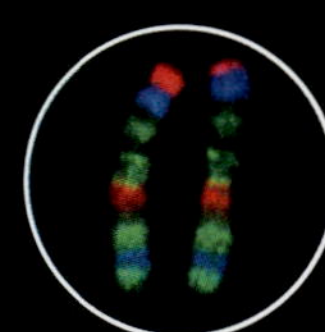

SEX AND ALLELES

In asexual reproduction, one parent transmits its genes to offspring. In sexual reproduction, offspring inherit genes from two parents who usually differ in some number of alleles.

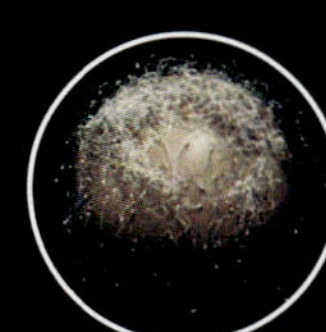

MEIOSIS IN REPRODUCTION

Meiosis is a nuclear division process that reduces the chromosome number. It occurs only in cells that play a role in sexual reproduction in eukaryotes.

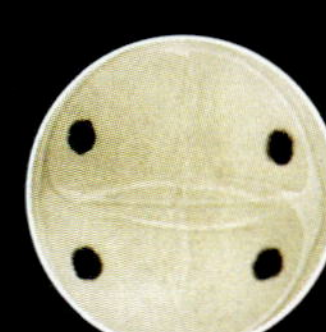

STAGES OF MEIOSIS

The chromosome number becomes reduced by the two nuclear divisions of meiosis. During this process, the chromosomes are sorted into four new nuclei.

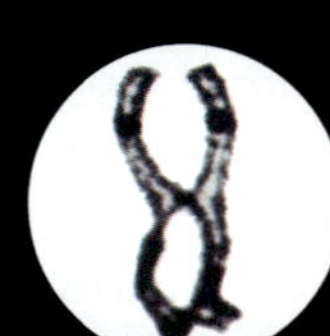

SHUFFLING PARENTAL DNA

During meiosis, homologous chromosomes swap segments, then are randomly sorted into separate nuclei. Both processes lead to novel combinations of alleles among offspring.

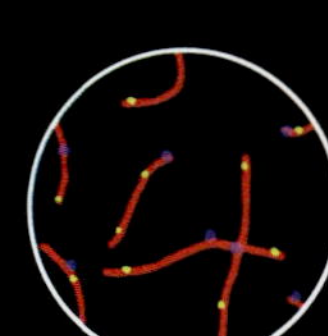

MITOSIS AND MEIOSIS COMPARED

Similarities between mitosis and meiosis suggest meiosis originated by evolutionary remodeling of mechanisms that already existed for mitosis and, before that, for repairing damaged DNA.

12.1 WHY SEX?

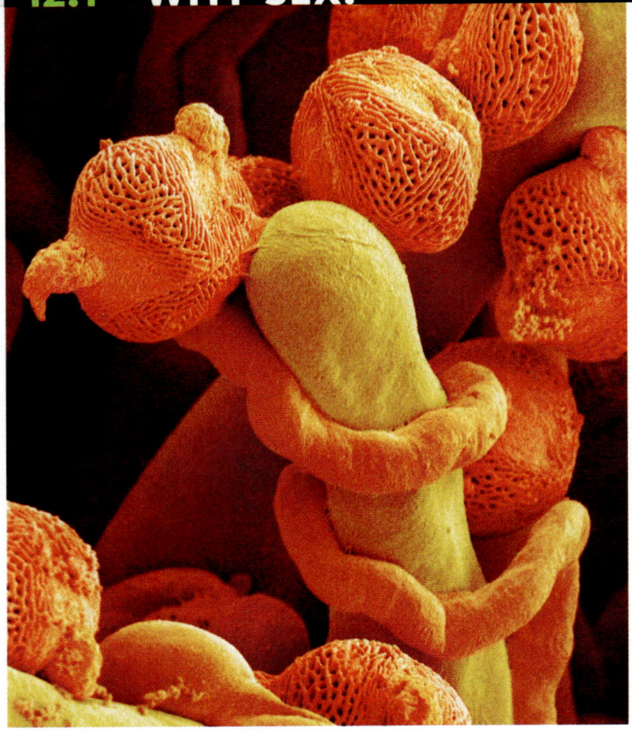

FIGURE 12.1 A moment in sexual reproduction. Sex mixes up the genetic material of two individuals. In flowering plants, pollen grains (orange) germinate on flower carpels (yellow). Pollen tubes with male gametes inside grow from the grains down into tissues of the ovary, which house the flower's female gametes.

INTRODUCING ALLELES

In asexual reproduction, a single individual gives rise to offspring that are identical to itself and to one another. By contrast, **sexual reproduction** involves two individuals and mixes their genetic material (**FIGURE 12.1**).

In your somatic (body) cells and in those of many other sexually reproducing eukaryotes, one chromosome of each homologous pair is maternal, and the other is paternal (Section 8.3). Homologous chromosomes carry the same genes (**FIGURE 12.2A**). However, the corresponding genes on maternal and paternal chromosomes often vary—just a bit—in DNA sequence. Over evolutionary time, unique mutations accumulate in separate lines of descent, and some of those mutations occur in genes. Thus, the DNA sequence of any gene may differ from the corresponding gene on the homologous chromosome (**FIGURE 12.2B**). Different forms of the same gene are called **alleles**.

Alleles may encode slightly different forms of the gene's product. Such differences influence the details of traits shared by a species. Consider that one of the approximately 20,000 genes in human chromosomes encodes beta globin (Section 9.5). Like most human genes, the beta globin gene has multiple alleles—more than 700 in this case. A few beta globin alleles cause sickle-cell anemia, several cause beta thalassemia, and so on. Allele differences among individuals are one reason that the members of a sexually reproducing species are not identical. Offspring of sexual reproducers inherit new combinations of alleles, which is the basis of new combinations of traits.

ON THE ADVANTAGES OF SEX

If the function of reproduction is the perpetuation of one's genes, then an organism that reproduces only by mitosis would seem to win the evolutionary race. When it reproduces, it passes all of its genes to every one of its offspring. Only about half of a sexual reproducer's genes are passed to each offspring. Yet most eukaryotes reproduce sexually, at least some of the time. Why?

Consider that all offspring of an asexual reproducer are clones. They have the same alleles—and the same traits—as their parent. Consistency is a good thing if an organism lives in a favorable, unchanging environment. Alleles that help it survive and reproduce do the same for its descendants. However, most environments are constantly changing. Offspring of asexual reproducers are equally vulnerable to unfavorable environmental change. In changing environments, sexual reproducers have the evolutionary edge. Their offspring inherit different combinations of alleles, so they vary in the details of their shared traits. Some may have a particular combination of traits that suits them perfectly to a change. As a group, their diversity offers them a better chance of surviving an environmental challenge than clones.

Perhaps the most important advantage of sexual reproduction involves the inevitable occurrence of harmful mutations. A population of sexual reproducers has a better chance of weathering the effects of such mutations. With asexual reproduction, individuals bearing a harmful mutation necessarily pass it to all of their offspring. This outcome would be rare in sexual reproduction, because each offspring of a sexual union has a 50 percent chance of inheriting a parent's mutation. Thus, all else being equal, harmful mutations accumulate in an asexually reproducing population more quickly than in a sexually reproducing one.

Environmental change is the norm. Consider, for example, the interaction between a predatory species and its prey. In each generation, prey individuals with traits that allow them to hide from, fend off, or escape the predator will leave the most young. However, the predator is constantly changing too: In each generation, predators best able to find, capture, and overcome prey leave the most descendants. Thus predators and prey are locked in a constant race, with each genetic improvement in one species

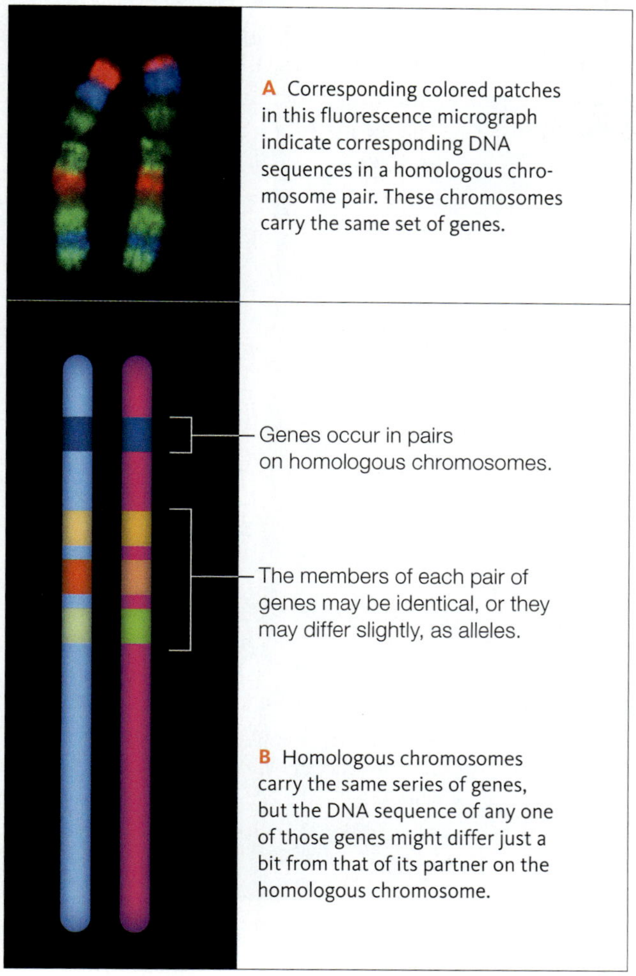

A Corresponding colored patches in this fluorescence micrograph indicate corresponding DNA sequences in a homologous chromosome pair. These chromosomes carry the same set of genes.

B Homologous chromosomes carry the same series of genes, but the DNA sequence of any one of those genes might differ just a bit from that of its partner on the homologous chromosome.

FIGURE 12.2 Genes on chromosomes. Different forms of a gene are called alleles.

countered by an improvement in the other. The Red Queen hypothesis holds that sexual reproduction is widespread because of the pressure for constant change resulting from such species interactions. The name of the hypothesis is a reference to Lewis Carroll's book *Through the Looking Glass*. In the book, the Queen of Hearts tells Alice, "It takes all the running you can do, to keep in the same place."

alleles Forms of a gene with slightly different DNA sequences; may encode slightly different versions of the gene's product.
sexual reproduction Reproductive mode by which offspring arise from two parents and inherit genes from both.

TAKE-HOME MESSAGE 12.1

Paired genes on homologous chromosomes may vary in DNA sequence as alleles. Alleles arise by mutation.

Alleles are the basis of differences in shared traits. Offspring of sexual reproducers inherit new combinations of parental alleles—thus new combinations of such traits.

PEOPLE MATTER

National Geographic Grantee
MAURINE NEIMAN

Animals that reproduce solely by asexual means are very rare. However, some populations of the New Zealand freshwater snail pictured in the chapter opener do just that. Maurine Neiman compares sexually reproducing and asexual populations of these snails to determine the costs and benefits of sex. Neiman and her collaborators Amy Krist and Adam Kay have discovered that sexual and asexual snails might differ in their need for phosphorus. Like most sexual organisms, the sexual snails have two chromosome sets; like most animals that cannot reproduce sexually, the asexual snails have at least three. DNA has a high phosphorus content, so having the extra sets of chromosomes multiplies each organism's requirement for this nutrient, which could add up to a big minus in the cost-of-asexuality equation.

Neiman's research has shown that extra chromosome sets translate into a disadvantage in low-phosphorus environments for the asexual snails, setting the stage for the possibility that the sexual snails—with the fewest chromosome sets of all—are likely to beat out asexuals when phosphorus is scarce. Follow-up experiments will extend beyond explaining the predominance of sex. These tiny snails have spread far beyond their native New Zealand, and huge populations of them are now disrupting ecosystems all over the world. The invasive populations are always asexual (and in fact many other invasive species have three or more sets of chromosomes). Fertilizers and detergents contain a high level of phosphorus, so agricultural runoff and other types of water pollution may be fueling population explosions of these species.

12.2 WHY IS MEIOSIS NECESSARY FOR SEXUAL REPRODUCTION?

FIGURE 12.3 Gametes. This illustration shows a human egg (female gamete) surrounded by sperm (male gametes).

MEIOSIS HALVES THE CHROMOSOME NUMBER

Sexual reproduction involves the fusion of mature reproductive cells—**gametes**—from two parents (**FIGURE 12.3**). Gametes have a single set of chromosomes, so they are **haploid** (n): Their chromosome number is half of the diploid ($2n$) number (Section 8.3). **Meiosis**, the nuclear division mechanism that halves the chromosome number, is necessary for gamete formation. Meiosis also gives rise to new combinations of parental alleles.

Gametes arise by division of **germ cells**, which are immature reproductive cells that form in special reproductive organs (**FIGURE 12.4**). Animals and plants make gametes somewhat differently. In animals, meiosis in diploid germ cells gives rise to eggs (female gametes) or sperm (male gametes). In plants, haploid germ cells form by meiosis. Gametes form when these cells divide by mitosis.

The first part of meiosis is similar to mitosis. A cell duplicates its DNA before either nuclear division process begins. As in mitosis, a spindle forms, and its microtubules move the duplicated chromosomes to opposite spindle poles. However, meiosis sorts the chromosomes into new nuclei not once, but twice, so it results in the formation of four haploid nuclei. The two consecutive nuclear divisions are called meiosis I and meiosis II:

Interphase	→	Meiosis I	→	Meiosis II
DNA replication		prophase I metaphase I anaphase I telophase I		prophase II metaphase II anaphase II telophase II

In some cells, meiosis II occurs immediately after meiosis I. In others, a period of protein synthesis—but no DNA replication—intervenes between the divisions.

During meoisis I, every duplicated chromosome aligns with its homologous partner (**FIGURE 12.5** ❶). Then the homologous chromosomes are pulled away from one another and packaged into separate nuclei ❷. At this stage of meiosis, the chromosome number has been reduced. Each

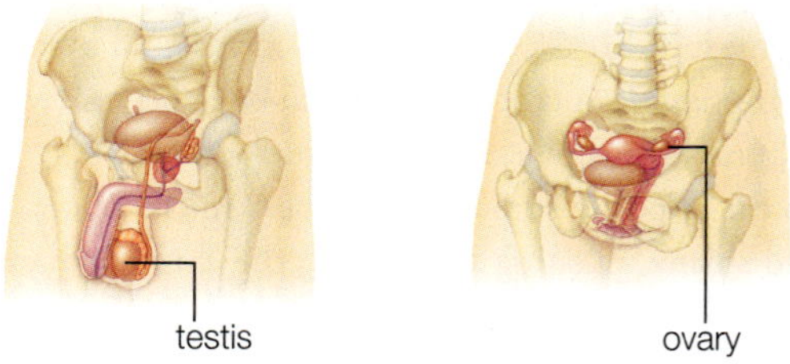

A Reproductive organs of humans. Meiosis in germ cells inside testes and ovaries produces gametes (sperm and eggs).

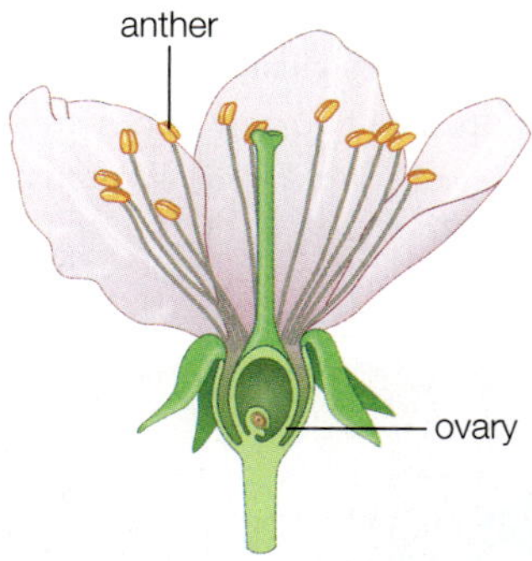

B Reproductive organs of a flowering plant. Meiosis produces haploid germ cells inside anthers and ovaries. These cells divide by mitosis to give rise to gametes (sperm and eggs).

FIGURE 12.4 {Animated} Examples of reproductive organs in **A** animals and **B** plants.

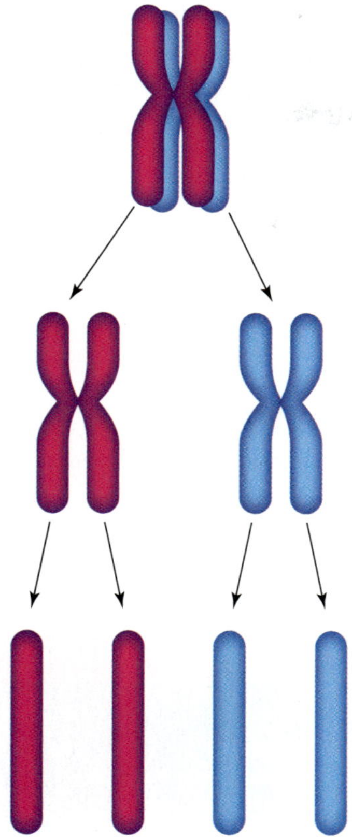

1 Chromosomes are duplicated before meiosis begins. During meiosis I, each chromosome in the nucleus pairs with its homologous partner. The nucleus contains two of each chromosome, so it is diploid (2*n*).

2 Homologous partners separate and are packaged into two new nuclei. Each new nucleus contains one of each chromosome, so it is haploid (*n*). The chromosomes are still duplicated.

3 Sister chromatids separate in meiosis II and are packaged into four new nuclei. Each new nucleus contains one of each chromosome, so it is haploid (*n*). The chromosomes are now unduplicated.

FIGURE 12.5 How meiosis halves the chromosome number.

of the two new nuclei has one copy of each chromosome, so it is haploid (*n*). The chromosomes are still duplicated (the sister chromatids remain attached to one another).

During meiosis II, the sister chromatids are pulled apart, and each becomes an individual, unduplicated chromosome **3**. The chromosomes are sorted into four new nuclei. Each new nucleus still has one copy of each chromosome, so it is haploid (*n*).

Thus, meiosis partitions the chromosomes of one diploid nucleus (2*n*) into four haploid (*n*) nuclei. The next section zooms in on the details of this process.

FERTILIZATION RESTORES THE CHROMOSOME NUMBER

Haploid gametes form by meiosis. The diploid chromosome number is restored at **fertilization**, when two haploid gametes fuse to form a **zygote**, the first cell of a new individual. Thus, meiosis halves the chromosome number, and fertilization restores it.

If meiosis did not precede fertilization, the chromosome number would double with every generation. As you will see in Chapter 14, chromosome number changes can have drastic consequences, particularly in animals. An individual's set of chromosomes is like a fine-tuned blueprint that must be followed exactly, page by page, in order to build a body that functions normally.

fertilization Fusion of two gametes to form a zygote.
gamete Mature, haploid reproductive cell; e.g., an egg or a sperm.
germ cell Immature reproductive cell that gives rise to haploid gametes when it divides.
haploid Having one of each type of chromosome characteristic of the species.
meiosis Nuclear division process that halves the chromosome number. Basis of sexual reproduction.
zygote Cell formed by fusion of two gametes at fertilization; the first cell of a new individual.

TAKE-HOME MESSAGE 12.2

The nuclear division process of meiosis is the basis of sexual reproduction in plants and animals.

Meiosis halves the diploid (2*n*) chromosome number, to the haploid number (*n*), for forthcoming gametes.

When two gametes fuse at fertilization, the diploid chromosome number is restored in the resulting zygote.

12.3 WHAT HAPPENS TO A CELL DURING MEIOSIS?

FIGURE 12.6 shows the stages of meiosis in a diploid (2*n*) cell, which contains two sets of chromosomes. DNA replication occurs before meiosis I, so each chromosome has two sister chromatids.

Meiosis I The first stage of meiosis I is prophase I ❶. During this phase, the chromosomes condense, and homologous chromosomes align tightly and swap segments (more about segment-swapping in the next section). The nuclear envelope breaks up. A spindle forms, and by the end of prophase I, microtubules attach one chromosome of each homologous pair to one spindle pole, and the other to the opposite spindle pole. These microtubules grow and shrink, pushing and pulling the chromosomes as they do. At metaphase I ❷, all of the microtubules are the same length, and the chromosomes are aligned in the middle of the cell. In anaphase I ❸, the spindle pulls the homologous chromosomes of each pair apart and toward opposite spindle poles. The two sets of chromosomes reach the spindle poles during telophase I ❹, and a new nuclear envelope forms around each cluster of chromosomes as the DNA loosens up. The two new nuclei are haploid (*n*); each contains one set of (duplicated) chromosomes. The

FIGURE 12.6 {Animated} Meiosis. Two pairs of chromosomes are illustrated in a diploid (2*n*) cell. Homologous chromosomes are indicated in blue and pink. Micrographs show meiosis in a lily plant cell (*Lilium regale*).

FIGURE IT OUT: During which stage of meiosis does the chromosome number become reduced?
Answer: Anaphase I

MEIOSIS I: ONE DIPLOID NUCLEUS TO TWO HAPLOID NUCLEI

① Prophase I
Homologous chromosomes condense, pair up, and swap segments. Spindle microtubules attach to them as the nuclear envelope breaks up.

② Metaphase I
Homologous chromosome pairs are aligned between spindle poles. Spindle microtubules attach the two chromosomes of each pair to opposite spindle poles.

③ Anaphase I
All of the homologous chromosomes separate and begin heading toward the spindle poles.

④ Telophase I
A complete set of chromosomes clusters at both ends of the cell. A nuclear envelope forms around each set, so two haploid (*n*) nuclei form.

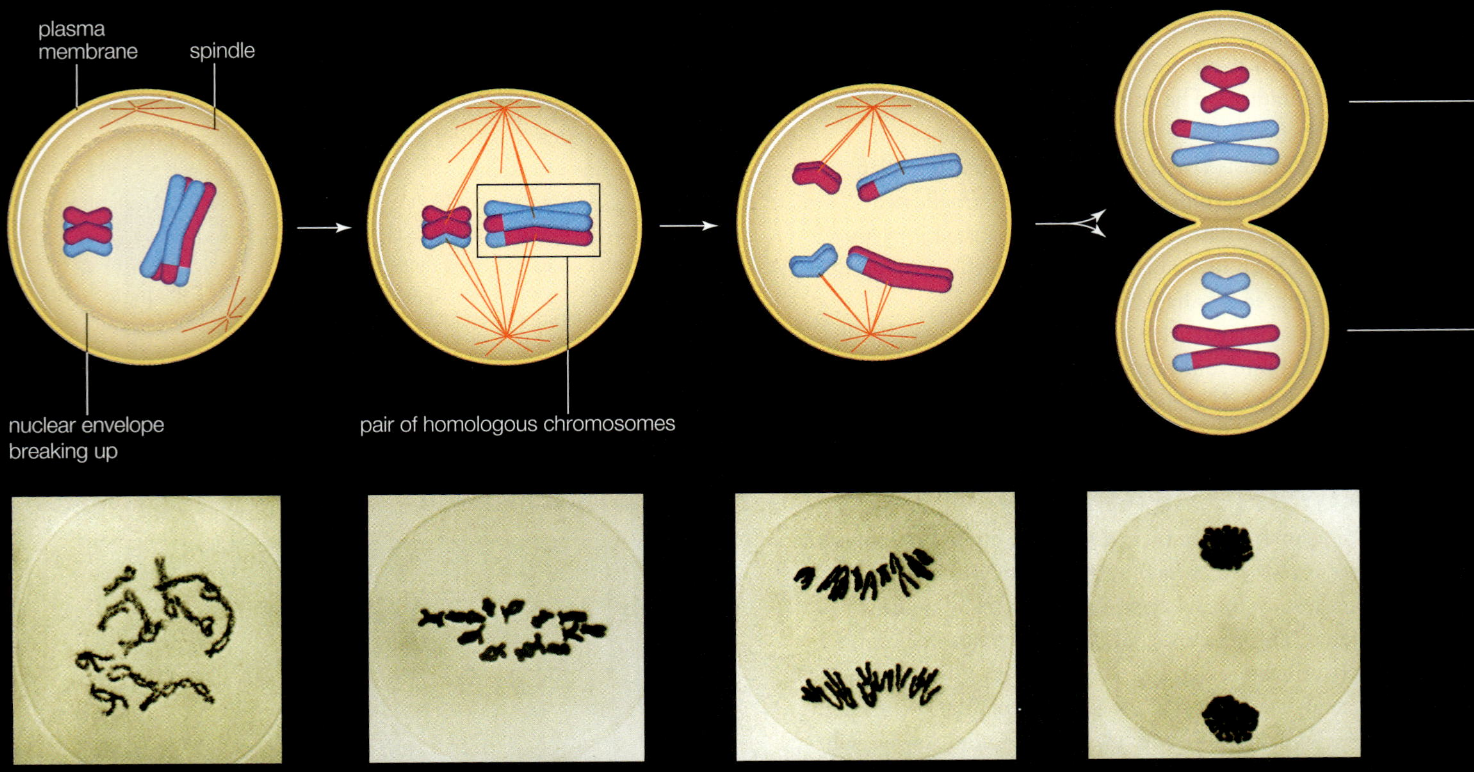

cytoplasm often divides at this point. Each chromosome is still duplicated (it consists of two sister chromatids).

Meiosis II DNA replication does not occur before meiosis II, which proceeds simultaneously in both nuclei that formed in meiosis I. In prophase II 5, the chromosomes condense and the nuclear envelope breaks up. A new spindle forms. By the end of prophase II, spindle microtubules attach each chromatid to one spindle pole, and its sister chromatid to the opposite spindle pole. These microtubules push and pull the chromosomes, aligning them in the middle of the cell at metaphase II 6. In anaphase II 7, the spindle microtubules pull the sister chromatids apart and toward opposite spindle poles. Each chromosome is now unduplicated (it consists of one molecule of DNA). During telophase II 8, these chromosomes reach the spindle poles. New nuclear envelopes form around the four clusters of chromosomes as the DNA loosens up. The cytoplasm often divides at this point to form four haploid (*n*) cells whose nuclei contain one set of (unduplicated) chromosomes.

TAKE-HOME MESSAGE 12.3

During meiosis, the nucleus of a diploid (2*n*) cell divides twice. Four haploid (*n*) nuclei form, each with a full set of chromosomes—one of each type.

MEIOSIS II: TWO HAPLOID NUCLEI TO FOUR HAPLOID NUCLEI

5 Prophase II
The chromosomes condense. Spindle microtubules attach to each sister chromatid as the nuclear envelope breaks up.

6 Metaphase II
The (still duplicated) chromosomes are aligned midway between spindle poles.

7 Anaphase II
Sister chromatids separate. The now unduplicated chromosomes head to the spindle poles.

8 Telophase II
A complete set of chromosomes clusters at both ends of the cell. A new nuclear envelope forms around each set, so four haploid (*n*) nuclei form.

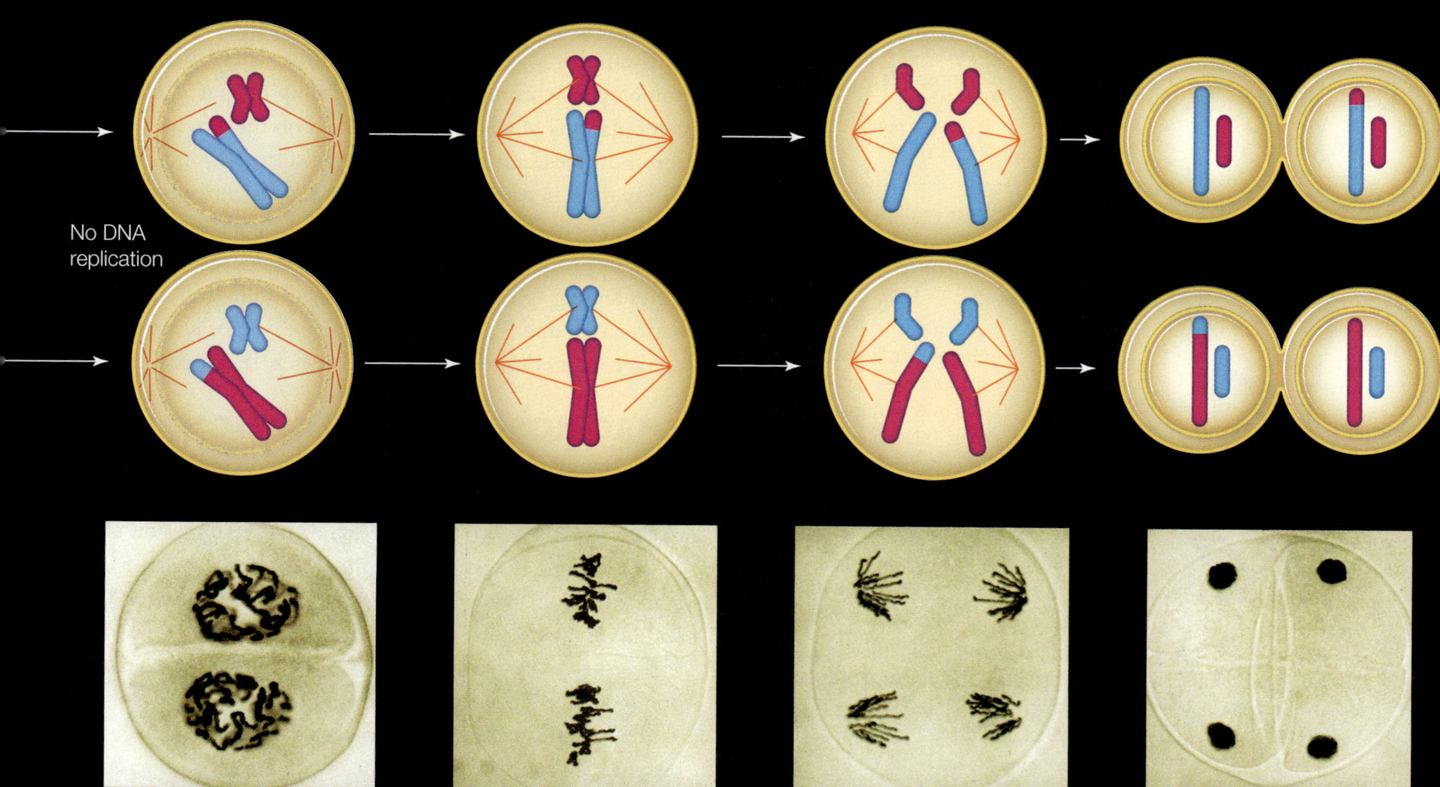

CREDITS: (4) Bottom photos, With thanks to the John Innes Foundation Trustees, computer enhanced by Gary Head; Top art, © Cengage Learning.

12.4 HOW DOES MEIOSIS GIVE RISE TO NEW COMBINATIONS OF PARENTAL ALLELES?

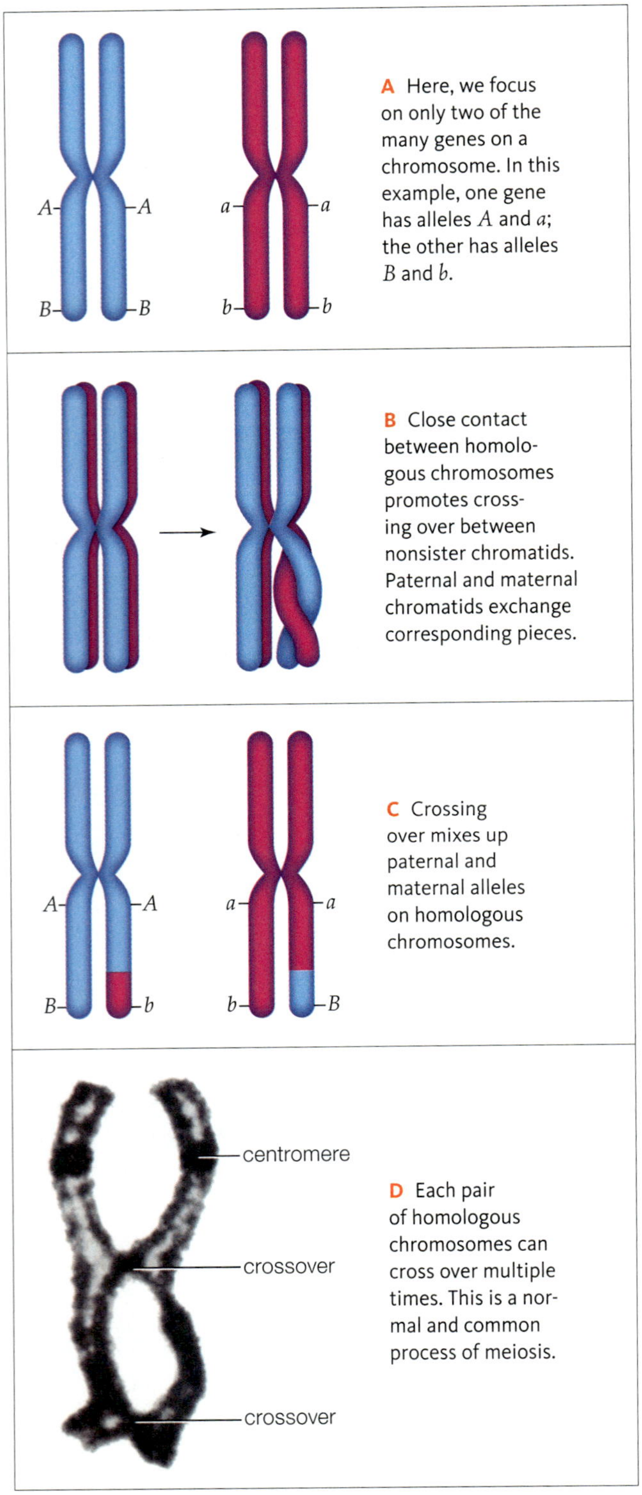

FIGURE 12.7 {Animated} Crossing over. Blue signifies a paternal chromosome, and pink, its maternal homologue. For clarity, we show only one pair of homologous chromosomes.

FIGURE IT OUT: In how many places is the chromosome pictured in D crossing over?

Answer: Two

The previous section mentioned briefly that duplicated chromosomes swap segments with their homologous partners during prophase I. It also showed how spindle microtubules align and then separate homologous chromosomes during anaphase I. These events, along with fertilization, contribute to the variation in combinations of traits among the offspring of sexually reproducing species.

CROSSING OVER IN PROPHASE I

Early in prophase I of meiosis, all chromosomes in the cell condense. When they do, each is drawn close to its homologous partner, so that the chromatids align along their length:

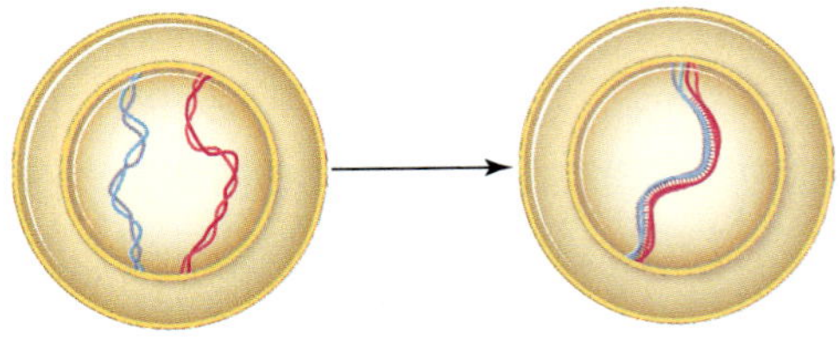

This tight, parallel orientation favors **crossing over**, a process by which a chromosome and its homologous partner exchange corresponding pieces of DNA during meiosis (**FIGURE 12.7**). Homologous chromosomes may swap any segment of DNA along their length, although crossovers tend to occur more frequently in certain regions.

Swapping segments of DNA shuffles alleles between homologous chromosomes. It breaks up the particular combinations of alleles that occurred on the parental chromosomes, and makes new ones on the chromosomes that end up in gametes. Thus, crossing over introduces novel combinations of traits among offspring. It is a normal and frequent process in meiosis, but the rate of crossing over varies among species and among chromosomes. In humans, between 46 and 95 crossovers occur per meiosis, so on average each chromosome crosses over at least once.

CHROMOSOME SEGREGATION

Normally, all of the new nuclei that form in meiosis I receive a complete set of chromosomes. However, whether a new nucleus ends up with the maternal or paternal version of a chromosome is entirely random. The chance that the maternal or the paternal version of any chromosome will end up in a particular nucleus is 50 percent. Why? The answer has to do with the way the spindle segregates the homologous chromosomes during meiosis I.

The process of chromosome segregation begins in prophase I. Imagine one of your own germ cells undergoing meiosis. Crossovers have already made genetic mosaics of

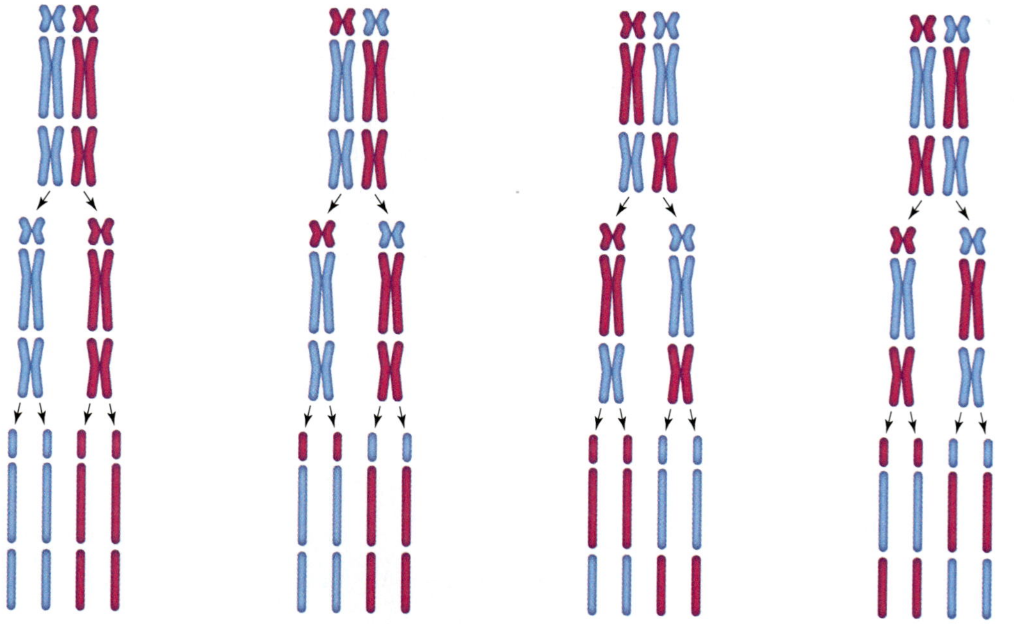

FIGURE 12.8 {Animated} Hypothetical segregation of three pairs of chromosomes in meiosis I. Maternal chromosomes are pink; paternal, blue. Which chromosome of each pair gets packaged into which of the two new nuclei that form at telophase I is random. For simplicity, no crossing over occurs in this example, so all sister chromatids are identical.

its chromosomes, but for simplicity let's put crossing over aside for a moment. Just call the twenty-three chromosomes you inherited from your mother the maternal ones, and the twenty-three from your father the paternal ones.

During prophase I, microtubules fasten your cell's chromosomes to the spindle poles. Chances are very low that all of the maternal chromosomes get attached to one pole and all of the paternal chromosomes get attached to the other. Microtubules extending from a spindle pole bind to the centromere of the first chromosome they contact, regardless of whether it is maternal or paternal. Though each homologous partner becomes attached to the opposite spindle pole, there is no pattern to the attachment of the maternal or paternal chromosomes to a particular pole.

Now imagine that your germ cell has just three pairs of chromosomes (**FIGURE 12.8**). By metaphase I, those three pairs of maternal and paternal chromosomes have been divided up between the two spindle poles in one of four ways **1**. In anaphase I, homologous chromosomes separate and are pulled toward opposite spindle poles. In telophase I, a new nucleus forms around the chromosomes that cluster at each spindle pole. Each nucleus contains one of eight possible combinations of maternal and paternal chromosomes **2**.

crossing over Process by which homologous chromosomes exchange corresponding segments of DNA during prophase I of meiosis.

In telophase II, each of the two nuclei divides and gives rise to two new haploid nuclei. The two new nuclei are identical because no crossing over occurred in our hypothetical example, so all of the sister chromatids were identical. Thus, at the end of meiosis in this cell, two (2) spindle poles have divvied up three (3) chromosome pairs. The resulting four nuclei have one of eight (2^3) possible combinations of maternal and paternal chromosomes **3**.

Cells that give rise to human gametes have twenty-three pairs of homologous chromosomes, not three. Each time a human germ cell undergoes meiosis, the four gametes that form end up with one of 8,388,608 (or 2^{23}) possible combinations of homologous chromosomes. That number does not even take into account crossing over, which mixes up the alleles on maternal and paternal chromosomes, or fusion with another gamete at fertilization.

TAKE-HOME MESSAGE 12.4

Crossing over—recombination between nonsister chromatids of homologous chromosomes—occurs during prophase I.

Homologous chromosomes can get attached to either spindle pole in prophase I, so each chromosome of a homologous pair can end up in either one of the two new nuclei.

Crossing over and random sorting of chromosomes into gametes give rise to new combinations of alleles—thus new combinations of traits—among offspring of sexual reproducers.

12.5 ARE THE PROCESSES OF MITOSIS AND MEIOSIS RELATED?

This chapter opened with hypotheses about evolutionary advantages of asexual and sexual reproduction. It seems like a giant evolutionary step from producing clones to producing genetically varied offspring, but was it really?

By mitosis and cytoplasmic division, one cell becomes two new cells that have the parental chromosomes. Mitotic (asexual) reproduction results in clones of the parent. Meiosis results in the formation of haploid gametes. Gametes of two parents fuse to form a zygote, which is a cell of mixed parentage. Meiotic (sexual) reproduction results in offspring that differ genetically from the parent, and from one another.

Though the end results differ, there are striking parallels between the four stages of mitosis and meiosis II (**FIGURE 12.9**). As one example, a spindle forms and separates chromosomes during both processes. There are many more similarities at the molecular level.

Long ago, the molecular machinery of mitosis may have been remodeled into meiosis. Evidence for this hypothesis includes a host of shared molecules, including the products

FIGURE 12.9 **{Animated}** Comparing meiosis II with mitosis.

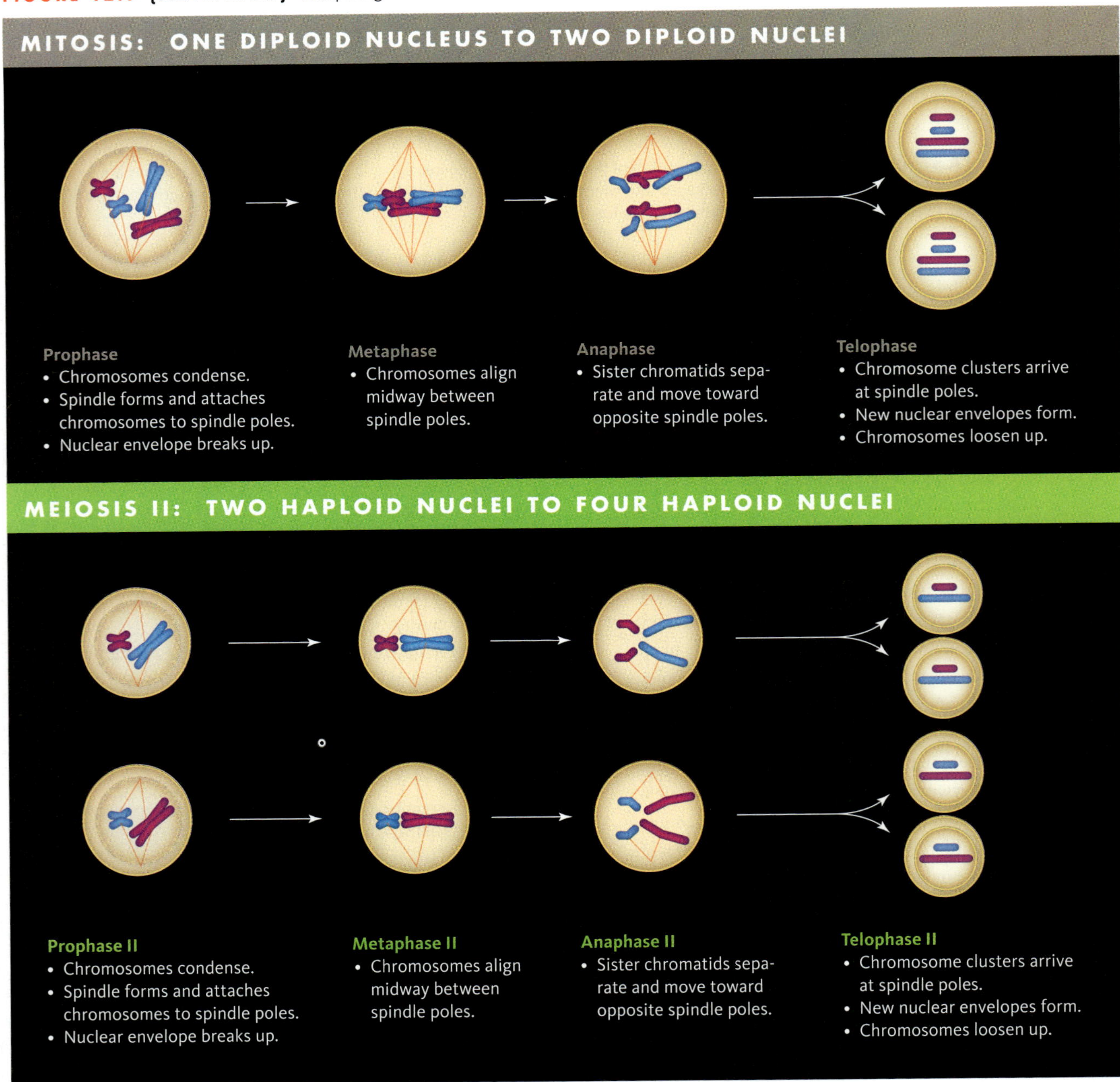

of the *BRCA* genes (Sections 10.6 and 11.5) that are made by all modern eukaryotes. By monitoring and fixing problems with the DNA—such as damaged or mismatched bases (Section 8.5)—these molecules actively maintain the integrity of a cell's chromosomes. It turns out that many of the same molecules help homologous chromosomes cross over in prophase I of meiosis (**FIGURE 12.10**). Some proteins function as part of checkpoints in both mitosis and meiosis, so mutations that affect them or the rate at which they are made can affect the outcomes of both nuclear division processes.

In anaphase of mitosis, sister chromatids are pulled apart. What would happen if the connections between the sisters did not break? Each duplicated chromosome would be pulled to one or the other spindle pole—which is exactly what happens in anaphase I of meiosis.

The shared molecules and mechanisms imply a shared evolutionary history; sexual reproduction probably originated with mutations that affected processes of mitosis. As you will see in later chapters, the remodeling of existing processes into new ones is a common evolutionary theme.

FIGURE 12.10 Example of a molecule that functions in mitosis and meiosis. This fluorescence micrograph shows homologous chromosome pairs (red) in the nucleus of a human cell during prophase I of meiosis. Centromeres are blue. Yellow pinpoints the location of a protein called MLH1 assisting with crossovers. MLH1 also helps repair mismatched bases during mitosis.

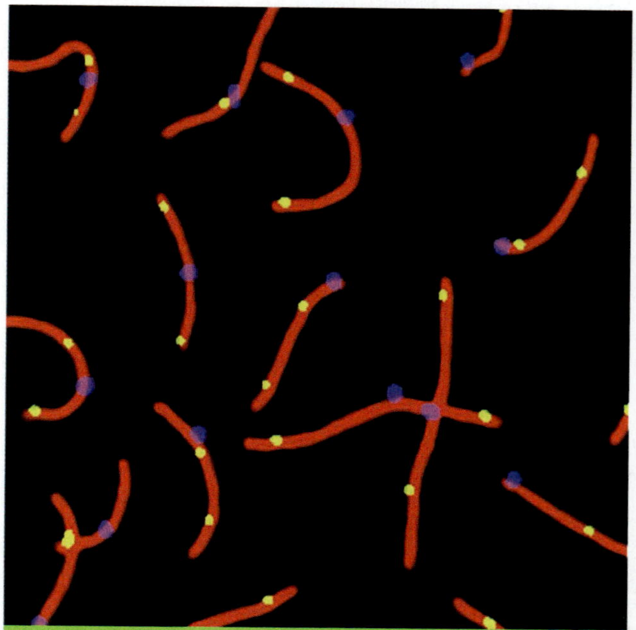

TAKE-HOME MESSAGE 12.5

Meiosis may have evolved by the remodeling of existing mechanisms of mitosis.

12.6 HOW TO SURVIVE FOR 80 MILLION YEARS WITHOUT SEX

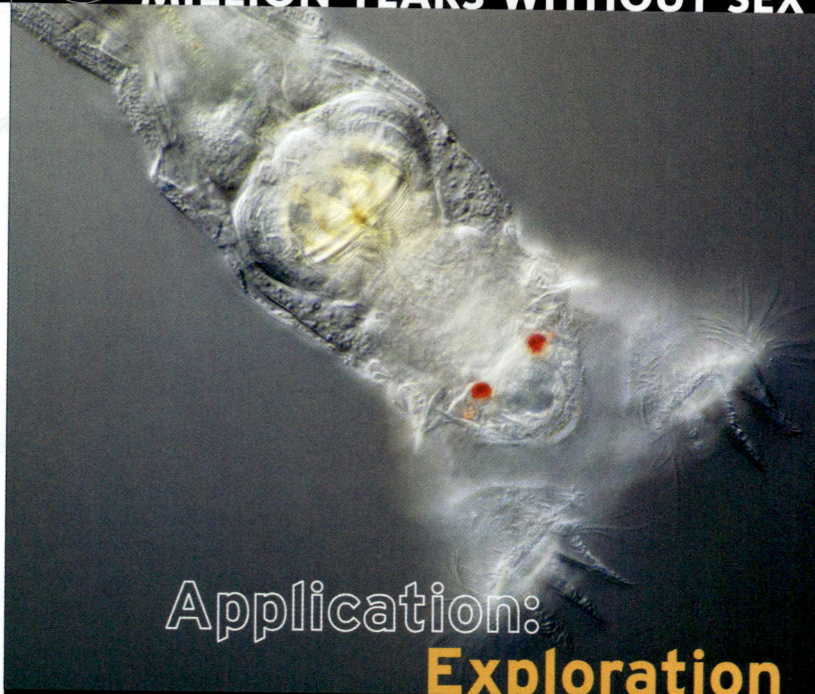

Application: Exploration

FIGURE 12.11 A bdelloid rotifer. All of these tiny animals are female.

WHY DO MALES EXIST? No male has ever been found among the tiny freshwater creatures called bdelloid rotifers (FIGURE 12.11). Females have been reproducing for 80 million years solely through cloning themselves. Bdelloids are one of the few groups of animals to have completely abandoned sex.

Compared to sex, asexual reproduction is often seen as a poor long-term strategy because it lacks crossing over—the chromosomal shuffling that brings about genetic diversity thought to give species an adaptive edge in the face of new challenges. Bdelloids have contradicted this theory by being very successful; over 360 species are alive today.

A newly discovered ability may help to explain the success of the bdelloids despite their rejection of sex. These rotifers can apparently import genes from bacteria, fungi, protists, and even plants. If the main advantage of sex is that it promotes genetic diversity, why worry about it when you have the genes of entire kingdoms available to you?

The direct swapping of genetic material is incredibly rare in animals, but bdelloids are bringing in external genes to an extent completely unheard of in complex organisms. Each rotifer is a genetic mosaic whose DNA spans almost all the major kingdoms of life: About 10 percent of its active genes have been pilfered from other organisms.

Summary

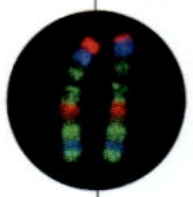

SECTION 12.1 **Sexual reproduction** mixes up the genetic information of two parents. The offspring of sexual reproducers typically vary in shared, inherited traits. This variation in traits can offer an evolutionary advantage over genetically identical offspring produced by asexual reproduction.

Sexual reproduction produces offspring with pairs of chromosomes, one of each homologous pair from the mother and the other from the father. The two chromosomes of a pair carry the same genes. The DNA sequence of paired genes often varies slightly, in which case they are called **alleles**. Alleles are the basis of differences in shared, heritable traits. They arise by mutation.

SECTION 12.2 **Meiosis**, the basis of sexual reproduction in eukaryotes, is a nuclear division mechanism that halves the chromosome number for forthcoming **gametes**. **Haploid** (n) gametes are mature reproductive cells that form from **germ cells**. The fusion of two haploid gametes during **fertilization** restores the diploid parental chromosome number in the **zygote**, the first cell of the new individual.

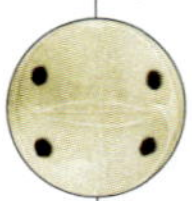

SECTION 12.3 DNA replication occurs before meiosis, so each chromosome consists of two molecules of DNA (sister chromatids). Two nuclear divisions (I and II) occur during meiosis. Meiosis I begins when the chromosomes condense and align tightly with their homologous partners during prophase I. Microtubules then extend from the spindle poles, penetrate the nuclear region, and attach to one or the other chromosome of each homologous pair. At metaphase I, all chromosomes are lined up at the spindle equator. During anaphase I, homologous chromosomes separate and move to opposite spindle poles. Two nuclear envelopes form around the two sets of chromosomes during telophase I. The cytoplasm may divide at this point. There may be a resting period before meiosis resumes, but DNA replication does not occur.

The second nuclear division, meiosis II, occurs in both haploid nuclei that formed in meiosis I. The chromosomes are still duplicated; each still consists of two sister chromatids. The chromosomes condense in prophase II, and align in metaphase II. Sister chromatids of each chromosome are pulled apart from each other in anaphase II, so at the end of meiosis each chromosome consists of one molecule of DNA. By the end of telophase II, four haploid nuclei have typically formed, each with a complete set of (unduplicated) chromosomes.

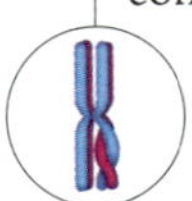

SECTION 12.4 Meiosis shuffles parental alleles, so offspring inherit non-parental combinations of them. During prophase I, homologous chromosomes exchange corresponding segments. This **crossing over** mixes up the alleles on maternal and paternal chromosomes, thus giving rise to combinations of alleles not present in either parental chromosome. The random segregation of maternal and paternal chromosomes into gametes also contributes to variation in traits among offspring of sexual reproducers. Microtubules can attach the maternal or the paternal chromosome of each pair to one or the other spindle pole. Either chromosome may end up in any new nucleus, and in any gamete.

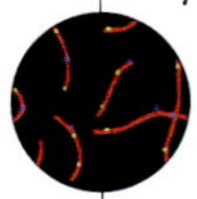

SECTION 12.5 Like mitosis, meiosis requires a spindle to move and sort duplicated chromosomes, but meiosis occurs only in cells that are involved in sexual reproduction. The process of meiosis resembles that of mitosis, and may have evolved from it. Many of the same molecules function the same way in both processes.

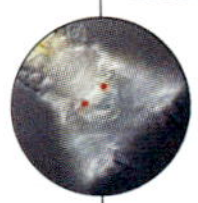

SECTION 12.6 A few groups of animals have survived for millions of years by reproducing only asexually, despite the lack of chromosome shufflings that bring about genetic diversity in offspring. Bdelloid rotifers may have offset this disadvantage by picking up new genes from organisms in other kingdoms.

Self-Quiz

Answers in Appendix VII

1. The main evolutionary advantage of sexual over asexual reproduction is that it produces ________ .
   a. more offspring per individual
   b. more variation among offspring
   c. healthier offspring

2. Meiosis is a necessary part of sexual reproduction because it ________ .
   a. divides two nuclei into four new nuclei
   b. reduces the chromosome number for gametes
   c. produces clones that can cross over

3. Meiosis ________ .
   a. occurs in all eukaryotes
   b. supports growth and tissue repair in multicelled species
   c. gives rise to genetic diversity among offspring
   d. is part of the life cycle of all cells

4. Sexual reproduction in animals requires ________ .
   a. meiosis c. germ cells
   b. fertilization d. all of the above

5. Meiosis ________ the parental chromosome number.
   a. doubles c. maintains
   b. halves d. mixes up

6. Dogs have a diploid chromosome number of 78. How many chromosomes do their gametes have?
   a. 39 c. 156
   b. 78 d. 234

7. The cell in the diagram to the *right* is in anaphase I, not anaphase II. I know this because ________ .

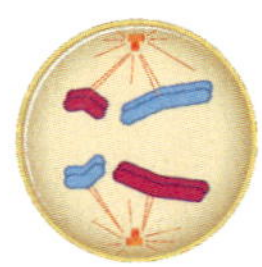

Data Analysis Activities

BPA and Abnormal Meiosis In 1998, researchers at Case Western University were studying meiosis in mouse oocytes when they saw an unexpected and dramatic increase of abnormal meiosis events (**FIGURE 12.12**). Improper segregation of chromosomes during meiosis is one of the main causes of human genetic disorders, which we will discuss in Chapter 14.

The researchers discovered that the spike in meiotic abnormalities began immediately after the mouse facility started washing the animals' plastic cages and water bottles in a new, alkaline detergent. The detergent had damaged the plastic, which as a result was leaching bisphenol A (BPA). BPA is a synthetic chemical that mimics estrogen, the main female sex hormone in animals. BPA is still widely used to manufacture polycarbonate plastic items (including water bottles) and epoxies (including the coating on the inside of metal cans of food).

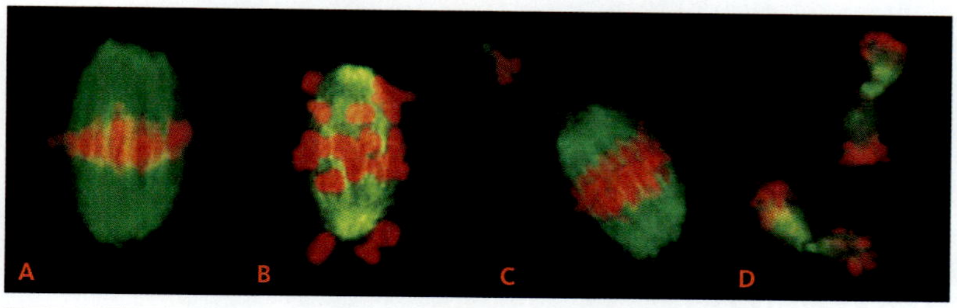

Caging materials	Total number of oocytes	Abnormalities
Control: New cages with glass bottles	271	5 (1.8%)
Damaged cages with glass bottles		
Mild damage	401	35 (8.7%)
Severe damage	149	30 (20.1%)
Damaged bottles	197	53 (26.9%)
Damaged cages with damaged bottles	58	24 (41.4%)

FIGURE 12.12 Meiotic abnormalities associated with exposure to damaged plastic caging. Fluorescent micrographs show nuclei of single mouse oocytes in metaphase I. **A** Normal metaphase; **B–D** examples of abnormal metaphase. Chromosomes appear red; spindle fibers, green.

1. What percentage of mouse oocytes displayed abnormalities of meiosis with no exposure to damaged caging?
2. Which group of mice showed the most meiotic abnormalities in their oocytes?
3. What is abnormal about metaphase I as it is occurring in the oocytes shown in the micrographs in **FIGURE 12.12B**, **C**, and **D**?

8. The cell pictured to the *right* is in which stage of nuclear division?
   a. anaphase c. anaphase II
   b. anaphase I d. none of the above

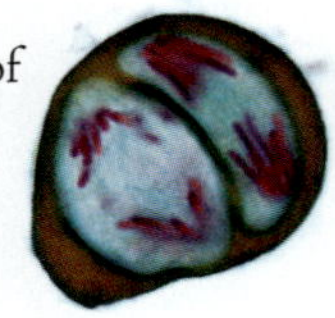

9. Crossing over mixes up ______ .
   a. chromosomes c. zygotes
   b. alleles d. gametes
10. Crossing over happens during which phase of meiosis?
   a. prophase I c. anaphase I
   b. prophase II d. anaphase II
11. ______ contributes to variation in traits among the offspring of sexual reproducers.
   a. Crossing over c. Fertilization
   b. Random attachment of chromosomes to spindle poles d. both a and b
   e. all are factors
12. Which of the following is one of the very important differences between mitosis and meiosis?
   a. Chromosomes align midway between spindle poles only in meiosis.
   b. Homologous chromosomes pair up only in meiosis.
   c. DNA is replicated only in mitosis.
   d. Sister chromatids separate only in meiosis.
   e. Interphase occurs only in mitosis.
13. Match each term with its description.

___ interphase		a. different forms of a gene
___ metaphase I		b. useful for varied offspring
___ alleles		c. may be none between meiosis I and meiosis II
___ zygotes		d. chromosomes lined up
___ gametes		e. haploid
___ males		f. form at fertilization
___ prophase I		g. mash-up time

Critical Thinking

1. In your own words, explain why sexual reproduction tends to give rise to greater genetic diversity among offspring in fewer generations than asexual reproduction.

2. Make a simple sketch of meiosis in a cell with a diploid chromosome number of 4. Now try it when the chromosome number is 3.

3. The diploid chromosome number for the body cells of a frog is 26. What would the frog chromosome number be after three generations if meiosis did not occur before gamete formation?

CREDITS: (12) Reprinted from *Current Biology*, Vol 13, (Apr 03), Authors Hunt, Koehler, Susiarjo, Hodges, Ilagan, Voigt, Thomas, Thomas and Hassold, Bisphenol A Exposure Causes Meiotic Aneuploidy in the Female Mouse, pp. 546–553, © 2003 Cell Press. Published by Elsevier Ltd. With permission from Elsevier; (in text S-Q 8) Michael Clayton/University of Wisconsin, Department of Botany.

Eye color, like many other human traits, does not occur in discrete forms. The continuous range of variation in color among individuals is the result of interactions among several genes involved in making and distributing melanins.

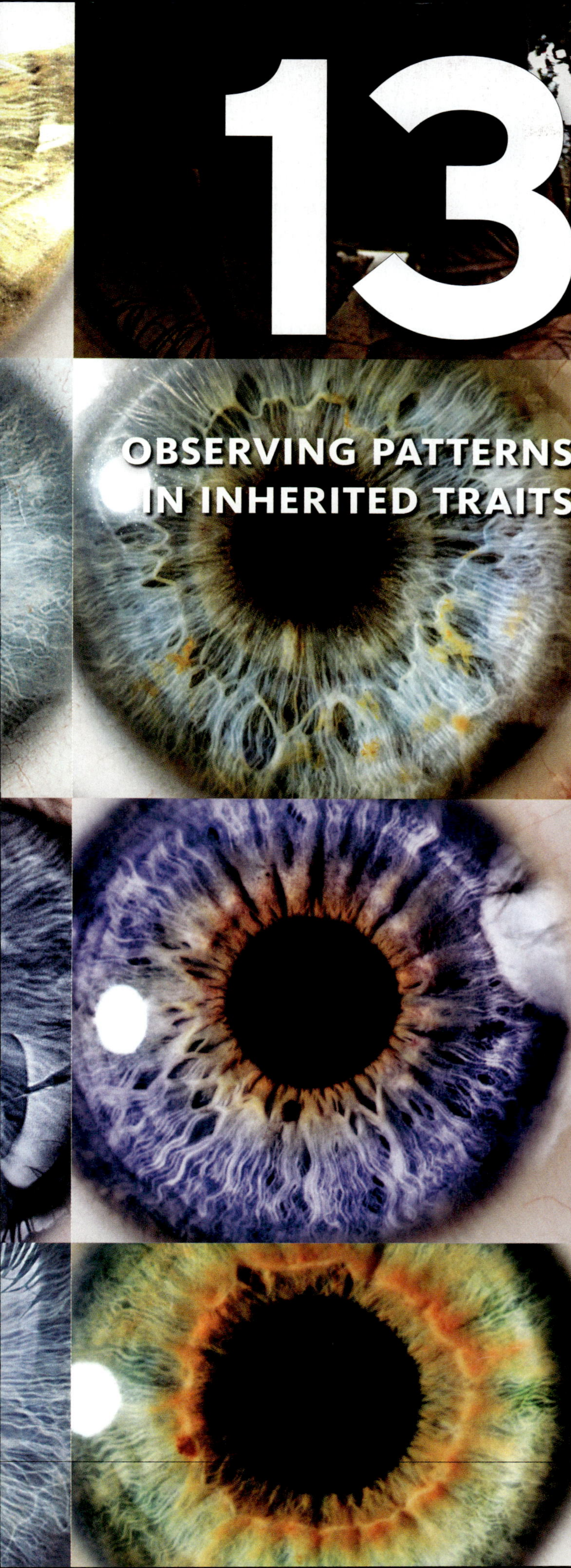

13 OBSERVING PATTERNS IN INHERITED TRAITS

Links to Earlier Concepts

You may want to review what you know about traits (Section 1.4), chromosomes (8.3), genes and gene expression (9.1), mutation (9.5), sexual reproduction and alleles (12.1), and meiosis (12.2–12.4). You will revisit probability and sampling error (1.7), laws of nature (1.8), protein structure (3.4,3.5), pigments (6.1), clones (8.6), gene control (10.1, 10.2, 11.5), and epigenetics (10.5).

KEY CONCEPTS

WHERE MODERN GENETICS STARTED

Gregor Mendel discovered that inherited traits are specified in units. The units, which are distributed into gametes in predictable patterns, were later identified as genes.

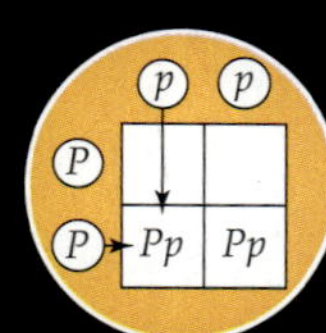

MONOHYBRID CROSSES

Tracking inheritance patterns of single traits led to the discovery that during meiosis, pairs of genes on homologous chromosomes separate and end up in different gametes.

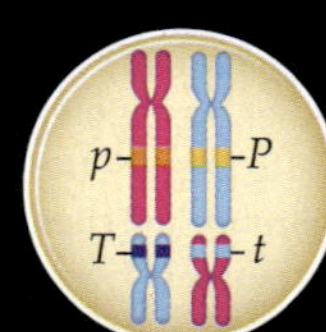

DIHYBRID CROSSES

Tracking inheritance patterns of two unrelated traits led to the discovery that in most cases, genes of a pair segregate into gametes independently of other gene pairs.

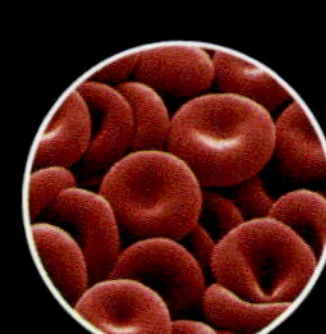

NON-MENDELIAN INHERITANCE

An allele may be partly dominant over a nonidentical partner, or codominant with it. Multiple genes may influence a trait; some genes influence many traits.

COMPLEX VARIATIONS IN TRAITS

Environmental factors can alter the expression of genes that influence a trait. Many traits appear in a continuous range of forms.

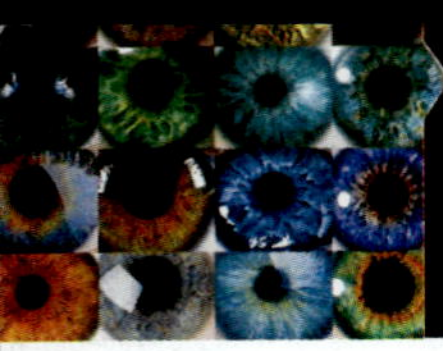

13.1 HOW DO ALLELES CONTRIBUTE TO TRAITS?

In the nineteenth century, people thought that hereditary material must be some type of fluid, with fluids from both parents blending at fertilization like milk into coffee. However, the idea of "blending inheritance" failed to explain what people could see with their own eyes. Children sometimes have traits such as freckles that do not appear in either parent. A cross between a black horse and a white one does not produce gray offspring.

The naturalist Charles Darwin did not accept the idea of blending inheritance, but he could not come up with an alternative hypothesis even though inheritance was central to his theory of natural selection. (We return to Darwin and his theory of natural selection in Chapter 16.) At the time, no one knew that hereditary information (DNA) is divided into discrete units (genes), an insight that is critical to understanding how traits are inherited. However, even before Darwin presented his theory, someone had been gathering evidence that would support it. Gregor Mendel (*above*), an Austrian monk, had been carefully breeding thousands of pea plants. By keeping detailed records of how traits passed from one generation to the next, Mendel had been collecting evidence of how inheritance works.

1 In the flowers of garden pea plants, pollen grains that form in anthers produce male gametes; female gametes form in carpels.

2 Experimenters can control the transfer of hereditary material from one pea plant to another by snipping off a flower's anthers (to prevent the flower from self-fertilizing), and then brushing pollen from another flower onto its carpel.

In this example, pollen from a plant that has purple flowers is brushed onto the carpel of a white-flowered plant.

3 Later, seeds develop inside pods of the cross-fertilized plant. An embryo in each seed develops into a mature pea plant.

4 Every plant that arises from the cross has purple flowers. Predictable patterns such as this are evidence of how inheritance works.

FIGURE 13.1 {Animated} Breeding garden pea plants.

MENDEL'S EXPERIMENTS

Mendel cultivated the garden pea plant (**FIGURE 13.1**). This species is naturally self-fertilizing, which means its flowers produce male and female gametes 1 that form viable embryos when they meet up. In order to study inheritance, Mendel had to carry out controlled matings between individuals with specific traits, then observe and document the traits of their offspring. To keep an individual pea plant from self-fertilizing, Mendel removed the pollen-bearing anthers from its flowers. He then cross-fertilized the plant by brushing its egg-bearing carpels with pollen from another plant 2. He collected the seeds 3 from the cross-fertilized individual, and recorded the traits of the new pea plants that grew from them 4.

Many of Mendel's experiments started with plants that "breed true" for particular traits such as white flowers or purple flowers. Breeding true for a trait means that, new mutations aside, all offspring have the same form of the trait as the parent(s), generation after generation. For example, all offspring of pea plants that breed true for white flowers also have white flowers. As you will see in the next section, Mendel cross-fertilized pea plants that breed true for different forms of a trait, and discovered that the traits of the offspring often appear in predictable patterns. Mendel's meticulous work tracking pea plant traits led him to conclude (correctly) that hereditary information passes from one generation to the next in discrete units.

INHERITANCE IN MODERN TERMS

DNA was not proven to be hereditary material until the 1950s (Section 8.1), but Mendel discovered its units, which we now call genes, almost a century before then. Today, we know that individuals of a species share certain traits because their chromosomes carry the same genes.

Each gene occurs at a specific location, or **locus** (plural, loci), on a particular chromosome (**FIGURE 13.2**). The somatic cells of humans and other animals are diploid,

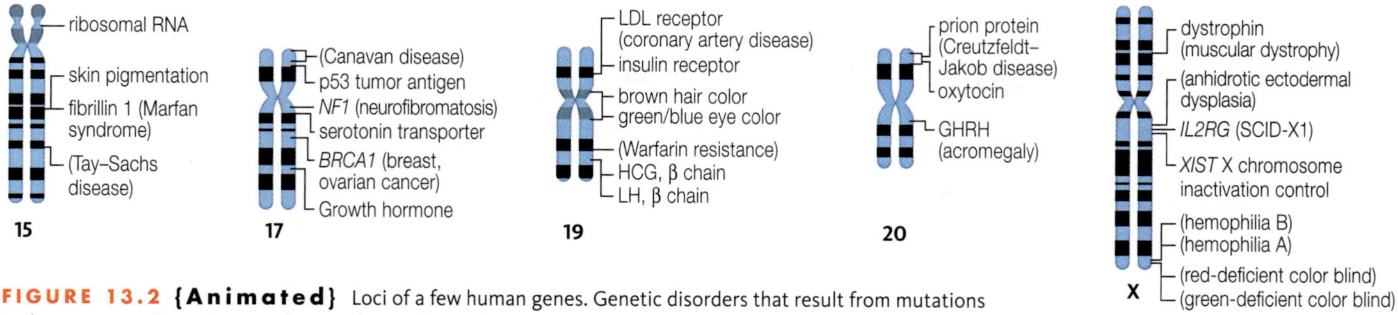

FIGURE 13.2 {Animated} Loci of a few human genes. Genetic disorders that result from mutations in the genes are shown in parentheses. The number or letter below each chromosome is its name; the characteristic banding patterns appear after staining. A similar map of all 23 human chromosomes is in Appendix IV.

so they have pairs of genes, on pairs of homologous chromosomes. In most cases, both genes of a pair are expressed. Genes at the same locus on a pair of homologous chromosomes may be identical, or they may vary as alleles (Section 12.1). Organisms breed true for a trait because they carry identical alleles of genes governing that trait. An individual with two identical alleles of a gene is **homozygous** for the allele. By contrast, an individual with two different alleles of a gene is **heterozygous** (*hetero*– means mixed).

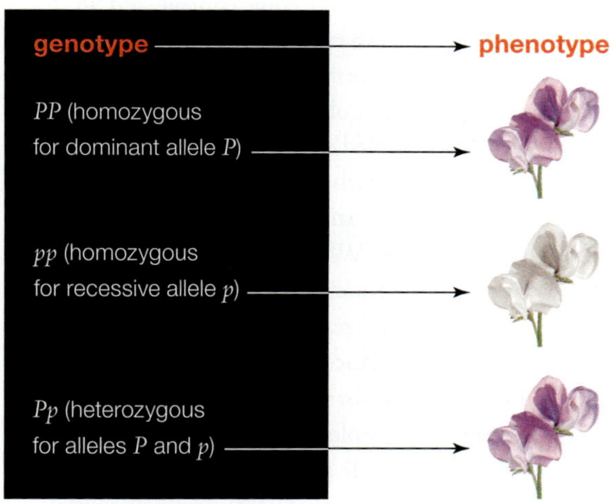

FIGURE 13.3 Genotype gives rise to phenotype. In this example, the dominant allele *P* specifies purple flowers; the recessive allele *p*, white flowers.

dominant Refers to an allele that masks the effect of a recessive allele paired with it in heterozygous individuals.
genotype The particular set of alleles that is carried by an individual's chromosomes.
heterozygous Having two different alleles of a gene.
homozygous Having identical alleles of a gene.
hybrid The heterozygous offspring of a cross or mating between two individuals that breed true for different forms of a trait.
locus Location of a gene on a chromosome.
phenotype An individual's observable traits.
recessive Refers to an allele with an effect that is masked by a dominant allele on the homologous chromosome.

Hybrids are heterozygous offspring of a cross or mating between individuals that breed true for different forms of a trait.

When we say that an individual is homozygous or heterozygous, we are discussing its **genotype**, the particular set of alleles it carries. Genotype is the basis of **phenotype**, which refers to the individual's observable traits. "White-flowered" and "purple-flowered" are examples of pea plant phenotypes that arise from differences in genotype.

The phenotype of a heterozygous individual depends on how the products of its two different alleles interact. In many cases, the effect of one allele influences the effect of the other, and the outcome of this interaction is reflected in the individual's phenotype. An allele is **dominant** when its effect masks that of a **recessive** allele paired with it. Usually, a dominant allele is represented by an italic capital letter such as *A*; a recessive allele, with a lowercase italic letter such as *a*. Consider the purple- and white-flowered pea plants that Mendel studied. In these plants, the allele that specifies purple flowers (let's call it *P*) is dominant over the allele that specifies white flowers (*p*). Thus, a pea plant homozygous for the dominant allele (*PP*) has purple flowers; one homozygous for the recessive allele (*pp*) has white flowers (**FIGURE 13.3**). A heterozygous plant (*Pp*) has purple flowers.

TAKE-HOME MESSAGE 13.1

Gregor Mendel indirectly discovered the role of alleles in inheritance by carefully breeding pea plants and tracking traits of their offspring.

Genotype refers to the particular set of alleles that an individual carries. Genotype is the basis of phenotype, which refers to the individual's observable traits.

A homozygous individual has two identical alleles of a gene. A heterozygous individual has two nonidentical alleles of the gene.

A dominant allele masks the effect of a recessive allele paired with it in a heterozygous individual.

13.2 HOW ARE ALLELES DISTRIBUTED INTO GAMETES?

When homologous chromosomes separate during meiosis (Section 12.3), the gene pairs on those chromosomes separate too. Each gamete that forms carries only one of the two genes of a pair (**FIGURE 13.4**). Thus, plants homozygous for the dominant allele (*PP*) can only make gametes that carry the dominant allele *P* ❶. Plants homozygous for the recessive allele (*pp*) can only make gametes that carry the recessive allele *p* ❷. If these homozygous plants are crossed (*PP* × *pp*), only one outcome is possible: A gamete carrying allele *P* meets up with a gamete carrying allele *p* ❸. All offspring of this cross will have both alleles—they will be heterozygous (*Pp*).

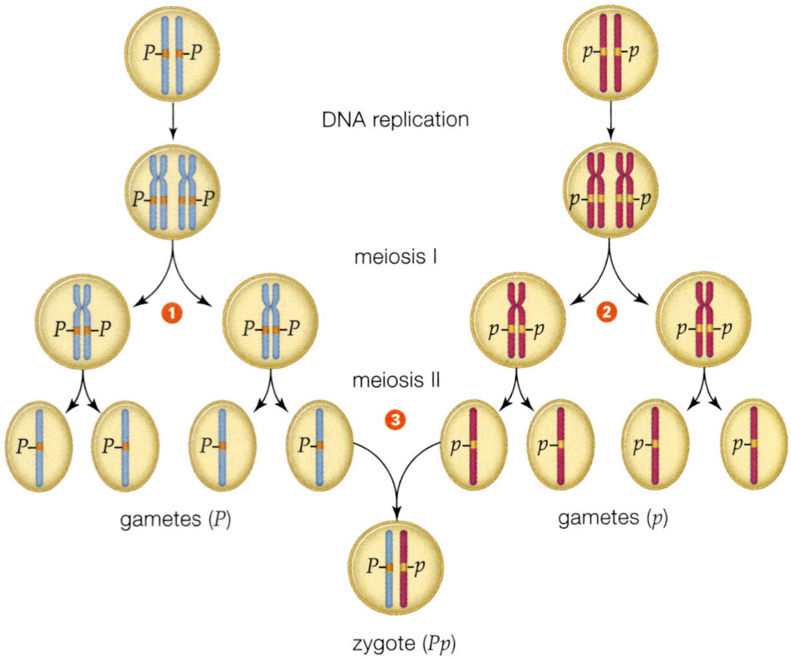

FIGURE 13.4 Segregation of genes on homologous chromosomes into gametes. Homologous chromosomes separate during meiosis, so the pairs of genes they carry separate too. Each of the resulting gametes carries one of the two members of each gene pair. For clarity, only one set of chromosomes is illustrated.

❶ All gametes made by a parent homozygous for a dominant allele carry that allele.

❷ All gametes made by a parent homozygous for a recessive allele carry that allele.

❸ If these two parents are crossed, the union of any of their gametes at fertilization produces a zygote with both alleles. All offspring of this cross will be heterozygous.

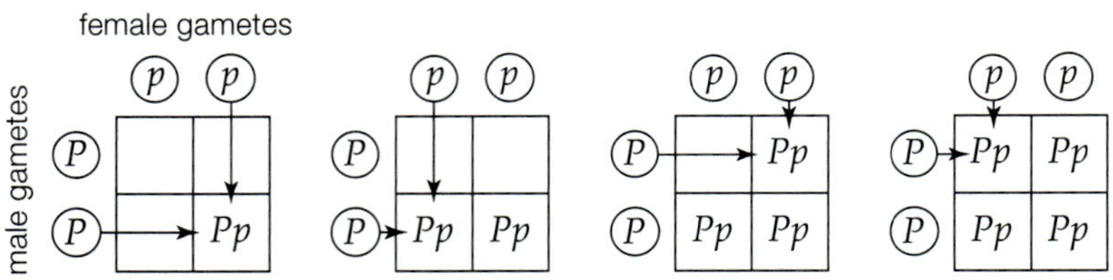

FIGURE 13.5 Making a Punnett square. Parental gametes are listed in circles on the top and left sides of a grid. Each square is filled with the combination of alleles that would result if the gametes in the corresponding row and column met up.

A grid called a **Punnett square** is helpful for predicting the outcomes of such crosses (**FIGURE 13.5**).

Our example illustrated a pattern so predictable that it can be used as evidence of a dominance relationship between alleles. In a **testcross**, an individual that has a dominant trait (but an unknown genotype) is crossed with an individual known to be homozygous for the recessive allele. The pattern of traits among the offspring of the cross can reveal whether the tested individual is heterozygous or homozygous. If all of the offspring of the testcross have the dominant trait, then the parent with the unknown genotype is homozygous for the dominant allele. If any of the offspring have the recessive trait, then it is heterozygous.

Dominance relationships between alleles determine the phenotypic outcome of a **monohybrid cross**, in which individuals that are identically heterozygous for one gene—*Pp* for example—are bred together or self-fertilized. The frequency at which traits associated with the alleles appear among the offspring depends on whether one of the alleles is dominant over the other.

To perform a monohybrid cross, we would start with two individuals that breed true for two different forms of a trait. In garden pea plants, flower color (purple and white) is one example of a trait with two distinct forms, but there are many others. Mendel investigated seven of them: stem length (tall and short), seed color (yellow and green), pod texture (smooth and wrinkled), and so on (**TABLE 13.1**). A cross between individuals that breed true for two forms of a trait yields offspring identically heterozygous for the alleles that govern the trait. When these F_1 (first generation) hybrids are crossed, the frequency at which the two traits appear in the F_2 (second generation) offspring offers information about a dominance relationship between the alleles. F is an abbreviation for filial, which means offspring.

A cross between two purple-flowered heterozygous individuals (*Pp*) is an example of a monohybrid cross. Each of these plants can make two types of gametes: ones that carry a *P* allele, and ones that carry a *p* allele (**FIGURE 13.6A**). So, in a monohybrid cross between two *Pp* plants (*Pp* × *Pp*), the two types of gametes can meet up in four possible ways at fertilization:

Possible Event		Probable Outcome
Sperm *P* meets egg *P*	⟶	zygote genotype is *PP*
Sperm *P* meets egg *p*	⟶	zygote genotype is *Pp*
Sperm *p* meets egg *P*	⟶	zygote genotype is *Pp*
Sperm *p* meets egg *p*	⟶	zygote genotype is *pp*

Three out of four possible outcomes of this cross include at least one copy of the dominant allele *P*. Each time

TABLE 13.1

Mendel's Seven Pea Plant Traits

Trait	Dominant Form	Recessive Form
Seed Shape	Round	Wrinkled
Seed Color	Yellow	Green
Pod Texture	Smooth	Wrinkled
Pod Color	Green	Yellow
Flower Color	Purple	White
Flower Position	Along Stem	At Tip
Stem Length	Tall	Short

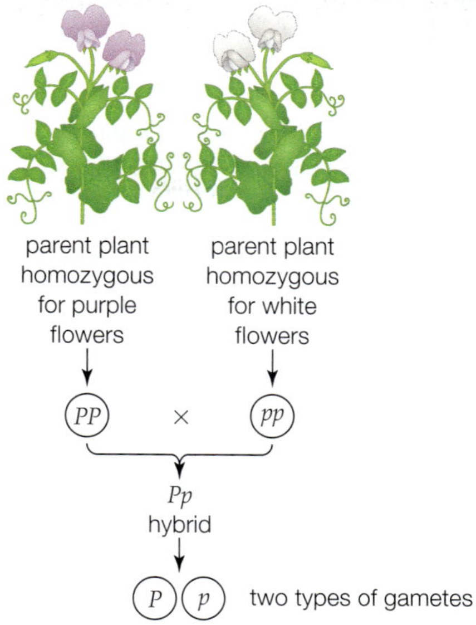

A All of the F_1 offspring of a cross between two plants that breed true for different forms of a trait are identically heterozygous (*Pp*). These offspring make two types of gametes: *P* and *p*.

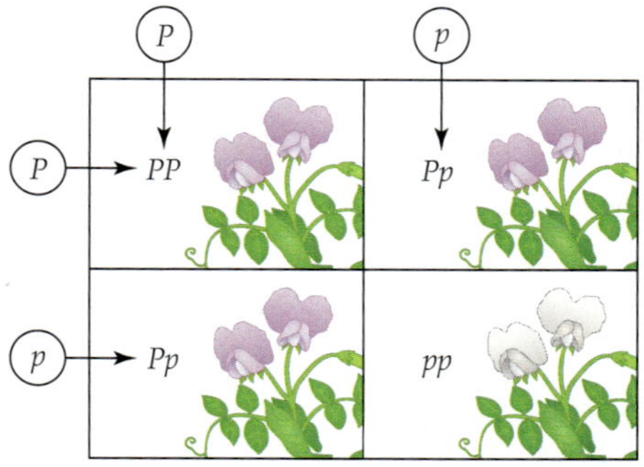

B A monohybrid cross is a cross between these F_1 offspring. In this example, the phenotype ratio in F_2 offspring is 3:1 (3 purple to 1 white).

FIGURE 13.6 {Animated} A monohybrid cross.

FIGURE IT OUT: How many genotypes are possible in the F_2 generation?

Answer: Three: *PP*, *Pp*, and *pp*

fertilization occurs, there are 3 chances in 4 that the resulting offspring will inherit a *P* allele, and have purple flowers. There is 1 chance in 4 that it will inherit two recessive *p* alleles, and have white flowers. Thus, the probability that a particular offspring of this cross will have purple or white flowers is 3 purple to 1 white, which we represent as a ratio of 3:1 (**FIGURE 13.6B**). The 3:1 pattern is an indication that purple and white flower color are specified by alleles with a clear dominance relationship: Purple is dominant; white, recessive. If the probability of one individual inheriting a particular genotype is difficult to imagine, think about probability in terms phenotypes of many offspring. In this example, there will be roughly three purple-flowered plants for every white-flowered one.

The phenotype ratios in the F_2 offspring of Mendel's monohybrid crosses were all close to 3:1. These results became the basis of his **law of segregation**, which we state here in modern terms: Diploid cells carry pairs of genes, on pairs of homologous chromosomes. The two genes of each pair are separated from each other during meiosis, so they end up in different gametes.

TAKE-HOME MESSAGE 13.2

Homologous chromosomes carry pairs of genes. The two genes of each pair are separated from each other during meiosis, so they end up in different gametes.

law of segregation The two members of each pair of genes on homologous chromosomes end up in different gametes during meiosis.
monohybrid cross Cross between two individuals identically heterozygous for one gene; for example $Aa \times Aa$.
Punnett square Diagram used to predict the genetic and phenotypic outcome of a cross.
testcross Method of determining genotype by tracking a trait in the offspring of a cross between an individual of unknown genotype and an individual known to be homozygous recessive.

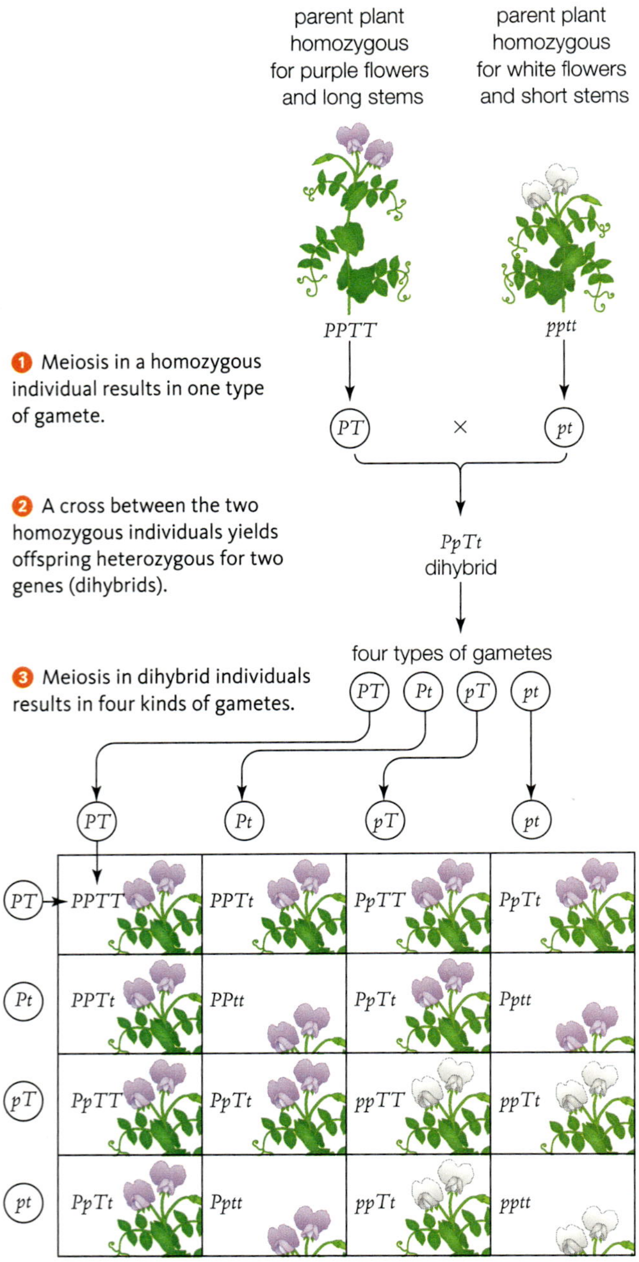

4 If two of the dihybrid individuals are crossed, the four types of gametes can meet up in 16 possible ways. Of 16 possible offspring genotypes, 9 will result in plants that are purple-flowered and tall; 3, purple-flowered and short; 3, white-flowered and tall; and 1, white-flowered and short. Thus, the ratio of phenotypes is 9:3:3:1.

FIGURE 13.7 {Animated} A dihybrid cross between plants that differ in flower color and plant height. In this example, *P* and *p* are dominant and recessive alleles for flower color; *T* and *t* are dominant and recessive alleles for height.

FIGURE IT OUT: What do the flowers inside the boxes represent?

Answer: Phenotypes of the F_2 offspring

A monohybrid cross allows us to track alleles of one gene pair. What about alleles of two gene pairs? In a **dihybrid cross**, individuals identically heterozygous for alleles of two genes (dihybrids) are crossed. As with a monohybrid cross, the pattern of traits seen in the offspring of the cross depends on the dominance relationships between alleles of the genes.

Let's use a gene for flower color (*P*, purple; *p*, white) and one for plant height (*T*, tall; *t*, short) in an example. **FIGURE 13.7** shows a dihybrid cross starting with one parent plant that breeds true for purple flowers and tall stems (*PPTT*), and one that breeds true for white flowers and short stems (*pptt*). The *PPTT* plant only makes gametes with the dominant alleles (*PT*); the *pptt* plant only makes gametes with the recessive alleles (*pt*) 1. So, all offspring from a cross between these parent plants (*PPTT* × *pptt*) will be dihybrids (*PpTt*) with purple flowers and tall stems 2.

Four combinations of alleles are possible in the gametes of *PpTt* dihybrids 3. If two *PpTt* plants are crossed (a dihybrid cross, *PpTt* × *PpTt*), the four types of gametes can combine in sixteen possible ways at fertilization 4. Nine of the sixteen genotypes would give rise to tall plants with purple flowers; three, to short plants with purple flowers; three, to tall plants with white flowers; and one, to short plants with white flowers. Thus, the ratio of phenotypes among the offspring of this dihybrid cross would be 9:3:3:1.

Mendel discovered the 9:3:3:1 ratio of phenotypes among the offspring of his dihybrid crosses, but he had no idea what it meant. He could only say that "units" specifying one trait (such as flower color) are inherited independently of "units" specifying other traits (such as plant height). In time, Mendel's hypothesis became known as the **law of independent assortment**, which we state here in modern terms: During meiosis, the two genes of a pair tend to be sorted into gametes independently of how other gene pairs are sorted into gametes.

Mendel published his results in 1866, but apparently his work was read by few and understood by no one at the time. In 1871 he was promoted, and his pioneering experiments ended. When he died in 1884, he did not know that his work with pea plants would be the starting point for modern genetics.

CONTRIBUTION OF CROSSOVERS

How two genes get sorted into gametes depends partly on whether they are on the same chromosome. When homologous chromosomes separate during meiosis, either member of the pair can end up in either of the two new nuclei that form. This random assortment happens

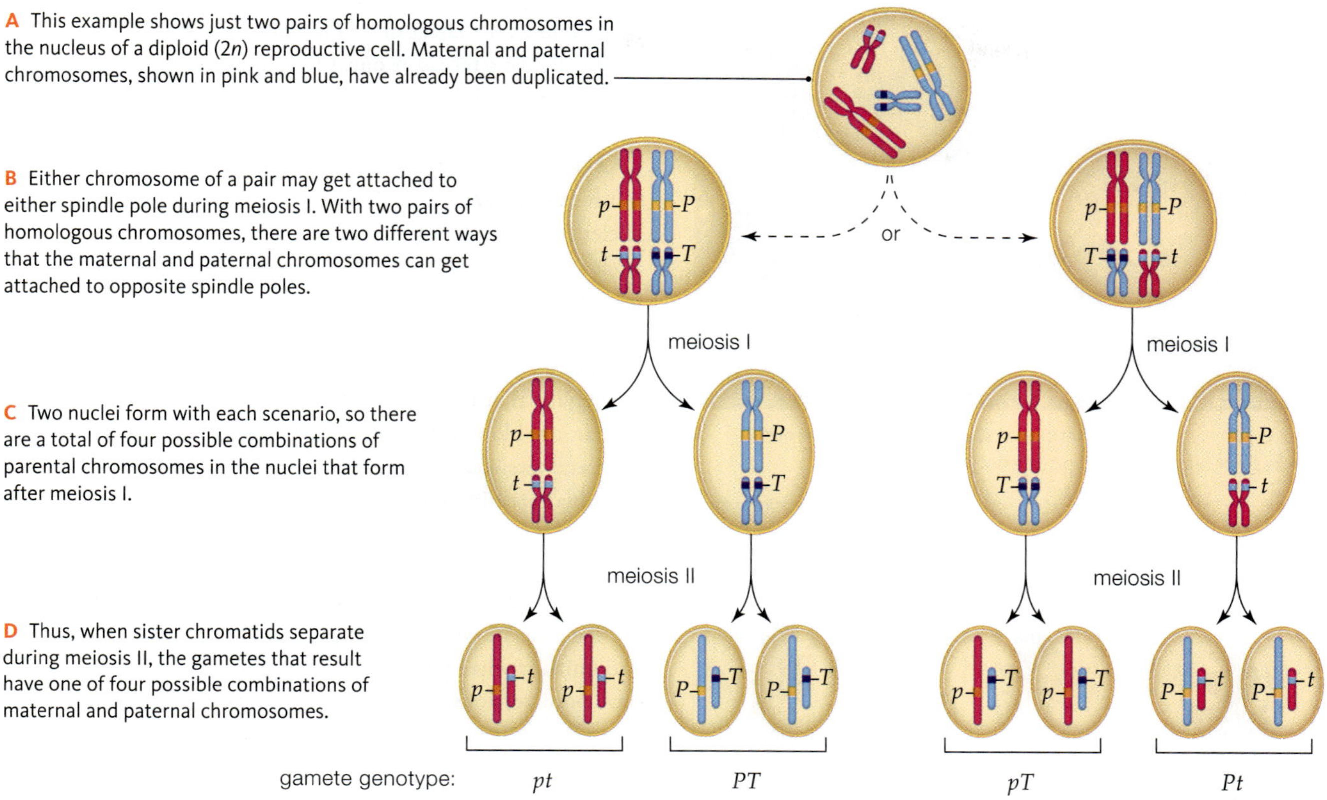

FIGURE 13.8 {Animated} Independent assortment of genes on different chromosomes. Genes that are far apart on the same chromosome usually assort independently too, because crossovers typically separate them.

independently for each pair of homologous chromosomes in the cell. Thus, genes on one chromosome assort into gametes independently of genes on the other chromosomes (**FIGURE 13.8**).

Pea plants have seven chromosomes. Mendel studied seven pea genes, and all of them assorted into gametes independently of one another. Was he lucky enough to choose one gene on each of those chromosomes? As it turns out, some of the genes Mendel studied *are* on the same chromosome. These genes are far enough apart that crossing over occurs between them very frequently—so frequently that they tend to assort into gametes independently, just as if they were on different chromosomes. By contrast, genes that are very close together on a chromosome usually do not assort independently into gametes, because crossing over does not happen between them very often. Thus, gametes usually end up with parental combinations of alleles of these genes.

Genes that do not assort independently into gametes are said to be linked. Linked genes were identified by tracking inheritance in human families over several generations. All of the genes on a chromosome are called a **linkage group**. Peas have 7 different chromosomes, so they have 7 linkage groups. Humans have 23 different chromosomes, so they have 23 linkage groups.

TAKE-HOME MESSAGE 13.3

During meiosis, gene pairs on homologous chromosomes tend to be distributed into gametes independently of how other gene pairs are distributed.

Independent assortment depends on proximity. Genes that are closer together on a chromosome get separated less frequently by crossovers, so gametes often receive parental combinations of alleles of these genes.

dihybrid cross Cross between two individuals identically heterozygous for two genes; for example *AaBb* × *AaBb*.
law of independent assortment During meiosis, members of a pair of genes on homologous chromosomes tend to be distributed into gametes independently of other gene pairs.
linkage group All genes on a chromosome.

13.4 ARE ALL GENES INHERITED IN A MENDELIAN PATTERN?

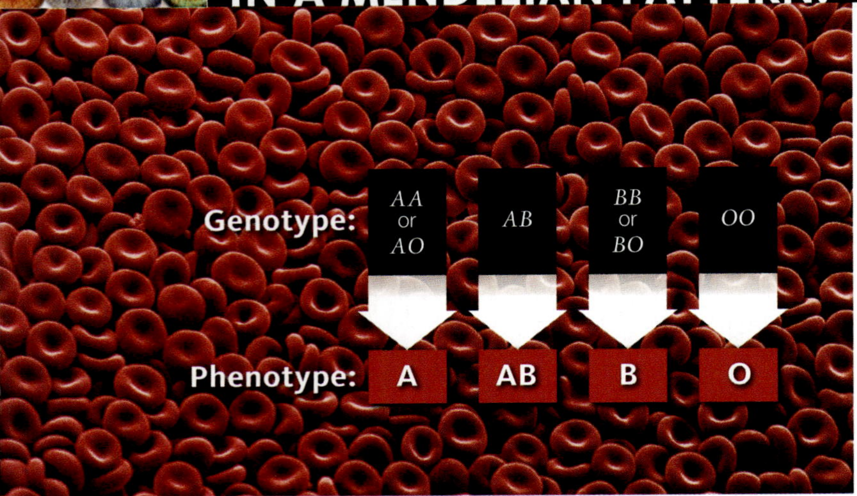

FIGURE 13.9 Combinations of alleles (genotype) that are the basis of human blood type (phenotype).

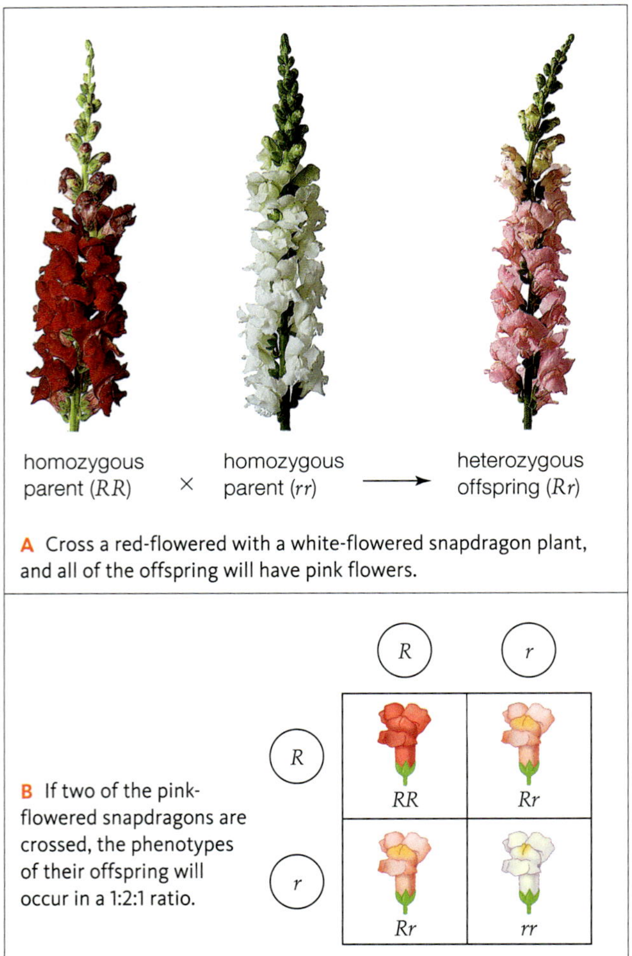

A Cross a red-flowered with a white-flowered snapdragon plant, and all of the offspring will have pink flowers.

B If two of the pink-flowered snapdragons are crossed, the phenotypes of their offspring will occur in a 1:2:1 ratio.

FIGURE 13.10 {Animated} Incomplete dominance in heterozygous (pink) snapdragons. One allele (*R*) results in the production of a red pigment; the other (*r*) results in no pigment.

FIGURE IT OUT: Is the experiment in **B** a monohybrid cross or a dihybrid cross?

Answer: A monohybrid cross

In the Mendelian inheritance patterns discussed in the last two sections, the effect of a dominant allele on a trait fully masks that of a recessive one. Other inheritance patterns are more common, but also more complex.

CODOMINANCE

With **codominance**, traits associated with two nonidentical alleles of a gene are equally apparent in heterozygotes; neither allele is dominant or recessive. Codominance may occur in **multiple allele systems**, in which three or more alleles of a gene persist at relatively high frequency among individuals of a population. Consider the *ABO* gene, which encodes an enzyme that modifies a carbohydrate on the surface of human red blood cells. The *A* and *B* alleles encode slightly different versions of this enzyme, which in turn modify the carbohydrate differently. The *O* allele has a mutation that prevents its enzyme product from becoming active at all.

The two alleles you carry for the *ABO* gene determine the form of the carbohydrate on your blood cells, so they are the basis of your blood type. The *A* and the *B* allele are codominant when paired. If your genotype is *AB*, then you have both versions of the enzyme, and your blood type is AB. The *O* allele is recessive when paired with either the *A* or *B* allele. If your genotype is *AA* or *AO*, your blood type is A. If your genotype is *BB* or *BO*, it is type B. If you are *OO*, it is type O (**FIGURE 13.9**).

Receiving incompatible blood in a transfusion is dangerous because the immune system attacks red blood cells bearing molecules that do not occur in one's own body. The attack can cause the blood cells to clump or burst, with potentially fatal consequences. The blood cells of people with type O blood do not carry the carbohydrate that can trigger this immune response. Thus, people with type O blood can donate blood to anyone; they are called universal donors. However, because their body is unfamiliar with the carbohydrates made by people with type A or B blood, they can receive type O blood only. People with type AB blood can receive a transfusion of any blood type, so they are called universal recipients.

INCOMPLETE DOMINANCE

With **incomplete dominance**, one allele is not fully dominant over the other, so the heterozygous phenotype is an intermediate blend of the two homozygous phenotypes. A gene that affects flower color in snapdragon plants is an example. One allele of the gene (*R*) encodes an enzyme that makes a red pigment. The enzyme encoded by a mutated allele (*r*) cannot make any pigment. Plants homozygous for the *R* allele (*RR*) make a lot of red pigment, so they have red flowers. Plants homozygous for the *r* allele (*rr*) make no pigment, so their flowers are white. Heterozygous plants (*Rr*) make only enough red pigment to color their flowers pink (**FIGURE 13.10**). A cross between two heterozygous plants yields red-, pink-, and white-flowered offspring in a 1:2:1 ratio.

EPISTASIS

Some traits are affected by multiple genes, an effect called polygenic inheritance or **epistasis**. Consider fur color in dogs, which depends on pigments called melanins. A dark brown melanin gives rise to brown or black fur; a reddish melanin is responsible for yellow fur. The production and deposition of melanin pigments in fur depends on several genes. The product of one gene (*TYRP1*) helps make the brown melanin. A dominant allele (*B*) of this gene results in a higher production of this melanin than the recessive allele (*b*). A different gene (*MC1R*) affects which type of melanin is produced. A dominant allele (*E*) of this gene triggers production of the brown melanin; its recessive partner (*e*) carries a mutation that results in production of the reddish form. Dogs homozygous for the *e* allele are yellow because they produce only the reddish melanin (**FIGURE 13.11**).

PLEIOTROPY

In many cases, a single gene influences multiple traits, an effect called **pleiotropy**. Mutations that alter the gene affect all of the traits at once. Many complex genetic disorders, including sickle-cell anemia (Section 9.5) and Marfan syndrome, arise as a result of mutations in pleiotropic genes. Marfan syndrome is caused by mutations that affect fibrillin. Long fibers of this protein impart elasticity to tissues of the heart, skin, blood vessels, tendons, and other body parts. Mutations cause tissues to form with defective fibrillin or none at all. The largest blood vessel leading from the heart, the aorta, is particularly affected. The aorta's thick wall is not as elastic as it should be, and it eventually stretches and becomes leaky. Thinned and weakened, the aorta can rupture during exercise—an abruptly fatal outcome.

About 1 in 5,000 people have Marfan syndrome, and there is no cure. Its effects—and risks—are manageable with early diagnosis, but symptoms are often missed. Many affected people die suddenly and early without ever knowing they had the disorder (**FIGURE 13.12**).

codominance Effect in which the full and separate phenotypic effects of two alleles are apparent in heterozygous individuals.
epistasis Polygenic inheritance, in which a trait is influenced by multiple genes.
incomplete dominance Effect in which one allele is not fully dominant over another, so the heterozygous phenotype is an intermediate blend between the two homozygous phenotypes.
multiple allele system Gene for which three or more alleles persist in a population at relatively high frequency.
pleiotropy Effect in which a single gene affects multiple traits.

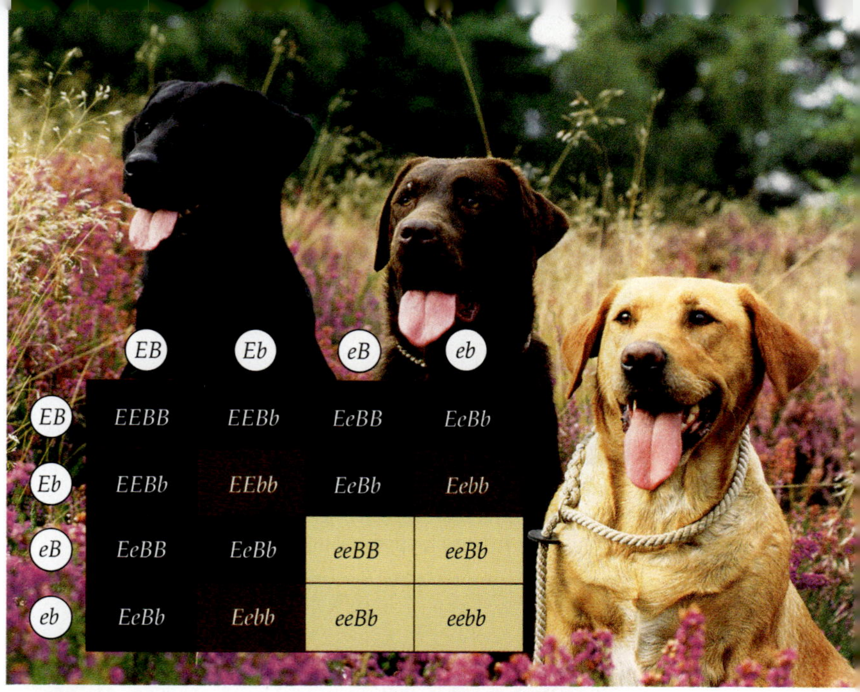

FIGURE 13.11 {Animated} An example of epistasis. Interactions among products of two gene pairs affect coat color in Labrador retrievers. Dogs with alleles *E* and *B* have black fur. Those with an *E* and two recessive *b* alleles have brown fur. Dogs homozygous for the recessive *e* allele have yellow fur.

FIGURE 13.12
A heartbreaker: Marfan syndrome. In 2006, 21-year-old basketball star Haris Charalambous collapsed and died suddenly during warm-up exercises. An autopsy revealed that his aorta had burst, an effect of the Marfan syndrome that Charalambous did not realize he had. Assistant trainer Brian Jones says, "Haris was just the nicest, funniest kid in the world. With his size, he was sort of lovably goofy. He was everybody's best friend."

TAKE-HOME MESSAGE 13.4

Some alleles are not dominant or recessive when paired.

With incomplete dominance, one allele is not fully dominant over another, so the heterozygous phenotype is an intermediate blend of the two homozygous phenotypes.

In codominance, two alleles have full and separate effect, so the phenotype of a heterozygous individual comprises both homozygous phenotypes.

In some cases, one gene influences multiple traits. In other cases, multiple genes influence the same trait.

13.5 DOES THE ENVIRONMENT AFFECT PHENOTYPE?

A The color of the snowshoe hare's fur varies by season. In summer, the fur is brown (*left*); in winter, white (*right*). Both forms offer seasonally appropriate camouflage from predators.

B The height of a mature yarrow plant depends on the elevation at which it grows.

C The body form of the water flea on the top develops in environments with few predators. A longer tail spine and a pointy head (*bottom*) develop in response to chemicals emitted by insects that prey on the fleas.

FIGURE 13.13 {Animated} Examples of environmental effects on phenotype.

The phrase "nature versus nurture" refers to a centuries-old debate about whether human behavioral traits arise from one's genetics (nature) or from environmental factors (nurture). It turns out that both play a role. The environment affects the expression of many genes, which in turn affects phenotype—including human behavioral traits. We can summarize this thinking with an equation:

genotype + **environment** ⟶ **phenotype**

Epigenetics research is revealing that the environment makes an even greater contribution to this equation than most biologists had suspected (Section 10.5).

Environmentally driven changes in gene expression patterns involve gene control (Section 10.1). For example, environmental cues trigger some cell-signaling pathways that end with methyl groups being removed from or added to particular regions of DNA. The change in methylation enhances or suppresses gene expression in those regions.

SOME ENVIRONMENTAL EFFECTS

Mechanisms that adjust phenotype in response to external cues are part of an individual's normal ability to adapt to its environment, as the following examples illustrate.

Seasonal Changes in Coat Color Seasonal changes in temperature and the length of day affect the production of melanin and other pigments that color the skin and fur of many animals. These species have different color phases in different seasons (**FIGURE 13.13A**). Hormonal signals triggered by the seasonal changes cause fur to be shed, and new fur grows back with different types and amounts of pigments deposited in it. The resulting change in phenotype provides these animals with seasonally appropriate camouflage from predators.

Effect of Altitude on Yarrow In plants, a flexible phenotype gives immobile individuals an ability to thrive in diverse habitats. For example, genetically identical yarrow plants grow to different heights at different altitudes (**FIGURE 13.13B**). More challenging temperature, soil, and water conditions are typically encountered at higher altitudes. Differences in altitude are also correlated with changes in the reproductive mode of yarrow: Plants at higher altitude tend to reproduce asexually, and those at lower altitude tend to reproduce sexually.

Alternative Phenotypes in Water Fleas Water fleas have different phenotypes depending on whether the aquatic insects that prey on them are present (**FIGURE 13.13C**). Individuals also switch between asexual and sexual modes of reproduction depending on environmental conditions. During the early spring, competition is scarce in their freshwater pond habitats. At that time, the fleas reproduce rapidly by asexual means, giving birth to large numbers of female offspring that quickly fill the ponds. Later in the season, competition for resources intensifies as the pond water becomes warmer, saltier, and more crowded. Under these conditions, some of the water fleas start giving birth to males, and then reproducing sexually. The increased genetic diversity of sexually produced offspring may offer the population an advantage in a more challenging environment.

Psychiatric Disorders Does the environment affect human genes? Researchers recently discovered that mutations in four human gene regions are associated with five psychiatric disorders: autism, depression, schizophrenia, bipolar disorder, and attention deficit hyperactivity disorder (ADHD). However, there must be an environmental component too, because one person with the mutations might get one type of disorder, while a relative with the same mutations might get another—two different results from the same genetic underpinnings. Moreover, the majority of people who carry these mutations never end up with a psychiatric disorder.

Recent discoveries in animal models are beginning to unravel some of the mechanisms by which environment can influence mental state in humans. For example, we now know that learning and memory are associated with dynamic and rapid DNA modifications in brain cells. Mood is, too. Stress-induced depression causes methylation-based silencing of a particular nerve growth factor gene; some antidepressants work by reversing this methylation. As another example, rats whose mothers are not very nurturing end up anxious and having a reduced resilience for stress as adults. The difference between these rats and ones who had nurturing maternal care is traceable to epigenetic DNA modifications that result in a lower than normal level of another nerve growth factor. Drugs can reverse these modifications—and their effects. We do not yet know all of the genes that influence human mental state, but the implication of such research is that future treatments for many disorders will involve deliberate modification of methylation patterns in an individual's DNA.

TAKE-HOME MESSAGE 13.5

The environment influences gene expression, and therefore can alter phenotype.

Cell-signaling pathways link environmental cues with changes in gene expression.

PEOPLE MATTER

DR. GAY BRADSHAW

The air explodes with the sound of high-powered rifles and the startled infant watches his family fall to the ground, the image seared into his memory. He and other orphans are then transported to distant locales to start new lives. Ten years later, the teenaged orphans begin a killing rampage, leaving more than a hundred victims.

A scene describing post-traumatic stress disorder (PTSD) in Kosovo or Rwanda? The similarities are striking—but the teenagers are young elephants, and the victims, rhinoceroses.

Gay Bradshaw, a psychologist and the director of the Kerulos Center in Oregon, has brought the latest insights from human neuroscience and psychology to bear on startling field observations of elephant behavior. She suspects that some threatened elephant populations might be suffering from chronic stress and trauma brought on by human encroachment and killing. "The loss of older elephants," says Bradshaw, "and the extreme psychological and physical trauma of witnessing the massacres of their family members interferes with a young elephant's normal development."

Under normal conditions, an early and healthy emotional relationship between an infant and its mother fosters the development of self-regulatory structures in the brain's right hemisphere. All mammals share this developmental attachment mechanism. With trauma, a malfunction can develop that makes the individual vulnerable to PTSD and predisposed to violence as an adult. Individuals who survive trauma often face a lifelong struggle with depression, suicide, or behavioral dysfunctions. In addition, children of trauma survivors can exhibit similar symptoms, an effect that is likely to be epigenetic at least in part.

As with humans, an intact, functioning social order helps buffer the effects of trauma in elephants. When park rangers introduced older males into the herd of marauding adolescent orphans, the orphans' violent behavior abruptly stopped.

13.6 DO ALL TRAITS OCCUR IN DISTINCT FORMS?

FIGURE 13.14 Face length varies continuously in dogs. A gene with 12 alleles influences this trait.

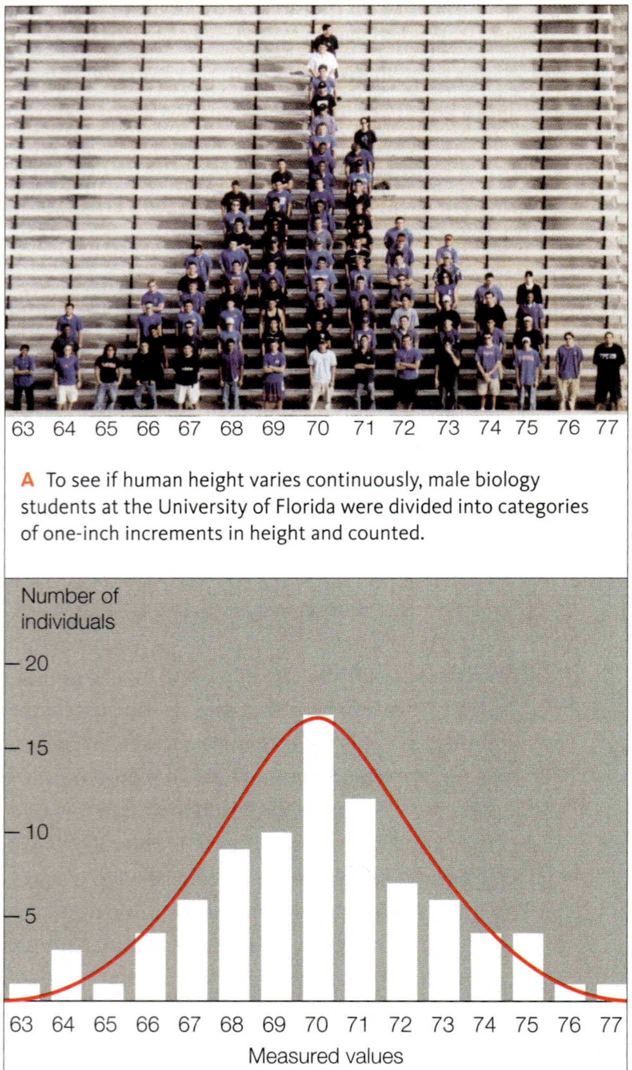

A To see if human height varies continuously, male biology students at the University of Florida were divided into categories of one-inch increments in height and counted.

B Graphing the data that resulted from the experiment in **A** produces a bell-shaped curve, an indication that height does vary continuously in humans.

FIGURE 13.15 {Animated} Continuous variation.

The pea plant phenotypes that Mendel studied appeared in two or three forms, which made them easy to track through generations. However, many other traits do not appear in distinct forms. Such traits are often the result of complex genetic interactions—multiple genes, multiple alleles, or both—with added environmental influences (we return to this topic in Chapter 17, as we consider some evolutionary consequences of variation in phenotype). Tracking traits with complex variation presents a special challenge, which is why the genetic basis of many of them has not yet been completely unraveled.

Some traits occur in a range of small differences that is called **continuous variation**. Continuous variation can be an outcome of epistasis, in which multiple genes affect a single trait. The more genes that influence a trait, the more continuous is its variation. Traits that arise from genes with a lot of alleles may also vary continuously. Some genes have regions of DNA in which a series of 2 to 6 nucleotides is repeated hundreds or thousands of times in a row. These **short tandem repeats** can spontaneously expand or contract very quickly compared with the rate of mutation, and the resulting changes in the gene's DNA sequence may be preserved as alleles. For example, short tandem repeats have given rise to 12 alleles of a homeotic gene that influences the length of the face in dogs, with longer repeats associated with longer faces (**FIGURE 13.14**).

Human skin color varies continuously (a topic that we return to in Chapter 14), as does human eye color (shown in the chapter opener). How do we determine whether a particular trait varies continuously? Let's use another human trait, height, as an example. First, the total range of phenotypes is divided into measurable categories—inches, in this case (**FIGURE 13.15A**). Next, the individuals in each category are counted; these counts reveal the relative frequencies of phenotypes across the range of values. Finally, the data is plotted as a bar chart (**FIGURE 13.15B**). A graph line around the top of the bars shows the distribution of values for the trait. If the line is a bell-shaped curve, or **bell curve**, the trait varies continuously.

bell curve Bell-shaped curve; typically results from graphing frequency versus distribution for a trait that varies continuously.
continuous variation Range of small differences in a shared trait.
short tandem repeat In chromosomal DNA, sequences of a few nucleotides repeated multiple times in a row.

TAKE-HOME MESSAGE 13.6

The more genes and other factors that influence a trait, the more continuous is its range of variation.

13.7 Application: MENACING MUCUS

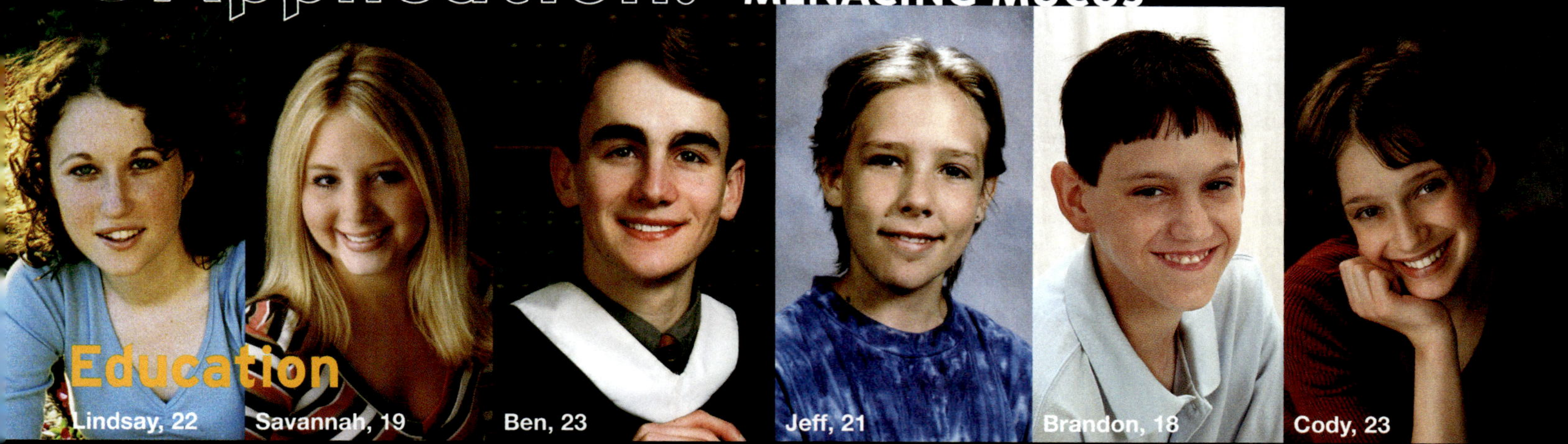

FIGURE 13.16 Cystic fibrosis. These are a few of the many young victims of cystic fibrosis, which occurs most often in people of northern European ancestry. At least one young person dies every day in the United States from complications of this disease.

CYSTIC FIBROSIS (CF) IS THE MOST COMMON FATAL GENETIC DISORDER IN THE UNITED STATES. It occurs in people homozygous for a mutated allele of the *CFTR* gene. This gene encodes an active transport protein that moves chloride ions out of epithelial cells. Sheets of these cells line the passageways and ducts of the lungs, liver, pancreas, intestines, reproductive system, and skin. When chloride ions leave these cells, water follows by osmosis. The process maintains a thin film of water on the surface of the epithelial sheets.

The allele most commonly associated with CF has a 3 base pair deletion. It is called *ΔF508* because the protein it encodes is missing the normal 508th amino acid, a phenylalanine (F). The deletion prevents proper membrane trafficking of newly assembled polypeptides, which are left stranded in endoplasmic reticulum. The altered protein can function properly, but it never reaches the cell surface to do its job.

Epithelial cell membranes that lack the CFTR protein cannot transport chloride ions. Too few chloride ions leave these cells. Not enough water leaves them either, so the surfaces of epithelial cell sheets are too dry. Mucus that normally slips through the body's tubes sticks to the walls of the tubes instead. This outcome has pleiotropic effects because thick globs of mucus accumulate and clog passageways and ducts throughout the body. Breathing becomes difficult as the mucus obstructs the smaller airways of the lungs. Digestive problems arise as ducts that lead to the gut get clogged with mucus. Males are typically infertile because their sperm flow is hampered.

CFTR also helps alert the immune system to the presence of disease-causing bacteria in the lungs. The CFTR protein functions as a receptor: It binds directly to bacteria and triggers endocytosis. Endocytosis of bacteria into epithelial cells lining the respiratory tract initiates an immune response. When the cells lack CFTR, this early alert system fails, so bacteria have time to multiply before being detected by the immune system. Thus, chronic bacterial infections of the lungs are a hallmark of cystic fibrosis. Antibiotics help control infections, but there is no cure. Most affected people die before age thirty, when their tormented lungs fail (**FIGURE 13.16**).

The *ΔF508* allele is at least 50,000 years old and very common: in some populations, 1 in 25 people are heterozygous for it. Why does the allele persist if it is so harmful? *ΔF508* is codominant with the normal allele. Heterozygous individuals typically have no symptoms of cystic fibrosis; their cells have plasma membranes with enough CFTR to transport chloride ions normally. The *ΔF508* allele may offer these individuals an advantage in surviving certain deadly infectious diseases. CFTR's receptor function is an essential part of the immune response to bacteria in the respiratory tract. However, the same function allows bacteria to enter cells of the gastrointestinal tract, where they can be deadly. For example, endocytosis of *Salmonella typhi* bacteria into epithelial cells lining the gut results in a dangerous infection called typhoid fever. Cells lacking CFTR do not take up these bacteria. Thus, people who carry *ΔF508* are probably less susceptible to typhoid fever and other bacterial diseases that begin in the intestinal tract.

Summary

SECTION 13.1 Gregor Mendel indirectly discovered the role of alleles in inheritance by breeding pea plants and tracking traits of the offspring. Each gene occurs at a **locus**, or location, on a chromosome. Individuals with identical alleles are **homozygous** for the allele. **Heterozygous** individuals, or **hybrids**, have two nonidentical alleles. A **dominant** allele masks the effect of a **recessive** allele on the homologous chromosome. **Genotype** (an individual's particular set of alleles) gives rise to **phenotype**, which refers to an individual's observable traits.

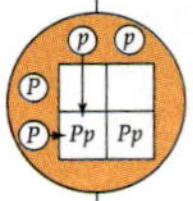

SECTION 13.2 Crossing individuals that breed true for two forms of a trait yields identically heterozygous offspring. A cross between such offspring is a **monohybrid cross**. The frequency at which the traits appear in offspring of such **testcrosses** can reveal dominance relationships among the alleles associated with those traits.

Punnett squares are useful for determining the probability of offspring genotype and phenotype. Mendel's monohybrid cross results led to his **law of segregation** (stated here in modern terms): Diploid cells have pairs of genes on homologous chromosomes. The two genes of a pair separate from each other during meiosis, so they end up in different gametes.

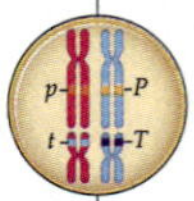

SECTION 13.3 Crossing individuals that breed true for two forms of two traits yields F_1 offspring identically heterozygous for alleles governing those traits. A cross between such offspring is a **dihybrid cross**. The frequency at which the two traits appear in F_2 offspring can reveal dominance relationships between alleles associated with those traits. Mendel's dihybrid cross results led to his **law of independent assortment** (stated here in modern terms): Paired genes on homologous chromosomes tend to sort into gametes independently of other gene pairs during meiosis. Crossovers can break up **linkage groups**.

SECTION 13.4 With **incomplete dominance**, the phenotype of heterozygous individuals is an intermediate blend of the two homozygous phenotypes. With **codominant** alleles, heterozygous individuals have both homozygous phenotypes. Codominance may occur in **multiple allele systems** such as the one underlying ABO blood typing. With **epistasis**, two or more genes affect the same trait. A **pleiotropic** gene affects two or more traits.

SECTION 13.5 An individual's phenotype is influenced by environmental factors. Environmental cues alter gene expression by way of cell signaling pathways that ultimately affect gene controls.

SECTION 13.6 A trait that is influenced by multiple genes often occurs in a range of small increments of phenotype called **continuous variation**. Continuous variation typically occurs as a **bell curve** in the range of values. Multiple alleles such as those that arise in regions of **short tandem repeats** can give rise to continuous variation.

SECTION 13.7 Cystic fibrosis occurs in people homozygous for a mutated allele of the *CFTR* gene. The allele persists at high frequency despite its devastating effects. Carrying the allele may offer heterozygous individuals protection from dangerous gastrointestinal tract infections.

Self-Quiz

Answers in Appendix VII

1. A heterozygous individual has a ______ for a trait being studied.
   a. pair of identical alleles
   b. pair of nonidentical alleles
   c. haploid condition, in genetic terms

2. An organism's observable traits constitute its ______ .
   a. phenotype
   b. variation
   c. genotype
   d. pedigree

3. In genetics, F stands for filial, which means ______ .
   a. friendly
   b. offspring
   c. final
   d. hairlike

4. The second-generation offspring of a cross between individuals who are homozygous for different alleles of a gene are called the ______ .
   a. F_1 generation
   b. F_2 generation
   c. hybrid generation
   d. none of the above

5. F_1 offspring of the cross $AA \times aa$ are ______ .
   a. all AA
   b. all aa
   c. all Aa
   d. 1/2 AA and 1/2 aa

6. Refer to question 5. Assuming complete dominance, the F_2 generation will show a phenotypic ratio of ______ .
   a. 3:1 b. 9:1 c. 1:2:1 d. 9:3:3:1

7. A testcross is a way to determine ______ .
   a. phenotype b. genotype c. both a and b

8. Assuming complete dominance, crosses between two dihybrid F_1 pea plants, which are offspring from a cross $AABB \times aabb$, result in F_2 phenotype ratios of ______ .
   a. 1:2:1 b. 3:1 c. 1:1:1:1 d. 9:3:3:1

9. The probability of a crossover occurring between two genes on the same chromosome ______ .
   a. is unrelated to the distance between them
   b. decreases with the distance between them
   c. increases with the distance between them

10. A gene that affects three traits is ______ .
    a. epistatic
    b. a multiple allele system
    c. pleiotropic
    d. dominant

11. The phenotype of individuals heterozygous for ______ alleles comprises both homozygous phenotypes.
    a. epistatic
    b. codominant
    c. pleiotropic
    d. hybrid

Data Analysis Activities

Carrying the Cystic Fibrosis Allele Offers Protection from Typhoid Fever Epithelial cells that lack the CFTR protein cannot take up bacteria by endocytosis. Endocytosis is an important part of the respiratory tract's immune defenses against common *Pseudomonas* bacteria, which is why *Pseudomonas* infections of the lungs are a chronic problem in cystic fibrosis patients. Endocytosis is also the way that *Salmonella typhi* enter cells of the gastrointestinal tract, where internalization of this bacteria can result in typhoid fever.

Typhoid fever is a common worldwide disease. Its symptoms include extreme fever and diarrhea, and the resulting dehydration causes delirium that may last several weeks. If untreated, it kills up to 30 percent of those infected. Around 600,000 people die annually from typhoid fever. Most of them are children.

In 1998, Gerald Pier and his colleagues compared the uptake of *S. typhi* by different types of epithelial cells: those homozygous for the normal allele, and those heterozygous for the $\Delta F508$ allele associated with CF. (Cells that are homozygous for the mutation do not take up any *S. typhi* bacteria.) Some of the results are shown in **FIGURE 13.17**.

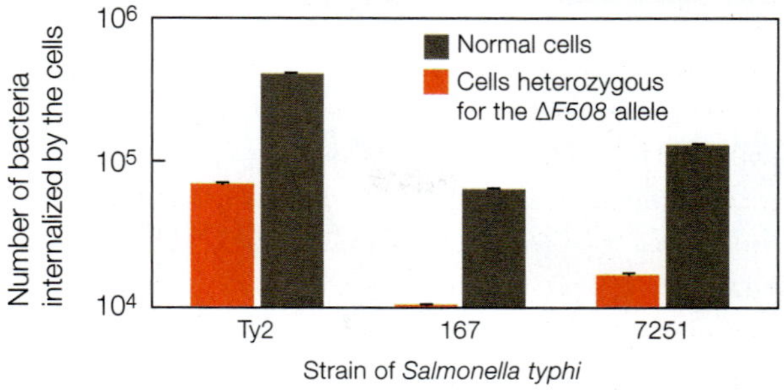

FIGURE 13.17 Effect of the $\Delta F508$ mutation on the uptake of three different strains of *Salmonella typhi* bacteria by epithelial cells.

1. Regarding the Ty2 strain of *S. typhi*, about how many more bacteria were able to enter normal cells (those heterozygous for the normal allele) than cells heterozygous for the $\Delta F508$ allele?
2. Which strain of bacteria entered normal epithelial cells most easily?
3. Entry of all three *S. typhi* strains into the heterozygous epithelial cells was inhibited. Is it possible to tell which strain was most inhibited?

12. ________ in a trait is indicated by a bell curve.

13. Match the terms with the best description.

___ dihybrid cross	a. *bb*
___ monohybrid cross	b. $AaBb \times AaBb$
___ homozygous condition	c. *Aa*
___ heterozygous condition	d. $Aa \times Aa$

Genetics Problems

Answers in Appendix VII

1. Mendel crossed a true-breeding pea plant with green pods and a true-breeding pea plant with yellow pods. All offspring had green pods. Which color is recessive?

2. Assuming that independent assortment occurs during meiosis, what type(s) of gametes will form in individuals with the following genotypes?

a. *AABB* b. *AaBB* c. *Aabb* d. *AaBb*

3. Determine the predicted genotype frequencies among the offspring of an $AABB \times aaBB$ mating.

4. Heterozygous individuals perpetuate some alleles that have lethal effects in homozygous individuals. A mutated allele (M^L) associated with taillessness in Manx cats *(left)* is an example. Cats homozygous for this allele (M^LM^L) typically die before birth due to severe spinal cord defects. In a case of incomplete dominance, cats heterozygous for the M^L allele and the normal, unmutated allele (M) have a short, stumpy tail or none at all. Two M^LM heterozygous cats mate. What is the probability that any of their kittens will be heterozygous (M^LM)?

5. People homozygous for a base-pair substitution in the beta-globin gene have sickle-cell anemia (Section 9.5). A couple who are planning to have children discover that both of the individuals are heterozygous for the mutated allele (Hb^S) and the normal allele (Hb^A). Calculate the probability that any one of their children will be born with sickle-cell anemia.

6. In sweet pea plants, an allele for purple flowers (P) is dominant to an allele for red flowers (p). An allele for long pollen grains (L) is dominant to an allele for round pollen grains (l). Bateson and Punnett crossed a plant having purple flowers/long pollen grains with one having white flowers/round pollen grains. All F_1 offspring had purple flowers and long pollen grains. Among the F_2 generation, the researchers observed the following phenotypes:

296 purple flowers/long pollen grains
19 purple flowers/round pollen grains
27 red flowers/long pollen grains
85 red flowers/round pollen grains

What is the best explanation for these results?

CREDITS: (17) © Cengage Learning; (in text) © Leslie Faltheisek.

This family lives in Tanzania, where exposure to intense sunlight is responsible for the skin cancer that kills almost everyone with the albino phenotype. An abnormally low amount of melanin leaves people with this trait defenseless against UV radiation in the sun's rays. Recessive alleles on an autosome give rise to albinism.

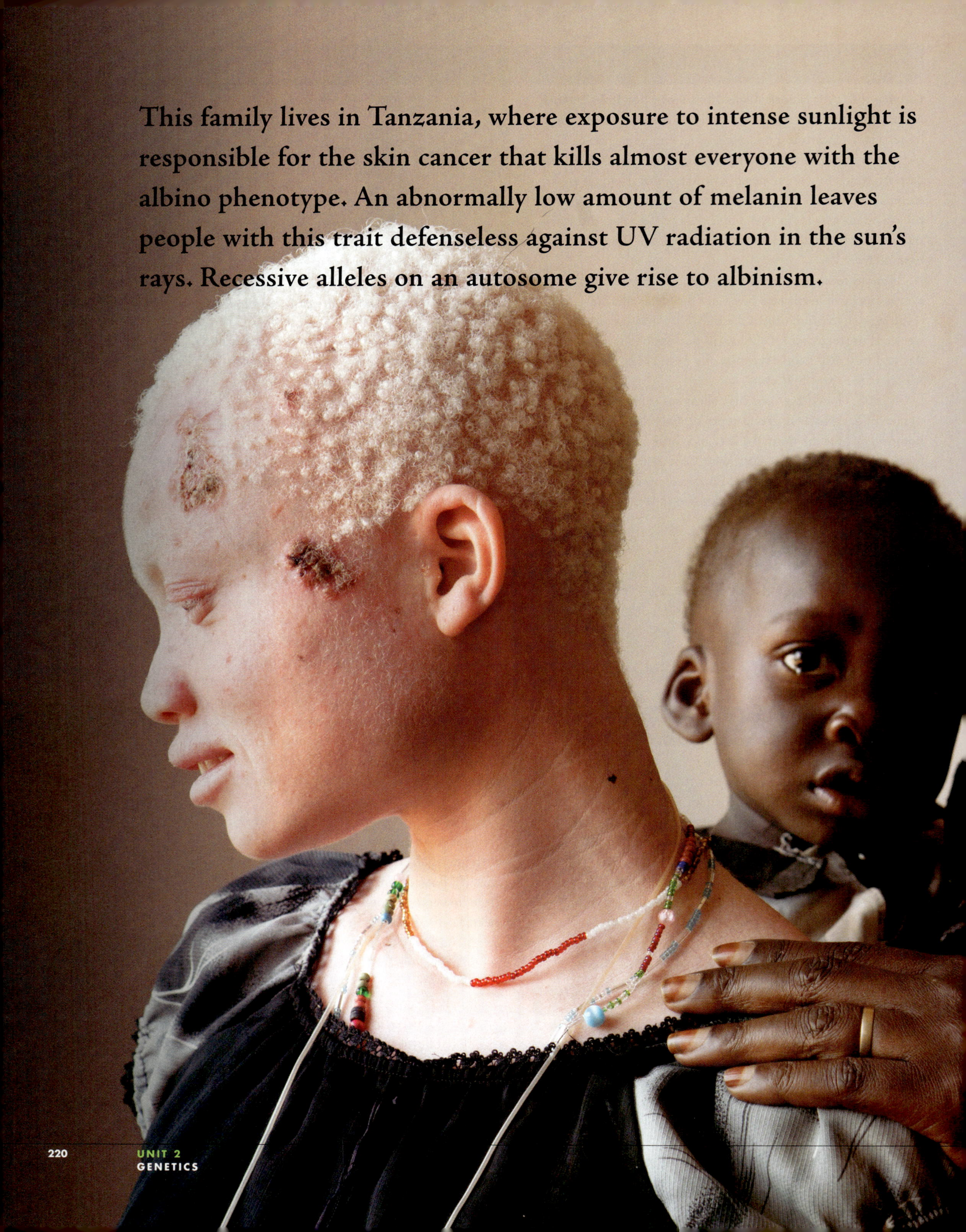

14 HUMAN INHERITANCE

Links to Earlier Concepts

Be sure you understand dominance relationships (Sections 13.1, 13.4, and 13.5), gene expression (9.1, 9.2), and mutations (9.5). You will use your knowledge of chromosomes (8.3), DNA replication and repair (8.4, 8.5), meiosis (12.2, 12.3), and sex determination (10.3). Sampling error (1.7), proteins (3.4), cell components (4.5, 4.9, 4.10), metabolism (5.4), pigments (6.1), telomeres (11.4), and oncogenes (11.5) will turn up in the context of genetic disorders.

KEY CONCEPTS

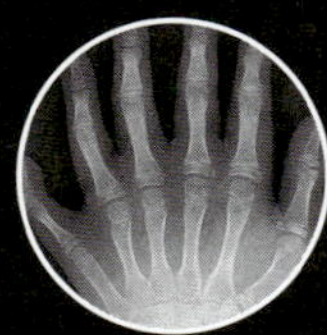

TRACKING TRAITS IN HUMANS

Inheritance patterns in humans are revealed by following traits through generations of a family. Tracked traits are often genetic abnormalities or syndromes associated with a genetic disorder.

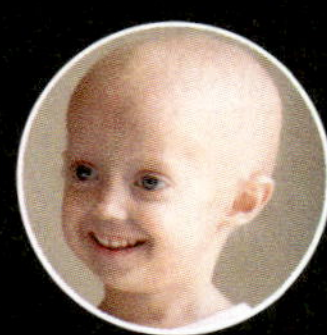

AUTOSOMAL INHERITANCE

Traits associated with dominant alleles on autosomes appear in every generation. Traits associated with recessive alleles on autosomes can skip generations.

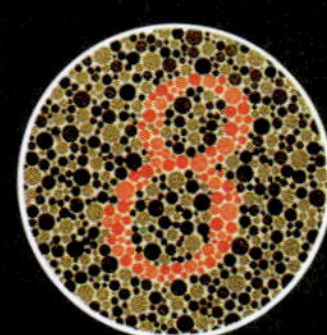

SEX-LINKED INHERITANCE

Traits associated with alleles on the X chromosome tend to affect more men than women. Men cannot pass such alleles to a son; carrier mothers bridge affected generations.

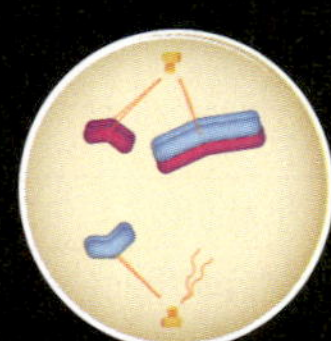

CHROMOSOME CHANGES

Some genetic disorders arise after large-scale change in chromosome structure. With few exceptions, a change in the number of autosomes is fatal in humans.

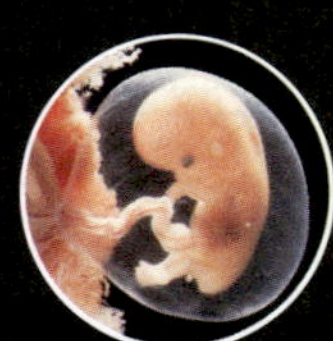

GENETIC TESTING

Genetic testing provides information about the risk of passing a harmful allele to offspring. Prenatal testing can reveal a genetic abnormality or disorder in a developing fetus.

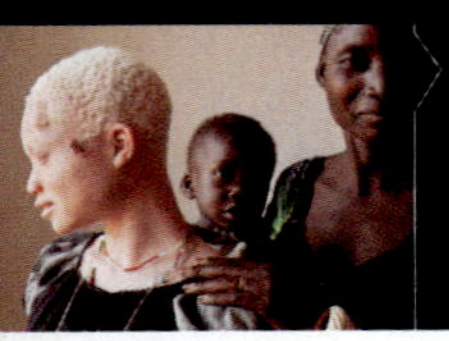

14.1 HOW DO WE STUDY INHERITANCE PATTERNS IN HUMANS?

TABLE 14.1

Examples of Genetic Abnormalities and Disorders in Humans

Disorder or Abnormality	Main Symptoms
Autosomal dominant inheritance pattern	
Achondroplasia	One form of dwarfism
Aniridia	Defects of the eyes
Camptodactyly	Rigid, bent fingers
Familial hypercholesterolemia	High cholesterol level; clogged arteries
Huntington's disease	Degeneration of the nervous system
Marfan syndrome	Abnormal or missing connective tissue
Polydactyly	Extra fingers, toes, or both
Progeria	Drastic premature aging
Neurofibromatosis	Tumors of nervous system, skin
Autosomal recessive inheritance pattern	
Albinism	Absence of pigmentation
Hereditary methemoglobinemia	Blue skin coloration
Cystic fibrosis	Difficulty breathing; chronic lung infections
Ellis–van Creveld syndrome	Dwarfism, heart defects, polydactyly
Fanconi anemia	Physical abnormalities, marrow failure
Galactosemia	Brain, liver, eye damage
Hereditary hemochromatosis	Joints, organs damaged by iron overload
Phenylketonuria (PKU)	Mental impairment
Sickle-cell anemia	Anemia, pain, swelling, frequent infections
Tay–Sachs disease	Deterioration of mental and physical abilities; early death
X-linked recessive inheritance pattern	
Androgen insensitivity syndrome	XY individual but having some female traits; sterility
Red–green color blindness	Inability to distinguish red from green
Hemophilia	Impaired blood clotting ability
Muscular dystrophies	Progressive loss of muscle function
X-linked anhidrotic dysplasia	Mosaic skin (patches with or without sweat glands); other ill effects
X-linked dominant inheritance pattern	
Fragile X syndrome	Intellectual, emotional disability
Incontinentia pigmenti	Abnormalities of skin, hair, teeth, nails, eyes; neurological problems
Changes in chromosome number	
Down syndrome	Mental impairment; heart defects
Turner syndrome (XO)	Sterility; abnormal ovaries, sexual traits
Klinefelter syndrome	Sterility; mild mental impairment
XXX syndrome	Minimal abnormalities
XYY condition	Mild mental impairment or no effect
Changes in chromosome structure	
Chronic myelogenous leukemia (CML)	Overproduction of white blood cells; organ malfunctions
Cri-du-chat syndrome	Mental impairment; abnormal larynx

Some organisms, including pea plants and fruit flies, are ideal for genetic analysis. They have relatively few chromosomes, they reproduce quickly under controlled conditions, and breeding them poses few ethical problems. It does not take long to follow a trait through many generations. Humans, however, are a different story. Unlike flies grown in laboratories, we humans live under variable conditions, in different places, and we live as long as the geneticists who study us. Most of us select our own mates and reproduce if and when we want to. Our families tend to be on the small side, so sampling error (Section 1.7) is a major factor in studying them.

Because of these and other challenges, geneticists often use historical records to track traits through many generations of a family. These researchers make standardized charts of genetic connections called **pedigrees** (**FIGURE 14.1**). Analysis of a pedigree can reveal whether a trait is associated with a dominant or recessive allele, and whether the allele is on an autosome or a sex chromosome. Pedigree analysis also allows geneticists to determine the probability that a trait will recur in future generations of a family or a population.

TYPES OF GENETIC VARIATION

Some easily observed human traits follow Mendelian inheritance patterns. Like the flower color of pea plants, these traits are controlled by a single gene with alleles that have a clear dominance relationship. (Appendix IV shows a map of human chromosomes with the locations of some of these alleles.) Consider how someone who is homozygous for two recessive alleles of the *MC1R* gene (Section 13.4) makes the reddish melanin but not the brown melanin, so this person has red hair.

Single genes on autosomes or sex chromosomes also govern more than 6,000 genetic abnormalities and disorders. **TABLE 14.1** lists a few examples. A genetic abnormality is a rare or uncommon version of a trait, such as having six fingers on a hand. By contrast, a genetic disorder sooner or later causes medical problems that may be severe. A genetic disorder is often characterized by a specific set of symptoms (a syndrome). Most research in the field of human genetics focuses on disorders, because what we learn may help us develop treatments for affected people.

The next two sections of this chapter focus on inheritance patterns of human single-gene disorders, which affect about 1 in 200 people. Keep in mind that these patterns are the least common. Most human traits are

pedigree Chart showing the pattern of inheritance of a trait through generations in a family.

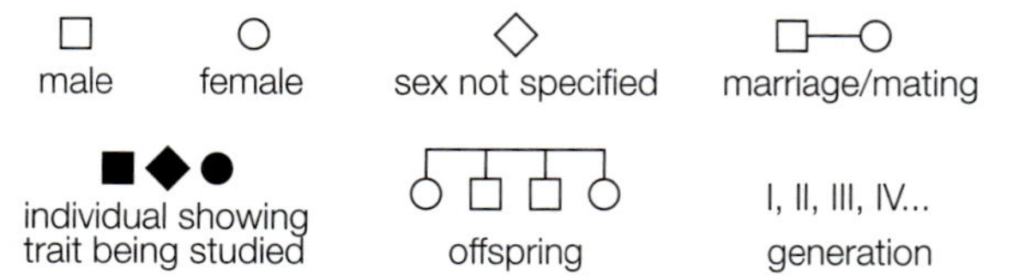

A Standard symbols used in pedigrees.

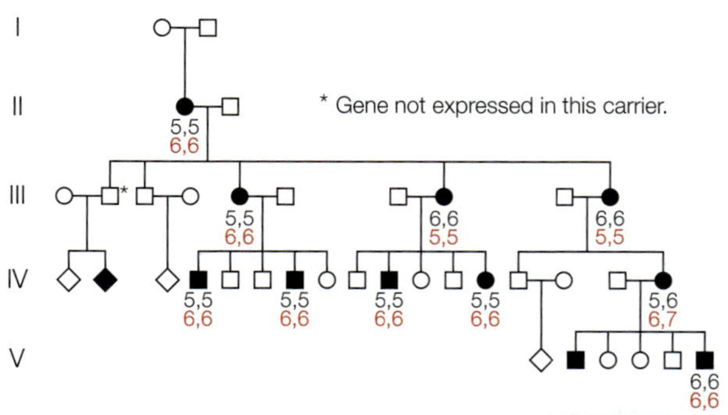

B A pedigree for polydactyly, which is characterized by extra fingers, toes, or both. The black numbers signify the number of fingers on each hand; the red numbers signify the number of toes on each foot. Though it occurs on its own, polydactyly is also one of several symptoms of Ellis–van Creveld syndrome.

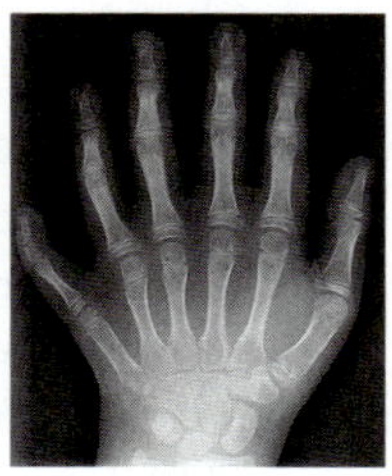

FIGURE 14.1 {Animated} Pedigrees.

polygenic (Section 13.4) and have epigenetic contributions (Section 13.5). Many genetic disorders are like this, including diabetes, asthma, obesity, cancers, heart disease, and multiple sclerosis. The inheritance patterns of these disorders are complex, and despite intense research our understanding of the genetics behind them remains incomplete.

Sections 14.4 and 14.5 explore some causes and effects of major changes in chromosomes—alterations in chromosome number or structure. Such changes occur in about 1 of every 100 births worldwide, and in many cases they have drastic consequences on health.

TAKE-HOME MESSAGE 14.1

Human inheritance patterns are often studied by tracking genetic abnormalities or disorders through family trees.

A genetic disorder is an inherited condition that causes medical problems. A genetic abnormality is a rare version of an inherited trait.

Some human genetic traits are governed by single genes and are inherited in a Mendelian fashion. Many others are influenced by multiple genes and epigenetics.

PEOPLE MATTER

DR. NANCY WEXLER

The village of Barranquitas, Venezuela, has the highest incidence of Huntington's disease in the world. A person with this incurable, fatal hereditary disorder gradually loses muscle control. Eventually, serious problems with swallowing cause many patients to die from choking or malnutrition. Beyond the physical symptoms, deep depression can often take hold.

Huntington's affects 1 in 10,000 people worldwide, but in Barranquitas the rate is more like 1 in 10. Some 1,000 villagers already have full-blown Huntington's; many more carry the gene. Such a high concentration of Huntington's patients made this region the backbone of Nancy Wexler's research. Wexler has been coming here for more than 30 years to study the genetics behind the disorder.

This is more than an academic pursuit or a career goal for her. "My mother died of Huntington's and she was a scientist. My father was a scientist too, and so we said 'Let's find a cure.' And we still say that. You can't get up in the morning without having hope and confidence that the cure is just around the corner."

Wexler and her colleagues collected DNA from and compiled an extended pedigree for nearly 10,000 Venezuelans. Her research was critical to the discovery that a dominant allele on human chromosome 4 causes Huntington's. As the daughter of a Huntington's sufferer, she has a one-in-two chance of carrying the fatal genetic flaw herself.

In 1999, Wexler cofounded a care home for Huntington's sufferers near Lake Maracaibo in Venezuela. The home, Casa Hogar, is a haven for more than 50 people whose families can no longer cope. Casa Hogar is facing a chronic lack of funding, possibly even closure. Still, Wexler remains confident that one day it won't be needed. "We never know when some miraculous discovery is going to be made," she said. "There are science breakthroughs on the horizon and happening now so I am very hopeful about the cure in the near future."

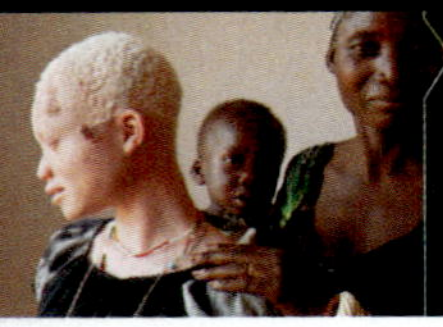

14.2 HOW DO WE KNOW WHEN A TRAIT IS AFFECTED BY AN ALLELE ON AN AUTOSOME?

THE AUTOSOMAL DOMINANT PATTERN

A trait associated with a dominant allele on an autosome appears in people who are heterozygous for it as well as those who are homozygous. Such traits appear in every generation of a family, and they occur with equal frequency in both sexes. When one parent is heterozygous, and the other is homozygous for the recessive allele, each of their children has a 50 percent chance of inheriting the dominant allele and having the associated trait (**FIGURE 14.2A**).

Achondroplasia A form of hereditary dwarfism called achondroplasia offers an example of an autosomal dominant disorder (one caused by a dominant allele on an autosome). Mutations associated with achondroplasia occur in a gene for a growth hormone receptor. The mutations cause the receptor, a regulatory molecule that slows bone development, to be overly active. About 1 in 10,000 people is heterozygous for one of these mutations. As adults, affected people are, on average, about four feet, four inches (1.3 meters) tall, with arms and legs that are short relative to torso size (**FIGURE 14.2B**). An allele that causes achondroplasia can be passed to children because its expression does not interfere with reproduction, at least in heterozygous people. The homozygous condition results in severe skeletal malformations that cause early death.

Huntington's Disease Alleles that cause Huntington's disease are also inherited in an autosomal dominant pattern. Mutations associated with this disorder alter a gene for a cytoplasmic protein whose function is still unknown. The mutations are insertions caused by expansion of a short tandem repeat (Section 13.6), in which the same three nucleotides become repeated many, many times in the gene's sequence. The altered gene encodes an oversized protein product that gets chopped into pieces inside nerve cells of the brain. The pieces accumulate in cytoplasm as large clumps that eventually prevent the cells from functioning properly. Brain cells involved in movement, thinking, and emotion are particularly affected. In the most common form of Huntington's, symptoms do not start until after the age of thirty, and affected people die during their forties or fifties. With this and other late-onset disorders, people tend to reproduce before symptoms appear, so the allele is often passed unknowingly to children.

Hutchinson–Gilford Progeria Hutchinson–Gilford progeria is an autosomal dominant disorder characterized by drastically accelerated aging. It is usually caused by a mutation in the gene for lamin A, a protein subunit of intermediate filaments that support the nuclear envelope (Section 4.9). Lamins also have roles in mitosis, DNA synthesis and repair, and transcriptional regulation. The

A A dominant allele on an autosome (red) affects heterozygous people.

B Achondroplasia affects Ivy Broadhead (*left*), as well as her brother, father, and grandfather.

C Five-year-old Megan is already showing symptoms of Hutchinson–Gilford progeria.

FIGURE 14.2 {Animated} Autosomal dominant inheritance.

mutation, a base-pair substitution, adds a signal for a splice site (Section 9.2). The resulting lamin A protein is too short and cannot be processed correctly after translation. Cells that carry this mutation have a nucleus that is grossly abnormal, with nuclear pore complexes that do not assemble properly and membrane proteins localized to the wrong side of the nuclear envelope. The function of the nucleus as protector of chromosomes and gateway of transcription is severely impaired, so DNA damage accumulates quickly. The effects are pleiotropic. Outward symptoms begin to appear before age two, as skin that should be plump and resilient starts to thin, muscles weaken, and bones soften. Premature baldness is inevitable (**FIGURE 14.2C**). Most people with the disorder die in their early teens as a result of a stroke or heart attack brought on by hardened arteries, a condition typical of advanced age. Progeria does not run in families because affected people do not live long enough to reproduce.

THE AUTOSOMAL RECESSIVE PATTERN

A recessive allele on an autosome is expressed only in homozygous people, so traits associated with the allele tend to skip generations. Both sexes are equally affected. People heterozygous for the allele are carriers, which means that they have the allele but not the trait. Any child of two carriers has a 25 percent chance of inheriting the allele from both parents (**FIGURE 14.3A**). Being homozygous for the allele, such children would have the trait.

Tay–Sachs Disease Alleles associated with Tay–Sachs disease are inherited in an autosomal recessive pattern. In the general population, about 1 in 300 people is a carrier for one of these allele, but the incidence is ten times higher in some groups, such as Jews of eastern European descent. The gene altered in Tay–Sachs encodes a lysosomal enzyme responsible for breaking down a particular type of lipid. Mutations cause this enzyme to misfold and become destroyed, so cells make the lipid but cannot break it down. Typically, newborns homozygous for a Tay–Sachs allele seem normal, but within three to six months they become irritable, listless, and may have seizures as the lipid accumulates in their nerve cells. Blindness, deafness, and paralysis follow. Affected children usually die by age five (**FIGURE 14.3B**).

Albinism Albinism, a genetic abnormality characterized by an abnormally low level of the pigment melanin, is also inherited in an autosomal recessive pattern. Mutations associated with the albino phenotype occur in genes involved in melanin synthesis. Skin, hair, or eye pigmentation may be reduced or missing, as shown in the chapter opening photo. In the most dramatic form of the phenotype, the skin is very white and does not tan, and the hair is white. The lack of pigment in the irises of the eyes allows underlying blood vessels to show through, so the irises appear red. Melanin plays a role in the retina, so people with the albino phenotype tend to have vision problems.

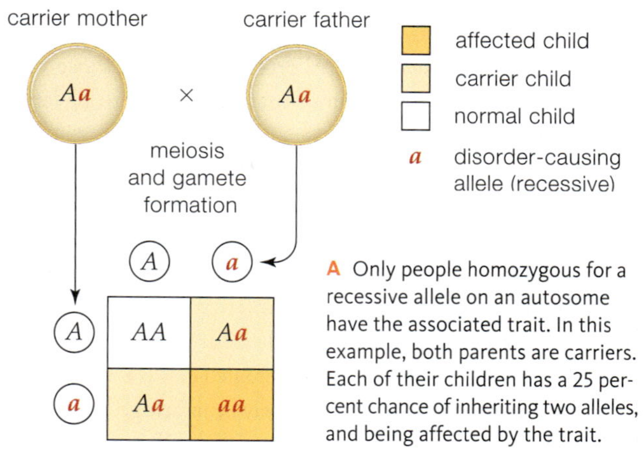

A Only people homozygous for a recessive allele on an autosome have the associated trait. In this example, both parents are carriers. Each of their children has a 25 percent chance of inheriting two alleles, and being affected by the trait.

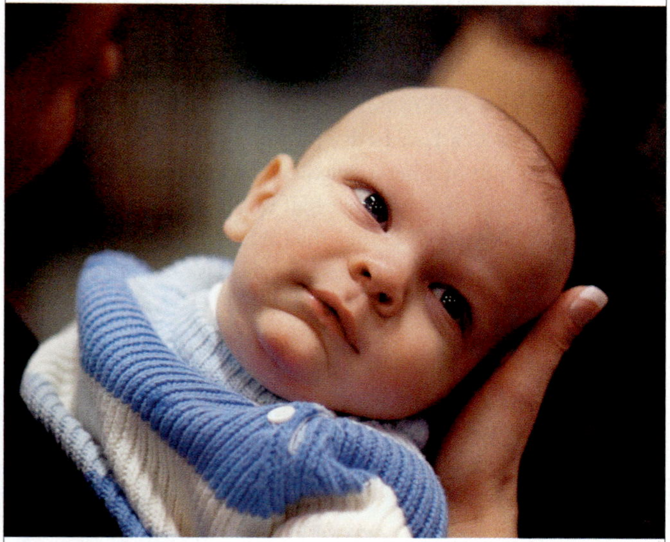

B Conner Hopf was diagnosed with Tay–Sachs disease at age 7½ months. He died before his second birthday.

FIGURE 14.3 {Animated} Autosomal recessive inheritance.

> **TAKE-HOME MESSAGE 14.2**
>
> With an autosomal dominant inheritance pattern, anyone with the allele, homozygous or heterozygous, has the associated trait. The trait tends to appear in every generation.
>
> With an autosomal recessive inheritance pattern, only persons who are homozygous for an allele have the associated trait. The trait tends to skip generations.

14.3 HOW DO WE KNOW WHEN A TRAIT IS AFFECTED BY AN ALLELE ON AN X CHROMOSOME?

Many genetic disorders are associated with alleles on the X chromosome (**FIGURE 14.4**). Almost all of them are inherited in a recessive pattern, probably because those caused by dominant X chromosome alleles tend to be lethal in male embryos.

THE X-LINKED RECESSIVE PATTERN

A recessive allele on the X chromosome (an X-linked recessive allele) leaves two clues when it causes a genetic disorder. First, an affected father never passes an X-linked recessive allele to a son, because all children who inherit their father's X chromosome are female (**FIGURE 14.5A**). Thus, a heterozygous female is always the bridge between an affected male and his affected grandson. Second, the disorder appears in males more often than in females. This is because all males who carry the allele have the disorder, but not all heterozygous females do. Remember that one of the two X chromosomes in each cell of a female is inactivated as a Barr body (Section 10.3). As a result, only about half of a heterozygous female's cells express the recessive allele. The other half of her cells express the dominant, normal allele that she carries on her other X chromosome, and this expression can mask the phenotypic effects of the recessive allele.

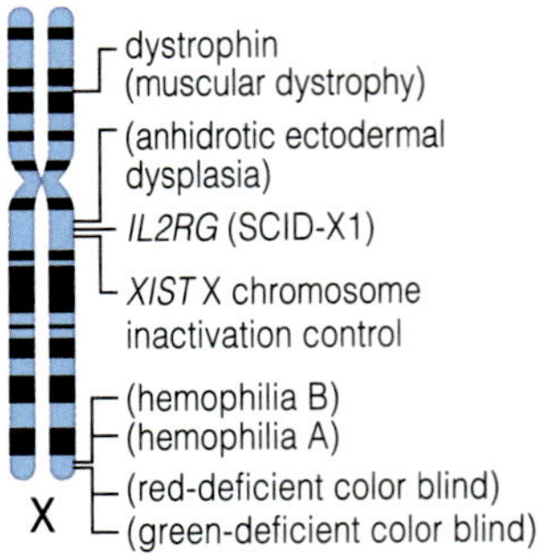

FIGURE 14.4 The human X chromosome. This chromosome carries about 2,000 genes—almost 10 percent of the total. Most X chromosome alleles that cause genetic disorders are inherited in a recessive pattern. A few disorders are listed (in parentheses).

Red–Green Color Blindness Color blindness refers to a range of conditions in which an individual cannot distinguish among some or all colors in the spectrum of visible light. These conditions are typically inherited in an X-linked recessive pattern, because most of the genes involved in color vision are on the X chromosome.

Humans can sense the differences among 150 colors, and this perception depends on pigment-containing receptors

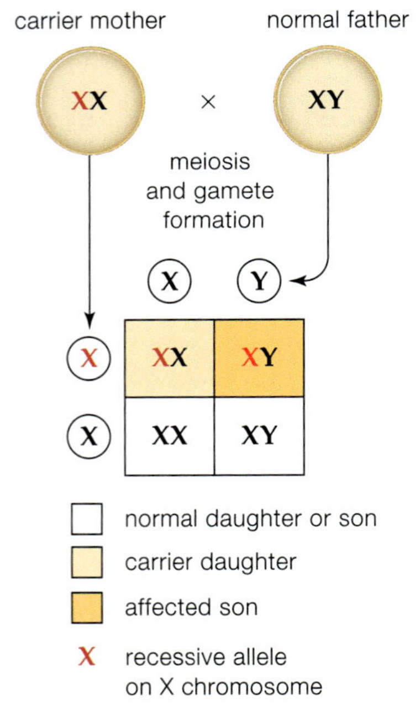

A In this example of X-linked inheritance, the mother carries a recessive allele on one of her two X chromosomes (red).

You may have one form of red–green color blindness if you see a 7 in this circle instead of a 29.

You may have another form of red–green color blindness if you see a 3 instead of an 8 in this circle.

B A view of color blindness. The photo on the *left* shows how a person with red–green color blindness sees the photo on the *right*. The perception of blues and yellows is normal; red and green appear similar. The circles are part of a standardized test for color blindness. A set of 38 of these circles is commonly used to diagnose deficiencies in color perception.

FIGURE 14.5 **{Animated}** X-linked recessive inheritance.

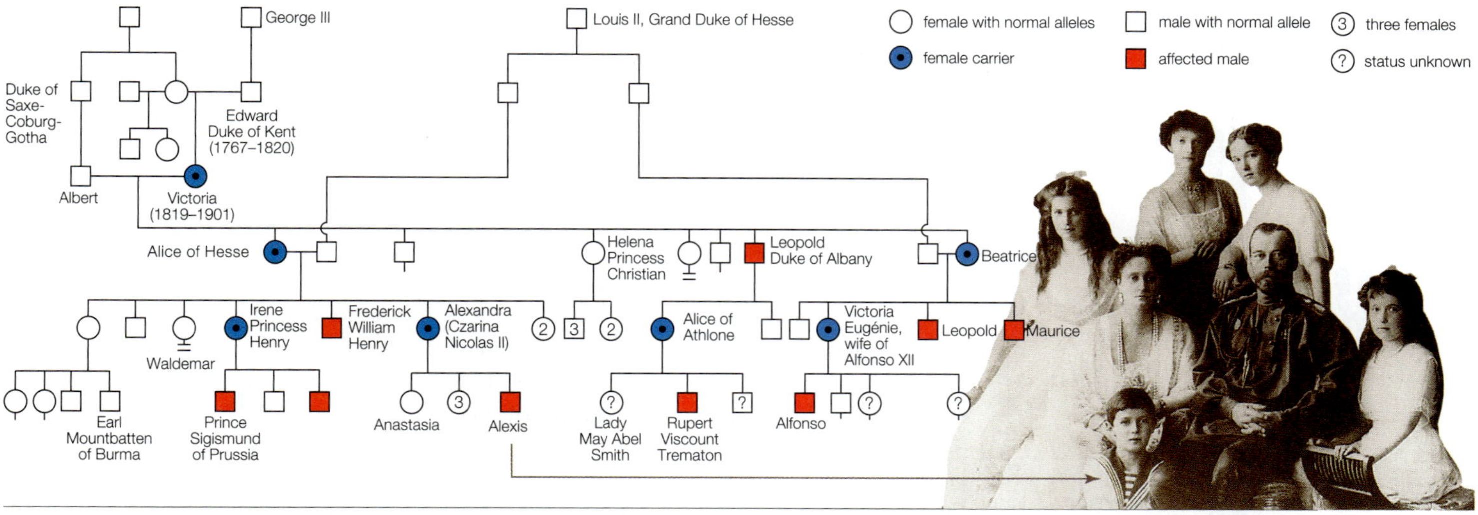

FIGURE 14.6 {Animated} A classic case of X-linked recessive inheritance: a partial pedigree of the descendants of Queen Victoria of England. At one time, the recessive X-linked allele that resulted in hemophilia was present in eighteen of Victoria's sixty-nine descendants, who sometimes intermarried. Of the Russian royal family members shown, the mother (Alexandra Czarina Nicolas II) was a carrier.

FIGURE IT OUT: How many of Alexis's siblings were affected by hemophilia A?

Answer: None

in the eyes. Mutations that result in altered or missing receptors affect color vision. For example, people who have red–green color blindness see fewer than 25 colors because receptors that respond to red and green wavelengths are weakened or absent (**FIGURE 14.5B**). Some confuse red and green; others see green as gray.

Duchenne Muscular Dystrophy An X-linked recessive disorder, Duchenne muscular dystrophy (DMD), is characterized by muscle degeneration. It is caused by mutations in the X chromosome gene for dystrophin, a cytoskeletal protein that links actin microfilaments in cytoplasm to a complex of proteins in the plasma membrane. This complex structurally and functionally links the cell to extracellular matrix. When dystrophin is absent, the entire protein complex is unstable. Muscle cells, which are subject to stretching, are particularly affected. Their plasma membrane is easily damaged, and they become flooded with calcium ions. Eventually, the muscle cells die and become replaced by fat cells and connective tissue.

DMD affects about 1 in 3,500 people, almost all of them boys. Symptoms begin between ages three and seven. Anti-inflammatory drugs can slow the progression of DMD, but there is no cure. When an affected boy is about twelve years old, he will begin to use a wheelchair and his heart muscle will start to fail. Even with the best care, he will probably die before the age of thirty, from a heart disorder or respiratory failure (suffocation).

Hemophilia A Hemophilia A is an X-linked recessive disorder that interferes with blood clotting. Most of us have a blood clotting mechanism that quickly stops bleeding from minor injuries. That mechanism involves factor VIII, a protein product of a gene on the X chromosome. Bleeding can be prolonged in males who carry a mutation in this gene. Females who have two mutated alleles are also affected (heterozygous females make enough factor VIII to have a clotting time that is close to normal). Affected people tend to bruise very easily, but internal bleeding is their most serious problem. Repeated bleeding inside the joints disfigures them and causes chronic arthritis.

In the nineteenth century, the incidence of hemophilia A was relatively high in royal families of Europe and Russia, probably because the common practice of inbreeding kept the allele in their family trees (**FIGURE 14.6**). Today, about 1 in 7,500 people in the general population is affected. That number may be rising because the disorder is now a treatable one. More affected people are living long enough to transmit the mutated allele to children.

TAKE-HOME MESSAGE 14.3

Men who have an X-linked allele have the associated trait, but not all heterozygous women do. Thus, the trait appears more often in men.

Men transmit an X-linked allele to their daughters, but not to their sons.

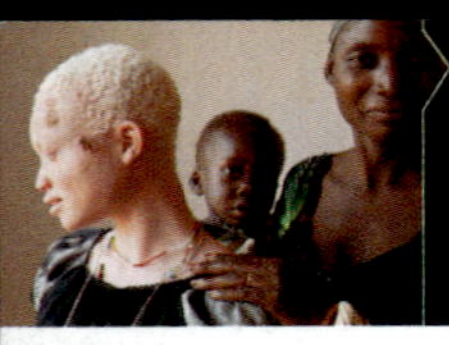

14.4 HOW DOES CHROMOSOME STRUCTURE CHANGE?

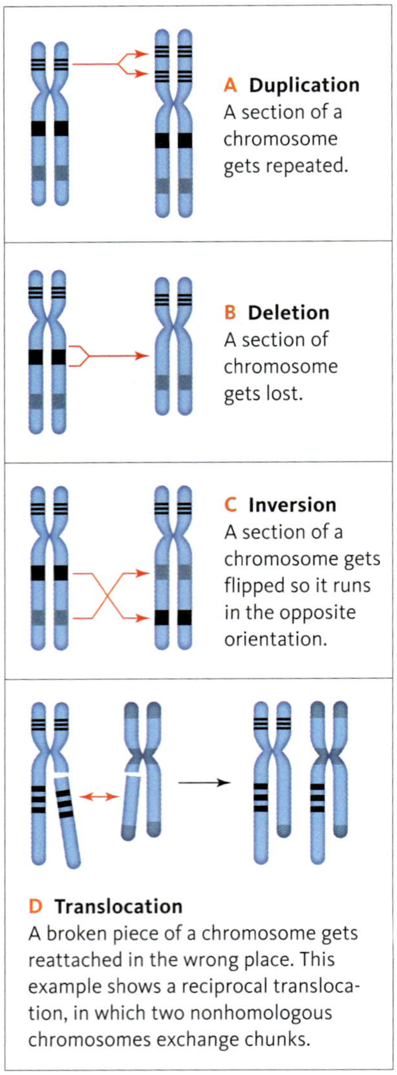

A Duplication A section of a chromosome gets repeated.

B Deletion A section of chromosome gets lost.

C Inversion A section of a chromosome gets flipped so it runs in the opposite orientation.

D Translocation A broken piece of a chromosome gets reattached in the wrong place. This example shows a reciprocal transloca-tion, in which two nonhomologous chromosomes exchange chunks.

FIGURE 14.7 {Animated} Major changes in chromosome structure.

Mutation is a term that generally refers to small-scale changes in DNA sequence—one or a few nucleotides. Chromosome changes on a larger scale also occur. Like mutations, these changes may be induced by exposure to chemicals or radiation. Others are an outcome of faulty crossing over during prophase I of meiosis. For example, nonhomologous chromosomes may align and swap segments at spots where the DNA sequence is similar. Homologous chromosomes sometimes misalign along their length. In both cases, crossing over results in the exchange of segments that are not equivalent.

Chromosome structural changes also result from the activity of **transposable elements**, which are segments of DNA, hundreds to thousands of nucleotides long, that can move spontaneously within or between chromosomes. Repeated DNA sequences at their ends allow the segments to move during mitosis or meiosis. Transposable elements are common in the DNA of all species; about 45 percent of human DNA consists of them or their remnants.

TYPES OF CHROMOSOMAL CHANGE

Regardless of the cause, large-scale changes in chromosome structure can be categorized into several groups (FIGURE 14.7). In most cases, these changes have drastic effects on health; about half of all miscarriages are due to chromosome abnormalities of the developing embryo.

Duplication Even normal chromosomes have DNA sequences that are repeated two or more times. These repetitions are called **duplications** (FIGURE 14.7A). Some newly occurring duplications, such as the expansion mutations that cause Huntington's disease, cause genetic abnormalities or disorders. Others, as you will see shortly, have been evolutionarily important.

Deletion Large-scale deletions in a chromosome (FIGURE 14.7B) often have severe consequences. Duchenne muscular dystrophy most often arises from deletions in the X chromosome. A different deletion in chromosome 5 shortens life span, impairs mental functioning, and results in an abnormally shaped larynx. This disorder, cri-du-chat (French for "cat's cry"), is named for the sound that affected infants make when they cry.

Inversion With an **inversion**, a segment of chromosomal DNA becomes oriented in the reverse direction, with no molecular loss (FIGURE 14.7C). An inversion may not affect a carrier's health if it does not interrupt a gene or gene control region, because the individual's cells still contain their full complement of genetic material. However, fertility may be compromised because a chromosome with an inversion does not pair properly with its homologous partner during meiosis. Crossovers between these mispaired chromosomes can produce other chromosome abnormalities that reduce the viability of forthcoming embryos. People who carry an inversion may not know about it until they are diagnosed with infertility and their karyotype is checked.

Translocation If a chromosome breaks, the broken part may get attached to a different chromosome, or to a different part of the same one. This type of structural change is called a **translocation**. Most translocations are reciprocal, in which two nonhomologous chromosomes exchange broken parts (FIGURE 14.7D). A reciprocal translocation between chromosomes 8 and 14 is the usual cause of Burkitt's lymphoma, an aggressive cancer of the immune system. This translocation moves a proto-oncogene to a region that is vigorously transcribed in immune cells, with the result being uncontrolled cell divisions that are characteristic of cancer (Section 11.5). Many other reciprocal translocations have no adverse effects on health, but, like inversions, they can compromise fertility. During meiosis, translocated chromosomes pair abnormally and segregate improperly; about half of the resulting gametes carry major duplications or deletions. If one of these gametes unites with a normal gamete at fertilization, the resulting embryo almost always dies. As with inversions, people who carry a translocation may not know about it until they have difficulty with fertility.

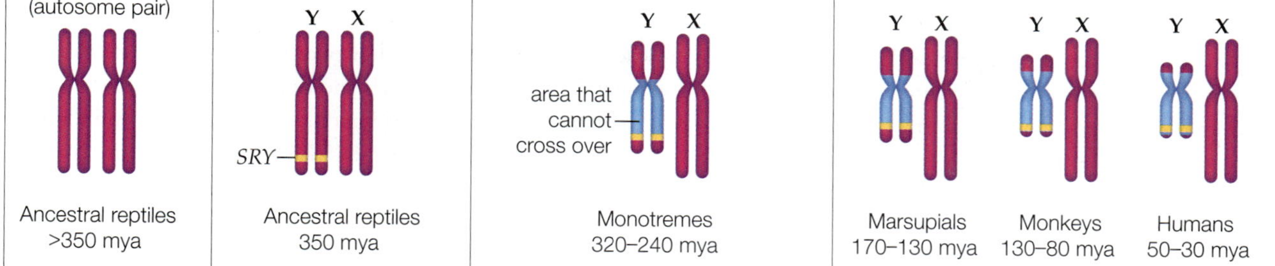

A Before 350 mya, sex was determined by temperature, not by chromosome differences.

B The *SRY* gene begins to evolve 350 mya. The DNA sequences of the chromosomes diverge as other mutations accumulate.

C By 320–240 mya, the DNA sequences of the chromosomes are so different that the pair can no longer cross over in one region. The Y chromosome begins to shorten.

D Three more times, the pair stops crossing over in yet another region. Each time, the DNA sequences of the chromosomes diverge, and the Y chromosome shortens. Today, the pair crosses over only at a small region near the ends.

FIGURE 14.8 Evolution of the Y chromosome. Today, the *SRY* gene determines male sex. Homologous regions of the chromosomes are shown in pink; mya, million years ago. Monotremes are egg-laying mammals; marsupials are pouched mammals.

CHROMOSOME CHANGES IN EVOLUTION

There is evidence of major structural alterations in the chromosomes of all known species. For example, duplications have often allowed a copy of a gene to mutate while the original carried out its unaltered function. The multiple and strikingly similar globin chain genes of mammals apparently evolved by this process. Globin chains, remember, associate to form molecules of hemoglobin (Section 9.5). Two identical genes for the alpha chain—and five other slightly different versions of it—form a cluster on chromosome 16. The gene for the beta chain clusters with four other slightly different versions on chromosome 11.

As another example, X and Y chromosomes were once homologous autosomes in ancient, reptilelike ancestors of mammals (**FIGURE 14.8**). Ambient temperature probably determined the gender of those organisms, as it still does in turtles and some other modern reptiles. About 350 million years ago, a gene on one of the two homologous chromosomes mutated. The change, which was the beginning of the male sex determination gene *SRY*, interfered with crossing over during meiosis. A reduced frequency of crossing over allowed the chromosomes to diverge around the changed region as mutations began to accumulate separately in the two chromosomes. Over evolutionary time, the chromosomes became so different that they no longer crossed over at all in the changed region, so they diverged even more. Today, the Y chromosome is much smaller than the X, and is homologous with it only in a tiny part. The Y crosses over mainly with itself—by translocating duplicated regions of its own DNA.

Some chromosome structure changes contributed to differences among closely related organisms, such as apes and humans. Human somatic cells have twenty-three pairs of chromosomes, but cells of chimpanzees, gorillas, and orangutans have twenty-four. Thirteen human chromosomes are almost identical with chimpanzee chromosomes. Nine more are similar, except for some inversions. One human chromosome matches up with two in chimpanzees and the other great apes (**FIGURE 14.9**). During human evolution, two chromosomes evidently fused end to end and formed our chromosome 2. How do we know? The region where the fusion occurred contains remnants of a telomere (Section 11.4).

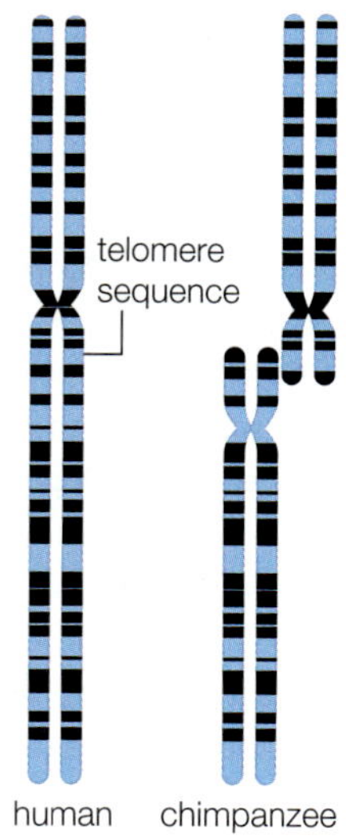

FIGURE 14.9 Human chromosome 2 compared with chimpanzee chromosomes 2A and 2B.

duplication Repeated section of a chromosome.
inversion Structural rearrangement of a chromosome in which part of the DNA becomes oriented in the reverse direction.
translocation Structural change of a chromosome in which a broken piece gets reattached in the wrong location.
transposable element Segment of DNA that can move spontaneously within or between chromosomes.

TAKE-HOME MESSAGE 14.4

A segment of a chromosome may be duplicated, deleted, inverted, or translocated. Any of these changes are usually harmful or lethal, but may be conserved in the rare circumstance that it has a neutral or beneficial effect.

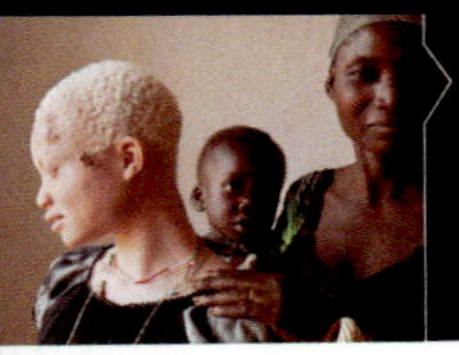

14.5 WHAT ARE THE EFFECTS OF CHROMOSOME NUMBER CHANGES IN HUMANS?

About 70 percent of flowering plant species, and some insects, fishes, and other animals, are **polyploid**, which means that they have three or more complete sets of chromosomes. Cells in some adult human tissues are normally polyploid, but inheriting more than two full sets of chromosomes is invariably fatal in humans.

An **aneuploid** individual inherited too many or too few copies of a particular chromosome. Less than 1 percent of children are born with a diploid chromosome number that differs from the normal 46.

Changes in chromosome number are usually the outcome of **nondisjunction**, the failure of chromosomes to separate properly during nuclear division—mitosis or meiosis. Nondisjunction during meiosis (FIGURE 14.10) can affect chromosome number at fertilization. For example, if a normal gamete (n) fuses with another gamete that has an extra chromosome ($n+1$), the resulting zygote will have three copies of one type of chromosome and two of every other type ($2n+1$), a type of aneuploidy called trisomy. If an $n-1$ gamete fuses with a normal n gamete, the new individual will be $2n-1$, a condition called monosomy.

AUTOSOMAL ANEUPLOIDY AND DOWN SYNDROME

In most cases, autosomal aneuploidy in humans is fatal before birth or shortly thereafter. An important exception is trisomy 21. A person born with three chromosomes 21 has a high likelihood of surviving infancy, and will develop Down syndrome. Mild to moderate mental impairment and health problems such as heart disease are hallmarks of this disorder. Other effects may include a somewhat flattened facial profile, a fold of skin that starts at the inner corner of each eyelid, white spots on the iris (FIGURE 14.11), and one deep crease (instead of two shallow creases) across each palm. The skeleton grows and develops abnormally, so older children have short body parts, loose joints, and misaligned bones of the fingers, toes, and hips. The muscles and reflexes are weak, and motor skills such as speech develop slowly. With medical care, affected individuals live about fifty-five years. Early training can help these individuals learn to care for themselves and to take part in normal activities. Down syndrome occurs once in about 700 births, and the risk increases with maternal age.

SEX CHROMOSOME ANEUPLOIDY

Nondisjunction also causes alterations in the number of X and Y chromosomes, with a frequency of about 1 in 400 live births. Most often, such alterations lead to mild difficulties in learning and impaired motor skills such as a speech delay. These problems may be very subtle.

Turner Syndrome Individuals with Turner syndrome have an X chromosome and no corresponding X or Y chromosome (XO). The syndrome is thought to arise most frequently as an outcome of inheriting an unstable Y chromosome from the father. The zygote starts out being genetically male, with an X and a Y chromosome. Sometime during early development, the Y chromosome breaks up and is lost, so the embryo continues to develop as a female.

There are fewer people affected by Turner syndrome than other chromosome abnormalities: Only about 1 in 2,500 newborn girls has it. XO individuals grow up well proportioned but short, with an average height of four feet, eight inches (1.4 meters). Their ovaries do not develop

FIGURE 14.10 **{Animated}** An example of nondisjunction during meiosis. Of the two pairs of homologous chromosomes shown here, one fails to separate during anaphase I. The chromosome number is altered in the resulting gametes.

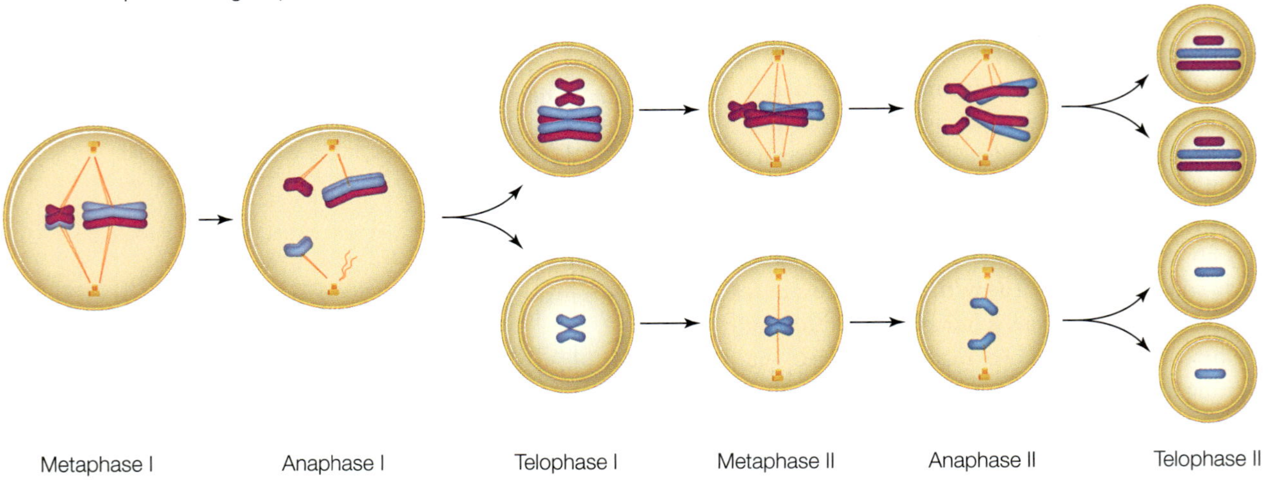

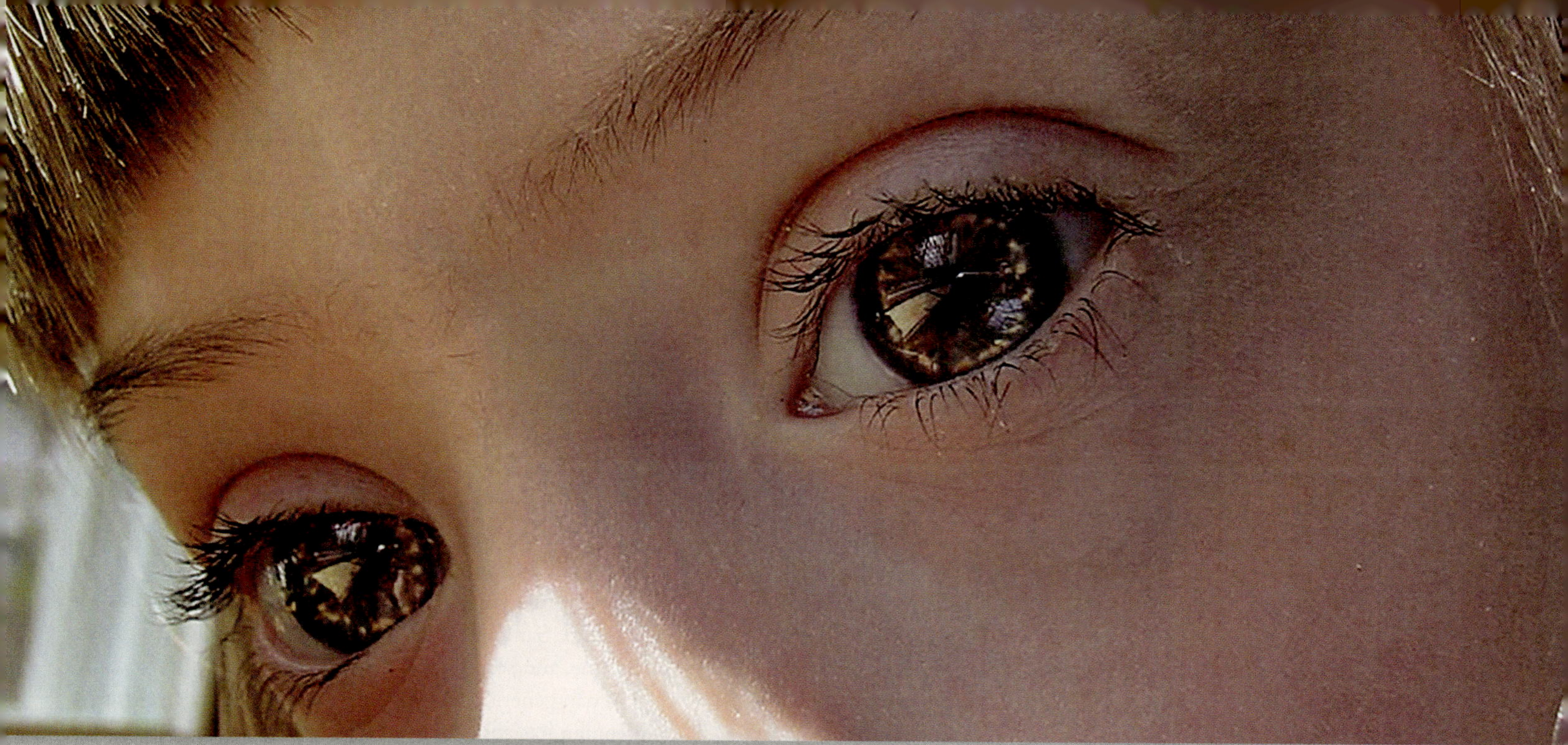

FIGURE 14.11 A Down syndrome phenotype. Excess tissue deposits on the colored part of the eye give rise to a ring of starlike white speckles, a lovely phenotypic effect of the chromosome number change that causes Down syndrome.

properly, so they do not make enough sex hormones to become sexually mature and do not develop secondary sexual traits such as enlarged breasts.

XXX Syndrome A female may inherit multiple X chromosomes, a condition called XXX syndrome. XXX syndrome occurs in about 1 of 1,000 births. As with Down syndrome, risk increases with maternal age. Only one X chromosome is typically active in female cells, so having extra X chromosomes usually does not cause physical or medical problems, but mild mental impairment may occur.

Klinefelter Syndrome About 1 out of every 500 males has an extra X chromosome (XXY). The resulting disorder, Klinefelter syndrome, develops at puberty. XXY males tend to be overweight, tall, and have mild mental impairment. They make more estrogen and less testosterone than normal males. This hormone imbalance causes affected men to have small testes and a small prostate gland, a low sperm count, sparse facial and body hair, a high-pitched voice, and enlarged breasts. Testosterone injections during puberty can minimize some of these traits.

XYY Syndrome About 1 in 1,000 males is born with an extra Y chromosome (XYY), a result of nondisjunction of the Y chromosome during sperm formation. Adults tend to be taller than average and have mild mental impairment, but most are otherwise normal. XYY men were once thought to be predisposed to a life of crime. This misguided view was based on sampling error (too few cases in narrowly chosen groups such as prison inmates) and bias (the researchers who gathered the karyotypes also took the personal histories of the participants). That view has since been disproven: Men with XYY syndrome are only slightly more likely to be convicted for crimes than unaffected men. Researchers now believe this slight increase can be explained by poor socioeconomic conditions related to the effects of the syndrome.

aneuploid Having too many or too few copies of a particular chromosome.
nondisjunction Failure of sister chromatids or homologous chromosomes to separate during nuclear division.
polyploid Having three or more of each type of chromosome characteristic of the species.

TAKE-HOME MESSAGE 14.5

Polyploidy is fatal in humans, but not in flowering plants and some other organisms.

Aneuploidy can arise from nondisjunction during meiosis. In humans, most cases of aneuploidy are associated with some degree of mental impairment.

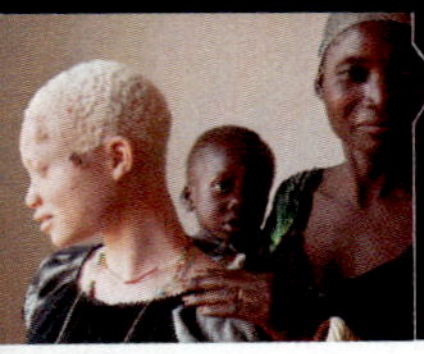

14.6 HOW DO WE USE WHAT WE KNOW ABOUT HUMAN INHERITANCE?

Studying human inheritance patterns has given us many insights into how genetic disorders arise and progress, and how to treat them. Some disorders can be detected early enough to start countermeasures before symptoms develop. For this reason, most hospitals in the United States now screen newborns for mutations that cause phenylketonuria, or PKU. The mutations affect an enzyme that converts the amino acid phenylalanine to tyrosine. Without this enzyme, the body becomes deficient in tyrosine, and phenylalanine accumulates to high levels. The imbalance inhibits protein synthesis in the brain, which in turn results in severe neurological symptoms. Restricting all intake of phenylalanine can slow the progression of PKU, so routine early screening has resulted in fewer individuals suffering from the symptoms of the disorder.

The probability that a child will inherit a genetic disorder can often be estimated by testing prospective parents for alleles known to be associated with genetic disorders. Karyotypes and pedigrees are also useful in this type of screening, which can help the parents make decisions about family planning. Genetic screening can also be done post-conception, in which case it is called prenatal diagnosis (prenatal means before birth). Prenatal diagnosis checks for physical and genetic abnormalities in an embryo or fetus. More than 30 conditions are detectable prenatally, including aneuploidy, hemophilia, Tay–Sachs disease, sickle-cell anemia, muscular dystrophy, and cystic fibrosis. If the disorder is treatable, early detection allows the newborn to receive prompt and appropriate treatment. A few defects are even surgically correctable before birth. Prenatal diagnosis also gives parents time to prepare for the birth of an affected child, and an opportunity to decide whether to continue with the pregnancy or terminate it.

As an example of how prenatal diagnosis works, consider a woman who becomes pregnant at age thirty-five. Her doctor will probably perform a procedure called obstetric sonography, in which ultrasound waves directed across the woman's abdomen form images of the fetus's limbs and internal organs. If the images reveal a physical defect that may be the result of a genetic disorder, a more invasive technique would be recommended for further diagnosis. With fetoscopy, sound waves pulsed from inside the mother's uterus yield images much higher in resolution than ultrasound. Samples of tissue or blood are often taken at the same time, and some corrective surgeries can be performed.

Human genetics studies show that our thirty-five-year-old woman has about a 1 in 80 chance that her baby will be born with a chromosomal abnormality, a risk more than six times greater than when she was twenty years old. Thus, even if no abnormalities are detected by ultrasound, she probably will be offered a more thorough diagnostic procedure, amniocentesis, in which a small sample of fluid is drawn from the amniotic sac enclosing the fetus (**FIGURE 14.12**). The fluid contains cells shed by the fetus, and those cells can be tested for genetic disorders. Chorionic villus sampling (CVS) can be performed earlier than amniocentesis. With this technique, a few cells from the chorion are removed and tested for genetic disorders. (The chorion is a membrane that surrounds the amniotic sac and helps form the placenta, an organ that allows substances to be exchanged between mother and embryo.)

An invasive procedure often carries a risk to the fetus. The risks vary by the procedure. Amniocentesis has improved so much that, in the hands of a skilled physician, the procedure no longer increases the risk of miscarriage. CVS occasionally disrupts the placenta's development and thus causes underdeveloped or missing fingers and toes in 0.3 percent of newborns. Fetoscopy raises the miscarriage risk by a whopping 2 to 10 percent.

FIGURE 14.12 **{Animated}** An 8-week-old fetus. With amniocentesis, fetal cells shed into the fluid inside the amniotic sac are tested for genetic disorders. Chorionic villus sampling tests cells of the chorion, which is part of the placenta.

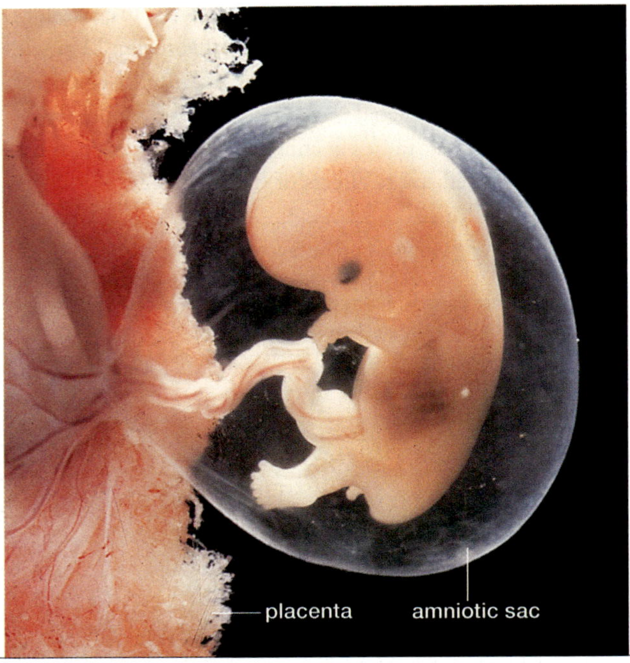

TAKE-HOME MESSAGE 14.6

Studying inheritance patterns for genetic disorders has helped researchers develop treatments for some of them.

Genetic testing can provide prospective parents with information about the health of their future children.

14.7 Application: SHADES OF SKIN

FIGURE 14.13 Fraternal twins Kian and Remee. Both of the children's grandmothers are of European descent, and have pale skin. Both of their grandfathers are of African descent, and have dark skin. The twins inherited different alleles of some genes that affect skin color from their parents, who, given the appearance of their children, must be heterozygous for those alleles.

SKIN COLOR, LIKE MOST OTHER HUMAN TRAITS, HAS A GENETIC BASIS. The color of human skin begins with melanosomes, which are organelles that make melanin pigments. Most people have about the same number of melanosomes in their skin cells. Variations in skin color arise from differences in formation and deposition of melanosomes in the skin, as well as in the kinds and amounts of melanins they make. More than 100 genes are involved in these processes.

Human skin color variation may have evolved as a balance between vitamin production and protection against harmful UV radiation in the sun's rays. Dark skin would have been beneficial under the intense sunlight of African savannas where humans first evolved. Melanin is a natural sunscreen: It prevents UV radiation from breaking down folate, a vitamin essential for normal sperm formation and embryonic development.

Early human groups that migrated to regions with cold climates were exposed to less sunlight. In these regions, lighter skin color is beneficial. Why? UV radiation stimulates skin cells to make a molecule the body converts to vitamin D. Where sunlight exposure is minimal, UV radiation is less of a risk than vitamin D deficiency, which has serious health consequences for developing fetuses and children.

The evolution of regional variations in human skin color began with mutations. Consider a gene on chromosome 15 that encodes a transport protein in melanosome membranes. Nearly all people of African, Native American, or east Asian descent carry the same allele of this gene. Between 6,000 and 10,000 years ago, a mutation gave rise to a different allele. The mutation, a single base-pair substitution, changed the 111th amino acid of the transport protein from alanine to threonine. The change results in less melanin—and lighter skin color—than the original African allele does. Today, nearly all people of European descent carry this mutated allele.

A person of mixed ethnicity may make gametes that contain different combinations of alleles for dark and light skin. It is fairly rare that one of those gametes contains mainly alleles for dark skin, or mainly alleles for light skin, but it happens (FIGURE 14.13). Skin color is only one of many human traits that vary as a result of single nucleotide mutations. The small scale of such changes offers a reminder that all of us share the genetic legacy of common ancestry.

CREDIT: (13) Gary Roberts/worldwidefeatures.com.

Summary

SECTION 14.1 Geneticists study inheritance patterns in humans by tracking genetic disorders and abnormalities through families. A genetic abnormality is an uncommon version of a heritable trait that does not result in medical problems. A genetic disorder is a heritable condition that sooner or later results in mild or severe medical problems. Geneticists make **pedigrees** to reveal inheritance patterns for alleles that can be predictably associated with specific phenotypes.

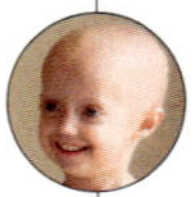

SECTION 14.2 An allele is inherited in an autosomal dominant pattern if the trait it specifies appears in everyone who carries it, and both sexes are affected with equal frequency. Such traits appear in every generation of families that carry the allele. An allele is inherited in an autosomal recessive pattern if the trait it specifies appears only in homozygous people. Such traits also appear in both sexes equally, but they can skip generations.

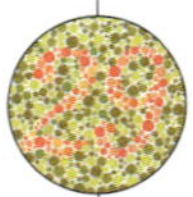

SECTION 14.3 An allele is inherited in an X-linked pattern when it occurs on the X chromosome. Most X-linked disorders are inherited in a recessive pattern, and these tend to appear in men more often than in women. Heterozygous women have a dominant, normal allele that can mask the effects of the recessive one; men do not. Men can transmit an X-linked allele to their daughters, but not to their sons. Only a woman can pass an X-linked allele to a son.

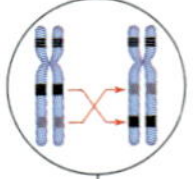

SECTION 14.4 Faulty crossovers and the activity of **transposable elements** can give rise to major changes in chromosome structure, including **duplications**, **inversions**, and **translocations**. Some of these changes are harmful or lethal in humans; others affect fertility. Even so, major structural changes have accumulated in the chromosomes of all species over evolutionary time.

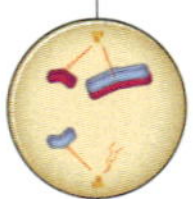

SECTION 14.5 Occasionally, abnormal events occur before or during meiosis, and new individuals end up with the wrong chromosome number. Consequences range from minor to lethal changes in form and function.

Chromosome number change is usually an outcome of **nondisjunction**, in which chromosomes fail to separate properly during meiosis. **Polyploid** individuals have three or more of each type of chromosome. Polyploidy is lethal in humans, but not in flowering plants, and some insects, fishes, and other animals.

Aneuploid individuals have too many or too few copies of a chromosome. In humans, most cases of autosomal aneuploidy are lethal. Trisomy 21, which causes Down syndrome, is an exception. A change in the number of sex chromosomes usually results in some degree of impairment in learning and motor skills.

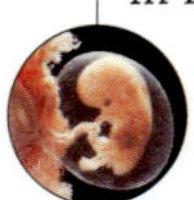

SECTION 14.6 Prospective parents can estimate their risk of transmitting a harmful allele to offspring with genetic screening, in which their pedigrees and genotype are analyzed by a genetic counselor. Prenatal genetic testing can reveal a genetic disorder before birth.

SECTION 14.7 Like most other human traits, skin color has a genetic basis. Minor differences in the alleles that govern melanin production and the deposition of melanosomes affect skin color. The differences probably evolved as a balance between vitamin production and protection against harmful UV radiation.

Self-Quiz

Answers in Appendix VII

1. Constructing a family pedigree is particularly useful when studying inheritance patterns in organisms that ______ .
   a. produce many offspring per generation
   b. produce few offspring per generation
   c. have a very large chromosome number
   d. reproduce asexually
   e. have a fast life cycle

2. Pedigree analysis is necessary when studying human inheritance patterns because ______ .
   a. humans have more than 20,000 genes
   b. of ethical problems with human experimentation
   c. inheritance in humans is more complicated than in other organisms
   d. genetic disorders occur in humans
   e. all of the above

3. A recognized set of symptoms that characterize a genetic disorder is a(n) ______ .
   a. syndrome b. disease c. abnormality

4. If one parent is heterozygous for a dominant allele on an autosome and the other parent does not carry the allele, any child of theirs has a ______ chance of having the associated trait.
   a. 25 percent c. 75 percent
   b. 50 percent d. no chance; it will die

5. Is this statement true or false? A son can inherit an X-linked recessive allele from his father.

6. A trait that is present in a male child but not in either of his parents is characteristic of ______ inheritance.
   a. autosomal dominant
   b. autosomal recessive
   c. X-linked recessive
   d. It is not possible to answer this question without more information.

7. Color blindness is inherited in an ______ pattern.
   a. autosomal dominant c. X-linked dominant
   b. autosomal recessive d. X-linked recessive

8. A female child inherits one X chromosome from her mother and one from her father. What sex chromosome does a male child inherit from each of his parents?

Data Analysis Activities

Skin Color Survey of Native Peoples In 2000, researchers measured the average amount of UV radiation received in more than fifty regions of the world, and correlated it with the average skin reflectance of people native to those regions (reflectance is a way to measure the amount of melanin pigment in skin). Some of the results of this study are shown in **FIGURE 14.14**.

1. Which country receives the most UV radiation?
2. Which country receives the least UV radiation?
3. People native to which country have the darkest skin?
4. People native to which country have the lightest skin?
5. According to these data, how does the skin color of indigenous peoples correlate with the amount of UV radiation incident in their native regions?

Country	Skin Reflectance	UVMED
Australia	19.30	335.55
Kenya	32.40	354.21
India	44.60	219.65
Cambodia	54.00	310.28
Japan	55.42	130.87
Afghanistan	55.70	249.98
China	59.17	204.57
Ireland	65.00	52.92
Germany	66.90	69.29
Netherlands	67.37	62.58

FIGURE 14.14 Skin color of indigenous peoples and regional incident UV radiation. Skin reflectance measures how much light of 685-nanometer wavelength is reflected from skin; UVMED is the annual average UV radiation received at Earth's surface.

9. Alleles for Tay–Sachs disease are inherited in an autosomal recessive pattern. Why would two parents with a normal phenotype have a child with Tay–Sachs?
   a. Both parents are homozygous for a Tay–Sachs allele.
   b. Both parents are heterozygous for a Tay–Sachs allele.
   c. A new mutation gave rise to Tay–Sachs in the child.
   d. b or c

10. The *SRY* gene gives rise to the male phenotype in humans (Sections 10.3 and 14.4). What do you think the inheritance pattern of *SRY* alleles is called?

11. Nondisjunction may occur during _______ .
   a. mitosis   c. fertilization
   b. meiosis   d. both a and b

12. Nondisjunction can result in _______ .
   a. duplications   c. crossing over
   b. aneuploidy   d. pleiotropy

13. Is this statement true or false? Inheriting three or more of each type of chromosome characteristic of the species results in a condition called polyploidy.

14. Klinefelter syndrome (XXY) can be easily diagnosed by _______ .
   a. pedigree analysis   c. karyotyping
   b. aneuploidy   d. phenotypic treatment

15. Match the chromosome terms appropriately.

___ polyploidy	a. symptoms of a genetic disorder
___ deletion	b. segment of a chromosome moves to a nonhomologous chromosome
___ aneuploidy	c. extra sets of chromosomes
___ translocation	d. gets around
___ syndrome	e. a chromosome segment lost
___ transposable element	f. one extra chromosome

Genetics Problems

Answers in Appendix VII

1. Does the phenotype indicated by the red circles and squares in this pedigree show an inheritance pattern that is autosomal dominant, autosomal recessive, or X-linked?

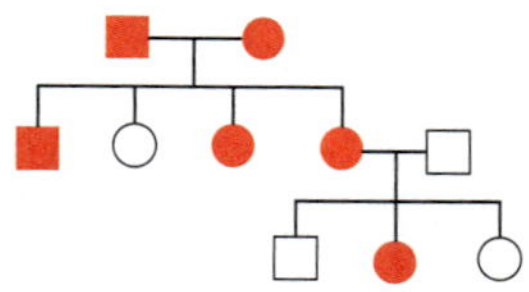

2. Human females have two X chromosomes (XX); males have one X and one Y chromosome (XY).

a. Does a male inherit an X chromosome from his mother or his father?

b. With respect to X-linked alleles, how many different types of gametes can a male produce?

c. A female homozygous for an X-linked allele can produce how many types of gametes with respect to that allele?

d. A female heterozygous for an X-linked allele can produce how many types of gametes with respect to that allele?

3. Somatic cells of individuals with Down syndrome usually have an extra chromosome 21; they contain forty-seven chromosomes. A few individuals with Down syndrome have forty-six chromosomes: two normal-appearing chromosomes 21, and a longer-than-normal chromosome 14. Speculate on how this chromosome abnormality arises.

4. An allele responsible for Marfan syndrome (Section 13.4) is inherited in an autosomal dominant pattern. What is the chance that a child will inherit the allele if one parent does not carry it and the other is heterozygous?

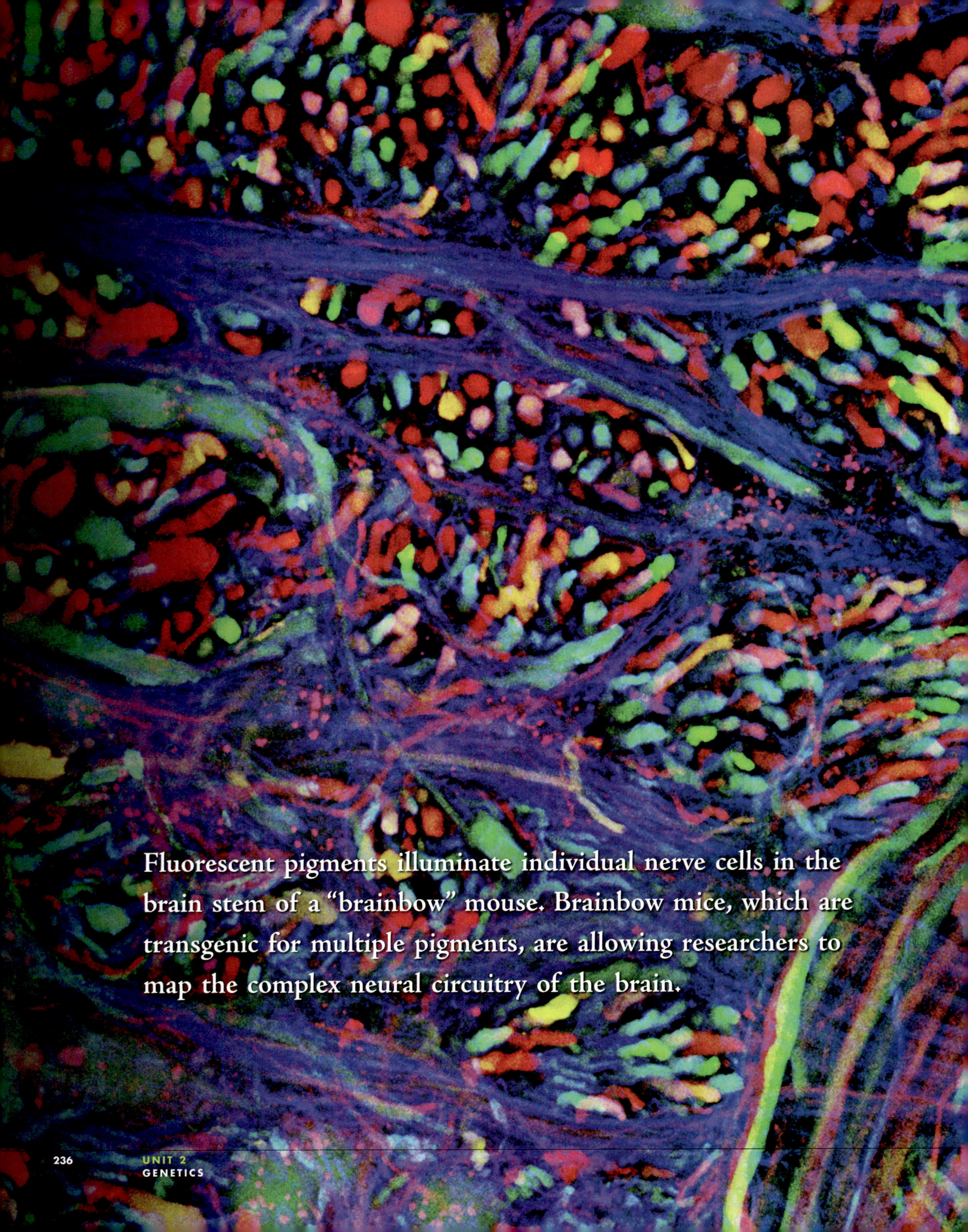

Fluorescent pigments illuminate individual nerve cells in the brain stem of a "brainbow" mouse. Brainbow mice, which are transgenic for multiple pigments, are allowing researchers to map the complex neural circuitry of the brain.

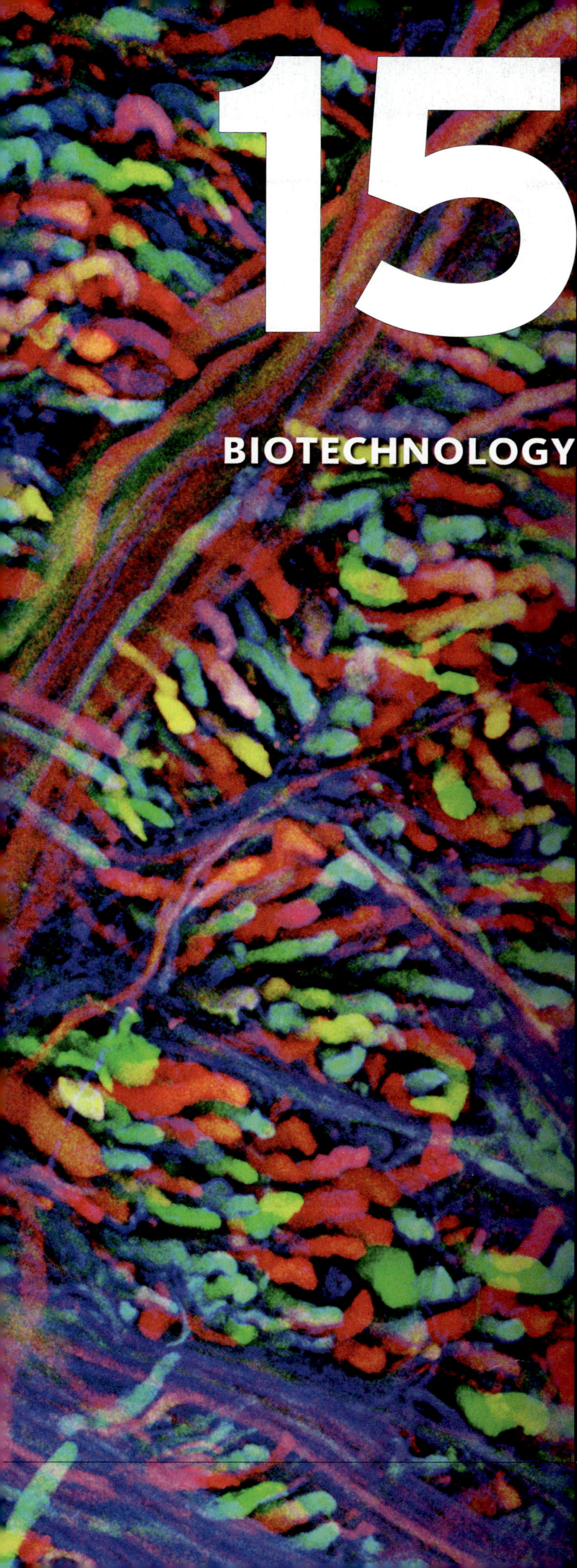

15 BIOTECHNOLOGY

Links to Earlier Concepts

This chapter builds on your understanding of DNA (Sections 8.2, 8.3, 13.1, 14.6) and DNA replication (8.4). Clones (8.6), gene expression (9.1, 9.2), and knockouts (10.2) are important in genetic engineering, particularly in research on human traits (13.6) and genetic disorders (Chapter 14). You will revisit tracers (2.1), triglycerides (3.3), denaturation (3.5), bacteria (4.4), β-carotene (6.1), mutations (9.5), the *lac* operon (10.4), cancer (11.5), and alleles (12.1).

KEY CONCEPTS

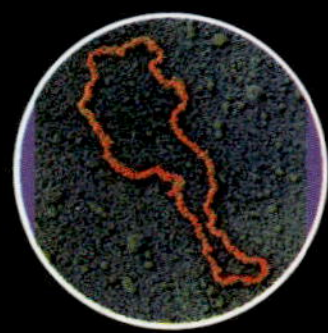

DNA CLONING

Researchers make recombinant DNA by cutting and pasting together DNA from different species. Plasmids and other vectors can carry foreign DNA into host cells.

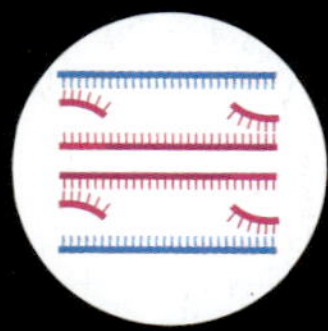

FINDING NEEDLES IN HAYSTACKS

Genetic engineering, the directed modification of an organism's genes, relies on laboratory techniques for isolating and identifying particular fragments of DNA.

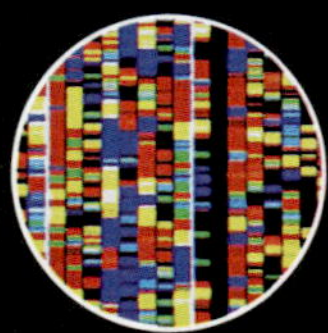

DNA SEQUENCING

Sequencing reveals the linear order of nucleotides in DNA. Comparing genomes offers insights into human genes and evolution. DNA sequence can be used to identify individuals.

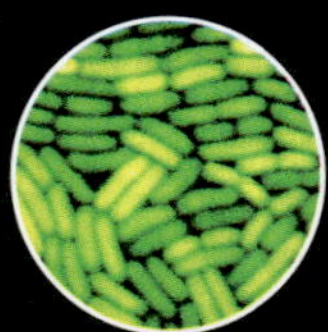

GENETIC ENGINEERING

Genetic engineering is now a routine part of research and industrial applications. Genetically modified organisms are used to produce food, medicines, and other products.

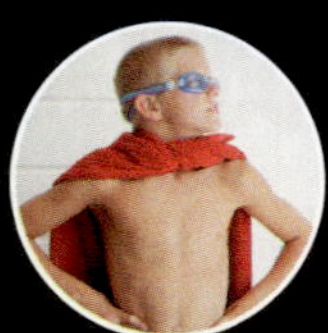

GENE THERAPY

The directed modification of human DNA continues to be tested in medical applications. It also continues to raise ethical questions about modifying the human genome.

Photograph courtesy of © Dr. Jean Levit. The Brainbow technique was developed in the laboratories of Jeff W. Lichtman and Joshua R. Sanes at Harvard University. This image has received the Bioscape imaging competition 2007 prize.

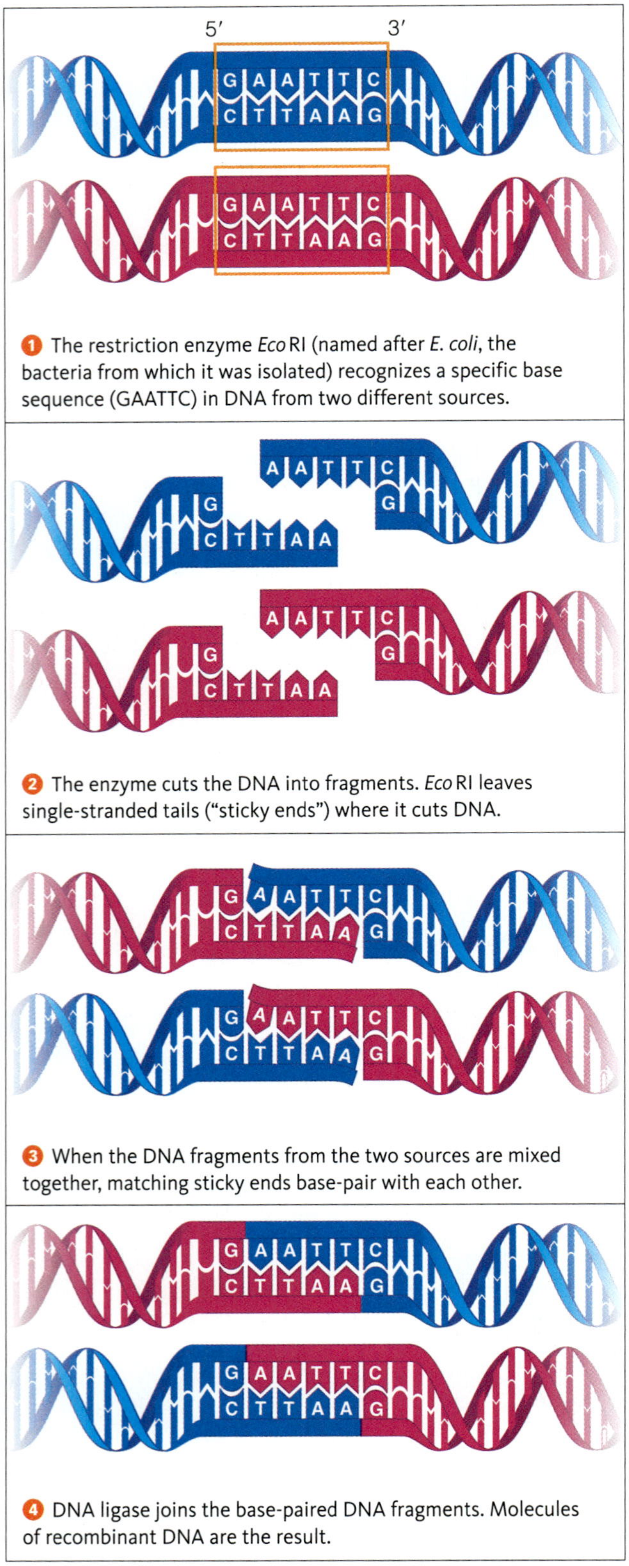

❶ The restriction enzyme *Eco* RI (named after *E. coli*, the bacteria from which it was isolated) recognizes a specific base sequence (GAATTC) in DNA from two different sources.

❷ The enzyme cuts the DNA into fragments. *Eco* RI leaves single-stranded tails ("sticky ends") where it cuts DNA.

❸ When the DNA fragments from the two sources are mixed together, matching sticky ends base-pair with each other.

❹ DNA ligase joins the base-paired DNA fragments. Molecules of recombinant DNA are the result.

FIGURE 15.1 {Animated} Making recombinant DNA.

FIGURE IT OUT: Why did the enzyme cut both strands of DNA?

Answer: Because the recognition sequence occurs on both strands.

CUT AND PASTE

In the 1950s, excitement over the discovery of DNA's structure (Section 8.2) gave way to frustration: No one could determine the order of nucleotides in a molecule of DNA. Identifying a single base among thousands or millions of others turned out to be a huge technical challenge.

Research in a seemingly unrelated field yielded a solution when Werner Arber, Hamilton Smith, and their coworkers discovered how some bacteria resist infection by bacteriophage (Section 8.1). These bacteria have enzymes that chop up any injected viral DNA before it has a chance to integrate into the bacterial chromosome. The enzymes restrict viral growth; hence their name, restriction enzymes. A **restriction enzyme** cuts DNA wherever a specific nucleotide sequence occurs (**FIGURE 15.1** ❶). The discovery of restriction enzymes allowed researchers to cut chromosomal DNA into manageable chunks. It also allowed them to combine DNA fragments from different organisms. How? Many restriction enzymes leave single-stranded tails on DNA fragments ❷. Researchers realized that complementary tails will base-pair, regardless of the source of DNA ❸. The tails are called "sticky ends," because two DNA fragments stick together when their matching tails base-pair. The enzyme DNA ligase (Section 8.4) can be used to seal the gaps between base-paired sticky ends, so continuous DNA strands form ❹. Thus, using appropriate restriction enzymes and DNA ligase, researchers can cut and paste DNA from different sources. The result, a hybrid molecule that consists of genetic material from two or more organisms, is called **recombinant DNA.**

Making recombinant DNA is the first step in **DNA cloning**, a set of laboratory methods that uses living cells to

FIGURE 15.2 Plasmid cloning vectors. **A** Micrograph of a plasmid. **B** A commercial plasmid cloning vector. Restriction enzyme recognition sequences are indicated on the right by the name of the enzyme that cuts them. Researchers insert foreign DNA into the vector at these sequences. Bacterial genes (gold) help researchers identify host cells that take up a vector with inserted DNA. This vector carries two antibiotic resistance genes and the *lac* operon (Section 10.4).

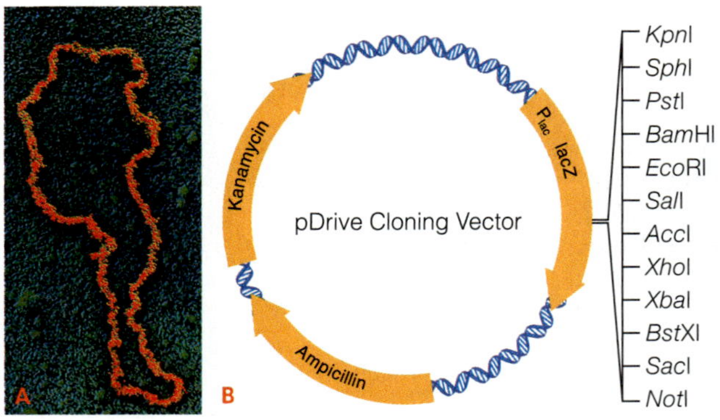

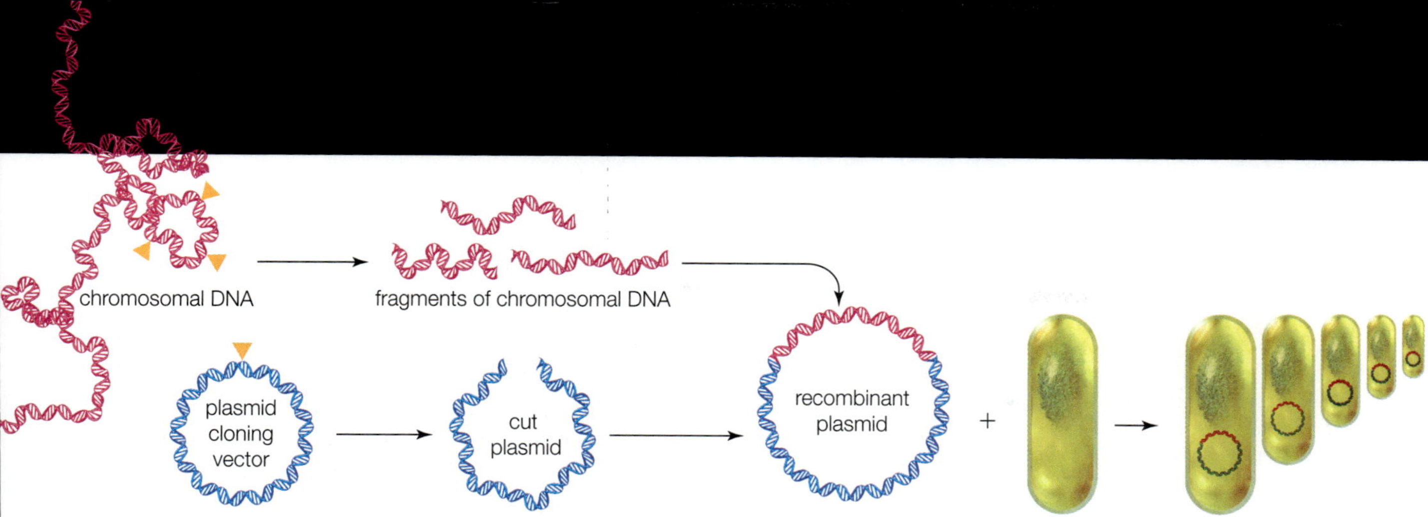

A A restriction enzyme (gold triangles) cuts a specific nucleotide sequence in chromosomal DNA and in a plasmid cloning vector.

B A fragment of chromosomal DNA and the cut plasmid base-pair at their sticky ends. DNA ligase joins the two pieces of DNA, so a recombinant plasmid forms.

C The recombinant plasmid is inserted into a host bacterial cell. When the cell reproduces, it copies the plasmid along with its chromosome. Each descendant cell receives a plasmid.

FIGURE 15.3 {Animated} An example of cloning. Here, a fragment of chromosomal DNA is inserted into a bacterial plasmid.

mass-produce specific DNA fragments. Researchers clone a fragment of DNA by inserting it into a **cloning vector**, which is a molecule that can carry foreign DNA into host cells. Bacterial plasmids (Section 4.4) may be used as cloning vectors (**FIGURE 15.2**). A bacterium copies all of its DNA before it divides, so its offspring inherit plasmids along with chromosomes. If a plasmid carries a fragment of foreign DNA, that fragment gets copied and distributed to descendant cells along with the plasmid (**FIGURE 15.3**).

A host cell into which a cloning vector has been inserted can be grown in the laboratory (cultured) to yield a huge population of genetically identical cells, or clones (Section 8.6). Each clone contains a copy of the vector and the inserted DNA fragment. The hosted DNA fragment can be harvested in large quantities from the clones.

cDNA CLONING

Remember from Section 9.2 that eukaryotic DNA contains introns. Unless you are a eukaryotic cell, it is not very easy to determine which parts of eukaryotic DNA encode gene products. Thus, researchers who study gene expression in eukaryotes often start with mature mRNA, because introns are removed during post-transcriptional processing.

An mRNA cannot be cut with restriction enzymes or pasted with DNA ligase, because these enzymes work only on double-stranded DNA. Thus, cloning with mRNA requires **reverse transcriptase**, a replication enzyme that uses an RNA template to assemble a strand of complementary DNA, or **cDNA**:

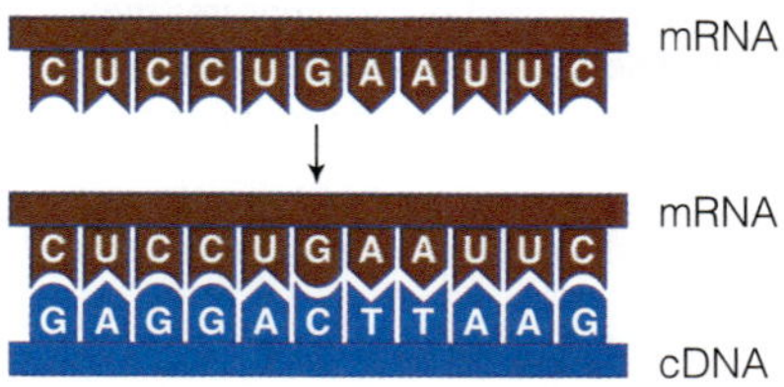

DNA polymerase is used to copy the cDNA into a second strand of DNA. The outcome is a double-stranded DNA version of the original mRNA:

Like any other double-stranded DNA, this fragment may be cut with restriction enzymes and pasted into a cloning vector using DNA ligase.

cDNA Complementary strand of DNA synthesized from an RNA template by the enzyme reverse transcriptase.
cloning vector A DNA molecule that can accept foreign DNA and be replicated inside a host cell.
DNA cloning Set of methods that uses living cells to make many identical copies of a DNA fragment.
recombinant DNA A DNA molecule that contains genetic material from more than one organism.
restriction enzyme Type of enzyme that cuts DNA at a specific nucleotide sequence.
reverse transcriptase An enzyme that uses mRNA as a template to make a strand of cDNA.

TAKE-HOME MESSAGE 15.1

DNA cloning uses living cells to mass-produce particular DNA fragments. Restriction enzymes cut DNA into fragments, then DNA ligase seals the fragments into cloning vectors. Recombinant DNA molecules result.

A cloning vector that holds foreign DNA can be introduced into a living cell. When the host cell divides, it gives rise to huge populations of genetically identical cells (clones), each with a copy of the foreign DNA.

15.2 HOW DO RESEARCHERS STUDY ONE GENE IN THE CONTEXT OF MANY?

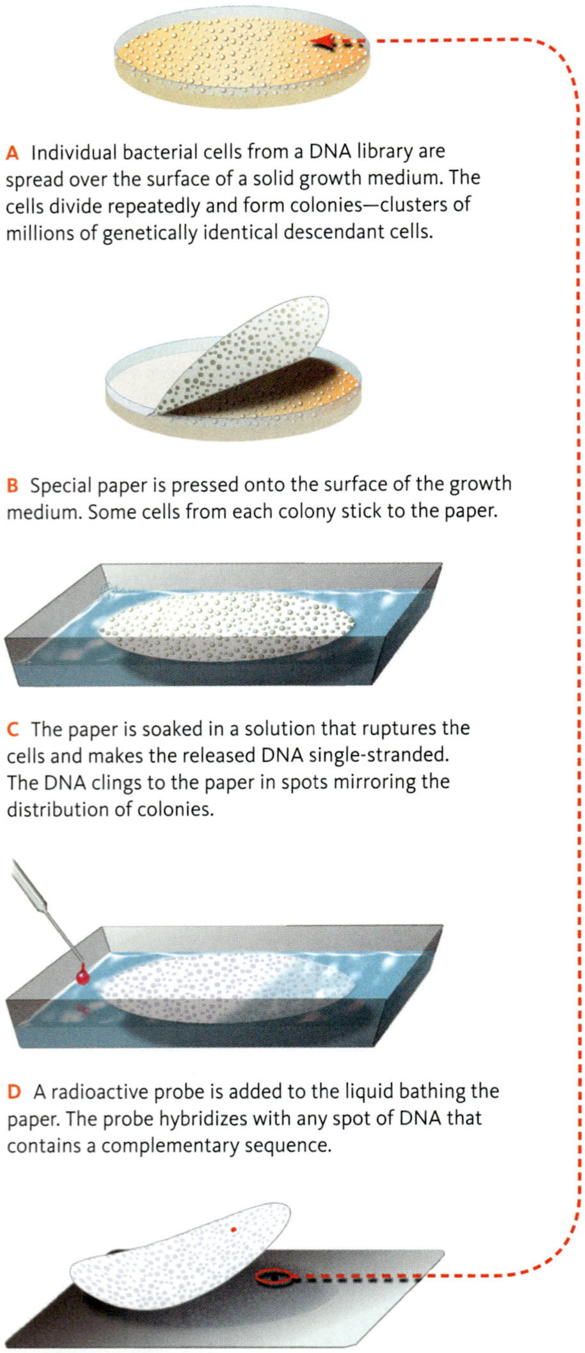

A Individual bacterial cells from a DNA library are spread over the surface of a solid growth medium. The cells divide repeatedly and form colonies—clusters of millions of genetically identical descendant cells.

B Special paper is pressed onto the surface of the growth medium. Some cells from each colony stick to the paper.

C The paper is soaked in a solution that ruptures the cells and makes the released DNA single-stranded. The DNA clings to the paper in spots mirroring the distribution of colonies.

D A radioactive probe is added to the liquid bathing the paper. The probe hybridizes with any spot of DNA that contains a complementary sequence.

E The paper is pressed against x-ray film. The radioactive probe darkens the film in a spot where it has hybridized. The spot's position is compared to the positions of the original bacterial colonies. Cells from the colony that corresponds to the spot are cultured, and their DNA is harvested.

FIGURE 15.4 {Animated} Nucleic acid hybridization. In this example, a radioactive probe helps identify a bacterial colony that contains a targeted sequence of DNA.

DNA LIBRARIES

The entire set of genetic material—the **genome**—of most organisms consists of thousands of genes. To study or manipulate a single gene, researchers must first separate it from all of the other genes in a genome. They often begin by cutting an organism's DNA into pieces, and then cloning all the pieces. The result is a genomic library, a set of clones that collectively contain all of the DNA in a genome. Researchers may also harvest mRNA, make cDNA copies of it, and then clone the cDNA. The resulting cDNA library represents only those genes being expressed at the time the mRNA was harvested.

Genomic and cDNA libraries are **DNA libraries**, sets of cells that host various cloned DNA fragments. In such libraries, a cell that contains a particular DNA fragment of interest is mixed up with thousands or millions of others that do not—a needle in a genetic haystack. One way to find that clone among the others involves the use of a **probe**, which is a fragment of DNA or RNA labeled with a tracer (Section 2.1). For example, to find a particular gene, researchers may use radioactive nucleotides to synthesize a short strand of DNA complementary in sequence to a similar gene. Because the nucleotide sequences of the probe and the gene are complementary, the two can hybridize. (Remember from Section 8.4 that nucleic acid hybridization is the establishment of base pairing between nucleic acid strands.) When the probe is mixed with DNA from a library, it will hybridize with the gene, but not with other DNA (**FIGURE 15.4**). Researchers can pinpoint a clone that hosts the gene by detecting the label on the probe. That clone is isolated and cultured, and DNA can be extracted in bulk from the cultured cells for research or other purposes.

PCR

The **polymerase chain reaction** (**PCR**) is a technique used to mass-produce copies of a particular section of DNA without having to clone it in living cells (**FIGURE 15.5**). The reaction can transform a needle in a haystack—that one-in-a-million fragment of DNA—into a huge stack of needles with a little hay in it.

The starting material for PCR is any sample of DNA with at least one molecule of a targeted sequence. It might be DNA from a mixture of 10 million different clones, a sperm, a hair left at a crime scene, or a mummy—essentially any sample that has DNA in it.

The PCR reaction is similar to DNA replication (Section 8.4). It requires two primers. Each base-pairs with one end of the section of DNA to be amplified,

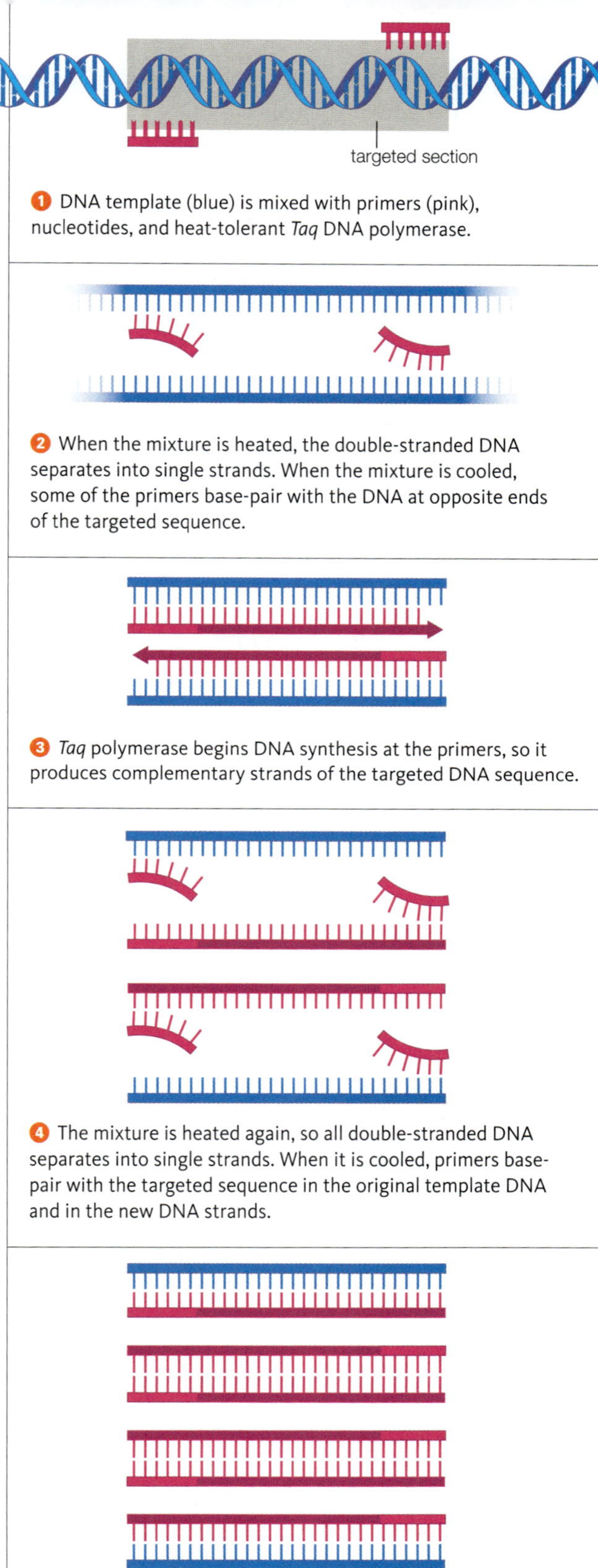

❶ DNA template (blue) is mixed with primers (pink), nucleotides, and heat-tolerant *Taq* DNA polymerase.

❷ When the mixture is heated, the double-stranded DNA separates into single strands. When the mixture is cooled, some of the primers base-pair with the DNA at opposite ends of the targeted sequence.

❸ *Taq* polymerase begins DNA synthesis at the primers, so it produces complementary strands of the targeted DNA sequence.

❹ The mixture is heated again, so all double-stranded DNA separates into single strands. When it is cooled, primers base-pair with the targeted sequence in the original template DNA and in the new DNA strands.

❺ Each cycle of heating and cooling can double the number of copies of the targeted DNA section.

FIGURE 15.5 {Animated} Two rounds of PCR. Each cycle of this reaction can double the number of copies of a targeted section of DNA. Thirty cycles can make a billion copies.

or mass-produced ❶. Researchers mix these primers with the starting (template) DNA, nucleotides, and DNA polymerase, then expose the reaction mixture to repeated cycles of high and low temperatures. A few seconds at high temperature disrupts the hydrogen bonds that hold the two strands of a DNA double helix together (Section 8.2), so every molecule of DNA unwinds and becomes single-stranded. As the temperature of the reaction mixture is lowered, the single DNA strands hybridize with the primers ❷.

The DNA polymerases of most organisms denature at the high temperature required to separate DNA strands. The kind that is used in PCR reactions, *Taq* polymerase, is from *Thermus aquaticus*. This bacterial species lives in hot springs and hydrothermal vents, so its DNA polymerase necessarily tolerates heat. *Taq* polymerase, like other DNA polymerases, recognizes hybridized primers as places to start DNA synthesis ❸. Synthesis proceeds along the template strand until the temperature rises and the DNA separates into single strands ❹. The newly synthesized DNA is a copy of the targeted section. When the mixture is cooled, the primers rehybridize, and DNA synthesis begins again. Each cycle of heating and cooling takes only a few minutes, but it can double the number of copies of the targeted section of DNA ❺. Thirty PCR cycles may amplify that number a billionfold.

DNA library Collection of cells that host different fragments of foreign DNA, often representing an organism's entire genome.
genome An organism's complete set of genetic material.
polymerase chain reaction (PCR) Method that rapidly generates many copies of a specific section of DNA.
probe Short fragment of DNA labeled with a tracer; designed to hybridize with a nucleotide sequence of interest.

TAKE-HOME MESSAGE 15.2

Researchers isolate one gene from the many other genes in a genome by making a DNA library or with PCR.

Probes may be used to identify one clone that hosts a particular DNA fragment of interest among many other clones in a DNA library.

PCR quickly mass-produces copies of a particular section of DNA.

15.3 HOW IS DNA SEQUENCE DETERMINED?

Researchers use a technique called **sequencing** to determine the order of nucleotides in a fragment of DNA that has been isolated by cloning or PCR. The most common method is similar to DNA replication (Section 8.4). The DNA to be sequenced (the template) is mixed with nucleotides, a primer, and DNA polymerase. Starting at the primer, the polymerase joins the nucleotides into a new strand of DNA, in the order dictated by the sequence of the template (**FIGURE 15.6**).

Remember that DNA polymerase can add a nucleotide only to the hydroxyl group on the 3′ carbon of a DNA strand. The sequencing reaction mixture includes four kinds of modified nucleotides that lack the hydroxyl group on their 3′ carbon ❶. Each kind (A, C, G, or T) is labeled with a different colored pigment. During the reaction, the polymerase randomly adds either a regular nucleotide or a modified nucleotide to the end of a growing DNA strand. If it adds a modified nucleotide, the 3′ carbon of the strand will not have a hydroxyl group, so synthesis of the strand ends there ❷. The reaction produces millions of DNA fragments of different lengths—incomplete, complementary copies of the starting DNA ❸. Each fragment of a given length ends with the same modified nucleotide. For example, if the tenth base in the template DNA was thymine, then any newly synthesized fragment that is 10 bases long ends with a modified adenine.

The DNA fragments are then separated by length. Using a technique called **electrophoresis**, an electric field pulls the fragments through a semisolid gel. Fragments of different sizes move through the gel at different rates. The shorter the fragment, the faster it moves, because shorter fragments slip through the tangled molecules of the gel faster than longer fragments do. All fragments of the same length move through the gel at the same speed, so they gather into bands. All fragments in a given band have the same modified nucleotide at their ends, and the pigment labels now impart distinct colors to the bands ❹. Each color designates one of the four modified nucleotides, so the order of colored bands in the gel represents the DNA sequence ❺.

FIGURE 15.6 {Animated} DNA sequencing, in which DNA polymerase is used to incompletely replicate a section of DNA.

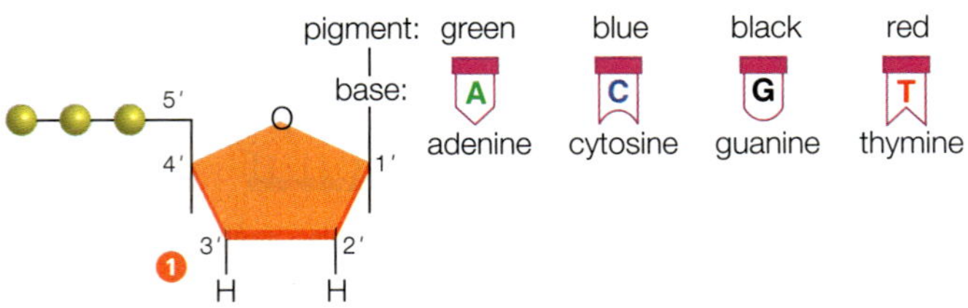

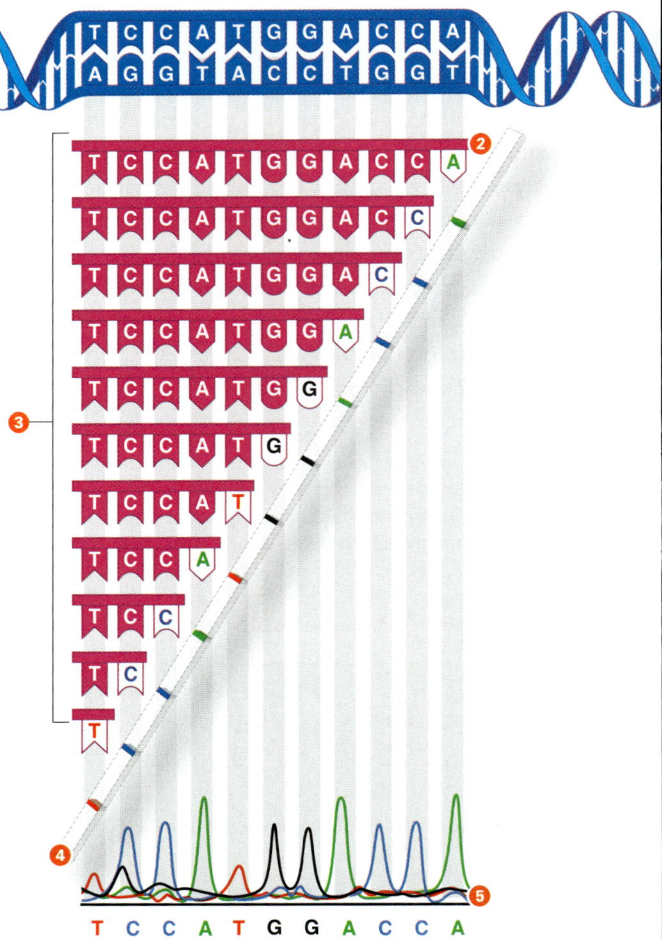

❶ Sequencing depends on modified nucleotides that have a hydrogen atom instead of a hydroxyl group on the 3′ carbon (compare the structure with those in **FIGURE 8.4**). Each is labeled with a colored pigment.

❷ DNA polymerase uses a section of DNA as a template to synthesize new strands of DNA. Synthesis of each new strand stops when a modified nucleotide is added.

❸ At the end of the reaction, there are many incomplete copies of the template DNA in the mixture.

❹ Electrophoresis separates the copied DNA fragments into bands according to their length. All of the DNA strands in each band end with the same nucleotide; thus, each band is the color of that nucleotide's tracer pigment.

❺ A computer detects and records the color of successive bands on the gel (see **FIGURE 15.7** for an example). The order of colors of the bands represents the sequence of the template DNA.

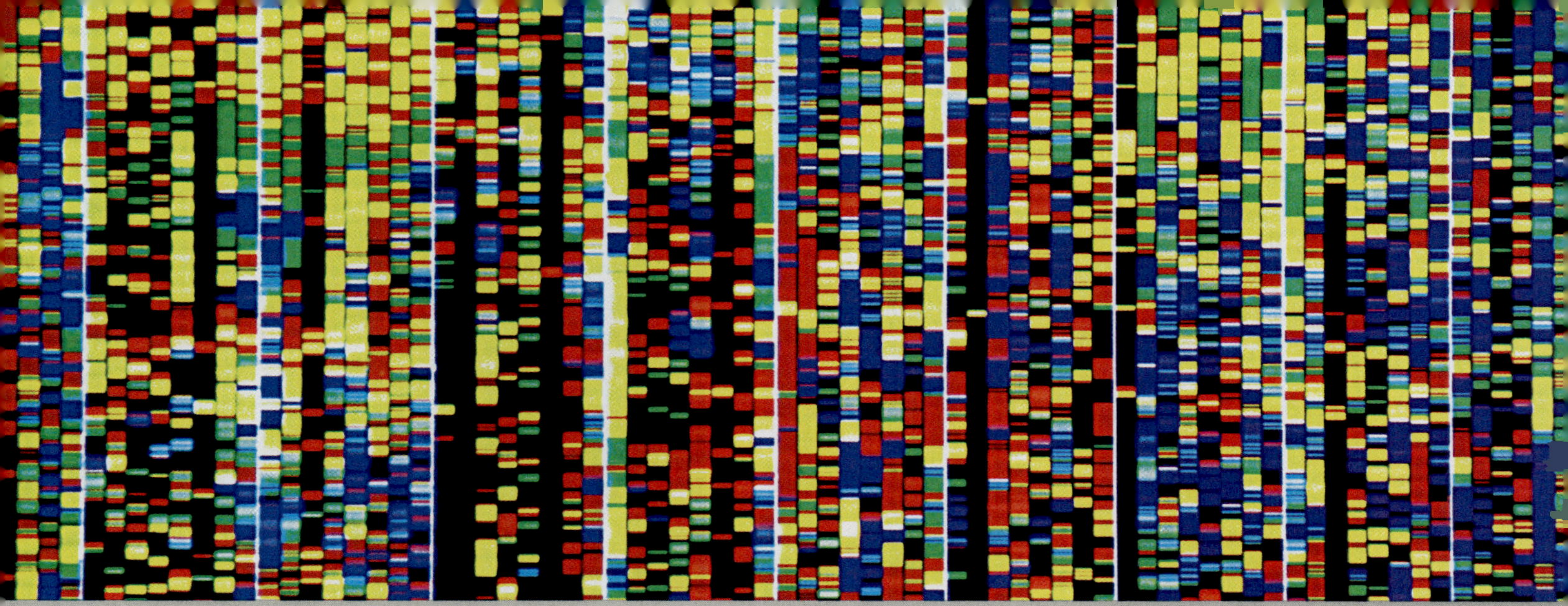

FIGURE 15.7 Human DNA sequence data. The order of colors in each vertical lane reveals one part of the DNA sequence.

THE HUMAN GENOME PROJECT

The sequencing method we have just described was invented in 1975. Ten years later, it had become so routine that scientists began to consider sequencing the entire human genome—all 3 billion nucleotides. Proponents of the idea said it could provide huge payoffs for medicine and research. Opponents said this daunting task would divert attention and funding from more urgent research. It would require 50 years to sequence the human genome given the techniques of the time. However, the techniques continued to improve rapidly, and with each improvement more nucleotides could be sequenced in less time. Automated (robotic) DNA sequencing and PCR had just been invented. Both were still too cumbersome and expensive to be useful in routine applications, but they would not be so for long. Waiting for faster technologies seemed the most efficient way to sequence the genome, but just how fast did they need to be before the project should begin?

A few privately owned companies decided not to wait, and started sequencing. One of them intended to determine the genome sequence in order to patent it. The idea of patenting the human genome provoked widespread outrage, but it also spurred commitments in the public sector. In 1988, the National Institutes of Health (NIH) essentially took over the project by hiring James Watson (of DNA structure fame) to head an official Human Genome Project, and providing $200 million per year to fund it.

electrophoresis Technique that separates DNA fragments by size.
sequencing Method of determining the order of nucleotides in DNA.

A partnership formed between the NIH and international institutions that were sequencing different parts of the genome. Watson set aside 3 percent of the funding for studies of ethical and social issues arising from the work. He later resigned over a patent disagreement, and geneticist Francis Collins took his place.

Amid ongoing squabbles over patent issues, Celera Genomics formed in 1998. With biologist Craig Venter at its helm, the company intended to commercialize human genetic information. Celera invented faster techniques for sequencing genomic DNA, because the first to have the complete sequence had a legal basis for patenting it. The competition motivated the international partnership to accelerate its efforts. Then, in 2000, U.S. President Bill Clinton and British Prime Minister Tony Blair jointly declared that the sequence of the human genome could not be patented. Celera kept sequencing anyway, and, in 2001, the competing governmental and corporate teams published about 90 percent of the sequence. In 2003, fifty years after the discovery of the structure of DNA, the sequence of the human genome was officially completed (**FIGURE 15.7**).

TAKE-HOME MESSAGE 15.3

With DNA sequencing, a strand of DNA is partially replicated. Electrophoresis is used to separate the resulting fragments by length.

Improved sequencing techniques and worldwide efforts allowed the human genome sequence to be determined.

CREDIT: (7) Patrick Landmann/Science Source.

15.4 HOW DO WE USE WHAT RESEARCHERS DISCOVER ABOUT GENOMES?

It took 15 years to sequence the human genome for the first time, but the techniques have improved so much that sequencing an entire genome now takes about a day. Anyone can now pay to have their genome sequenced (a cost of $5,000 to $8,000, at this writing).

Despite our ability to determine the sequence of an individual's genome, however, it will be a long time before we understand all the information coded within that sequence. The human genome contains a massive amount of seemingly cryptic data. One way to decipher it is by comparing it to the genomes of other species, the premise being that all organisms are descended from shared ancestors, so all genomes are related to some extent. We see evidence of such genetic relationships simply by comparing the raw sequence data, which, in some regions, is extremely similar across many species (**FIGURE 15.8**).

The study of genomes is called **genomics**, a broad field that encompasses whole-genome comparisons, structural analysis of gene products, and surveys of small-scale variations in sequence. Genomics is providing powerful insights into evolution, and it has many medical benefits. We have learned the function of many human genes by studying their counterpart genes in other species. For instance, researchers comparing human and mouse genomes discovered a human version of a mouse gene, *APOA5*, that encodes a lipoprotein (Section 3.4). Mice with an *APOA5* knockout have four times the normal level of triglycerides in their blood. The researchers then looked for—and found—a correlation between *APOA5* mutations and high triglyceride levels in humans. High triglycerides are a risk factor for coronary artery disease.

DNA PROFILING

About 99 percent of your DNA is exactly the same as everyone else's. The shared part is what makes you human; the differences make you a unique member of the species. If you compared your DNA with your neighbor's, about 2.97 billion nucleotides of the two sequences would be identical; the remaining 30 million or so nonidentical nucleotides are sprinkled throughout your chromosomes. The sprinkling is not entirely random because some regions of DNA vary less than others. Such conserved regions are of particular interest to researchers because they are the ones most likely to have an essential function. When a conserved sequence does vary among people, the variation tends to be in nucleotides at a particular location. A base-pair substitution carried by a measurable percentage of a population, usually above 1 percent, is called a **single-nucleotide polymorphism**, or **SNP** (pronounced "snip").

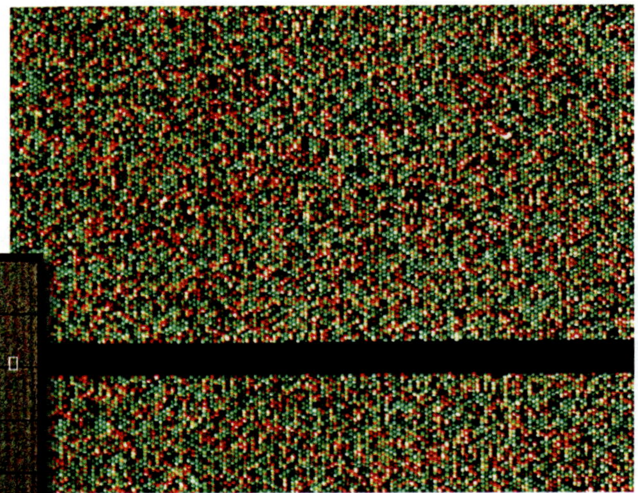

A *Left*, a SNP chip, shown actual size, that tests 550,000 SNPs. The small white box indicates the magnified portion shown in **B**.

B *Above*, each spot is a region where the individual's genomic DNA has hybridized with one SNP sequence. A red or green dot means that the individual is homozygous for a SNP; a combined signal (yellow dot) indicates heterozygosity.

FIGURE 15.9 A SNP–chip analysis.

Alleles of most genes differ by single nucleotides, and differences in alleles are the basis of the variation in human traits that makes each individual unique (Section 12.1). In fact, those differences are so unique that they can be used to identify you. Identifying an individual by his or her DNA is called **DNA profiling**.

One type of DNA profiling involves SNP-chips (**FIGURE 15.9**). A SNP-chip is a tiny glass plate with

FIGURE 15.8 Genomic DNA alignment. This is a region of the gene for a DNA polymerase. Nucleotides that differ from those in the human sequence are highlighted. The chance that any two of these sequences would randomly match is about 1 in 10^{46}.

```
758 GATAATCCTGTTTTGAACAAAAGGTCAAATTGCTGAATAGAAA-GTCTTGATTAACTAAAAGATGTACAAAGTGGAATTA 836  Human
752 GATAATCCTGTTTTGAACAAAAGGTCAAATTGCTGAATAGAAA-GTCTTGATTAACTAAAAGATGTACAAAGTGGAATTA 830  Mouse
751 GATAATCCTGTTTTGAACAAAAGGTCAAATTGCTGAATAGAAA-GTCTTGATTAACTAAAAGATGTACAAAGTGGAATTA 829  Rat
754 GATAATCCTGTTTTGAACAAAAGGTCAAATTGCTGAATAGAAA-GTCTTGATTAACTAAAAGATGTACAAAGTGGAATTA 832  Dog
782 GATAATCCTGTTTTGAACAAAAGGTCAAATTGCTGAATAGAAA-GTCTTGATTAACTAAAAGATGTACAAAGTGGAATTA 860  Chicken
758 GATAATCCTGTTTTGAACAAAAGGTCAAATTGCTGAATAGAAA-GTCTTGATTAAGTAAAAGATGTACAAAGTGGAATTA 836  Frog
823 GATAATCCTGTTTTGAACAAAAGGTCAGATTGCTGAATAGAAAAGGCTTGATTAAAGCAGAGATGTACAAAGTGGACGCA 902  Zebrafish
763 GATAATCCTGTTTTGAACAAAAGGTCAAATTGTTGAATAGAGACGCTTTGATAAAGCGGAGGAGGTACAAAGTGGGACC- 841  Pufferfish
```

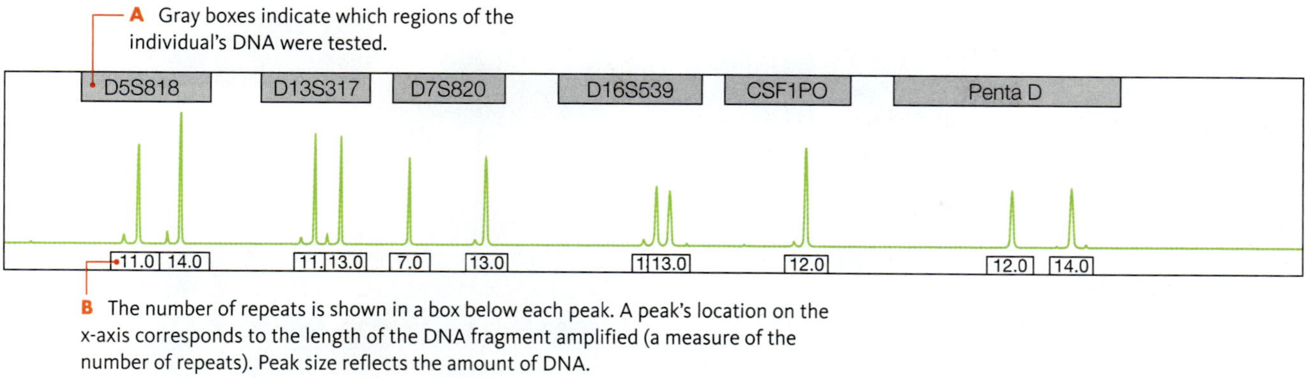

FIGURE 15.10 {Animated} An individual's (partial) short tandem repeat profile. Remember, human body cells are diploid. Double peaks appear on a profile when the two members of a chromosome pair carry a different number of repeats.

FIGURE IT OUT: How many repeats does this individual have at the Penta D region?

Answer: 12 on one chromosome; 14 on the other

microscopic spots of DNA stamped on it. The DNA sample in each spot is a short, synthetic single strand with a unique SNP sequence. When an individual's genomic DNA is washed over a SNP-chip, it hybridizes only with DNA spots that have a matching SNP sequence. Probes reveal where the genomic DNA has hybridized—and which SNPs are carried by the individual.

Another method of DNA profiling involves analysis of short tandem repeats in an individual's chromosomes (Section 13.6). Short tandem repeats tend to occur in predictable spots, but the number of times a sequence is repeated in each spot differs among individuals. For example, one person's DNA may have fifteen repeats of the nucleotides TTTTC at a certain spot on one chromosome. Another person's DNA may have this sequence repeated only twice in the same location. Such repeats slip spontaneously into DNA during replication, and their numbers grow or shrink over generations. Unless two people are identical twins, the chance that they have identical short tandem repeats in even three regions of DNA is 1 in a quintillion (10^{18}), which is far more than the number of people who have ever lived. Thus, an individual's array of short tandem repeats is, for all practical purposes, unique.

Analyzing a person's short tandem repeats begins with PCR, which is used to copy ten to thirteen particular regions of chromosomal DNA known to have repeats. The lengths of the copied DNA fragments differ among most individuals, because the number of tandem repeats in those regions also differs. Thus, electrophoresis can be used to reveal an individual's unique array of short tandem repeats (**FIGURE 15.10**).

Short tandem repeat analysis will soon be replaced by full genome sequencing, but for now it continues to be a common DNA profiling method. Geneticists compare short tandem repeats on Y chromosomes to determine relationships among male relatives, and to trace an individual's ethnic heritage. They also track mutations that accumulate in populations over time by comparing DNA profiles of living humans with those of ancient ones. Such studies are allowing us to reconstruct population dispersals that happened in the ancient past.

Short tandem repeat profiles are routinely used to resolve kinship disputes, and as evidence in criminal cases. Within the context of a criminal or forensic investigation, DNA profiling is called DNA fingerprinting. As of January 2013, the database of DNA fingerprints maintained by the Federal Bureau of Investigation (the FBI) contained the short tandem repeat profiles of 10.1 million convicted offenders, and had been used in over 190,000 criminal investigations. DNA fingerprints have also been used to identify the remains of over 470,000 people, including the individuals who died in the World Trade Center on September 11, 2001.

DNA profiling Identifying an individual by analyzing the unique parts of his or her DNA.
genomics The study of genomes.
single-nucleotide polymorphism (SNP) One-nucleotide DNA sequence variation carried by a measurable percentage of a population.

TAKE-HOME MESSAGE 15.4

Analysis of the human genome sequence is yielding new information about our genes and how they work.

DNA profiling identifies individuals by the unique parts of their DNA.

CREDIT: (10) Raw STR data courtesy of © Orchid Cellmark, www.orchidcellmark.com.

15.5 WHAT IS GENETIC ENGINEERING?

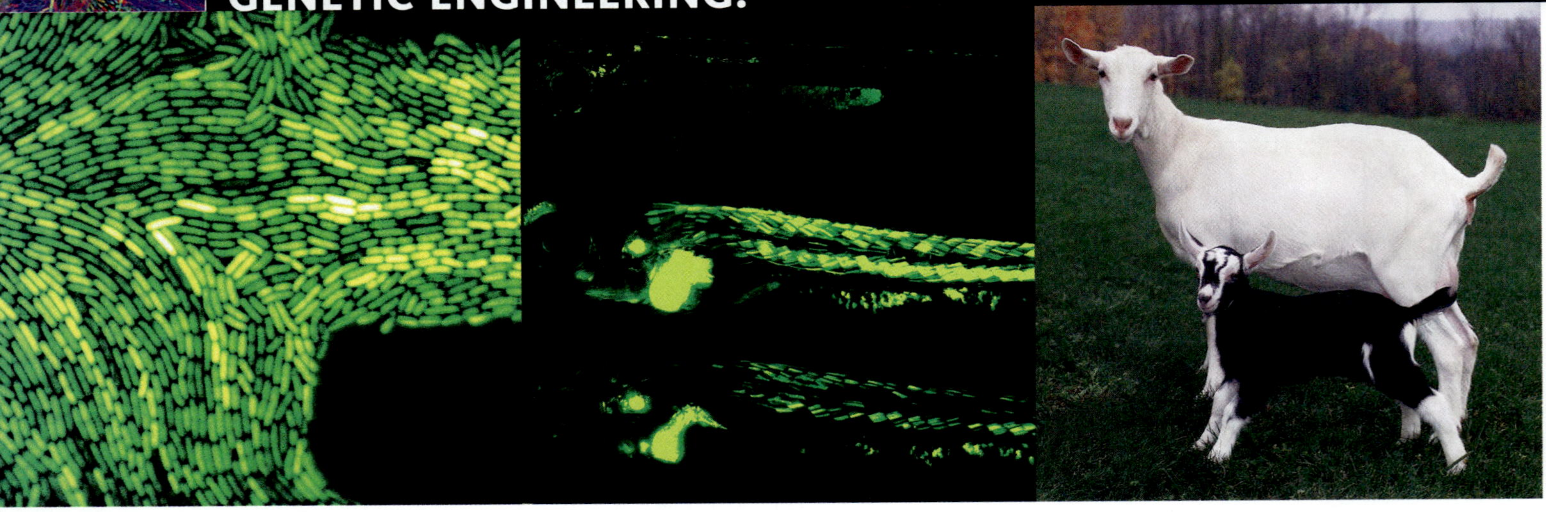

A *E. coli* bacteria transgenic for a fluorescent jellyfish protein. Variation in fluorescence among the genetically identical cells reveals differences in gene expression that may help us understand why some bacteria become dangerously resistant to antibiotics, and others do not.

B Zebrafish engineered to glow in places where BPA, an endocrine-disrupting chemical, is present. The fish are literally illuminating where this pollutant acts in the body—and helping researchers discover what it does when it gets there.

C Transgenic goats produce human antithrombin, an anti-clotting protein. Antithrombin harvested from their milk is used as a drug during surgery or childbirth to prevent blood clotting in people with hereditary antithrombin deficiency. This genetic disorder carries a high risk of life-threatening clots.

FIGURE 15.11 Examples of GMOs.

Genetic engineering is a process by which an individual's genome is deliberately modified. A gene from one species may be transferred to another to produce an organism that is **transgenic**, or a gene may be altered and reinserted into an individual of the same species. Both methods result in a **genetically modified organism**, or **GMO**.

The most common GMOs are bacteria (**FIGURE 15.11A**) and yeast. These cells have the metabolic machinery to make complex organic molecules, and they are easily modified to produce, for example, medically important proteins. People with diabetes were among the first beneficiaries of such organisms. Insulin for their injections was once extracted from animals, but it provoked an allergic reaction in some people. Human insulin, which does not provoke allergic reactions, has been produced by transgenic *E. coli* since 1982. Slight modifications of the gene have yielded fast-acting and slow-release forms of human insulin.

Genetically engineered microorganisms also produce proteins used in foods. For example, cheese is traditionally made with an extract of calf stomachs, which contain the enzyme chymotrypsin. Most cheese manufacturers now use chymotrypsin produced by genetically engineered bacteria. Other enzymes produced by GMOs improve the taste and clarity of beer and fruit juice, slow bread staling, or modify certain fats.

The first genetically modified animals were mice. Today, engineered mice are commonplace, and they are invaluable in research (an example is shown in the chapter opener). We have discovered the function of human genes (including the *APOA5* gene discussed in Section 15.4) by inactivating their counterparts in mice. Genetically modified mice are also used as models of human diseases. For example, researchers inactivated the molecules involved in the control of glucose metabolism, one by one. Studying the effects of the knockouts in mice has resulted in much of our current understanding of how diabetes works in humans.

Other genetically modified animals are useful in research (**FIGURE 15.11B**), and some make molecules that have medical and industrial applications. Various transgenic goats produce proteins used to treat cystic fibrosis, heart attacks, blood clotting disorders (**FIGURE 15.11C**), and even nerve gas exposure. Goats transgenic for a spider silk gene produce the silk protein in their milk; researchers can spin this protein into nanofibers that are useful in medical and electronics applications. Rabbits make human interleukin-2, a protein that triggers divisions of immune cells. Genetic engineering has also given us pigs with heart-healthy fat and environmentally friendly low-phosphate feces, muscle-bound trout, chickens that do not transmit bird flu, and cows that do not get mad cow disease.

As crop production expands to keep pace with human population growth, many farmers have begun to rely on genetically modified crop plants. Genes are often introduced into plant cells by way *Agrobacterium tumefaciens*. These bacteria carry a plasmid with genes that cause tumors to form on infected plants; hence the name Ti plasmid (for

Tumor-inducing). Researchers replace the tumor-inducing genes with foreign or modified genes, then use the plasmid as a vector to deliver the genes into plant cells. Whole plants can be grown from cells that integrate a recombinant plasmid into their chromosomes (**FIGURE 15.12**).

Many genetically modified crops carry genes that impart resistance to devastating plant diseases. Others offer improved yields. GMO crops such as Bt corn and soy help farmers use smaller amounts of toxic pesticides (**FIGURE 15.13**). Organic farmers often spray their crops with spores of Bt (*Bacillus thuringiensis*), a bacterial species that makes a protein toxic only to some insect larvae. Researchers transferred the gene encoding the Bt protein into plants. The engineered plants produce the Bt protein, and larvae die shortly after eating their first and only GMO meal.

Transgenic crop plants are also being developed for impoverished regions of the world. Genes that confer drought tolerance, insect resistance, and enhanced nutritional value are being introduced into plants such as corn, rice, beans, sugarcane, cassava, cowpeas, banana, and wheat. The resulting GMO crops may help people who rely on agriculture for food and income.

At this writing, ninety-two crops have been approved for unrestricted use in the United States. Worldwide, more than 330 million acres are currently planted in GMO crops, the majority of which are corn, sorghum, cotton, soy, canola, and alfalfa engineered for resistance to the herbicide glyphosate. Rather than tilling the soil to control weeds, farmers spray their fields with glyphosate, which kills the weeds but not the engineered crops.

Many people worry that our ability to tinker with genetics has surpassed our ability to understand the impact of the tinkering. Controversy raised by GMO use invites you to read the research and form your own opinions. The alternative is to be swayed by media hype (the term "Frankenfood," for instance), or by reports from potentially biased sources (such as herbicide manufacturers).

---

genetic engineering Process by which deliberate changes are introduced into an individual's genome.
genetically modified organism (GMO) Organism whose genome has been modified by genetic engineering.
transgenic Refers to a genetically modified organism that carries a gene from a different species.

TAKE-HOME MESSAGE 15.5

Genetic engineering is the deliberate alteration of an individual's genome, and it results in a genetically modified organism (GMO).

A transgenic organism carries a gene from a different species.

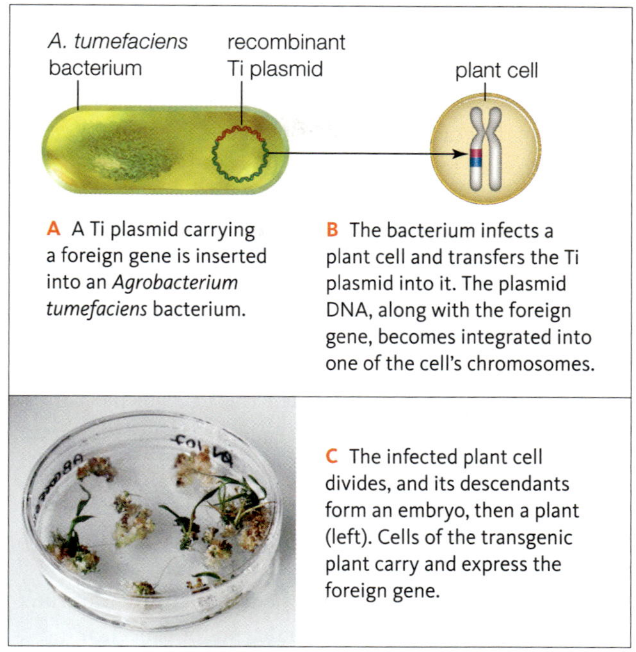

A A Ti plasmid carrying a foreign gene is inserted into an *Agrobacterium tumefaciens* bacterium.

B The bacterium infects a plant cell and transfers the Ti plasmid into it. The plasmid DNA, along with the foreign gene, becomes integrated into one of the cell's chromosomes.

C The infected plant cell divides, and its descendants form an embryo, then a plant (left). Cells of the transgenic plant carry and express the foreign gene.

FIGURE 15.12 {Animated} Using the Ti plasmid to make transgenic plants.

FIGURE 15.13 Farmers can use much less pesticide on crops that make their own. The genetically modified plants that produced the row of corn on the *top* carry a gene from the bacteria *Bacillus thuringensis* (Bt) that conferred insect resistance. Compare the corn from unmodified plants, *bottom*. No pesticides were used on either crop.

15.6 CAN WE GENETICALLY MODIFY PEOPLE?

GENE THERAPY

We know of more than 15,000 serious genetic disorders. Collectively, they cause 20 to 30 percent of infant deaths each year, and account for half of all mentally impaired patients and a fourth of all hospital admissions. They also contribute to many age-related disorders, including cancer, Parkinson's disease, and diabetes. Drugs and other treatments can minimize the symptoms of some genetic disorders, but gene therapy is the only cure. **Gene therapy** is the transfer of recombinant DNA into an individual's body cells, with the intent to correct a genetic defect or treat a disease. The transfer, which occurs by way of lipid clusters or genetically engineered viruses, inserts an unmutated gene into an individual's chromosomes.

Human gene therapy is a compelling reason to embrace genetic engineering research. It is now being tested as a treatment for AIDS, muscular dystrophy, heart attack, sickle-cell anemia, cystic fibrosis, hemophilia A, Parkinson's disease, Alzheimer's disease, several types of cancer, and inherited diseases of the eye, the ear, and the immune system. Results have been encouraging. For example, gene therapy has recently been used to treat acute lymphoblastic leukemia, a typically fatal cancer of bone marrow cells. A viral vector was used to insert a gene into immune cells extracted from patients. When the engineered cells were reintroduced into the patients' bodies, the inserted gene directed the destruction of the cancer cells. The therapy worked astonishingly well: In one patient, all traces of the leukemia vanished in eight days. However, the outcome of manipulating a gene in a living individual can be unpredictable. In an early trial, for example, twenty boys were treated with gene therapy for a severe X-linked genetic disorder called SCID-X1. Five of them developed leukemia, and one died. The researchers had wrongly predicted that cancer related to the therapy would be rare. Research now implicates the gene targeted for repair, especially when combined with the virus that delivered it. Integration of the modified viral DNA activated nearby proto-oncogenes (Section 11.5) in the children's chromosomes.

EUGENICS

The idea of selecting the most desirable human traits, **eugenics**, is an old one. It has been used as a justification for some of the most horrific episodes in human history, including the genocide of 6 million Jews during World War II. Thus, it continues to be a hotly debated social issue. For example, using gene therapy to cure human genetic disorders seems like a socially acceptable goal to most people, but imagine taking this idea a bit further. Would it also be acceptable to engineer the genome of an individual who is within a normal range of phenotype in order to modify a particular trait? Researchers have already produced mice that have improved memory, enhanced learning ability, bigger muscles, and longer lives. Why not people?

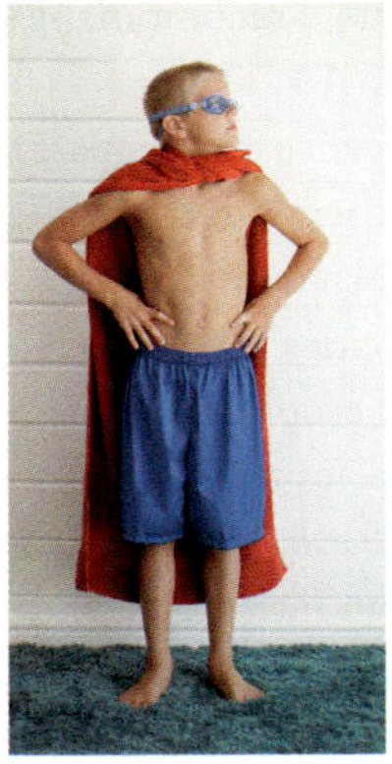

Given the pace of genetics research, the debate is no longer about how we would engineer desirable traits, but how we would choose the traits that are desirable. Realistically, cures for many severe but rare genetic disorders will not be found, because the financial return would not cover the cost of the research. Eugenics, however, may be profitable. How much would potential parents pay to be sure that their child will be tall or blue-eyed? Would it be okay to engineer "superhumans" with breathtaking strength or intelligence? How about a treatment that can help you lose that extra weight, and keep it off permanently? The gray area between interesting and abhorrent can be very different depending on who is asked. In a survey conducted in the United States, more than 40 percent of those interviewed said it would be fine to use gene therapy to make smarter and cuter babies. In one poll of British parents, 18 percent would be willing to use it to keep a child from being aggressive, and 10 percent would use it to keep a child from growing up to be homosexual.

Some people are concerned that gene therapy puts us on a slippery slope that may result in irreversible damage to ourselves and to the biosphere. We as a society may not have the wisdom to know how to stop once we set foot on that slope; one is reminded of our peculiar human tendency to leap before we look. And yet, something about the human experience allows us to dream of such things as wings of our own making, a capacity that carried us into space. In this brave new world, the questions before you are these: What do we stand to lose if serious risks are not taken? And, do we have the right to impose the potential consequences on people who would choose not to take those risks?

eugenics Idea of deliberately improving the genetic qualities of the human race.

gene therapy Treating a genetic defect or disorder by transferring a normal or modified gene into the affected individual.

TAKE-HOME MESSAGE 15.6

Genes can be transferred into a person's cells to correct a genetic defect or treat a disease. However, the outcome of altering a person's genome has been unpredictable.

We as a society must continue to work our way through the ethical implications of applying DNA technologies.

15.7 Application: PERSONAL GENETIC TESTING

Actress Angelina Jolie discovered via genetic testing that she carries a *BRCA1* mutation associated with an 87% lifetime risk of developing breast cancer. Even though she did not yet have cancer, Jolie underwent a double mastectomy. By doing so, she reduced her risk of breast cancer to 5%.

Education

FIGURE 15.14 Celebrity Angelina Jolie chose preventive treatment after genetic testing showed she had a very high risk of breast cancer.

DO YOU WANT TO KNOW YOUR SNP PROFILE? Finding out which of about 1 million SNPs you carry has never been easier. Genetic testing companies can extract your DNA from a few drops of spit, then analyze it using a SNP-chip. Results typically include estimated risks of developing conditions associated with your particular set of SNPs. For example, the test will probably determine whether you are homozygous for one allele of the *MC1R* gene (Section 13.4). If you are, the company's report will tell you that you have red hair. Few SNPs have such a clear effect, however. Consider the lipoprotein particles that carry fats and cholesterol through our bloodstreams. These particles consist of variable amounts and types of lipids and proteins, one of which is specified by the gene *APOE*. About one in four people carries an allele of this gene, *ε4*, that increases one's risk of developing Alzheimer's disease later in life. If you are heterozygous for this allele, a DNA testing company will report that your lifetime risk of developing the disease is about 29 percent, as compared with about 9 percent for someone who has no *ε4* allele.

What, exactly, does a 29 percent lifetime risk of Alzheimer's disease mean? The number is a probability statistic; it means, on average, 29 of every 100 people who have the *ε4* allele eventually get the disease. However, a risk is just that. Not everyone who has the *ε4* allele develops Alzheimer's, and not everyone who develops the disease has the allele. Other unknown factors, including epigenetic modifications of DNA, contribute to the disease. We still have a limited understanding of how genes contribute to many health conditions, particularly age-related ones such as Alzheimer's disease

Geneticists believe that it will be at least five to ten more years before genotyping can be used to accurately predict an individual's future health problems. Nonetheless, we are at a tipping point; personalized genetic testing is already beginning to revolutionize medicine. Cancer treatments are now being tailored to fit the genetic makeup of individual patients. People who discover they carry alleles associated with a heightened risk of a medical condition are being encouraged to make lifestyle changes that could delay the condition's onset or prevent it entirely. Preventive treatments based on personal genetics are becoming more common—and more mainstream (FIGURE 15.14).

Summary

SECTION 15.1 In **DNA cloning**, researchers use **restriction enzymes** to cut a sample of DNA into pieces, and then use DNA ligase to splice the fragments into plasmids or other **cloning vectors**. The resulting molecules of **recombinant DNA** are inserted into host cells such as bacteria. Division of host cells produces huge populations of genetically identical descendant cells (clones), each with a copy of the cloned DNA fragment.

The enzyme **reverse transcriptase** is used to transcribe RNA into **cDNA** for cloning.

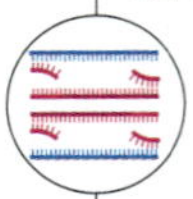

SECTION 15.2 A **DNA library** is a collection of cells that host different fragments of DNA, often representing an organism's entire **genome**. Researchers can use **probes** to identify cells in a library that carry a specific fragment of DNA. The **polymerase chain reaction** (**PCR**) uses primers and a heat-resistant DNA polymerase to rapidly increase the number of copies of a targeted section of DNA.

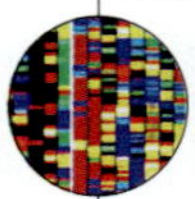

SECTION 15.3 Advances in **sequencing**, which reveals the order of nucleotides in DNA, allowed the DNA sequence of the entire human genome to be determined. DNA polymerase is used to partially replicate a DNA template. The reaction produces a mixture of DNA fragments of all different lengths; **electrophoresis** separates the fragments by length into bands.

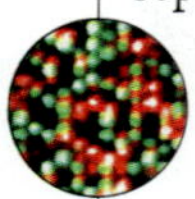

SECTION 15.4 **Genomics** provides insights into the function of the human genome. Similarities between genomes of different organisms are evidence of evolutionary relationships, and can be used as a predictive tool in research.

DNA profiling identifies a person by the unique parts of his or her DNA. An example is the determination of an individual's array of short tandem repeats or **single-nucleotide polymorphisms** (**SNPs**). Within the context of a criminal investigation, a DNA profile is called a DNA fingerprint.

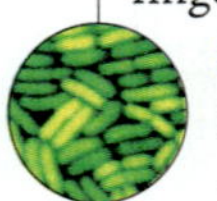

SECTION 15.5 Recombinant DNA technology is the basis of **genetic engineering**, the directed modification of an organism's genetic makeup with the intent to modify its phenotype. A gene from one species is inserted into an individual of a different species to make a **transgenic** organism, or a gene is modified and reinserted into an individual of the same species. The result of either process is a **genetically modified organism** (**GMO**).

Bacteria and yeast, the most common genetically engineered organisms, produce proteins that have medical value. The majority of the animals that are being created by genetic engineering are used for medical applications or research. Most transgenic crop plants, which are now in widespread use worldwide, were created to help farmers produce food more efficiently. Some have enhanced nutritional value.

SECTION 15.6 With **gene therapy**, a gene is transferred into body cells to correct a genetic defect or treat a disease. Potential benefits of genetically modifying humans must be weighed against potential risks. The practice raises ethical issues such as whether **eugenics** is desirable in some circumstances.

SECTION 15.7 Personal genetic testing, which reveals a person's unique array of SNPs, is beginning to revolutionize the way medicine is practiced.

Self-Quiz Answers in Appendix VII

1. _______ cut(s) DNA molecules at specific sites.
   a. DNA polymerase c. Restriction enzymes
   b. DNA probes d. DNA ligase

2. A _______ is a molecule that can be used to carry a fragment of DNA into a host organism.
   a. cloning vector c. GMO
   b. chromosome d. cDNA

3. Reverse transcriptase assembles a(n) _______ on a(n) _______ template.
   a. mRNA; DNA c. DNA; ribosome
   b. cDNA; mRNA d. protein; mRNA

4. For each species, all _______ in the complete set of chromosomes is/are the _______ .
   a. genomes; phenotype c. mRNA; start of cDNA
   b. DNA; genome d. cDNA; start of mRNA

5. A set of cells that host various DNA fragments collectively representing an organism's entire set of genetic information is a _______ .
   a. genome c. genomic library
   b. clone d. GMO

6. _______ is a technique to determine the order of nucleotide bases in a fragment of DNA.
   a. PCR c. Electrophoresis
   b. Sequencing d. Nucleic acid hybridization

7. Fragments of DNA can be separated by electrophoresis according to _______ .
   a. sequence b. length c. species

8. PCR can be used _______ .
   a. to increase the number of specific DNA fragments
   b. in DNA fingerprinting
   c. to modify a human genome
   d. a and b are correct

9. An individual's set of unique _______ can be used as a DNA profile.
   a. DNA sequences c. SNPs
   b. short tandem repeats d. all of the above

10. True or false? Some humans are genetically modified.

Data Analysis Activities

Enhanced Spatial Learning Ability in Mice with an Autism Mutation Autism is a neurobiological disorder with symptoms that include impaired social interactions and stereotyped patterns of behavior. Around 10 percent of autistic people have an extraordinary skill or talent such as greatly enhanced memory.

Mutations in neuroligin 3, an adhesion protein that connects brain cells to one another, have been associated with autism. One mutation changes amino acid 451 from arginine to cysteine. In 2007, Katsuhiko Tabuchi and his colleagues genetically modified mice to carry the same arginine-to-cysteine substitution in their neuroligin 3. Mice with the mutation had impaired social behavior. Spatial learning ability was tested in a water maze, in which a platform is submerged a few millimeters below the surface of a deep pool of warm water. The platform is not visible to swimming mice. Mice do not particularly enjoy swimming, so they locate a hidden platform as fast as they can. When tested again, they can remember its location by checking visual cues around the edge of the pool. How quickly they remember the platform's location is a measure of spatial learning ability (**FIGURE 15.15**).

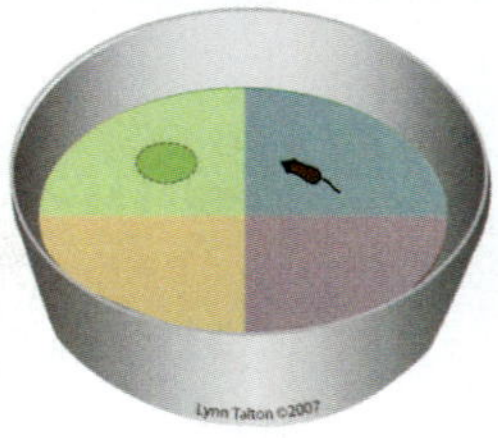

a water maze

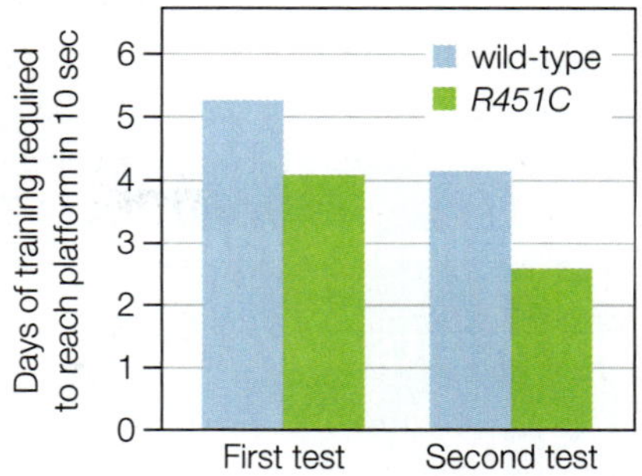

FIGURE 15.15 Spatial learning ability in mice with a mutation in neuroligin 3 (*R451C*), compared with unmodified (wild-type) mice.

1. In the first test, how many days did unmodified mice need to learn to find the location of a hidden platform within 10 seconds?
2. Did the modified or the unmodified mice learn the location of the platform faster in the first test?
3. Which mice learned faster the second time around?
4. Which mice showed the greatest improvement in memory between the first and the second test?

11. Which of the following can be used to carry foreign DNA into host cells? Choose all correct answers.
    a. RNA d. lipid clusters
    b. viruses e. bacteria
    c. PCR f. plasmids

12. Transgenic _______ can pass a foreign gene to offspring.
    a. plants c. bacteria
    b. animals d. all of the above

13. _______ can correct a genetic defect in an individual.
    a. Cloning vectors c. Eugenics
    b. Gene therapy d. a and b

14. Match the recombinant DNA method with the appropriate enzyme.
    ___ PCR a. *Taq* polymerase
    ___ cutting DNA b. DNA ligase
    ___ cDNA synthesis c. reverse transcriptase
    ___ DNA sequencing d. restriction enzyme
    ___ pasting DNA e. DNA polymerase (not *Taq*)

15. Match each term with the most suitable description.
    ___DNA profile a. GMO with a foreign gene
    ___Ti plasmid b. alleles commonly contain them
    ___eugenics c. a person's unique collection of short tandem repeats
    ___SNP d. selecting "desirable" traits
    ___transgenic e. genetically modified
    ___GMO f. used in plant gene transfers

Critical Thinking

1. The results of a paternity test using short tandem repeats are shown in the table below. Who's the daddy? How sure are you?

	Mother	Baby	Alleged Father #1	Alleged Father #2
D3S1358	15, 17	17, 23	23, 27	17, 15
TH01	9, 9	9, 9	9, 12	12, 12
D21S11	29, 29	29, 27	27, 28	29, 28
D18S51	14, 18	18, 20	15, 20	17, 22
Penta E	14, 14	14, 14	14, 14	15, 16
D5D818	11, 14	14, 16	12, 16	14, 20
D13S317	11, 13	10, 13	8, 10	18, 18
D7S820	7, 13	13, 13	13, 19	13, 13
D16S539	13, 13	13, 15	12, 15	10, 12
CSF1PO	12, 12	10, 12	8, 10	12, 17
Penta D	12, 14	5, 12	14, 14	18, 25
amelogenin	X, X	X,Y	X,Y	X,Y
vWa	15, 17	17, 22	15, 22	22, 22
D8S1179	13, 13	8, 13	8, 13	15, 15
TPOX	11, 11	11, 11	10, 11	17, 22
FGA	23, 23	23, 25	18, 25	23, 23

High in the Andes, a scientist infers the stride of an extinct dinosaur by measuring the distance between its fossilized footprints. We often reconstruct history by studying physical evidence of events that took place long ago. This practice relies on a foundational premise of science: Natural phenomena that occurred in the past can be explained by the same physical, chemical, and biological processes operating today.

16 EVIDENCE OF EVOLUTION

Links to Earlier Concepts

You may wish to review critical thinking (Section 1.5) before reading this chapter, which explores a clash between belief and science (1.7). What you know about alleles (12.1) will help you understand natural selection. The chapter revisits radioisotopes (2.1), the effect of photosynthesis on Earth's early atmosphere (7.2), the genetic code and mutations (9.3, 9.5), master genes (10.2, 10.3), and evolution by gene duplication (14.4).

KEY CONCEPTS

EMERGENCE OF EVOLUTIONARY THOUGHT

Nineteenth-century naturalists investigating the global distribution of species discovered patterns that could not be explained within the framework of traditional belief systems.

A THEORY TAKES FORM

Evidence of evolution, or change in lines of descent, led Charles Darwin and Alfred Wallace to develop a theory of how traits that define each species change over time.

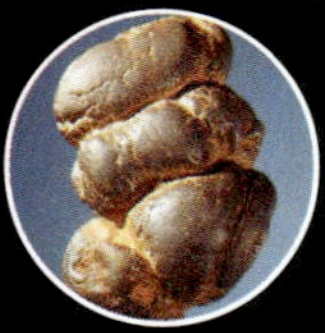

EVIDENCE FROM FOSSILS

The fossil record provides physical evidence of past changes in many lines of descent. We use the property of radioisotope decay to determine the age of rocks and fossils.

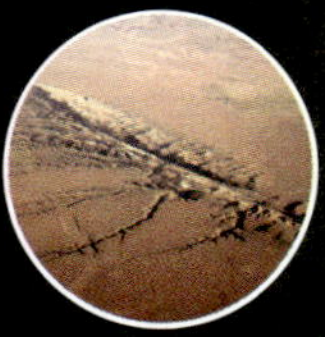

INFLUENTIAL GEOLOGIC FORCES

By studying rock layers and fossils in them, we can correlate geologic events with evolutionary events. Such correlations help explain the distribution of species, past and present.

EVIDENCE IN BODY FORM

Comparisons of genes, developmental patterns, and body form provide information about how organisms are related to one another. Lineages with recent common ancestry are most similar.

16.1 HOW DID OBSERVATIONS OF NATURE CHANGE OUR THINKING IN THE 19TH CENTURY?

A Emu, native to Australia. B Rhea, native to South America. C Ostrich, native to Africa.

FIGURE 16.1 Similar-looking, related species native to distant geographic realms. These birds are unlike most others in several unusual features, including long, muscular legs and an inability to fly. All are native to open grassland regions about the same distance from the equator.

People have long been curious about the natural world and our place in it. About 2,300 years ago, the Greek philosopher Aristotle described nature as a continuum of organization, from lifeless matter through complex plants and animals. Aristotle's work greatly influenced later European thinkers, who adopted his view of nature and modified it in light of their own beliefs. By the fourteenth century, Europeans generally believed that a "great chain of being" extended from the lowest form (snakes), up through humans, to spiritual beings. Each link in the chain was a species, and each was said to have been forged at the same time, in one place and in a perfect state. The chain itself was complete and continuous. Because everything that needed to exist already did, there was no room for change.

European naturalists who embarked on globe-spanning survey expeditions brought back tens of thousands of plants and animals from Asia, Africa, North and South America, and the Pacific Islands. Each newly discovered species was carefully catalogued as another link in the chain of being. By the late 1800s, naturalists were seeing patterns in where species live and similarities in body plans, and had started to think about the natural forces that shape life. These naturalists were pioneers in **biogeography**, the study of patterns in the geographic distribution of species and communities. Some of the patterns raised questions that could not be answered within the framework of prevailing belief systems. For example, globe-trotting explorers had discovered plants and animals living in extremely isolated places. The isolated species looked suspiciously similar to species living across vast expanses of open ocean, or on the other side of impassable mountain ranges. The three birds in **FIGURE 16.1** live on different continents, but they share a set of unusual features. These flightless birds sprint about on long, muscular legs in flat, open grasslands about the same distance from the equator. All raise their long necks to watch for predators. Alfred Wallace, an explorer who was particularly interested in the geographical distribution of animals, thought that the shared set of unusual traits might mean that these three birds descended from a common ancestor (and he was right), but he had no idea how they could have ended up on different continents.

Naturalists of the time also had trouble classifying organisms that are very similar in some features, but different in others. For example, the plants in **FIGURE 16.2** are native to different continents. Both live in hot

FIGURE 16.2 Similar-looking, unrelated species. On the *left*, an African milk barrel cactus (*Euphorbia horrida*), native to the Great Karoo desert of South Africa. On the *right*, saguaro cactus (*Carnegiea gigantea*), native to the Sonoran Desert of Arizona.

deserts where water is seasonally scarce. Both have rows of sharp spines that deter herbivores, and both store water in their thick, fleshy stems. However, their reproductive parts are very different, so these plants cannot be as closely related as their outward appearance might suggest.

Observations such as these are examples of **comparative morphology**, the study of anatomical patterns—similarities and differences among the body plans of organisms. Today, comparative morphology is part of taxonomy (Section 1.4), but in the nineteenth century it was the only way to distinguish species. In some cases, comparative morphology revealed anatomical details—body parts with no apparent function, for example—that added to the mounting confusion. If every species had been created in a perfect state, then why were there useless parts such as wings in birds that do not fly, eyes in moles that are blind, or remnants of a tail in humans (**FIGURE 16.3**)?

Fossils were puzzling too. A **fossil** is physical evidence—remains or traces—of an organism that lived in the ancient past. Geologists mapping rock formations exposed by erosion or quarrying had discovered identical sequences of rock layers in different parts of the world. Deeper layers held fossils of simple marine life. Layers above those held similar but more complex fossils (**FIGURE 16.4**). In higher layers, fossils that were similar but even more complex resembled modern species. What did these sequences mean?

Fossils of many animals unlike any living ones were also being unearthed. If these animals had been perfect at the time of creation, then why had they become extinct?

Taken as a whole, the accumulating findings from biogeography, comparative morphology, and geology did not fit with prevailing beliefs of the nineteenth century. If species had not been created in a perfect state (and extinct species, fossil sequences, and "useless" body parts implied that they had not), then perhaps species had indeed changed over time.

---

biogeography Study of patterns in the geographic distribution of species and communities.
comparative morphology The scientific study of similarities and differences in body plans.
fossil Physical evidence of an organism that lived in the ancient past.

TAKE-HOME MESSAGE 16.1

Increasingly extensive observations of nature in the nineteenth century did not fit with prevailing belief systems.

Cumulative findings from biogeography, comparative morphology, and geology led naturalists to question traditional ways of interpreting the natural world.

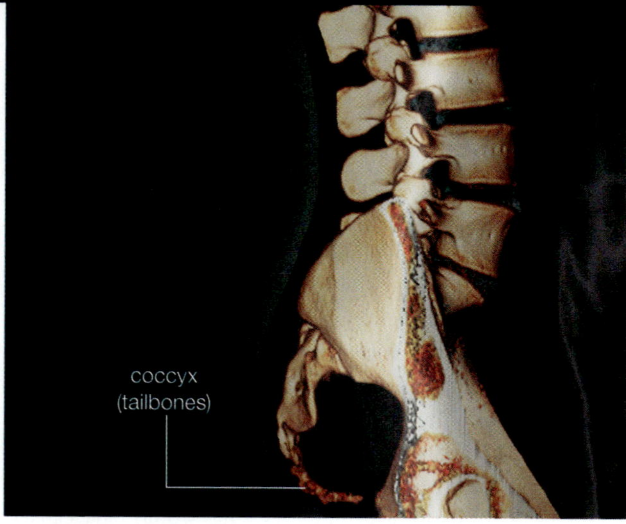

FIGURE 16.3 A vestigial structure: human tailbones. Nineteenth-century naturalists were well aware of—but had trouble explaining—body structures such as human tailbones that had apparently lost most or all function.

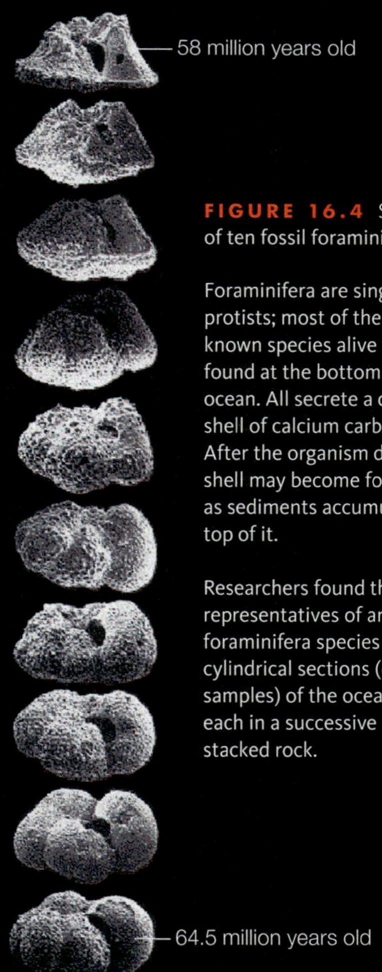

FIGURE 16.4 Sequence of ten fossil foraminifera.

Foraminifera are single-celled protists; most of the 4,000 known species alive today are found at the bottom of the ocean. All secrete a durable shell of calcium carbonate. After the organism dies, the shell may become fossilized as sediments accumulate on top of it.

Researchers found these representatives of ancient foraminifera species in cylindrical sections (core samples) of the ocean floor, each in a successive layer of stacked rock.

16.2 WHAT IS NATURAL SELECTION?

Around 1800, naturalists were trying to explain the mounting evidence that life on Earth, and even Earth itself, had changed over time. Georges Cuvier (*left*), an expert in zoology and paleontology, proposed an idea startling for the time: Many species that had once existed were now extinct. Cuvier knew about evidence that Earth's surface had changed. For example, he had seen fossilized seashells on mountainsides far from modern seas. Like most others of his time, he assumed Earth's age to be in the thousands, not billions, of years. He reasoned that geologic forces unlike any known at the time would have been necessary to raise seafloors to mountaintops in this short time span. Catastrophic geological events would have caused extinctions, after which surviving species repopulated Earth.

Jean-Baptiste Lamarck (*left*) was thinking about processes that drive **evolution**, or change in a line of descent. A line of descent is also called a **lineage**. Lamarck thought that a species gradually improved over generations because of an inherent drive toward perfection, up the chain of being. By Lamarck's hypothesis, environmental pressures cause an internal need for change in an individual's body, and the resulting change is inherited by offspring. (Lamarck was correct in thinking that environmental factors affect traits, but his understanding of how traits are passed to offspring was incomplete.)

Charles Darwin (*left*) had earned a theology degree from Cambridge after an attempt to study medicine. All through school, however, he had spent most of his time with faculty members and other students who embraced natural history. In 1831, when he was 22, Darwin joined a 5-year survey expedition to South America on the ship *Beagle*, and he quickly became an enthusiastic naturalist. During the *Beagle*'s voyage, Darwin found many unusual fossils, and saw diverse species living in environments that ranged from the sandy shores of remote islands to plains high in the Andes. Along the way, he read the first volume of a new and popular book, Charles Lyell's *Principles of Geology*. Lyell (*left*) was a proponent of what became known as the theory of uniformity, the idea that gradual, everyday geological processes such as erosion could have sculpted Earth's current landscape over great spans of time. The theory challenged the prevailing belief that Earth was 6,000 years old. By Lyell's calculations, it must have taken millions of years to sculpt Earth's surface. Darwin's exposure to Lyell's ideas gave him insights into the history of the regions he would encounter on his journey.

A Fossil of a glyptodon, an automobile-sized mammal that existed from 2 million to 15,000 years ago.

B A modern armadillo, about a foot long.

FIGURE 16.5 Ancient relatives: glyptodon and armadillo. Though widely separated in time, these animals share a restricted distribution and unusual traits, including a shell and helmet of keratin-covered bony plates—a material similar to crocodile and lizard skin. (The fossil in **A** is missing its helmet.)

Among the thousands of specimens Darwin collected during the *Beagle*'s voyage were fossil glyptodons. These armored mammals are extinct, but they have many unusual traits in common with modern armadillos (**FIGURE 16.5**). Armadillos also live only in places where glyptodons once lived. Could the odd shared traits and restricted distribution mean that glyptodons were ancient relatives of armadillos? If so, perhaps traits of their common ancestor had changed in the line of descent that led to armadillos. But why would such changes occur?

Economist Thomas Malthus (*left*) had correlated increases in the size of human populations with episodes of famine, disease, and war. He proposed the idea that humans run out of food, living space, and other resources because they tend to reproduce beyond the capacity of their environment to sustain them. When that happens, the individuals of a population must either compete with one another for the limited resources, or develop technology to increase productivity. Darwin realized that Malthus's ideas had wider application: All populations, not just human ones, must have the capacity to produce more individuals than their environment can support.

TABLE 16.1

Principles of Natural Selection, in Modern Terms

Observations About Populations

- Natural populations have an inherent capacity to increase in size over time.
- As population size increases, resources that are used by its individuals (such as food and living space) eventually become limited.
- When resources are limited, individuals of a population compete for them.

Observations About Genetics

- Individuals of a species share certain traits.
- Individuals of a natural population vary in the details of those shared traits.
- Shared traits have a heritable basis, in genes. Slightly different forms of those genes (alleles) give rise to variation in shared traits.

Inferences

- A certain form of a shared trait may make its bearer better able to survive.
- Individuals of a population that are better able to survive tend to leave more offspring.
- Thus, an allele associated with an adaptive trait tends to become more common in a population over time.

Darwin started thinking about how individuals of a species often vary a bit in the details of shared traits such as size, coloration, and so on. He saw such variation among finch species on isolated islands of the Galápagos archipelago. This island chain is separated from South America by 900 kilometers (550 miles) of open ocean, so most species living on the islands did not have the opportunity for interbreeding with mainland populations. The Galápagos island finches resembled finch species in South America, but many of them had unique traits that suited them to their particular island habitat.

Darwin was familiar with dramatic variations in traits of pigeons, dogs, and horses produced through selective breeding. He recognized that a natural environment could similarly select traits that make individuals of a population suited to it. It dawned on Darwin that having a particular form of a shared trait might give an individual an advantage over competing members of its species. In any population, some individuals have forms of shared traits that make them better suited to their environment than others. In other words, individuals of a natural population vary in fitness. Today, we define **fitness** as the degree of adaptation to a specific environment, and measure it by relative genetic contribution to future generations. A trait that enhances an individual's fitness is called an evolutionary **adaptation**, or **adaptive trait**.

Over many generations, individuals with the most adaptive traits tend to survive longer and reproduce more than their less fit rivals. Darwin understood that this process, which he called **natural selection**, could be a mechanism by which evolution occurs. If an individual has a form of a trait that makes it better suited to an environment, then it is better able to survive. If an individual is better able to survive, then it has a better chance of living long enough to produce offspring. If individuals with an adaptive, heritable trait produce more offspring than those that do not, then the frequency of that trait will tend to increase in the population over successive generations. **TABLE 16.1** summarizes this reasoning.

Darwin wrote out his ideas about natural selection, but let ten years pass without publishing them. In the meantime, Alfred Wallace (*left*), who had been studying wildlife in the Amazon basin and the Malay Archipelago, sent an essay to Darwin for advice. Wallace's essay outlined evolution by natural selection—the very same hypothesis as Darwin's. Wallace had written earlier letters to Darwin and Lyell about patterns in the geographic distribution of species, and had come to the same conclusion. In 1858, the idea of evolution by natural selection was presented at a scientific meeting, with Darwin and Wallace credited as authors. Wallace was in the field and knew nothing about the meeting, which Darwin did not attend. The next year, Darwin published *On the Origin of Species*, which laid out detailed evidence in support of natural selection. Many people had already accepted the idea of descent with modification (evolution). However, there was a fierce debate over the idea that evolution occurs by natural selection. Decades would pass before experimental evidence from the field of genetics led to its widespread acceptance as a theory by the scientific community.

As you will see in the remainder of this chapter, the theory of evolution by natural selection is supported by and helps explain the fossil record as well as similarities in the form, function, and biochemistry of living things.

adaptation (adaptive trait) A heritable trait that enhances an individual's fitness in a particular environment.
evolution Change in a line of descent.
fitness Degree of adaptation to an environment, as measured by an individual's relative genetic contribution to future generations.
lineage Line of descent.
natural selection Differential survival and reproduction of individuals of a population based on differences in shared, heritable traits. Driven by environmental pressures.

TAKE-HOME MESSAGE 16.2

Evidence that Earth and the species on it had changed over very long spans of time led to the theory of evolution by natural selection.

Natural selection is a process in which individuals of a population survive and reproduce with differing success depending on the details of their shared, heritable traits.

Traits favored in a particular environment are adaptive.

16.3 WHY DO BIOLOGISTS STUDY ROCKS AND FOSSILS?

A Fossil skeleton of an ichthyosaur that lived about 200 million years ago. These marine reptiles were about the same size as modern porpoises, breathed air like them, and probably swam as fast, but the two groups are not closely related.

B Extinct wasp encased in amber, which is ancient tree sap. This 9-mm-long insect lived about 20 million years ago.

C Fossilized imprint of a leaf from a 260-million-year-old *Glossopteris*, a type of plant called a seed fern.

D Fossilized footprint of a theropod, a name that means "beast foot." This group of carnivorous dinosaurs, which includes the familiar *Tyrannosaurus rex*, arose about 250 million years ago.

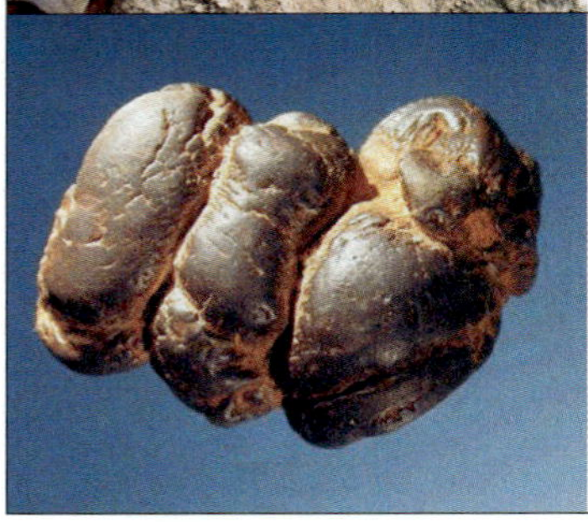

E Coprolite (fossilized feces). Fossilized food remains and parasitic worms inside coprolites offer clues about the diet and health of extinct species. A foxlike animal excreted this one.

FIGURE 16.6 Examples of fossils.

Even before Darwin's time, fossils were recognized as stone-hard evidence of earlier forms of life (**FIGURE 16.6**). Most fossils consist of mineralized bones, teeth, shells, seeds, spores, or other hard body parts. Trace fossils such as footprints and other impressions, nests, burrows, trails, eggshells, or feces are evidence of an organism's activities.

The process of fossilization typically begins when an organism or its traces become covered by sediments, mud, or ash. Groundwater then seeps into the remains, filling spaces around and inside of them. Minerals dissolved in the water gradually replace minerals in bones and other hard tissues. Mineral particles that crystallize and settle out of the groundwater inside cavities and impressions form detailed imprints of internal and external structures. Sediments that slowly accumulate on top of the site exert increasing pressure, and, after a very long time, extreme pressure transforms the mineralized remains into rock.

Most fossils are found in layers of sedimentary rock that forms as rivers wash silt, sand, volcanic ash, and other materials from land to sea. Mineral particles in the materials settle on the seafloor in horizontal layers that vary in thickness and composition. After many millions of years, the layers of sediments become compacted into layered sedimentary rock. Even though most sedimentary rock forms at the bottom of a sea, geologic processes can tilt the rock and lift it far above sea level, where the layers may become exposed by the erosive forces of water and wind (the chapter opening photo shows an example).

Biologists study sedimentary rock formations in order to understand life's historical context. Features of the formations can provide information about conditions in the environment in which they formed. Consider banded iron, a unique formation named after its distinctive striped appearance (*left*). Huge deposits of this sedimentary rock are the source of most iron we mine for steel today, but they also hold a record of how the evolution of the noncyclic pathway of photosynthesis changed the chemistry of Earth. Banded iron started forming about 2.4 billion years ago, right after photosynthesis evolved (Section 7.2). At that time, Earth's atmosphere and ocean contained very little oxygen, so almost all of the iron on Earth was in a reduced form (Section 5.4). Reduced iron dissolves in water, and ocean water contained a lot of it. Oxygen released into the ocean by early photosynthetic bacteria quickly combined with the dissolved iron. The resulting oxidized iron compounds are completely insoluble in water, and they began to rain down on the ocean floor in massive quantities. These compounds

accumulated in sediments that would eventually become compacted into banded iron formations.

The massive sedimentation of oxidized iron continued for about 600 million years. After that, ocean water no longer contained very much dissolved iron, and oxygen gas bubbling out of it had oxidized the iron in rocks exposed to the atmosphere.

THE FOSSIL RECORD

We have fossils for more than 250,000 known species. Considering the current range of biodiversity, there must have been many millions more, but we will never know all of them. Why not? The odds are against finding evidence of an extinct species, because fossils are relatively rare. When an organism dies, its remains are often obliterated quickly by scavengers. Organic materials decompose in the presence of moisture and oxygen, so remains that escape scavenging can endure only if they dry out, freeze, or become encased in an air-excluding material such as sap, tar, or mud. Remains that do become fossilized are often deformed, crushed, or scattered by erosion and other geologic assaults.

In order for us to know about an extinct species that existed long ago, we have to find a fossil of it. At least one specimen had to be buried before it decomposed or something ate it. The burial site had to escape destructive geologic events, and it had to be accessible for us to find.

Most ancient species had no hard parts to fossilize, so we do not find much evidence of them. For example, there are many fossils of bony fishes and mollusks with hard shells, but few fossils of the jellyfishes and soft worms that were probably much more common. Also think about relative numbers of organisms. Fungal spores and pollen grains are typically released by the billions. By contrast, the earliest humans lived in small bands and few of their offspring survived. The odds of finding even one fossilized human bone are much smaller than the odds of finding a fossilized fungal spore. Finally, imagine two species, one that existed only briefly and the other for billions of years. Which is more likely to be represented in the fossil record? Despite these challenges, the fossil record is substantial enough to help us reconstruct large-scale patterns in the history of life.

TAKE-HOME MESSAGE 16.3

Fossils are evidence of organisms that lived in the remote past, a stone-hard historical record of life.

The fossil record will never be complete. Geologic events have obliterated much of it. The rest of the record is slanted toward species that had hard parts, lived in dense populations with wide distribution, and persisted for a long time.

PEOPLE MATTER

National Geographic Explorer-in-Residence
DR. PAUL SERENO

A real-life Indiana Jones, paleontologist Paul Sereno blends his background as an artist with a love for science and history. Sereno's passion carries him to the remote corners of the world to discover new species under the harshest of conditions.

Sereno's fieldwork began in 1988 in the Andes, where his team discovered the first dinosaurs to roam the Earth, including the most primitive of all: *Eoraptor*. This work culminated in the most complete picture yet of the dawn of the dinosaur era, some 225 million years ago. In the 1990s, Sereno's expeditions shifted to the Sahara. There, his teams have since excavated more than 70 tons of fossils representing organisms such as the huge-clawed fish-eater *Suchomimus*, the gigantic *Carcharodontosaurus* (its jaws pictured above, with Sereno), and a series of crocs including the 40-foot-long "SuperCroc" *Sarcosuchus*, the world's largest crocodile. In 2001, a trip to India yielded the Asian continent's first dinosaur skull—a new species of predator, *Rajasaurus*. Also in 2001, Sereno began an ongoing series of expeditions to China, first exploring remote areas of the Gobi in Inner Mongolia and discovering a herd of more than 20 dinosaurs that had died in their tracks. In 2012, he reported the discovery of *Pegomastax*, a bizarre, cat-sized dinosaur with a parrotlike beak and sharp fangs.

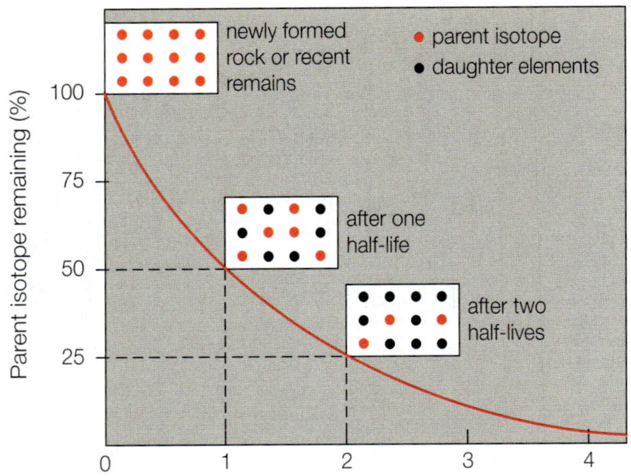

FIGURE 16.7 {Animated} Half-life.

FIGURE IT OUT: How much of any radioisotope remains after two of its half-lives have passed? Answer: 25 percent

A Long ago, ^{14}C and ^{12}C were incorporated into the tissues of a nautilus. The carbon atoms were part of organic molecules in the animal's food. ^{12}C is stable and ^{14}C decays, but the proportion of the two isotopes in the nautilus's tissues remained the same. Why? The nautilus continued to gain both types of carbon atoms in the same proportions from its food.

B The nautilus stopped eating when it died, so its body stopped gaining carbon. The ^{12}C atoms in its tissues were stable, but the ^{14}C atoms (represented as red dots) were decaying into nitrogen atoms. Thus, over time, the amount of ^{14}C decreased relative to the amount of ^{12}C. After 5,730 years, half of the ^{14}C had decayed; after another 5,730 years, half of what was left had decayed, and so on.

C Fossil hunters discover the fossil and measure its content of ^{14}C and ^{12}C. They use the ratio of these isotopes to calculate how many half-lives passed since the organism died. For example, if its ^{14}C to ^{12}C ratio is one-eighth of the ratio in living organisms, then three half-lives $(½)^3$ must have passed since it died. Three half-lives of ^{14}C is 17,190 years.

FIGURE 16.8 {Animated} Example of how radiometric dating is used to find the age of a carbon-containing fossil. Carbon 14 (^{14}C) is a radioisotope of carbon that decays into nitrogen. It forms in the atmosphere and combines with oxygen to become CO_2, which enters food chains by way of photosynthesis.

Remember from Section 2.1 that a radioisotope is a form of an element with an unstable nucleus. Atoms of a radioisotope become atoms of other elements—daughter elements—as their nucleus disintegrates. This radioactive decay is not influenced by temperature, pressure, chemical bonding state, or moisture; it is influenced only by time. Thus, like the ticking of a perfect clock, each type of radioisotope decays at a constant rate. The time it takes for half of the atoms in a sample of radioisotope to decay is called a **half-life** (**FIGURE 16.7**).

Half-life is a characteristic of each radioisotope. For example, radioactive uranium 238 decays into thorium 234, which decays into something else, and so on until it becomes lead 206. The half-life of the decay of uranium 238 to lead 206 is 4.5 billion years.

The predictability of radioactive decay can be used to find the age of a volcanic rock (the date it solidified). Rock deep inside Earth is hot and molten, so atoms swirl and mix in it. Rock that reaches the surface cools and hardens. As the rock cools, minerals crystallize in it. Each kind of mineral has a characteristic structure and composition. For example, the mineral zircon (*left*) consists mainly of ordered arrays of zirconium silicate molecules ($ZrSiO_4$). Some of the molecules in a newly formed zircon crystal have uranium atoms substituted for zirconium atoms, but never lead atoms. However, uranium decays into lead at a predictable rate. Thus, over time, uranium atoms disappear from a zircon crystal, and lead atoms accumulate in it. The ratio of uranium atoms to lead atoms in a zircon crystal can be measured precisely. That ratio can be used to calculate how long ago the crystal formed (its age).

zircon

We have just described **radiometric dating**, a method that can reveal the age of a material by measuring its content of a radioisotope and daughter elements. The oldest known terrestrial rock, a tiny zircon crystal from the Jack Hills in Western Australia, is 4.404 billion years old.

Recent fossils that still contain carbon can be dated by measuring their carbon 14 content (**FIGURE 16.8**). Most of the ^{14}C in a fossil will have decayed after about 60,000 years. The age of fossils older than that can be estimated by dating volcanic rocks in lava flows above and below the fossil-containing layer of sedimentary rock.

FINDING A MISSING LINK

The discovery of intermediate forms of cetaceans (an order of animals that includes whales, dolphins, and porpoises) provides an example of how scientists use fossil finds and radiometric dating to piece together evolutionary history. For some time, evolutionary biologists predicted that the ancestors of modern cetaceans walked on land, then took up life in the water. Evidence in support of this line of thinking includes a set of distinctive features of the skull and lower jaw that cetaceans share with some kinds of ancient carnivorous land animals. DNA sequence comparisons indicate that the ancient land animals were probably artiodactyls, hooved mammals with an even number of toes (two or four) on each foot (**FIGURE 16.9A**). Modern representatives of the artiodactyl lineage include camels, hippopotamuses, pigs, deer, sheep, and cows.

Until recently, we had no fossils demonstrating gradual changes in skeletal features that accompanied a transition of whale lineages from terrestrial to aquatic life. Researchers knew there were intermediate forms because they had found a representative fossil skull of an ancient whalelike animal, but without a complete skeleton the rest of the story remained speculative.

Then, in 2000, Philip Gingerich and his colleagues unearthed complete skeletons of two ancient whales: a fossil *Rodhocetus kasrani* excavated from a 47-million-year-old rock formation in Pakistan, and a fossil *Dorudon atrox*, from 37-million-year-old rock in Egypt (**FIGURE 16.9B,C**). Both fossil skeletons had whalelike skull bones, as well as intact ankle bones. The ankle bones of both fossils have distinctive features in common with those of extinct and modern artiodactyls. Modern cetaceans do not have even a remnant of an ankle bone (**FIGURE 16.9D**).

Rodhocetus and *Dorudon* were not direct ancestors of modern whales, but their telltale ankle bones mean they are long-lost relatives. Both whales were offshoots of the ancient artiodactyl-to-modern-whale lineage as it transitioned from life on land to life in water. The proportions of limbs, skull, neck, and thorax indicate *Rodhocetus* swam with its feet, not its tail. Like modern whales, the 5-meter (16-foot) *Dorudon* was clearly a fully aquatic tail-swimmer: The entire hindlimb was only about 12 centimeters (5 inches) long, much too small to have supported the animal's tremendous body out of water.

half-life Characteristic time it takes for half of a quantity of a radioisotope to decay.
radiometric dating Method of estimating the age of a rock or fossil by measuring the content and proportions of a radioisotope and its daughter elements.

TAKE-HOME MESSAGE 16.4

The predictability of radioisotope decay can be used to estimate the age of rock layers and fossils in them.

Radiometric dating helps evolutionary biologists retrace changes in ancient lineages.

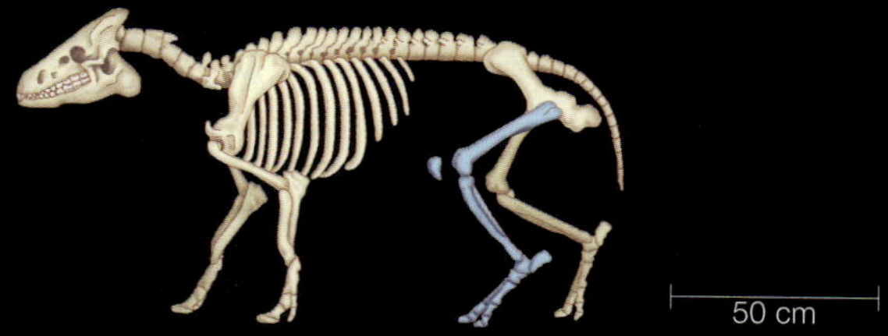

A *Elomeryx*, a small terrestrial animal that lived about 30 million years ago. This is a member of the same artiodactyl group (even-toed hooved mammals) that gave rise to modern representatives, including hippopotamuses. *Elomeryx* shares a common ancestor with whales, and is thought to resemble this ancestor.

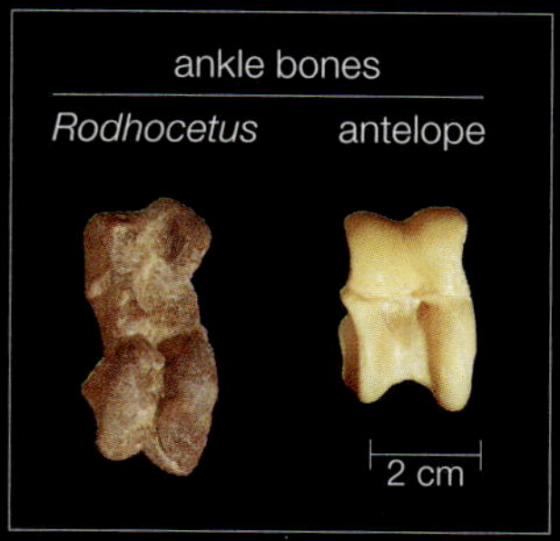

B *Rodhocetus kasrani*, an ancient whale that lived about 47 million years ago. Its distinctive ankle bones are evidence of a close evolutionary connection to artiodactyls. Artiodactyls are defined by the unique "double-pulley" shape of the bone (right) that forms the lower part of their ankle joint.

C *Dorudon atrox*, an ancient whale that lived about 37 million years ago. Its tiny, artiodactyl-like ankle bones were much too small to have supported the weight of its huge body on land, so this mammal had to be fully aquatic.

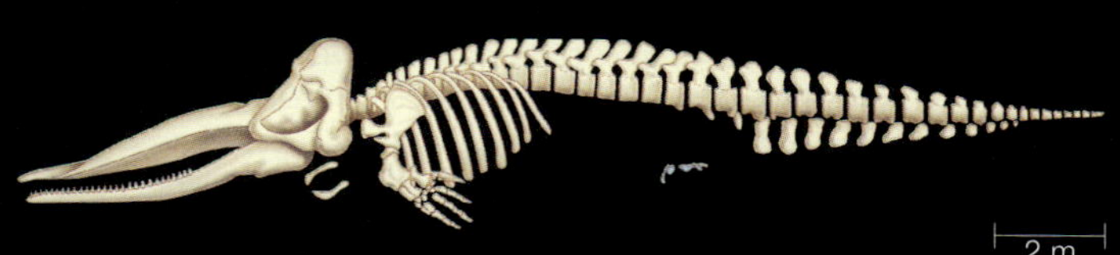

D Modern cetaceans such as the sperm whale have remnants of a pelvis and leg, but no ankle bones.

FIGURE 16.9 Ancient relatives of whales. The ancestor of whales and other cetaceans was an artiodactyl that walked on land. The lineage transitioned from life on land to life in water over millions of years, and as it did, the animals' limb bones became smaller and smaller. Comparable hindlimb bones are highlighted in blue.

16.5 HOW HAS EARTH CHANGED OVER GEOLOGIC TIME?

Wind, water, and other forces continuously sculpt Earth's surface, but they are only part of a much bigger picture of geological change. All continents that exist today were once part of a supercontinent—**Pangea**—that split into fragments and drifted apart.

The idea that continents move around, originally called continental drift, was proposed in the early 1900s to explain why the Atlantic coasts of South America and Africa seem to "fit" like jigsaw puzzle pieces, and why the same types of fossils occur in identical rock formations on both sides of the Atlantic Ocean. It also explained why the magnetic poles of gigantic rock formations point in different directions on different continents. Rock forms when molten lava solidifies on Earth's surface. Some iron-rich minerals become magnetic as they solidify, and their magnetic poles align with Earth's poles when they do. If the continents never moved, then all of these ancient rocky magnets should be aligned north-to-south, like compass needles. Indeed, the magnetic poles of each rock formation are aligned—but they do not always point north-to-south. Either Earth's magnetic poles veer dramatically from their north–south axis, or the continents must wander.

The concept of moving continents was initially greeted with skepticism because there was no known mechanism capable of causing such movements. Then, in the late 1950s, deep-sea explorers found immense ridges and trenches stretching thousands of kilometers across the seafloor (**FIGURE 16.10**). The discovery led to the **plate tectonics theory**, which explains how continents move: Earth's outer layer of rock is cracked into immense plates, like a huge cracked eggshell. Molten rock streaming from an undersea ridge ❶ or continental rift at one edge of a plate pushes old rock at the opposite edge into a trench ❷. The movement is like that of a colossal conveyor belt that transports continents on top of it to new locations. The plates move no more than 10 centimeters (4 inches) a year—about half as fast as your toenails grow—but it is enough to carry a continent all the way around the world after 40 million years or so (**FIGURE 16.11**).

The San Andreas Fault, which extends 800 miles through California, marks the boundary between two tectonic plates.

FIGURE 16.10 Plate tectonics. Huge pieces of Earth's outer layer of rock slowly drift apart and collide. As these plates move, they convey continents around the globe.

❶ At oceanic ridges, plumes of molten rock welling up from Earth's interior drive the movement of tectonic plates. New crust spreads outward as it forms on the surface, forcing adjacent tectonic plates away from the ridge and into trenches elsewhere.

❷ At trenches, the advancing edge of one plate plows under an adjacent plate and buckles it.

❸ Faults are ruptures in Earth's crust where plates meet. The diagram shows a rift fault, in which plates move apart. The photo above shows a strike-slip fault, in which two abutting plates slip against one another in opposite directions.

❹ Plumes of molten rock rupture a tectonic plate at what are called "hot spots." The Hawaiian Islands have been forming from molten rock that continues to erupt from a hot spot under the Pacific Plate. This and other tectonic plates are shown in Appendix V.

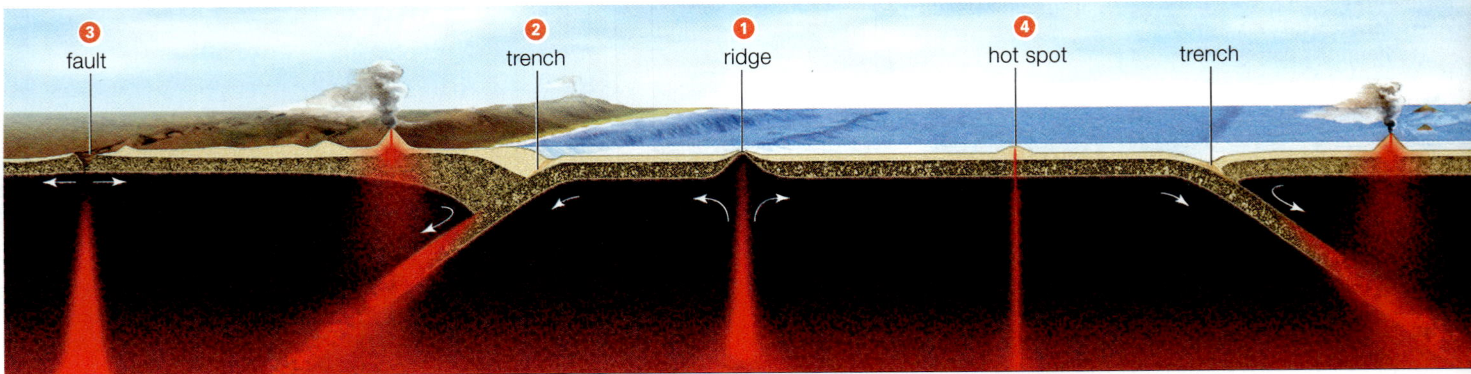

CREDITS: (10) top, Courtesy of SanAndreasFault.org; bottom, © Cengage Learning.

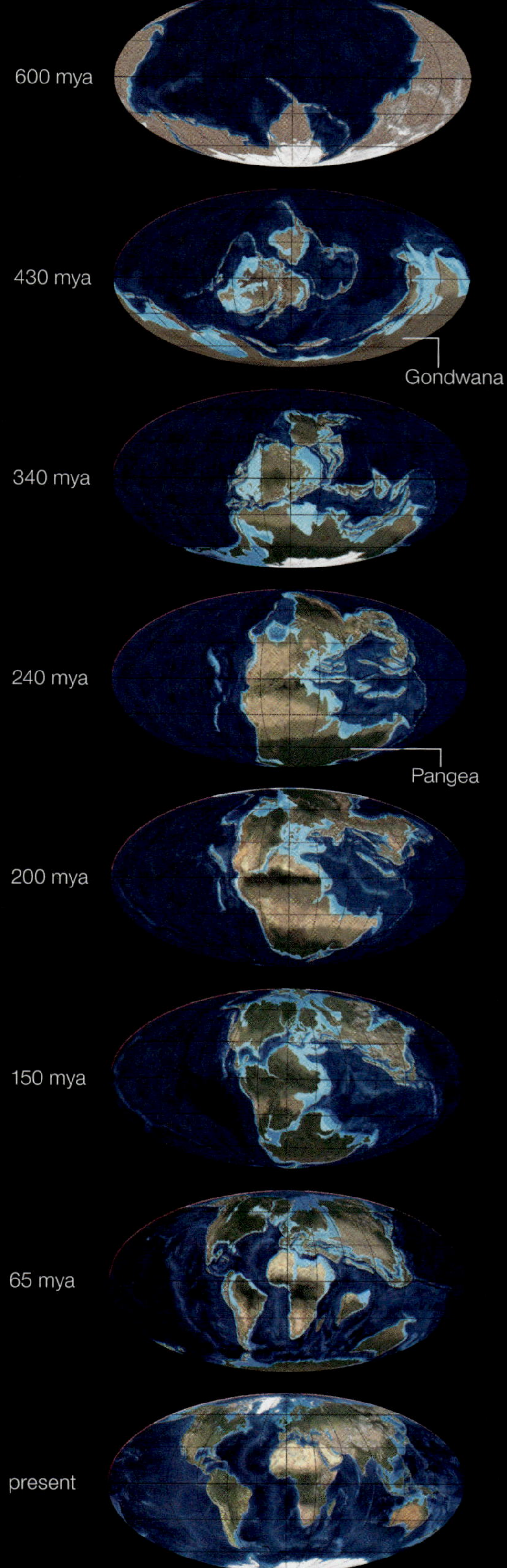

FIGURE 16.11 A series of reconstructions of the drifting continents. mya: million years ago.

Evidence of tectonic movement is all around us, in faults ③ and other geological features of our landscapes. For example, volcanic island chains (archipelagos) form as a plate moves across an undersea hot spot. These hot spots are places where a plume of molten rock wells up from deep inside Earth and ruptures a tectonic plate ④.

The fossil record also provides evidence in support of plate tectonics. Consider an unusual rock formation that exists in a huge belt across Africa. The sequence of rock layers in this formation is so complex that it is quite unlikely to have formed more than once, but identical sequences of layers also occur in huge belts that span India, South America, Madagascar, Australia, and Antarctica. Across all of these continents, the layers are the same ages. They also hold fossils found nowhere else, including imprints of the seed fern *Glossopteris* (pictured in **FIGURE 16.6C**). The most probable explanation for these observations is that the layered rock formed in one long belt on a single continent, which later broke up.

We now know that at least five times since Earth's outer layer of rock solidified 4.55 billion years ago, supercontinents formed and then split up again. One called **Gondwana** formed about 500 million years ago. Over the next 230 million years, this supercontinent wandered across the South Pole, then drifted north until it merged with other landmasses to form Pangea (**FIGURE 16.11**). Most of the landmasses currently in the Southern Hemisphere as well as India and Arabia were once part of Gondwana. Many modern species, including the birds pictured in **FIGURE 16.1**, live only in these places.

As you will see in later chapters, the changes brought on by plate tectonics have had a profound impact on life. Colliding continents have physically separated organisms living in oceans, and brought together those that had been living apart on land. As continents broke up, they separated organisms living on land, and brought together ones that had been living in separate oceans. Such changes have been a major driving force of evolution, a topic that we return to in the next chapter.

Gondwana Supercontinent that existed before Pangea, more than 500 million years ago.
Pangea Supercontinent that formed about 270 million years ago.
plate tectonics theory Theory that Earth's outer layer of rock is cracked into plates, the slow movement of which rafts continents to new locations over geologic time.

TAKE-HOME MESSAGE 16.5

Over geologic time, movements of Earth's crust have caused dramatic changes in continents and oceans. These changes profoundly influenced the course of life's evolution.

16.6 WHAT IS THE GEOLOGIC TIME SCALE?

Eon	Era	Period	Epoch	mya	Major Geologic and Biological Events
Phanerozoic	Cenozoic	Quaternary	Recent	0.01	Modern humans evolve. Major extinction event is now under way.
			Pleistocene	2.5	
		Neogene	Pliocene	5.3	Tropics, subtropics extend poleward. Climate cools; dry woodlands and grasslands emerge. Adaptive radiations of mammals, insects, birds.
			Miocene	23.0	
		Paleogene	Oligocene	33.9	
			Eocene	56.0	
			Paleocene	66.0 ◄	Major extinction event
	Mesozoic	Cretaceous	Upper	100.5	Flowering plants diversify; sharks evolve. All dinosaurs and many marine organisms disappear at the end of this epoch.
			Lower	145.0	Climate very warm. Dinosaurs continue to dominate. Important modern insect groups appear (bees, butterflies, termites, ants, and herbivorous insects including aphids and grasshoppers). Flowering plants originate and become dominant land plants.
		Jurassic		201.3 ◄	Age of dinosaurs. Lush vegetation; abundant gymnosperms and ferns. Birds appear. Pangea breaks up. Major extinction event
		Triassic		252 ◄	Recovery from the major extinction at end of Permian. Many new groups appear, including turtles, dinosaurs, pterosaurs, and mammals. Major extinction event
	Paleozoic	Permian		299	Supercontinent Pangea and world ocean form. Adaptive radiation of conifers. Cycads and ginkgos appear. Relatively dry climate leads to drought-adapted gymnosperms and insects such as beetles and flies.
		Carboniferous		359 ◄	High atmospheric oxygen level fosters giant arthropods. Spore-releasing plants dominate. Age of great lycophyte trees; vast coal forests form. Ears evolve in amphibians; penises evolve in early reptiles (vaginas evolve later, in mammals only). Major extinction event
		Devonian		419	Land tetrapods appear. Explosion of plant diversity leads to tree forms, forests, and many new plant groups including lycophytes, ferns with complex leaves, seed plants.
		Silurian		443 ◄	Radiations of marine invertebrates. First appearances of land fungi, vascular plants, bony fishes, and perhaps terrestrial animals (millipedes, spiders). Major extinction event
		Ordovician		485	Major period for first appearances. The first land plants, fishes, and reef-forming corals appear. Gondwana moves toward the South Pole and becomes frigid.
		Cambrian		541	Earth thaws. Explosion of animal diversity. Most major groups of animals appear (in the oceans). Trilobites and shelled organisms evolve.
Proterozoic				2,500	Oxygen accumulates in atmosphere. Origin of aerobic metabolism. Origin of eukaryotic cells, then protists, fungi, plants, animals. Evidence that Earth mostly freezes over in a series of global ice ages between 750 and 600 mya.
Archaean and earlier					3,800–2,500 mya. Origin of bacteria and archaea. 4,600–3,800 mya. Origin of Earth's crust, first atmosphere, first seas. Chemical, molecular evolution leads to origin of life (from protocells to anaerobic single cells).

FIGURE 16.12 {Animated} The geologic time scale (above) correlated with sedimentary rock exposed by erosion in the Grand Canyon (opposite). Orange triangles mark times of great mass extinctions. "First appearance" refers to appearance in the fossil record, not necessarily the first appearance on Earth. mya: million years ago. Dates are from the International Commission on Stratigraphy, 2013.

Similar sequences of sedimentary rock layers occur around the world. Transitions between the layers mark boundaries between great intervals of time in the **geologic time scale**, which is a chronology of Earth's history (**FIGURE 16.12**). Each layer's composition offers clues about conditions on Earth during the time the layer was deposited. Fossils in the layers are a record of life during that period of time.

geologic time scale Chronology of Earth's history.

TAKE-HOME MESSAGE 16.6

The geologic time scale correlates geological and evolutionary events of the ancient past.

CREDITS: (12) art, © Cengage Learning; photo, © Michael Pancier.

Each rock layer has a composition and set of fossils that reflect events during its deposition. For example, Coconino Sandstone, which stretches from California to Montana, is mainly weathered sand. Ripple marks and reptile tracks are the only fossils in it. Many think it is the remains of a vast sand desert, similar to the modern Sahara.

FIGURE IT OUT: Which formation is marked with the ? ? Answer: Tapeats sandstone

16.7 WHAT EVIDENCE DOES EVOLUTION LEAVE IN BODY FORM?

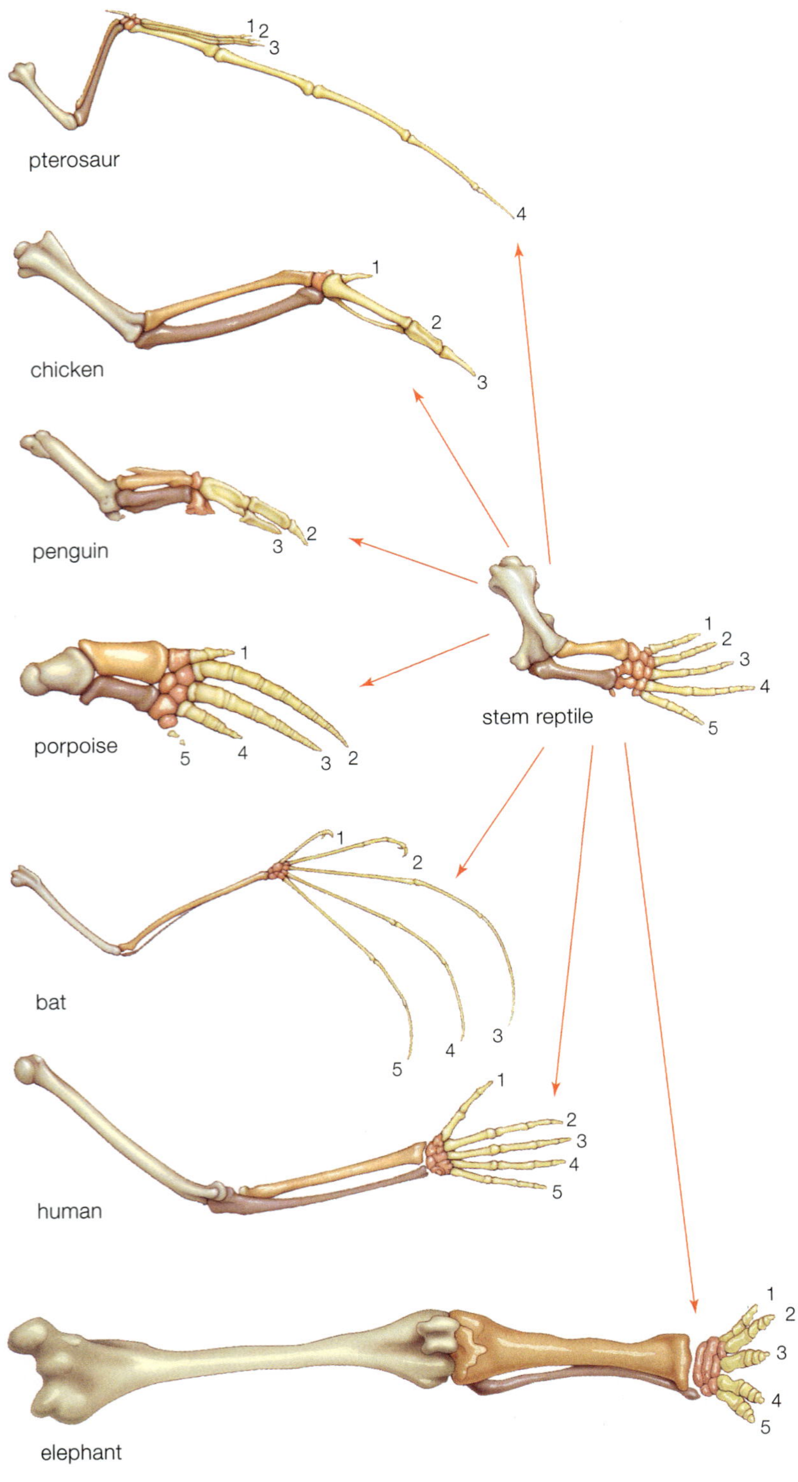

FIGURE 16.13 {Animated} Morphological divergence among vertebrate forelimbs, starting with the bones of an ancient stem reptile. The number and position of many skeletal elements were preserved when these diverse forms evolved; notice the bones of the forearms. Certain bones were lost over time in some of the lineages (compare the digits numbered 1 through 5). Drawings are not to scale.

To biologists, remember, evolution means change in a line of descent. How do they reconstruct evolutionary events that occurred in the ancient past? Evolutionary biologists are a bit like detectives, using clues to piece together a history that they did not witness in person. Fossils provide some clues. The body form and function of organisms that are alive today provide others.

MORPHOLOGICAL DIVERGENCE

Body parts that appear similar in separate lineages because they evolved in a common ancestor are called **homologous structures** (*hom–* means "the same"). Homologous structures may be used for different purposes in different groups, but the very same genes direct their development.

A body part that outwardly appears very different in separate lineages may be homologous in underlying form. Vertebrate forelimbs, for example, vary in size, shape, and function. However, they clearly are alike in the structure and positioning of bony elements, and in their internal patterns of nerves, blood vessels, and muscles.

As you will see in the next chapter, populations that are not interbreeding diverge genetically, and in time these divergences give rise to changes in body form. Change from the body form of a common ancestor is an evolutionary pattern called **morphological divergence**. Consider the limb bones of modern vertebrate animals. Fossil evidence suggests that many vertebrates are descended from a family of ancient "stem reptiles" that crouched low to the ground on five-toed limbs. Descendants of this ancestral group diversified over millions of years, and eventually gave rise to modern reptiles, birds, and mammals. A few lineages that had become adapted to walking on land even returned to life in the seas. During this time, the limbs became adapted for many different purposes (**FIGURE 16.13**). They became modified for flight in extinct reptiles called pterosaurs and in bats and most birds. In penguins and porpoises, the limbs are now flippers useful for swimming. In humans, five-toed forelimbs became arms and hands with four fingers and an opposable thumb. Among elephants, the limbs are now strong and pillarlike, capable of supporting a great deal of weight. Limbs degenerated to nubs in pythons and boa constrictors, and they disappeared entirely in other snakes.

MORPHOLOGICAL CONVERGENCE

Body parts that appear similar in different species are not always homologous; they sometimes evolve independently in lineages subject to the same environmental pressures. The independent evolution of similar body parts in different lineages is **morphological convergence**. Structures that are similar as a result of morphological convergence are called **analogous structures**. Analogous structures look alike but did not evolve in a shared ancestor; they evolved independently after the lineages diverged.

For example, bird, bat, and insect wings all perform the same function, which is flight. However, several clues tell us that the wing surfaces are not homologous. All of the wings are adapted to the same physical constraints that govern flight, but each is adapted in a different way. In the case of birds and bats, the limbs themselves are homologous, but the adaptations that make those limbs useful for flight differ. The surface of a bat wing is a thin, membranous extension of the animal's skin. By contrast, the surface of a bird wing is a sweep of feathers, which are specialized structures derived from skin. Insect wings differ even more. An insect wing forms as a saclike extension of the body wall. Except at forked veins, the sac flattens and fuses into a thin membrane. The sturdy, chitin-reinforced veins structurally support the wing. Unique adaptations for flight are evidence that wing surfaces of birds, bats, and insects are analogous structures that evolved after the ancestors of these modern groups diverged (**FIGURE 16.14**).

As another example of morphological convergence, the similar external structures of American cacti and African euphorbias (see **FIGURE 16.2**) are adaptations to similarly harsh desert environments where rain is scarce. Distinctive accordion-like pleats allow the plant body to swell with water when rain does come. Water stored in the plants' tissues allows them to survive long dry periods. As the stored water is used, the plant body shrinks, and the folded pleats provide it with some shade in an environment that typically has none. Despite these similarities, a closer look reveals many differences that indicate the two types of plants are not closely related. For example, cactus spines have a simple fibrous structure; they are modified leaves that arise from dimples on the plant's surface. Euphorbia spines project smoothly from the plant surface, and they are not modified leaves: In many species the spines are actually dried flower stalks (*below*).

FIGURE 16.14 Morphological convergence in animals. The surfaces of an insect wing, a bat wing, and a bird wing are analogous structures. The diagram shows how the evolution of wings (yellow dots) occurred independently in the three separate lineages that led to bats, birds, and insects. You will read more about diagrams that show evolutionary relationships in Section 17.12.

analogous structures Similar body structures that evolved separately in different lineages.
homologous structures Body structures that are similar in different lineages because they evolved in a common ancestor.
morphological convergence Evolutionary pattern in which similar body parts evolve separately in different lineages.
morphological divergence Evolutionary pattern in which a body part of an ancestor changes in its descendants.

TAKE-HOME MESSAGE 16.7

Body parts are often modified differently in different lines of descent.

Some body parts that appear alike evolved independently in different lineages.

16.8 HOW DO SIMILARITIES IN DNA AND PROTEIN REFLECT EVOLUTION?

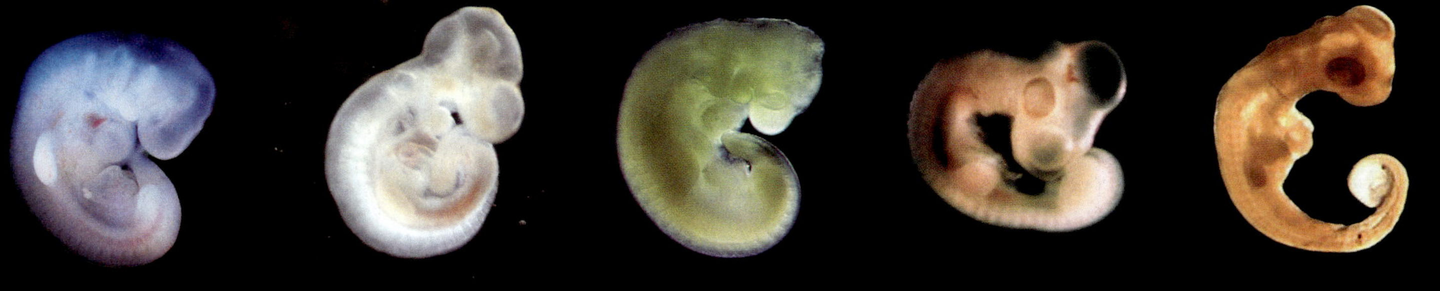

FIGURE 16.15 Visual comparison of vertebrate embryos. All vertebrates go through an embryonic stage in which they have four limb buds, a tail, and divisions called somites along their back. From *left* to *right*: human, mouse, bat, chicken, alligator.

Evolution also leaves clues in patterns of embryonic development: In general, the more closely related animals are, the more similar is their development. For example, all vertebrates go through a stage during which a developing embryo has four limb buds, a tail, and a series of somites—divisions of the body that give rise to the backbone and associated skin and muscle (**FIGURE 16.15**).

Animals have similar patterns of embryonic development because the very same master genes direct the process. Remember from Section 10.2 that the development of an embryo into the body of a plant or animal is orchestrated by layer after layer of master gene expression. The failure of any single master gene to participate in this symphony of expression can result in a drastically altered body plan, typically with devastating consequences. Because a mutation in a master gene typically unravels development completely, these genes tend to be highly conserved. Even among lineages that diverged a very long time ago, such genes often retain similar sequences and functions.

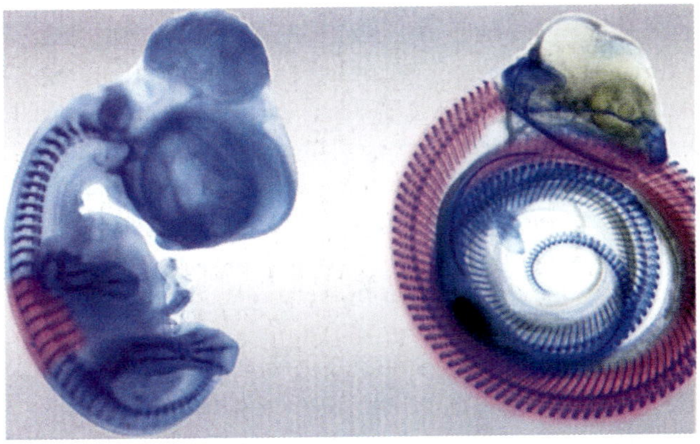

FIGURE 16.16 How differences in body form arise from differences in master gene expression. Expression of the *Hoxc6* gene is indicated by purple stain in two vertebrate embryos, chicken (*left*) and garter snake (*right*). Expression of this gene causes a vertebra to develop ribs as part of the back. Chickens have 7 vertebrae in their back and 14 to 17 vertebrae in their neck; snakes have upwards of 450 back vertebrae and essentially no neck.

Consider homeotic genes called *Hox*. Like other homeotic genes, *Hox* gene expression helps sculpt details of the body's form during embryonic development. Vertebrate animals have multiple sets of the same ten *Hox* genes that occur in insects and other arthropods. You have already read about one of these genes, *antennapedia*, which determines the identity of the thorax (the body part with legs) in fruit flies. One vertebrate version of *antennapedia* is called *Hoxc6*,

FIGURE 16.17 Example of a protein comparison. Here, part of the amino acid sequence of mitochondrial cytochrome *b* from 20 species is aligned. This protein is a crucial component of mitochondrial electron transfer chains. The honeycreeper sequence is identical in ten species of honeycreeper; amino acids that differ in the other species are shown in red. Dashes are gaps in the alignment.

FIGURE IT OUT: Based on this comparison, which species is the most closely related to the honeycreepers?

Answer: The song sparrow

honeycreepers (10)	...CRDVQFGWLIRNLHANGASFFFICIYLHIGRGIYYGSYLNK--ETWNIGVILLLTLMATAFVGYVLPWGQMSFWG...
song sparrow	...CRDVQFGWLIRNLHANGASFFFICIYLHIGRGIYYGSYLNK--ETWNVGIILLLALMATAFVGYVLPWGQMSFWG...
Gough Island finch	...CRDVQFGWLIRNIHANGASFFFICIYLHIGRGLYYGSYLYK--ETWNVGVILLLTLMATAFVGYVLPWGQMSFWG...
deer mouse	...CRDVNYGWLIRYMHANGASMFFICLFLHVGRGMYYGSYTFT--ETWNIGIVLLFAVMATAFMGYVLPWGQMSFWG...
Asiatic black bear	...CRDVHYGWIIRYMHANGASMFFICLFMHVGRGLYYGSYLLS--ETWNIGIILLFTVMATAFMGYVLPWGQMSFWG...
bogue (a fish)	...CRDVNYGWLIRNLHANGASFFFICIYLHIGRGLYYGSYLYK--ETWNIGVVLLLVMGTAFVGYVLPWGQMSFWG...
human	...TRDVNYGWIIRYLHANGASMFFICLFLHIGRGLYYGSFLYS--ETWNIGIILLLATMATAFMGYVLPWGQMSFWG...
thale cress (a plant)	...MRDVEGGWLLRYMHANGASMFLIVVYLHIFRGLYHASYSSPREFVWCLGVVIFLLMIVTAFIGYVLPWGQMSFWG...
baboon louse	...ETDVMNGWMVRSIHANGASWFFIMLYSHIFRGLWVSSFTQP--LVWLSGVIILFLSMATAFLGYVLPWGQMSFWG...
baker's yeast	...MRDVHNGYILRYLHANGASFFFMVMFMHMAKGLYYGSYRSPRVTLWNVGVIIFTLTIATAFLGYCCVYGQMSHWG...

and it determines the identity of the back (as opposed to the neck or tail). Expression of the *Hoxc6* gene causes ribs to develop on a vertebra. Vertebrae of the neck and tail normally develop with no *Hoxc6* expression, and no ribs (**FIGURE 16.16**).

Given that the very same genes direct development in all of the vertebrate lineages, how do the adult forms end up so different? Part of the answer is that there are differences in the onset, rate, or completion of early steps in development brought about by variations in master gene expression. The variation has arisen at least in part as a result of gene duplications followed by mutation, the same way that multiple globin genes evolved in primates (Section 14.4).

Genes that are not conserved are the basis of major phenotypic differences that define species. Over time, inevitable mutations change the DNA sequence of a lineage's genome. The more recently two lineages diverged, the less time there has been for unique mutations to accumulate in the DNA of each one. That is why the genomes of closely related species tend to be more similar than those of distantly related ones—a general rule that can be used to estimate relative times of divergence. Two species with very few similar genes probably have not shared an ancestor for a long time—long enough for many mutations to have accumulated in the DNA of their separate lineages. Consider that about 88 percent of the mouse genome sequence is identical with the human genome, as is 73 percent of the zebrafish genome, 47 percent of the fruit fly genome, and 25 percent of the rice genome.

Getting useful information from comparing DNA requires a lot more data than comparing proteins. This is because coincidental homologies are statistically more likely to occur with DNA comparisons—there are only four nucleotides in DNA versus twenty amino acids in proteins. Thus, proteins are more commonly compared. By comparing the amino acid sequence of a protein among several species, the number of amino acid differences can be used as a measure of relative relatedness (**FIGURE 16.17**).

TAKE-HOME MESSAGE 16.8

Similarities in patterns of animal development occur because the same genes direct the process. Similar developmental patterns—and shared genes—are evidence of common ancestry, which can be ancient.

Mutations change the nucleotide sequence of each lineage's DNA over time. There are generally fewer differences between the DNA of more closely related lineages.

Similar genes give rise to similar proteins. Fewer differences occur among the proteins of more closely related lineages.

CREDITS: (18) © David A. Kring, NASA/Univ. Arizona Space Imagery Center; (in text) © U.S. Geological Survey.

16.9 REFLECTIONS OF A DISTANT PAST

Application: Exploration

FIGURE 16.18 The K–Pg boundary sequence, an unusual, worldwide sedimentary rock formation that formed 66 million years ago.

WHAT KILLED THE DINOSAURS? Most scientists now think that the dinosaurs perished in the aftermath of a catastrophic meteorite impact. No human witnesses were around at the time, so how do they know what happened? The event is marked by an unusual, worldwide formation of sedimentary rock (FIGURE 16.18). There are plenty of dinosaur fossils below this formation, which is called the K–Pg boundary sequence (formerly known as the K–T boundary). Above it, there are none, anywhere. The rock consists of an unusual clay rich in iridium, an element much more abundant in asteroids than in Earth's crust. It also contains shocked quartz (*left*) and small glass spheres called tektites, minerals that form when quartz or sand undergoes a sudden, violent application of extreme pressure. The only processes on Earth that produce these minerals are atomic bomb explosions and meteorite impacts.

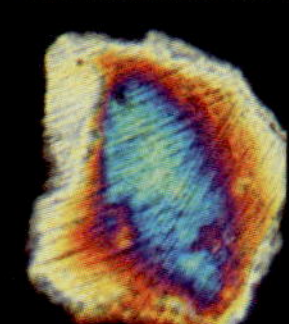

Geologists concluded that the K–Pg boundary layer must have originated with extraterrestrial material, and began looking for evidence of a meteorite that hit Earth 66 million years ago—one big enough to cover the entire planet with its debris. Twenty years later, they found it: an impact crater the size of Ireland off the coast of the Yucatán Peninsula. To make a crater this big, a meteorite 20 km (12 miles) wide would have slammed into Earth with the force of 100 trillion tons of dynamite—enough to cause an ecological disaster of sufficient scale to wipe out almost all life on Earth.

Summary

SECTION 16.1 Expeditions by nineteenth-century explorers yielded increasingly detailed observations of nature. Geology, **biogeography**, and **comparative morphology** of organisms and their **fossils** led to new ways of thinking about the natural world.

SECTION 16.2 Prevailing belief systems may influence interpretation of the underlying cause of a natural event. Nineteenth-century naturalists tried to reconcile traditional belief systems with physical evidence of **evolution,** or change in a **lineage** over time.

Humans select desirable traits in animals by selective breeding. Charles Darwin and Alfred Wallace independently came up with a theory of how environments also select traits, stated here in modern terms: A population tends to grow until it exhausts environmental resources. As that happens, competition for those resources intensifies among the population's members. Individuals with forms of shared, heritable traits that give them an advantage in this competition tend to produce more offspring. Thus, **adaptive traits** (**adaptations**) that impart greater **fitness** to an individual become more common in a population over generations. The process in which environmental pressures result in the differential survival and reproduction of individuals of a population is called **natural selection**. It is one of the processes that drives evolution.

SECTION 16.3 Fossils are typically found in stacked layers of sedimentary rock. Younger fossils usually occur in layers deposited more recently, on top of older fossils in older layers. Fossils are relatively scarce, so the fossil record will always be incomplete.

SECTION 16.4 A radioisotope's characteristic **half-life** can be used to determine the age of rocks and fossils. This technique, **radiometric dating**, helps us understand the ancient history of many lineages.

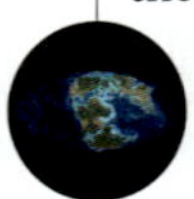

SECTION 16.5 According to the **plate tectonics theory**, Earth's crust is cracked into giant plates that carry landmasses to new positions as they move. Earth's landmasses have periodically converged as supercontinents such as **Gondwana** and **Pangea**.

SECTION 16.6 Transitions in the fossil record are the boundaries of great intervals of the **geologic time scale**, a chronology of Earth's history that correlates geologic and evolutionary events.

SECTION 16.7 Comparative morphology is one way to study evolutionary connections among lineages. **Homologous structures** are similar body parts that, by **morphological divergence**, became modified differently in different lineages. Such parts are evidence of a common ancestor. **Analogous structures** are body parts that look alike in different lineages but did not evolve in a common ancestor. By the process of **morphological convergence**, they evolved separately after the lineages diverged.

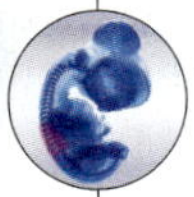

SECTION 16.8 We can discover and clarify evolutionary relationships through comparisons of DNA and protein sequences, because lineages that diverged recently tend to share more sequences than ones that diverged long ago. Master genes that affect development tend to be highly conserved, so similarities in patterns of embryonic development reflect shared ancestry that can be evolutionarily ancient.

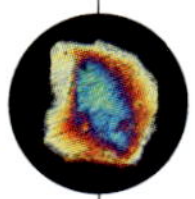

SECTION 16.9 A mass extinction 66 million years ago may have been caused by an asteroid impact that left traces in a worldwide sedimentary rock formation.

Self-Quiz

Answers in Appendix VII

1. The number of species on an island usually depends on the size of the island and its distance from a mainland. This statement would most likely be made by ______ .
   a. an explorer
   b. a biogeographer
   c. a geologist
   d. a philosopher

2. The bones of a bird's wing are similar to the bones in a bat's wing. This observation is an example of ______ .
   a. uniformity
   b. evolution
   c. comparative morphology
   d. a lineage

3. Evolution ______ .
   a. is natural selection
   b. is change in a line of descent
   c. can occur by natural selection
   d. b and c are correct

4. A trait is adaptive if it ______ .
   a. arises by mutation
   b. increases fitness
   c. is passed to offspring
   d. occurs in fossils

5. In which type of rock are you more likely to find a fossil?
   a. basalt, a dark, fine-grained volcanic rock
   b. limestone, composed of calcium carbonate sediments
   c. slate, a volcanically melted and cooled shale
   d. granite, which forms by crystallization of molten rock below Earth's surface

6. If the half-life of a radioisotope is 20,000 years, then a sample in which three-quarters of that radioisotope has decayed is ______ years old.
   a. 15,000
   b. 26,667
   c. 30,000
   d. 40,000

7. Did Pangea or Gondwana form first?

8. Forces that cause geologic change include ______ (select all that are correct).
   a. erosion
   b. natural selection
   c. volcanic activity
   d. tectonic plate movement
   e. wind
   f. meteorite impacts

9. Through ______ , a body part of an ancestor is modified differently in different lines of descent.

Data Analysis Activities

Discovery of Iridium in the K–Pg Boundary Layer In the late 1970s, geologist Walter Alvarez was investigating the composition of the K–Pg boundary sequence in different parts of the world. He asked his father, Nobel Prize–winning physicist Luis Alvarez, to help him analyze the elemental composition of the layer (*right*, Luis and Walter Alvarez with a section of the boundary sequence).

The Alvarezes and their colleagues tested the K–Pg boundary sequence in Italy and Denmark. They discovered that it contains a much higher iridium content than the surrounding rock layers. Some of their results are shown in **FIGURE 16.19**.

1. What was the iridium content of the K–Pg boundary sequence?
2. How much higher was the iridium content of the boundary layer than the sample taken 0.7 meter above the sequence?

Sample Depth	Average Abundance of Iridium (ppb)
+ 2.7 m	< 0.3
+ 1.2 m	< 0.3
+ 0.7 m	0.36
boundary layer	41.6
− 0.5 m	0.25
− 5.4 m	0.30

FIGURE 16.19 Abundance of iridium in and near the K-Pg boundary sequence in Stevns Klint, Denmark. Many rock samples taken from above, below, and at the boundary were tested for iridium content. Depths are given as meters above or below the boundary.

The iridium content of an average Earth rock is 0.4 parts per billion (ppb) of iridium. An average meteorite contains about 550 parts per billion of iridium.

10. Homologous structures among major groups of organisms may differ in _______ .
    a. size b. shape c. function d. all of the above

11. By altering steps in the program by which embryos develop, a mutation in a _______ may lead to major differences in body form between related lineages.
    a. derived trait c. homologous structure
    b. homeotic gene d. all of the above

12. The dinosaurs died _______ million years ago.

13. All of the following data types can be used as evidence of shared ancestry except similarities in _______ .
    a. amino acid sequence d. embryonic development
    b. DNA sequence e. form due to convergence
    c. fossil morphology f. all are appropriate

14. Match the terms with the most suitable description.

___ fitness	a. line of descent
___ fossils	b. measured by relative genetic contribution to future generations
___ natural selection	c. human arm and bird wing
___ half-life	d. evidence of ancient life
___ homologous structures	e. characteristic of radioisotope
___ analogous structures	f. insect wing and bird wing
___ lineage	g. survival of the fittest
___ sedimentary rock	h. good for finding fossils

CENGAGE brain .com **To access course materials, please visit www.cengagebrain.com.**

Critical Thinking

1. Radiometric dating does not measure the age of an individual atom. It is a measure of the age of a quantity of atoms—a statistic. As with any statistical measure, its values may deviate around an average (see sampling error, Section 1.7). Imagine that one sample of rock is dated ten different ways. Nine of the tests yield an age close to 225,000 years. One test yields an age of 3.2 million years. Do the nine consistent results imply that the one that deviates is incorrect, or does the one odd result invalidate the nine that are consistent?

2. If you think of geologic time spans as minutes, life's history might be plotted on a clock such as the one shown *below*. According to this clock, the most recent epoch started in the last 0.1 second before noon. Where does that put you?

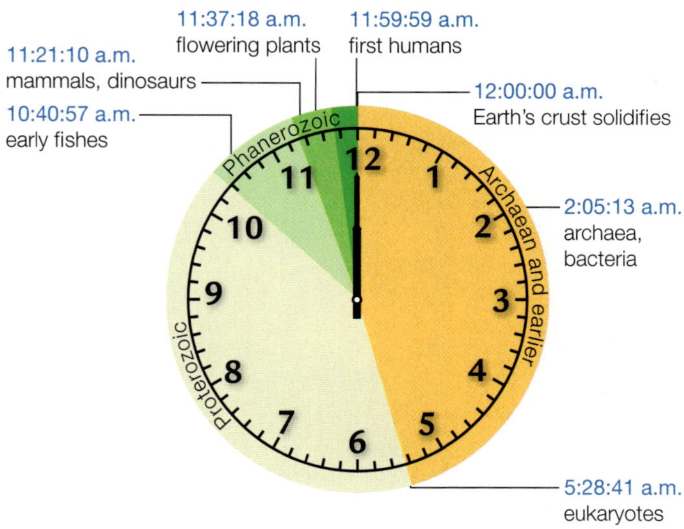

A male peacock spider approaches a female, signaling his intent to mate with her by raising and waving colorful flaps, and gesturing his legs in time with abdominal vibrations. If his courtship display fails to impress her, she will kill him.

17 PROCESSES OF EVOLUTION

Links to Earlier Concepts

A review of species (Section 1.4), mutations (8.5, 9.5), sexual reproduction (12.1, 12.2), inheritance (13.1, 13.2), traits (13.4–13.6, 14.1), and natural selection (16.2) will be helpful. This chapter also revisits populations (1.1), experiments (1.5), sampling error (1.7), homeotic genes (10.2), *BRCA* genes (10.6, 15.7), phenotype (13.1), polyploidy (14.5), genetic screening (14.6), transgenic plants (15.5), plate tectonics (16.5), and the geologic time scale (16.6).

KEY CONCEPTS

MICROEVOLUTION

Members of a population inherit different alleles, which are the basis of differences in phenotype. An allele may increase or decrease in frequency in a population, a change called microevolution.

PROCESSES OF MICROEVOLUTION

Natural selection can shift or maintain the range of variation in heritable traits. Change in allele frequency can occur by chance alone. Gene flow between populations keeps them similar.

HOW SPECIES ARISE

Speciation typically starts after gene flow ends. Microevolution leads to genetic divergences, which are reinforced as mechanisms evolve that prevent interbreeding.

MACROEVOLUTION

Macroevolutionary patterns include the origin of major groups, one species giving rise to many, two species evolving jointly, and mass extinctions.

PHYLOGENY

Understanding a group's evolutionary history can help us protect endangered species. It also helps us understand the origin and transmission of infectious diseases.

17.1 WHAT IS MICROEVOLUTION?

VARIATION IN SHARED TRAITS

Section 1.1 introduced a **population** as a group of interbreeding individuals of the same species in some specified area. The individuals of a population (and a species) share morphological, physiological, and behavioral traits because they have the same genes. Almost every shared trait varies a bit among members of a sexually reproducing species (**FIGURE 17.1**); this variation arises mainly because different individuals inherit different combinations of alleles (Sections 13.1 and 13.2).

Some traits occur in distinct forms, or morphs. A trait with only two forms is dimorphic (*di–* means two). The pea plants that Gregor Mendel studied are dimorphic for flower color—their flowers are either white or purple. In this case, dimorphic flower color arises from two alleles with a clear dominance relationship. Traits with more than two distinct forms are polymorphic (*poly–*, many). Human blood type, which is determined by the codominant *ABO* alleles, is an example (Section 13.4). The genetic basis of traits that vary continuously (Section 13.6) is often quite complex. Any of the genes that influence such traits may have multiple alleles.

In earlier chapters, you learned about the processes that introduce and maintain variation in traits among individuals of a species (**TABLE 17.1**). Mutation is the original source of new alleles. Other events shuffle alleles into different combinations, and what a shuffle that is! There are $10^{116,446,000}$ possible combinations of human alleles. Not even 10^{10} people are living today. Unless you have an identical twin, it is unlikely that another person with your precise genetic makeup has ever lived, or ever will.

TABLE 17.1

Sources of Variation in Traits Among Individuals of a Species

Genetic Event	Effect
Mutation	Source of new alleles
Crossing over at meiosis I	Introduces new combinations of alleles into chromosomes
Independent assortment at meiosis I	Mixes maternal and paternal chromosomes
Fertilization	Combines alleles from two parents
Changes in chromosome number or structure	Often dramatic changes in structure and function

AN EVOLUTIONARY VIEW OF MUTATIONS

Being the original source of new alleles, mutations are worth another look, this time in context of their impact on populations. We cannot predict when or in which individual a particular gene will mutate. We can, however, predict the average mutation rate of a species, which is the probability that a mutation will occur in a given interval. In the human species, that rate is about 2.2×10^{-9} mutations per base pair per year. In other words, what we define as the human genome sequence changes by about 70 nucleotides every decade.

Many mutations give rise to structural, functional, or behavioral alterations that reduce an individual's chances of surviving and reproducing. Even one biochemical change may be devastating. Consider collagen, a fibrous protein that structurally supports tissues that compose skin, bones, tendons, lungs, blood vessels, and other parts of the vertebrate body. If one of the genes for collagen mutates in a way that changes the protein's structure, the entire body may be affected. A mutation such as this can change phenotype (Section 13.1) so drastically that it results in death, in which case it is called a **lethal mutation**.

A **neutral mutation** is one that has no effect on survival or reproduction. For instance, a mutation that results in your earlobes being attached to your head instead of swinging freely should not in itself stop you from surviving and reproducing as well as anybody else. So, natural selection does not affect the frequency of this particular mutation in a population.

Occasionally, a change in the environment favors a mutation that had previously been neutral or even somewhat harmful. Even if a beneficial mutation bestows only a slight advantage, its frequency tends to increase in a population over time. This is because natural selection operates on traits with a genetic basis.

Mutations have been altering genomes for billions of years, and they are still at it. Cumulatively, they have given rise to Earth's staggering biodiversity. Think about it: The

allele frequency Abundance of a particular allele among members of a population.
gene pool All the alleles of all the genes in a population; a pool of genetic resources.
lethal mutation Mutation that alters phenotype so drastically that it causes death.
microevolution Change in an allele's frequency in a population.
neutral mutation A mutation that has no effect on survival or reproduction.
population A group of organisms of the same species who live in a specific location and breed with one another more often than they breed with members of other populations.

FIGURE 17.1 Sampling morphological variation among zigzag Nerite snails and (insets) humans.

reason you do not look like an apple or an earthworm or even your next-door neighbor began with mutations that occurred in different lines of descent.

ALLELE FREQUENCIES

Together, all the alleles of all the genes of a population comprise a pool of genetic resources—a **gene pool**. The members of a population breed with one another more often than they breed with members of other populations, so their gene pool is more or less isolated.

We refer to the abundance of any particular allele among members of a population as its **allele frequency**. Any change in allele frequency in the gene pool of a population (or a species) is called **microevolution**. Microevolution is always occurring in natural populations because, as you will see in the next sections, processes that drive it—mutation, natural selection, and genetic drift—are always operating. Remember, even though we can recognize patterns of evolution, none of them are purposeful. Evolution simply fills the nooks and crannies of opportunity.

TAKE-HOME MESSAGE 17.1

Individuals of a natural population share morphological, physiological, and behavioral traits characteristic of the species. Alleles are the main basis of differences in the details of those shared traits.

All alleles of all individuals in a population make up the population's gene pool. An allele's abundance in the gene pool is called its allele frequency.

Microevolution is change in allele frequency. It is always occurring in natural populations because processes that drive it are always operating.

CREDITS: (1) background, © David McIntyre/Photographer's Direct; inset far left, © Roderick Hulsbergen/ http://www.photography.euweb.nl; all others, © JupiterImages Corporation.

17.2 HOW DO WE KNOW WHEN A POPULATION IS EVOLVING?

GENETIC EQUILIBRIUM

Early in the twentieth century, Godfrey Hardy (a mathematician) and Wilhelm Weinberg (a physician) independently applied the rules of probability to population genetics. They realized that, under certain theoretical conditions, allele frequencies in a sexually reproducing population's gene pool would remain stable from one generation to the next. The population would remain in this stable state, called **genetic equilibrium**, as long as all of the following five conditions are met:

1. Mutations never occur.
2. The population is infinitely large.
3. The population is isolated from all other populations of the species—no individual enters or leaves.
4. Mating is random.
5. All individuals survive and produce the same number of offspring.

Thus, if we see an allele's frequency change in a shared gene pool, at least one of the five conditions is not being met. As you can imagine, all five of these conditions never occur in natural populations.

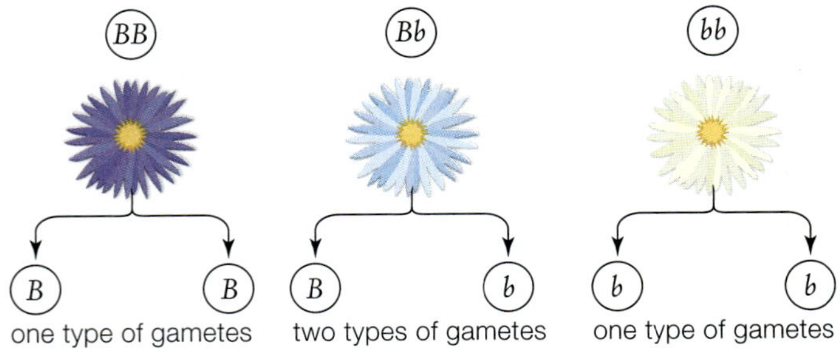

A In this two-allele system, *B* specifies dark blue flowers; *b*, white. Plants that are homozygous (*B* or *b*) make one kind of gamete. Heterozygous plants (*Bb*) have light blue flowers and make two kinds of gametes (*B* and *b*).

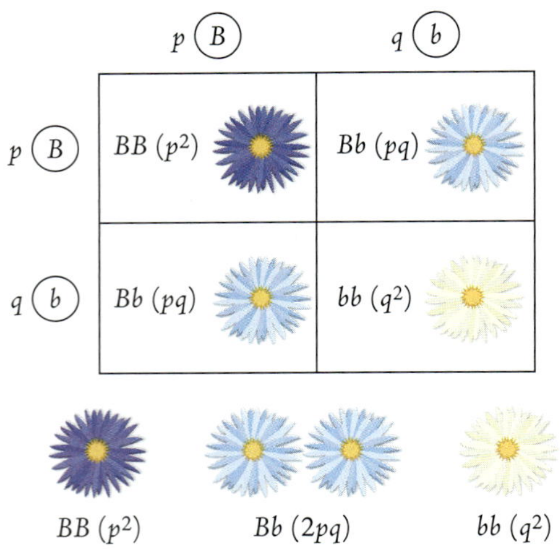

B Say p is the proportion of *B* alleles in the gene pool, and q is the proportion of *b* alleles. This Punnett square shows that in each generation, the predicted proportion of offspring that will inherit two *B* alleles is $p \times p$, or p^2. Likewise, the proportion that will inherit both alleles is $2pq$, and the proportion that will inherit two *b* alleles is q^2.

FIGURE 17.2 {Animated} Calculating Hardy–Weinberg frequencies. In this example, two alleles of a gene show incomplete dominance over flower color.

FIGURE IT OUT: If ¼ of this population has dark blue flowers and ¼ has white flowers, what proportion of the next generation will have light blue flowers (assuming genetic equilibrium)?

Answer: Half of the gametes have a *B* allele; the other half have a *b* allele. Both p and $q = 0.5$, so $2pq = 50$ percent.

APPLYING THE HARDY–WEINBERG LAW

The concept of genetic equilibrium under ideal conditions is called the Hardy–Weinberg law. To see how it works, consider a hypothetical gene that encodes a blue pigment in daisies. A plant homozygous for one allele (*BB*) has dark blue flowers. A plant homozygous for the other allele (*bb*) has white flowers. These two alleles are inherited in a pattern of incomplete dominance, so a heterozygous plant (*Bb*) has medium-blue flowers (**FIGURE 17.2A**).

Start with the concept that allele frequencies always add up to one. For a gene with two alleles, the following equation is true:

$$p + q = 1.0$$

where p is the frequency of one allele in the population, and q is the frequency of the other. Remember from Section 13.2 that paired alleles assort into different gametes during meiosis. **FIGURE 17.2B** shows the predictable proportions in which those gametes meet up at fertilization in our example. The predicted fraction of offspring that inherit two *B* alleles (*BB*) is $p \times p$, or p^2; the fraction that inherit two *b* alleles (*bb*) is q^2; and the fraction that inherit one *B* allele and one *b* allele (*Bb*) is $2pq$. Note that the frequencies of the three genotypes, whatever they may be, add up to 1.0:

$$p^2 + 2pq + q^2 = 1.0$$

Suppose our hypothetical population consists of 1,000 plants: 490 homozygous (*BB*), 420 heterozygous (*Bb*), and 90 homozygous (*bb*), and each of these individuals

genetic equilibrium Theoretical state in which an allele's frequency never changes in a population's gene pool.

makes just two gametes. All 980 gametes made by the *BB* individuals will have the *B* allele, as will half of the gametes made by the 420 *Bb* individuals. Thus, the frequency of the *B* allele among the pool of gametes is:

$$B\ (p) = \frac{980 + 420}{2{,}000 \text{ alleles}} = \frac{1{,}400}{2{,}000} = 0.7$$

$$b\ (q) = \frac{180 + 420}{2{,}000 \text{ alleles}} = \frac{600}{2{,}000} = 0.3$$

Using our equation, the proportion of individuals in the next generation is predicted to be:

$$BB\ (p^2) = (0.7)^2 = 0.49$$
$$Bb\ (2pq) = 2\,(0.7 \times 0.3) = 0.42$$
$$bb\ (q^2) = (0.3)^2 = 0.09$$

These proportions are the same as the ones in the parent population. As long as the five conditions required for genetic equilibrium are met, traits specified by the alleles should show up in the same proportions in each generation. If they do not, the population is evolving.

REAL-WORLD SITUATIONS

Genetic equilibrium is often used as a benchmark. As an example, researchers used it to determine the carrier frequency of an allele that causes hereditary hemochromatosis (HH), the most common genetic disorder among people of Irish ancestry. This autosomal recessive disorder causes affected individuals to absorb too much iron from food. As a result, they can have liver problems, fatigue, and arthritis. The recessive allele's frequency (q) was found to be 0.14. If $q = 0.14$, then p, the frequency of the normal allele, must be 0.86. Thus, the carrier frequency, $2pq$, was calculated to be 0.24 (24 percent of the population).

As another example, consider that mutations in the *BRCA* genes have been linked to breast cancer in adults (Sections 10.6 and 15.7). A deviation from predicted allele frequencies suggested that these mutations also have effects even before birth, so researchers investigated the frequency of mutated alleles of these genes among newborn girls. They found fewer individuals homozygous for these alleles than expected, based on the number of heterozygous individuals. Thus, in homozygous form, *BRCA* mutations impair the survival of female embryos.

TAKE-HOME MESSAGE 17.2

Researchers measure genetic change by comparing it with a theoretical baseline of genetic equilibrium.

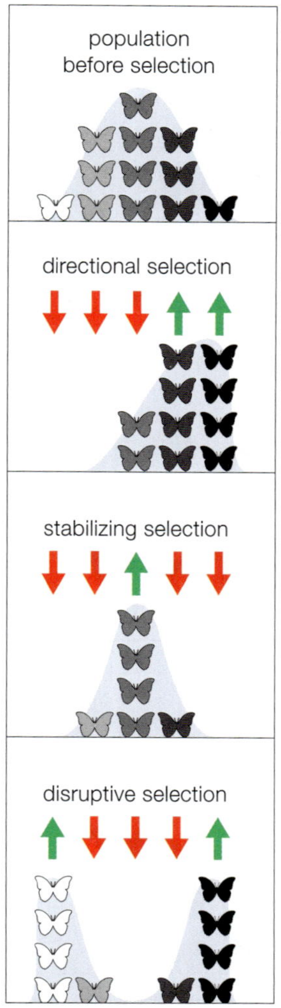

FIGURE 17.3 Overview of three modes of natural selection.

The remainder of this chapter explores the mechanisms and effects of processes that drive evolution, including natural selection. Remember from Section 16.2 that natural selection is a process in which environmental pressures result in the differential survival and reproduction of individuals of a population. It influences the frequency of alleles in a population by operating on phenotypes with a heritable, genetic basis.

We observe different patterns of natural selection, depending on the selection pressures and the organisms involved. Sometimes, individuals with a trait at one extreme of a range of variation are selected against, and those at the other extreme are adaptive. We call this pattern directional selection. With stabilizing selection, midrange forms of a trait are adaptive, and the extremes are selected against. With disruptive selection, forms at the extremes of the range of variation are adaptive, and the intermediate forms are selected against. We will discuss these three modes of natural selection, which **FIGURE 17.3** summarizes, in the following two sections.

Section 17.6 explores sexual selection, a mode of natural selection that operates on a population by influencing mating success. This section also discusses balanced polymorphism, a particular case in which natural selection maintains a relatively high frequency of multiple alleles in a population.

Natural selection and other processes of evolution can alter a population so much that it becomes a new species. We discuss mechanisms of speciation in the final sections.

TAKE-HOME MESSAGE 17.3

Natural selection, one of the most influential processes of evolution, can increase, decrease, or maintain the frequency of specific traits in a population.

17.4 WHAT IS THE EFFECT OF DIRECTIONAL SELECTION?

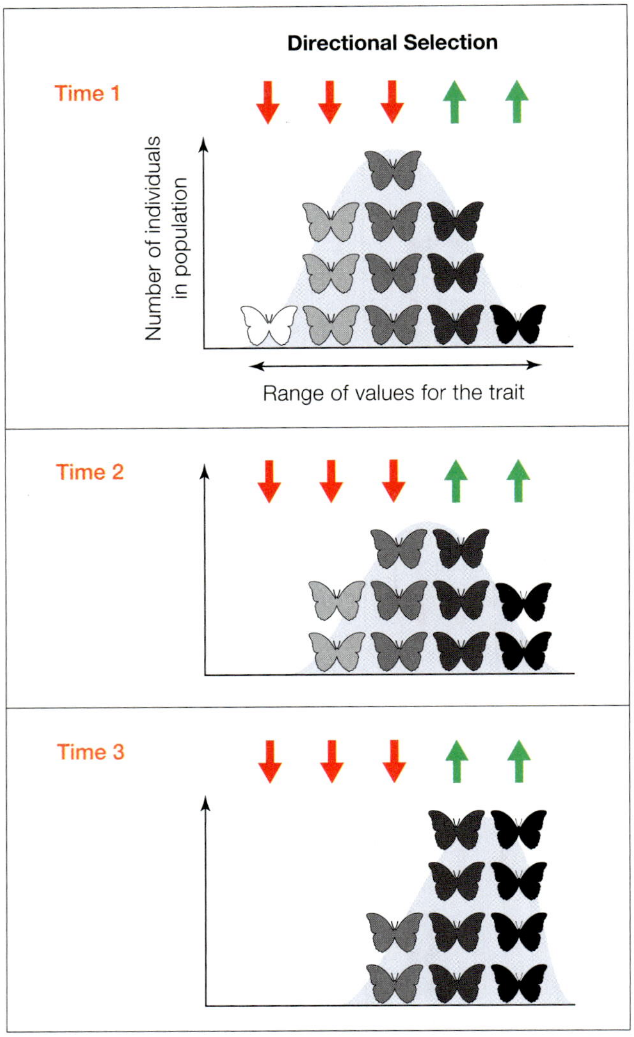

FIGURE 17.4 {Animated} With directional selection, a form of a trait at one end of a range of variation is adaptive. Bell-shaped curves indicate continuous variation. Red arrows indicate which forms are being selected against; green, forms that are adaptive.

A Light-colored moths on a nonsooty tree trunk (*top*) are hidden from predators; dark ones (*bottom*) stand out.

B In places where soot darkens tree trunks, the dark color (*bottom*) provides more camouflage than the light color (*top*).

FIGURE 17.5 {Animated} Adaptive value of two color forms of the peppered moth.

Directional selection shifts allele frequencies in a consistent direction, so forms at one end of a range of phenotypic variation become more common over time (**FIGURE 17.4**). The following examples show how field observations provide evidence of directional selection.

THE PEPPERED MOTH

A well-documented change in the coloration of peppered moths illustrates how environmental change influences directional selection. Peppered moths feed and mate at night, then rest on trees during the day. In preindustrial England, the vast majority of peppered moths were white with black speckles, and a small number were much darker. At this time, the air was clean, and light-gray lichens grew on the trunks and branches of most trees. When light-colored moths rested on lichen-covered trees, they were well camouflaged, whereas darker moths were not (**FIGURE 17.5A**). By the 1850s, dark moths had become much more common than light moths. The industrial revolution had begun, and smoke emitted by coal-burning factories was killing lichens. Dark moths were better camouflaged on lichen-free, soot-darkened trees (**FIGURE 17.5B**).

Scientists suspected that predation by birds was the selective pressure that shaped moth coloration, and in the 1950s, H. B. Kettlewell set out to test this hypothesis. He bred dark and light moths in captivity, marked them for easy identification, then released them in several areas. His team recaptured more of the dark moths in the polluted areas and more light ones in the less polluted ones. The researchers also observed predatory birds eating more light-colored moths in soot-darkened forests, and more dark-colored moths in cleaner, lichen-rich forests. Dark-colored moths were clearly at a selective advantage in industrialized areas.

Pollution controls went into effect in 1952. As a result of improved environmental standards, tree trunks gradually became free of soot, and lichens made a comeback. Kettlewell observed that moth phenotypes shifted too: Wherever pollution decreased, the frequency of dark moths decreased as well. Recent research has confirmed Kettlewell's results implicating birds and soot as selective agents of peppered moth coloration. It has also shown that coloration in peppered moths is determined by a single gene. Individuals that carry a dominant allele of this gene are black; those homozygous for a recessive allele are lighter.

WARFARIN-RESISTANT RATS

Human attempts to control the environment can result in directional selection. Consider that the average city in the United States sustains about one rat for every ten people. Rats thrive in urban centers, where garbage is plentiful and

FIGURE 17.6 Rats thrive wherever people do. Spreading poisons around buildings and soil does not usually exterminate rat populations, which recover quickly. Rather, the practice exerts directional selection favoring resistant rats.

natural predators are not (**FIGURE 17.6**). Part of their success stems from an ability to reproduce very quickly: Rat populations can expand within weeks to match the amount of garbage available for them to eat.

For decades, people have been fighting back with dogs, traps, ratproof storage facilities, and poisons that include arsenic and cyanide. Baits laced with warfarin, an organic compound that interferes with blood clotting, became popular in the 1950s. Rats that ate the poisoned baits died within days after bleeding internally or losing blood through cuts or scrapes. Warfarin was extremely effective, and its impact on harmless species was much lower than that of other rat poisons. It quickly became the rat poison of choice. By 1980, however, about 10 percent of rats in urban areas were resistant to warfarin. What happened?

Warfarin exposure had exerted directional selection on rat populations, favoring a particular allele in the rats' gene pool. Warfarin inhibits the gene's product, an enzyme that recycles vitamin K after it has been used to activate blood clotting factors. A mutation had made the enzyme less active, but also insensitive to warfarin. "What happened" was evolution by natural selection. Rats with the normal allele died after eating warfarin. The lucky ones with a warfarin-resistance allele survived and passed it to their offspring. The rat populations recovered quickly, and a higher proportion of rats in the next generation carried the mutated allele. With each onslaught of warfarin, the frequency of this allele in the rat populations increased.

When warfarin resistance increased in rat populations, people stopped using this poison. The frequency of the warfarin-resistance allele in rat populations declined, probably because rats that carry the allele are not as healthy as ones that do not. Now, savvy exterminators in urban areas know that the best way to control a rat infestation is to exert another kind of selection pressure: Remove their source of food, which is usually garbage. Then the rats will eat each other.

directional selection Mode of natural selection in which phenotypes at one end of a range of variation are favored.

TAKE-HOME MESSAGE 17.4

Directional selection causes allele frequencies underlying a range of variation to shift in a consistent direction.

17.5 WHAT TYPES OF NATURAL SELECTION FAVOR INTERMEDIATE OR EXTREME FORMS OF TRAITS?

Natural selection does not always result in a directional shift in a population's range of phenotypes. In some cases, a midrange form of a trait is adaptive; in others, a midrange form is eliminated and the most extreme forms are adaptive.

STABILIZING SELECTION

Stabilizing selection tends to preserve midrange phenotypes in a population. With this mode of natural selection, an intermediate form of a trait is adaptive, and extreme forms are not (**FIGURE 17.7**).

Stabilizing selection maintains an intermediate body mass in populations of sociable weavers (**FIGURE 17.8**). These birds live in the African savanna, where they build

large communal nests (*left*), and their body mass has a genetic basis. Between 1993 and 2000, Rita Covas and her colleagues investigated selection pressures that operate on sociable weaver body mass by capturing and weighing thousands of birds before and after the breeding seasons. The results indicated that optimal body mass in sociable weavers is a trade-off between the risks of starvation and predation. Birds that carry less fat are more likely to starve than fatter birds. However, birds that carry more fat spend more time eating,

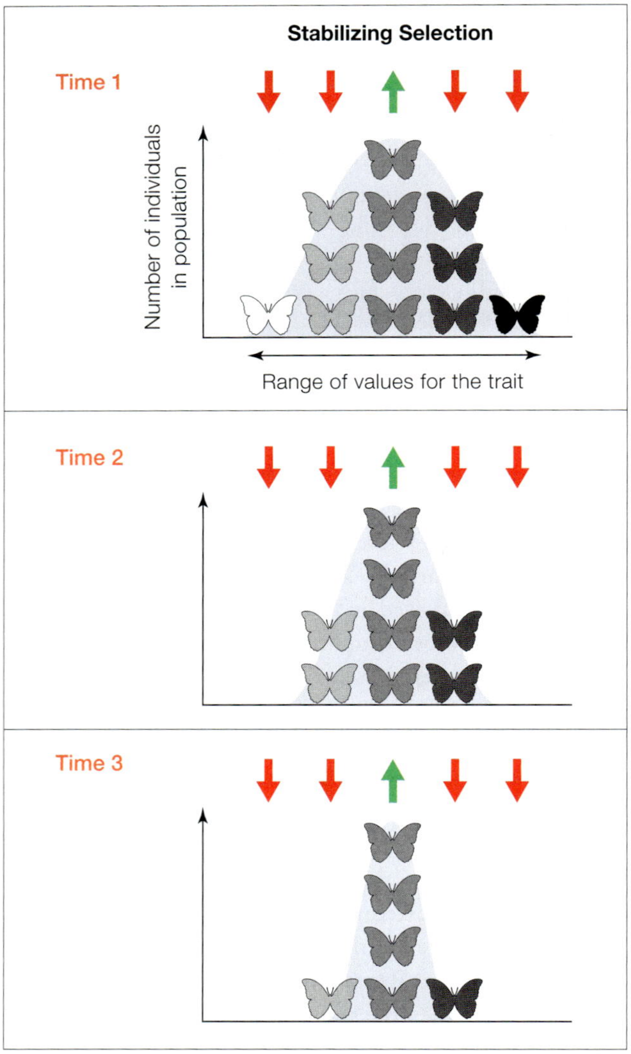

FIGURE 17.7 {Animated} Stabilizing selection eliminates extreme forms of a trait, and maintains an intermediate form. Red arrows indicate which forms are being selected against; green, the form that is adaptive. Compare the data set from a field experiment in **FIGURE 17.8**.

FIGURE 17.8 Stabilizing selection in sociable weavers (*top*). The graph (*bottom*) shows the number of birds (out of 977) that survived a breeding season.

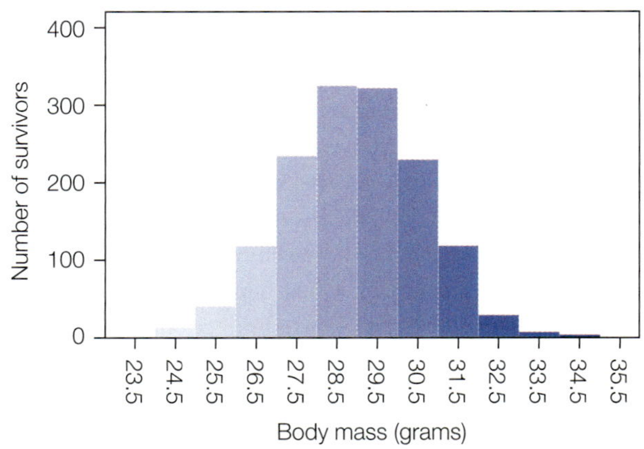

FIGURE IT OUT: What is the optimal mass of a sociable weaver?

Answer: About 29 grams

which in this species means foraging in open areas where they are easily accessible to predators. Fatter birds are also more attractive to predators, and not as agile when escaping. Thus, predators are agents of selection that eliminate the fattest individuals. Birds of intermediate weight have the selective advantage, and they make up the bulk of sociable weaver populations.

DISRUPTIVE SELECTION

With **disruptive selection**, forms of a trait at both ends of a range of variation are adaptive, and intermediate forms are not (**FIGURE 17.9**).

Consider the black-bellied seedcracker, a colorful finch species native to Cameroon, Africa. In these birds, there is a genetic basis for bill size. The bill of a typical black-bellied seedcracker, male or female, is either 12 millimeters wide, or wider than 15 millimeters (**FIGURE 17.10**). Birds with a bill size between 12 and 15 millimeters are uncommon. It is as if every human adult were 4 feet or 6 feet tall, with no one of intermediate height. Seedcrackers with the large and small bill forms inhabit the same geographic range, and they breed randomly with respect to bill size.

Environmental factors that affect feeding performance maintain the dimorphism in seedcracker bill size. The finches feed mainly on the seeds of two types of sedge, a grasslike plant. One sedge produces hard seeds; the other, soft seeds. Small-billed birds are better at opening the soft seeds, but large-billed birds are better at cracking the hard ones. During Cameroon's semiannual wet seasons, when both hard and soft sedge seeds are abundant, all seedcrackers feed on both types. During the region's dry seasons, when seeds become scarce and competition for food intensifies, each bird focuses on eating the seeds that it opens most efficiently: Small-billed birds feed mainly on soft seeds, and large-billed birds feed mainly on hard seeds. Birds with intermediate-sized bills cannot open either type of seed as efficiently as the other birds, so they are less likely to survive the dry seasons.

disruptive selection Mode of natural selection in which traits at the extremes of a range of variation are adaptive, and intermediate forms are not.

stabilizing selection Mode of natural selection in which an intermediate form of a trait is adaptive, and extreme forms are not.

TAKE-HOME MESSAGE 17.5

With stabilizing selection, an intermediate phenotype is adaptive, and extreme forms are selected against.

With disruptive selection, an intermediate form of a trait is selected against, and extreme phenotypes are adaptive.

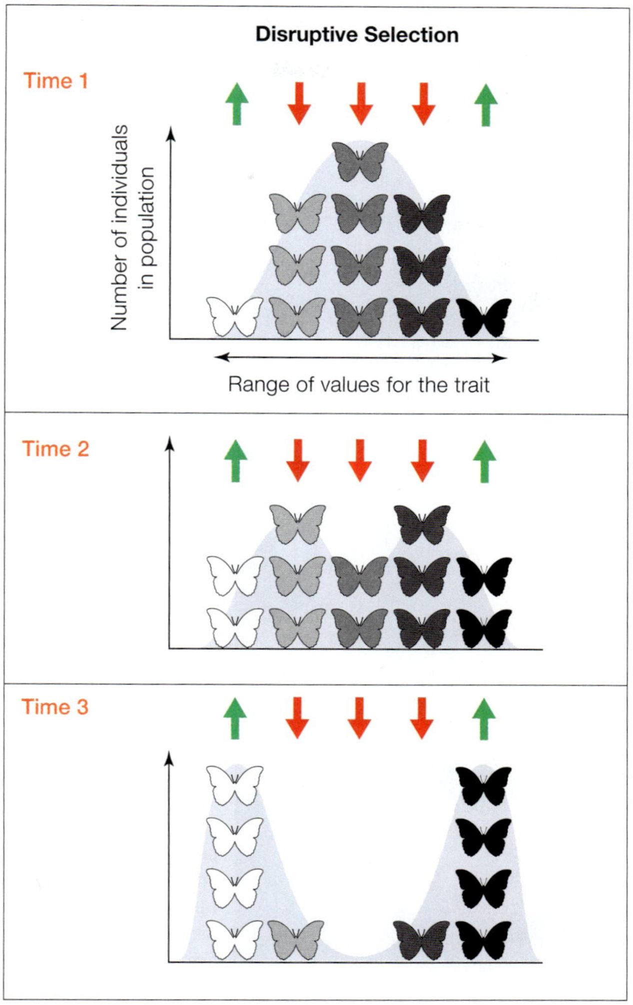

FIGURE 17.9 {Animated} Disruptive selection eliminates midrange forms of a trait, and maintains extreme forms. Red arrows indicate which forms are being selected against; green, the form that is adaptive.

FIGURE 17.10 {Animated} Disruptive selection in African seedcracker populations maintains a distinct dimorphism in bill size. Competition for scarce food during dry seasons favors birds with bills that are either 12 millimeters wide (*left*) or 15 to 20 millimeters wide (*right*). Birds with bills of intermediate size are selected against.

17.6 HOW DOES NATURAL SELECTION MAINTAIN DIVERSITY?

A Male elephant seals fight for sexual access to a cluster of females. Males of this species typically compete for access to the clusters of females.

B A male bird of paradise engaged in a flashy courtship display has caught the eye (and, perhaps, the sexual interest) of a female. Female birds of paradise are choosy; a male mates with any female that accepts him.

C Female stalk-eyed flies prefer to mate with males that have the longest eyestalks, a trait that provides no obvious selective advantage other than sexual attractiveness.

FIGURE 17.11 Sexual selection in action.

Selection pressures that operate on natural populations are often not as clear-cut as the examples in the previous sections might suggest. A particular form of a trait may be adaptive in one circumstance but harmful in another. Even individuals of the same species can play a role.

SURVIVAL OF THE SEXIEST

Not all evolution is driven by selection for traits that influence survival. Competition for mates is a selective pressure that shapes form and behavior in many species. Consider how males and females of many sexually reproducing species are different in size or another aspect of their appearance—a **sexual dimorphism**. Individuals of one sex (often males) are more colorful, larger, or more aggressive than individuals of the other sex. These traits can seem puzzling because they take energy and time away from activities that enhance survival, and some actually hinder an individual's ability to survive. Why, then, do they persist?

The answer is **sexual selection**, in which the evolutionary winners outreproduce others of a population because they are better at securing mates. With this mode of natural selection, the most adaptive forms of a trait are those that help individuals defeat rivals for mates, or are most attractive to the opposite sex.

For example, the females of some species cluster in defensible groups when they are sexually receptive, and males compete for sole access to the groups. Competition for the ready-made harems favors brawny, combative males (**FIGURE 17.11A**).

As another example, males or females that are choosy about mates act as selective agents on their own species. The females of some species shop for a mate among males that display species-specific cues such as a highly specialized appearance or courtship behavior (**FIGURE 17.11B**). The cues often include flashy body parts or movements, traits that tend to attract predators and in some cases are a physical hindrance. However, to a female member of the species, a flashy male's survival despite his obvious handicap may imply health and vigor, two traits that are likely to improve her chances of bearing healthy, vigorous offspring. Selected males pass alleles for their attractive traits to the

---

balanced polymorphism Maintenance of two or more alleles of a gene at high frequency in a population.
frequency-dependent selection Natural selection in which a trait's adaptive value depends on its frequency in a population.
sexual dimorphism Difference in appearance between males and females of a species.
sexual selection Mode of natural selection in which some individuals outreproduce others of a population because they are better at securing mates.

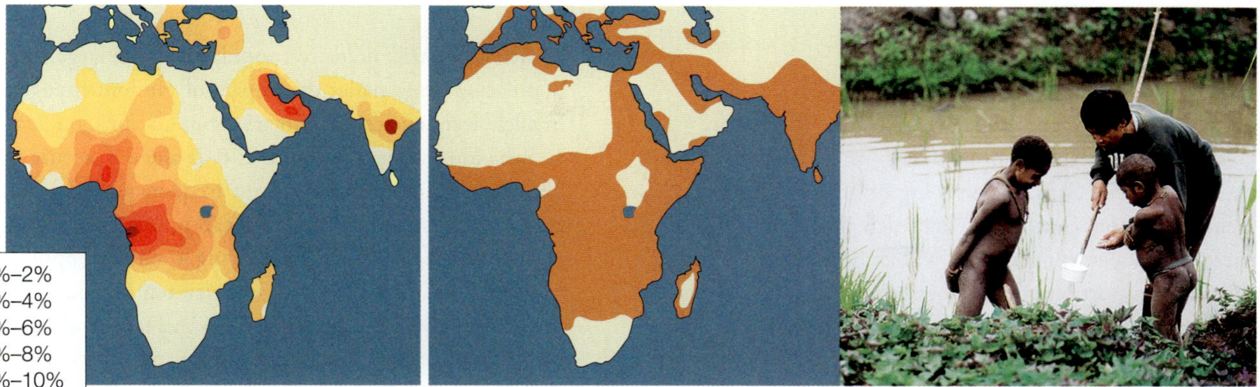

A Distribution (by percentage) of people who carry the sickle-cell allele.

B Distribution of malaria cases (orange) in Africa, Asia, and the Middle East in the 1920s, before the start of programs to control mosquitoes, which transmit the parasitic protist that causes the disease. Notice the correlation with the distribution of the sickle-cell allele in **A**. The photo shows a physician searching for mosquito larvae in Southeast Asia.

FIGURE 17.12 Malaria and sickle-cell anemia.

next generation of males, and females pass alleles that influence mate preference to the next generation of females. This type of sexual selection can result in highly exaggerated traits (**FIGURE 17.11C**).

MAINTAINING MULTIPLE ALLELES

Any mode of natural selection may keep two or more alleles of a gene circulating at relatively high frequency in a population's gene pool, a state called **balanced polymorphism**. For example, sexual selection maintains multiple alleles that govern eye color in populations of *Drosophila* fruit flies. Female flies prefer to mate with rare white-eyed males, until the white-eyed males become more common than red-eyed males, at which point the red-eyed flies are again preferred. This is also an example of **frequency-dependent selection**, in which the adaptive value of a particular form of a trait depends on its frequency in a population.

Balanced polymorphism can also arise in environments that favor heterozygous individuals (Section 13.1). Consider the gene that encodes the beta globin chain of hemoglobin. *HbA* is the normal allele; the codominant *HbS* allele carries a mutation that causes sickle-cell anemia (Section 9.5). Even with medical care, about 15 percent of individuals homozygous for the *HbS* allele die by age 18 from complications of the disorder.

Despite being so harmful, the *HbS* allele persists at very high frequency among the human populations in tropical and subtropical regions of Asia, Africa, and the Middle East. Why? Populations with the highest frequency of the *HbS* allele also have the highest incidence of malaria (**FIGURE 17.12**). Mosquitoes transmit the parasitic protist that causes malaria, *Plasmodium*, to human hosts (more about this in Section 20.5). *Plasmodium* multiplies in the liver and then in red blood cells. The cells rupture and release new parasites during recurring bouts of severe illness.

It turns out that people who make both normal and sickle hemoglobin are more likely to survive malaria than people who make only normal hemoglobin. In *HbA/HbS* heterozygous individuals, *Plasmodium*-infected red blood cells sometimes sickle. The abnormal shape brings the cells to the attention of the immune system, which destroys them along with the parasites they harbor. By contrast, *Plasmodium*-infected red blood cells of individuals homozygous for the normal *HbA* allele do not sickle, so the parasite may remain hidden from the immune system.

In areas where malaria is common, the persistence of the *HbS* allele is a matter of relative evils. Malaria and sickle-cell anemia are both potentially deadly. Heterozygous individuals are not completely healthy, but they do have a better chance of surviving malaria than those who carry two normal alleles. With or without malaria, people with both alleles are more likely to live long enough to reproduce than individuals who are homozygous for the *HbS* allele. The result is that nearly one-third of the people living in the most malaria-ridden regions of the world are heterozygous for the *HbS* allele.

TAKE-HOME MESSAGE 17.6

With sexual selection, adaptive forms of a trait are those that give an individual an advantage in securing mates.

Sexual selection can reinforce phenotypic differences between males and females, and sometimes it results in exaggerated traits.

Balanced polymorphism can be an outcome of frequency-dependent selection, or of environmental pressures that favor heterozygous individuals.

17.7 WHAT MECHANISMS OTHER THAN NATURAL SELECTION AFFECT ALLELE FREQUENCIES?

GENETIC DRIFT

Genetic drift is random change in an allele's frequency over time, brought about by chance alone. We explain genetic drift in terms of probability—the chance that some event will occur. Sample size is important in probability. Remember from Section 1.7 that each time you flip a coin, there is a 50 percent chance it will land heads up. With 10 flips, the proportion of times heads actually land up may be very far from 50 percent. With 1,000 flips, that proportion is more likely to be near 50 percent.

A The flour beetle (*Tribolium castaneum*), shown here on a flake of cereal.

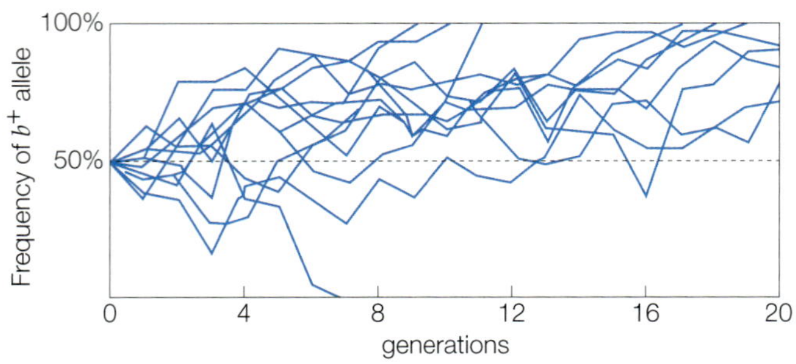

B The size of these populations was maintained at 10 breeding individuals. Allele b^+ was lost in one population (one graph line ends at 0).

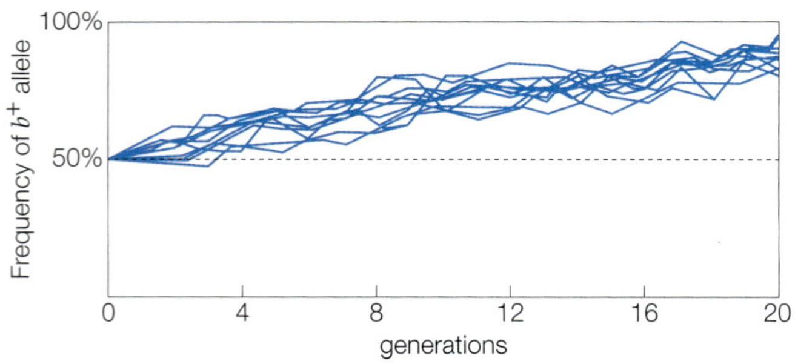

C The size of these populations was maintained at 100 individuals. Drift in these populations was less than the small populations in **B**.

FIGURE 17.13 {Animated} Genetic drift experiment in flour beetles (shown in **A**). Beetles heterozygous for alleles b^+ and b were maintained in populations of **B** 10 individuals or **C** 100 individuals for 20 generations. Graph lines in **C** are smoother than in **B**, indicating that drift was greatest in the sets of 10 beetles and least in the sets of 100. Notice that the average frequency of allele b^+ rose at the same rate in both groups, an indication that natural selection was at work too: Allele b^+ was weakly favored.

FIGURE IT OUT: In how many populations did allele b^+ become fixed?

Answer: Six

The same rule holds for populations: the larger the population, the smaller the impact of random changes in allele frequencies. Imagine two populations, one with 10 individuals, the other with 100. If allele *X* occurs in both populations at a 10 percent frequency, then only one person carries the allele in the small population. If that individual dies without reproducing, then the population's gene pool will lose allele *X*. However, ten individuals in the large population carry the allele. All ten would have to die without reproducing for the allele to be lost. Thus, the chance that the small population will lose allele *X* is greater than that for the large population. This is a general effect: The loss of genetic diversity is possible in all populations, but it is more likely in small ones (**FIGURE 17.13**). When all individuals of a population are homozygous for an allele, we say that the allele is **fixed**. The frequency of a fixed allele will not change unless a new mutation occurs, or an individual bearing another allele enters the population.

BOTTLENECKS AND THE FOUNDER EFFECT

A drastic reduction in population size, which is called a **bottleneck**, can greatly reduce genetic diversity. For example, northern elephant seals (shown in **FIGURE 17.11A**) underwent a bottleneck during the late 1890s, when hunting reduced their population size to about twenty individuals. Hunting restrictions have since allowed the population to recover, but genetic diversity among its members has been greatly reduced. The bottleneck and subsequent genetic drift eliminated many alleles that had been previously present in the population.

A loss of genetic diversity can also occur when a small group of individuals establishes a new population. If the founding group is not representative of the original population in terms of allele frequencies, then the new population will not be representative of it either. This outcome is called the **founder effect** (**FIGURE 17.14A**). Consider that all three *ABO* alleles for blood type (Section 13.4) are common in most human populations. Native Americans are an exception, with the majority of individuals

bottleneck Reduction in population size so severe that it reduces genetic diversity.

fixed Refers to an allele for which all members of a population are homozygous.

founder effect After a small group of individuals found a new population, allele frequencies in the new population differ from those in the original population.

gene flow The movement of alleles into and out of a population.

genetic drift Change in allele frequency due to chance alone.

inbreeding Mating among close relatives.

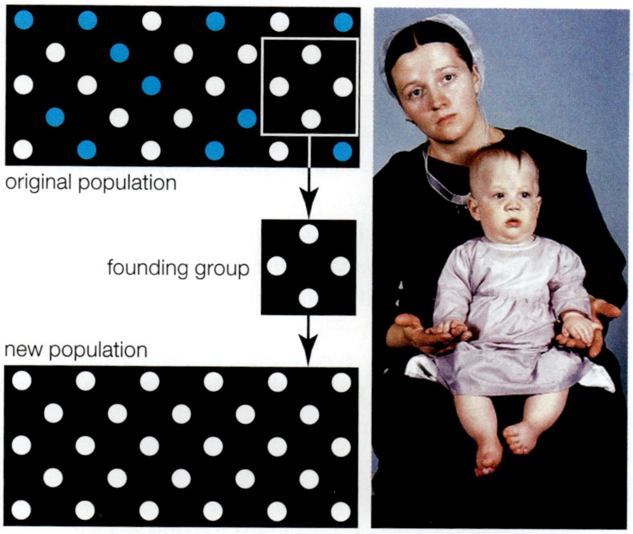

A The founder effect: a group that founds a new population is not representative of the original population. so allele frequencies differ between the new and the old populations.

B A high frequency of an allele that causes Ellis–van Creveld syndrome among the Lancaster Amish began with the founder effect.

FIGURE 17.14 The founder effect: mechanism and example.

being homozygous for the O allele. Native Americans are descendants of early humans who migrated from Asia between 14,000 and 21,000 years ago, across a narrow land bridge that once connected Siberia and Alaska. Analysis of DNA from ancient skeletal remains reveals that most early Americans were also homozygous for the O allele. Modern Siberian populations have all three alleles. Thus, the first humans in the Americas were probably members of a small group that had reduced genetic diversity compared with the general population.

Founding populations are often necessarily inbred. **Inbreeding** is breeding between close relatives. Closely related individuals tend to share more alleles than nonrelatives do, so inbred populations often have unusually high numbers of individuals homozygous for recessive alleles, some of which are harmful. This outcome is minimized in human populations where incest (mating between parents and children or between siblings) is discouraged or forbidden.

The Old Order Amish in Lancaster County, Pennsylvania, offer an example of the effects of inbreeding within human populations. Amish people marry only within their community. Intermarriage with other groups is not permitted, and no "outsiders" are allowed to join the community. As a result, Amish populations are moderately inbred, and many of their individuals are homozygous for harmful recessive alleles. The Lancaster population has an unusually high frequency of a recessive allele that causes Ellis–van Creveld syndrome, a genetic disorder characterized by dwarfism, polydactyly, and heart defects, among other symptoms. This allele has been traced to a man and his wife, two of a group of 400 Amish people who immigrated to the United States in the mid-1700s. As a result of the founder effect and inbreeding since then, about 1 of 8 people in the Lancaster population is now heterozygous for the allele, and 1 in 200 is homozygous for it (**FIGURE 17.14B**).

GENE FLOW

Individuals of natural populations tend to mate or breed most frequently with other members of their own population. However, not all populations of a species are completely isolated from one another, and nearby populations may occasionally interbreed. Also, individuals sometimes leave one population and join another. **Gene flow**, the movement of alleles between populations, occurs in both cases. Gene flow can change or stabilize allele frequencies, thus countering the evolutionary effects of mutation, natural selection, and genetic drift.

Gene flow is typical among populations of animals, but it also occurs in less mobile organisms. Consider the acorns

that jays disperse when they gather nuts for the winter (*left*). Every fall, these birds visit acorn-bearing oak trees repeatedly, then bury the acorns in the soil of territories that may be as much as a mile away. The jays transfer acorns (and the alleles carried by these seeds) among populations of oak trees that may otherwise be genetically isolated.

Gene flow also occurs when wind or animals transfer pollen from one plant to another, often over great distances. Many opponents of genetic engineering cite gene flow from transgenic crop plants into wild populations via pollen transfer. For example, herbicide-resistance genes and the *Bt* gene (Section 15.5) are now commonly found in weeds and unmodified crop plants. Long-term effects of this gene flow are currently unknown.

TAKE-HOME MESSAGE 17.7

Genetic drift can reduce a population's genetic diversity. Its effect is greatest in small populations.

A population's genetic diversity may be lowered after a bottleneck, or as a result of the founder effect.

Gene flow tends to oppose the evolutionary effects of mutation, natural selection, and genetic drift in a population.

17.8 HOW DO SPECIES ATTAIN AND MAINTAIN SEPARATE IDENTITIES?

Mutation, natural selection, and genetic drift operate on all natural populations, and they do so independently in populations that are not interbreeding. When gene flow does not keep two populations alike, different genetic changes accumulate in each one. Over time, the populations may become so different that we call them different species. The evolutionary process in which new species arise is called **speciation**.

Evolution is a dynamic, extravagant, messy, and ongoing process that can be challenging for people who like their categories neat. Speciation offers a perfect example, because it rarely occurs at a precise moment in time: Individuals often continue to interbreed even as populations are diverging, and populations that have already diverged may come together and interbreed again.

FIGURE 17.15 {Animated} Reproductive isolating mechanisms that can prevent interbreeding.

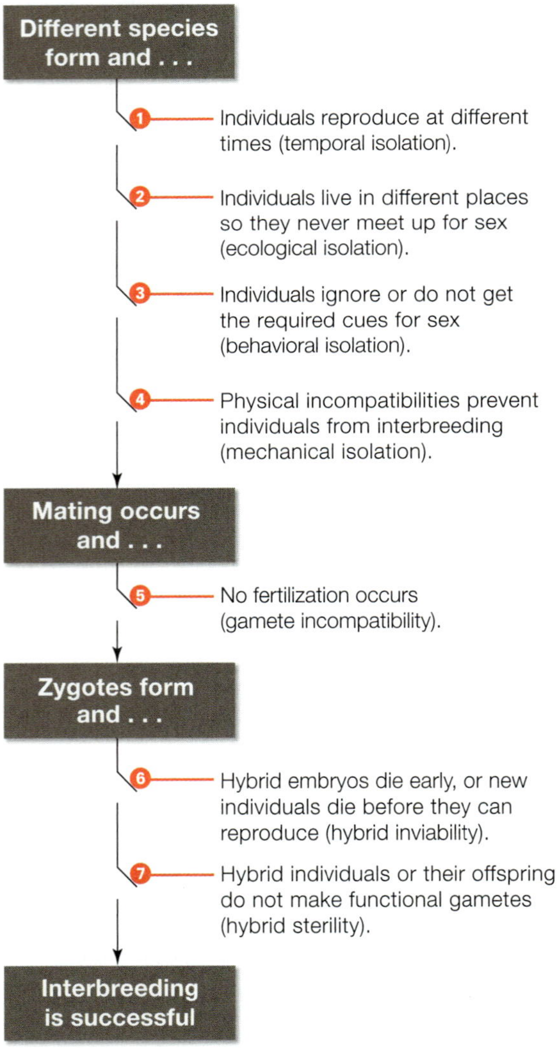

Every time speciation happens, it happens in a unique way, and each species is a product of its own unique evolutionary history. However, there are recurring patterns. For example, reproductive isolation is always part of speciation. **Reproductive isolation**, the end of gene flow between populations, is part of the process by which sexually reproducing species attain and maintain their separate identities. Mechanisms that prevent successful interbreeding reinforce differences between diverging populations (**FIGURE 17.15**).

MECHANISMS OF REPRODUCTIVE ISOLATION

1 Temporal Isolation Some closely related species cannot interbreed because the timing of their reproduction differs. The periodical cicada (*left*) offers an example. Cicada larvae feed on roots as they mature underground, then the adults emerge to reproduce. Three cicada species reproduce every 17 years. Each has a sibling species with nearly identical form and behavior, except that the siblings emerge on a 13-year cycle instead of a 17-year cycle. Sibling species have the potential to interbreed, but they can only get together once every 221 years!

2 Ecological Isolation Closely related species that are adapted to different microenvironments in the same region may be ecologically isolated. For example, two species of manzanita, a plant native to the Sierra Nevada mountain range, rarely hybridize. One species that lives high up on dry, rocky hillsides is better adapted for conserving water. The other, less drought-adapted species lives on lower slopes where water stress is not as intense. The physical separation makes cross-pollination unlikely.

3 Behavioral Isolation In animals, behavioral differences can stop gene flow between related species. Males and females of many species engage in courtship displays before sex (the chapter opener photo shows an example). In a typical pattern, the female recognizes the sounds and movements of a male of her species as an overture to sex, but females of different species do not.

❹ Mechanical Isolation The size or shape of an individual's reproductive parts may prevent it from mating with members of closely related species. For example, plants called black sage and white sage grow in the same areas, but hybrids rarely form because the flowers of these two related species have become specialized for different pollinators (**FIGURE 17.16**).

❺ Gamete Incompatibility Even if gametes of different species do meet up, they often have molecular incompatibilities that prevent a zygote from forming. For example, the molecular signals that trigger pollen germination in flowering plants are species-specific. Gamete incompatibility may be the primary speciation route among animals that release their eggs and free-swimming sperm into water.

❻ Hybrid Inviability Genetic changes are the basis of divergences in form, function, and behavior. Even chromosomes of species that diverged relatively recently may be different enough that a hybrid zygote ends up with extra or missing genes, or genes with incompatible products. Such outcomes typically disrupt embryonic development. Hybrids that do survive embryonic development often have reduced fitness. For example, hybrid offspring of lions and tigers have more health problems and a shorter life expectancy than individuals of either parent species.

❼ Hybrid Sterility Some interspecies crosses produce robust but sterile offspring. For example, mating between a female horse (64 chromosomes) and a male donkey (62 chromosomes) produces a mule. Mules are healthy, but their 63 chromosomes cannot pair up evenly during meiosis, so this animal makes few viable gametes.

If hybrids are fertile, their offspring usually have lower and lower fitness with each successive generation. Incompatible nuclear and mitochondrial DNA may be the cause (mitochondrial DNA is inherited from the mother only).

reproductive isolation The end of gene flow between populations.
speciation Evolutionary process in which new species arise.

TAKE-HOME MESSAGE 17.8

Speciation is an evolutionary process in which new species form. It varies in its details and duration.

Reproductive isolation, which occurs by one of several mechanisms, is always a part of speciation.

A Black sage is pollinated mainly by honeybees and other small insects.

B The flowers of black sage are too delicate to support larger insects. Big insects access the nectar of small sage flowers only by piercing from the outside, as this carpenter bee is doing. When they do so, they avoid touching the flower's reproductive parts.

C The reproductive parts (anthers and stigma) of white sage flowers are too far away from the petals to be brushed by honeybees, so honeybees are not efficient pollinators of this species. White sage is pollinated mainly by larger bees and hawkmoths, which brush the flower's stigma and anthers as they pry apart the petals to access nectar.

FIGURE 17.16 Mechanical isolation in sage.

17.9 WHAT IS ALLOPATRIC SPECIATION?

Genetic changes that lead to a new species can begin with physical separation between populations. With **allopatric speciation**, a physical barrier arises and separates two populations, ending gene flow between them (*allo*– means different; *patria*, fatherland). Then, reproductive isolating mechanisms evolve that prevent interbreeding even if the diverging populations do meet up again later.

Gene flow between populations separated by distance is often inconsistent. Whether a geographic barrier can completely block that gene flow depends on how the species travels (such as by swimming, walking, or flying), and how it reproduces (for example, by internal fertilization or by pollen dispersal).

A geographic barrier can arise in an instant, or over an eon. The Great Wall of China is an example of a barrier that arose abruptly. As it was being built, the wall interrupted gene flow among nearby populations of insect-pollinated plants; DNA sequence comparisons show that trees, shrubs, and herbs on either side of the wall are diverging genetically. Geographic isolation usually occurs much more slowly. For example, it took millions of years of tectonic plate movements (Section 16.5) to bring the two continents of North and South America close enough to collide. The land bridge where the two continents now connect is called the Isthmus of Panama. When this isthmus formed about 4 million years ago, it cut off the flow of water—and gene flow among populations of aquatic organisms—as it separated one large ocean into what are now the Pacific and Atlantic Oceans (**FIGURE 17.17**).

SPECIATION IN ARCHIPELAGOS

New species rarely form on island chains such as the Florida Keys that are in close proximity to a mainland. Being close to a mainland means gene flow is essentially unimpeded between island and mainland populations. By contrast, allopatric speciation is common on archipelagos (island chains) such as the Hawaiian and Galápagos Islands. These island chains are so geographically isolated that, for most species, gene flow does not occur between island and mainland populations.

The Hawaiian archipelago includes 19 islands and more than 100 atolls stretching 1,500 miles in the Pacific Ocean. These islands arose from hot spots on the ocean floor (Section 16.5). Because they were the tops of volcanoes, we can assume that their fiery surfaces were initially barren and inhospitable to life. Later, winds and currents carried a few individuals of mainland species to them. The individuals reproduced, and their descendants established populations. The lack of gene flow with mainland populations allowed the island populations to diverge. Today, thousands of

FIGURE 17.17 An example of allopatric speciation. When the Isthmus of Panama formed 4 million years ago, it cut off gene flow among ocean-dwelling populations of snapping shrimp. Today, shrimp species on opposite sides of the isthmus are so similar that they might interbreed, but they are behaviorally isolated: Instead of mating when they are brought together, they snap their claws at one another aggressively. The photos show two of the many closely related species that live on opposite sides of the isthmus.

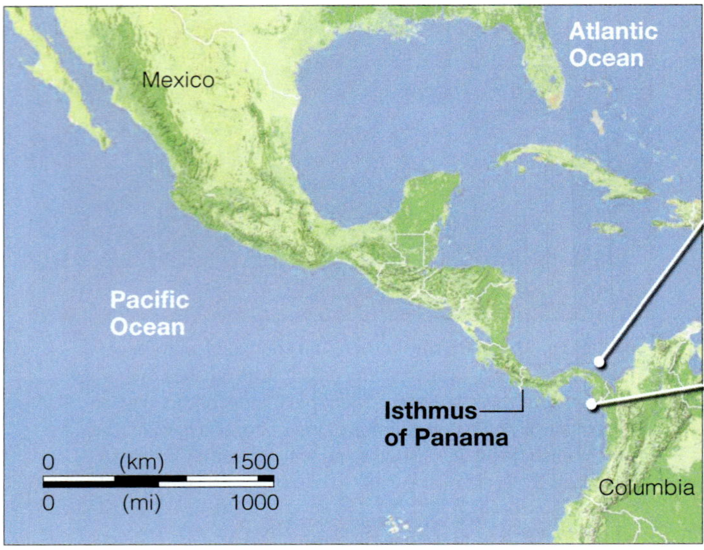

CREDITS: (17) left, © Cengage Learning 2011; right, © Arthur Anker.

FIGURE 17.18 {Animated} Allopatric speciation gave rise to the Hawaiian honeycreepers. The Hawaiian Islands (*right*) are separated from mainland continents by thousands of miles of open ocean—a geographic barrier that prevented gene flow between island colonizers and mainland populations. DNA sequence comparisons suggest that the ancestor of all Hawaiian honeycreepers resembled the housefinch (*Carpodacus*, *left*).

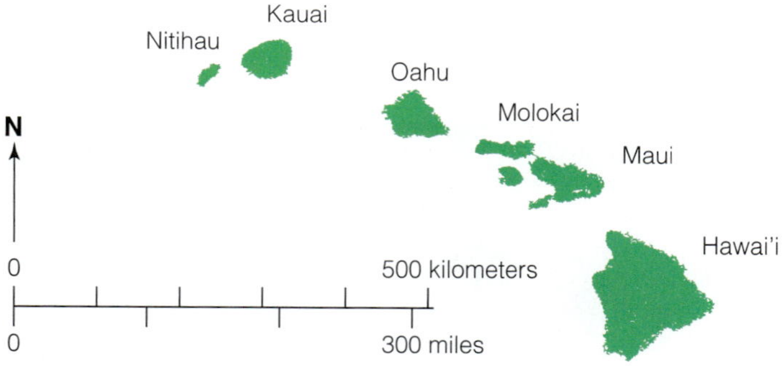

species are unique to this island chain. Consider Hawaiian honeycreepers, birds that are descendants of a mainland finch species that arrived on the islands about 3.5 million years ago. A buffet of fruits, seeds, nectars, tasty insects, and the near absence of competitors and predators allowed the finch's descendants to thrive. In the absence of gene flow, the island finch population diverged from the ancestral mainland species. Further divergences occurred because habitats on the landmasses of the Hawaiian archipelago vary dramatically—from lava beds, rain forests, and grasslands to dry woodlands and snow-capped peaks. Selection pressures differ within and between these habitats. The cumulative result of all these divergences is a spectacular array of honeycreeper species (**FIGURE 17.18**).

allopatric speciation Speciation pattern in which a physical barrier ends gene flow between populations.

TAKE-HOME MESSAGE 17.9

A physical barrier that intervenes between populations or subpopulations of a species prevents gene flow among them. When gene flow ends, genetic divergences give rise to new species. This process is allopatric speciation.

17.10 CAN SPECIATION OCCUR WITHOUT A PHYSICAL BARRIER TO GENE FLOW?

SYMPATRIC SPECIATION

In **sympatric speciation**, populations inhabiting the same geographic region speciate in the absence of a physical barrier between them (*sym–* means together).

Sympatric speciation can occur in a single generation when the chromosome number multiplies. Polyploidy (having three or more sets of chromosomes, Section 14.5) typically arises when an abnormal nuclear division during meiosis or mitosis doubles the chromosome number. For example, if the nucleus of a somatic cell in a flowering plant fails to divide during mitosis, the resulting polyploid cell may proliferate and give rise to shoots and flowers. If the flowers can self-fertilize, a new polyploid species may be the result. Common bread wheat originated after related species hybridized, and then the chromosome number of the hybrid offspring doubled (**FIGURE 17.19**). Today, about 95 percent of ferns and 70 percent of flowering plant species are polyploid, as well as a few conifers, insects and other arthropods, mollusks, fishes, amphibians, and reptiles.

Sympatric speciation can also occur with no change in chromosome number. The mechanically isolated sage plants you just learned about speciated with no physical barrier to gene flow. As another example, more than 500 species of cichlid fishes arose by sympatric speciation in the shallow waters of Lake Victoria. This large freshwater lake sits isolated from river inflow on an elevated plain in Africa's Great Rift Valley. Since Lake Victoria formed about 400,000 years ago, it has dried up three times. DNA sequence comparisons indicate that almost all of the cichlid species in this lake arose since the last dry spell, which was 12,400 years ago. How could hundreds of species arise so quickly? In this case, the answer begins with differences in the water—color of ambient light and clarity—in different parts of the lake. The light in the lake's shallower, clear water is mainly blue; the light that penetrates the deeper, muddier water is mainly red. The cichlids vary in color and in patterning (**FIGURE 17.20**). Outside of captivity, female cichlids rarely mate with males of other species. Given a choice, they prefer to mate with brightly colored males of their own species. Their preference has a basis in genes that encode light-sensitive pigments of the retina (part of the eye). Retinal pigments made by species that live mainly in shallow areas of the lake are more sensitive to blue light. The males of these species are also the bluest. Retinal pigments made by species that live mainly in deeper areas of the lake are more sensitive to red light. Males of these species are redder. In other words, the colors that a female cichlid sees best are the same colors displayed by males of her species. Thus, mutations in genes that affect color perception are likely to affect a female's choice of mates. Such mutations are probably the way sympatric speciation occurs in these fish.

FIGURE 17.19 {Animated} Sympatric speciation in wheat. The wheat genome, which consists of seven chromosomes, occurs in slightly different forms called A, B, C, D, and so on. Many wheat species are polyploid, carrying more than two copies of the genome. For example, modern bread wheat (*Triticum aestivum*) is hexaploid, with six copies of the wheat genome: two each of genomes A, B, and D (or 42 AABBDD).

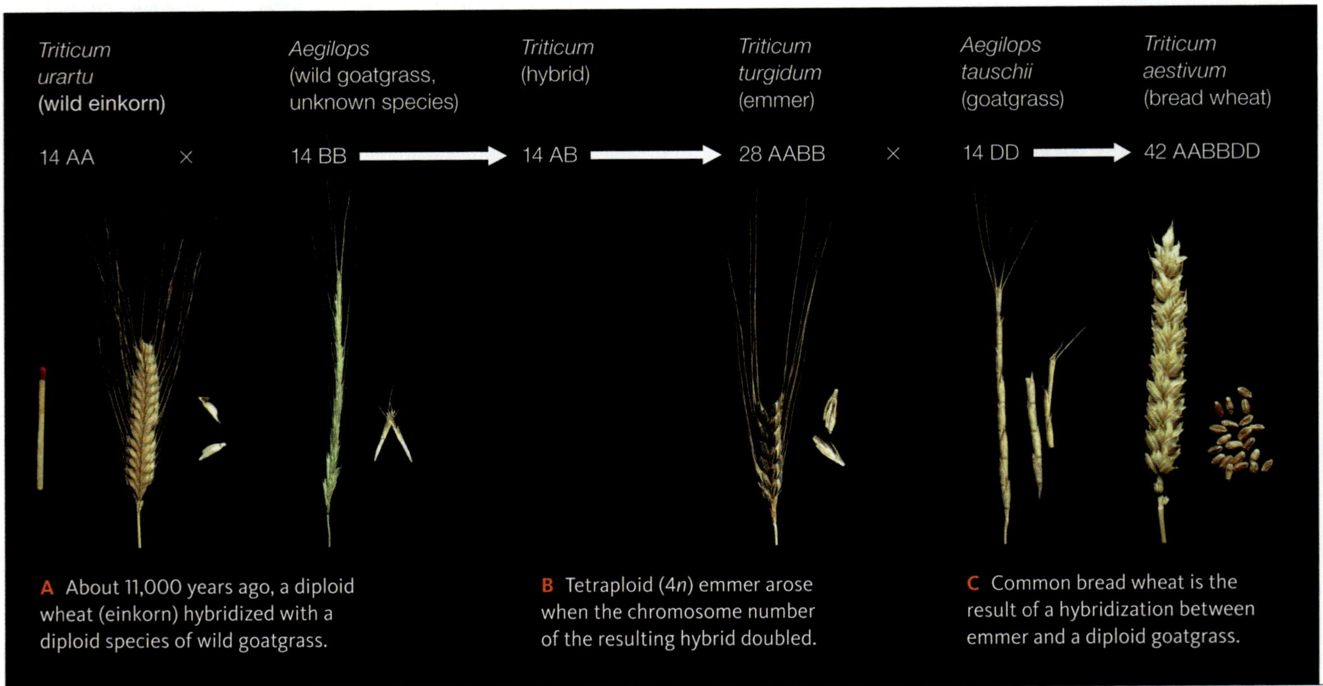

A About 11,000 years ago, a diploid wheat (einkorn) hybridized with a diploid species of wild goatgrass.

B Tetraploid (4*n*) emmer arose when the chromosome number of the resulting hybrid doubled.

C Common bread wheat is the result of a hybridization between emmer and a diploid goatgrass.

CREDIT: (19) Photos by © J. Honegger, courtesy of S. Stamp, E. Merz, www.sortengarten/ethz.ch.

FIGURE 17.20 Red fish, blue fish: Males of four closely related species of cichlid native to Lake Victoria, Africa. Hundreds of cichlid species arose by sympatric speciation in this lake. Mutations in genes that affect females' perception of the color of ambient light in deeper or shallower regions of the lake also affect their choice of mates. Female cichlids prefer to mate with brightly colored males of their own species.

FIGURE IT OUT: What form of natural selection has been driving sympatric speciation in Lake Victoria cichlids?

Answer: Sexual selection

PARAPATRIC SPECIATION

With **parapatric speciation**, adjacent populations speciate despite being in contact across a common border. Divergences spurred by local selection pressures are reinforced because hybrids that form in the contact zone are less fit than individuals on either side of it.

Consider velvet walking worms, which resemble caterpillars but may be more related to spiders: They are predatory, and shoot streams of glue from their head to entangle insect prey. Two rare species of velvet walking worm are native to the island of Tasmania. The giant velvet walking worm and the blind velvet walking worm can interbreed, but they only do so in a tiny area where their habitats overlap. Hybrid offspring are sterile, which may be the main reason the two species are maintaining separate identities in the absence of a physical barrier between their adjacent populations.

parapatric speciation Speciation pattern in which populations speciate while in contact along a common border.
sympatric speciation Speciation pattern in which speciation occurs within a population, in the absence of a physical barrier to gene flow.

TAKE-HOME MESSAGE 17.10

With sympatric speciation, divergence within an interbreeding population leads to new species that inhabit the same area, with no physical barrier to gene flow.

With parapatric speciation, populations maintaining contact along a common border evolve into distinct species.

PEOPLE MATTER

National Geographic Grantee
DR. JULIA J. DAY

Understanding processes that lead to speciation is fundamental to explaining the diversity of life. National Geographic grantee Julia Day studies why some environments give rise to and maintain higher species richness than others. Her research focuses on comparing the staggering biodiversity of three East African great lakes: Tanganyika, Victoria, and Malawi. Cichlid fishes in these lakes evolved into "flocks" of many hundred species, most of which are close relatives. Each species flock displays astonishing levels of ecological, phenotypic, and behavioral diversity. In addition to being generally very colorful, these fish have a stunning variety of adaptations: unusual dietary habits, clever feeding behaviors, and sophisticated reproductive behaviors and parental care strategies.

Lake Tanganyika, despite being larger and older than Lakes Victoria and Malawi, harbors only about one-third the number of cichlid species in the other two lakes. Why the difference? Day and her team analyzed mitochondrial DNA samples from almost all of the Lake Tanganyika cichlids. Her results indicate that species in this lake diversified from several distinct lineages, rather than from a single ancestor (as occurred in Lakes Victoria and Malawi). Cichlid diversifications in Lake Tanganyika occurred several times, possibly coinciding with periods of changing water levels in the lake, or with successive invasions of new ancestral cichlid species. Speciation also occurred about six times more slowly than in the other lakes—a rate more in line with typical speciation rates of plants and animals in other environments.

Something about Lake Tanganyika was less conducive for cichlid speciation than the other lakes. Day thinks that diversification may have been relatively inhibited by the presence of older species already filling Lake Tanganyika's niches. Unlike Lakes Victoria and Malawi, Lake Tanganyika has not completely dried up in the past.

As these unique ecosystems are threatened from increasing populations and climate change, understanding and assessing the origins and maintenance of their biodiversity are of high priority.

CREDITS: (20) Kevin Bauman, www.african-cichlid.com; (in text) Courtesy of Julia J. Day.

17.11 WHAT IS MACROEVOLUTION?

FIGURE 17.21 An example of stasis. *Left*, compare a 320-million-year-old coelacanth fossil found in Montana with a live coelacanth. *Right*, a few of the coelacanth's unusual ancestral features that have been lost in almost all other fish lineages over evolutionary time.

Microevolution is change in allele frequencies within a single species or population. **Macroevolution** is our name for evolutionary patterns on a larger scale: trends such as land plants evolving from green algae, the dinosaurs disappearing in a mass extinction, a burst of divergences from a single species, and so on.

PATTERNS OF MACROEVOLUTION

Stasis

With the simplest macroevolutionary pattern, **stasis**, a lineage persists for millions of years with little evolutionary change. Consider coelacanths, an order of ancient lobe-finned fish that had been assumed extinct for at least 70 million years until a fisherman caught one in 1938. The modern coelacanth species are very similar to fossil specimens hundreds of millions of years old (**FIGURE 17.21**).

Exaptation

Major evolutionary novelties often stem from the adaptation of an existing structure for a completely new purpose. This macroevolutionary pattern is called **exaptation**. For example, the feathers that allow modern birds to fly are derived from feathers that first evolved in some dinosaurs. Those dinosaurs could not have used their feathers for flight, but they probably did use them for insulation. Thus, we say that flight feathers in birds are an exaptation of insulating feathers in dinosaurs.

Mass Extinctions

By current estimates, more than 99 percent of all species that ever lived are now **extinct**, which means they no longer have living members. In addition to continuing small-scale extinctions, the fossil record indicates that there have been more than twenty mass extinctions, which are simultaneous losses of many lineages. These include five catastrophic events in which the majority of species on Earth disappeared (Section 16.6).

Adaptive Radiation

With **adaptive radiation**, one lineage rapidly diversifies into several new species. Adaptive radiation can occur after individuals colonize a new environment that has a variety of different habitats with few or no competitors. The adaptation of populations to different regions of the new environment produces many new species. The Hawaiian honeycreepers arose this way, as did the Lake Victoria cichlids.

Adaptive radiation may also occur after a key innovation evolves. A **key innovation** is a new trait that allows its bearer to exploit a habitat more efficiently or in a novel way. The evolution of lungs offers an example, because lungs were a key innovation that opened the way for an adaptive radiation of vertebrates on land.

Adaptive radiation can also occur after geologic or climatic events eliminate some species from a habitat. The surviving species then have access to resources from

adaptive radiation A burst of genetic divergences from a lineage gives rise to many new species.
coevolution The joint evolution of two closely interacting species; each species is a selective agent for traits of the other.
exaptation Evolutionary adaptation of an existing structure for a completely new purpose.
extinct Refers to a species that no longer has living members.
key innovation An evolutionary adaptation that gives its bearer the opportunity to exploit a particular environment much more efficiently or in a new way.
macroevolution Large-scale evolutionary patterns and trends.
stasis Evolutionary pattern in which a lineage persists with little or no change over evolutionary time.

which they had previously been excluded. This is the way mammals were able to undergo an adaptive radiation after the dinosaurs disappeared.

Coevolution The process by which close ecological interactions between two species cause them to evolve jointly is called **coevolution**. One species acts as an agent of selection on the other, and each adapts to changes in the other. Over evolutionary time, the two species may become so interdependent that they can no longer survive without one another.

Relationships between coevolved species can be quite intricate. Consider the large blue butterfly (*Maculinea arion*), a parasite of ants (**FIGURE 17.22**). After hatching, the larvae (caterpillars) feed on wild thyme flowers and then drop to the ground. An ant that finds a caterpillar strokes it, which makes the caterpillar exude honey. The ant eats the honey and continues to stroke the caterpillar, which secretes more honey. This interaction continues for hours, until the caterpillar suddenly hunches itself up into a shape that appears (to an ant) very much like an ant larva. The deceived ant then picks up the caterpillar and carries it back to the ant nest, where, in most cases, other ants kill it—except, however, if the ants are of the species *Myrmica sabuleti*. Secretions of the caterpillar fool these ants into treating it just like a larva of their own. For the next 10 months, the caterpillar lives in the nest and grows to gigantic proportions by feeding on ant larvae. After it metamorphoses into a butterfly, the insect emerges from the ground to mate. Eggs are deposited on wild thyme near another *M. sabuleti* nest, and the cycle starts anew. This relationship between ant and butterfly is typical of coevolved relationships in that it is extremely specific. Any increase in the ants' ability to identify a caterpillar in their nest selects for caterpillars that better deceive the ants, which in turn select for ants that can better identify the caterpillars. Each species exerts directional selection on the other.

A To a *Myrmica sabuleti* ant, a honey-exuding, hunched-up *Maculinea arion* caterpillar appears to be an ant larva. This deceived ant is preparing to carry the caterpillar back to its nest, where the caterpillar will eat ant larvae for the next 10 months until it pupates.

B A *Maculinea arion* butterfly emerges from the pupa and lays its eggs on wild thyme flowers. Larvae that emerge from the eggs will survive only if a colony of *Myrmica sabuleti* ants adopts them.

FIGURE 17.22 An example of coevolved species.

EVOLUTIONARY THEORY

Biologists do not doubt that macroevolution occurs, but many disagree about how it occurs. However we choose to categorize evolutionary processes, the very same genetic change may be at the root of all evolution—fast or slow, large-scale or small-scale. Dramatic jumps in morphology, if they are not artifacts of gaps in the fossil record, may be the result of mutations in homeotic or other regulatory genes. Macroevolution may include more processes than microevolution, or it may not. It may be an accumulation of many microevolutionary events, or it may be an entirely different process. Evolutionary biologists may disagree about these and other hypotheses, but all of them are trying to explain the same thing: how all species are related by descent from common ancestors.

TAKE-HOME MESSAGE 17.11

Macroevolution comprises large-scale patterns of evolutionary change such as adaptive radiation, the origin of major groups, and mass extinctions.

17.12 WHY DO WE STUDY EVOLUTIONARY HISTORY?

TABLE 17.2

Examples of Characters

	Bird	Bat	Lion
Warm-blooded	Y	Y	Y
Hair	N	Y	Y
Milk	N	Y	Y
Teeth	N	Y	Y
Wings	Y	Y	N
Feathers	Y	N	N

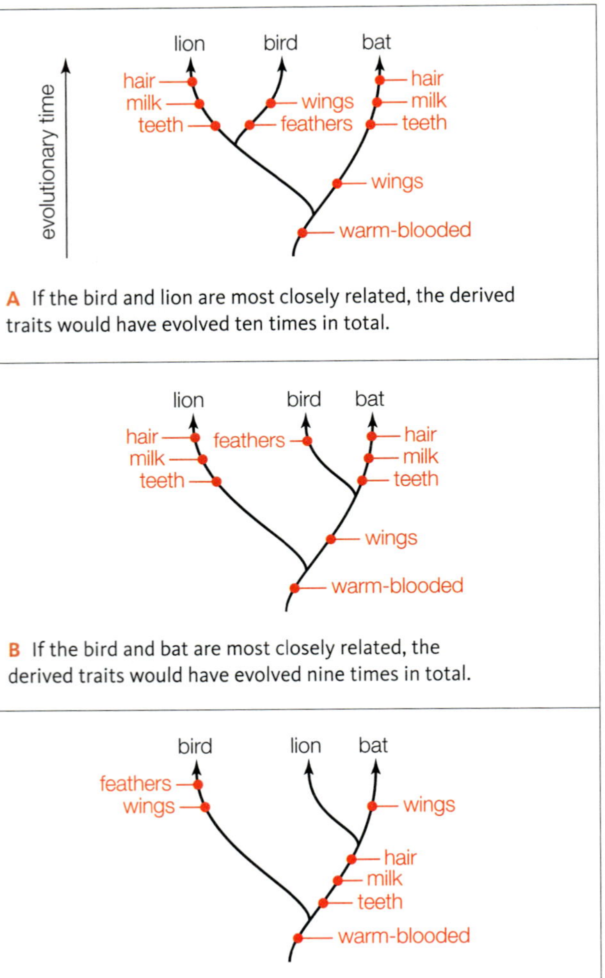

A If the bird and lion are most closely related, the derived traits would have evolved ten times in total.

B If the bird and bat are most closely related, the derived traits would have evolved nine times in total.

C If the lion and bat are most closely related, the derived traits would have evolved seven times in total.

FIGURE 17.23 An example of cladistics, using parsimony analysis with characters in **TABLE 17.2**.

A, **B**, and **C** show the three possible evolutionary pathways that could connect birds, bats, and lions; red indicates the evolution of a derived trait. The pathway most likely to be correct (**C**) is the simplest—the one in which the derived traits would have had to evolve the fewest number of times in total.

Classifying life's tremendous diversity into a series of taxonomic ranks (Section 1.4) is a useful endeavor, in the same way that it is useful to organize a telephone book or contact list in alphabetical order. Today, however, reconstructing evolutionary relationships among organisms has become at least as important as classifying them. Thus, biologists often focus on unraveling **phylogeny**, the evolutionary history of a species or a group of species. Phylogeny is a kind of genealogy that follows a lineage's evolutionary relationships through time.

Humans were not around to witness the evolution of most species, but we can use evidence to understand events in the ancient past (Chapter 16). For example, each species bears traces of its own unique evolutionary history in its characters. A **character** is a quantifiable, heritable trait, such as the nucleotide sequence of ribosomal RNA or the presence of wings (**TABLE 17.2**).

Traditional classification schemes group organisms based on shared characters: Birds have feathers, cacti have spines, and so on. By contrast, evolutionary biology tries to fit each species into a bigger picture of evolution: Every living thing is related if you just go back far enough in time. Evolutionary biologists pinpoint what makes the organisms share the characters in the first place: a common ancestor. They determine common ancestry by identifying derived traits. A **derived trait** is a character present in a group under consideration, but not in any of the group's ancestors. A group whose members share one or more defining derived traits is called a **clade**. By definition, a clade is a **monophyletic group**: one that consists of an ancestor (in which a derived trait evolved) together with any and all of its descendants.

Each species is a clade. Many higher taxonomic rankings are also equivalent to clades—flowering plants, for example, are both a phylum and a clade—but some are not. For example, the traditional Linnaean class Reptilia ("reptiles") includes crocodiles, alligators, tuataras, snakes, lizards, turtles, and tortoises. While it is convenient to classify these animals together, they would not constitute a clade unless birds are also included, as you will see in Chapter 24.

It is the recent nature of a derived trait that defines a clade. Consider how alligators look a lot more like lizards than birds. In this case, the similarity in appearance does indicate shared ancestry, but it is a more distant relationship than alligators have with birds. Evolutionary biologists discovered that alligators and birds share a more recent common ancestor than alligators and lizards do. Derived traits—a gizzard and a four-chambered heart—evolved in the lineage that gave rise to alligators and birds, but not in the one that gave rise to lizards.

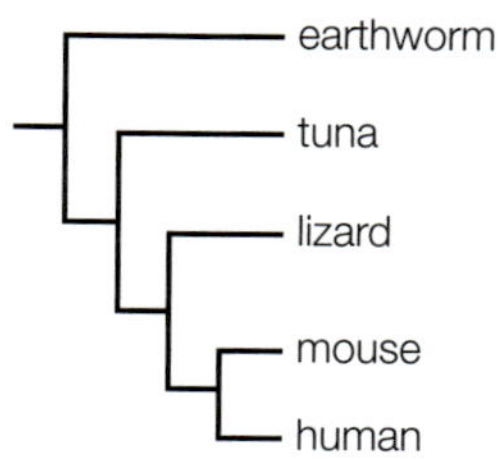

A Evolutionary connections among clades are represented as lines on a cladogram. Sister groups emerge from a node, which represents a common ancestor.

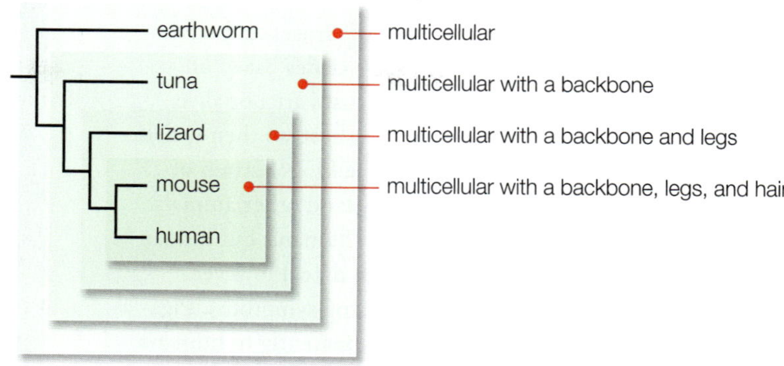

B A cladogram can be viewed as "sets within sets" of derived traits. Each set (an ancestor together with all of its descendants) is a clade.

FIGURE 17.24 {Animated} An example of a cladogram.

A species' ancestry remains the same no matter how it evolves. However, as with traditional taxonomy, we can make mistakes grouping organisms into clades if the information we have is incomplete. A clade is necessarily a hypothesis, and which organisms it includes may change when new discoveries are made. As with all hypotheses, the more data that support a cladistic grouping, the less likely it is to require revision.

CLADISTICS

In the big picture of evolution, all clades are interconnected; an evolutionary biologist's job is to figure out where the connections are. Making hypotheses about evolutionary relationships among clades is called **cladistics**. One way of doing this involves the logical rule of simplicity: When there are several possible ways that a group of clades can be connected, the simplest evolutionary pathway is probably the correct one. By comparing all of the possible connections among the clades, we can identify the simplest—the one in which the defining derived traits evolved the fewest number of times. The process of finding the simplest pathway is called parsimony analysis (**FIGURE 17.23**).

The result of a cladistic analysis is an **evolutionary tree**—a diagram of evolutionary connections—called a cladogram. **Cladograms** visually summarize our best data-supported hypotheses about how a group of clades are related (**FIGURE 17.24**). Data from an outgroup (a species not closely related to any member of the group under study) may be included in order to "root" the tree. Each line in a cladogram represents a lineage, which may branch into two lineages at a node. The node represents a common ancestor of two lineages. Every branch of a cladogram is a clade; the two lineages that emerge from a node on a cladogram are called **sister groups**.

character Quantifiable, heritable characteristic or trait.
clade A group whose members share one or more defining derived traits.
cladistics Making hypotheses about evolutionary relationships among clades.
cladogram Evolutionary tree diagram that shows evolutionary connections among a group of clades.
derived trait A novel trait present in a clade but not in the clade's ancestors.
evolutionary tree Diagram showing evolutionary connections.
monophyletic group An ancestor in which a derived trait evolved, together with all of its descendants.
phylogeny Evolutionary history of a species or group of species.
sister groups The two lineages that emerge from a node on a cladogram.

TAKE-HOME MESSAGE 17.12

Evolutionary biologists study phylogeny in order to understand how all species are connected by shared ancestry.

A clade is a monophyletic group whose members share one or more derived traits. Cladistics is a method of making hypotheses about evolutionary relationships among clades.

Cladograms and other evolutionary tree diagrams are hypotheses based on our best understanding of the evolutionary history of a group of organisms.

17.13 HOW CAN WE USE WHAT WE LEARN ABOUT EVOLUTIONARY HISTORY?

Researchers often study the evolution of viruses and other infectious agents by grouping them into clades based on biochemical characters. Even though viruses are not alive, they can mutate every time they infect a host, so their genetic material changes over time. Consider the H5N1 strain of influenza (flu) virus, which infects birds and other animals. H5N1 has a very high mortality rate in humans, but human-to-human transmission has been rare to date. However, the virus replicates in pigs without causing symptoms. Pigs transmit the virus to other pigs—and apparently to humans too. A phylogenetic analysis of H5N1 isolated from pigs showed that the virus "jumped" from birds to pigs at least three times since 2005, and that one of the isolates had acquired the potential to be transmitted among humans. An increased understanding of how this virus adapts to new hosts is helping researchers design more effective vaccines for it.

The story of the Hawaiian honeycreepers offers an example of how finding ancestral connections can help species that are still living. The first Polynesians arrived on the Hawaiian Islands sometime before 1000 A.D., and Europeans followed in 1778. Hawaii's rich ecosystem was hospitable to these newcomers and their domestic animals and crops. Escaped livestock began to eat and trample rain forest plants that had provided honeycreepers with food and shelter. Entire forests were cleared to grow imported crops, and plants that escaped cultivation began to crowd out native plants. Mosquitoes accidentally introduced in 1826 spread diseases such as avian malaria from imported chickens to native bird species. Stowaway rats and snakes ate their way through populations of native birds and their eggs. Mongooses deliberately imported to eat the rats and snakes preferred to eat birds and bird eggs.

The very isolation that had spurred adaptive radiations also made honeycreepers vulnerable to extinction. Divergence from the ancestral species had led to the loss of unnecessary traits such as defenses against mainland predators and diseases. Specializations such as extravagantly elongated beaks became hindrances when the birds' habitats suddenly changed or disappeared. Thus, at least 43 honeycreeper species that had thrived on the islands before humans arrived were extinct by 1778. Conservation efforts began in the 1960s, but 26 more species have since disappeared. Today, 35 of the remaining 68 species are endangered (**FIGURE 17.25**). They are still pressured by established populations of invasive, nonnative species of plants and animals. Rising global temperatures are also allowing mosquitoes to invade high-altitude habitats that had previously been too cold for the insects, so honeycreeper species remaining in these habitats are now succumbing to avian malaria and other mosquito-borne diseases.

A The palila has an adaptation that allows it to feed on the seeds of a native Hawaiian plant, which are toxic to most other birds. The one remaining palila population is declining because these plants are being trampled by cows and gnawed to death by goats and sheep. Only about 1,200 palila remained in 2010.

B The unusual lower bill of the akekee points to one side, allowing this bird to pry open buds that harbor insects. Avian malaria carried by mosquitoes to higher altitudes is decimating the last population of this species. Between 2000 and 2007, the number of akekee plummeted from 7,839 birds to 3,536.

C This poouli—rare, old, and missing an eye—died in 2004 from avian malaria. There were two other poouli alive at the time, but neither has been seen since then.

FIGURE 17.25 Three honeycreeper species: going, going, gone.

As more and more honeycreeper species become extinct, the group's reservoir of genetic diversity dwindles. The lowered diversity means the group as a whole is less resilient to change, and more likely to suffer catastrophic losses. Deciphering their phylogeny can tell us which honeycreeper species are most different from the others—and those are the ones most valuable in terms of preserving the group's genetic diversity. Such research allows us to concentrate our resources and conservation efforts on those species that hold the best hope for the survival of the entire group. For example, we now know the poouli (**FIGURE 17.25C**) to be the most distant relative in the honeycreeper family. Unfortunately, the knowledge came too late; the poouli is probably extinct. Its extinction means the loss of a large part of evolutionary history of the group: One of the longest branches of the honeycreeper family tree is gone forever.

TAKE-HOME MESSAGE 17.13

Among other applications, phylogeny research can help us understand the spread of infectious diseases, and also to focus our conservation efforts.

17.14 Application: DIRECTIONAL SELECTION: SUPERBAD SUPERBUGS

Sustainability

FIGURE 17.26 Abundant use of antibiotics in livestock has contributed to the worldwide prevalence of resistant bacteria.

WE ARE NOW PAYING THE PRICE FOR OVERUSE OF ANTIBIOTICS. Prior to the 1940s, scarlet fever, tuberculosis, and pneumonia caused one-fourth of the annual deaths in the United States. Since the 1940s, we have been relying on antibiotics such as penicillin to fight these and other dangerous bacterial diseases. We have also been using antibiotics in other, less dire circumstances. For an as-yet unknown reason, antibiotics promote growth in cattle, pigs, poultry (FIGURE 17.26), and even fish. The agricultural industry uses antibiotics mainly for this purpose. In addition, antibiotics that are used to prevent or treat infection are given to entire flocks or herds at once. In 2011, the U.S. agricultural industry used 13.7 million kilograms of antibiotics—more than four times the amount used to treat people.

Bacteria evolve at a much accelerated rate compared with humans, in part because they reproduce much more quickly. The common intestinal bacteria *E. coli* can divide every 17 minutes. Each new generation is an opportunity for mutation, so a bacterial gene pool can diversify very fast. Different species of bacteria can also share DNA, further adding to the diversity of their respective gene pools.

As you learned in Section 12.1, having a diverse gene pool is an advantage in a changing environment. When a natural bacterial population is exposed to an antibiotic, some cells in it are likely to survive because they carry an allele that offers resistance. As susceptible cells die and the survivors reproduce, the frequency of these antibiotic-resistance alleles increases in the population (an example of directional selection). A typical two-week course of antibiotics can potentially exert selection pressure on over a thousand generations of bacteria, and antibiotic-resistant strains are the outcome.

Antibiotic-resistant bacteria bred inside treated animals and humans end up in the environment, where they can easily spread to other individuals. These bacteria have plagued hospitals for years, and now we find them everywhere. They are common in day-care centers, schools, gyms, prisons, and other places where people are in close contact. We find them in pets and wildlife such as rabbits, mongooses, sharks, rodents, reptiles, and even in Arctic penguins. Drinking water supplies and Antarctic seawater contain them. A recent report found antibiotic-resistant bacteria in more than half of the samples of supermarket ground beef and pork chops, and over 80 percent of ground turkey. Some of these bacteria are resistant to most available antibiotics.

In humans, an infection with antibiotic resistant bacteria tends to be longer, more severe, and more likely to be deadly. The Centers for Disease Control and Prevention says that these "superbugs" contribute to death in up to 50 percent of patients who become infected.

Summary

SECTIONS 17.1, 17.2 All alleles of all genes in a **population** constitute a **gene pool**. Mutations may be **neutral**, **lethal**, or adaptive. **Microevolution** is change in **allele frequency** of a population. Deviations from **genetic equilibrium** indicate that a population is evolving.

SECTIONS 17.3–17.6 In **directional selection**, a phenotype at one end of a range of variation is adaptive. An intermediate form of a trait is adaptive in **stabilizing selection**; extreme forms are adaptive in **disruptive selection**. **Sexual dimorphism** is one outcome of **sexual selection**. **Frequency-dependent selection** or any other mode of natural selection can give rise to a **balanced polymorphism**.

SECTION 17.7 **Genetic drift**, which is most pronounced in small or **inbreeding** populations, can cause alleles to be **fixed**. The **founder effect** may occur after an evolutionary **bottleneck**. **Gene flow** can counter the effects of mutation, natural selection, and genetic drift.

SECTIONS 17.8–17.10 **Reproductive isolation** is always a part of **speciation** (**TABLE 17.3**). With **allopatric speciation**, a geographic barrier arises and ends gene flow between populations. **Sympatric speciation** occurs with no barrier to gene flow. With **parapatric speciation**, populations in contact along a common border speciate.

SECTION 17.11 **Macroevolution** refers to large-scale patterns of evolution. In **exaptation**, a lineage uses a structure for a different purpose than its ancestor. With **stasis**, a lineage changes little over evolutionary time. A **key innovation** can result in an **adaptive radiation**. **Coevolution** occurs when two species act as agents of selection upon one another. A lineage with no more living members is **extinct**.

SECTION 17.12 Evolutionary biologists reconstruct evolutionary history (**phylogeny**) by comparing physical, behavioral, and biochemical traits, or **characters**, among species. A **clade** is a **monophyletic group** that consists of an ancestor in which one or more **derived traits** evolved, together with all of its descendants. Making hypotheses about the evolutionary history of a group of clades is called **cladistics**. In **evolutionary tree** diagrams such as **cladograms**, each line represents a lineage. A lineage branches into two **sister groups** at a node, which represents a shared ancestor.

SECTION 17.13 Reconstructing phylogeny is part of our efforts to preserve endangered species. It also allows us to understand the spread of some infectious diseases.

SECTION 17.14 Overuse of antibiotics has resulted in directional selection for dangerous antibiotic-resistant bacteria that are now common in the environment.

TABLE 17.3

Comparison of Speciation Models

	Allopatric	Parapatric	Sympatric
Original population(s)			
Initiating event:	physical barrier arises	selection pressures differ	genetic change
Reproductive isolation occurs			
New species arises:	in isolation	in contact along common border	within existing population

Self-Quiz

Answers in Appendix VII

1. _______ is the original source of new alleles.
   a. Mutation
   b. Natural selection
   c. Genetic drift
   d. Gene flow
   e. All are original sources of new alleles

2. Evolution can only occur in a population when _______ .
   a. mating is random
   b. there is selection pressure
   c. (neither is necessary)

3. Match the modes of natural selection with their best descriptions.

   ___ stabilizing — a. eliminates extreme forms of a trait
   ___ directional — b. eliminates midrange forms of a trait
   ___ disruptive — c. shifts allele frequency in one direction

4. Sexual selection frequently influences aspects of body form and can lead to _______ .
   a. sexual dimorphism
   b. male aggression
   c. exaggerated traits
   d. all of the above

5. The persistence of the sickle allele at high frequency in a population is an example of _______ .
   a. bottlenecking
   b. inbreeding
   c. natural selection
   d. balanced polymorphism

6. _______ tends to keep populations of a species similar to one another.
   a. Genetic drift
   b. Gene flow
   c. Mutation
   d. Natural selection

7. The theory of natural selection does not explain _______ .
   a. genetic drift
   b. the founder effect
   c. gene flow
   d. how mutations arise
   e. inheritance
   f. any of the above

8. A fire devastates all trees in a wide swath of forest. Populations of a species of tree-dwelling frog on either side of the burned area diverge to become separate species. This is an example of _______ .

Data Analysis Activities

Resistance to Rodenticides in Wild Rat Populations Beginning in 1990, rat infestations in northwestern Germany started to intensify despite continuing use of rat poisons. In 2000, Michael H. Kohn and his colleagues analyzed the genetics of wild rat populations around Munich. For part of their research, they trapped wild rats in five towns, and tested those rats for resistance to warfarin and the more recently developed poison bromadiolone. The results are shown in **FIGURE 17.27**.

1. In which of the five towns were most of the rats susceptible to warfarin?
2. Which town had the highest percentage of poison-resistant wild rats?
3. What percentage of rats in Olfen were resistant to warfarin?
4. In which town do you think the application of bromadiolone was most intensive?

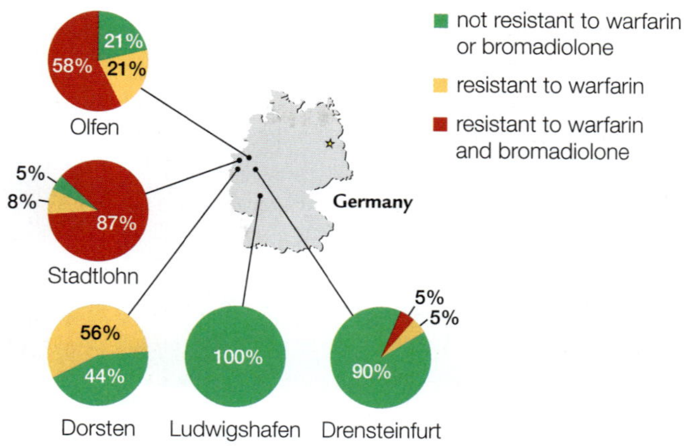

FIGURE 17.27 Resistance to rat poisons in wild populations of rats in Germany, 2000.

9. Sex in many birds is typically preceded by an elaborate courtship dance. If a male's movements are unrecognized by the female, she will not mate with him. This is an example of ______ .
   a. reproductive isolation
   b. behavioral isolation
   c. sexual selection
   d. all of the above

10. Cladistics ______ .
    a. is a way of reconstructing evolutionary history
    b. may involve parsimony analysis
    c. is based on derived traits
    d. all of the above are correct

11. In cladistics, the only taxon that is always correct as a clade is the ______ .
    a. genus b. family c. species d. kingdom

12. In evolutionary trees, each node represents a(n) ______ .
    a. single lineage
    b. extinction
    c. point of divergence
    d. adaptive radiation

13. In cladograms, sister groups are ______ .
    a. inbred
    b. the same age
    c. represented by nodes
    d. in the same family

14. Match the evolutionary concepts.

___ gene flow	a. can lead to interdependent species
___ sexual selection	b. changes in a population's allele frequencies due to chance alone
___ extinct	c. alleles enter or leave a population
___ genetic drift	d. evolutionary history
___ phylogeny	e. genetic winners are the sexiest
___ adaptive radiation	f. burst of divergences
___ coevolution	g. no more living members
___ cladogram	h. evolutionary diagram

Critical Thinking

1. Species have been traditionally characterized as "primitive" and "advanced." For example, mosses were considered to be primitive, and flowering plants advanced; crocodiles were primitive and mammals were advanced. Why do most biologists of today think it is incorrect to refer to any modern species as primitive?

2. Rama the cama, a llama–camel hybrid, was born in 1997. The idea was to breed an animal that has the camel's strength and endurance, and the llama's gentle disposition. However, instead of being large, strong, and sweet, Rama is smaller than expected and has a camel's short temper. The breeders plan to mate him with Kamilah, a female cama. What potential problems with this mating should the breeders anticipate?

3. Two species of antelope, one from Africa, the other from Asia, are put into the same enclosure in a zoo. To the zookeeper's surprise, individuals of the different species begin to mate and produce healthy, hybrid baby antelopes. Explain why a biologist might not view these offspring as evidence that the two species of antelope are in fact one.

4. Some people think that many of our uniquely human traits arose by sexual selection. Over thousands of years, women attracted to charming, witty men perhaps prompted the development of human intellect beyond what was necessary for mere survival. Men attracted to women with juvenile features may have shifted the species as a whole to be less hairy and softer featured than any of our simian relatives. Can you think of a way to test these hypotheses?

A low tide exposes rocky structures called stromatolites in Australia's Shark Bay. Bacteria have been building structures like these for billions of years. Fossil stromatolites are among our earliest evidence of life on Earth.

18 LIFE'S ORIGIN AND EARLY EVOLUTION

Links to Earlier Concepts

This chapter explains how scientists investigate life's origin and lays out what we know about life's early history. It draws on your knowledge of scientific method (Section 1.5), plate tectonics (16.5), and the geologic time scale (16.6). We discuss the origin of archaea and bacteria (4.4), and of eukaryotic organelles (4.5). We also return to the link between photosynthesis and aerobic respiration (7.2).

KEY CONCEPTS

BUILDING BLOCKS OF LIFE

Simulations show how organic monomers could have formed on the early Earth. Such compounds also form in space and could have been delivered by meteorites.

THE FIRST CELLS FORM

Experiments provide insight into how cells could have arisen from nonliving material through known physical and chemical processes.

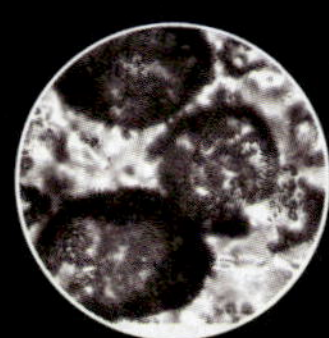

LIFE'S EARLY EVOLUTION

The first cells were anaerobic and prokaryotic. Evolution of oxygen-producing photosynthesis altered Earth's atmosphere, creating selection pressure that favored aerobic organisms.

EUKARYOTIC ORGANELLES

The nucleus and endomembrane system are likely derived from infoldings of the plasma membrane. Bacteria are the ancestors of mitochondria and chloroplasts.

Photograph by Frans Lanting, National Geographic Creative.

18.1 WHAT WAS THE BUILDING MATERIAL FOR EARTH'S FIRST LIFE?

CONDITIONS ON THE EARLY EARTH

To understand how life on Earth could have begun, it helps to know a bit about our planet's history. Scientists estimate that Earth formed by about 4.6 billion years ago through the aggregation of dust and rock bits that were orbiting our sun. Although no one has yet discovered rocks that date that far back, geologists have discovered crystals of zircon (a type of mineral) that formed about 4.3 billion years ago. As Section 16.4 explains, zircon crystals form when molten rock cools, and the timing of their formation can be determined by radiometric dating.

Geologists can learn what conditions were like on the early Earth by analyzing ancient crystals and rocks. For example, the composition of some 4.3-billion-year-old zircon crystals indicates that they formed in the presence of water. This suggests that water had begun to pool on Earth's surface by 4.3 billion years ago. The presence of water is important because water is essential to life. All metabolic reactions that take place in cells require water.

We know that Earth's early atmosphere contained little or no oxygen gas, because the oldest existing rocks show no evidence of iron oxidation (rusting). If oxygen had been present, it would have interfered with assembly of simple organic compounds.

FORMATION OF SIMPLE ORGANIC COMPOUNDS

All life consists of the same simple organic compounds (amino acids, fatty acids, nucleotides, and simple sugars). Thus, reactions that formed these compounds were probably the first step on the road to life. There are several ways in which this initial step could have taken place.

In the early 1950s, Stanley Miller and Harold Urey showed that lightning-fueled atmospheric reactions can produce simple organic compounds. They simulated Earth's early atmosphere by filling a reaction chamber with a mix of gases and water vapor, then zapped it with sparks to simulate lightning (**FIGURE 18.1**). Within a week, amino acids and other organic molecules formed. We now know that the mix of gases in the Miller–Urey experiment probably did not accurately represent those present on the early Earth. However, later experiments carried out using more accurate gas mixtures also produced amino acids.

Discovery of organic compounds inside meteorites that fell to Earth suggests another source for life's building blocks. Organic materials that formed in interstellar clouds of ice, dust, and gases could been carried to Earth on meteorites. Keep in mind that during Earth's early years, meteorites fell to Earth thousands of times more frequently than they do today.

Reactions deep in the sea near hydrothermal vents could also have produced simple organic compounds. A **hydrothermal vent** is an underwater hot spring—a place where superheated, mineral-rich water streams out through a rocky opening. Scientists have simulated conditions near a hydrothermal vent by combining hot water with carbon monoxide (CO), potassium cyanide (KCN), and metal ions like those in rocks near the vents. Under these conditions, amino acids form within a week.

Note that the three mechanisms of organic compound formation discussed above are not mutually exclusive. Most likely all three operated simultaneously and contributed to an accumulation of simple organic compounds in Earth's early seas. It was in these seas that life first began.

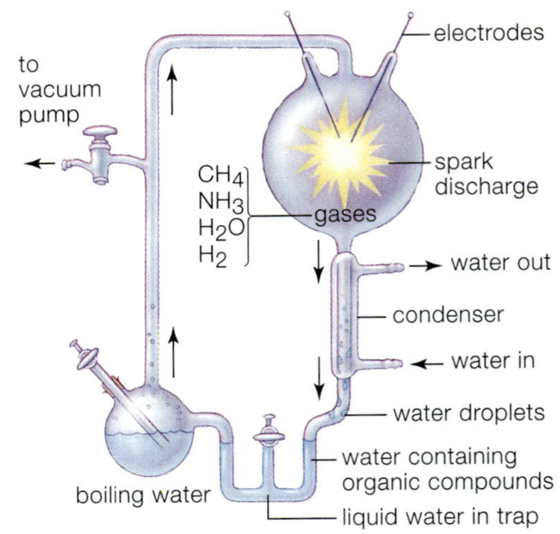

FIGURE 18.1 {Animated} Diagram of an apparatus designed by Stanley Miller and Harold Urey to test whether organic compounds could have formed by chemical interactions in Earth's early atmosphere. Water, hydrogen gas (H_2), methane (CH_4), and ammonia (NH_3) circulated through the apparatus as sparks from an electrode simulated lightning.

FIGURE IT OUT: Which gas in this mixture provided the nitrogen for the amino group in the amino acids?

Answer: Ammonia

TAKE-HOME MESSAGE 18.1

- Earth formed by 4.6 billion years ago. By 4.3 billion years ago, it had seas and an atmosphere with little or no oxygen.
- Simulations of conditions on the early Earth show organic compounds could have formed in the atmosphere or seas.
- Organic compounds also form in space and could have been delivered to Earth by meteorites.

hydrothermal vent Underwater opening where hot, mineral-rich water streams out.

PEOPLE MATTER

National Geographic Explorer-in-Residence DR. ROBERT BALLARD

Undersea explorer Bob Ballard is best known for finding wrecks of historic ships such as the RMS *Titanic*. His proudest accomplishment, however, is discovering that life thrives in the deep, dark ocean depths near hydrothermal vents. In 1977, Ballard led a team that used a submersible vehicle to explore a region of the seafloor where tectonic plates are moving apart (as described in Section 16.5). As Ballard and others had predicted, the team found openings (vents) in the seafloor. Mineral-rich water heated by geothermal energy streamed from the vents into the frigid waters of the deep ocean. To Ballard's astonishment, the team also found a wealth of previously unknown life. He says, "We discovered that this whole life system was living not off the energy from the sun, but from the energy of the Earth." Dr. Ballard's discovery led other scientists to reconsider life's origins. Many scientists now think that Earth's earliest life arose deep in the sea near hydrothermal vents.

Left, a black smoker, a type of hydrothermal vent.

CREDITS: From the IMAX film "Volcanoes of the Deep Sea" produced by The Stephen Low Company in association with Rutgers University; (top) O. Louis Mazzatenta/National Geographic Creative.

18.2 WHAT STEPS LED TO THE FORMATION OF THE FIRST CELLS?

ON THE ROAD TO LIFE

In addition to sharing the same molecular components, all cells have a plasma membrane with a lipid bilayer. As Chapter 9 explained, cells have a genome of DNA that enzymes transcribe into RNA, and ribosomes that translate RNA into proteins. All cells replicate, and pass on copies of their genetic material to their descendants. The similarities in structure, metabolism, and replication processes among all known organisms are considered evidence that they all are descendants of the same cellular ancestor.

Time has erased all evidence of the earliest cells, but scientists can still investigate this first chapter in life's history. They use their knowledge of chemistry to design experiments that test whether a particular hypothesis about how life began is plausible. Results of such studies support the hypothesis that cells arose as a result of a stepwise process, beginning with inorganic materials (**FIGURE 18.2**). Each step on the road to life can be explained by chemical and physical mechanisms that still operate today.

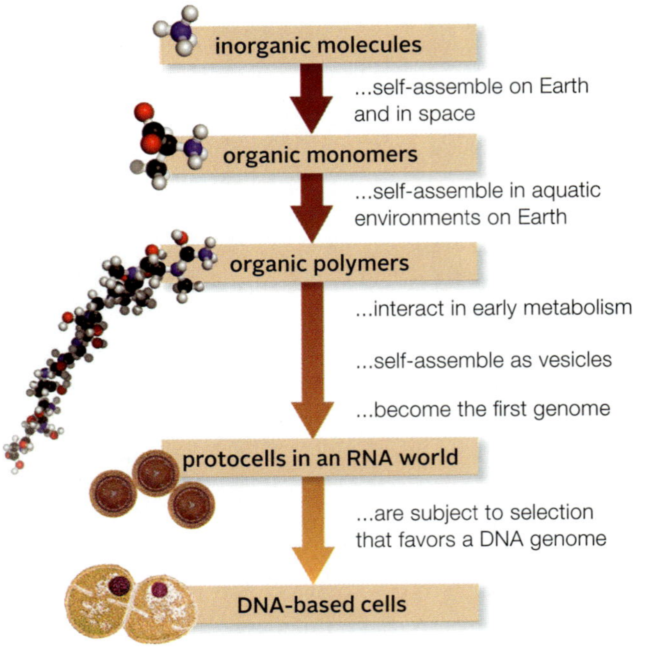

FIGURE 18.2 **{Animated}** Proposed sequence for the evolution of cells. Scientists investigate this process by carrying out experiments and simulations that test hypotheses about feasibility of individual steps.

ORIGIN OF METABOLISM

Modern cells take up organic subunits, concentrate them, and assemble them into organic polymers (Section 3.1). Before there were cells, a nonbiological process that concentrated organic subunits in one place would have increased the chance that the subunits would combine.

By one hypothesis, this process occurred on clay-rich tidal flats. Clay particles have a slight negative charge, so positively charged molecules in seawater stick to them. At low tide, evaporation would have concentrated the subunits even more, and energy from the sun could have caused them to bond together as polymers. In simulations of tidal flat conditions, amino acids assemble in short chains.

By another hypothesis, metabolic reactions began in the high-temperature, high-pressure environment near a hydrothermal vent. Rocks around the vents contain iron sulfide (pyrite) and are porous, with many tiny chambers about the size of cells (**FIGURE 18.3**). Metabolism may have begun when iron sulfide in the rocks donated electrons to dissolved carbon monoxide (CO), forming organic compounds. Under simulated vent conditions, organic compounds such as pyruvate do form and accumulate in rocky chambers. In addition, all modern organisms have proteins that use iron–sulfur clusters as cofactors (Section 5.5). The universal requirement for these cofactors may be a legacy of life's rocky beginnings.

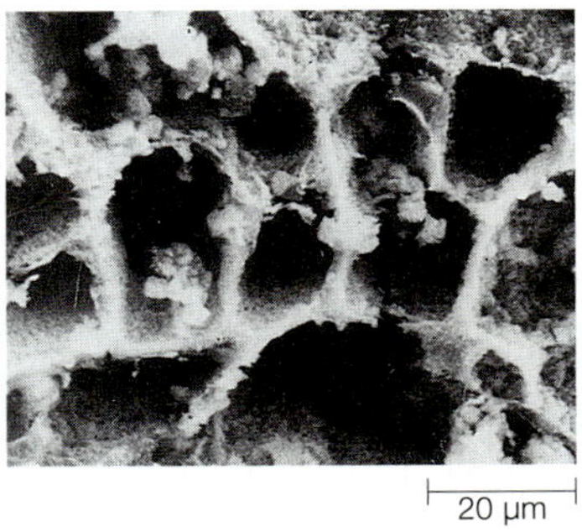

FIGURE 18.3 Cell-sized chambers in iron-sulfide-rich rocks formed by simulations of conditions near hydrothermal vents. Similar chambers could have served as protected environments in which the first metabolic reactions took place.

ORIGIN OF THE CELL MEMBRANE

Molecules formed by early synthetic reactions would have floated away from one another unless something enclosed them. In modern cells, a plasma membrane serves this role. If the first reactions took place in tiny rock chambers, the rock would have acted as a boundary. Over time, lipids produced by reactions in a chamber could have accumulated and lined the chamber wall. Such lipid-enclosed collections of interacting molecules may have been the first protocells. A **protocell** is a membrane-enclosed collection of molecules that takes up material and replicates itself.

Experiments by Jack Szostak and others have shown that rock chambers are not necessary for protocell formation. **FIGURE 18.4A** is a computer model of one type of protocell that Szostak investigates. **FIGURE 18.4B** is a photo of a protocell that formed in his laboratory. It has a lipid bilayer membrane enclosing strands of RNA. Such a protocell "grows" by incorporating additional fatty acids

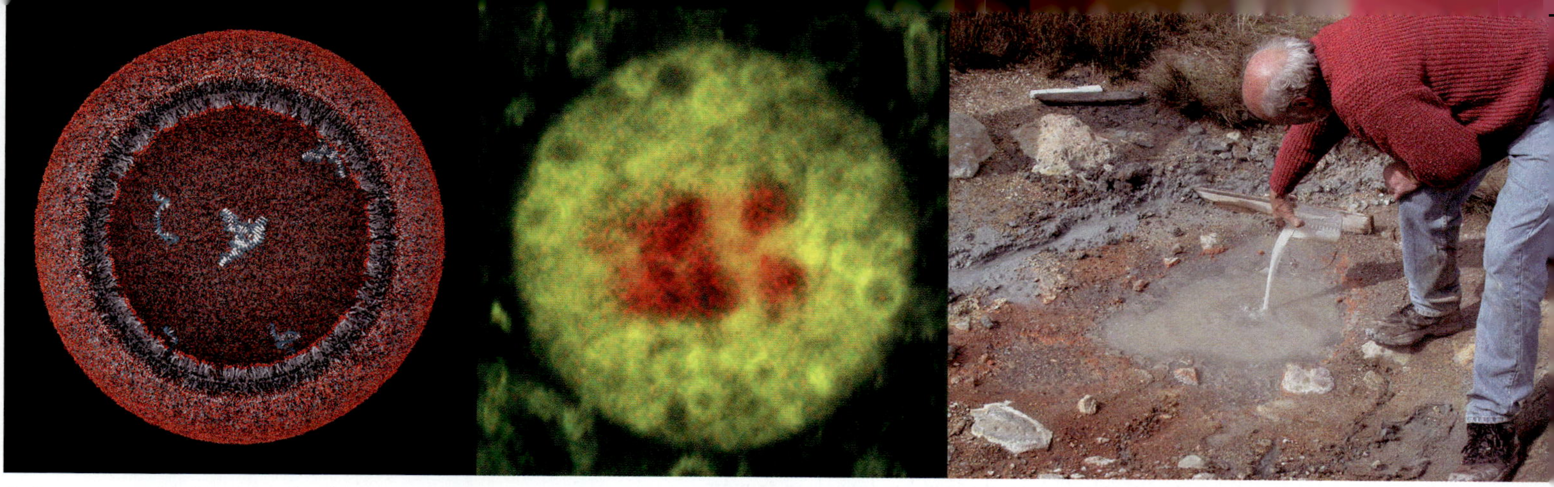

A Computer model of a protocell with a bilayer membrane of fatty acids around strands of RNA.

B Laboratory-formed protocell consisting of RNA-coated clay (red) surrounded by fatty acids and alcohols.

C Field-testing a hypothesis about protocell formation. David Deamer pours a mix of small organic molecules and phosphates into a hot acidic pool in Russia.

FIGURE 18.4 Protocells. Scientists test hypotheses about protocell formation through laboratory simulations and field experiments.

into its membrane and additional nucleotides into its RNA. Mechanical force causes division.

David Deamer studies protocell formation in both the laboratory and the field. In the lab, he has shown that the small organic molecules carried to Earth on meteorites can react with minerals and seawater to form vesicles with a lipid bilayer membrane. In the field, he tests whether specific environmental conditions can facilitate this process. In one experiment, he added a mix of organic subunits to the acidic waters of a clay-rich volcanic pool in Russia (**FIGURE 18.4C**). Although organic subunits bound tightly to the clay, no vesicle-like structures formed. Deamer concluded that hot acidic waters of volcanic springs do not provide conditions that favor protocell formation. He continues to carry out experiments to determine what naturally occurring conditions do favor this process.

ORIGIN OF THE GENOME

All modern cells have a genome of DNA. They pass copies of their DNA to descendant cells, which use instructions encoded in the DNA to build proteins. Some of these proteins are enzymes that synthesize new DNA, which is passed along to descendant cells, and so on. Thus, protein synthesis depends on DNA, which is built by proteins. How did this cycle begin?

In the 1960s, Francis Crick and Leslie Orgel addressed this dilemma by suggesting that RNA may have been the first molecule to encode genetic information. Since then, evidence for an early **RNA world**—a time when RNA both stored genetic information and functioned like an enzyme in protein synthesis—has accumulated. Scientists discovered **ribozymes**, RNAs that function as enzymes, in living cells. For example, the rRNA in ribosomes speeds formation of peptide bonds during protein synthesis (Section 9.3). Other ribozymes cut noncoding bits (introns) out of newly formed RNAs (Section 9.2). In addition, researchers have produced self-replicating ribozymes that copy themselves by assembling free nucleotides.

If the earliest self-replicating genetic systems were RNA-based, then why do all organisms have a genome of DNA? The structure of DNA may hold the answer. Compared to a double-stranded DNA molecule, single-stranded RNA breaks apart more easily and mutates more often. Thus, a switch from RNA to DNA would make larger, more stable genomes possible.

TAKE-HOME MESSAGE 18.2

- All living cells carry out metabolic reactions, are enclosed within a plasma membrane, and can replicate themselves.
- Concentration of molecules on clay particles or in tiny rock chambers near hydrothermal vents may have helped start metabolic reactions.
- Vesicle-like structures with outer membranes can form spontaneously.
- An RNA-based system of inheritance may have preceded DNA-based systems.

protocell Membranous sac that contains interacting organic molecules; hypothesized to have formed prior to the earliest life forms.
ribozyme RNA that functions as an enzyme.
RNA world Hypothetical early interval when RNA served as the genetic information.

18.3 WHAT DO WE KNOW ABOUT EARLY CELLS?

TRAITS OF THE UNIVERSAL COMMON ANCESTOR

The processes described in Section 18.2 may have produced cellular life more than once. If so, all but one of those early cell lineages became extinct. We know from analysis of modern genomes that all species on Earth today are descended from a cell that lived as early as 4 billion years ago.

Given what scientists know about relationships among modern species, most assume that this common ancestor was prokaryotic, meaning it did not have a nucleus. Oxygen was scarce on the early Earth, so the ancestral cell must also have been anaerobic (capable of living without oxygen). Other aspects of this cell's metabolism are less clear. It may have been a heterotroph that fermented organic compounds. Alternatively, it may have been an autotroph that stripped electrons from inorganic material and assembled its own food from carbon dioxide. It was most likely not photosynthetic. Photosynthesis is a complicated process that requires the evolution of much specialized metabolic machinery.

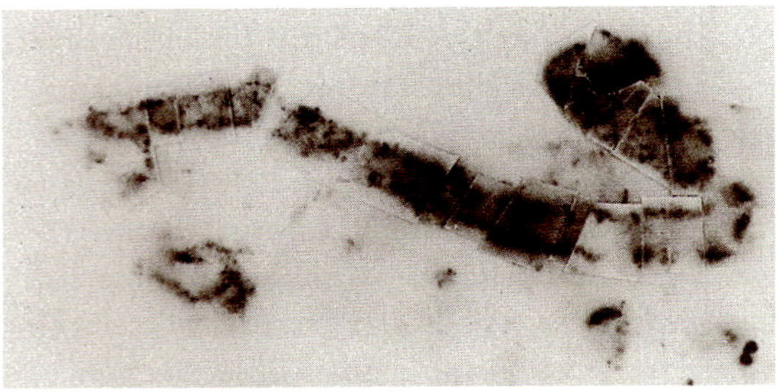

A Micrograph of a filament that may be a 3.5-billion-year-old fossilized chain of photosynthetic bacteria.

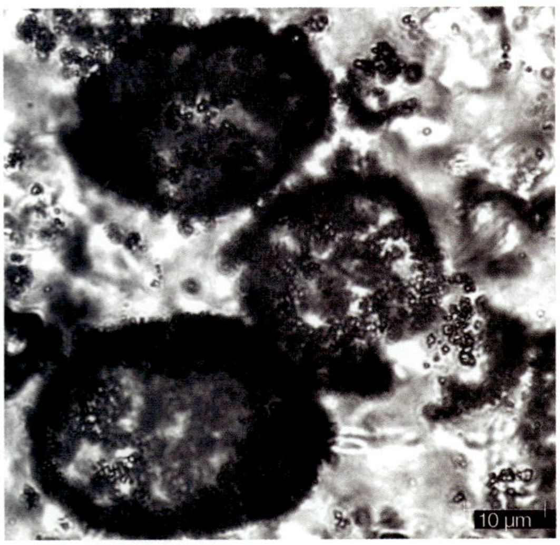

B Micrograph showing structures that may be fossilized 3.4-billion-year-old sulfur-reducing bacteria.

FIGURE 18.5 Possible microfossils from western Australia.

FIGURE 18.6 Fossil stromatolite. It consists of layered remains of countless photosynthetic bacteria and sediment they captured.

Finding and identifying signs of early cells is a challenge. Cells are microscopic and most have no hard parts to fossilize. In addition, few ancient rocks that might hold early fossils still exist. Tectonic plate movements have destroyed nearly all rocks older than about 4 billion years, and slightly younger rocks have often been subject to heating and other processes that destroy traces of biological material. To add to the difficulty, structures formed by nonbiological processes sometimes resemble fossils. To avoid mistakenly accepting such material as a genuine fossil, scientists constantly reanalyze purported fossil finds and they often question one another's conclusions.

EVIDENCE OF EARLY PROKARYOTIC CELLS

The divergence that led to domains Bacteria and Archaea occurred very early in the history of life, and we have no fossils from before this divergence. At present, the oldest proposed cell microfossils (microscopic fossils) come from rock formations in Western Australia. In 1993, William Schopf proposed that 3.5-billion-year-old filaments in these rocks are fossilized chains of photosynthetic bacteria (**FIGURE 18.5A**). This interpretation has been challenged by other scientists, including Martin Brasier, who notes that mineral deposits that resemble these "fossil" filaments can form by geologic processes.

Brasier has put forward his own proposed contender for the title of oldest fossil cells (**FIGURE 18.5B**). Like Schopf's find, Brasier's cells come from Western Australia, but they are a bit younger, dating back 3.4 billion years. Pyrite in and around these cells suggests they may have been similar to the sulfate-reducing bacteria that live in present-day mud flats. These anaerobic bacteria use sulfur in the same way aerobic organisms use oxygen, that is, as the final electron acceptor in an energy-producing pathway. As a by-product of their metabolism, the bacteria produce hydrogen sulfide gas that combines with iron to form pyrite.

Additional evidence of early cellular life comes from fossil stromatolites (**FIGURE 18.6**). A **stromatolite** is a rocky conical or dome-shaped structure composed of layers of cells, cell remains, and sediment. It forms when photosynthetic bacteria living in shallow sunlit water form a mat that traps sediments. Once a layer of sediment covers the bacteria, they grow up through it and trap more sediment. Scientists can observe this process in modern stromatolites such as those of Australia's Shark Bay, shown in the chapter's opening photo. The stromatolites in Shark Bay began growing an estimated 2,000 years ago and now stand up to 1.5 meters high. They consist of sediments and cyanobacteria, a group of bacteria that produce oxygen as a by-product of photosynthesis.

The oldest structures that might be fossil stromatolites date to 3.5 billion years ago. They are, however, highly degraded and may have formed by geologic rather than biological processes. Fossils that are undisputed stromatolites first appear about 2.8 billion years ago. Stromatolites reached their peak abundance about 1.25 billion years ago, when they were common worldwide.

EVIDENCE OF EARLY EUKARYOTES

It is difficult to find a nucleus in a fossilized cell, but other traits can indicate that the cell was probably eukaryotic. Eukaryotic cells are generally larger than prokaryotic ones. A cell wall with complex patterns, spines, or spikes likely housed a eukaryote. Researchers also look for biomarkers specific to eukaryotes. A **biomarker** is a substance that occurs only in or predominantly in cells of a specific type. For example, steroids are mainly found in eukaryotes, so traces of steroids are biomarkers for this group.

At present, the oldest widely accepted eukaryote microfossils date to about 1.8 billion years ago. The fossils resemble resting stage cells (cysts) that form during the life cycle of some present-day marine protists. However, we do not know whether these early fossil eukaryotes are related to any modern groups.

biomarker Substance found only or mainly in cells of one type.
stromatolite Rocky structures composed of layers of bacterial cells and sediments.

TAKE-HOME MESSAGE 18.3

- All present-day life descended from the same anaerobic prokaryotic cell.
- Microfossils show prokaryotic cells were present by 3.4 billion years ago and eukaryotic cells by 1.8 billion years ago.
- Fossil stromatolites provide evidence of photosynthetic prokaryotes.

Many types of bacteria carry out photosynthesis, but only cyanobacteria do so by an oxygen-producing pathway (**FIGURE 18.7**). Evolution of oxygen-producing photosynthesis had a dramatic effect on early life. By about 2.5 billion years ago, oxygen released by cyanobacteria began to accumulate in Earth's seas and air (Section 7.2).

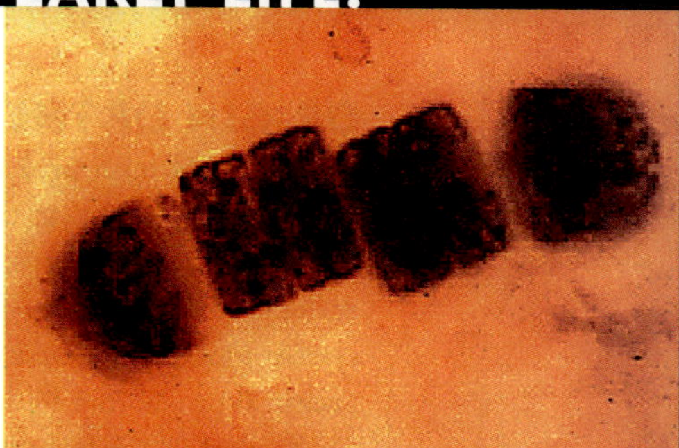

FIGURE 18.7 Fossilized chain of cyanobacteria from about 850 million years ago. Such cells produce oxygen as a by-product of photosynthesis. Activity of such cells added oxygen to Earth's seas and air.

In the seas, increased oxygen created a new selective pressure. Oxygen was toxic to many species that had evolved in its absence. It reacted with metal ions in their cells, forming free radicals (Section 2.2) that damaged essential cell components. Species that could not detoxify free radicals either went extinct or became restricted to the low-oxygen environments that remained.

By contrast, species with metabolic machinery capable of detoxifying oxygen thrived. Some cells began to carry out aerobic respiration, in which oxygen serves as the final electron acceptor (Section 7.5). As you will learn, one group of aerobic bacteria later evolved into mitochondria, the organelles that power eukaryotic cells.

An increase in atmospheric oxygen also led to formation of the **ozone layer**, a region of the upper atmosphere that contains a high concentration of ozone gas (O_3). The ozone layer absorbs ultraviolet (UV) radiation from the sun, preventing it from reaching Earth's surface. As Section 8.5 explained, such radiation is a dangerous mutation. Before the ozone layer formed, life existed only in the seas, where water shielded organisms from incoming UV radiation. Without an ozone layer to screen out some of this radiation, life could not have moved onto land.

ozone layer Atmospheric region with a high concentration of ozone (O_3) that screens out incoming UV radiation.

TAKE-HOME MESSAGE 18.4

- Oxygen released by photosynthetic cyanobacteria began accumulating in the air and seas about 2.5 billion years ago.
- As oxygen levels rose, species capable of detoxifying oxygen had a selective advantage.
- Oxygen reacted in the upper atmosphere to form an ozone layer that shielded Earth's surface from UV radiation, making life on land possible.

18.5 HOW DID EUKARYOTIC ORGANELLES ARISE?

ORIGIN OF THE NUCLEUS

The DNA of most prokaryotes lies unenclosed in the cell's cytoplasm. By contrast, the DNA of a eukaryotic cell is always enclosed within a nucleus and associated with an endomembrane system. The nucleus and endomembrane system probably evolved when the plasma membrane of an ancestral prokaryote folded inward (**FIGURE 18.8**).

Studies of the few types of bacteria that have internal membranes illustrates that such infolding does occur and how it can be advantageous. For example, some modern marine bacteria have membrane infoldings that increase the surface area available to hold membrane-associated enzymes.

FIGURE 18.8 {Animated} One hypothesis for the steps in organelle evolution.

Membrane infolding produced the endomembrane system. Then, an early endosymbiosis produced mitochondria, which occur in almost all eukaryotes. Later, photosynthetic bacteria entered a eukaryotic cell and, over many generations, evolved into chloroplasts.

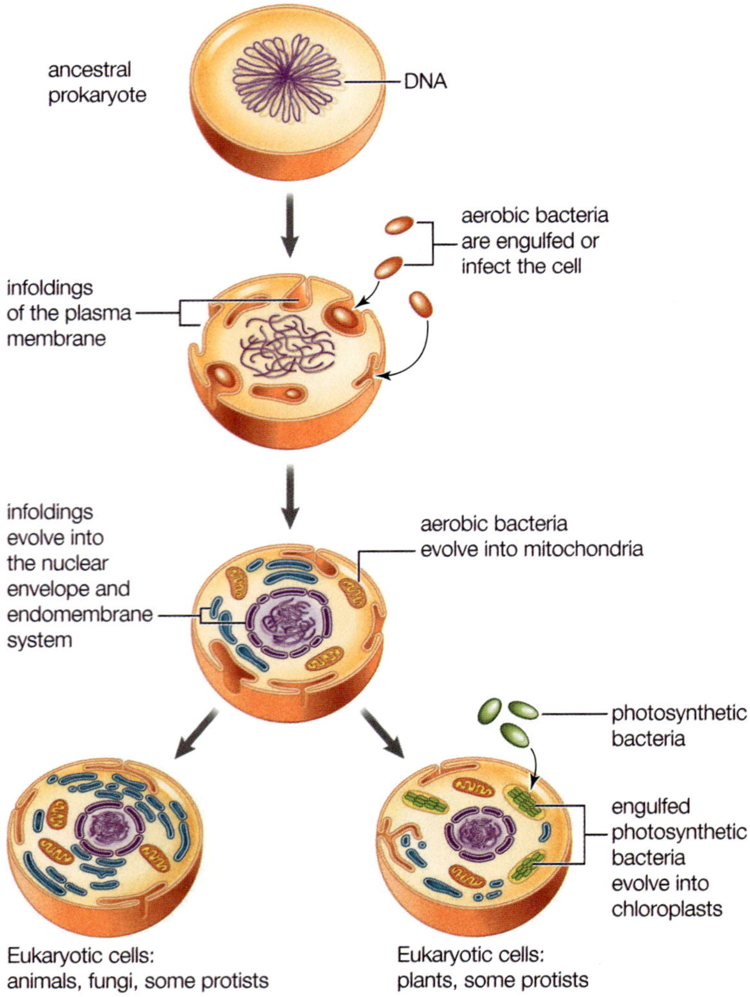

Internal membranes also protect a genome from physical or biological threats. Consider *Gemmata obscuriglobus*, one of the few bacteria that houses its DNA inside a membrane (**FIGURE 18.9**). Like a eukaryotic nuclear envelope (Section 4.5), this membrane consists of a two lipid bilayer, but it does not have the equivalent nuclear pores. Compared to typical bacteria, *Gemmata obscuriglobus* withstands much higher levels of mutation-causing radiation. Researchers attribute this species' radiation resistance to the tighter packing, and therefore higher shielding, of its DNA within the membrane-enclosed compartment. The DNA of other bacteria is more vulnerable to radiation because it is more spread out. The membrane also helps prevent the cells' DNA from integrating any foreign genetic material. For example, it provides protection from viruses that inject their genetic material into bacterial cells.

ORIGIN OF MITOCHONDRIA AND CHLOROPLASTS

The **endosymbiont hypothesis** proposes that mitochondria and chloroplasts are descendants of aerobic bacteria that entered and replicated inside a host cell. Whether this host was a prokaryotic cell or an early eukaryote remains unclear. When the host cell divided, it passed some "guest" cells, referred to as endosymbionts, along to its offspring. As the two species lived together over many generations, genes that both partners carried were free to mutate. A gene could lose its function in one partner if the duplicate gene carried by the other partner still worked. Eventually, the host and endosymbiont became incapable of living independently. The endosymbionts had evolved into organelles.

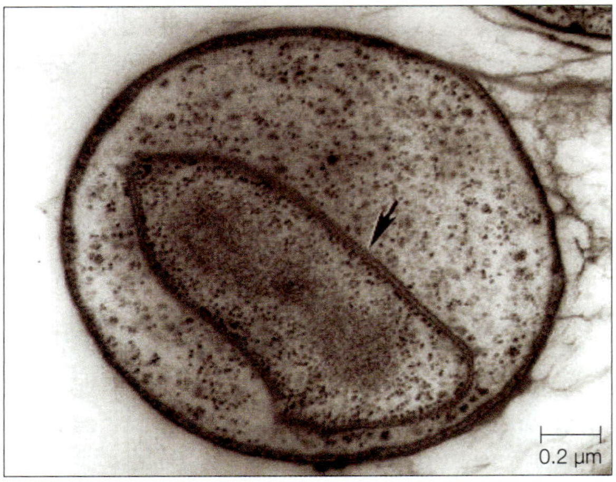

FIGURE 18.9 Nucleus-like structure in a prokaryote. The DNA of *Gemmata obscuriglobus*, a species of bacteria, is enclosed by a double lipid bilayer membrane (indicated by the arrow).

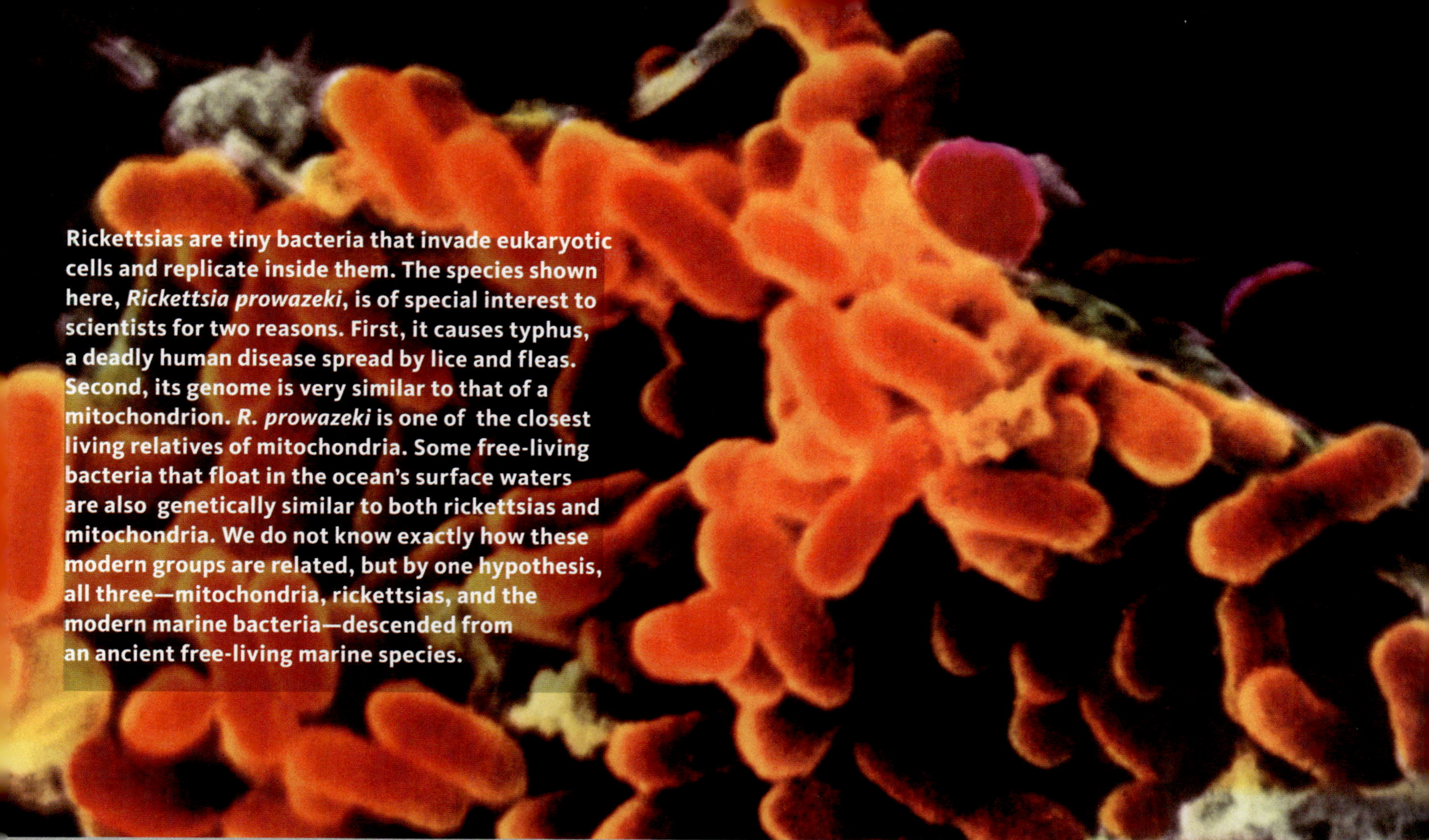

FIGURE 18.10 Modern relatives of mitochondria.

As evidence in support of the endosymbiont hypothesis, mitochondria and chloroplasts resemble bacteria in their size and shape, and they replicate independently of the cell that holds them. Like bacteria, they have their own DNA in the form of a single circular chromosome. They also have at least two outer membranes, with the innermost membrane structurally similar to a bacterial plasma membrane.

Metabolic and genetic similarities between organelles and specific bacterial groups are taken as evidence of shared ancestry. So far, two groups of bacteria have been identified as possible close relatives of mitochondria (**FIGURE 18.10**). Chloroplasts are thought to have evolved from cyanobacteria. Cyanobacteria are the only modern bacteria that, like chloroplasts, carry out photosynthesis by a pathway that produces oxygen.

Nearly all eukaryotic lineages have mitochondria or mitochondria-like organelles, but only some have chloroplasts. Thus, biologists postulate that the two types of organelles were acquired independently in the sequence illustrated in **FIGURE 18.8**.

The endosymbiont hypothesis assumes that cells can enter and live inside other cells. It also assumes that such a relationship can, over time, become essential to the partners. Studies of modern-day cell partnerships lend support to both assumptions. One such study was carried out by microbiologist Kwang Jeon. In 1966, Jeon was studying *Amoeba proteus*, a species of single-celled protist. By accident, his amoebas became infected by a rod-shaped bacterium. Most infected amoebas died right away. A few, however, survived and reproduced despite their infection. Intrigued, Jeon maintained these infected cultures to see what would happen. Five years later, the descendant amoebas were host to many bacterial cells, yet they seemed healthy. In fact, when these amoebas were treated with bacteria-killing drugs that usually do not harm amoebas, they died. Apparently the amoebas had come to require the bacteria for some life-sustaining function. Further investigations revealed that the amoebas had lost the ability to make an essential enzyme. They now depended on their bacterial partners to make that enzyme for them.

endosymbiont hypothesis Theory that mitochondria and chloroplasts evolved from bacteria that entered and lived in a host cell.

TAKE-HOME MESSAGE 18.5

The nucleus and endomembrane system may have evolved from infoldings of the plasma membrane.

Mitochondria and chloroplasts may have evolved when bacterial endosymbionts and their hosts became mutually dependent.

CREDIT: (10) CNRI/Science Source.

18.6 WHAT HAPPENED DURING THE PRECAMBRIAN?

The evolutionary events discussed in earlier sections of this chapter took place during a very long interval of time commonly called the Precambrian (**FIGURE 18.11**). The **Precambrian** encompasses almost all of Earth's history, from its origin 4.6 billion years ago to the beginning of the Cambrian period (542 million years ago). The Cambrian was the first period in our current geologic eon.

The Precambrian opened on a lifeless planet. By 4.3 billion years ago, water had pooled in seas and allowed protocells to form (1). Cells may have evolved more than once, but all present-day life descends from the same single-celled prokaryotic anaerobe (2). An early divergence separated the domains Bacteria and Archaea (3). Later, an evolutionary branching from the Archaea became the domain Eukarya (4). As oxygen released by cyanobacteria began to accumulate, aerobic cells thrived. One type of aerobic bacteria evolved into the mitochondria of early eukaryotes (5). Similarly, cyanobacteria that partnered with one lineage of eukaryotes evolved into chloroplasts (6).

Multicellularity evolved in several eukaryotic lineages (7). A **multicellular organism** consists of interdependent cells that differ in their structure and function. The oldest fossil of a multicellular organism that we can assign to a modern group is a filamentous red alga that lived in the ocean about 1.2 billion years ago.

Two of the most familiar eukaryotic groups, fungi and animals, arose in the sea during the late Precambrian. The oldest fossil fungi are single celled flagellates. One group of modern fungi (chytrids) retains this form. The oldest animal fossils that scientists can assign to a modern group are sponges.

multicellular organism Organism that consists of interdependent cells of multiple types.

Precambrian Period from 4.6 billion to 542 million years ago.

TAKE-HOME MESSAGE 18.6

Life arose and diversified during the Precambrian. By the close of this period, bacteria, archaea, and eukaryotes—including early fungi and animals—lived in the seas.

FIGURE 18.11 **{Animated}** Important Precambrian events.

(1) The first protocells form in Earth's seas.

(2) Single-celled prokaryotic ancestor of all modern life evolves.

(3) Bacteria and Archaea diverge.

(4) Eukarya diverge from Archaea.

(5) Aerobic bacteria enter an early eukaryotic cell and evolve into mitochondria.

(6) Oxygen-producing photosynthetic bacteria enter a eukaryotic cell, evolve into chloroplasts.

(7) Multicellularity evolves independently in several eukaryotic lineages.

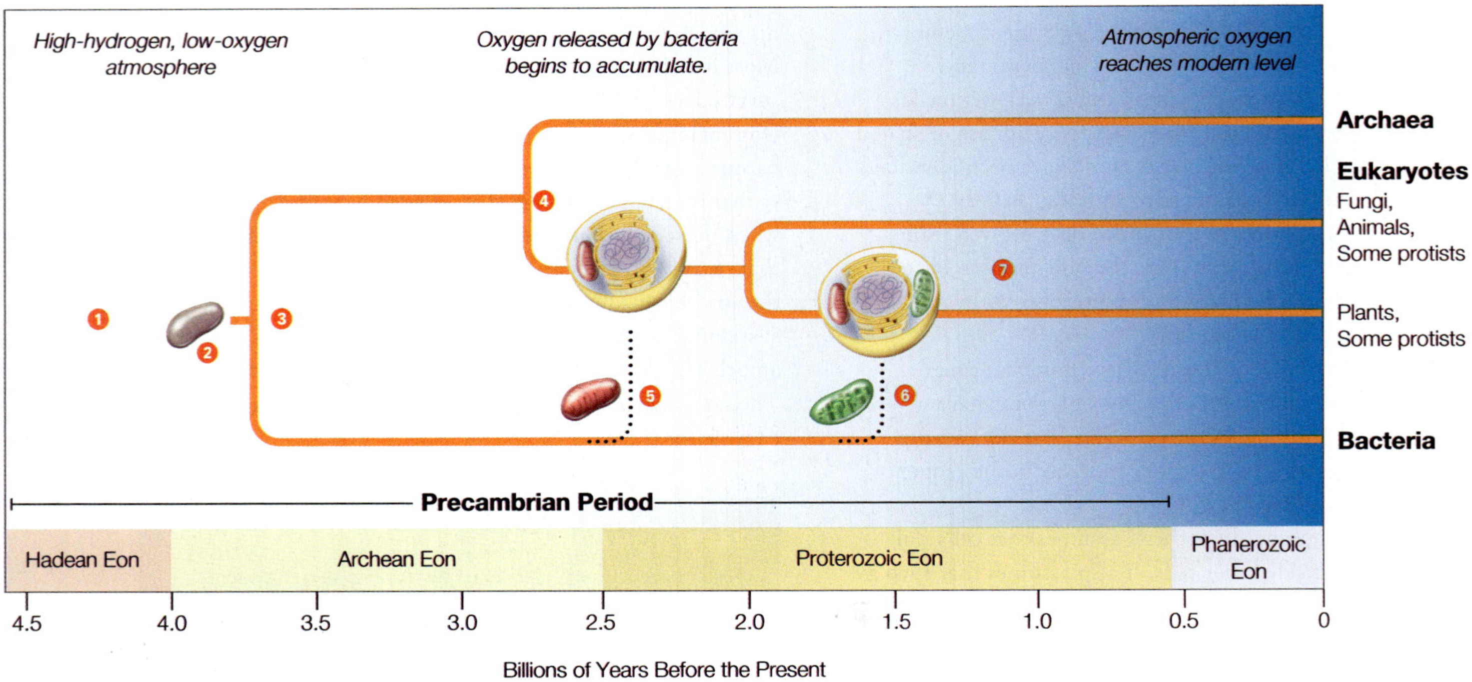

18.7 Application: LIFE IN EXTREME HABITATS

Exploration

FIGURE 18.12 Mars-like landscape of Chile's Atacama Desert. Scientist Jay Quade, visible in the distance at the right, was a member of a team that found bacteria living beneath this desert soil. The inset photo shows the relative sizes of Earth and Mars.

WE LIVE IN A VAST UNIVERSE THAT WE HAVE ONLY BEGUN TO EXPLORE. So far, we know of only one planet that has life—Earth. In addition, biochemical, genetic, and metabolic similarities among Earth's species imply that all evolved from a common ancestor that lived billions of years ago. What properties of the ancient Earth allowed life to arise, survive, and diversify? Could similar processes occur on other planets? These are some of the questions posed by astrobiology, the study of life's origins and distribution in the universe.

Astrobiologists study Earth's extreme habitats to determine the range of conditions that living things can tolerate. One group found bacteria living about 30 centimeters (1 foot) below the soil surface of Chile's Atacama Desert, a place said to be the driest on Earth (FIGURE 18.12). Another group drilled 3 kilometers (almost 2 miles) beneath the soil surface in Virginia, and found bacteria thriving at high pressure and temperature. They named their find *Bacillus infernus*, or "bacterium from hell."

Knowledge gained from studies of life on Earth inform the search for extraterrestrial life. On Earth, all metabolic reactions involve interactions among molecules in aqueous solution (dissolved in water). We assume that the same physical and chemical laws operate throughout the universe, so liquid water is considered an essential requirement for life. Thus, scientists were excited when a robotic lander discovered ice in the soil of Mars, our closest planetary neighbor. If there is life on Mars, it is likely to be underground. Mars has no ozone layer, so ultraviolet radiation would fry organisms at the planet's surface. However, Martian life may exist in deep rock layers just as it does on Earth.

Any Martian life is also almost certainly anaerobic. Mars is only about half the size of Earth. As a result of its smaller size, Mars has less gravity than Earth and is less able to keep atmospheric gases from drifting off into space. The relatively small amount of atmosphere that remains consists mainly of carbon dioxide, some nitrogen, and only traces of oxygen.

Suppose scientists do find evidence of microbial life on Mars or another planet. Why would it matter? Such a discovery would support the hypothesis that life on Earth arose as a consequence of physical and chemical processes that occur throughout the universe. The discovery of extraterrestrial microbes would also make the possibility of nonhuman intelligent life in the universe more likely. The more places life exists, the more likely it is that complex, intelligent life evolved on other planets in the same manner that it did on Earth.

Summary

SECTION 18.1 Earth formed by about 4.6 billion years ago, and by 4.3 billion years ago water had begun to pool in shallow seas. Earth's early seas and atmosphere were free of oxygen gas, as demonstrated by the lack of rust in the oldest remaining rocks.

Laboratory simulations demonstrate that the simple organic building blocks of life could have formed by lightning-fueled reactions in the atmosphere or as a result of reactions in hot, mineral-rich water around **hydrothermal vents**. Such compounds also form in deep space and can be carried to Earth by meteorites.

SECTION 18.2 Researchers carry out experiments to determine whether their hypotheses about steps on the road to life are plausible. Such experiments show that metabolic reactions could have begun after proteins formed spontaneously on an clay-rich tidal flat. Clay attracts amino acids that can bond under the heat of the sun. Metabolic reactions could also have begun in tiny cavities in rocks near deep-sea hydrothermal vents.

Laboratory-produced vesicles serve as a model for the **protocells** that likely preceded cells.

DNA now serves as the molecule of inheritance for all life. However, an **RNA world** may have preceded the current DNA-based system. RNA still is a part of ribosomes that carry out protein synthesis in all organisms. Existence of **ribozymes**, RNAs that act as enzymes, lends support to the RNA world hypothesis. A later switch from RNA to DNA would have made the genome more stable.

SECTION 18.3 Genetic studies indicate that all modern life descended from an anaerobic prokaryotic cell that lived about 4 billion years ago. The divergence of bacteria and archaea occurred soon thereafter.

The small size of cells, lack of hard parts, and destruction and alteration of rocks by geologic processes make it difficult to find and identify evidence of very early life. The oldest microfossils of prokaryotic cells date to about 3.5 billion years ago. The oldest stromatolite fossils date to about the same time. **Stromatolites** are dome-shaped rocky structures that form over thousands of years as colonies of photosynthetic bacteria trap sediment. In some locations, stromatolites are still forming as a result of the activity of oxygen-releasing cyanobacteria.

Fossils of large cells and cells with elaborate cell walls are considered likely eukaryotes, as are fossils associated with certain **biomarkers**. The earliest such fossil cells date to about 1.8 billion years ago.

SECTION 18.4 After photosynthesis evolved, oxygen released by cyanobacteria accumulated in Earth's air and seas. The increased oxygen in air and water favored cells that carried out aerobic respiration. This ATP-forming metabolic pathway was a key innovation in the evolution of eukaryotic cells.

Oxygen molecules react in Earth's upper atmosphere to form an **ozone layer**. Formation of the ozone layer provided protection from UV radiation and may have opened the way for life to move onto land.

SECTION 18.5 Internal membranes typical of eukaryotic cells may have evolved through infoldings of the plasma membrane of prokaryotic ancestors. The existence of internal membranes in some present-day bacteria supports this hypothesis.

Mitochondria and chloroplasts resemble bacteria in their size, shape, genome, and membrane structure. The **endosymbiont hypothesis** holds that these organelles are in fact modified bacteria. Presumably, bacteria entered a host cell and, over generations, host and guest cells came to depend on one another for essential metabolic processes. Mitochondria are descended from aerobic bacteria similar to modern rickettsias; chloroplasts, from cyanobacteria.

Support for the endosymbiont hypothesis comes from observation of modern instances in which bacterial cells have taken up permanent residence inside eukaryotes.

SECTION 18.6 The interval from 4.6 billion years ago to 542 million years ago is called the **Precambrian**. All three domains of life (Bacteria, Archaea, and Eukarya) evolved during the Precambrian. By the close of the Precambrian, **multicellular organisms** such as red algae and sponges lived in the seas.

SECTION 18.7 Astrobiology is the study of life's origin and distribution in the universe. The discovery of cells in deserts and deep below Earth's surface suggests that life may exist in similar settings on other planets. If life exists on Mars, it must be anaerobic, because the thin Martian atmosphere contains little oxygen.

Self-Quiz

Answers in Appendix VII

1. An abundance of ______ in Earth's early atmosphere would have prevented the spontaneous assembly of organic compounds on early Earth.
   a. hydrogen b. methane c. oxygen d. nitrogen

2. Miller and Urey created a reaction chamber that simulated conditions in Earth's early atmosphere to test the hypothesis that ______.
   a. lightning-fueled atmopheric reactions could have produced organic compounds
   b. meteorites contain organic compounds
   c. organic compounds form at hydrothermal vents
   d. oxygen prevents formation of organic compounds

3. The prevalence of iron–sulfur cofactors in organisms supports the hypothesis that life arose ______.
   a. in outer space c. near deep-sea vents
   b. on tidal flats d. in the upper atmosphere

Data Analysis Activities

A Changing Earth Modern conditions on Earth are unlike those when life first evolved. **FIGURE 18.13** shows how the frequency of asteroid impacts and the composition of the atmosphere have changed over time. Use this figure and information in the chapter to answer the following questions.

1. Which occurred first a decline in asteroid impacts, or a rise in the atmospheric level of oxygen?
2. How do modern levels of carbon dioxide and oxygen compare to those at the time when the first cells arose?
3. Which is now more abundant, oxygen or carbon dioxide?

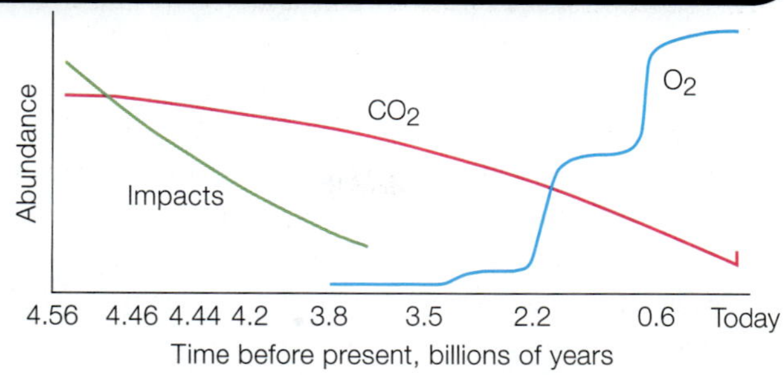

FIGURE 18.13 How asteroid impacts (green), atmospheric carbon dioxide concentration (pink), and oxygen concentration (blue) changed over geologic time.

4. RNA in ribosomes catalyzes formation of peptide bonds in all organisms. This supports the hypothesis that _______ .
   a. an RNA world existed prior to the rise of DNA
   b. RNA can hold more information than DNA
   c. only protists use RNA as their genetic material
   d. all of the above

5. By one hypothesis, clay _______ .
   a. facilitated assembly of early polypeptides
   b. was present at hydrothermal vents
   c. provided energy for early metabolism
   d. served as an early genome

6. The evolution of _______ resulted in an increase in the levels of atmospheric oxygen.
   a. protocells
   b. the ozone layer
   c. oxygen-producing photosynthesis
   d. multicellular eukaryotes

7. Mitochondria most resemble _______ .
   a. archaea c. cyanobacteria
   b. aerobic bacteria d. early eukaryotes

8. The presence of ozone in the upper atmosphere protects life from _______ .
   a. free radicals c. viruses
   b. ultraviolet (UV) radiation d. ionizing radiation

9. A ribozyme consists of _______ .
   a. clay b. DNA c. RNA d. lipids

10. A rise in oxygen in Earth's air and seas put organisms that engaged in _______ at a selective advantage.
    a. aerobic respiration c. photosynthesis
    b. fermentation d. sexual reproduction

11. Which of the following was not present on Earth when chloroplasts first evolved?
    a. archaea c. protists
    b. bacteria d. animals

12. Chloroplasts most resemble _______ .
    a. archaea c. cyanobacteria
    b. aerobic bacteria d. early eukaryotes

13. During the Precambrian, _______ .
    a. protocells formed
    b. some bacteria evolved into mitochondria
    c. multicellular eukaryotes evolved
    d. all of the above

14. Arrange these events in order of occurrence, with 1 being the earliest and 6 the most recent.

___ 1	a. water pools to form Earth's first seas
___ 2	b. origin of mitochondria
___ 3	c. first protocells form
___ 4	d. Precambrian ends
___ 5	e. origin of chloroplasts
___ 6	f. first animals appear

Critical Thinking

1. Researchers looking for fossils of the earliest life forms face many hurdles. For example, few sedimentary rocks date back more than 3 billion years. Review what you learned about plate tectonics (Section 16.5). Explain why so few remaining samples of these early rocks remain.

2. Rickettsias always live as parasites inside eukaryotic cells, and their genomes are much smaller than those of free-living bacteria. Organisms that live only as parasites often have reduced genomes compared to their free-living relatives. How could a parasitic lifestyle contribute to a reduction in genome size?

A population explosion of photosynthetic cyanobacteria colors water of California's Klamath River green.

19

VIRUSES, BACTERIA, AND ARCHAEA

Links to Earlier Concepts

This chapter's discussion of viruses and viroids reverse transcription (Section 15.1), ribozymes (repair (8.5), and cancer (11.5). The chapter also and archaea (4.4). You will learn more about th classification system (1.4), bacteria that gave ri (18.5), and antibiotic resistance (17.14).

KEY CONCEPTS

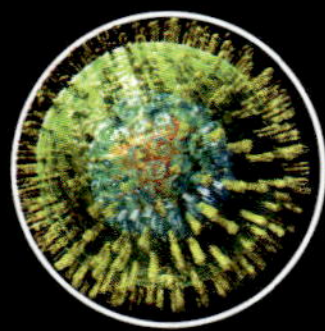

VIRUSES AND VIROIDS

Viruses are noncellular, with a p a genome of nucleic acid. They r to replicate. Viroids are RNAs th

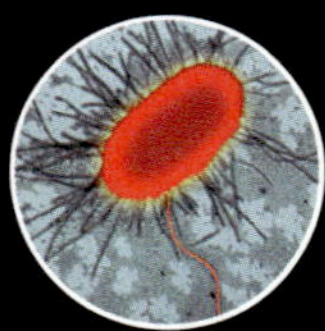

FEATURES OF "PROKAR

Bacteria and archaea are small, s simple, and metabolically diverse binary fission and can exchange existing individuals.

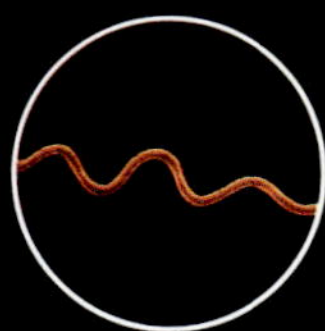

BACTERIAL DIVERSITY

Bacteria put oxygen into the air, plants with essential nitrogen, ar materials. Some that live in or o benefit us, but others are pathog

ARCHAEAL DIVERSITY

The archaea were discovered rela Some species live in very hot or places; others live in human or a Few contribute to human disease

Photograph by David McLain/National Geographic Creative.

19.1 WHAT ARE VIRUSES?

VIRAL CHARACTERISTICS

A **virus** is a noncellular infectious particle that can replicate only in a living cell. A typical virus is so tiny that you would need an electron microscope to see it. A virus does not have ribosomes or other metabolic machinery, but it does have a genome of RNA or DNA. By many definitions, viruses are not alive. However, because they are replicators with a genome that can mutate, they do evolve by natural selection.

A free viral particle (a virus outside a cell) always has a genome enclosed within a protein coat. In some cases, the coat also encloses viral enzymes. Coat proteins protect the viral genome and facilitate its delivery into a host cell. In all viruses, some proteins of the viral coat must bind to proteins at the surface of a host cell during infection.

Viruses are either enveloped or nonenveloped. Most plant viruses are non-enveloped, meaning their protein coat is their outermost layer (**FIGURE 19.1A**). By contrast, most animal viruses have an envelope, a layer of cell membrane derived from the cell in which the virus formed (**FIGURE 19.1B**).

VIRAL REPLICATION

Viral replication begins when a virus attaches to membrane proteins of an appropriate host cell. After attachment, viral genetic material enters the cell. The result is a cellular hijacking. An infected cell puts aside normal tasks and instead produces viral components. These components self-assemble as new viral particles and are released from the cell.

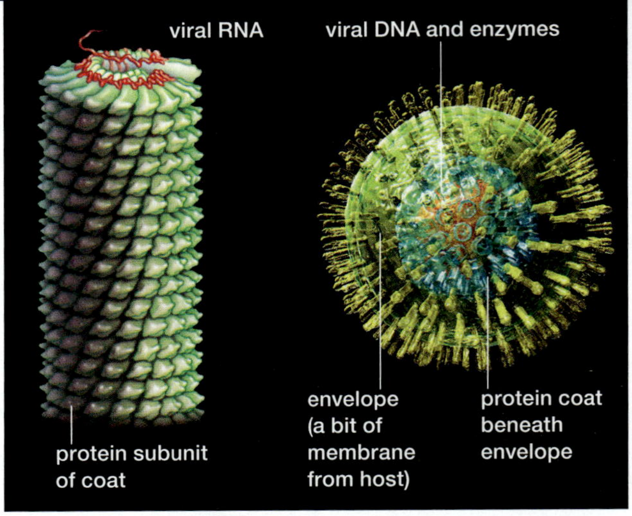

A Tobacco mosaic virus, a nonenveloped plant virus. **B** Herpesvirus, an enveloped animal virus.

FIGURE 19.1 {Animated} Examples of viral structure.

Bacteriophage replication Viruses known as **bacteriophages** (or phages for short) infect bacteria and archaea. Phages have two replication pathways (**FIGURE 19.2**). Both begin when the phage attaches to a bacterium and injects DNA. In the **lytic pathway** (steps A–E), this DNA is put to use immediately, causing the host cell to produce new viral DNA and proteins. The viral components assemble as virus particles, which escape when the cell breaks open (lyses) and dies.

FIGURE IT OUT: What does the blue circle in A represent?

Answer: The bacterial chromosome

FIGURE 19.2 {Animated} Bacteriophage replication pathways.

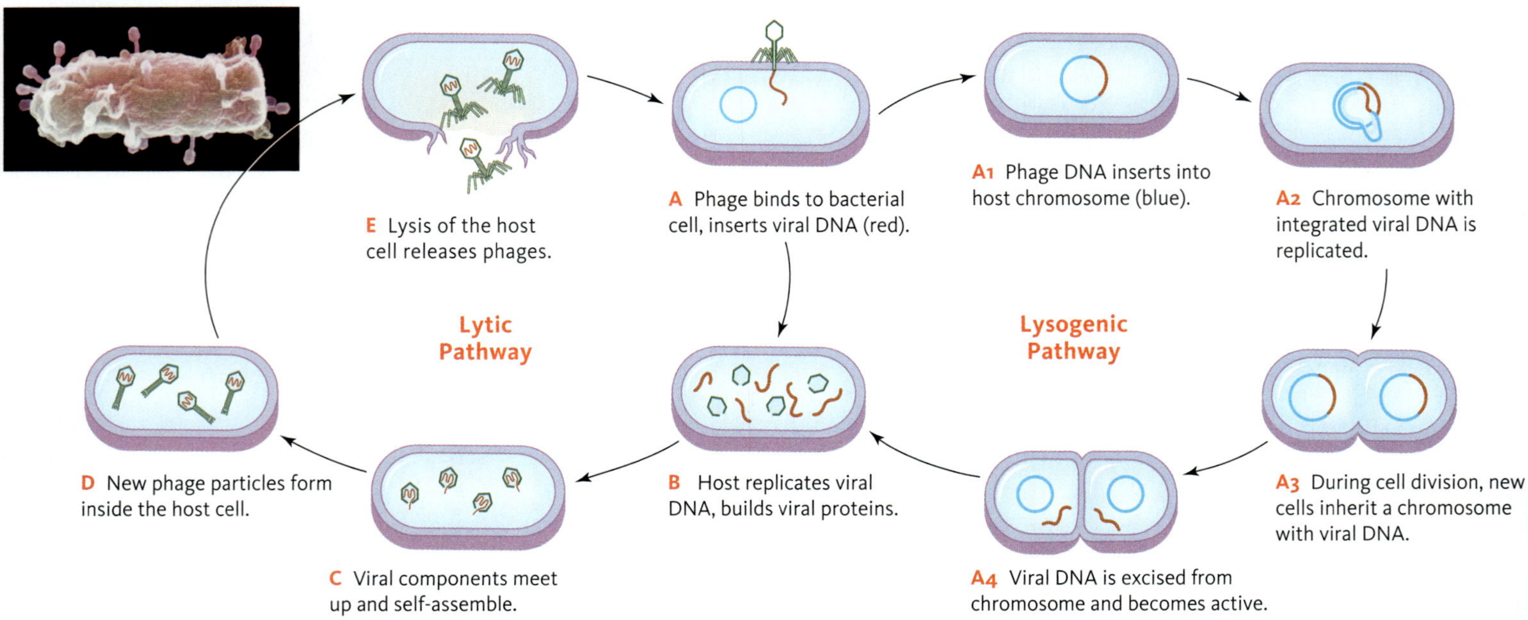

FIGURE 19.3 Structure of HIV, an enveloped retrovirus.

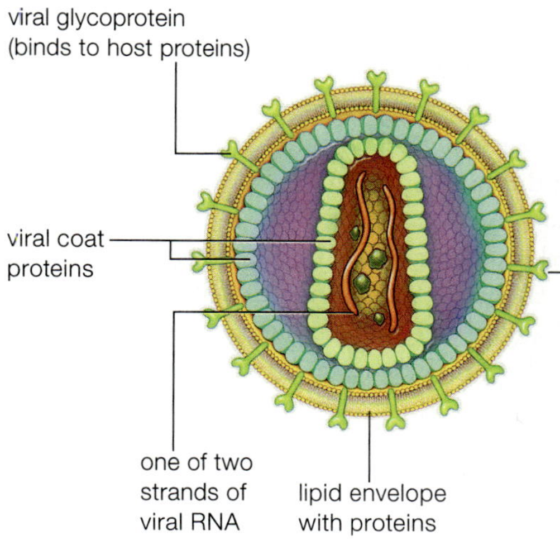

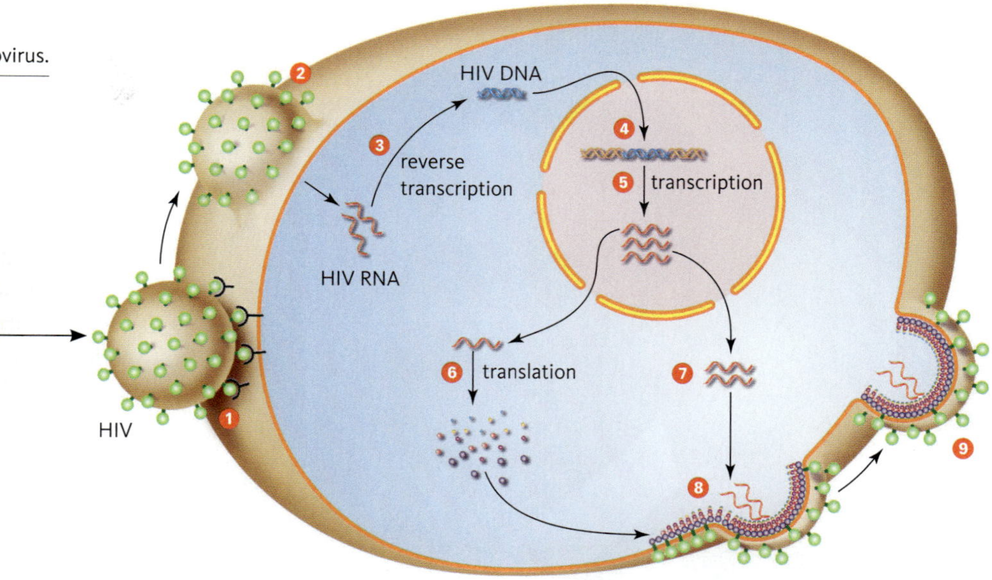

FIGURE 19.4 {Animated} HIV replication.

1. Viral protein binds to proteins at the surface of a white blood cell.
2. Viral RNA and enzymes enter the cell.
3. Viral reverse transcriptase uses viral RNA to make double-stranded viral DNA.
4. Viral DNA enters the nucleus and is integrated into the host genome.
5. Transcription produces viral RNA.
6. Some viral RNA is translated to produce viral proteins.
7. Other viral RNA forms the new viral genome.
8. Viral proteins and viral RNA self-assemble at the host plasma membrane.
9. New virus buds from the host cell, with an envelope of host plasma membrane.

Viral DNA also becomes integrated into the host cell's genome in the **lysogenic pathway** (steps A1–A4). However, viral genes are not expressed, so the host cell remains healthy. When this cell reproduces, a copy of the viral DNA is passed on along with the host's genome. Like miniature time bombs, the viral DNA in descendant cells awaits a signal to enter the lytic pathway.

Some bacteriophages always replicate by the lytic pathway. They kill their host cell quickly and are not passed from one bacterial generation to the next. Others replicate by either the lytic or lysogenic pathway, depending on conditions in the host cell.

HIV replication **HIV (human immunodeficiency virus)** (**FIGURE 19.3**) is an enveloped RNA virus that causes the disease known as AIDS (acquired immunodeficiency syndrome). HIV replicates inside human white blood cells (**FIGURE 19.4**). During infection, spikes of viral protein attach to proteins in the cell's plasma membrane 1. The viral envelope and the membrane fuse, allowing viral enzymes and RNA to enter the cell 2. One of these enzymes, reverse transcriptase, converts viral RNA into double-stranded DNA that the cell can transcribe 3. RNA viruses that use reverse transcriptase to produce viral DNA in a host are called **retroviruses**. Once viral DNA is made, it is moved into the nucleus, along with another viral enzyme that integrates it into a host chromosome 4. The viral DNA is then transcribed along with host genes 5. Some of the resulting viral RNA is translated into viral proteins 6, and some becomes the genetic material of new HIV particles 7. These particles self-assemble at the plasma membrane 8 and a bit of the blood cell's plasma membrane becomes their viral envelope 9.

Drugs that fight HIV interfere with viral binding to the host, reverse transcription, integration of DNA, or processing of viral polypeptides to form viral proteins.

bacteriophage Virus that infects bacteria.
HIV (human immunodeficiency virus) Virus that causes AIDS.
lysogenic pathway Bacteriophage replication path in which viral DNA becomes integrated into the host's chromosome and is passed to the host's descendants.
lytic pathway Bacteriophage replication pathway in which a virus immediately replicates in its host and kills it.
retrovirus RNA virus that uses the enzyme reverse transcriptase to produce viral DNA in a host cell.
virus Noncellular, infectious particle of protein and nucleic acid; replicates only in a host cell.

TAKE-HOME MESSAGE 19.1

A virus is a noncellular infectious particle that consists of nucleic acid enclosed in a protein coat and sometimes an outer envelope.

A virus replicates by binding to a specific type of host cell, taking over the host's metabolic machinery, and using that machinery to produce viral components. These components self-assemble to form new viral particles.

19.2 HOW DO VIRUSES AFFECT HUMAN HEALTH?

PEOPLE MATTER

Virologist
DR. NATHAN WOLFE

Where will the next global health threat arise and what can we do to head it off? National Geographic Explorer Dr. Nathan Wolfe is working to create an early warning system that can forecast and contain new plagues before they kill millions. He says, "The way that pandemics come to us is through our interaction with animals. That's true whether it's swine flu (H1N1), bird flu, Ebola, or HIV. These are all animal diseases that jumped to humans." Wolfe monitors hunters and other individuals who have high levels of contact with wild animals. When these individuals butcher an animal, they collect a spot of the animal's blood that is analyzed for viruses. "We watch as the viruses are pinging at us, and when something takes hold in humans, we sound the alarm." Wolfe recalls, "When I started this work in 1999, we were particularly interested in retroviruses because we knew they had the potential to cause devastating pandemics like HIV. But we didn't have a good idea of the frequency with which they were crossing over from animals to humans. Our results were shocking. We discovered that cross-species transmission wasn't rare; it was happening on a regular basis." Despite the many threats, Wolfe remains optimistic. "It's really the dawn of a new scientific era. As we try to detect viruses that can do great harm, we could also discover the next generation of vaccines and cures. If we can provide even a few months of early warning for just one pandemic, the benefits will outweigh all the time and energy we're devoting. Imagine preventing health crises, not just responding to them."

COMMON VIRAL DISEASES

Some viruses live in our body without any ill effects, but other are **pathogens**, meaning they cause disease. Viral diseases usually produce mild symptoms and sicken us only briefly. For example, some viruses infect cells in our upper respiratory system and cause common colds. Viruses that infect cells in the lining of our gut often cause a brief bout of vomiting and diarrhea.

Other viral pathogens persist in our body for long periods. Typically, an initial infection causes symptoms that subside quickly. However, the virus remains present in a latent state and can reawaken later on. For example, herpes simplex virus 1 (HSV-1) remains dormant in nerve cells for years. When activated, it replicates and causes painful "cold sores" at the edge of the lips (**FIGURE 19.5**). Another herpesvirus causes similar sores on the genitals.

A few types of viral infections increase the risk of cancer. Infection by certain strains of sexually transmitted human papillomaviruses (HPV) is the main cause of cervical cancer. Similarly, infection by some hepatitis viruses raises the risk of liver cancer.

EMERGING VIRAL DISEASES

An **emerging disease** is a disease that has only recently been detected in humans, or has recently expanded its range. AIDS is an emerging viral disease that was first identified in humans in 1981. Since then, it has caused about 30 million deaths. AIDS is a communicable disease, meaning HIV spreads from one infected person to another. Currently, about 34 million people are infected by HIV.

By contrast, the virus that causes West Nile fever cannot spread directly from person to person. Mosquitoes carry the virus from host to host, so we say they are the **vector** for this virus. Birds serve as a "reservoir" for West Nile virus, which means the virus can replicate in birds and birds serve as a source of virus for new human infections. West Nile virus was unknown in North America until 1999, when it emerged in New York. It has since spread across the continent. In 2012, there were more than 4,000 cases of West Nile fever in the United States and 163 deaths.

Ebola virus causes the emerging disease Ebola hemorrhagic fever. Symptoms of this incurable disease include fever, muscle pain, and massive internal and external bleeding. The death rate is close to 90 percent. Ebola virus

emerging disease A disease that was previously unknown or has recently begun spreading to a new region.
pathogen Disease-causing agent.
vector Animal that carries a pathogen from one host to the next.
viral recombination Multiple strains of virus infect a host simultaneously and swap genes.

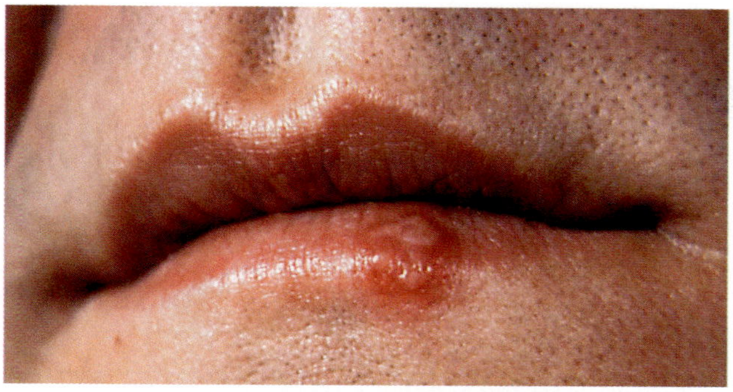

FIGURE 19.5 Sign of an active herpes simplex virus 1 infection. Fluid rich in viral particles leaks from the open sore.

is highly contagious in humans, but so far it has been seen only in Africa, where fruit bats are its most likely reservoir.

VIRAL RECOMBINATION

Like living organisms, viruses have genomes that can mutate. RNA viruses such as HIV and influenza viruses (the viruses that cause the flu) mutate especially quickly. Viral genomes can also change when two viruses infect a host at the same time and exchange genes, a common process called **viral recombination**. A new subtype of H1N1 influenza virus that appeared in Asia in 2009 arose by recombination. The media called this virus a "swine flu," but it had a composite genome, with genes from a human flu virus, bird flu virus, and two different swine flu viruses. Fortunately, H1N1 proved less deadly than anticipated and rapid development of a vaccine put a stop to its spread by August 2010.

Another strain of influenza, H5N1, occasionally infects people who have direct contact with birds. From 2003 to 2012, there have been 608 reported human cases of influenza H5N1, mainly in Asia. Of these, 359 (about 60 percent) were fatal. Fortunately, person-to-person transmission of the H5N1 virus is exceedingly rare.

Health officials continue to carefully monitor H5N1 and H1N1 influenza. Either virus could mutate, and their coexistence raises the possibility of a potentially disastrous gene exchange. If H1N1 picked up genes from avian H5N1, the result could be a flu virus that is both easily transmissible and deadly.

TAKE-HOME MESSAGE 19.2

Viruses cause many widespread, familiar diseases such as the common cold and cold sores, and some cause cancers.

Other viruses cause emerging diseases such as AIDS and West Nile virus that have only recently become global threats.

19.3 HOW DO VIRUSES AND VIROIDS AFFECT PLANTS?

Plant viruses usually enter a plant through a wound made by an insect or by pruning. Symptoms of viral infection typically include stunting and curling, yellowing, or spotting of leaves (**FIGURE 19.6A**). Viral infections can dramatically decrease crop yields, so plant viruses are of great commercial importance. Infected plants cannot be rid of a virus, so farmers focus on preventing viral infections. They grow plants that have been bred for viral resistance and try to fend off insects that transmit viral diseases.

Viroids, another type of plant pathogen, also enter plants through wounds. They consist solely of single-stranded RNA. Viroids are remarkably small, with a genome comprising fewer than 400 nucleotides. By comparison, even the smallest viral genome has thousands of nucleotides. Unlike the genetic material of a virus, viroid RNA does not encode proteins. The viroid causes disease by interfering with normal gene expression within the plant cell. A viroid infection typically stunts a plant and deforms its parts (**FIGURE 19.6B**). The viroid replicates with assistance of a plant enzyme (RNA polymerase) that normally transcribes the plant's DNA to RNA.

FIGURE 19.6 Symptoms of virus or viroid infection in plants.

A Mottled leaves of a plant with tobacco mosaic virus.

B Potato tuber deformed by a viroid infection.

viroid Small noncoding RNA that can infect plants.

TAKE-HOME MESSAGE 19.3

Viruses and viroids (small RNAs) can enter a plant through a wound and cause disease.

19.4 HOW ARE BACTERIA AND ARCHAEA SIMILAR?

The two lineages of prokaryotes (cells without a nucleus) belong to the domains Bacteria and Archaea. Later sections in this chapter describe unique features of each of these lineages. In this section, we focus on the structural and functional traits they have in common (**TABLE 19.1**).

TABLE 19.1

Prokaryotic Traits

1.	No nucleus, endoplasmic reticulum, Golgi bodies
2.	Single chromosome (a circular DNA molecule); many species also contain plasmids
3.	Cell wall (in most species)
4.	Ribosomes distributed in the cytoplasm
5.	Asexual reproduction by binary fission
6.	Capacity for gene exchange among existing cells through conjugation, transduction, and transformation

CELL SIZE AND STRUCTURE

All bacteria and archaea are single-celled and nearly all are too small to be observed without a light microscope. Three cell shapes are common. Spherical cells are referred to as cocci (singular, coccus); rod-shaped cells as bacilli (singular, bacillus); and spiral cells as spirilli (singular, spirillum). **FIGURE 19.7** shows the typical structure of a prokaryotic cell. Most bacteria and archaea secrete a porous cell wall around their plasma membrane. Many also secrete a layer of slime or a capsule onto this wall. Within the cell, a single circular chromosome (a ring of double-stranded DNA) typically lies exposed in the cytoplasm. Some bacteria have internal membranes, but no known prokaryote has mitochondria, a nuclear envelope, or an endomembrane system like that of eukaryotes. Ribosomes scattered through the cytoplasm assemble proteins.

Bacteria and archaea often have one or more flagella that rotate like a propeller. By contrast, a eukaryotic flagellum whips from side to side. Many bacteria and archaea also have hairlike projections called pili that function in adhesion or locomotion. A type of retractable pilus can draw cells together for gene exchanges.

FIGURE 19.7 {Animated} Structural features of a prokaryote.

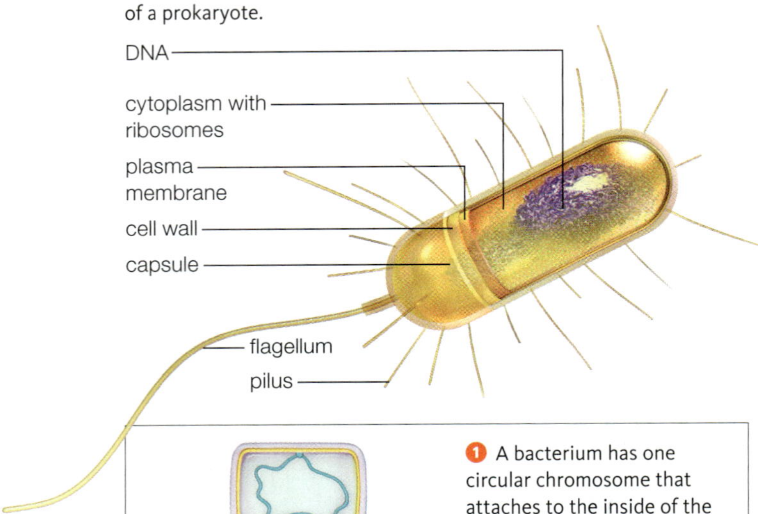

REPRODUCTION AND GENE EXCHANGE

Bacteria and archaea usually reproduce by **binary fission**, a type of asexual reproduction (**FIGURE 19.8**). The process begins when the cell replicates its single chromosome, which is attached to the inside of the plasma membrane ❶. The DNA replica attaches to the plasma membrane adjacent to the parent molecule. Addition of new membrane and wall material elongates the cell and moves the two DNA molecules apart ❷. Then, membrane and cell wall material is deposited across the cell's midsection ❸. This material partitions the cell, yielding two identical descendant cells ❹.

In addition to inheriting DNA "vertically" from a parent cell, prokaryotes engage in **horizontal gene transfer**: a transfer of genetic material between existing individuals. **Conjugation** involves transfer of genes on a plasmid. Recall that a plasmid is a small circle of DNA separate from the chromosome (Section 15.2). During conjugation, a cell that contains a plasmid extends a special sex pilus to another cell, draws it close, then puts a copy of the plasmid into its partner (**FIGURE 19.9**). Both bacteria and archaea have plasmids and conjugation can sometimes transfer genes between the two groups.

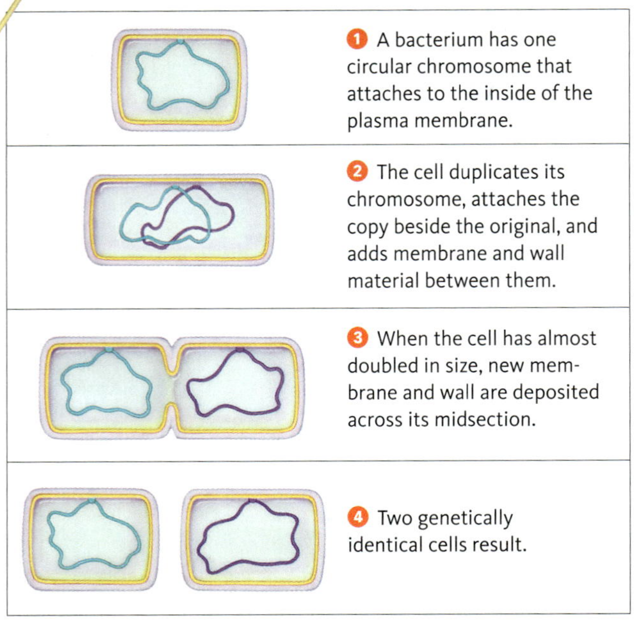

FIGURE 19.8 {Animated} Binary fission in bacteria.

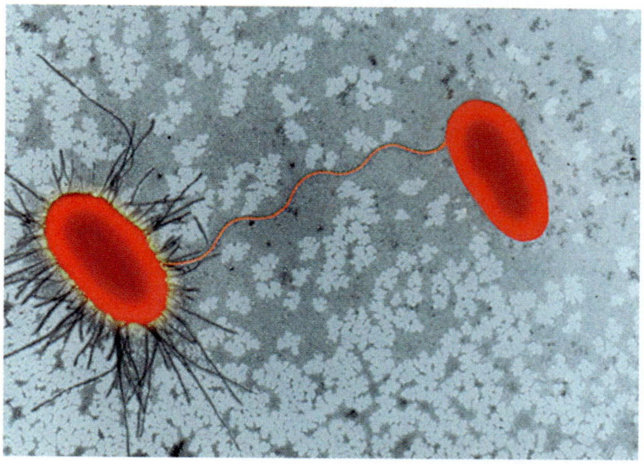

FIGURE 19.9 {Animated} Prokaryotic conjugation. One cell extends a sex pilus out to another, draws it close, and gives it a copy of a plasmid.

FIGURE IT OUT: Does conjugation increase the number of cells?
Answer: No. It is not a mode of reproduction.

CARBON SOURCE	ENERGY SOURCE: Light	ENERGY SOURCE: Chemicals
Inorganic source such as CO_2	**Photoautotrophs** bacteria, archaea, photosynthetic protists, plants	**Chemoautotrophs** bacteria, archaea
Organic source such as glucose	**Photoheterotrophs** bacteria, archaea	**Chemoheterotrophs** bacteria, archaea, fungi, animals, nonphotosynthetic protists

FIGURE 19.10 {Animated} Modes of obtaining energy and carbon, and the groups that can use them.

FIGURE IT OUT: Which group of organisms can build their own food from CO_2 in the dark?
Answer: Chemoautotrophs

Genes also move between prokaryotes by transduction and tranformation. Transduction occurs when a virus picks up a bit of DNA from one prokaryotic host cell, then transfers the DNA to its next host. Transformation occurs when bacteria or archaea take up DNA from their environment and integrate it into their genome.

METABOLIC DIVERSITY

Collectively, prokaryotes use all four possible mechanisms of obtaining energy and nutrients (**FIGURE 19.10**). Metabolic diversity gives these cells the ability to thrive in a vast variety of environments. Many bacteria and most archaea are anaerobic: They can (or must) live where there is no oxygen. Others are aerobic, which means they require (or at least tolerate) oxygen.

As Section 6.1 explained, autotrophs build their own food from an inorganic source of carbon such as carbon dioxide (CO_2). Many bacteria and some archaea are **photoautotrophs**, which carry out photosynthesis. Like plants and photosynthetic protists, these organisms use light energy to fuel assembly of their food. No eukaryotes are **chemoautotrophs**, organisms that fuel the assembly of their food by oxidizing (removing electrons from) inorganic substances such as hydrogen sulfide or methane. Chemoautotrophic bacteria and archaea are the main producers in dark places such as the seafloor.

Heterotrophs obtain carbon by breaking down organic molecules from their environment. Some bacteria and archaea are **photoheterotrophs** that use light energy to fuel this process. No eukaryotes are photoheterotrophs. **Chemoheterotrophs** obtain both energy and carbon through the breakdown of organic compounds. This group includes fungi, animals, and nonphotosynthetic protists, as well as some archaea and most bacteria. All pathogenic bacteria are chemoheterotrophs that extract the organic compounds they need to live from their host. Other bacterial chemoheterotrophs serve as decomposers.

binary fission Method of asexual reproduction that divides one bacterial or archaeal cell into two identical descendant cells.
chemoautotroph Organism that uses carbon dioxide as its carbon source and obtains energy by oxidizing inorganic molecules.
chemoheterotroph Organism that obtains energy and carbon by breaking down organic compounds.
conjugation Mechanism of horizontal gene transfer in which one prokaryote passes a plasmid to another.
horizontal gene transfer Transfer of genetic material between existing individuals.
photoautotroph Organism that obtains carbon from carbon dioxide and energy from light.
photoheterotroph Organism that obtains its carbon from organic compounds and its energy from light.

TAKE-HOME MESSAGE 19.4

Prokaryotes (bacteria and archaea) are small, typically walled cells with no nuclear membrane, mitochondria, or endomembrane system.

Both groups of prokaryotes reproduce asexually by binary fission. They exchange genes by conjugation and other means of horizontal gene transfer.

Prokaryotes are metabolically diverse, collectively making use of all four modes of obtaining energy and nutrients. Many are anaerobic.

19.5 WHAT ARE SOME MAJOR BACTERIAL LINEAGES?

There are many bacterial lineages and new ones are constantly being discovered. Here we consider a few major groups to provide insight into bacterial diversity.

OXYGEN-PRODUCING CYANOBACTERIA

Photosynthesis evolved in many bacterial lineages, but only **cyanobacteria** release oxygen as a by-product. If, as evidence suggests, chloroplasts evolved from ancient cyanobacteria, we have cyanobacteria and their chloroplast descendants to thank for nearly all the oxygen in Earth's air.

Some cyanobacteria also carry out **nitrogen fixation**: They incorporate nitrogen from the air into ammonia (NH_3). Nitrogen fixation is an important ecological service. Photosynthetic eukaryotes need nitrogen but they cannot use the gaseous form ($N \equiv N$) because they do not have an enzyme that can break the molecule's triple bond. Plants and photosynthetic protists can, however, take up ammonia released by nitrogen-fixing bacteria.

Cyanobacteria can partner with fungi to form lichens (which we discuss in Section 22.3) and can grow on the surface of soils, but most are aquatic. The photo that opens this chapter shows a river tinted green by cyanobacteria. Aquatic cyanobacteria grow as single cells or as long filaments of cells arranged end to end (**FIGURE 19.11**). When conditions become unfavorable for growth, some filamentous cyanobacteria produce thick-walled resting cells. These cells are easily dispersed and remain dormant until environmental conditions improve.

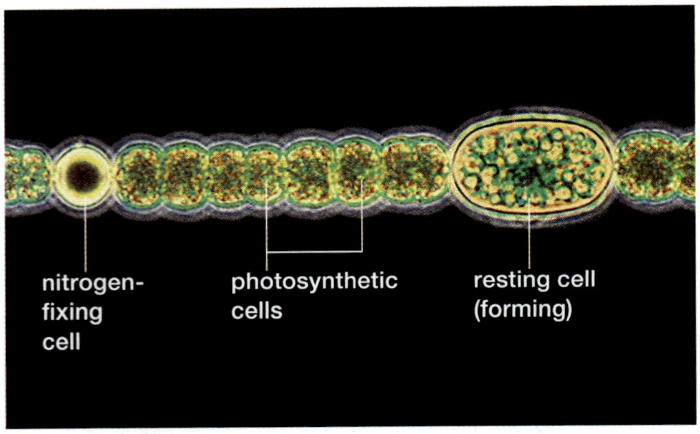

FIGURE 19.11 Chain of aquatic cyanobacterial cells.

HIGHLY DIVERSE PROTEOBACTERIA

Proteobacteria, the most diverse bacterial group, are defined on the basis of their ribosomal RNA sequence rather than any structural or metabolic feature. Most are aerobic. Some, such as the purple bacteria, carry out photosynthesis but do not produce oxygen as a by-product. Other proteobacteria are chemoautotrophs or chemoheterotrophs. One chemoautotrophic species, *Thiomargarita namibiensis*, is among the largest prokaryotes and can be seen even without a microscope (**FIGURE 19.12A**). A giant vacuole containing nitrate and sulfur makes up the bulk of the cell. The cell uses nitrate in its energy-releasing reactions and forms sulfur as a by-product of these reactions.

Many chemoheterotrophic proteobacteria live in other organisms. For example, members of the genus *Rhizobium* live inside the roots of legumes such as peas and beans. The bacteria receive sugars from the plant, and in turn provide the plant with ammonia from nitrogen fixation. Rickettsias live as parasites inside animals and are considered close relatives of mitochondria (Section 18.5).

Scientists use some proteobacteria in biological research and biotechnology. The most well-studied prokaryote is *Escherichia coli*, a rod-shaped species of proteobacteria that normally lives in the mammalian gut (**FIGURE 19.12B**). It is easily grown in the laboratory, so researchers often investigate genetic and metabolic processes by studying *E. coli*. It is also widely used in industrial biotechnology. Recombinant *E. coli* now produce many synthetic hormones and other proteins for medical use.

When biotechnologists want to alter a plant's genome, they may turn to *Agrobacterium*. These soil proteobacteria have a plasmid that gives them the capacity to infect plants and cause a tumor. As Section 15.5 explained, scientists can produce recombinant plants by inserting the foreign genes into the tumor-inducing plasmid, then infecting a plant with recombinant bacteria.

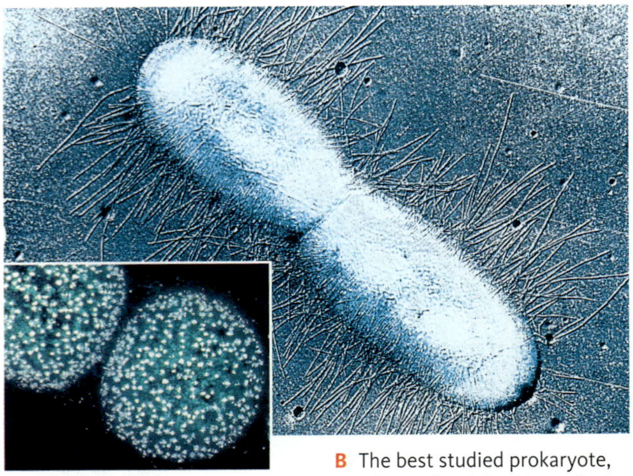

A The largest prokaryote, *Thiomargarita namibiensis*. Cells can be up to 0.75 mm wide.

B The best studied prokaryote, *E. coli*, can divide by binary fission every 20 minutes.

FIGURE 19.12 Proteobacteria.

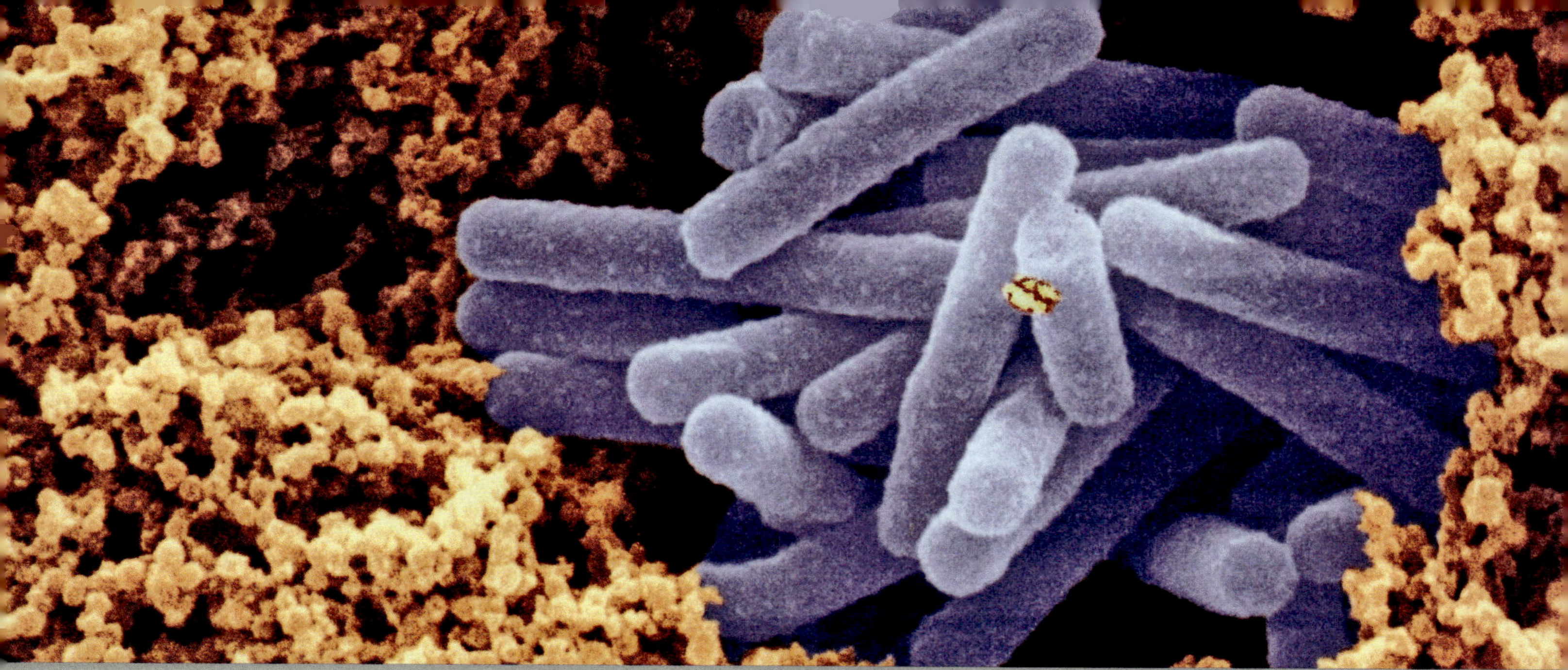

FIGURE 19.13 Lactobacilli in yogurt. The cells ferment milk sugars to produce lactate that gives yogurt its tangy taste.

THICK-WALLED GRAM POSITIVES

Gram-positive bacteria have thick cell walls that stain purple when prepared for microscopy by a process known as Gram staining. Most are chemoheterotrophs and many serve as decomposers. **Decomposers** break down complex organic molecules in wastes and remains into inorganic components that plants can take up and use. Decomposers also break down pesticides and pollutants, thus improving the environment for other organisms.

Actinomycetes are Gram-positive decomposers that grow through soil as long, branching chains of cells. Their presence gives freshly exposed soil its distinctive "earthy" smell. Many antibiotics, including streptomycin and vancomycin, were first isolated from actinomycetes.

Soil also contains members of the Gram-positive genus *Lactobacillus*. These cells ferment sugars and produce lactate. Lactobacilli sometimes spoil milk, but they are also used to produce sour foods such as sauerkraut and yogurt (**FIGURE 19.13**). Streptococci, which are close relatives of lactobacilli, include species that cause strep throat and the skin disease impetigo. Streptococci are also the usual cause of bacterial pneumonias.

Soil bacteria in the genera *Clostridium* and *Bacillus* form endospores when conditions are unfavorable. An **endospore** contains the cell's DNA and a bit of cytoplasm in a protective coat. It is functionally similar to a cyanobacterial resting cell, but much tougher. Endospores survive drying, boiling, and radiation. If they enter a human body and germinate, the result can be deadly. In 2001, distribution of *Bacillus anthracis* endospores through the United States mail resulted in five deaths from the disease anthrax.

SPIRAL-SHAPED SPIROCHETES

Spirochetes are very small bacteria with a shape like a stretched-out spring (**FIGURE 19.14**). Some live in the cattle gut and help their host by breaking down cellulose. Others are aquatic decomposers and some fix nitrogen. Still others are pathogens. One pathogenic spirochete causes the sexually transmitted disease syphilis, and another causes Lyme disease, which we discuss in the next section.

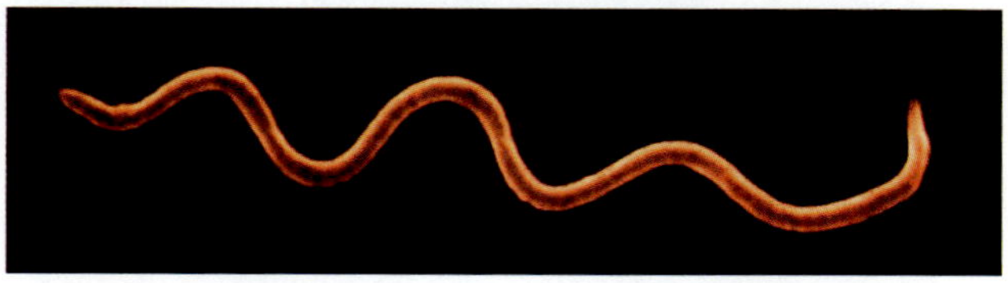

FIGURE 19.14 **A spirochete.**

TAKE-HOME MESSAGE 19.5

- Bacteria play important ecological roles. They put oxygen into the air, fix nitrogen, and act as decomposers.
- Bacteria play an important role in scientific research, biotechnology, and food production.
- Many bacterial groups include species that adversely affect human health.

cyanobacteria Photosynthetic, oxygen-producing bacteria.
decomposer Organism that breaks down organic compounds in wastes and remains into their inorganic subunits.
endospore Resistant resting stage of some soil bacteria.
Gram-positive bacteria Bacteria with thick cell walls that are colored purple when prepared for microscopy by Gram staining.
nitrogen fixation Incorporation of nitrogen gas into ammonia.
proteobacteria Most diverse bacterial lineage.
spirochetes Bacteria that resemble a stretched-out spring.

CREDITS: (13) SciMAT/Science Source; (14) Stem Jems/Science Source.

19.6 HOW DO BACTERIA AFFECT HUMAN HEALTH?

NORMAL FLORA

Bacteria and other microorganisms that normally live in or on our body are our **normal flora**. The normal flora serves as a first defense against pathogens. For example, lactobacilli that normally live in the human mouth, gut, and vagina produce lactate, an acid that helps keep acid-intolerant pathogenic bacteria from becoming established. Some side effects of antibiotics result from the effects of these drugs on our normal flora. Antibiotics kill beneficial bacteria as well as harmful ones. When this disruption takes place in the gut, it causes diarrhea. When it happens in the vagina, the resulting overgrowth of yeast (fungal cells) can cause vaginitis.

In addition to their protective role, gut bacteria also benefit us by producing some essential vitamins. Our intestines are home to lactobacilli that synthesize some B vitamins and to *E. coli* that produce vitamin K.

TOXINS AND DISEASE

Pathogenic bacteria produce toxins that cause the symptoms of disease. A bacterial toxin may be a substance that bacteria release into their environment (an exotoxin), or a molecule integral to the cell's wall (an endotoxin). Exotoxins directly harm human cells. For example, botulinum toxin released by *Clostridium botulinum* causes a dangerous paralyzing food poisoning by disrupting nerve cell function. Another bacterial exotoxin, Shiga toxin, is a ribosome-inhibiting protein (Section 9.6) that interferes with protein synthesis by intestinal cells. Endotoxins do not directly harm cells. Rather, they elicit an immune response that typically includes fever and aches. If this immune response spirals out of control, the resulting loss of blood pressure and organ failure can be deadly.

Bacteria in our normal skin flora can act as pathogens when a cut, scrape, or other wound allows them into our body. For example, *Staphylococcus aureus* that can live harmlessly on the skin sometimes infects deep skin layers. An antibiotic-resistant strain—methicillin-resistant *Staphylococcus aureus* (MRSA)—is especially dangerous, because it does not respond to most commonly used drugs.

Bacterial pathogens also enter the body in tainted food or water. In the United States, most cases of bacterial food poisoning involve *Salmonella* or *Campylobacter*, both proteobacteria. These bacteria live in the gut of animals such as chickens, and bacteria-rich animal feces sometimes get into food. Symptoms of food poisoning typically include fever, diarrhea, and abdominal cramps. Usually an affected

FIGURE 19.15 Signs of tuberculosis. A doctor at an Ethiopian hospital points out signs of TB on a chest x-ray.

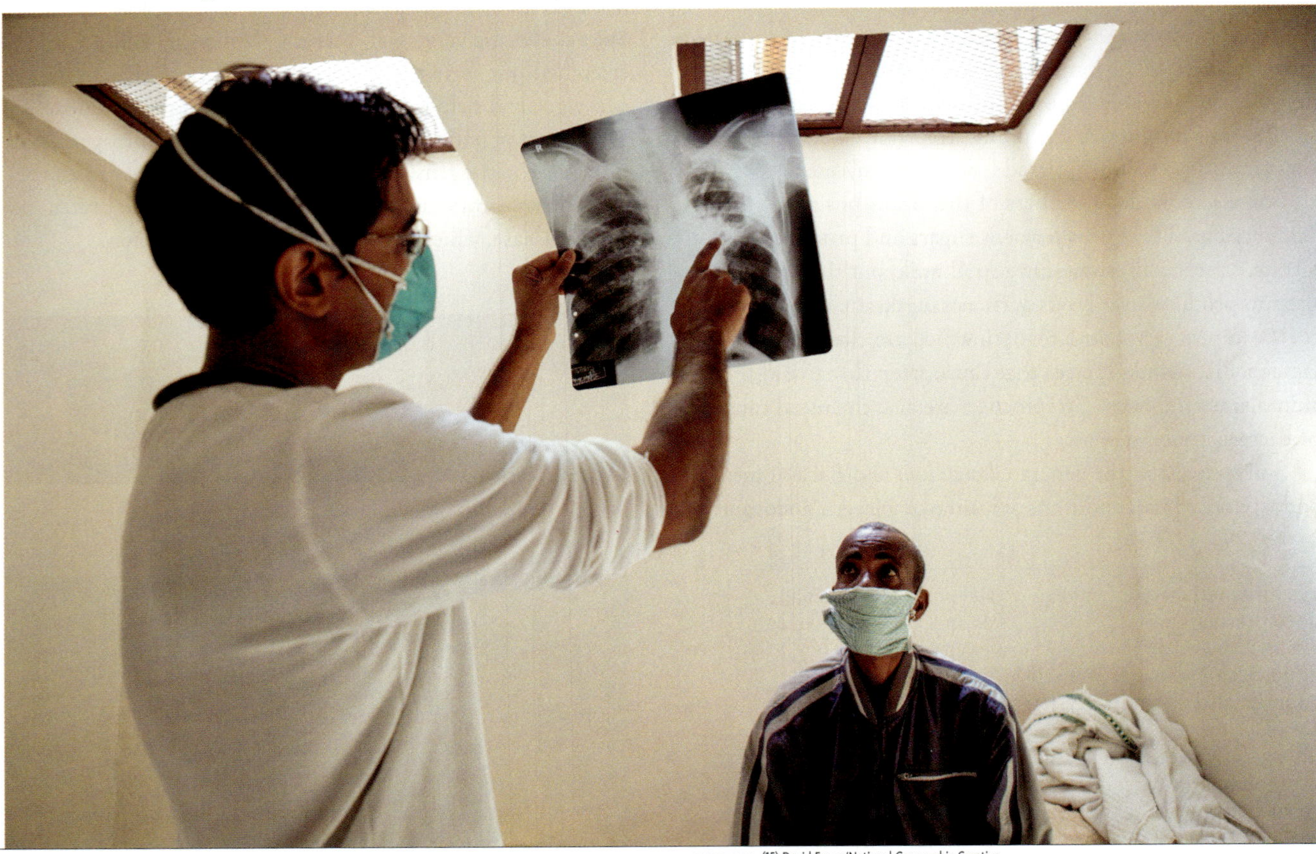

CREDITS: (15) David Evans/National Geographic Creative.

person's immune system rids the body of bacteria and recovery occurs within a few days to a week.

In regions where safe drinking water is not readily available, cholera sickens millions of people each year and kills about 100,000. Cholera-causing bacteria (*Vibrio cholerae*) produce an exotoxin that poisons intestinal cells. The resulting massive diarrhea causes solute and water losses that can be fatal. Tainted water also spreads *Helicobacter pylori*, the only bacterial species known to cause cancer. A long-term *H. pylori* infection increases the risk of ulcers and stomach cancer.

Communicable diseases caused by bacteria include strep throat, whooping cough, bacterial pneumonia, and the sexually transmitted diseases gonorrhea and syphilis. *Mycobacterium tuberculosis* causes the communicable disease tuberculosis (TB), which infects an estimated one-third of the world population. Fortunately, in most otherwise healthy people, a TB infection remains inactive; bacteria survive, but the immune system keeps them in check and the affected person cannot infect others. In a minority of people, the infection becomes active; bacteria grow in the lungs and cause coughing, chest pain, fever, and weight loss (**FIGURE 19.15**). A person with active TB spreads the disease by expelling tiny bacteria-laden droplets in coughs and sneezes. In 2011, about 2 million people died as a result of active TB. Both active and inactive TB infections can be cured by antibiotics, but killing all the bacteria in the body requires months of treatment. To further complicate matters, strains of *M. tuberculosis* resistant to most commonly used antibiotics are increasingly prevalent.

The most common vector-borne bacterial disease in the United States is Lyme disease. It occurs most frequently in northeastern and upper midwestern states. Ticks transmit the spirochete (*Borrelia burgdorferi*) that causes this disease and mice serve as its reservoir. Some infected people develop a rash at the site where the tick bite introduced the bacteria. More often, flulike symptoms and fever are the first indications of disease. Untreated, Lyme disease can damage joints, the heart, and the nervous system.

normal flora Normally harmless or beneficial microorganisms that typically live in or on a body.

TAKE-HOME MESSAGE 19.6

- Bacteria of our normal flora protect us against infections and provide us with vitamins.
- Pathogenic bacteria release toxins or have toxins in their cell wall.
- Bacteria cause wound infections, food poisoning, and communicable diseases, and one species causes cancer.

19.7 HOW WERE ARCHAEA DISCOVERED?

Science is a self-correcting process in which new discoveries can overturn even long-held ways of thinking. For example, biologists historically divided all life into two domains, prokaryotes and eukaryotes. These groups were very different in size and structure, so scientists thought they represented distinct lineages (clades) that parted ways early in the history of life (**FIGURE 19.16A**).

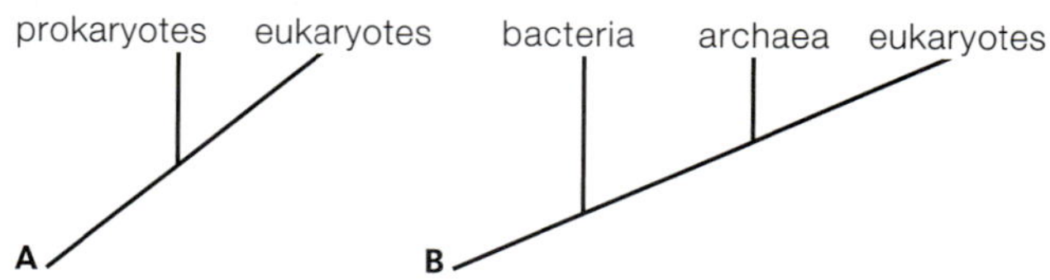

FIGURE 19.16 The **A** two-domain and **B** three-domain trees of life. The two-domain model was widely accepted until new evidence revealed previously unknown differences between bacteria and archaea. The three-domain model is now in wide use.

In the late 1970s Carl Woese began investigating evolutionary relationships among prokaryotes. By comparing ribosomal RNA gene sequences, Woese found that some methane-making cells were as similar to eukaryotes as they were to typical bacteria. Woese proposed that the methane makers were not bacteria but rather a previously unrecognized branch on the tree of life. To accommodate this branch, he proposed a new classification system with three domains of life: bacteria, archaea, eukaryotes (**FIGURE 19.16B**).

Woese's ideas were initially greeted with skepticism, but as years went by, evidence in support of them mounted. Archaea and bacteria have different cell wall and membrane components. Archaea and eukaryotes organize their DNA around histone proteins, which bacteria do not have. Sequencing the genome of the archaeon *Methanococcus jannaschii* (**FIGURE 19.17**) provided definitive support for Woese's proposal. Most of this archaeon's genes have no counterpart in bacteria. Today, the hypothesis that bacteria and archaea constitute a single lineage has been discarded, and the three-domain classification system is in widespread use. Woese compares discovery of archaea to the discovery of a new continent, which he and others are exploring.

FIGURE 19.17 An archaeon, *Methanococcus jannaschii*.

TAKE-HOME MESSAGE 19.7

Genetic studies of prokaryotes revealed that the cells now known as archaea constitute a lineage distinct from bacteria.

19.8 WHERE DO ARCHAEA LIVE?

Many archaea live in places where few or no other cells survive. Archaea were first discovered in hot springs, and many are **extreme thermophiles**, meaning they grow only at a very high temperature (**FIGURE 19.18A**). Some archaea that live near deep-sea hydrothermal vents grow even at 110°C (230°F). These organisms are chemoautotrophs that reduce sulfur to obtain energy.

Other archaea are **extreme halophiles**, organisms that live in highly salty water. Salt-loving archaea live in the Dead Sea, the Great Salt Lake, and smaller brine-filled lakes (**FIGURE 19.18B**). Most are photoheterotrophs that capture light energy with a red pigment (bacteriorhodopsin).

Many archaea, including some extreme thermophiles and extreme halophiles, are **methanogens**, or methane makers. These chemoautotrophs form ATP by pulling electrons from hydrogen gas or acetate. The reactions produce methane (CH_4), an odorless gas. Methanogenic archaea are strict anaerobes, meaning they cannot live in the presence of oxygen. They abound in sewage, marsh sediments, and the animal gut (**FIGURE 19.18C**). Cattle have methanogens in their stomach and release methane gas by belching.

About a third of people have significant numbers of methanogens in their intestine, so their flatulence (farts) contains methane. Archaea have also been discovered in the human mouth and vagina. So far, no archaea have been proven to be human pathogens. However, some methanogens that live in the mouth contribute to gum disease. By their presence, they may improve the environment for pathogenic bacteria that cause this disease.

Researchers continue to investigate the diversity of and evolutionary relationships among the archaea. Two major lineages have been identified. Most of the extreme thermophiles belong to the lineage Crenarchaeota, and most methanogenic and salt-loving archaea belong to the lineage Euryarchaeota.

extreme halophile Organism adapted to life in a highly salty environment.
extreme thermophile Organism adapted to life in a very high-temperature environment.
methanogen Organism that produces methane gas (CH_4) as a metabolic by-product.

TAKE-HOME MESSAGE 19.8

Archaea live in extremely salty and extremely high-temperature environments, as well as in oxygen-free sediments and the animal gut.

Few archaea contribute to human disease.

A Thermally heated waters. Archaeal extreme thermophiles live in the waters of this hot spring in Nevada.

B Highly salty waters. Pigmented extreme halophiles color the brine in this California lake.

C The gut of many animals. Cows belch frequently to release the methane produced by archaea in their stomach.

FIGURE 19.18 Examples of archaeal habitats.

CREDITS: (18A) © Savannah River Ecology Laboratory; (18B) Courtesy of Benjamin Brunner; (18C) Dr. John Brackenbury/Science Source.

19.9 Application: SHARING VIRUSES

FIGURE 19.19 Wild chimpanzees from a population infected by SIV, the chimpanzee equivalent of HIV.

WE SHARE MANY TRAITS WITH CHIMPANZEES, INCLUDING SUSCEPTIBILITY TO SOME VIRUSES. HIV is similar to simian immunodeficiency virus (SIV), which infects wild chimpanzees in Africa (**FIGURE 19.19**). In chimpanzees, as in humans, the virus harms the immune system. Researchers think HIV evolved after SIV entered and survived inside a human. Most likely, the person was exposed to the virus while butchering an infected animal. Chimpanzees and other primates are hunted and butchered as "bushmeat" in many parts of Africa.

To find out when HIV first infected humans, researchers looked for HIV in old tissue samples stored at hospitals. The earliest HIV samples known date to about 1960 and come from two people who lived in Africa's Democratic Republic of the Congo. Given what scientists know about the mutation rates for viruses, they estimate that HIV first infected humans in the early 1900s.

Comparing genes of HIV in stored and modern blood samples has allowed researchers to trace the movement of the virus out of Africa. In 1966, HIV was introduced from Africa to Haiti, where new mutations arose. In about 1969, someone infected by HIV with those new mutations introduced the virus to the United States. From there, it spread quietly for 12 years until AIDS was identified as a threat in 1981. Two years later, scientists showed that HIV causes AIDS.

Could other viruses also make their way from nonhuman primates into the human population? Some already are trickling in. For example, people exposed to the blood or saliva of a wild Old World primate can become infected by simian foamy virus (SFV). So far as we know, SFV infection has no adverse effect on human health. However, some viruses, such as HIV, often do not cause symptoms until years after an initial infection. There is also no evidence that SFV can currently spread from person to person. However, viruses can evolve. In addition, opportunities for SFV and other currently unknown viruses to infect humans are likely to increase as human populations expand into previously remote areas of primate habitat.

CREDIT: (19) Michael Nichols/National Geographic Creative.

Summary

SECTION 19.1 A **virus** is a noncellular infectious agent with a protein coat around a core of DNA or RNA. In enveloped viruses, the coat is enclosed by a bit of plasma membrane derived from a previous host. A virus lacks ribosomes and other metabolic machinery, so it must replicate inside a host cell. Viruses attach to a host cell, then viral genes and enzymes direct the host to replicate viral genetic material and make viral proteins. New viral particles self-assemble and are released.

Bacteriophages can replicate in bacteria by two different pathways. Replication by the **lytic pathway** is rapid, and the new viral particles are released by lysis. During the **lysogenic pathway**, the virus enters a latent state that extends the cycle.

HIV (human immunodeficiency virus), the virus that causes AIDS, is an enveloped **retrovirus**. Its RNA must be reverse transcribed to DNA to begin the process of viral replication. The virus acquires its envelope as it buds from the plasma membrane of a host cell.

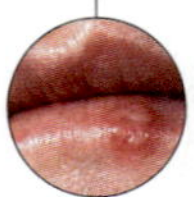

SECTION 19.2 Some viruses are human **pathogens**, meaning they cause disease. Viruses cause common colds, flus, and some cancers. Many viral diseases are communicable, but others require a **vector**, an animal that carries them between hosts. An **emerging disease** is newly recognized in humans (like Ebola) or is spreading to a new region (like West Nile virus). Changes to viral genomes occur as a result of mutation and **viral recombination**.

SECTION 19.3 Plants can be infected by viruses or **viroids**, which are tiny bits of RNA. Unlike a viral genome, viroid RNA does not encode proteins. A viral infection can stunt plants and deform plant parts.

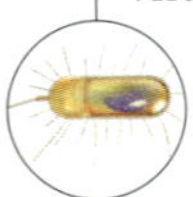

SECTION 19.4 Bacteria and archaea are small, structurally simple cells. They do not have a nucleus or cytoplasmic organelles typical of eukaryotes. The single circular chromosome of double-stranded DNA resides in the cytoplasm. Most have a cell wall and many have flagella and hairlike pili.

Bacteria and archaea reproduce by **binary fission**: replication of a single, circular chromosome and division of a parent cell into two genetically equivalent descendants. **Horizontal gene transfers** move genes between existing cells, as when **conjugation** moves a plasmid with a few genes from one cell into another.

As a group, prokaryotes are metabolically diverse, with both anaerobes and aerobes. Some are **photoautotrophs** like plants and others are **chemoheterotrophs** like fungi and animals. Two modes of nutrition occur only among prokaryotes: **chemoautotrophs** extract energy from inorganic chemicals, and **photoheterotrophs** capture light energy but get their carbon from organic compounds.

SECTIONS 19.5, 19.6 Bacteria have essential ecological roles. **Cyanobacteria** produce oxygen during photosynthesis and are relatives of chloroplasts. Some also carry out **nitrogen fixation**, producing ammonia that algae and plants require. **Proteobacteria**, the most diverse bacterial lineage, also include nitrogen-fixers, as well as relatives of mitochondria. **Gram-positive bacteria** have thick walls and many live in soil, where they act as **decomposers**. Formation of **endospores** allows some Gram-positive bacteria to survive adverse conditions. **Spirochetes** resemble a stretched-out spring. They may be free-living or pathogens.

Bacteria that normally live in or on our body are part of our protective **normal flora**. Pathogenic bacteria cause disease by producing toxins.

SECTIONS 19.7, 19.8 Archaea are now recognized as members of a domain distinct from bacteria. They include **methanogens** (methane makers), **extreme halophiles** (salt lovers), and **extreme thermophiles** (heat lovers). Archaea coexist with bacteria in many habitats, including human and animal bodies. Few are pathogens.

SECTION 19.9 Scientists use their knowledge of evolution to investigate how pathogens such as HIV arise, function, and spread. HIV is a descendant of a virus that infects wild chimpanzees. Other viruses that infect wild primates have also been detected in humans.

Self-Quiz

Answers in Appendix VII

1. _______ can have a genome of either RNA or DNA.
   a. Bacteria b. Viroids c. Viruses d. Archaea

2. Which is smallest?
   a. bacterium b. viroid c. virus d. archaeon

3. In _______, viral DNA becomes integrated into a bacterial chromosome and is passed to descendant cells.
   a. binary fission c. the lysogenic pathway
   b. the lytic pathway d. both b and c

4. During HIV replication, reverse trancriptase reads viral _______ to produce viral _______.
   a. DNA; RNA c. RNA; proteins
   b. RNA; DNA d. DNA; proteins

5. Viral genomes can be altered by _______.
   a. binary fission c. mutation
   b. recombination d. both b and c

6. Prokaryotic conjugation is a type of _______.
   a. sexual reproduction c. horizontal gene transfer
   b. asexual reproduction d. both b and c

7. How many chromosomes do bacteria have?

8. _______ are oxygen-releasing photoautotrophs.
   a. Spirochetes c. Cyanobacteria
   b. Archaeans d. Viroids

Data Analysis Activities

Maternal Transmission of HIV Since the AIDS pandemic began, there have been more than 8,000 cases of mother-to-child HIV transmission in the United States. In 1993, American physicians began giving antiretroviral drugs to HIV-positive women during pregnancy and treating both mother and infant in the months after birth. Only about 10 percent of mothers were treated in 1993, but by 1999 more than 80 percent got antiviral drugs.

FIGURE 19.20 shows the number of AIDS diagnoses among children in the United States. Use the information in this graph to answer the following questions.

1. How did the number of children diagnosed with AIDS change during the late 1980s?
2. What year did new AIDS diagnoses in children peak, and how many children were diagnosed that year?
3. How did the number of AIDS diagnoses change as the use of antiretrovirals in mothers and infants increased?

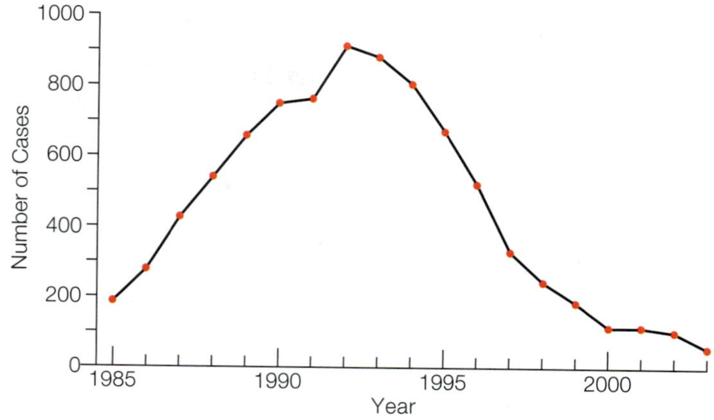

FIGURE 19.20 Number of new AIDS diagnoses in the United States per year among children exposed to HIV during pregnancy, birth, or by breast-feeding.

9. Nitrogen-fixation by bacteria produces ______ .
   a. nitrogen gas b. methane c. oxygen d. ammonia

10. Vitamin-producing *E. coli* cells in your gut are ______ .
    a. normal flora c. proteobacteria
    b. chemoheterotrophs d. all of the above

11. A chemical that is released into the environment by bacteria and causes symptoms of disease is a(n) ______ .
    a. endospore b. endotoxin c. exotoxin d. vector

12. Infection by some ______ can cause cancer.
    a. viruses c. archaea
    b. bacteria d. both a and b

13. A Gram-positive coccus is ______ .
    a. spherical b. rod-shaped c. spiral-shaped

14. Match each disease with the type of pathogen that causes it. Choices may be used more than once.
    ___ tuberculosis a. virus
    ___ AIDS b. bacteria
    ___ influenza (flu)
    ___ Lyme disease
    ___ cholera

15. Match each group with the most suitable description.
    ___ cyanobacteria a. contain no protein
    ___ methanogens b. live in salty places
    ___ viroids c. live in hot places
    ___ retroviruses d. release oxygen
    ___ extreme halophiles e. release methane
    ___ extreme thermophiles f. RNA genome inside a protein coat

Critical Thinking

1. Rhinoviruses, the most common cause of colds, do not have a lipid envelope. Compared to enveloped viruses, such nonenveloped viruses tend to remain infectious outside the body longer, are more likely to be spread by contact with surfaces, and are less likely to be rendered harmless by exposure to hand sanitizer or handwashing. Explain how the lack of an envelope contributes to these characteristics.

2. Methanogens have been found in the human gut and deep-sea sediments, but not on human skin or in the surface waters of the ocean. What physiological trait of methanogens could explain this distribution?

3. Review the description of Fred Griffith's experiments with *Streptococcus pneumoniae* in Section 8.1. Using your knowledge of bacterial biology, explain the process by which the harmless bacteria became dangerous after being exposed to components of harmful bacterial cells.

4. The antibiotic penicillin acts by interfering with the production of new bacterial cell wall. Cells treated with penicillin do not die immediately, but they cannot reproduce. Explain how penicillin halts binary fission.

5. Many compounds secreted by soil bacteria have been isolated and are now produced synthetically for use as antibiotics. What function do you think these compounds play in the bacteria themselves? Devise an experiment that will test your hypothesis about the natural function of these substances.

A stand of giant kelp, the largest protists, off the coast of California. Like trees in a forest, kelp shelter and feed a wide variety of organisms.

20 THE PROTISTS

Links to Earlier Concepts

This chapter describes protists, a collection of eukaryotic lineages introduced in Section 1.3. We reexamine cell structures such as chloroplasts (6.3,6.4), pseudopods, and flagella (4.9); reconsider the effects of osmosis on cells (5.6); and return to photosynthetic pigments (6.1). We also delve again into how organelles evolve by endosymbiosis (18.5).

KEY CONCEPTS

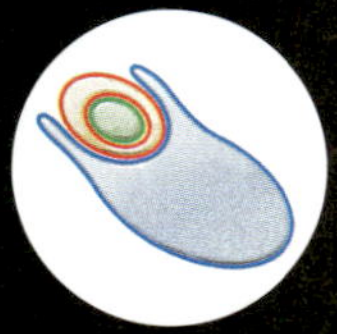

A COLLECTION OF LINEAGES

The protist lineages include single-celled, colonial, and multicelled organisms. Photosynthetic protists have chloroplasts derived from either bacteria or other protists.

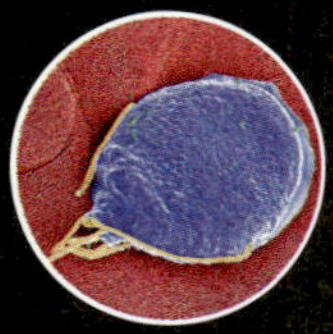

SINGLE-CELLED LINEAGES

Most protist lineages are entirely single-celled. Such groups include flagellated protozoans, shelled cells called foraminifera and radiolaria, and ciliates, dinoflagellates, and apicomplexans.

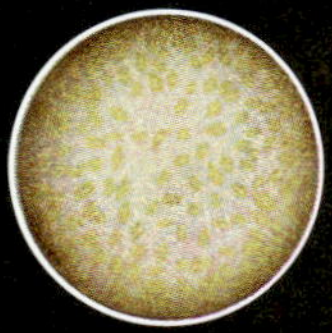

BROWN ALGAE AND RELATIVES

Multicellular brown algae and single-celled photosynthetic diatoms belong to the same lineage as filamentous heterotrophs called water molds.

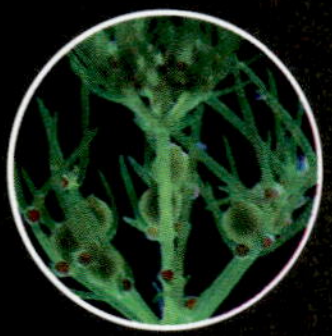

RED ALGAE, GREEN ALGAE

Red algae and green algae include single-celled and multicelled aquatic producers. Red algae have pigments that allow them to live in deep waters. Green algae are the closest relatives of land plants.

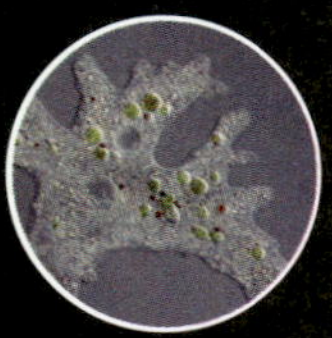

AMOEBAS, SLIME MOLDS, AND CHOANOFLAGELLATES

Shape-shifting heterotrophic amoebas live in aquatic habitats. The related slime molds feed on forest floors. Choanoflagellates are the closest living protist relatives of animals.

Photograph by Brian J. Skerry, National Geographic Creative.

20.1 WHAT ARE PROTISTS?

A **protist** is a eukaryotic organism that is not a fungus, plant, or animal. Protists are not a clade (Section 17.12), because they include all members of some lineages, plus some members of other lineages. As **FIGURE 20.1** illustrates, some protists are more closely related to plants, fungi, or animals than to other protists. Like other eukaryotes, protists undergo mitosis and, if they reproduce sexually, meiosis.

Most protists live as single cells, but some lineages include colonial or multicellular members. Cells of a **colonial organism** live together, but remain self-sufficient. Each cell retains the traits required to survive and reproduce on its own. By contrast, cells of a **multicellular organism** have a division of labor and are interdependent.

Though typically small, protists have a huge ecological impact. Heterotrophic protists decompose organic material, prey on smaller organisms such as bacteria, or live inside the bodies of larger species. Photosynthetic protists use the same oxygen-producing photosynthetic pathway as cyanobacteria, and their activity produces much of the oxygen we breathe. They also take up huge amounts of carbon dioxide, and build organic compounds that are the basis for aquatic food chains.

All protist chloroplasts evolved by endosymbiosis (Section 18.5 and **FIGURE 20.2**). **Primary endosymbiosis** occurs when a bacterium enters a cell and its descendants evolve into an organelle ❶. Red algae, green algae, and land plants share an ancestor in which chloroplasts evolved by primary endosymbiosis. Their chloroplasts have two membranes, one from the bacteria, and one from the vacuole in which it was engulfed. **Secondary endosymbiosis** occurs when a photosynthetic protist engulfed by a heterotrophic protist evolves into a chloroplast ❷. In this case, the chloroplast often has more than two membranes. For example, some red algae evolved into chloroplasts in other protists.

colonial organism Organism composed of many similar cells, each capable of living and reproducing on its own.
multicellular organism Organism composed of interdependent cells that vary in their structure and function.
primary endosymbiosis Evolution of an organelle from bacteria that entered a host cell and lived inside it.
protist Eukaryote that is not a plant, fungus, or animal.
secondary endosymbiosis Evolution of an organelle from a protist that itself contains organelles that arose by primary endosymbiosis.

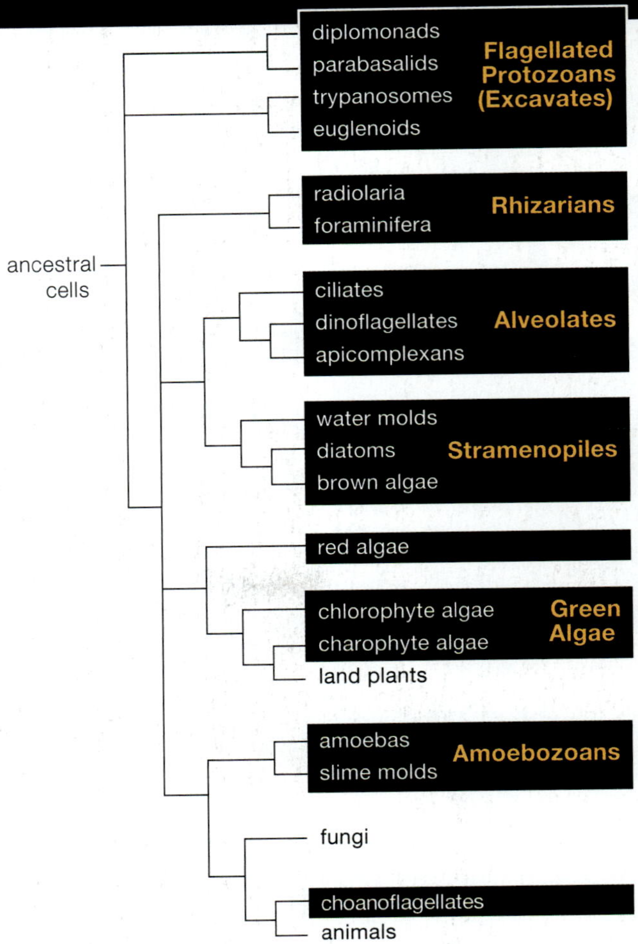

FIGURE 20.1 Phylogenetic tree for the eukaryotes discussed in this book. Like all such trees, it is a hypothesis. Protist groups are indicated by black boxes.

FIGURE IT OUT: Are land plants more closely related to red algae or brown algae?
Answer: Red algae

TAKE-HOME MESSAGE 20.1

Protists include members of multiple eukaryotic lineages.

Most protists are single-celled, but multicellular and colonial forms evolved in some lineages.

Some protists are heterotrophs; others are photosynthetic with chloroplasts that evolve from bacteria or other protists.

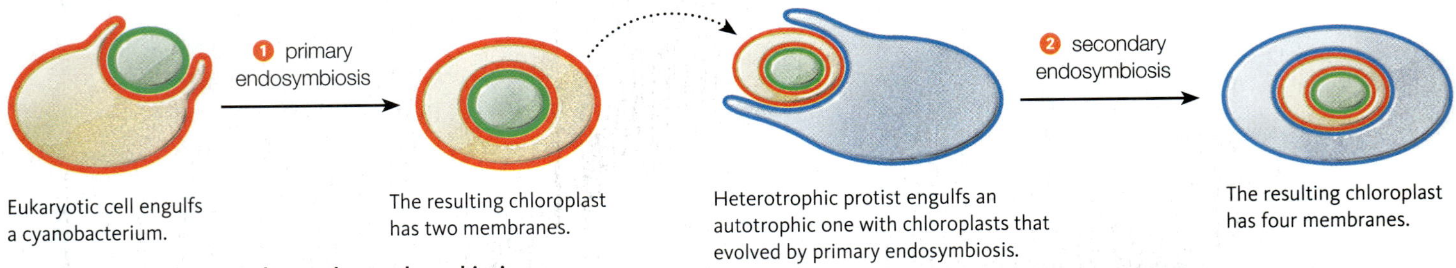

FIGURE 20.2 Primary and secondary endosymbiosis.

20.2 WHAT ARE FLAGELLATED PROTOZOANS?

Single-celled, unwalled heterotrophic protists are commonly called protozoans. **Flagellated protozoans** (clade Excavata) have one or more flagella and a flagellated feeding groove. Food that enters this groove is taken up by phagocytosis. A **pellicle**, a layer of elastic proteins beneath the plasma membrane, helps these unwalled cells retain their shape.

ANAEROBIC FLAGELLATES

Diplomonads and parabasalids are two closely related groups of flagellates. Both can live where oxygen is low because they have modified mitochondria that make ATP anaerobically.

Diplomonads are unusual in that they have two nuclei. Most are parasitic. One species attaches to the human intestinal lining and causes the waterborne disease giardiasis (**FIGURE 20.3A**). Symptoms of giardiasis include cramps, nausea, and severe diarrhea. Infected people or animals excrete cysts (a hardy resting stage) of the protist in feces.

Parabasalids have a single nucleus and four flagella. One species causes trichomoniasis, a sexually transmitted disease. This parasite does not make cysts, so it cannot survive very long outside the human body. Fortunately for the parasite, sexual intercourse delivers it directly into hosts.

TRYPANOSOMES AND EUGLENOIDS

Trypanosomes are tapered cells that have a single large mitochondrion. Their flagellum is attached to the cell body by a membrane (**FIGURE 20.3B**). All trypanosomes are parasites. Biting insects transmit trypanosomes that cause human diseases such as sleeping sickness.

Euglenoids are a group of primarily freshwater protists closely related to trypanosomes. They have multiple mitochondria and a pellicle composed of protein strips (**FIGURE 20.4**). The interior of a euglenoid is saltier than its freshwater habitat, so water tends to diffuse into the cell. Like many other freshwater protists, euglenoids have one or more **contractile vacuoles**, organelles that collect excess water, then contract and expel it to the outside through a pore. Many euglenoids have chloroplasts descended from a green alga. They can carry out photosynthesis in the light, but become heterotrophs in the dark. An eyespot near the base of a long flagellum helps these cells detect light.

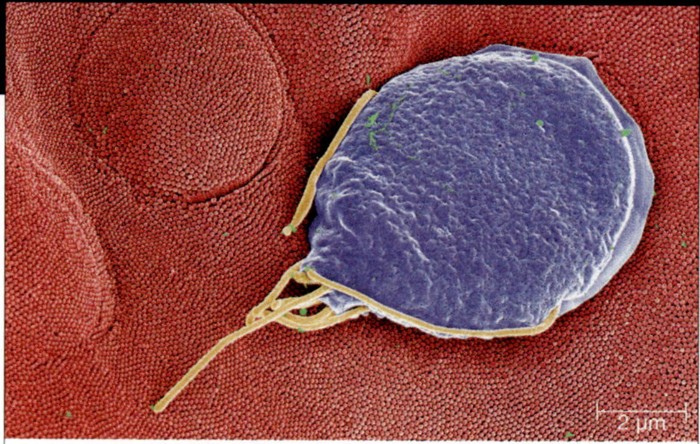

A Diplomonad (*Giardia lamblia*) attached to the lining of the human intestine.

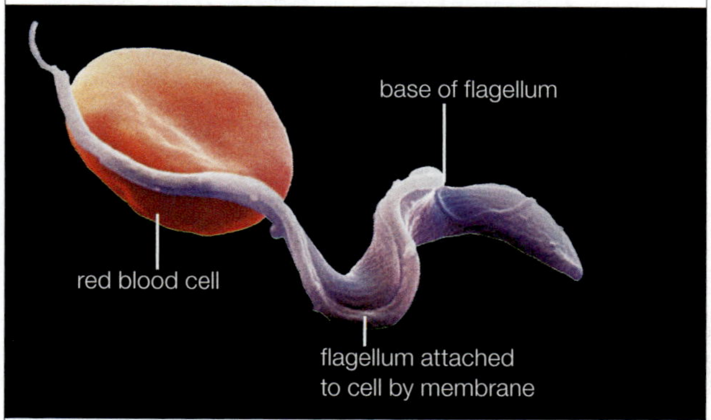

B Trypanosome with a human red blood cell. Wavelike movement of the membrane-enclosed flagellum propels the trypanosome.

FIGURE 20.3 Examples of flagellated protozoans. "Protozoan" is a general term used to describe members of the primarily heterotrophic groups of unicellular protists.

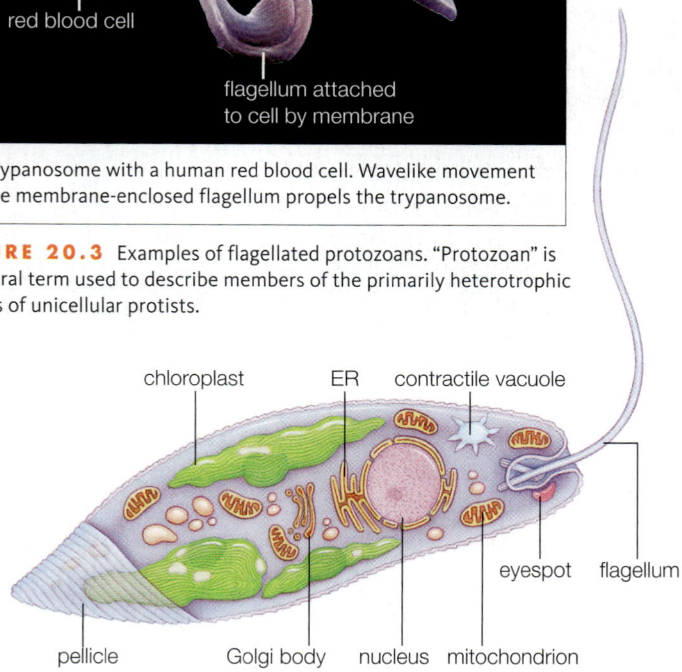

FIGURE 20.4 {Animated} Body plan of a euglenoid (*Euglena*), a freshwater species with chloroplasts that evolved from a green alga.

TAKE-HOME MESSAGE 20.2

Flagellated protozoans are unwalled unicellular organisms with one or more flagella.

Parabasalids and diplomonads are heterotrophs that have modified mitochondria and can live in anaerobic habitats. Some are important human pathogens.

Trypanosomes are parasites with one large mitochondrion. Some cause human disease.

Euglenoids are freshwater protists. They feed as heterotrophs or use chloroplasts that evolved from a green alga.

contractile vacuole In freshwater protists, an organelle that collects and expels excess water.
euglenoid Flagellated protozoan with multiple mitochondria; may be heterotrophic or have chloroplasts descended from green algae.
flagellated protozoan Protist belonging to an entirely or mostly heterotrophic lineage with no cell wall and one or more flagella.
pellicle Layer of proteins that gives shape to many unwalled, single-celled protists.
trypanosome Parasitic flagellated protist with a single mitochondrion and a flagellum that runs along the back of the cell.

20.3 WHAT ARE RHIZARIANS?

Rhizarians include two groups of single-celled marine protists—foraminifera and radiolaria—that secrete a sieve-like shell. Like amoebas, which we discuss in Section 20.8, rhizarians are heterotrophs that use cytoplasmic extensions to capture prey. However, this mechanism of feeding evolved independently in the two groups; genetic comparisons show rhizarians and amoebas are not close relatives.

FIGURE 20.5 Living foraminiferan. This is one of the planktonic species, and the tiny gold specks are algae that live in its cytoplasm.

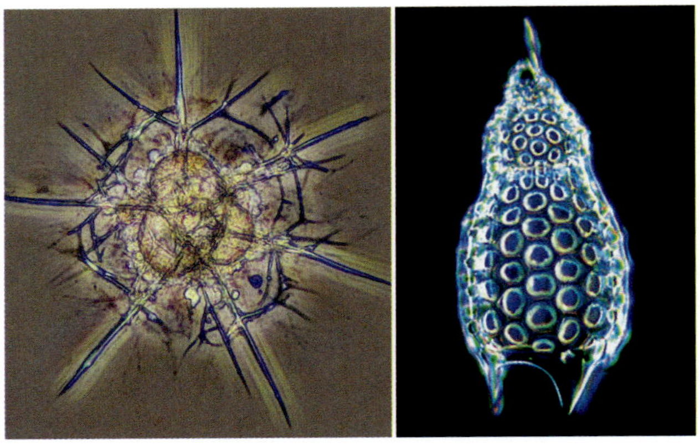

A Living radiolarian. B Radiolarian shell.

FIGURE 20.6 Two silica-shelled radiolarians.

CHALKY OR SILICA-SHELLED CELLS

Foraminifera secrete a calcium carbonate ($CaCO_3$) shell. Including its shell, an individual foraminiferal cell can be as big as a grain of sand. Most species live on the seafloor, where they probe the water and sediments for prey. Others are members of the marine plankton. **Plankton** is a community of mostly microscopic organisms that drift or swim in open waters. Planktonic foraminifera often have smaller photosynthetic protists such as diatoms or algae living inside them (**FIGURE 20.5**).

By taking up carbon dioxide (CO_2) and incorporating it into their calcium carbonate shells, foraminifera help the sea to absorb more carbon dioxide from the air. Lowering the concentration of CO_2 in the air is important because excess atmospheric CO_2 is one cause of global climate change.

Radiolaria secrete a glassy silica (SiO_2) shell (**FIGURE 20.6**) and have two cytoplasmic layers. The inner layer holds all the typical eukaryotic organelles and the outer one contains numerous gas-filled vacuoles that keep the cell afloat. Radiolaria are planktonic and they are most abundant in nutrient-rich tropical waters.

RHIZARIAN REMAINS

Foraminifera and radiolaria have lived and died in the oceans for more than 500 million years, so remains of countless cells have fallen to the seafloor. Over time, geologic processes transformed some accumulations of foraminiferal shells into chalk and limestone, two types of calcium carbonate–rich sedimentary rock. The giant blocks of limestone used to build the great pyramids of Egypt consist largely of the shells of foraminifera that fell to the seafloor about 50 million years ago. Similarly, geologic processes transformed silica-rich deposits of radiolarian shells into a rock called chert.

Fossil foraminifera and radiolaria help geologists locate deposits of oil (petroleum). Oil forms over millions of years from lipids in the remains of marine protists. To locate rock layers likely to contain oil deposits, geologists drill into rock and look for fossil foraminifera or radiolaria. Observing which fossil species are present allows the geologists to determine how old the rocks are and whether they were laid down in an environment likely to favor oil formation.

foraminifera Heterotrophic single-celled protists with a porous calcium carbonate shell and long cytoplasmic extensions.
plankton Community of tiny drifting or swimming organisms.
radiolaria Heterotrophic single-celled protists with a porous shell of silica and long cytoplasmic extensions.

TAKE-HOME MESSAGE 20.3

Two related lineages of heterotrophic marine cells have porous secreted shells. Foraminifera have chalk shells and live in sediments or drift as part of plankton. Radiolarians have silica shells and are planktonic.

20.4 WHAT ARE ALVEOLATES?

Ciliates, dinoflagellates, and apicomplexans belong to the clade Alveolata. "Alveolus" means sac, and the defining trait of alveolates is a layer of sacs beneath the plasma membrane. In this section, we discuss alveolates that are primarily aquatic: the ciliates and dinoflagellates. The section that follows covers apicomplexans, which are parasites.

CILIATED CELLS

Ciliates, or ciliated protozoans, are unwalled heterotrophic cells that use their many cilia to move and feed. Most are predators in seawater, fresh water, or damp soil. Members of the genus *Paramecium* are common in ponds (**FIGURE 20.7**). Their cilia sweep food into an oral groove, from which the particles are taken up by phagocytosis, then digested in food vacuoles. Exocytosis expels digestive waste, and contractile vacuoles squirt out excess water.

A few types of ciliates have adapted to life in the animal gut. Some gut ciliates help cattle, sheep, and related grazers break down cellulose in plant material. Others help termites digest wood. Only one ciliate (*Balantidium coli*) is a known human pathogen. It lives in the pig gut; human infections occur when pig feces get into drinking water.

WHIRLING DINOFLAGELLATES

Dinoflagellates are single-celled protists that typically have two flagella, one at the cell's tip and the other running through a groove around the cell's middle like a belt (**FIGURE 20.8**). Dinoflagellate means "whirling flagellate," and the combined actions of the two flagella cause the cell to spin as it moves forward. Many dinoflagellates deposit cellulose in the sacs beneath their plasma membrane, and these deposits form thick protective plates.

The vast majority of dinoflagellates are part of the marine plankton, and they are especially abundant in warm climates. In tropical seas, dinoflagellates are the most common source of **bioluminescence**, light produced by a living organism. When disturbed, dinoflagellates produce a blue glow by an oxidation reduction reaction (Section 5.4).

Some planktonic dinoflagellates prey on bacteria; others are photosynthetic, with chloroplasts that evolved from a red alga. Photosynthetic dinoflagellates that live in the tissues of reef-building corals help maintain coral reefs. The protists receive shelter within the coral, which is a type of invertebrate animal. In return, they provide the coral with sugars and oxygen. Without its dinoflagellate partners, a reef-building coral will die.

bioluminescence Light emitted by a living organism.
ciliate Single-celled, heterotrophic protist with many cilia.
dinoflagellate Single-celled, aquatic protist that moves with a whirling motion; may be heterotrophic or photosynthetic.

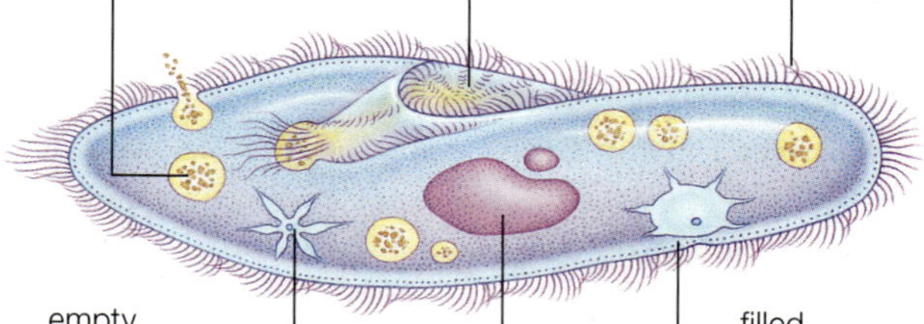

FIGURE 20.7 Structure of *Paramecium*, a freshwater ciliate.

Many free-living dinoflagellates are an important food source for marine animals, but others make toxins that poison animals that eat them. These toxins can move up food chains, so animals and people who eat fish or shellfish that ate toxic dinoflagellates can also be poisoned.

A few types of dinoflagellates are parasites of aquatic animals. For example, some dinoflagellates infect and sicken crabs and other crustaceans.

FIGURE 20.8 A photosynthetic dinoflagellate (*Karenia brevis*).

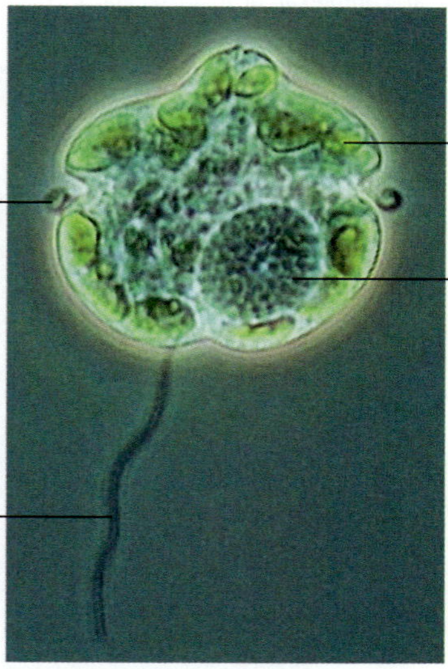

TAKE-HOME MESSAGE 20.4

Ciliates and dinoflagellates are two closely related lineages of single-celled, primarily aquatic protists.

Most ciliates are predators, but some live inside animals. Cilia on the cell's surface are used in locomotion and feeding.

Dinoflagellates have two flagella and move with a whirling motion. Most are planktonic predators or producers, but some are animal parasites. Some photosynthetic species live as partners inside reef-building corals.

20.5 HOW DOES MALARIA AFFECT HUMAN HEALTH?

Apicomplexans are parasitic alveolates that reproduce only inside cells of their hosts. Their name refers to a complex of microtubules at their apical (upper) end that allows them to pierce and enter a host cell.

Malaria is a deadly disease caused by apicomplexans of the genus *Plasmodium*. A bite from a female mosquito introduces the parasite into a human body, where it reproduces in the liver and in red blood cells (**FIGURE 20.9**). Malaria symptoms usually start a week or two after a mosquito bite, when infected liver cells rupture and release *Plasmodium* cells, metabolic wastes, and cellular debris into the blood. Shaking, chills, a burning fever, and sweats result. After the initial episode, symptoms may subside for weeks or even months. However, an ongoing infection damages the liver, spleen, kidneys, and brain. If untreated, malaria nearly always results in death.

Plasmodium cannot survive at low temperatures, so malaria is mainly a tropical disease. Today, it remains a threat in parts of Mexico, Central America, South America, Asia, and most especially in Africa. In 2011, 91 percent of the 670,000 malarial deaths reported to the World Health Organization occurred in Africa.

FIGURE 20.9 **{Animated}** Artist's depiction of a *Plasmodium*, the apicomplexan protist that causes malaria, emerging from a human red blood cell.

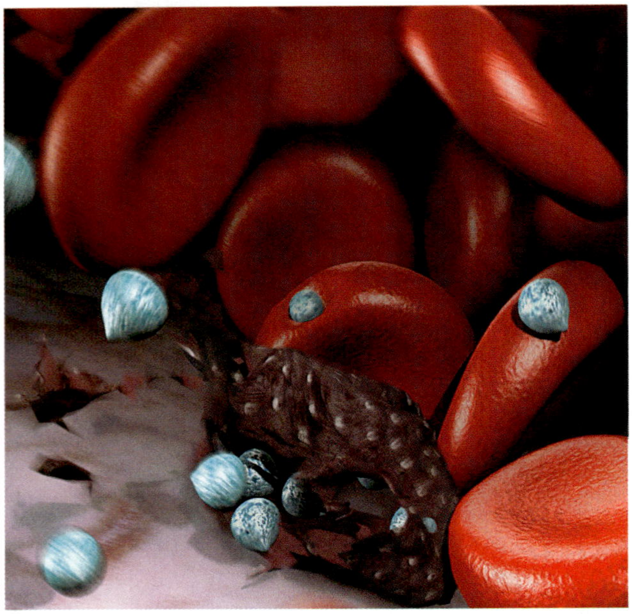

apicomplexan Parasitic protist that reproduces in cells of its host.

TAKE-HOME MESSAGE 20.5

Malaria, a disease caused by an apicomplexan protist, causes many human deaths in the tropics, especially in Africa.

PEOPLE MATTER

National Geographic Explorer
KEN BANKS

Ken Banks develops low-cost, low-tech innovations that improve the lives of people in developing nations. While involved in conservation work in Africa, Banks saw a huge unmet need for technology that could send information between groups in remote areas with no Internet access. Banks created a text messaging application called FrontlineSMS, which does just that. It is now used to deliver vital information in more than 130 nations.

In Cambodia, health officials use the system to pinpoint and contain malaria outbreaks. Community volunteers in rural areas are trained to diagnose malaria and are given a cell phone to report their findings. A volunteer who diagnoses a new malaria case gives the affected person an antimalarial drug, and reports the case to health officials via a text message. Three days later, the volunteer rechecks the patient's blood to be sure the drug has worked and the parasites are gone. If parasites remain, the person probably has a drug-resistant strain of malaria. If so, this too is immediately reported to health officials. By monitoring incoming reports from hundreds of volunteers, health officials can deploy antimalarial drugs and insecticide-treated nets where they will have the greatest impact.

20.6 WHAT ARE STRAMENOPILES?

DIATOMS AND BROWN ALGAE

Diatoms and brown algae are photosynthetic members of the clade Stramenopila. Their chloroplasts, which evolved from a red alga, contain chlorophylls *a* and *c*. An accessory pigment (fucoxanthin) gives them a brown color.

Brown algae are multicelled "seaweeds" of temperate or cool seas (**FIGURE 20.10A**). Most grow along rocky seashores. They range in size from microscopic filaments to the largest protists—giant kelp that can reach 30 meters (100 feet) in height. The photo that opened this chapter shows a stand of giant kelp off the coast of California. Like trees in a forest, kelp shelter a wide variety of other organisms. Kelp are also the source of alginates, polysaccharides commonly used to thicken foods, beverages, cosmetics, and body lotions.

Diatoms (**FIGURE 20.10B,C**) are cells with a two-part silica shell. The upper part of the shell overlaps the lower part, like a shoe box. Diatoms live in lakes, seas, and damp soils. When marine diatoms die, their shells accumulate on the seafloor. In places where deposits of diatom shells have been lifted onto land by geological processes, people mine the resulting "diatomaceous earth." This silica-rich material is used in filters and cleaners, and as an insecticide that is nontoxic to vertebrates. Diatoms also contributed the bulk of the organic material to the seafloor deposits from which most oil formed. Like dinoflagellates, diatoms serve as food for many animals. Also like dinoflagellates, some diatoms produce a toxin that accumulates in fish, crabs, and shellfish. Eating seafood tainted by this toxin (domoic acid) can cause a potentially life-threatening form of food poisoning.

WATER MOLDS

Colorless water molds are also among the stramenopiles. **Water molds** are heterotrophic protists that grow as a mesh of nutrient-absorbing filaments. Fungi have a similar growth pattern, and water molds were once considered a type of fungus. More recently, genetic studies revealed their close relationship to the brown algae and diatoms.

Most water molds are decomposers in aquatic habitats or moist soil, but some are important plant pathogens. For example, in the mid-1800s, an outbreak of the water mold *Phytophthora infestans* destroyed Ireland's potato crop. The resulting famine led to hundreds of thousands of human deaths. In recent years, the spread of a related water mold (*P. ramorum*) has decimated oak forests in California and southern Oregon (**FIGURE 20.11A**). Water molds also infect animals. Freshwater fish serve as hosts for *Saprolegnia*, and infections often break out on fish farms and in aquariums. Affected fish develop cottony patches on their body (**FIGURE 20.11B**) and few survive infection.

A A brown alga (*Fucus*), common along rocky shores.

B Scanning electron micrograph of a diatom shell.

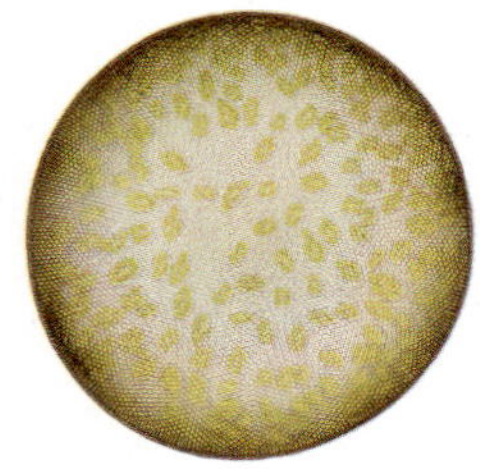

C Light micrograph of a live diatom with its brownish chloroplasts.

FIGURE 20.10 **Photosynthetic stramenopiles.**

A Trunk of an oak infected by *Phytophthora ramorum*.

B *Saprolegnia* growing on a fish.

FIGURE 20.11 **Pathogenic water molds.**

brown alga Multicelled marine protist with a brown accessory pigment in its chloroplasts.
diatom Single-celled photosynthetic protist with a brown accessory pigment in its chloroplasts and a two-part silica shell.
water mold Heterotrophic protist that grows as a mesh of nutrient-absorbing filaments.

TAKE-HOME MESSAGE 20.6

Brown algae are multicellular seaweeds. Diatoms are silica-shelled cells. Both stramenopile groups have chloroplasts that contain a brown accessory pigment.

Water molds are filamentous heterotrophs. Most are decomposers, but some are parasites of plants or animals.

CREDITS: (10A) Garo/ Science Source; (10B) Science Museum of Minnesota; (10C) © Wim van Egmond/ Visuals Unlimited; (11A) Pavel Svihra; (11B) Heather Angel.

20.7 WHICH PROTISTS ARE CLOSEST TO PLANTS?

Archaeplastids are organisms whose cellulose-walled cells contain chloroplasts with two membranes. Scientists think these chloroplasts evolved from cyanobacteria by primary endosymbiosis. Archaeplastids include two protist groups (red algae and green algae) that we discuss here, as well as land plants, which we consider in detail in the next chapter.

FIGURE 20.12 Branching and sheetlike red algae growing 75 meters (250 ft) beneath the sea surface in the Gulf of Mexico.

RED ALGAE

Red algae (clade Rhodophyta) are photosynthetic, generally multicelled protists that live in clear, warm seas. Most grow either as thin sheets or in a branching pattern (**FIGURE 20.12**). One subgroup, the coralline algae, deposits calcium carbonate in its cells. The rigid material produced by these algae contributes bulk to some tropical reefs.

Chloroplasts of red algae contain chlorophyll *a*, as well as red accessory pigments (phycobilins). Phycobilins absorb blue–green light, which penetrates deeper into water than light of other wavelengths. Thus, red algae can thrive at greater depths than other algae.

Two gelatinous polysaccharides, agar and carrageenan, are extracted from red algae for use as thickeners or stabilizers in foods, cosmetics, and personal care products. Agar is also important in microbiology. When mixed with appropriate nutrients, it forms a semisolid culture medium for growing bacteria or fungi.

Nori, the edible seaweed used to wrap sushi, is a red alga (*Porphyra*). Like many multicelled algae and all plants, *Porphyra* has a life cycle that alternates between haploid and diploid bodies—an **alternation of generations** (**FIGURE 20.13**). The nori **gametophyte**, or haploid body, is a sheetlike seaweed (1) that forms gametes by mitosis (2). When gametes join at fertilization, they form a diploid zygote (3) that grows and develops into the diploid body, or **sporophyte** (4). The nori sporophyte is a tiny, branching filament. The sporophyte produces haploid spores by meiosis (5). The spore germinates, then grows and develops into a new gametophyte to complete the cycle (6).

GREEN ALGAE

Green algae are single-celled, colonial, or multicellular autotrophs whose chloroplasts contain chlorophylls *a* and *b*. Some single-celled green algae live in the soil, others in

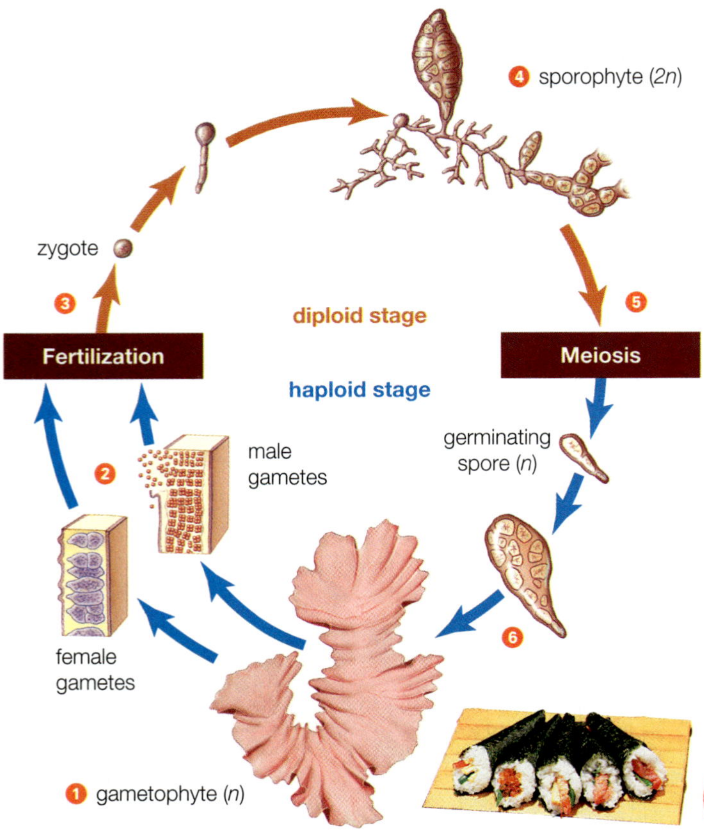

FIGURE 20.13 {Animated} Life cycle of a red alga (*Porphyra*), commonly known as nori.

(1) The haploid gametophyte is sheetlike.
(2) Gametes form at its edges.
(3) Fertilization produces a diploid zygote.
(4) The zygote develops into a diploid sporophyte.
(5) Haploid spores form by meiosis on the sporophyte, and are released.
(6) Spores germinate and develop into a new gametophyte.

FIGURE IT OUT: Are the cells in the sheets of algae used to wrap sushi haploid or diploid?

Answer: Haploid

the sea, and still others partner with a fungus to form a composite organism known as a lichen. However, the vast majority of green algae live in fresh water.

Chlamydomonas is a single-celled green alga common in ponds. Flagellated haploid cells (**FIGURE 20.14A**) reproduce asexually when conditions favor growth. If nutrients become scarce, gametes form by mitosis and fuse to form a diploid zygote with a thick wall. When conditions once again become favorable, the zygote undergoes meiosis, yielding four haploid, flagellated cells.

Volvox is a colonial freshwater species (**FIGURE 20.14B**). A colony consists of hundreds to thousands of flagellated cells that resemble *Chlamydomonas*. Thin cytoplasmic strands join the cells together as a whirling sphere. Daughter colonies form inside the parental colony, which eventually ruptures and releases them.

Wispy sheets of the multicelled green alga *Ulva* cling to coastal rocks worldwide. In some species, these sheets grow longer than your arm, but are less than 40 microns thick (**FIGURE 20.14C**). *Ulva* is commonly known as sea lettuce and is harvested for use in salads and soups.

Like the chloroplasts of green algae, those of land plants have two membranes and contain chlorophylls *a* and *c*. This and other similarities indicate that plants descended from a green alga. More specifically, land plants descended from a lineage of mostly freshwater green algae known as the charophyte algae (clade Charophyta). Modern members of this clade share unique structural and functional features with land plants (**FIGURE 20.15**).

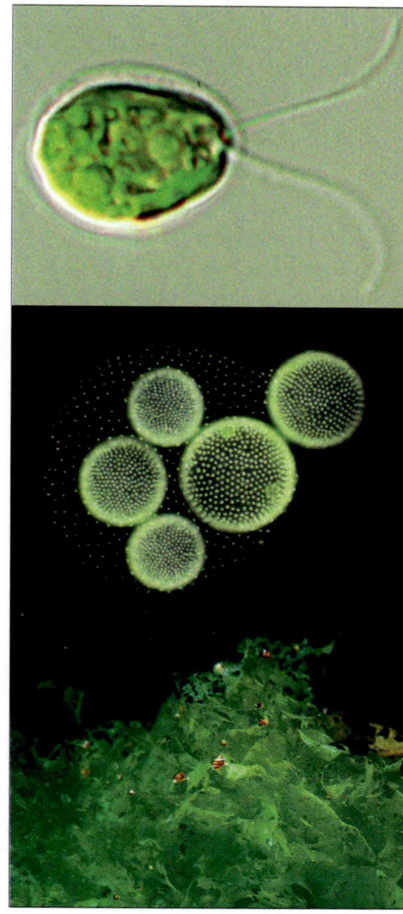

A Single celled organism. *Chlamydomonas*, a single celled freshwater alga, uses its two flagella to swim.

B Colony. *Volvox* colonies consist of flagellated cells connected by thin strands of cytoplasm. New colonies form inside a parent colony and are released.

C Multicellular organism. Long, thin sheets of sea lettuce (*Ulva*) are common along coasts where there is little wave action.

FIGURE 20.14 Three levels of organization among green algae.

A *Coleochaete orbicularis* grows as a multicellular disk attached to a surface.

B *Chara*, commonly known as musk-grass or stinkweed for its strong odor.

FIGURE 20.15 Charophyte algae. Like plants, and unlike most other green algae, charophyte cells divide their cytoplasm by cell plate formation (Section 11.3) and have plasmodesmata (cytoplasmic connections between neighboring cells, Section 4.10).

alternation of generations Of land plants and some algae, a life cycle that includes haploid and diploid multicelled bodies.
gametophyte Gamete-producing haploid body that forms in the life cycle of land plants and some multicelled algae.
green alga Single-celled, colonial, or multicelled photosynthetic protist that has chloroplasts containing chlorophylls *a* and *b*.
red alga Photosynthetic protist; typically multicelled, with chloroplasts containing red accessory pigments (phycobilins).
sporophyte Spore-forming diploid body that forms in the life cycle of land plants and some multicelled algae.

TAKE-HOME MESSAGE 20.7

Red algae and green algae are protists that belong to the same clade as the land plants. All members of this clade deposit cellulose in their cell wall.

Red algae are mostly multicellular and marine. They have red accessory pigments that allow them to live at greater depths than other algae.

Green algae include single-celled, colonial, and multicelled species. One subgroup, the charophyte algae, includes the closest living relatives of land plants.

20.8 WHAT ARE AMOEBOZOANS?

Amoebozoans are shape-shifting heterotrophic cells that typically lack a wall or pellicle. They move about and capture smaller cells by extending lobes of cytoplasm called pseudopods (Section 4.9).

AMOEBAS

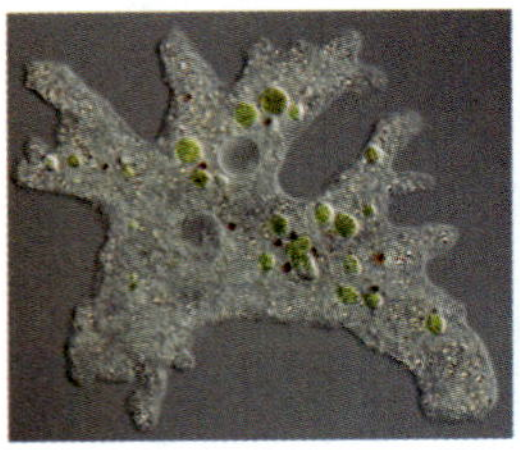

FIGURE 20.16 Freshwater amoeba.

Amoebas spend their entire lives as single cells (**FIGURE 20.16**). Most live in marine or freshwater sediments or in damp soil, where they are important predators of bacteria. A few affect human health. *Entamoeba* infects the human gut, causing cramps, aches, and bloody diarrhea. Infections are most common in developing nations where amoebic cysts contaminate drinking water. Another freshwater amoeba, *Naegleria fowleri*, can enter the nose and cause a rare but deadly brain infection. Most people become infected while swimming in a warm pond or lake, but people have also died after using tap water to irrigate their sinuses.

SLIME MOLDS

Slime molds are sometimes described as "social amoebas." Two types are common on the floor of temperate forests.

Cellular slime molds spend most of their existence as individual amoeba-like cells (**FIGURE 20.17**). Each cell eats bacteria and reproduces by mitosis ❶. If food runs out, thousands of cells stream together ❷ to form a cohesive multicelled unit referred to as a "slug" ❸. The slug moves to a suitable spot, where its component cells differentiate, forming a fruiting body with resting spores atop a stalk ❹. After disperal, each spore will release an amoeba-like cell.

Plasmodial slime molds spend most of their life cycle as a plasmodium, a multinucleated mass that forms when a diploid cell undergoes mitosis many times without cytoplasmic division. A plasmodium streams along surfaces, feeding on microbes and organic matter (**FIGURE 20.18**). When food runs out, the plasmodium develops into many spore-bearing fruiting bodies.

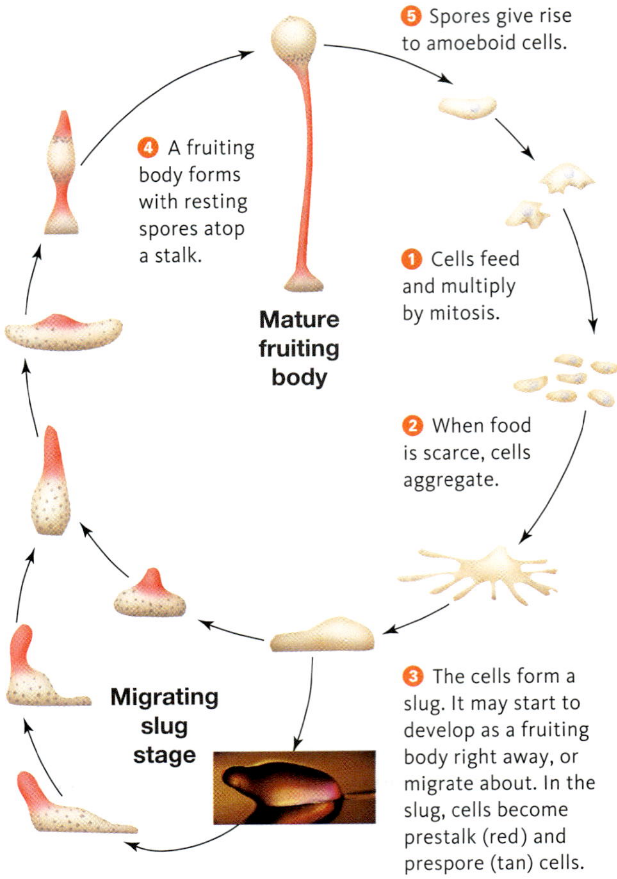

FIGURE 20.17 {Animated} Life cycle of *Dictyostelium discoideum*, a cellular slime mold.

FIGURE 20.18 Plasmodial slime mold streaming across a log.

amoeba Single-celled protist that extends pseudopods to move and to capture prey.
amoebozoan Shape-shifting heterotrophic protist with no pellicle or cell wall; an amoeba or slime mold.
cellular slime mold Amoeba-like protist that feeds as a single predatory cell; joins with others to form a multicellular spore-bearing structure under unfavorable conditions.
plasmodial slime mold Protist that feeds as a multinucleated mass; forms a spore-bearing structure when environmental conditions become unfavorable.

TAKE-HOME MESSAGE 20.8

Amoebozoans are heterotrophic protists with cells that lack a wall or pellicle, so they can constantly change shape.

Amoebas live as single cells, usually in fresh water.

Slime molds live on forest floors. Plasmodial slime molds feed as a big multinucleated mass. Cellular slime molds feed as single cells, but come together as a multicelled mass when conditions are unfavorable. Both types of slime molds form fruiting bodies that release spores.

20.9 WHICH PROTISTS ARE CLOSEST TO ANIMALS?

Aquatic, heterotrophic protists called **choanoflagellates** are the closest known protistan relatives of animals. A choanoflagellate cell has a flagellum surrounded by a "collar" of threadlike projections (**FIGURE 20.19**). Movement of the flagellum sets up a current that draws food-laden water through the collar. As you will learn in Chapter 23, some sponge cells have a similar structure and function.

Most choanoflagellates live as single cells, but some form colonies. The colonies arise by mitosis, when descendant cells do not separate after division. Instead, cellular offspring stick together with the help of adhesion proteins. Choanoflagellate adhesion proteins are similar to those of animals, and researchers have discovered that even solitary choanoflagellates have these proteins. By one hypothesis, the common ancestor of animals and choanoflagellates was a single-celled protist with adhesion proteins that helped it capture prey. Later, these proteins were put to use in a new context, helping choanoflagellates stick together to form multicelled colonies. Later still, the proteins allowed animal cells to adhere to one another in multicelled bodies. This modification in the use of adhesion proteins is an example of exaptation, an evolutionary process by which a trait that evolves with one function later takes on a different function (Section 17.11).

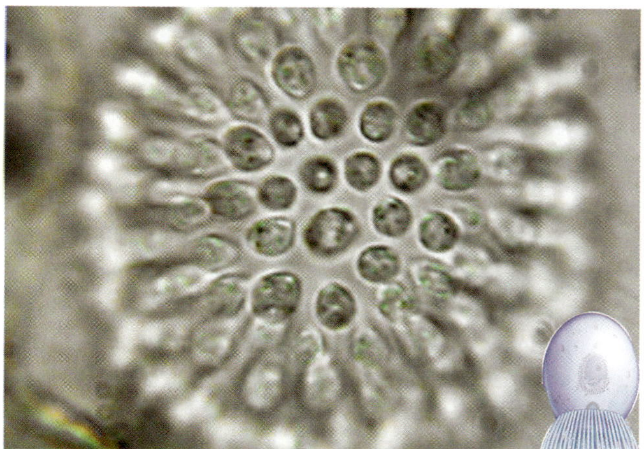

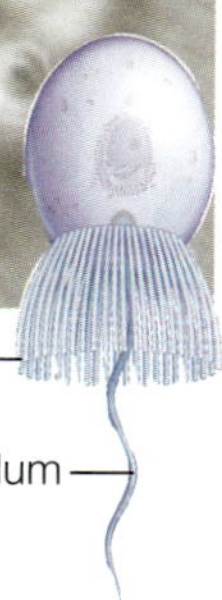

FIGURE 20.19 Choanoflagellate colony. It is composed of genetically identical cells like that illustrated at right.

choanoflagellate Heterotrophic freshwater protist with a flagellum and a food-capturing "collar." May be solitary or colonial.

TAKE-HOME MESSAGE 20.9

Choanoflagellates are close relatives of animals. Studies of colonial choanoflagellates may help us discover how the first animals evolved.

20.10 ALGAL BLOOMS

FIGURE 20.20 A bloom of dinoflagellate cells colors seawater along Florida's Gulf coast red.

LIKE LAND PLANTS, ALGAE GROW FASTER WITH THE ADDITION OF CERTAIN NUTRIENTS. Fertilizing a houseplant or a lawn results in a spurt of growth. Similarly, adding nutrients to an aquatic habitat encourages photosynthetic protists to reproduce. We do not fertilize our waters on purpose, but fertilizers drain from croplands and lawns, and sewage and animal wastes from factory farms get into rivers and flow to the sea. The result is an "algal bloom," a population explosion of aquatic protists.

If the protist involved secretes toxins, the bloom can poison other species, including humans. Even nontoxic algal blooms can have negative environmental effects. The huge number of cells eventually die. Bacteria that decompose the remains deplete the water of oxygen, causing fish and other aquatic animals to smother.

Toxic algal blooms affect every coastal region of the United States. Blooms of toxin-producing dinoflagellates are common in the Gulf of Mexico and along the Atlantic coast (**FIGURE 20.20**). Along the Pacific coast, population explosions of toxic diatoms are most common.

Keeping harmful algal toxins out of the human food supply requires constant vigilance. These toxins have no color or odor, and are unaffected by heating or freezing. Government agencies use laboratory tests to detect algal toxins in water samples and shellfish. When the toxins reach a threatening level, a shore is closed to shellfishing.

Summary

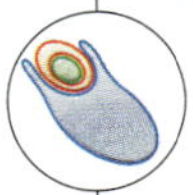

SECTION 20.1 **Protists** are a diverse collection that includes members of many eukaryotic groups, some only distantly related to one another. Most protists are single cells, but some are **colonial organisms** or **multicellular organisms**. Some protists are heterotrophs and others have chloroplasts that evolved from cyanobacteria by **primary endosymbiosis**. Still others have chloroplasts that evolved from a red or green alga by **secondary endosymbiosis**.

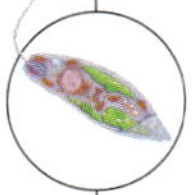

SECTION 20.2 **Flagellated protozoans** are single unwalled cells. A protein covering, or **pellicle**, helps maintain the cell's shape. Diplomonads and parabasalids have modified mitochondria and are anaerobic. Members of both groups include species that infect humans.

Trypanosomes are parasites with a single mitochondrion and a flagellum that attaches to the side of the body. The related **euglenoids** typically live in fresh water. They have a **contractile vacuole** that squirts out excess water. Some have chloroplasts that evolved from a green alga.

SECTION 20.3 **Foraminifera** are single cells with a calcium carbonate shell. **Radiolaria** are single-celled and have a silica shell. Both groups are marine heterotrophs and part of **plankton**. Cytoplasmic extensions that stick out through their porous shell capture prey.

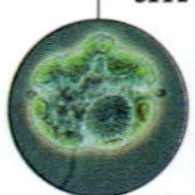

SECTIONS 20.4, 20.5 Tiny sacs (alveoli) beneath the plasma membrane characterize alveolates. All are single-celled. **Dinoflagellates** are whirling aquatic cells. They may be autotrophs or heterotrophs. Some are capable of **bioluminescence**. The **ciliates** have many cilia. They include aquatic predators and parasites. **Apicomplexans** live as parasites in the cells of animals. Mosquitoes transmit the apicomplexan that causes malaria.

SECTION 20.6 Stramenopiles include two photosynthetic groups and the colorless water molds. **Diatoms** are silica-shelled photosynthetic cells. Deposits of ancient diatom shells are mined as diatomaceous earth. **Brown algae** include microscopic strands and giant kelp, the largest protists. Brown algae are the source of algins, compounds used as thickeners and emulsifiers. Diatoms and brown algae share a brown accessory pigment. **Water molds** grow as a mesh of absorptive filaments. They include decomposers and parasites.

SECTION 20.7 Most **red algae** are multicelled and marine. Accessory pigments called phycobilins allow them to capture light even in deep waters. Red algae are commercially important as the source of agar, carrageenan, and as dry sheets (nori) used for wrapping sushi. **Green algae** may be single cells, colonial, or multicelled. They are the closest relatives of land plants.

Like land plants, some algae have an **alternation of generations**, a life cycle in which two kinds of multicelled bodies form: a diploid, spore-producing **sporophyte** and a haploid, gamete-producing **gametophyte**.

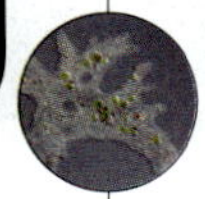

SECTION 20.8 **Amoebozoans** include solitary **amoebas** and slime molds, both heterotrophic. The **plasmodial slime molds** feed as a multinucleated mass. Amoeba-like cells of **cellular slime molds** aggregate when food is scarce. Both types of slime molds form multicelled fruiting bodies that disperse resting spores.

SECTION 20.9 **Choanoflagellates** are heterotrophic solitary or colonial protists that are close relatives of the animals.

SECTION 20.10 An algal bloom is a population explosion of aquatic protists that results from nutrient enrichment of an aquatic habitat. Algal blooms endanger other organisms.

Self-Quiz Answers in Appendix VII

1. All flagellated protozoans ________ .
   a. lack mitochondria c. live as single cells
   b. are photosynthetic d. cause disease

2. Deposits of shells from ancient ________ are mined as chalk and limestone.
   a. dinoflagellates c. radiolaria
   b. diatoms d. foraminifera

3. The presence of a contractile vacuole indicates that a single-celled protist ________ .
   a. is marine c. is photosynthetic
   b. lives in fresh water d. secretes a toxin

4. Cattle benefit when ________ in their gut help them digest plant material.
   a. red algae c. ciliates
   b. diatoms d. foraminifera

5. An insect bite can transmit a malaria-causing ________ to a human host.
   a. trypanosome c. ciliate
   b. apicomplexan d. diplomonad

6. ________ are the closest relatives of the land plants.
   a. Green algae c. Brown algae
   b. Red algae d. Euglenoids

7. Accessory pigments of ________ allow them to carry out photosynthesis at greater depths than other algae.
   a. euglenoids c. brown algae
   b. green algae d. red algae

8. The most common bioluminescent protists in tropical seas are ________ .
   a. red algae b. diatoms c. dinoflagellates d. radiolaria

9. The ________ are important plant pathogens.
   a. dinoflagellates c. water molds
   b. ciliates d. slime molds

Data Analysis Activities

Tracking Changes in Algal Blooms Reports of fish kills along Florida's southwest coast date to the mid-1800s, suggesting algal blooms are a natural phenomenon in this region. However, University of Miami researchers suspected that increased nutrient delivery from land contributed to an increase in the abundance of the dinoflagellates. Since the 1950s, the population of coastal cities in southwestern Florida has soared and the amount of agriculture increased. To find out if nearshore nutrients are affecting dinoflagellate numbers, the researchers looked at records for coastal waters that have been monitored for more than 50 years. **FIGURE 20.21** shows the average abundance and distribution of dinoflagellates during two time periods.

1. How did the average concentration of dinoflagellates in waters less than 5 kilometers from shore change between the two time periods?
2. How did the average concentration of dinoflagellates in waters more than 25 kilometers from shore change between the two time periods?
3. Do these data support the hypothesis that human activity increased the abundance of dinoflagellates by adding nutrients to coastal waters?

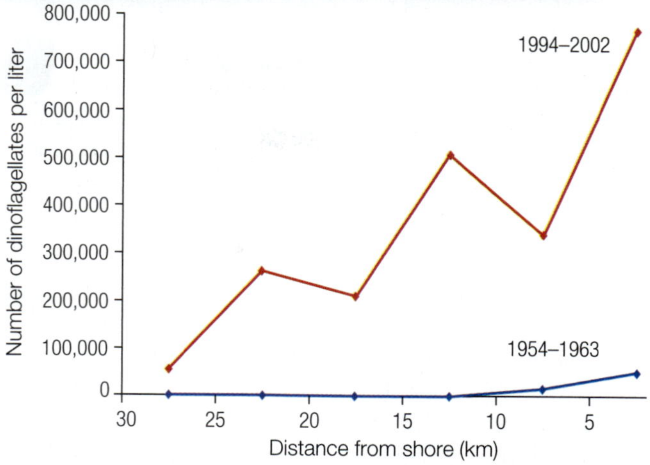

FIGURE 20.21 Average concentration of dinoflagellate cells detected at various distances from the shore during two time periods: 1954–1963 (blue line) and 1994–2002 (red line). Samples were collected from offshore waters between Tampa Bay and Sanibel Island.

4. If the two graph lines became farther apart as distance from the shore increased, what would that suggest about the nutrient source?

10. Silica-rich shells of ancient diatoms are the source of diatomaceous earth that can be used _______ .
    a. to thicken foods c. as a fertilizer
    b. as a gelatin substitute d. as an insecticide
11. The sporophyte of a multicellular alga _______ .
    a. is haploid c. produces spores
    b. is a single cell d. produces gametes
12. Cellular slime molds most often live _______ .
    a. on the forest floor c. in an animal gut
    b. in a tropical sea d. in a mountain lake
13. The protist that causes the sexually transmitted disease trichomoniasis is a(n) _______ .
    a. flagellated protozoan c. ciliate
    b. radiolaria d. apicomplexan
14. All green algae _______ .
    a. have a cell wall c. are multicellular
    b. are marine d. all of the above
15. Match each item with its description.

____	choanoflagellate	a. chalky-shelled heterotroph
____	apicomplexan	b. silica-shelled autotroph
____	foraminiferan	c. deep dweller with phycobilins
____	diatom	d. close relative of animals
____	brown alga	e. close relative of land plants
____	red alga	f. multicelled, with fucoxanthin
____	green alga	g. cause of malaria

Critical Thinking

1. Imagine you are in a developing country where sanitation is poor. Having read about parasitic flagellates in water and damp soil, what would you consider safe to drink? What foods might be best to avoid or which food preparation methods might make them safe to eat?

2. Water in abandoned swimming pools often turns green. If you examined a drop of this water with a microscope, how could you tell whether the water contains protists and, if so, which group they might belong to?

3. The "snow alga" *Chlamydomonas nivalis* (*right*) lives on glaciers. It is a green alga, but it has so many carotenoid pigments that it appears red. Besides their role in photosynthesis, what other function might these light-absorbing carotenoids serve in the alga's bright, arctic environment?

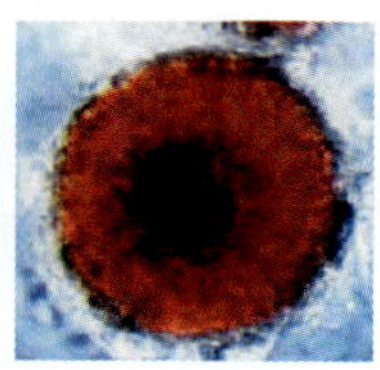

4. The protist that causes malaria evolved from a photosynthetic ancestor and has the remnant of a chloroplast. The organelle no longer functions in photosynthesis, but it remains essential to the protist. Why might targeting this organelle yield an antimalarial drug that produces minimal side effects in humans?

CREDITS: (21) Based on research by Larry Brand; (inset) Brian Duval.

Ancient bristlecone pines in California's White Mountains. These plants can live more than 4,500 years.

21 PLANT EVOLUTION

Links to Earlier Concepts

Section 20.7 introduced the algae that are the closest relative of plants. Section 12.2 introduced gamete formation in plants and here you will see specific examples. You will learn about evolution of cell walls strengthened with lignin (4.10) and a waxy cuticle perforated by stomata (6.5). We also discuss plant fossils (16.3) and coevolution with pollinators (17.11).

KEY CONCEPTS

ADAPTIVE TRENDS AMONG PLANTS

Plants evolved from a green alga. Over time, changes in structure, life cycle, and reproductive processes adapted plants to life in increasingly drier climates.

THE BRYOPHYTES

Bryophytes include the oldest plant lineages such as mosses. All are low-growing plants that disperse by releasing spores, and have flagellated sperm that swim to eggs.

SEEDLESS VASCULAR PLANTS

Internal pipelines that characterize vascular plants allow them to stand taller than bryophytes. Seedless vascular plants such as ferns release spores. Their sperm swim through water to eggs.

GYMNOSPERMS

Gymnosperms are seed plants—vascular plants that make pollen and release seeds, rather than spores. They do not require water for sperm to reach eggs. Gymnosperm seeds are not enclosed within a fruit.

ANGIOSPERMS

Angiosperms are seed plants that make flowers and form seeds inside fruits. They are the most widely dispersed and diverse plant group. Many enlist animals to move pollen or disperse seeds.

Photograph by Marc Moritsch, National Geographic Creative.

21.1 HOW DID PLANTS ADAPT TO LIFE ON LAND?

Plants are multicelled, photosynthetic eukaryotes that typically live on land. They first appeared about 500 million years ago. Green algae grew at the water's edge, and one lineage, the charophytes, gave rise to the first land plants. Like green algae, plants have cells with walls made of cellulose and chloroplasts that contain chlorophylls *a* and *b*.

A developmental trait defines plants. Their clade is called the embryophytes (meaning embryo-bearing plants), because their embryos form inside a chamber of parental tissues and receive nourishment from the parent during early development.

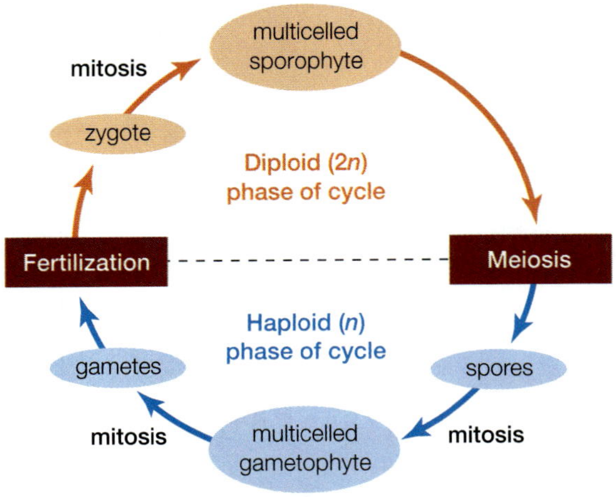

FIGURE 21.2 {Animated} Generalized plant life cycle.

STRUCTURAL ADAPTATIONS

Moving onto land posed challenges to a previously aquatic lineage. A multicelled green alga absorbs all the water, dissolved nutrients, and gases it needs across its body surface. Water also buoys algal parts, helping an alga stand upright. By contrast, land plants face the threat of drying out, must take up water and nutrients from soil, and have to stand upright on their own.

A variety of structural features allow land plants to meet these challenges. Most secrete a waxy **cuticle** that reduces water loss (**FIGURE 21.1**). Closable pores called **stomata** (singular, stoma) extend across the cuticle. Depending on conditions in the environment and inside the plant, stomata can open to allow gas exchange or close to conserve water.

Early plants were held in place by threadlike structures, but these structures did not deliver water or dissolved minerals to the rest of the plant body. True roots contain vascular tissue: internal pipelines that transport water and nutrients among plant parts. **Xylem** is the vascular tissue that distributes water and mineral ions taken up by roots. **Phloem** is the vascular tissue that distributes sugars made by photosynthetic cells. Of the 295,000 or so modern plant species, more than 90 percent are **vascular plants**, meaning plants that have xylem and phloem. Mosses and other ancient plant lineages are referred to as **nonvascular plants**, meaning they do not have these vascular tissues.

An organic compound called **lignin** stiffens cell walls in vascular tissue. Lignin-reinforced vascular tissue not only distributes materials, it also helps vascular land plants stand upright and supports their branches. Most vascular plants have branches with many leaves that increase the surface area for capturing light and for gas exchange.

FIGURE 21.1 Diagram of a vascular plant leaf in cross-section, showing some traits that contribute to the success of this group. The micrograph shows a stoma, a pore at the surface of a squash plant's leaf. Photosynthetic cells on either side of the stoma control its opening and closing to regulate water loss and gas exchange.

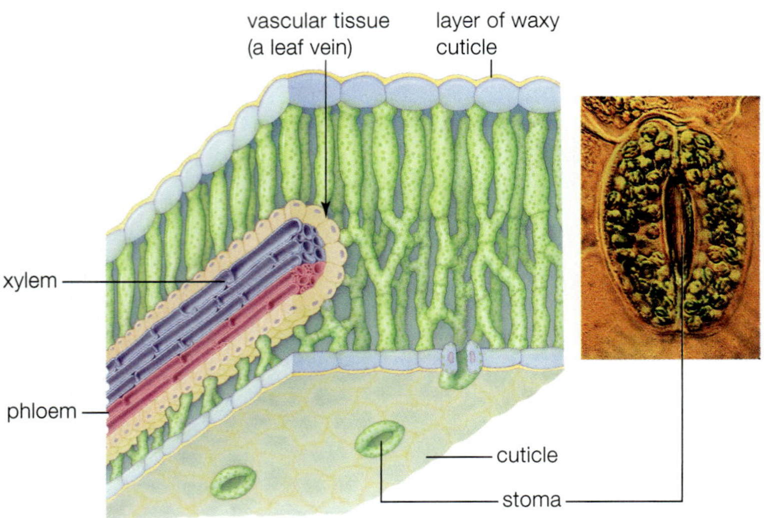

LIFE CYCLE CHANGES

Adapting to life in dry habitats also involved life cycle changes. Like some algae, land plants have an alternation of generations (**FIGURE 21.2**). Meiosis of cells in a diploid, multicelled **sporophyte** produces spores, which in plants are walled haploid cells. A spore germinates, divides by mitosis, and develops into a haploid, multicelled **gametophyte**. Mitosis of cells in a gametophyte produces gametes that combine at fertilization to form a diploid zygote. The zygote grows and develops into a new sporophyte.

In nonvascular plants, the gametophyte is the largest and longest-lived part of the life cycle. By contrast, the sporophyte dominates the life cycle of vascular plants. Living in dry conditions favors an increased emphasis on spore production, because spores withstand drying out better than gametes do.

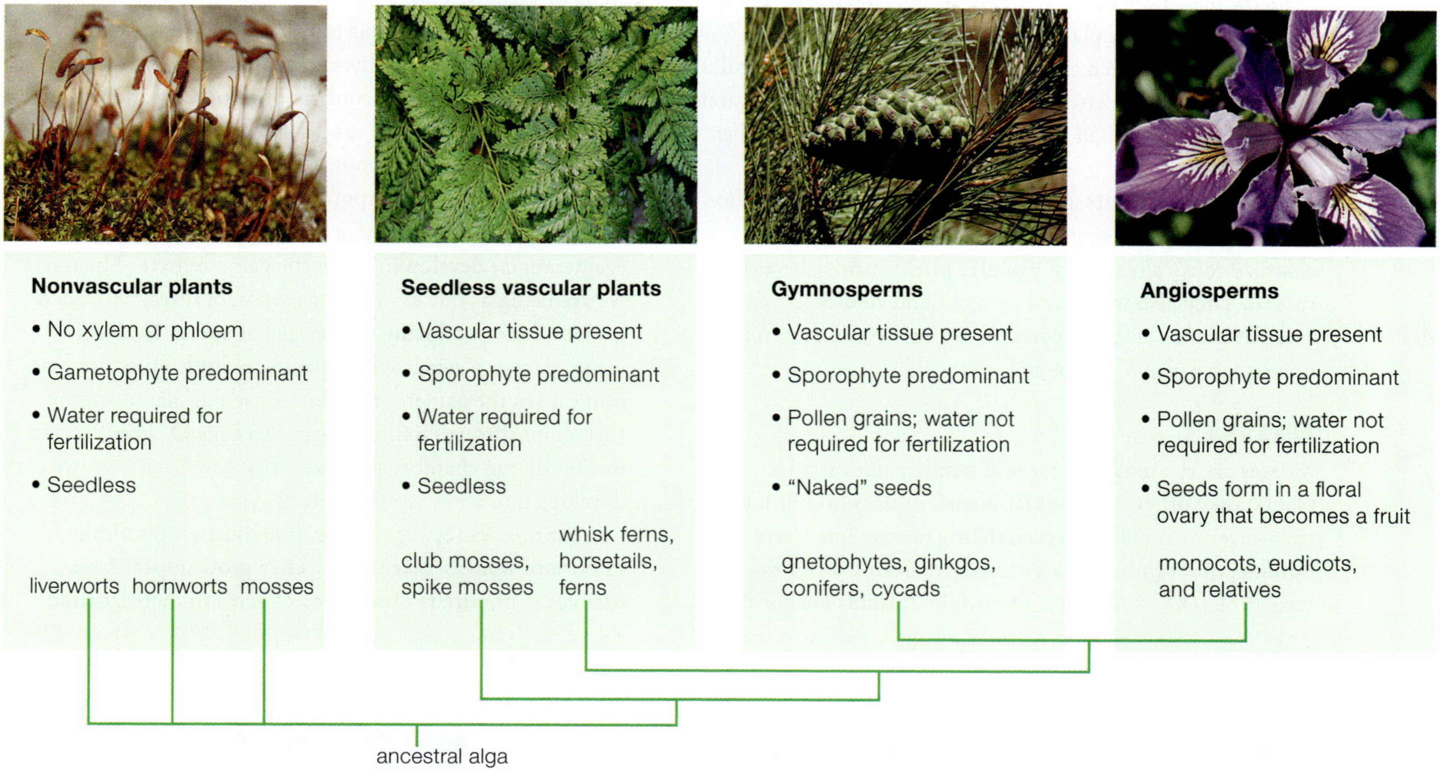

FIGURE 21.3 **{Animated}** Traits of the modern plant groups and relationships among them. Only the gymnosperms and angiosperms are clades.

POLLEN AND SEEDS

Evolution of new reproductive traits in seed plants gave this vascular plant lineage a competitive edge in dry habitats. Seed plants do not release spores. Instead, their spores give rise to gametophytes inside specialized structures on the sporophyte body.

A **pollen grain** is a walled, immature male gametophyte of a seed plant. Once released, it can be transported to another plant by wind or animals even in the driest of times. By contrast, plants that do not make pollen (nonvascular plants and seedless vascular plants) can only reproduce when a film of water allows their sperm to swim to eggs.

Fertilization of a seed plant takes place on the sporophyte body. It results in development of a **seed**—an embryo sporophyte packaged with a supply of nutritive tissue inside a protective seed coat. Seed plants disperse a new generation by releasing seeds.

There are two modern seed plant lineages. Gymnosperms include nonflowering seed producers such as pine trees. Angiosperms, which make flowers and package their seeds inside a fruit, are the most widely distributed and diverse plant group.

FIGURE 21.3 summarizes the relationships among the major plant groups and the traits of each group.

cuticle Secreted covering at a body surface.
gametophyte Multicelled, haploid, gamete-producing body.
lignin Material that stiffens cell walls of vascular plants.
nonvascular plant Plant that does not have xylem and phloem; a bryophyte such as a moss.
phloem Plant vascular tissue that distributes sugars.
plant Multicelled photosynthetic organism in which embryos form on and are nurtured by the parent.
pollen grain Walled, immature male gametophyte of a seed plant.
seed Embryo sporophyte of a seed plant packaged with nutritive tissue inside a protective coat.
sporophyte Multicelled, diploid, spore-producing body.
stoma Opening across a plant's cuticle and epidermis; can be opened for gas exchange or closed to prevent water loss.
vascular plant Plant with xylem and phloem.
xylem Plant vascular tissue that distributes water and dissolved mineral ions.

TAKE-HOME MESSAGE 21.1

Nonvascular plants such as mosses have a life cycle dominated by a haploid gametophyte.

Vascular plants have a life cycle dominated by a diploid sporophyte with vascular tissue. Like nonvascular plants, the oldest vascular plant lineages disperse by releasing spores.

Seed plants, the most recently evolved plant lineage, can reproduce even in dry times because they make pollen. They disperse by releasing seeds, not spores.

21.2 WHAT ARE NONVASCULAR PLANTS?

Nonvascular plants, also known as **bryophytes**, include three modern lineages: mosses, hornworts, and liverworts. These are the only modern plants in which the gametophyte is larger and longer-lived than the sporophyte. All nonvascular plants produce flagellated sperm that require a film of water to swim to eggs, and all disperse by releasing spores, rather than seeds.

Nonvascular plants absorb nutrients across their surface, rather than withdrawing them from soil, so they can colonize rocky sites where vascular plants cannot become rooted. They also withstand drought and cold better than vascular plants. In some parts of the Arctic and Antarctic, nonvascular plants are the only plant life.

MOSSES

Mosses are the most diverse and familiar nonvascular plants. Like other nonvascular plants, mosses do not have true leaves or roots. However, their gametophytes have leaflike photosynthetic parts arranged around a central stalk (**FIGURE 21.4** ❶). Threadlike **rhizoids** anchor the gametophyte, but do not take up water and nutrients as the roots of vascular plants do.

The moss sporophyte is not photosynthetic, so it depends on the gametophyte for nourishment even when mature. The sporophyte consists of a spore-producing structure (a sporangium) atop a stalk ❷. Meiosis of cells inside the sporangium produces haploid spores ❸. After dispersal by the wind, a spore germinates and grows into a gametophyte. Multicellular gamete-producing structures (gametangia) develop in or on the gametophyte. The moss we are using as our example has separate sexes, with each gametophyte producing either eggs or sperm ❹, but in some species, a gametophyte produces both. In either case, rain causes the gametophyte to release flagellated sperm that swim through a film of water to eggs ❺. Fertilization inside the egg chamber produces a zygote ❻ that grows and develops into a new sporophyte ❼.

Peat mosses (*Sphagnum*) are the most economically important nonvascular plants. They grow in peat bogs that cover hundreds of millions of acres in high-latitude

FIGURE 21.4 {Animated} Life cycle of a common moss (*Polytrichum*).

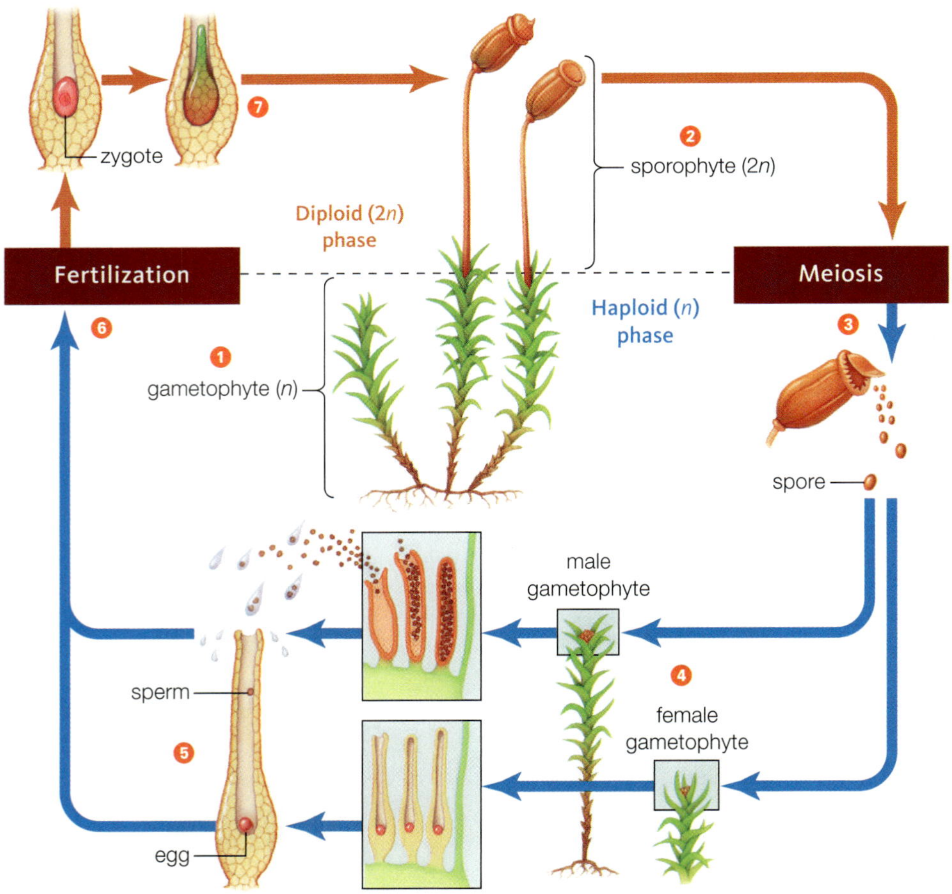

❶ The moss gametophyte has photosynthetic leaflike parts. Rhizoids hold it in place.
❷ The nonphotosynthetic moss sporophyte consists of a stalk and a capsule.
❸ Spores form by meiosis in the capsule, are released, and drift with the winds.
❹ Spores develop into gametophytes that produce eggs or sperm in gametangia at their tips.
❺ Sperm released from tips of sperm-producing gametophytes swim through water to eggs at tips of egg-producing gametophytes.
❻ Fertilization produces a zygote.
❼ The zygote grows and develops into a sporophyte while remaining attached to and nourished by its egg-producing parent.

regions of Europe, Asia, and North America. Many peat bogs have persisted for thousands of years, and layer upon layer of plant remains have become compressed as carbon-rich material called peat. Blocks of peat are cut, dried, and burned as fuel, especially in Ireland (**FIGURE 21.5**). Freshly harvested peat moss is also an important commercial product. The moss is dried and added to planting mixes to help soil retain moisture.

LIVERWORTS AND HORNWORTS

Liverworts may be the most ancient of the surviving plant lineages. The oldest known fossils of land plants are spores that resemble those of modern liverworts. In addition, genetic comparisons put liverworts near the base of the plant family tree.

Some liverwort gametophytes look leafy and others are flattened sheets. In the widespread liverwort genus *Marchantia*, eggs and sperm form on separate plants. The gametangia are elevated above the main gametophyte body on stalks (**FIGURE 21.6**). Members of this genus also reproduce asexually by producing small clumps of cells in cups on the gametophyte surface. Some *Marchantia* species can be pests in commercial greenhouses. Liverwort infestations are difficult to eradicate because the tiny spores can persist even after all plants are killed.

Hornworts have a flat, ribbonlike or rosette-shaped gametophyte and a pointy, hornlike sporophyte (**FIGURE 21.7**). The base of the sporophyte is embedded in the gametophyte, and spores form in an upright capsule at its tip. When spores mature, the tip of the capsule splits, releasing them.

Unlike the sporophyte of a moss or liverwort, that of a hornwort grows continually from its base. It also has chloroplasts and, in some cases, can survive even after the gametophyte dies. These traits, together with genetic similarities, suggest that hornworts are the closest living relatives of the vascular plants. As you will see, vascular plants have a sporophyte-dominated life cycle.

---

bryophyte Nonvascular plant; a moss, liverwort, or hornwort.
rhizoid Threadlike structure that holds a nonvascular plant in place.

TAKE-HOME MESSAGE 21.2

Mosses, liverworts, and hornworts are three lineages of low-growing plants that do not have lignin-reinforced vascular tissues. All have flagellated sperm that require a film of water to swim to eggs, and all disperse by releasing spores.

Nonvascular plants (bryophytes) are the only plants in which the gametophyte is largest and longest-lived. The sporophyte remains attached to the gametophyte even when mature.

FIGURE 21.5 Cutting blocks of peat in Ireland for use as fuel. Peat is the compressed, carbon-rich remains of *Sphagnum* moss.

FIGURE 21.6 {Animated} Liverwort (*Marchantia*). Tiny sporophytes with yellow capsules form on umbrella-shaped, female gametangia.

FIGURE 21.7 Hornwort. Photosynthetic hornlike sporophytes grow from a flattened gametophyte body.

21.3 WHAT ARE SEEDLESS VASCULAR PLANTS?

Only vascular plants have lignin-strengthened vascular tissue. This innovation allowed evolution of larger, branching sporophytes with roots, stems, and leaves. Such sporophytes are the predominant generation in all vascular plants. Like nonvascular plants, **seedless vascular plants** have flagellated sperm that swim to eggs, and they disperse by releasing spores directly into the environment. The spores develop into tiny, short-lived gametophytes.

Two lineages of seedless vascular plants survived to the present. One (monilophytes) includes ferns and related groups, the other (lycophytes) includes club mosses and their relatives. These lineages diverged from a common ancestor before leaves and roots evolved, so each developed these features in a different way.

FERNS AND CLOSE RELATIVES

With 12,000 or so species, ferns are the most diverse and familiar seedless vascular plants. Most live in the tropics. A typical fern sporophyte has fronds (leaves) and roots that grow from a **rhizome**, a horizontal underground stem (FIGURE 21.8 ❶). Fiddleheads, the young, tightly coiled fronds of some ferns, are harvested from the wild as food.

Fern spores form in **sori** (singular, sorus), clusters of capsules (sporangia) that develop on the lower surface of fronds ❷. When these capsules pop open, spores disperse on the wind. After a spore germinates, it grows into a photosynthetic, heart-shaped gametophyte just a few centimeters wide ❸. Eggs and sperm form in chambers (gametangia) on the underside of the tiny gametophyte. Rain stimulates release of sperm, which swim through a film of water to reach and fertilize an egg ❹. The resulting zygote develops into a new sporophyte ❺, and its parental gametophyte dies.

Fern sporophytes vary in size and form. Some floating ferns have fronds only 1 millimeter long. Tree ferns can be 25 meters (80 feet) high (FIGURE 21.9A). Many tropical ferns are epiphytes, meaning they attach to and grow on a trunk or branch of another plant but do not withdraw any nutrients from it.

Whisk ferns and horsetails are close relatives of ferns. Most whisk ferns are in the genus *Psilotum*. Their sporophyte has leafless photosynthetic stems that grow from rhizomes (FIGURE 21.9B). Spores form in fused sporangia at the tips of short lateral branches. Florists use these whisk ferns to add visual interest to mixed bouquets.

The sporophyte of a horsetail (*Equisetum*) has hollow stems with tiny nonphotosynthetic leaves at the joints. Photosynthesis occurs in stems and leaflike branches

FIGURE 21.8 {Animated} Life cycle of a fern.

❶ The familiar leafy form is the diploid sporophyte.
❷ Meiosis in cells on the underside of fronds (leaves) produces haploid spores.
❸ After spores are released, they germinate and grow into tiny gametophytes that produce eggs and sperm.
❹ Sperm swim to eggs and fertilize them, forming a zygote.
❺ The sporophyte begins its development attached to the gametophyte, but it continues to grow and live independently after the gametophyte dies.

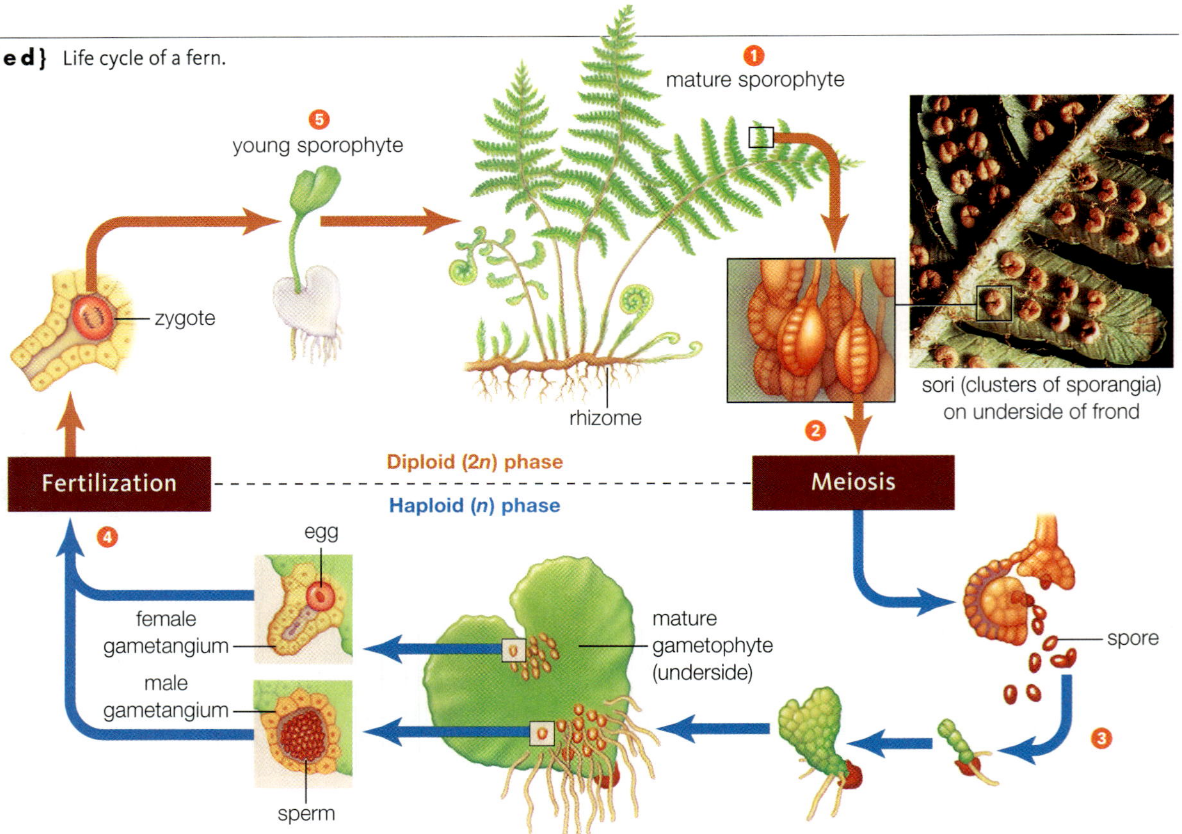

A Tree ferns in Australia.

B Whisk fern (*Psilotum*)

C Horsetail (*Equisetum*)

D Club moss (*Lycopodium*)

FIGURE 21.9 Seedless vascular plant sporophytes. These plants disperse by releasing spores.

(**FIGURE 21.9C**). Deposits of silica in the stem support the plant and give stems a sandpaper-like texture that helps fend off herbivores. Before scouring powders and pads became widely available, people used stems of horsetails as pot scrubbers. Horsetail spores form on a **strobilus**, a soft cone-shaped structure composed of modified leaves.

LYCOPHYTES

Lycophytes include plants commonly known as quillworts, spikemosses, and club mosses. Club mosses of the genus *Lycopodium* are common in temperate forests (**FIGURE 21.9D**). Their sporophyte has a rhizome from which roots and upright stems with tiny leaves grow. When a *Lycopodium* plant is several years old, it begins to make spore-producing strobili seasonally. *Lycopodium* is gathered from the wild for a variety of uses. The stems are used in wreaths and bouquets. Spores, which have a waxy, flammable coating, are sold as "flash powder" for special effects. When the spores are sprayed out as a fine mist and ignited, they produce a bright flame.

rhizome Stem that grows horizontally along or under the ground.
seedless vascular plant Plant that disperses by releasing spores and has xylem and phloem. For example, a club moss or fern.
sorus Cluster of spore-producing capsules on a fern leaf.
strobilus Of some nonflowering plants, a spore-forming, cone-shaped structure composed of modified leaves.

TAKE-HOME MESSAGE 21.3

Sporophytes with vascular tissue in their roots, stems, and leaves dominate the life cycle of seedless vascular plants.

Tiny, short-lived gametophytes produce flagellated sperm that swim to eggs through films of water.

Ferns, the most diverse seedless vascular plants, belong to the same lineage as whisk ferns and horsetails. Club mosses and related groups belong to another seedless vascular lineage.

21.4 HOW HAVE VASCULAR PLANTS CHANGED OVER TIME?

A Fossil of an early vascular plant (*Cooksonia*). Its stems always divided into two equal branches.

B Painting of a Carboniferous "coal forest." An understory of ferns is shaded by tree-sized relatives of modern horsetails and lycophytes.

C Early angiosperms such as magnolias (*foreground*) evolved while dinosaurs walked on Earth.

FIGURE 21.10 Early vascular plants.

FROM TINY BRANCHERS TO COAL FORESTS

The oldest fossils of vascular plants are spores that date to about 450 million years ago, during the late Ordovician period. Early vascular plants such as *Cooksonia* stood only a few centimeters high and had a simple branching pattern, with no leaves or roots (**FIGURE 21.10A**). Spores formed at branch tips. By the early Devonian, taller species with a more complex branching pattern were common worldwide.

As the Devonian continued, a taller structure evolved in some seedless vascular plants. The oldest forest we know about existed about 385 million years ago in what is now upstate New York. Fossil stumps and fronds discovered at this site indicate that some plants of this forest stood about 8 meters (26 feet) high and resembled modern tree ferns in their structure.

During the Carboniferous period (359–299 million years ago), ancient relatives of horsetails and lycophytes evolved into massively stemmed giants (**FIGURE 21.10B**). Some stood 40 meters (more than 130 feet) high. After these forests first formed, climates changed, and the sea level repeatedly rose and fell. Over time, pressure and heat transformed the compacted organic remains of Carboniferous forests into **coal**. Often you will hear about annual production rates for coal or some other fossil fuel. In fact, we do not "produce" these materials. Coal is a nonrenewable source of energy, and one that we are on the way to depleting.

RISE OF THE SEED PLANTS

Seed plants evolved late in the Devonian period (about 365 million years ago) and gymnosperms lived beside seedless plants in Carboniferous forests. As Earth became cooler and drier during the Permian period (299–251 million years ago), gymnosperms replaced seedless plants in many habitats. Angiosperms, the flowering plants, first appear in the fossil record about 125 million years ago, during the Cretaceous period (**FIGURE 21.10C**).

Reproductive traits of seed-bearing plants put them at an advantage in dry habitats. Gametophytes of seedless vascular plants develop from spores that were released into the environment. By contrast, gametophytes of seed plants form protected inside reproductive parts on a sporophyte body (**FIGURE 21.11**). Pollen grains that give rise to male gametophytes develop from haploid spores that form inside **pollen sacs**. These spores are called **microspores**. Egg-producing gametophytes develop from haploid **megaspores** that form inside a protective chamber called an **ovule**.

A seed-bearing plant releases pollen grains, but holds onto its eggs. Wind or animals can deliver pollen from one seed plant to the ovule of another, a process called **pollination**. Because sperm of seed plants do not need to swim through a film of water to reach eggs, these plants can reproduce even during dry times.

After fertilization, an ovule develops into a seed, which contains an embryo sporophyte and food that nourishes it until it can make its own. By contrast, seedless plants release single-celled spores that do not have a store of food.

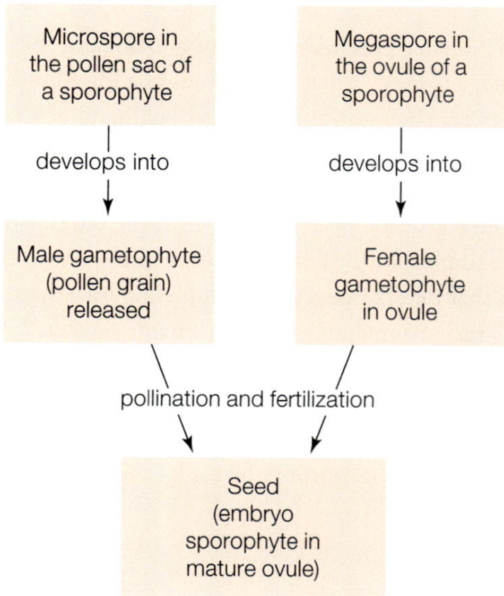

FIGURE 21.11 How a seed forms.

Seed plants are also the only plants that produce true wood. As a woody plant grows, addition of lignin-stiffened xylem cells widens its parts. Wood may have first evolved as a means of improving water delivery through the plant, but it also provided structural advantages. Wood makes a plant sturdier and more resistant to mechanical stress, so it can grow taller and overshadow its competitors.

coal Fossil fuel formed over millions of years by compaction and heating of plant remains.
megaspore Haploid spore formed in ovules of seed plants; gives rise to an egg-producing gametophyte.
microspore Haploid spore formed in pollen sacs of seed plants; gives rise to a sperm-producing male gametophyte.
ovule Of seed plants, reproductive structure in which egg-bearing gametophyte develops; after fertilization, it matures into a seed.
pollen sac Of seed plants, reproductive structure in which pollen grains develop.
pollination Delivery of a pollen grain to the egg-bearing part of a seed plant.

TAKE-HOME MESSAGE 21.4

Seedless vascular plants arose before the seed-bearing lineages. They were widespread during the Carboniferous.

Seed plants produce pollen grains that wind or animals can carry to eggs. As a result, seed plants can reproduce under drier conditions than seedless plants.

The egg-bearing gametophyte of a seed plant forms in an ovule that becomes a seed after fertilization. Dispersing seeds rather than spores increases reproductive success.

CREDITS: (11) © Cengage Learning; Jeff Benca.

PEOPLE MATTER

National Geographic Explorer
JEFF BENCA

Like many children, Jeff Benca was fascinated by dinosaurs. As a teen he transferred his interest to another ancient group—the lycophytes. By the time he graduated from college, he had one of the world's largest lycophyte collections. Now a graduate student, he investigates what these early-diverging vascular plants can tell us about conditions in Earth's deep past.

Paleobotanists have long used fossil leaf shape of woody flowering plants as a tool for inferring ancient climatic conditions. For example, botanists know that woody angiosperms with more serrated leaf margins (like maples) are now prominent in temperate regions with cool, wet climates. By contrast woody angiosperms with smooth leaf margins (such as magnolias) now predominate in tropical regions. If ferns and lycophytes also show climate-influenced changes in leaf shape, their fossils could yield a wealth of information about Earth's distant past. The fossil record of angiosperms extends back 125 million years, but ferns and lycophytes have fossil records exceeding 360 million years.

To investigate how climate affects leaves of ferns and lycophytes, Benca grows members of these groups under a variety of conditions in climate-controlled growth chambers and compares their leaf shapes. He is also looking at how the shape of fern and lycophyte leaves varies with elevation (and climate) on a tropical mountain.

Aside from his research, Benca advocates conservation of lycophytes. Although this ancient lineage survived several mass extinctions, many species are now declining swiftly due to habitat loss. As a result, lycophytes are in need of increased research attention and protection.

21.5 WHAT ARE GYMNOSPERMS?

Gymnosperms are vascular seed plants whose seeds are "naked," meaning that unlike the seeds of angiosperms they are not within a fruit. (*Gymnos* means naked and *sperma* is taken to mean seed.) However, many gymnosperms enclose their seeds in a fleshy or papery covering.

CONIFERS

Conifers, the most diverse and familiar gymnosperms, include 600 or so species of trees and shrubs that have needlelike or scalelike leaves, and produce woody seed cones. Conifers are typically more resistant to drought and cold than flowering plants, and they abound in cool forests of the Northern Hemisphere. Conifers include the tallest trees in the Northern Hemisphere (redwoods), and the most abundant (pines). They also include the long-lived bristlecone pines shown in the chapter-opening photo.

Conifers are of great commercial importance. We use fir bark to mulch gardens, use oils from cedar in cleaning products, and eat seeds, or "pine nuts," of some pines. Pines also provide lumber for building homes, and some make a sticky resin that deters insects from tunneling into them. We use this resin to make turpentine, a paint solvent.

FIGURE 21.12 illustrates the life cycle of one conifer, the ponderosa pine. Seed cones and pollen cones develop on the same tree. Ovules form on cone scales of seed cones ①. Inside the ovule, a megaspore forms by meiosis ② and

FIGURE 21.12 {Animated} Conifer (pine) life cycle.

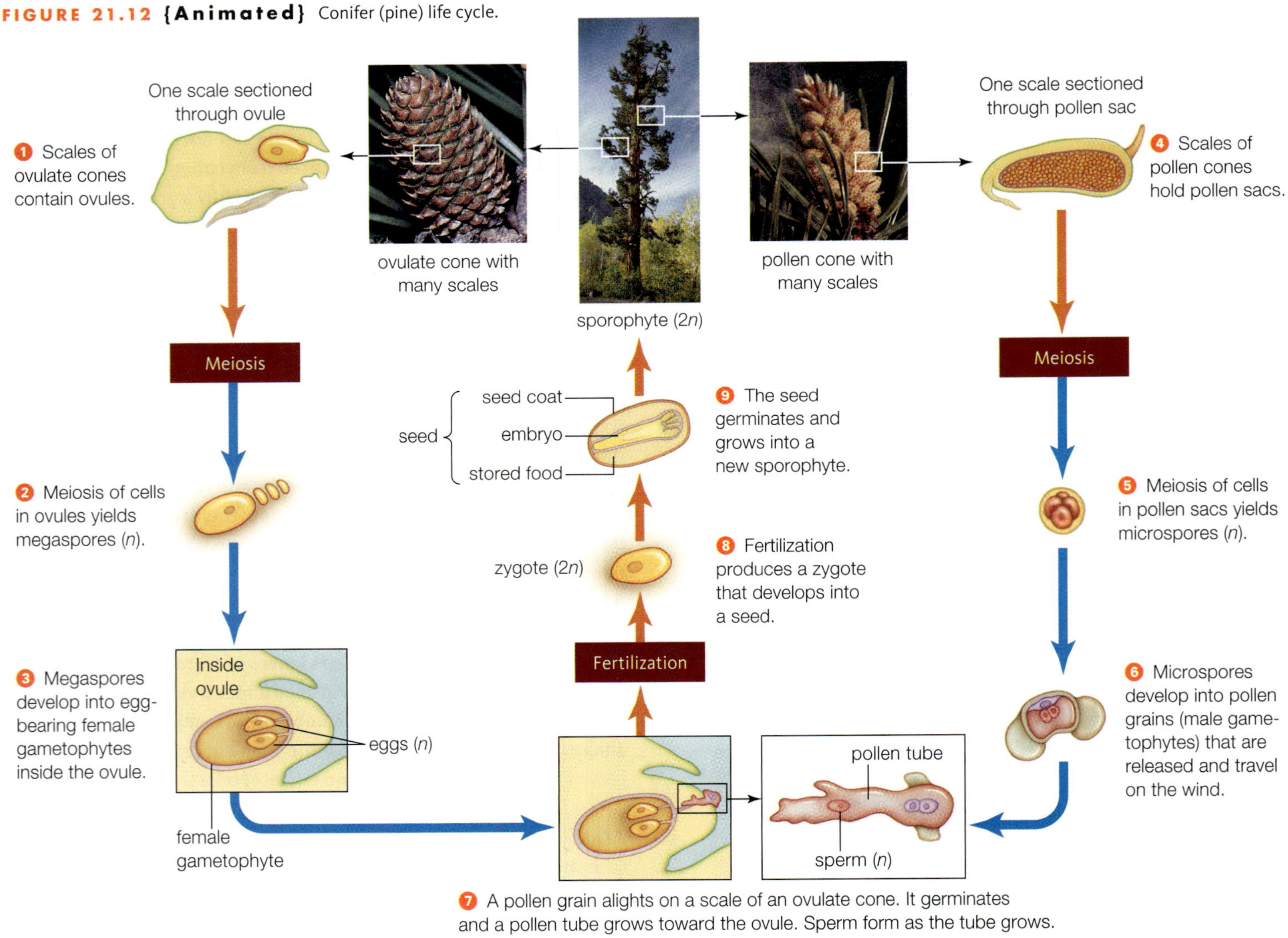

FIGURE IT OUT: Does the pollen tube grow through tissue of a male cone or a female cone?

Answer: A female cone. It grows after a pollen grain alights on a female cone.

develops into a female gametophyte ❸. Male cones hold pollen sacs ❹, where microspores form ❺ and develop into pollen grains ❺. The pollen grains are released and drift with the winds. Pollination occurs when one lands on the scale of a seed cone ❻. The pollen grain then germinates: Some cells develop into a pollen tube that grows through the ovule tissue and delivers sperm to the egg ❼. Pollen tube growth in gymnosperms is an astonishingly leisurely process. It typically takes about a year for the tube to grow through the ovule to the egg. When fertilization occurs, it produces a zygote ❽. Over about six months, the zygote develops into an embryo sporophyte that, along with tissues of the ovule, becomes a seed ❾. The seed is released, germinates, then grows and develops into a new sporophyte.

LESSER KNOWN LINEAGES

Cycads and ginkgos were most diverse about 200 million years ago. They are the only modern seed plants that have flagellated sperm. Sperm emerge from pollen grains, then swim in fluid produced by the plant's ovule.

The 130 species of modern cycads live mainly in the dry tropics and subtropics (**FIGURE 21.13A**). Cycads often resemble palms but the two groups are not close relatives. The "sago palms" commonly used in landscaping and as houseplants are actually cycads. Cycad seeds have a fleshy covering and were traditionally used as food and medicine in Guam and other regions. However, they contain toxins that increase the risk of neurodegenerative disease and cancers.

The only living ginkgo species is *Ginkgo biloba*, the maidenhair tree (**FIGURE 21.13B**). It is a deciduous native of China. Deciduous plants drop all their leaves at once seasonally. The ginkgo's pretty fan-shaped leaves and resistance to insects, disease, and air pollution make it a popular tree along city streets. Female trees produce fleshy plum-sized seeds with a strong unpleasant odor, so male trees are preferred for urban landscaping. Extracts of *G. biloba* have been touted as a possible memory aid and a treatment for Alzheimer's disease, but recent well-designed studies have found no such beneficial effect.

Gnetophytes include woody vines, tropical trees, and shrubs. Members of the genus *Ephedra* are evergreen desert shrubs with broomlike green stems and inconspicuous leaves (**FIGURE 21.13C**). Native Americans brewed tea using one species common in semiarid regions of the American West. Some Eurasian *Ephedra* species produce ephedrine, a stimulant similar to amphetamines. Another gnetophyte, *Welwitschia*, grows only in Africa's Namib desert (**FIGURE 21.13D**). It has a taproot that grows deep into the soil, a short woody stem, and two long, straplike leaves. These leaves split lengthwise repeatedly, giving the plant a strange shaggy appearance. *Welwitschia* is very long-lived; some plants are more than a thousand years old.

A Cycad with fleshy seeds.

B Fan-shaped leaves and fleshy seeds of *Ginkgo biloba*.

C Pollen cones of *Ephedra*, a gnetophye.

D *Welwitschia*, a gnetophyte, with seed cones and two long, wide leaves.

FIGURE 21.13 Gymnosperm diversity.

gymnosperm Seed plant whose seeds are not enclosed within a fruit; a conifer, cycad, ginkgo, or gnetophyte.

TAKE-HOME MESSAGE 21.5

Gymnosperms are seed-bearing plants. Their eggs and seeds form on the surface of an ovule, not inside ovaries as occurs in flowering plants.

All gymnosperms produce pollen. In cycads and ginkgos, sperm emerge from pollen grains and swim through fluid released by the ovule. Other groups have nonmotile sperm.

21.6 WHAT ARE ANGIOSPERMS?

FIGURE 21.14 Structure of a typical flower.

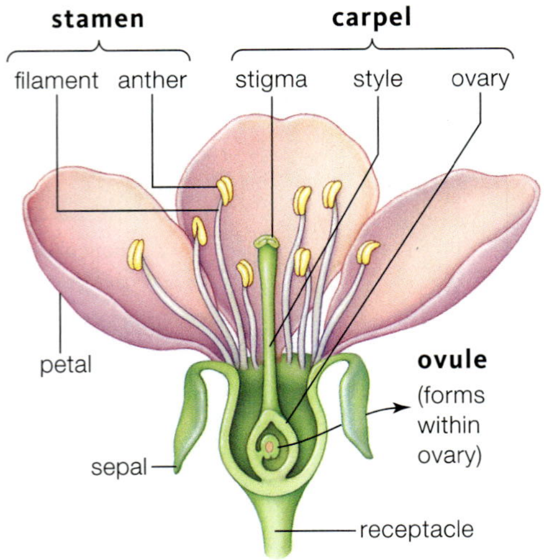

FLOWERS AND FRUITS

Angiosperms, the most diverse seed plant lineage, are the only plants that make flowers and fruits. A **flower** is a specialized reproductive shoot. Floral structure varies, but most flowers include the parts shown in **FIGURE 21.14**. Sepals, which usually have a green leaflike appearance, ring the base of a flower and enclose it until it opens. Inside the sepals are petals, which are often brightly colored. The petals surround the stamens. **Stamens** are organs that produce pollen. Typically a stamen consists of a tall stalk (the filament), topped by an anther that holds two pollen sacs. The innermost part of the flower is the **carpel**, the organ that captures pollen and produces eggs. The carpel consists of a stigma, style, and ovary. The sticky or hairy stigma, which is specialized for receiving pollen, sits atop a stalk called the style. At the base of the style is an **ovary**,

FIGURE 21.15 {Animated} Life cycle of a typical angiosperm.

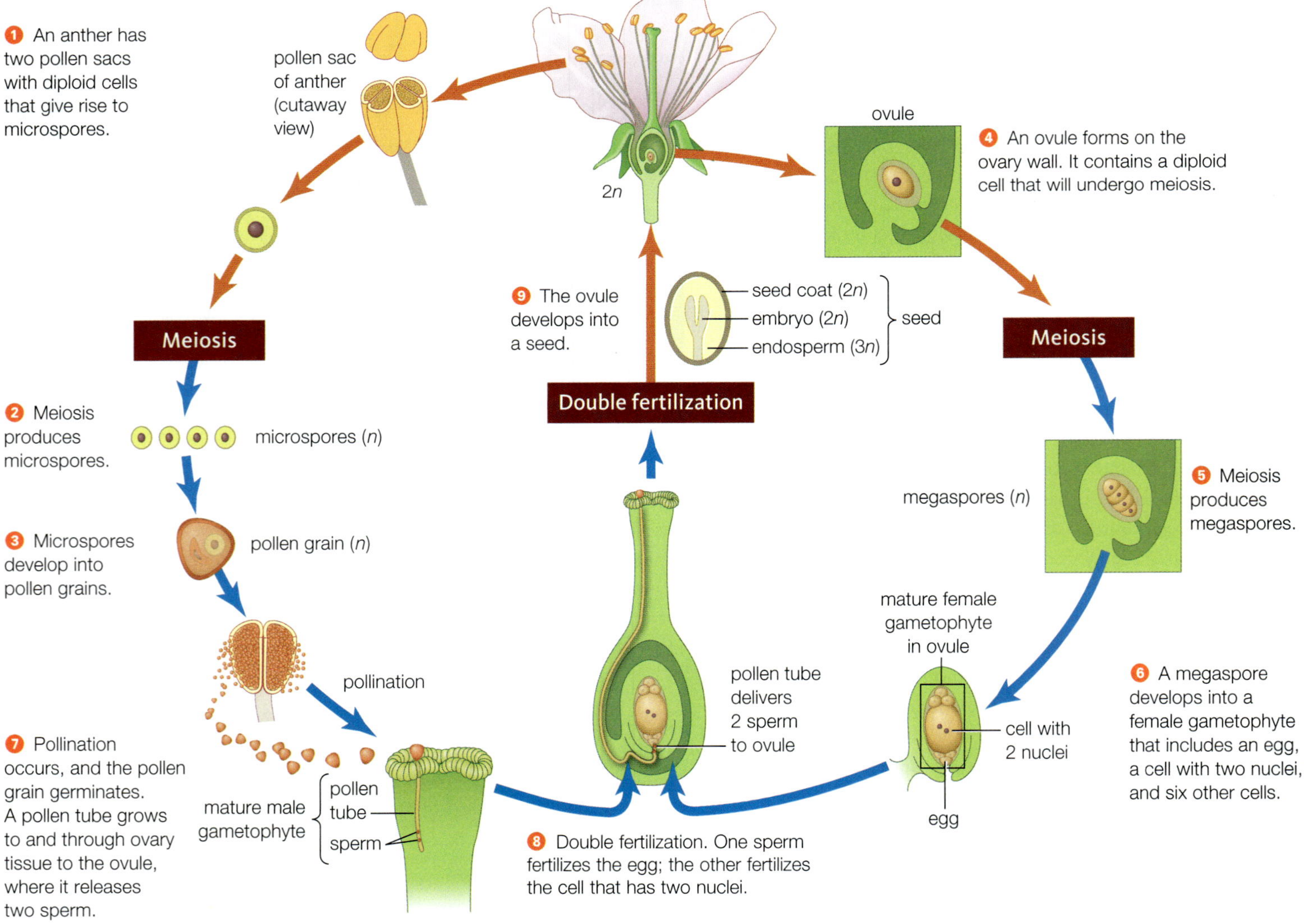

a chamber that contains one or more ovules. The name angiosperm refers to the fact that seeds form within an ovary. (*Angio*– means enclosed chamber, and *sperma*, seed.) After fertilization, an ovule matures into a seed and the ovary becomes a **fruit**.

FIGURE 21.15 shows a typical flowering plant life cycle. Inside pollen sacs in the flower's anthers ❶, diploid cells produce microspores by meiosis ❷. The microspores develop into pollen grains (immature male gametophytes) ❸. At the same time, ovules form in an ovary at the base of a carpel ❹. Meiosis of cells in an ovule yields haploid megaspores ❺. The megaspores undergo mitosis to produce a female gametophyte that includes a haploid egg, a cell with two nuclei, and other cells ❻.

Pollination occurs when a pollen grain arrives on a receptive stigma ❼. The pollen grain germinates, and a pollen tube grows through the style to the ovary at the base of the carpel. Two nonflagellated sperm form inside the pollen tube as it grows.

Double fertilization occurs when a pollen tube delivers the two sperm into the ovule ❽. One sperm fertilizes the egg to create a zygote. The other sperm fuses with the cell that has two nuclei to form a triploid ($3n$) cell. After double fertilization, the ovule matures into a seed ❾. The zygote develops into an embryo sporophyte, and the triploid cell gives rise to the **endosperm**, a nutritious tissue that will serve as a source of food for the developing embryo.

FIGURE 21.16 Representatives of four angiosperm lineages. The vast majority of angiosperms are monocots or eudicots.

MAJOR LINEAGES

Gene comparisons have identified the oldest angiosperm lineages among modern plants. Water lilies (**FIGURE 21.16A**) are among the basal angiosperms, meaning they belong to a lineage that branched off before the three major angiosperm lineages evolved. These major lineages are magnoliids, eudicots (true dicots), and monocots (**FIGURE 21.16B–D**). The 9,200 magnoliids include magnolias as well as avocados. The 80,000 **monocots** include palms, lilies, grasses, orchids, and irises. The 170,000 **eudicots** include most familiar broadleaf plants such as tomatoes, cabbages, poppies, and roses, as well as the cacti and all flowering shrubs and trees.

Monocots and eudicots derive their group names from the number of seed leaves, or **cotyledons**, in the embryo. Monocots have one cotyledon and eudicots have two. The two groups also differ in the arrangement of their vascular tissues, number of flower petals, and other traits. Some eudicots are woody, but no monocots produce true wood.

angiosperms Highly diverse seed plant lineage; only plants that make flowers and fruits.
carpel Floral reproductive organ that produces female gametophytes; typically consists of a stigma, style, and ovary.
cotyledon Seed leaf of a flowering plant embryo.
endosperm Nutritive tissue in the seeds of flowering plants.
eudicot Flowering plant in which the embryo has two cotyledons.
flower Specialized reproductive shoot of a flowering plant.
fruit Mature ovary of a flowering plant; encloses a seed or seeds.
monocot Flowering plant in which the embryo has one cotyledons.
ovary In flowering plants, the enlarged base of a carpel, inside which one or more ovules form and eggs are fertilized.
stamen Floral reproductive organ that produces male gametophytes; typically consists of an anther on the tip of a filament.

TAKE-HOME MESSAGE 21.6

Angiosperms are flowering plants. A flower is a special reproductive shoot that makes pollen in stamens and eggs in carpels.

Angiosperm seeds are enclosed within a fruit. The fruit forms after fertilization from tissue of the ovary, the chamber in which ovules formed.

Angiosperm seeds contain endosperm, a nutritive tissue.

Magnoliids, monocots, and eudicots are the main lineages of flowering plants.

CREDITS: (16A) Smithsonian Institution Department of Botany, G.A. Cooper @ USDA-NRCS PLANTS Database; (16B) © Donald Johansson/iStockphoto.com; (16C) ©WendyTownrow/iStockphoto.com; (16D) Courtesy of Dr. Thomas L. Rost.

21.7 WHY ARE ANGIOSPERMS SO DIVERSE AND WIDESPREAD?

It would be nearly impossible to overestimate the ecological importance of angiosperms. As the most abundant plants in the majority of land habitats, they provide food and shelter for a variety of animals. Their ecological significance is a reflection of their diversity, which stems from factors discussed below.

ACCELERATED LIFE CYCLE

Compared to gymnosperms, most angiosperms have a shorter life cycle. A dandelion or grass can grow from a seed, mature, and produce seeds of its own within a month or so. In contrast, gymnosperms such as pine trees take years to mature. Producing and dispersing seeds fast allows an angiosperm to expand its range faster than a gymnosperm.

PARTNERSHIPS WITH POLLINATORS

Flowers give angiosperms an edge by facilitating animal-assisted pollination. As seed plants evolved, some animals began feeding on their protein-rich pollen. The plants lost a bit of pollen, but benefited when the insects inadvertently transferred pollen from one plant to another of the same species. An animal that facilitates pollination by transferring pollen between plants is a **pollinator**. Most pollinators are insects (**FIGURE 21.17A**), but birds, bats, and other animals fulfill this role for some plants.

Over time, many flowering plants coevolved with their pollinators. (As Section 17.11 explained, coevolution refers to the joint evolution of two or more species as a result of their close ecological interactions.) Producing sugary nectar encourages more pollinator visits, which improves pollination rates and enhances seed production. Conspicuous petals and distinctive scents that attract pollinator attention also provide a selective advantage. So do mutations that alter floral shapes in ways that maximize the likelihood of pollen transfer onto a pollinator or from a pollinator to a receptive stigma.

ANIMAL-DISPERSED FRUITS

Many angiosperms also benefit by having animals disperse their seeds (**FIGURE 21.17B**). Some plants make seeds with hooks or spines that can stick to animal fur. Others make brightly colored, fleshy fruits that attract fruit-eating birds or mammals. Such plants benefit when the fruit eater later regurgitates their seeds or passes them unharmed in its feces. Some fruits have features that fend off animals that destroy seeds in the fruits they eat. For example, rodents tend to chew up seeds, so chile pepper fruits contain a compound (capsaicin) that tastes "hot" to rodents and other mammals. Birds do not chew seeds, cannot taste the compound, and so serve as the chile's main dispersers.

pollinator Animal that moves pollen, thus facilitating pollination.

TAKE-HOME MESSAGE 21.7

Accelerated life cycles and animal assistance in pollination and seed dispersal contribute to angiosperm success.

FIGURE 21.17 Animal assistants of flowering plants.

A Bees commonly serve as pollinators.

B Birds often eat fruits and disperse seeds.

21.8 Application: SAVING SEEDS

FIGURE 21.18 The Svalbard Global Seed Vault. To learn more about the vault and the seeds stored inside it, visit www.croptrust.org.

PLANT DIVERSITY IS DECLINING. We know of threats to about 12,000 wild plant species, and the actual number of threatened plants is no doubt much higher. As a result, many valuable sources of food, medicine, and other products could disappear before we ever learn their value.

A decline in the diversity of cultivated plants raises additional concerns. In the past, food crops were locale specific. People developed and planted plant varieties that did well in their region, and they saved seeds from their crops to plant the following year. Now, many traditional varieties of crop plants are disappearing as farmers worldwide turn to the same few large companies for seeds. Different varieties of plants are resistant to different diseases, so widespread planting of one variety increases the risk that a disease could decimate the global supply of a crop. The more varieties of a crop we plant, the more likely it is that some will resist a particular disease.

Sustaining the wild relatives of crop plants provides a form of insurance. Such plants provide a reservoir of genetic diversity that plant breeders can draw upon to meet future challenges.

One way to ensure the survival of potentially useful plants is by storing their seeds in seed banks. Today, there are more than 1,500 seed banks around the world. The most ambitious, the Svalbard Global Seed Vault, was built in 2008 on a Norwegian island about 700 miles from the North Pole. This so-called doomsday vault serves as a backup for other seed banks (**FIGURE 21.18**). Here, deep inside a mountain, seeds are stored in a permanently chilled, earthquake-free zone 400 feet above sea level. The location was chosen to ensure that the seeds will remain high and dry even if global climate change causes the polar ice caps to melt. The vault also has an advanced security system and has been engineered to withstand any nearby explosions.

The Svalbard vault now holds the world's most diverse collection of seeds, with more than 750,000 samples and material from nearly every nation. The United States has stored seeds of its native crops such as chile peppers from New Mexico, as well as seeds that American scientists collected elsewhere in the world. Under the deep-freeze conditions in Svalbard, the seeds are expected to survive about 100 years.

Summary

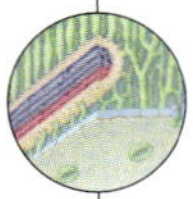

SECTION 21.1 **Plants** evolved from green algae. They are embryophytes; they form a multicelled embryo on the parental body. Key adaptations that allowed plants to move into dry habitats include a waterproof **cuticle** with **stomata**, and internal pipelines of vascular tissue (**xylem** and **phloem**) reinforced by **lignin**.

Plant life cycles involve an alternation of generations. Two types of two multicelled bodies form: a haploid **gametophyte** and a diploid **sporophyte**. The gametophyte predominates in the **nonvascular plants**, but in **vascular plants**, the sporophyte is larger and longer-lived. **Seeds** and **pollen grains** that can be dispersed without water are adaptations that contribute to the success of seed plants.

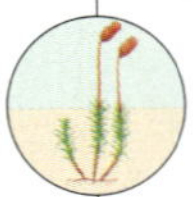

SECTION 21.2 **Bryophytes** are low-growing, nonvascular plants that disperse by releasing spores. They include three modern lineages: mosses, liverworts, and hornworts. All have a gametophye-dominated life cycle and produce flagellated sperm that swim through a film of water to eggs. Mosses are the most diverse bryophytes. **Rhizoids** attach them to soil or a surface. Remains of some mosses form peat, which is dried and burned as fuel.

SECTION 21.3 In **seedless vascular plants**, sporophytes have vascular tissues, and they are the larger, longer-lived phase of the life cycle. Typically the sporophyte's roots and shoots grow from a horizontal stem, or **rhizome**. Tiny free-living gametophytes make flagellated sperm that require water for fertilization. Ferns, the most diverse group of seedless vascular plants, produce spores in **sori**. Many ferns grow as epiphytes. Club mosses and horsetails produce spores in conelike **strobili**.

SECTION 21.4 Forests of giant seedless vascular plants thrived during the Carboniferous period. Later, heat and pressure transformed remains of these forests to **coal**. Seed plants rose to dominance during the Permian, as the climate became cooler and drier. Seed plant sporophytes have **pollen sacs**, where **microspores** form and develop into immature male gametophytes (pollen grains). They also have **ovules**, where **megaspores** form and develop into female gametophytes. **Pollination** unites the egg and sperm of a seed plant.

SECTION 21.5 **Gymnosperms** are seed plants that do not enclose their seeds within a fruit. Conifers have woody seed cones and include commercially important groups such as the pines. They are more resistant to drought and to cold than flowering plants. Cycads and ginkgos are gymnosperm lineages that have flagellated sperm and fleshy seeds. Gnetophytes include desert shrubs.

SECTION 21.6 **Angiosperms** are seed plants that make **flowers** and **fruits**. The **stamens** of a flower produce pollen inside pollen sacs. The **carpel** has a stigma specialized for receiving pollen, atop a stalk called the style. An **ovary** at the base of the carpel holds one or more ovules. After pollination, double fertilization occurs. One sperm delivered by the pollen tube fertilizes the egg and the other fertilizes a cell that has two nuclei. After fertilization, the flower's ovary becomes a fruit that contains one or more seeds. A flowering plant seed includes an embryo sporophyte and **endosperm**, a nutritious tissue. The two main angiosperm lineages, **eudicots** and **monocots**, differ in their number of **cotyledons** and other traits.

SECTION 21.7 Angiosperms are the most widely distributed and diverse plant group. Accelerated life cycles and partnerships with animal **pollinators** and seed dispersers contributed to their success.

SECTION 21.8 Sustaining many varieties of crop plants and their wild relatives ensures that plant breeders will have a reservoir of genetic diversity to tap into if widely planted varieties fail. Seed banks can help us maintain a wide variety of potentially valuable plant species.

Self-Quiz

Answers in Appendix VII

1. The first plants were ______ .
   a. ferns
   b. flowering plants
   c. bryophytes
   d. conifers

2. Which of the following statements is false?
   a. Ferns produce seeds inside sori.
   b. Bryophytes do not have xylem or phloem.
   c. Gymnosperms and angiosperms produce seeds.
   d. Only angiosperms produce flowers.

3. In bryophytes, eggs are fertilized in a chamber on the ______ and a zygote develops into a ______ .
   a. sporophyte; gametophyte
   b. gametophyte; sporophyte
   c. sorus; cone

4. Horsetails and ferns are ______ plants.
   a. multicelled aquatic
   b. nonvascular seed
   c. seedless vascular
   d. seed-bearing vascular

5. Coal consists primarily of compressed remains of the ______ that dominated Carboniferous swamp forests.
   a. seedless vascular plants
   b. conifers
   c. flowering plants
   d. mosses

6. The ______ produce flagellated sperm.
   a. mosses
   b. ferns
   c. conifers
   d. monocots
   e. a and b
   f. a through c

7. A seed is a ______ .
   a. female gametophyte
   b. mature ovule
   c. mature pollen grain
   d. modified microspore

8. True or false? Both spores and sperm of a seedless vascular plant are haploid.

Data Analysis Activities

Insect-Assisted Fertilization in Moss Plant ecologist Nils Cronberg suspected that crawling insects facilitate fertilization of mosses. To test his hypothesis, he carried out an experiment. He placed patches of male and female moss gametophytes in dishes, either next to one another or with water-absorbing plaster between them so sperm could not swim between plants. He then looked at how the presence or absence of insects affected the number of sporophytes formed. **FIGURE 21.19** shows his results.

1. Why is sporophyte formation a good way to determine if fertilization occurred?
2. How close did the male and female patches have to be for sporophytes to form in the absence of insects?
3. Does this study support the hypothesis that insects aid moss fertilization?
4. How might a crawling insect aid moss fertilization?

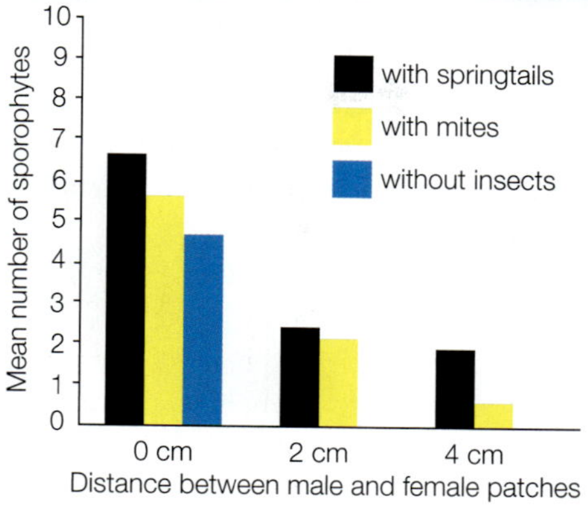

FIGURE 21.19 Sporophyte production in female moss patches with and without two types of crawling insects (springtails and mites). No sporophytes formed in the insect-free dishes when moss patches were 2 or 4 centimeters apart.

9. Only angiosperms produce _______ .
   a. pollen c. fruits
   b. seeds d. all of the above

10. The _______ do not have xylem or phloem.
    a. mosses b. ferns c. monocots d. a and b

11. A waxy cuticle helps land plants _______ .
    a. conserve water c. reproduce
    b. take up carbon dioxide d. stand upright

12. Pollinators aid many _______ .
    a. conifers c. angiosperms
    b. mosses d. ferns

13. _______ produce seeds on woody cones.
    a. Cycads c. Ginkgos
    b. Conifers d. Hornworts

14. Match the terms appropriately.

___ bryophyte	a. seeds, but no fruit
___ seedless vascular plant	b. flowers and fruits
___ gymnosperm	c. no xylem or phloem
___ angiosperm	d. xylem and phloem, but no ovule

15. Match the terms appropriately.

___ ovule	a. gamete-producing body
___ monocot	b. spore-producing body
___ gametophyte	c. becomes seed
___ sporophyte	d. horizontal stem
___ fruit	e. mature ovary
___ endosperm	f. nutritive tissue in seed
___ rhizome	g. where fern spores form
___ sorus	h. single haploid cell
___ microspore	i. one type of angiosperm

Critical Thinking

1. Early botanists admired ferns but found their life cycle perplexing. In the 1700s, they learned to propagate ferns by sowing what appeared to be tiny dustlike "seeds" that they collected from the undersides of fronds. Despite many attempts, the botanists could not locate the pollen source, which they assumed must stimulate these "seeds" to develop. Imagine you could write to these botanists. Compose a note that explains the fern life cycle and clears up their confusion.

2. In most plants the largest, longest-lived body is a diploid sporophyte. By one hypothesis, diploid dominance was favored because it allowed a greater level of genetic diversity. Suppose that a recessive mutation arises. It is mildly disadvantageous now, but it will be useful in some future environment. Explain why such a mutation would be more likely to persist in a fern than in a moss.

3. The photo at the *right* is a micrograph of a longitudinal section through the stem of a squash plant. The stem has been dyed with a substance that tints lignin red. Can you identify the red-ringed structures? Would you expect to find similar structures in the stem of a monocot such as a corn plant? Would you find them in the leafy green part of a moss?

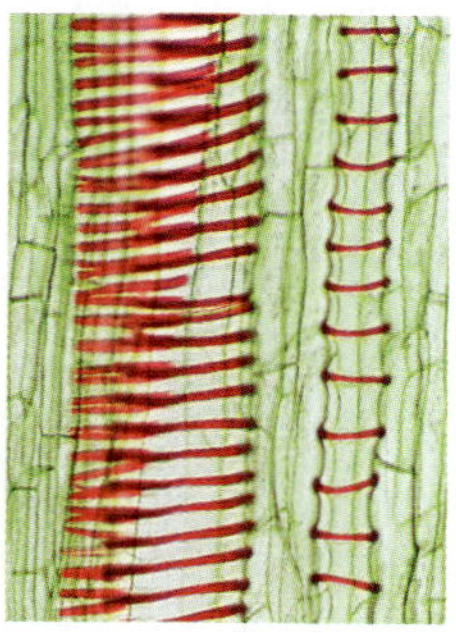

CREDIT: (19) *Science* 1 September 2006: Vol. 313 no. 5791 p. 1255 DOI: 10.1126/science.1128707; (in text) © M.I. Walker/ Wellcome Images.

A spore-producing fruiting body of a stinkhorn fungus on a forest floor.

22 FUNGI

Links to Earlier Concepts

Before starting, review Figure 20.1 to get a sense of where fungi fit in the eukaryotic family tree. In this chapter you will learn about fungal interactions with cyanobacteria (19.5) and green algae (20.7). You will also draw on your knowledge of the organic molecules chitin (3.2) and lignin (21.1), and your understanding of the processes of nutrient cycling (1.2) and fermentation (7.6).

KEY CONCEPTS

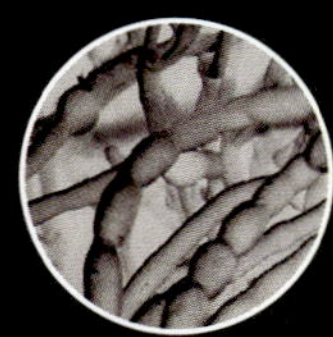

ABSORPTIVE FEEDERS

Fungi include single-celled yeasts and multicelled molds and mushrooms. They secrete digestive enzymes onto organic matter, then absorb released nutrients.

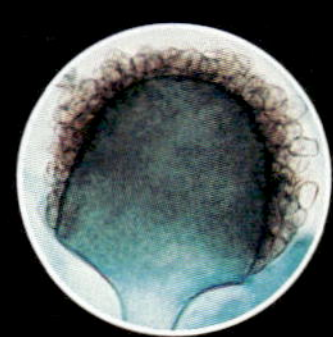

SPORE PRODUCTION

Fungi produce spores sexually and asexually. The major fungal groups are defined by how they produce spores by meiosis.

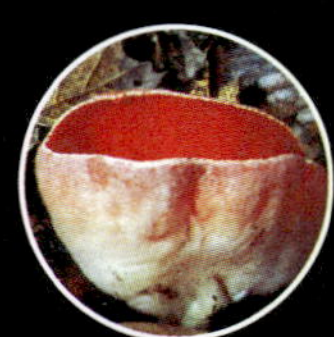

FRUITING BODIES

Some sac fungi and most club fungi produce spores on multicelled fruiting bodies during sexual reproduction. Mushrooms are fungal fruiting bodies.

FUNGAL ECOLOGY

Most fungi act as decomposers. Some form mutually beneficial partnerships with photosynthetic cells, or with plant roots. Still others are parasites of plants or animals.

HUMAN USES OF FUNGI

We eat fungi, enlist them to produce foods and drinks, use them in genetic studies and biotechnology, and extract medicinal or psychoactive compounds from them.

Photograph by Vilainecrevette/Shutterstock.com

22.1 WHAT IS A FUNGUS?

FIGURE 22.1 A multicelled club fungus growing on a tree. The inset micrograph shows the filaments (hyphae) that make up the body of fungi such as this one.

ABSORPTIVE FEEDERS

A **fungus** (plural, fungi) is a eukaryote that secretes digestive enzymes onto its food, then absorbs the resulting breakdown products. Most fungi are decomposers that feed on organic wastes and remains. A lesser number live on or in other living organisms, and some of these are parasites.

Fungal digestive enzymes can break down many sturdy structural proteins that animal digestive enzymes cannot. For example, fungi can digest the cellulose and lignin in plant cell walls, as well as keratin, the main protein in animal skin, claws, hair, and fur. The capacity to digest these tough structural materials makes fungi important both as decomposers and as parasites.

FILAMENTOUS STRUCTURE

Some fungi live as single cells, and these are informally called **yeasts**. Multicelled fungi live as a mesh of threadlike filaments collectively called a **mycelium** (plural, mycelia). Each filament in the mycelium is a **hypha** (plural, hyphae). A hypha consists of walled cells attached end to end (**FIGURE 22.1**). It grows by adding cells to it tips.

Fungal cell walls contain chitin, a polysaccharide also found in the body covering of crabs and insects. Depending on the fungal group, there may or may not be cross-walls between cells of a hypha. When cross-walls do exist, they are porous, so materials can flow between adjacent cells. Thus nutrients or water taken up in one part of a mycelium are shared with cells in other parts of the fungal body.

SPORE PRODUCERS

Fungi disperse by releasing microscopic spores. A fungal spore is typically one or more haploid (n) cells enclosed within a thick coat. When the spore germinates (becomes active), it grows into a new haploid mycelium.

Fungi produce spores both asexually and sexually (**FIGURE 22.2**). During asexual reproduction, multicelled fungi form spores by mitosis at the tips of specialized hyphae ❶. Sexual reproduction begins when two hyphae meet and the cytoplasm of cells at their tips fuses to

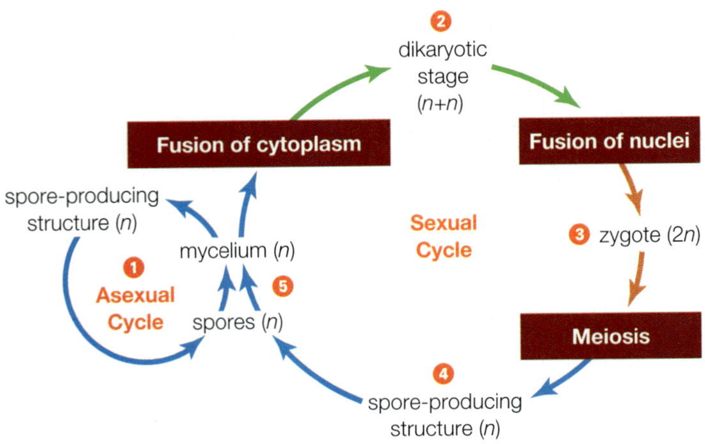

FIGURE 22.2 Generalized life cycle of a fungus.
❶ Asexual reproduction occurs when a haploid mycelium produces spores by mitosis at the tips of specialized hyphae.
❷ Sexual reproduction begins when haploid hyphae of two individuals meet and cells at their tips fuse. This cytoplasmic fusion produces a dikaryotic cell (a cell with two nuclei).
❸ Fusion of nuclei in a dikaryotic cell creates a diploid zygote.
❹ The zygote undergoes meiosis and produces a haploid spore-bearing structure.
❺ Haploid spores form by mitosis and germinate to form a new haploid mycelium.

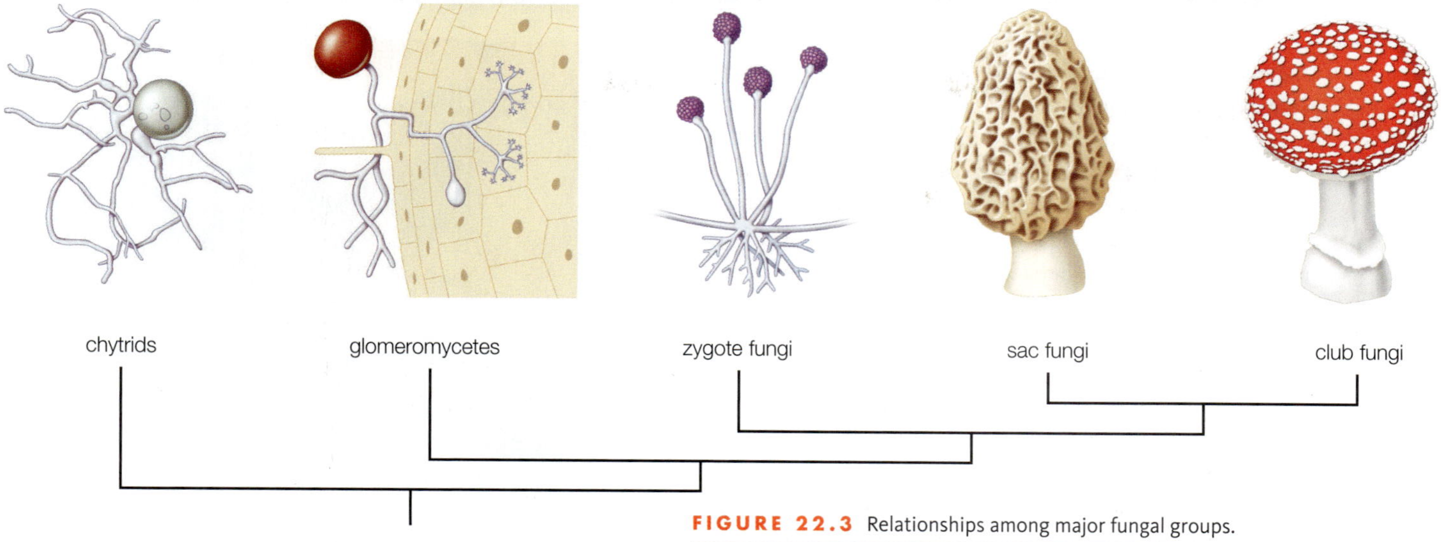

FIGURE 22.3 Relationships among major fungal groups.

produce a dikaryotic cell ❷. **Dikaryotic** means having two genetically distinct types of nuclei ($n+n$). The fungal zygote forms when the two nuclei inside a dikaryotic cell fuse ❸. This diploid zygote undergoes meiosis and gives rise to a structure that produces haploid spores by mitosis ❹. Those spores germinate, releasing cells that divide by mitosis to form a new haploid mycelium ❺.

FIVE MAJOR SUBGROUPS

More than 70,000 fungal species have been named, and there may be a million yet to be discovered. Most fungi belong to one of five groups (**FIGURE 22.3**). **Chytrids** include the oldest fungal lineages, and are the only living fungi that make flagellated spores. Most are aquatic decomposers, but some live in the gut of herbivores, such as cattle and sheep, where they help their host digest cellulose. Others are parasites of plants or animals. **Glomeromycetes** are soil fungi with hyphae that grow into plant roots and branch inside the cell walls of plant root cells. Most **zygote fungi** (zygomycetes) are molds that live in damp places. **Molds** grow as a mass of hyphae and reproduce asexually as long as food is plentiful. During sexual reproduction, zygote fungi produce a thick-walled structure called a zygospore.

Unlike the lineages mentioned thus far, sac fungi and club fungi have cross-walls between cells of their hyphae. Cross walls reinforce hyphae, so sac fungi and club fungi include larger-bodied species than other fungal groups.

Sac fungi (ascomycetes) reproduce sexually by producing spores in sac-shaped structures. They are the most diverse fungal group. Sac fungi include yeasts, molds, parasites of plants and animals, species that partner with photosynthetic cells to form lichens, and species such as morels that form large fruiting bodies. A fungal fruiting body is a spore-producing sexual organ made of intertwined hyphae.

Club fungi (basidiomycetes) reproduce sexually by producing spores in club-shaped structures. Most familiar mushrooms are fruiting bodies of club fungi. This group also includes some parasites.

chytrid Fungus with flagellated spores.
club fungi Fungi that produce spores in club-shaped structures during sexual reproduction.
dikaryotic Having two genetically distinct nuclei.
fungus Unicellular or multicellular eukaryotic heterotroph that digests food outside the body, then absorbs the resulting breakdown products. Has chitin-containing cell walls.
glomeromycete Fungus that partners with plant roots; fungal hyphae grow inside the cell walls of root cells.
hypha Component of a fungal mycelium; a filament made up of cells arranged end to end.
mold Fungus that grows as a mass of asexually reproducing hyphae.
mycelium Mass of threadlike filaments (hyphae) that make up the body of a multicelled fungus.
sac fungi Fungi that form spores in a sac-shaped structure during sexual reproduction.
yeast Fungus that lives as single cell.
zygote fungi Fungi that live in damp places and form a thick-walled zygospore during sexual reproduction.

TAKE-HOME MESSAGE 22.1

Fungi secrete digestive enzymes onto organic matter and absorb the released nutrients. They live as single cells or, more often, as a multicelled mycelium. Most are decomposers, but some live in or on other organisms.

Fungi disperse by releasing spores that form by either mitosis or meiosis. Only the oldest lineage, the chytrids, produces flagellated spores.

Glomeromycete fungi are soil fungi whose hyphae branch inside cell walls of root cells. Zygote fungi are typically molds.

Sac fungi and club fungi have hyphae with cross-walls. Some members of both groups form large fruiting bodies.

22.2 HOW DO LIFE CYCLES DIFFER AMONG FUNGAL GROUPS?

The black bread mold (*Rhizopus stolinifera*) has a life cycle typical of zygote fungi (**FIGURE 22.4**). As long as food is plentiful, it grows as a haploid mycelium and produces spores by mitosis ❶. When the food supply dwindles, lack of food and proximity of a compatible sexual partner cause development of special side branches (gametangia) ❷. Zygote fungi do not have walls between their cells, so many haploid nuclei from within a hypha can flow into each gametangium. When the two gametangia come into contact, their walls break down, and their cytoplasm fuses. The result is an immature zygospore containing multiple nuclei from each parent ❸. As the zygospore matures, the haploid nuclei within it pair up and fuse, forming diploid nuclei. A mature zygospore contains multiple diploid nuclei and has a thick, protective wall ❹. It is the only diploid stage in the zygote fungus life cycle. When the zygospore germinates, a hypha emerges and cells at its tip undergo meiosis to produce haploid spores ❺.

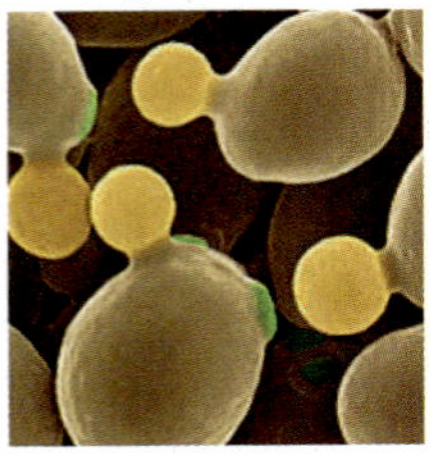

Sac fungi include yeasts, molds, and species with large fruiting bodies. Yeasts usually reproduce asexually by budding (*left*), a process by which a descendant cell with a nucleus and a small amount of cytoplasm pinches off from a parental cell. When sac fungi that grow as molds reproduce asexually, they produce spores by mitosis at the tips of specialized hyphae (**FIGURE 22.5A**). When they or any other sac fungi reproduce sexually, spores form in a sac-shaped structure called an ascus. In some sac fungi, such as the cup fungi, the ascus forms on a large fruiting body called an ascocarp (**FIGURE 22.5B**).

A Asexual reproduction by the mold *Penicillium* (*left*) produces spores atop specialized hyphae (*right*).

B Cup fungi reproduce sexually by producing an ascocarp (*left*). Spores (*right*) form by meiosis of cells on the cup's concave surface.

FIGURE 22.5 Sac fungus reproduction.

FIGURE 22.4 {Animated} Life cycle of black bread mold, a zygote fungus.

❶ As long as food is plentiful, a haploid mycelium grows in size and produces spores by mitosis on specialized hyphae.
❷ When nutrients are limited and hyphae of two compatible individuals come into close proximity, they produce branches (gametangia) that grow toward one another. As these branches grow, haploid nuclei stream into them and accumulate at their tips.
❸ Cytoplasmic fusion of the gametangia produces a zygospore that contains many haploid nuclei from each parent.
❹ Nuclei within the zygospore pair up and fuse to produce a mature zygospore with many diploid nuclei.
❺ The zygospore germinates, and an aerial hypha emerges. Meiosis of cells within this hypha gives rise to haploid spores which are released from its tip.

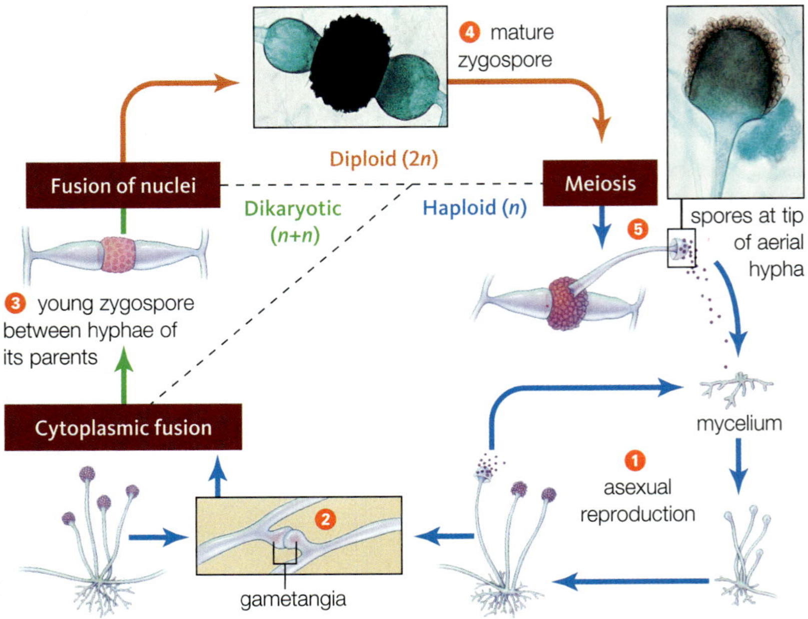

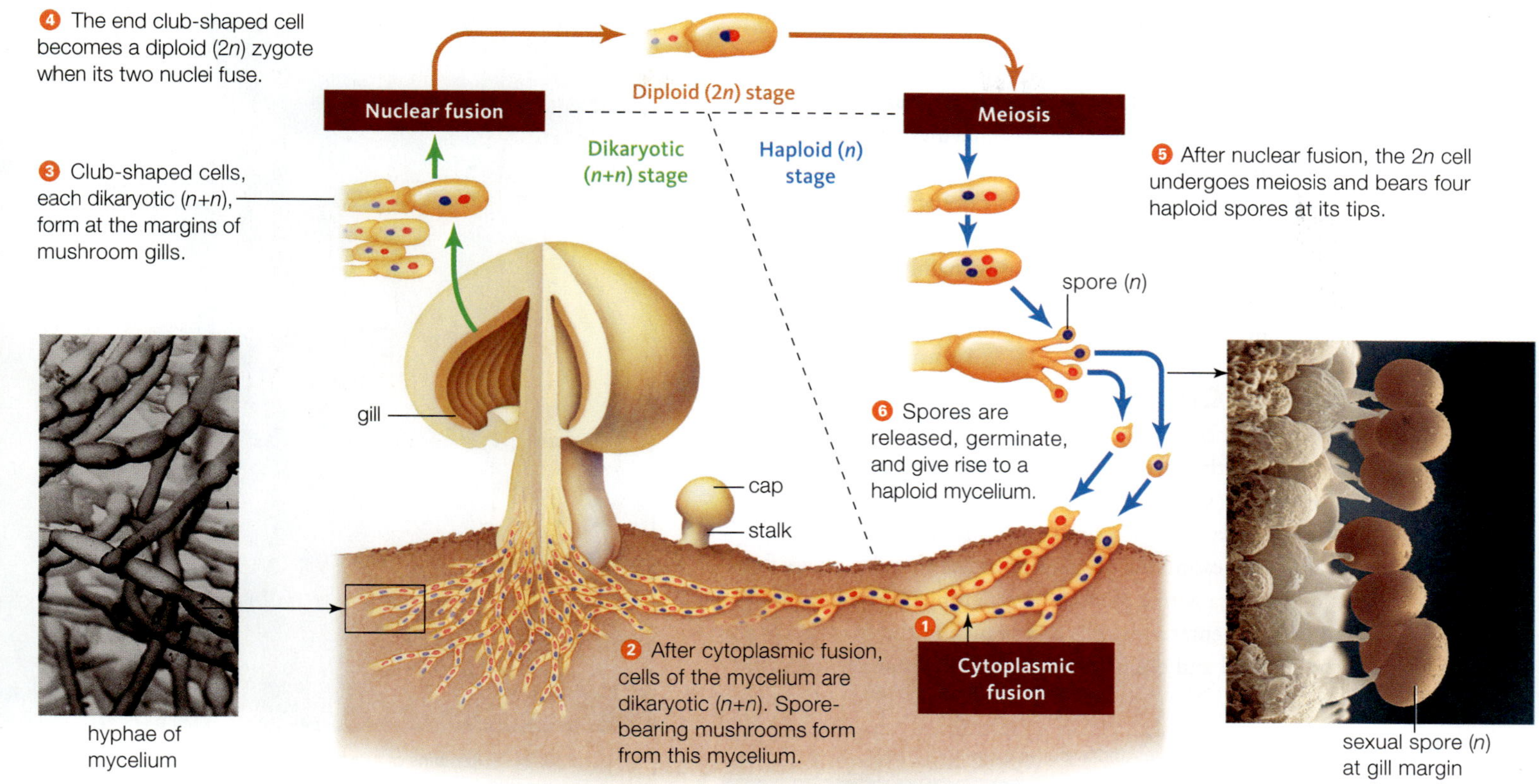

FIGURE 22.6 {Animated} Life cycle of a club fungus, the button mushroom *Agaricus bisporus*. The red and blue dots represent genetically distinct nuclei.

FIGURE IT OUT: What process produces a dikaryotic cell from two haploid cells?

Answer: Cytoplasmic fusion

When club fungi reproduce sexually, they form spores by meiosis in club-shaped cells called basidia. These cells form on a fruiting body (a basidiocarp) composed of dikaryotic hyphae. A button mushroom is a basidiocarp (**FIGURE 22.6**). Haploid hyphae of a such fungi grow in the soil and feed by decomposing organic material. Unlike the haploid hyphae of molds, those of the fungi that form fruiting bodies do not produce spores asexually.

When hyphae of two mushroom-forming club fungi meet, they fuse and form a dikaryotic mycelium 1. This mycelium continues to grow through the soil unseen and can reach great size. Mycelia of some mushroom-forming fungi extend for miles. Embryonic mushrooms form on the mycelium. When it rains, hyphae soak up water, and these tiny mushrooms expand and break through the soil surface 2. Producing fruiting bodies after a rain ensures that spores will disperse when conditions favor their survival.

The underside of a mushroom's cap has thin tissue sheets (gills) fringed with club-shaped, dikaryotic cells 3. Fusion of the nuclei in a dikaryotic cell forms a diploid zygote 4. The zygote undergoes meiosis, forming four haploid spores 5. After dispersal, these spores germinate and the life cycle begins again 6.

Fruiting bodies of club fungi come in a dazzling variety of shapes, colors, and sizes. Some look like a coral or a rubbery seaweed. Roughly spherical puffballs can be more than a meter around. Stinkhorns, such as the one in the chapter opening photo, have a phallic shape and smell like rotting meat or feces. Flies and beetles attracted by this odor alight on the tip of the fruiting body, where they unknowingly pick up spores.

TAKE-HOME MESSAGE 22.2

Zygote fungi produce spores asexually until food runs out. Then they make a diploid zygospore that undergoes meiosis.

Sac fungi have diverse body forms and reproductive mechanisms. Some form fruiting bodies called ascocarps; others are yeast and molds that usually reproduce asexually.

Most club fungi produce fruiting bodies called basidiocarps and do not reproduce asexually.

CREDITS: (6) left, © Micrograph, Garry T. Cole, University of Texas, Austin/BPS; right, © Eye of Science/Science Source; art, © Cengage Learning.

22.3 WHAT ECOLOGICAL ROLES DO FUNGI PLAY?

NATURE'S RECYCLERS

Fungi provide an important ecological service by breaking down complex compounds in organic wastes and remains. When a fungus secretes digestive enzymes onto these materials, some soluble nutrients escape into nearby soil or water. Plants and other producers can then take up these substances to meet their own needs. Bacteria also serve as decomposers, but they tend to grow mainly on surfaces. By contrast, fungal hyphae can extend deep into a dead log or other bulky food source.

BENEFICIAL PARTNERS

Many fungi take part in a mutualism, an interspecific interaction that benefits both participants. For example, a **mycorrhiza** (plural, mycorrhizae) is a partnership between a soil fungus and the root of a vascular plant (**FIGURE 22.7**). An estimated 80 percent of the vascular plants form a mycorrhiza with a glomeromycete fungus. Hyphae of these fungi enter root cells and branch in the space between the cell wall and the plasma membrane. Some sac fungi and club fungi form mycorrhizae in which hyphae surround a root and grow between its cells. Most forest mushrooms are fruiting bodies of mycorrhizal sac fungi or club fungi.

Hyphae of all mycorrhizal fungi functionally increase the absorptive surface area of their plant partner. Hyphae are thinner than even the smallest roots and can grow between soil particles. The fungus shares water and nutrients taken up by its hyphae with root cells. In return, the plant supplies the fungus with sugars.

A **lichen** is a composite organism that consists of a sac fungus and either cyanobacteria or green algae (**FIGURE 22.8A**). The fungus makes up the bulk of a lichen's mass, with its hyphae surrounding the photosynthetic cells. These cells provide the fungus with sugars and, if the cells are cyanobacteria, with fixed nitrogen. Lichens play an important ecological role by colonizing places too hostile for most organisms, such as exposed rocks (**FIGURE 22.8B**). They break down the rock and produce soil by releasing acids and by holding water that freezes and thaws. When soil forms, plants can move in and take root. Long ago, lichens may have preceded plants onto land.

FIGURE 22.9 Remains of a "zombie fly" infected by the zygote fungus *Entomophthora muscae*. An infected fly ceases normal activities, climbs to the top of stem, and clings there with outstretched wings until it dies. After it does, spore-bearing hyphae erupt through its abdominal wall and shed spores that can cause new infections.

FIGURE 22.7 Mycorrhiza; a fungus and a root of a hemlock tree.

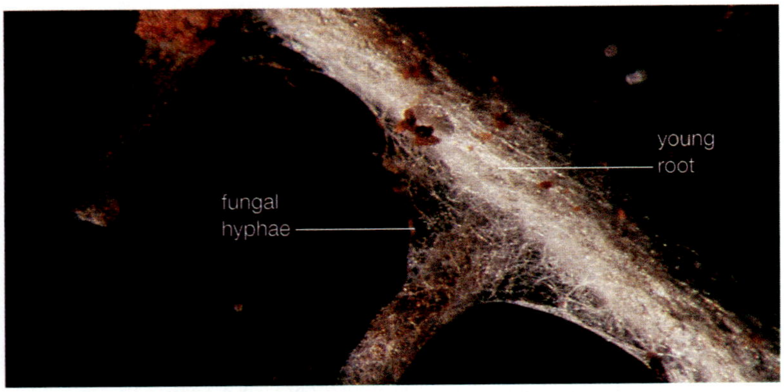

FIGURE 22.8 {Animated} Lichens. A fungus partnered with green algal cells or cyanobacteria.

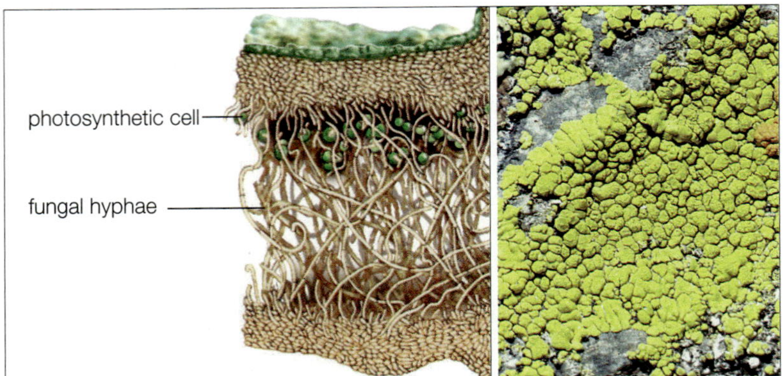

A Structure of a leafy lichen. **B** Lichen growing on a rock.

Fungal partners also enhance the nutrition of some animals. Chytrid fungi that live in the stomachs of grazing hoofed mammals such as cattle, deer, and moose aid their hosts by breaking down otherwise indigestible cellulose. Similarly, fungal partners of some ants and termites serve as an external digestive system. Leaf-cutter ants gather bits of leaf to sustain the fungus that lives in their colony. The ants cannot digest leaves, but they do eat the fungus.

PARASITES AND PATHOGENS

Many sac fungi and club fungi are plant parasites. Powdery mildews (sac fungi) and rusts and smuts (club fungi) are parasites that grow only in living plants. Their hyphae grow into cells of stems and leaves, where they suck up photosynthetically produced sugars. The resulting loss of nutrients stunts the plant, prevents it from producing seeds, and may eventually kill it. However, the plant usually does not die before the fungus has produced spores on the surface of its infected parts.

Other pathogenic fungi produce toxins that kill plant tissues, then feed on the resulting remains. The club fungus *Armallaria* causes root rot by infecting trees and woody shrubs in forests worldwide. Once an infected tree dies, the fungus decomposes the stumps and logs left behind. In one Oregon forest, the mycelium of a single honey mushroom (*A. ostoyae*) extends across nearly 4 square miles (10 km^2). It has been growing for an estimated 2,400 years.

Many more fungi infect plants than animals. Among animals, those that do not maintain a high body temperature are most vulnerable to fungal infections. Hundreds of fungal species infect insects, and some turn their hosts into seeming zombies, the better to disperse their spores (**FIGURE 22.9**).

Most human fungal infections involve body surfaces. Infected areas become raised, red, and itchy. For example, several species of fungus infect skin between the toes and on the sole of the foot, causing "athlete's foot." Fungal vaginitis (a vaginal yeast infection) occurs when a yeast (*Candida*) that normally lives in the vagina in low numbers undergoes a population explosion. Fungi also cause skin infections misleadingly known as "ringworm." No worm is involved. Rather, a ring-shaped lesion forms as fungal hyphae grow outward from the initial infection site. Life-threatening fungal infections are rare and usually occur only in people whose immune response is weak as a result of other factors.

---

lichen Composite organism consisting of a fungus and green algae or cyanobacteria.

mycorrhiza Mutually beneficial partnership between a fungus and a plant root.

PEOPLE MATTER

National Geographic Grantee
DR. DEEANN REEDER

DeeAnn Reeder is an expert on bats, with a particular interest in species that hibernate. As a National Geographic Society Grantee, she has explored bat diversity in the new country of South Sudan. She is also part of a team investigating white nose syndrome, a bat disease first described in New York in 2006. By early 2013, it had been discovered in 21 states and four Canadian provinces, and it had killed millions of North America's bats.

Bats with white nose syndrome fly when they should be hibernating, lose weight, and have fuzzy white filaments of the fungus *Geomyces destructans* on their wings, ears, and muzzle. Most fungal infections do not kill mammals, and fungi often infect animals weakened by other pathogens. Thus, Reeder and other biologists initially assumed the fungus was a symptom of a disease, rather than its cause. However, when they experimentally infected healthy bats with *G. destructans*, they found the fungus alone is sufficient to cause the disease. Reeder and others are now working to understand why the fungus has such deadly effects, which types of bat colonies are most susceptible to infection, and what can be done to halt the spread of the fungus.

TAKE-HOME MESSAGE 22.3

- The decomposition of organic wastes and remains by fungal decomposers releases nutrients that other organisms can use.
- Fungi that partner with plant roots or live in the animal gut enhance the nutrition of their hosts. Fungi also partner with cyanobacteria or green algae to form a lichen.
- Parasitic fungi that infect plants either steal sugars from living cells or kill cells and digest their remains.
- Fungi cause disease in plants more often than in animals. Animals with a high body temperature are least susceptible to fungal diseases.
- Human fungal infections usually involve body surfaces and are rarely life-threatening.

CREDIT: (in text) Stephen Alvarez/National Geographic Creative.

22.4 HOW DO WE USE FUNGI?

Many fungal fruiting bodies serve as human food. Button mushrooms, shiitake mushrooms, and oyster mushrooms decompose organic matter, so these species are easily cultivated. By contrast, edible mycorrhizal fungi such as chanterelles, porcini mushrooms, morels, and truffles need a living plant partner, so they are typically gathered from the wild. Picking wild mushrooms is best done with the aid of someone experienced in identifying species. Each year thousands of people become ill after eating poisonous mushrooms they mistook for edible ones.

Truffle-producing fungi, which are highly prized by gourmets, have evolved an interesting dispersal strategy. Truffles form underground near their host trees and, when mature, produce an odor similar to that of an amorous male wild pig. Female wild pigs detect the scent and root through the soil in search of their seemingly subterranean suitor. When they unearth the truffles, they eat them, then disperse fungal spores in their feces. Human truffle hunters usually rely on trained dogs to sniff out the fungi.

FIGURE 22.10 Fungi as food. Sac fungi help us produce breads, wine, and blue cheeses.

Fermentation by fungi helps us make a variety of products. Fermentation by one mold (*Aspergillus*) helps make soy sauce. Another mold (*Penicillium*) produces the tangy blue veins in cheeses such as Roquefort (**FIGURE 22.10**). A packet of baker's yeast holds spores of the sac fungus *Saccharomyces cerevisiae*. When these spores germinate in bread dough, they ferment sugar and produce carbon dioxide that causes the dough to rise. Other strains of *S. cerevisiae* are used to make wine and beer, and also to produce ethanol biofuel.

Geneticists and biotechologists also make use of yeasts. Like *E. coli* bacteria, yeasts grow readily in laboratories and they have the added advantage of being eukaryotes like us. Checkpoint genes that regulate the eukaryotic cell cycle (Section 11.1) were first discovered in the yeast *S. cerevisiae*. This discovery was the first step toward our current understanding of how mutations of these genes cause human cancers. Genetically engineered *S. cerevisiae* and other yeasts are now used to produce proteins that serve as vaccines or other medicines.

Some naturally occurring fungal compounds have medicinal or psychoactive properties. The initial source of the antibiotic penicillin was a soil fungus (*Penicillium chrysogenum*). Another soil fungus gave us cyclosporin, an immune suppressant used to prevent rejection of transplanted organs. Ergotamine, a compound used to relieve migraines, was first isolated from ergot (*Claviceps purpurea*), a club fungus that infects rye plants. Ergotamine is also used in synthesis of the hallucinogen LSD. Another hallucinogen, psilocybin, is the active ingredient in so-called "magic mushrooms."

In Asia, dried remains of a fungus-infected caterpillar have been prized as a medicine for thousands of years. The fungus (*Ophiocordyceps sinensis*) infects a soil-dwelling moth larva. After the insect dies, a fungal fruiting body about the size of a blade of grass emerges above ground. When gathered from the wild and dried, the fungal fruiting bodies and attached insect remains are more expensive than gold. A compound (cordycepin) first extracted from this fungus boosts production of testosterone (the male sex hormone) in mice and shows promise as an anticancer drug.

TAKE-HOME MESSAGE 22.4

We eat fungal fruiting bodies and use yeasts to produce foods and beverages. Yeasts are also used in genetic studies, in production of recombinant proteins, and to synthesize ethanol as a biofuel.

Some fungi make compounds that are prized for their medicinal or psychoactive properties.

22.5 Application: SPREAD OF FUNGAL PATHOGENS

Conservation

FIGURE 22.11 Frogs killed by the chytrid fungus Bd at a high-elevation lake in California's Sierra Nevada Mountains.

THE DISPERSAL OF FUNGAL PATHOGENS BY GLOBAL TRADE AND TRAVEL CAN HAVE DEVASTATING EFFECTS ON ECOSYSTEMS.

In the early twentieth century, a plant-infecting sac fungus native to China was introduced to North America in imported plants. The fungus caused a chestnut blight that eliminated all mature American chestnut trees. Some American chestnuts persist as root systems, but the blight always kills them before they mature and reproduce.

Today, human-facilitated spread of a fungal pathogen is among the foremost causes of an amphibian extinction crisis. The fungus, *Batrachochytrium dendrobatidis*, is a chytrid known as Bd. It is native to Africa, where it infects African clawed frogs without sickening them.

International trade in African clawed frogs began in 1934, when scientists discovered the frogs could be used in pregnancy tests. If a female frog is injected with blood from a pregnant woman, hormones in her blood cause the frog to lay eggs. Until the 1950s, this frog-based test was the main method of determining pregnancy, so millions of African clawed frogs were exported for this purpose. The Bd chytrid traveled with them. It became established in new regions when people released infected frogs or dumped water with Bd spores into the environment. Bd now occurs on all continents except Antarctica, and trade in amphibians continues to introduce it into previously uninfected habitats.

The effect of Bd varies among amphibian species. Many frogs outside of Africa have little or no resistance to Bd, so their death rate from infection is extremely high (FIGURE 22.11). The fungus grows on an amphibian's skin, causing the skin to thicken. Thickened skin prevents the frog from absorbing water properly, so it eventually dies of dehydration.

CREDIT: (11) Joel Sartore/National Geographic Creative.

Summary

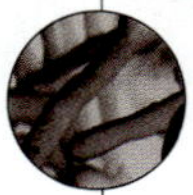

SECTION 22.1 **Fungi** are heterotrophs that secrete digestive enzymes onto organic matter and absorb released nutrients. Their cell walls include chitin and they disperse by releasing spores. In a multicelled species, spores germinate and give rise to filaments called **hyphae**. The filaments typically grow as an extensive mesh called a **mycelium**. Depending on the group, the cells of a hypha may be haploid (n) or **dikaryotic** ($n+n$). Water and nutrients move freely between cells of a hypha.

The oldest fungal lineage, the **chytrids**, are a mostly aquatic group and the only fungi with flagellated spores. **Zygote fungi** include many familiar **molds** that grow on fruits, breads, and other foods. **Glomeromycetes** live in soil and extend their hyphae into plant roots. **Sac fungi** are the most diverse fungal group. They include single-celled **yeasts**, molds, and species that produce multicelled fruiting bodies. Most familiar mushrooms are fruiting bodies of **club fungi**. Multicelled sac fungi and club fungi can form large fruiting bodies because their hyphae, unlike those of other fungi, have porous cross walls between the cells. These walls reinforce the structure of a hypha.

SECTION 22.2 Mechanisms of spore formation differ among fungal groups. Zygote fungi such as bread molds usually reproduce asexually by producing spores atop specialized hyphae until food runs out. Then hyphae fuse to produce a diploid zygospore that undergoes meiosis and releases haploid spores.

Sac fungal yeasts reproduce asexually by budding. During sexual reproduction, a sac fungus produces spores in a sac-shaped structure (an ascus). In some sac fungi, including the cup fungi, the asci form on a fruiting body called an ascocarp.

Club fungi produce spores in club-shaped structures (basidia) on fruiting bodies called basidiocarps. Typically, a dikaryotic mycelium grows by mitosis. When conditions favor reproduction, a basidiocarp, also made of dikaryotic hyphae, develops. A mushroom is an example. Fusion of nuclei in dikaryotic club-shaped cells at the edges of gills produces diploid cells. Meiosis in these cells produces haploid spores.

SECTION 22.3 Fungi have many important roles in ecosystems. Most are decomposers. By breaking down organic wastes and remains, they make nutrients that were tied up in these materials available to producers.

A **lichen** is a composite organism composed of a fungus and photosynthetic cells of a green alga or cyanobacterium. The fungus, which makes up the bulk of the lichen body, obtains a supply of nutrients from its photosynthetic partner. Lichens are important pioneers in new habitats because they facilitate the breakdown of rock to form soil.

A **mycorrhiza** is an interaction between a fungus and a plant root. Fungal hyphae penetrate roots and supplement their absorptive surface area. The fungus shares absorbed mineral ions with the plant and obtains some photosynthetic sugars in return.

Fungi that parasitize plants either insert their hyphae into plant cells to steal sugars, or kill plant cells and suck up the released nutrients. Fungi parasitize animals less frequently than they do plants. Animals that have a low body temperature are most often infected. Some fungi infect insects and alter their behavior.

Fungi that infect humans tend to affect body surfaces, as when they cause athlete's foot or a vaginal yeast infection. Typically, fungi cause life-threatening infections only in people whose immune respone is impaired.

SECTION 22.4 Many fungal fruiting bodies are edible, although some produce dangerous toxins. Fungi that carry out fermentation reactions help us produce food products, alcoholic beverages, and ethanol for use as a biofuel. Study of yeasts can provide insights into eukaryotic genetics. Recombinant yeasts produce vaccines and other desired proteins. Some compounds extracted from fungi are useful as medicines or psychoactive drugs.

SECTION 22.5 Human activities have introduced some fungal pathogens to new environments where they have negative effects on the health of other organisms. One sac fungus introduced to North America wiped out American chestnuts. Today, a chytrid fungus native to Africa threatens amphibian species worldwide.

Self-Quiz

Answers in Appendix VII

1. All fungi _______ .
   a. are multicelled
   b. form flagellated spores
   c. are heterotrophs
   d. all of the above

2. Most fungi obtain nutrients from _______ .
   a. wastes and remains
   b. living plants
   c. living animals
   d. photosynthesis

3. A fungus that usually lives as a mass of hyphae and reproduces asexually is called a _______ .
   a. lichen
   b. mold
   c. yeast
   d. mushroom

4. A _______ steals sugars from a living plant cell.
   a. rust or smut
   b. mushroom
   c. lichen
   d. mold

5. In many _______ , an extensive dikaryotic mycelium is the most conspicuous phase of the life cycle.
   a. chytrids
   b. zygote fungi
   c. sac fungi
   d. club fungi

6. A _______ produces spores by meiosis in an ascus.
   a. chytrid
   b. zygote fungus
   c. sac fungus
   d. club fungus

Data Analysis Activities

Fighting a Forest Fungus The club fungus *Armillaria ostoyae* infects living trees and acts as a parasite, withdrawing nutrients from them. After the tree dies, the fungus continues to feed on its remains. Fungal hyphae grow out from the roots of infected trees as well as the roots of dead stumps. If these hyphae contact roots of a healthy tree, they can invade and cause a new infection.

Canadian forest pathologists hypothesized that removing fungus-infected stumps after logging could help prevent tree deaths. To test this hypothesis, they carried out an experiment. In one region of a forest they removed stumps after logging. In another, they left stumps behind as a control. For more than 20 years, they recorded tree deaths and whether *A. ostoyae* caused them. **FIGURE 22.12** shows the results.

1. Which tree species was most often killed by *A. ostoyae* in control forests? Which was least affected by the fungus?
2. For the most affected species, what percentage of deaths did *A. ostoyae* cause in control and in experimental regions?
3. Looking at the overall results, do the data support the hypothesis? Does stump removal reduce tree mortality from *A. ostoyae*?

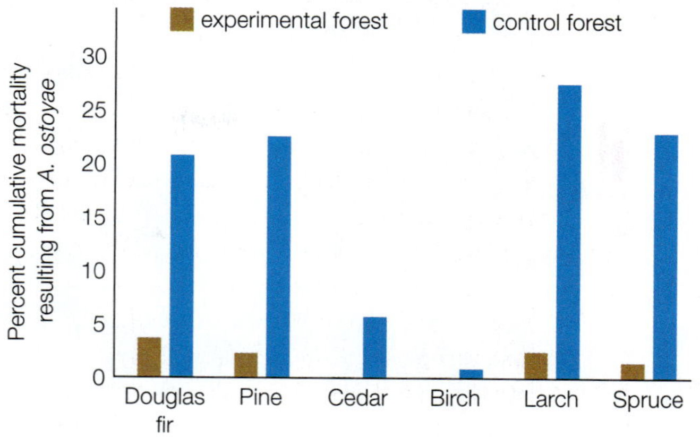

FIGURE 22.12 Results of a long-term study of how logging practices affect tree deaths caused by the fungus *A. ostoyae*. In the experimental forest, whole trees—including stumps—were removed (brown bars). The control half of the forest was logged conventionally, with the stumps left behind (blue bars).

7. A mushroom is _______ .
   a. the digestive organ of a club fungus
   b. the only part of the fungal body made of hyphae
   c. a reproductive structure that releases sexual spores
   d. the only diploid phase in the club fungus life cycle

8. Spores released from a mushroom's gills are _______ .
   a. dikaryotic
   b. diploid
   c. flagellated
   d. produced by meiosis

9. _______ by yeast helps us produce bread, soy sauce, and ethanol as a biofuel.
   a. Photosynthesis
   b. Fermentation
   c. Decomposition
   d. Budding

10. A _______ helps to break down rocks and form soil.
    a. chytrid
    b. glomeromycete
    c. mycorrhiza
    d. lichen

11. _______ are mycorrhizal fungi with hyphae that grow into a root cell and branch inside it.
    a. Glomeromycetes
    b. Chytrids
    c. Zygote fungi
    d. Club fungi

12. A truffle is an example of a _______ .
    a. spore
    b. lichen
    c. mycorrhizal fungus
    d. plant pathogen

13. Frogs worldwide are now threatened by a pathogenic _______ native to Africa.
    a. glomeromycete
    b. chytrid
    c. zygote fungus
    d. club fungus

14. Human fungal infections _______ .
    a. usually involve the skin
    b. typically produce hallucinations
    c. are often life-threatening
    d. all of the above

15. Match the terms appropriately.

___ hypha	a. produces flagellated spores
___ chitin	b. component of fungal cell walls
___ chytrid	c. partnership between a fungus and photosynthetic cells
___ zygote fungus	d. filament of a mycelium
___ club fungus	e. fungus–root partnership
___ lichen	f. bread mold is an example
___ mycorrhiza	g. many form mushrooms

Critical Thinking

1. Developing antifungal drugs is more difficult than developing antibacterial drugs because compounds that harm fungi more frequently cause side effects in humans than compounds that harm bacteria. Explain why this is the case, given the evolutionary relationships among bacteria, fungi, and humans.

2. Bakers who want to be sure their yeast is alive "proof" it. They test the yeast's viability by putting a bit of it in warm water with some sugar. A few minutes later, they look for a sign the yeast has been active. What do they look for?

3. Molds usually reproduce sexually when food is running low. Why might sexual reproduction be more advantageous at this time than when food remains plentiful?

CREDIT: (12) After graph from www.pfc.forestry.ca.

Sea stars, sea anemones, and pink encrusting sponges in a Pacific Northwest tide pool.

23

Links to Earlier Concepts

This chapter draws on your knowledge of animal tissues and organs (Section 1.1) and of homeotic genes (10.2), fossils (16.3, 16.4), analogous structures (16.7), speciation (17.8), and biomarkers (18.3). You will see another use of chitin (3.2) and more effects of osmosis (5.6). You will learn how animals interact with dinoflagellates (20.4), protists that cause malaria (20.5), and flowering plants (21.7).

ANIMALS I: JOR INVERTEBRATE GROUPS

KEY CONCEPTS

INTRODUCING THE ANIMALS

Animals digest food in their body, and most move about. They evolved from a colonial protist. The overwhelming majority of modern animals are invertebrates.

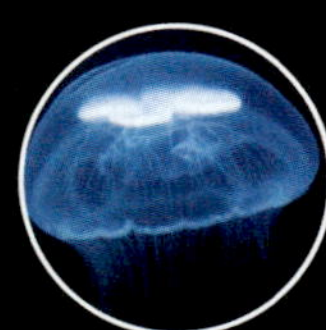

SPONGES AND CNIDARIANS

Sponges filter food from water and have an asymmetrical body. Cnidarians are predators with a radially symmetrical body that has two tissue layers.

BILATERAL INVERTEBRATES

Most animals are bilaterally symmetrical and have organ systems. In flatworms, organs are hemmed in by tissue. In other bilateral animals, the organs reside in a body cavity.

THE MOST SUCCESSFUL ANIMALS

Arthropods are the most diverse animals. Crustaceans abound in seas, and insects on land. Insects play essential roles in ecosystems and have important economic and health effects.

RELATIVES OF CHORDATES

Echinoderms are on the same branch of the animal family tree as animals with backbones. Adults have a spiny skin and a radially symmetrical body.

Photograph by George Grall, National Geographic Creative.

23.1 HOW HAVE ANIMAL BODY PLANS EVOLVED?

Animals are multicelled heterotrophs whose unwalled body cells are typically diploid. Most animals ingest food (take it in) and digest it inside their body. Nearly all are motile (able to move from place to place) during all or part of their life.

This chapter describes major groups of **invertebrates**, animals that do not have a backbone. The next chapter focuses on vertebrates (animals with a backbone) and their closest invertebrate relatives. Keep in mind that although invertebrates have a simpler structure than vertebrates, they are not evolutionarily inferior. Some invertebrate lineages have been around for more than 500 million years, and about 95 percent of all animals are invertebrates.

FIGURE 23.1 shows an evolutionary tree for the major animal groups covered in this book, and we will use it as a guide to discuss evolutionary trends. All animals are multicellular ❶ and constitute the clade Metazoa. The earliest animals were probably aggregations of cells, and this level of organization persists in sponges. However, most other modern animals have cells organized as tissues ❷.

Tissue organization begins in animal embryos. Embryos of jellies and other cnidarians have two tissue layers: an outer **ectoderm** and an inner **endoderm**. In other modern animals, embryonic cells typically rearrange themselves to form a middle tissue layer called **mesoderm** (**FIGURE 23.2**). Evolution of a three-layer embryo allowed an important increase in structural complexity. Most internal organs in animals develop from embryonic mesoderm.

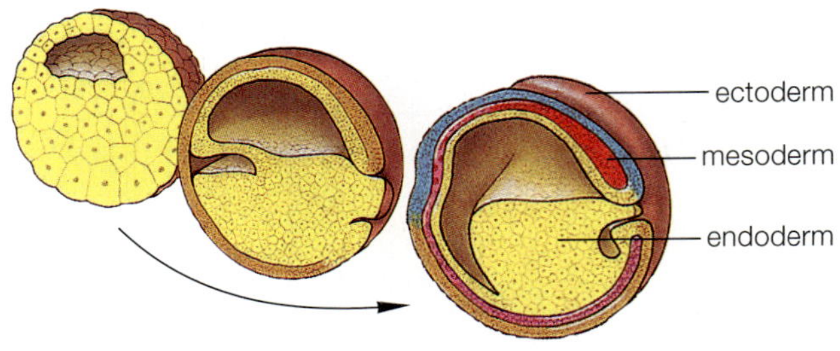

FIGURE 23.2 How a three-layer animal embryo forms. Most animals have this type of embryo.

Animals with the simplest structural organization are asymmetrical; you cannot divide their body into halves that are mirror images. Jellies, sea anemones, and other cnidarians have **radial symmetry**: Their body parts are repeated around a central axis, like spokes of a wheel ❸. Radial animals usually attach to an underwater surface or drift along. A radial body plan allows them to capture food that can arrive from any direction. Animals with a three-layer body plan typically have **bilateral symmetry**: The body's left and right halves are mirror images ❹. Such lineages typically undergo **cephalization**, an evolutionary process whereby many nerve cells and sensory structures become concentrated at the front of the body. These structures help the animal find food or avoid threats as it moves head-first through its environment.

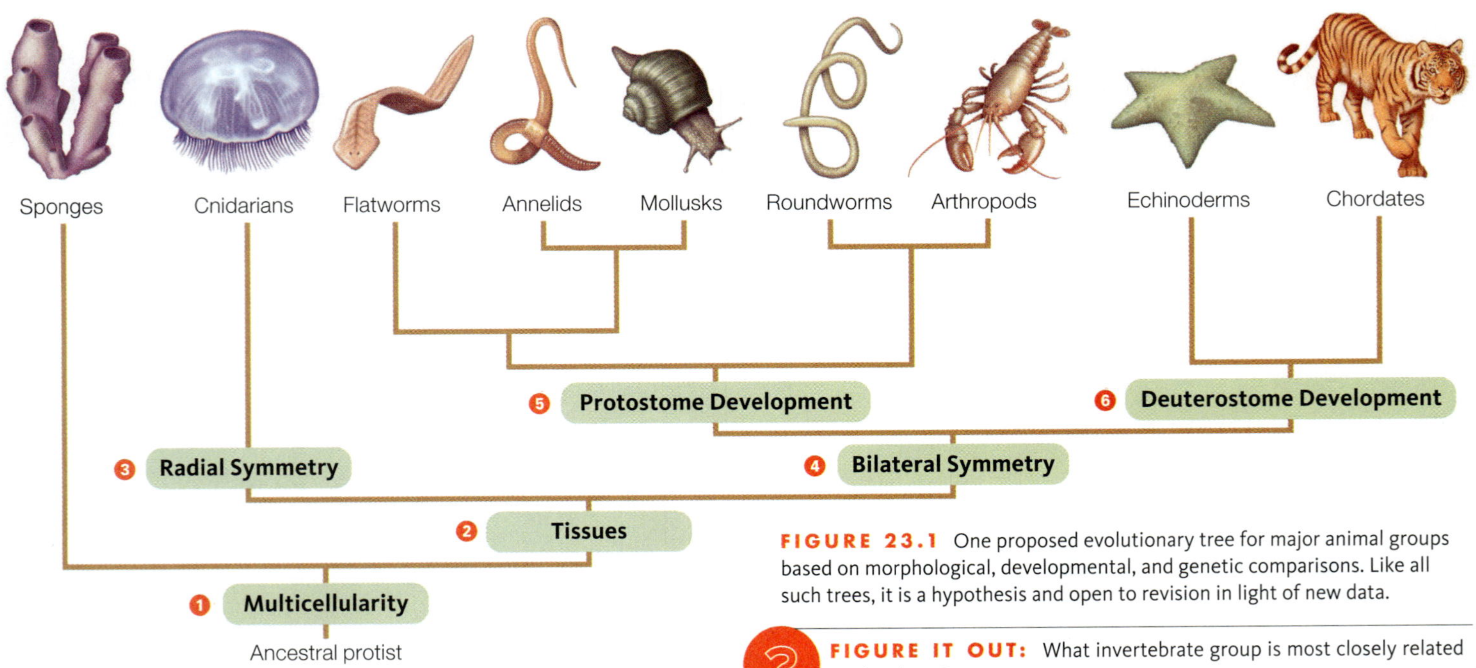

FIGURE 23.1 One proposed evolutionary tree for major animal groups based on morphological, developmental, and genetic comparisons. Like all such trees, it is a hypothesis and open to revision in light of new data.

FIGURE IT OUT: What invertebrate group is most closely related to chordates?
Answer: Echinoderms

Developmental differences define the two clades of bilaterally symmetrical animals. Among **protostomes**, the first opening that appears on the embryo becomes the mouth ❺. *Proto–* means first and *stoma* means opening. In **deuterostomes**, the mouth develops from the second embryonic opening ❻.

Some animals digest food in a saclike cavity with a single opening. Most animals have a tubular gut, with a mouth at one end and an anus at the other. Parts of the tube are typically specialized for taking in food, digesting food, absorbing nutrients, or compacting the wastes. A tubular gut can carry out all of these tasks simultaneously, whereas a saclike cavity cannot.

FIGURE 23.3 {Animated} Animal body cavities. Most animals, and all vertebrates, have a coelomate body plan, as in C.

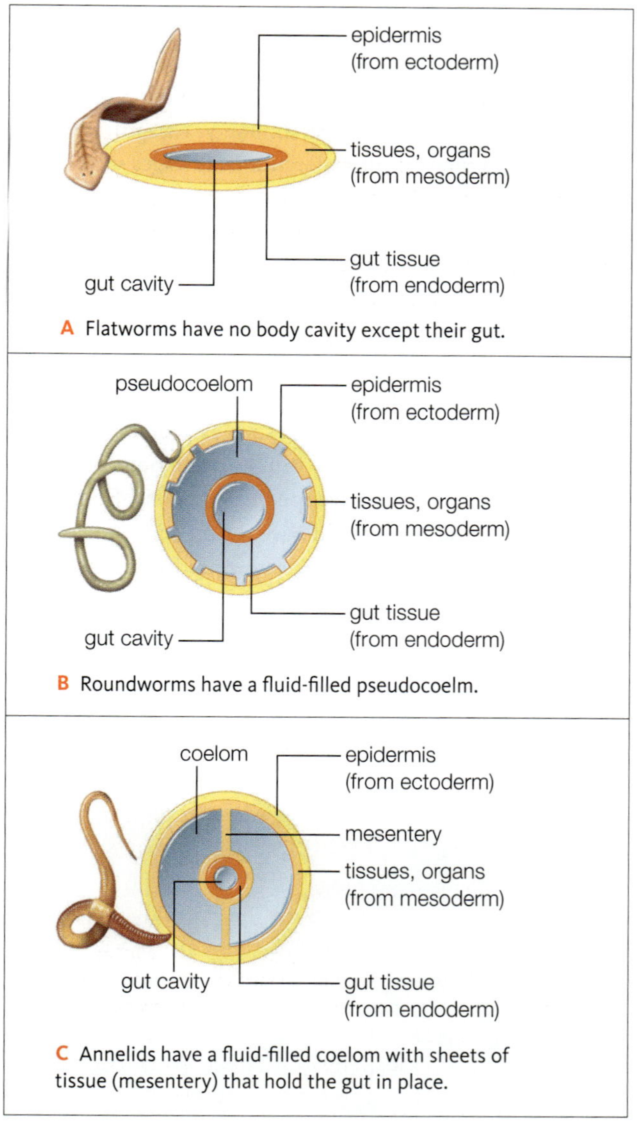

A Flatworms have no body cavity except their gut.

B Roundworms have a fluid-filled pseudocoelm.

C Annelids have a fluid-filled coelom with sheets of tissue (mesentery) that hold the gut in place.

A mass of tissues and organs surrounds the flatworm gut (**FIGURE 23.3A**). However, most bilateral animals have a fluid-filled body cavity around their gut. In a few animals such as roundworms this body cavity is only partially lined, in which case it is called a **pseudocoelom** (**FIGURE 23.3B**). More typically, bilateral animals have a **coelom**, a body cavity lined with a tissue derived from mesoderm (**FIGURE 23.3C**). In such animals, sheets of tissue called mesentery suspend the gut in the center of a fluid-filled cavity. Coelomic fluid cushions the gut and keeps it from being distorted by body movements. The fluid also helps distribute material through the body, and in some animals it plays a role in locomotion.

Most bilaterally symmetrical animals have some degree of **segmentation**, a division of a body into similar units repeated one after the other along the main axis. Evolution of segmentation allowed other innovations in body form to evolve. When many segments carry out the same function, some segments can become modified over evolutionary time, while others retain their original function.

animal Multicelled heterotroph with unwalled cells. Most ingest food and are motile during at least part of the life cycle.
bilateral symmetry Having paired structures so the right and left halves are mirror images.
cephalization Evolutionary trend toward having a concentration of nerve and sensory cells at the head end.
coelom Body cavity lined with tissue derived from mesoderm.
deuterostomes Lineage of bilateral animals in which the second opening on the embryo surface develops into a mouth.
ectoderm Outermost tissue layer of an animal embryo.
endoderm Innermost tissue layer of an animal embryo.
invertebrate Animal that does not have a backbone.
mesoderm Middle tissue layer of a three-layered animal embryo.
protostomes Lineage of bilateral animals in which the first opening on the embryo surface develops into a mouth.
pseudocoelom Unlined body cavity around the gut.
radial symmetry Having parts arranged around a central axis, like the spokes of a wheel.
segmentation Having a body composed of similar units that repeat along its length.

TAKE-HOME MESSAGE 23.1

- Early animals had no body symmetry or tissues.
- The next evolutionary step was a radially symmetrical body that develops from an embryo with two tissue layers.
- Most modern animals have a bilaterally symmetrical body that develops from an embryo with three tissue layers.
- Most bilaterally symmetrical animals have a coelom.

23.2 HOW DID ANIMALS ARISE AND DIVERSIFY?

COLONIAL ORIGINS

According to the **colonial theory of animal origins**, the first animals evolved from a colonial protist. At first, all cells in the colony were similar. Each could reproduce and carry out all other essential tasks. Later, mutations produced cells that specialized in some tasks and did not carry out others. Perhaps these cells captured food more efficiently but did not make gametes, whereas others made gametes but did not catch food. The division of labor among interdependent cells made them more efficient, allowing the colonies with mutations to obtain more food and produce more offspring. Over time, additional types of specialized cell types evolved.

What was the protist ancestral to animals like? Choanoflagellates, the modern protists most closely related to animals, provide some clues. As Section 20.9 explains, choanoflagellates are flagellated cells that live either as single cells or as a colony of genetically identical cells. In their structure, choanoflagellate cells closely resemble some cells in the bodies of modern sponges.

EARLY EVOLUTION

FIGURE 23.4 The only named placozoan, *Trichoplax adhaerans*. It is about 2 millimeters wide and two micrometers thick. This one is colored red by the red algae it fed on.

Early animals may have been similar to **placozoans**, a little-known group of tiny marine organisms that have the simplest body and smallest genome of all modern animals. A placozoan has an asymmetrical body with four types of cells (**FIGURE 23.4**). Ciliated cells on its surface allow it to glide along the seafloor, where it ingests bacteria and algae. Genetic comparisons among living species suggest that placozoans may be the oldest surviving animal lineage.

Sponges are another ancient group. In Oman, sedimentary rocks laid down 635 million years ago contain traces of complex steroids that today are made only by marine sponges. Similarly aged rocks from Australia contain what appear to be fossils of sponge bodies.

A collection of 570-million-year-old fossils from Australia provide evidence of an early animal diversification. The fossil species, collectively known as Ediacarans, include a variety of soft-bodied organisms that may have been early marine invertebrates. Many Ediacarans are unlike any modern animals, whereas others appear to be related to some modern invertebrates (**FIGURE 23.5**). In general, the connections between Ediacarans and modern animals remain unclear. One biologist has even suggested some Ediacaran fossils are the remains of land-dwelling microbes, lichens, or fungi.

FIGURE 23.5 An Ediacaran fossil. *Spriggina* was about 3 centimeters (1 inch) long. By one hypothesis, it was a soft-bodied ancestor of arthropods, a group that includes modern crabs and insects.

AN EXPLOSION OF DIVERSITY

Animals underwent a dramatic adaptive radiation during the Cambrian period (542–488 million years ago). By the end of this interval, all major animal lineages were present in the seas. What caused this Cambrian explosion in diversity? Rising oxygen levels and changes in global climate may have played a role. Also, supercontinents were breaking up. As landmasses moved, they could have isolated different populations, creating new opportunities for allopatric speciation (Section 17.9). Biological interactions also encouraged speciation. Once the first predators arose, mutations that produced defenses such as protective hard parts would have been favored. Mutations in homeotic genes (Section 10.2) may have sped things along. Some mutations in these genes could have resulted in changes in body form that proved adaptive as new predators arose or the habitat changed.

colonial theory of animal origins Hypothesis that the first animals evolved from a colonial protist.

placozoans Group of tiny marine animals having a simple asymmetrical body and a small genome; considered an ancient lineage.

TAKE-HOME MESSAGE 23.2

The ancestor of all animals was probably a colonial protist. It may have resembled choanoflagellates, the modern protist group most closely related to animals.

Animals probably originated more than 600 million years ago. Placozoans and sponges are ancient animal lineages that survived to the present day.

Environmental and biological factors encouraged a great adaptive radiation during the Cambrian period.

Sponges (phylum Porifera) are aquatic animals with a porous body that does not have tissues. The vast majority live in tropical seas, and nearly all are sessile, meaning they live fixed in place. Sponge bodies vary in size and shape. Some fit on a fingertip, whereas others stand meters tall. Asymmetrical vaselike or columnar forms are most common, but some sponges grow as a thin crust.

Flat, nonflagellated cells cover a sponge's outer surface, flagellated collar cells line the inner surface, and a jellylike extracellular matrix lies in between (**FIGURE 23.6**). In many sponges, cells in this matrix secrete fibrous proteins, glassy silica spikes, or both. These materials structurally support the body, and help fend off predators. Some protein-rich sponges are harvested from the sea, dried, cleaned, and bleached. Their rubbery protein remains (*left*) are then sold as bathing and cleaning supplies.

The typical sponge is a **suspension feeder** that eats material it filters from the surrounding water. Action of flagella on collar cells draws food-laden water through pores in a sponge's body wall. The collar cells filter food from the water and engulf it by phagocytosis. Digestion is intracellular. Amoeba-like cells in the matrix receive the breakdown products of digestion from collar cells and distribute them to other cells in the sponge body.

Most sponges are **hermaphrodites**, meaning each individual can produce both eggs and sperm. Typically, the sponge releases its sperm into the water (**FIGURE 23.7**) but holds onto its eggs. Fertilization produces a zygote that develops into a ciliated larva. A **larva** (plural, larvae) is a young, sexually immature stage in an animal life cycle. Sponge larvae swim briefly, then settle and become adults. Many sponges also reproduce asexually when small buds or fragments break away and grow into new sponges.

Some freshwater species can survive dry conditions by producing gemmules, tiny clumps of resting cells encased in a hardened coat. Gemmules are often dispersed by the wind. Those that land in a hospitable habitat become active and grow into new sponges.

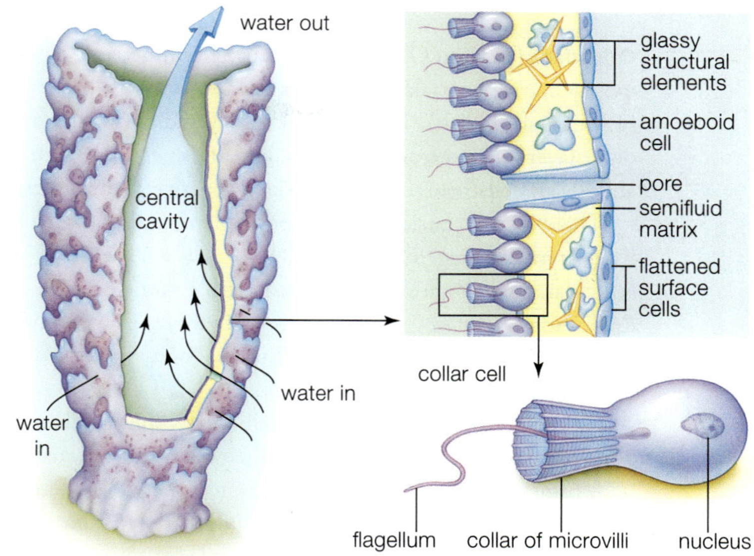

FIGURE 23.6 {**Animated**} Body plan of a simple sponge.

FIGURE 23.7 Barrel sponge releasing sperm.

TAKE-HOME MESSAGE 23.3

Sponges are typically suspension feeders with a porous, asymmetrical body. Fibers and glassy spikes in the body wall support the body and help deter predators.

Sponges are hermaphrodites. Their ciliated larvae swim about briefly before settling and developing into adults.

hermaphrodite Animal that makes both eggs and sperm.
larva Sexually immature stage in some animal life cycles.
sponge Aquatic invertebrate that has no tissues or organs and filters food from the water.
suspension feeder Animal that filters food from water around it.

23.4 WHAT ARE CNIDARIANS?

Cnidarians (phylum Cnidaria) are radially symmetrical, mostly marine animals, such as sea anemones and jellies (also called jellyfish). They have a two-layered body, with an outer layer derived from ectoderm, and an inner layer from endoderm. Jellylike secreted material lies between the layers.

There are two basic cnidarian body plans, and both have a tentacle-ringed mouth (**FIGURE 23.8**). Medusae (singular, medusa) are dome-shaped, with a mouth on the dome's lower surface. Most swim or drift about. Polyps have an upward-facing mouth atop a cylindrical body that is typically attached to a surface.

Cnidarian life cycles vary. Most jellies have a life cycle in which polyps that reproduce asexually by budding alternate with gamete-producing medusae. Sea anemones, corals, and the hydras (a freshwater group) exist only as polyps that can both bud and produce gametes. In all cnidarians, the zygote produced by sexual reproduction develops into a bilaterally symmetrical ciliated larvae called a planula.

The name Cnidaria is from *cnidos*, the Greek word for the stinging nettle plant, and refers to the animals' mechanism of feeding. Cnidarians are predators. Their tentacles have **cnidocytes**, specialized stinging cells that help them capture prey. A cnidocyte contains a nematocyst, an organelle that functions like a jack-in-the-box (**FIGURE 23.9**). When prey brushes a nematocyst's trigger, a barbed thread pops out and delivers a dose of venom. People that brush against a jellyfish can trigger this same response and end up with a painful sting. Some box jellies that live near Australia produce a venom so powerful that their sting occasionally kills people.

FIGURE 23.8 {Animated} Cidarian body plans and representative examples.

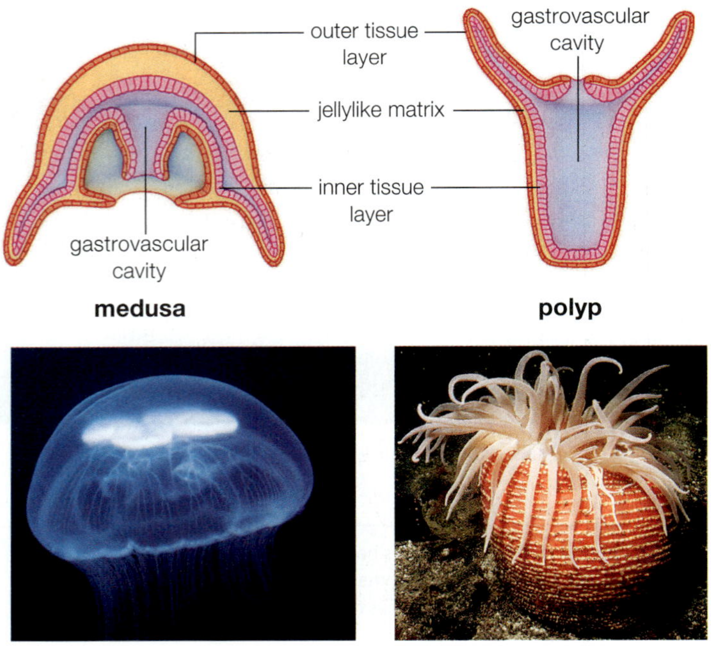

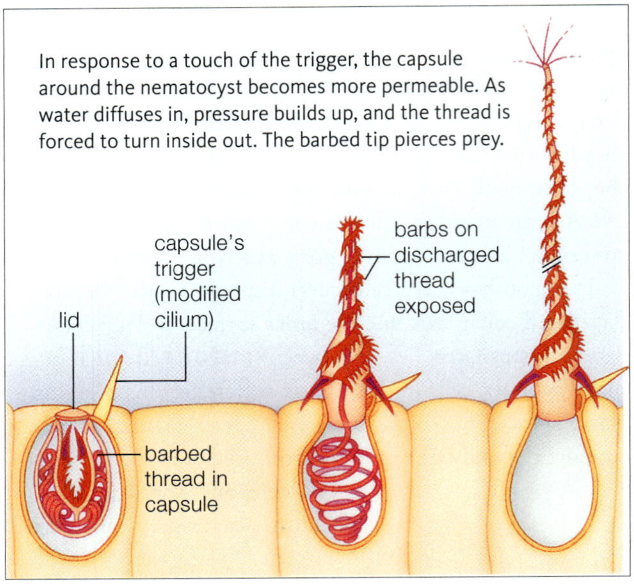

FIGURE 23.9 {Animated} Nematocyst action.

Tentacles move captured food to the mouth, which opens to a **gastrovascular cavity**. This saclike region functions in digestion and gas exchange. Unlike sponges, cnidarians digest their food extracellularly. Enzymes secreted into the gastrovascular cavity break down prey. Released nutrients are then absorbed and are distributed through the body by diffusion. Digestive waste exits through the mouth.

Cnidarians are brainless, but interconnecting nerve cells extend through their tissues as a **nerve net**. Body parts move when these nerve cells signal contractile cells to shorten. Contraction alters the shape of the animal by redistributing fluid trapped within the gastrovascular cavity. By analogy, think of what happens when you squeeze a water-filled balloon. A fluid-filled cavity that contractile cells exert force against is a **hydrostatic skeleton**. Many soft-bodied invertebrates have this type of skeleton.

cnidarian Radially symmetrical invertebrate with two tissue layers; uses tentacles with stinging cells to capture food.
cnidocyte Stinging cell unique to cnidarians.
gastrovascular cavity A saclike gut that also functions in gas exchange.
hydrostatic skeleton Of soft-bodied invertebrates, a fluid-filled chamber that contractile cells exert force on.
nerve net Decentralized mesh of nerve cells that allows movement in cnidarians.

FIGURE 23.10 A coral reef. The corals are colonies of polyps. Dinoflagellates in their tissues give them their color.

Most cnidarians are solitary, but colonial groups exist. Siphonophores such as the Portuguese man-of-war (*Physalia*) look like a single animal, but are colonies of many polyps and medusae. Coral reefs are built by colonies of polyps enclosed in a skeleton of secreted calcium carbonate. In a mutually beneficial relationship, photosynthetic dinoflagellates live inside each polyp's tissues (**FIGURE 23.10**). The protists receive shelter and carbon dioxide from the coral, and give it sugars and oxygen in return. If a reef-building coral loses its protist partners, an event called "coral bleaching," it may die.

TAKE-HOME MESSAGE 23.4

Cnidarians are radially symmetrical, mostly marine predators with two tissue layers. There are two body plans: medusa and polyp. In both, tentacles with stinging cells surround a mouth that opens onto a gastrovascular cavity. A nerve net interacts with the hydrostatic skeleton to allow movement.

Cnidarians include sea anemones, jellies, and corals. Reef-building corals secrete calcium carbonate that forms reefs.

PEOPLE MATTER

National Geographic Explorer
DR. DAVID GRUBER

Biofluorescence is the process by which organisms absorb light of one wavelength and emit light of another. The stunning photograph *below*, taken by Dr. David Gruber, shows a coral polyp illuminated by blue light. The coral has proteins that absorb violet wavelengths, then emit green, orange, and red wavelengths.

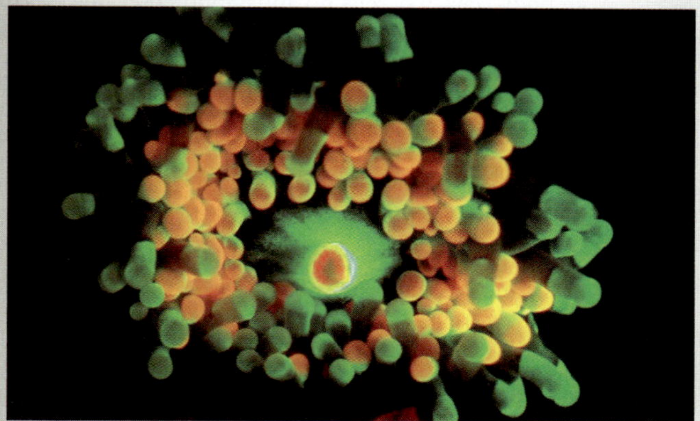

Nearly all known biofluorescent animals are cnidarians, so Gruber searches the world's reefs for species that glow. He and his research team have discovered over 30 new fluorescent proteins from corals, including the brightest one found to date. Recently, he has also been finding biofluorescence in several species of sharks, rays, and bony fishes. Fluorescent proteins are useful in science and medicine. They allow researchers to visualize specific cells and structures using fluorescence microscopy (Section 4.2). Among other applications, this technique can be used in early detection of cancers and to study how signals travel through the brain. We do not know how having biofluorescent proteins benefits cnidarians, but they are linked to coral health. Gruber also suspects that other reef-dwelling organisms recognize the fluorescence as a signal.

23.5 WHAT IS A FLATWORM?

With this section, we begin our survey of protostomes, one of the two lineages of bilaterally symmetrical animals. All of these animals develop from a three-layered embryo.

Flatworms (phylum Platyhelminthes) are the simplest protostomes. They have a flattened body with an array of organ systems, but no body cavity other than a gastrovascular cavity. Like cnidarians, they rely entirely on diffusion to move nutrients and gases through their body. Some flatworms are free-living and others are parasites. Nearly all are hermaphrodites.

FIGURE 23.11 A marine flatworm. Brightly colored flatworms like this one are common inhabitants of coral reef ecosystems.

FREE-LIVING FLATWORMS

Most free-living flatworms live in tropical seas, and many of these are brilliantly colored (FIGURE 23.11). A lesser number live in fresh water, and a few live in damp places on land. Free-living flatworms typically glide along, propelled by the action of cilia that cover the body surface. Some marine species can also swim with an undulating motion.

Planarians are free-living flatworms common in ponds. They have a highly branched gastrovascular cavity (FIGURE 23.12A), and nutrients diffuse from the fine branches to all body cells. There is no anus; food enters and wastes leave through the mouth. The mouth is not on the planarian's head, but rather at the tip of a muscular tube (called the pharynx) that extends from the animal's lower surface.

A planarian's head has chemical receptors and eyespots that detect light. These sensory structures send messages to a simple brain that consists of paired groupings of nerve cell bodies (ganglia). A pair of nerve cords extend from the brain and run the length of the body (FIGURE 23.12B).

A planarian's body fluid has a higher solute concentration than the fresh water around it, so water tends to move into the body by osmosis. A system of tubes regulates internal water and solute levels by driving excess fluids out through a pore at the body surface (FIGURE 23.12C).

A planarian has female and male sex organs (FIGURE 23.12D), but cannot fertilize its own eggs. Freshwater planarians typically swap sperm. By contrast, some marine flatworms battle over who will assume the male role. In

FIGURE 23.12 {Animated} Organ systems of a planarian, a free-living, freshwater flatworm about a centimeter long.

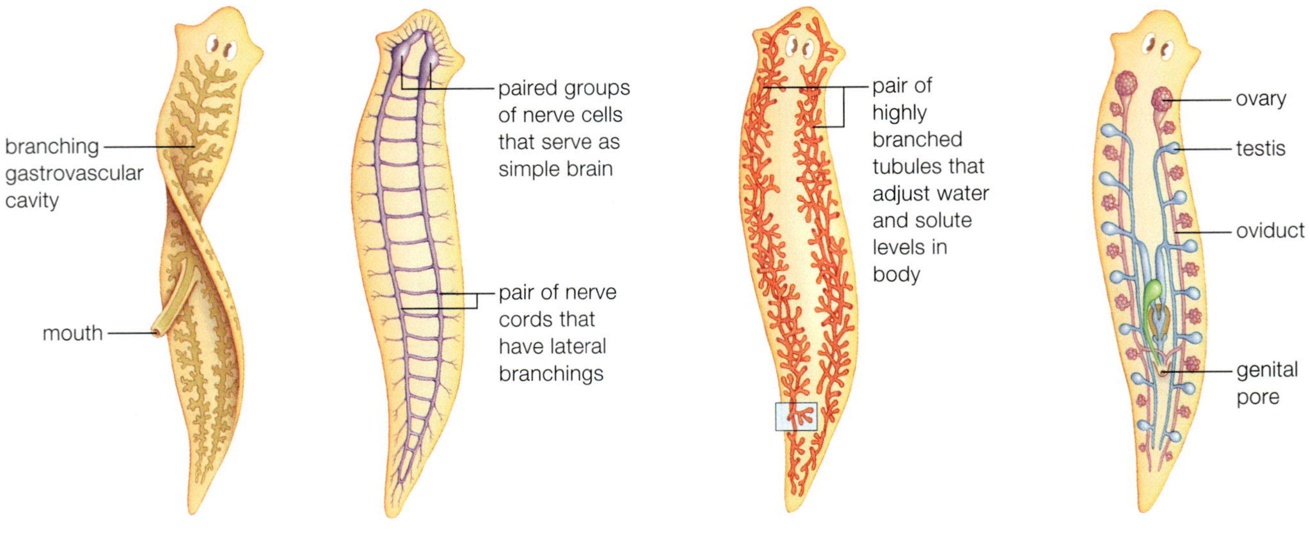

A Digestive system. B Nervous system. C Solute-regulating system. D Reproductive system.

CREDITS: (11) © Cory Gray; (12) © Cengage Learning.

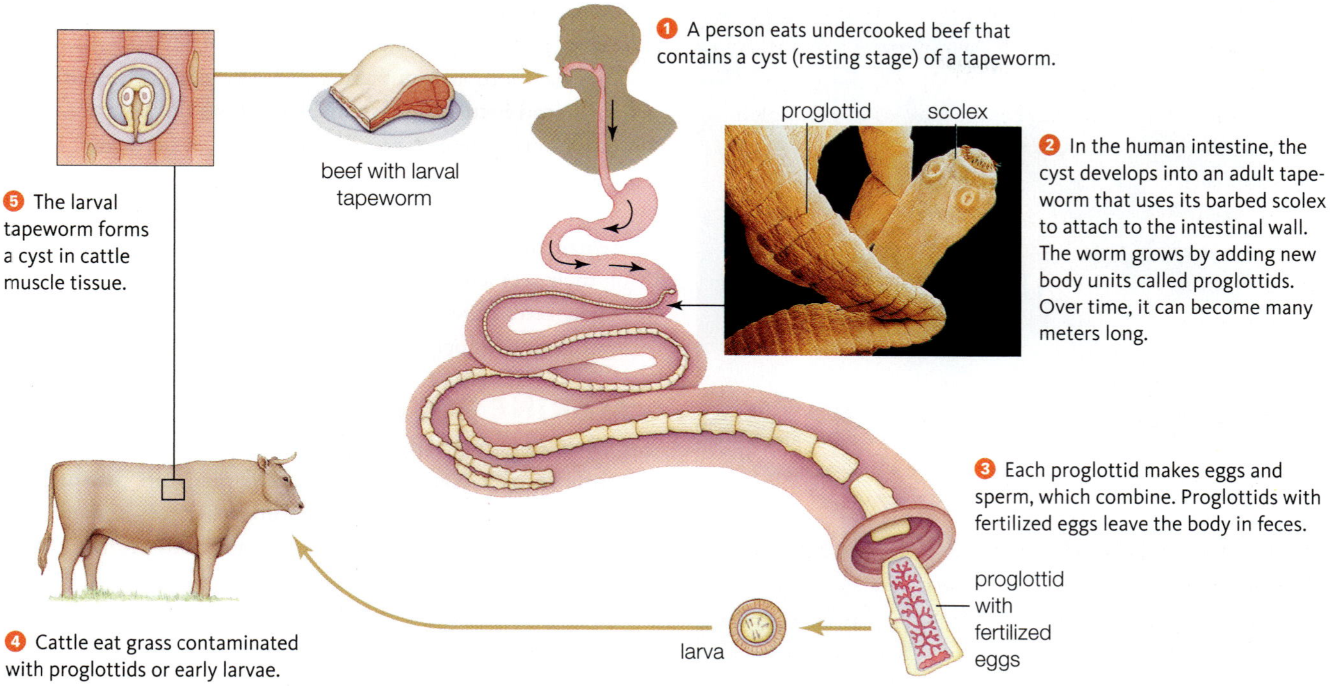

FIGURE 23.13 {Animated} Life cycle of a beef tapeworm.

a behavior described as "penis fencing," each flatworm attempts to stab its penis into a partner's body and squirt in some sperm, while fending off its partner's attempts to do the same.

Planarians also reproduce asexually and some have an amazing capacity for regeneration. During asexual reproduction, the body splits in two near the middle, then each piece regrows the missing parts. This capacity for regrowth is also put to use when a planarian is injured, as by a predator. A planarian can regenerate itself even when more than 99 percent of its body is gone.

PARASITIC FLATWORMS

Flukes and tapeworms are parasitic flatworms whose life cycle often involves multiple hosts. Typically, larvae reproduce asexually in one or more intermediate hosts before developing into adults. Adults reproduce sexually in a final or definitive host. For example, aquatic snails are the intermediate host for blood fluke larvae (*Schistosoma*), but the adults can only reproduce sexually inside a mammal, such as a human. Humans become infected when they swim or stand in water where infected snails live. The infectious larval form of the fluke crosses human skin and migrates into internal organs, where it develops into a sexually reproducing adult. The resulting disease, schistosomiasis, affects about 200 million people, with most cases in Southeast Asia and northern Africa.

Tapeworms are parasites that live and reproduce in the vertebrate gut. The head has a scolex, a structure with hooks or suckers that allow the worm to attach to the gut wall. Behind the head are body units called proglottids. Unlike planarians and flukes, tapeworms do not have a gastrovascular cavity. Instead, the worm absorbs nutrients from the food in the host's gut. **FIGURE 23.13** shows the life cycle of a beef tapeworm.

flatworm Bilaterally symmetrical invertebrate with organs but no body cavity; for example, a planarian or tapeworm.

TAKE-HOME MESSAGE 23.5

Flatworms are bilateral animals with a simple nervous system, and a system for regulating water and solutes in internal body fluids. They develop from a three-layer embryo and do not have a coelom. Most groups have a gastrovascular cavity.

Free-living flatworms include many tropical marine species and the freshwater planarians.

Flukes and tapeworms are parasitic flatworms. Some are human pathogens.

23.6 WHAT ARE ANNELIDS?

A The bristleworm uses its many appendages to burrow into sediment on marine mudflats. It is an active predator with hard jaws.

B The feather duster worm secretes and lives inside a hard tube. Feathery extensions on the head capture food from the water. Cilia move captured food to the mouth.

FIGURE 23.14 Polychaete worms.

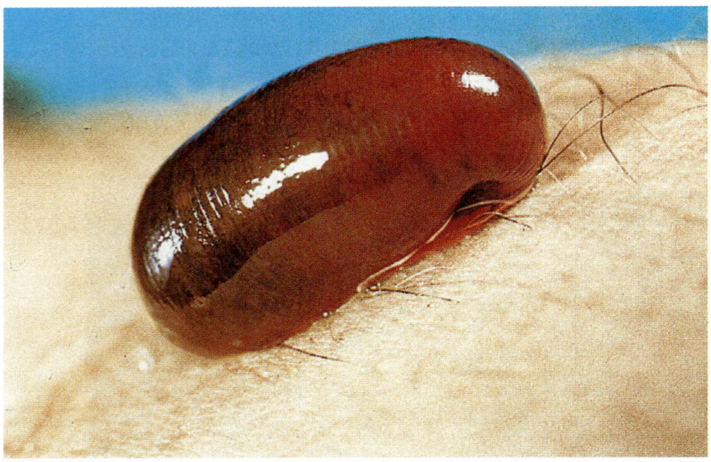

FIGURE 23.15 A leech feeding on human blood.

Annelids (phylum Annelida) are bilateral worms with a coelom and conspicuous segmentation, inside and out. They have a tubular gut and a closed circulatory system. In a **closed circulatory system**, blood flows through a continuous system of vessels. All exchanges between blood and the tissues take place across a vessel wall.

Two annelid subgroups, the polychaetes and oligochaetes, are named for the chitin-reinforced bristles called chaetae on their segments. Polychaetes have many bristles and oligochaetes have few. (*Poly*– means many; *oligo*– means few.) A third subgroup, the leeches, lacks bristles entirely.

MARINE POLYCHAETES

Polychaetes are mostly marine. Those commonly known as bristleworms or sandworms are active predators that use their chitin-hardened jaws to capture other soft-bodied invertebrates (**FIGURE 23.14A**). Each body segment has a pair of bristle-tipped paddlelike appendages that the worm uses to burrow in sediments and pursue prey.

Fan worms and feather duster worms live in a tube that they make from sand grains and mucus. The worm's head end protrudes from the tube and has elaborate tentacles to capture any food that drifts by (**FIGURE 23.14B**). Some other tube-dwelling polychaete worms live near deep-sea hydrothermal vents (Section 18.1). These worms can be up to 2.4 meters (7 feet) long. They do not have tentacles. In fact, as adults they do not even have a mouth. Rather than ingesting food, they rely entirely on the activity of chemoautotrophic bacteria that live inside them to provide nutrients. The bacteria take up dissolved materials from the water and use them to assemble organic compounds that the worm breaks down for energy.

LEECHES

A leech lacks bristles and has a sucker at either end of its body. Most leeches live in fresh water, but some are marine and others live in damp places on land. A typical leech is a scavenger or preys on small invertebrates. An infamous minority attach to a vertebrate, pierce its skin, and suck blood (**FIGURE 23.15**). Saliva of bloodsucking leeches contains a protein that prevents blood from clotting while the leech feeds. For this reason, doctors who reattach a severed finger or ear sometimes apply leeches to the reattached body part. Action of the leeches prevents clots from forming inside the newly reconnected blood vessels.

EARTHWORMS

Oligochaetes include marine and freshwater species, but the land-dwelling earthworms are most familiar. In the United States, most earthworms seen in backyards, bait

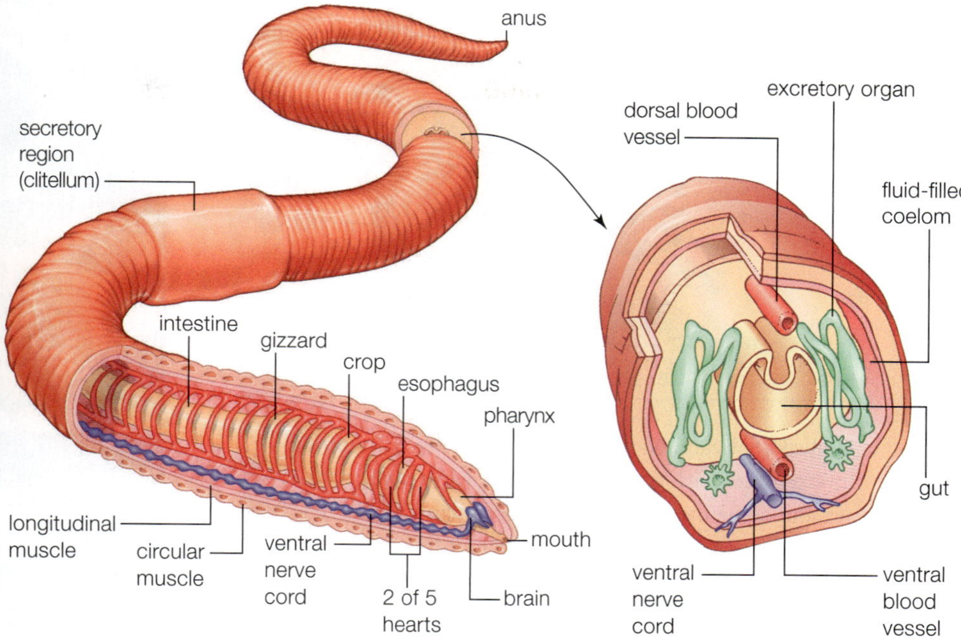

FIGURE 23.16 {Animated} Body plan of an earthworm. Each segment contains a coelomic chamber full of organs. A gut, ventral nerve cord, and dorsal and ventral blood vessels run through all coelomic chambers.

FIGURE IT OUT: Why do people sometimes describe the body of a coelomate animal such as an earthworm as "a tube inside a tube"?

Answer: One tube, the gut, is inside another tube, the body wall. The fluid-filled coelom lies between the tubes.

shops, and compost heaps are introduced species native to Europe. However, some undisturbed habitats do retain native earthworms, including a species in Oregon that can reach more than a meter in length.

A cuticle of secreted proteins coats an earthworm's body (**FIGURE 23.16**). Visible grooves at the body surface correspond to internal partitions. Gas exchange occurs across the body surface, and the closed circulatory system helps distribute oxygen. Hearts in the anterior of the worm provide the pumping power to move the blood.

A fluid-filled coelom runs the length of the body and is divided into chambers, one per segment. A tubular gut extends through all coelomic chambers. Earthworms are scavengers that eat their way through the soil and digest organic debris. The worms improve soil by loosening its particles and by excreting tiny bits of organic matter that decomposers can easily break down. Excreted earthworm "castings" are sold as a natural fertilizer.

Most body segments have a pair of excretory organs that regulate the solute composition and volume of coelomic fluid. The organs collect coelomic fluid, adjust its composition, and expel waste through a pore in the body wall.

A simple brain connects to a pair of nerve cords that run the length of the body. The brain coordinates locomotion and receives information from sensory cells, such as light-detecting cells in the worm's body wall.

The earthworm has two sets of muscles. Longitudinal muscles parallel the body's long axis, and circular muscles ring the body. The worm's coelomic fluid is a hydrostatic skeleton. Action of the muscles puts pressure on the fluid trapped inside body segments, causing them to change shape. When a segment's longitudinal muscles contract, the segment gets shorter and fatter. When circular muscles contract, a segment gets longer and thinner. Coordinated waves of contraction that run along the body propel the worm through soil.

Earthworms are hermaphrodites, but they cannot fertilize themselves. During mating, a secretory organ (the clitellum) produces mucus that glues two worms together while they swap sperm. Later, the same organ secretes a silky case that protects the fertilized eggs.

annelid Segmented worm with a coelom, complete digestive system, and closed circulatory system.

closed circulatory system Circulatory system in which blood flows through a continuous network of vessels; all materials are exchanged across the walls of those vessels.

TAKE-HOME MESSAGE 23.6

Annelids are bilateral, coelomate, segmented worms. They include aquatic worms, earthworms, and leeches.

A closed circulatory system distributes materials through the annelid body.

CREDITS: (16) left, Courtesy of © Christine Evers; right, After Solomon, 8th edition, p. 624, figure 29-4.

23.7 WHAT ARE MOLLUSKS?

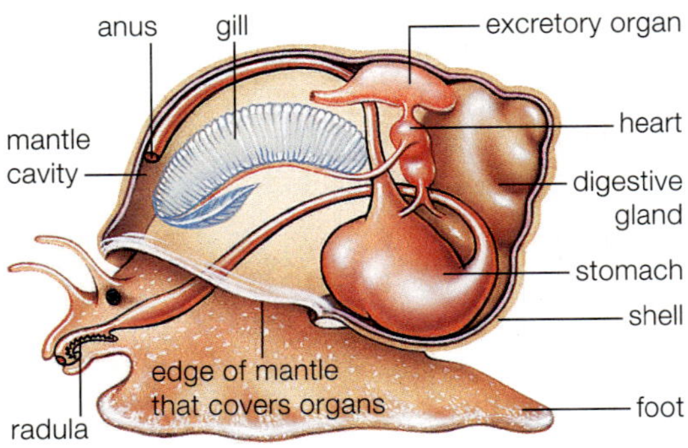

FIGURE 23.17 Body plan of an aquatic snail.

Mollusks (phylum Mollusca) are bilaterally symmetrical invertebrates with a reduced coelom. Most dwell in the sea, but some live in fresh water or on land. All have a mantle, which is a skirtlike extension of the upper body wall that covers a mantle cavity (**FIGURE 23.17**). In shelled mollusks, the shell consists of a calcium-rich, bonelike material secreted by the mantle. Aquatic mollusks typically have one or more respiratory organs called gills inside their fluid-filled mantle cavity. All mollusks have a tubular gut. In most mollusks, the mouth contains a radula, a tonguelike organ hardened with chitin.

Aquatic mollusks typically have separate sexes and produce a ciliated swimming larva called a trochophore. Marine polychaetes have the same sort of larva, and this similarity is taken as evidence that annelids and mollusks shared a common trochophore-producing ancestor.

With more than 100,000 living species, mollusks are second only to arthropods in level of diversity. There are three main groups: gastropods, bivalves, and cephalopods (**FIGURE 23.18**).

A Gastropods (belly footed).

B Bivalves (two-part shell). The blue dots are this scallop's eyes.

C Cephalopods (jet-propelled).

FIGURE 23.18 Representatives of the main mollusk groups.

GASTROPODS

Gastropods are the most diverse mollusk lineage. Their name means "belly foot," and most species glide about on the broad muscular foot that makes up most of the lower body mass (**FIGURE 23.18A**). A gastropod shell, when present, is one-piece and often coiled.

Gastropods have a distinct head that usually has eyes and sensory tentacles. In many aquatic species, a part of the mantle forms an inhalant siphon, a tube that the animal uses to draw water into its mantle cavity. The circulatory system of gastropods and bivalves is open. In an **open circulatory system**, vessels do not form a continuous loop. Instead, fluid (called hemolymph) leaves vessels and seeps around tissues before returning to the heart. Cells exchange substances with hemolymph while it is outside of vessels.

Gastropods include the only terrestrial mollusks. In snails and slugs that live on land, a lung replaces the gill. Mucus secreted by glands on the foot protects the animal as it moves across dry, abrasive surfaces. Most mollusks have separate sexes, but land snails and slugs tend to be hermaphrodites. Unlike other mollusks, which produce a swimming larva, embryos of these groups develop directly into adults.

Nudibranchs, commonly called sea slugs, are marine gastropods with a greatly reduced shell. Nudibranch means "naked gills." Many nudibranchs defend themselves from predation using defensive weapons taken from their prey. Some eat sponges and store sponge toxins, secreting them when disturbed. Others eat cnidarians and store cnidocytes in outpouchings on their body (**FIGURE 23.19**).

BIVALVES

Bivalves include many mollusks that end up on dinner plates, including mussels, oysters, clams, and scallops (**FIGURE 23.18B**). All have a hinged, two-part shell. A bivalve has no obvious head, but many have simple

bivalve Mollusk with a hinged two-part shell.
cephalopod Predatory mollusk with a closed circulatory system; moves by jet propulsion.
gastropod Mollusk in which the lower body is a broad "foot."
mollusk Invertebrate with a reduced coelom and a mantle.
open circulatory system System in which hemolymph leaves vessels and seeps through tissues before returning to the heart.

FIGURE 23.19 Spanish shawl nudibranchs.

eyes around the edge of the mantle. Most bivalves feed by drawing water into their mantle cavity and trapping food in mucus on the gills. Waving cilia on the gills direct particle-laden mucus to the mouth.

CEPHALOPODS

Cephalopods include squids (**FIGURE 23.18C**), nautiluses, octopuses (**FIGURE 23.20**), and cuttlefish. All are predators and most have beaklike, biting mouthparts in addition to a radula. Cephalopods move by jet propulsion. They draw water into their mantle cavity, then force it out through a funnel-shaped siphon. The foot has been modified into arms and/or tentacles that extend from the head. Cephalopod means "head-footed." Cephalopods include the fastest invertebrates (squids), the biggest (giant squid), and the smartest (octopuses). Of all invertebrates, octopuses have the largest brain relative to body size.

Five hundred million years ago, large cephalopods with a long, conelike shell were the top predators in the seas. Today, nautiluses have a coiled external shell, but all other cephalopods have a highly reduced shell or none. Shell loss may have been driven by competition with jawed fishes, which evolved 400 million years ago. Cephalopods with the smallest shell could be fastest and most agile. A speedier lifestyle required other changes as well. Of all mollusks, only cephalopods have a closed circulatory system. Competition with fishes also favored improved eyesight. Like vertebrates, cephalopods have eyes with a lens that focuses light. As mollusks and vertebrates are not closely related, this type of eye is thought to have evolved independently in these groups. This is an example of convergent evolution.

FIGURE 23.20 Octopus.

TAKE-HOME MESSAGE 23.7

Mollusks are bilateral, soft-bodied, coelomate animals, the only ones with a mantle over the body mass.

Mollusks include snails and slugs that glide about on a huge foot, bivalves with a hinged two-part shell, and the jet-propelled cephalopods.

CREDITS: (19) © Alex Kirstitch; (20) NURC/UNCW and NOAA/FGBNMS.

23.8 WHAT ARE ROUNDWORMS?

Roundworms, or nematodes (phylum Nematoda) are cylindrical worms with a pseudocoelom (**FIGURE 23.21**). They have a tubular digestive system, excretory organs, and a nervous system, but no circulatory or respiratory organs. Like arthropods such as insects, they have a cuticle that they periodically **molt** (shed and replace) as they grow. Most roundworms feed on organic debris in soil or water and are less than a millimeter long. Parasitic species tend to be larger: Some roundworms that infect whales can be meters long.

The soil roundworm *Caenorhabditis elegans* is frequently used in scientific studies. It is useful as a model for developmental processes because it has the same tissue types as more complex organisms, but it is transparent, has fewer than 1,000 body cells, and reproduces fast. In addition, its genome is about 1/30 the size of the human genome.

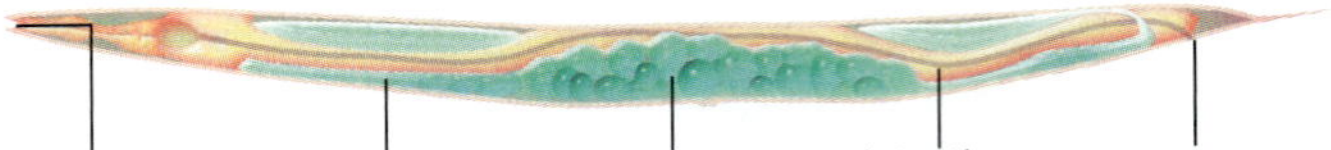

FIGURE 23.21 {Animated} Body plan of a free-living roundworm (*Caenorhabditis elegans*).

FIGURE 23.22 Parasitic roundworms.

A The grossly enlarged leg of the man on the left is a sign of lymphatic filariasis.

B Intestinal parasite, *Ascaris*, passed by a child.

C Plant-infecting roundworm entering a root.

Several kinds of parasitic roundworms infect humans. Mosquito-transmitted roundworms cause a disfiguring tropical disease called lymphatic filariasis. The worms travel in the body's lymph vessels and destroy the vessels' valves so lymph pools in the lower limbs (**FIGURE 23.22A**). Elephantiasis, the common name for this disease, refers to fluid-filled, "elephant-like" legs. The swelling is permanent because lymph vessels remain damaged even after the worms have been eliminated.

The intestinal parasite *Ascaris lumbricoides* (**FIGURE 23.22B**) currently infects more than 1 billion people. Most of those affected live in developing tropical nations, but occasional infections occur in rural parts of the American Southeast. Typically an infected person has no symptoms, but a heavy infection can block the intestine.

Pinworms (*Enterobius vermicularis*) commonly infect children in the United States. The small worms, about the size of a staple, live in the rectum. At night, females crawl out and lay eggs on the skin around the anus. Their movement produces an itching sensation. Scratching to relieve the itch puts eggs onto fingertips and under fingernails. If swallowed, the eggs start a new infection.

Parasitic roundworms also infect our livestock, pets, and crop plants. A roundworm that lives in pigs can also infect humans who eat undercooked pork, causing trichinosis. Dogs are susceptible to heartworms transmitted by mosquitoes. Cats usually become infected by eating an infected rodent that serves as an intermediate host. Many crop plants are susceptible to nematode parasites. In some cases, the worms suck on plant roots and in others they actually enter the plant (**FIGURE 23.22C**). Either way, the infection stunts the plant's growth and lowers crop yields.

molting Periodic shedding of an outer body layer or part.
roundworm Cylindrical worm with a pseudocoelom.

TAKE-HOME MESSAGE 23.8

Roundworms are bilateral worms with a cuticle that they molt as they grow. Organ systems almost fill their false coelom. Some are agricultural pests or human parasites.

23.9 WHAT ARE ARTHROPODS?

Arthropods (phylum Arthropoda) are bilaterally symmetrical invertebrates with a tubular gut, an open circulatory system, and a reduced coelom. Modern representatives include spiders, lobsters, barnacles, centipedes, and insects. Trilobites, a now-extinct arthropod lineage, were the most abundant and diverse animal group in Cambrian seas (**FIGURE 23.23**). They disappeared about 425 million years ago, and we do not know why. Despite this loss, arthropods remain the most diverse animal phylum. In this section, we consider traits that contribute to their success.

A tough cuticle stiffened by chitin protects and supports the arthropod body. The cuticle is an **exoskeleton**, an external skeleton to which muscles attach. The cuticle does not inhibit movement because it is thin at the joints where two body parts connect. In fact, *arthropod* means jointed leg. The cuticle consists of secreted materials and does not grow, so, like roundworms, arthropods must periodically molt their cuticle and replace it with a larger one.

When some arthropods moved onto land, their exoskeleton supported their body against the force of gravity and helped them conserve water. Thus, arthropods thrive even in dry environments. The largest land invertebrate, the coconut crab, is an arthropod. It can weigh as much as 4 kilograms (about 9 pounds).

Aquatic arthropods have gills, but life on land required a different respiratory apparatus. The evolution of lungs or gas-delivering tracheal tubes allowed arthropods to breathe air as they moved onto land.

Arthropods have a segmented body plan. In most groups, some segments have fused to form distinct body regions. In crustaceans such as lobsters, segments of the head and thorax (midbody) form a unit called the cephalothorax (**FIGURE 23.24**). Modified segments and appendages give arthropods the ability to exploit resources in new ways. For example, in American lobsters, the appendages on one segment have evolved into large, muscular claws that can crush food. In flying insects, one or two body segments have wings that evolved from extensions of the body wall.

Well-developed sensory structures allow arthropods to monitor their environment. All arthropod lineages have one or more pairs of eyes. Some groups have **compound eyes**, which consist of many image-forming units that are highly sensitive to motion. Most arthropods also have paired **antennae** (singular, antenna), sensory structures that can detect chemicals in water or air.

FIGURE 23.23 Fossil trilobite. Over 20,000 species of this now-extinct group have been identified from fossils.

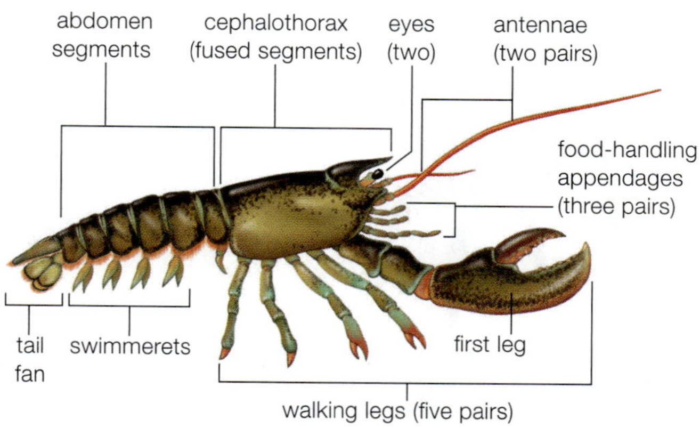

FIGURE 23.24 Body plan of an American lobster.

Metamorphosis, a dramatic change in form between the larval and adult stages, occurs in most lineages. For instance, wingless caterpillars that chew leaves metamorphose into winged butterflies that fly, sip nectar, mate, and then deposit eggs. Evolution of metamorphosis allowed arthropods to, in essence, have two different bodies, each adapted to utilize a different set of resources and complete a different task. Having different bodies also keeps adults and juveniles from competing for the same resources.

antenna Of some arthropods, sensory structure on the head that detects touch and odors.
arthropod Invertebrate with jointed legs and a hard exoskeleton that is periodically molted.
compound eye Of some arthropods, a motion-sensitive eye made up of many image-forming units.
exoskeleton Hard external parts that muscles attach to and move.
metamorphosis Dramatic remodeling of body form during the transition from larva to adult.

TAKE-HOME MESSAGE 23.9

Arthropods are the most diverse animal phylum. A hardened, jointed exoskeleton, modified body segments, and specialized appendages, respiratory structures, and sensory structures contribute to their evolutionary success.

In many groups, larvae and adults have dramatically different body forms.

23.10 WHAT ARE THE MAIN KINDS OF ARTHROPODS?

FIGURE 23.25 Horseshoe crabs.

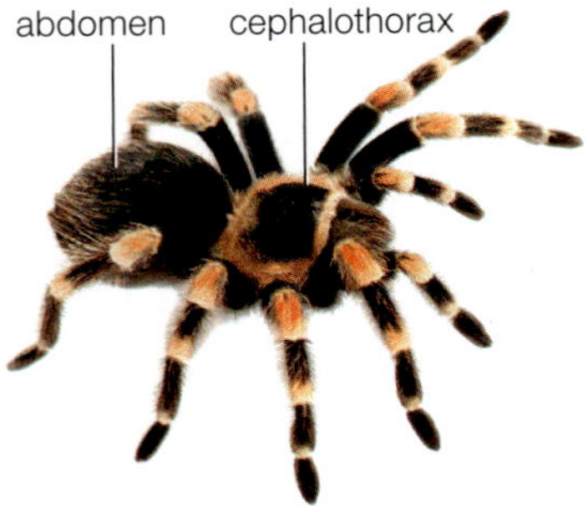

A Spider (tarantula), a predator.

B Scorpion, a predator.

C Tick, a parasite.

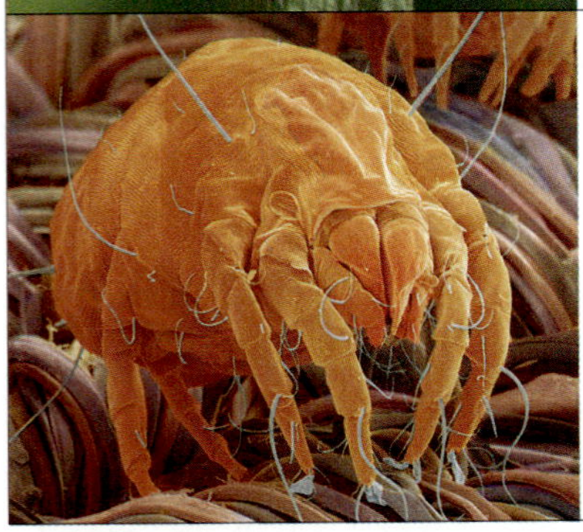

D Dust mite, a scavenger (micrograph).

FIGURE 23.26 Arachnids. All have eight walking legs and no antennae.

CHELICERATES

A pair of feeding appendages (chelicerae) in front of the mouth distinguishes the **chelicerates**. Their head also has one or more pairs of eyes, but no antennae. The body is usually divided into a cephalothorax and an abdomen.

Horseshoe crabs are an ancient chelicerate lineage of marine bottom-feeders. A horseshoe-shaped shield covers their cephalothorax, and their last abdominal segment has evolved into a long spine that serves as a rudder (FIGURE 23.25). Each spring, eggs laid along Atlantic coasts by horseshoe crabs feed millions of migratory shorebirds.

Arachnids have four pairs of walking legs attached to their cephalothorax. They include spiders, scorpions, ticks, and mites, nearly all of which live on land (FIGURE 23.26). Spiders and scorpions are venomous predators. Spiders dispense venom through their chelicerae, which have been modified into fangs. Some catch prey in a web made of silk ejected by glands on their abdomen. Others such as tarantulas actively hunt and capture prey. Scorpions are active hunters too. They capture prey with pincers and dispense venom through a stinger on their abdomen.

Ticks are parasites that suck blood from vertebrates. Some can transmit human pathogens, such as the bacteria that cause Lyme disease (Section 19.5).

Mites include predators, scavengers, and parasites. Most are less than a millimeter long. Dust mites are scavengers that can cause discomfort to people who are allergic to their feces. Other mites that burrow into skin cause scabies in humans and mange in dogs.

MYRIAPODS

Myriapods have a head with one pair of antennae attached to an elongated body with many similar segments. Myriapod means "many footed," and centipedes and millipedes do have a lot of feet. Centipedes have a low-slung, flattened body with one pair of legs per segment, for a total of 30 to 50 legs (FIGURE 23.27A). These fast-moving predators use fangs to inject paralyzing venom. Most centipedes prey on insects, but some big tropical species eat small vertebrates. Millipedes are slower-moving animals that feed on decaying vegetation. Their cylindrical body has a cuticle hardened by calcium carbonate. There are two pairs of legs per segment, for a total of a few hundred (FIGURE 23.27B).

CRUSTACEANS

Most marine arthropods are **crustaceans**. Their great diversity and abundance are reflected in their nickname, "insects of the seas." Crustaceans have two pairs of antennae and a distinct cephalothorax and abdomen. In many, the exoskeleton is hardened by calcium.

A Centipede, a speedy predator.

B Millipede, a scavenger of decaying plant material.

FIGURE 23.27 {Animated} Myriapods.

You are probably familiar with some decapod crustaceans. This group of bottom-feeding scavengers includes the lobsters, crayfish, crabs, and shrimps. Decapod means ten-legged; members of this group typically have five pairs of walking legs on their cephalothorax.

Krill and copepods are small swimmers that eat plankton. They in turn serve as an important food source for larger animals. Krill, which have a shrimplike body a few centimeters long, swim in upper ocean waters (**FIGURE 23.28A**). They are so plentiful and nutritious that a 100-ton blue whale can subsist almost entirely on the krill it filters from seawater. Copepods live in oceans and lakes. Most are only millimeters long (**FIGURE 23.28B**).

Larval barnacles swim, but adults secrete a thick calcified shell and live fixed in place. They filter food from the water with feathery legs (**FIGURE 23.28C**). Adults cannot move about, so you would think that mating might be tricky. But barnacles tend to settle in groups, and most are hermaphrodites. An individual extends a penis, often several times its body length, out to neighbors.

A few crustaceans, including some crabs and the animals commonly called pill bugs (*right*), have adapted to life on land.

INSECTS

Insects are the most diverse arthropods. Until recently, they were thought to be close relatives of myriapods because of structural similarities. Then, gene comparisons made scientists rethink the connections. The currently favored hypothesis holds that insects descended from freshwater crustaceans. We consider insect diversity and ecology in detail in the next section.

A Antarctic krill.

B Copepod.

C Barnacle.

FIGURE 23.28 Examples of crustaceans. See also **FIGURE 23.24**.

TAKE-HOME MESSAGE 23.10

Horseshoe crabs are an ancient lineage of marine bottom-feeders. They are most closely related to arachnids (spiders, scorpions, ticks, and mites), which live on land. Arachnids have four pairs of walking legs on their cephalothorax.

Centipedes and millipedes live on land. Members of this group have elongated bodies with many similar segments.

Crustaceans are mostly marine arthropods with two pairs of antennae and an exoskeleton hardened with calcium. They include bottom-feeding crabs and lobsters, tiny swimmers such as copepods and krill, filter-feeding barnacles, and the terrestrial pill bugs.

arachnids Land-dwelling arthropods with no antennae and four pairs of walking legs; spiders, scorpions, mites, and ticks.

chelicerates Arthropod group with specialized feeding structures (chelicerae) and no antennae; arachnids and horseshoe crabs.

crustaceans Mostly marine arthropods with a calcium-hardened cuticle and two pairs of antennae; for example lobsters, crabs, krill, and barnacles.

myriapods Long-bodied terrestrial arthropods with one pair of antennae and many similar segments; centipedes and millipedes.

CREDITS: (27) © Eric Isselée/Shutterstock; (in text) © Chris Howey/Shutterstock.com; (28A) © David Tipling/Photographer's Choice/Getty Images; (28B) Herve Chaumeton/Agence Nature; (28C) Courtesy of Dave & Debra Wrobel.

23.11 WHAT MAKES INSECTS SO DIVERSE AND IMPORTANT?

With more than a million named species, **insects** are the most diverse class of animals. They are also breathtakingly abundant. By some estimates, ants alone make up about 10 percent of the total weight of all living land animals. Insects live on every continent, including Antarctica. Here we consider the traits that contribute to insect diversity and the ways that they interact with us and with other organisms.

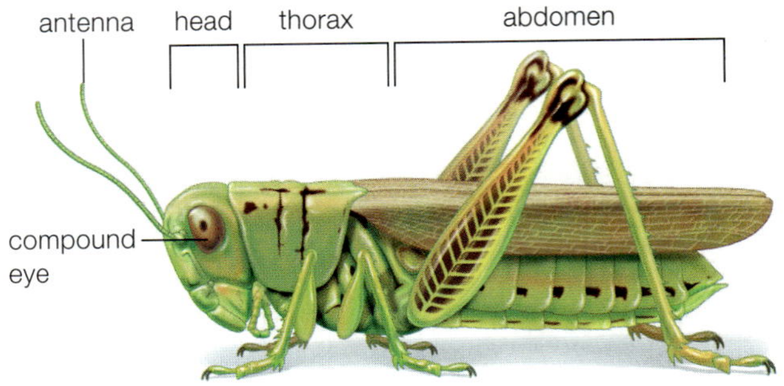

FIGURE 23.29 Body plan of an insect, a grasshopper.

CHARACTERISTICS OF INSECTS

Insects have a three-part body plan, with a distinct head, thorax, and abdomen (FIGURE 23.29). The head has one pair of antennae and two compound eyes. Three pairs of legs attach to the thorax. In winged insects, the thorax also has one or two pairs of wings. Like other arthropods, insects have a tubular gut and an open circulatory system. The insect respiratory system consists of tracheal tubes that carry air from openings on the insect's surface deep into the body where gas exchange occurs. The insect abdomen contains digestive organs, sex organs, and water-conserving excretory organs (Malpighian tubules) that serve the same function as vertebrate kidneys.

The earliest insects were wingless, ground-dwelling scavengers that did not undergo metamorphosis. A few modern insects such as bristletails and silverfish retain this type of body form and development. When they hatch from an egg, they look like a tiny adult, and they simply grow bigger with each molt.

Most modern insects have wings and undergo metamorphosis. Insects are the only winged invertebrates, and evolving an ability to fly gave them dispersal abilities unrivaled among other land invertebrates. Evolution of metamorphosis allowed insects to utilize multiple sets of resources. With incomplete metamorphosis, an egg hatches into a nymph that gradually takes on the adult form over the course of several molts. Cockroaches, grasshoppers, and dragonflies undergo incomplete metamorphosis. A dragonfly lives as an aquatic nymph for one to three years before emerging from the water and undergoing a final molt to the winged adult form.

FIGURE 23.30 Complete metamorphosis in a butterfly.

With complete metamorphosis, a larva grows and molts without altering its form, then undergoes pupation. A pupa is a nonfeeding body in which larval tissues are remodeled into the adult form (FIGURE 23.30). Members of the four most diverse insect orders have wings and undergo complete metamorphosis. There are about 150,000 species of flies (Diptera), and at least as many beetles (Coleoptera). The order Hymenoptera includes 130,000 species of wasps, ants, and bees. Moths and butterflies are Lepidoptera, a group of about 120,000 species. As a comparison, consider that there are about 4,500 species of mammals.

INSECT ECOLOGY

Insects play essential roles in just about every land ecosystem. The vast majority of flowering plants are pollinated by members of the four diverse insect orders discussed above. Other insect orders contain few or no pollinators. Close interactions between pollinator and flowering plant lineages likely contributed to an increased rate of speciation that increased the diversity of both.

Insects serve as food for a variety of wildlife. Most songbirds nourish their nestlings on an insect diet. Those that migrate often travel long distances in order to nest and raise their young in areas where insect abundance is seasonally high. Aquatic larvae of insects such as dragonflies and mayflies serve as food for trout and other freshwater fish. Most amphibians and reptiles feed mainly on insects. Even humans eat insects. In many cultures, they are considered a tasty source of protein.

Insects dispose of wastes and remains. Flies and beetles quickly discover an animal corpse or a pile of feces (FIGURE 23.31). They lay their eggs in or on this organic material, which serves as food for their larvae. By their

FIGURE 23.31 Dung beetles. Nature's clean-up crew.

actions, these insects keep wastes and remains from piling up, and help distribute nutrients through ecosystems.

HEALTH AND ECONOMIC EFFECTS

Plant-eating insects are our main competitors for plant products. Each year, they devour as much as one-third of all crops grown in the United States. Pine beetles and other insects bore into living trees and reduce lumber harvests. Other wood-boring beetles damage wooden buildings, as do some termites and ants.

Parasitic insects can pose a threat to human health and well-being. Mosquitoes cause more than 1 million deaths

per year by transmitting malaria, West Nile fever, and other deadly diseases. Fleas transmit bubonic plague, and body lice spread typhus. Bedbugs (*left*) do not cause disease, but infestations are stressful and eliminating them can be costly. On the other hand, some insects help us keep pest species in check. For example, larvae of ladybird beetles devour aphids that would otherwise suck juices from valuable crops.

Economically valuable products produced by insects include honey, silk, and shellac. Honey is floral nectar concentrated and stored by honeybees. Larval silk moths produce silk as they spin a cocoon in which to pupate. Asian bark-feeding bugs excrete shellac. We use it to give a shiny coating to wood products and to candies such as jelly beans.

insects Most diverse arthropod group; members have six legs, two antennae, and, in some groups, wings.

TAKE-HOME MESSAGE 23.11

Insects include the only winged invertebrates.

Most insects undergo metamorphosis, which allows them to exploit resources in two different ways over the course of their life.

Insects pollinate plants, serve as food for wildlife and humans, dispose of wastes and remains, and produce products such as silk. Some harm us by eating crops and spreading diseases.

CREDITS: (31) Joel Sartore, National Geographic Creative; (in text) CDC/ Piotr Naskrecki.

23.12 WHAT ARE ECHINODERMS?

In Section 23.1 we introduced the two major lineages of animals, protostomes and deuterostomes. Here we begin our survey of deuterostome lineages. Echinoderms are the most diverse invertebrate deuterostomes. We discuss other invertebrate deuterostomes and the vertebrates (also deuterostomes) in the next chapter.

Echinoderms (phylum Echinodermata) have interlocking spines and plates of calcium carbonate in their body wall. Their phylum name means "spiny-skinned." All begin life as bilateral free-swimming larvae, then develop into radially symmetrical adults with five parts (or multiples of five) arrayed around a central axis. The fact that the larva is bilateral suggests that echinoderms had a bilateral ancestor.

Sea stars (also called starfish) are the most familiar echinoderms (**FIGURE 23.32**). They do not have a brain, but do have a decentralized nervous system. Eyespots at the tips of the arms detect light and movement. A typical sea star is an active predator that moves about on tiny, fluid-filled tube feet. Tube feet are part of a **water–vascular system** unique to echinoderms. Fluid-filled canals extend into each arm, and side canals deliver coelomic fluid into muscular ampullae that function like the bulb on a medicine dropper. When an ampulla contracts, it forces fluid into a tube foot. The increase in internal fluid pressure extends the foot.

A sea star eats bivalve mollusks by forcing its stomach out through its mouth and into the bivalve's shell. Acid and enzymes secreted by the stomach kill the mollusk and begin to digest it. The partially digested food is then taken into the stomach, where digestion is completed with the aid of digestive glands that extend into the arms. Arms also hold reproductive organs. Each sea star is either male or female.

Gas exchange occurs by diffusion across the tube feet and tiny skin projections at the body surface. There are no specialized excretory organs.

In addition to sea stars, echinoderms include brittle stars, sea urchins, and sea cucumbers. Brittle stars are the most diverse and abundant echinoderms (**FIGURE 23.33A**). These scavengers generally live in deep water. They have a central disk and flexible arms that move in a snakelike way. Sea urchins have a stiff, rounded cover of calcium carbonate plates through which movable spines protrude (**FIGURE 23.33B**). Most graze on algae. Sea cucumbers are, as their name implies, cucumber-shaped, and have microscopic plates embedded in a soft body (**FIGURE 23.33C**). Most feed like earthworms, by burrowing through sediments and ingesting them. If attacked, a sea cucumber expels organs through its anus to distract the predator. If this defense succeeds and the sea cucumber escapes, its missing parts grow back. Sea stars also have great powers of regeneration; an animal can grow from one arm and a bit of central disk.

FIGURE 23.32 Sea star body plan.

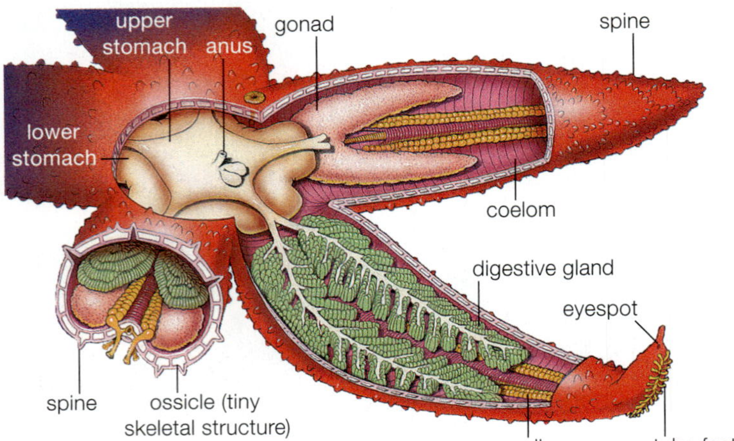

A Brittle star.

B Sea urchin.

C Sea cucumber.

FIGURE 23.33 Echinoderm diversity.

echinoderms Invertebrates with a water–vascular system and hardened plates and spines embedded in the skin or body.

water–vascular system Of echinoderms, a system of fluid-filled tubes and tube feet that function in locomotion.

TAKE-HOME MESSAGE 23.12

Echinoderms are spiny-skinned invertebrates that are radial as adults. They have a decentralized nervous system and a unique water–vascular system that functions in locomotion.

23.13 Application: MEDICINES FROM THE SEA

FIGURE 23.34 A cone snail eating a fish that it has injected with venom.

THERE ARE ABOUT A THOUSAND SPECIES OF CONE SNAILS. Each species of these predatory snails produces a venom that contains around a hundred different peptides. Cone snail peptides typically act on a prey animal's nervous system. The snail catches prey such as a small fish by harpooning it with a modified radula, injecting it with venom, then engulfing it (**FIGURE 23.34**). The venom sedates the fish and dulls its pain, so it does not struggle and possibly damage the snail's harpoon.

The human nervous system uses the same chemical signals a fish nervous system does, so compounds used by a snail to sedate or anesthetize fish can be useful as human medicines. For example, an injectable painkiller sold under the brand name Prialt is a synthetic version of a cone snail peptide. It is far more powerful than morphine, yet not addictive. Another synthetic snail peptide is being tested as a treatment for epilepsy. Researchers have only begun to explore the potential of cone snail venom. Given the large number of cone snails and complexity of their venoms, there is plenty to do.

Nonvenomous invertebrates can also be a source of new medicines. Many invertebrates produce chemicals that have antibacterial, antiprotozoal, or antifungal activity. To develop new drugs with these properties, researchers extract compounds from invertebrates, then test the ability of the compounds to kill pathogens cultured in the laboratory. Researchers also test the effect of each compound on cultured human cells. An ideal candidate drug is one that kills pathogens, but does no harm to human cells. For example, some compounds made by marine sponges kill disease-causing protists (such as trypanosomes and apicomplexans), but have little effect on human cells growing in culture.

A new appreciation of the potential medicinal value of compounds produced by marine invertebrates has many researchers worried about the declining state of our oceans. Pollution, destructive harvesting methods, overharvesting, and climate change are likely to drive many potentially valuable species to extinction before we have time to discover the beneficial compounds they make.

CREDIT: (34) © K.S. Matz.

Summary

SECTION 23.1 **Animals** are multicelled heterotrophs that digest food inside their body. Most have an embryo with three layers: **ectoderm**, **endoderm**, and **mesoderm**. Most animals are **invertebrates**. Early invertebrates had no body symmetry. Cnidarians have **radial symmetry**, but most animals have **bilateral symmetry** and underwent **cephalization**. Bilateral animals typically have a body cavity, either a **pseudocoelom** or a **coelom**. Many have **segmentation**, with repeating body units. There are two lineages of bilateral animals, **protostomes** and **deuterostomes**. **TABLE 23.1** summarizes the traits of major animal lineages discussed in this chapter.

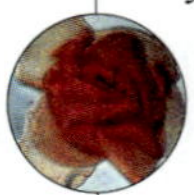

SECTION 23.2 The **colonial theory of animal origins** states that animals evolved from a colonial protist. Early animals may have resembled modern **placozoans**, the simplest living animals. Sponges are the animals for which we have the earliest fossil evidence. They evolved more than 600 million years ago. A great adaptive radiation during the Cambrian gave rise to most modern animal lineages.

SECTION 23.3 **Sponges** have a porous body with no tissues or organs. These **suspension feeders** filter food from water. Each is a **hermaphrodite**, producing both eggs and sperm. The ciliated **larva** is the only motile stage.

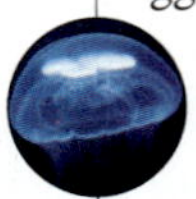

SECTION 23.4 **Cnidarians** such as jellies, corals, and sea anemones are carnivores with two tissue layers. Only cnidarians have **cnidocytes**, which they use to capture prey. A **gastrovascular cavity** functions in both respiration and digestion. Cnidarians have a **hydrostatic skeleton**. A **nerve net** gives commands to contractile cells that redistribute fluid and change the body's shape.

TABLE 23.1

Comparative Summary of Animal Body Plans

Group	Adult Symmetry	Embryonic Layers	Digestive Cavity	Circulatory System
Sponges	None	None	None	None
Cnidarians	Radial	2	Gastrovascular cavity	None
Protostomes				
Flatworms	Bilateral	3	Gastrovascular cavity	None
Annelids	Bilateral	3	Tubular gut	Closed
Mollusks	Bilateral	3	Tubular gut	Most open; cephalopods closed
Roundworms	Bilateral	3	Tubular gut	None
Arthropods	Bilateral	3	Tubular gut	Open
Deuterostomes				
Echinoderms	Radial	3	Tubular gut	Open
Vertebrates	Bilateral	3	Tubular gut	Closed

SECTION 23.5 **Flatworms**, the simplest animals with organ systems, include marine species and the freshwater planarians, as well as parasitic tapeworms and flukes. Some tapeworms and flukes infect humans.

SECTION 23.6 **Annelids** are segmented worms. Their **closed circulatory system** and digestive, solute-regulating, and nervous systems extend through coelomic chambers. Annelids move when muscles exert force on coelomic fluid, altering the shape of segments in a coordinated manner. Oligochaetes include aquatic species and the familiar earthworms. Polychaetes are predatory marine worms. Leeches are scavengers, predators, or bloodsucking parasites.

SECTION 23.7 **Mollusks** have a reduced coelom and a sheetlike mantle that drapes back over itself. They include **gastropods** (such as snails), **bivalves** (such as scallops), and **cephalopods** (such as squids and octopuses). Except in the cephalopods, which are adapted to a speedy, predatory lifestyle, blood flows through an **open circulatory system**.

SECTION 23.8 **Roundworms** (nematodes) have an unsegmented, cylindrical body covered by a cuticle that is **molted** as the animal grows. Most are decomposers in soil, but some are parasites of plants or animals, including humans. One free-living soil roundworm is commonly used in studies of development.

SECTIONS 23.9–23.11 There are more than 1 million **arthropod** species. Their diversity is attributed to structural traits such as a hardened **exoskeleton**, jointed appendages, specialized segments, and sensory structures such as **antennae** and **compound eyes**. Dramatic body changes during **metamorphosis** allow some arthropods to exploit multiple resources during their lifetime.

Chelicerates include marine horseshoe crabs and the mostly terrestrial **arachnids** (spiders, scorpions, ticks, and mites). **Crustaceans** are the most abundant arthropods in the seas. They include bottom-feeding crabs, free-swimming krill and copepods, and the sessile shelled barnacles. Centipedes and millipedes are **myriapods**.

Insects include the only winged invertebrates. Some serve as decomposers and pollinators; others harm crops and transmit diseases.

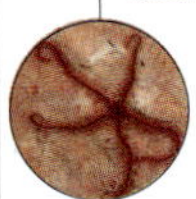

SECTION 23.12 **Echinoderms** such as sea stars, brittle stars, sea urchins, and sea cucumbers belong to the deuterostome lineage. Spines and other hard parts embedded in their skin support the body. There is no central nervous system. A **water–vascular system** with tube feet functions in locomotion. Adults are radial, but larvae are bilateral, implying a bilateral ancestry.

SECTION 23.13 Invertebrate species are a largely untapped source of medicines. Compounds that predatory snails use to subdue their fish prey can sometimes be used as drugs because all vertebrates have similar nervous systems.

Data Analysis Activities

Use of Horseshoe Crab Blood Horseshoe crab blood clots immediately upon exposure to bacterial endotoxins (Section 19.6), so it can be used to test injectable drugs for the presence of dangerous bacteria. To keep horseshoe crab populations stable, blood is extracted from captured animals, which are then returned to the wild. Concerns about the survival of animals after bleeding led researchers to do an experiment. They compared survival of animals captured and maintained in a tank with that of animals captured, bled, and kept in a similar tank. **FIGURE 23.35** shows the results.

1. In which trial did the most control crabs die? In which did the most bled crabs die?
2. Looking at the overall results, how did the mortality of the two groups differ?
3. Based on these results, would you conclude that bleeding harms horseshoe crabs more than capture alone does?

	Control Animals		Bled Animals	
Trial	Number of crabs	Number that died	Number of crabs	Number that died
1	10	0	10	0
2	10	0	10	3
3	30	0	30	0
4	30	0	30	0
5	30	1	30	6
6	30	0	30	0
7	30	0	30	2
8	30	0	30	5
Total	200	1	200	16

FIGURE 23.35 Mortality of young male horseshoe crabs kept in tanks during the two weeks after their capture. Blood was taken from half the animals on the day of their capture. Control animals were handled, but not bled. This procedure was repeated eight times with different sets of horseshoe crabs.

Self-Quiz Answers in Appendix VII

1. True or false? Animal cells have chitin walls.

2. A coelom is a ______ .
   a. type of bristle
   b. resting stage
   c. sensory organ
   d. lined body cavity

3. Cnidarians alone have ______ .
   a. cnidocytes
   b. a mantle
   c. a hydrostatic skeleton
   d. a radula

4. Flukes are most closely related to ______ .
   a. tapeworms
   b. roundworms
   c. arthropods
   d. echinoderms

5. Which group has six legs and two antennae?
   a. crustaceans
   b. insects
   c. spiders
   d. horseshoe crabs

6. The ______ are mollusks with a hinged shell.
   a. bivalves
   b. barnacles
   c. gastropods
   d. cephalopods

7. ______ have the smallest genome of all living animals.
   a. Sponges
   b. Placozoans
   c. Jellies
   d. Flatworms

8. Which of these groups includes the most species?
   a. protostomes
   b. roundworms
   c. arthropods
   d. mollusks

9. The ______ include the only winged invertebrates.
   a. cnidarians
   b. echinoderms
   c. arthropods
   d. mollusks

10. The ______ move about on tube feet.
    a. cnidarians
    b. echinoderms
    c. annelids
    d. flatworms

11. Annelids and cephalopods have a(n) ______ circulatory system.
    a. open
    b. closed

12. A pupa forms during the life cycle of some ______ .
    a. crustaceans
    b. echinoderms
    c. annelids
    d. insects

13. Match the organisms with their descriptions.

___ echinoderms	a. complete gut, pseudocoelom
___ mollusks	b. tube feet, spiny skin
___ sponges	c. simplest organ systems
___ cnidarians	d. body with lots of pores
___ flatworms	e. jointed exoskeleton
___ roundworms	f. mantle over body mass
___ annelids	g. segmented worms
___ arthropods	h. tentacles with stinging cells

Critical Thinking

1. Most hermaphrodites cannot fertilize their own eggs, but tapeworms can. Explain the advantages and disadvantages of such self-fertilization.

2. Acidity makes it harder for invertebrates to build structures containing calcium carbonate. List the types of animals that an increase in ocean acidity could harm.

3. Why are pesticides designed to kill insects more likely to harm lobsters and crabs than fish?

CREDIT: (35) left, Data: Walls, E., Berkson, J., *Fish. Bull.* 101:457–459 (2003); right, Jane Burton/ Bruce Coleman, Ltd.

Excavation of fossil dinosaur bones in northern Wyoming. Much of our information about vertebrate evolution comes from studies of such fossil skeletons.

24

ANIMALS II: THE CHORDATES

Links to Earlier Concepts

As we continue our survey of animals, you may wish to review the discussion of vertebrates and the animal evolutionary tree (Sections 23.1, 23.2). The geologic time scale (16.6) will help put events discussed here in perspective. You will see more examples of fossils (16.3) and of homologous structures (16.7). You will draw on your understanding of speciation, adaptive radiation, extinction (17.10–17.13), and cladistics (17.12).

KEY CONCEPTS

CHARACTERISTICS OF CHORDATES

Four traits characterize chordate embryos: a supporting rod (notochord), a dorsal nerve cord, a pharynx with gill slits, and a tail that extends past the anus.

THE FISHES

Fishes were the first vertebrates. Most modern fishes have jaws. They include cartilaginous fishes such as sharks, and two lineages of bony fishes.

FROM WATER TO LAND

Tetrapods, animals that walk on four legs, evolved from a lineage of bony fishes. Amphibians, the first tetrapods, live on land, but their eggs must develop in water.

THE AMNIOTES

Amniotes are adapted to life on land. Reptiles (including birds) and mammals are the two modern amniote lineages. Dinosaurs are an extinct lineage of reptiles.

EARLY PRIMATES TO HUMANS

Primates have grasping hands with nails rather than claws. Our closest extinct relatives are primates that walked upright. Fossils and genetics provide information about our history.

Photograph by Lynn Johnson, National Geographic Creative.

24.1 WHAT ARE CHORDATES?

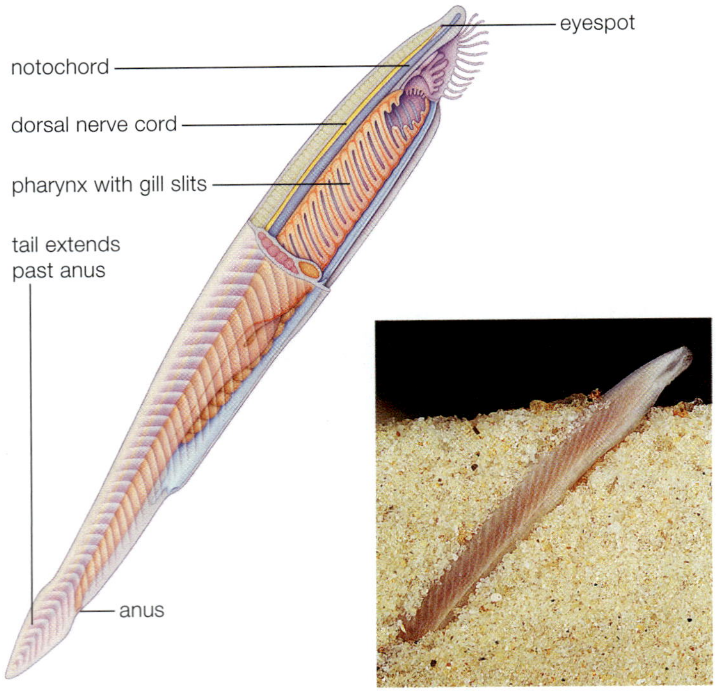

FIGURE 24.1 **{Animated}** Body plan and photo of a lancelet. Both adults and larvae have all four defining chordate traits.

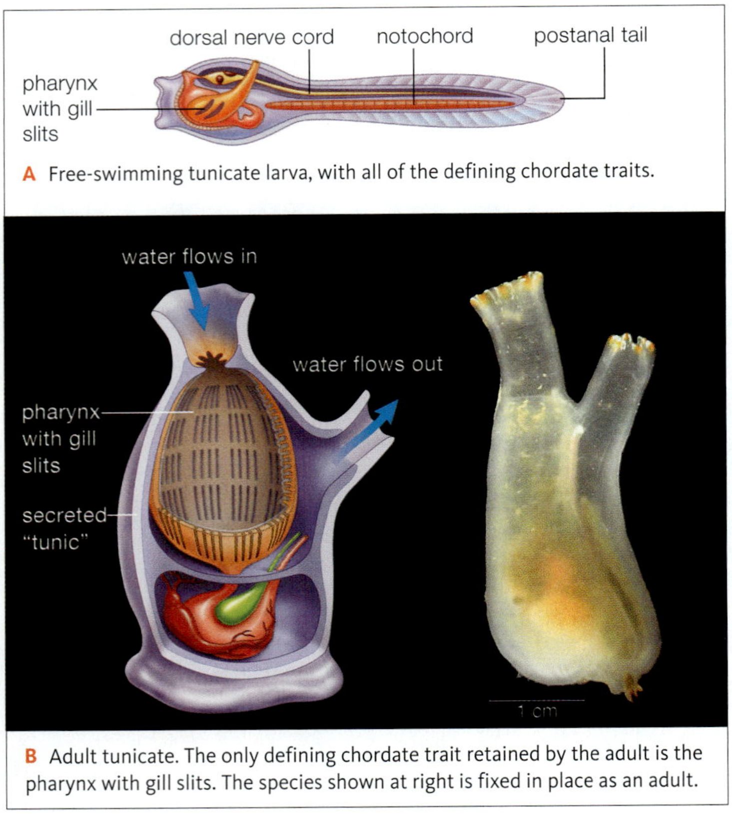

A Free-swimming tunicate larva, with all of the defining chordate traits.

B Adult tunicate. The only defining chordate trait retained by the adult is the pharynx with gill slits. The species shown at right is fixed in place as an adult.

FIGURE 24.2 **{Animated}** Tunicates.

CHORDATE CHARACTERISTICS

The previous chapter ended with a discussion of echinoderms, one of the deuterostome lineages. The other major deuterostome lineage is the **chordates** (phylum Chordata). These bilaterally symmetrical, coelomate animals typically have a complete digestive system and a closed circulatory system. Four traits of chordate embryos define the lineage:

1. A **notochord**, a rod of stiff but flexible connective tissue, extends the length of the body and supports it.
2. A dorsal, hollow nerve cord parallels the notochord.
3. Narrow gill slits open across the wall of the pharynx (the throat region).
4. A muscular tail extends beyond the anus.

Depending on the subgroup, some, none, or all of these embryonic traits persist in the adult.

Most chordate species are **vertebrates** (subphylum Vertebrata), animals that have a backbone. However, the chordates also include two lineages of marine invertebrates.

INVERTEBRATE CHORDATES

Lancelets (Cephalochordata) are invertebrates with an elongated body that retains all characteristic chordate features into adulthood (**FIGURE 24.1**). A lancelet is about 5 centimeters (2 inches) long. Its dorsal nerve cord connects to a simple brain. An eyespot at the end of the nerve cord detects light, but there are no paired sensory organs like those of fishes. Lancelets are suspension feeders. They usually bury themselves so only their head sticks out, then draw water in through their mouth. Food particles are filtered out as the water exits through the gill slits. Cilia on the gills move food particles to the gut. Lancelets are either male or female.

Tunicates (Urochordata) larvae have all the characteristic chordate features (**FIGURE 24.2A**). They swim about briefly, then undergo metamorphosis. The tail breaks down and other parts become rearranged. Adults retain only the pharynx with gill slits (**FIGURE 24.2B**). They secrete a carbohydrate-rich covering, or "tunic," that gives the group its common name. An adult feeds by drawing water in through an oral opening, capturing food on its gill slits, then expelling water through a second opening. Most tunicates are sessile, but those called salps drift or swim in the sea. Tunicates are hermaphrodites; an individual releases both eggs and sperm.

OVERVIEW OF CHORDATE EVOLUTION

Until recently, lancelets were considered the closest invertebrate relatives of vertebrates. An adult lancelet looks more like a fish than an adult tunicate does, but such apparent similarities can be deceiving. Studies of developmental processes and gene sequences have revealed that tunicates are closer to the vertebrates, as illustrated in **FIGURE 24.3**.

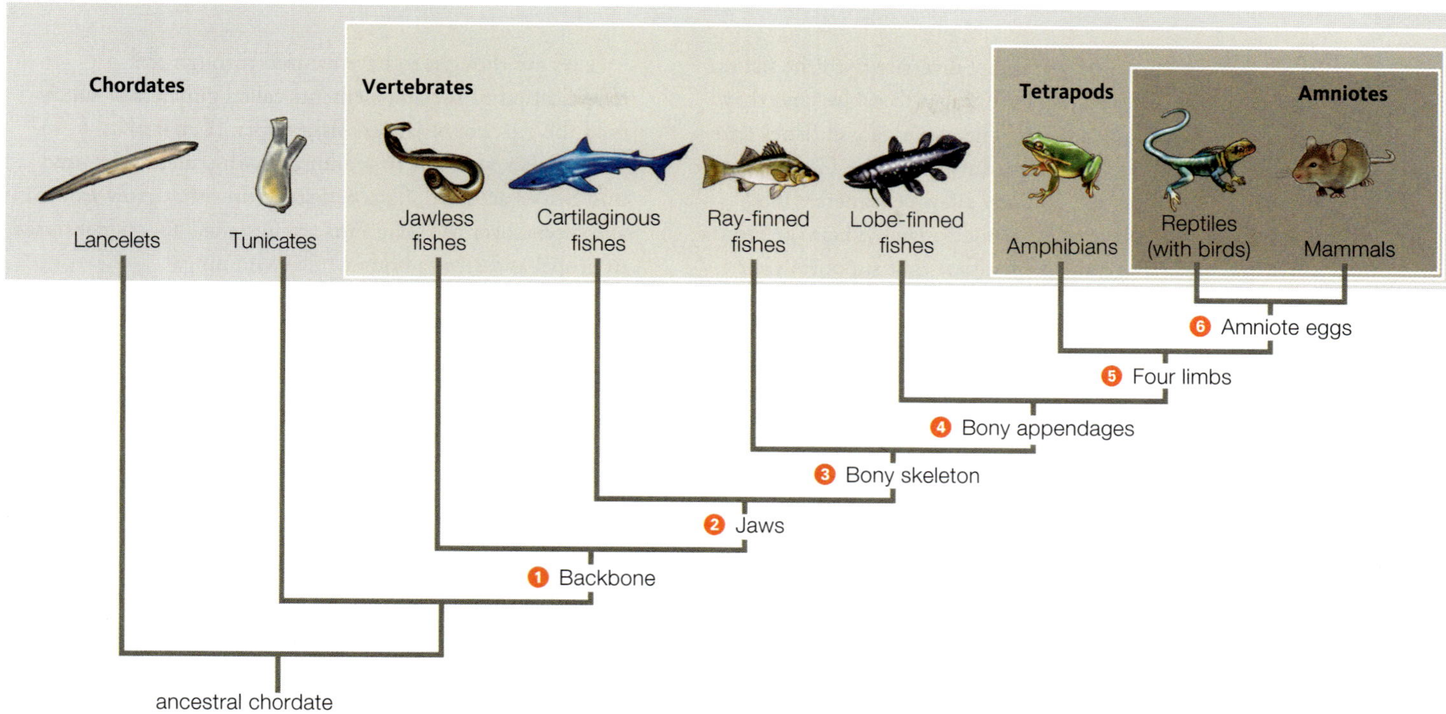

FIGURE 24.3 {Animated} Evolutionary tree diagram for the vertebrates with key traits for each clade. Boxes denote some of the clades that nest within the chordate clade and will be discussed again within the chapter.

Most chordates are vertebrates ①. Their backbone and other skeletal elements are components of the vertebrate **endoskeleton**, or internal skeleton. The vertebrate endoskeleton consists of living cells and grows with an animal, so it does not have to be molted.

The first vertebrates were jawless fishes that sucked up or scraped up food. Later, hinged skeletal elements called jaws evolved ②. Fishes with jaws were able to exploit new strategies for feeding. The vast majority of fishes and other modern vertebrates have jaws.

Jawless fishes had a skeleton of cartilage, and one group of jawed fishes (cartilaginous fishes) retains this trait. In other jawed animals, bone replaces cartilage as the major component of the skeleton ③. Sturdy fins supported by bones evolved in one group of bony fishes (lobe-finned fishes) ④. A branch from this lineage gave rise to the **tetrapods**—vertebrates that have four limbs or are descendants of a four-limbed ancestor ⑤.

Early tetrapods spent some time on land, but laid their eggs in water. Eggs that enclosed an embryo within a series of waterproof membranes evolved later, in **amniotes** ⑥. These specialized eggs and other adaptations to a dry habitat allowed amniotes to disperse widely on land. Modern amniotes include the animals traditionally known as reptiles (lizards, snakes, and turtles), as well as birds and mammals.

amniote Vertebrate whose egg has waterproof membranes that allow it to develop away from water; a reptile, bird, or mammal.
chordate Animal with an embryo that has a notochord, dorsal nerve cord, pharyngeal gill slits, and a tail that extends beyond the anus. For example, a lancelet or a vertebrate.
endoskeleton Internal skeleton made up of hardened components such as bones.
lancelet Invertebrate chordate that has a fishlike shape and retains the defining chordate traits into adulthood.
notochord Stiff rod of connective tissue that runs the length of the body in chordate larvae or embryos.
tetrapod Vertebrate with four legs, or a descendant thereof.
tunicate Invertebrate chordate that loses most of its defining chordate traits during the transition to adulthood.
vertebrate Animal with a backbone.

TAKE-HOME MESSAGE 24.1

All chordate embryos have a notochord, a dorsal hollow nerve cord, a pharynx with gill slits in its wall, and a tail that extends past the anus.

Most chordates are vertebrates, which have an internal skeleton that includes a backbone. The first vertebrates were jawless fishes.

A group of jawed fishes with a bony skeleton gave rise to four-legged animals (tetrapods) that colonized the land.

Amniotes, a tetrapod subgroup, have specialized eggs that allow them to develop away from water.

24.2 WHAT ARE FISHES?

We begin our survey of vertebrate diversity with the fishes. These were the first vertebrate lineages to evolve, and they remain the most fully aquatic. The earliest fossil fishes date to about 530 million years ago, during the late Cambrian period. They had a tapered body a few centimeters long, and a head with a pair of eyes, but no jaws. Their skeleton consisted of cartilage, the same tissue that supports your ears and nose.

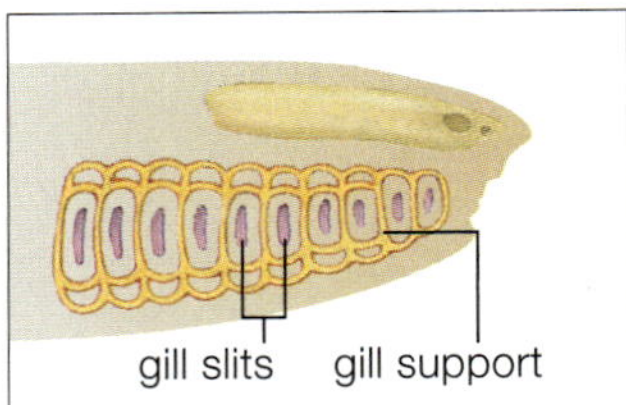

A Jawless fish

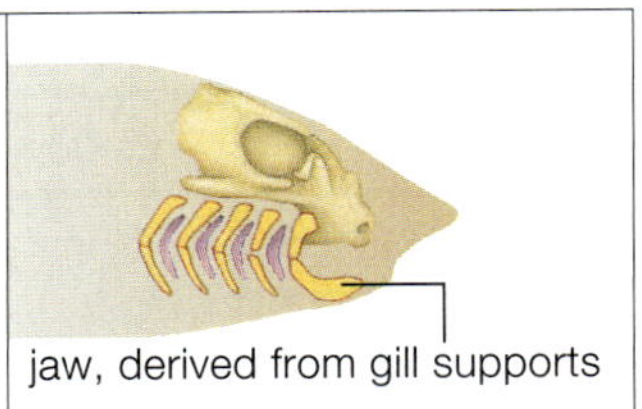

B Jawed fish

FIGURE 24.4 {Animated} Proposed mechanism for the evolution of jaws. Modification of gill supports of a jawless fish A produced a hinged, two-part jaw in early jawed fishes B.

A Hagfish. It feeds on worms and scavenges on the seafloor.

B Parasitic lamprey. It attaches to another fish with its oral disk and scrapes off bits of flesh.

FIGURE 24.5 Modern jawless fishes.

Jaws are thought to have evolved through the modification of skeletal elements called gill arches, which hold the gill slits of jawless fishes open (**FIGURE 24.4**). Jawed fishes were the first animals with scales and paired fins. Scales are hard, flattened structures that grow from and often cover the skin. Fins are flattened appendages used to propel and steer a body while swimming.

JAWLESS FISHES

Hagfishes and lampreys are modern **jawless fishes**. Relationships among lampreys, hagfishes, and jawed fishes have long inspired debate. The most recent research indicates that hagfishes and lampreys are monophyletic; both belong to a lineage that branched off before jawed fishes evolved. Lampreys and hagfishes have a cartilage skeleton and a cylindrical body about a meter long. They do not have fins and, like a lancelet, they move with a wiggling motion. Although they have no jaws, they do have hard mouthparts made of keratin, the main protein in hair.

Hagfishes (**FIGURE 24.5A**) feed on soft-bodied invertebrates and dead or dying fish. When threatened, they secrete a slimy mucus. Although the slime deters most predators, humans still harvest hagfish. Most products sold as "eelskin" are really hagfish skin.

Many lampreys parasitize other fish. A parasitic lamprey (**FIGURE 24.5B**) attaches to a host fish, then uses its hard mouthparts to scrape off bits of flesh. In the early 1900s, parasitic Atlantic lampreys that entered the Great Lakes via newly built canals decimated populations of native fish such as trout. Fishery managers now lower lamprey numbers with dams, nets, and poisons.

CARTILAGINOUS FISHES

The mostly marine **cartilaginous fishes** (Chondrichthyes) are jawed fishes with a skeleton of cartilage. Multiple gill slits are uncovered and visible at the body surface, as they are in jawless fishes. On the lower surface of the body, a single multipurpose opening called the **cloaca** serves as the exit for digestive and urinary waste. The cloaca also functions in reproduction.

Most cartilaginous fishes are sharks, rays, or skates. Streamlined predatory sharks chase down and tear apart prey (**FIGURE 24.6A**). Other sharks are bottom-feeders that suck up invertebrates and act as scavengers. Still others strain plankton from the seawater. Rays have a flattened body with large pectoral fins and a long, thin tail. Manta rays glide through seawater and filter out plankton (**FIGURE 24.6B**). Stingrays are bottom-feeders. A sharp spearlike point at the tip of their tail defends them against predators. In some stingrays, the tail can deliver a dose of venom. Skates are bottom-feeders that resemble rays but have a shorter, thicker tail.

BONY FISHES

There are two modern lineages of bony fishes: ray-finned fishes and lobe-finned fishes. In both groups, bone replaces cartilage in some or most of the skeleton, and gill slits are typically hidden beneath a gill cover.

Ray-finned fishes have thin, weblike fins with flexible supports derived from skin. **FIGURE 24.7A** shows the internal anatomy of a typical ray-finned fish. Like all fishes, it has a closed circulatory system with one heart, and a urinary system with paired kidneys that filter blood, adjust its composition, and eliminate wastes. In most ray-finned fishes, urinary waste, digestive waste, and gametes exit the body through three separate openings. Ray-finned fishes typically have a gas-filled swim bladder. By adjusting the volume of gas in its swim bladder, a fish can change its buoyancy so that it remains suspended at the desired depth. With more than 21,000 freshwater and marine species, ray-finned fishes are the most diverse vertebrate group.

Lobe-finned fishes include coelacanths (Section 17.11) and lungfishes (**FIGURE 24.7B**). Their pelvic and pectoral fins are fleshy, with supporting bones inside them. Lungfishes have gills and one or two lungs. Their lungs are air-filled sacs with an associated network of tiny blood vessels. A lungfish gulps air into its lungs, then oxygen diffuses from its lungs into the blood. As the next section explains, bony fins and simple lungs proved advantageous when descendants of lobe-finned fishes ventured onto land.

cartilaginous fish Jawed fish with a skeleton of cartilage; a shark, ray, or skate.
cloaca Body opening that serves as the exit for digestive waste and urine; also functions in reproduction.
jawless fish Fish with a skeleton of cartilage, no fins or jaws; a lamprey or hagfish.
lobe-finned fish Jawed fish with fleshy fins that contain bones; a coelacanth or lungfish.
ray-finned fish Jawed fish with fins supported by thin rays derived from skin; member of most diverse lineage of fishes.

TAKE-HOME MESSAGE 24.2

Fishes are the most diverse vertebrates. Like other vertebrates, they have a closed circulatory system with one heart, and a urinary system with paired kidneys.

Lampreys and hagfishes are jawless fishes. Most fishes have jaws that evolved from gill supports.

Cartilaginous fishes such as sharks have a skeleton of cartilage. Bony fishes include the highly diverse ray-finned fishes and the lobe-finned fishes, which are the closest relatives of tetrapods.

A Galápagos shark, a fast-swimming predator.

B Manta ray, a slow-moving plankton feeder.

FIGURE 24.6 {Animated} Two cartilaginous fishes. Note the gill slits visible at the body surface of both fishes.

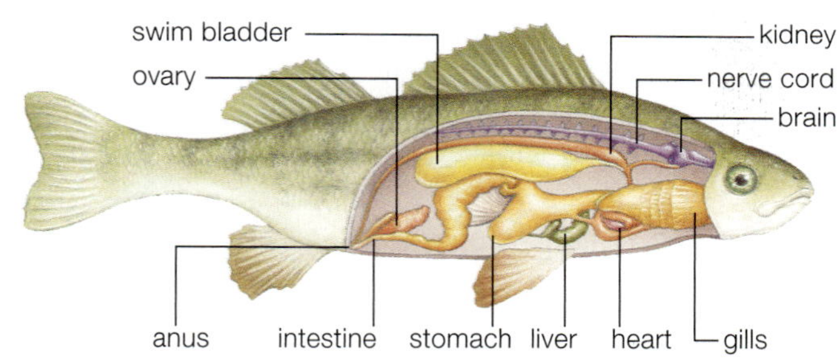

A Anatomy of a ray-finned fish (a perch).

B Lungfish, a lobe-finned fish, with bony pelvic and pectoral fins.

FIGURE 24.7 {Animated} Bony fishes. Both the ray-finned and lobe-finned lineages have covered gills and a skeleton that consists mainly of bone.

CREDITS: (6A) Jonathan Bird/Oceanic Research Group, Inc.; (6B) © Gido Braase/Deep Blue Productions; (7A) © Cengage Learning; (7B) © Wernher Krutein/photovault.com.

24.3 WHAT ARE AMPHIBIANS?

THE FIRST TETRAPODS

Amphibians are scaleless, land-dwelling vertebrates that typically breed in water. All are carnivores. Amphibians were the first tetrapods. Recently discovered fossil footprints from Poland show that amphibians were walking on land by about 395 million years ago, in the Devonian. The animal that left the footprints was about 2.5 meters (8 feet) long.

A variety of fossils demonstrate how fishes adapted to swimming evolved into four-legged walkers (**FIGURE 24.8**). Bones of a lobe-finned fish's pectoral fins and pelvic fins are homologous with those of an amphibian's front and hind limbs. During the transition to land, these bones became larger and better able to bear weight. Ribs enlarged and a distinct neck emerged, allowing the head to move independently of the rest of the body.

The transition to land was not simply a matter of skeletal changes. Lungs, which had previously served an accessory purpose, became larger and more complex. Division of the previously two-chambered heart into three chambers allowed blood to flow in two circuits, one to the body and one to those increasingly important lungs. Changes to the inner ear improved detection of airborne sounds. Eyes became protected from drying out by eyelids.

What drove the move onto land? An ability to spend time out of water would have been favored in seasonally dry places. In addition, it would have allowed individuals to escape aquatic predators and to access a new source of food—insects—which also evolved during the Devonian.

FIGURE 24.9 Salamander, with equal-sized forelimbs and hindlimbs.

MODERN AMPHIBIANS

Salamanders and newts have a body form similar to early tetrapods. Their tail is long, and their forelimbs are about the same size as their back limbs (**FIGURE 24.9**). When a salamander walks, the movements of its limbs and bending of its trunk resemble movements made by a fish swimming through water. Early tetrapods probably used this motion to walk in the water before one lineage ventured onto land.

Larval salamanders look like small versions of adults, except for the presence of gills. In a few species (axolotls), gills are retained into adulthood. More typically, gills disappear as the animal matures and lungs develop.

Frogs and toads belong to the most diverse amphibian lineage, with more than 5,000 species. The long, muscular

FIGURE 24.8 Fossil species from the late Devonian illustrate how a fish body became adapted for life on land.

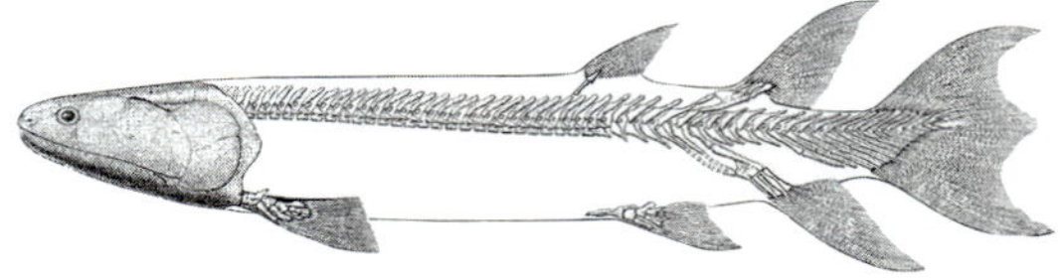

1 Fish (*Eusthenopteron*) with bony fins.

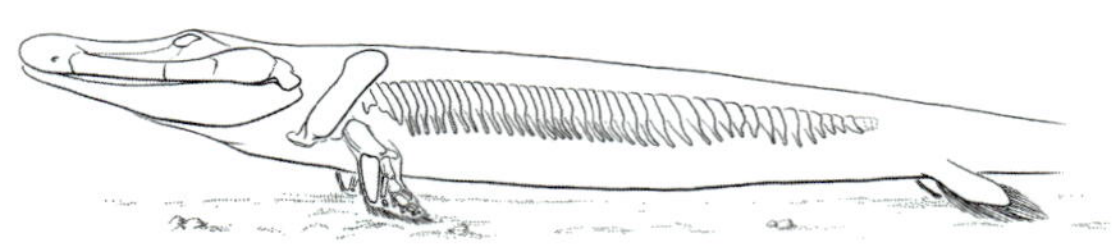

2 Fish (*Tiktaalik*) with sturdier weight-bearing pectoral fins, wristlike bones, and enlarged ribs.

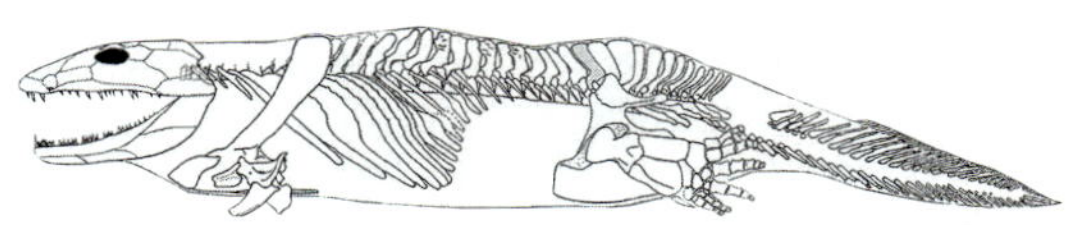

3 Early amphibian (*Icthyostegna*) with well-developed ribs, and thick limbs with distinct digits.

A Long, muscular hindlimbs allow an adult frog to leap. Frogs spend much of their time in water.

B Toads have a thicker skin and can live away from water.

C Aquatic larvae (tadpoles) have gills, a long tail, and no limbs.

FIGURE 24.10 Frogs and toads. Both frogs and toads have longer hindlimbs than forelimbs and are capable of hopping or jumping.

hindlimbs of an adult frog allow it to swim, hop, and make spectacular leaps (**FIGURE 24.10A**). The much smaller forelimbs help absorb the impact of landings. Toads can hop, but they usually walk, and they have somewhat shorter hind legs than frogs (**FIGURE 24.10B**). Toads are also better adapted to dry conditions. Both frogs and toads undergo metamorphosis, during which the gilled, tailed larva (**FIGURE 24.10C**) transforms itself into an adult with lungs and no tail.

DECLINING DIVERSITY

Amphibian populations throughout the world are declining or disappearing. Researchers correlate many declines with shrinking or deteriorating habitats. For example, nearly all amphibians need to deposit their eggs and sperm in water, and their larvae must develop in water. Thus, they prefer to breed in low-lying ground where rain collects and forms pools of standing water. Humans inadvertently destroy these appropriate breeding spots by leveling the ground.

Other factors that contribute to amphibian declines include introduction of new species in amphibian habitats, long-term shifts in climate, increases in ultraviolet radiation, water pollution, and the spread of pathogens and parasites such as the chytrid fungus discussed in Section 22.1.

amphibian Tetrapod with scaleless skin; it typically develops in water, then lives on land as a carnivore with lungs. For example, a frog or salamander.

TAKE-HOME MESSAGE 24.3

- The tetrapod lineage branched off from the lobe-finned fishes. Amphibians are the oldest tetrapod lineage.
- Lungs and a three-chambered heart adapt amphibians to life on land, but they typically lay their eggs in water.
- Frogs and toads undergo metamorphosis from a gilled, tailed, limbless larva to an adult with four limbs and no tail.

24.4 WHAT ARE AMNIOTES?

FIGURE 24.11 Artist's depiction of the largest feathered dinosaur yet discovered. A human figure is shown for scale.

Amniotes branched off from an amphibian ancestor about 300 million years ago, during the Carboniferous. A variety of traits adapt them to life in dry places. They have lungs, rather than gills, throughout their life. Their skin is rich in keratin, a protein that makes it waterproof. A pair of well-developed kidneys help conserve water, and fertilization usually takes place inside the female's body. Amniotes produce eggs in which an embryo develops bathed in fluid, so amniotes can develop on dry land. A series of membranes within the egg function in gas exchange, nutrition, and waste removal.

An early branching of the amniote lineage separated ancestors of mammals from the common ancestor of all modern **reptiles**. You probably do not think of birds as reptiles, but the reptile clade includes turtles, lizards, snakes, crocodilians, and birds:

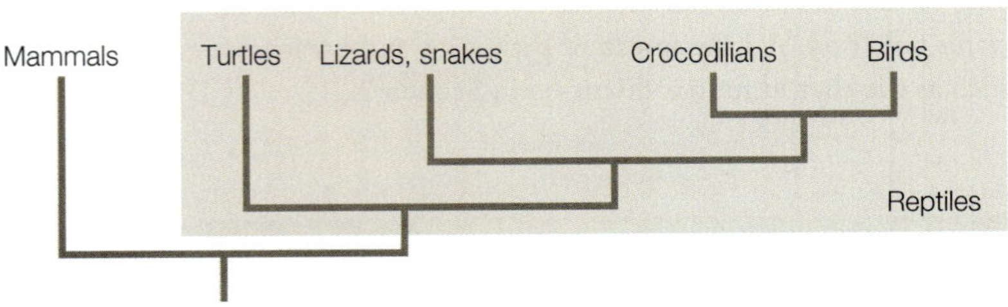

Dinosaurs are extinct members of the reptile clade. Biologists define them by skeletal features, such as the shape of their pelvis and hips. Like other reptiles, dinosaurs produced amniote eggs. One dinosaur group, the theropods, includes many feathered species (**FIGURE 24.11**). Most likely, the feathers benefited the dinosaurs by providing insulation. Like modern birds and mammals, these feathered dinosaurs may have been **endotherms**, animals that maintain their body temperature by adjusting their production of metabolic heat. Endotherm means "heated from within." Most nonbird reptiles are **ectotherms**, animals whose internal temperature varies with that of their environment.

Birds branched off from a theropod dinosaur lineage during the Jurassic, and are the only surviving descendants of dinosaurs. All dinosaurs became extinct by the end of the Cretaceous, probably as a result of an asteroid impact (Section 16.9).

dinosaur Group of reptiles that include the ancestors of birds; became extinct at the end of the Cretaceous.
ectotherm Animals whose body temperature varies with that of its environment.
endotherm Animal that maintains its temperature by adjusting its production of metabolic heat; for example, a bird or mammal.
reptile Amniote subgroup that includes lizards, snakes, turtles, crocodilians, and birds.

TAKE-HOME MESSAGE 24.4

Amniotes are animals that produce eggs in which the young can develop away from water. They have waterproof skin and highly efficient kidneys. Fertilization is typically internal.

An early divergence separated the ancestors of mammals from the ancestors of reptiles—a group in which biologists include turtles, lizards, snakes, crocodilians, and birds.

Birds evolved from a group of feathered dinosaurs.

24.5 WHAT ARE THE NONBIRD REPTILES?

All modern reptiles have scales and a cloaca. With the exception of birds, which we consider next, all are ectotherms. They are often seen basking in the sun to warm their bodies.

Lizards and snakes constitute the most diverse group of modern reptiles. Their body is covered with overlapping scales. Most lizards and snakes lay eggs (**FIGURE 24.12A**), but females of some species brood eggs in their body and give birth to well-developed young. The live-bearers do not provide any nourishment to the offspring inside their body, as some mammals do.

The smallest lizard can fit on a dime (*left*). The largest, the Komodo dragon, grows up to 3 meters (10 feet) long. Most lizards are predators, although iguanas are herbivores.

The first snakes evolved during the Cretaceous, from short-legged, long-bodied lizards. Some modern snakes have bony remnants of hindlimbs, but most lack limb bones entirely. All are predators. Many have flexible jaws that help them swallow prey whole. All snakes have teeth, but not all have fangs. Rattlesnakes and other fanged types bite and subdue prey with venom they produce in modified salivary (saliva-producing) glands. Each year in the United States there are about 7,000 reported snake bites, five of them fatal. The bites are a defense response.

Turtles have a bony, keratin-covered shell attached to their skeleton. They lack teeth, but a thick layer of keratin covers their jaws, forming a horny beak (**FIGURE 24.12B**). Most turtles that live in the sea feed on invertebrates such as sponges or jellies; others feed mainly on sea grass. Freshwater turtles prey on fish and invertebrates. Land-dwelling turtles, which are commonly called tortoises, feed on plants.

Crocodilians—crocodiles, alligators, and caimans—are stealthy predators with a long snout and many sharp peglike teeth (**FIGURE 24.12C**). They spend much of their time in water and a long, powerful tail propels them when they swim. Crocodilians are the closest living relatives of birds. Like most birds, crocodilians are highly vocal and engage in complex parental behavior. During courtship males and females grunt and bellow. After a female mates, she digs a nest, lays eggs, then buries and guards them. The young call when they are ready to hatch, and their mother helps them dig their way out.

Lizards, snakes, and turtles have a three-chambered heart, but the crocodilians have a heart with four chambers. A four-chambered heart improves efficiency of oxygen delivery by preventing oxygen-poor blood from ever mixing with oxygen-rich blood.

A Hognose snakes emerging from leathery amniote eggs.

B A turtle in a defensive posture. Note the horny beak.

C Spectacled caiman, a crocodilian. All crocodilians are predators that spend much of their time in water.

FIGURE 24.12 {Animated} Examples of nonbird reptiles.

TAKE-HOME MESSAGE 24.5

All modern nonbird reptiles are ectotherms that have a scale-covered body.

Lizards and snakes have an elongated body with overlapping scales. Most lizards and all snakes are predators.

Turtles have a bony shell and a horny beak.

Crocodiles and alligators are the closest living relatives of birds. Like birds, they have a four-chambered heart and care for their eggs and hatchlings.

24.6 WHAT ARE BIRDS?

FIGURE 24.13 An owl in flight. Muscles that connect the wing to the breastbone power flight.

Birds are the only living animals with feathers. Bird feathers have roles in flight, insulation, and courtship displays. They also shed water and thus help keep the skin beneath dry.

Most modern birds have a body well adapted to flight (**FIGURE 24.13**). Like humans, birds stand upright. In birds, the forelimbs have evolved into wings. Each wing is covered with long feathers that extend outward and increase its surface area. The feathers give the wing a shape that helps lift the bird. Flight muscles connect an enlarged sternum (breastbone) with a protruding keel to bones of the upper limb. When flight muscles contract, the resulting powerful downstroke lifts the bird.

Most birds are surprisingly lightweight, which helps them become and remain airborne. Air cavities inside a bird's bones keep its body weight low, as does the lack of a bladder (an organ that stores urinary waste in many other vertebrates). Rather than having heavy, bony teeth, birds have a beak made of keratin, which is lighter in weight.

Birds use a lot of energy in flight and to maintain their body temperature. Like mammals, birds are endotherms.

FIGURE 24.14 {Animated} Bird's egg. The amnion, chorion, and allantois are membranes characteristic of amniote eggs.

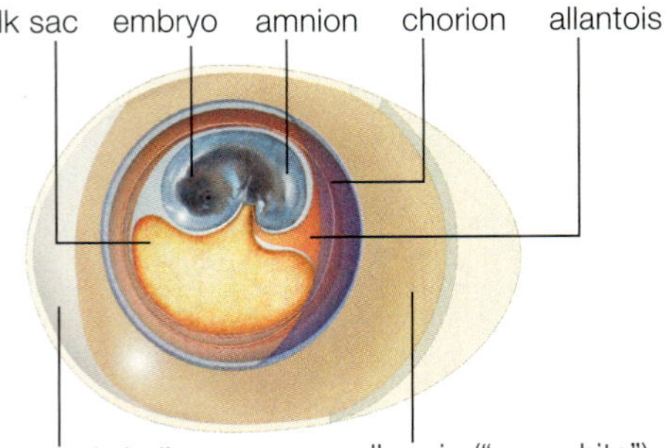

A unique system of air sacs keeps air flowing continually through a bird's lungs, ensuring an adequate oxygen supply to flight muscles. A four-chambered heart pumps blood quickly from the lungs to wing muscles and back. A bird also has a large heart for its size and a rapid heartbeat.

Flying requires good eyesight and a great deal of coordination. Compared to a lizard of a similar body weight, a bird has much larger eyes and a bigger brain. A bird's eyes are largely immobile. Thus, to alter its view, it must turn its highly flexible neck.

As in other reptiles, fertilization is internal. However, most male birds do not have a penis. To inseminate a female, a male must press his cloaca against hers, a maneuver poetically described as a cloacal kiss. After fertilization occurs in the female's body, she lays an egg with the characteristic amniote membranes (**FIGURE 24.14**). Nutrients from the yolk and water from the albumin in the egg sustain the developing embryo. Like crocodiles, birds encase their eggs in a hard calcium carbonate shell.

More than half of all bird species belong to a subgroup called perching birds. Sparrows, finches, jays, and other birds seen at backyard feeders are in this group. The next most diverse group, the hummingbirds, includes the most agile fliers and the only birds capable of flying backward. At the other extreme, penguins and the ratite birds such as ostriches can no longer become airborne. Penguins "fly" through water by flapping their wings in the same manner that other birds do when they fly through air.

bird Feathered reptile with a body adapted for flight.

TAKE-HOME MESSAGE 24.6

Birds are endothermic amniotes with feathers. Lightweight bones and highly efficient respiratory and circulatory systems adapt a bird to flight.

FURRY OR HAIRY MILK MAKERS

Mammals are furry or hairy animals in which females nourish their offspring with milk secreted from mammary glands. The group name is derived from the Latin *mamma*, meaning breast. Like birds, mammals have a four-chambered heart and are endotherms. Fur or hair helps them retain body heat.

In other vertebrates, an individual's teeth may vary in size, but they are all the same shape. Only mammals have teeth of four different shapes (**FIGURE 24.15**). This assortment of teeth allows mammals to process a wider variety of foods than other vertebrates.

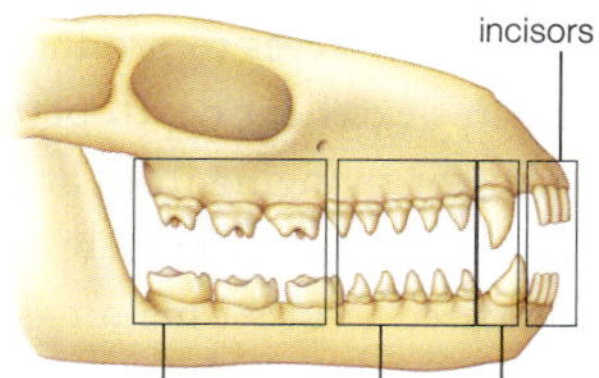

FIGURE 24.15 {Animated} Four types of mammalian teeth.

MODERN SUBGROUPS

Monotremes, the oldest mammalian lineage, lay eggs with a leathery shell. Only three monotreme species survive, the duck-billed platypus (**FIGURE 24.16A**) and two kinds of spiny anteater (echidna). Monotreme eggs hatch while young are still tiny, hairless, and blind. The young then cling to the mother or are held in a skin fold on her belly. A monotreme has mammary glands but no nipples. Young lap up milk that oozes from openings on their mother's skin.

Marsupials are pouched mammals. Young marsupials develop briefly in their mother's body, nourished by egg yolk and by nutrients that diffuse from the mother's tissues. They are born at an early developmental stage, and crawl to a permanent pouch on the surface of their mother's belly. A nipple in the pouch supplies milk that sustains continued development. Most of the 240 modern marsupials live in Australia and on nearby islands. Kangaroos and koalas are the best known. Opossums (**FIGURE 24.16B**) are the only marsupials native to North America.

In **placental mammals** (eutherians), maternal and embryonic tissues combine as an organ called the placenta. A placenta allows maternal and embryonic bloodstreams to transfer materials without mixing. Placental embryos grow faster than those of other mammals, and the offspring are born more fully developed. After birth, young suck milk from nipples on their mother's surface (**FIGURE 24.16C**). Monotremes and marsupials have a cloaca, but placental mammals have separate openings that service the urinary, reproductive, and excretory systems.

Placental mammals tend to outcompete other mammals and are now dominant on all continents except Australia. Of the approximately 4,000 species, nearly half are rodents such as rats and mice. The next most diverse group is bats, with 1,000 species. Humans are primates, the lineage we discuss in the next section.

A Platypuses are monotremes.

B Opossums are marsupials.

C Hamsters are placental mammals.

FIGURE 24.16 {Animated} Mammalian mothers and their young. All female mammals produce milk to feed offspring.

mammal Animal with hair or fur; females secrete milk from mammary glands.
marsupial Mammal in which young are born at an early stage and complete development in a pouch on the mother's surface.
monotreme Egg-laying mammal.
placental mammal Mammal in which maternal and embryonic bloodstreams exchange materials by means of a placenta.

TAKE-HOME MESSAGE 24.7

Mammals are endothermic amniotes that have hair or fur. Young are nourished by milk secreted from the female's mammary glands.

Egg-laying monotremes, pouched marsupials, and placental mammals are subgroups. Placental mammals are now the dominant lineage in most regions.

CREDITS: (15) © Cengage Learning; (16A) Jean Phillipe Varin/Jacana/Science Source; (16B) Jack Dermid; (16C) © Minden Pictures/SuperStock.

24.8 WHAT ARE PRIMATES?

PRIMATE CHARACTERISTICS

Primates are an order of placental mammals that includes humans, apes, monkeys, and their close relatives. Primates first evolved in tropical forests, and many of the group's characteristic traits arose as adaptations to life among the branches. Primate shoulders have an extensive range of motion (**FIGURE 24.17**). Unlike most mammals, a primate can extend its arms out to its sides, reach above its head, and rotate its forearm at the elbow. With the exception of humans, all living primates have both hands and feet capable of grasping. Mammals often have claws or hooves, but tips of primate fingers and toes typically have touch-sensitive pads protected by flat nails.

FIGURE 24.17 Adapted to climbing. An orangutan (an Asian ape) demonstrates her wide range of shoulder motion and her ability to grasp with her hands and feet.

Most mammals have eyes set toward the side of their head, but primate eyes tend to face forward. As a result, both eyes view the same area, each from a slightly different vantage point. The brain integrates the signals it receives from the two eyes to produce a three-dimensional image. A primate's excellent depth perception adapts it to a life spent leaping or swinging from limb to limb.

Primates have a large brain for their body size. Compared to other mammals, primates devote more brain area to vision and information processing, and less to smell.

Primates have a varied diet, and their teeth reflect this lack of specialization. They have teeth for tearing flesh, as well as for grinding material.

Most primates spend their life in a social group that includes adults of both sexes. Female primates usually give birth to only one or two young at a time and provide care for an extended period after birth.

MODERN SUBGROUPS

FIGURE 24.18 shows relationships among living primates. The oldest surviving primate lineage includes the lemurs, agile climbers that live in the forests of Madagascar, an island off the coast of Africa. Lemurs are active during the day and eat mainly fruits, leaves, and flowers. They have poor color vision. Like most other mammals, lemurs have a prominent wet nose and a vertically split upper lip that attaches tightly to the underlying gum (**FIGURE 24.19A**).

FIGURE 24.18 Evolutionary tree for some modern primates.

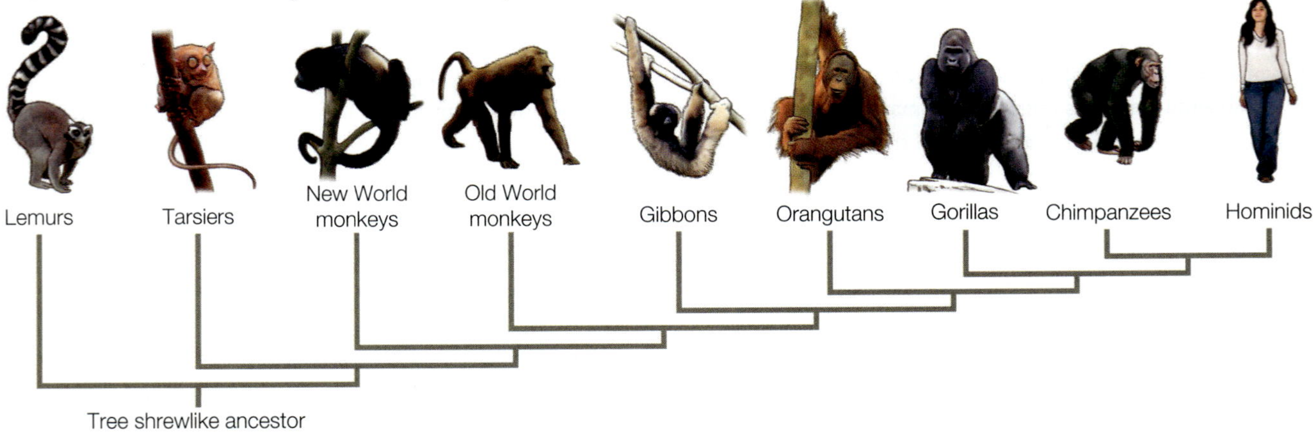

FIGURE IT OUT: Which modern primates are human ancestors?

Answer: None of them. All species ancestral to humans are extinct. We share a common ancestor with chimpanzees and bonobos.

A Lemur with a wet nose, cleft upper lip.

B Tarsier with a dry nose, uncleft upper lip.

C Squirrel monkey (New World monkey) with flat face, prehensile tail.

D Baboon (Old World monkey) with long nose, short tail.

E Gorilla, the largest ape.

F Chimpanzee, one of our two closest living relatives.

FIGURE 24.19 Primate diversity.

Tarsiers are small, nocturnal insect eaters that live in South Asia (**FIGURE 24.19B**). Like monkeys, apes, and humans, they are dry-nosed primates, with nostrils set in dry skin. Their upper lip is not cleft and has a reduced attachment to the gum. Evolution of a movable upper lip allowed a wider range of facial expressions and vocalizations.

Anthropoid primates—monkeys, apes, and humans—are typically active during the day and have good color vision (*anthropoid* means humanlike). New World monkeys (**FIGURE 24.19C**) climb through forests of Central and South America. They have a flat face and a nose with widely separated nostrils. A long tail helps them maintain balance. In many species, the tail is prehensile, meaning it can grasp things. Old World monkeys live in Africa, the Middle East, and Asia. They tend to be larger than New World monkeys and have a longer nose with closely set nostrils. Some are tree-climbing forest dwellers. Others, such as baboons (**FIGURE 24.19D**), spend most of their time on the ground in grasslands or deserts. Not all Old World monkeys have a tail, but in those that do it is short and never prehensile.

Most modern, tailless, nonhuman primates belong to the group commonly known as **apes**. About 24 species of small apes called gibbons inhabit Southeast Asian forests. They are sometimes referred to as "lesser apes," in comparison with the larger apes, or "great apes." The forest-dwelling orangutan of Sumatra and Borneo is the only surviving Asian great ape. All African great apes (gorillas, chimpanzees, and bonobos) live in social groups, and spend most of their time on the ground. Gorillas (**FIGURE 24.19E**), the largest living primates, live in forests and feed mainly on leaves. Chimpanzees (**FIGURE 24.19F**) and the bonobos, or pygmy chimpanzees, are our closest living relatives. The chimpanzee/bonobo lineage and the lineage leading to humans parted ways between 6 and 8 million years ago. Chimpanzees and bonobos eat fruit, but also catch insects and cooperatively hunt small mammals, including monkeys. The two species differ in their behavior, with chimpanzees engaging in more intraspecific aggression and bonobos spending more time in nonreproductive sex acts.

TAKE-HOME MESSAGE 24.8

Primates include lemurs, tarsiers, monkeys, apes, and humans.

Primate traits such as a flexible shoulder joint and grasping hands with nails are adaptations to climbing. Compared to other mammals, primates have a larger brain with a greater area devoted to vision and less to smell.

Humans are most closely related to apes, and the chimpanzees and bonobos are our closest living relatives.

anthropoid primate Humanlike primate; monkey, ape, or human.
ape Common name for a tailless nonhuman primate; a gibbon, orangutan, gorilla, chimpanzee, or bonobo.
primate Mammal having grasping hands with nails and a body adapted to climbing; for example, a lemur, monkey, ape, or human.

CREDITS: (19A) © iStockphoto.com/Catharina van den Dikkenberg; (19B) © iStockphoto.com/Robin O'Connel; (19C) primates.com; (19D) © iStockphoto.com/JasonRWarren; (19E) © Dallas Zoo, Robert Cabello; (19F) Kenneth Garrett/ National Geographic Creative.

24.9 HOW DO HUMANS AND APES DIFFER?

Humans share an ancestor with great apes, but differ from them in many aspects. Compared to a chimpanzee, a human has a flatter face, a smaller jaw with reduced canine teeth, and a larger brain (**FIGURE 24.20**). Our bodies are similarly sized, but our brains are three times bigger.

We also walk differently. When gorillas and chimpanzees walk, they lean forward and support their weight on their knuckles. By contrast, humans walk upright. Habitual upright walking, or **bipedalism**, is the defining trait of hominins. **Hominins** include modern humans and all extinct species more closely related to us than to any other primate.

Evolution of bipedalism involved many skeletal changes (**FIGURE 24.21**). A knuckle-walking ape has a backbone with a C-shape, whereas the human backbone has an S-shaped curve that keeps our head centered over our feet. Apes have flat feet capable of grasping, but human feet have a pronounced arch and a non-opposable big toe. An ape's spinal cord enters near the rear of skull, whereas the spinal cord of a human enters at the skull's base.

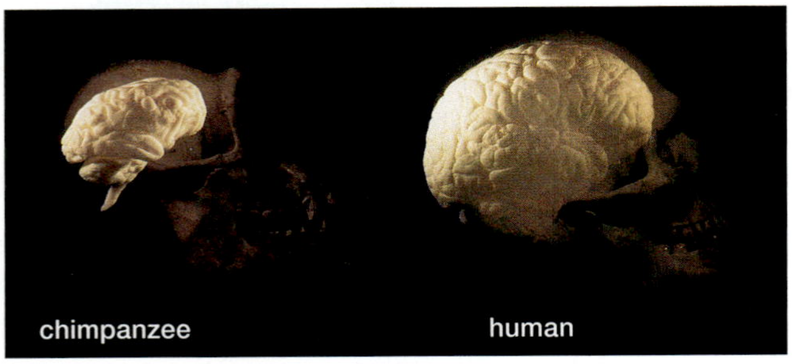

FIGURE 24.20 Comparison of chimpanzee and human skulls.

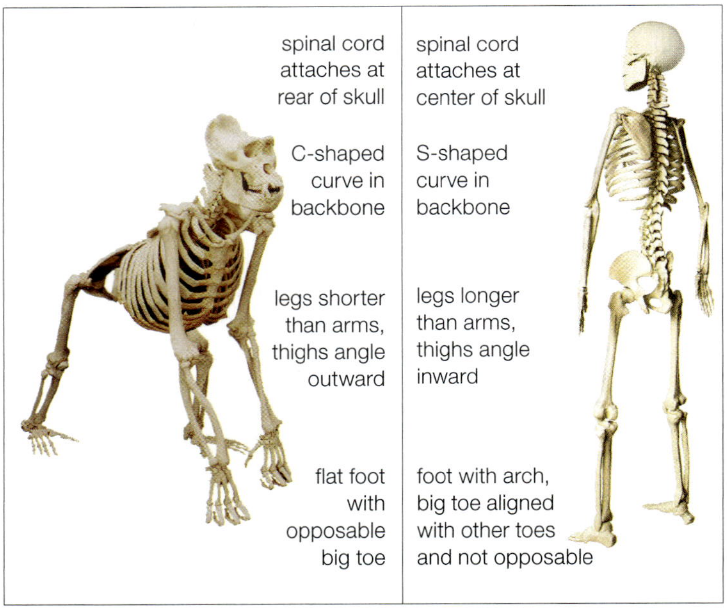

FIGURE 24.21 {Animated} Some skeletal differences between a knuckle-walking gorilla (*left*) and a bipedal human (*right*).

What favored the transition from a four-legged gait to bipedalism? A change in climate may have been a factor. Hominins evolved in Africa at a time when lush rain forests were giving way to woodlands interspersed with grassy plains. In this altered habitat, an ability to move efficiently across open ground would have been favored, and human walkers use less energy to move than chimpanzees do. Bipedalism also keeps a body cooler. A bipedal animal gains less heat from the ground than a four-legged walker and also intercepts less warming sunlight. In addition, an upright stance makes it easier to scan the horizon for predators, and it frees hands to gather and carry food or other resources.

Human hands also differ from those of apes. A human thumb is longer, stronger, and more maneuverable than that of a chimpanzee. Many monkeys and apes grasp objects in a power grip, but only humans routinely use a precision grip, pinching together the tips of the thumb and forefinger for fine manipulation of objects:

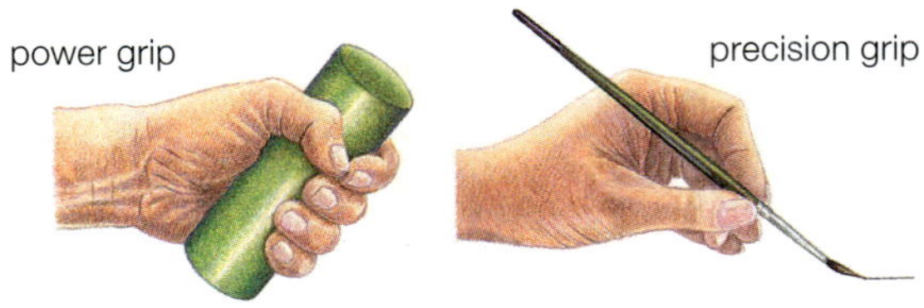

Humans are the only living primates without a thick coat of body hair. We have about the same number of body hairs as other primates, but ours are shorter and finer, giving our skin a more naked appearance. Most likely the decrease in hominin hairiness was driven by a need to stay cool. Sweat evaporates more quickly from bare skin than from under thick hair, and an improved cooling ability would have reduced the risk of overheating while walking or running across hot, sunny grasslands.

The many differences between humans and apes did not arise all at once. Rather, as hominins evolved, different traits changed at different rates. Hominins walked upright for millions of years before their brain size began to increase.

bipedalism Habitual upright walking.
hominin Human or an extinct primate species more closely related to humans than to any other primates.

TAKE-HOME MESSAGE 24.9

Compared to apes, humans have a larger brain, more flexible hands, and less insulating body hair.

Unlike apes, humans are bipedal, and have a skeleton that adapts them to this upright posture.

Differences between apes and humans arose over millions of years as evolutionary forces altered our hominin ancestors.

24.10 WHAT WERE EARLY HOMININS LIKE?

FIGURE 24.23 Reconstruction of *Australopithecus sediba*, a possible human ancestor that lived in South Africa 2 million years ago.

Fossil skull fragments from *Sahelanthropus tchadensis* may be the oldest evidence of hominins. This species, which lived about 7 million years ago in west-central Africa, is known only from fossil skulls. The skulls opened to the spinal cord at their base, rather than at the rear, suggesting that this species walked upright. *S. tchadensis* had a hominid-like flat face, prominent brow, small canines, and a chimpanzee-sized brain. Another proposed early hominin, *Orrorin tugenensis*, lived about 6 million years ago in East Africa. Two sturdy fossilized femurs (thighbones) suggest that it stood upright; bipedal species have thicker femurs than four-legged walkers. In other traits, this species was apelike.

Ardipithecus ramidus was clearly a hominin. It lived in East Africa between 5.8 and 5.2 million years ago and left a wealth of fossil remains, including a largely complete skeleton of a female informally known as Ardi (**FIGURE 24.22A**). Features of the *A. ramidus* pelvis suggest that this species walked upright when on the ground. However, elongated arms, curved fingers, and an outwardly splayed big toe indicate it also spent time climbing along branches.

The best known early hominins, **australopiths**, belong to the genus *Australopithecus*, which lived in Africa from about 4 million to 1.2 million years ago. Australopith fossils reveal a trend toward smaller teeth and improvements in the ability to walk upright, but little increase in brain size.

Some *Australopithecus* species are considered likely human ancestors. One of these, *A. afarensis*, left behind fossils of more than 300 individuals, among them a nearly complete female skeleton known as Lucy (**FIGURE 24.22B**). Two *A. afarensis* individuals walking across a layer of newly deposited volcanic ash may also have left the trail of 3.6-million-year-old footprints discovered in Tanzania. *A. afarensis* males stood about 1.5 to 1.8 meters (5 to 5.5 feet) tall, and females 1 meter (3 feet) tall. They had a pelvis and legs suited to upright walking, although their gait may not have been as fluid as that of modern humans. They also retained the long arms and curved fingers of a climber.

Australopithecus sediba, a species recently discovered in South Africa, is the most humanlike australopith discovered thus far (**FIGURE 24.23**). *A. sediba* lived about 2 million years ago. Its hands were remarkably humanlike and were probably capable of a precision grip. In addition, its brain, although chimpanzee-sized and largely apelike, has a humanlike frontal region.

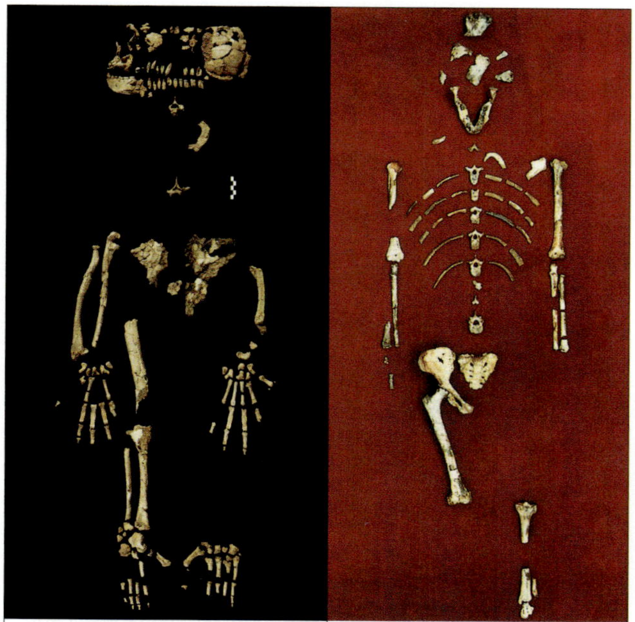

A Ardi, an *Ardipithecus ramidus* (4.4 million years ago).

B Lucy, an *Australopithicus afarensis* (3.5 million years ago).

FIGURE 24.22 Fossil skeletons of two early female hominins.

australopith Extinct African hominins in the genus *Australopithecus*; some are considered likely human ancestors.

TAKE-HOME MESSAGE 24.10

Some African fossil species from as far back as 7 million years ago have features that suggest they were bipedal.

Ardipithecus ramidus, an East African hominin known from many fossils, walked upright on the ground, but retained many traits related to climbing.

Australopiths, a group of African hominins, show a trend toward improved upright walking, smaller teeth, and greater hand dexterity. Like other early hominins, they retained a chimpanzee-sized brain.

24.11 WHAT SPECIES ARE IN THE GENUS *HOMO*?

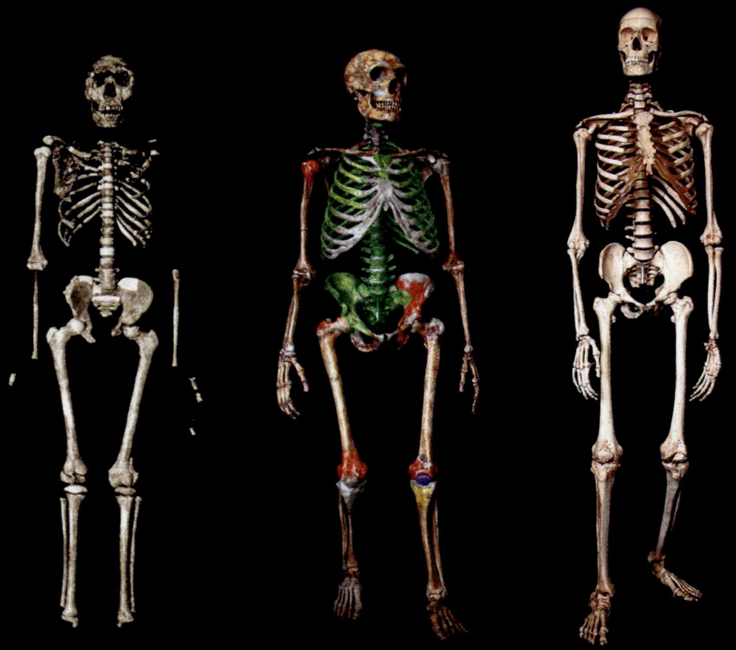

A *Homo erectus.* Fossil of Turkana boy, who lived 1.6 million years ago in what is now Kenya.

B *Homo neanderthalensis.* Reconstruction of an adult male based on multiple fossils, each denoted by a different color.

C *Homo sapiens.* (Modern human.) Skeletal model of a typical adult male.

FIGURE 24.24 Some members of the genus *Homo*.

Modern humans, *Homo sapiens*, are the only surviving member of the genus *Homo*. Fossil species are typically assigned to this genus on the basis of their relatively large brain, although other factors such as tool use can play a role. Researchers disagree over how many extinct *Homo* species there were, with some lumping many fossils into a few species and others splitting the same fossils among a greater number of groups. From a scientific standpoint, both approaches are equally valid. Both lumpers and splitters formulate hypotheses about how a particular fossil relates to other known fossils. They then seek evidence (in the form of additional fossil discoveries) that will confirm or refute their hypothesis.

EARLY *HOMO* SPECIES

The most ancient named species in our genus, ***Homo habilis***, lived in East Africa between 2.3 million and 1.4 million years ago. *Homo habilis* means "handy man" and is a reference to stone tools found in the same area as the fossil. *H. habilis* had a somewhat larger brain and smaller teeth than australopiths, but its size and limb proportions were quite apelike. Some researchers assign a few of the largest-brained fossils to a different species, *Homo rudolfensis*.

A dramatically different species of hominin appears in the fossil record beginning about 1.8 million years ago. Called ***Homo erectus***, which means "upright man," it was taller than *H. habilis*, with body proportions more like those of modern humans. It also had smaller teeth and a larger brain. The most complete *H. erectus* fossil known is a skeleton of a juvenile male from Kenya, commonly referred to as Turkana boy (**FIGURE 24.24A**). Although this individual apparently died at about age nine, his brain was already twice the size of a chimpanzee's. *H. erectus* used a variety of stone tools to cut up animal carcasses, scrape meat from bones, and extract marrow. Most likely the meat came from scavenging, rather than hunting. Bits of burnt plant material and bone left behind by a million-year-old campfire inside a South African cave suggest that *H. erectus* may have also been the first species to cook.

By 1.75 million years ago, a population of *Homo* had become established in what is now the Republic of Georgia. They are informally known as the Dmanisi hominins for the place where they were discovered. Most researchers think the Dmanisi hominins are descended from *H. erectus* and place them in this species. Although the Dmanisi hominins have the smallest brains of any known *H. erectus* population, they were sophisticated tool makers.

H. erectus populations became established in Indonesia by 1.6 million years ago, and China by 1.15 million years ago. *H. erectus* is also considered a likely ancestor of Neanderthals and modern humans.

NEANDERTHALS

In the mid-1800s, scientists discovered 40,000-year-old humanlike fossils in Germany's Neander Valley and named them ***Homo neanderthalensis***. Since then, many fossils with similar features have been unearthed, including a few largely complete skeletons. Scientists now consider Neanderthals our closest extinct relatives.

Neanderthals lived in Africa, the Middle East, Europe, and in central Asia as far east as Siberia. The last known population lived in seaside caves in Gibraltar until perhaps as recently as 28,000 years ago. Compared to modern humans, Neanderthals had a shorter, stockier build, with thicker bones and bulkier muscles (**FIGURE 24.24B**). A stocky body minimizes the surface area available for heat loss, and modern Arctic peoples have a similar body shape. Neanderthals had a braincase that was longer and lower than that of modern humans, but their brain was as big as ours or bigger. Their face had pronounced brow ridges and a large nose with widely spaced nostrils. Unlike modern humans, they typically lacked a protruding chin, an area of thickened bone in the middle of the lower jawbone. Fossils of Neanderthal individuals who survived despite disabilities such as the loss of a limb testify to a compassionate social structure. Some simple burials suggest possible symbolic thought. The anatomy of the voice box suggests Neanderthals were able to speak.

FLORES HOMININS

In 2003, scientists discovered 18,000-year-old hominin fossils on the Indonesian island of Flores. The fossilized individuals stood about a meter tall, and had a heavy brow. Their brain was australopith-sized, but shaped more like that of *H. erectus*. The media nicknamed the diminutive individuals "hobbits," a reference to small-bodied beings in the fiction of J.R.R. Tolkien. The scientists who discovered the fossils labeled them a new species, *Homo floresiensis*. Other scientists think that the fossils are modern humans with a genetic or nutritional disorder. Additional study of the existing fossils and of any new fossils discovered will determine which hypothesis is correct.

HOMO SAPIENS

Section 24.9 described the traits that characterize our species (*Homo sapiens*). To date, the oldest *H. sapiens* fossils discovered are two partial male skulls from Ethiopia, in East Africa. Known as Omo I and Omo II, they date to 195,000 years ago. The skulls were found in the same region where investigators also unearthed *H. sapiens* remains (two adult males and a child) from about 160,000 years. By 115,000 years ago, *H. sapiens* had extended their range into South Africa. The more recent global dispersal of this species is the topic of our next section.

Homo erectus Extinct hominin that arose about 1.8 million years ago in East Africa; migrated out of Africa.
Homo habilis Extinct hominin; earliest named *Homo* species; known only from Africa, where it arose 2.3 million years ago.
Homo neanderthalensis Extinct hominin; closest known relative of *H. sapiens*; lived in Africa, Europe, Asia.

TAKE-HOME MESSAGE 24.11

The earliest *Homo* species, *Homo habilis*, lived in Africa and resembled australopiths, but had a somewhat larger brain and may have made tools.

Homo erectus were taller, had a still larger brain, and were proportioned like modern humans. Some *H. erectus* ventured out of Africa and became established in Europe and Asia.

Homo neanderthalensis (Neanderthals), our closest extinct relatives, were a widespread group of stocky, large-brained hominins. Small-brained, short hominins discovered on an Indonesian island may be members of another *Homo* species.

The oldest fossils of our own species, *Homo sapiens*, come from East Africa.

PEOPLE MATTER

National Geographic Explorers-in-Residence
DR. LOUISE LEAKEY AND DR. MEAVE LEAKEY

Paleontologist Louise Leakey is a third-generation fossil-finder. Her grandfather, Louis Leakey, discovered and named *Homo habilis*, and her father, Richard Leakey, excavated the *Homo erectus* fossil known as "Turkana boy." She remembers the latter discovery well. She says, "At age 12, the discovery of the *Homo erectus* from the west side of Turkana was a very exciting time... We were able to engage and help and excavate it. There was a real sense of excitement about that excavation."

Louise now carries out her own research in the Turkana Basin, searching for fossils in collaboration with her mother, Meave. Both women are National Geographic Explorers. Among their accomplishments, they have recently shown that *Homo habilis* survived until at least 1.4 million years ago, and that it coexisted with *Homo erectus* for about half a million years. This long overlap between the two species argues against a widely held hypothesis that *H. habilis* was the ancestor of *H. erectus*. Louise and Meave Leakey have also discovered a 3.5-million-year-old skull that may belong to a previously unknown genus, and could be a human ancestor. They have tentatively named their find *Kenyanthropus platyops*, and continue to look for additional fossils that will confirm its identity as a new species.

CREDIT: (in text) Kenneth Garrett/National Geographic Creative.

24.12 WHAT DO OUR GENES REVEAL ABOUT HUMAN ORIGIN AND DISPERSAL?

We are all African. Regardless of where your recent ancestors lived, look far enough back and you'll discover your African roots. Both fossil finds and genetic studies indicate that our species originated in Africa. The oldest *Homo sapiens* fossils come from Ethiopia, and modern Africans are more genetically diverse than people of any other region. The large number of differences among African populations indicates that these populations have existed and have been accumulating random mutations for a very long time. Furthermore, most genetic variation seen in non-African populations is a subset of the variation found inside Africa. This is evidence that founder effects (Section 17.7) occurred when subsets of the African population left to colonize the rest of the world.

The ancestors of modern non-African populations began their journey out of Africa sometime between 80,000 and 60,000 years ago. Early in this journey, they encountered and successfully bred with Neanderthals. As a result, modern non-African peoples have a bit of Neanderthal DNA in their genome, whereas Africans do not. Modern humans would later live alongside Neanderthals in Europe and Asia, but to date there is no genetic evidence that matings in these regions contributed to our gene pool.

Our species expanded its range over tens of thousands of years as small groups ventured away from their homelands. As they did, unique mutations arose in different lineages and were passed along to descendants. Today, the distribution of these mutations among different ethnic groups provides evidence of the routes taken by the ancient travelers. By mapping the frequency of maternal and paternal genetic markers in modern peoples, geneticists have created a picture of when and where ancient humans moved around the world (**FIGURE 24.25**).

The first people to colonize Eurasia probably crossed from the Horn of Africa onto the Arabian peninsula. From there, they moved along the coastline to India, and reached Southeast Asia and Australia by 50,000 years ago. Shortly after this, another group traveled through the Middle East and into South Central Asia. Offshoots from this lineage colonized North Asia and Europe. Then, about 15,000 years ago, some people ventured across a land bridge that temporarily connected Siberia to North America. Over the next thousand years, their descendants spread to the tip of South America.

TAKE-HOME MESSAGE 24.12

- Modern humans originated in Africa, and expanded their range worldwide over tens of thousands of years.
- As humans left Africa, they interbred with Neanderthals, so modern non-African humans have some Neanderthal DNA.
- Genetic differences among modern ethnic groups can be used to reconstruct the routes of early human migrations.

FIGURE 24.25 Human migration routes, as determined by genetic analysis. To learn more about how geneticists use genes to trace journeys and determine our ancestry, visit National Geographic's Genographic Project site at *https://genographic.nationalgeographic.com*.

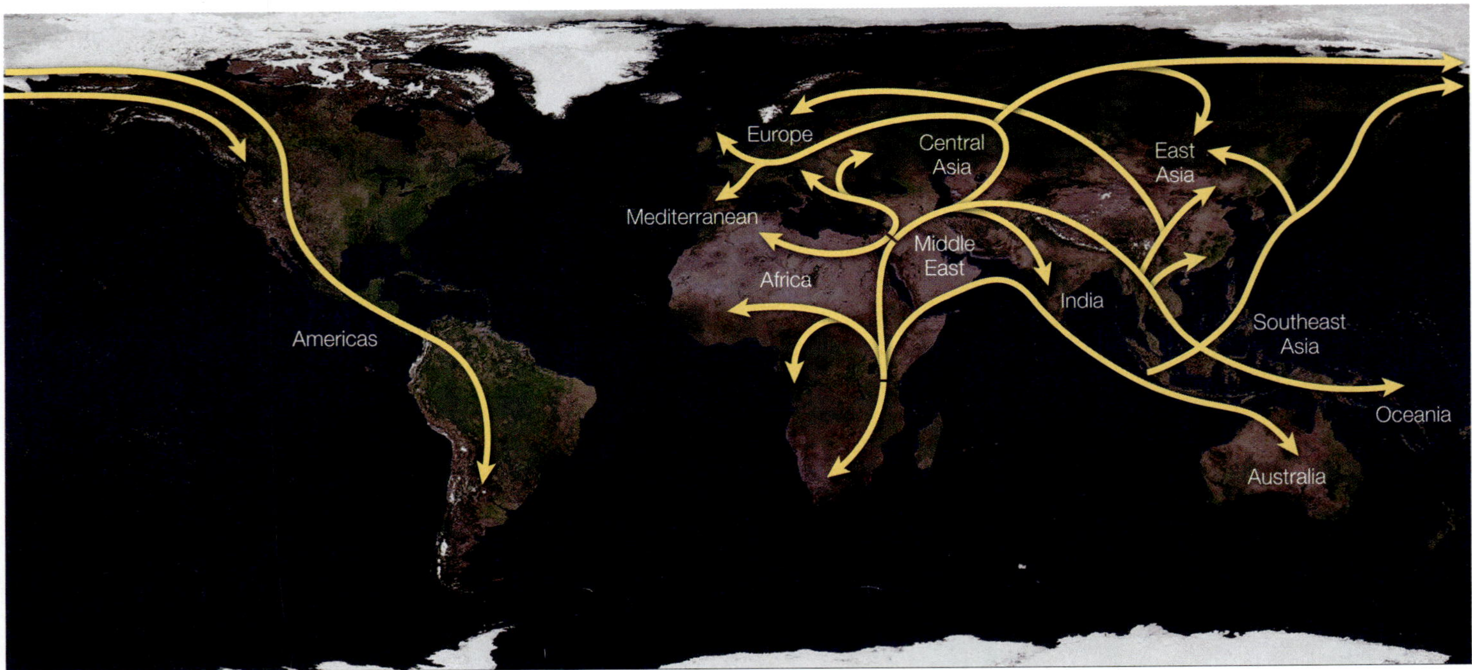

CREDIT: (25) photo, NASA; data, National Geographic.

24.13 Application: THE DOWNSIDE OF WALKING UPRIGHT

ACHING BACKS ARE AMONG THE PRICES WE PAY FOR WALKING UPRIGHT. In most vertebrates, the backbone is oriented horizontally, so it does not have weight pressing down on it. Our upright stance forced our lower back to support the weight of our upper body and put us at a heightened risk for lower back pain and injuries.

Walking upright also puts tremendous pressure on our knees. When you walk, you place all of your body weight on first one leg then the other. By contrast, four-legged walkers typically distribute their weight across at least two limbs. In addition, we fully extend our legs, whereas apes maintain a bent-legged stance. Fully extending your leg while walking decreases the effort expended by your thigh muscles, but also increases the risk you will injure your knee.

The combination of a relatively large brain and a pelvis adapted to an upright stance also makes for a tight fit during labor (FIGURE 24.26). Giving birth is typically more difficult for humans than it is for other primates.

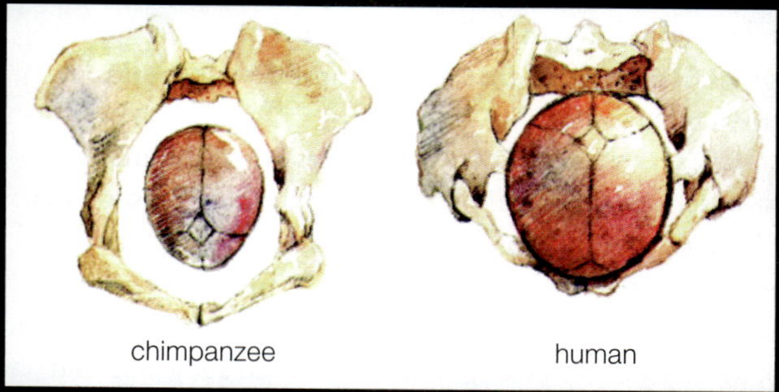

FIGURE 24.26 Passage of the chimpanzee or human skull through the birth canal during labor. Both are shown from below.

Summary

SECTION 24.1 Four features define **chordate** embryos: a **notochord**, a dorsal hollow nerve cord, a pharynx with gill slits, and a tail extending past the anus. Some or all of these features persist in adults. Chordates include two groups of marine invertebrates, the **tunicates** and **lancelets**, but most are **vertebrates**, which have an **endoskeleton** that includes a backbone. The first vertebrates were jawless fishes. **Tetrapods** are vertebrates with four limbs. **Amniotes** are tetrapods that produce eggs that allow embryos to develop away from water.

SECTION 24.2 Fishes are fully aquatic vertebrates. The earliest fishes had no jaws, fins, or scales. Modern **jawless fishes**, the hagfishes and lampreys, also lack these traits. Jaws evolved by a modification of structures that support gill slits. **Cartilaginous fishes** such as sharks and rays have a skeleton of cartilage and a **cloaca** that functions in excretion and reproduction. There are two lineages of bony fishes. **Ray-finned fishes**, the most diverse vertebrate group, have fins reinforced with thin rays derived from skin, whereas bones support the fins of **lobe-finned fishes**.

SECTION 24.3 **Amphibians**, the first lineage of tetrapods, evolved from a lobe-finned fish ancestor. A variety of modifications including a three-chamber heart adapt amphibians to life on land. Modern amphibians such as frogs, toads, and salamanders are carnivores. Most spend some time on land, but return to the water to reproduce. Many amphibian species face the threat of extinction.

SECTION 24.4 A variety of traits adapt amniotes to life away from water. Their skin and kidneys help minimize water loss. Fertilization usually takes place inside the female's body. Amniote eggs encase a developing embryo in fluid, so they can develop away from water.

Reptiles (which include birds and **dinosaurs**) belong to one amniote lineage; mammals to another. Birds descended

Summary continued

from dinosaurs. Some dinosaurs may have been **endotherms**, like modern birds and mammals. Nonbird reptiles are **ectotherms** that rely on environmental sources of heat.

SECTION 24.5 Lizards and snakes belong to the most diverse reptile lineage. All are carnivores with flattened scales. Turtles have a bony, keratin-covered shell and a horny beak rather than teeth. Crocodilians are aquatic carnivores and the closest relatives of birds. Like birds, they have a four-chamber heart and assist young.

SECTION 24.6 **Birds** are reptiles with feathers. Most have a body adapted to flight, with lightweight bones, a four-chamber heart, a beak rather than teeth, and a highly efficient respiratory system.

SECTION 24.7 **Mammals** nourish young with milk secreted by mammary glands, have fur or hair, and have more than one kind of tooth. Three lineages are egg-laying mammals (**monotremes**), pouched mammals (**marsupials**), and **placental mammals**, the most diverse group. Placental mammals develop faster than other mammals and are born at a later stage of development. They have become the primary mammal group in most regions. Rodents and bats are the most diverse groups of placental mammals.

SECTIONS 24.8, 24.9 **Primates** are a mammalian order adapted to climbing. They have flexible shoulder joints, grasping hands tipped by nails, and good depth perception. They rely on vision more than smell. Lemurs and their relatives are wet-nosed primates with a fixed upper lip. Most primates belong to the dry-nosed subgroup and have a movable upper lip. Old World monkeys, New World monkeys, **apes**, and humans share a common ancestor and are grouped as **anthropoids**. Apes and humans do not have a tail. The lineage leading to humans and the chimpanzee/bonobo lineage diverged from a common ancestor an estimated 6 to 8 million years ago.

Humans and our closest extinct ancestors are **hominins**, a group defined by **bipedalism**. We also have larger brains, a coat of finer hair, and more flexible hands than other primates. These traits evolved at different times.

SECTION 24.10 The earliest fossils that may be hominins date to 7 million years ago (**FIGURE 24.27**). *Ardipithecus ramidus* lived about 5 million years ago and walked upright on the ground, but also climbed among tree branches. **Australopiths** are a genus of chimpanzee-sized hominins that likely include some human ancestors. They walked upright, but had small brains.

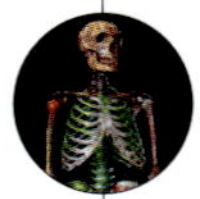

SECTIONS 24.11, 24.12 The human genus (*Homo*) arose by 2 million years ago. The oldest named species, ***H. habilis***, resembled australopiths, but had a slightly larger brain and may have made tools. They lived in Africa.

H. erectus had a much larger brain and a body form like that of modern humans. *H. erectus* made tools and may have used fire. Some members of this species left Africa and became established in Europe and Asia. *H. erectus* is thought to be the ancestor of both Neanderthals (***Homo neanderthalensis***) and modern humans (*H. sapiens*). Neanderthals, our closest extinct relatives, had a big brain and a stocky body. They lived in Africa, Europe, and Asia, then disappeared about 28,000 years ago. Fossils on an Indonesian island may be another species, *H. floresiensis*.

Fossil evidence and genetic comparisons indicate that our species, *H. sapiens*, arose by about 195,000 years ago in East Africa. As humans moved out of Africa, they met and interbred with Neanderthals, so modern non-African populations contain some Neanderthal DNA. Geneticists can determine the paths taken by early humans by looking at differences in the frequency of genetic markers among modern ethnic groups.

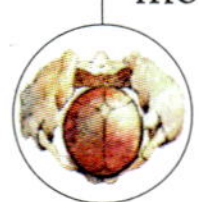

SECTION 24.13 Modifications of the skeleton that adapt our body to bipedalism put us at risk for knee and back problems. Coupled with our large skull size, they make childbirth more difficult.

FIGURE 24.27 Estimated dates for the origin and extinction of some hominins discussed in this chapter. Relationships among them remain a matter of debate. However, *H. erectus* is considered a likely ancestor of *H. neanderthalensis* and *H. sapiens*.

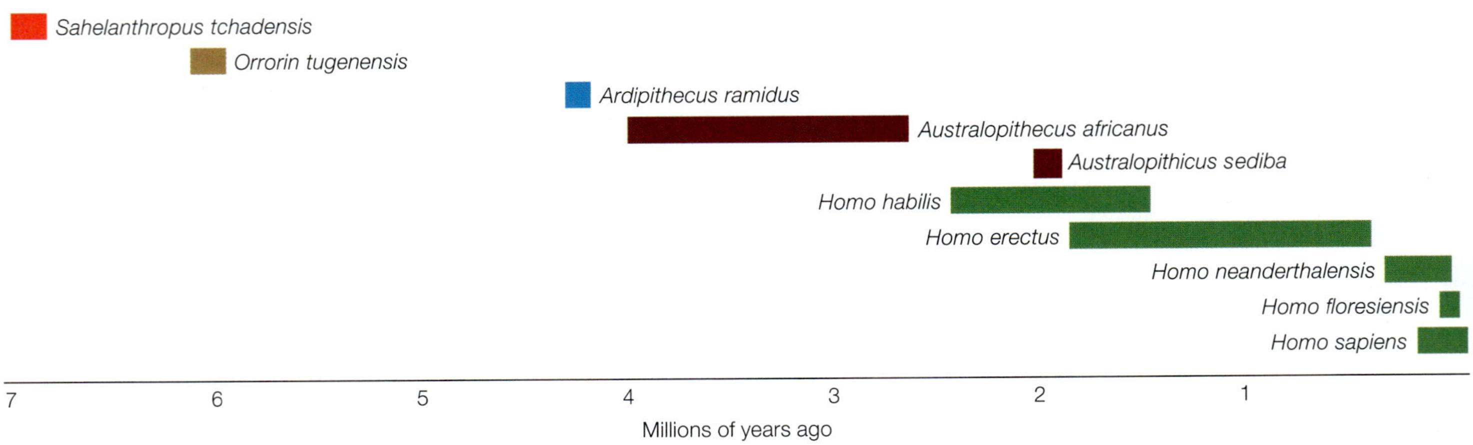

Data Analysis Activities

Neanderthal Hair Color The *MC1R* gene regulates pigmentation in humans (Sections 14.1 and 15.1 revisited), so loss-of-function mutations in this gene affect hair and skin color. A person with two mutated alleles for this gene makes more of a reddish pigment than a brownish one, resulting in red hair and pale skin. DNA extracted from two Neanderthal fossils has an *MC1R* mutation that has not yet been found in humans. To see how this mutation affects activity of the *MC1R* gene, Carles Lalueza-Fox and her team introduced the mutant allele into cultured monkey cells (**FIGURE 24.28**).

1. How did *MC1R* activity in monkey cells with the mutant allele differ from that in cells with the normal allele?
2. What does this imply about the mutation's effect on Neanderthal hair color?
3. What purpose do the cells with the jellyfish gene for green fluorescent protein serve in this experiment?

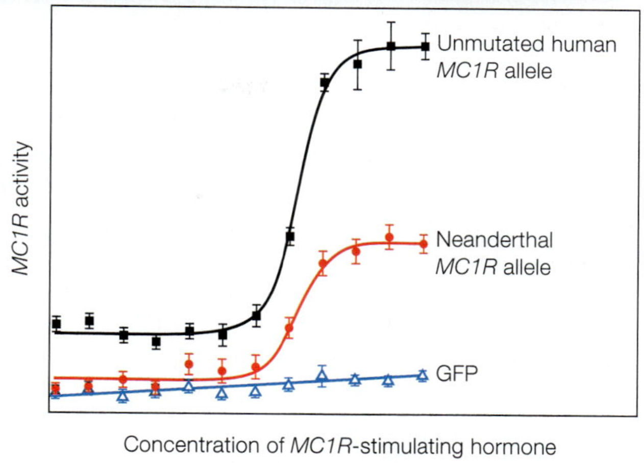

FIGURE 24.28 *MC1R* activity in monkey cells transgenic for an unmutated *MC1R* gene, the Neanderthal *MC1R* allele, or the gene for green fluorescent protein (GFP). GFP is not related to *MC1R*.

Self-Quiz Answers in Appendix VII

1. All chordates have (a) _______ as embryos.
   a. backbone b. jaws c. notochord d. both b and c

2. The _______ are invertebrate chordates.
   a. echinoderms c. lancelets
   b. hagfishes d. lampreys

3. Vertebrate jaws evolved from _______ .
   a. gill supports b. ribs c. scales d. teeth

4. Sharks and rays are _______ .
   a. ray-finned fishes c. jawless fishes
   b. cartilaginous fishes d. lobe-finned fishes

5. A divergence from _______ gave rise to tetrapods.
   a. ray-finned fishes c. cartilaginous fishes
   b. lizards d. lobe-finned fishes

6. Reptiles, including birds, belong to one major lineage of amniotes, and _______ belong to another.
   a. sharks c. mammals
   b. frogs and toads d. salamanders

7. Reptiles are adapted to life on land by _______ .
   a. tough skin d. amniote eggs
   b. internal fertilization e. both a and c
   c. good kidneys f. all of the above

8. The closest modern relatives of birds are _______ .
   a. crocodilians b. mammals c. turtles d. lizards

9. The defining trait of hominins is _______ .
   a. tool use b. bipedalism c. a large brain d. endothermy

10. Among living animals, only birds have _______ .
    a. a cloaca c. feathers
    b. a four-chamber heart d. amniote eggs

11. *Homo erectus* _______ .
    a. was the earliest member of the genus *Homo*
    b. was one of the australopiths
    c. evolved in Africa and dispersed to many regions
    d. disappeared as the result of an asteroid impact

12. Match the organisms with the appropriate description.

___ tunicates	a. pouched mammals
___ fishes	b. invertebrate chordates
___ amphibians	c. feathered amniotes
___ primates	d. egg-laying mammals
___ birds	e. extinct hominins
___ monotremes	f. have grasping hands with nails
___ marsupials	g. first land tetrapods
___ placental mammals	h. most diverse mammal lineage
___ australopiths	i. oldest vertebrate lineage

Critical Thinking

1. Researchers recently compared mitochondrial DNA from a fossil finger found in Siberia to mitochondrial DNA (mitoDNA) from modern humans. The sequence of the finger DNA differed from modern human DNA at an average of 385 sites. Neanderthal mitoDNA differs from modern human mitoDNA at an average of 202 sites, and chimpanzee mitoDNA differs from human mitoDNA at an average of 1,462 sites. Using this information, draw a tree that shows the relationship between chimpanzees, humans, Neanderthals, and the owner of this finger.

A male tungara frog inflates his vocal sac to make his species-specific mating call. Female frogs are attracted by the sound of this call.

25 ANIMAL BEHAVIOR

Links to Earlier Concepts

This chapter builds on your understanding of the concepts of sexual selection (17.6) and adaptation (16.2). The chapter provides many examples of scientific experiments (1.6).

KEY CONCEPTS

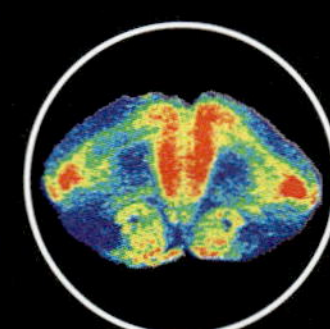

GENETIC FOUNDATIONS
Genes affecting the ability to detect stimuli or to respond to nervous or hormonal signals influence behavior. Genetic differences within and among species cause behavioral differences.

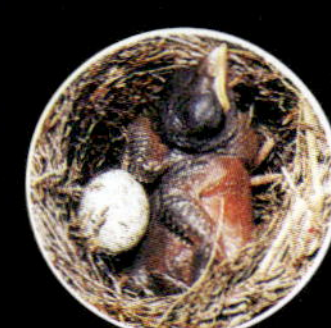

INSTINCT AND LEARNING
Instinctive behavior can be performed without practice, but most behavior has a learned component. Some types of learning can only occur during a certain portion of the lifetime.

ANIMAL COMMUNICATION
Animal communication signals arise only if communication benefits both signal senders and receivers. Predators sometimes take advantage of the communication behavior of their prey.

MATING AND PARENTING
In most cases, females are choosier about mates than males. Monogamy is rare in most animals. Whether there is parental care, and who delivers it, varies among animal groups.

FORMING GROUPS
Grouping together provides benefits such as protection, but also has costs such as increased competition. Self-sacrificing behavior evolved in some animals that live in large family groups.

Photograph by George Grall, National Geographic Creative.

25.1 HOW DO GENES INFLUENCE BEHAVIOR?

SENSING AND RESPONDING

Like other traits, behavioral traits have a genetic basis. The structure of the nervous system determines the types of stimuli that an animal can detect and the types of responses it can make. Gene differences that affect the structure and activity of the nervous system cause many differences in behavior. Hormones interact closely with the nervous system, so genes that influence hormone action also affect behavior. Genes with influence on metabolism and structural traits can also have behavioral effects. For example, singing requires both sound-making structures and sufficient energy to do so.

Like other genetically determined traits, behaviors evolve. In considering why animals behave as they do, biologists often discuss "proximate" and "ultimate" causes of a behavior. The proximate causes of a behavior are the genetic and physiological mechanisms that bring about the behavior. The ultimate cause of a behavior is the specific selective pressure that caused it to evolve. Keep in mind that not all behavior is adaptive all the time. A behavior that evolved in one context can be maladaptive in another.

FIGURE 25.1 Coastal garter snake dining on a banana slug. The snake is genetically predisposed to recognize slugs as prey. By contrast, garter snakes from inland regions where there are no banana slugs do not recognize them as food.

FIGURE 25.2 Genetic polymorphism for foraging behavior in fruit fly larvae. When a larva is placed in the center of a yeast-filled plate, its genotype at the *foraging* locus influences whether it moves a little or a lot while it feeds. Black lines show a representative larva's path.

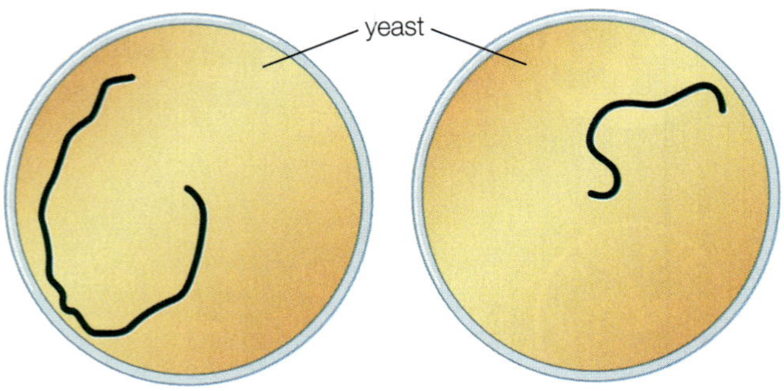

A Rovers (genotype *FF* or *Ff*) move often as they feed. When a rover's movements on a petri dish filled with yeast are traced for 5 minutes, the trail is relatively long.

B Sitters (genotype *ff*) move little as they feed. When a sitter's movements on a petri dish filled with yeast are traced for 5 minutes, the trail is relatively short.

VARIATION WITHIN A SPECIES

One way to investigate the genetic basis of behavior is to examine behavioral and genetic differences among members of a single species. For example, Stevan Arnold studied feeding behavior in two populations of garter snakes of the Pacific Northwest. One population lives in coastal forests, where it hunts the banana slugs common on the forest floor (**FIGURE 25.1**). The other population lives inland, where there are no banana slugs. Inland snakes eat fishes and tadpoles. The two populations differ in their inborn food preferences. When Arnold offered a slug to newborn garter snakes, offspring of coastal snakes ate it, but offspring of inland snakes ignored it.

Arnold hypothesized that inland snakes lack the genetically determined ability to associate the scent of slugs with food. He predicted that if coastal garter snakes were crossed with inland snakes, the resulting offspring would make an intermediate response to slugs. Results from his experimental crosses confirmed this prediction. We do not know which allele or alleles underlie this difference.

We know more about the genetic basis of differences in foraging behavior among fruit fly larvae. The wingless, wormlike larvae crawl about eating yeast that grows on decaying fruit. In wild fruit fly populations, about 70 percent of the fly larvae are "rovers," which means they tend to move around a lot as they feed. Rovers often leave one patch of food to seek another (**FIGURE 25.2A**). The remaining 30 percent of larvae are "sitters"; they tend to move little once they find a patch of yeast (**FIGURE 25.2B**). When food is absent, rovers and sitters move the same amount, so both are equally energetic.

The proximate cause of the difference in larval behaviors is a difference in alleles of a gene called *foraging*. Flies with a dominant allele of this gene have the rover phenotype. Sitters are homozygous for the recessive allele. The *foraging* gene encodes an enzyme involved in learning about olfactory cues. Rovers make more of the enzyme than sitters.

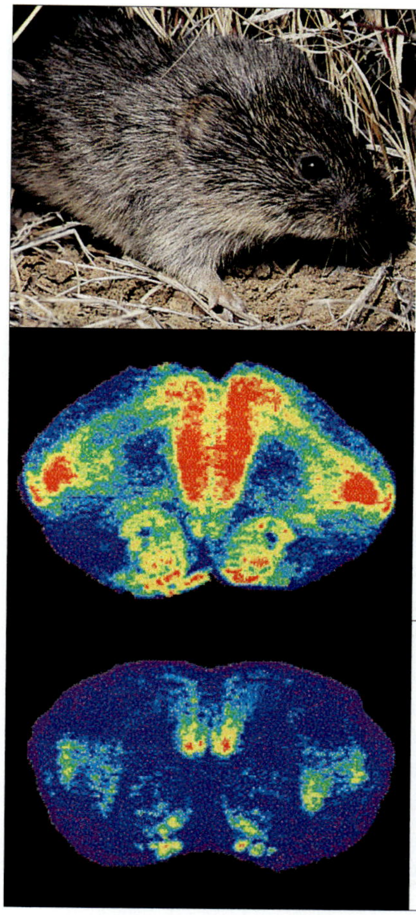

A Vole (*Microtus*), a type of small rodent.

B PET scan of a monogamous prairie vole's brain with many oxytocin receptors (red).

C PET scan of a promiscuous mountain vole's brain with fewer oxytocin receptors.

FIGURE 25.3 Genetic roots of mating and bonding behavior in voles. Closely related vole species vary in their behavior, and in the number and distribution of receptors for the hormone oxytocin.

The ultimate cause of the behavioral variation in larval foraging behavior is competition for food. In experimental populations with limited food, both rovers and sitters are most likely to survive to adulthood when their foraging type is rare. A rover does best when surrounded by sitters, and vice versa. Presumably, when there are lots of larvae of one type they all compete for food in the same way. Under these circumstances, a fly that behaves differently than the majority is at an advantage. As a result, natural selection maintains both alleles of the *foraging* gene.

VARIATION AMONG SPECIES

A behavior's genetic basis can sometimes be clarified by comparing related species. For example, studies of voles (**FIGURE 25.3A**) reveal that inherited differences in the number and distribution of certain hormone receptors influence mating and bonding behavior. Differences in mating behavior between closely related species of vole make these animals ideal subjects for study of this behavior. Most voles, like most mammals, are promiscuous; they have multiple mates. However, some voles form lifelong, largely monogamous relationships. For example, in prairie voles, a permanent social bond forms after a night of repeated matings. The hormone oxytocin plays a central role in a female prairie vole's bonding behavior. When females who are part of an established pair are injected with a chemical that interferes with oxytocin action, they dump their long-term partners in favor of other males.

Females of promiscuous vole species are less influenced by oxytocin than prairie voles. When researchers compared the brains of promiscuous and monogamous species, they found a striking difference in the number and distribution of oxytocin receptors (**FIGURE 25.3B,C**). Monogamous prairie voles have many oxytocin receptors in the part of the brain associated with social learning. Promiscuous mountain voles have far fewer of these receptors.

In male voles, variations in the distribution of receptors for another hormone (arginine vasopressin, or AVP) correlate with bonding tendency. Compared to males of promiscuous vole species, males of monogamous species have more AVP receptors. Scientists isolated the prairie vole gene for the AVP receptor and used genetic engineering to insert it into the brain of male mice, which are naturally promiscuous. After this treatment, the genetically modified mice preferred a female with whom they had already mated over an unfamiliar female. These results confirm the role of AVP in fostering monogamy among male rodents.

EPIGENETIC EFFECTS

Epigenetic mechanisms (Section 10.5) sometimes cause heritable changes in behavior. For example, a female rat who did not receive much licking and grooming from her mother early in life will fail to lick and groom her own pups. The lack of tactile stimulation affects methylation patterns in a rat pup's DNA, and the resulting changes in gene expression disrupt oxytocin's influence in the adult. The methylation patterns are inherited, so this negative effect on female parental behavior can persist for generations.

TAKE-HOME MESSAGE 25.1

Many genes shape the nervous system. Such genes affect traits such as an animal's ability to sense and respond to specific stimuli.

Genetic differences affect species-specific behavior, and also differences in behavior among individuals of a species.

Epigenetic mechanisms can result in inherited behavioral tendencies.

25.2 WHAT IS THE DIFFERENCE BETWEEN INSTINCTIVE AND LEARNED BEHAVIORS?

INSTINCTIVE BEHAVIOR

All animals are born with the capacity for **instinctive behavior**—an innate response to a specific and usually simple stimulus. For example, a newborn coastal garter snake behaves instinctively when it eats a banana slug in response to its smell.

The life cycle of the cuckoo bird provides several examples of instinct at work. This European bird is a "brood parasite," meaning it lays its eggs in the nest of other birds. A newly hatched cuckoo is blind, but contact with an object beside it stimulates an instinctive response. That hatchling maneuvers the object, which is usually one of its foster parents' eggs, onto its back, then shoves it out of the nest (**FIGURE 25.4A**). The cuckoo will also shove its foster parents' chicks over the side, if they happen to hatch before it does. Instinctively shoving objects out of the nest ensures the cuckoo has its foster parents' undivided attention.

Instinctive responses are advantageous only if the triggering stimulus almost always signals the same situation. Doing away with an egg benefits a cuckoo chick because the egg always houses a future competitor for food. However, instinctive responses can open the way to exploitation. Cuckoos often look nothing like the chicks they displace, yet their foster parents feed them all the same (**FIGURE 25.4B**). The cuckoo chick exploits its foster parents' instinctive urge to fill any gaping mouth in their nest with food.

A Young cuckoo shoving its foster parents' egg from the nest. It pushes out any object beside it.

B A foster parent (*left*) feeds a cuckoo chick (*right*) in response to a simple cue: a gaping mouth.

FIGURE 25.4 Instinctive responses to simple cues.

TIME-SENSITIVE LEARNING

A **learned behavior** is one that is altered by experience. Instinctive behavior is sometimes modified with learning. A garter snake's initial strikes at prey are instinctive, but the snake learns to avoid dangerous or unpalatable prey. Learning may occur throughout an animal's life, or be restricted to a critical period.

Imprinting is a form of learning that occurs during a genetically determined time period early in life. For example, baby geese learn to follow the large object that looms over them after they hatch (**FIGURE 25.5A**). With rare exceptions, this object is their mother. When mature, the geese will seek out a sexual partner that appears similar to the imprinted object.

A genetic capacity to learn, combined with actual experiences in the environment, shapes most forms of behavior. For example, a male songbird has an inborn capacity to recognize his species' song when he hears older males singing it. The young male uses these overheard songs as a model for his own song. Males reared with no model or exposed only to songs of other species often sing a simplified version of their species' song.

Many birds can only learn the details of their species-specific song during a limited period early in life. For example, a male white-crowned sparrow will not sing normally if he does not hear a male "tutor" of his own species during his first month. Hearing a same-species tutor later in life will not normalize his singing.

Most birds must also practice their song to perfect it. In one experiment, researchers temporarily paralyzed throat muscles of zebra finches who were beginning to sing. After being temporarily unable to practice, these birds never mastered their song. In contrast, temporary paralysis of throat muscles in very young birds or adults did not impair later song production. Thus, in this species, there is a critical period for song practice, as well as for song learning.

CONDITIONED RESPONSES

Nearly all animals are lifelong learners. Most learn to associate certain stimuli with rewards and others with negative consequences. With classical conditioning, an animal's involuntary response to a stimulus becomes associated with a stimulus that accompanies it. In the most famous example, Ivan Pavlov rang a bell whenever he fed a dog. Eventually, the dog's reflexive response—increased salivation—was elicited by the sound of the bell alone. Conditioned taste aversion is a type of classical conditioning in which an animal learns to avoid a food that made it sick at an earlier time. This learned response protects the animal from repeated ingestion of toxic substances. Some plants and fungi take advantage of animals' capacity for this type of learning by producing nausea-inducing substances.

With operant conditioning, an animal modifies a voluntary behavior in response to consequences of that behavior. This type of learning was first described in the context of learning experiments in the lab. A rat that presses a lever and is rewarded with a food pellet becomes more likely to press the lever again. A rat that receives a shock when it enters a particular area will quickly learn to avoid that area. In nature as well, animals learn to repeat behaviors that provide food or mating opportunities and to avoid those that cause pain.

OTHER TYPES OF LEARNING

With **habituation**, an animal learns by experience not to respond to a stimulus that has neither positive nor negative effects. Pigeons in cities often become habituated to people: The birds learned not to flee from the throngs who walk past them.

Many animals learn the landmarks in their environment. A mental map of the landmarks can be put to use when an animal needs to return home. For example, a fiddler crab foraging up to 10 meters (30 feet) away from its burrow is able to scurry straight home when it perceives a threat.

Animals also learn the details of their social landscape. They learn to recognize mates, offspring, or competitors by their appearance, calls, odor, or some combination of cues. For example, two male lobsters that meet up for the first time will fight (**FIGURE 25.5B**). Later, they will recognize one another by scent and behave accordingly, with the loser actively avoiding the winner.

With imitation learning, an animal copies behavior it observes in another individual. For example, Ludwig Huber and Bernhard Voelkl allowed marmoset monkeys to watch another marmoset open a container to get the treat inside. Some marmosets used their hands to open the container; others used their teeth. When the observing monkeys were later given a similar container, they used either their hands or their teeth depending on which technique they had observed (**FIGURE 25.5C**).

A Imprinting. Nobel laureate Konrad Lorenz with geese that imprinted on him. Under normal circumstances, newborn geese imprint on and follow their mother.

B Social learning. Captive male lobsters fight at their first meeting. The loser remembers the winner's scent and avoids him. Without another meeting, memory of the defeat lasts up to two weeks.

C Imitative learning. Marmosets given an opportunity to observe another marmoset opening a container filled with food later opened the container the same way. The marmoset above had observed another opening a container with its teeth.

FIGURE 25.5 Experimental demonstrations of animal learning.

habituation Learning not to respond to a repeated neutral stimulus.
imprinting Learning that can occur only during a specific interval in an animal's life.
instinctive behavior An innate response to a simple stimulus.
learned behavior Behavior that is modified by experience.

TAKE-HOME MESSAGE 25.2

Instinctive behavior can initially be performed without any prior experience, as when a simple cue triggers a response. Even instinctive behavior may be modified by experience.

Certain types of learning can only occur at particular times in the life cycle.

Learning can affect both voluntary and involuntary behaviors.

25.3 HOW AND WHY DO ANIMALS COMMUNICATE?

Communication signals are evolved cues that transmit information from one member of a species to another. A communication signal arises and persists only if it benefits both the signal sender and the signal receiver. If signaling is disadvantageous for either party, then natural selection will tend to favor individuals that do not send or respond to it.

Pheromones are chemical signals that convey information among members of a species. Signal pheromones cause a rapid shift in the receiver's behavior. Sex attractants that help males and females of many species find each other are one example. The alarm pheromone a honeybee emits when she perceives a threat to her hive is another. Alarm pheromone causes other honeybees to rush out of the hive and attack the potential intruder. Priming pheromones cause longer-term responses, as when a chemical in the urine of male mice triggers ovulation in females. Producing a pheromone requires less energy than calling or gesturing. However, the amount of information a pheromone can convey is limited; it is either released or not. Properties of acoustical, visual, and tactile signals vary continuously, and so can convey more information.

A Courtship display in albatrosses. The display involves a series of coordinated movements, some accompanied by calls.

B Threat display of a male collared lizard. This display allows rival males to assess one another's strengths without engaging in a potentially damaging fight.

FIGURE 25.6 Examples of visual communication.

FIGURE 25.7 **{Animated}** Honeybee waggle dance. Orientation of dancer's straight run (middle of the figure 8) conveys where food is, relative to the direction of the sun.

Acoustical signals often advertise the presence of an animal or group of animals. Many male vertebrates, including songbirds, whales, frogs, some fish, and many insects, make sounds to attract prospective mates. In many cases, sounds made by males also function in territoriality. Some birds and mammals give alarm calls that inform others of potential threats. A prairie dog makes one kind of call when it sees an eagle and another when it sees a coyote. Upon hearing the call, other prairie dogs respond appropriately: They either dive into burrows (to escape an eagle's attack) or stand erect (to spot the coyote).

Properties of acoustical signals vary depending on whether or not the sender benefits by revealing its position. Sounds that lure mates or offspring are typically easily localized, whereas alarm calls are not.

Visual communication is most widespread in animals that have good eyesight and are active during the day. Most bird courtship involves coordinated visual signaling (FIGURE 25.6A). Selection favors clear signals, so movements that serve as signals often become exaggerated. Body form may evolve in concert with movements, as when bright-colored feathers enhance a courtship display. A courtship display assures a prospective mate that the displayer is of the correct species and is in good health.

Threat displays advertise good health too, but they serve a different purpose. When two potential rivals meet, a threat display demonstrates each individual's strength and how well armed it is (FIGURE 25.6B). If the rivals are not evenly matched, the weaker individual retreats, and both benefit by avoiding a fight that could lead to injury.

With tactile displays, touch transmits information. For example, after discovering food, a foraging honeybee worker returns to the hive and dances in the dark, surrounded by a crowd of fellow workers. The speed and orientation of

pheromone Chemical that serves as a communication signal between members of an animal species.

CREDITS: (6A) G. Ziesler/ZEFA; (6B) © Eitan Grunwald, www.herptrips.com; (7) © Cengage Learning.

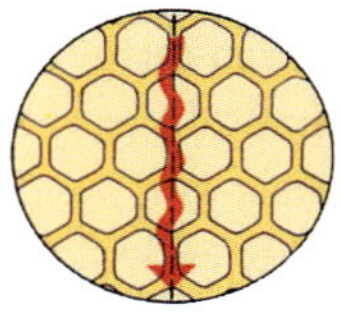

When bee moves straight down comb, recruits fly to source directly away from the sun.

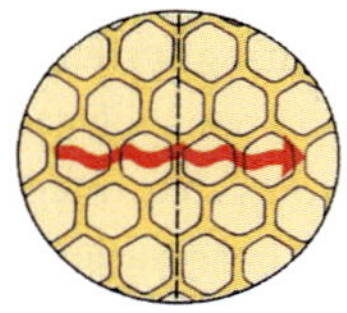

When bee moves to right of vertical, recruits fly at 90° angle to right of the sun.

her dance convey information about the distance of and direction to the food (**FIGURE 25.7**).

Predators sometimes tap into signaling systems of prey. For example, calls made by a male tungara frog (shown in the chapter opening photo) attract females. Unfortunately, such calls also attract frog-eating bats (**FIGURE 25.8**). As another example, fireflies attract mates by producing flashes of light in a characteristic pattern. Some female fireflies prey on males of other species. When a predatory female sees the flash from a male of the prey species, she flashes back as if she were a female of his own species. When he approaches ready to mate, she captures and eats him.

FIGURE 25.8 **Intercepted signals.** A frog-eating bat with a male tungara frog. Both female frogs and frog-eating bats locate male frogs by listening for the males' courtship call.

TAKE-HOME MESSAGE 25.3

A chemical, visual, acoustical, or tactile communication signal transfers information from one individual to another of the same species. Both signaler and receiver benefit from the transfer.

Signals can draw attention of predators as well as intended receivers. Features of signals reflect a balance between the benefit of sending information and the potential cost of signaling.

PEOPLE MATTER

National Geographic Explorer
DR. ISABELLE CHARRIER

Some animals are able to recognize individuals by their voice. Isabelle Charrier says, "My scientific interest is in understanding how animals are able to identify each other individually by voice, especially in species showing strong environmental and ecological constraints. Colonial birds and mammals are good models for this problem: Vocal recognition between mates or between parents and offspring is effective and reliable in spite of the high background noise and the high risk of confusion between individuals."

Much of Charrier's research focuses on seals and related marine mammals. These animals spend most of their life at sea, but come ashore to give birth and nurse their young. Charrier's field studies span the globe, from walrus colonies in the Arctic to sea lion colonies in Australia. She has studied the features of vocal signals and their role in recognition in interactions between mothers and pups, between males and females, and between territorial males. She says "My favorite field experience was when I did my Ph.D.: I stayed nearly nine months on Amsterdam Island [French Subantarctic island] to study Subantarctic fur seals. This was so great, to study mother–pup vocal recognition from birth until weaning. You can really develop special interactions with the animals by being on the colony for hours every day." Her greatest challenge has been studying walruses in the Arctic. She says, "You fight every day with the weather and sea conditions, and walruses are quite difficult to find and approach. It results in a great amount of frustration, but walruses are amazing to study."

As for why she does her research, she says, "Learning more about animal communication systems is not only important for our general knowledge, but also an essential approach and a powerful tool to protect some species and their environment."

25.4 WHAT FACTORS AFFECT REPRODUCTIVE BEHAVIOR?

A Male hangingfly dangling a moth as a gift for a potential mate.

B Male fiddler crabs (*top*) wave their one enlarged claw to attract a female (*bottom*).

C Male sage grouse gather on a communal display ground, where they dance, puff out their neck, and make booming calls.

FIGURE 25.9 How to impress a female.

MATING SYSTEMS

Animal mating systems have traditionally been categorized as promiscuous, polygamous, or monogamous. In recent years, studies that integrate paternity analysis with behavioral observations have shown that species do not always fit cleanly into these categories. Members of a species often vary in their behavior.

Some animals such as prairie voles form a pair bond, which means two individuals mate, preferentially spend time together, and cooperate in rearing offspring. However, the participants in a pair bond may also seize opportunities for matings with other individuals. A paternity study of the prairie voles discussed in Section 25.1 found that 7 percent of pair-bonded females produced litters of mixed male parentage; and 15 percent of pair-bonded males sired offspring with females other than their partner. Thus, prairie voles are now described as socially monogamous, but genetically promiscuous. As a species, they form an exclusive social relationship with a single partner. However, the genetic composition of their offspring reflects a tendency to mate with multiple partners.

Even social monogamy is rare in most animals, with the exception of birds. An estimated 90 percent of birds are socially monogamous. Among this group, the vast majority of species that have been examined for paternity patterns are genetically promiscuous. This is not surprising as promiscuity benefits both males and females by increasing the genetic diversity among the offspring they produce.

Multiple matings can also provide another benefit: more offspring. This benefit usually applies most strongly to males. Sperm are energetically inexpensive to produce, so a male's reproductive success is usually limited by access to mates. By contrast, the main limit on a female's reproductive success is her capacity to produce large, yolk-rich eggs or, in mammals, to carry developing young.

When one sex exerts selection pressure on the other, we expect sexual selection to occur (Section 17.6). In many species, females choose among males on the basis of their ability to provide necessary resources. For example, a female hangingfly will only mate with a male while feeding on prey that he has provided as a "nuptial gift" (**FIGURE 25.9A**).

In other species, males establish a mating territory, an area that includes resources that females require for reproduction. A **territory** is an area from which an animal or group of animals actively exclude others. A male fiddler crab's territory is a stretch of shoreline in which he excavates a burrow. Fiddler crabs have **sexual dimorphism**, in which body forms of males and females differ. A male has one oversized claw (**FIGURE 25.9B**) and during mating season, he stands outside his burrow waving it. The claw is used both in the courtship display and in territorial disputes with other males. Females attracted by a male's display check out the location and dimensions of his burrow before mating with him. Burrow location and size are important because they affect development of the larvae.

In mammals such as elephant seals and elk, the strongest males hold large territories, and mate with all of the females inside the area. Other males may never get to mate at all unless they sneak into an area under another male's control.

In some birds such as sage grouse, males converge at a communal display ground called a **lek**. Males stamp their feet and emit booming calls by puffing and deflating their large neck pouches for female onlookers (**FIGURE 25.9C**). The most popular males mate with many females.

PARENTAL CARE

Parental care requires time and energy that an individual could otherwise invest in reproducing again. It arises only if the genetic benefit of providing care (increased offspring survival now) offsets the cost (reduced opportunity to produce offspring later). If the parental care provided by a single parent is sufficient to successfully rear an offspring, individuals who spare themselves this cost by leaving parental duties to their mate are at a selective advantage.

Which sex cares for the young varies by species. In about 90 percent of mammals, the female rears young (**FIGURE 25.10A**). In the remaining 10 percent, both sexes participate. No mammal relies solely on male parental care. Female mammals sustain developing young in their body, so they have a greater investment in newborns than males. Also, most male mammals do not lactate, although in two fruit bat species, both males and females nurse the young.

Most fishes provide no parental care, but in those that do, this duty usually falls to the male. Males guard eggs until they hatch (**FIGURE 25.10B**) or, in the case of sea horses, protect them inside their body. The prevalence of male parental care in fishes may be related to sex differences in how fertility changes with age. A female fish's fertility increases dramatically with age, whereas a male's does not. Thus, a female fish who invests energy in care forgoes more future reproduction than a male fish does.

In birds, two-parent care is most common (**FIGURE 25.10C**). Chicks of birds that cooperate in care of their young tend to hatch while in a relatively helpless state. Chicks of sage grouse and other birds in which females alone provide care tend to hatch when more fully developed.

lek Of some birds, a communal mating display area for males.
sexual dimorphism Distinct male and female body forms.
territory Area an animal or group occupies and defends.

A Female grizzly bears care for their cub for as long as two years.

B Male clownfish guard eggs (*right*) until they hatch.

C Male and female terns cooperate in the care of their chick.

FIGURE 25.10 Who cares for the young?

TAKE-HOME MESSAGE 25.4

Genetic monogamy is rare. Mating with multiple partners benefits both males and females by increasing the genetic diversity among their offspring.

In many species, a few males monopolize opportunities to mate either by enticing females to choose them or by defending a territory with many females inside it.

Whether parental care occurs in a species and who delivers it depends on the benefits and costs of care to each parent.

25.5 WHY DO SOME ANIMALS GROUP TOGETHER?

FIGURE 25.11 A selfish herd. Sardines attempt to hide behind one another as a predatory sailfish attacks.

Most animals live solitary lives, coming together only to mate. However, some spend time in groups. A group may be a temporary, like a herd of zebras or a school of fish, or it may be more permanent, like a prairie dog colony or a wolf pack. A tendency to group together, whether briefly or for the longer term, will evolve only if the benefits of being near others outweigh the costs.

BENEFITS OF GROUPS

In many species, grouping together reduces the risk of predation. There is safety in numbers. Whenever animals cluster, individuals at the margins of the group inadvertently shield others from predators. A **selfish herd** is a temporary aggregation that arises when animals hide behind one another to escape a threat. Small fish form a selfish herd when under attack by a predator (**FIGURE 25.11**).

In a group, multiple individuals can be on the alert for predators. In some cases, an animal that spots a threat will warn others of its approach. Birds, monkeys, meerkats, and prairie dogs are among the animals that make alarm calls in response to a predator. Even in species that do not make alarm calls, individuals can benefit from the vigilance of others. Often, when an animal notices another group member beginning to flee, it will do likewise. In what is called the "confusion effect," a predator has a more difficult time picking out a specific individual to pursue when a group of prey is scattering.

Not all prey groups flee from predators. Some present a united defense. For example, when threatened by wolves, musk oxen stand back to back, presenting an imposing display of sharp horns. Sawfly caterpillars also gather in a group. When a hungry bird approaches, the caterpillars rear up and vomit partly digested eucalyptus leaves. In experiments, birds presented with such a slimy cluster eat fewer caterpillars than birds that were offered caterpillars one at a time.

Social animals form more permanent multigenerational groups, in which members benefit by cooperating in some tasks. Members of a social group are usually relatives. Wolves and lions are social animals who cooperate in catching prey. Cooperative hunting allows a group of predators to capture larger or faster prey. Even so, cooperative hunting is often no more efficient than hunting alone. In one study, researchers observed that a solitary

CREDIT: (11) Paul Nicklen/National Geographic Creative.

lion catches prey about 15 percent of the time. Two lions hunting together catch prey twice as often as a solitary lion, but having to share the spoils of the hunt means the amount of food per lion is the same. When more lions join a hunt, the success rate per lion falls. Wolves have a similar pattern. Among carnivores that hunt cooperatively, hunting success is usually not the major advantage of group living. Individuals hunt together, but also cooperate in fending off scavengers (**FIGURE 25.12**), caring for one another's young, and defending their territory.

Group living can also facilitate learning by imitation. Imitative tool production is an example. Chimpanzees strip leaves from branches to make "fishing sticks," simple tools that they use to capture insects as food. Different groups of chimpanzees use slightly different methods of tool-shaping and insect-fishing. In each group, individuals learn by imitating the behavior of others.

FIGURE 25.12 Lionesses enjoy the results of a communal hunt, while spotted hyenas try to steal a share.

COSTS OF GROUP LIVING

Grouping together has costs. Individuals that group together are easier for predators to locate and for parasites and contagious diseases to infect. Members of a group also compete more for resources. Consider how many seabirds form dense breeding colonies in which competition for space and food is intense (**FIGURE 25.13**). Given the opportunity, a pair of breeding herring gulls will cannibalize the eggs and even the chicks of their neighbors.

FIGURE 25.13 A colony of nesting penguins. Parasites spread easily through such colonies and the colony's large size makes it easy for predators to find.

DOMINANCE HIERARCHIES

Cost and benefits are often not equally distributed among members of a group. Many animals that live in permanent groups form a **dominance hierarchy**. In this type of social system, dominant animals get a greater share of resources and breeding opportunities than subordinate ones. Typically, dominance is established by physical confrontation. In most wolf packs, one dominant male breeds with one dominant female. The other members of the pack are nonbreeding brothers and sisters, aunts and uncles. All hunt and carry food back to individuals that guard the young in their den.

Why would a subordinate give up resources and often breeding privileges? Challenging a strong individual can be dangerous, as is living on one's own. Subordinates get their chance to reproduce by outliving a dominant peer.

TAKE-HOME MESSAGE 25.5

- Individual animals sometimes form a temporary group that reduces their risk of predation.
- Animals that live in permanent social groups cooperate in tasks such as hunting and rearing young.
- Costs of grouping include increased risk of disease and parasitism, and increased competition for resources.

dominance hierarchy Social system in which resources and mating opportunities are unequally distributed within a group.
selfish herd Temporary group that forms when individuals cluster to minimize their individual risk of predation.
social animal Animal that lives in a multigenerational group in which members, who are usually relatives, cooperate in some tasks.

25.6 HOW DID EUSOCIAL ANIMALS EVOLVE?

A Honeybee queen surrounded by sterile worker females.

B Naked mole-rat worker in its burrow.

FIGURE 25.14 Examples of eusocial animals.

From a genetic standpoint, the greatest cost an individual can pay for living in a group is the failure to breed. Yet, in some animals, a social system has evolved in which permanently sterile workers care cooperatively for the offspring of just a few breeding individuals. Such animals are said to be eusocial. **Eusocial animals** live in a multigenerational family group in which sterile workers carry out all tasks essential to the group's welfare, while other members of the group produce offspring.

Most eusocial animals are members of the order Hymenoptera, which includes bees, wasps, and ants. Some bees and wasps are eusocial, as are all ants. In all eusocial hymenoptera, the workers that maintain and defend a colony are sterile females. For example, the only egg-laying female in a honeybee colony is the queen (**FIGURE 25.14A**). Termites are also eusocial, but a termite colony has both male and female workers.

Only two species of eusocial vertebrates are known. Both are African mole-rats, mouse-sized rodents that live in underground burrows (**FIGURE 25.14B**). A clan of eusocial mole-rats includes a reproductive female, one or two reproductive males, and their worker offspring. Workers of both sexes dig the burrows, care for the young, provide the clan with food, and defend it.

Workers in a eusocial species engage in **altruistic behavior**, which in biology means behavior that enhances another individual's reproductive success at the altruist's expense. According to William Hamilton's **theory of inclusive fitness**, genes that promote altruism can evolve if those helped are relatives. If an individual with an allele that promotes altruism helps enough others with same allele, the frequency of that allele will increase, despite the reduced reproductive success of the altruist.

It is easy to see the genetic advantage of caring for one's own offspring. They have copies of some of your genes. However, remember that in a sexually reproducing diploid species, each offspring typically inherits half its genes from its mother and half from its father. Thus each individual shares only 50 percent of its genes with each of its parents. It also shares 50 percent of its genes with any of its siblings (brothers and sisters). By sacrificing its own reproductive success to help rear its siblings, a sterile worker promotes copies of its own "self-sacrifice" genes in these close relatives.

In all eusocial species, sterile workers assist fertile relatives with whom they share genes. Honeybees and ants may have an added incentive to make sacrifices. Hymenoptera have a haplodiploid sex determination system, in which diploid females develop from fertilized eggs and haploid males from unfertilized ones. As a result, female hymenopterans share 75 percent of their genes with their sisters. This may help explain why eusociality has arisen many times in this order.

In mole-rats, eusocial behavior is thought to have evolved mainly as a result of ecological factors. Mole-rats live in a very dry habitat where burrowing is difficult and food supplies are patchy. Dispersing from an existing clan and starting a new one is unlikely to be successful. Thus, most individuals do better by staying and assisting their relatives.

altruistic behavior Behavior that benefits others at the expense of the individual.
eusocial animal Animal that lives in a multigenerational group in which many sterile workers cooperate in all tasks essential to the group's welfare, while a few members of the group produce offspring.
theory of inclusive fitness Alleles associated with altruism can be advantageous if the expense of this behavior to the altruist is outweighed by increases in the reproductive success of relatives.

TAKE-HOME MESSAGE 25.6

- Eusocial animals live in family groups in which many sterile workers care for the young of a few breeding individuals.
- In eusocial bees and ants, all workers are female. In termites and mole-rats, there are both male and female workers.
- The theory of inclusive fitness explains how genes that favor altruism can evolve if individuals help close relatives.
- Ecological factors can also favor evolution of eusociality. In mole-rats, a harsh habitat makes it difficult for individuals to leave an established clan and breed elsewhere.

25.7 Application: ALARMING BEE BEHAVIOR

FIGURE 25.15 Africanized honeybee workers stand guard at their hive entrance. If a threat appears, they will release an alarm pheromone.

HONEYBEES STING ONLY IN DEFENSE. Stinging is an evolved response to the threat of animals that raid hives for honey. The European bees used as pollinators of domestic crops and in commercial honey production have been selectively bred to have a high threat threshold. Africanized bees, known in the popular press as "killer bees," are more easily put on the defensive (**FIGURE 25.15**).

Africanized honeybees arose in Brazil in the 1950s. Bee breeders there had imported African bees in the hope of breeding an improved pollinator for this region's tropical orchards. Some of the African imports escaped and mated with European honeybees that had already become established there. Descendants of the resulting hybrids expanded northward and have now become established in much of the United States.

All honeybees can sting only once, and all make the same kind of venom, but Africanized bees sting with less provocation, respond to threats in greater numbers, and are more persistent in pursuit of perceived threats. Vibrations, as from a lawn mower or construction equipment, have triggered several attacks.

What makes Africanized honeybees so testy? One factor is a greater response to the alarm pheromone that workers release in response to a threat. Workers inside the hive detect this chemical signal and rush out to drive off the intruder. Researchers tested the response of Africanized honeybees and European honeybees to alarm pheromone by positioning a cloth near the entrance of hives and releasing an artificial pheromone. Africanized bees flew out of a hive and attacked the cloth faster than European bees, and they plunged six to eight times as many stingers into the cloth.

Human deaths from Africanized bee stings remain rare: There have been about 20 since the bees arrived in the United States in 1990. However, even a single sting can be fatal to someone allergic to honeybee venom. In addition, bee stings are highly painful and people who receive a large number of stings often require hospitalization. Livestock and pet animals have also been stung, in some cases fatally. Thus, the spread of Africanized bees through the United States is a matter of concern.

Summary

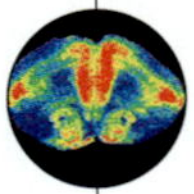

SECTION 25.1 Behavior is a response to stimuli. Genes that influence the nervous or endocrine systems often affect behavior. Studies of behavioral differences within a species or among closely related species can shed light on the proximate (genetic and physiological) and ultimate (evolutionary) causes of a behavior. Epigenetic effects can also influence behavior.

SECTION 25.2 **Instinctive behavior** is inborn; it occurs without any prior experience. Instinctive responses are often triggered by a simple stimulus.

Learned behavior arises in response to experience. Most behavior has a learned component. **Imprinting** is time-sensitive learning. With **habituation**, an animal learns to disregard certain frequently encountered stimuli. Animals also form mental maps, learn the identity of other individuals, develop conditioned responses, and imitate observed behaviors.

SECTION 25.3 Animals communicate by means of chemical, visual, acoustical, or tactile signals. Chemical signals called **pheromones** may elicit an immediate change in behavior, or cause a physiological change that affects behavior over the longer term.

Signaling systems are adaptive only when they benefit both the sender and the receiver. Evolution influences properties of signals, as when courtship calls are easily localized, but alarm calls are not. Predators sometimes take advantage of the communication system of their prey.

SECTION 25.4 Animal mating systems vary among species, and often within species. Monogamy is rare. Even among species that form pair bonds, either or both partners may mate with others. Both males and females benefit by mating with multiple partners, but males tend to be less choosy about mates than females. Sexual selection favors traits that give an individual a competitive edge in attracting mates. Some male birds display for females at a **lek**. In other animals, males establish and defend a **territory** that contains resources that females need. With territorial systems, some males have many mates, whereas other do not mate at all. Territorial animals may have **sexual dimorphism.**

Parental care evolves when the cost in terms of lost opportunities to produce additional offspring later is offset by the benefit of increased survival among current offspring. Which parent or parents care for offspring varies among animal groups. In mammals, maternal care is the rule and no cases of sole paternal care are known. In fishes that provide care, the care-giver is usually male. In birds, both parents usually cooperate in rearing young.

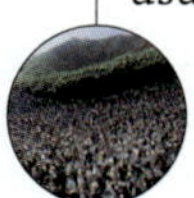

SECTION 25.5 Most animals are solitary, but some form temporary or long-term groups. Animals may temporarily come together as a **selfish herd** in response to the threat of predation. **Social animals** form multigenerational groups and benefit by cooperating in tasks such as defense or rearing the young. With a **dominance hierarchy**, resource and mating opportunities are distributed unequally among group members. Species that live in large groups incur costs, including increased disease and parasitism, and more intense competition for resources.

SECTION 25.6 Honeybees, ants, termites, and some rodents called mole-rats are **eusocial animals**: They live in colonies with overlapping generations and have a reproductive division of labor. Most colony members do not reproduce; they assist their relatives instead.

The **theory of inclusive fitness** states that **altruistic behavior** can be perpetuated when altruistic individuals help their reproducing relatives. Altruistic individuals help perpetuate alleles that lead to their altruism by promoting reproductive success of close relatives who are likely to share these alleles.

SECTION 25.7 "Killer bees" are a hybrid of African and European honeybee strains. Their defensive stinging behavior is triggered more easily than that of other honeybees and they sting in greater numbers.

Self-Quiz

Answers in Appendix VII

1. Genes affect the behavior of individuals by ________ .
   a. influencing the development of nervous systems
   b. affecting how individuals respond to hormones
   c. determining which stimuli can be detected
   d. all of the above

2. Stevan Arnold offered slug meat to newborn garter snakes from different populations to test his hypothesis that the snakes' response to slugs ________ .
   a. was shaped by indirect selection
   b. is an instinctive behavior
   c. is based on pheromones
   d. is adaptive

3. The proximate cause of feeding differences between sitter and rover fruit flies is ________ .
   a. a difference in alleles at one locus that affects learning
   b. natural selection related to competition for food
   c. a difference in alleles for oxytocin receptors
   d. a difference in their ability to find food

4. The honeybee dance transmits information about ________ by way of tactile signals.
   a. predators
   b. mating opportunities
   c. location of food
   d. amount of honey

5. A ________ is a chemical that conveys information between individuals of the same species.
   a. pheromone
   b. neurotransmitter
   c. hormone
   d. all of the above

6. In what group of vertebrates is monogamy and cooperative care of the young by two parents most common?

Data Analysis Activities

Nestling Responses to Alarm Calls As adults, the Australian songbirds known as scrubwrens give different alarm calls depending on whether a predatory bird is on the ground or in flight. The length of the call indicates the urgency of the threat. Scrubwrens make well-concealed nests on the ground, so predators on the ground pose more of a threat to nestlings than those flying above. By contrast, flying predators pose more of a threat to adults. To gauge the response of scrubwren nestlings to parental alarm calls, Dirk Platzen recorded the calls and played them to 5- and 11-day-old nestlings. **FIGURE 25.16** shows how playback of the recordings affected the rate at which chicks made peeping sounds.

1. Which type of alarm call reduced peeping most in 11-day-old nestlings? Which had the least effect?
2. Given that a predator on the ground poses the greatest threat to young birds, which response would you expect to be greatest in 5-day-old nestlings? Is it?
3. Which types of calls did chicks become increasingly responsive to as they aged?

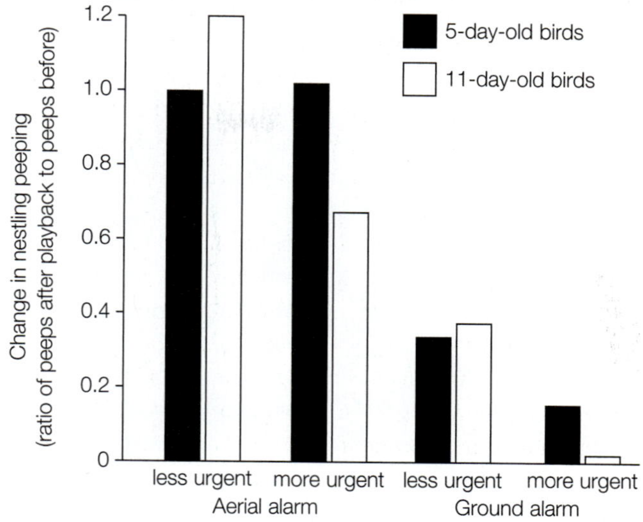

FIGURE 25.16 Change in peeping behavior of 5-day-old nestlings (black bars) and 11-day-old nestlings (white bars) in response to playback of alarm calls by their parent. A value of 0 indicates complete suppression of nestling peeps, 1 indicates no change, and a value above 1 indicates an increase in peeping.

7. List two possible benefits of living in a group.

8. All honeybee workers are _______ .
   a. male
   b. sterile
   c. female
   d. b and c

9. Eusocial insects _______ .
   a. live in extended family groups
   b. include termites, honeybees, and ants
   c. show a reproductive division of labor
   d. all of the above

10. In fishes that provide parental care, that care is most often delivered by _______ .
    a. the male parent
    b. the female parent
    c. siblings
    d. both parents

11. Match the terms with their most suitable description.

___	altruistic behavior	a. time-dependent form of learning
___	selfish herd	b. communal display area
___	habituation	c. learning to ignore a stimulus
___	lek	d. defended area with resources
___	territory	e. assisting another individual at one's own expense
___	imprinting	f. unequal distribution of benefits
___	dominance hierarchy	g. involuntary response becomes tied to a stimulus
___	classical conditioning	h. individuals hide behind others

Critical Thinking

1. For billions of years, the only bright objects in the night sky were stars or the moon. Night-flying moths use them to navigate a straight line. Today, the instinct to fly toward bright objects causes moths to exhaust themselves fluttering around streetlights and banging against brightly lit windowpanes. This behavior clearly is not adaptive, so why does it persist?

2. A female chimpanzee engages in sex only during her fertile period, which is advertised by a swelling of her external genitals. Bonobos, the closest relatives of the chimpanzees, are more sexually active than chimpanzees. Female bonobos mate even when they are not fertile. In both species, mating is promiscuous; each female mates with many males. By one hypothesis, female promiscuity may help prevent male infanticide. Chimpanzee males have been observed to kill infants in the wild, but this behavior has never been observed among bonobos.

Explain why it would be disadvantageous for a male to kill the infant of a female with whom he had sex. How might the bonobo female's increased sexual activity lower likelihood of male infanticide?

CREDIT: (16) Platzen, D. & Magrath, R., "Adaptive differences in response to two types of parental alarm call in altricial nestlings," Proceedings of the Royal Society of London 2005, Series B; Biological Sciences, vol. 272. pp. 1101–1106, by permission of The Royal Society.

A red knot sandpiper, marked with leg bands for an ongoing study of its rapidly declining population.

26 POPULATION ECOLOGY

Links to Earlier Concepts

Earlier chapters explored the evolutionary history and genetic nature of populations, including those of humans (Sections 1.1, 17.1, 24.12). Now you will consider ecological factors that limit population growth. Again, science does not address social issues—in this case, how exponential growth affects human lives (1.7). It can only help explain how ecological conditions and events sustain a population's growth, or put a stop to it.

KEY CONCEPTS

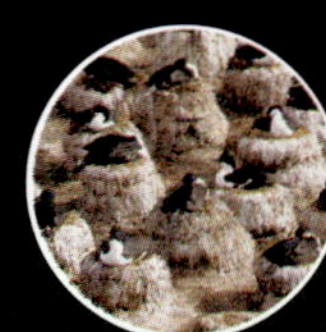

THE VITAL STATISTICS

Ecologists describe a population in terms of its size, density, and the distribution of its members. They often use sampling methods to determine these statistics.

EXPONENTIAL GROWTH

With unlimited resources, a population will grow exponentially. Although the growth rate remains constant, the number of individuals added increases with each generation.

LIMITS ON INCREASES IN SIZE

Density-dependent factors such as increased competition for resources slow population growth. Density-independent factors such as extreme weather can also affect population size.

LIFE HISTORY PATTERNS

Natural selection affects how quickly organisms mature, how often they reproduce, and the number of young per breeding event. Different environments favor different patterns of investment in offspring.

THE HUMAN POPULATION

Technological innovations have allowed human populations to sidestep some limits on growth and expand into new habitats. We are now using resources at an unsustainable rate.

Photograph by © Kevin Elsby/Alamy.

26.1 HOW DO WE DESCRIBE POPULATIONS?

A Clumped distribution of hippopotamuses. B Near-uniform distribution of nesting seabirds. C Random distribution of dandelions.

FIGURE 26.1 Population distribution patterns.

A population, remember, is a group of interbreeding individuals of the same species in a specified area (Section 17.1). Members of a population breed with one another more than they breed with members of other populations. Many factors affect the size and structure of a population, and the study of those factors is called population ecology. **Ecology** is a branch of biology that deals with how organisms interact with one another and the environment. Ecology is not the same as environmentalism, which is advocacy for protection of the environment. However, environmentalists often cite results of ecological studies when drawing attention to environmental concerns.

SIZE, DENSITY, AND DISTRIBUTION

Demographics are statistics that describe a population. These include population size, density, and distribution, which we discuss here, as well as age structure and fertility rates, which we consider in Section 26.6.

Population size is the total number of individuals in a population. **Population density** is the number of individuals per unit area or volume. Examples of population density include the number of dandelions per square meter of lawn or the number of amoebas per milliliter of pond water. **Population distribution** describes the location of individuals relative to one another. Members of a population may be clumped together, be an equal distance apart, or be distributed randomly. The most common population distribution pattern is a clumped one, in which members of the population are closer to one another than would be predicted by chance alone. A patchy distribution of resources encourages clumping. Hippopotamuses clump in muddy river shallows (**FIGURE 26.1**). A cool, damp, north-facing slope may be covered with ferns, whereas an adjacent drier south-facing slope has none. Limited dispersal ability increases the likelihood of a clumped distribution: As the saying goes, the nut does not fall far from the tree. Asexual reproduction is another source of clusters. It produces colonies of coral and vast stands of some trees. Finally, as Section 25.5 explained, some animals benefit by grouping together.

Intense competition for limited resources can produce a near uniform distribution, with individuals more evenly spaced than would be expected by chance. Creosote bushes in deserts of the American Southwest grow in this pattern. Competition for limited water among the root systems keeps the plants from growing in close proximity. Similarly, seabirds in breeding colonies often show a near uniform distribution. Each nesting bird aggressively repels others that get within reach of its beak (**FIGURE 26.1B**).

A random distribution occurs when resources are distributed uniformly through the environment, and proximity to others neither benefits nor harms individuals. For example, when the wind-dispersed seeds of dandelions land on the uniform environment of a suburban lawn, dandelion plants grow in a random pattern (**FIGURE 26.1C**). Wolf spider burrows are also randomly distributed relative to one another. When seeking a burrow site, the spiders neither avoid one another nor seek one another out.

Often, the scale of the area sampled and the timing of a study influence the observed demographics. For example, seabirds are spaced almost uniformly at a nesting site, but the nesting sites are clumped along a shoreline. The birds crowd together during the breeding season, but disperse when breeding is over.

SAMPLING A POPULATION

It is often impractical to count all members of a population, so biologists frequently use sampling techniques to estimate population size. **Plot sampling** estimates the total number of individuals in an area on the basis of direct counts in a small portion of the area. For example, ecologists might estimate the number of grass plants in a grassland, or the number of clams in a mudflat, by measuring the number of individuals in several 1-meter by 1-meter square plots. To estimate total population size, scientists first determine the average number of individuals per sample plot. They then multiply that average by the number of plots that would fit in the population's range. Estimates derived from plot sampling are most accurate when species are not very mobile and conditions across their habitat are uniform.

Mark–recapture sampling is used to estimate the population size of mobile animals. With this technique, animals are captured, marked with a unique identifier of some sort, then released. After marked individuals return to the population, scientists capture another sample of the population. The proportion of marked animals in the second sample is taken to be representative of the proportion marked in the population as a whole. Suppose 100 deer are captured, marked, and released. Later, 50 of these deer are recaptured along with 50 unmarked deer. Marked deer constitute half the recaptured group, so the group previously caught and marked (100 deer) must have been half of the population. Thus the total population is estimated at 200.

Information about the traits of individuals in a sample plot or capture group can be used to infer properties of the population as a whole. For example, if half the recaptured deer are of reproductive age, half of the population is assumed to share this trait.

demographics Statistics that describe a population.
ecology Study of interactions among organisms, and among organisms and their environment.
mark–recapture sampling Method of estimating population size of mobile animals by marking individuals, releasing them, then checking the proportion of marks among individuals recaptured at a later time.
plot sampling Using demographics observed in sample plots to estimate demographics of a population as a whole.
population density Number of individuals per unit area.
population distribution Location of population members relative to one another; clumped, uniformly dispersed, or randomly dispersed.
population size Total number of individuals in a population.

TAKE-HOME MESSAGE 26.1

Demographics such as size, density, and distribution pattern are used to describe populations. Demographics of a population are often inferred on the basis of a study of a smaller subgroup within that population.

Environmental conditions and interactions among individuals can influence a population's demographics, which often change over time.

PEOPLE MATTER

National Geographic Grantee
DR. KAREN DeMATTEO

A Chesapeake Bay retriever named Train helps Karen DeMatteo study jaguars, puma, ocelots, oncillas, and bush dogs. Individuals of these species occur at low density in the dense South American forest, so locating them is a challenge. DeMatteo could have used standard survey technology (such as radio telemetry or camera traps) to get information, but those methods have some weaknesses. For example, an animal could avoid the part of the forest where a camera is set up. So, DeMatteo decided to look instead for what the animals leave behind in the forest—their scat (feces).

This is where Train comes in. Like other dogs, he has a terrific sense of smell, and he has been trained to use it to find specific types of scat. He behaves like a drug-sniffing dog, except instead of alerting his handler when he finds a drug, he alerts DeMatteo when he finds predator scat. DeMatteo collects the scat that Train finds, and later analyzes its DNA in the laboratory. Results of this analysis, along with GPS and spatial mapping locations of where scat was found, allow DeMatteo to determine the predators' distribution and understand differences in habitat use. DeMatteo says, "The goal is to expand our knowledge of how these species move through the landscape so we can determine locations for biological corridors/wildlife crossings that maximize animal movement and minimize human–wildlife conflict."

CREDIT: (in text) Photo courtesy Karen DeMatteo.

26.2 WHY DOES THE SIZE OF A POPULATION CHANGE?

Starting Size of Population		Net Monthly Increase	New Size of Population
2,000	× r =	800	2,800
2,800	× r =	1,120	3,920
3,920	× r =	1,568	5,488
5,488	× r =	2,195	7,683
7,683	× r =	3,073	10,756
10,756	× r =	4,302	15,058
15,058	× r =	6,023	21,081
21,081	× r =	8,432	29,513
29,513	× r =	11,805	41,318
41,318	× r =	16,527	57,845
57,845	× r =	23,138	80,983
80,983	× r =	32,393	113,376
113,376	× r =	45,350	158,726
158,726	× r =	63,490	222,216
222,216	× r =	88,887	311,103
311,103	× r =	124,441	435,544
435,544	× r =	174,218	609,762
609,762	× r =	243,905	853,667
853,667	× r =	341,467	1,195,134

A Increases in size over time. Note that the net increase becomes larger with each generation.

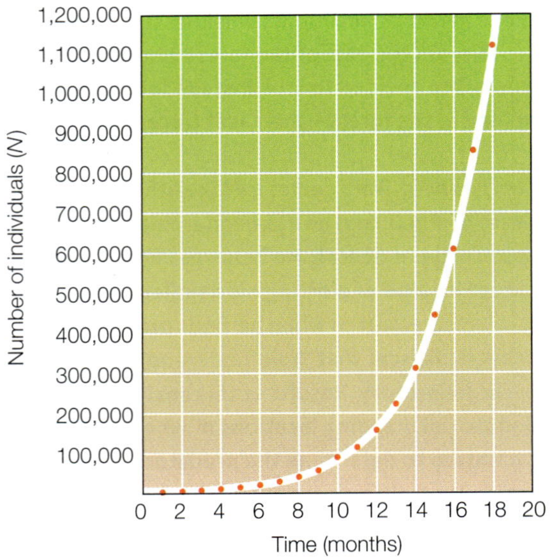

B Graphing numbers over time produces a J-shaped curve.

FIGURE 26.2 {Animated} Exponential growth in a hypothetical population of mice with a per capita rate of growth (r) of 0.4 per mouse per month and a population size of 2,000.

IMMIGRATION AND EMIGRATION

In nature, populations continually change in size. Individuals are added to a population by births and **immigration**, the arrival of new residents that previously belonged to another population. Individuals are removed from it by deaths and **emigration**, the departure of individuals who take up permanent residence elsewhere.

In many animal species, young of one or both sexes leave the area where they were born to breed elsewhere. For example, young freshwater turtles typically emigrate from their parental population and become immigrants at another pond some distance away. By contrast, seabirds typically breed where they were born. However, some individuals may emigrate and end up at breeding sites more than a thousand kilometers away. The tendency of individuals to emigrate to a new breeding site is usually related to resource availability and crowding. As resources decline and crowding increases, the likelihood of emigration rises.

ZERO TO EXPONENTIAL GROWTH

If we set aside the effects of immigration and emigration, we can define **zero population growth** as an interval during which the number of births is balanced by an equal number of deaths. As a result, population size remains unchanged, with no net increase or decrease in the number of individuals.

We can measure births and deaths in terms of rates per individual, or per capita. *Capita* means head, as in a head count. Subtract a population's per capita death rate (d) from its per capita birth rate (b) and you have the **per capita growth rate**, or r:

b	$-$	d	$=$	r
(per capita birth rate)		(per capita death rate)		(per capita growth rate)

Imagine 2,000 mice living in the same field. If 1,000 mice are born each month, then the birth rate is 0.5 births per mouse per month (1,000 births/2,000 mice). If 200 mice die one way or another each month, then the death rate is 200/2,000 or 0.1 deaths per mouse per month. Thus, r is 0.5 – 0.1, or 0.4 per mouse per month.

As long as r remains constant and greater than zero, **exponential growth** will occur, which means that the population's size will increase by the same proportion of its total in every successive time interval. We can calculate population growth (G) for each interval based on the number of individuals (N) and the per capita growth rate:

N	$\times$	r	$=$	G
(number of individuals)		(per capita growth rate)		(population growth per unit time)

Our hypothetical population of field mice consists of 2,800 after one month (**FIGURE 26.2A**). A net increase of 800 fertile mice has increased the number of breeders. If all of the fertile mice reproduce, the population size will expand by 1,120 individuals (2,800 × 0.4). The total population size is now 3,920. At this growth rate, the number of mice would rise from 2,000 to more than 1 million in under two years! Graphing the increases against time results in a J-shaped curve, which is characteristic of exponential population growth (**FIGURE 26.2B**).

With exponential growth, the number of new individuals added increases each generation, although the per capita growth rate stays the same. Exponential population growth is analogous to compound interest in a bank account. The annual interest *rate* stays fixed, yet every year the *amount* of interest paid increases. The annual interest paid into the account adds to the size of the balance, so the next interest payment will be based on the increased balance. Similarly, with exponential growth, the number of individuals added increases in each generation.

Imagine a single bacterium in a culture flask. After thirty minutes, the cell divides in two. Those two cells divide, and so on every thirty minutes. If no cells die between divisions, then the population size will double in every interval—from 1 to 2, then 4, 8, 16, 32, and so on. After 9–1/2 hours, there have been nineteen doublings, so the population now consists of more than 500,000 cells. Ten hours (twenty doublings) later, there are more than a million. Curve 1 in **FIGURE 26.3** is a plot of this increase.

Now suppose that 25 percent of the descendant cells die every thirty minutes in our hypothetical population of bacteria. In this scenario, it takes seventeen hours, not ten, for that population to reach 1 million. Thus, deaths slow the rate of increase but do not stop exponential growth (curve 2 in **FIGURE 26.3**). Exponential growth will occur in any population in which the birth rate exceeds the death rate—in other words, as long as *r* is greater than zero.

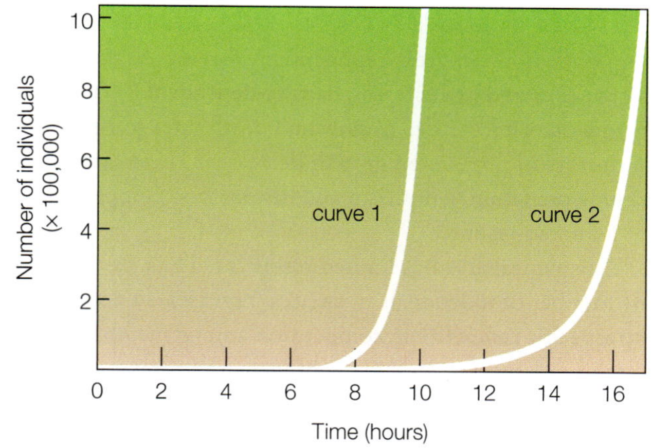

FIGURE 26.3 Effect of deaths on the rate of increase for two hypothetical populations of bacteria, both starting with one cell. Plot the population growth for bacterial cells that reproduce every half hour and you get growth curve 1. Next, plot the population growth of bacterial cells that divide every half hour, with 25 percent dying between divisions, and you get growth curve 2. Deaths slow the rate of increase, but as long as the birth rate exceeds the death rate, exponential growth will continue.

BIOTIC POTENTIAL

The growth rate for a population under ideal conditions is its **biotic potential**. This is a theoretical rate at which the population would grow if shelter, food, and other essential resources were unlimited and there were no predators or pathogens. Factors that affect biotic potential include the age at which reproduction typically begins, how long individuals remain reproductive, and the number of offspring that are produced each time an individual reproduces. Microbes such as bacteria have some of the highest biotic potentials, whereas large-bodied mammals have some of the lowest.

Regardless of the species, populations seldom reach their biotic potential because of the effects of limiting factors, a topic we discuss in detail in the next section.

biotic potential Maximum possible population growth rate under optimal conditions.
emigration Movement of individuals out of a population.
exponential growth A population grows by a fixed percentage in successive time intervals; the size of each increase is determined by the current population size.
immigration Movement of individuals into a population.
per capita growth rate (*r*) Of a population, the change in individuals added over some time interval, divided by the number of individuals in the population.
zero population growth Interval in which births equal deaths.

TAKE-HOME MESSAGE 26.2

The size of a population depends on its birth rate, death rate, immigration, and emigration.

Exponential growth occurs when the per capita growth rate (*r*) of a population is stable and greater than one. With exponential growth, more and more individuals are added with each successive generation.

Most populations never achieve the maximal possible growth rate for their species, because of limiting factors.

26.3 WHAT CONSTRAINS POPULATION GROWTH?

DENSITY-DEPENDENT FACTORS

No population can grow exponentially forever. As the degree of crowding increases, **density-dependent limiting factors** cause birth rates to slow and death rates to rise, so the rate of population growth decreases. These factors include predation, parasitism and disease, and competition for a limited resource.

Any natural area has limited resources. Thus, as the number of individuals in an area increases, so does **intraspecific competition**: competition among members of the same species. As a result of increased competition, some individuals fail to secure what they need to survive and reproduce. Competition has a detrimental effect even on winners, because energy they use in competition for resources is not available for reproduction. Essential resources for which animals might compete include food, water, hiding places, and nesting sites (**FIGURE 26.4**). Plants compete for nutrients, water, and access to sunlight.

Parasitism and contagious disease increase with crowding because the closer individuals are to one another, the more easily parasites and pathogens can spread. Predation increases with density too, because predators often concentrate their efforts on the most abundant prey.

FIGURE 26.4 Example of a limiting factor. Wood ducks build nests only inside tree hollows of specific dimensions (*left*). In some places, lack of access to appropriate hollows now limits the size of the wood duck population. Adding artificial nesting boxes (*right*) to these environments can help increase the size of the duck populations.

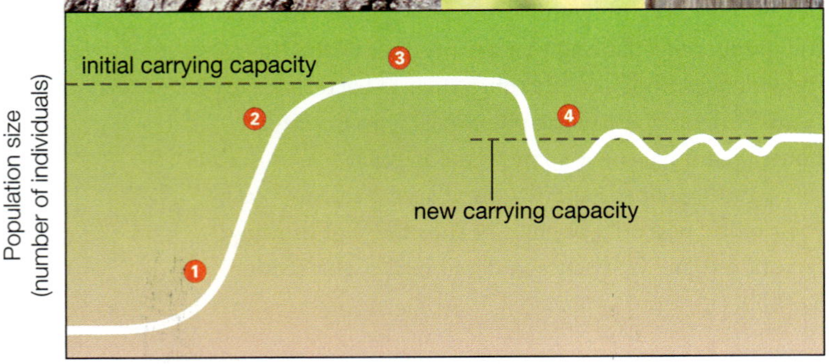

FIGURE 26.5 Logistic growth. Note the initial S-shaped curve.
1 At low density, the population grows exponentially.
2 As crowding increases, density-dependent limiting factors slow the rate of growth.
3 Population size levels off at the carrying capacity for that species.
4 Any change in the carrying capacity will result in a corresponding shift in population size.

FIGURE IT OUT: How does adding nest boxes affect the carrying capacity for wood duck populations? Answer: It increases carrying capacity.

LOGISTIC GROWTH

Logistic growth occurs when density-dependent factors affect population size over time (**FIGURE 26.5**). When the population is small, density-dependent limiting factors have little effect and the population grows exponentially 1. Then, as population size and the degree of crowding rise, these factors begin to limit growth 2. Eventually, the population size levels off at the environment's carrying capacity 3. **Carrying capacity (*K*)** is the maximum number of individuals that a population's environment can support indefinitely. The result is an S-shaped curve.

The equation that describes logistic growth is:

$$(r \times N)\frac{(K - N)}{K} = G$$

As with the exponential growth equation, G is growth per unit time, *r* is the per capita growth rate, and *N* is current population size. *K* is carrying capacity. The $(K-N)/K$ part of the equation represents the proportion of carrying capacity not yet used. As a population grows, this proportion decreases, so G becomes smaller and smaller. At carrying capacity, the equation become $G=rN(0)$, which means no individuals are added.

Carrying capacity is species-specific, environment-specific, and can change over time 4. For example, the carrying capacity for a plant species decreases when nutrients in the soil become depleted. Human activities also affect carrying capacity. For example, human harvest of horseshoe crabs has decreased the carrying capacity for red knot sandpipers (shown in the chapter opening photo). Horseshoe crab eggs are these birds' main food during their long-distance migration.

DENSITY-INDEPENDENT FACTORS

Sometimes, natural disasters or weather-related events affect population size. A volcanic eruption, hurricane, or flood can decrease population size. So can human-caused events such

FIGURE 26.6 Overshoot and crash. A reindeer herd introduced to an island in 1944 increased in size exponentially, then crashed after an especially cold and snowy winter in 1963–64.

as an oil spill. These events are called **density-independent limiting factors**, because crowding does not influence the likelihood of their occurrence or the magnitude of their effect.

In nature, density-dependent and density-independent factors often interact to determine a population's size. Consider what happened after the 1944 introduction of 29 reindeer to St. Matthew Island, an uninhabited island off the coast of Alaska. When biologist David Klein visited the island in 1957, he found 1,350 well-fed reindeer (**FIGURE 26.6**). Klein returned in 1963 and counted 6,000 reindeer. The population had soared far above the island's carrying capacity. A population can temporarily overshoot an environment's carrying capacity, but the high density cannot be sustained. Klein observed that some effects of density-dependent limiting factors were already apparent. For example, the average body size of the reindeer had decreased. When Klein returned in 1966, only 42 reindeer survived. The single male had abnormal antlers, and was thus unlikely to breed. There were no fawns. Klein figured out that thousands of reindeer had starved to death during the winter of 1963–1964. That winter was unusually harsh, with low temperatures, high winds, and 140 inches of snow. Most reindeer, already in poor condition as a result of increased competition, starved when deep snow covered their food. A population decline had been expected—a population that exceeds its carrying capacity usually shrinks and falls below that capacity—but bad weather magnified the extent of the crash. By the 1980s, there were no reindeer left on the island.

carrying capacity (*K*) Maximum number of individuals of a species that a particular environment can sustain; can change over time.
density-dependent limiting factor Factor that limits population growth and has a greater effect in dense populations; for example, competition for a limited resource.
density-independent limiting factor Factor that limits population growth and arises regardless of population size; for example, a flood.
intraspecific competition Competition for resources among members of the same species.
logistic growth A population grows exponentially at first, then growth slows as population size approaches the environment's carrying capacity for that species.

TAKE-HOME MESSAGE 26.3

- Carrying capacity is the maximum number of individuals of a population that can be sustained indefinitely by the resources in a given environment.
- With logistic growth, the growth of a population is fastest when its individuals are at low density, then it slows as the population size approaches the carrying capacity of its environment.
- The effects of density-dependent factors such as disease cause a logistic growth pattern. Density-independent factors such as natural disasters also affect population growth.

26.4 HOW DO LIFE HISTORY PATTERNS VARY?

A population's growth rate is affected by its members' **life history**, which is the schedule of how individuals allocate resources to reproduction over the course of their lifetime. Life history traits include the age at which reproduction begins, the frequency of reproduction, and the number of offspring produced during each reproductive event.

TABLE 26.1

Life Table for an Annual Plant Cohort*

Age Interval (days)	Survivorship (number surviving at start of interval)	Number Dying During Interval	Death Rate (number dying/ number surviving)	"Birth" Rate During Interval (number of seeds from each plant)
0–63	996	328	0.329	0
63–124	668	373	0.558	0
124–184	295	105	0.356	0
184–215	190	14	0.074	0
215–264	176	4	0.023	0
264–278	172	5	0.029	0
278–292	167	8	0.048	0
292–306	159	5	0.031	0.33
306–320	154	7	0.045	3.13
320–334	147	42	0.286	5.42
334–348	105	83	0.790	9.26
348–362	22	22	1.000	4.31
362–	0	0	0	0
		996		

* Phlox drummondii; *data from W. J. Leverich and D. A. Levin, 1979.*

TIMING OF BIRTHS AND DEATHS

One way to investigate life history traits is to focus on a **cohort**—a group of individuals born during the same interval—from their time of birth until the last one dies. Ecologists often divide a natural population into age classes and record the age-specific birth rates and mortality. The resulting data is summarized in a life table (**TABLE 26.1**).

Information about age-specific death rates can also be illustrated by a **survivorship curve**, a plot that shows how many members of a cohort remain alive over time. Ecologists have described three types of curves. A type I curve is convex, indicating that the death rate remains low until relatively late in life (**FIGURE 26.7A**). Humans and other large mammals that produce and care for one or two offspring at a time show this pattern. A diagonal type II curve indicates that the death rate of the population does not vary much with age (**FIGURE 26.7B**). In lizards, small mammals, and large birds, old individuals are about as likely to die of disease or predation as young ones. A type III curve is concave, indicating that the death rate for a population peaks early in life (**FIGURE 26.7C**). Marine animals that release eggs into water have this type of curve, as do plants that release enormous numbers of tiny seeds.

r-SELECTION AND *K*-SELECTION

To produce offspring, an individual must invest resources that it could otherwise use to grow and maintain itself. Species differ in the manner in which they distribute parental investment among offspring and over the course

FIGURE 26.7 {Animated} Survivorship curves. Gray lines are theoretical curves. Red dots are data from field studies.

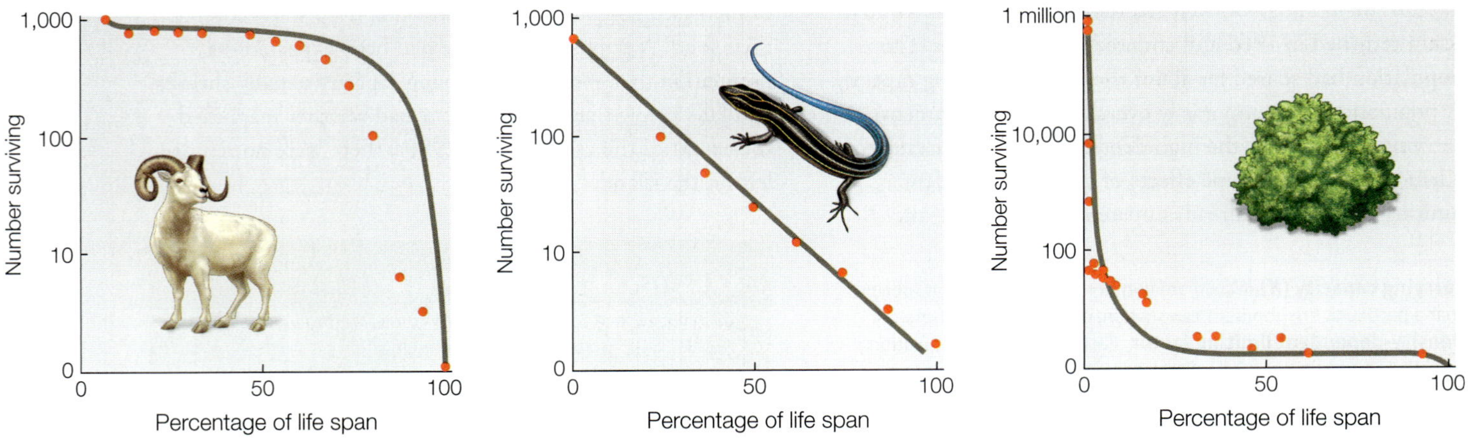

A Type I curve. Mortality is highest late in life. Data for Dall sheep (*Ovis dalli*).

B Type II curve. Mortality varies little with age. Data for five-lined skink (*Eumeces fasciatus*).

C Type III curve. Mortality is highest early in life. Data for a desert shrub (*Cleome droserifolia*).

FIGURE IT OUT: Based on the death rates listed for the annual plant in Table 26.1, what type of survivorship curve does that plant have?

Answer: A type III curve, with mortality highest early in life.

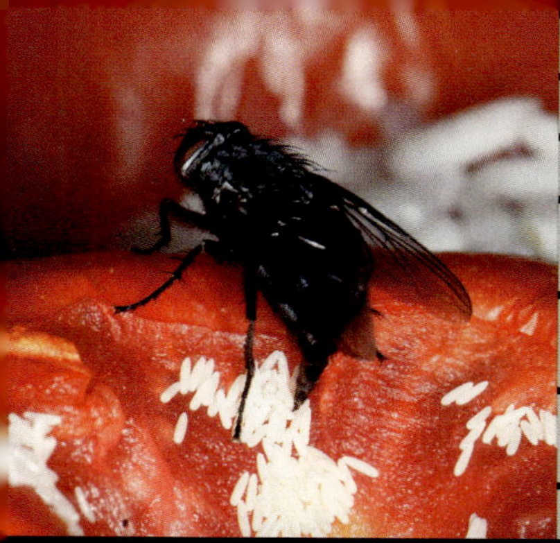

Opportunistic life history	Equilibrial life history
shorter development	longer development
early reproduction	later reproduction
fewer breeding episodes, many young per episode	more breeding episodes, few young per episode
less parental investment per young	more parental investment per young
higher early mortality, shorter life span	low early mortality, longer life span
result of *r*-selection	result of *K*-selection

A Fly laying many eggs.

B Whale with its single calf.

FIGURE 26.8 Two types of life history. Most species have a mix of opportunistic and equilibrial life history traits.

of their lifetime. Life history patterns vary continuously among species, but ecologists have described two theoretical extremes at either end of this continuum. Both maximize the number of offspring that will be produced and survive to adulthood, but they do so under very different environmental conditions.

When a species lives where conditions vary in an unpredictable manner, its populations seldom reach the carrying capacity of their environment. As a result, there is little competition for resources and deaths occur mainly as a result of density-independent factors. Such conditions favor an opportunistic life history, in which individuals produce as many offspring as possible, as quickly as possible. Opportunistic species are said to be subject to ***r*-selection**, because they maximize *r*, the per capita growth rate. They tend to have a short generation time and small body size. Opportunistic species usually have a type III survivorship curve, with mortality heaviest early in life. For example, weedy plants such as dandelions have an opportunistic life history. They mature within weeks, produce many tiny seeds, then die. Flies are opportunistic animals. A female fly can lay hundreds of small eggs in a temporary food source such as a rotting tomato (**FIGURE 26.8A**).

When a species lives in a more stable environment, its populations often approach carrying capacity. Under these circumstances, the ability to successfully compete for resources has a major influence on reproductive success. Thus, an equilibrial life history, in which parents produce a few, high-quality offspring, is adaptive. Equilibrial species are shaped by ***K*-selection**, in which adaptive traits provide a competitive advantage when population size is near carrying capacity (*K*). Such species tend to have a large body and a long generation time. This type of life history is typical of large mammals that take years to reach adulthood and begin reproducing. For example, a female blue whale reaches maturity at the age of 6 to 10 years. She then produces only one large calf at a time, and continues to invest in the calf by nursing it after its birth (**FIGURE 26.8B**). Similarly, a coconut palm grows for years before beginning to produce a few coconuts at a time. In both whales and coconut palms, a mature individual produces young for many years.

Some species have mixes of traits that cannot be explained by *r*-selection or *K*-selection alone. For example, century plants (a type of agave) and bamboo are large and long-lived, but they reproduce only once. Atlantic eels and Pacific salmon are unusual among vertebrates in also having a one-shot reproductive strategy. Such a strategy can evolve when opportunities for reproduction are unlikely to be repeated. In the century plant and bamboo, climate conditions that favor reproduction occur only rarely. In the eels and salmon, physiological changes related to migration between fresh water and salt water make a repeat journey impossible.

cohort Group of individuals born during the same interval.
***K*-selection** Selection that favors traits that allow their bearers to outcompete others for limited resources; occurs when a population is near its environment's carrying capacity.
life history A set of traits related to growth, survival, and reproduction such as life span, age-specific mortality, age at first reproduction, and number of breeding events.
***r*-selection** Selection that favors traits that allow their bearers to produce the most offspring the most quickly; occurs when population density is low and resources are abundant.
survivorship curve Graph showing how many members of a cohort remain alive over time.

TAKE-HOME MESSAGE 26.4

Tracking a group of same-aged individuals from birth to death reveals patterns of reproduction and survival. These data can be summarized in life tables or survivorship curves.

Different life history patterns are adaptive under different environmental conditions and population densities.

26.5 HOW DOES PREDATION AFFECT LIFE HISTORY TRAITS?

FIGURE 26.9 {Animated} How predation affects life history traits in guppies. David Reznik (shown above) and John Endler carried out an experiment using wild guppies that had evolved in the presence of a predator (pike cichlid) that preferentially preys on large guppies. Control group guppies remained in their home pool with pike cichlids. Other guppies were moved to a guppy-free pool with a predator (killifish) that preferentially eats small guppies. Data at left show the average age and weight at maturity for descendants of both groups 11 years after the study began.

Many species are subject to different types of predation at different stages in their life cycle. In some cases, this predation pressure can influence life history traits.

AN EXPERIMENTAL STUDY

A long-term study by the evolutionary biologists John Endler and David Reznick illustrates the effect of predation on life history traits. Endler and Reznick studied populations of guppies, small fishes that are native to shallow freshwater streams in the mountains of Trinidad (**FIGURE 26.9**). The scientists focused their attention on a region where many small waterfalls prevent guppies in one part of a stream from moving easily to another. As a result of these natural barriers, each stream holds several populations of guppies that have very little gene flow between them (Section 17.7).

The waterfalls also keep guppy predators from moving from one part of the stream to another. The main guppy predators, killifishes and cichlids, differ in size and prey preferences. The relatively small killifish preys mostly on immature guppies, and ignores the larger adults. The cichlids are bigger fish. They tend to pursue mature guppies and ignore small ones.

Some parts of the streams hold one type of predator but not the other. Thus, different guppy populations face different predation pressures. Reznick and Endler discovered that guppies in regions with cichlids grow faster and are smaller at maturity than guppies in regions with killifish. Guppies in populations hunted by cichlids also reproduce earlier, have more offspring at a time, and breed more frequently.

Were these differences in life history traits genetic, or did some environmental variation cause them? To find out, the biologists collected guppies from both cichlid- and killifish-dominated streams. They reared the groups in separate aquariums under identical predator-free conditions. Two generations later, the groups continued to show the differences observed in natural populations. The researchers concluded that differences between guppies preyed on by different predators have a genetic basis.

Reznick and Endler hypothesized that the predators act as selective agents on guppy life history patterns. They made a prediction: If life history traits evolve in response to predation, then these traits will change when a population is exposed to a new predator that favors different prey traits. To test their prediction, they found a stream region above a waterfall that had killifish but no guppies or cichlids. To this region, they introduced guppies from a site below the waterfall, where there were cichlids but no killifish. Thus, at the experimental site, guppies that had previously lived only with cichlids (which eat large guppies) were now exposed to killifish (which eat smaller ones). The control site was the downstream region below the waterfall, where relatives of the transplanted guppies still coexisted with cichlids.

Reznick and Endler revisited the stream over the course of eleven years and thirty-six generations of guppies. They monitored traits of guppies above and below the waterfall. The recorded data showed that guppies at the upstream experimental site were evolving. Exposure to a previously unfamiliar predator caused changes in the guppies' rate of growth, age at first reproduction, and other life history traits. By contrast, guppies at the control site showed no such changes. Reznick and Endler concluded that life history traits in guppies can evolve rapidly in response to the selective pressure exerted by predation.

COLLAPSE OF A FISHERY

The evolution of life history traits in response to predation is not merely of theoretical interest. It has economic importance. Just as guppies evolved in response to predators, a population of Atlantic codfish (*Gadus morhua*) evolved in response to human fishing pressure. From the mid-1980s to early 1990s, the number of fishing boats targeting the North Atlantic population of codfish increased. As the yearly catch rose, the age of sexual maturity shifted, and fishes that reproduce while young and small became a larger component of the population. These early-reproducing individuals were at an advantage because both commercial fisherman and sports fishermen preferentially caught and kept larger fish (**FIGURE 26.10**). Fishing pressure continued to rise until 1992, when declining cod numbers caused the Canadian government to ban cod fishing in some areas. That ban, and later restrictions, came too late to stop the Atlantic cod population from crashing. In some areas, the population declined by 97 percent and still shows no signs of recovery.

FIGURE 26.10 Fishermen with a prized catch, a large Atlantic codfish. Both sport fishermen and commercial fisherman preferentially harvested the largest codfish.

Looking back, it is clear that life history changes were an early sign that the North Atlantic cod population was in trouble. Had biologists recognized what was happening, they might have been able to save the fishery and protect the livelihood of more than 35,000 fishers and associated workers. Ongoing monitoring of the life history data for other economically important fishes may help prevent similar disastrous crashes in the future.

TAKE-HOME MESSAGE 26.5

The action of a predator that preferentially focuses on prey individuals of a particular size can exert directional selection on life history traits of a prey species.

When humans are the predator, monitoring life history changes in the prey population can help prevent the overharvesting of that species.

26.6 HOW HAS THE SIZE OF THE HUMAN POPULATION CHANGED OVER TIME?

EXPANSIONS AND INNOVATIONS

For most of its history, the human population grew very slowly (**FIGURE 26.11**). The growth rate began to pick up about 10,000 years ago, and it soared during the past two centuries. Three trends promoted the large increases. First, humans expanded into new habitats. Second, they developed technologies that increased the carrying capacity of their habitats. Third, they sidestepped limiting factors that typically restrain population growth.

Modern humans evolved in Africa by about 200,000 years ago, and by 43,000 years ago, their descendants were established in much of the world (Section 24.12). The invention of agriculture about 11,000 years ago provided a more dependable food supply than traditional hunting and gathering. In the middle of the eighteenth century, people learned to harness energy in fossil fuels to operate machinery. This innovation opened the way to high-yielding mechanized agriculture and improved food distribution systems. Food production was further enhanced in the early 1900s, when the invention of synthetic nitrogen fertilizers increased crop yields. The invention of synthetic pesticides in the mid-1900s also contributed to increased food production.

Disease has historically dampened human population growth. During the mid-1300s, one-third of Europe's population was lost in a pandemic known as the Black Death. Beginning in the mid-1800s, an increased understanding of the link between microorganisms and illness led to improvements in food safety, sanitation, and medicine. People began to pasteurize foods and drinks, heating them to kill harmful bacteria. They also began to protect their drinking water.

Advances in sanitation also lowered the death rate associated with medical treatment. In the mid-1800s, Ignaz Semmelweis, a physician in Vienna, began urging doctors to wash their hands between patients. His advice was largely ignored until after his death, when Louis Pasteur popularized the idea that unseen organisms cause disease. Acceptance of this idea also revolutionized surgery, which had been carried out without regard for cleanliness.

A worldwide decline in death rates without an equivalent drop in birth rates is responsible for the ongoing explosion in human population size. The population is now more than 7 billion and is expected to reach 9 billion by 2050.

FERTILITY AND FUTURE GROWTH

Birth rates have begun to slow as a result of contraception use. The **total fertility rate** of a population is the number of offspring a woman would be expected to have during her reproductive years given the current age-specific birth

FIGURE 26.11 Growth curve (red) for the world human population.

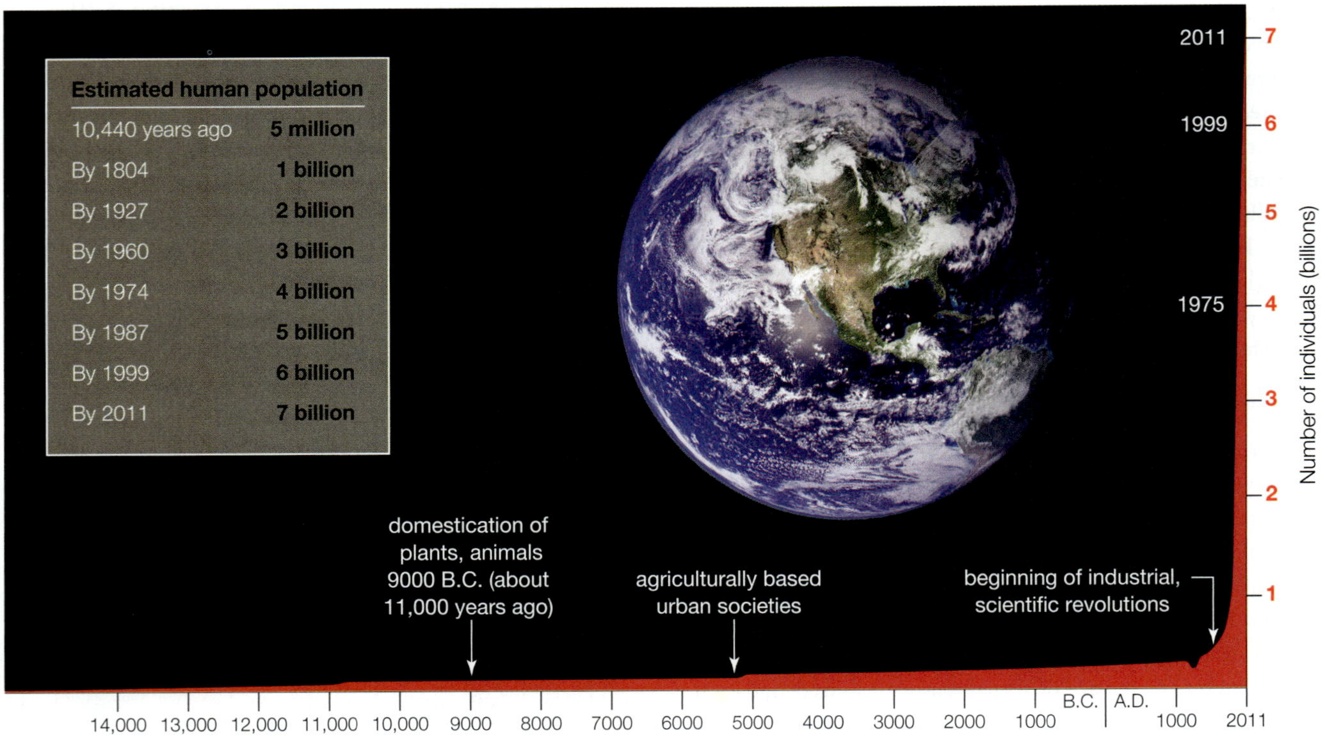

rate. In 1950, the total fertility rate for humans averaged 6.5 worldwide. By 2012, it had declined to 2.6, but it still remains above the **replacement fertility rate**—the number of children a woman must bear to replace herself with one daughter of reproductive age. At present, the replacement fertility rate is 2.1 for developed countries, and as high as 2.5 in some developing countries. (It is higher in developing countries because more daughters die before reaching the age of reproduction.) As long as the total fertility rate exceeds the replacement rate, the human population will continue to grow.

Age structure, which is the distribution of individuals among age groups, affects the rate of population growth. Individuals are often categorized as pre-reproductive, reproductive, or post-reproductive. Members of the pre-reproductive category have a capacity to produce offspring when mature. Along with reproductive individuals, they constitute a population's **reproductive base**.

FIGURE 26.12 shows age structure diagrams for the world's three most populous countries. China and India already have more than one billion people apiece; together, they hold 38 percent of the world population. Next in line is the United States, with more than 315 million. Notice the size of the reproductive base in each diagram. The broader the base of an age structure diagram, the greater the proportion of young people, and the greater the expected growth. Government policies that favor couples who have only one child have helped China to narrow its pre-reproductive base.

Even if every couple now living decides to have no more than two children, two factors will keep the world population increasing for many years. First, longevity continues to increase. Second, the population has a broad pre-reproductive base; about a quarter of the world population has not yet begun to reproduce.

---

age structure Distribution of population members among various age categories.
replacement fertility rate Number of children a woman must bear to replace herself with one daughter of reproductive age.
reproductive base Of a population, members of the reproductive and pre-reproductive age categories.
total fertility rate Expected number of children a women will bear over the course of a lifetime.

TAKE-HOME MESSAGE 26.6

Agricultural, technological, and medical innovations have decreased human death rates.

Worldwide, human population growth has slowed. However, a large number of young people are approaching reproductive age, so our population is predicted to continue growing.

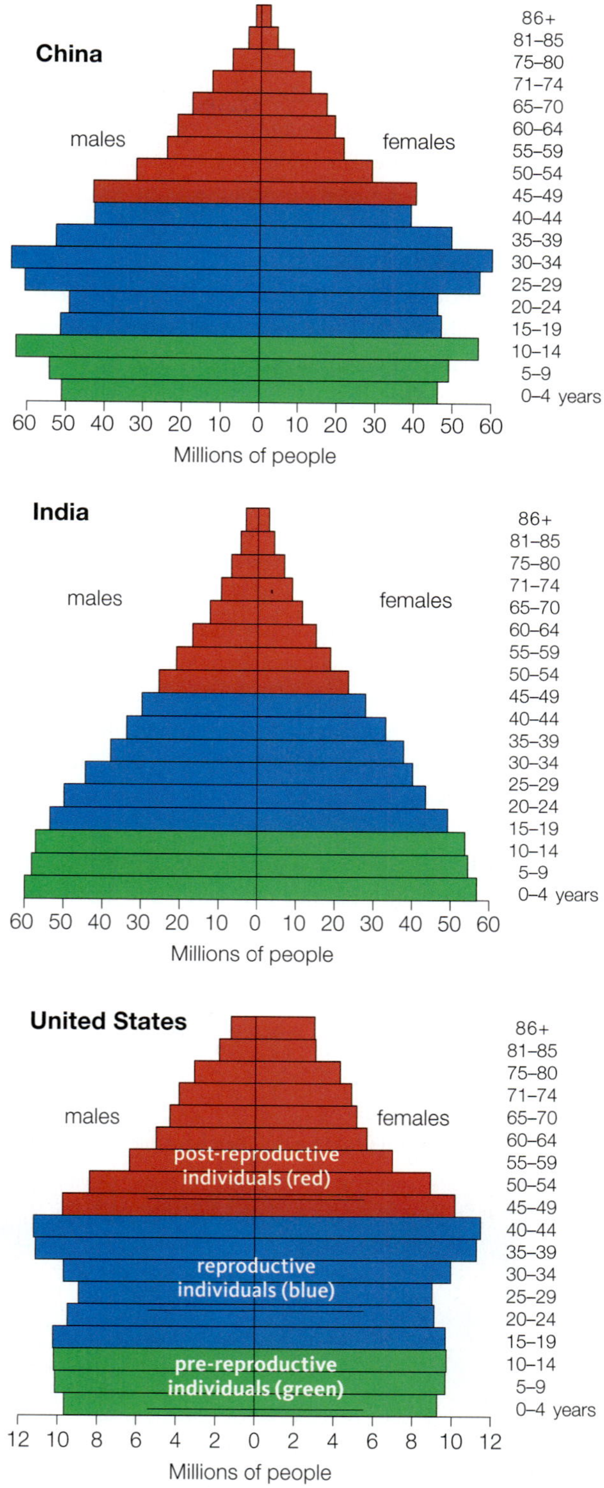

FIGURE 26.12 Age structure diagrams for the world's three most populous countries. The width of each bar represents the number of individuals in a 5-year age group. The *left* side of each chart indicates males; the *right* side, females.

FIGURE IT OUT: Which country has the largest number of men in the 45 to 49 age group?

Answer: China

26.7 HOW DOES INDUSTRIALIZATION AFFECT HUMAN POPULATION GROWTH?

A DEMOGRAPHIC TRANSITION

Demographic factors vary among countries, with the most highly developed countries having the lowest birth rates and infant mortality, and the highest life expectancy. The **demographic transition model** describes how changes in population growth often unfold in four stages of economic development (**FIGURE 26.13**).

Living conditions are harshest during the preindustrial stage, before technological and medical advances become widespread. Birth and death rates are both high, so the growth rate is low ❶.

In the transitional stage, food production and health care improve as industrialization begins. The death rate drops fast, but the birth rate declines more slowly ❷. As a result, the rate of population growth increases. Many of the world's least developed countries are in this stage. Examples include Afghanistan, Haiti, Cambodia, and Ethiopia.

During the industrial stage, when industrialization is in full swing, the birth rate declines. People move from rural areas to towns and cities, where birth control is often more easily available and couples tend to want smaller families. The birth rate moves closer to the death rate, and the population grows less rapidly ❸. The United States is in this stage.

In the postindustrial stage, a population's growth rate becomes negative. The birth rate falls below the death rate, and population size slowly decreases ❹. Japan and some European countries are in this stage.

The demographic transition model was developed based on analysis of what happened when western Europe and North America industrialized in the late 1800s. Whether it can accurately predict changes in modern developing countries remains to be seen. Less developed countries now receive aid from highly developed countries, but must also compete against these countries in a global market.

TABLE 26.2

Ecological Footprints*

Country	Hectares per Capita
United States	8.0
Canada	7.0
France	5.0
United Kingdom	4.9
Japan	4.7
Mexico	3.0
Brazil	2.9
China	2.2
India	0.9

* *Global Footprint Network, 2010 data*

DEVELOPMENT AND CONSUMPTION

What is Earth's carrying capacity for humans? There is no simple answer to this question. For one thing, we cannot predict what new technologies may arise or the effects they will have. For another, different types of societies require different amounts of resources to sustain them. On a per capita basis, people in highly developed countries use far

FIGURE 26.13 **{Animated}** Demographic transition model. The model correlates changes in population growth with industrialization.

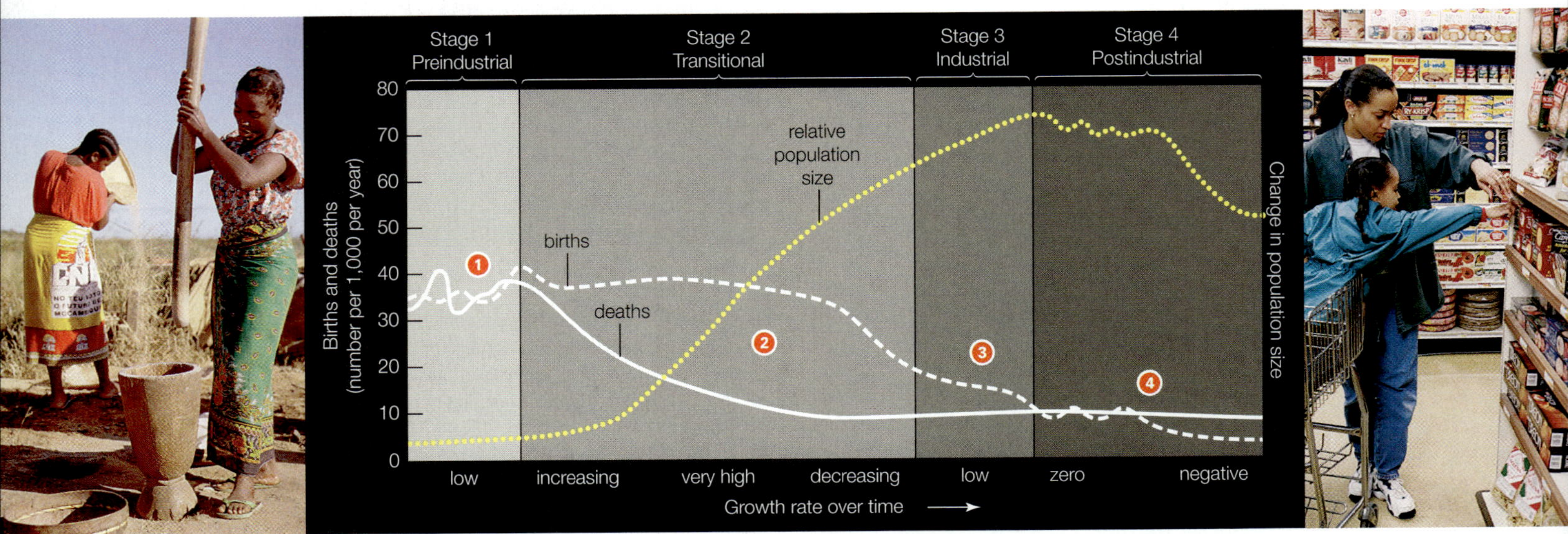

more resources than those in less developed countries, and they also generate more waste and pollution.

Ecological footprint analysis is one widely used method of measuring and comparing resource use. An **ecological footprint** is the amount of Earth's surface required to support a particular level of development and consumption in a sustainable fashion. It includes the amount of area required to grow crops, graze animals, produce forest products, catch fish, hold buildings, and take up any carbon emitted by burning fossil fuels.

In 2010, the per capita global footprint for the human population was 2.7 hectares, or about 6.5 acres (**TABLE 26.2**). The world's two most populous countries, China and India, were below that average, whereas the per capita footprint of the United States was more than three times the average. In other words, the lifestyle of an average person in the United States requires three times as much of Earth's sustainable resources as the lifestyle of an average world citizen. It requires more than eight times the resources of a person in India.

The United States is unlikely to lower its resource consumption to match that of India. In fact, billions of people in India, China, and other less developed nations dream that one day they or their offspring will enjoy the same type of lifestyle as the average American.

Ecological footprint analysis tells us that, with current technology, Earth may not have enough resources to make those dreams come true. The Global Footprint Network estimates that for everyone now alive to live like an average person in the United States would require the resources of five Earths. Such analysis suggests that the human population is currently living well beyond its ecological means and is in the process of racking up a deficit that will take its toll on future generations.

demographic transition model Model describing the changes in human birth and death rates that occur as a region becomes industrialized.

ecological footprint Area of Earth's surface required to sustainably support a particular level of development and consumption.

TAKE-HOME MESSAGE 26.7

- Historically, population growth has slowed after countries have industrialized. The same trend may eventually occur in currently developing countries.
- The most highly developed countries have relatively low growth rates, but they consume a disproportionately large share of Earth's resources.
- The current global level of resource use is unsustainable, and the requirement for resources is expected to rise.

26.8 A HONKIN' MESS

FIGURE 26.14 California park overrun by Canada geese.

CANADA GEESE WERE HUNTED TO NEAR EXTINCTION IN THE LATE 1800S. In the early 1900s, federal laws and international treaties were put in place to protect them and other migratory birds. In recent decades, the number of geese in the United States has soared. For example, in 1970, Michigan had about 9,000 geese; today, it has 300,000. These plant-eating birds often congregate at golf courses and parks (**FIGURE 26.14**). They are considered pests because they produce large amounts of slimy, green feces that soil shoes, feet, and clothing.

Controlling the number of Canada geese poses a challenge because several different populations spend time in the United States. In the past, nearly all Canada geese seen in the United States were migratory. The geese nested in northern Canada, flew to the United States to spend the winter, then returned to Canada. Most Canada geese still migrate, but some populations have lost this trait. During the winter, migratory birds often mingle with nonmigratory ones.

Life is more difficult for migratory geese than for nonmigratory ones. Flying hundreds of miles to and from a northern breeding area takes lots of energy and is dangerous. Compared to a migratory bird, one that stays put can devote more energy to producing young. If the nonmigrant lives in a suburban or urban area, it also benefits from an unnatural abundance of food (grass) and an equally unnatural lack of predators. Not surprisingly, the biggest increases in Canada geese have been among nonmigratory birds that live where humans are plentiful. We have increased the carrying capacity for these birds.

In 2006, increasing complaints about Canada geese led the U.S. Fish and Wildlife Service to encourage wildlife managers to look for ways to reduce nonmigratory Canada goose populations, without harming migratory ones.

CREDIT: (14) Courtesy of Joel Peter.

Summary

SECTION 26.1 **Ecology** is the study of interactions among organisms and their living and nonliving environment. Ecologists describe a population in terms of its **demographics**. They estimate **population size** by using a sampling method such as **plot sampling** or **mark–recapture sampling**. Other demographics include **population density** and **population distribution**. Most populations have a clumped distribution.

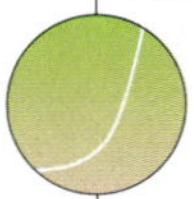

SECTION 26.2 A population's per capita birth rate minus its per capita death rate gives us *r*, the **per capita growth rate**. When birth rate and death rate are equal, there is **zero population growth**.

A population in which *r* is a value greater than zero undergoes **exponential growth**. The increase in size in any interval is determined by the equation $G = r \times N$, where G is population growth and *N* is the number of individuals. With exponential growth, a graph of population size against time produces a J-shaped growth curve. The maximum possible exponential growth rate under optimal conditions is the population's **biotic potential**.

Emigration and **immigration** of individuals can also affect population size.

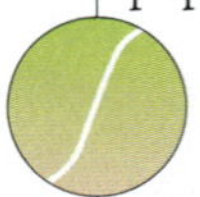

SECTION 26.3 The effects of **density-dependent limiting factors** such as disease and **intraspecific competition** lead to **logistic growth**: a population begins growing exponentially, then growth slows as the population size approaches its environment's **carrying capacity**. With logistic growth, population size over time plots out as an S-shaped curve. Carrying capacity varies by species, among environments, and over time. A population may temporarily overshoot carrying capacity, then crash.

Density-independent limiting factors such as extreme weather events can also influence the growth rate of a population, but their effect does not vary with crowding.

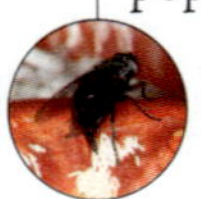

SECTIONS 26.4, 26.5 The time to maturity, number of reproductive events, number of offspring per event, and life span are aspects of a **life history** pattern. Such patterns can be studied by following a **cohort**, a group of individuals born at the same time. Three types of **survivorship curves** are common: a high death rate late in life, a constant death rate at all ages, or a high death rate early in life. Life histories have a genetic basis and are subject to natural selection. At low population density, ***r*-selection** favors quickly producing as many offspring as possible. At a higher population density, ***K*-selection** favors investing more time and energy in fewer, higher-quality offspring. Most populations have a mixture of both *r*-selected and *K*-selected traits. Predation can also affect life history traits.

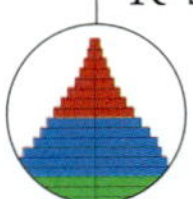

SECTION 26.6 The human population has surpassed 7 billion. Expansion into new habitats and the invention of agriculture allowed early increases. Medicine and technology have allowed greater increases. Today, the global **total fertility rate** is declining, but it remains above the **replacement fertility rate.** The **age structure** varies among countries, with some having a broad **reproductive base** that will cause numbers to increase for at least sixty years.

SECTION 26.7 The **demographic transition model** predicts that population growth slows with advancing economic development. Some of the world's least developed countries remain in the earliest stages of their demographic transition. World resource consumption will probably continue to rise because nations continue to develop and a highly developed nation has a much larger **ecological footprint** than a developing one. However, with current technology, Earth does not have enough resources to support the existing population in the style of developed nations.

SECTION 26.8 Human activities have caused shifts in the size of some Canada goose populations in the United States. Most recently, there has been a dramatic rise in nonmigratory populations of these birds because we have increased their carrying capacity.

Self-Quiz

Answers in Appendix VII

1. Most commonly, individuals of a population show a ________ distribution through their habitat.

2. The rate at which population size grows or declines depends on the rate of ________ .
   a. births c. immigration e. a and b
   b. deaths d. emigration f. all of the above

3. Suppose 200 fish are marked and released in a pond. The following week, 200 fish are caught and 100 of them have marks. How many fish are in this pond?

4. A population of worms is growing exponentially in a compost heap. Thirty days ago there were 400 worms and now there are 800. How many worms will there be thirty days from now, assuming conditions remain constant?

5. For a given species, the maximum rate of increase per individual under ideal conditions is its ________ .
   a. biotic potential c. environmental resistance
   b. carrying capacity d. density control

6. ________ is a density-independent factor that influences population growth.
   a. Resource competition c. Predation
   b. Infectious disease d. Harsh weather

7. A life history pattern for a population is a set of adaptive traits such as ________ .
   a. longevity c. age at reproductive maturity
   b. fertility d. all of the above

8. The human population is now over 7 billion. It reached 6 billion in ________ .
   a. 2007 b. 1999 c. 1802 d. 1350

Data Analysis Activities

Monitoring Iguana Populations In 1989, Martin Wikelski began a long-term study of marine iguana populations in the Galápagos Islands. He marked the iguanas on two islands—Genovesa and Santa Fe—and collected data on how their body size, survival, and reproductive rates varied over time. The iguanas eat algae and have no predators, so deaths typically result from food shortages, disease, or old age. His studies showed that the iguana populations decline during El Niño events, when water surrounding the islands heats up.

In January 2001, an oil tanker ran aground and leaked a small amount of oil into the water near Santa Fe. **FIGURE 26.15** shows the number of marked iguanas that Wikelski and his team counted just before the spill and about a year later.

1. Which island had more marked iguanas at the time of the first census?
2. How much did the population size on each island change between the first and second census?
3. Wikelski concluded that changes on Santa Fe were the result of the oil spill, rather than a factor common to both islands. How would the census numbers be different from those he observed if an adverse event had affected both islands?

FIGURE 26.15 Shifting numbers of marked marine iguanas on two Galápagos islands. An oil spill occurred near Santa Fe just after the January 2001 census (orange bars). A second census was carried out in December 2001 (green bars).

9. Compared to the less developed countries, the highly developed ones have a higher ______.
   a. death rate
   b. birth rate
   c. total fertility rate
   d. resource consumption rate

10. An increase in infant mortality will ______ a population's replacement fertility rate.
   a. raise
   b. lower
   c. not affect

11. Species that usually colonize empty habitats are more likely to have traits that are favored by ______.
   a. *r*-selection
   b. *K*-selection

12. All members of a cohort are the same ______.
   a. sex
   b. size
   c. age
   d. weight

13. Match each term with its most suitable description.

___ carrying capacity	a. maximum rate of increase per individual under ideal conditions
___ exponential growth	b. population growth plots out as an S-shaped curve
___ biotic potential	c. maximum number of individuals sustainable by the resources in a given environment
___ limiting factor	d. population growth plots out as a J-shaped curve
___ logistic growth	e. essential resource that restricts population growth when scarce

Critical Thinking

1. When researchers moved guppies from pools with cichlids that eat large guppies to pools with killifish that eat small ones, life history traits were not the only traits that changed. Over generations, male guppies became more colorful. Why do you think this change occurred?

2. Each summer, a giant saguaro cactus produces tens of thousands of tiny black seeds. Most die, but the few that land in a sheltered spot sprout the following spring. The saguaro is a slow-growing CAM plant (Section 6.5). After fifteen years, it may be only knee high, and it will not flower for another fifteen years. It may live for 200 years. Saguaros share their habitat with annuals such as poppies, which sprout, form seeds, and die in just a few weeks. Speculate on how these different life histories can both be adaptive in the same desert environment.

3. Age structure diagrams for two hypothetical populations are shown below. Describe the growth rate of each population and discuss the current and future social and economic problems that each is likely to face.

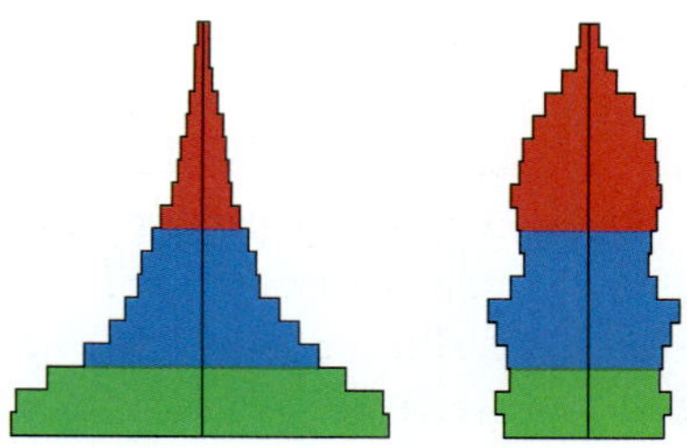

A red-billed oxpecker perches on a cape buffalo. Oxpeckers benefit buffalo and other grazing animals by removing and eating the ticks that parasitize them.

27

Links to Earlier Concepts

In this chapter you will revisit biogeography and take a closer look at global patterns in species distribution (Section 16.1). You will be reminded of effects of pathogens (19.2) and how species interactions lead to natural selection (17.11). You will see examples of field experiments (1.6) and apply what you know about populations and factors affecting their growth (26.1–26.3).

COMMUNITY ECOLOGY

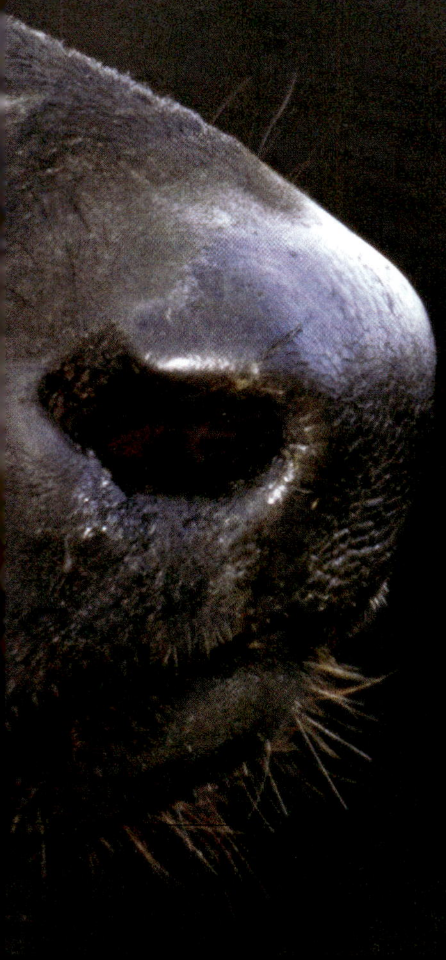

KEY CONCEPTS

COMMUNITY CHARACTERISTICS

Physical characteristics of the environment and interactions among species affect the number of species in a community and their relative abundance.

FORMS OF SPECIES INTERACTIONS

Interspecific interactions such as commensalism, mutualism, competition, predation, and parasitism can influence the population size of participating species.

CHANGES IN COMMUNITIES

The composition of a community changes over time, but the exact outcome of future changes is difficult to predict. Random factors and disturbances can influence community structure.

EFFECTS OF A SINGLE SPECIES

The addition or loss of a single species can have a dramatic effect on the population size of other species within a community. Some species foster diversity by their presence.

LESSONS FROM ISLANDS

The number of species on an island depends on immigration and extinction rates. A model based on studies of islands helps scientists predict how many species will live in a particular community.

Photograph by © Kim Wolhuter, National Geographic Creative.

27.1 WHAT FACTORS AFFECT COMMUNITY STRUCTURE?

The type of place where a species normally lives is its **habitat**. In any specific habitat, all the species that live there constitute a **community**. Communities often nest one inside another. For example, we find a community of microbial organisms inside the gut of a termite. That termite is part of a larger community of organisms living on a fallen log. The log-dwellers are part of a larger forest community.

Even communities that are similar in scale differ in their species diversity. There are two components to species diversity. The first, species richness, refers to the number of species. The second is species evenness, or the relative abundance of each species. For example, a pond that has five fish species in nearly equal numbers has a higher species diversity than a pond with one abundant fish species and four rare ones.

Community structure is dynamic, which means that in any community, the array of species and their relative abundances tends to change over time. Communities change over a long time span as they form and then age. They also change over the short term as a result of disturbances.

Geography and climate affect community structure. Factors such as soil quality, sunlight intensity, rainfall, and temperature vary with latitude and elevation. Tropical regions receive the most sunlight energy and have the most even temperature. For most plants and animal groups, the number of species is greatest in the tropical regions near the equator, and declines as you move toward the poles. Tropical forest communities have more types of trees than temperate ones. Similarly, tropical reef communities are more diverse than comparable communities farther from the equator.

Species interactions also influence community structure. In some cases, the effect is indirect. For example, when songbirds eat caterpillars, the birds indirectly benefit trees that the caterpillars feed on, while directly reducing the abundance of caterpillars.

TABLE 27.1

Direct Two-Species Interactions

Type of Interaction	Effect on Species 1	Effect on Species 2
Commensalism	Helpful	None
Mutualism	Helpful	Helpful
Interspecific competition	Harmful	Harmful
Predation, herbivory, parasitism, parasitoidism	Helpful	Harmful

Biologists categorize direct species interactions by their effects on both participating species (**TABLE 27.1**). For example, **commensalism** helps one species and has no effect on the other. Commensal orchids live attached to the trunk or branches of a tree (**FIGURE 27.1**). Having a perch in the light benefits the orchid, and the tree is unaffected. Relationships are considered commensal only when one species benefits, and the other neither benefits nor is harmed by the relationship. Should evidence of either effect come to light, the relationship is reclassified.

Species interactions may be fleeting or long term. **Symbiosis**, which means "living together," refers to a relationship in which two species have a prolonged close association. Two species that interact closely for generations often coevolve, regardless of whether one species helps or harms another. As Section 17.11 explained, coevolution is an evolutionary process in which each species acts as a selective agent that shifts the range of variation in the other.

FIGURE 27.1 Commensal orchids on a tree trunk. The orchids benefit by growing on the tree, which is unaffected by their presence.

commensalism Species interaction that benefits one species and neither helps nor harms the other.
community All species that live in a particular area.
habitat Type of environment in which a species typically lives.
symbiosis One species lives in or on another in a commensal, mutualistic, or parasitic relationship.

TAKE-HOME MESSAGE 27.1

The types and abundances of species in a community are affected by physical factors such as climate and by biological factors such as interactions among species.

With commensalism, one species benefits from an interaction that has no effect on its partner species. A commensalism in which one species lives on or in another is a type of symbiosis.

27.2 HOW DO SPECIES COOPERATE?

FIGURE 27.2 Obligate mutualists: a yucca moth on a yucca flower. Yucca moths mate on such flowers. Afterward, the female gathers pollen from one flower and carries it with her to another flower that has not yet begun to produce pollen. She lays her eggs in the second flower's ovary, then uses the pollen she brought along to fertilize it. By pollinating the flower, she ensures that seeds will form to feed her larvae. Moth larvae eat some seeds, but plenty survive to give rise to new yucca plants.

Mutualism is an interspecific interaction that benefits both participants. Flowering plants and their pollinators are a familiar example. In some cases, two species coevolve a mutual dependence. For example, each species of yucca plant is pollinated by one species of yucca moth, whose larvae develop only on that plant (**FIGURE 27.2**). More often, mutualistic relationships are less exclusive. Most flowering plants have more than one pollinator, and most pollinators service more than one species of plant.

Photosynthetic organisms often supply sugars to their nonphotosynthetic partners, as when plants lure pollinators with nectar. In addition, many plants make sugary fruits that attract seed-dispersing animals. Plants also provide sugars to mycorrhizal fungi and nitrogen-fixing bacteria. The plants' fungal or bacterial symbionts return the favor by supplying their host with other essential nutrients. Similarly, photosynthetic dinoflagellates provide sugars to reef-building corals, and photosynthetic bacteria or algae in a lichen feed their fungal partner.

Animals often share ingested nutrients with mutualistic microorganisms that live in their gut. For example, *Escherichia coli* bacteria in your colon provide you with vitamin K in return for a steady food supply and a nice, warm place to live.

mutualism Species interaction that benefits both species.

TAKE-HOME MESSAGE 27.2

A mutualism is a species interaction in which each species benefits by associating with the other.

In each species, individuals who maximize the benefit they receive while minimizing their costs will be favored.

Other mutualisms involve protection. For example, an anemonefish and a sea anemone fend off one another's predators (**FIGURE 27.3**). Ants protect bull acacia trees from leaf-eating insects, and in return the tree houses the ant in special hollow thorns and provides them with sugar-rich foods. The oxpecker bird shown in the chapter's opening photo protects its partner from ticks, which it eats.

From an evolutionary standpoint, mutualism is best described as reciprocal exploitation. Each individual increases its fitness by extracting a resource, such as protection or food, from its partner. If taking part in the mutualism has a cost, such as the cost of producing nectar, selection favors individuals who minimize that cost. For example, a flower that produces the minimum amount of nectar necessary to attract pollinators will do better than one that expends more energy to produce additional nectar.

FIGURE 27.3 Mutualism between a sea anemone and a pink anemone fish. The tiny but aggressive fish chases away predatory butterfly fishes that would like to bite off tips of the anemone's stinging tentacles. The fish cannot survive and reproduce without the protection of an anemone. The anemone does not need a fish to protect it, but it does better with one.

27.3 HOW DO SPECIES COMPETE?

TYPES OF COMPETITION

Interspecific competition, which is competition among members of different species, is not usually as intense as intraspecific competition. The requirements of two species might be similar, but they are never as close as they are for members of one species.

Each species has a unique set of ecological requirements and roles that we refer to as its **ecological niche**. Both physical and biological factors define the niche. Aspects of an animal's niche include the temperature range it can tolerate, the species it eats, and the places it can breed. A description of a flowering plant's niche would include its soil, water, light, and pollinator requirements. The more similar the niches of two species are, the more intensely those species will compete.

Competition takes two forms. With interference competition, one species actively prevents another from accessing some resource. As an example, one species of scavenger will often chase another away from a carcass (**FIGURE 27.4**). As another example, some plants use chemicals to fend off potential competition. Aromatic compounds that ooze from tissues of sagebrush plants, black walnut trees, and eucalyptus trees seep into the soil around these plants. The chemicals prevent other kinds of plants from germinating or growing.

In exploitative competition, species do not interact directly; by using the same resource, each reduces the amount of the resource available to the other. For example, deer and blue jays both eat acorns in oak forests. The more acorns the deer eat, the fewer are available for the jays.

FIGURE 27.4 Interference competition among scavengers. A golden eagle attacks a fox with its talons to drive it away from a resource—the moose carcass in the foreground.

COMPETITIVE EXCLUSION

Deer and blue jays share a fondness for acorns, but each also has other sources of food. Any two species usually differ at least a bit in their resource requirements. Species compete most intensely when the supply of a shared resource is an important limiting factor for both.

In the 1930s, G. Gause conducted experiments with two species of ciliated protozoans (*Paramecium*) that compete for the same prey: bacteria. He cultured the *Paramecium* species separately and together (**FIGURE 27.5**). Within weeks, population growth of one species outpaced the other, which went extinct. This and other experiments are the basis for the concept of **competitive exclusion**: Whenever two species require the same limited resource to survive or reproduce, the better competitor will drive the less competitive species to extinction in that habitat.

When resource needs of competitors are not exactly the same, competing species continue to coexist, but the presence of each reduces the carrying capacity (Section 26.3) of the habitat for the other. For example, the reproductive success of a flowering plant is decreased by competition from other species that flower at the same time and rely on the same pollinators.

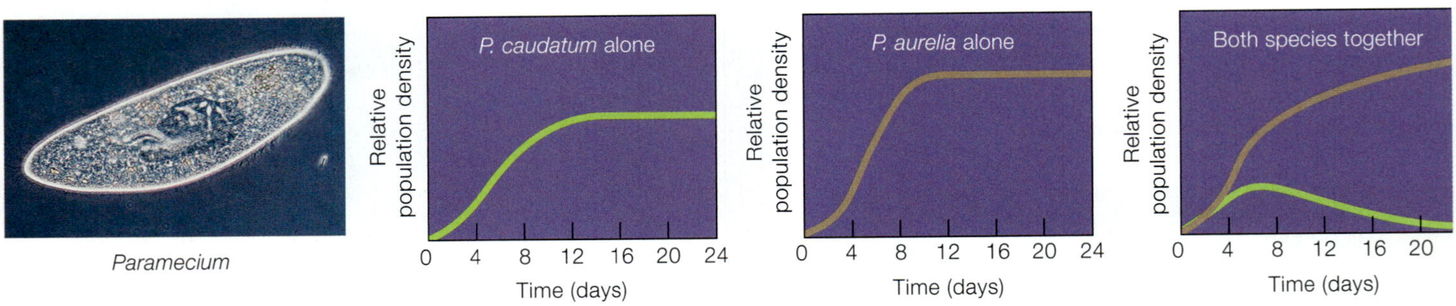

FIGURE 27.5 **{Animated}** Results of competitive exclusion between two *Paramecium* species that compete for the same food. When the two species were grown together, *P. aurelia* drove *P. caudatum* to extinction.

Even strikingly different organisms can influence one another through competition. Wolf spiders and carnivorous plants called sundews compete for insect prey in Florida swamps. Sundews grown in the presence of wolf spiders make fewer flowers than sundews in spider-free enclosures. Presumably, competition for insect prey reduces the energy the plants could devote to flowering.

RESOURCE PARTITIONING

Resource partitioning is an evolutionary process by which species become adapted to use a shared limiting resource in a way that minimizes competition. Consider three species of annual plants that commonly coexist in abandoned fields. All require water and nutrients, but the roots of each species take resources up from a different depth. This variation allows the plants to coexist. Similarly, eight species of woodpecker coexist in Oregon forests. All feed on insects and nest in hollow trees, but the details of their foraging behavior and nesting preferences vary. Differences in nesting time also reduce competitive interactions.

Resource partitioning arises as a result of the directional selection that occurs when species with similar requirements share a habitat and compete for a limiting resource. In each species, those individuals who differ most from the competing species have the least competition and thus leave the most offspring. Over generations, directional selection leads to **character displacement**: The range of variation for one or more traits is shifted in a direction that lessens the intensity of competition for a limiting resource. For example, competing seed-eating birds might evolve greater differences in bill size so they can eat different seeds.

character displacement As a result of competition between two species, the species become less similar in their resource requirements.
competitive exclusion Process whereby two species compete for a limiting resource, and one drives the other to local extinction.
ecological niche All of a species' requirements and roles in an ecosystem.
interspecific competition Competition between members of two species.
resource partitioning Evolutionary process whereby species become adapted in different ways to access different portions of a limited resource; allows species with similar needs to coexist.

TAKE-HOME MESSAGE 27.3

In some competitive interactions, one species controls or blocks access to a resource, regardless of whether it is scarce or abundant. In other interactions, one species is better than another at exploiting a shared resource.

When two species compete, individuals whose needs are least like those of the competing species are favored.

PEOPLE MATTER

National Geographic Grantee
DR. NAYUTA YAMASHITA

In animals, resource partitioning often involves evolved differences in feeding behavior. Nayuta Yamashita (above left) and her colleagues Chia Tan (center) and Chris Vinyard (right) investigate how differences in body form influence food usage among primates. They find three species of bamboo lemurs (Section 24.8) that coexist in Madagascar of particular interest. Yamashita says, "The lemurs are all bamboo specialists that feed on the same species of giant bamboo, though they eat different parts."

To find out the mechanisms by which the lemurs divvy up their food supply, Yamashita measured physical properties of the bamboo they eat and compared the shapes of the lemurs' teeth and jaws. She also measured how much force their jaws could exert. Results of these studies revealed that differences in bite force reflect resource partitioning. The greater bamboo lemur (shown in the photo above) has the most powerful bite, and is the only bamboo lemur that can access the toughest part of the bamboo plant—the pith inside mature shoots.

Yamashita is also investigating another interesting aspect of bamboo lemur feeding: their ability to ingest cyanide. The bamboo that the lemurs eat contains a lot of this poison, yet the lemurs eat it with no apparent ill effect.

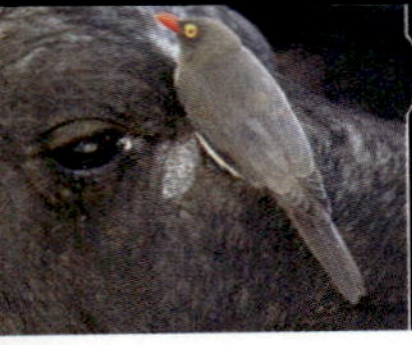

27.4 HOW DO PREDATORS AND THEIR PREY INTERACT?

PREDATOR AND PREY ABUNDANCE

With **predation**, one species (the predator) captures, kills, and digests another species (the prey). The abundance of prey species in a community affects how many predators it can support. The number of predators reduces the number of prey, but the extent of this effect depends partly on how the predator species responds to changes in prey density. With some predators, such as web-spinning spiders, the proportion of prey killed is constant, so the number killed in any given interval depends solely on prey density. As the number of flies in an area increases, more and more become caught in a web. More often, the number of prey killed depends in part on the time it takes predators to capture, eat, and digest prey. As prey density increases, the rate of kills rises steeply at first because there are more prey to catch. Eventually, the rate of increase slows, because a predator is exposed to more prey than it can handle at one time. A wolf that just killed a caribou will not hunt another until it has eaten and digested the first one.

A Wasp that can inflict a painful sting. Like many stinging bees and wasps, it has a yellow and black pattern.

B Fly, which lacks a stinger, mimics the color pattern of stinging insects.

FIGURE 27.7 **Mimicry.**

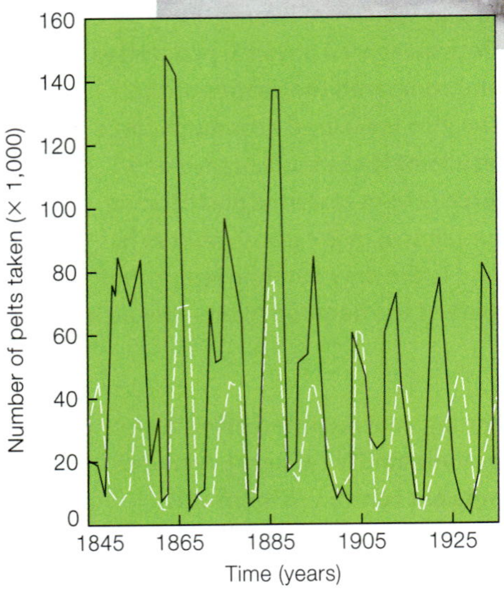

FIGURE 27.6 Graph of the number of Canadian lynx (*dashed* line) and snowshoe hares (*solid* line), based on counts of pelts sold by trappers to Hudson's Bay Company during a ninety-year period.

Predator and prey populations sometimes rise and fall in a cyclical fashion. **FIGURE 27.6** shows historical data for the numbers of lynx and their main prey, the snowshoe hare. Both populations rise and fall over an approximately ten-year cycle, with predator abundance lagging behind prey abundance. Field studies indicate that lynx numbers fluctuate mainly in response to hare numbers. However, the size of the hare population is affected by the abundance of the hare's food as well as the number of lynx. Hare populations continued to rise and fall even when predators were experimentally excluded from areas.

PREDATOR–PREY ARMS RACES

Predator and prey exert selection pressure on one another. Suppose a mutation arises that gives a prey species a more effective defense. Over generations, directional selection will cause this mutation to spread through the prey population. If some members of a predator population have a trait that makes them better at thwarting the improved defense, they and their descendants will be at an advantage. Thus, predators exert selection pressure that favors improved prey defense, which in turn exerts selection pressure on predators, and so it goes over many generations.

You have already learned about some defensive adaptations. Many prey species have hard or sharp parts that make them difficult to eat. Think of a snail's shell or a sea urchin's spines. Others contain chemicals that taste bad or sicken predators. Most defensive compounds in animals come from the plants they eat. For example, a monarch butterfly caterpillar takes up chemicals from the milkweed that it feeds on. A bird that later eats the butterfly will be sickened by these chemicals and avoid similar butterflies.

Prey animals use a variety of mechanisms to fend off a predator. Section 1.6 described how eyespots and a hissing sound protect some butterflies. A lizard's tail may detach from the body and wiggle a bit as a distraction. Many animals, including skunks, exude or squirt a foul-smelling, irritating repellent when frightened. Bees and wasps sting.

Well-defended prey often have warning coloration, a conspicuous color pattern that predators learn to avoid. For example, many species of stinging wasps and bees have a pattern of black and yellow stripes (**FIGURE 27.7A**). The similar appearance of bees and wasps is an example of one type of **mimicry**, an evolutionary pattern in which one species comes to resemble another. Bees and wasps benefit from their similar appearance. The more often a predator is stung by a black-and-yellow-striped insect, the less likely it is to attack a similar-looking insect.

In another type of mimicry, prey masquerade as a species that has a defense they lack. For example, some flies that cannot sting resemble bees or wasps that can (**FIGURE 27.7B**). The fly benefits when predators avoid it after an encounter with the better-defended look-alike species.

Camouflage is a body form and coloration pattern that allows an animal to blend into its surroundings, and thus avoid detection. For example, snowshoe hares such as the one in **FIGURE 27.6** turn white in winter, making it harder for predators to spot them in a snowy landscape.

Predators can benefit from camouflage too (**FIGURE 27.8**). Other predator adaptations include sharp teeth and claws that can pierce protective hard parts. Speedy prey select for faster predators. For example, the cheetah, the fastest land animal, can run 114 kilometers per hour (70 mph). Its preferred prey, Thomson's gazelles, run 80 kilometers per hour (50 mph).

PLANT–HERBIVORE ARMS RACE

With **herbivory**, an animal eats a plant or plant parts. The number and type of plants in a community can influence the number and type of herbivores present. Two types of defenses have evolved in response to herbivory. Some plants have adapted to withstand and recover quickly from herbivory. For example, prairie grasses are seldom killed by native grazers such as bison. The grasses store enough resources in roots to grow back lost shoots. Other plants have traits that deter herbivory. Such traits include spines or thorns, leaves that are hard to chew or digest, and chemicals that taste bad or sicken herbivores. Ricin, the toxin made by castor bean plants (Section 9.6), sickens many herbivores. Caffeine in coffee beans and nicotine in tobacco leaves are evolved defenses against insects.

Existence of plant defenses favors herbivores capable of overcoming those defenses. For example, a capacity to withstand cyanide evolved in the bamboo-eating lemurs discussed in the prior section. Similarly, eucalyptus leaves contain toxins that make them poisonous to most mammals, but not to koalas. Specialized liver enzymes allow koalas to break down toxins made by a few eucalyptus species.

A Frilly pink body parts of a flower mantis help hide it from insect prey attracted to the real flowers.

B Fleshy protrusions give a scorpionfish the appearance of an algae-covered rock. Fish that come close for a nibble end up as prey.

FIGURE 27.8 Camouflage.

TAKE-HOME MESSAGE 27.4

Predation benefits predators and harms prey. Predator numbers may fluctuate in response to prey availability.

Predators can coevolve with their prey. Herbivores coevolve with the plants that they eat.

With mimicry, one species benefits by its resemblance to another species.

camouflage Coloration or body form that helps an organism blend in with its surroundings and escape detection.
herbivory An animal feeds on plants or plant parts.
mimicry An evolutionary pattern in which one species becomes more similar in appearance to another.
predation One species captures, kills, and eats another.

CREDITS: (8A) © Bob Jensen Photography; (8B) David Burdick, NOAA.

27.5 HOW DO PARASITES AND THEIR HOSTS INTERACT?

PARASITISM

With **parasitism**, one species (the parasite) benefits by feeding on another (the host), without immediately killing it. Endoparasites such as parasitic roundworms live and feed inside their host (**FIGURE 27.9A**). An ectoparasite such as a tick feeds while attached to a host's external surface (**FIGURE 27.9B**).

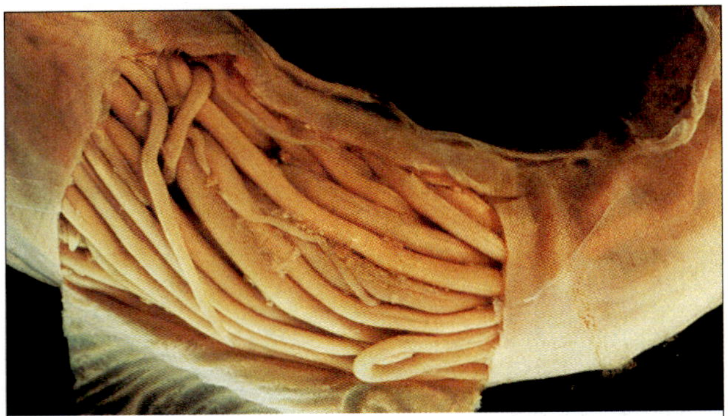

A Endoparasitic roundworms in the intestine of a host pig.

B Ectoparasitic ticks attached to and sucking blood from a finch.

C Ectoparasitic dodder (*Cuscuta*), also known as strangleweed or devil's hair. This parasitic flowering plant has almost no chlorophyll. Leafless stems wrap around a host plant, and modified roots absorb water and nutrients from the host plant's vascular tissue.

FIGURE 27.9 Parasites inside and out.

A parasitic way of life has evolved in members of a diverse variety of groups. Bacterial, fungal, protistan, and invertebrate parasites feed on vertebrates. Lampreys (Section 24.2) attach to and feed on other fish. There are even a few parasitic plants that withdraw nutrients from other plants (**FIGURE 27.9C**).

Parasites usually do not kill their host immediately. In terms of evolutionary fitness, killing a host too fast is bad for the parasite. Ideally, a host will live long enough to give a parasite time to produce some offspring. The longer the host survives, the more parasite offspring can be produced. Thus, parasites with less-than-fatal effects on hosts are at a selective advantage.

Although parasites typically do not kill their hosts, many still have an important impact on a host population. Most pathogens are parasites that cause disease in their hosts. Even when a parasite does not cause obvious symptoms, its presence can weaken a host, making it more vulnerable to predation or less attractive to potential mates. Some parasites cause their host to become sterile. Others shift the sex ratio among their host's offspring.

Adaptations to a parasitic lifestyle include traits that allow the parasite to locate hosts and to feed undetected. For example, ticks that feed on mammals or birds move toward a source of heat and carbon dioxide, which may be a potential host. A chemical in tick saliva acts as a local anesthetic, preventing the host from noticing the feeding tick. Endoparasites often have traits that help them evade a host's immune defenses.

Among hosts, traits that fend off parasites or reduce the toll that a parasite takes on fitness are favored. Many animals groom or preen themselves to remove ectoparasites. One type of seabird (the crested auklet) secretes a citrus-scented compound that it spreads on its feathers as a tick repellent. Chimpanzees and other primates sometimes fold up tough, indigestible leaves and swallow them whole, a practice that helps rid their gut of parasitic worms.

PARASITOIDS

As many as 15 percent of all insects may be **parasitoids**, which lay their eggs in other insects. Larvae that hatch from the eggs develop in the host's body, eat its tissue, and eventually kill it. Parasitoids reduce the size of a host population in two ways. First, as the parasitoid larvae grow inside their host, they withdraw nutrients and prevent it from reproducing. Second, the presence of these larvae eventually leads to the death of the host.

FIGURE 27.10 Biological pest control. A commercially raised parasitoid wasp about to deposit a fertilized egg in an aphid. This wasp is used to reduce aphid populations. After the egg hatches, a wasp larva devours the aphid from the inside.

FIGURE 27.11 Brood parasitism. Ants tend a caterpillar of the Alcon blue butterfly, which smells like an ant and imitates sounds made by the queen ant. Ant workers feed the caterpillar more eagerly than their own larvae and will even feed it ant larvae if food runs short.

BIOLOGICAL PEST CONTROL

Parasites and parasitoids are commercially-raised and released in target areas as a form of **biological pest control** in which a pest's natural enemies are used to reduce its numbers (**FIGURE 27.10**). Biological pest control has some advantages over pesticides. Most chemical insecticides kill a wide variety of insects, including some that help control pests. Insecticides also have negative effects on human health. By contrast, the parasites and parasitoids used as biological control agents usually target only a limited number of species.

For a species to be an effective biological control agent, it must be adapted to take advantage of a specific host species and to survive in that species' habitat. The ideal biological control agent excels at finding the target host species, has a population growth rate comparable to the host's, and has offspring that disperse widely.

Introducing a species into a community as a biological control agent always entails some risks. The introduced parasites sometimes attack nontargeted species in addition to, or instead of, those that they were expected to control. For example, parasitoid wasps were introduced to Hawaii to control stinkbugs that feed on some Hawaiian crops. Instead, the parasitoids decimated the population of koa bugs, Hawaii's largest native bug. Introduced parasitoids have also been implicated in ongoing declines of many native Hawaiian butterfly and moth populations.

BROOD PARASITISM

With **brood parasitism**, one egg-laying species benefits by having another raise its offspring. The host species suffers by squandering care on unrelated individuals. The European cuckoos are brood parasites, as are North American cowbirds. Not having to invest in parental care allows a female cowbird to produce a large number of eggs, as many as thirty in a single reproductive season.

The presence of brood parasites decreases the reproductive rate of the host species and favors host individuals that detect and eject foreign young. Some avian brood parasites counter this host defense by producing eggs that closely resemble those of their host species. In cuckoos, different subpopulations have different host preferences and egg coloration. Females of each subpopulation lay eggs that closely resemble those of their preferred host.

As another example, some butterflies such as the Alcon blue butterfly outsource care of their young to ants (**FIGURE 27.11**). Caterpillars of these butterflies smell like an ant and make sounds similar to those made by ants. Worker ants, fooled by these false cues, carry the caterpillars into their nest, where they care for them as if they were members of the colony, feeding them and protecting them from predators.

biological pest control Use of a pest's natural enemies to control its population size.
brood parasitism One egg-laying species benefits by having another raise its offspring.
parasitism Relationship in which one species withdraws nutrients from another species, without immediately killing it.
parasitoid An insect that lays eggs in another insect, and whose young devour their host from the inside.

TAKE-HOME MESSAGE 27.5

Parasites feed on living host individuals, harming them by withdrawing nutrients from them.

Parasitoids reduce the number of host organisms by preventing reproduction and eventually killing the host.

Brood parasites reduce the reproductive rate of hosts by tricking them into caring for young that are not their own.

CREDITS: (10) © Peter J. Bryant/Biological Photo Service; (11) © Darlyne A. Murawski/National Geographic Creative.

27.6 HOW DO COMMUNITIES CHANGE OVER TIME?

A The 1980 Mount Saint Helens eruption wiped out the biological community at the base of this Cascade volcano.

B In less than a decade, pioneer species had arrived and the process of primary succession had begun.

C Twelve years later, seedlings of Douglas firs had taken hold.

FIGURE 27.12 **{Animated}** An example of succession.

SUCCESSIONAL CHANGE

Species composition of a community changes over time. Often, some species alter the habitat in ways that allow others to come in and replace them. This type of change, which takes place over a long interval, is referred to as ecological succession.

Primary succession begins when **pioneer species** colonize a barren habitat that lacks soil, such as land exposed by the retreat of a glacier, a newly formed volcanic island, or a region where volcanic material has buried existing soil (**FIGURE 27.12**). The earliest pioneers to colonize such environments are often mosses and lichens (Sections 21.2 and 22.3), which are small, have a brief life cycle, and can tolerate intense sunlight, extreme temperature changes, and little or no soil. Some hardy annual flowering plants with wind-dispersed seeds are also frequent pioneers. Pioneer species help build and improve the soil. In doing so, they often set the stage for their own replacement. Many pioneer species partner with nitrogen-fixing bacteria, so they can grow in nitrogen-poor habitats. Seeds of later species find shelter inside mats of the pioneers. Organic wastes and remains accumulate and, by adding volume and nutrients to soil, this material helps other species take hold. Later successional species often shade and eventually displace earlier ones.

In **secondary succession**, a disturbed area within a community recovers. If improved soil is still present, secondary succession can occur fast. It commonly occurs in abandoned agricultural fields and burned forests.

When the concept of ecological succession was first developed in the late 1800s, it was thought to be a predictable and directional process that culminates in a "climax community," an array of species that persists over time and is reconstituted in the event of a disturbance. Ecologists now realize that the species composition of a community changes frequently, and in unpredictable ways. Communities do not journey along a well-worn path to some predetermined climax state. Random events can determine the order in which species arrive in a habitat and thus affect the course of succession in unpredictable ways.

Ecologists had an opportunity to investigate these factors after the 1980 eruption of Mount Saint Helens leveled about 600 square kilometers (235 square miles) of forest in Washington State (**FIGURE 27.12**). They recorded the natural pattern of colonization and carried out experiments in plots inside the blast zone. The results of these and other studies showed that the presence of some pioneers helped certain other, later-arriving plants become established, whereas the presence of other pioneers kept the same late successional species out.

EFFECTS OF DISTURBANCE

The type and rate of disturbances also influence species composition in communities. According to the **intermediate disturbance hypothesis**, species richness is greatest in habits where disturbances are moderate in their intensity or frequency. In such habitats, there is enough time for new colonists to arrive and become established but not enough for competitive exclusion to cause extinctions:

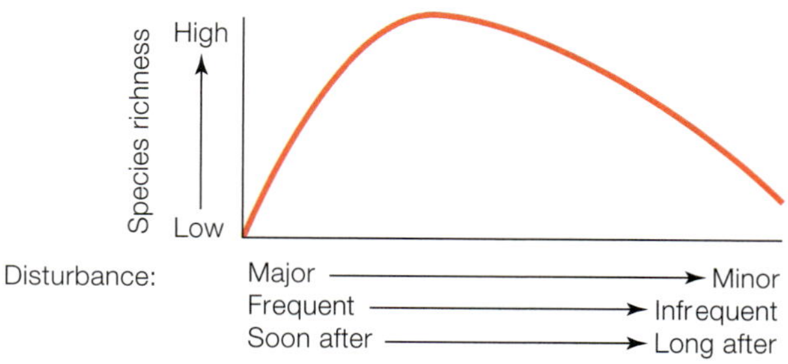

In communities repeatedly subjected to a particular type of physical disturbance, individuals that withstand or benefit from that disturbance have a selective advantage. For example, some plants in areas subject to periodic fires produce seeds that germinate only after a fire. Seedlings of these plants benefit from the lack of competition for resources in newly burned areas. Other plants have an ability to resprout quickly after a fire (**FIGURE 27.13**). Because different species respond differently to fire, the frequency of this disturbance affects competitive interactions. Human suppression of naturally occurring fires can change the composition of a biological community by allowing species that are not fire-adapted to outcompete fire-adapted ones that previously predominated.

Some species are especially intolerant of physical disturbance of their environment. These **indicator species** are the first to decline or disappear when conditions change, so they can provide an early warning of environmental degradation. For example, a decline in a trout population can be an early sign of problems in a stream, because trout are highly sensitive to pollutants and cannot tolerate low oxygen levels. Some lichens are intolerant of air pollution and serve as indicators of air quality (**FIGURE 27.14**).

indicator species Species whose presence and abundance in a community provides information about conditions in the community.
intermediate disturbance hypothesis Species richness is greatest in communities with moderate levels of disturbance.
pioneer species Species that can colonize a new habitat.
primary succession A new community becomes established in an area where there was previously no soil.
secondary succession A new community develops in a site where a community previously existed.

FIGURE 27.13 Adapted to disturbance. Some woody shrubs, such as this toyon, resprout from their roots after a fire. In the absence of occasional fire, toyons are outcompeted and displaced by species that grow faster but are less fire resistant.

FIGURE 27.14 Indicator species. Lichens take up nutrients—and pollutants—from airborne dust, so many cannot survive where the level of air pollution is high. The U.S. Forest Service monitors the types and numbers of lichens on tree trunks as part of its program to assess the health of forest communities.

TAKE-HOME MESSAGE 27.6

Succession is a process in which one array of species replaces another over time. It can occur in a barren, soilless habitat (primary succession), or in a habitat where a community previously existed (secondary succession).

Physical factors affect succession, but so do species interactions and disturbances. As a result, the course that succession will take in a community is difficult to predict.

Some species are especially intolerant of certain physical changes to the environment, and the size of their populations can serve as an indicator of the state of an environment.

27.7 HOW CAN A SINGLE SPECIES ALTER COMMUNITY STRUCTURE?

The loss or addition of one species sometimes alters the abundance of many other species in a community.

KEYSTONE SPECIES

A **keystone species** is a species that has a disproportionately large effect on a community relative to its abundance. It is named for the central, wedge-shaped stone in a stone arch; a keystone keeps other stones in place. Robert Paine coined the term "keystone species" as a result of his experiments on the rocky shores of California's coast. Species in the rocky intertidal zone withstand pounding surf by clinging to rocks. A rock to cling to is a limiting factor. Paine set up control plots with the sea star *Pisaster ochraceus* and its main prey—chitons, limpets, barnacles, and mussels. Then he removed all sea stars from his experimental plots.

Sea stars prey mainly on mussels. With sea stars gone from experimental plots, mussels took over, crowding out seven other species of invertebrates (**FIGURE 27.15**).

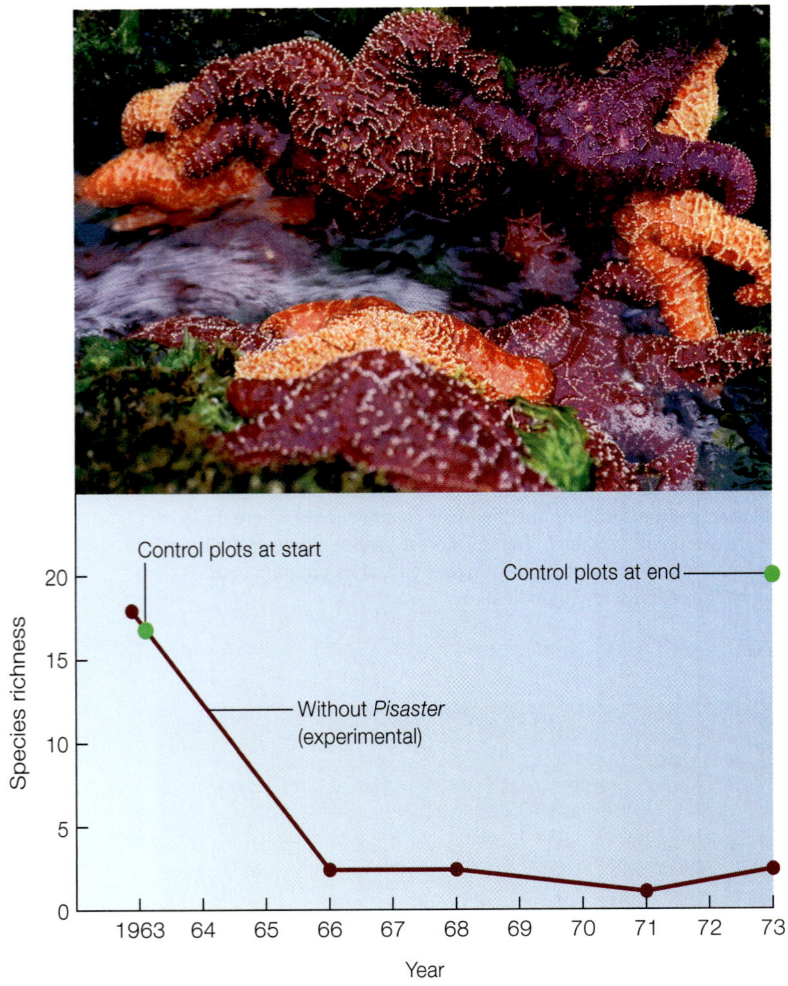

FIGURE 27.15 Effect of removal of the sea star *Pisaster* on the number of species in tide pools. Sea stars were not removed from control plots.

Paine concluded that sea stars are a keystone species. They normally keep the number of prey species in the intertidal zone high by preventing competitive exclusion by mussels.

Keystone species need not be predators. For example, beavers are keystone species in some communities. These large, herbivorous rodents cut down trees by gnawing through their trunks. The beaver then uses the felled trees to build a dam, thus creating a deep pool where a shallow stream would otherwise exist. By altering the physical conditions in a section of the stream, the beaver affects the types of fish and aquatic invertebrates that can live there.

EXOTIC SPECIES

The arrival of a new species in a community can also cause dramatic changes. When you hear someone speaking enthusiastically about exotic species, you can safely bet the speaker is not an ecologist. An **exotic species** is a nonnative species, one that evolved in one community, then dispersed from its home and became established elsewhere.

As the pace of global travel and trade have picked up, so has the rate of species introductions. More than 4,500 exotic species have become established in the United States. An estimated 25 percent of Florida's plant and animal species are exotics. In Hawaii, 45 percent are exotic. Some species were brought in for use as food crops, to brighten gardens, or to provide textiles. Other species arrived as stowaways along with cargo from distant regions. Increased dispersal rates among pest species are an unanticipated side effect of increased global trade and improvements in shipping. Speedier ships make quicker trips, increasing the likelihood that pests hidden in cargo holds will survive a journey.

Many exotic species have little impact on their adopted community, but others are invasive. An invasive species harms an existing community. The water mold that is currently killing off oaks in the Pacific Northwest (Section 20.6) is an invasive species, as are the fungi that currently threaten North America's bats (Section 22.3) and frogs (Section 22.5). Sea lampreys are considered an invasive species in the Great Lakes (Section 24.2).

As another example, consider the vine called kudzu (*Pueraria lobata*). Native to Asia, it was introduced to the American Southeast as a food for grazers and to control erosion, but it quickly became an invasive weed. Kudzu overgrows trees, telephone poles, houses, and almost everything else in its path (**FIGURE 27.16A**).

Gypsy moths (*Lymantria dispar*) are an invasive species too. They entered the northeastern United States in the mid-1700s and now range into the Southeast, Midwest, and Canada. Gypsy moth caterpillars (**FIGURE 27.16B**) preferentially feed on oaks. Loss of leaves to gypsy moths

A Kudzu native to Asia is overgrowing trees across the southeastern United States.

B Gypsy moth caterpillars native to Europe and Asia now feed on oaks in much of the United States.

C Nutrias native to South America now abound in freshwater marshes of the Gulf States.

FIGURE 27.16 Three exotic species that are altering natural communities in the United States. To learn more about invasive species in the United States, visit the National Invasive Species Information Center online at www.invasivespeciesinfo.gov.

weakens the trees, making them less efficient competitors and more susceptible to parasites.

Large semiaquatic rodents called nutrias (*Myocastor coypus*) were imported from South America for their fur. Today, descendants of animals that either escaped or were intentionally released thrive in freshwater marshes and along rivers in twenty states (**FIGURE 27.16C**). Their appetite for plants threatens native vegetation and crop plants, and their burrowing contributes to marsh erosion and damages levees, increasing the risk of flooding.

Most invasive species have a far greater impact in their new home than they did in the community in which they evolved. When a species leaves its community of origin behind, it also leaves behind the competitors, predators, and parasites with which it coevolved and which helped to keep its numbers in check. If the invasive species is a parasite, predator, or herbivore, it also leaves behind hosts or prey that had coevolved with it and had defenses against it. As a result, an invasive species often reaches a higher population density in its new home than it achieved in its old one.

exotic species A species that evolved in one community and later became established in a different one.

keystone species A species that has a disproportionately large effect on community structure relative to its abundance.

TAKE-HOME MESSAGE 27.7

Keystone species greatly influence the diversity and abundance of other species within their community.

An exotic species evolved in one community and now lives in another. It is considered invasive if it threatens other members of its new community.

When an invasive species enters a new community, it encounters species that have not evolved defenses against it. Thus invasive species often have a greater impact in their new community than they did in their community of origin.

27.8 HOW MANY SPECIES WILL A COMMUNITY HOLD?

Islands are natural laboratories for population studies. They have also been laboratories for community studies. Consider Surtsey, an island that appeared in the mid-1960s when an undersea volcano erupted 33 kilometers (21 miles) from the coast of Iceland (**FIGURE 27.17**). Bacteria and fungi were early colonists. The first vascular plant became established on the island in 1965. Mosses appeared two years later and thrived. The first lichens were found five years after that. The rate at which vascular plants were introduced to the island picked up dramatically after a seagull colony became established in 1986.

FIGURE 27.17 Vascular plant colonization of Surtsey, a volcanic island shown forming in the inset photo. Seagulls first began nesting on the island in 1986.

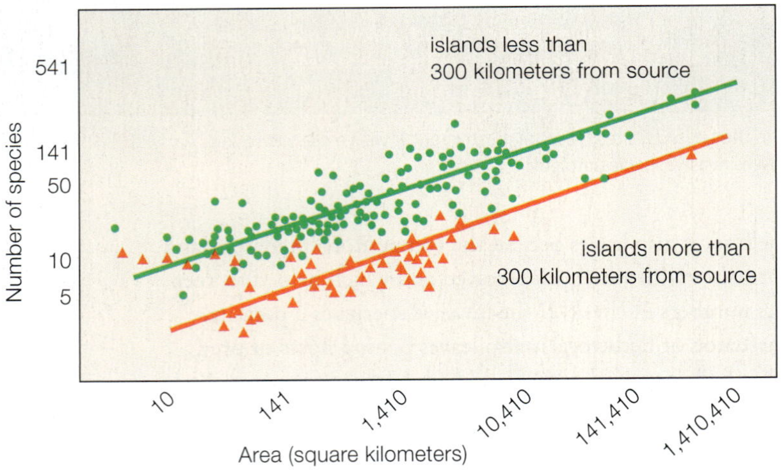

FIGURE 27.18 {Animated} Island biodiversity patterns.

Distance effect: Species richness on islands of a given size declines as distance from a source of colonists rises. Green circles are values for islands less than 300 kilometers from the colonizing source. Orange triangles are values for islands more than 300 kilometers (190 miles) from a source of colonists.

Area effect: Among islands the same distance from a source of colonists, larger islands tend to support more species than smaller ones.

FIGURE IT OUT: Which is likely to have more species, a 100-km² island more than 300 km from a colonizing source or a 500-km² island less than 300 km from a colonist source?

Answer: The 500-km² island

The number of species on Surtsey will not continue increasing forever. How many species will there be when the number levels off? According to the **equilibrium model of island biogeography**, the number of species living on any island reflects a balance between immigration rates for new species and extinction rates for established ones.

Two factors determine the equilibrium number of species (**FIGURE 27.18**). First, there is a **distance effect**; islands far from a source of colonists receive fewer immigrants than those closer to a source. Most species cannot disperse very far, so they will not turn up far from a mainland. Second, there is an **area effect**; an island's size affects both immigration rates and extinction rates. More colonists will happen upon a larger island simply by virtue of its size. Also, big islands are more likely to offer a variety of habitats, such as high and low elevations. These options make it more likely that a new arrival will find a suitable habitat. Finally, big islands can support larger populations of species than small islands. The larger a population, the less likely it is to become locally extinct as the result of some random event. For example, a fire large enough to wipe out the population of a small island might leave survivors on a larger island.

Robert H. MacArthur and Edward O. Wilson developed the equilibrium model of island biogeography in the late 1960s. Since then, the model has been modified and its use has been expanded to help scientists think about habitat islands, which are natural settings surrounded by a "sea" of habitat that has been disturbed by humans. Many parks and wildlife preserves fit this description. Island-based models can help ecologists estimate the size of an area needed to ensure survival of a species that inhabits it.

area effect Larger islands have more species than small ones.
distance effect Islands close to a mainland have more species than those farther away.
equilibrium model of island biogeography Model that predicts the number of species on an island based on the island's area and distance from the mainland.

TAKE-HOME MESSAGE 27.8

Studies of islands helped biologists understand how the area of a community and its distance from a source of colonist species influences the species richness of the community.

27.9 Application: FIGHTING FOREIGN FIRE ANTS

Sustainability

FIGURE 27.19 Nest mounds of red imported fire ants in a Texas field. The inset photo shows the ants attacking a quail egg.

RED IMPORTED FIRE ANTS (*SOLENOPSIS INVICTA*) ARRIVED IN THE SOUTHEASTERN UNITED STATES IN THE 1930s. These natives of South America probably arrived as stowaways on a cargo ship. They are invasive and have established colonies as far west as California and as far north as Kansas and Delaware.

The spread of red imported fire ants poses a threat to wildlife. Competition from these invasive ants typically causes a region's native ant populations to decline, and the resulting change in species composition can harm ant-eating animals. For example, the Texas horned lizard feeds mainly on native harvester ants, and cannot eat the red imported fire ants that have largely replaced its natural prey. Red imported fire ants also harm native species by feeding on their eggs and feeding on or stinging their young. Ground-nesting animals such as quail are especially vulnerable to fire ant predation (FIGURE 27.19).

The presence of red imported fire ants can even affect native plants. The ants interfere with pollination by displacing or preying on native pollinators such as ground-nesting bees. They also impede dispersal of native plants whose seeds would normally be spread by native ants that the fire ants have replaced.

Given the problems the imported ants are causing in the United States, you might wonder what things are like in their native South America. The ants are not considered much of a concern there, in part because they are far less common. In South America, parasites, predators, and diseases keep the ants' numbers in check.

Invicta means "invincible" in Latin, and *S. invicta* lives up to its name. Pesticides do not slow the spread of this species. To fight the ants, scientists have now turned to biological controls, enlisting the help of the ants' natural enemies.

Phorid flies are parasitoids that target red imported fire ants. The flies kill their host in a rather gruesome way. A female fly pierces the cuticle of an adult ant, then lays an egg in the ant's soft tissues. The egg hatches into a larva, which grows and eats its way through the tissues to the ant's head. When the larva is ready to undergo metamorphosis, it secretes an enzyme that makes the ant's head fall off. The fly larva then develops into an adult within the shelter of the detached ant head.

Several phorid fly species have been introduced in various southern states. The flies seem to be surviving, reproducing, and increasing their range. They are not expected to kill off all *S. invicta* in affected areas. Rather, the hope is that the flies will reduce the density of invading colonies.

Ecologists are also exploring other options. They are testing effects of imported pathogenic fungi or protists that infect *S. invicta* but not native ants. Another idea is to introduce a parasitic South American ant that invades *S. invicta* colonies and kills the egg-laying queens.

Summary

SECTION 27.1 Each species in a **community** occupies a certain **habitat** characterized by physical and chemical features and by the array of other species living in it. The number of species in a community tends to be greatest in the tropics. Species interactions affect community structure. With **commensalism**, one species benefits and the other is unaffected. A **symbiosis** is an interaction in which one species lives in or on another.

SECTION 27.2 In a **mutualism**, both species benefit from an interaction. Some mutualists cannot complete their life cycle without the interaction. Mutualists who maximize their own benefits while limiting the cost of cooperating are at a selective advantage.

SECTION 27.3 A species' roles and requirements define its unique **ecological niche**. With **interspecific competition**, two species with similar resource requirements are harmed by one another's presence. When one competitor drives the other to local extinction, we call the process **competitive exclusion**. Competing species with similar requirements become less similar when directional selection causes **character displacement**. This change in traits allows **resource partitioning**.

SECTION 27.4 **Predation** benefits a predator at the expense of the prey it captures, kills, and eats. Predators and their prey exert selective pressures on one another. Evolved prey defenses include **camouflage** and **mimicry**. Predators have traits that allow them to overcome prey defenses. With **herbivory**, an animal eats a plant or plant parts. Plants have traits that discourage herbivory or that allow a quick recovery from the loss of shoots.

SECTION 27.5 **Parasitism** involves feeding on a host without killing it. **Parasitoids** are insects that lay eggs on a host insect, then larvae devour the host. Parasites and parasitoids are often used in **biological pest control**. A **brood parasite** steals parental care from another species.

SECTION 27.6 Ecological succession is the sequential replacement of one array of species by another over time. **Primary succession** happens in new habitats. **Secondary succession** occurs in disturbed ones. The first species of a community are **pioneer species**. Their presence may help, hinder, or not affect later colonists.

Modern models of succession emphasize the unpredictability of the outcome as a result of chance events, ongoing changes, and disturbance. The **intermediate disturbance hypothesis** predicts that a moderate level of disturbance keeps a community diverse. An **indicator species** is highly sensitive to disturbance and can provide information about the health of the environment.

SECTION 27.7 **Keystone species** play a major role in determining the composition of a community. Removal of a keystone species or introduction of an **exotic species** can dramatically alter community structure.

SECTION 27.8 A community supports a finite number of species. The **equilibrium model of island biogeography** predicts the number of species that an island will sustain based on the **area effect** and the **distance effect**. Scientists can use this model to predict the number of species that habitat islands such as parks can sustain.

SECTION 27.9 Arrival of an invasive exotic species such as the red imported fire ant can decrease populations of native species that compete for the same resources. Natural enemies of the ants are now being used to reduce their numbers.

Self-Quiz

Answers in Appendix VII

1. The type of place where a species typically lives is called its ________ .
   a. niche
   b. habitat
   c. community
   d. population

2. Which cannot be a symbiosis?
   a. mutualism
   b. parasitism
   c. commensalism
   d. interspecific competition

3. Lizards that eat flies they catch on the ground and birds that eat flies they catch in the air are engaged in ________ competition.
   a. exploitative
   b. interference
   c. intraspecific
   d. interspecific
   e. both a and d
   f. both b and c

4. ________ can lead to resource partitioning.
   a. Mutualism
   b. Parasitism
   c. Commensalism
   d. Interspecific competition

5. Match the terms with the most suitable descriptions.
   ___ mutualism
   ___ parasitism
   ___ commensalism
   ___ predation
   ___ interspecific competition
   a. one free-living species feeds on another and usually kills it
   b. two species interact and both benefit by the interaction
   c. two species interact and one benefits while the other is neither helped nor harmed
   d. one species feeds on another but usually does not kill it
   e. two species access a resource

6. A tick is a(n) ________ .
   a. brood parasite
   b. ectoparasite
   c. endoparasite

7. By a currently favored hypothesis, species richness of a community is greatest between physical disturbances of ________ intensity or frequency.
   a. low
   b. intermediate
   c. high
   d. variable

8. ________ species are the first to colonize a new habitat.

Data Analysis Activities

Biological Control of Fire Ants Ant-decapitating phorid flies are just one of the biological control agents used to battle imported fire ants. Researchers have also enlisted the help of a fungal parasite that infects the ants and slows their production of offspring. An infected colony dwindles in numbers and eventually dies out. Are these biological controls useful against imported fire ants? To find out, USDA scientists treated infested areas with either traditional pesticides or pesticides plus biological controls (both flies and the parasite). The scientists left some plots untreated as controls. **FIGURE 27.20** shows the results.

1. How did population size in the control plots change during the first four months of the study?
2. How did population size in the two types of treated plots change during this same interval?
3. If this study had ended after the first year, would you conclude that biological controls had a major effect?
4. How did the two types of treatment (pesticide alone versus pesticide plus biological controls) differ in their longer-term effects? Which is most effective?

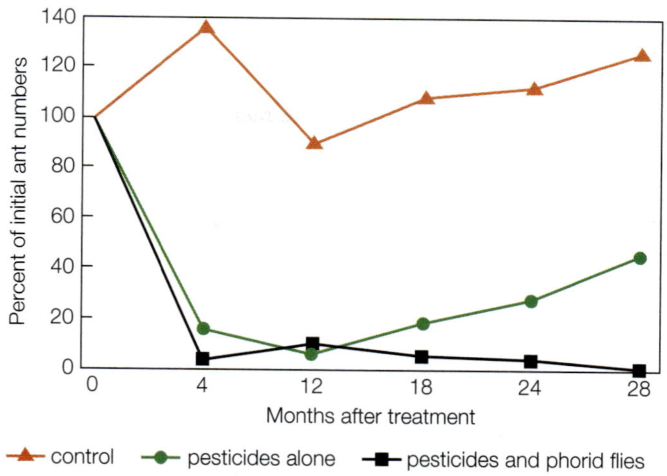

FIGURE 27.20 Effects of two methods of controlling red imported fire ants. The graph shows the numbers of red imported fire ants over a 28-month period. Orange triangles represent untreated control plots. Green circles are plots treated with pesticides alone. Black squares are plots treated with pesticide and biological control agents (phorid flies and a fungal parasite).

9. Growth of a forest in an abandoned corn field is an example of _______ .
   a. primary succession c. secondary succession
   b. resource partitioning d. competitive exclusion

10. Species richness is greatest in communities _______ .
   a. near the equator c. near the poles
   b. in temperate regions d. that recently formed

11. If you remove a species from a community, the population size of its main _______ is likely to increase.
   a. parasite b. competitor c. predator

12. A _______ has another species rear its young.
   a. mutualist c. brood parasite
   b. pioneer species d. exotic species

13. Herbivory benefits _______ .
   a. the herbivore c. both a and b
   b. the plant d. neither a nor b

14. Match the terms with the most suitable descriptions.

___ area effect	a. greatly affects other species
___ pioneer species	b. first species established in a new habitat
___ indicator species	c. more species on large islands than small ones at same distance from the source of colonists
___ keystone species	d. species that is especially sensitive to changes in the environment
___ exotic species	e. allows competitors to coexist
___ resource partitioning	f. often outcompete, displace native species of established community

Critical Thinking

1. With antibiotic resistance rising, researchers are looking for ways to reduce the use of antibiotics. Some cattle once fed antibiotic-laced food now get probiotic feed instead, with cultured bacteria that can establish or bolster populations of helpful bacteria in the animal's gut. The idea is that if a large population of beneficial bacteria is in place, then harmful bacteria cannot become established or thrive. Which ecological principle is guiding this research?

2. Phasmids are plant-eating insects that mimic leaves or sticks, as shown at the *left*. Most rest during the day, and feed at night. If they do move in daytime, they do so very slowly. If disturbed, a phasmid will fall to the ground and lie motionless. Speculate on selective pressures that could have shaped phasmid morphology and behavior. Suggest an experiment to test a hypothesis about how its appearance or behavior may be adaptive.

An energy and nutrient transfer. The caterpillar is tapping into the sunlight energy that this leaf captured and stored in the chemical bonds of sugars. The caterpillar also obtains nutrients such as nitrogen that the plant took up from the soil and used to build its leaves.

28 ECOSYSTEMS

Links to Earlier Concepts

This chapter returns to the concept of a one-way flow of energy in nature (Sections 1.2, 5.1) and looks at nitrogen-fixing bacteria (19.5) in the context of nitrogen cycling. You will be reminded of properties of chemical bonds (2.3) and the pathways of photosynthesis (6.3) and aerobic respiration (7.1). You will see how slow movements of Earth's crust (16.5) influence the cycling of some nutrients.

KEY CONCEPTS

ORGANIZATION OF ECOSYSTEMS

A one-way flow of energy and cycling of nutrients among species maintain an ecosystem. Nutrients and energy are transferred through food chains that interconnect as food webs.

BIOGEOCHEMICAL CYCLES

In a biogeochemical cycle, a nutrient moves slowly among its environmental reservoirs (air, water, and rocks). Nutrients move more quickly into, through, and out of food webs.

THE WATER CYCLE

Most of Earth's water is in its oceans. Evaporation, condensation, precipitation, and flow of rivers and streams move water. Water also transports soluble forms of some nutrients.

THE CARBON CYCLE

Most of Earth's carbon is in rocks, but organisms take carbon up from water or the air. Carbon dioxide is one of the atmospheric greenhouse gases that help keep Earth's surface warm.

NITROGEN AND PHOSPHORUS CYCLES

Plants take up soluble forms of nitrogen and phosphorus from soil. Nitrogen is abundant in air, but only certain bacteria can use this form. Phosphorus has no major gaseous form; most of it is in rocks.

Photograph by Mateusz Sulecki/National Geographic My Shot.

28.1 HOW DO ENERGY AND NUTRIENTS MOVE THROUGH AN ECOSYSTEM?

An **ecosystem** is a community of organisms together with the nonliving components of their environment. It is an open system, because it requires ongoing inputs of energy to persist. Most ecosystems also gain nutrients from and lose nutrients to other ecosystems.

Features of ecosystems vary widely. In climate, soil type, array of species, and other features, prairies differ from forests, which differ from tundra and deserts. Reefs differ from the open ocean, which differs from streams and lakes. Yet, despite these differences, ecologists have found that all systems are alike in many aspects.

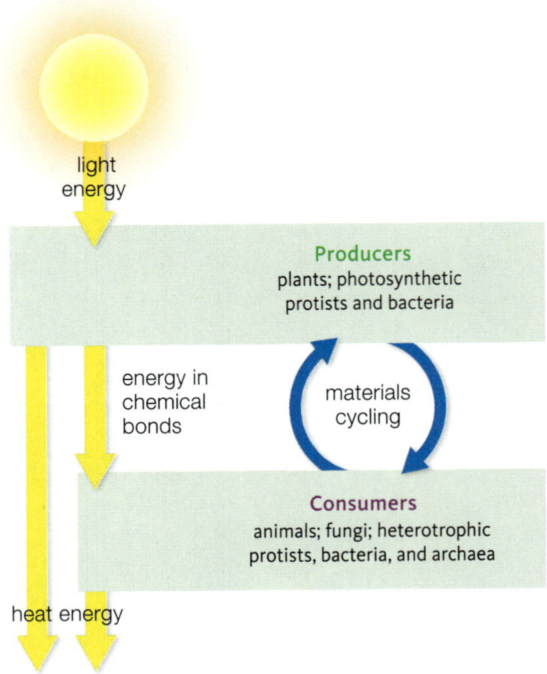

FIGURE 28.1 {Animated} One-way flow of energy (yellow arrows) and nutrient cycling (blue arrows) in the most common type of ecosystem. All light energy that enters the system eventually returns to the environment as heat energy that is not reused. By contrast, nutrients are continually recycled.

PRODUCERS AND CONSUMERS

All ecosystems run on energy that **producers** capture from the environment (**FIGURE 28.1**). Producers are autotrophs (Section 6.1), meaning they obtain energy directly from the environment and use an inorganic form of carbon to build sugars. In most ecosystems, photoautotrophs such as plants, algae, and photosynthetic protists and bacteria are the main producers. In some dark environments such as deep-sea hydrothermal vent ecosystems, chemoautotrophs fill this role. As Section 19.4 explained, chemoautotrophs are bacteria or archaea that fuel the assembly of their food by oxidizing (removing electrons from) inorganic substances such as hydrogen sulfide.

Primary production is the rate at which an ecosystem's producers capture and store energy. Day length, temperature, and the availability of nutrients, including nitrogen and phosphorus, affect the rate of photosynthesis and so influence primary production. As a result, primary production can vary seasonally within an ecosystem and also differs among ecosystems. Per unit area, primary production on land tends to be higher than that in the oceans. However, because oceans cover about 70 percent of Earth's surface, marine producers—photosynthetic bacteria and protists—contribute nearly half of the global net primary production.

As Section 1.2 explained, **consumers** obtain energy and carbon by feeding on tissues and remains of producers and one another. We describe consumers by their diets. Herbivores eat plants. Carnivores eat the flesh of animals. Parasites live inside or on a living host and feed on its tissues. Omnivores eat both animals and plants. **Detritivores** such as earthworms eat small bits of decaying organic matter, or detritus. **Decomposers** feed on wastes and remains, breaking them down into inorganic building blocks. Bacteria, archaea, and fungi serve as decomposers.

ENERGY FLOW, NUTRIENT CYCLING

Energy captured by producers is converted to bond energy in organic molecules. This energy is released by metabolic reactions that give off heat. Energy flow through living organisms is a one-way process because producers cannot capture the energy of heat in chemical bonds, so it escapes from the ecosystem (Section 5.1). In contrast, nutrients cycle within an ecosystem. Producers take up hydrogen, oxygen, and carbon from inorganic sources such as the air and water. They also take up dissolved nitrogen, phosphorus, and other necessary minerals. Nutrients that producers used to build their bodies are used in turn to build the bodies of the consumers who eat them. When producers or consumers die, decomposition returns nutrients to the environment, from which producers take them up again.

TROPHIC STRUCTURE

Ecologists often look at energy transfers in terms of who eats whom. The hierarchy of feeding relationships in an ecosystem is the ecosystem's trophic structure. *Troph* refers to feeding, as in autotroph, which means self-feeding. In any ecosystem, all organisms at the same **trophic level** are the same number of transfers away from that system's source of energy. Producers are at the first trophic level. The primary consumers that eat them are at the second trophic level. The second-level consumers eat primary consumers, and so on.

A **food chain** is a sequence of steps by which some energy captured by primary producers is transferred to higher trophic levels. For example, in one tallgrass prairie food chain, energy flows from grasses to grasshoppers, to sparrows, and finally to bird-eating hawks (**FIGURE 28.2**). At the first trophic level in this food chain, grasses and other plants are the producers. At the second trophic level, grasshoppers are primary consumers. At the third trophic level, sparrows that eat grasshoppers are second-level consumers. At the fourth trophic level, hawks that eat sparrows are third-level consumers.

Energy captured by producers usually passes through no more than four or five trophic levels. Even in ecosystems with many species, the number of participants in each food chain is limited. The inefficiency of energy transfers constrains the length of food chains. Only 5 to 30 percent of the energy in tissues of an organism at one trophic level ends up in tissues of an organism at the next trophic level.

Several factors limit the efficiency of transfers. All organisms lose energy as metabolic heat, and this energy is not available to organisms at the next trophic level. Also, some energy gets stored in molecules that most consumers cannot break down. For example, most carnivores cannot access energy tied up in bones, scales, hair, feathers, or fur. Many herbivores cannot digest cellulose and lignin that reinforce the tissues of plants.

---

consumer Organism that obtains energy and carbon by feeding on tissues, wastes, or remains of other organisms.
decomposer Organism that feeds on biological remains and breaks organic material down into its inorganic subunits.
detritivore Consumer that feeds on small bits of organic material.
ecosystem A biological community and its environment.
food chain Description of who eats whom in one path of energy flow through an ecosystem.
primary production The rate at which an ecosystem's producers capture and store energy.
producer An organism that obtains energy directly from the environment and carbon from inorganic sources; an autotroph.
trophic level Position of an organism in a food chain.

TAKE-HOME MESSAGE 28.1

An ecosystem is a community of autotrophic producers and heterotrophic consumers that interact with one another and with their nonliving environment.

Nutrients cycle within an ecosystem, but energy flows through in one direction only.

Ecosystems vary in their primary production. Marine ecosystems generally have a lower primary production per unit area than land ecosystems.

Energy and nutrients are passed up food chains. Inefficiency of energy transfers limits the number of steps in food chains.

FIGURE 28.2 {Animated} Food chain. An example of who eats whom in a tallgrass prairie ecosystem in Kansas. Yellow arrows indicate energy flow. Plants are the main producers. Sunlight energy they capture and store in their tissues supplies the energy that the ecosystem's consumers require.

FIGURE IT OUT: Which of these organisms are heterotrophs?

Answer: All of the consumers are heterotrophs. They cannot make their own food.

28.2 HOW DO FOOD CHAINS INTERCONNECT?

An organism that participates in one food chain usually has a role in many others as well. The food chains of an ecosystem cross-connect as a **food web**. **FIGURE 28.3** shows some participants in an arctic food web.

Most food webs include both **grazing food chains**, in which herbivores eat producers, and **detrital food chains**, in which producers die and are then consumed by detritivores. Herbivores tend to be relatively large animals such as mammals, whereas detritivores tend to be smaller animals such as worms or insects.

In most land ecosystems, the bulk of the energy that gets stored in producer tissues moves through detrital food chains. For example, in an arctic ecosystem, voles, lemmings, and hares eat some living plant parts. However, far more

FIGURE 28.3 **{Animated}** Some organisms in an arctic food web. Yellow arrows indicate path of energy flow (point from eaten to eater).

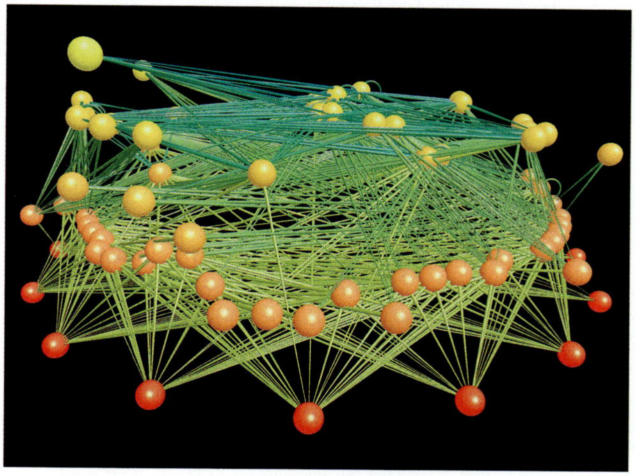

FIGURE 28.4 Computer model for a land food web in East River Valley, Colorado. Balls signify species. Their colors identify trophic levels, with producers (coded red) at the bottom and top predators (yellow) at top. The connecting lines thicken, as they go from an eaten species to the eater.

plant material ends up as detritus. Bits of dead plant material sustain detritivores such as nematodes and soil-dwelling insects.

Understanding food webs helps ecologists predict how ecosystems will respond to change. Neo Martinez and his colleagues constructed the food web diagram shown in **FIGURE 28.4**. By comparing different food webs, Martinez realized that trophic interactions connect species more closely than people thought. On average, each species in a food web was typically two links away from all other species. Ninety-five percent of species were within three links of one another, even in large communities with many species. As Martinez concluded in his paper about these findings, "Everything is linked to everything else." He cautioned that the extinction of any species in a food web has a potential impact on many other species.

detrital food chain Food chain in which energy is transferred directly from producers to detritivores.
food web Set of cross-connecting food chains.
grazing food chain Food chain in which energy is transferred from producers to grazers (herbivores).

TAKE-HOME MESSAGE 28.2

Two types of food chains connect in most food webs. Tissues of living producers are the base for grazing food chains. Producer remains are the base for detrital food chains.

Even in complex ecosystems, trophic interactions link each species in a food web with many others.

28.3 WHAT ARE ECOLOGICAL PYRAMIDS?

A food web diagram is one way of depicting the trophic relationships of species in a particular ecosystem. Ecological pyramid diagrams are another. A biomass pyramid shows the amount of organic material in the bodies of organisms at each trophic level at a specific time. An energy pyramid shows the amount of energy that flows through each trophic level in a given interval. **FIGURE 28.5** shows ecological pyramids for one freshwater spring ecosystem in Florida.

Most commonly, producers account for most of the biomass in a pyramid, and top carnivores contribute relatively little. The Florida ecosystem has lots of aquatic plants but very few gars (a top predator in this ecosystem). Similarly, if you walk through a prairie, you will see more grams of grass than of coyote.

An energy pyramid is always broadest at the bottom. This is why people promote a vegetarian diet by touting the ecological benefits of "eating lower on the food chain." They are referring to the energy losses in transfers between plants, livestock, and humans. When a person eats a plant, he or she gets most of the calories in that food. When a plant is used to feed livestock, only a small percentage of that food's calories ends up in meat a person can eat. Thus, feeding a population of meat-eaters requires far greater crop production than sustaining a population of vegetarians.

FIGURE 28.5 {Animated} Ecological pyramids for Silver Springs, an aquatic ecosystem in Florida.

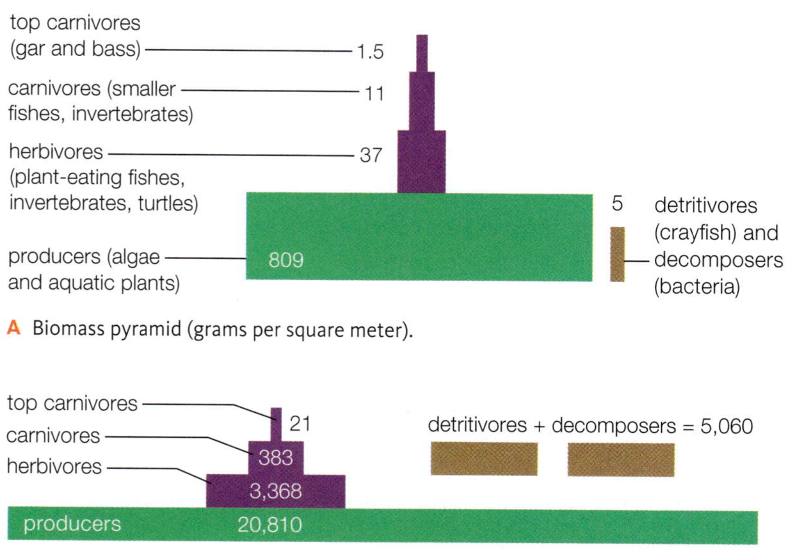

A Biomass pyramid (grams per square meter).

B Energy flow pyramid (kilocalories per square meter per year).

TAKE-HOME MESSAGE 28.3

Ecological pyramids depict the distribution of materials and energy among the trophic levels of an ecosystem.

28.4 WHAT IS A BIOGEOCHEMICAL CYCLE?

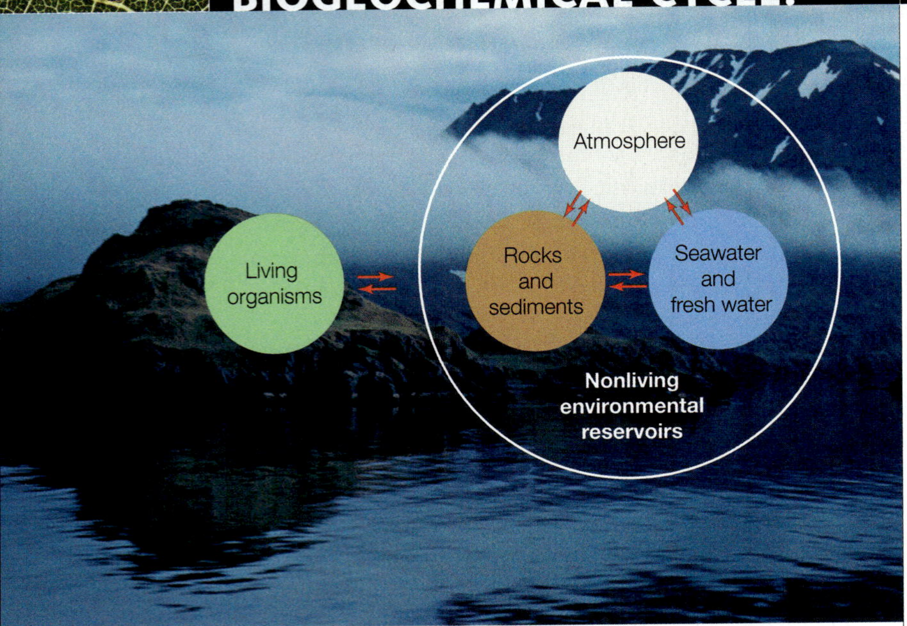

FIGURE 28.6 Generalized biogeochemical cycle. For any nutrient, the cumulative amount in all environmental reservoirs far exceeds the amount in living organisms.

In a **biogeochemical cycle**, an essential element moves from one or more environmental reservoirs, through the biological component of an ecosystem, and then back to the reservoirs (**FIGURE 28.6**). Depending on the element, environmental reservoirs may include Earth's rocks and sediments, waters, and atmosphere.

Chemical and geologic processes move elements to, from, and among environmental reservoirs. For example, elements locked in rocks can become part of the atmosphere as a result of volcanic activity. As one of Earth's crustal plates moves under another, rocks on the seafloor can be uplifted so they become part of a landmass. On land, the rocks are exposed to the erosive forces of wind and rain. As the rocks are slowly broken down, elements in them enter rivers, and eventually seas. Compared to the movement of elements among organisms of an ecosystem, the movement of elements among nonbiological reservoirs is far slower. Processes such as erosion and uplifting operate over thousands or millions of years.

biogeochemical cycle A nutrient moves among environmental reservoirs and into and out of food webs.

TAKE-HOME MESSAGE 28.4

A biogeochemical cycle is the slow movement of a nutrient among environmental reservoirs and into, through, and out of food webs.

28.5 WHAT IS THE WATER CYCLE?

The **water cycle** moves water from the ocean to the atmosphere, onto land, and back to the oceans (**FIGURE 28.7**). Sunlight energy causes evaporation, the conversion of liquid water to water vapor. Water vapor that enters the cool upper layers of the atmosphere condenses into droplets, forming clouds. When droplets get large and heavy enough, they fall as precipitation—as rain, snow, or hail.

Oceans cover about 70 percent of Earth's surface, so most rainfall returns water directly to the oceans. Most precipitation that falls on land seeps into the ground. Some of this water remains between soil particles as **soil water**. Plant roots can tap into this water source. Water that drains through soil layers often collects in **aquifers**, which are natural underground reservoirs consisting of porous rock layers. **Groundwater** is water in soil and aquifers. Water that falls on impermeable rock or on saturated soil becomes **runoff**: It flows over the ground into streams. The flow of groundwater and surface water returns water to oceans.

Movement of water results in movement of other nutrients. Carbon, nitrogen, and phosphorus all have soluble forms that flowing water can carry from place to place. As water trickles through soil, it brings nutrients from topsoil into deeper soil layers. As a stream flows over limestone, water slowly dissolves the rock and carries carbonates back to the seas where the limestone formed.

The vast majority of Earth's water (97 percent) is in oceans, and most fresh water is frozen as ice (**TABLE**

FIGURE 28.7 {Animated} The water cycle. Water moves from the ocean to the atmosphere, land, and back. The arrows identify processes that move water.

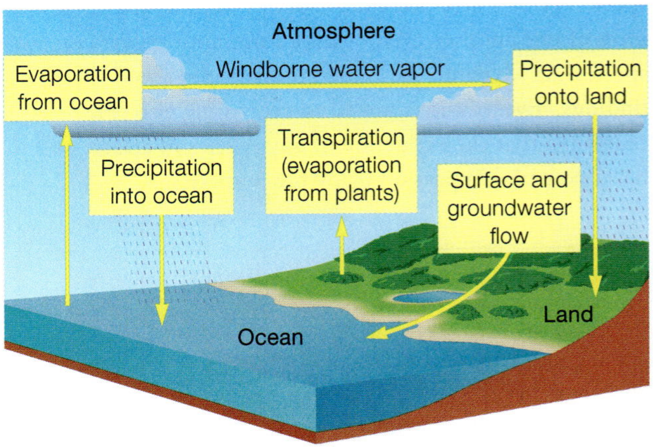

aquifer Porous rock layer that holds some groundwater.
groundwater Soil water and water in aquifers.
runoff Water that flows over soil into streams.
soil water Water between soil particles.
water cycle Movement of water among Earth's oceans, atmosphere, and the freshwater reservoirs on land.

28.1). Thus, the fresh water available to meet human needs and sustain land ecosystems is limited. Water overdrafts are now common—humans remove water from aquifers, lakes, or rivers faster than natural processes replenish it.

Aquifers supply about half of the drinking water in the United States. Overdrawing water from an aquifer can lower the water table, which is the topmost level at which the rock is saturated with water. When the water table falls, wells that tap an aquifer can run dry.

Consider what has happened to the largest aquifer in the United States, the Ogallala aquifer. This aquifer stretches from South Dakota to Texas and supplies irrigation water for 27 percent of the nation's crops. For the past thirty years, withdrawals have exceeded replenishment by a factor of ten. As a result, the water table has dropped as much as 50 meters (150 feet) in some regions.

Water in many rivers is currently over-allocated, meaning the amount of water promised to various stakeholders such as cities and farmers exceeds the amount that currently flows through the river. In such rivers, diversion of water for human use results in lowered or nonexistent flow in some portion of the river. The lack of water can have a devastating effect on biological communities that depend on the river. Rivers convey sediment and nutrients as well as water, so decreased flow alters ecosystems by slowing delivery of these materials to the river's delta, the region where the river approaches the sea.

TABLE 28.1

Environmental Water Reservoirs

Reservoir	Volume (10^3 cubic kilometers)
Ocean	1,370,000
Polar ice, glaciers	29,000
Groundwater	4,000
Surface water (lakes, rivers)	230
Atmosphere (water vapor)	14

TAKE-HOME MESSAGE 28.5

Water moves slowly from its main reservoir—the oceans—through the atmosphere, onto land, then back to the oceans.

Fresh water constitutes only a tiny portion of Earth's water, and most fresh water is frozen as ice.

Excessive withdrawal of water from aquifers and rivers depletes sources of drinking water and endangers ecosystems.

PEOPLE MATTER

National Geographic Grantee
JONATHAN WATERMAN

For million of years, the Colorado River has flowed from its source high in the Rocky Mountains to the Gulf of Mexico. Today, the river often does not reach the sea; its waters instead irrigate cropland and flow into pipes that supply cities such as Los Angeles, Las Vegas, Phoenix, and Denver.

In 2008, author Jonathan Waterman set out to raft the length of the Colorado River, intending to document the extent and causes of its decline. He says, "I saw a river being both depleted and salted thick by farms." In Mexico, the river essentially disappeared. "Fifty miles from the sea, 1.5 miles south of the Mexican border, I saw the river evaporate into a scum of phosphates and discarded water bottles." (See the photo above.) Waterman ended up walking for ten days through what should have been the river's delta region. Eventually he reached a tributary, the Rio Hardy, which he rafted until it too ran dry about 12 miles from the sea.

Waterman detailed the story of his journey and of the Colorado River's plight in his book *Running Dry: A Journey from Source to Sea Down the Colorado River*. He says of his books, "All were underlain by the premise that we can affect change in human behavior, to protect and preserve wild places, or human life." With regard to the Colorado River, Waterman's optimism about the possibility of change may be well-founded. In 2012, the governments of the United States and Mexico agreed to increase water flow through the Colorado's delta region through at least 2017.

28.6 HOW DOES CARBON MOVE THROUGH ECOSYSTEMS?

THE CARBON CYCLE

In the **carbon cycle**, natural processes move carbon among Earth's atmosphere, oceans, soils, and into and out of food webs (**FIGURE 28.8**). It is an **atmospheric cycle**, a biogeochemical cycle in which a gaseous form of the element plays a significant role.

On land, plants take up carbon dioxide from the atmosphere and incorporate it into their tissues when they carry out photosynthesis ❶. Plants and most other land organisms release carbon dioxide into the atmosphere by the process of aerobic respiration ❷.

The greatest flow of carbon between nonbiological reservoirs takes place between the atmosphere and the oceans. The air holds about 750 gigatons of carbon, mainly in the form of carbon dioxide (CO_2). Seawater holds 38,000–40,000 gigatons of dissolved carbon, primarily in bicarbonate (HCO_3^-) and carbonate (CO_3^{2-}) ions. Bicarbonate ions form when atmospheric carbon dioxide dissolves in water ❸. Aquatic producers take up bicarbonate and convert it to CO_2 that they use in photosynthesis. As on land, aquatic organisms carry out aerobic respiration and release carbon dioxide ❹.

Earth's rocks and sediments form its single greatest reservoir of carbon, with more than 65 million gigatons. Limestone is a sedimentary rock that forms over millions of years when sediments containing calcium carbonate shells of marine organisms such as foraminifera (Section 20.3) become compacted ❺. Limestone and other rocks derived from marine sediments can be uplifted onto land by movements of tectonic plates (Section 16.5). Producers do not take up carbon from rocks and sediment, so carbon in these reservoirs has little effect on these ecosystems.

Soil contains about 1,600 gigatons of carbon, more than twice as much as the atmosphere. The carbon in soil resides in humus and in living soil organisms. Over time, bacteria and fungi in the soil decompose humus and release carbon dioxide into the air. The speed of decomposition increases with temperature. In a tropical forest, decomposition and

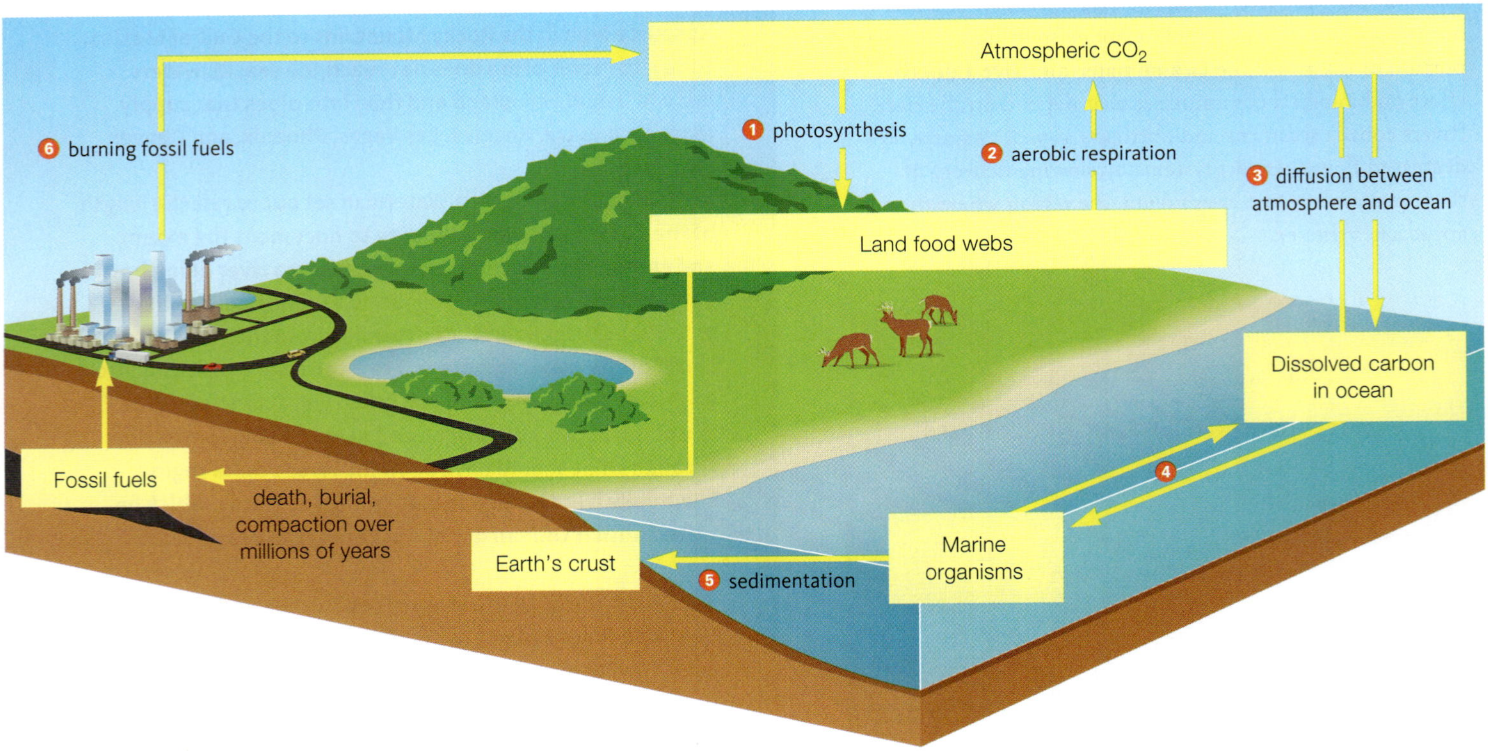

FIGURE 28.8 {Animated} The carbon cycle.

❶ Carbon enters land food webs when plants take up carbon dioxide from the air for use in photosynthesis.

❷ Carbon returns to the atmosphere as carbon dioxide when plants and other land organisms carry out aerobic respiration.

❸ Carbon diffuses between the atmosphere and the ocean. Bicarbonate forms when carbon dioxide dissolves in seawater.

❹ Marine producers take up bicarbonate for use in photosynthesis, and marine organisms release carbon dioxide from aerobic respiration.

❺ Many marine organisms incorporate carbon into their shells. After they die, these shells become part of the sediments. Over time, the sediments become carbon-rich rocks such as limestone and chalk in Earth's crust.

❻ Burning of fossil fuels derived from the ancient remains of plants puts additional carbon dioxide into the atmosphere.

FIGURE 28.9 {Animated} Greenhouse effect.

1 Earth's atmosphere reflects some sunlight energy back into space.

2 More light energy reaches and warms Earth's surface.

3 Earth's warmed surface emits heat energy. Some of this energy escapes through the atmosphere into space. But some is absorbed and then emitted in all directions by greenhouse gases. The emitted heat warms Earth's surface and lower atmosphere.

FIGURE IT OUT: Do greenhouse gases reflect heat energy toward the Earth?

Answer: No. The gases absorb heat energy, then reemit it in all directions.

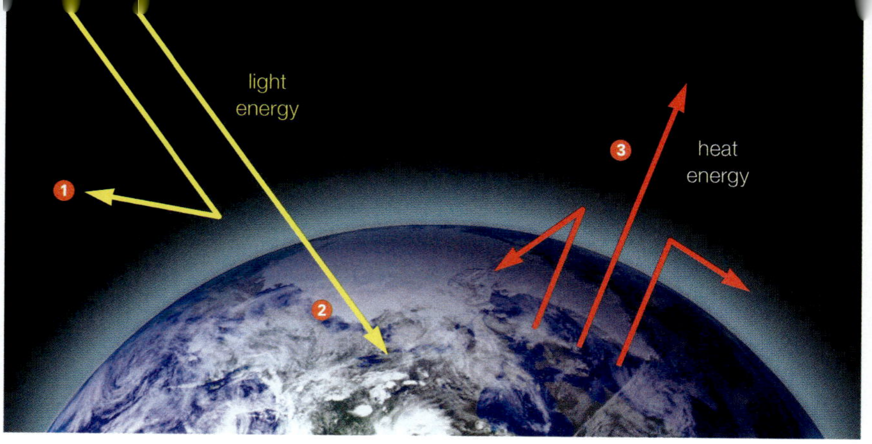

nutrient uptake proceed rapidly, so most of the forest's carbon is stored in living plants, rather than in soil. By contrast, in temperate zone forests and grasslands, soil holds more carbon than the plants do. The most carbon-rich soils are in the arctic, where low temperature hampers decomposition, and in peatbogs (Section 21.2), where acidic, anaerobic conditions do the same.

Fossil fuels such as coal, oil, and natural gas are another carbon reservoir. Such fuels are carbon-rich remains of ancient photosynthesizers. Fossil fuels hold an estimated 5,000 gigatons of carbon. Until the Industrial Revolution, this carbon, like that in rocks, had little impact on ecosystems. However, we now withdraw 4 to 5 gigatons of carbon from fossil fuel reservoirs each year 6. Our use of this fuel puts more carbon dioxide into the air than can dissolve in the oceans. Each year, about 2 percent of the extra carbon we release by burning fossil fuels dissolves in seawater. The other 98 percent increases the carbon dioxide level of the atmosphere.

THE GREENHOUSE EFFECT

The ongoing rise in atmospheric carbon dioxide is a matter of concern. Atmospheric carbon dioxide helps keep Earth warm enough for life. In what is known as the **greenhouse effect**, sunlight heats Earth's surface, then carbon dioxide and other "greenhouse gases" absorb some heat radiating from the surface and reradiate it toward Earth (**FIGURE 28.9**). Without the greenhouse effect, heat from Earth's surface would escape into space, leaving the planet cold and lifeless.

Given the greenhouse effect, we would predict that increases in the atmospheric concentration of carbon dioxide and other greenhouse gases would raise the temperature of Earth's surface. Evidence supports this prediction. In 2013, the carbon dioxide content of Earth's atmosphere rose above 400 parts per million for the first time in several million years. The result is **global climate change**, a trend toward rising temperature and shifts in other climate patterns.

Earth's climate has always varied over time. During ice ages, much of the planet was covered by glaciers. Other periods were warmer than the present, and tropical plants and coral reefs thrived at now-cool latitudes. Scientists can correlate past large-scale temperature changes with shifts in Earth's orbit, which varies in a regular fashion over 100,000 years, and Earth's tilt, which varies over 40,000 years. Changes in solar output and volcanic eruptions also affect Earth's temperature. However, there is a consensus among scientists that these factors do not play a major role in the current temperature rise, whereas the rise in greenhouse gases does. We discuss the consequences of global climate change in more detail in Section 30.6, when we consider human effects on the biosphere.

TAKE-HOME MESSAGE 28.6

Most of Earth's carbon is in rocks. Not much carbon moves out of this reservoir into the world of life.

Oceans and soils hold more carbon than the air. Carbon flows continually between these reservoirs and into and out of food webs. Producers take up carbon dioxide for photosynthesis, and all organisms release carbon dioxide as a result of aerobic respiration.

At present, burning of fossil fuels is releasing carbon into the air faster than the ocean can absorb it. As a result, the atmospheric concentration of carbon dioxide is increasing.

Carbon dioxide is a greenhouse gas. The presence of such gases in the atmosphere is essential to keeping Earth's surface warm enough to support life. However, the increasingly high level of these gases is causing global climate change.

atmospheric cycle Biogeochemical cycle in which a gaseous form of an element plays a significant role.
carbon cycle Movement of carbon, mainly between the oceans, atmosphere, and living organisms.
global climate change A rise in average temperature that is altering climate patterns around the world.
greenhouse effect Warming of Earth's lower atmosphere and surface as a result of heat trapped by greenhouse gases.

CREDITS: (9) photo, NASA; art, © Cengage Learning.

28.7 HOW DOES NITROGEN MOVE THROUGH ECOSYSTEMS?

THE NITROGEN CYCLE

Nitrogen moves in an atmospheric cycle known as the **nitrogen cycle** (**FIGURE 28.10**). The main nitrogen reservoir is the atmosphere, which is about 80 percent nitrogen gas. Nitrogen gas consists of two atoms of nitrogen held together by a triple covalent bond as N_2, or $N{\equiv}N$. Recall from Section 2.3 that a triple bond holds atoms together more strongly than a single or double bond would.

All organisms use nitrogen to build ATP, nucleic acids, and proteins. Photosynthetic organisms also use it to build chlorophyll. Despite the universal need for nitrogen and the abundance of atmospheric nitrogen, no eukaryote can make use of nitrogen gas. Eukaryotes do not have an enzyme that can break the strong bond between the two nitrogen atoms.

Certain types of bacteria and archaea can break the triple bond. They carry out **nitrogen fixation**, combining nitrogen atoms in nitrogen gas with hydrogen to produce ammonia (NH_3). Ammonia dissolves to form ammonium (NH_4^+) ❶. Biological nitrogen fixation has a high activation energy (Section 5.2); it requires an input of 16 molecules of ATP to convert one molecule of nitrogen to ammonia.

You have already learned about two major groups of nitrogen-fixers. Nitrogen-fixing cyanobacteria live in aquatic habitats, soil, and as components of lichens (Sections 19.5 and 22.3). Other nitrogen-fixing bacteria live on their own in soil, or inside plant parts such as nodules on roots of peas and other legumes. Some deep-sea archaea also fix nitrogen.

An additional small amount of ammonium forms as a result of lightning-fueled reactions in the atmosphere. Energy from the lightning causes nitrogen gas to react with atmospheric water vapor.

Plants can take up ammonium from soil water ❷ and use it in metabolic reactions. Animals meet their nitrogen needs by eating plants or one another. Bacterial and fungal decomposers return ammonium to the soil when they break down organic wastes and remains, a process called ammonification ❸.

Nitrification is a two-step, oxygen-requiring process that converts ammonium to nitrates (NO_3^-) ❹. First, ammonia-oxidizing bacteria or archaea convert ammonium to nitrite (NO_2^-), then nitrite-oxidizing bacteria convert nitrites to nitrates. Like ammonium, nitrates can be taken up and used by plants ❺. Nitrification is essential to ecosystem health because it prevents ammonium from accumulating to toxic concentrations. We make use of the bacteria that carry out this process in sewage treatment plants. Sewage contains large amounts of ammonium formed from urea excreted in urine.

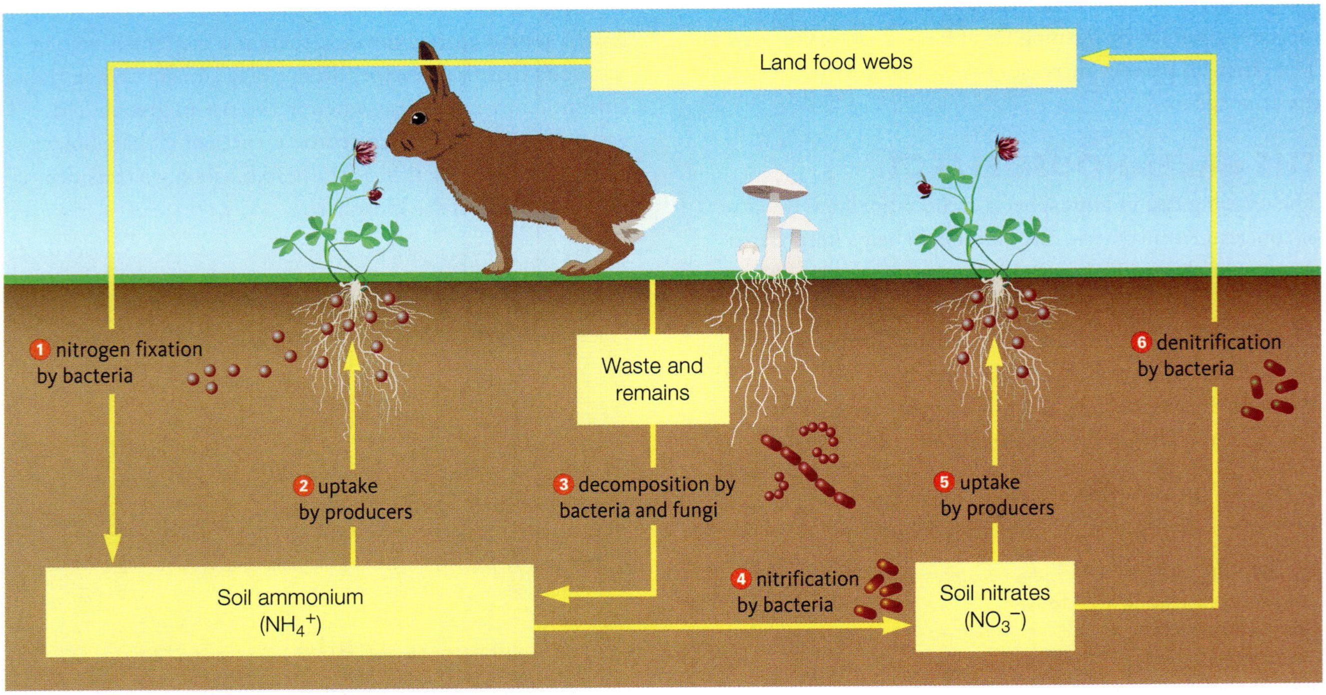

FIGURE 28.10 **{Animated}** Nitrogen cycle in a land ecosystem.

FIGURE 28.11 Tractor applying industrially produced, nitrogen-rich fertilizer to a cornfield. Nitrogen is the nutrient that most commonly limits corn growth. The inset photo shows corn grown with adequate nitrogen (*left*) and in nitrogen-deficient soil (*right*).

Conversion of nitrates to nitrogen gas is called **denitrification** ⑥. This anaerobic reaction is carried out mainly by bacteria. In sewage treatment plants, denitrifying bacteria are used to remove nitrates from wastewater before the water is released into the environment. In ecosystems, denitrification results in a decline in the amount of soluble nitrogen available to producers.

ALTERATIONS TO THE CYCLE

In the early 1900s, scientists invented a method of fixing atmospheric nitrogen and producing ammonium on an industrial scale. This process allowed production of synthetic nitrogen fertilizers that have boosted crop yields (**FIGURE 28.11**). These fertilizers have helped feed a rapidly increasing human population. However, their use has also disrupted the biological portion of the nitrogen cycle. A lot of nitrate from synthetic fertilizers is carried away from croplands in runoff and nitrate contamination of aquatic ecosystems encourages algal blooms (Section 20.10). When nitrate contaminates drinking water, it can pose a threat to human health. Among other effects, ingested nitrate inhibits iodine uptake by the thyroid gland and may increase the risk of thyroid cancer. The U.S. Environmental Protection Agency (EPA) has set a maximum standard for nitrate in public drinking water and requires periodic testing to ensure this standard is met.

Humans also interfere with the nitrogen cycle by burning wood and fossil fuels. Combustion of these materials releases nitrous oxide gas (N_2O) into the atmosphere. An increase in atmospheric nitrous oxide is a matter of concern for two reasons. First, nitrous oxide is a greenhouse gas, and a highly persistent and effective one. It can remain in the atmosphere for more than 100 years, and it traps 300 times as much heat as an equivalent amount of CO_2. Second, nitrous oxide contributes to destruction of the ozone layer. As Section 18.4 explains, ozone high in the atmosphere protects life at Earth's surface from damaging effects of ultraviolet radiation. We discuss ozone destruction in more detail in Section 30.5.

denitrification Conversion of nitrates or nitrites to nitrogen gas.
nitrification Conversion of ammonium to nitrate.
nitrogen cycle Movement of nitrogen among the atmosphere, soil, and water, and into and out of food webs.
nitrogen fixation Conversion of nitrogen gas to ammonia.

TAKE-HOME MESSAGE 28.7

Nitrogen-fixing prokaryotes convert gaseous nitrogen to ammonium that plants take up and use. Other prokaryotes convert ammonium to nitrites and nitrates. Still others change nitrates into nitrogen gas, which can escape from an ecosystem.

Use of synthetic nitrogen fertilizer and burning of fossil fuels add nitrogen-containing pollutants to ecosystems.

28.8 HOW DOES PHOSPHORUS MOVE THROUGH ECOSYSTEMS?

Atoms of phosphorus are highly reactive, so phosphorus does not occur naturally in its elemental form. Most of Earth's phosphorus is bonded to oxygen as phosphate (PO_4^{3-}), an ion that occurs in rocks and sediments. In the **phosphorus cycle**, phosphorus passes quickly through food webs as it moves from land to ocean sediments, then slowly back to land (**FIGURE 28.12**). Because little phosphorus exists in a gaseous form and its major reservoir is sedimentary rock, the phosphorus cycle is called a **sedimentary cycle**.

In the geochemical portion of the phosphorus cycle, weathering and erosion move phosphates from rocks into soil, lakes, and rivers ①. Leaching and runoff carry dissolved phosphates to the ocean ②. Here, most phosphorus comes out of solution and settles as rocky deposits along continental margins ③. Slow movements of Earth's crust can uplift these deposits onto land ④, where weathering releases phosphates from rocks once again.

All organisms require phosphorus as a component of nucleic acids and phospholipids. The biological portion of the phosphorus cycle begins when producers take up phosphate. Land plants take up dissolved phosphate from the soil water ⑤. Land animals get phosphates by eating the plants or one another. Phosphorus returns to the soil in the wastes and remains of organisms ⑥.

Like nitrogen, phosphorus is often a limiting factor for plant growth, so fertilizers usually contain both phosphorus and nitrogen. Phosphate-rich rock is mined for this purpose. Guano, which is phosphate-rich droppings from seabird or bat colonies, is also mined and used as fertilizer. Also like nitrogen, phosphates often run off from the site where they are applied and enter aquatic habitats. The influx of phosphorus can encourage the growth of aquatic producers, resulting in an algal bloom.

phosphorus cycle Movement of phosphorus among Earth's rocks and waters, and into and out of food webs.
sedimentary cycle Biochemical cycle in which the atmosphere plays little role and rocks are the major reservoir.

TAKE-HOME MESSAGE 28.8

Rocks are the main phosphorus reservoir. Weathering and erosion of rock move phosphates into soil and water. Plants take up dissolved phosphates from the soil.

Phosphate can be a limiting nutrient for plants, so phosphate-rich rock and guano are mined for use as fertilizer. However, excess phosphate from these or other sources can result in an algal bloom when it enters aquatic habitats.

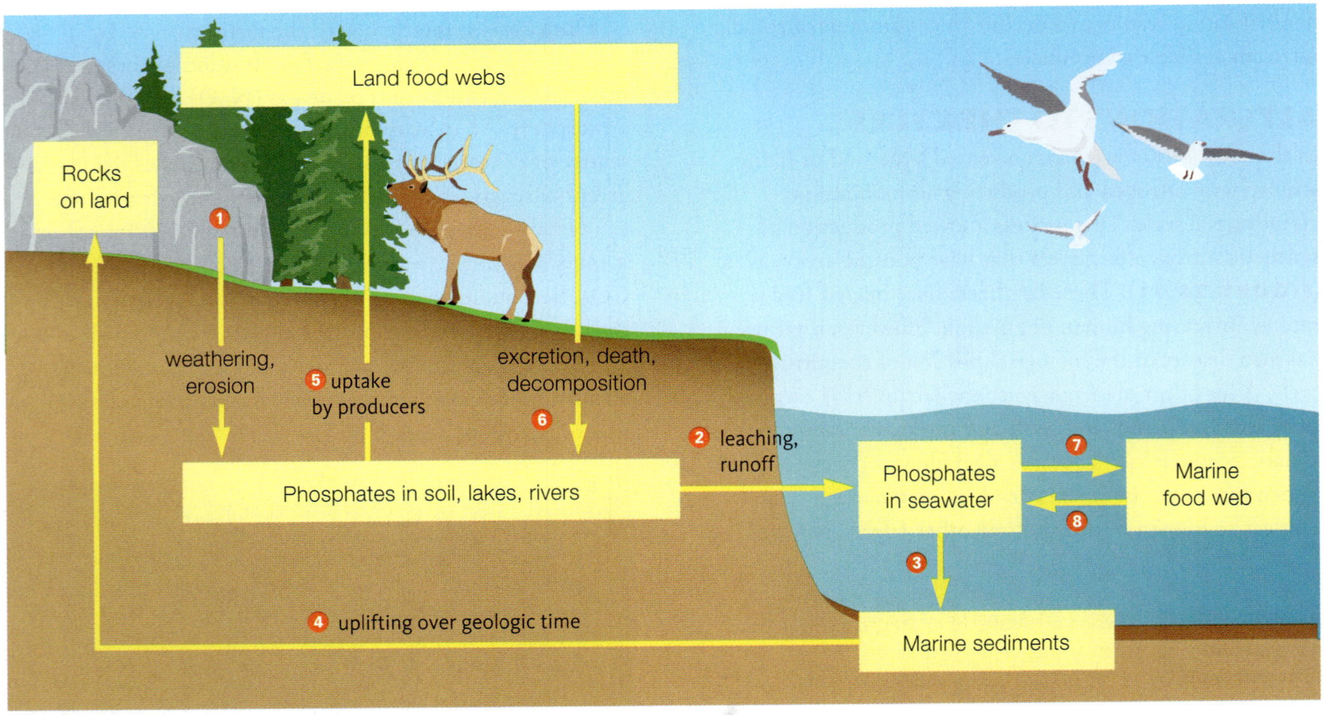

FIGURE 28.12 **{Animated}** The phosphorus cycle.

28.9 Application: TOXIC TRANSFERS

FIGURE 28.13 Loon with its fish prey. Along with nutrients, the loon takes in any toxic pollutants that accumulated in the fish's body.

NUTRIENTS ARE NOT THE ONLY THINGS THAT MOVE UP FOOD CHAINS. Pollutants enter food chains and pass from one trophic level to the next.

By the process of **bioaccumulation**, an organism's tissues store a pollutant taken up from the environment, causing the amount in the body to increase over time. The ability of some plants to bioaccumulate toxic substances makes them useful in phytoremediation of polluted soils.

In animals, hydrophobic chemical pollutants ingested or absorbed across skin tend to accumulate in fatty tissues. The amount of pollutants in an animal's body increases over time, so longer-lived species tend to be more affected by fat-soluble pollutants than shorter-lived ones. Within a species, old individuals tend to have a higher pollutant load than younger ones.

The concentration of a chemical in organisms increases as the pollutant moves up a food chain, a process known as **biological magnification.** As a result of bioaccumulation and biological magnification, even seemingly low concentrations of pollutants in an environment can cause harm, with long-lived animals at the top of food chains being most affected.

For example, tissues of a fish-eating bird such as a loon contain contaminants that the bird took in from the bodies of the fish it ate (FIGURE 28.13). Loons in many regions contain high levels of methylmercury, a neurotoxin produced by coal-burning power plants (Section 2.6). The higher the mercury level in a lake, the lower the bird's reproductive success. Among other ill effects, loons with a high mercury level show impaired parental behavior.

bioaccumulation The concentration of a chemical pollutant in the tissues of an organism rises over the course of the organism's lifetime.

biological magnification A chemical pollutant becomes increasingly concentrated as it moves up through food chains.

Summary

SECTION 28.1 There is a one-way flow of energy into and out of an **ecosystem**, and a cycling of materials among the organisms within it. All ecosystems have inputs and outputs of energy.

Producers convert energy from an inorganic source (usually light) into chemical bond energy. **Primary production**, the rate at which producers capture and store energy, can vary over time and between locations.

Consumers feed on producers or one another. For example, **detritivores** eat small bits of organic remains; **decomposers** break wastes and remains down into their inorganic components.

A **food chain** shows one path of energy and nutrient flow among organisms. Each organism in a food chain is at a different **trophic level**, with the primary producer being the first level and consumers at higher levels. The inefficiency of energy transfers from one trophic level limits most food chains to four or five links.

SECTION 28.2 The many food chains within an ecosystem interconnect to form a **food web**. Most food webs include both **grazing food chains**, in which herbivores eat producers, and **detrital food chains**, in which producers die and are then consumed by detritivores. As a result of the multiple connections through food webs, a change that affects one species in an ecosystem will have effects on many others.

SECTION 28.3 Energy pyramids and biomass pyramids show how energy and organic compounds are distributed among organisms within an ecosystem. All energy pyramids are largest at their base.

SECTION 28.4 In the nonbiological portion of a **biogeochemical cycle**, an element moves among environmental reservoirs such as Earth's atmosphere, rocks, and waters. In the biological portion of the cycle, elements move through an ecosystem's food web, then return to the environment.

SECTION 28.5 In the **water cycle**, evaporation, condensation, and precipitation move water from its main reservoir—the oceans—into the atmosphere, onto land, then back to oceans. Water that falls onto land may become part of the **groundwater**, which means it may be become **soil water** or be stored in an **aquifer**. Alternatively, it may become **runoff**. The water cycle helps move soluble forms of other nutrients.

Earth has a limited amount of freshwater, most of which is frozen as ice. Overwithdrawing water from rivers or aquifers to meet human needs can harm other species.

SECTION 28.6 The main reservoir for carbon is rocks, but the **carbon cycle** moves carbon mainly among seawater, the air, soils, and living organisms in an **atmospheric cycle**. Carbon dioxide contributes to the **greenhouse effect**. Greenhouse gases keep Earth's surface warm enough to support life. However, as a result of fossil fuel consumption and other human activities, the levels of these gases are increasing. The increase correlates with and is considered the most likely cause of the ongoing **global climate change**.

SECTION 28.7 The **nitrogen cycle** is an atmospheric cycle. Air is the main reservoir for N_2, a gaseous form of nitrogen that plants cannot use. Plants can take up and use ammonium that bacteria produce by **nitrogen fixation**. Fungi and bacteria that act as decomposers add ammonium derived from remains to the soil. Plants also use nitrates that some bacteria produce from ammonium through **nitrification**. Nitrogen is returned to the air by bacteria that carry out **denitrification** of nitrates. Humans add extra nitrogen to ecosystems by using synthetic fertilizer and by burning fossil fuels, which releases nitrous oxide.

SECTION 28.8 The **phosphorus cycle** is a **sedimentary cycle** with no significant atmospheric component. Phosphorus from rocks dissolves in water and is taken up by producers. Phosphate-rich rocks and deposits of bird droppings are mined for use as fertilizer.

SECTION 28.9 Toxic substances move up through food chains in the same way that nutrients do. Such toxins accumulate in the bodies of organisms, a process called **bioaccumulation**. Predators end up with high levels of these toxins because of **biological magnification**; when they eat prey, they ingest all the toxins that accumulated in the prey's body over its lifetime, and each predator typically eats many prey.

Self-Quiz

Answers in Appendix VII

1. In most ecosystems, producers use energy from _______ to build organic compounds.
   a. inorganic chemicals c. heat
   b. sunlight d. lower trophic levels

2. Decomposers are commonly _______ .
   a. fungi c. plants
   b. bacteria d. a and b

3. Organisms at the first trophic level _______ .
   a. capture energy from a nonliving source
   b. are eaten by organisms at higher trophic levels
   c. are shown at the bottom of an energy pyramid
   d. all of the above

4. Primary productivity on land is affected by _______ .
   a. nutrient availability c. temperature
   b. amount of sunlight d. all of the above

5. A(n) _______ is an autotroph.
   a. producer c. detritivore
   b. herbivore d. top carnivore

Data Analysis Activities

Rising Atmospheric Carbon To assess the impact of human activity on the carbon dioxide level in Earth's atmosphere, it helps to take a long view. One useful data set comes from deep core samples of Antarctic ice. The oldest ice core that has been fully analyzed dates back a bit more than 400,000 years. Air bubbles trapped in the ice provide information about the gas content in Earth's atmosphere at the time the ice formed. Combining ice core data with more recent direct measurements of atmospheric carbon dioxide—as in **FIGURE 28.14**—can help scientists put current changes in the atmospheric carbon dioxide into historical perspective.

1. What was the highest carbon dioxide level between 400,000 B.C. and 0 A.D.?
2. During this period, how many times did carbon dioxide reach a level comparable to that measured in 1980?
3. The industrial revolution occurred around 1800. How much did carbon dioxide levels change in the 800 years prior to the event? In 175 years after it?
4. Did carbon dioxide levels rise more between 1800 and 1975 or between 1980 and 2013?

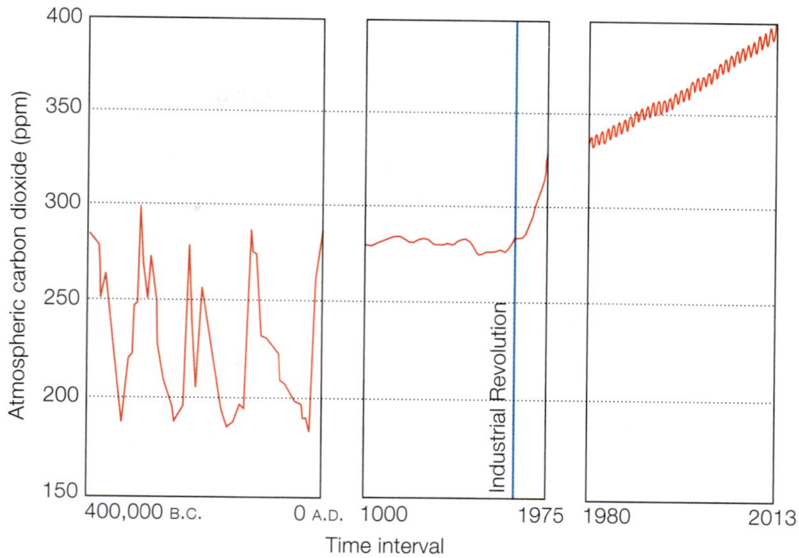

FIGURE 28.14 Changes in atmospheric carbon dioxide levels. Direct measurements began in 1980. Earlier data are based on air bubbles in ice cores. The oscillations in the directly measured data result from seasonal differences in photosynthesis.

6. Most of Earth's fresh water is ______ .
   a. in lakes and streams c. frozen as ice
   b. in aquifers and soil d. in bodies of organisms

7. Earth's largest carbon reservoir is ______ .
   a. the atmosphere c. seawater
   b. sediments and rocks d. living organisms

8. Carbon is released into the atmosphere by ______ .
   a. photosynthesis c. burning fossil fuels
   b. aerobic respiration d. b and c

9. Greenhouse gases ______ .
   a. help keep Earth's surface warm enough for life
   b. are released by natural and human activities
   c. include nitrous oxide and carbon dioxide
   d. all of the above

10. The ______ cycle is a sedimentary cycle.
   a. phosphorus c. nitrogen
   b. carbon d. water

11. Earth's largest phosphorus reservoir is ______ .
   a. the atmosphere c. sediments and rocks
   b. bird droppings d. living organisms

12. Most plants obtain ______ by taking it up from the air.
   a. nitrogen c. phosphorus
   b. carbon d. a and b

13. Nitrogen fixation converts ______ to ______ .
   a. nitrogen gas; ammonia c. ammonia; nitrates
   b. nitrates; nitrites d. nitrites; nitrogen oxides

14. Soil in ______ is richest in carbon.
   a. the arctic b. the tropics c. temperates zones

15. Match each term with its most suitable description.

___	carbon dioxide	a. contains triple bond
___	bicarbonate	b. product of nitrogen fixation
___	ammonium	c. marine carbon source
___	nitrogen gas	d. greenhouse gas

Critical Thinking

1. Marguerite has a vegetable garden in Maine. Eduardo has one in Florida. List the variables that could cause differences in the primary production of these gardens.

2. A watershed is an area in which all rainfall drains into a particular river. Find out which watershed you live in at the *Science in Your Watershed* site at http://water.usgs.gov/wsc.

3. The sulfur cycle is another important biogeochemical cycle. Rocks are the main reservoir for sulfur, but some sulfur compounds are dissolved in water, and sulfur oxide occurs in the atmosphere. Which of these sources do you think plants tap to meet their need for sulfur? Which essential biological compounds contain sulfur?

4. Rather than using fertilizer, a farmer may rotate crops, planting legumes one year, then another crop, then legumes again. Explain how crop rotation keeps soil fertile.

The Congo rain forest in central Africa, Earth's second largest tropical rain forest. Year-round warmth and rainfall support a diverse variety of evergreen, broadleaf trees. Their continual photosynthesis removes an enormous amount of carbon dioxide from the atmosphere.

29 THE BIOSPHERE

Links to Earlier Concepts

In this chapter, you will consider the biosphere (Section 1.1) as a whole, revisit properties of water (2.4), and see how carbon-fixing pathways (6.5), herbivory (27.4), and fire (27.6) affect plant distribution. We compare regional primary production (28.1) and learn more about the effects of global climate change (28.6), succession (27.6), and morphological convergence (16.7).

KEY CONCEPTS

AIR CIRCULATION PATTERNS

Latitudinal differences in the amount of solar energy reaching the ground cause air to move away from the equator, giving rise to major surface winds and latitudinal patterns in rainfall.

CURRENTS AND CLIMATES

Heating of the tropical seas sets ocean waters in motion. The circulating water affects climate on land. Interactions between oceans, air, and land influence coastal climates.

LAND BIOMES

A biome consists of geographically separated regions that have a similar climate and soils, and so support similar types of vegetation. Biomes vary in their productivity and species-richness.

FRESHWATER ECOSYSTEMS

Lakes undergo succession. In temperate zones, seasonal temperature changes determine when water within a lake mixes. Properties of a river, and the life it supports, vary along its length.

COASTAL AND MARINE ECOSYSTEMS

Life thrives in coastal wetlands, on coral reefs, and in the ocean's upper, sunlit water. Organisms also live in the ocean's deeper, darker waters and on the seafloor.

29.1 WHAT DETERMINES AIR CIRCULATION PATTERNS?

The biosphere includes all places where life exists on Earth (Section 1.1). The geographical distribution of species within the biosphere depends largely on climate. **Climate** refers to average weather conditions, such as cloud cover, temperature, humidity, and wind speed, over time. Regional climates differ because many factors that influence winds and ocean currents vary from place to place.

SEASONAL EFFECTS

Each year, Earth rotates around the sun in an elliptical path (**FIGURE 29.1**). Seasonal changes in day length and temperature arise because Earth's axis is not perpendicular to the plane of this ellipse, but rather tilts at a 23-degree angle. In June, when the Northern Hemisphere is angled toward the sun, it receives more intense sunlight and has longer days than the Southern Hemisphere (**FIGURE 29.1A**). In December, the opposite occurs (**FIGURE 29.1C**). Twice a year—on spring and autumn equinoxes—Earth's axis is perpendicular to incoming sunlight. On these days, every place on Earth has 12 hours of daylight and 12 hours of darkness (**FIGURE 29.1B,D**).

In each hemisphere, the extent of seasonal change in day length increases with latitude. At 25° north or south of the equator, the longest day length is a bit less than 14 hours. By contrast, 60° north or south of the equator, the longest day length is nearly 19 hours.

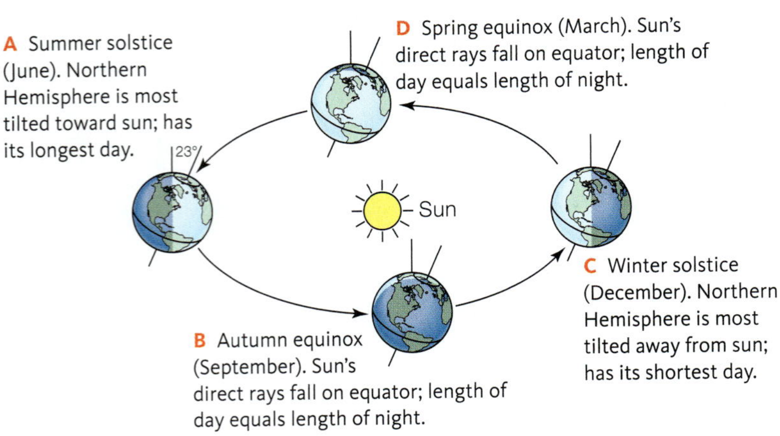

FIGURE 29.1 Effects of Earth's tilt and yearly rotation around the sun. The 23° tilt of Earth's axis causes the Northern Hemisphere to receive more intense sunlight and have longer days in summer than in winter.

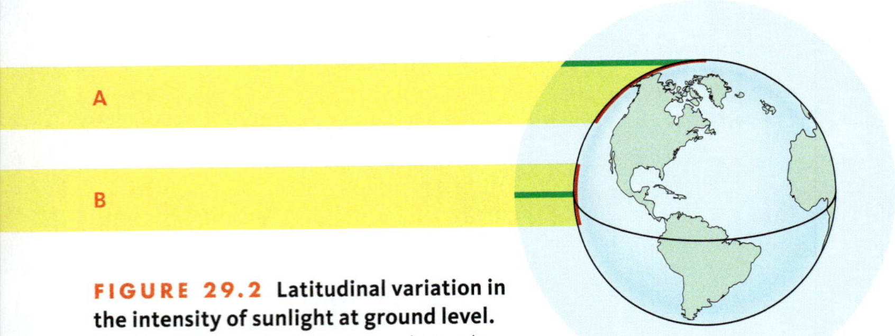

FIGURE 29.2 Latitudinal variation in the intensity of sunlight at ground level. For simplicity, we depict two equal parcels of incoming radiation on an equinox, a day when incoming rays are perpendicular to Earth's axis. Rays that fall on high latitudes **A** have passed through more atmosphere (blue) than those that fall near the equator **B**. Compare the length of the green lines. (Atmosphere is not to scale.)

In addition, energy in the rays that fall at the high latitude is spread over a greater area than energy that falls on the equator. Compare the length of the red lines.

As a result of these two factors, the amount of sunlight energy that reaches the Earth's surface decreases with increasing latitude.

AIR CIRCULATION AND RAINFALL

On any given day, equatorial regions receive more sunlight energy than higher latitudes for two reasons. First, fine particles of dust, water vapor, and greenhouse gases absorb some solar radiation or reflect it back into space. Sunlight traveling to high latitudes passes through more atmosphere to reach Earth's surface than light traveling to the equator, so less energy reaches the ground at high latitudes (**FIGURE 29.2A**). Second, energy in an incoming parcel of sunlight is spread out over a smaller surface area at the equator than at the higher latitudes (**FIGURE 29.2B**). As a result, Earth's surface warms more at the equator than at the poles.

Knowing about two properties of air can help you understand how regional differences in surface warming give rise to global air circulation and rainfall patterns. First, as air warms, it becomes less dense and rises. Hot-air balloonists take advantage of this effect when they take off from the ground by heating the air inside their balloon. Second, warm air can hold more water than cooler air. This is why you can "see your breath" in cold weather; water vapor in warm exhaled air condenses into droplets when exposed to the cold outside your body.

The global air circulation pattern begins at the equator, where intense sunlight heats air and causes evaporation from the ocean. The result is an upward movement of warm, moist air (**FIGURE 29.3A**). As this air rises to higher altitudes, it cools and flows north and south, releasing moisture as rain. This rain supports tropical rain forests such as the one shown in the photo that opens this chapter.

By the time the air reaches 30° north or south of the equator, it has given up most moisture, so little rain falls here. Many of the world's great deserts are about 30° from the equator. The air has also cooled, so it sinks toward Earth's surface (**FIGURE 29.3B**).

Idealized Pattern of Air Circulation | **Major Winds Near Earth's Surface**

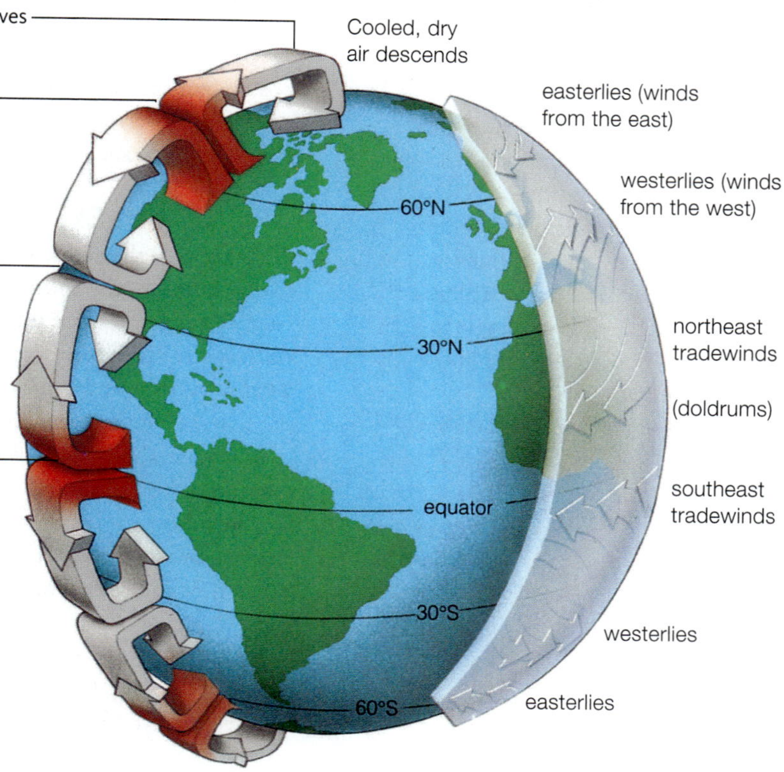

FIGURE 29.3 {Animated} Global air circulation patterns.

As air continues flowing along Earth's surface toward the poles, it again picks up heat and moisture. At a latitude of about 60°, warm, moist air rises again, losing moisture as it does so (**FIGURE 29.3C**). The resulting rains support temperate zone forests.

Cold, dry air descends near the poles (**FIGURE 29.3D**). Precipitation is sparse, and polar deserts form.

SURFACE WIND PATTERNS

Major wind patterns arise as air in the lower atmosphere moves continually from latitudes where air is sinking toward those where air is rising. Earth's rotation affects the direction of these winds. Air masses are not attached to Earth's surface, so the Earth spins beneath them, moving fastest at the equator and most slowly at the poles. Thus, as an air mass moves away from the equator, the speed at which the Earth rotates beneath it continually slows. As a result, major winds trace a curved path relative to the Earth's surface (**FIGURE 29.3E**). In the Northern Hemisphere, winds curve toward the right of their initial direction; in the Southern Hemisphere, they curve toward the left. For example, between 30° north and 60° north, surface air traveling toward the North Pole is deflected right, or toward the east. Winds are named for the direction from which they blow, so prevailing winds in the United States are westerlies—they blow from west to east.

Winds blow most consistently from one region where air is rising to another such location. Where air actually rises, winds are intermittent, as in the doldrums near the equator.

climate Average weather conditions in a region.

TAKE-HOME MESSAGE 29.1

Equatorial regions receive more sunlight energy than higher latitudes.

Sunlight-induced heating drives the rise of moisture-laden air at the equator. This air cools as it moves north and south, releasing rains that support tropical forests. Deserts form where cool, dry air sinks. Sunlight energy also drives moisture-laden air aloft at 60° north and south latitude. This air gives up moisture as it flows toward the equator or the pole.

Major surface winds arise as air in the lower atmosphere moves toward latitudes where air rises and away from latitudes where it sinks. These winds trace a curved path relative to Earth's surface because of Earth's rotation.

29.2 HOW DO CURRENTS AND LANDFORMS AFFECT CLIMATE?

OCEAN CURRENTS

Latitudinal variations in sunlight affect ocean temperature and set major currents in motion. At the equator, vast volumes of water warm and expand, making the sea level about 8 centimeters (3 inches) higher than it is at either pole. The existence of this "slope" starts sea surface water moving toward the poles. As the water moves toward cooler latitudes, it gives up heat to the air above it.

Enormous volumes of water flow as ocean currents. Directional movement of surface currents is influenced by the major winds, Earth's rotation, and the distribution of land masses. Surface currents circulate clockwise in the Northern Hemisphere and counterclockwise in the Southern Hemisphere (**FIGURE 29.4**).

Swift, deep, and narrow currents of water flow away from the equator along the east coast of continents. Along the east coast of North America, warm water flows north, as the Gulf Stream. Slower, shallower, broader currents of cold water parallel the west coast of continents and flow toward the equator.

Ocean currents affect climate. For example, Pacific Northwest coasts are cool and foggy in summer because the cold California current chills the air, and water condenses out of the cooled air as droplets. As another example, Boston and Baltimore are warm and muggy in summer because air masses pick up heat and moisture from the warm Gulf Stream, then flow over these cities.

REGIONAL EFFECTS

Differences in the ability of water and land to absorb and release heat give rise to coastal breezes. In the daytime, land warms faster than water. As air over land warms and rises, cooler offshore air moves in to replace it (**FIGURE 29.5A**). After sundown, land cools more quickly than the water, so the breezes reverse direction (**FIGURE 29.5B**).

Differential heating of water and land also causes **monsoons**, which are winds that change direction seasonally. Consider how the continental interior of Asia heats up in the summer, so air rises above it. Moist air from over the warm Indian Ocean, which is to the south, moves in to

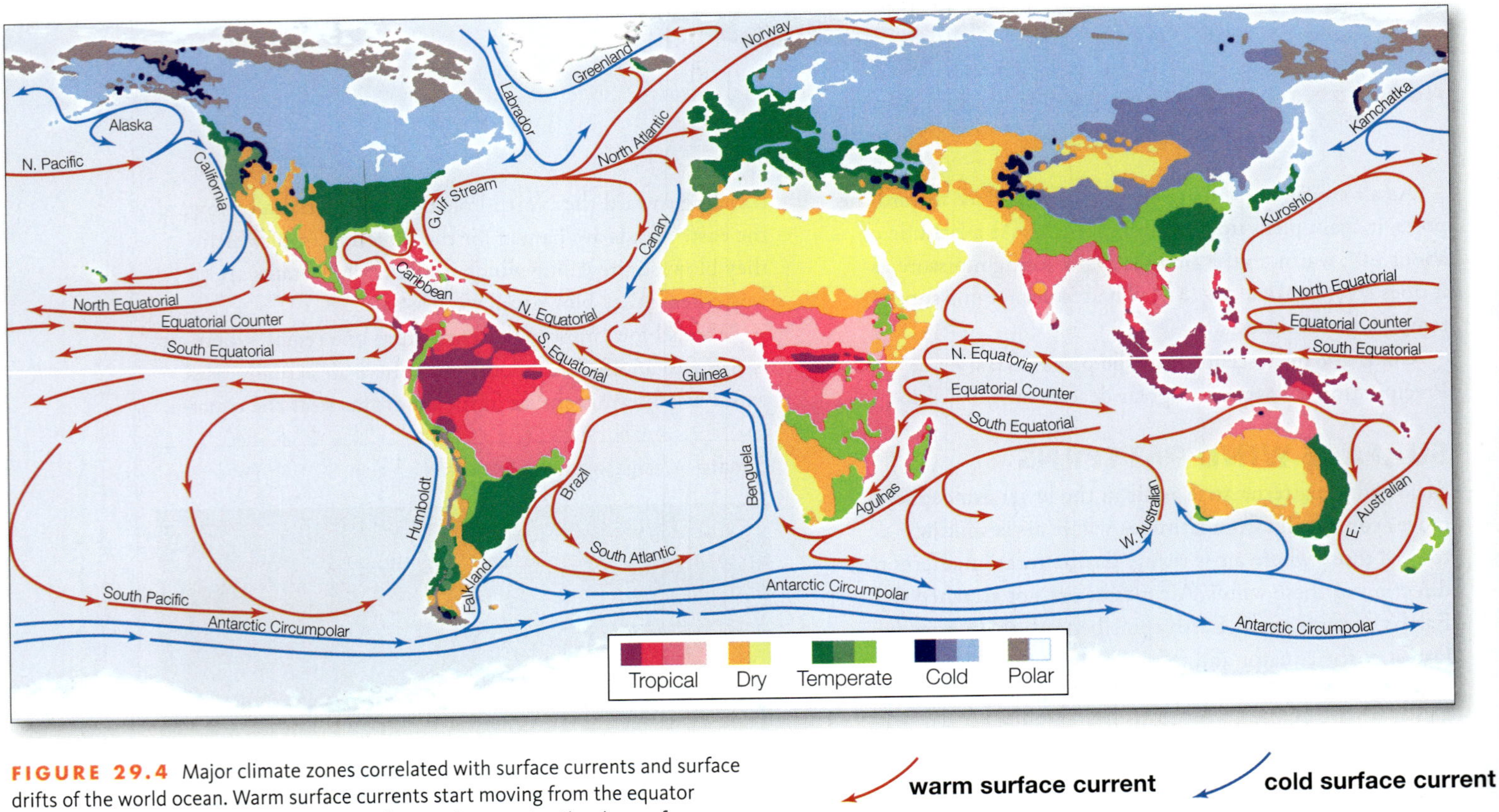

FIGURE 29.4 Major climate zones correlated with surface currents and surface drifts of the world ocean. Warm surface currents start moving from the equator toward the poles, but prevailing winds, Earth's rotation, gravity, the shape of ocean basins, and landforms influence the direction of flow. Water temperatures, which differ with latitude and depth, contribute to regional differences in air temperature and rainfall.

CREDIT: (4) NASA.

replace the rising air, and this north-blowing wind delivers heavy rains. In the winter, the continental interior is cooler than the ocean. As a result, cool, dry wind blowing from the north toward southern coasts causes a seasonal drought.

Proximity to an ocean moderates climate. Seattle, Washington has much milder winters than Minneapolis, Minnesota, even though Seattle is slightly farther north. Air over Seattle draws heat from the adjacent Pacific Ocean, a heat source not available to Minneapolis. Mountains, valleys, and other surface features of the land affect climate too. Suppose you track a warm air mass after it picks up moisture off California's coast. It moves inland as wind from the west, and piles up against the Sierra Nevada, a high mountain range that parallels the coast. The air cools as it rises in altitude and loses moisture as rain (**FIGURE 29.6**). The result is a **rain shadow**, a semiarid or arid region of sparse rainfall on the leeward side of high mountains. *Leeward* is the side facing away from the wind. The Himalayas, Andes, Rockies, and other great mountain ranges cast similar rain shadows.

monsoon Wind that reverses direction seasonally.
rain shadow Dry region downwind of a coastal mountain range.

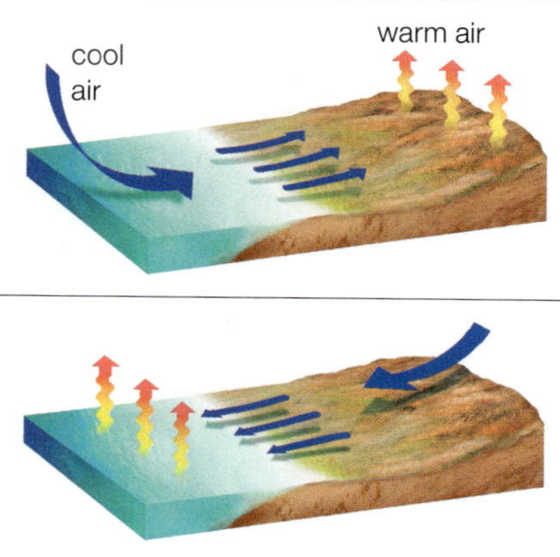

FIGURE 29.5 **Coastal breezes.**

TAKE-HOME MESSAGE 29.2

Surface ocean currents are set in motion by latitudinal differences in solar radiation. Currents are affected by winds and by Earth's rotation.

The collective effects of winds and ocean currents around landforms determine regional climate patterns.

FIGURE 29.6 {Animated} Rain shadow effect. On the side of mountains facing away from prevailing winds, rainfall is light. Black numbers signify annual precipitation, in centimeters, averaged on both sides of the Sierra Nevada, a mountain range. White numbers signify elevations, in meters.

29.3 WHAT ARE THE MAJOR BIOMES?

FIGURE 29.7 Major biomes. Climate is the main factor determining the distribution of biomes.

DIFFERENCES AMONG BIOMES

Biomes are areas of land characterized by their climate and type of vegetation (**FIGURE 29.7**). Most biomes consist of widely separated areas on different continents. For example, the temperate grassland biome includes areas of North American prairie, South African veld, South American pampa, and Eurasian steppe. Grasses and other nonwoody flowering plants constitute the bulk of the vegetation in all of these regions.

Rainfall and temperature are the main determinants of the type of biome in a given region. Desert biomes get the least annual rainfall, grasslands and shrublands get more, and forests get the most. Deserts occur where temperatures soar the highest and tundra where they drop the lowest.

Soils also influence biome distribution. Soils consist of a mixture of mineral particles and varying amounts of humus. Water and air fill spaces between soil particles. Soil properties vary depending on the types, proportions, and compaction of particles. Deserts have sandy or gravelly, fast-draining soil with little topsoil. Topsoil tends to be deepest in natural grasslands, where it can be more than one meter thick. For this reason, grasslands are often converted to agricultural uses.

Climate and soils affect primary production, so primary production varies among biomes.

SIMILARITIES WITHIN A BIOME

Unrelated species living in widely separated parts of a biome often have similar body structures that arose by the process of morphological convergence (Section 16.7). For example, cacti with water-storing stems that live in North American deserts are similar to euphorbs with water-storing stems that live in African deserts. Cacti and euphorbs do not share an ancestor with a water-storing stem. Rather, this feature evolved independently in the two groups as a result of living in similar environments. Similarly, an ability to carry out C4 photosynthesis evolved independently in grasses growing in warm grasslands on different continents. Under hot, dry conditions, C4 photosynthesis is more efficient than the more common C3 pathway (Section 6.5).

biome A region (often discontinuous) characterized by its climate and dominant vegetation.

TAKE-HOME MESSAGE 29.3

Biomes are vast expanses of land dominated by distinct kinds of plants that support characteristic communities. They vary in their primary productivity.

Evolution often produces similar solutions to environmental challenges in different regions of a biome.

29.4 WHAT IS THE MOST PRODUCTIVE BIOME?

FIGURE 29.8 Tropical rain forest. The graphic to the right shows the soil profile.

Tropical rain forests dominated by evergreen broadleaf trees form mainly between latitudes 10° north and south in equatorial Africa, the East Indies, Southeast Asia, South America, and Central America. Rain falls throughout the year and sums to an annual total of 130 to 200 centimeters (50 to 80 inches). The regular rains, combined with an average warm temperature of 25°C (77°F) and little variation in day length, allows photosynthesis to continue year-round. Of all land biomes, tropical forests have the greatest primary production. Per unit area, they remove more carbon from the atmosphere than other forests or grasslands.

Tropical rain forest is also the most structurally complex and species-rich biome. The forest has a multilayer structure (**FIGURE 29.8**). Its broadleaf trees can stand 30 meters (100 feet) tall. The trees often form a closed canopy that prevents most sunlight from reaching the forest floor. Vines and epiphytes (plants that grow on another plant, but do not withdraw nutrients from it) thrive in the shade beneath the canopy. Compared to other land biomes, tropical rain forests have the greatest variety and numbers of insects, as well as the most diverse collection of birds and primates. Age is key to this diversity: Tropical rain forest is the oldest modern biome. Some rain forests have existed for more than 50 million years, so there has been plenty of time for many evolutionary branchings to occur.

Trees in tropical rain forests shed leaves continually, but decomposition and mineral cycling happen so fast in this warm, moist environment that litter does not accumulate. The soil is highly weathered and heavily leached. Being a poor nutrient reservoir, this soil is not well suited to agriculture. Nevertheless, deforestation is an ongoing threat to tropical rain forests. Tropical forests are located in developing countries with fast-growing human populations who look to the forest as a source of lumber, fuel, and potential cropland. As human populations expand, more and more trees fall to the ax.

Deforestation in any region leaves fewer trees to remove carbon dioxide from the atmosphere. In rain forests, it also causes the extinction of species found nowhere else in the world. Among the potential losses are plants that make potentially life-saving chemicals. Two chemotherapy drugs, vincristine and vinblastine, were extracted from the rosy periwinkle, a low-growing plant native to Madagascar's rain forests. Today, these drugs help fight leukemia, lymphoma, breast cancer, and testicular cancer.

tropical rain forest Highly productive and species-rich biome in which year-round rains and warmth support continuous growth of evergreen broadleaf trees.

TAKE-HOME MESSAGE 29.4

Near the equator, year-round warmth and rains support tropical rain forests, the most productive, structurally complex, and species-rich biome.

29.5 WHAT TYPES OF FORESTS OCCUR IN COOLER CLIMATES?

A North American temperate deciduous forest in fall.

B Boreal forest (taiga) in Siberia.

FIGURE 29.9 **Cool forest biomes.**

TEMPERATE DECIDUOUS FORESTS

Temperate deciduous forests are dominated by broadleaf trees that lose all their leaves seasonally before a cold winter. Leaves often turn color before dropping (**FIGURE 29.9A**). Trees remain dormant while water is locked in snow and ice. In the spring, when conditions again favor growth, deciduous trees flower and put out new leaves. Also during the spring, leaves that were shed the prior autumn decay to form a rich humus. Rich soil and a somewhat open canopy that lets some sunlight through allows shorter understory plants to flourish.

Temperate deciduous forests are limited to the Northern Hemisphere, occurring in parts of eastern North America, western and central Europe, and areas of Asia, including Japan. In all these regions, 50 to 150 centimeters (about 20–60 inches) of precipitation falls throughout the year. Winters are cool and summers are warm.

North America has the most species-rich examples of this biome, with different tree species characterizing forests in different regions. For example, Appalachian forests include mainly oaks, whereas beeches and maples dominate Ohio's forests.

CONIFEROUS FORESTS

Conifers (trees with seed-bearing cones) are the main plants in coniferous forests. Although conifers do shed and replace their leaves, they do so continually, not all at once like deciduous trees. Conifer leaves are typically needle-shaped, with a thick cuticle and stomata that are sunk below the leaf surface. These adaptations help conifers conserve water during drought or times when the ground is frozen. As a group, conifers tolerate poorer soils and drier habitats than most broadleaf trees.

Conifers also withstand cold better than other trees. The most extensive land biome is the coniferous forest that sweeps across northern Asia, Europe, and North America (**FIGURE 29.9B**). It is known as **boreal forest**, or taiga, which means "swamp forest" in Russian. Pine, fir, and spruce predominate. Most rain falls in the summer. Winters are long, cold, and dry. Also in the Northern Hemisphere, montane coniferous forests extend southward through the great mountain ranges. Spruce and fir dominate at the highest elevations. At lower elevations, the mix becomes firs and pines. Conifers also dominate temperate lowlands along the Pacific coast from Alaska into northern California. These forests hold the world's tallest trees: Sitka spruce to the north, and coast redwoods to the south.

We find other conifer-dominated ecosystems in the eastern United States. About a quarter of New Jersey is pine barrens, a mixed forest of pitch pines and scrub oaks that grow in sandy, acidic soil. Pine forest covers about one-third of the Southeast. Fast-growing loblolly pines that dominate these forests are a major source of lumber.

boreal forest Extensive high-latitude forest of the Northern Hemisphere; conifers are the predominant vegetation.

temperate deciduous forest Northern Hemisphere biome in which the main plants are broadleaf trees that lose their leaves in fall and become dormant during cold winters.

TAKE-HOME MESSAGE 29.5

Temperate broadleaf forests grow in the Northern Hemisphere where cold winters prevent year-round growth.

Conifers dominate high-latitude boreal forests (taiga), which are Earth's most extensive biome.

Plants adapted to periodic lightning-ignited fires dominate grasslands, savanna, and chaparral.

Grasslands form in the interior of continents between deserts and temperate forests. Their soils are rich, with a deep layer of topsoil, so they are often converted to cropland. Annual rainfall is enough to keep desert from forming, but not enough to support woodlands. Low-growing grasses and other nonwoody plants tolerate strong winds, sparse and infrequent rain, and intervals of drought. Growth tends to be seasonal. Constant trimming by grazers, along with periodic fires, keeps trees and most shrubs from taking hold. When fire is suppressed by human activity, the low-growing grasses may be overgrown by woody plants.

North America's temperate grasslands, called prairies, once covered much of the continent's interior, where summers are hot and winters are cold and snowy. The prairies supported herds of elk, pronghorn antelope, and bison (**FIGURE 29.10A**) that were prey to wolves. Today, these predators and prey are largely absent from most of their former range. Nearly all prairies have been plowed under and now sustain production of wheat and other crops.

Savannas are broad belts of grasslands with a few scattered shrubs and trees. Savannas lie between the tropical forests and hot deserts of Africa, India, and Australia. In these regions, temperature remains high year-round, but rain falls seasonally. Africa's savannas are famous for their abundant wildlife. Herbivores include giraffes, zebras, a variety of antelopes, and immense herds of wildebeests (**FIGURE 29.10B**).

Chaparral is a biome dominated by drought-resistant, fire-adapted shrubs whose small, leathery leaves help them withstand drought. Chaparral occurs along the western coast of continents, between 30 and 40 degrees north or south latitude. Mild winters bring a moderate amount of rain, and the summer is hot and dry. Chaparral is California's most extensive ecosystem (**FIGURE 29.10C**). It also occurs in regions that border the Mediterranean, and in Chile, Australia, and South Africa. Many chaparral plants produce aromatic oils that help fend off insects, but also make them highly flammable. After a fire, plants resprout from roots and fire-resistant seeds germinate.

A In Kansas, tallgrass prairie with grazing bison.

B African savanna with grazing wildebeest.

C California chaparral dominated by shrubby plants with leathery leaves. Deer are the main grazers here.

FIGURE 29.10 Fire-adapted biomes.

chaparral Biome of dry shrubland in regions with hot, dry summers and cool, rainy winters.
grassland Biome in the interior of continents; perennial grasses and other nonwoody plants adapted to grazing and fire predominate.
savanna Biome dominated by perennial grasses with a few scattered shrubs and trees.

TAKE-HOME MESSAGE 29.6

Plants in grasslands, savanna, and chapparal have adaptations that allow them to withstand or recover after periodic fires.

Grasslands have a deep layer of topsoil and are often converted to farmland.

CREDITS: (10A) Joel Sartore/National Geographic Creative; (10B) Jonathan Scott/Planet Earth Pictures; (10C) Jack Wilburn/Animals Animals.

29.7 WHAT IS THE DRIEST BIOME?

Deserts receive an average of less than 10 centimeters (4 inches) of rain per year. They cover about one-fifth of Earth's land surface and many are located at about 30° north and south latitude, where dry air sinks. Rain shadows also reduce rainfall. For example, Chile's Atacama Desert is on the leeward side of the Andes, and the Himalayas prevent rain from falling in China's Gobi desert.

Lack of rainfall keeps the humidity in deserts low. With little water vapor to block the sun's rays, intense sunlight reaches and heats the ground. At night, the lack of insulating water vapor in the air allows the temperature to fall fast. As a result, deserts tend to have larger daily temperature shifts than other biomes. Desert soils have very little topsoil. They also tend to be somewhat salty, because rain that falls usually evaporates before seeping into the ground. Rapid evaporation allows any salt in rainwater to accumulate at the soil surface.

ADAPTED TO DROUGHT

Despite their harsh conditions, most deserts support some plant life. Desert plants often have spines or fuzz at their surface. In addition to deterring herbivory, these structures reduce water loss by trapping some water and thus keeping the humidity around the stomata high. Where rains fall seasonally, some plants reduce water loss by producing leaves only after a rain, then shedding them when dry conditions return. Other desert-adapted plants store water in their tissues. For example, the stem of a barrel cactus has a spongy pulp that holds water. The stem swells after a rain, then shrinks as the plant uses stored water.

Woody desert shrubs such as mesquite and creosote have extensive, efficient root systems that take up the little water that is available. Mesquite roots can tap into water that lies as much as 60 meters (197 feet) beneath the soil surface.

Alternative carbon-fixing pathways also help desert plants minimize water loss. Cacti, agaves, and euphorbs are CAM plants, which open their stomata only at night when temperature declines (Section 6.5). This reduces the water they lose to evaporation.

Most deserts contain a mix of annuals and perennials (**FIGURE 29.11**). The annuals are adapted to desert life by a life cycle that allows them to sprout and reproduce in the short time that the soil is moist.

DESERT CRUST

A desert crust forms at the surface of many desert soils. The crust is a community that can include cyanobacteria, lichens, mosses, and fungi (**FIGURE 29.12**). These organisms secrete organic molecules that glue them and the surrounding soil particles together. The resulting crust benefits members of the larger desert community in several ways. Bacteria in the crust fix nitrogen (Section 19.5), making this nutrient available to plants. The crust also holds soil particles in place. When the fragile connections within the desert crust are broken, the soil can blow away. Negative effects of such disturbance increase when windblown soil buries healthy crust in an undisturbed area, killing more crust organisms and allowing more soil to take flight.

FIGURE 29.11 Sonoran Desert after the rains. Perennial cacti are surrounded by annual wildflowers.

FIGURE 29.12 Desert crust.

desert Biome with little rain and low humidity; plants that have water-storing and water-conserving adaptations predominate.

TAKE-HOME MESSAGE 29.7

- The desert biome has low rainfall; poor, salty soil; and large swings in daily temperature.
- Desert plants have water-conserving adaptations.
- A diverse community of plants, fungi, and microorganisms holds desert soil particles together, forming a crustlike structure at the soil surface.

Arctic tundra extends between the ice cap of the North Pole and the belts of boreal forests in the Northern Hemisphere. Most of this biome in northern Russia and Canada. Arctic tundra is Earth's youngest modern biome; it first appeared about 10,000 years ago when glaciers retreated at the end of the last ice age.

Conditions in this biome are harsh; snow blankets the ground for as long as nine months of the year. Annual precipitation is usually less than 25 centimeters (10 inches), and cold temperature keeps the snow that does fall from melting. During a brief summer, plants grow fast under the nearly continuous sunlight (**FIGURE 29.13**). Lichens and shallow-rooted, low-growing plants are the main producers.

Even at midsummer, only the surface layer of tundra soil thaws. Below that lies **permafrost**, a layer of frozen soil that is as much as 500 meters (1,600 feet) thick in places. Permafrost acts as a barrier that prevents drainage, so the soil above it remains perpetually waterlogged. The cool, anaerobic conditions in this soil slow decay, so organic remains can build up. Organic matter in permafrost makes the arctic tundra one of Earth's greatest stores of carbon.

FIGURE 29.13 Arctic tundra in the summer.

---

arctic tundra Highest-latitude Northern Hemisphere biome, where low, cold-tolerant plants survive with only a brief growing season.
permafrost Continually frozen soil layer that lies beneath arctic tundra and prevents water from draining.

TAKE-HOME MESSAGE 29.8

Arctic tundra prevails at high latitudes, where short, cool summers alternate with long, cold winters.

Arctic tundra is the youngest biome, and organic matter frozen in the permafrost makes it one of Earth's greatest stores of carbon.

PEOPLE MATTER

National Geographic Explorer
DR. KATEY WALTER ANTHONY

Thawing permafrost releases methane, the same flammable gas used to heat homes and cook meals. In arctic lakes, the released methane bubbles up from the depths in a way that is hard to quantify—until the first clear ice of fall captures a snapshot of emissions from the lake in bubbles at its surface. To test whether a frozen bubble contains methane, ecologist Katey Walter Anthony has an assistant plunge a pick into a bubble while she holds a match near it. If the bubble is methane, the gas ignites, as shown in the photo above.

Walter Anthony says the amount of methane bubbling up out of arctic lakes is troubling, because some of it seems to be coming not from bottom mud but from deeper geologic reservoirs that had previously been securely capped by permafrost—and that contain hundreds of times more methane than is in the atmosphere now. By venting methane into the atmosphere, the lakes are amplifying the global warming that created them: Methane is a greenhouse gas that traps 25 times more heat than carbon dioxide. Increased Arctic methane release could amplify global warming over the next few centuries.

"If we could only capture the gas, it would make a great energy source," Walter Anthony says. Unlike coal, methane burns without spewing sulfur dioxide or mercury, or leaving ash behind.

29.9 WHAT ARE THE MAIN TYPES OF FRESHWATER ECOSYSTEMS?

With this section, we turn our attention to Earth's waters. We begin here with freshwater systems, continue to coasts in the next section, then dive into the oceans.

LAKES

A lake is a body of standing fresh water. If sufficiently deep, it will have zones that differ in their physical characteristics and species composition (**FIGURE 29.14**). Nearest shore is the littoral zone. Here, sunlight penetrates all the way to the lake bottom and aquatic plants are the main producers. A lake's open waters include an upper, well-lit limnetic zone, and a profundal zone where light does not penetrate. The main producers in the limnetic zone are members of the phytoplankton, a group of photosynthetic microorganisms that includes green algae, diatoms, and cyanobacteria. They serve as food for zooplankton, which are tiny consumers such as copepods. In the profundal zone, there is not enough light for photosynthesis, so consumers depend on food produced above. Debris that drifts down feeds detritivores and decomposers.

FIGURE 29.14 **Lake zonation.** A lake's littoral zone extends all around the shore to a depth where rooted aquatic plants stop growing. Its limnetic zone is the open water where light penetrates and photosynthesis occurs. Below the limnetic zone is the cooler, dark water of the profundal zone.

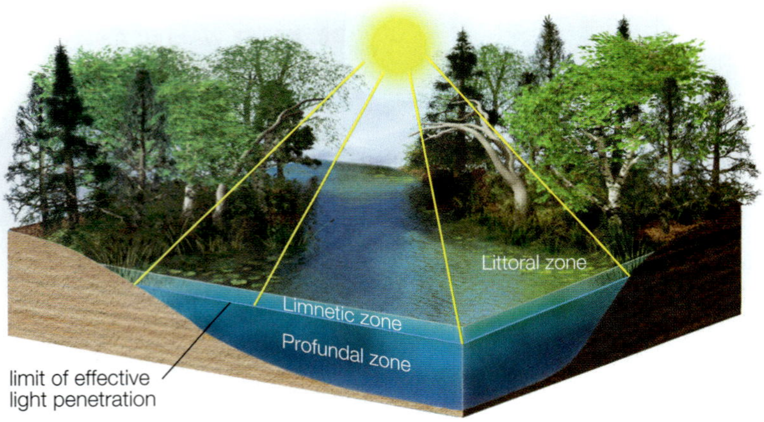

FIGURE 29.15 **An oligotrophic lake.** Crater Lake in Oregon is a collapsed volcano that filled with snowmelt. It began filling about 7,700 years ago; from a geologic standpoint, it is a young lake.

Nutrient Content and Succession Like a habitat on land, a lake undergoes succession; it changes over time. A newly formed lake is oligotrophic: deep, clear, and nutrient-poor, with low primary productivity (**FIGURE 29.15**). Over time, the lake becomes eutrophic. **Eutrophication** refers to processes, either natural or artificial, that enrich a body of water with nutrients. The increase in nutrients allows more producer growth, and primary productivity rises.

Seasonal Changes Temperate zone lakes undergo seasonal changes that affect primary productivity. Unlike most substances, water is not most dense in its solid state (ice). As water cools, its density increases until it reaches 4°C (39°F). Below this temperature, any additional cooling decreases water's density—which is why ice floats on water (Section 2.4). In an ice-covered lake, water just under the ice is near freezing and at its lowest density. The densest (4°C) water is at the bottom (**FIGURE 29.16A**).

In spring, winds cause currents that lead to a spring overturn, during which oxygen-rich water in the surface layers moves down and nutrient-rich water from the lake's depths moves up (**FIGURE 29.16B**). After the spring overturn, longer days and the dispersion of nutrients through the water encourage primary productivity.

In summer, a lake has three layers (**FIGURE 29.16C**). The upper layer is warm and oxygen-rich. Below this is a thermocline, a thin layer where temperature falls rapidly. Beneath the thermocline is the coolest water. The upper and lower waters on either side of this boundary do not mix. As a result, decomposers deplete oxygen dissolved near the lake bottom, and nutrients near the lake bottom cannot escape into surface waters. Nutrient shortages limit growth and production declines.

In autumn, the lake's upper waters cool, the thermocline vanishes, and a fall overturn occurs (**FIGURE 29.16D**). Oxygen-rich water moves down while nutrient-rich water moves up. This overturn brings nutrients to the surface and favors a brief burst of primary productivity. However, unlike the spring overturn, it does not lead to sustained production because decreasing light and temperature slow photosynthesis. Primary productivity will not peak again until after the next spring overturn.

STREAMS AND RIVERS

Flowing-water ecosystems start as freshwater springs or seeps. As water flows downslope, streams grow and merge. Rainfall, snowmelt, geography, altitude, and shade cast by plants affect flow volume and temperature. Minerals in rocks beneath the flowing water dissolve in it, affecting the water's solute concentrations. Because water in different parts of a river moves at different speeds, contains different solutes, and differs in temperature, the species composition of a river varies along its length (**FIGURE 29.17**).

THE ROLE OF DISSOLVED OXYGEN

The amount of oxygen dissolved in water is one of the most important factors affecting aquatic organisms. More oxygen dissolves in cooler, fast-flowing water than in warmer, still water. Thus, an increase in water temperature or decrease in its flow rate can cause aquatic species with high oxygen needs to suffocate.

In freshwater habitats, aquatic larvae of mayflies and stoneflies are the first invertebrates to disappear when the oxygen content of the water decreases. These insect larvae are active predators that demand considerable oxygen, so they serve as indicator species. Gilled snails disappear, too. Declines in populations of invertebrates can have cascading effects on the fishes that feed on them. Fishes can also be more directly affected. Trout and salmon are especially intolerant of low oxygen. Carp (including goldfish) are among the most tolerant; they survive tepid, stagnant water in ponds and tiny fish bowls.

No fishes can survive when the oxygen content of water falls below 4 parts per million. Leeches thrive as most competing invertebrates disappear. In waters with the lowest oxygen concentration, annelids called sludge worms (*Tubifex*) often are the only animals. The worms are colored red by their large amount of hemoglobin, which allows them to exploit low-oxygen habitats where predators and competition for food are scarce.

eutrophication Nutrient enrichment of an aquatic habitat.

TAKE-HOME MESSAGE 29.9

Within a lake, amounts of light, dissolved oxygen, and nutrients vary with depth. Primary productivity varies with a lake's age and—in temperate zones—with the season.

Rivers move nutrients into and out of ecosystems. Characteristics such as temperature and nutrient content usually vary along the length of a river.

Species differ in their dissolved oxygen needs. Cold, fast-moving water holds more oxygen than still, warm water.

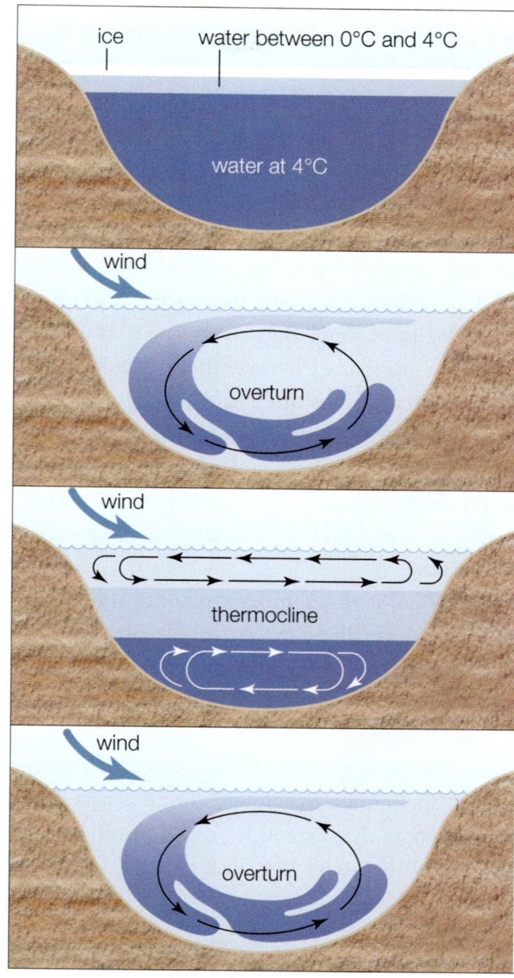

A Winter. Ice covers a thin layer of slightly warmer water just below it. Densest (4°C) water is at the bottom. Winds do not affect water under the ice, so there is little circulation.

B Spring. Ice thaws. Upper water warms to 4°C and sinks. Winds blowing across water create currents that help overturn water, bringing nutrients up from the bottom.

C Summer. Sun-warmed water floats on a thermocline, a layer across which temperature changes abruptly. Upper and lower water do not mix because of this thermal boundary.

D Fall. Upper water cools and sinks downward, eliminating the thermocline. Vertical currents mix water that was separated during the summer.

FIGURE 29.16 {Animated} Seasonal changes in a temperate zone lake.

FIGURE IT OUT: Which overturn results in the greatest rise in productivity?

Answer: The spring overturn, because increased nutrient availability is accompanied by increased light.

FIGURE 29.17 Effect of turbulence. As water flows over rocks, it picks up soluble minerals and becomes aerated, so it contains more oxygen.

29.10 WHAT ARE COASTAL ECOSYSTEMS?

COASTAL WETLANDS

An **estuary** is a partly enclosed body of water where fresh water from a river or rivers mixes with seawater. Seawater is denser than fresh water, so fresh water floats on top of the seawater where they meet. The size and shape of the estuary, and the rate at which freshwater flows into it, determine how quickly the saltwater and fresh water mix and the effects of tides. In all estuaries, an influx of water from upstream continually replenishes nutrients and allows a high level of productivity. Incoming fresh water also carries silt. Where the velocity of water flow slows, the silt falls to the bottom, forming mudflats. Photosynthetic bacteria and protists in biofilms on mudflats often account for a large portion of an estuary's primary production. Plants adapted to withstand changes in water level and salinity also serve as producers. The high salt content of estuary plants makes them unpalatable to most herbivores, so detrital food webs typically predominate.

Cordgrass (*Spartina*) is the dominant plant in the salt marshes of many estuaries along the Atlantic coast (**FIGURE 29.18A**). It is adapted to life in estuaries by an ability to withstand immersion during high tides and to tolerate salty, waterlogged, anaerobic soil.

"Mangrove" is the common term for salt-tolerant woody plants common in sheltered areas along tropical coasts. Prop roots, adventitious roots that extend from the trunk, help the plant stay upright (**FIGURE 29.18B**).

A Cordgrass (*Spartina*) in a South Carolina salt marsh. Salt taken up in water by roots is excreted by glands on the leaves.

B Mangroves along the shore in Florida. Specialized cells at the surface of some exposed prop roots allow gas exchange with air.

FIGURE 29.18 Two types of coastal wetlands.

FIGURE 29.19 Vertical zonation in the intertidal zone.

ROCKY AND SANDY SEASHORES

As with lakes, an ocean's shoreline is the littoral zone. This zone can be divided into three vertical regions that differ in their physical characteristics and species diversity (**FIGURE 29.19**). The upper littoral zone, which is also called the splash zone, regularly receives ocean spray but is submerged only during the highest of high tides. This zone gets the most sun, but has the fewest species. The midlittoral zone is covered by water during an average high tide and dry during a low tide. The lower littoral zone, exposed only during the lowest tide of the lunar cycle, is home to the most species.

You can easily see the zonation along a rocky shore. Multicelled algae ("seaweeds") that cling to rocks are the main producers, and grazing food chins predominate. Primary consumers include snails and sea urchins. Zonation is less obvious on sandy shores where detrital food chains start with material washed ashore. Some crustaceans eat detritus in the upper littoral zone. Nearer to the water, other invertebrates feed as they burrow through the sand.

estuary A highly productive ecosystem where nutrient-rich water from a river mixes with seawater.

TAKE-HOME MESSAGE 29.10

- Estuaries are highly productive areas where fresh water and seawater mix.
- Mangrove wetlands form along sheltered tropical coasts.
- Grazing food chains predominate on rocky shores, and detrital food chains on sandy shores.

WHAT ARE CORAL REEFS?

Coral reefs are wave-resistant formations that consist primarily of calcium carbonate secreted by many generations of coral polyps. Reef-forming corals live mainly in shallow, clear, warm waters between latitudes 25° north and 25° south. About 75 percent of all coral reefs are in the Indian and Pacific Oceans. A healthy reef is home to living corals and a huge number of other species (**FIGURE 29.20**). Biologists estimate that about a quarter of all marine fish species are associated with coral reefs.

Australia's Great Barrier Reef parallels Queensland for 2,500 kilometers (1,550 miles). This is the largest reef in the world, and it is also the largest example of biological architecture. Scientists estimate that it began forming about 600,000 years ago. Today, the Great Barrier Reef supports about 500 coral species, 3,000 fish species, 1,000 kinds of mollusks, and 40 kinds of sea snakes.

Photosynthetic dinoflagellates live inside the tissues of all reef-building corals (Section 23.4). Dinoflagellates live protected with the coral's tissues, where they receive plenty of carbon dioxide. In return, they provide the coral with oxygen and sugars.

Stress can cause a coral to expel its dinoflagellates. Because dinoflagellates give the coral its color, expelling these protists turns the coral white, an event called **coral bleaching**. When a coral is stressed for more than a short time, the dinoflagellate population in the coral's tissues cannot rebound and the coral dies, leaving its bleached hard parts behind (**FIGURE 29.21**).

The incidence of coral bleaching events has been increasing. Rising sea temperatures and sea level associated with global climate change most likely play a role. People also stress reefs by discharging sewage and other pollutants into coastal waters, by causing erosion that clouds water with sediments, and by destructive fishing practices. Fishing nets break pieces off corals. Fishermen hoping to capture reef fishes for the pet trade use explosives or sodium cyanide to stun the fishes, and destroy corals in the process. Invasive species also threaten reefs. Hawaiian reefs are threatened by exotic algae, including several species imported for cultivation during the 1970s.

Human-induced damage to reefs is taking a huge toll. For example, the Indo-Pacific region, the global center for reef diversity, lost about 3,000 square kilometers (1,160 square miles) of living coral reef each year between 1997 and 2003.

coral bleaching A coral expels its photosynthetic dinoflagellate symbionts in response to stress and becomes colorless.
coral reef Highly diverse marine ecosystem centered around reefs built by living corals that secrete calcium carbonate.

FIGURE 29.20 Healthy coral reef near Fiji. The coral gets its color from pigments of symbiotic dinoflagellates that live in its tissues and supply it with sugars.

FIGURE 29.21 "Bleached" reef near Australia. The coral skeletons shown here belong mainly to staghorn coral (*Acropora*), a genus especially likely to undergo coral bleaching.

TAKE-HOME MESSAGE 29.11

Coral reefs form by the action of living corals that lay down a calcium carbonate skeleton. Photosynthetic dinoflagellates in the coral's tissues are necessary for the coral's survival.

Rising water temperature, pollutants, fishing, and exotic species contribute to loss of reefs.

Declines in coral reefs will affect the enormous number of fishes and invertebrate species that make their home on or near the reefs.

29.12 WHERE IS LIFE IN THE OPEN OCEAN?

FIGURE 29.22 Life at a hydrothermal vent on the seafloor. Giant tube worms are the most conspicuous members of this deep sea community. These annelids can grow more than 2 meters in length. The bright red plume that extends from the worm's tube gets its color from hemoglobin, the same pigment in your red blood cells. The red plume captures oxygen and dissolved sulfur from the seawater around it. As an adult, the worm never eats. Rather, sulfur absorbed by the worm serves as the energy source for chemoautotrophic bacteria that live inside it and provide it with sugars. The worms in turn serve as food for crabs that nibble on their exposed plumes.

The ocean's open waters are the **pelagic province**. This province includes the water over continental shelves and the more extensive waters farther offshore. In the ocean's upper, bright waters, phytoplankton such as single-celled algae and bacteria are the primary producers, and grazing food chains predominate. Depending on the region, some light may penetrate as far as 1,000 meters (more than a half mile) beneath the sea surface. Below that, organisms live in continual darkness, and organic material that drifts down from above serves as the basis of detrital food chains.

The **benthic province** is the ocean bottom, its rocks, and sediments. Species richness is greatest on continental shelves (the underwater edges of continents). The benthic province also includes largely unexplored species-rich regions, including seamounts and hydrothermal vents.

Seamounts are undersea mountains that stand 1,000 meters or more tall, but are still below the sea surface. They attract large numbers of fishes and are home to many marine invertebrates. Like islands, seamounts often are home to species that evolved there and live nowhere else.

At **hydrothermal vents**, hot water rich in dissolved minerals spews out from an opening on the ocean floor. The water is seawater that seeped into cracks in the ocean floor at the margins of tectonic plates and was heated by heat energy from within the Earth. Minerals in this water settle out when it mixes with the cold deep-sea water. Chemoautotrophic bacteria and archaea that obtain energy by removing electrons from minerals are the main producers in food webs that include diverse invertebrates, including large numbers of tube worms (**FIGURE 29.22**).

Life exists even in the deepest sea. A remote-controlled submersible that sampled sediments in the deepest part of the ocean (the Mariana Trench) brought up foraminifera that live 11 kilometers (7 miles) below the surface. Sediment samples from deep in the Mediterranean Sea turned up another surprise, tiny animals that do not use oxygen. The animals, called loriciferans, live between sand grains and are distant relatives of insects and nematodes. They are the only animals known to live their lives entirely without oxygen.

benthic province The ocean's sediments and rocks.
hydrothermal vent Place where hot, mineral-rich water streams out from an underwater opening in Earth's crust.
pelagic province The ocean's open waters.
seamount An undersea mountain.

TAKE-HOME MESSAGE 29.12

In the pelagic province's upper waters, photosynthesis supports grazing food chains. In deeper, darker waters of this province, organisms feed mainly on detritus that drifts down from above.

The benthic province has pockets of high species diversity at undersea mountains (seamounts) and near hydrothermal vents. A hydrothermal vent ecosystem does not run on energy from the sun; the producers are chemoautotrophs rather than photoautotrophs.

29.13 Application: EFFECTS OF EL NIÑO

FIGURE 29.23 Satellite photos showing the effect of El Niño on primary productivity in the Pacific Ocean.

FLUCTUATIONS IN CLIMATE INFLUENCE THE DISTRIBUTION AND ABUNDANCE OF ORGANISMS. Consider the effects of El Niño, a recurring climate event in which equatorial waters of the eastern and central Pacific Ocean warm above their average temperature. The term El Niño means "baby boy" and refers to Jesus; it was first used by Peruvian fishermen to describe local weather changes and a shortage of fishes that occurred in some years around Christmas. Scientists now know that during an El Niño, marine currents interact with the atmosphere in ways that influence weather patterns worldwide.

During an El Niño, unusually warm water flows toward eastern Pacific coasts, displacing currents that would otherwise bring up nutrients from the deep. Without these nutrients, marine primary producers decline in numbers (FIGURE 29.23). The dwindling producer populations and warming water cause a decrease in populations of small, cold-water fishes, as well as the larger consumers that rely on them, such as seals and sea lions.

An El Niño causes rainfall patterns to shift worldwide. In the El Niño winter of 1997–1998, torrential rains caused flooding and landslides along eastern Pacific coasts, while Australia and Indonesia suffered from drought-driven crop failures and raging wildfires. An El Niño typically brings cooler, wetter weather to the American Gulf states, and reduces the likelihood of hurricanes.

An El Niño usually persists for 6 to 18 months. It may be followed by an interval in which the temperature of the eastern Pacific remains near its average, or by a La Niña. During a La Niña, eastern Pacific waters become cooler than average. As a result, the west coast of the United States gets little rainfall and the likelihood of hurricanes in the Atlantic increases.

Outbreaks of human disease often occur during an El Niño. For example, the increased ocean temperature in the Pacific leads to an increased incidence of cholera. Copepods, a type of small crustacean, serve as a reservoir for cholera-causing bacteria between disease outbreaks. During an El Niño, the rise in the temperature of the ocean's surface results in a rise in the number of cholera-carrying copepods. An El Niño also brings an increase in malaria to coastal communities in South Asia and Latin America.

The United States National Oceanographic and Atmospheric Administration (NOAA) monitors sea surface temperature and studies El Niño events. NOAA's goal is to determine how El Niño affects global weather patterns and the extent of its effects. Such studies could help us develop a method of predicting when an El Niño or La Niña event is likely to occur and which regions are at a heightened risk for flooding, drought, hurricanes, or epidemics as a result. Predicting and planning for such occurrences could help prevent or minimize their harmful effects. The current data on sea surface temperature, as well as information about the monitoring program, are available on NOAA's website at www.elnino.noaa.gov.

Summary

SECTIONS 29.1, 29.2 Global air circulation patterns affect **climate** and the distribution of communities within the biosphere. Air is set in motion when sunlight heats tropical regions more than higher latitudes. Ocean currents distribute heat worldwide and influence weather patterns. Interactions between ocean currents, air currents, and landforms determine where regional phenomena such as **rain shadows** or **monsoons** occur.

SECTION 29.3 **Biomes** are characterized by a particular type of vegetation. Many biomes include multiple discontinuous areas. Climate and soil properties affect the distribution of biomes.

SECTIONS 29.4, 29.5 The broadleaf evergreen trees that predominate in **tropical rain forests** grow all year. These enormously productive forests are home to a large number of species. Tropical rain forest is the oldest existing biome. Trees in **temperate deciduous forests** shed their leaves all at once just before a cold winter that prevents growth. Conifers that dominate Northern Hemisphere high-latitude boreal forests withstand cold and drought better than broadleaf trees.

SECTION 29.6 **Grasslands** dominated by plants adapted to fire and grazing form in the somewhat moist interior of midlatitude continents. **Savannas** include fire-adapted grasses and scattered shrubs. Shrubby, fire-adapted **chaparral** is common in California and other coastal regions with hot, dry summers and cool, wet winters.

SECTION 29.7 Deserts form in regions with little precipitation and widely fluctuating temperature. Drought-adapted plants dominate. Desert crust holds soil particles in place and provides plants with nutrients.

SECTION 29.8 The Northern Hemisphere's **arctic tundra** is dominated by short plants that grow only during the brief, cool summer when daylight is abundant. Tundra is the youngest biome, and its **permafrost** is a great reservoir of carbon.

SECTION 29.9 Gradients in light, temperature, dissolved gases, and nutrients affect the distribution of species in aquatic ecosystems. Lakes undergo a natural process of **eutrophication**, becoming more productive over time. In temperate zone lakes, a spring overturn and a fall overturn cause vertical mixing of waters.

SECTIONS 29.10–29.12 Nutrient-rich fresh water mixes with seawater in an **estuary**. Mangrove wetlands form on sheltered tropical coasts. Along rocky shores, multicellular algae form the base for grazing food chains. On sandy shores, detrital food chains predominate.

Accumulated calcium carbonate skeletons of coral polyps form **coral reefs**. Photosynthetic dinoflagellates in the corals are producers for this ecosystem. Many marine species associate with reefs. When stressed, a coral may eject its photosynthetic symbionts, an event called **coral bleaching**.

Life exists throughout the ocean. In the **pelagic province**, grazing food chains predominate in sunlit waters. Detritus forms the base for food chains in deeper, darker waters. **Seamounts** are regions of high diversity in the **benthic province**. At **hydrothermal vents**, chemoautotrophic bacteria and archaea are the producers.

SECTION 29.13 Interactions among Earth's air and waters affect weather worldwide and have effects on human health, as when the Pacific warms during an El Niño and then cools during a La Niña.

Self-Quiz

Answers in Appendix VII

1. The Northern Hemisphere is most tilted toward the sun in ________ .
   a. spring b. summer c. autumn d. winter

2. Which latitude will have the most hours of daylight on the summer solstice?
   a. 0° (the equator) c. 45° north
   b. 30° north d. 60° north

3. Warm air ________ and it holds ________ water than cold air.
   a. sinks; less c. sinks; more
   b. rises; less d. rises; more

4. A rain shadow is a reduction in rainfall ________ .
   a. on the inland side of a coastal mountain range
   b. during an El Niño event
   c. that results from global warming

5. The Gulf Stream is a current that flows ________ along the eastern coast of the United States.
   a. north to south b. south to north

6. ________ have a deep layer of humus-rich topsoil.
   a. Deserts c. Rain forests
   b. Grasslands d. Seamounts

7. Biomes differ in their ________ .
   a. climate c. soils
   b. dominant plants d. all of the above

8. Grasslands most often are found ________ .
   a. at 30° north and south c. in interior of continents
   b. at high altitudes d. all of the above

9. Permafrost underlies ________ .
   a. arctic tundra c. boreal forest
   b. temperate forest d. all of the above

10. The warmer water is, the ________ oxygen it can hold.
    a. more b. less

11. Chemoautotrophic bacteria and archaea are the primary producers for food webs ________ .
    a. in mangrove wetlands c. on coral reefs
    b. at seamounts d. at hydrothermal vents

Data Analysis Activities

Changing Sea Temperatures In an effort to predict El Niño or La Niña events in the near future, the National Oceanographic and Atmospheric Administration collects information about sea surface temperature (SST) and atmospheric conditions. Scientists compare monthly temperature averages in the eastern equatorial Pacific Ocean to historical data and calculate the difference (degree of anomaly) to see if El Niño conditions, La Niña conditions, or neutral conditions are developing. El Niño is a rise in the average SST above 0.5°C. A decline of the same amount is La Niña. **FIGURE 29.24** shows data for nearly 39 years.

1. When did the greatest positive temperature deviation occur during this time period?
2. What type of event, if any, occurred during the winter of 1982–1983? What about the winter of 2001–2002?
3. During a La Niña event, less rain than normal falls in the American West and Southwest. In the time interval shown, what was the longest interval without a La Niña event?

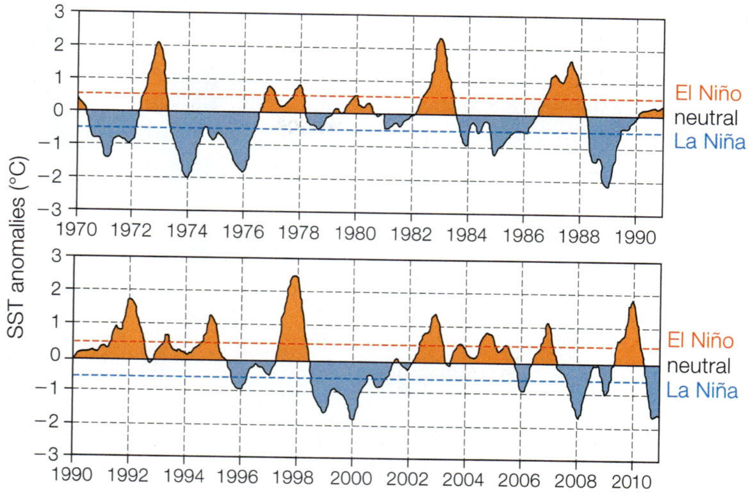

FIGURE 29.24 Sea surface temperature anomalies (differences from the historical mean) in the eastern equatorial Pacific Ocean. A rise above the dashed red line is an El Niño event; a decline below the blue line is La Niña.

4. What type of conditions were in effect in the fall of 2007 when California suffered severe wildfires?

12. Corals rely on symbiotic ________ for sugars.
    a. fungi
    b. bacteria
    c. dinoflagellates
    d. green algae

13. Which of the following biomes borders on boreal forest?
    a. savanna
    b. taiga
    c. tundra
    d. chaparral

14. Unrelated species living in geographically separated parts of a biome may resemble one another as a result of ________ .
    a. competitive interactions
    b. morphological convergence
    c. morphological divergence
    d. coevolution

15. Match the terms with the most suitable description.

___ tundra	a. broadleaf forest near equator
___ chaparral	b. partly enclosed by land; where fresh water and seawater mix
___ desert	c. African grassland with trees
___ savanna	d. low-growing plants at high latitudes or elevations
___ estuary	e. dry shrubland
___ boreal forest	f. at latitudes 30° north and south
___ prairie	g. mineral-rich, superheated water supports communities
___ tropical rain forest	h. conifers dominate
___ hydrothermal vents	i. North American grassland

Critical Thinking

1. On April 26, 1986, a meltdown occurred at the Chernobyl nuclear power plant in Ukraine. Nuclear fuel burned for nearly ten days and released 400 times more radioactive material than the atomic bomb that dropped on Hiroshima. Winds carried radioactive fallout around the globe. By 1998, the rate of thyroid abnormalities in children living downwind from the site was nearly seven times as high as for those upwind; their thyroid gland concentrated the iodine radioisotopes. Chernoboyl is at 51° north latitude. In what direction did the major winds carry the fallout after the accident?

2. Owners of off-road recreational vehicles would like increased access to government-owned deserts. Some argue that it is the perfect place for off-roaders because "There's nothing there." Do you agree?

3. Rita Colwell, the scientist who discovered why cholera outbreaks often occur during an El Niño, is concerned that global climate change could increase the incidence of this disease. By what mechanism might global warming cause an increase in cholera outbreaks?

Young elephants and their keepers at David Sheldrick Wildlife Trust in Kenya. Poaching and human–wildlife conflicts left the elephants motherless at an early age. They will need human care for eight to ten years before they can survive in the wild.

30 HUMAN EFFECTS ON THE BIOSPHERE

Links to Earlier Concepts

This chapter considers the causes of an ongoing mass extinction (Section 16.6) in light of human population growth (26.6). We look again at effects of species introductions and will draw on your knowledge of pH (2.5), the ozone layer (18.4), and water and nutrient cycles (28.5–28.8).

KEY CONCEPTS

AN EXTINCTION CRISIS

Humans have increased the frequency of extinctions by overharvesting, and by habitat degradation and fragmentation. The extent of species losses is not fully known.

HARMFUL LAND USES

Plowing grasslands and cutting forests have long-term and long-range effects. By allowing soil erosion and affecting rainfall patterns, these practices make it difficult to restore plant cover.

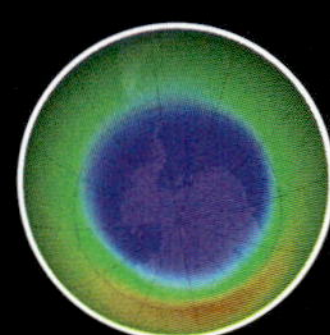

EFFECTS OF POLLUTANTS

Some airborne pollutants fall to Earth in acid rain. Others harm the protective ozone layer or contribute to global climate change.

CONSERVING BIODIVERSITY

Earth's biodiversity is the product of billions of years of evolution. Conservation biologists prioritize which areas to protect by assessing which are most threatened and most biodiverse.

REDUCING NEGATIVE IMPACTS

Extraction of fuel and other nonrenewable resources harms the environment. Using such resources carefully can help reduce threats to biodiversity.

Photograph by Michael Nichols , National Geographic Creative.

30.1 WHY ARE SO MANY SPECIES THREATENED OR ENDANGERED?

A White abalone. B Pyne's ground plum. C Texas blind salamander. D Florida perforate reindeer lichen.

FIGURE 30.1 Examples of endangered species native to the United States. To learn more about these and other threatened and endangered species, visit the United States Fish and Wildlife Service's endangered species website at www.fws.gov/endangered/.

Extinction, like speciation, is a natural process. Species arise and become extinct on an ongoing basis. Scientists estimate that 99 percent of all species that have ever lived are now extinct. The rate of extinction picks up dramatically during a mass extinction, when many kinds of organisms in many different habitats become extinct in a relatively short period. We are currently in the midst of such an event. Unlike most previous mass extinctions, this one is not the inevitable result of a physical catastrophe such as a volcanic eruption or asteroid impact. Humans are the driving force behind the current rise in extinctions and our actions will determine the extent of the losses.

An **endangered species** is a species that faces extinction in all or part of its range. A **threatened species** is one that is likely to become endangered in the near future. Keep in mind that not all rare species are threatened or endangered. Some species have always been uncommon. A species is considered endangered when one or more of its populations have declined or are declining.

CAUSES OF SPECIES DECLINE

When European settlers first arrived in North America, they found between 3 and 5 billion passenger pigeons. In the 1800s, commercial hunting caused a steep decline in the bird's numbers. The last time anyone saw a wild passenger pigeon was 1900, and he shot it. The last captive member of the species died in 1914.

We continue to overharvest species. The crash of the Atlantic codfish population, described in Section 26.5, is one recent example. Another is the fate of the white abalone, a gastropod mollusk native to kelp forests off the coast of California (**FIGURE 30.1A**). Heavy harvesting of this species during the 1970s reduced the population to about 1 percent of its original size. In 2001, it became the first invertebrate to be listed as endangered by the United States Fish and Wildlife Service. Although some white abalone remain in the wild, population density remains too low for effective reproduction. The species' only hope for survival is a program of captive breeding. If this program succeeds, individuals will be reintroduced to the wild.

Species are overharvested not only as food, but also for use in traditional medicine, for the pet trade, and for ornamentation. Some orchids prized by collectors have become nearly extinct in the wild. Most of the orphan elephants shown in the chapter opening photo lost their mother to poachers who kill to obtain ivory tusks. The majority of elephant tusks harvested this way end up in China, in the form of decorative carved objects.

Overharvest directly reduces population size, but humans also affect species indirectly by altering their habitat. Many species requires a highly specific type of habitat, and any degradation, fragmentation, or destruction of that habitat reduces population numbers.

An **endemic species** remains confined to the area in which it evolved. Such species are more likely to go extinct as a result of habitat degradation than species with a more widespread distribution. Consider Pyne's ground plum (**FIGURE 30.1B**), a flowering plant that lives only in cedar glades near a rapidly growing city in Tennessee. The plant is threatened by conversion of its habitat to homes and industrial use. Texas blind salamanders (**FIGURE 30.1C**) are among the species endemic to Edwards Aquifer, a series of water-filled, underground limestone formations. Excessive withdrawals of water, along with water pollution, threaten the salamander and other species in the aquifer. A lichen endemic to Florida is endangered by development of its scrubland habitat (**FIGURE 30.1D**).

Deliberate or accidental species introductions can also pose a threat (Section 27.7). Rats that reached islands by stowing away on ships attack and endanger many ground-nesting birds that evolved in the absence of egg-eating ground predators. Exotic species also cause problems by outcompeting native ones. For example, California's native golden trout declined after European brown trout and eastern brook trout were introduced into California's mountain streams for sport fishing.

CREDITS: (1A) John Butler, NOAA; (1B, D) Joel Sartore/National Geographic Creative; (1C) Joe Fries, U.S. Fish & Wildlife Service.

Decline or extinction of one species can endanger others. Consider running buffalo clover and the buffalo that graze on it. Both were once common in the Midwest. The plants thrived in the open woodlands where soil was enriched by buffalo droppings and periodically disturbed by the animals' hooves. Buffalo helped to disperse the clover's seeds. When buffalo were hunted to near extinction, buffalo clover populations declined as well.

THE UNKNOWN LOSSES

The International Union for Conservation of Nature and Natural Resources (IUCN) monitors threats to species worldwide. Its species listings have historically focused on vertebrates. Scientists have only recently begun to consider the threats to invertebrates and to plants. Our impact on protists and fungi is largely unknown, and the IUCN does not address threats to bacteria or archaea.

Microbiologist Tom Curtis is among those making a plea for increased research on microbial ecology and microbial diversity. He argues that we have barely begun to comprehend the vast number of microbial species and to understand their importance. Curtis writes, "I make no apologies for putting microorganisms on a pedestal above all other living things. For if the last blue whale choked to death on the last panda, it would be disastrous but not the end of the world. But if we accidentally poisoned the last two species of ammonia-oxidizers, that would be another matter. It could be happening now and we wouldn't even know . . ." Ammonia-oxidizing bacteria play an essential role in the nitrogen cycle by converting ammonia in wastes and remains to nitrites (Section 28.7). Without them, wastes would pile up and plants would not have access to the nitrogen they need to grow.

endangered species Species that faces extinction in all or a part of its range.
endemic species Species that remains restricted to the area where it evolved.
threatened species Species likely to become endangered in the near future.

TAKE-HOME MESSAGE 30.1

- Species often decline when humans destroy or fragment natural habitat by converting it to human use, or degrade it through pollution or withdrawal of an essential resource.
- Humans also directly cause declines by overharvesting species.
- Global travel and trade can introduce exotic species that harm native ones.
- The number of endangered species remains largely unknown.

PEOPLE MATTER

National Geographic Explorer
DR. PAULA KAHUMBU

Linking conservationists with members of the public and with one another is one of Paula Kahumbu's main goals. She says, "Conservationists do crucial work on a shoestring, cut off from the rest of the world. They're in remote, isolated places, some even risking their lives, with no chance of getting on the international radar screen. Meanwhile, millions of people who care about the catastrophic loss of wildlife and habitats aren't sure how to help." Kahumbu is executive director of WildlifeDirect (http://wildlifedirect.org), an online platform that gives conservationists a way to share day-by-day challenges and victories via blogs, diaries, videos, photos, and podcasts. Thanks to her efforts, people concerned about wildlife and wild places can view problems in real time and track the impact of their own contributions. The site attracts thousands of visitors daily, with online donations going directly to projects across Africa, Asia, and South America.

Information from conservationists involved with WildlifeDirect helped Kahumbu recognize an important conservation issue in Africa—misuse of the insecticide carbofuran. This chemical, which is banned for use on food crops in the United States and for all uses in Europe, remains widely available in Africa. Reports flowing into Wildlife Direct revealed that some farmers were using carbofuran-laced carcasses to poison lions that they believed threatened livestock. This practice caused declines in endangered lions, hyenas, and vultures. Others put carbofuran into irrigated rice fields to deliberately kill water birds for consumption as human food. Although both harmful practices have declined as a result of the efforts of Kahumbu and other conservationists, she continues to work toward a ban on the sale of carbofuran in Africa.

Kahumbu began her conservation work with research on African elephants and she remains a strong advocate for their protection. She favors a global ban on all trade in ivory, increased efforts to detect smuggled ivory, and stiffer penalties for those who take part in any aspect of the ivory trade.

CREDIT: (in text) Rebecca Hale/National Geographic Creative.

30.2 WHAT ARE THE EFFECTS OF DESERTIFICATION AND DEFORESTATION?

With this section, we begin a survey of some of the ways that human activities threaten species by destroying or degrading habitats.

DESERTIFICATION

Deserts naturally expand and contract over geological time as climate conditions vary. However, human activities sometimes result in the rapid conversion of a grassland or woodland to desert, a process called **desertification**. As human populations increase, greater numbers of people are forced to farm in areas that are ill-suited to agriculture. Others allow livestock to overgraze in grasslands. In both cases, desertification can occur.

A well-documented instance of desertification occurred in the United States during the mid-1930s, when large portions of prairie on the southern Great Plains were plowed under to plant crops. This plowing exposed deep prairie topsoil to the force of the region's constant winds. Then came a drought, and the result was an economic and ecological disaster. Winds carried more than a billion tons of topsoil aloft as sky-darkening dust clouds turned the region into what came to be known as the Dust Bowl (**FIGURE 30.2A**). Tons of displaced soil fell to earth as far away as New York City and Washington, D.C.

Today, Africa's Sahara desert is expanding south into the Sahel region. Overgrazing in this region strips grasslands of their vegetation and allows winds to erode the soil. Winds carry the soil aloft and westward (**FIGURE 30.2B**). Soil particles land as far away as the southern United States and the Caribbean. In China's northwestern regions, overplowing and overgrazing have expanded the Gobi desert so that dust clouds periodically darken skies above Beijing. Winds carry some of the dust across the Pacific to the west coast of the United States.

Drought encourages desertification, which results in more drought in a positive feedback cycle. Plants cannot thrive in a region where the topsoil has blown away. With less transpiration, less water enters the atmosphere, so local rainfall decreases.

The best way to prevent desertification is to avoid farming in areas subject to high winds and periodic drought. If these areas must be used, methods that do not repeatedly disturb the soil can minimize risk of desertification.

DEFORESTATION

The amount of forested land is currently stable or increasing in North America, Europe, and China, but tropical forests continue to disappear at an alarming rate. In Brazil, increases in the export of soybeans and free-range beef have helped make the country the world's seventh-largest economy. However, this economic expansion has come at the expense of the country's woodlands and forests (**FIGURE 30.3**).

Deforestation has detrimental effects beyond the immediate destruction of forest organisms. For example, deforestation encourages flooding because water runs off into streams, rather than being taken up by tree roots. Deforestation also raises risk of landslides in hilly areas. Tree roots tend to stabilize the soil. When they are removed, waterlogged soil becomes more likely to slide.

Soils of deforested areas become nutrient-poor because of the increased loss of nutrient ions in runoff. **FIGURE**

FIGURE 30.2 Dust storms, one outcome of desertification.

A Dust cloud in the Great Plains during the 1930s.

B Dust blows across the Atlantic from North Africa.

FIGURE 30.3 Deforestation. Cropland that once was rainforest extends right up to the edge of Iguaçu National Park, a World Heritage Site that straddles the border between Brazil and Argentina. Threatened primates and large cats are among the species that rely on this forest habitat.

30.4 shows results of an experiment in which scientists deforested a region in New Hampshire and monitored the nutrient content of runoff. Deforestation caused a spike in loss of essential soil nutrients such as calcium.

Like desertification, deforestation affects local climate. The loss of plants means reduced transpiration, so the amount of local rainfall declines. In shady forests, transpiration also results in evaporative cooling. When a forest is cut down, shade disappears and the evaporative cooling ceases. Thus, the temperature in a deforested area is typically higher than in an adjacent forested area.

Once a tropical forest has been logged, the resulting nutrient losses and drier, hotter conditions can make it impossible for tree seeds to germinate or for seedlings to survive. Thus, deforestation can be difficult to reverse.

Because forests take up and store huge amounts of carbon dioxide, ongoing forest losses also contribute to global climate change by increasing the atmospheric concentration of this greenhouse gas.

desertification Conversion of dry grassland to desert.

TAKE-HOME MESSAGE 30.2

Desertification occurs when excess plowing or grazing of a grassland causes soil to blow away. With fewer plants, rainfall declines.

Deforestation increases flooding and loss of soil nutrients, raises the local temperature, and decreases rainfall. Changes in soil and temperature produced by deforestation make it difficult for new trees to become established.

FIGURE 30.4 {Animated} Effect of experimental deforestation on nutrient losses from soil. After deforestation, calcium (Ca) levels in runoff increased sixfold (gray). An undisturbed plot in the same forest showed no increase during this time (green).

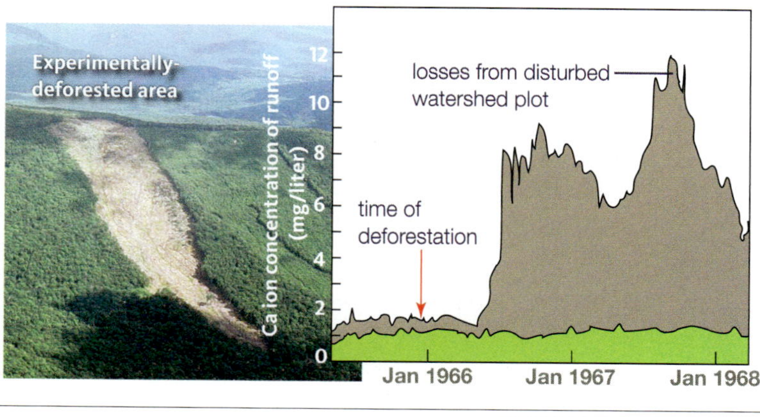

A Juvenile sea lion with ring of discarded plastic around its neck. As the animal grows, the plastic will cut into its neck, causing a wound and impairing its ability to feed.

B Recently deceased Laysan albatross chick, dissected to reveal the contents of its gut. Scientists found more than 300 pieces of plastic inside the bird. One of the pieces had punctured its gut wall, resulting in its death. The chick was fed the plastic by its parents, who gathered the material from the ocean surface, mistaking it for food.

FIGURE 30.5 Perils of plastic.

Seven billion people use and discard a lot of stuff. Where does all the waste go? Historically, unwanted material was buried in the ground or dumped out at sea. Trash was out of sight, and also out of mind.

We now know that burying garbage can contaminate groundwaters, as when lead from discarded batteries seeps into the ground. We also know that solid waste dumped into oceans harms marine life (**FIGURE 30.5**). In the United States, solid municipal waste can no longer legally be dumped at sea. Nevertheless, plastic constantly enters our coastal waters. Foam cups and containers from fast-food outlets, plastic shopping bags, plastic water bottles, and other litter ends up in storm drains. From there it is carried to streams and rivers that can convey it to the sea. A seawater sample taken near the mouth of the San Gabriel River in southern California had 128 times as much plastic as plankton by weight.

Once in the ocean, trash can persist for a surprisingly long time. Components of a disposable diaper will last for more than 100 years, as will fishing line. A plastic bag will be around for more than 50 years, and a cigarette filter for more than 10.

Ocean currents can carry bits of plastic for thousands of miles. These plastic bits can end up accumulating in some areas of the ocean. Consider the Great Pacific Garbage Patch, a region of the north central Pacific that the media often describes as an "island of trash." In fact, the plastic is not easily visible. Rather, the garbage patch is a region where a high concentration of confetti-like plastic particles swirl slowly around an area as large as the state of Texas. The small bits of plastic absorb and concentrate toxic compounds such as pesticides and industrial chemicals from the seawater around them, making the plastic all the more harmful to marine organisms that mistakenly eat it. Scientists recently estimated that fish living in mid-depth water of the north central Pacific consume as much as 240,000 tons of this chemically tainted plastic each year.

You can help reduce the impact of plastic trash by choosing more durable objects over disposable ones, and avoiding plastic products when other, less environmentally harmful alternatives exist. If you use plastic, be sure to recycle or dispose of it properly.

TAKE-HOME MESSAGE 30.3

Plastic trash often ends up in the ocean, where it harms marine life.

You can minimize your environmental impact by avoiding disposable plastic goods and by recycling.

CREDITS: (5A) Frans Lanting/National Geographic Creative; (5B) Clair Fackler/NOAA.

Plastic trash is one example of a pollutant. **Pollutants** are natural or man-made substances released into soil, air, or water in greater than natural amounts. The presence of a pollutant disrupts the physiological processes of organisms that evolved in its absence, or that are adapted to lower levels of it. Some pollutants come from a few distinct sites, or point sources. Pollutants that come from point sources are usually the easiest to control: Identify the few sources of the pollutant, and you can take action there. It is more difficult to deal with pollution from nonpoint sources, which are more numerous and widely dispersed.

Sulfur dioxide and nitrogen oxide gases are common air pollutants. Most sulfur dioxide pollution comes from point sources—coal-burning power plants and smelters (factories that extract metals from ore). Nitrogen oxides come largely from nonpoint sources such as cars and other vehicles that burn gas and oil, and the agricultural use of synthetic, nitrogen-rich fertilizers.

Sulfur dioxide and nitrogen oxides coat dust particles when the weather is dry. Dry acid deposition occurs when this coated dust falls to the ground. Wet acid deposition, or **acid rain**, occurs when pollutants react with gases and water vapor in air and fall as acidic precipitation. The pH of unpolluted rainwater is about 5.6 (Section 2.5). The pH of acid rain can be as low as 2. In the United States, federal regulations limiting sulfur dioxide emissions have helped reduce the acidity of precipitation (**FIGURE 30.6A**). The world's main sulfur dioxide emitters are now China and India, where industrialization and coal use continue to rise.

Acid rain that falls on or drains into waterways, ponds, and lakes affects aquatic organisms. For example, heightened acidity impairs the growth of diatoms, which serve as producers in many lakes. The high acidity of some streams in the northeastern United States and Canada has contributed to a decline in populations of Atlantic salmon. These salmon breed and lay their eggs in streams, but spend most of their life in the ocean. A young salmon exposed to acidic water cannot make the physiological changes necessary for a transition to life in saltwater.

Acid rain that falls on forests burns tree leaves and increases the pH of soil water. As acidic water drains through soil, positively charged hydrogen ions displace positively charged nutrient ions such as calcium, so soil loses these nutrients. The acidity also causes soil particles to release metals such as aluminum that can harm plants. The combination of nutrient-poor soil and exposure to toxic metals weakens trees, making them more susceptible to insects and pathogens, and thus more likely to die (**FIGURE 30.6B**). Effects are most pronounced at high elevations where trees are frequently exposed to clouds of acidic fog.

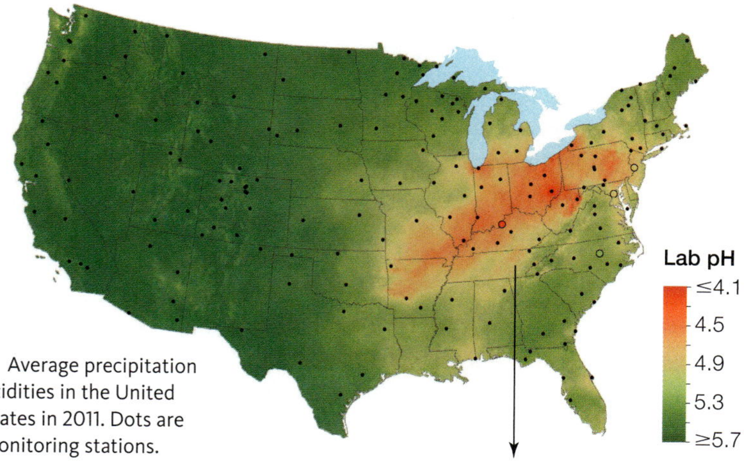

A Average precipitation acidities in the United States in 2011. Dots are monitoring stations.

B Dying trees in Great Smoky Mountains National Park, where acid rain harms leaves and causes loss of nutrients from soil.

FIGURE 30.6 {Animated} Acid rain in the United States.

FIGURE IT OUT: Is rain more acidic on the East Coast or the West Coast?

Answer: The East Coast

acid rain Low-pH rain that forms when sulfur dioxide and nitrogen oxides mix with water vapor in the atmosphere.

pollutant A substance that is released into the environment by human activities and interferes with the function of organisms that evolved in the absence of the substance or with lower levels.

TAKE-HOME MESSAGE 30.4

Burning fossil fuels, metal production, and use of synthetic nitrogen fertilizer contribute to the acidification of rain and to dry acid deposition.

Increased acidity in water and soil impairs the health and development of many types of organisms.

30.5 WHAT ARE THE EFFECTS OF OZONE DEPLETION AND POLLUTION?

DEPLETION OF THE OZONE LAYER

In the upper layers of the atmosphere, between 17 and 27 kilometers (10.5 and 17 miles) above sea level, the ozone (O_3) concentration is so great that scientists refer to this region as the **ozone layer**. The ozone layer benefits living organisms by absorbing most ultraviolet (UV) radiation from incoming sunlight. UV radiation, remember, damages DNA and causes mutations (Section 8.5).

In the mid-1970s, scientists noticed that Earth's ozone layer was thinning. Its thickness had always varied a bit with the season, but now the average level was declining steadily from year to year. By the mid-1980s, the spring ozone thinning over Antarctica was so pronounced that people began referring to the lowest-ozone region as an "ozone hole" (**FIGURE 30.7A**).

Declining ozone quickly became an international concern. With a thinner ozone layer, people would be exposed to more UV radiation, the main cause of skin cancers. Higher UV levels also harm wildlife, which do not have the option of avoiding sunlight. In addition, exposure to higher-than-normal UV levels affects plants and other producers, slowing the rate of photosynthesis and the release of oxygen into the atmosphere.

Chlorofluorocarbons, or CFCs, are the main ozone destroyers. These odorless gases were once widely used as propellants in aerosol cans, as coolants, and in solvents and plastic foam. In response to the potential threat posed by the thinning ozone layer, countries worldwide agreed in 1987 to phase out the production of CFCs and other ozone-destroying chemicals. As a result of that agreement (the Montreal Protocol), the concentrations of CFCs in the atmosphere are no longer rising dramatically (**FIGURE 30.7B**). However, CFCs break down quite slowly, so scientists expect them to remain at a level that significantly thins the ozone layer for several decades.

NEAR-GROUND OZONE POLLUTION

Near the ground, where ozone levels are naturally low, ozone is considered a pollutant. Ground-level ozone forms when nitrogen oxides and volatile organic compounds released by burning or evaporating fossil fuels are exposed to sunlight. Warm temperature speeds the reaction. Thus, ground-level ozone tends to vary daily (being higher in the daytime) and seasonally (being higher during the summer).

Ozone does not persist for long, so ozone emitted into the lower atmosphere never makes it to the ozone layer where would be useful. In the lower atmosphere, this strong oxidizing agent irritates the eyes and respiratory tracts of people and wildlife and interferes with plant growth.

You can help reduce ozone pollution by avoiding actions that put fossil fuels or their combustion products into the air at times that favor ozone production. For example, on hot, sunny, still days, postpone filling your gas tank or using gasoline-powered appliances such as lawn mowers until the evening, when there is less sunlight to power the conversion of pollutants to ozone.

A Ozone levels in the upper atmosphere in September 2012, the Antarctic spring.

Purple indicates the least ozone, with blue, green, and yellow indicating increasingly higher levels.

Check the current status of the ozone hole at NASA's website (http://ozonewatch.gsfc.nasa.gov/).

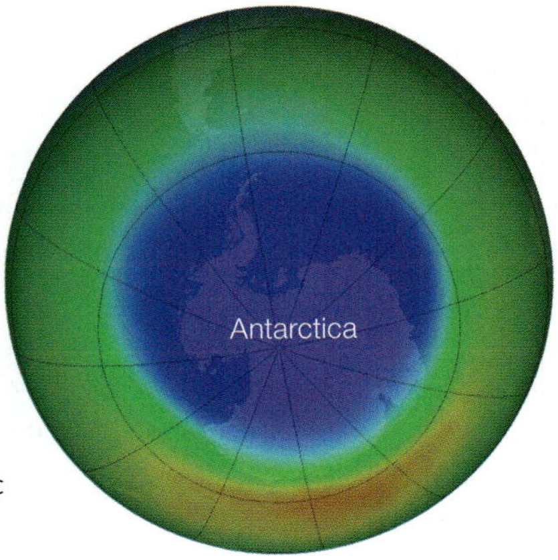

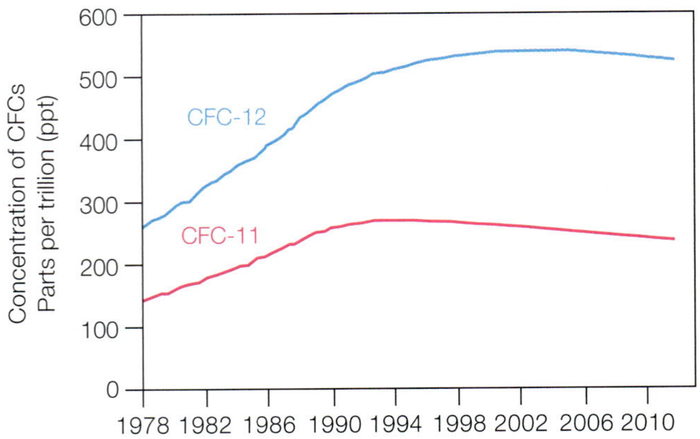

B Concentration of two CFCs in the upper atmosphere. These pollutants destroy ozone. A worldwide ban on CFCs has successfully halted the rise in atmospheric CFC concentration.

FIGURE 30.7 Destruction of the ozone layer.

ozone layer High atmospheric layer with a high concentration of ozone (O_3) that prevents much ultraviolet radiation from reaching Earth's surface.

TAKE-HOME MESSAGE 30.5

Certain synthetic chemicals destroy ozone in the upper atmosphere's protective ozone layer. This layer shields us and other life from UV radiation in sunlight.

Evaporation and burning of fossil fuels increase the amount of ozone in the lower atmosphere, where the gas is considered a harmful pollutant.

CREDITS: (7A) NASA Ozone Watch; (7B) From www.esrl.noaa.gov.

30.6 WHAT ARE THE EFFECTS OF GLOBAL CLIMATE CHANGE?

FIGURE 30.9 Melting glaciers, one sign of a warming world. Water from melting glaciers contributes to rising sea level.

Ongoing climate change affects ecosystems worldwide. How is the global climate changing? Most notably, average temperatures are increasing (**FIGURE 30.8**). Warming is more pronounced at temperate and polar latitudes than at the equator.

Rising temperature elevates sea level by two mechanisms. Water expands as it is heated, and heating also melts sea ice and glaciers (**FIGURE 30.9**). Together, thermal expansion and the addition of meltwater from glaciers cause sea level to rise. In the past century, the sea level has risen about 20 centimeters (8 inches). As a result, some coastal wetlands are disappearing underwater.

The warming climate is already having widespread effects on biological systems. Temperature changes are important cues for many temperate zone species. Abnormally warm spring temperatures are causing deciduous trees to put leaves out earlier, and spring-blooming plants to flower earlier. Animal migration times and breeding seasons are also shifting. Species arrays in biological communities are changing as warmer temperatures allow some species to expand their range to higher latitudes or elevations. Of course, not all species can move or spread quickly, and warmer temperatures are expected to drive some of these species to extinction. For example, warming of tropical waters is already stressing reef-building corals and increasing the frequency of coral bleaching events.

FIGURE 30.8 Rising temperature. The temperature anomaly is the deviation from average temperature in the period 1951–1980.

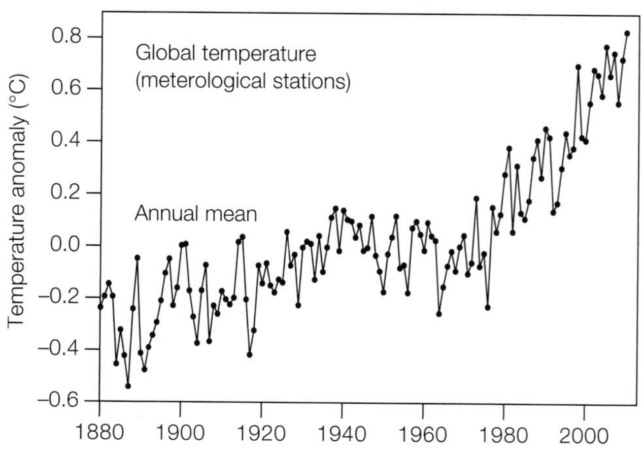

Global warming is just one aspect of global climate change. Temperature affects evaporation, winds, and currents, so many weather patterns are expected to change as global temperature continues to rise. For example, warmer temperatures are correlated with extremes in rainfall patterns: periods of drought interrupted by unusually heavy rains. In addition, warmer seas tend to increase the intensity of hurricanes.

As Section 28.6 explained, the most widely accepted explanation for global climate change is the rising levels of greenhouse gases such as carbon dioxide. Fossil fuel combustion is the single biggest source of greenhouse gas emissions, and the use of these fuels is still rising as large nations such as China and India become increasingly industrialized. Reducing greenhouse gas emissions will be a challenge, but efforts are under way to increase the efficiency of processes that require fossil fuels, to shift to alternative energy sources such as solar and wind power, and to develop innovative ways to store carbon dioxide.

TAKE-HOME MESSAGE 30.6

The rise in global temperature is causing the sea level to rise, and is affecting weather patterns.

These changes are altering the range of some species, threatening others with extinction, and disrupting the structure of biological communities.

CREDITS: (8) NASA Goddard Institute for Space Studies; (9) National Snow and Ice Data Center, W. O. Field; (9B) National Snow and Ice Data Center, B. F. Molnia.

30.7 WHAT IS CONSERVATION BIOLOGY?

THE VALUE OF BIODIVERSITY

Every nation has some amount of biological wealth, which we call **biodiversity**. A region's biodiversity is measured at three levels: the genetic diversity within species, species diversity, and ecosystem diversity. Biodiversity is currently declining at all three levels, in all regions.

Conservation biology addresses these declines. The goals of this relatively new field of biology are (1) to survey the range of biodiversity, and (2) to find ways to maintain and use biodiversity to benefit human populations by encouraging people to value their region's natural resources and use those resources in nondestructive ways.

Why should we protect biodiversity? From a selfish standpoint, doing so is an investment in our future. Healthy ecosystems are essential to the survival of our species. Other organisms produce the oxygen we breathe and the food we eat. They remove waste carbon dioxide from the air and decompose and detoxify wastes. Plants take up rain and hold soil in place, preventing erosion and reducing the risk of flooding. We are still discovering medically valuable compounds produced by wild species. Wild relatives of crop plants are reservoirs of genetic diversity that plant breeders draw on to protect and improve crops.

There are ethical reasons to preserve biodiversity too. All living species are the result of an ongoing evolutionary process that stretches back billions of years. Each species has a unique combination of traits, and extinction removes that collection of traits from the world forever.

SETTING PRIORITIES

Protecting biological diversity is often a tricky proposition. Even in developed countries, people often oppose environmental protections because they fear such measures will have adverse economic consequences. However, taking care of the environment can make good economic sense. With a bit of planning, people can both preserve and profit from their biological wealth.

The resources available for conserving areas are limited, so conservation biologists must often make difficult choices about which areas should be targeted for protection first. These biologists identify **hot spots**, places that are home to species found nowhere else and are under great threat of destruction. Once identified, hot spots can take priority in worldwide conservation efforts.

On a broader scale, conservation biologists define ecoregions, which are land or aquatic regions characterized by climate, geography, and the species found within them. The most widely used ecoregion system was developed by scientists of the World Wildlife Fund and defines 867 distinctive land ecoregions that they hope to maintain. **FIGURE 30.10** shows the locations and conservation status of ecoregions. Those that are critical or endangered are considered the top priority for conservation efforts.

The Klamath–Siskiyou forest in southwestern Oregon and northwestern California is one of North America's endangered ecoregions (**FIGURE 30.11**). It is home to many rare conifers. Two endangered birds, the northern

FIGURE 30.10 The location and conservation status of the land ecoregions deemed most important by the World Wildlife Fund.

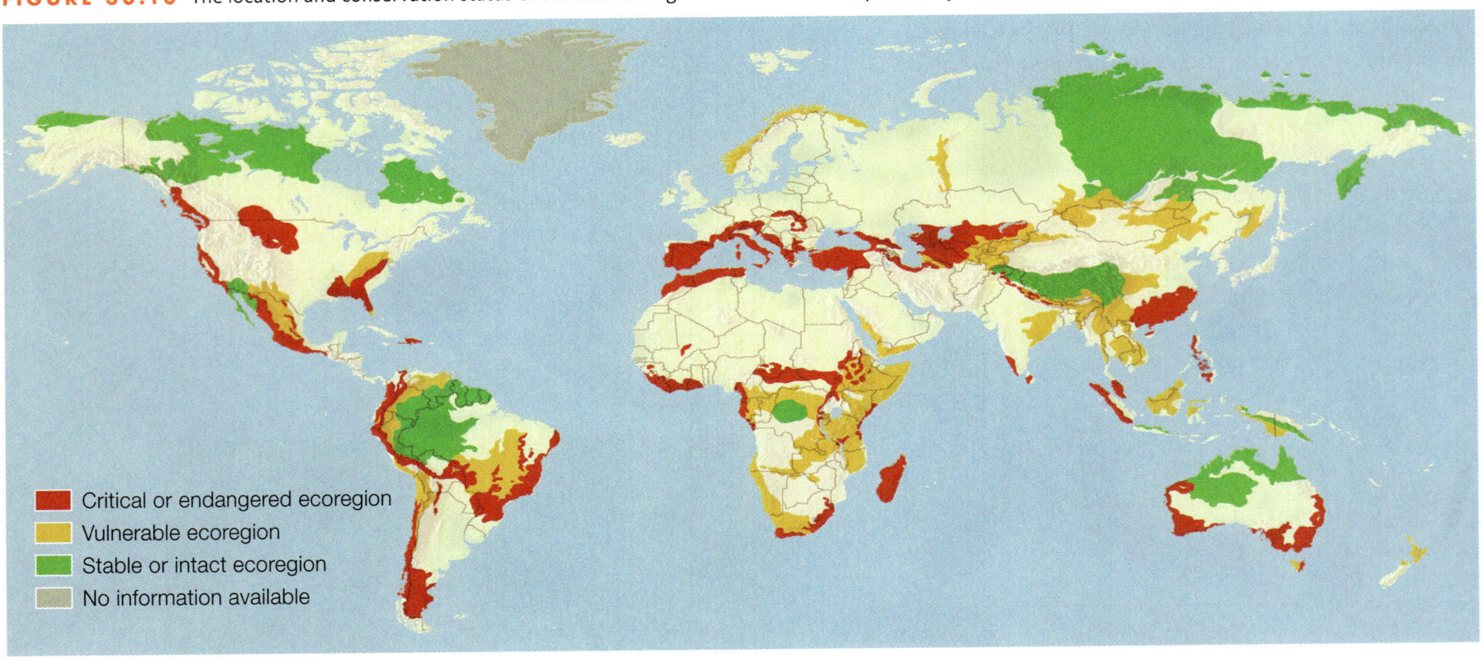

spotted owl and the marbled murrelet, nest in old-growth parts of the forest, and endangered coho salmon breed in streams that run through the forest. Logging threatens all of these species.

By focusing on hot spots and critical ecoregions rather than on individual endangered species, scientists hope to maintain ecosystem processes that naturally sustain biological diversity.

PRESERVATION AND RESTORATION

Worldwide, many ecologically important regions have been protected in ways that benefit local people. The Monteverde Cloud Forest in Costa Rica is one example. During the 1970s, George Powell was studying birds in this forest, which was rapidly being cleared. Powell decided to buy part of the forest as a nature sanctuary. His efforts inspired individuals and conservation groups to donate funds, and much of the forest is now protected as a private nature reserve. The reserve's plants and animals include more than 100 mammal species, 400 bird species, and 120 species of amphibians and reptiles. It is one of the few habitats left for jaguars and ocelots. A tourism industry centered on the reserve provides economic benefits to local people.

Sometimes an ecosystem is so damaged, or there is so little of it left, that conservation alone is not enough to sustain biodiversity. **Ecological restoration** is work designed to bring about the renewal of a natural ecosystem that has been degraded or destroyed.

For example, ecological restoration is occurring in Louisiana's coastal wetlands. More than 40 percent of the coastal wetlands in the United States are in Louisiana. These marshes are an ecological and economic treasure, but they are in trouble. Dams and levees built upstream of the marshes keep back sediments that would normally replenish sediments lost to the sea. Channels cut through the marshes for oil exploration and extraction have encouraged erosion, and the rising sea level threatens to flood the existing plants. Since the 1940s, Louisiana has lost an area of marshland the size of Rhode Island. Restoration efforts now under way aim to reverse some of those losses (**FIGURE 30.12**).

FIGURE 30.11 Klamath–Siskiyou forest, one of North America's critical ecoregions. Endangered northern spotted owls (inset) are endemic to this coniferous forest.

FIGURE 30.12 Ecological restoration in Louisiana's Sabine National Wildlife Refuge. Where marshland has become open water, sediments are barged in and marsh grasses are planted on them. The squares are new sediment with marsh grass.

biodiversity Of a region, the genetic variation within species, variety of species, and variety of ecosystems.
conservation biology Field of applied biology that surveys biodiversity and seeks ways to maintain and use it nondestructively.
ecological restoration Actively altering an area in an effort to restore an ecosystem that has been damaged or destroyed.
hot spot Threatened region that is habitat for species not found elsewhere and is considered a high priority for conservation efforts.

TAKE-HOME MESSAGE 30.7

Biodiversity is the genetic diversity of individuals of a species, the variety of species, and the variety of ecosystems.

Conservation biologists identify threatened regions that contain species not found elsewhere and prioritize which should be first to receive protection.

Through ecological restoration, we restore a biologically diverse ecosystem that has been destroyed or degraded.

CREDITS: (11) main, David Patte, USFWS; inset, USFWS; (12) Diane Borden-Bilot, U.S. Fish and Wildlife Service.

30.8 HOW CAN WE REDUCE OUR IMPACT?

A Bingham copper mine near Salt Lake City, Utah. This open pit mine is 4 kilometers (2.5 miles) wide and 1,200 meters (0.75 miles) deep, the largest man-made excavation on Earth.

B Pelican covered with oil that accidentally escaped from a deep sea-drilling platform in the Gulf of Mexico.

FIGURE 30.13 Environmental costs of resource extraction.

Ultimately, the health of life on Earth depends on our ability to recognize that the principles of energy flow and of resource limitation, which govern the survival of all systems of life, do not change. We must take note of these principles and find a way to live within our limits. The goal is living sustainably, which means meeting the needs of the present generation without reducing the ability of future generations to meet their own needs.

Promoting sustainability begins with recognizing the environmental consequences of one's own lifestyle. People in industrial nations use enormous quantities of resources, and the extraction, delivery, and use of these resources has negative effects on biodiversity. In the United States, the size of the average family has declined since the 1950s, while the size of the average home has doubled. All of the materials used to build and furnish those larger homes come from the environment. For example, an average new home contains about 500 pounds of copper in its wiring and plumbing.

Where does copper come from? Like most other nonrenewable mineral elements used in manufacturing, most copper is mined from the ground (**FIGURE 30.13A**). Surface mining strips an area of vegetation and soil, creating an ecological dead zone. Mining puts dust into the air, creates mountains of rocky waste, and can contaminate nearby waterways.

Minerals are mined worldwide and globalization makes it difficult to know the source of the raw materials in products you buy. Keep in mind that resource extraction in developing countries is often carried out under regulations that are less strict or less stringently enforced than those in the United States. As a result, the environmental impact of mining is even greater in these countries.

Nonrenewable mineral resources are used in electronic devices such as phones, computers, televisions, and MP3 players. Constantly trading up to the newest device may be good for the ego and the economy, but it is bad for the environment. Reducing consumption by fixing existing products is a sustainable resource use, as is recycling. Obtaining nonrenewable materials by recycling reduces the need for extraction of those resources, and it also helps keep materials out of landfills.

Reducing your energy use is another way to promote sustainability. Fossil fuels such as petroleum, natural gas, and coal supply most of the energy used by developed countries. You already know that burning these nonrenewable fuels contributes to global warming and acid rain. In addition, extracting and transporting these fuels can have negative impacts, as when oil spills harm aquatic species (**FIGURE 30.13B**).

Renewable energy sources have their own drawbacks. For example, dams in rivers of the Pacific Northwest generate renewable hydroelectric power, but they also prevent endangered salmon from returning to streams above the dam to breed. Similarly, wind turbines can harm birds and bats. Manufacture of panels used to collect solar energy requires using nonrenewable mineral resources, and production of the panels generates pollutants.

In short, all commercially produced energy has negative environmental impacts, so the best way to minimize your impact is to use less energy.

If you want to make more of a difference, learn about the threats to ecosystems in your own area. Support efforts to preserve and restore local biodiversity. Many ecological restoration projects are supervised by trained biologists but carried out primarily through the efforts of volunteers (**FIGURE 30.14**).

FIGURE 30.14 Volunteers restoring the Little Salmon River in Idaho so that salmon can migrate upstream to their breeding grounds.

TAKE-HOME MESSAGE 30.8

Extraction of material and energy for usage has effects that threaten biodiversity. Reducing energy consumption and recycling and reusing materials help minimize our impact.

30.9 Application: A LONG REACH

FIGURE 30.15 Polar bears investigate an American submarine that surfaced in ice-covered Arctic waters.

WE BEGAN THIS BOOK WITH A STORY OF BIOLOGISTS WHO VENTURED INTO A REMOTE NEW GUINEA FOREST, AND THEIR EXCITEMENT AT THE MANY PREVIOUSLY UNKNOWN SPECIES THAT THEY DISCOVERED (SECTION 1.9). At the far end of the globe, a U.S. submarine surfaced in Arctic waters and found polar bears on the ice-covered sea. The polar bears were about 445 kilometers (270 miles) from the North Pole and 805 kilometers (500 miles) from the nearest land (**FIGURE 30.15**).

Even such seemingly remote regions are no longer beyond the reach of human explorers—and human influence. In the Arctic, unusually warm temperatures are affecting the seasonal cycle of sea ice melting and formation. In recent years, sea ice has begun to thin and to break up earlier in the spring and to form later in the fall. A decrease in the persistence of sea ice is bad news for polar bears. They can only reach their main prey—seals—by traveling across ice. A longer ice-free period means less time for bears to feed.

We have only recently come to realize the effects that our actions can have on other species. A century ago, Earth's biological resources seemed inexhaustible. Now we know that many practices we began while largely ignorant of how natural systems operate take a heavy toll on the biosphere. It would be presumptuous to think that we alone have had a profound impact on the world of life. As long ago as the Proterozoic, photosynthetic cells were irrevocably changing the course of evolution by enriching the atmosphere with oxygen. Over life's existence, the evolutionary success of some groups ensured the decline of others. What is new is the increasing pace of change and the capacity of our own species to recognize and perhaps to moderate its role in this increase.

CREDITS: (14) Mountain Visions/NOAA; (15) U.S. Navy photo by Chief Yeoman Alphanso Braggs.

Summary

SECTION 30.1 Extinction is a natural process, but we are in the midst of a human-caused mass extinction, with numbers of **threatened species** and **endangered species** rising. **Endemic species** are especially likely to be at risk. Overharvesting, species introductions, and habitat destruction, degradation, and fragmentation can push species toward extinction. We know about only a tiny fraction of the species that are currently under threat.

SECTION 30.2 Overplowing or overgrazing of grassland can cause **desertification**. Both desertification and deforestation affect soil properties and can alter rainfall patterns. Changes caused by deforestation are especially difficult to reverse.

SECTION 30.3 Human populations discard large amounts of trash. Chemicals that seep out of buried trash can contaminate groundwater. Trash, especially plastic, that washes into or is dumped into oceans poses a threat to marine life.

SECTION 30.4 Burning coal releases sulfur dioxide, and burning oil or gas releases nitrogen oxides. These **pollutants** react with gases and water vapor in air, then fall to earth as **acid rain**. The resulting increase in the acidity of soils and waters can sicken or kill organisms.

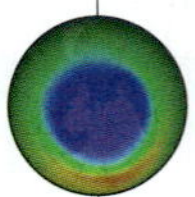

SECTION 30.5 The **ozone layer** in the upper atmosphere protects against incoming UV radiation. Chemicals known as CFCs were banned when they were found to cause thinning of the ozone layer. Near the ground, where the ozone concentration is naturally low, ozone emitted as a result of fossil fuel use is considered a pollutant. It irritates animal respiratory tracts and interferes with photosynthesis by plants.

SECTION 30.6 Global climate change resulting from an increase in greenhouse gases is causing glaciers to melt, thus raising the sea level. It also is affecting the range of species, allowing some to move into higher elevations or latitudes. Other species such as corals are showing signs of temperature-related stress. In addition, global climate change is expected to alter rainfall patterns and the intensity of hurricanes.

SECTION 30.7 Genetic diversity, species diversity, and ecosystem diversity are components of **biodiversity**, which is declining in all regions. **Conservation biology** involves surveying biodiversity and developing strategies to protect it and allow its sustainable use. Because resources are limited, biologists often focus on **hot spots**, where many unique species are under threat. They also attempt to ensure that portions of all ecoregions are protected. When an ecosystem has been totally or partially degraded, **ecological restoration** can help restore biodiversity.

SECTION 30.8 Extraction of nonrenewable mineral resources and fuels have detrimental effects on an ecosystem. Individuals can help sustain biodiversity by limiting their energy use and by reducing resource consumption through reuse and recycling.

SECTION 30.9 Human activities affect other species, even in remote places such as the Arctic. For example, polar bears in the Arctic are threatened by thinning sea ice, which is one effect of global climate change. Although humans are not the first species to change conditions on Earth in a way that harms other species, we are the first to do so with an understanding of our effects and while having the option of altering our behavior.

Self-Quiz

Answers in Appendix VII

1. A(n) ______ species has population levels so low it is at great risk of extinction in the near future.
   a. endemic
   b. endangered
   c. threatened
   d. indicator

2. Species are threatened by habitat ______ .
   a. fragmentation
   b. degradation
   c. destruction
   d. all of the above

3. Deforestation ______ .
   a. increases mineral runoff from soil
   b. decreases local temperature
   c. increases local rainfall
   d. all of the above

4. Sulfur dioxide released by coal-burning power plants contributes to ______ .
   a. ozone destruction
   b. sea level rise
   c. acid rain
   d. desertification

5. The "hole" in the ozone layer is most pronounced in ______ over ______ .
   a. fall; the Arctic
   b. fall; the equator
   c. spring; Antarctica
   d. spring; the equator

6. An increase in the size of the ozone hole would be expected to ______ .
   a. increase skin cancers
   b. reduce respiratory disorders
   c. both a and b

7. A large amount of plastic has accumulated in a region of the north central Pacific as a result of ______ .
   a. global warming
   b. ocean currents
   c. ozone depletion
   d. acid rain

8. Global climate change is causing ______ .
   a. a decrease in sea level
   b. glacial melting
   c. acid rain
   d. all of the above

9. The Montreal Protocol banned use of ______ , which contribute(s) to ozone depletion.
   a. DDT
   b. CFCs
   c. fossil fuels
   d. sulfur dioxides

Data Analysis Activities

Arctic PCB Pollution Winds carry chemical contaminants produced and released at temperate latitudes to the Arctic, where the chemicals enter food webs. As a result of biological magnification, top carnivores in arctic food webs, including people and polar bears, end up with high concentrations of these chemicals. Arctic people who eat a lot of local wildlife tend to have unusually high levels of industrial chemicals called polychlorinated biphenyls, or PCBs, in their body. The Arctic Monitoring and Assessment Programme studies the effects of these industrial chemicals on the health and reproduction of Arctic people. **FIGURE 30.16** shows how sex ratio at birth varies with average maternal PCB levels among people native to the Russian Arctic.

1. Which sex was more common in offspring of women with less than 1 microgram per milliliter of PCB in serum?
2. At what PCB concentrations were women more likely to have daughters?
3. In some villages in Greenland, nearly all recent newborns are female. Would you expect PCB levels in those villages to be above or under 4 micrograms per milliliter?

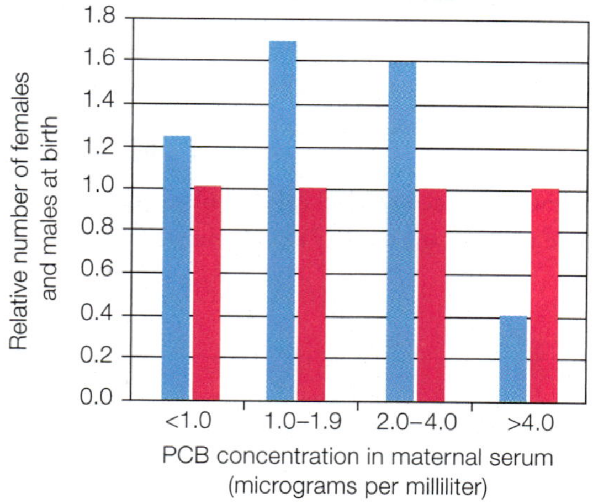

FIGURE 30.16 Effect of maternal PCB concentration on sex ratio of newborns in human populations native to the Russian Arctic. Blue bars indicate the relative number of males born per one female (pink bars).

10. A highly threatened region that is home to many unique species is a(n) _______ .
    a. ecoregion b. biome c. hot spot d. community

11. Biodiversity refers to _______ .
    a. genetic diversity c. ecosystem diversity
    b. species diversity d. all of the above

12. Restoring a marsh that has been damaged by human activities is an example of _______ .
    a. biological magnification c. ecological restoration
    b. bioaccumulation d. globalization

13. Individuals help sustain biodiversity by _______ .
    a. reducing resource consumption
    b. reusing materials
    c. recycling materials
    d. all of the above

14. Match the terms with the most suitable description.

___ hot spot	a. good up high; bad nearby
___ ozone	b. tree loss alters rainfall pattern and is difficult to reverse
___ biodiversity	c. can increase dust storms
___ acid rain	d. evolved in one region and remains there
___ endemic species	e. coal-burning is major cause
___ nonpoint source of pollution	f. involves release of pollutant in many areas
___ global climate change	g. has unique threatened species
___ deforestation	h. cause of rising sea level
___ desertification	i. genetic, species, and ecosystem diversity

Critical Thinking

1. In one seaside community in New Jersey, the U.S. Fish and Wildlife Service suggested trapping and removing feral cats (domestic cats that live in the wild). The goal was to protect some endangered wild birds (plovers) that nested on the town's beaches. Many residents were angered by the proposal, arguing that the cats have as much right to be there as the birds. Do you agree? Why or why not?

2. Burning fossil fuel puts excess carbon dioxide into the atmosphere, but deforestation and desertification also affect the atmospheric carbon dioxide concentration. Explain why a global decrease in the amount of vegetation is contributing to the rise in carbon dioxide.

3. The magnitude of acid rain's effects can be influenced by the properties of the rock that the rain runs over. Acid rain is least likely to significantly acidify lakes in regions where the bedrock consists of calcium carbonate–rich limestone or marble. How does the presence of these rocks mitigate the effects of acid rain?

4. Some of the bits of plastic that end up in the ocean contain phthalates, which are known endocrine disrupters. How might consumption of such plastic affect the survival and reproduction of vertebrates that accidentally eat the plastic?

Appendix I. Periodic Table of the Elements

The symbol for each element is an abbreviation of its name. Some symbols for elements are abbreviations for their Latin names. For instance, Pb (lead) is short for *plumbum*; the word "plumbing" is related—ancient Romans made their water pipes with lead.

Elements in each vertical column of the table behave in similar ways. For instance, all of the elements in the far right column of the table are inert gases; they do not interact with other atoms. In nature, such elements occur only as solitary atoms.

Atomic number → 11
Symbol → Na
Atomic mass → 22.99

Atomic masses are based on carbon12. Numbers in parentheses are mass numbers of most stable or best known isotopes of radioactive elements.

Period \ Group	1	2	3	4	5	6	7	8	9	10	11	12	13	14	15	16	17	18
1	1 H 1.008																	2 He 4.0026
2	3 Li 6.94	4 Be 9.0122											5 B 10.81	6 C 12.011	7 N 14.007	8 O 15.999	9 F 18.998	10 Ne 20.180
3	11 Na 22.990	12 Mg 24.305											13 Al 26.982	14 Si 28.085	15 P 30.974	16 S 32.06	17 Cl 35.45	18 Ar 39.948
4	19 K 39.098	20 Ca 40.078	21 Sc 44.956	22 Ti 47.867	23 V 50.942	24 Cr 51.996	25 Mn 54.938	26 Fe 55.845	27 Co 58.933	28 Ni 58.693	29 Cu 63.546	30 Zn 65.38	31 Ga 69.723	32 Ge 72.63	33 As 74.922	34 Se 78.96	35 Br 79.904	36 Kr 83.798
5	37 Rb 85.468	38 Sr 87.62	39 Y 88.906	40 Zr 91.224	41 Nb 92.906	42 Mo 95.96	43 Tc (97.91)	44 Ru 101.07	45 Rh 102.91	46 Pd 106.42	47 Ag 107.87	48 Cd 112.41	49 In 114.82	50 Sn 118.71	51 Sb 121.76	52 Te 127.60	53 I 126.90	54 Xe 131.29
6	55 Cs 132.91	56 Ba 137.33	* 71 Lu 174.97	72 Hf 178.49	73 Ta 180.95	74 W 183.84	75 Re 186.21	76 Os 190.23	77 Ir 192.22	78 Pt 195.08	79 Au 196.97	80 Hg 200.59	81 Tl 204.38	82 Pb 207.2	83 B 208.98	84 Po (208.98)	85 At (209.99)	86 Rn (222.02)
7	87 Fr (223.02)	88 Ra (226.03)	** 103 Lr (262.11)	104 Rf (265.12)	105 Db (268.13)	106 Sg (271.13)	107 Bh (270)	108 Hs (277.15)	109 Mt (276.15)	110 Ds (281.16)	111 Rg (280.16)	112 Cn (285.17)	113 Uut (284.18)	114 Fl (289.19)	115 Uup (288.19)	116 Lv (293)	117 Uus (294)	1 Uuo (294)

*Lanthanoids	57 La 138.91	58 Ce 140.12	59 Pr 140.91	60 Nd 144.24	61 Pm (144.91)	62 Sm 150.36	63 Eu 151.96	64 Gd 157.25	65 Tb 158.93	66 Dy 162.50	67 Ho 164.93	68 Er 167.26	69 Tm 168.93	70 Tb 173.05
**Actinoids	89 Ac (227.03)	90 Th 232.04	91 Pa 231.04	92 U 238.03	93 Np (237.05)	94 Pu (244.06)	95 Am (243.06)	96 Cm (247.07)	97 Bk (247.07)	98 Cf (251.08)	99 Es (252.08)	100 Fm (257.10)	101 Md (258.10)	102 No (259.10)

Appendix II. The Amino Acids

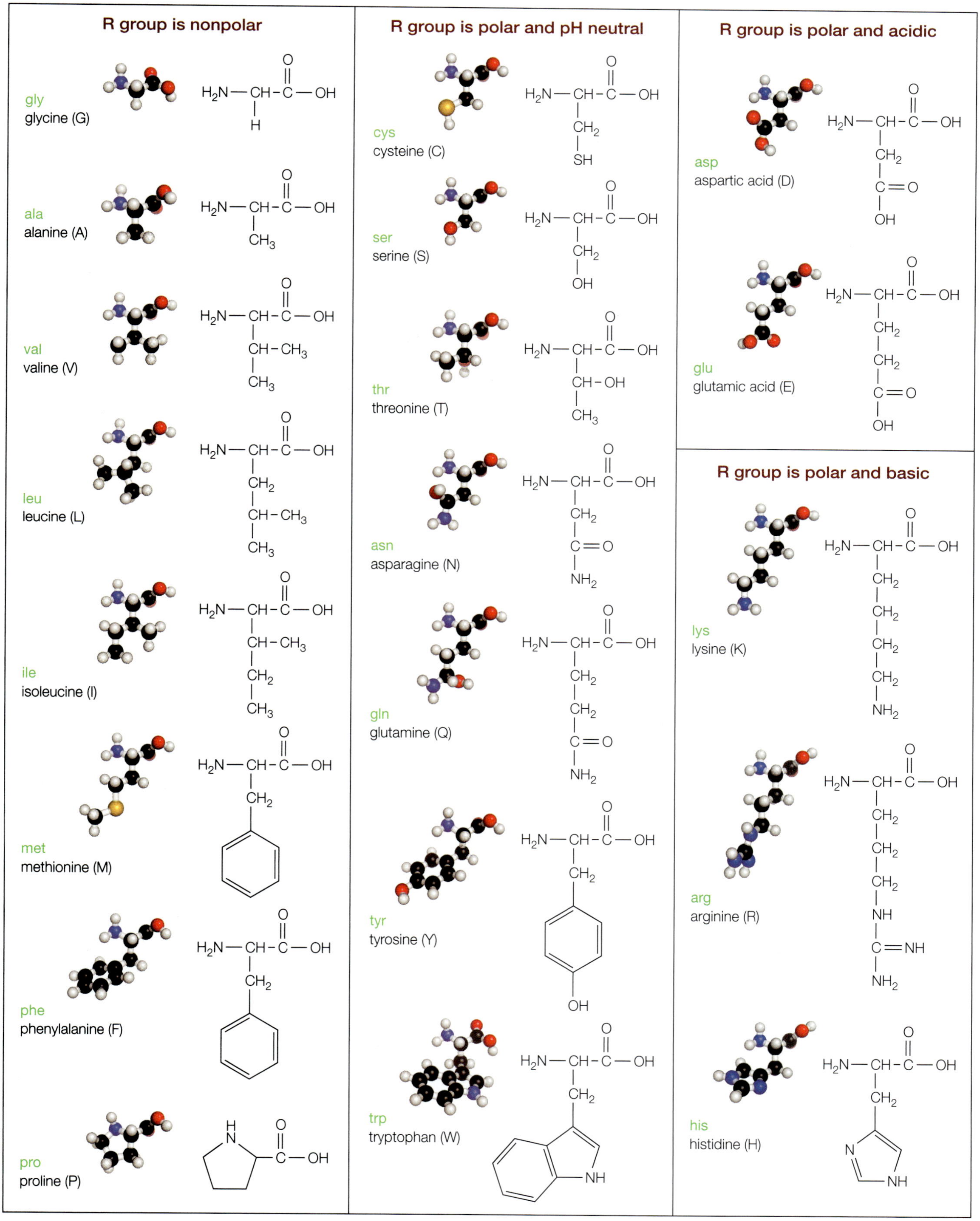

Appendix III. A Closer Look at Some Major Metabolic Pathways

Glycolysis

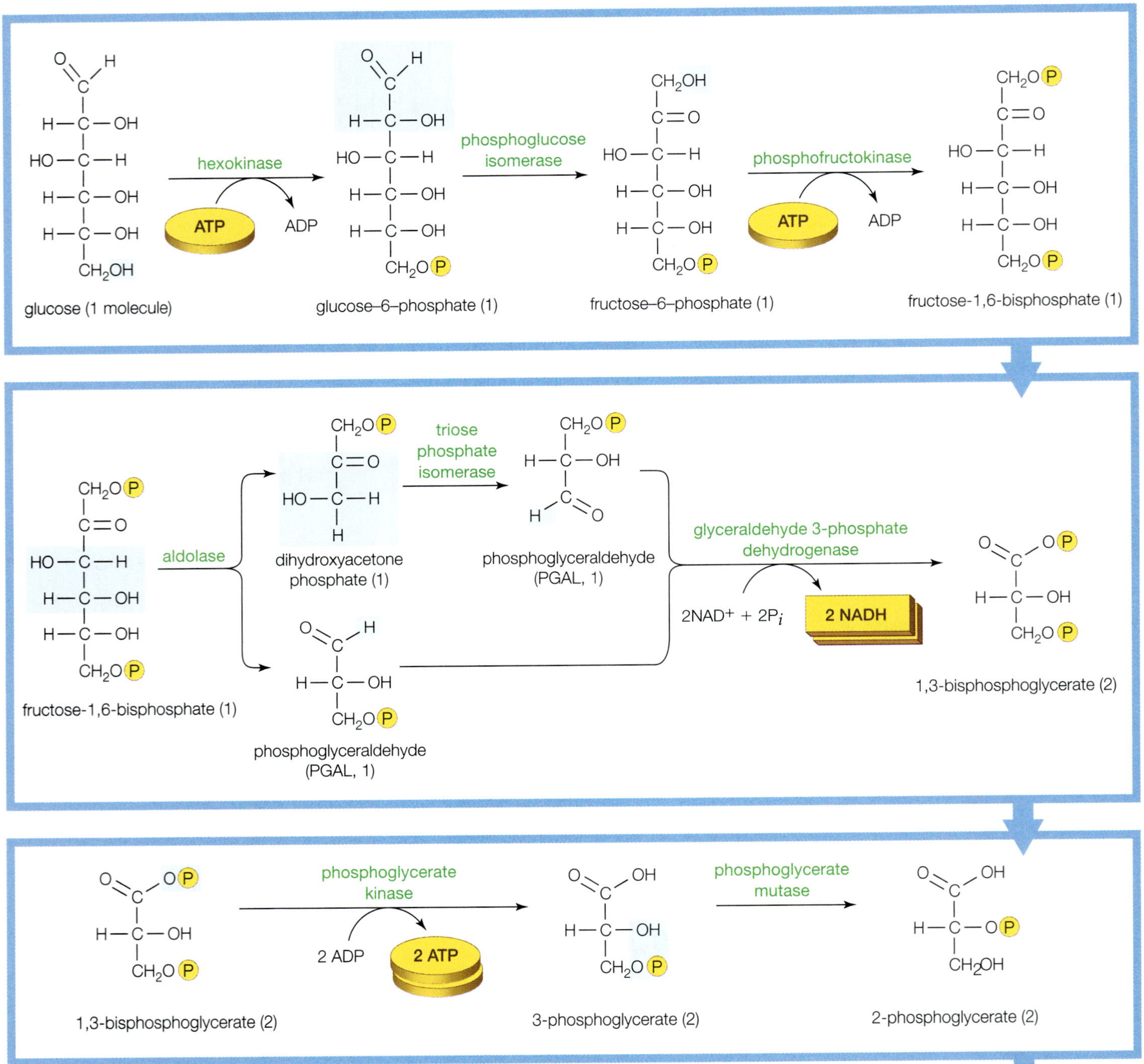

FIGURE A Glycolysis breaks down one glucose molecule into two 3-carbon pyruvate molecules for a net yield of two ATP. Enzyme names are indicated in green; parts of substrate molecules undergoing chemical change are highlighted blue.

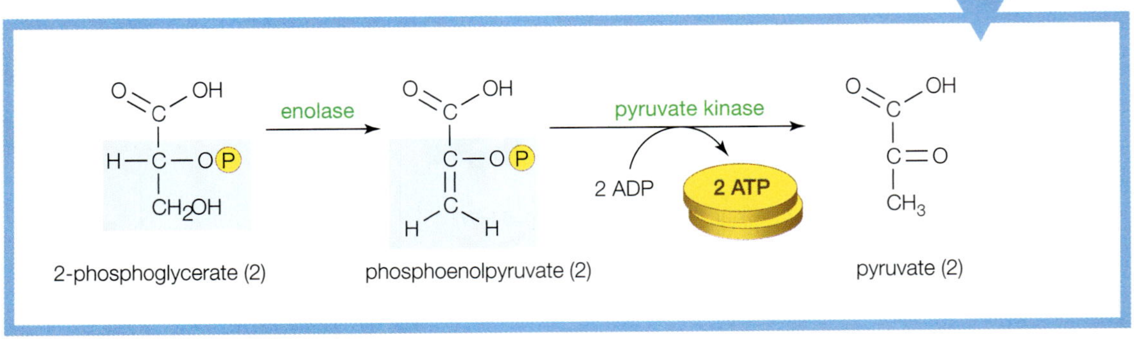

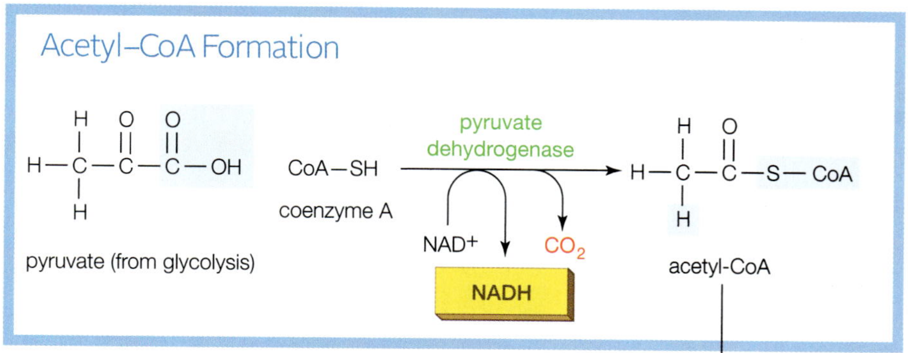

FIGURE B Second stage of aerobic respiration: acetyl–CoA formation and the Krebs cycle (also known as the citric acid cycle). Krebs cycle intermediates are shown in acid form for simplicity, but all are ionized in the watery fluid of a cell.

Two sets of these reactions break down two pyruvate (from glycolysis) for a net yield of two ATP and ten reduced coenzymes. At this stage of aerobic respiration, the carbon backbone of one glucose molecule has been completely oxidized: Six carbon atoms have left the cell, in six CO_2 (red).

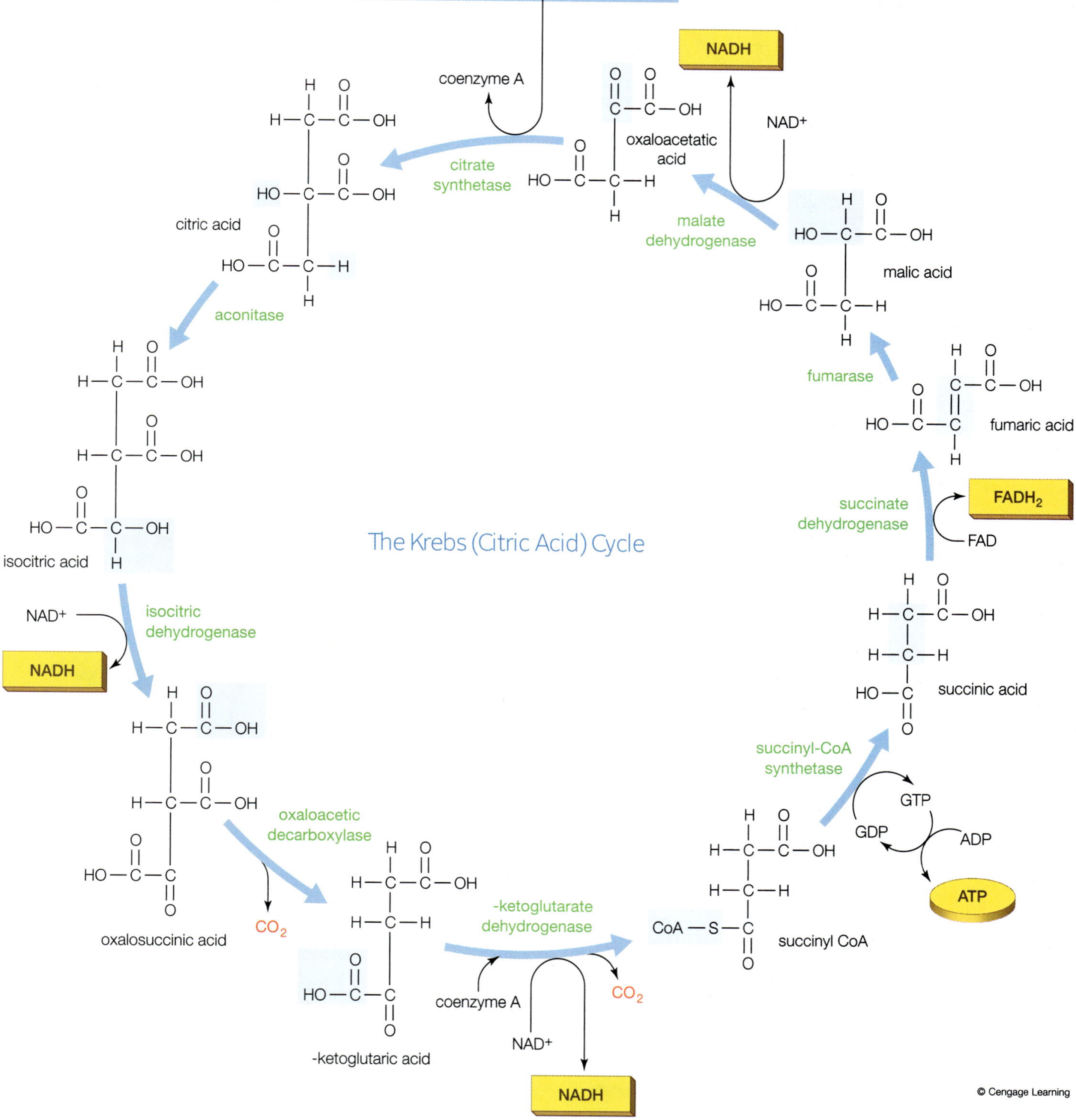

FIGURE C Details of the Calvin–Benson cycle. These light-independent reactions of photosynthesis use ATP and NADPH to fix carbon from carbon dioxide. The enzyme rubisco catalyzes the attachment of CO_2 to RuBP. The resulting PGA molecules are converted to PGAL, and the complex series of reactions that follow shuffle carbon atoms among sugar molecules to regenerate RuBP. One molecule of glucose is produced for six CO_2 molecules that enter the reactions. Water and some of the molecular participants are not shown, for clarity.

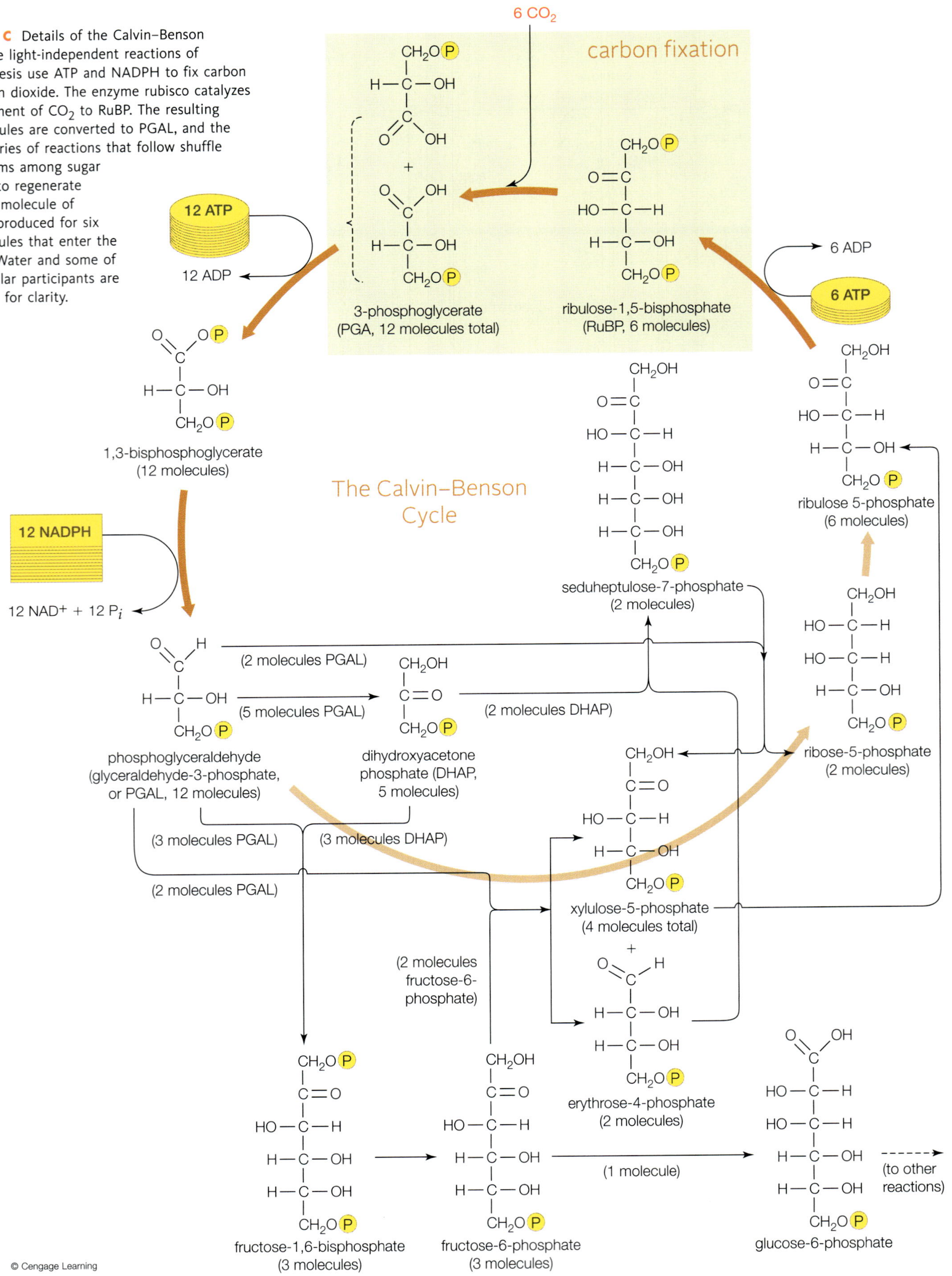

Appendix IV. A Plain English Map of the Human Chromosomes

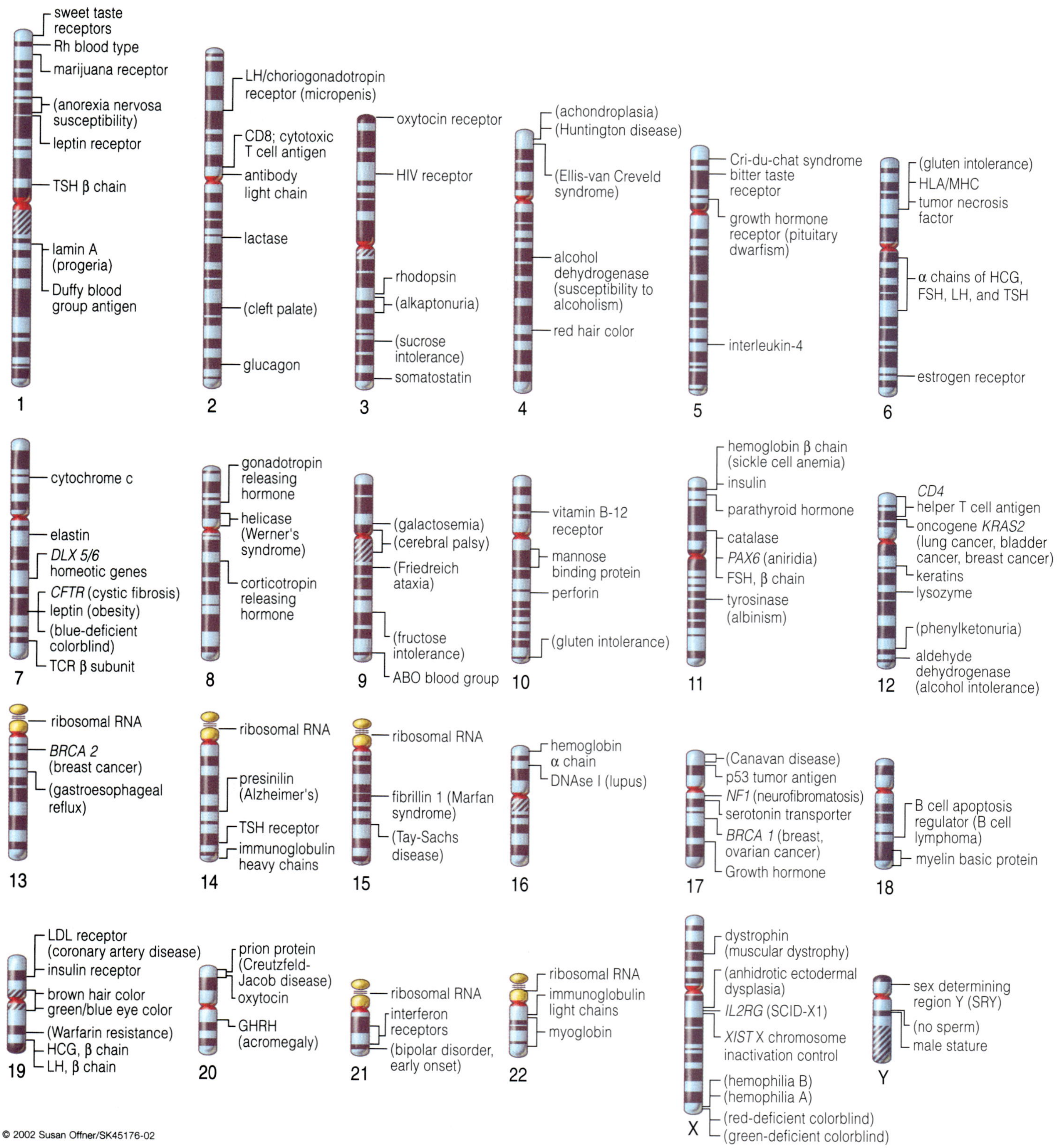

Haploid set of human chromosomes. The banding patterns characteristic of each type of chromosome appear after staining with a reagent called Giemsa. The locations of some of the 20,065 known genes (as of November, 2005) are indicated. Also shown are locations that, when mutated, cause some of the genetic diseases discussed in the text.

Appendix V. Restless Earth—Life's Changing Geologic Stage

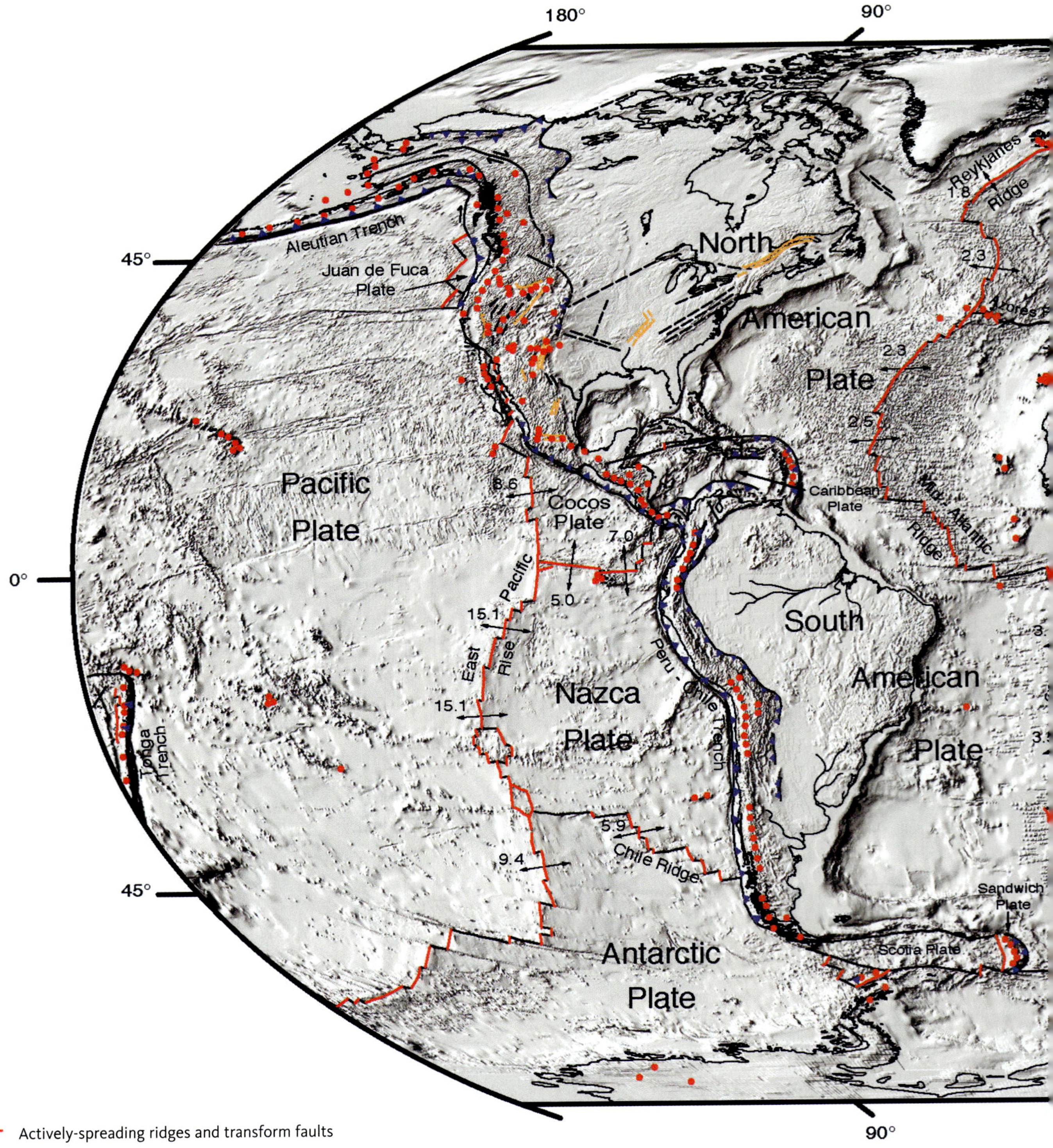

Actively-spreading ridges and transform faults

Total spreading rate, cm/year

Major active fault or fault zone; dashed where nature, location, or activity uncertain

Normal fault or rift; hachures on downthrown side

Reverse fault (overthrust, subduction zones); generalized; barbs on upthrown side

Volcanic centers active within the last one million years; generalized. Minor basaltic centers and seamounts omitted.

This NASA map summarizes the tectonic and volcanic activity of Earth during th past 1 million years. The reconstructions at far right indicate positions of Earth's major land masses through time.

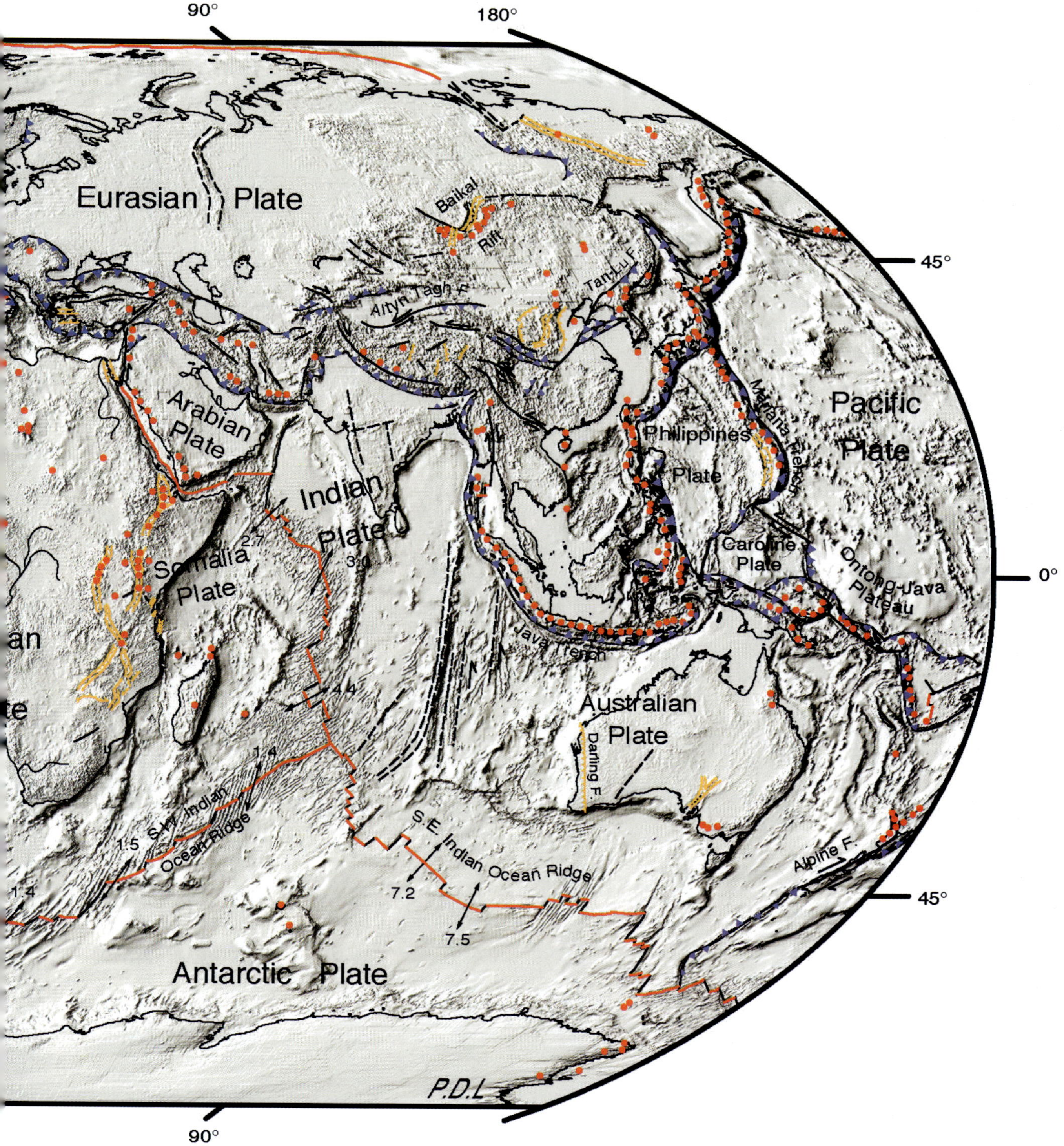

90°
180°
45°
0°
45°
90°
Eurasian Plate
Baikal Rift
Tan-Lu F.
Altyn Tagh F.
Arabian Plate
Indian Plate
Somalia Plate
Philippines Plate
Mariana Trench
Pacific Plate
Caroline Plate
Ontong-Java Plateau
Java Trench
Australian Plate
Darling F.
S.W. Indian Ocean Ridge
S.E. Indian Ocean Ridge
Alpine F.
Antarctic Plate
2.7
3.0
4.4
1.4
1.5
1.4
7.2
7.5
P.D.L.

Appendix VI. Units of Measure

LENGTH

1 kilometer (km) = 0.62 miles (mi)
1 meter (m) = 39.37 inches (in)
1 centimeter (cm) = 0.39 inches

To convert	*multiply by*	*to obtain*
inches	2.25	centimeters
feet	30.48	centimeters
centimeters	0.39	inches
millimeters	0.039	inches

AREA

1 square kilometer = 0.386 square miles
1 square meter = 1.196 square yards
1 square centimeter = 0.155 square inches

VOLUME

1 cubic meter = 35.31 cubic feet
1 liter = 1.06 quarts
1 milliliter = 0.034 fluid ounces = 1/5 teaspoon

To convert	*multiply by*	*to obtain*
quarts	0.95	liters
fluid ounces	28.41	milliliters
liters	1.06	quarts
milliliters	0.03	fluid ounces

WEIGHT

1 metric ton (mt) = 2,205 pounds (lb) = 1.1 tons (t)
1 kilogram (kg) = 2.205 pounds (lb)
1 gram (g) = 0.035 ounces (oz)

To convert	*multiply by*	*to obtain*
pounds	0.454	kilograms
pounds	454	grams
ounces	28.35	grams
kilograms	2.205	pounds
grams	0.035	ounces

TEMPERATURE

Celcius (°C) to Fahrenheit (°F): $°F = 1.8\,(°C) + 32$

Fahrenheit (°F) to Celsius: $°C = \dfrac{(°F - 32)}{1.8}$

	°C	°F
Water boils	100	212
Human body temperature	37	98.6
Water freezes	0	32

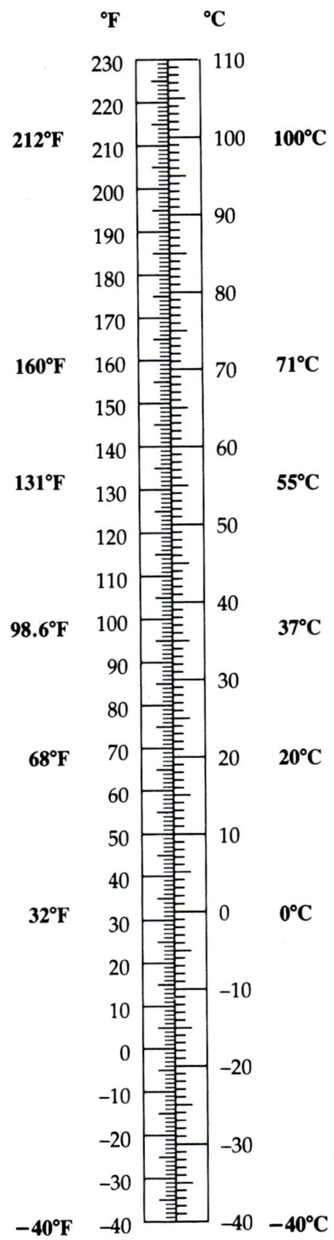

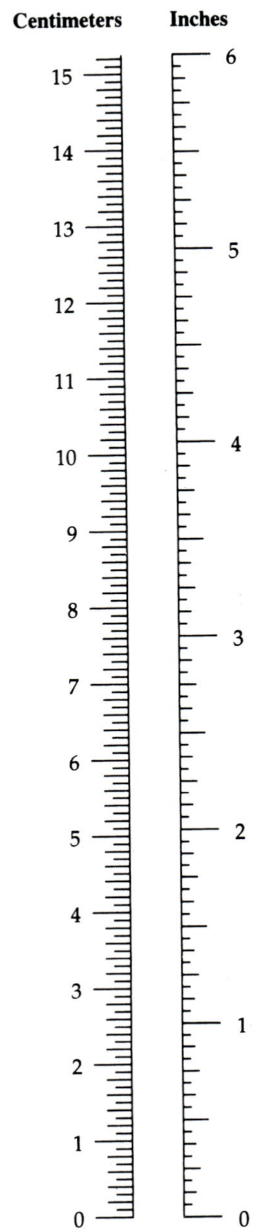

Appendix VII. Answers to Self-Quizzes and Genetics Problems

CHAPTER 1

1. a	1.1
2. c	1.1
3. energy, nutrients	1.2
4. homeostasis	1.2
5. d	1.2
6. reproduction	1.2
7. d	1.2
8. a, d, e	1.1–1.4
9. Animals	1.3
10. a, b	1.1, 1.3
11. domains	1.4
12. b	1.5
13. b	1.7
14. b	1.7
15. c	1.1
e	1.7
b	1.4
d	1.5
a	1.5
f	1.2

CHAPTER 2

1. a	2.1
2. b	2.1
3. d	2.2
4. d	2.1
5. b	2.2
6. a	2.3
7. a	2.3
8. c	2.3
9. c	2.4
10. c	2.4
11. d	2.2, 2.5
12. a	2.5
13. c	2.5
14. b	2.4
15. c	2.4
b	2.1
d	2.4
a	2.1, 2.2
f	2.4
e	2.1, 2.2

CHAPTER 3

1. c	3.1
2. four	3.1
3. b	3.1, 3.3, 3.4
4. e	3.2, 3.6
5. c	3.3
6. False	3.3, 3.7
7. b	3.3
8. starch, cellulose, glycogen	3.2
9. e	3.3
10. d	3.4, 3.6
11. d	3.5
12. d	3.6
13. a amino acid	3.4
b carbohydrate	3.2
c polypeptide	3.4
d fatty acid	3.3
14. c	3.3
a	3.2
b	3.3
15. g	3.4
a	3.3
b	3.4
c	3.3
d	3.6
j	3.3
f	3.6
i	3.4
h	3.2
e	3.2

CHAPTER 4

1. c	4.1
2. c	4.1
3. c	4.1, 4.4
4. b	4.1
5. b	4.3
6. c	4.3
7. a	4.3
8. b	4.5
9. a	4.6
10. c	4.7
11. a	4.6
12. c, b, d, a	4.6
13. d	4.10
14. a	4.10
15. c	4.7
g	4.8
e	4.4, 4.6
d	4.5
a	4.10
b	4.4, 4.9
f	4.1, 4.5

CHAPTER 5

1. c	5.1
2. b	5.1
3. d	5.1
4. a	5.2
5. c	5.2
6. c	5.2
7. temperature, pH, salt, pressure	5.3
8. d	5.4
9. a	5.5
10. more/less	5.6
11. c	5.6, 5.7
12. b	5.7
13. a	5.6
14. d	5.8
15. c	5.2
e	5.8
f	5.1
b	5.2
a	5.5
g	5.6
h	5.7
d	5.7

CHAPTER 6

1. autotroph: weed; heterotrophs: cat, bird, caterpillar	6.1
2. c	6.1
3. a	6.3
4. d	6.4
5. b	6.3, 6.4
6. b	6.4
7. c	6.4
8. c	6.3, 6.5
9. b	6.5
10. b	6.5
11. a	6.1
12. b	6.5, 6.6
13. f	6.5
h	6.5
g	6.4
d	6.4
e	6.5
b	6.1, 6.3
a	6.1
c	6.1

CHAPTER 7

1. False	7.1
2. d	7.1, 7.3
3. a	7.1, 7.6
4. c	7.3
5. b	7.1, 7.4, 7.5
6. d	7.1, 7.6
7. e	7.4
8. b	7.4
9. c	7.5
10. c	7.5
11. c	7.6
12. d	7.7
13. f	7.6
14. c	7.5
15. b	7.4
d	7.3
a	7.3, 7.6
c	7.4, 7.7
e	7.3
f	7.2

CHAPTER 8

1. c	8.2
2. c	8.2
3. b	8.2
4. b	8.2
5. a	8.3
6. b	8.3
7. b	8.3
8. a	8.4
9. d	8.4
10. c	8.2, 8.4
11. b	8.4
12. d	8.3, 8.4
13. d	8.5
14. d	8.6
15. d	8.1
b	8.6
a	8.2
g	8.3
e	8.4
h	8.4
c	8.3
f	8.5

CHAPTER 9

1. c	9.1
2. b	9.2
3. a	9.1
4. c	9.1
5. a	9.1
6. b	9.1, 9.3
7. b	9.2
8. c	9.3
9. a	9.3
10. a	9.3
11. a	9.2
12. a	9.2, 9.4
13. b	9.2, 9.4
14. c	9.4
15. c	9.3
b	9.2
e	9.4
a	9.2
f	9.3
d	9.2

CHAPTER 10

1. d	10.1
2. d	10.1, 10.2
3. b	10.1
4. b	10.1
5. h	10.1
6. b	10.1
7. c	10.2
8. c	10.2
9. d	10.2
10. b	10.2
11. b	10.3
12. b	10.3
13. c	10.3
14. b	10.4
15. f	10.2
a	10.3
b	10.1, 10.4
e	10.3
c	10.1
d	10.5

CHAPTER 11

1. e	11.1
2. b	11.1
3. d	11.1
4. e	11.1
5. c	11.2
6. c	11.1
7. a	11.1
8. c	11.1
9. a	11.1
10. d	11.3
11. interphase, prophase, metaphase, anaphase, telophase	11.2
12. d	11.5
13. a	11.5
14. c	11.3
f	11.2
a	11.5
g	11.3
b	11.3
e	11.5
d	11.2
h	11.4
15. d	11.2
b	11.2
c	11.2
e	11.1
a	11.2
f	11.3

CHAPTER 12

1. b	12.1
2 b	12.2
3. c	12.1, 12.4, 12.6
4. d	12.2
5. b	12.2
6. a	12.2
7. Sister chromatids are still attached	12.3
8. c	12.3
9. b	12.4
10. a	12.4
11. e	12.1, 12.4
12. b	12.4, 12.5
13. c	12.3
d	12.3
a	12.1
f	12.2
e	12.2
b	12.1, 12.6
g	12.4

CHAPTER 13

1. b	13.1
2. a	13.1
3. b	13.2
4. b	13.2
5. c	13.2
6. a	13.2, 13.3
7. b	13.2
8. d	13.3
9. c	13.3
10. c	13.4
11. b	13.4
12. Continuous variation	13.6
13. b	13.3
d	13.2
a	13.1
c	13.1

CHAPTER 14

1. b	14.1
2. b	14.1
3. a	14.1
4. b	14.2
5. False	14.3
6. d (could be due to both parents carrying an autosomal recessive allele, or the mom carrying an x-linked recessive allele)	14.2, 14.3
7. d	14.3
8. X from mom, Y from dad	14.3
9. d	14.2
10. Y-linked inheritance (this is a critical thinking question)	14.3
11. d	14.5
12. b	14.5
13. True	14.5
14. c	14.5
15. c	14.5
e	14.4
f	14.5
b	14.4
a	14.1
d	14.4

CHAPTER 15

1. c	15.1
2. a	15.1
3. b	15.1
4. b	15.2
5. c	15.2
6. b	15.3
7. b	15.3
8. d	15.2, 15.4
9. d	15.4
10. True	15.6
11. b	15.6
d	15.6
e	15.5
f	15.1
12. d	15.5
13. b	15.6
14. a	15.2
d	15.1
c	15.1
e	15.3
b	15.1
15. c	15.4
f	15.5
d	15.6
b	15.4
a	15.5
e	15.5

Appendix VII. Answers to Self-Quizzes and Genetics Problems *(continued)*

CHAPTER 16

Answer	Section
1. b	16.1
2. c	16.1, 16.7
3. d	16.2
4. b	16.2
5. b	16.3
6. d	16.4
7. Gondwana	16.6
8. a	16.1–16.3
c	16.6
d	16.5
e	16.3, 16.5
f	16.9
9. morphological divergence	16.7
10. d	16.7
11. b	16.8
12. 66	16.9
13. e	16.7, 16.8
14. b	16.2
d	16.1, 16.3
g	16.2
e	16.4
c	16.7
f	16.7
a	16.3

CHAPTER 17

Answer	Section
1. a	17.1
2. c	17.1, 17.7
3. a, c, b	17.3–17.5
4. d	17.6
5. d	17.6
6. b	17.7
7. f	17.1, 17.2, 17.7
8. allopatric speciation	17.9
9. d	17.6, 17.8
10. d	17.12
11. c (this is a critical thinking question)	17.12
12. c	17.12
13. b	17.12
14. c	17.7
e	17.6
g	17.11
b	17.7
d	17.12
f	17.11
a	17.11
h	17.12

CHAPTER 18

Answer	Section
1. c	18.1
2. a	18.1
3. c	18.2
4. a	18.2
5. a	18.2
6. c	18.4
7. b	18.5
8. b	18.4
9. c	18.2
10. a	18.4
11. d	18.5, 18.6
12. c	18.5
13. d	18.6
14. a	18.1
c	18.5
b	18.3
e	18.5
f	18.6
d	18.6

CHAPTER 19

Answer	Section
1. c	19.1
2. b	19.3
3. c	19.1
4. b	19.1
5. d	19.2
6. c	19.3
7. one	19.4
8. c	19.5
9. d	19.5
10. d	19.5
11. c	19.6
12. d	19.2, 19.7
13. a	19.4
14. b	19.6
a	19.1
a	19.2
b	19.6
b	19.6
15. d	19.5
e	19.7
a	19.3
f	19.1
b	19.8
c	19.8

CHAPTER 20

Answer	Section
1. c	20.2
2. d	20.3
3. b	20.2
4. c	20.4
5. b	20.5
6. a	20.7
7. d	20.7
8. c	20.4
9. c	20.6
10. d	20.6
11. c	20.7
12. a	20.8
13. a	20.2
14. a	20.7
15. d	20.9
g	20.5
a	20.3
b	20.6
f	20.6
c	20.7
e	20.6

CHAPTER 21

Answer	Section
1. c	21.1, 21.2
2. a (ferns produce spores, not seeds)	21.1, 21.3
3. b	21.2
4. c	21.3
5. a	21.4
6. e	21.2, 21.3, 21.5
7. b	21.5, 21.6
8. True	21.3
9. c	21.6
10. a	21.2
11. a	21.1
12. c	21.7
13. b	21.5
14. c	21.2
d	21.3
a	21.5
b	21.6
15. c	21.4
i	21.1
a	21.1
b	21.1
e	21.1
f	21.6
d	21.3
g	21.3
h	21.4

CHAPTER 22

Answer	Section
1. c	22.1
2. a	22.1
3. b	22.1
4. a	22.3
5. d	22.2
6. c	22.2
7. c	22.2
8. d	22.2
9. b	22.4
10. d	22.3
11. a	22.3
12. c	22.4
13. b	22.5
14. a	22.3
15. d	22.1
b	22.1
a	22.1
f	22.2
g	22.2
c	22.3
e	22.3

CHAPTER 23

Answer	Section
1. False	23.1
2. d	23.1
3. a	23.4
4. a	23.5
5. b	23.11
6. a	23.7
7. b	23.2
8. a	23.1, 23.9
9. c	23.9, 23.11
10. b	23.12
11. b	23.6, 23.7
12. d	23.11
13. b	23.12
f	23.7
d	23.3
h	23.4
c	23.5
a	23.8
g	23.6
e	23.9

CHAPTER 24

Answer	Section
1. c	24.1
2. c	24.1
3. a	24.2
4. b	24.2
5. d	24.3
6. c	24.4
7. f	24.4, 24.5
8. a	24.5
9. b	24.9
10. c	24.6
11. c	24.11
12. b	24.1
i	24.2
g	24.3
f	24.8
c	24.6
d	24.7
a	24.7
h	24.7
e	24.10

CHAPTER 25

Answer	Section
1. d	25.1
2. b	25.1
3. a	25.1
4. c	25.3
5. a	25.3
6. birds	25.4
7. cooperation in predator detection, defense, rearing young, learning by imitation, defending territory	25.5
8. d	25.6
9. d	25.6
10. a	25.4
11. e	25.6
h	25.5
c	25.2
b	25.4
d	25.4
a	25.2
f	25.5
g	25.2

CHAPTER 26

Answer	Section
1. clumped	26.1
2. f	26.2
3. 400	26.1
4. 1,600	26.2
5. a	26.2
6. d	26.3
7. d	26.4
8. b	26.6
9. d	26.7
10. a	26.6
11. a	26.4
12. c	26.4
13. c	26.3
d	26.2
a	26.2
e	26.3
b	26.3

CHAPTER 27

Answer	Section
1. b	27.1
2. d	27.1, 27.3
3. e	27.3
4. d	27.3
5. b	27.2
d	27.5
c	27.1
a	27.4
e	27.3
6. b	27.5
7. b	27.6
8. Pioneer	27.6
9. c	27.6
10. a	27.1
11. b	27.3
12. c	27.5
13. a	27.4
14. c	27.8
b	27.6
d	27.6
a	27.7
f	27.7
e	27.3

CHAPTER 28

Answer	Section
1. b	28.1
2. d	28.1
3. d	28.1
4. d	28.1
5. a	28.1
6. c	28.5
7. b	28.5
8. d	28.6
9. d	28.6, 28.7
10. a	28.8
11. c	28.8
12. b	28.6
13. a	28.7
14. d	28.6
15. d	28.6
c	28.6
b	28.7
a	28.7

CHAPTER 29

Answer	Section
1. b	29.1
2. d	29.1
3. d	29.1
4. a	29.2
5. b	29.2
6. b	29.2
7. d	29.3
8. c	29.6
9. a	29.8
10. b	29.9
11. d	29.12
12. c	29.11
13. c	29.3
14. b	29.3
15. d	29.8
e	29.6
f	29.7
c	29.6
b	29.10
h	29.5
i	29.6
a	29.4
g	29.12

CHAPTER 30

Answer	Section
1. b	30.1
2. d	30.1
3. a	30.2
4. c	30.4
5. c	30.5
6. a	30.5
7. b	30.3
8. b	30.6
9. b	30.5
10. c	30.7
11. d	30.7
12. c	30.7
13. d	30.8
14. g	30.7
a	30.5
i	30.7
e	30.4
d	30.1
f	30.4
h	30.6
b	30.2
c	30.2

CHAPTER 13: GENETICS PROBLEMS

1. Yellow is recessive. Because F_1 plants have a green phenotype and must be heterozygous, green must be dominant over the recessive yellow.

2. a. *AB*
   b. *AB*, *aB*
   c. *Ab*, *ab*
   d. *AB*, *Ab*, *aB*, *ab*

3. a. All offspring will be *AaBB*.

4. A mating of two M^L cats yields 1/4 *MM*, 1/2 M^LM, and 1/4 M^LM^L. Because M^LM^L is lethal, the probability that any one kitten among the survivors will be heterozygous is 2/3.

5. Because both parents are heterozygous (Hb^AHb^S), each child has a 1 in 4 possibility of inheriting two Hb^S alleles and having sickle cell anemia.

6. The data reveal that these genes do not assort independently because the observed ratio is very far from the 9:3:3:1 ratio expected with independent assortment. Instead, the results can be explained if the genes are located close to each other on the same chromosome.

CHAPTER 14: GENETICS PROBLEMS

1. Autosomal recessive. If the allele was inherited in a dominant pattern, individuals in the last generation would all have the phenotype. If it was X-linked, offspring of the first generation would all have the phenotype.

2. a. Human males (XY) inherit their X chromosome from their mother.
   b. A male with an X-linked allele produces two kinds of gametes: one with an X chromosome (and the X-linked allele), and the other with a Y chromosome.
   c. A female homozygous for an X-linked allele produces one type of gamete, which carries the X-linked allele.
   d. A female heterozygous for an X-linked allele produces two types of gametes: one that carries the X-linked allele, and another that carries the partnered allele on the homologous chromosome.

3. As a result of translocation, chromosome 21 may get attached to the end of chromosome 14. The new individual's chromosome number would still be 46, but its somatic cells would have the translocated chromosome 21 in addition to two normal chromosomes 21.

4. 50 percent

Glossary

acid Substance that releases hydrogen ions when it dissolves in water. **32**

acid rain Low-pH rain that forms when sulfur dioxide and nitrogen oxides mix with water vapor in the atmosphere. **515**

activation energy Minimum amount of energy required to start a reaction. **80**

activator Transcription factor that increases the rate of transcription. **164**

active site Of an enzyme, pocket in which substrates bind and a reaction occurs. **82**

active transport Energy-requiring mechanism in which a transport protein pumps a solute across a cell membrane against its concentration gradient. **90**

adaptation (adaptive trait) A heritable trait that enhances an individual's fitness in a particular environment. **257**

adaptive radiation Macroevolutionary pattern in which a burst of genetic divergences from a lineage gives rise to many new species. **292**

adaptive trait *See* adaptation.

adhering junction Cell junction composed of adhesion proteins that connect to cytoskeletal elements. Fastens animal cells to each other and basement membrane. **69**

adhesion protein Plasma membrane protein that helps cells stick to one another and (in animals) to extracellular matrix. **57**

aerobic Involving or occurring in the presence of oxygen. **115**

aerobic respiration Oxygen-requiring metabolic pathway that breaks down sugars to produce ATP. Includes glycolysis, acetyl–CoA formation, the Krebs cycle, and electron transfer phosphorylation. **114**

age structure Of a population, the distribution of its members among various age categories. **449**

alcoholic fermentation Anaerobic sugar breakdown pathway that produces ATP, CO_2, and ethanol. **122**

allele frequency Abundance of a particular allele in a population's gene pool. **275**

alleles Forms of a gene with slightly different DNA sequences; may encode slightly different versions of the gene's product. **192**

allopatric speciation Speciation pattern in which a physical barrier ends gene flow between populations. **288**

allosteric regulation Control of enzyme activity by a regulatory molecule or ion that binds to a region outside the enzyme's active site. **84**

alternation of generations Of land plants and some algae, a life cycle that includes haploid and diploid multicelled bodies. **338**

alternative splicing Post-translational RNA modification in which some exons are removed or joined in various combinations. **151**

altruistic behavior Behavior that benefits others at the expense of the individual. **432**

amino acid Small organic compound that is a subunit of proteins. Consists of a carboxyl group, an amine group, and a characteristic side group (R), all typically bonded to the same carbon atom. **44**

amniote Vertebrate whose egg has waterproof membranes that allow it to develop away from water; a reptile, bird, or mammal. **401**

amoeba Single-celled protist that extends pseudopods to move and to capture prey. **340**

amoebozoan Shape-shifting heterotrophic protist with no pellicle or cell wall; an amoeba or slime mold. **340**

amphibian Tetrapod with scaleless skin; develops in water, then lives on land as a carnivore with lungs. For example, a frog or salamander. **404**

anaerobic Occurring in the absence of oxygen. **115**

analogous structures Similar body structures that evolved separately in different lineages (by morphological convergence). **267**

anaphase Stage of mitosis during which sister chromatids separate and move toward opposite spindle poles. **181**

aneuploid Having too many or too few copies of a particular chromosome. **230**

angiosperms Highly diverse seed plant lineage; only plants that make flowers and fruits. **356**

animal A eukaryotic heterotroph that is made up of unwalled cells and develops through a series of stages. Most ingest food, reproduce sexually, and move. **9, 376**

annelid Segmented worm with a coelom, complete digestive system, and closed circulatory system. **384**

antenna Of some arthropods, sensory structure on the head that detects touch and odors. **389**

anthropoid primate Humanlike primate; monkey, ape, or human. **411**

anticodon In a tRNA, set of three nucleotides that base-pairs with an mRNA codon. **153**

antioxidant Substance that prevents oxidation of other molecules. **86**

ape Common name for a tailless nonhuman primate; a gibbon, orangutan, gorilla, chimpanzee, or bonobo. **411**

apicomplexan Parasitic protist that reproduces inside cells of its host; for example, the protist that causes malaria. **336**

aquifer Porous rock layer that holds some groundwater. **478**

arachnids Land-dwelling arthropods with no antennae and four pairs of walking legs; spiders, scorpions, mites, and ticks. **390**

archaea Singular **archaean**. Group of single-celled organisms that lack a nucleus but are more closely related to eukaryotes than to bacteria. **8**

arctic tundra Highest-latitude Northern Hemisphere biome, where low, cold-tolerant plants survive with only a brief growing season. **499**

area effect Larger islands have more species than small ones. **468**

arthropod Invertebrate with jointed legs and a hard exoskeleton that is periodically molted. **389**

asexual reproduction Reproductive mode of eukaryotes by which offspring arise from a single parent only. **178**

atmospheric cycle Biogeochemical cycle in which a gaseous form of an element plays a significant role. For example, the carbon cycle. **480**

atom Fundamental building block of all matter. Consists of varying numbers of protons, neutrons, and electrons. **4**

atomic number Number of protons in the atomic nucleus; determines the element. **24**

ATP Adenosine triphosphate. Nucleotide that consists of an adenine base, a ribose sugar, and three phosphate groups. Functions as a subunit of RNA and as a coenzyme in many reactions. Important energy carrier in cells. **46**

ATP/ADP cycle Process by which cells regenerate ATP. ADP forms when a phosphate group is removed from ATP, then ATP forms again as ADP gains a phosphate group. **87**

australopith Extinct African hominins in the genus *Australopithecus*; some are considered likely human ancestors. **413**

autosome A chromosome that is the same in males and females. **137**

autotroph Producer. An organism that makes its own food using energy from the environment and carbon from inorganic molecules such as CO_2. **100**

bacteria Singular bacterium. The most diverse and well-known group of prokaryotes. **8**

bacteriophage Virus that infects bacteria. **133, 316**

balanced polymorphism Maintenance of two or more alleles for a trait at high frequency in a population. **283**

Barr body Inactivated X chromosome in a cell of a female mammal. The other X chromosome is active. **168**

basal body Organelle that develops from a centriole; occurs at base of cilium or flagellum. **67**

base Substance that accepts hydrogen ions when it dissolves in water. **32**

base-pair substitution Type of mutation in which a single base pair changes. **156**

bell curve Bell-shaped curve; typically results from graphing frequency versus distribution for a trait that varies continuously. **216**

benthic province The ocean's sediments and rocks. **504**

bilateral symmetry Having paired structures so the right and left halves are mirror images. **376**

binary fission Method of asexual reproduction that divides one bacterial or archaeal cell into two identical descendant cells. **320**

bioaccumulation The concentration of a chemical pollutant in the tissues of an organism rises over the course of the organism's lifetime. **485**

biodiversity Scope of variation among living organisms; the genetic variation within species, variety of species, and variety of ecosystems. **518**

biofilm Community of microorganisms living within a shared mass of secreted slime. **59**

biogeochemical cycle A nutrient moves among environmental reservoirs and into and out of food webs. **478**

biogeography Study of patterns in the geographic distribution of species and communities. **254**

biological magnification A chemical pollutant becomes increasingly concentrated as it moves up through food chains. **485**

biological pest control Use of a pest's natural enemies to control its population size. **463**

biology The scientific study of life. **4**

bioluminescence Light emitted by a living organism. **335**

biomarker Substance found only or mainly in cells of one type. **307**

biome A region (often discontinuous) characterized by its climate and dominant vegetation. **494**

biosphere All regions of Earth where organisms live. **5**

biotic potential Maximum possible population growth rate under optimal conditions. **441**

bipedalism Habitual upright walking. **412**

bird Feathered reptile of a lineage in which the body became adapted for flight. **408**

bivalve Mollusk with a hinged two-part shell. For example, a clam. **386**

boreal forest Extensive high-latitude forest of the Northern Hemisphere; conifers are the predominant vegetation. **496**

bottleneck Reduction in population size so severe that it reduces genetic diversity. **284**

brood parasitism One egg-laying species benefits by having another raise its offspring. **463**

brown alga Multicelled marine protist with a brown accessory pigment in its chloroplasts. **337**

bryophyte Nonvascular plant; a moss, liverwort, or hornwort. **348**

buffer Set of chemicals that can keep the pH of a solution stable by alternately donating and accepting ions that contribute to pH. **32**

C3 plant Type of plant that uses only the Calvin–Benson cycle to fix carbon. **106**

C4 plant Type of plant that minimizes photorespiration by fixing carbon twice, in two cell types. **107**

Calvin–Benson cycle Cyclic carbon-fixing pathway that builds sugars from CO_2; the light-independent reactions of photosynthesis. **106**

camouflage Coloration or body form that helps an organism blend in with its surroundings and escape detection. **461**

CAM plant Type of plant that conserves water by fixing carbon twice, at different times of day. **107**

cancer Disease that occurs when a malignant neoplasm physically and metabolically disrupts body tissues. **185**

carbohydrate Molecule that consists primarily of carbon, hydrogen, and oxygen atoms in a 1:2:1 ratio. Complex kinds (e.g., cellulose, starch, glycogen) are polymers of simple kinds (sugars). **40**

carbon cycle Movement of carbon, mainly between the oceans, atmosphere, and living organisms. **480**

carbon fixation Process by which carbon from an inorganic source such as carbon dioxide becomes incorporated (fixed) into an organic molecule. **106**

carpel Floral reproductive organ that produces female gametophytes; consists of an ovary, stigma, and often a style. **356**

carrying capacity (*K*) Of a species, the maximum number of individuals that a particular environment can sustain; can change over time. **442**

cartilaginous fish Jawed fish with a skeleton of cartilage; a shark, ray, or skate. **402**

catalysis The acceleration of a reaction rate by a molecule that is unchanged by participating in the reaction. **82**

cDNA Complementary strand of DNA synthesized from an RNA template by the enzyme reverse transcriptase. **239**

cell Smallest unit of life; at minimum, consists of plasma membrane, cytoplasm, and DNA. **4**

cell cortex Reinforcing mesh of microfilaments under a plasma membrane. **66**

cell cycle A series of events from the time a cell forms until its cytoplasm divides. **178**

cell junction Structure that connects a cell to another cell or to extracellular matrix; e.g., tight junction, adhering junction, or gap junction (of animals); plasmodesmata (of plants). **69**

cell plate A disk-shaped structure that forms during cytokinesis in a plant cell; matures as a cross-wall between the two new nuclei. **182**

cell theory Theory that all organisms consist of one or more cells, which are the basic unit of life; all cells come from division of preexisting cells; and all cells pass hereditary material to offspring. **52**

cellular slime mold Amoeba-like protist that feeds as a single predatory cell; under unfavorable conditions, it joins with others to form a multicellular spore-bearing structure. **340**

cell wall Rigid but permeable structure that surrounds the plasma membrane of some cells. **59**

cellulose Tough, insoluble carbohydrate that is the major structural material in plants. **40**

central vacuole Large, fluid-filled vesicle in many plant cells. **62**

centriole Barrel-shaped organelle from which microtubules grow. **67**

centromere Of a duplicated eukaryotic chromosome, constricted region where sister chromatids attach to each other. **136**

cephalization Evolutionary trend whereby nerve cells and sensory structures become concentrated in the head of a bilateral animal. **376**

cephalopod Predatory mollusk that has a closed circulatory system and moves by jet propulsion. For example an octopus or squid. **387**

chaparral Biome of dry shrubland in regions with hot, dry summers and cool, rainy winters. **497**

character Quantifiable, heritable characteristic or trait. **294**

character displacement Evolutionary process in which two competing species become less similar in their resource requirements over time. **459**

charge Electrical property; opposite charges attract, and like charges repel. **24**

chelicerates Arthropod group with specialized feeding structures (chelicerae) and no antennae; arachnids and horseshoe crabs. **390**

chemical bond An attractive force that arises between two atoms when their electrons interact; joins atoms as molecules. *See* covalent bond, ionic bond. **28**

chemoautotroph Organism that uses carbon dioxide as its carbon source and obtains energy by oxidizing inorganic molecules. **321**

chemoheterotroph Organism that obtains energy and carbon by breaking down organic compounds. **321**

chlorophyll *a* Main photosynthetic pigment in plants. **100**

chloroplast Organelle of photosynthesis in the cells of plants and photosynthetic protists. Has two outer membranes enclosing semifluid stroma. Light-dependent reactions occur at its inner thylakoid membrane; light-independent reactions, in the stroma. Stores excess sugars as starch. **65**

choanoflagellate Heterotrophic freshwater protist with a flagellum and a food-capturing "collar." May be solitary or colonial. **341**

chordate Animal with an embryo that has a notochord, dorsal nerve cord, pharyngeal gill slits, and a tail that extends beyond the anus. A lancelet, tunicate, or vertebrate. **400**

chromosome A structure that consists of DNA and associated proteins; carries part or all of a cell's genetic information. **136**

chromosome number The total number of chromosomes in a cell of a given species. **136**

chytrid Fungus that makes flagellated spores. **365**

ciliate Single-celled, heterotrophic protist with many cilia. **335**

cilium Plural, **cilia** Short, movable structure that projects from the plasma membrane of some eukaryotic cells. **66**

clade A group whose members share one or more defining derived traits. **294**

cladistics Making hypotheses about evolutionary relationships among clades. **295**

cladogram Evolutionary tree diagram that shows evolutionary connections among a group of clades. **295**

cleavage furrow In a dividing animal cell, the indentation where cytoplasmic division will occur. **182**

climate Average weather conditions in a region. **490**

cloaca Body opening that serves as the exit for digestive waste and urine; also functions in reproduction. **402**

cloning vector A DNA molecule that can accept foreign DNA and be replicated inside a host cell. **239**

closed circulatory system Circulatory system in which blood flows through a continuous network of vessels; all materials are exchanged across the walls of those vessels. **384**

club fungus Fungus that produces spores in club-shaped structures during sexual reproduction. **365**

cnidarian Radially symmetrical invertebrate with two tissue layers; uses tentacles with stinging cells to capture food. For example, a jelly or a sea anemone. **380**

cnidocyte Stinging cell unique to cnidarians. **380**

coal Fossil fuel formed over millions of years by compaction and heating of plant remains. **352**

codominance Effect in which the full and separate phenotypic effects of two alleles are apparent in heterozygous individuals. **212**

codon In an mRNA, a nucleotide base triplet that codes for an amino acid or stop signal during translation. **152**

coelom A fluid-filled body cavity between the gut and body wall; it is lined with tissue derived from mesoderm. **377**

coenzyme An organic molecule that functions as a cofactor; e.g., NAD. **86**

coevolution The joint evolution of two closely interacting species; macroevolutionary pattern in which each species is a selective agent for traits of the other. **293**

cofactor A metal ion or organic molecule that associates with an enzyme and is necessary for its function. **86**

cohesion Property of a substance that arises from the tendency of its molecules to resist separating from one another. **31**

cohort Group of individuals born during the same interval. **444**

colonial organism Organism composed of many similar cells, each capable of living and reproducing on its own. **332**

colonial theory of animal origins Hypothesis that the first animals evolved from a colonial protist. **378**

commensalism Species interaction that benefits one species and neither helps nor harms the other. **456**

community All populations of all species in a given area. **5, 456**

comparative morphology The scientific study of similarities and differences in body plans. **255**

competitive exclusion Process whereby two species compete for a limiting resource, and one drives the other to local extinction. **458**

compound Molecule that has atoms of more than one element. **28**

compound eye Of some arthropods, a motion-sensitive eye made up of many image-forming units, each with its own lens. **389**

concentration Amount of solute per unit volume of a solution. **32**

condensation Chemical reaction in which an enzyme builds a large molecule from smaller subunits; water also forms. **39**

conjugation Mechanism of horizontal gene transfer in which one prokaryote passes a plasmid to another. **320**

conservation biology Field of applied biology that surveys biodiversity and seeks ways to maintain and use it nondestructively. **518**

consumer Organism that obtains energy and carbon by feeding on tissues, wastes, or remains of other organisms; a heterotroph. **6, 474**

continuous variation Range of small differences in a shared trait. **216**

contractile vacuole In freshwater protists, an organelle that collects and expels excess water. **333**

control group In an experiment, group of individuals identical to an experimental group except for the independent variable under investigation. **13**

coral bleaching A coral expels its photosynthetic dinoflagellate symbionts in response to stress and becomes colorless. **503**

coral reef Highly diverse marine ecosystem centered around reefs built by living corals that secrete calcium carbonate. **503**

cotyledon Seed leaf of a flowering plant embryo. **357**

covalent bond Chemical bond in which two atoms share a pair of electrons. **28**

critical thinking The act of judging information before accepting it. **12**

crossing over Process by which homologous chromosomes exchange corresponding segments of DNA during prophase I of meiosis. **198**

crustaceans Mostly marine arthropods with a calcium-hardened cuticle and two pairs of antennae; for example lobsters, crabs, krill, and barnacles. **390**

cuticle Secreted covering at a body surface. **69, 346**

cyanobacteria Photosynthetic, oxygen-producing bacteria. **322**

cytokinesis Cytoplasmic division; process in which a eukaryotic cell divides in two after mitosis or meiosis. **182**

cytoplasm Semifluid substance enclosed by a cell's plasma membrane. **52**

cytoskeleton Network of interconnected protein filaments that support, organize, and move eukaryotic cells and their parts. *See* microtubules, microfilaments, intermediate filaments. **66**

data Experimental results. **13**

decomposer Organism that feeds on wastes and remains; breaks organic material down into its inorganic subunits. **323, 474**

deductive reasoning Using a general idea to make a conclusion about a specific case. **12**

deletion Mutation in which one or more nucleotides are lost from DNA. **156**

demographics Statistics that describe a population. **438**

demographic transition model Model describing the changes in human birth and death rates that occur as a region becomes industrialized. **450**

denature To unravel the shape of a protein or other large biological molecule. **46**

denitrification Conversion of nitrates or nitrites to nitrogen gas. **483**

density-dependent limiting factor Factor that limits population growth and has a greater effect in dense populations; for example, competition for a limited resource. **442**

density-independent limiting factor Factor that limits population growth and acts regardless of population size; for example a flood. **443**

dependent variable In an experiment, a variable that is presumably affected by an independent variable being tested. **13**

derived trait A novel trait present in a clade but not in the clade's ancestors. **294**

desert Biome with little rain and low humidity; plants that have water-storing and water-conserving adaptations predominate. **448**

desertification Conversion of dry grassland to desert. **512**

detrital food chain Food chain in which energy is transferred directly from producers to detritivores. **476**

detritivore Consumer that feeds on small bits of organic material. **474**

deuterostomes Lineage of bilateral animals in which the second opening on the embryo surface develops into a mouth; includes echinoderms and chordates. **377**

development Multistep process by which the first cell of a new multicelled organism gives rise to an adult. **7**

diatom Single-celled photosynthetic protist with a brown accessory pigment in its chloroplasts and a two-part silica shell. **337**

differentiation Process by which cells become specialized during development; occurs as different cells in an embryo begin to use different subsets of their DNA. **142**

diffusion Spontaneous spreading of molecules or ions. **88**

dihybrid cross Cross between two individuals identically heterozygous for two genes; for example $AaBb \times AaBb$. **210**

dikaryotic Having two genetically distinct nuclei in a cell ($n + n$). **365**

dinoflagellate Single-celled, aquatic protist that moves with a whirling motion; may be heterotrophic or photosynthetic. **335**

dinosaur Group of reptiles that includes the ancestors of birds; became extinct at the end of the Cretaceous. **406**

diploid Having two of each type of chromosome characteristic of the species ($2n$). **137**

directional selection Mode of natural selection that shifts an allele's frequency in a consistent direction, so phenotypes at one end of a range of variation are favored. **278**

disruptive selection Mode of natural selection in which traits at the extremes of a range of variation are adaptive, and intermediate forms are not. **281**

distance effect Islands close to a mainland have more species than those farther away. **468**

DNA Deoxyribonucleic acid. Carries hereditary information that guides development and other activities; consists of two chains of nucleotides (adenine, guanine, thymine, and cytosine) twisted into a double helix. **7, 46**

DNA cloning Set of methods that uses living cells to make many identical copies of a DNA fragment. **238**

DNA library Collection of cells that host different fragments of foreign DNA, often representing an organism's entire genome. **240**

DNA ligase Enzyme that seals gaps in double-stranded DNA. **138**

DNA polymerase DNA replication enzyme. Uses one strand of DNA as a template to assemble a complementary strand of DNA from nucleotides. **138**

DNA profiling Identifying an individual by analyzing the unique parts of his or her DNA. **244**

DNA replication Process by which a cell duplicates its DNA before it divides. **138**

DNA sequence Order of nucleotides in a strand of DNA. **135**

DNA sequencing *See* sequencing.

dominance hierarchy Social system in which resources and mating opportunities are unequally distributed within a group. **431**

dominant Refers to an allele that masks the effect of a recessive allele paired with it in heterozygous individuals. **207**

dosage compensation Mechanism in which X chromosome inactivation equalizes gene expression between males and females. **168**

duplication Repeated section of a chromosome. **228**

echinoderms Invertebrates with a water-vascular system and hardened plates and spines embedded in the skin or body. Radials as adults, but bilateral as larvae. For example, a sea star. **394**

ECM *See* extracellular matrix.

ecological footprint Area of Earth's surface required to sustainably support a particular level of development and consumption. **451**

ecological niche All of a species' requirements and roles in an ecosystem. **458**

ecological restoration Actively altering an area in an effort to restore an ecosystem that has been damaged or destroyed. **519**

ecology Study of interactions among organisms, and among organisms and their environment. **438**

ecosystem A community interacting with its environment through a one-way flow of energy and cycling of materials. **5, 474**

ectoderm Outermost tissue layer of an animal embryo. **376**

ectotherm Animal whose body temperature varies with that of its environment; controls its internal temperature by altering its behavior; for example, a fish or a lizard. **406**

electron Negatively charged subatomic particle. **24**

electronegativity Measure of the ability of an atom to pull electrons away from other atoms. **28**

electron transfer chain Array of enzymes and other molecules in a cell membrane that accept and give up electrons in sequence, thus releasing the energy of the electrons in small, usable steps. **85**

electron transfer phosphorylation Process in which electron flow through electron transfer chains sets up a hydrogen ion gradient that drives ATP formation. **104**

electrophoresis Laboratory technique that separates DNA fragments by size. **242**

element A pure substance that consists only of atoms with the same number of protons. **24**

emergent property A characteristic of a system that does not appear in any of the system's component parts. **4**

emerging disease A disease that was previously unknown or has recently begun spreading to a new region. **318**

emigration Movement of individuals out of a population. **440**

endangered species A species that faces extinction in all or a part of its range. **510**

endemic species Species that remains restricted to the area where it evolved. **510**

endergonic Describes a reaction that requires a net input of free energy to proceed. **80**

endocytosis Process by which a cell takes in a small amount of extracellular fluid (and its contents) by the ballooning inward of the plasma membrane. **92**

endoderm Innermost tissue layer of an animal embryo. **376**

endomembrane system Series of interacting organelles (endoplasmic reticulum, Golgi bodies, vesicles) between nucleus and plasma membrane; produces lipids, proteins. **62**

endoplasmic reticulum (ER) Organelle that is a continuous system of sacs and tubes extending from the nuclear envelope. Smooth ER makes lipids and breaks down carbohydrates and fatty acids; rough ER modifies polypeptides made by ribosomes on its surface. **63**

endoskeleton Internal skeleton made up of hardened components such as bones. **401**

endosperm Nutritive tissue in the seeds of flowering plants. **357**

endospore Resistant resting stage of some soil bacteria. **323**

endosymbiont hypothesis Theory that mitochondria and chloroplasts evolved from bacteria that entered and lived in a host cell. **308**

endotherm Animal that maintains its temperature by adjusting its production of metabolic heat; for example, a bird or mammal. **406**

energy The capacity to do work. **78**

enhancer In eukaryotic cells, a binding site in DNA for an activator. **164**

entropy Measure of how much the energy of a system is dispersed. **78**

enzyme Protein or RNA that speeds up a chemical reaction without being changed by it. **39**

epigenetic Refers to heritable changes in gene expression that are not the result of changes in DNA sequence. **172**

epistasis Polygenic inheritance, in which a trait is influenced by multiple genes. **213**

equilibrium model of island biogeography Model that predicts the number of species on an island based on the island's area and distance from the mainland. **468**

ER *See* endoplasmic reticulum.

estuary A highly productive ecosystem where nutrient-rich water from a river mixes with seawater. **502**

eudicot Flowering plant in which the embryo has two seed leaves (cotyledons). For example, a tomato, cherry, or cactus. **357**

eugenics Idea of deliberately improving the genetic qualities of the human race. **248**

euglenoid Flagellated protozoan with multiple mitochondria; may be heterotrophic or have chloroplasts descended from green algae. **333**

eukaryote Organism whose cells characteristically have a nucleus; a protist, fungus, plant, or animal. **8**

eusocial animal Animal that lives in a multigenerational group in which many sterile workers cooperate in all tasks essential to the group's welfare, while a few members of the group produce offspring. **432**

eutrophication Nutrient enrichment of an aquatic habitat. **500**

evaporation Transition of a liquid to a vapor. **31**

evolution Change in a line of descent. **256**

evolutionary tree Diagram showing evolutionary connections. **295**

exaptation Evolutionary adaptation of an existing structure for a completely new purpose. **292**

exergonic Describes a reaction that ends with a net release of free energy. **80**

exocytosis Process by which a cell expels a vesicle's contents to extracellular fluid. **92**

exon Nucleotide sequence that remains in an RNA after post-transcriptional modification. **151**

exoskeleton Of some invertebrates, hard external parts that muscles attach to and move. **389**

exotic species A species that evolved in one community and later became established in a different one. **466**

experiment A test designed to support or falsify a prediction. **12**

experimental group In an experiment, a group of individuals who have a certain characteristic or receive a certain treatment as compared with a control group. **13**

exponential growth A population grows by a fixed percentage in successive time intervals; the size of each increase is determined by the current population size. **440**

extinct Refers to a species that no longer has living members. **292**

extracellular matrix (ECM) Complex mixture of cell secretions; its composition and function vary by cell type. E.g., basement membrane of epithelial tissue. **68**

extreme halophile Organism adapted to life in a highly salty environment. **326**

extreme thermophile Organism adapted to life in a very high-temperature environment. **326**

facilitated diffusion Passive transport mechanism in which a solute follows its concentration gradient across a membrane by moving through a transport protein. **90**

fat Lipid that consists of a glycerol molecule with one, two, or three fatty acid tails. *See* saturated fat, unsaturated fat. **42**

fatty acid Organic compound that consists of a chain of carbon atoms with an acidic carboxyl group at one end. Carbon chain of saturated types has single bonds only; that of unsaturated types has one or more double bonds. **42**

feedback inhibition Regulatory mechanism in which a change that results from some activity decreases or stops the activity. **84**

fermentation A metabolic pathway that breaks down sugars to produce ATP and does not require oxygen. E.g., lactate fermentation. **114**

fertilization Fusion of two gametes to form a zygote; part of sexual reproduction. **195**

first law of thermodynamics Energy cannot be created or destroyed. **78**

fitness Degree of adaptation to an environment, as measured by an individual's relative genetic contribution to future generations. **257**

fixed Refers to an allele for which all members of a population are homozygous. **284**

flagellated protozoan Protist belonging to an entirely or mostly heterotrophic lineage with no cell wall and one or more flagella. **333**

flagellum Long, slender cellular structure used for locomotion through fluid surroundings. **59**

flatworm Bilaterally symmetrical invertebrate with organs but no body cavity; for example, a planarian or tapeworm. **382**

flower Specialized reproductive structure of a flowering plant. **356**

fluid mosaic Model of a cell membrane as a two-dimensional fluid of mixed composition. **56**

food chain Description of who eats whom in one path of energy flow through an ecosystem. **475**

food web Set of cross-connecting food chains. **476**

foraminifera Heterotrophic single-celled protists with a porous calcium carbonate shell and long cytoplasmic extensions. **334**

fossil Physical evidence of an organism that lived in the ancient past. **255**

founder effect After a small group of individuals found a new population, allele frequencies in the new population differ from those in the original population. **284**

free radical Atom with an unpaired electron; most are highly reactive and can damage biological molecules. **27**

frequency-dependent selection Mode of natural selection in which a trait's adaptive value depends on its frequency in a population. **283**

fruit Mature ovary of a flowering plant; often with accessory parts; encloses a seed or seeds. **357**

functional group An atom (other than hydrogen) or a small molecular group bonded to a carbon of an organic compound; imparts a specific chemical property. **39**

fungus Single-celled or multicellular eukaryotic consumer that digests food outside its body, then absorbs the resulting breakdown products. Has chitin-containing cell walls. **9, 364**

gamete Mature, haploid reproductive cell; e.g., an egg or a sperm. **194**

gametophyte Multicelled, haploid, gamete-producing body that forms in the life cycle of land plants and some multicelled algae. **338, 346**

gap junction Cell junction that forms a closable channel across the plasma membranes of adjoining animal cells. **69**

gastropod Mollusk in which the lower body consists of a broad "foot"; for example, a snail or slug. **386**

gastrovascular cavity Saclike gut that also functions in gas exchange. **380**

gene A part of a chromosome that encodes an RNA or protein product in its DNA sequence. Unit of hereditary information. **148**

gene expression Process by which the information in a gene guides assembly of an RNA or protein product. **149**

gene flow The movement of alleles into and out of a population. **285**

gene pool All the alleles of all the genes in a population; a pool of genetic resources. **275**

gene therapy Treating a genetic defect or disorder by transferring a normal or modified gene into the affected individual. **248**

genetically modified organism (GMO) Organism whose genome has been modified by genetic engineering. **246**

genetic code Complete set of sixty-four mRNA codons. **152**

genetic drift Change in allele frequency due to chance alone. **284**

genetic engineering Process by which deliberate changes are introduced into an individual's genome. **246**

genetic equilibrium Theoretical state in which an allele's frequency never changes in a population's gene pool. **276**

genome An organism's complete set of genetic material. **240**

genomics The study of genomes. **244**

genotype The particular set of alleles that is carried by an individual's chromosomes. **207**

genus plural **genera** A group of species that share a unique set of traits; first part of a species name. **10**

geologic time scale Chronology of Earth's history; correlates geologic and evolutionary events. **264**

germ cell Immature reproductive cell that gives rise to haploid gametes when it divides. **194**

global climate change A currently ongoing rise in average temperature that is altering climate patterns around the world. **481**

glomeromycete Fungus that partners with plant roots; fungal hyphae grow inside the cell walls of root cells. **365**

glycolysis Set of reactions in which a six-carbon sugar (such as glucose) is broken down to two pyruvate for a net yield of two ATP. First part of carbohydrate-breakdown pathways. **116**

GMO *See* genetically modified organism.

Golgi body Membrane-enclosed organelle that modifies proteins and lipids, then packages the finished products into vesicles. **63**

Gondwana Supercontinent that existed before Pangea, more than 500 million years ago. **263**

Gram-positive bacteria Bacteria with thick cell walls that are colored purple when prepared for microscopy by Gram staining. **323**

grassland Biome in the interior of continents; perennial grasses and other nonwoody plants adapted to grazing and fire predominate. **497**

grazing food chain Food chain in which energy is transferred from producers to grazers (herbivores). **476**

green alga Single-celled, colonial, or multicelled photosynthetic protist that has chloroplasts containing chlorophylls *a* and *b*. **338**

greenhouse effect Warming of Earth's lower atmosphere and surface as a result of heat trapped by greenhouse gases. **481**

groundwater Soil water and water in aquifers. **478**

growth In multicelled species, an increase in the number, size, and volume of cells. **7**

growth factor Molecule that stimulates mitosis and differentiation. **184**

gymnosperm Seed plant whose seeds are not enclosed within a fruit; a conifer, cycad, ginkgo, or gnetophyte. **354**

habitat Type of environment in which a species typically lives. **456**

Glossary *(continued)*

habituation Learning not to respond to a repeated neutral stimulus. **425**

half-life Characteristic time it takes for half of a quantity of a radioisotope to decay. **260**

haploid Having one of each type of chromosome characteristic of the species. **194**

herbivory An animal feeds on plants or plant parts. **461**

hermaphrodite Animal that has both male and female gonads, either simultaneously or at different times in its life. **379**

heterotroph Consumer. An organism that obtains carbon from organic compounds assembled by other organisms. **100**

heterozygous Having two different alleles of a gene; describes genotype of a diploid organism. **207**

histone Type of protein that structurally organizes eukaryotic chromosomes. **136**

HIV (human immunodeficiency virus) Virus that causes AIDS. **317**

homeostasis Process in which an organism keeps its internal conditions within tolerable ranges by sensing and responding to change. **7**

homeotic gene Type of master gene; its expression controls formation of specific body parts during development. **166**

hominin Human or an extinct primate species more closely related to humans than to any other primates. **412**

Homo erectus Extinct hominin that arose about 1.8 million years ago in East Africa; migrated out of Africa. **414**

Homo habilis Extinct hominin; earliest named *Homo* species; known only from Africa, where it arose 2.3 million years ago. **414**

homologous chromosomes Chromosomes with the same length, shape, and genes. In sexual reproducers, one member of a homologous pair is paternal and the other is maternal. **179**

homologous structures Body structures that are similar in different lineages because they evolved in a common ancestor. **266**

Homo neanderthalensis Extinct hominin; closest known relative of *H. sapiens*; lived in Africa, Europe, Asia. **414**

homozygous Having identical alleles of a gene; describes genotype of a diploid organism. **207**

horizontal gene transfer Transfer of genetic material between existing individuals. **320**

hot spot Threatened region that is habitat for species not found elsewhere and is considered a high priority for conservation efforts. **518**

human immunodeficiency virus *See* HIV.

hybrid The heterozygous offspring of a cross or mating between two individuals that breed true for different forms of a trait. **207**

hydrocarbon Compound or region of one that consists only of carbon and hydrogen atoms. **38**

hydrogen bond Attraction between a covalently bonded hydrogen atom and another atom taking part in a separate covalent bond. Collectively, they impart special properties to liquid water and stabilize the structure of biological molecules. **30**

hydrolysis Water-requiring chemical reaction in which an enzyme breaks a molecule into smaller subunits. **39**

hydrophilic Describes a substance that dissolves easily in water. **30**

hydrophobic Describes a substance that resists dissolving in water. **30**

hydrostatic skeleton Of soft-bodied invertebrates, a fluid-filled chamber that muscles exert force against, redistributing the fluid. **380**

hydrothermal vent Underwater opening where hot, mineral-rich water streams out from an underwater opening in Earth's crust. **302, 504**

hypertonic Describes a fluid that has a high solute concentration relative to another fluid separated by a semipermeable membrane. **88**

hypha Component of a fungal mycelium; a filament made up of cells arranged end to end. **364**

hypothesis Testable explanation of a natural phenomenon. **12**

hypotonic Describes a fluid that has a low solute concentration relative to another fluid separated by a semipermeable membrane. **88**

immigration Movement of individuals into a population. **740**

inbreeding Mating among close relatives. **285**

incomplete dominance Effect in which one allele is not fully dominant over another, so the heterozygous phenotype is an intermediate blend between the two homozygous phenotypes. **212**

independent variable In an experiment, variable that is controlled by an experimenter in order to explore its relationship to a dependent variable. **13**

indicator species Species whose presence and abundance in a community provides information about conditions in the community. **465**

induced-fit model Substrate binding to an active site improves the fit between the two. **82**

inductive reasoning Drawing a conclusion based on observation. **12**

inheritance Transmission of DNA to offspring. **7**

insect Most diverse arthropod group; members have six legs, two antennae, and, in some groups, wings. **392**

insertion Mutation in which one or more nucleotides become inserted into DNA. **156**

instinctive behavior An innate response to a simple stimulus. **424**

intermediate disturbance hypothesis Species richness is greatest in communities with moderate levels of disturbance. **465**

intermediate filament Stable cytoskeletal element that structurally supports cell membranes and tissues. **66**

interphase In a eukaryotic cell cycle, the interval between mitotic divisions when a cell grows, roughly doubles the number of its cytoplasmic components, and replicates its DNA. **178**

interspecific competition Competition between members of different species. **458**

intraspecific competition Competition for resources among members of the same species. **442**

intron Nucleotide sequence that intervenes between exons and is removed during post-transcriptional modification. **151**

inversion Structural rearrangement of a chromosome in which part of the DNA becomes oriented in the reverse direction. **228**

invertebrate Animal that does not have a backbone. **376**

ion Charged atom. **27**

ionic bond Type of chemical bond in which a strong mutual attraction links ions of opposite charge. **28**

isotonic Describes two fluids with identical solute concentrations and separated by a semipermeable membrane. **88**

isotopes Forms of an element that differ in the number of neutrons their atoms carry. **24**

jawless fish Fish with a skeleton of cartilage, no fins or jaws; a lamprey or hagfish. **402**

karyotype Image of an individual's set of chromosomes arranged by size, length, shape, and centromere location. **137**

key innovation An evolutionary adaptation that gives its bearer the opportunity to exploit a particular environment much more efficiently or in a new way. **292**

keystone species A species that has a disproportionately large effect on community structure relative to its abundance. **466**

kinetic energy The energy of motion. **78**

knockout An experiment in which a gene is deliberately inactivated in a living organism; also, an organism that carries a knocked-out gene. **166**

Krebs cycle Cyclic pathway that, along with acetyl–CoA formation, breaks down pyruvate to carbon dioxide in aerobic respiration's second stage. **118**

***K*-selection** Selection favoring traits that allow their bearers to outcompete others for limited resources; occurs when a population is near its environment's carrying capacity. **445**

lactate fermentation Anaerobic sugar breakdown pathway that produces ATP and lactate. **122**

lancelet Invertebrate chordate that has a fishlike shape and retains the defining chordate traits into adulthood. **400**

larva Sexually immature stage in some animal life cycles. **379**

law of independent assortment During meiosis, members of a pair of genes on homologous chromosomes tend to be distributed into gametes independently of other gene pairs. **210**

law of nature Generalization that describes a consistent natural phenomenon for which there is incomplete scientific explanation. **18**

law of segregation The two members of each pair of genes on homologous chromosomes end up in different gametes during meiosis. **209**

lek Of some birds, a communal mating display area for males. **429**

lethal mutation Mutation that alters phenotype so drastically that it causes death. **274**

lichen Composite organism consisting of a fungus and green algae or cyanobacteria. **368**

life history A set of traits related to growth, survival, and reproduction such as life span, age-specific mortality, age at first reproduction, and number of breeding events. **444**

light-dependent reactions First stage of photosynthesis; metabolic pathway that converts light energy to chemical energy. A noncyclic pathway produces oxygen; a cyclic pathway does not. **103**

light-independent reactions Second stage of photosynthesis; metabolic pathway that uses ATP and NADPH to assemble sugars from water and CO_2. E.g., Calvin–Benson cycle in C3 plants. **103**

lignin Material that strengthens the cell walls of vascular plants. **68, 346**

lineage Line of descent. **256**

linkage group All genes on a chromosome. **211**

lipid A fat, steroid, or wax. **42**

lipid bilayer Double layer of lipids (mainly phospholipids) arranged tail-to-tail; structural foundation of all cell membranes. **43**

lobe-finned fish Jawed fish with fleshy fins that contain bones; a coelacanth or lungfish. **403**

locus Location of a gene on a chromosome. **206**

logistic growth A population grows exponentially at first, then growth slows as population size approaches the environment's carrying capacity for that species. **442**

lysogenic pathway Bacteriophage replication path in which viral DNA becomes integrated into the host's chromosome and is passed to the host's descendants. **317**

lysosome Enzyme-filled vesicle that breaks down cellular wastes and debris. **62**

lytic pathway Bacteriophage replication pathway in which a virus immediately replicates in its host and kills it. **316**

macroevolution Large-scale evolutionary patterns and trends; e.g., adaptive radiation, exaptation. **292**

mammal Animal with hair or fur; females secrete milk from mammary glands. **409**

mark–recapture sampling Method of estimating population size of mobile animals by marking individuals, releasing them, then checking the proportion of marks among individuals recaptured at a later time. **439**

marsupial Mammal in which young are born at an early stage and complete development in a pouch on the mother's surface. **409**

mass number Of an isotope, the total number of protons and neutrons in the atomic nucleus. **24**

master gene Gene encoding a product that affects the expression of many other genes. **166**

megaspore Of seed plants, haploid spore that forms in an ovule and gives rise to an egg-producing gametophyte. **352**

meiosis Nuclear division process that halves the chromosome number. Basis of sexual reproduction. **194**

mesoderm Middle tissue layer of a three-layered animal embryo. **376**

messenger RNA (mRNA) RNA that has a protein-building message. **148**

metabolic pathway Series of enzyme-mediated reactions by which cells build, remodel, or break down an organic molecule. **84**

metabolism All of the enzyme-mediated chemical reactions by which cells acquire and use energy as they build and break down organic molecules. **39**

metamorphosis Dramatic remodeling of body form during the transition from larva to adult. **389**

metaphase Stage of mitosis at which all chromosomes are aligned midway between spindle poles. **181**

metastasis The process in which malignant cells of a neoplasm spread from one part of the body to another. **185**

methanogen Organism that produces methane gas (CH_4) as a metabolic by-product. **326**

microevolution Change in allele frequency. **275**

microfilament Cytoskeletal element composed of actin subunits. Reinforces cell membranes; functions in movement and muscle contraction. **66**

microspore Of seed plants, a haploid spore formed in pollen sacs; gives rise to a pollen grain. **352**

microtubule Hollow cytoskeletal element composed of tubulin subunits. Involved in movement of a cell or its parts. **66**

mimicry An evolutionary pattern in which one species becomes more similar in appearance to another. **461**

mitochondrion Double-membraned organelle that produces ATP by aerobic respiration in eukaryotes. **64**

mitosis Nuclear division mechanism that maintains the chromosome number. Basis of body growth and tissue repair in multicelled eukaryotes; also asexual reproduction in some multicelled eukaryotes and many single-celled ones. **178**

model Analogous system used to test an object or event that cannot be tested directly. **12**

mold Fungus that grows as a mass of asexually reproducing hyphae. **365**

molecule Two or more atoms joined by chemical bonds. **4**

mollusk Invertebrate with a reduced coelom and a mantle. For example, a bivalve, gastropod, or cephalopod. **386**

molting Periodic shedding of an outer body layer or part. **388**

monocot Flowering plant with one seed leaf (cotyledon). For example, a grass, orchids, or palm. **357**

monohybrid cross Cross between two individuals identically heterozygous for one gene; e.g., $Aa \times Aa$. **208**

monomers Molecules that are subunits of polymers. **39**

monophyletic group An ancestor in which a derived trait evolved, together with all of its descendants. **294**

monotreme Egg-laying mammal. **409**

monsoon Wind that reverses direction seasonally. **492**

morphological convergence Evolutionary pattern in which similar body parts (analogous structures) evolve separately in different lineages. **266**

morphological divergence Evolutionary pattern in which a body part of an ancestor changes in its descendants. **266**

motor protein Type of energy-using protein that interacts with cytoskeletal elements to move the cell's parts or the whole cell. **66**

mRNA *See* messenger RNA.

multicellular organism Organism that consists of interdependent cells of multiple types. **310, 332**

multiple allele system Gene for which three or more alleles persist in a population at relatively high frequency. **212**

mutation Permanent change in the nucleotide sequence of DNA. **140**

mutualism Species interaction that benefits both species. **457**

mycelium Mass of threadlike filaments (hyphae) that make up the body of a multicelled fungus. **364**

mycorrhiza Mutually beneficial partnership between a fungus and a plant root. **368**

myriapod Long-bodied terrestrial arthropod with one pair of antennae and many similar segments; a centipede or millipede. **390**

natural selection Differential survival and reproduction of individuals of a population based on differences in shared, heritable traits. Driven by environmental pressures. **257**

neoplasm An accumulation of abnormally dividing cells. **184**

nerve net Of cnidarians, a mesh of interacting neurons with no central control organ. **380**

neutral mutation A mutation that has no effect on survival or reproduction. **274**

neutron Uncharged subatomic particle. **24**

niche *See* ecological niche.

nitrification Conversion of ammonium to nitrate. **482**

nitrogen cycle Movement of nitrogen among the atmosphere, soil, and water, and into and out of food webs. **482**

nitrogen fixation Incorporation of nitrogen gas (N_2) into ammonia (NH_3). **322, 482**

nondisjunction Failure of sister chromatids or homologous chromosomes to separate during nuclear division. **230**

nonvascular plant Plant that does not have xylem and phloem; a bryophyte such as a moss. **346**

normal flora Microorganisms that typically live on human surfaces, including the interior tubes and cavities of the digestive and respiratory tracts. **324**

notochord Stiff rod of connective tissue that runs the length of the body in chordate larvae or embryos. **400**

nuclear envelope A double membrane that constitutes the outer boundary of the nucleus. Pores in the membrane control which substances can cross. **61**

nucleic acid Polymer of nucleotides; DNA or RNA. **46**

nucleic acid hybridization Convergence of complementary nucleic acid strands. Arises because of base-pairing interactions. **138**

nucleoid Of a bacterium or archaeon, region of cytoplasm where the DNA is concentrated. **59**

nucleolus In a cell nucleus, a dense, irregularly shaped region where ribosomal subunits are assembled. **61**

nucleoplasm Viscous fluid enclosed by the nuclear envelope. **61**

nucleosome A length of chromosomal DNA wound twice around a spool of histone proteins. **136**

nucleotide Small organic compound that is a subunit of nucleic acids. Consists of a five-carbon sugar, nitrogen-containing base, and one or more phosphate groups. E.g., adenine, guanine, cytosine, thymine, uracil. **46**

nucleus Of a eukaryotic cell, organelle with a double membrane that holds, protects, and controls access to the cell's DNA. **8, 52** Of an atom, core region occupied by protons and neutrons. **24**

nutrient Substance that an organism needs for growth and survival but cannot make for itself. **6**

oncogene Gene that helps transform a normal cell into a tumor cell. **184**

open circulatory system Circulatory system in which hemolymph leaves vessels and flows among tissues before returning to the heart. **386**

operator In prokaryotes, a binding site in DNA for a repressor. **164**

operon Group of genes together with a promoter–operator DNA sequence that controls their transcription. **170**

organ In multicelled organisms, a structure that consists of tissues engaged in a collective task. **5**

organelle Structure that carries out a specialized metabolic function inside a cell; e.g., a mitochondrion. **52**

organic Describes a molecule that consists mainly of carbon and hydrogen atoms. **38**

organism Individual that consists of one or more cells. **4**

organ system In multicelled organisms, set of organs that interact closely in a collective task. **5**

osmosis Diffusion of water across a selectively permeable membrane; occurs when the fluids on either side of the membrane are not isotonic. **89**

osmotic pressure Amount of turgor that prevents osmosis into cytoplasm or other hypertonic fluid. **89**

ovary In flowering plants, the enlarged base of a carpel, inside which one or more ovules form and eggs are fertilized. Matures as a fruit. **356**

ovule Of seed plants, reproductive structure in which egg-bearing gametophyte develops; after fertilization, it matures into a seed. **352**

ozone layer Upper atmospheric region with a high concentration of ozone (O_3) that screens out incoming UV radiation. **307, 516**

Pangea Supercontinent that formed about 270 million years ago. **262**

parapatric speciation Speciation pattern in which populations speciate while in contact along a common border. **291**

parasitism Relationship in which one species withdraws nutrients from another species, without immediately killing it. **462**

parasitoid An insect that lays eggs in another insect, and whose young devour their host from the inside. **462**

passive transport Membrane-crossing mechanism that requires no energy input. **90**

pathogen Disease-causing agent. **318**

PCR *See* polymerase chain reaction.

pedigree Chart showing the pattern of inheritance of a trait through generations in a family. **222**

pelagic province The ocean's open waters. **504**

pellicle Layer of proteins that gives shape to many unwalled, single-celled protists. **333**

peptide Short chain of amino acids linked by peptide bonds. **44**

peptide bond A bond between the amine group of one amino acid and the carboxyl group of another. Joins amino acids in proteins. **44**

per capita growth rate (*r*) Of a population, the change in individuals added over some time interval, divided by the number of individuals in the population. **440**

periodic table Tabular arrangement of all known elements by their atomic number. **24**

permafrost Continually frozen soil layer that lies beneath arctic tundra and prevents water from draining. **499**

peroxisome Enzyme-filled vesicle that breaks down amino acids, fatty acids, and toxic substances. **62**

pH Measure of the number of hydrogen ions in a fluid. Decreases with increasing acidity. **32**

phagocytosis "Cell eating"; an endocytic pathway by which a cell engulfs particles such as microbes or cellular debris. **92**

phenotype An individual's observable traits. **207**

pheromone Chemical that serves as a communication signal among members of an animal species. **426**

phloem Complex vascular tissue of plants; its living sieve elements compose sieve tubes that distribute sugars. Each sieve element has an associated companion cell that provides it with metabolic support. **346**

phospholipid A lipid with a phosphate group in its hydrophilic head, and two nonpolar tails typically derived from fatty acids. Major component of cell membranes. **43**

phosphorus cycle Movement of phosphorus among Earth's rocks and waters, and into and out of food webs. **484**

phosphorylation A phosphate-group transfer. **87**

photoautotroph Organism that obtains carbon from carbon dioxide and energy from light. **321**

photoheterotroph Organism that obtains carbon from organic compounds and energy from light. **321**

photolysis Process by which light energy breaks down a molecule. **104**

photorespiration Reaction in which rubisco attaches oxygen instead of carbon dioxide to ribulose bisphosphate. **106**

photosynthesis Metabolic pathway by which most autotrophs use light energy to make sugars from carbon dioxide and water. Converts light energy into chemical bond energy. **6**

photosystem Cluster of pigments and proteins that converts light energy to chemical energy in photosynthesis. **104**

phylogeny Evolutionary history of a species or group of species. **294**

pigment An organic molecule that selectively absorbs light of certain wavelengths. Reflected light imparts a characteristic color. E.g., chlorophyll. **100**

pilus A protein filament that projects from the surface of some bacterial cells. **59**

pioneer species Species that can colonize a new habitat. **464**

placental mammal Mammal in which maternal and embryonic bloodstreams exchange materials by means of a placenta. **409**

placozoans Group of tiny marine animals having a simple asymmetrical body and a small genome; considered an ancient lineage. **378**

plankton Community of tiny drifting or swimming organisms. **334**

plant Multicelled, typically photosynthetic eukaryote; develops from an embryo that forms on the parent and is nourished by it. **9, 346**

plasma membrane A cell's outermost membrane; controls movement of substances into and out of the cell. **52**

plasmid Of many prokaryotes, a small ring of nonchromosomal DNA. **59**

plasmodesmata Cell junctions that form an open channel between the cytoplasm of adjacent plant cells. **69**

plasmodial slime mold Protist that feeds as a multinucleated mass and forms a spore-bearing structure when environmental conditions become unfavorable. **340**

plastid One of several types of double-membraned organelles in plants and algal cells; for example, a chloroplast or amyloplast. **65**

plate tectonics theory Theory that Earth's outer layer of rock is cracked into plates, the slow movement of which rafts continents to new locations over geologic time. **262**

pleiotropy Effect in which a single gene affects multiple traits. **213**

plot sampling Using demographics observed in sample plots to estimate demographics of a population as a whole. **438**

polarity Separation of charge into positive and negative regions. **28**

pollen grain Walled, immature male gametophyte of a seed plant. Forms in an anther. **347**

pollen sac Of seed plants, reproductive structure in which pollen grains develop. **352**

pollination Arrival of pollen on a receptive stigma of a seed plant. **352**

pollinator An animal that facilitates pollination by moving pollen from one plant to another. **358**

pollutant A substance that is released into the environment by human activities and interferes with the function of organisms that evolved in the absence of the substance or with lower levels. **515**

polymer Molecule that consists of multiple monomers. **39**

polymerase chain reaction (PCR) Method that rapidly generates many copies of a specific section of DNA. **240**

polypeptide Long chain of amino acids linked by peptide bonds. **44**

polyploid Having three or more of each type of chromosome characteristic of the species. **230**

population A group of organisms of the same species who live in a specific location and breed with one another more often than they breed with members of other populations. **5, 274**

population density Number of individuals per unit area. **438**

population distribution Location of population members relative to one another; clumped, uniformly dispersed, or randomly dispersed. **438**

population size Total number of individuals in a population. **438**

potential energy Stored energy. **79**

Precambrian Period from 4.6 billion to 542 million years ago. **310**

predation One species captures, kills, and eats another. **460**

prediction Statement, based on a hypothesis, about a condition that should exist if the hypothesis is correct. **12**

primary endosymbiosis Over generations, evolution of an organelle from bacteria that entered a host cell and lived inside it. **332**

primary production The rate at which an ecosystem's producers capture and store energy. **474**

primary succession A new community becomes established in an area where there was previously no soil. **464**

primary wall The first cell wall of young plant cells. **68**

primate Mammal having grasping hands with nails and a body adapted to climbing; for example, a lemur, monkey, ape, or human. **410**

primer Short, single strand of DNA that base-pairs with a targeted DNA sequence. **138**

prion Infectious protein. **46**

probability The chance that a particular outcome of an event will occur; depends on the total number of outcomes possible. **16**

probe Short fragment of DNA labeled with a tracer; designed to hybridize with a nucleotide sequence of interest. **240**

producer Organism that makes its own food using energy and nonbiological raw materials from the environment; an autotroph. **6, 474**

product A molecule that is produced by a reaction. **80**

prokaryote Informal name for a single-celled organism without a nucleus; a bacterium or archaean. **8**

promoter In DNA, a sequence to which RNA polymerase binds. **150**

prophase Stage of mitosis during which chromosomes condense and become attached to a newly forming spindle. **181**

protein Organic molecule that consists of one or more polypeptides. **44**

proteobacteria Most diverse bacterial lineage. **322**

protist Eukaryote that is not a plant, fungus, or animal. **8, 332**

protocell Membranous sac that contains interacting organic molecules; hypothesized to have formed prior to the earliest life forms. **304**

proton Positively charged subatomic particle. **24**

proto-oncogene Gene that, by mutation, can become an oncogene. **184**

protostomes Lineage of bilateral animals in which the first opening on the embryo surface develops into a mouth. **377**

pseudocoelom Unlined body cavity around the gut. **377**

pseudopod A temporary protrusion that helps some eukaryotic cells move and engulf prey. **67**

Punnett square Diagram used to predict the genetic and phenotypic outcome of a breeding or mating. **208**

pyruvate Three-carbon end product of glycolysis. **116**

radial symmetry Having parts arranged around a central axis, like the spokes of a wheel. **376**

radioactive decay Process by which atoms of a radioisotope emit energy and/or subatomic particles when their nucleus spontaneously breaks up. **25**

radioisotope Isotope with an unstable nucleus. **24**

radiolaria Heterotrophic single-celled protists with a porous shell of silica and long cytoplasmic extensions. **334**

radiometric dating Method of estimating the age of a rock or fossil by measuring the content and proportions of a radioisotope and its daughter elements. **260**

rain shadow Dry region downwind of a coastal mountain range. **493**

ray-finned fish Jawed fish with fins supported by thin rays derived from skin; member of most diverse lineage of fishes. **403**

reactant A molecule that enters a reaction and is changed by participating in it. **80**

reaction Process of molecular change, in which reactants become products. **39**

receptor protein Plasma membrane protein that triggers a change in cell activity after binding to a particular substance. **57**

recessive Refers to an allele with an effect that is masked by a dominant allele on the homologous chromosome. **207**

recognition protein Plasma membrane protein that identifies a cell as belonging to self (one's own body or species). **57**

recombinant DNA A DNA molecule that contains genetic material from more than one organism. **238**

red alga Photosynthetic protist; typically multicelled, with chloroplasts containing red accessory pigments (phycobilins). **338**

redox reaction Oxidation–reduction reaction, in which one molecule accepts electrons (it becomes reduced) from another molecule (which becomes oxidized). Also called electron transfer. **85**

replacement fertility rate Number of children a woman must bear to replace herself with one daughter of reproductive age. **449**

repressor Transcription factor that reduces the rate of transcription. **164**

reproduction Processes by which parents produce offspring. *See* sexual reproduction, asexual reproduction. **7**

reproductive base Of a population, members of the reproductive and pre-reproductive age categories. **449**

reproductive cloning Technology that produces genetically identical individuals. **142**

reproductive isolation The end of gene flow between populations. **286**

reptile Amniote subgroup that includes lizards, snakes, turtles, crocodilians, and birds. **406**

resource partitioning Evolutionary process whereby species become adapted in different ways to access different portions of a limited resource; allows species with similar needs to coexist. **459**

restriction enzyme Type of enzyme that cuts DNA at a specific nucleotide sequence. **238**

retrovirus RNA virus that uses the enzyme reverse transcriptase to produce viral DNA in a host cell. **317**

reverse transcriptase An enzyme that uses mRNA as a template to make a strand of cDNA. **239**

rhizoid Threadlike structure that holds a nonvascular plant in place. **348**

rhizome Stem that grows horizontally along or under the ground. **350**

ribosomal RNA (rRNA) RNA that is part of ribosomes. **148**

ribosome Organelle of protein synthesis. An intact ribosome has two subunits, each composed of rRNA and proteins. Ribosomes are not enclosed by membranes. **59**

ribozyme RNA that functions as an enzyme. **305**

RNA Ribonucleic acid. Nucleic acid with roles in gene expression; consists of a single-stranded chain of nucleotides (adenine, guanine, cytosine, and uracil). *See* messenger RNA, transfer RNA, ribosomal RNA. **46**

RNA polymerase Enzyme that carries out transcription. **150**

RNA world Hypothetical early interval when RNA served as the genetic information. **305**

roundworm Cylindrical worm with a pseudocoelom. **388**

rRNA *See* ribosomal RNA.

***r*-selection** Selection that favors traits that allow their bearers to produce many offspring quickly; occurs when population density is low and resources are abundant. **445**

rubisco Ribulose bisphosphate carboxylase. Carbon-fixing enzyme of the Calvin–Benson cycle. **106**

runoff Water that flows over soil into streams. **478**

sac fungi Fungi that form spores in a sac-shaped structure during sexual reproduction. **365**

salt Compound that releases ions other than H+ and OH− when it dissolves in water. **30**

sampling error Difference between results derived from testing an entire group of events or individuals, and results derived from testing a subset of the group. **16**

saturated fat Triglyceride that has three saturated fatty acid tails. **42**

savanna Biome dominated by perennial grasses with a few scattered shrubs and trees. **497**

science Systematic study of the observable world. **12**

scientific method Systematically making, testing, and evaluating hypotheses about the natural world. **13**

scientific theory Hypothesis that has not been disproven after many years of rigorous testing. **18**

SCNT *See* somatic cell nuclear transfer.

seamount An undersea mountain. **504**

secondary endosymbiosis Evolution of an organelle from a protist that itself contains organelles that arose by primary endosymbiosis. **332**

secondary succession A new community develops in a site where a community previously existed. **464**

secondary wall Lignin-reinforced wall that forms inside the primary wall of a plant cell. **68**

second law of thermodynamics Energy disperses spontaneously. **78**

sedimentary cycle Biochemical cycle in which the atmosphere plays little role and rocks are the major reservoir. **484**

seed Embryo sporophyte of a seed plant packaged with nutritive tissue inside a protective coat. **347**

seedless vascular plant Plant that disperses by releasing spores and has xylem and phloem. For example, a club moss or fern. **350**

segmentation Having a body composed of similar units that repeat along its length. **377**

selfish herd Temporary group that forms when individuals cluster to minimize their individual risk of predation. **430**

semiconservative replication Describes the process of DNA replication, which produces two copies of a DNA molecule: one strand of each copy is new, and the other is parental. **139**

sequencing Method of determining the order of nucleotides in DNA. **242**

sex chromosome Member of a pair of chromosomes that differs between males and females. **137**

sexual dimorphism Difference in appearance between males and females of a species. **282, 428**

sexual reproduction Reproductive mode by which offspring arise from two parents and inherit genes from both. **192**

sexual selection Mode of natural selection in which some individuals outreproduce others because they are better at securing mates. **282**

shell model Model of electron distribution in an atom. **26**

short tandem repeat In chromosomal DNA, sequences of a few nucleotides repeated multiple times in a row. Used in DNA profiling. **216**

single-nucleotide polymorphism (SNP) One-nucleotide DNA sequence variation carried by a measurable percentage of a population. **244**

sister chromatids The two attached DNA molecules of a duplicated eukaryotic chromosome; attachs at the centromere. **136**

sister groups The two lineages that emerge from a node on a cladogram. **295**

SNP *See* single-nucleotide polymorphism.

social animal Animal that lives in a multigenerational group in which members, who are usually relatives, cooperate in some tasks. **430**

soil water Water between soil particles. **478**

solute A substance dissolved in a solvent. **30**

solution Uniform mixure of solute completely dissolved in solvent. **30**

solvent Liquid that can dissolve other substances. **30**

somatic cell nuclear transfer (SCNT) Reproductive cloning method in which the DNA of an adult donor's body cell is transferred into an unfertilized egg. **142**

sorus Cluster of spore-producing capsules on a fern leaf. **350**

speciation Evolutionary process in which new species arise. **286**

species Unique type of organism designated by genus name and specific epithet. Of sexual reproducers, often defined as one or more groups of individuals that can potentially interbreed, produce fertile offspring, and do not interbreed with other groups. **10**

specific epithet Second part of a species name. **10**

spindle Temporary structure that moves chromosomes during nuclear division; consists of microtubules. **181**

spirochetes Bacteria that resemble a stretched-out spring. **323**

sponge Aquatic invertebrate that has no tissues or organs and filters food from the water. **379**

sporophyte Spore-forming diploid body that forms in the life cycle of land plants and some multicelled algae. **338, 346**

stabilizing selection Mode of natural selection in which an intermediate form of a trait is adaptive, and extreme forms are not. **280**

stamen Floral reproductive organ that produces male gametophytes; typically consists of an anther on the tip of a filament. **356**

stasis Evolutionary pattern in which a lineage persists with little or no change over evolutionary time. **292**

statistically significant Refers to a result that is statistically unlikely to have occurred by chance. **16**

steroid Type of lipid with four carbon rings and no fatty acid tails. **43**

stoma Plural **stomata**. Opening across a plant's cuticle and epidermis; can be opened for gas exchange or closed to prevent water loss. **346**

strobilus Of some nonflowering plants, a spore-forming, cone-shaped structure composed of modified leaves. **351**

stroma The cytoplasm-like fluid between the thylakoid membrane and the two outer membranes of a chloroplast. Site of light-independent reactions of photosynthesis. **103**

Glossary *(continued)*

stromatolite Rocky structures composed of layers of bacterial cells and sediments. **307**

substrate Of an enzyme, a reactant that is specifically acted upon by the enzyme. **82**

substrate-level phosphorylation The formation of ATP by the direct transfer of a phosphate group from a substrate to ADP. **116**

surface-to-volume ratio A relationship in which the volume of an object increases with the cube of the diameter, and the surface area increases with the square. Limits cell size. **53**

survivorship curve Graph showing how many members of a cohort remain alive over time. **444**

suspension feeder Animal that filters food from water around it. **379**

symbiosis One species lives in or on another in a commensal, mutualistic, or parasitic relationship. **456**

sympatric speciation Speciation pattern in which speciation occurs within a population, in the absence of a physical barrier to gene flow. **290**

taiga *See* boreal forest.

taxonomy The science of naming and classifying species. **10**

taxon, plural taxa Group of organisms that share a unique set of traits. **10**

telomere Noncoding, repetitive DNA sequence at the end of eukaryotic chromosomes; protects the coding sequences from degradation. **183**

telophase Stage of mitosis during which chromosomes arrive at opposite spindle poles and decondense, and two new nuclei form. **181**

temperate deciduous forest Northern Hemisphere biome in which the main plants are broadleaf trees that lose their leaves in fall and become dormant during cold winters. **496**

temperature Measure of molecular motion. **31**

territory Area an animal or group of animals occupies and defends. **428**

testcross Method of determining genotype by tracking a trait in the offspring of a cross between an individual of unknown genotype and an individual known to be homozygous recessive. **208**

tetrapod Vertebrate with four legs, or a descendant thereof. **401**

theory of inclusive fitness Alleles associated with altruism can be advantageous if the expense of this behavior to the altruist is outweighed by increases in the reproductive success of relatives. **432**

therapeutic cloning The use of SCNT to produce human embryos for research purposes. **142**

threatened species Species likely to become endangered in the near future. **510**

thylakoid membrane A chloroplast's highly folded inner membrane system; forms a continuous compartment in the stroma. Site of light reactions of photosynthesis. **103**

tight junctions In animals, arrays of adhesion proteins that join epithelial cells and collectively prevent fluids from leaking between them. **69**

tissue In multicelled organisms, specialized cells organized in a pattern that allows them to perform a collective function. **4**

total fertility rate Expected number of children a women will bear over the course of a lifetime. **448**

tracer A molecule with a detectable component; researchers can track it after delivery into a body or other system. **25**

trait An observable characteristic of an organism or species. **10**

transcription Process by which enzymes assemble an RNA using the nucleotide sequence of a gene as a template. **148**

transcription factor Protein that influences transcription by binding directly to DNA; for example, an activator or repressor. **164**

transfer RNA (tRNA) RNA that delivers amino acids to a ribosome during translation. **148**

transgenic Refers to a genetically modified organism that carries a gene from a different species. **246**

translation Process by which a polypeptide chain is assembled from amino acids in the order specified by an mRNA. **149**

translocation Of a chromosome, major structural change in which a broken piece gets reattached in the wrong location. In a plant, movement of organic compounds through phloem. **228**

transport protein Protein that allows specific ions or molecules to cross a membrane. Those that function in active transport require an energy input, as from ATP; those that function in facilitated difusion require no energy input. **57**

transposable element Segment of DNA that can move spontaneously within or between chromosomes. **228**

triglyceride A fat with three fatty acid tails. **42**

tRNA *See* transfer RNA.

trophic level Position of an organism in a food chain. **474**

tropical rain forest Highly productive and species-rich biome in which year-round rains and warmth support continuous growth of evergreen broadleaf trees. **495**

trypanosome Parasitic flagellated protist with a single mitochondrion and a flagellum that runs along the back of the cell. **333**

tumor A neoplasm that forms a lump. **184**

tunicate Invertebrate chordate that loses most of its defining chordate traits during the transition to adulthood. **400**

turgor Pressure that a fluid exerts against a wall, membrane, or other structure that contains it. **89**

unsaturated fat Triglyceride that has one or more unsaturated fatty acid tails. **42**

vacuole A membrane-enclosed, fluid-filled organelle that isolates or disposes of waste, debris, or toxic materials. **62**

variable In an experiment, a characteristic or event that differs among individuals or over time. **12**

vascular plant Plant with xylem and phloem. **346**

vector Of a disease, an animal that carries the pathogen from one host to the next. **318**

vertebrate Animal with a backbone. **400**

vesicle Small, membrane-enclosed organelle; different kinds store, transport, or break down their contents; e.g., a peroxisome or lysosome. **62**

viral recombination Multiple strains of virus infect a host simultaneously and swap genes. **319**

viroid Small noncoding RNA that can infect plants. **319**

virus Noncellular, infectious particle of protein and nucleic acid; replicates only in a host cell. **316**

water cycle Movement of water among Earth's oceans, atmosphere, and the freshwater reservoirs on land. **478**

water mold Heterotrophic protist that grows as a mesh of nutrient-absorbing filaments. **337**

water–vascular system Of echinoderms, a system of fluid-filled tubes and tube feet that function in locomotion. **394**

wavelength Distance between the crests of two successive waves. **100**

wax Water-repellent mixture of lipids with long fatty acid tails bonded to long-chain alcohols or carbon rings. **43**

X chromosome inactivation Developmental shutdown of one of the two X chromosomes in the cells of female mammals. **168**

xylem Complex vascular tissue of plants; its dead tracheids and vessel elements distribute water and dissolved minerals through the plant body. **346**

yeast Fungus that lives as single cell. **364**

zero population growth Interval in which births equal deaths. **440**

zygote Cell formed by fusion of two gametes at fertilization; the first cell of a new individual. **195**

zygote fungi Fungi that live in damp places and form a thick-walled zygospore during sexual reproduction. **365**

Index

Figures and tables are indicated by f and t. Glossary terms are indicated with bold page numbers. Applications related to human health are indicated with green bullets and environmental applications with red bullets.

A

D

MindTap Student User Guide

Welcome to MindTap's Student User Guide! This guide will help you get started with MindTap by providing in depth, step-by-step instructions, created specifically for the student user.

Contents

CENGAGE LEARNING
DIGITAL COURSE SUPPORT

Benefits of Using MindTap

MindTap from Cengage Learning represents a new approach to a highly personalized, online learning platform. A cloud-based learning solution, MindTap combines all of your learning tools - readings, multimedia, activities and assessments into a singular Learning Path that guides you through the curriculum. Instructors personalize the experience by customizing the presentation of these learning tools to their students; even seamlessly introducing their own content into the Learning Path via "apps" that integrate into the MindTap platform.

Where to Buy

- Bookstore: **MindTap is available bundled with your textbook for a valuable savings at the bookstore.**

First Time Login

To get started, navigate to: login.cengagebrain.com.
➔If this is your first time using a Cengage Learning Product, you will need to create a new account.

Create Profile

➔ Now you will need to enter the MindTap access code which came with your text.

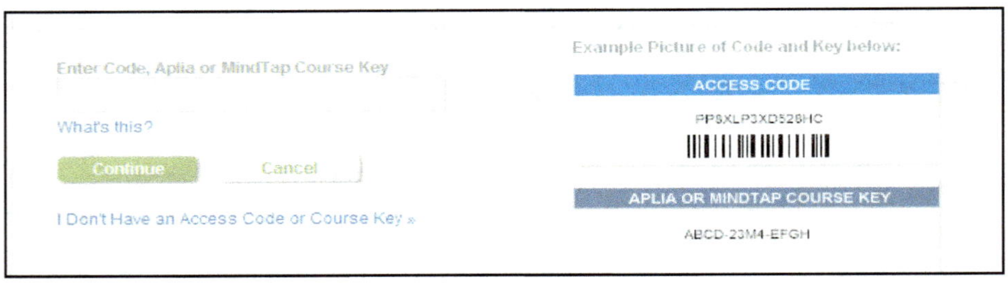

➔ Follow the directions and create a Cengage Learning student account.

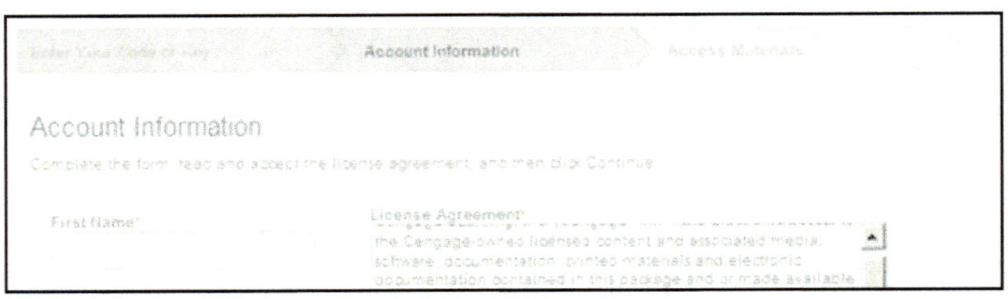

➔ Once completed, you will be directed to your dashboard.

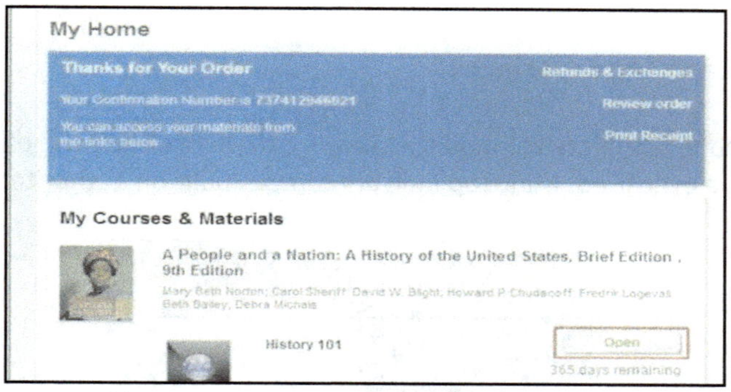

Enter Your Course Key

➔ To access the MindTap product which you just registered, press 

➔ Enter your instructor-provided MindTap CourseKey.

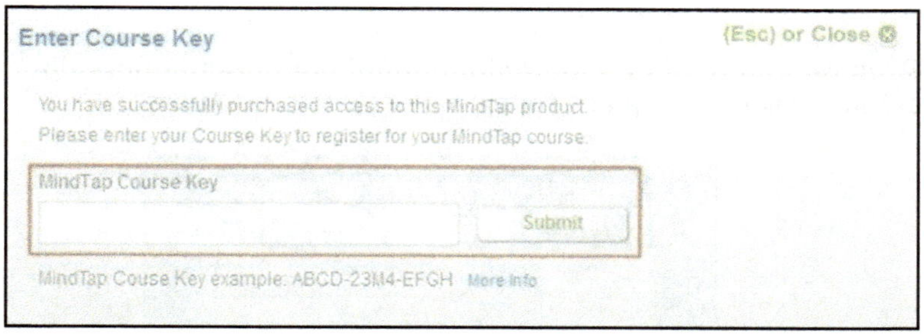

➔ Confirm that the Course Key corresponds to the correct course.

➔ Choose Continue and you are bought directly into your MindTap Course!

Returning User

To get started, navigate to: login.cengagebrain.com. Please log in to your CengageBrain account to activate your product. If you have forgotten your password please select the "Forgot Password" link to retrieve your password.

Walkthrough

Home Page – The Personal Learning Experience

➔ MindTap is a customizable, complete solution that is more than an eBook and different from any learning management system you've ever used. MindTap delivers courses built on authoritative Cengage Learning content.

Learning Path

➔ The MindTap experience begins with the **Learning Path**, which lists Pre-Lecture and Post-Lecture resources for every chapter. For example, click on Chapter 2 for examples of both resources.

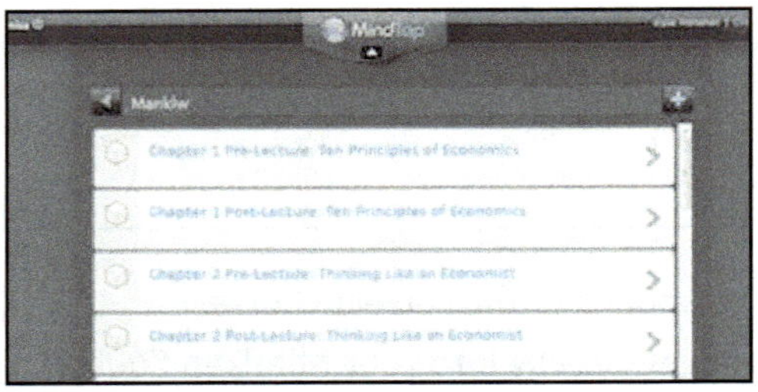

Pre-Lecture

Overview

➔ Each chapter presents the same consistent set of **Pre-Lecture** resources.

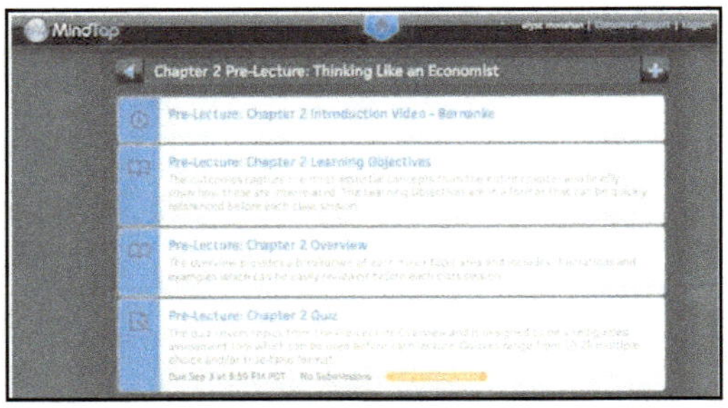

Introduction Video

➔ Every MindTap Pre-Lecture section begins with a relevant **Introduction Video** that highlights key topics—and engages interest.

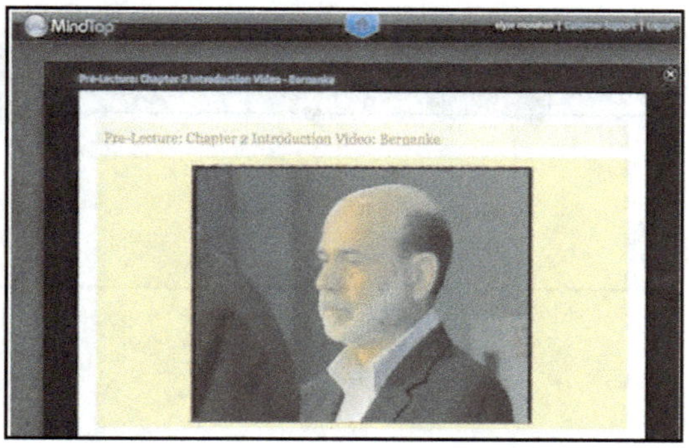

Learning Objectives

➔ After watching the video, you can review chapter **Learning Objectives**, which capture essential concepts in an easy-to-read format.

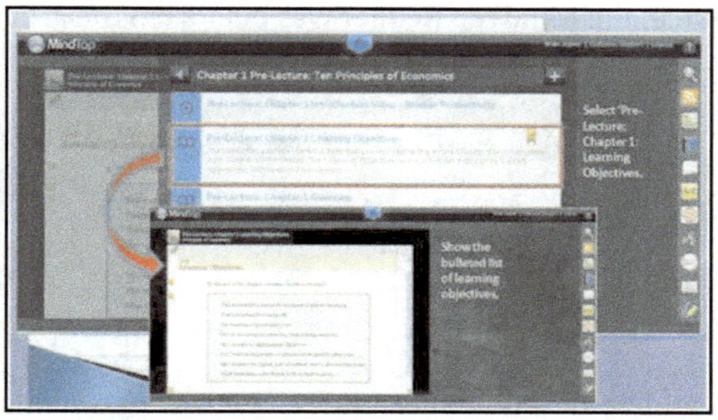

Chapter Overview

➔ Next, read the **Overview**. It provides a breakdown of major topic areas and includes illustrations and examples that students can review before class.

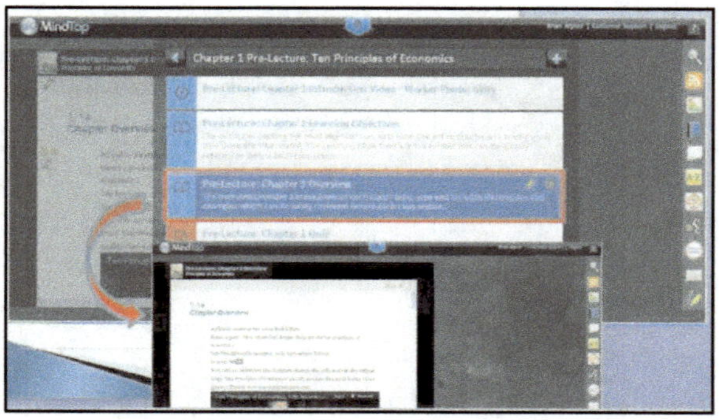

Chapter Quiz

→ The **Pre-Lecture Quiz** provides you with a clear understanding of their strong and weak areas with respect to the topic, so you can tailor your studying accordingly.

Post-Lecture

Overview

→ Within the **Post-Lecture**, students easily access their reading and homework assignments.

Reading Activity

→ An interactive learning resource, the **MindTap Reader** was built from the ground up to create a unique digital reading experience.

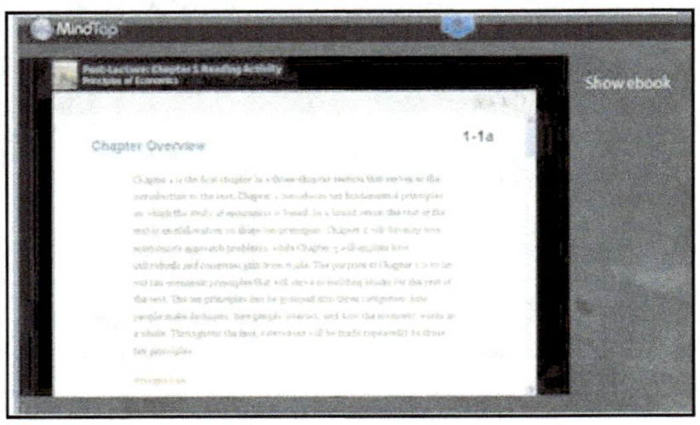

Homework

→ After completing the Reading Activity, move on to your **Homework**. You can complete and submit homework online—and receive instant feedback.

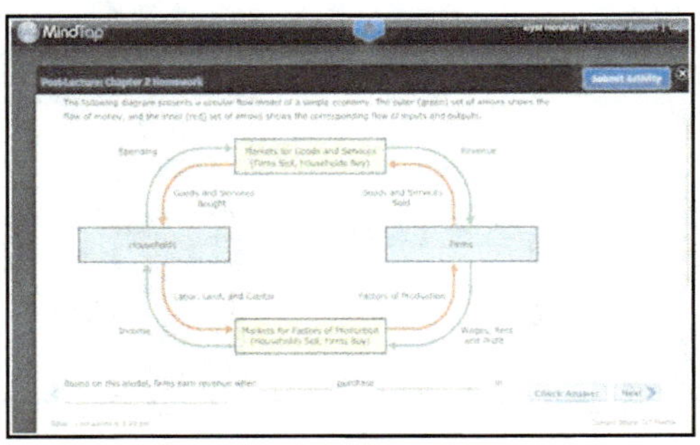

➔ When you answered incorrectly, you can read an explanation that clarifies the correct response.

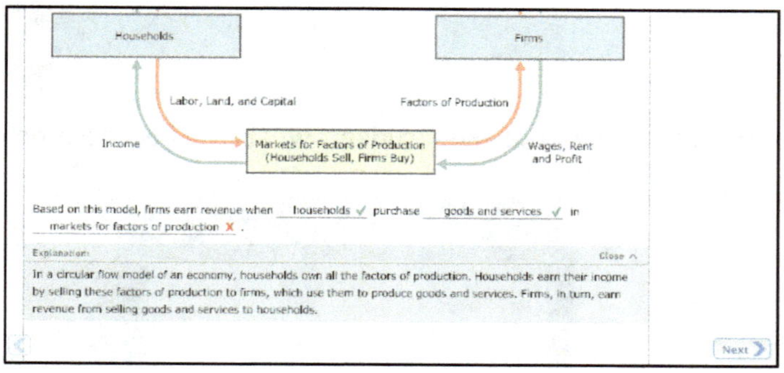

Features

Easy Navigation

➔ Within the reader, users can navigate from section to section or page to page with guidance from page thumbnails and arrows.

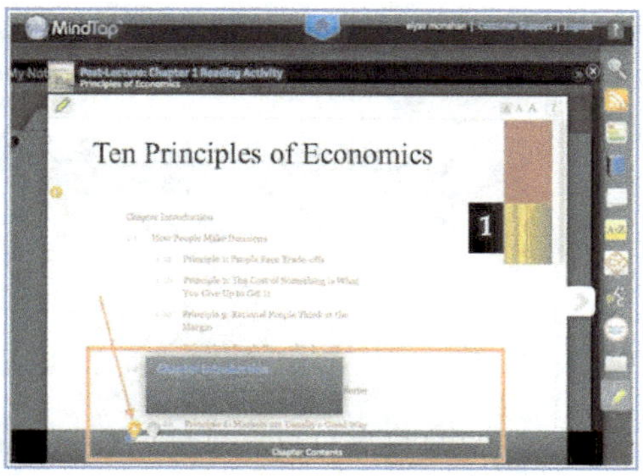

Convenient Tools

➔ The robust functionality of the MindTap Reader allows students to make notes, highlight text, and even find a definition right from the page.

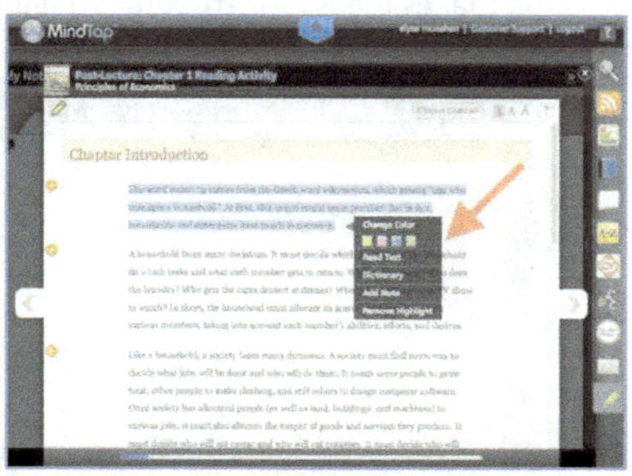

Learning Apps

Overview

➔ A comprehensive library of **Learning Apps** gives students what they need to prepare for a course or exam—all from a single platform.

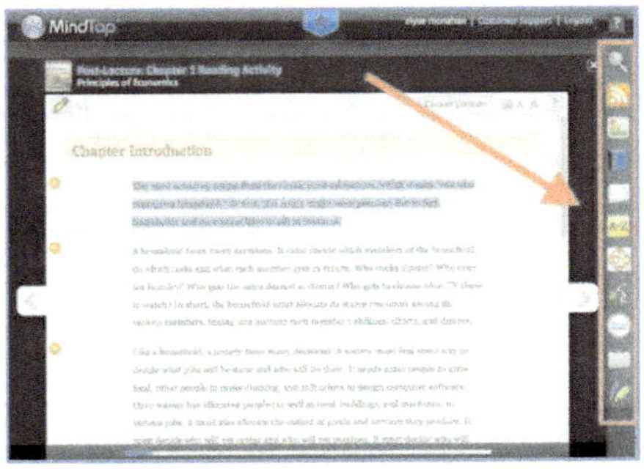

Notebook

➔ By clicking the **My Notes** icon, students can access the **Notebook**, which allows them to write, print, and share their notes. They can also access their notes outside the MindTap environment if they have an EverNote account.

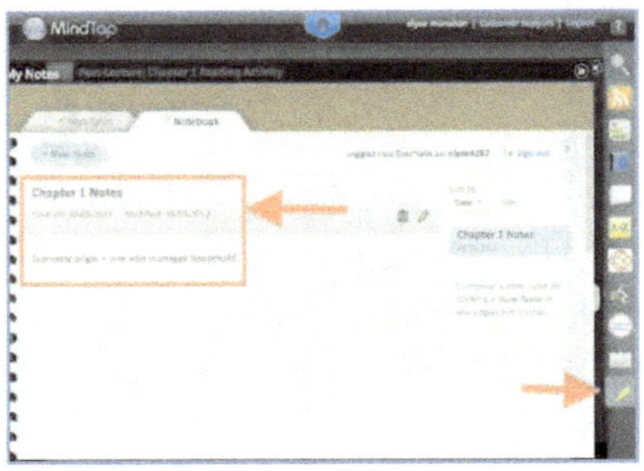

Dictionary

➔ The **Merriam-Webster dictionary app** makes it easy for students to find definitions of new or unfamiliar words. In addition, a click on a word that appears in bold font delivers the definition via the same app.

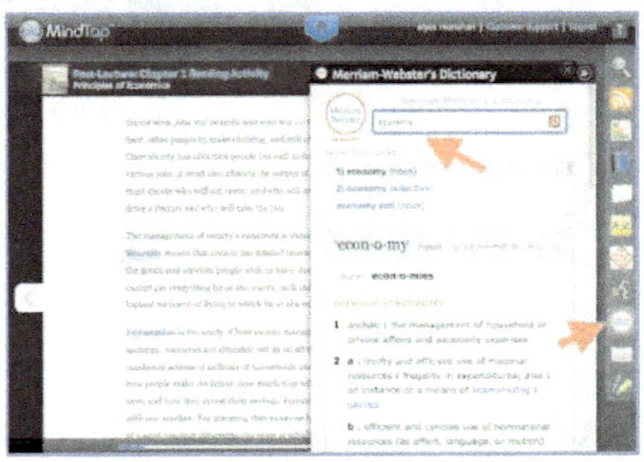

Flashcards

➔ Students can either click through MindTap Reader's **Flashcard app** or create their own flashcards to review key terms and concepts.

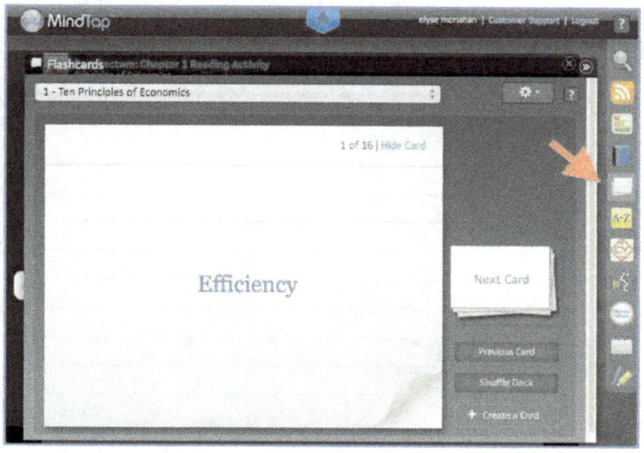

Text-to-Speech ReadSpeaker

➔ If students retain more information by hearing instead of reading, they can select a portion of the assignment and listen to it using the **ReadSpeaker text-to-speech app**.

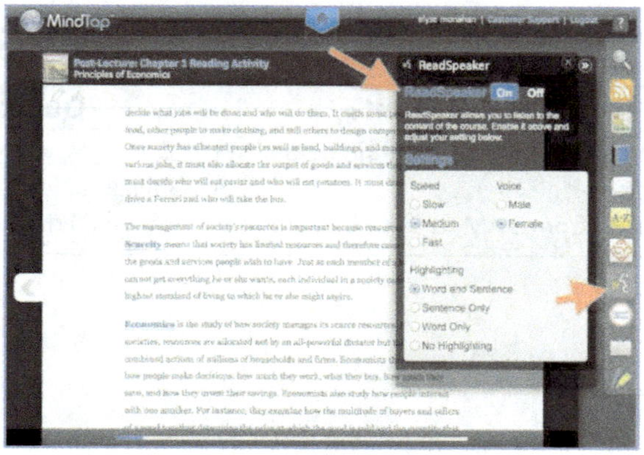

ConnectYard

➔ MindTaps boosts engagement by encouraging students to interact with their content, as well as with their peers, instructors, and tutors via the **ConnectYard social app**.

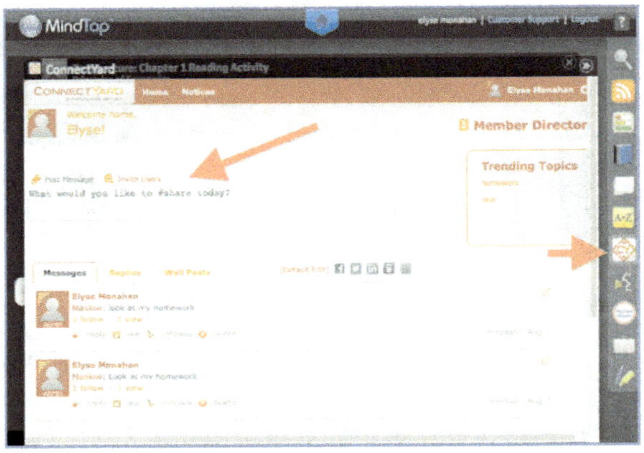

View Progress

➔ The **View Progress App** archives grades and makes them easily available to you, keeping you apprised of personal and current class standings.

Product Support

Questions with your CengageBrain account?	Questions regarding MindTap?
• Check the FAQs in the Support area of your CengageBrain home. • Write to cengagebrain.support@cengage.com • Call 866.994.2427 Mon. through Fri. from 8 AM to 6 PM EST	• Go to www.cengage.com/support for 24/7 live chat! • Call 800.354.9706 Mon. through Thurs. 8:30 AM to 9 PM EST and Fri. 8:30 AM to 6 PM EST

CHAPTER

25 The Dynamic Plant

Learning Outcomes

LO[1] Explain how plants transport and process water

LO[2] Explain how plants transport and process nutrients

LO[3] Identify plant macro- and micronutrients

LO[4] Explain how plants defend themselves

LO[5] Describe methods of weed control

"When yellow star thistle takes over an area, you see one or two plants every inch."

Weeds and Super Weeds

If you've ever done yard work and maintained a grassy backyard, then you've probably learned firsthand why people use the expression "growing like a weed." You pull a dandelion from your lawn this week and it is back next week just as big and hard to pull out as before. What makes weeds so vigorous, so expansive, so—exasperating? Ask America, or close to it.

> **What do you know?**
>
> **True or False? "In plants, water moves only in a one-way path from roots to leaves, and it does so only in the xylem." Explain.**

Dr. Joseph Di Tomaso is an ecologist who studies weeds at the University of California at Davis, and a major part of his job is fighting yellow star thistle. This hairy, spindly, gray-green weed grows to nearly a meter tall and produces a dozen or more bright yellow flowers, each guarded by a whorl of vicious spines. Finding it an attractive ornamental, California settlers first imported yellow star thistle from Europe or Asia before 1870. "Today," says Di Tomaso, "it's probably the single most common plant in California, including all the native species."

The Golden State has its own indigenous thistles, which grow one or two plants per acre. But when yellow star thistle takes over an area, you see one or two plants every *inch*. Each year, yellow star thistle infests 15 to 20 million acres of California's non-agricultural land. That's a collective area equivalent to the entire states of Connecticut, Maryland, and Delaware combined!

How does Dr. Di Tomaso explain the rambunctious nature of weeds in general? And knowing that, how does he then deal with California's colossal overgrowth of yellow star thistle?

The secret of any weed's success, he says, is an age-old question that has been asked a thousand times but never answered. Weeds, however, do tend to share certain traits: Rapid growth. Abundant seed production. The ability to spread through stems and roots as well as seeds. And, tremendous genetic variability, allowing them to occupy different kinds of habitats.

Dr. Di Tomaso uses a broad knowledge of plant physiology in his work to eradicate weeds. One area of plant physiology that he finds quite helpful is plant transport: how plants move the products of photosynthesis around internally, and how they transfer water, minerals, and other needed substances in a dynamic pattern that sustains life. Plant transport—our focus in the current chapter—is important to understanding why yellow star thistle can survive California's dry season. Along the way, you'll find answers to these central questions:

- How does water reach the top of tall trees?
- How does a plant avoid losing too much water on a hot day?
- Which nutrients must a plant have to grow?
- What is the worry over super weeds all about?

LO[1] Water Balance in Plants

When Dr. Di Tomaso pulls up a yellow star thistle, he is invariably impressed with the length of its taproot. This structure is ten times longer than the short, fibrous roots of the native grasses that occupied California prairies before pioneers introduced the noxious yellow weed. In late spring, a star thistle plant may stand only three inches tall, but its taproot already extends three feet down into the soil! These long roots are a key to yellow star thistle's success, because for nearly every land plant, water is the most important need. Over the course of one growing season, a single tall plant—corn or sunflower, let's say—consumes 17 times more water than a person does! But the water essentially passes right through; all but two percent of the imbibed moisture evaporates from the plant's leaves.

A plant's insatiable "thirst" is based on a fundamental difference in the way plant and animal bodies work. The fluids in most animal bodies travel in a recirculating pathway, maintained in a circulatory system that recycles water by and through organs with little water loss and relatively little input. In contrast, plant fluids travel in a flow-through pathway, a generally one-way path from roots through stems to leaves, then out into the environment (Fig. 25.1). A steady supply of water enables a plant to carry out photosynthesis (discussed in Chapter 3). Water also stops the plant from wilting, allowing it to remain erect with turgid or swollen cells. Furthermore, water helps a plant to transport substances internally and to stay cool despite a baking sun on outstretched leaves and stems. In California, in the desiccating heat of late summer, when shallow-rooted annual plants are withering and browning, the star thistle continues to suck moisture from the deep soil—and grow like a weed. Star thistle, in fact, draws up water so efficiently through its taproot and grows so luxuriantly that it can starve its competitors of light—even those that don't die off from drought alone.

Let's trace the one-way flow of water through a plant and see how the physical properties of water and plant architecture propel this life-sustaining movement.

How Roots Draw Water from the Soil

Rains come mostly in the winter and early spring in California, and a yellow star thistle absorbs this rainwater from the soil into its roots (Fig. 25.2). Millions of root hairs project from the cells of lateral roots into the spaces between soil particles. These hairs have a huge combined surface area for absorption. In many plants, root-associated fungi called *mycorrhizae* contribute to water uptake (review Chapter 12). After the water enters the roots, the plant stores some of it in special parenchyma cells in the root's innermost cylinder of tissue.

Water normally enters roots by simple diffusion, with no energy expenditure by the plant. Water diffuses into root hairs and then passes through the root cortex. Some of the fluid moves through the cytoplasm of root cells, but most of the water passes within the spaces between cell wall fibers (Fig. 25.2, Steps (1) and (2)). Next, water reaches the endodermis, the cell layer that separates the root cortex from the central cylinder, containing the plant's fluid conducting tubes, the xylem and phloem (review

Figure 25.1

Water Flows Through Plants in a One-Way Path

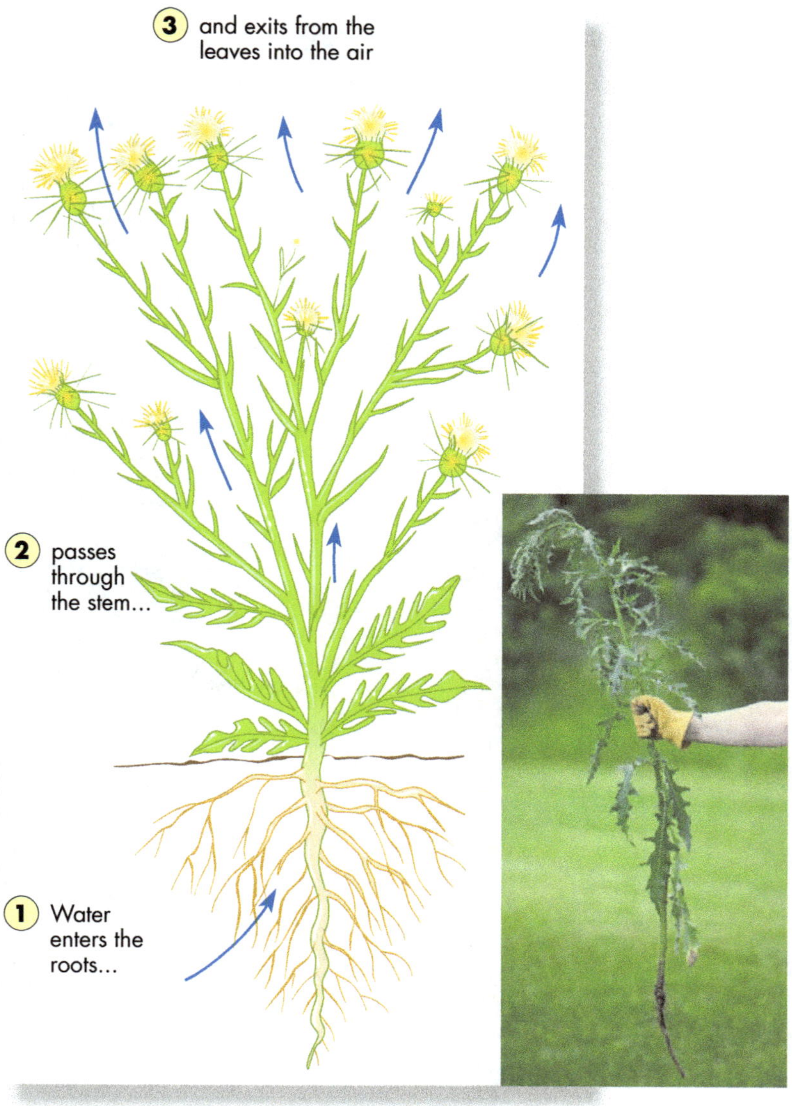

© SHERWIN MCGEHEE/ISTOCKPHOTO.COM

Figs. 23.7 and 23.8). Water can no longer move along and through the cell walls because the Casparian strip, the waxy belt surrounding each endodermal cell, acts like a gasket that prevents water from flowing in the spaces between cells. The Casparian strip forces water to pass through the cytoplasm of endodermal cells before it enters the hollow, water-conducting tubes of the xylem (Fig. 25.2, Step 3). Some water enters the xylem by diffusion, and some is drawn in as water leaves the root xylem and moves upward toward the stems and leaves.

Figure 25.2
How Water Enters Root Hairs and Passes into the Vascular Cylinder

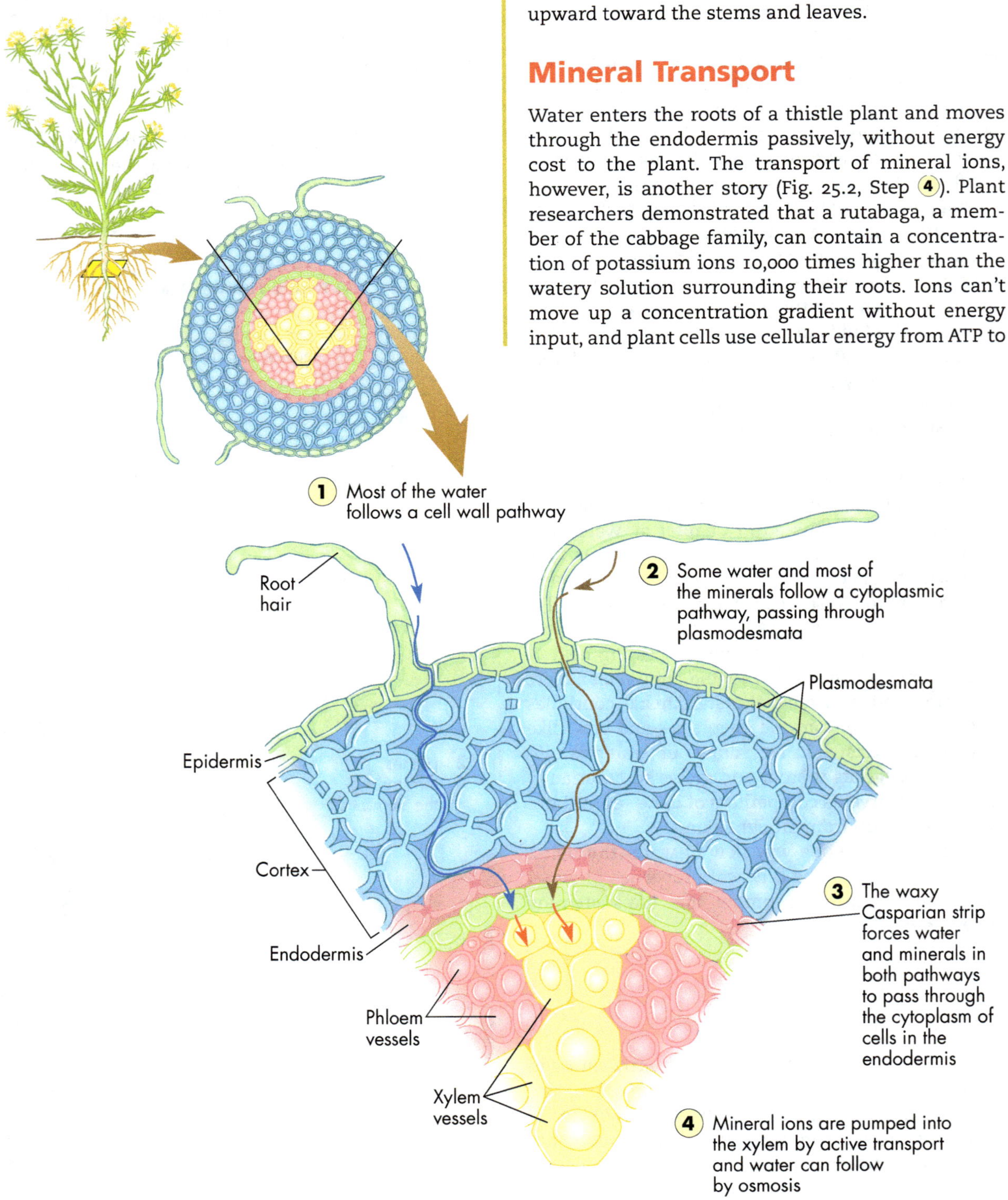

Mineral Transport

Water enters the roots of a thistle plant and moves through the endodermis passively, without energy cost to the plant. The transport of mineral ions, however, is another story (Fig. 25.2, Step 4). Plant researchers demonstrated that a rutabaga, a member of the cabbage family, can contain a concentration of potassium ions 10,000 times higher than the watery solution surrounding their roots. Ions can't move up a concentration gradient without energy input, and plant cells use cellular energy from ATP to

guttation upward pressure from plant roots producing tiny droplets of water that rim leaves early in the morning

root pressure force that is generated when fluid moves up xylem cells; root pressure can force water out of the leaves, where it accumulates in tiny droplets

capillary action the tendency of water to move upward in a thin tube; studies show that capillary action occurs in a plant's narrow xylem vessels

accumulate minerals in two ways. First, root epidermal cells have plasma membrane "transporters" or special proteins that actively transport minerals across the membrane barrier. Second, the root's parenchyma cells also have membrane proteins that actively pump ions into the adjacent xylem. Water then enters the xylem via osmosis and ions move throughout the plant dissolved in the water.

Water Transport: Root to Leaf

So far, we've seen that water moves into a plant's roots by simple diffusion and ions enter the xylem by active transport. But how does water reach the crown of a 1-meter-tall yellow star thistle, or more impressively, the top of a California redwood tree, Earth's tallest living thing, towering more than 100 m (328 ft) above the forest floor? By the same simple mechanisms?

Plant scientists have proposed four possible hypotheses about water transport to the tops of plants:

1. The plant root might *push* water upward.
2. Pumps distributed along the xylem might convey water upward like a bucket brigade.
3. Water might move upward by capillary action in the xylem.
4. Evaporation of water from the leaves (as water changes from liquid to gas and moves into the atmosphere) might *pull* water up from below.

But which is correct?

Root Pressure, Xylem Pumps, and Capillary Action

Farmers have long noticed that sometimes, in the early morning, the leaves of grass, tomatoes, strawberries, and numerous other plants are rimmed with tiny droplets of water. Plant researchers call this guttation, and it results from upward pressure from the roots. Could guttation force water all the way to the tops of trees?

Guttation droplets appear at special openings on leaf edges when the soil is nearly saturated with water and the leaves are not losing much through evaporation—in early morning, for example. Water and minerals leave the water-soaked soil and enter the vascular cylinder. Membrane proteins in the root's parenchyma cells expend energy pumping ions into the xylem; water then follows by the process of osmosis. The volume of fluid in the xylem increases, and this fluid must move somewhere. It tends not to flow back between the endodermal cells because the waxy Casparian strips prevent both the solutes and the water from leaking back out of the vascular cylinder (see Fig. 25.2). The fluid *can* move up the xylem, however, generating a force called root pressure. Root pressure can force water out of the leaves, where it can accumulate in tiny droplets.

After measuring the force of root pressure, botanists determined that it's much lower than the force needed to move water to the tops of tall trees. Such a force would have to be about 150 pounds per square inch (psi), roughly equivalent to the force you would exert on the floor by balancing all your weight on the tip of one big toe. What's more, many plants, including pines, don't develop root pressure at all, but still grow hundreds of feet tall; this rules out hypothesis 1.

As for hypothesis 2, you can perform a simple test of it yourself. To determine whether water is pushed along via little pumps in xylem tubes, you can pierce a plant stem with a needle. If the fluid was under pressure like the water in a garden hose or the blood in an artery, it would spew out from the hole you made. In fact, no fluid comes out—rather, air is sucked in. This rules out hypothesis 2.

©SCOTT CAMAZINE/PHOTO RESEARCHERS, INC.

Guttation

© DAVID FRANKLIN/ISTOCKPHOTO.COM

Hypothesis 3, capillary action, is the tendency of water to move upward in a thin tube, and indeed, studies show that capillary action does occur in narrow xylem vessels. However, water can cling to and creep up the inside of narrow tubes only to a height of about 0.5 m (20 in). While this would be enough for young yellow star thistle plants, it's not enough for mature

Figure 25.3
Transpiration-Cohesion-Tension Mechanism of Water Transport

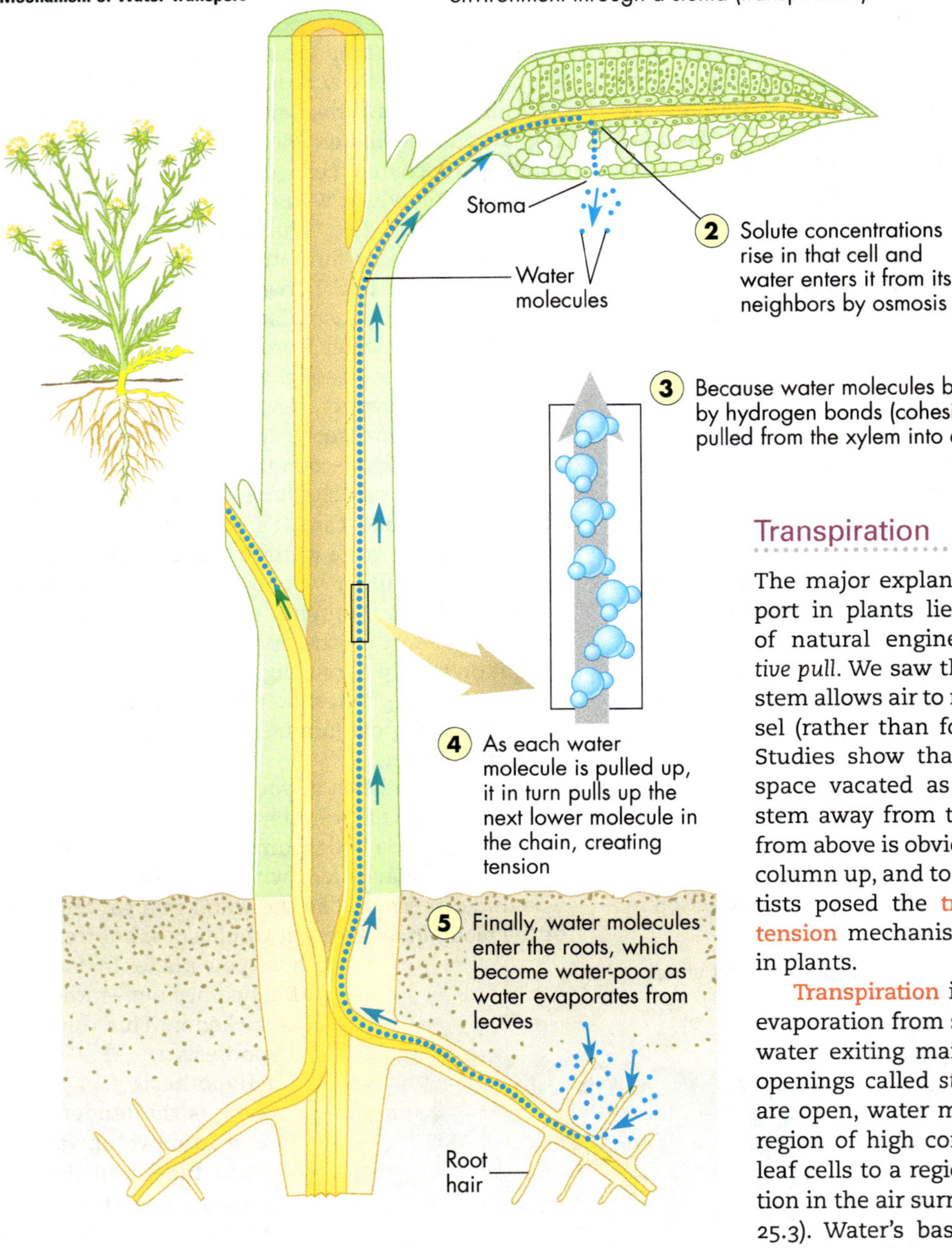

transpiration-cohesion-tension the mechanism scientists propose for water transport in plants, which explains how a column of water molecules is "pulled" up the stem

transpiration the loss of water from plants by evaporation, mainly through the stomata on stems and leaves

cohesion the tendency of like molecules to cling to each other

thistle plants and far too little for redwood trees. This rules out hypothesis 3.

Plant physiologists were left with hypothesis 4—that water is somehow *pulled* upward to the tops of tall plants. And this explanation turned out to be correct, as you'll read next.

Transpiration

The major explanation for water transport in plants lies in a fascinating bit of natural engineering called *evaporative pull*. We saw that puncturing a plant stem allows air to move into a xylem vessel (rather than for water to spew out). Studies show that air, in fact, fills the space vacated as water moves up the stem away from the wound. Something from above is obviously pulling the water column up, and to explain it, plant scientists posed the transpiration-cohesion-tension mechanism for water transport in plants.

Transpiration is the loss of water by evaporation from stems and leaves, with water exiting mainly through the little openings called *stomata*. When stomata are open, water molecules move from a region of high concentration inside the leaf cells to a region of lower concentration in the air surrounding the plant (Fig. 25.3). Water's basic physical properties allow it to form a strong, "sticky" chain that helps explain transpiration from roots to leaves. Water molecules tend both to *cohere* strongly to other water molecules and to *adhere* to unlike molecules. In a xylem tube, hydrogen bonds (review Fig. 2.3b) link water molecules to each other in a long, unbroken liquid chain (cohesion), while additional hydrogen bonding causes water to adhere to the cellulose lining of the xylem vessels (adhesion).

When a water molecule evaporates from an open stoma in a leaf, the water molecule below it in the column tends to move up and replace it, and, in turn, create tension and pull along the next water molecule. As a result, the entire liquid column moves upward in the xylem tube, and a new water molecule is drawn into the roots below. Because water molecules pull each other up from above, the entire column is under constant tension. As long as the water column remains unbroken and solar energy causes evaporation, water keeps moving in the plant's vascular tissues. This force also causes air to enter the xylem vessels of cut flowers in a bouquet. That's why florists recommend you submerge a fresh flower stem and remove an additional inch or two while it's still underwater. This eliminates the section containing the air bubbles, and allows the water column to reestablish itself so transpiration can continue and the flower will stay fresh longer.

Water Stress: A Break in the Chain

On a hot, dry day, evaporation from the leaves of a tree can outstrip water absorption by the roots, and tension on the chain of water molecules in each of the millions of xylem tubes becomes greater and greater. When this occurs, the adhesive forces of water molecules pulling inward against the xylem walls can grow so strong that the diameter of a tree trunk literally shrinks, a bit like a soda straw sucked flat. In summer, the upward tension on the chain of water molecules in a plant can become so intense that the column can snap. In fact, plant physiologists with sensitive microphones can actually hear the snapping and popping inside plants on a hot, dry day!

Once the water column breaks, an air bubble forms inside the xylem tube, and transpiration can no longer pull water from above. If the temperature drops, evaporation slows, or the soil becomes damp again fairly quickly, root pressure from below (as well as capillary action within the narrow xylem tubes) can rejoin the ends of the broken water column. If conditions aren't reversed soon enough, however, wilting can destroy the plant.

Stomata and Regulation of Water Loss

Transpiration rate is crucial to a plant's daily survival, and plants carefully regulate their water loss. In the leaf epidermis, each stoma is enclosed by two kidney-shaped guard cells (Fig. 25.4, Step 1). Like curved water balloons, guard cells arch away from each other when they swell with water, increasing the size of the opening between them (Step 2). When guard cells lose water and become flaccid, they slump together and close off the stoma, blocking water loss via transpiration (Step 3). In a plant

Figure 25.4

Stomata: Regulators of Water Loss

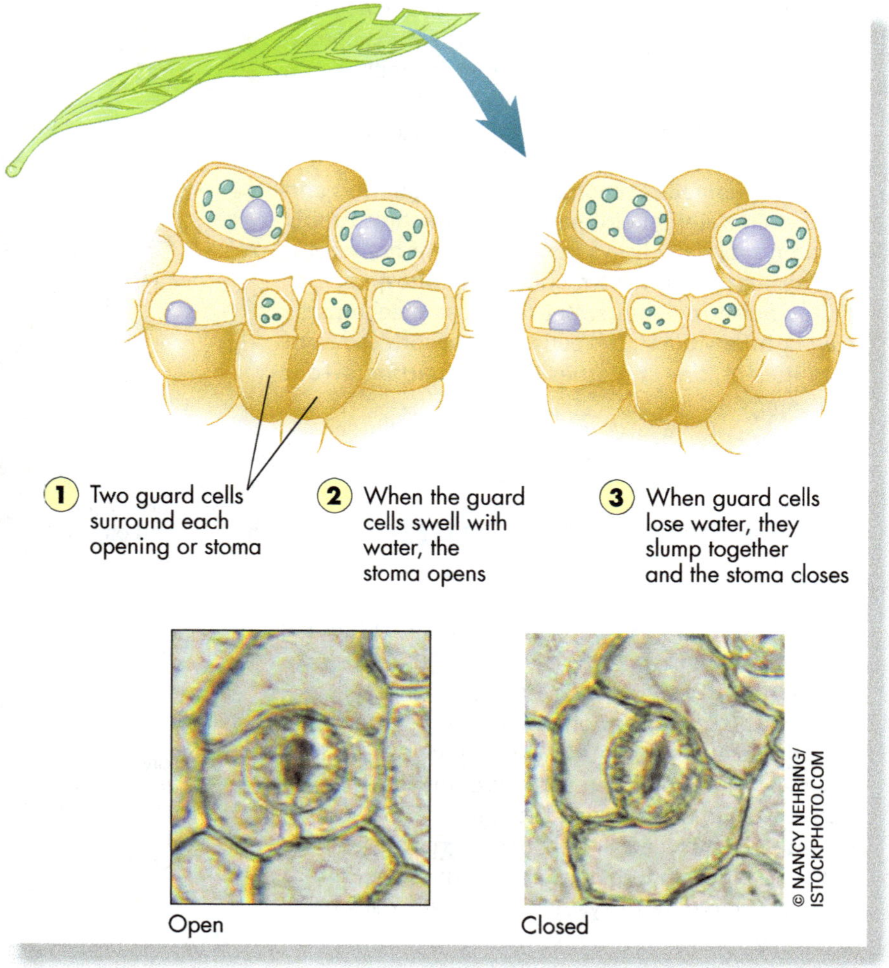

such as the yellow star thistle, the leaves are covered with gray, cottony hairs and these may also help decrease evaporation from the stomata.

What causes the guard cells flanking the stoma to swell or deflate and thus open or close the plant's "portholes"? Experiments show that when a plant begins to dry out, guard cells pump out potassium ions and water follows, leaving passively by osmosis. Loss of water from guard cells allows them to slump together, close the pore, and stop further water loss. After a rain, when the plant has plenty of water, potassium ions are pumped into guard cells, water follows by osmosis, guard cells swell, the stoma opens, and carbon dioxide can enter the leaf once more, allowing photosynthesis to continue.

In addition to drying, several other environmental agents affect guard cells. These include carbon dioxide, light, temperature, and daily rhythms in the plant. Together, they cause stomata to stay open, in general, when gas exchange can take place without threat of excess water loss, but to stay closed when there's a danger of dehydration. Plants such as the yellow star thistle that live in arid environments often close their stomata for much of the day. Instead, the stomata open and the plant absorbs carbon dioxide only at night, when the air is cooler. This prevents the plant from losing too much water. Photosynthesis is then completed during the day when light is available, but it takes place with the stomata closed.

The combined mechanisms of transpiration and guard cell swelling help maintain a constant supply of water to plant cells—a major aspect of homeostasis. These basic principles hold for gymnosperms as well as flowering plants. But for these mechanisms to work in any plant, the cells must absorb the right quantities of ions and other mineral nutrients—our next topic.

LO2 Movement of Nutrients Throughout the Plant

Star thistle has a built-in distribution problem: In the rainy spring, it generates sugars in its leaves but stores organic nutrients in its roots. Then in the late summer, when photosynthesis is difficult because stomata must stay closed to conserve water, the thistle must move nutrients stored in its roots upward to power the construction of flowers and seeds (Fig. 25.5). How does a plant distribute and redistribute photosynthetic products without a pump and circulatory system like an animal's heart and blood vessels?

Translocation: Movement of Sugar

translocation in plants, the transport of solutes in phloem cells

Plant scientists devised some clever methods for studying translocation, the movement of nutrients in the phloem of plants. Phloem is so delicate that puncturing the tubes even with a fine glass needle stops the flow of solutes. Tiny insects called *aphids*, however, can suck plant sap from phloem cells without disrupting them. Plant physiologists found that if they allow an aphid to insert its sharp feeding tube into the phloem, then cut away the insect's body, the phloem's contents ooze out of the little tube, and they can study this plant sap. Using the aphid technique, they discovered that the plant sap in phloem

Figure 25.5
Seasonal Sugar Transport

pressure flow hypothesis the best current hypothesis for phloem transport, which asserts that phloem transports substances from "sources," or areas of production, to "sinks," or areas of use, down a pressure gradient established via osmosis

tubes contains about 30 percent sugars (mostly sucrose) and 70 percent water.

Plant scientists also discovered that they can expose a plant's leaves to radioactive carbon dioxide, and once the carbon is incorporated into sugars by photosynthesis, they can trace its path through the phloem. With this radioactive procedure, they found that sugar often flows downward from leaves to other plant parts and this flow can be 10,000 times faster than normal diffusion. This convinces them that translocation involves a mass movement of phloem fluid based on bulk flow.

The best current hypothesis for phloem transport is the **pressure flow hypothesis**, the idea that phloem transports substances from "sources," or areas of production, to "sinks," or areas of use, down a pressure gradient set up by osmosis. Let's see how this works in a flowering plant such as the yellow star thistle. Photosynthesizing cells in the leaves (the source) produce sugars, and the companion cells load the sugars into the phloem's sieve tube elements by an active transport process (Fig. 25.6, Step ①). This energy-costly transport increases the solute concentration locally in the phloem, and water follows by osmosis from the xylem's nearby conducting elements (Step ②). The influx of water into the phloem cells plumps them up, increasing their internal pressure like little water balloons. This force is called *turgor pressure* (Latin, *turgere* = to be swollen). This pressure in the phloem forces the sugary solution through sieve plates at the ends of each sieve tube member and carries it away from the leaf (Step ③).

Meanwhile, cells in the root and other parts of the plant (the sinks) remove sugary solutes from the phloem (Step ④), and the solute content drops, decreasing the osmotic pressure in those regions. Now water flows out of the phloem and back into the xylem tubes, where transpirational pull carries it upward again (Step ⑤). Ordinary water pressure, the active loading activities of companion cells, and the unloading of sugars by root cells are thus behind the movement of sugars, amino acids, and a few mineral ions from one part of a yellow star thistle plant to another.

In the early spring, a similar mechanism probably causes plant sap to rise in the stems of biennials and perennials. As the weather warms and the days lengthen, the roots (now the sources) begin breaking down stored starch into sugar and loading it into the phloem's sieve tube members. Water from the xylem follows by osmosis, and as the roots bring in more water from the soil, enough pressure is created to push the sap up the phloem and into the shoot, where the emerging leaves (now the sink) take sugar from the

Figure 25.6
Translocation: How Plants Transport Sugars

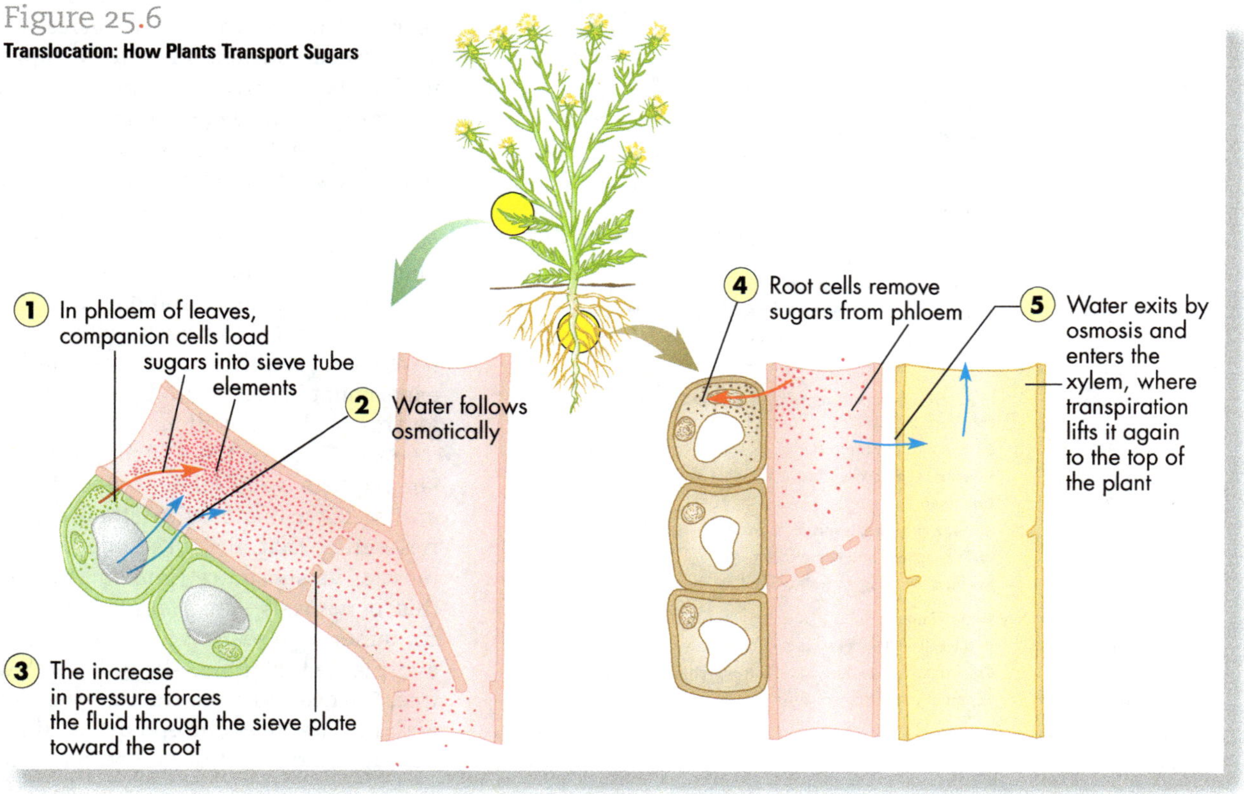

phloem and use it for energy and raw materials. Ever hear the phrase, "the sap's rising"? New Englanders carefully tap into the phloem of maple trees in early spring, drain off some of the sweet rising maple sap, and boil it down to make maple syrup.

In a sense, sugar translocates itself, since its production and use create osmotic pressure and mass flow. Other nutrients, however, move passively by the flow of fluid in both the xylem and the phloem.

Transport of Inorganic Substances in Xylem and Phloem

Plants transport a long list of nutrients (in addition to carbon, oxygen, and hydrogen in sugars) (Table 25.1) and they do it in either of two ways:

1. Some nutrients, such as calcium, move only in the xylem; they're carried along in the column of water that moves upward from the roots to the leaves via transpiration. These nutrients are deposited permanently wherever they leave the xylem and enter a living cell, and they can't be redistributed to the other parts of the plant.
2. Other nutrients, such as sulfur or phosphorus, can be redistributed to new tissues as needed, carried along passively in the phloem with the sugars and other phloem contents.

Plants can have mineral deficiencies just like animals can, and the two different modes of transport explain why different mineral deficiencies affect plants in different ways. Calcium deficiency, for example, always produces symptoms first in new leaves. The old leaves contain enough calcium, but the mineral can't move from these old leaves to the new leaves. In contrast, phosphorus deficiency always produces symptoms first in old leaves; as the plant's environment becomes depleted of the element, the plant mobilizes phosphorus in older leaves and transports it through the phloem to areas of new growth. Plants with phosphorus deficiency therefore have the characteristic yellow veins in their old leaves long before new leaves show this sign.

Table 25.1

Plant Nutrients and Their Functions

Nutients (Percent of Dry Weight)*	Location/Function
Macronutrients	
Carbon (45.0)	In all organic compounds
Oxygen (45.0)	In most organic compounds, including all sugars and carbohydrates
Hydrogen (6.0)	In all organic compounds
Nitrogen (1.0–4.0)	In proteins, nucleic acids, chlorophyll, and coenzymes
Potassium (1.0)	Involved in activating enzymes, protein synthesis, and regulation of osmotic balance
Calcium (0.5)	In cell walls and starch-digesting enzymes; regulates cell membrane permeability
Magnesium (0.2)	In chlorophyll and many cofactors
Phosphorus (0.2)	In nucleic acids, some coenzymes, ATP, and some lipids
Sulfur (0.1)	In proteins, some lipids, and coenzyme A
Micronutrients	
Iron (0.01)	In electron transport molecules; involved in synthesis of chlorophyll
Chlorine (0.01)	Essential to photosynthesis
Manganese (0.005)	Essential to photosynthesis; activates many enzymes
Boron (0.002)	Unknown
Zinc (0.002)	In some enzymes; involved in protein synthesis
Copper (0.0006)	In chloroplasts and some enzymes
Molybdenum (0.0001)	Essential to nitrogen fixation and assimilation
Elements Essential to Only Some Plants	
Silicon (0.25–2.0)	In the cell walls of grasses and horsetails
Sodium (trace)	Essential to a few desert and salt-marsh species
Cobalt (trace)	Essential to nitrogen fixation

* Concentration of nutrients expressed as percentage of dry weight in a typical plant. Actual proportions vary greatly from species to species.

LO3 Plant Nutrients

Wild daisies can share a meadow with dozens of other plant species, but when yellow star thistle spreads through a field, it chokes out virtually all the other plants. Why? We've already seen that in late summer, the aggressive, spiky weed can take in water through its deep taproot, while shallower-rooted plants dry up and die. But what about the role of nutrients? Another possibility is that yellow star thistle is exceptionally good at absorbing some plant nutrient and effectively starves the neighboring plants for that mineral. To see if this is true, we have to understand more about plant nutrients. Curiously, what may have been the very first quantitative experiment in biology probed the mechanisms of plant nutrition.

Around 1600, the Dutch physician Jan Baptista van Helmont questioned an assertion made by Aristotle centuries earlier that a plant's body derives most of its substance from soil. To test this hypothesis, van Helmont filled a container with 91 kg (200 lb) of dry soil and planted in it a 2.2-kg (5-lb) willow tree. For five years, he watered the tree with rain water. At the end of that time, he dug the tree out of its pot, removed

> "Fully 96 percent of a plant's dry weight is made up of carbon, oxygen, and hydrogen."

macronutrient
(Gr., *makros* = large + L., *nutrire* = to nourish) an inorganic chemical element, such as nitrogen, potassium, calcium, phosphorus, magnesium, or sulfur, required in large amounts for plant growth

micronutrient
(Gr., *mikros* = small + L., *nutrire* = to nourish) a nutrient, such as the minerals iron, copper, zinc, chlorine, manganese, molybdenum, and boron, required in small amounts for plant growth

nitrogen fixation
the assimilation of atmospheric nitrogen by certain prokaryotes into biologically usable nitrogenous compounds

legume
member of the pea family, including peas, beans, alfalfa, clover, and peanuts

nodule
a swelling on the roots of legumes and some other plants that houses nitrogen-fixing bacteria

the soil and dried it, weighed both soil and tree, and discovered that while the tree had gained 75 kg (165 lb), the soil had lost a mere 27.5 g (about an ounce). With this, van Helmont succeeded in showing quite clearly that the main mass of the willow tree came not from the soil, as Aristotle had suggested, but from some other source.

Today, plant scientists know that plants synthesize organic compounds from the carbon and oxygen in air, from the hydrogen in water, and from small amounts of minerals in soil (review Fig. 3.10). Fully 96 percent of a plant's dry weight is made up of carbon, oxygen, and hydrogen. Nevertheless, a plant can't live without the dozen or so chemical elements that make up the remaining four percent.

Plants need at least 16 different chemical elements, the so-called *essential elements*, which different species require in different amounts. All plants require the macronutrients in relatively large amounts, and the micronutrients in much smaller amounts. By drying many types of plants, weighing the dried plant matter, and analyzing its chemical content, plant physiologists were able to tabulate information on the nine macronutrients—carbon, oxygen, hydrogen, nitrogen, potassium, calcium, magnesium, phosphorus, and sulfur—and the seven micronutrients—iron, chlorine, manganese, boron, zinc, copper, and molybdenum. Plants use both macro- and micronutrients in amounts roughly comparable to the levels animals need. Just as animals suffer vitamin and mineral deficiency diseases if they lack essential nutrients, plants also show specific symptoms relating to specific deficiencies.

Macronutrients

In Chapters 2 and 3, we discussed the roles of carbon, oxygen, and hydrogen in biological molecules and reactions. Here, we'll concentrate on the other six macronutrients that make up most of the rest of the atoms in carbohydrates, proteins, lipids, and nucleic acids.

Nitrogen Fixation

After carbon, oxygen, and hydrogen, nitrogen is the most important of the macronutrients because it's an essential component of amino acids and proteins, chlorophyll, coenzymes, and nucleic acids. Nitrogen is often the most important nutrient that limits growth—the less nitrogen available, the slower the plant grows. Although gaseous nitrogen (N_2) makes up 78 percent of Earth's atmosphere, plants can't use it directly. Instead, the nitrogen must be fixed, or converted from the simple gas N_2 into some other form, such as ammonia (NH_3) or nitrate ions (NO_3^-), by a process called nitrogen fixation. Plant cells can then modify these fixed nitrogen compounds and incorporate them into plant proteins.

Farmers and home gardeners often add fixed nitrogen to the soil in the form of industrially produced nitrate-containing fertilizers. Wild-growing plants such as star thistle, of course, don't have the benefit of added fertilizers and must depend instead on nitrogen-fixing bacteria, including cyanobacteria, which can convert molecular nitrogen to usable forms. Many of these microorganisms live independently in the soil and reduce nitrogen gas to ammonia. Some of the most interesting ones, though, live in the root cells of certain vascular plants (Fig. 25.7). Peas, beans, alfalfa, clover, peanuts, and other legumes, or members of the pea family, have swellings, or nodules, on their roots that house nitrogen-fixing bacteria. Certain nonlegumes such as alder trees also form associations with nitrogen-fixing bacteria. Most other plants, including star thistle, don't form nitrogen-fixing nodules and get their nitrogen from soil-living bacteria, instead. Figure 25.7 shows how two kinds of plants, a soybean (a legume) and a yellow star thistle (a nonlegume) get enough nitrogen.

Nature is full of exceptions, and it turns out that soil-living, nitrogen-fixing bacteria can't survive in the acidic, waterlogged environment of bogs. Carnivorous plants such as the Venus flytrap and the pitcher plant have evolved the ability to gain fixed nitrogen from a different source: the proteins of the insects and other small animals they trap and digest.

Root nodules in legumes are a classic example of symbiosis, a close association of two different species: The plant supplies the bacteria with high-energy carbohydrates, while the bacteria provide biologically useful nitrogen. For centuries, farmers have rotated crops to take advantage of this symbiosis. Early farmers may have been unaware of microbes, but they observed that if they raised clover or alfalfa one year, the following year's crop of corn or wheat would grow more luxuriantly. Likewise, rice farmers have encouraged the growth of water ferns in their flooded rice paddies; cyanobacteria living symbiotically in the ferns fix nitrogen, which enriches the soil for the rice plants.

Other Macronutrients

Plants can often require 4 to 40 times more nitrogen than they do the other 5 macronutrients, but these other elements are still essential to normal plant growth and development. Calcium is an element that controls cell membrane permeability. Because of this, it plays a role in the opening and closing of stomata, directional growth in plant cells, responses to changing day lengths, and the tendency of roots to grow down and shoots to grow up. Calcium is also an important component of pectin, a substance that

Figure 25.7

Nitrogen-Fixing Bacteria in Root Nodules Help Nourish Some Plants

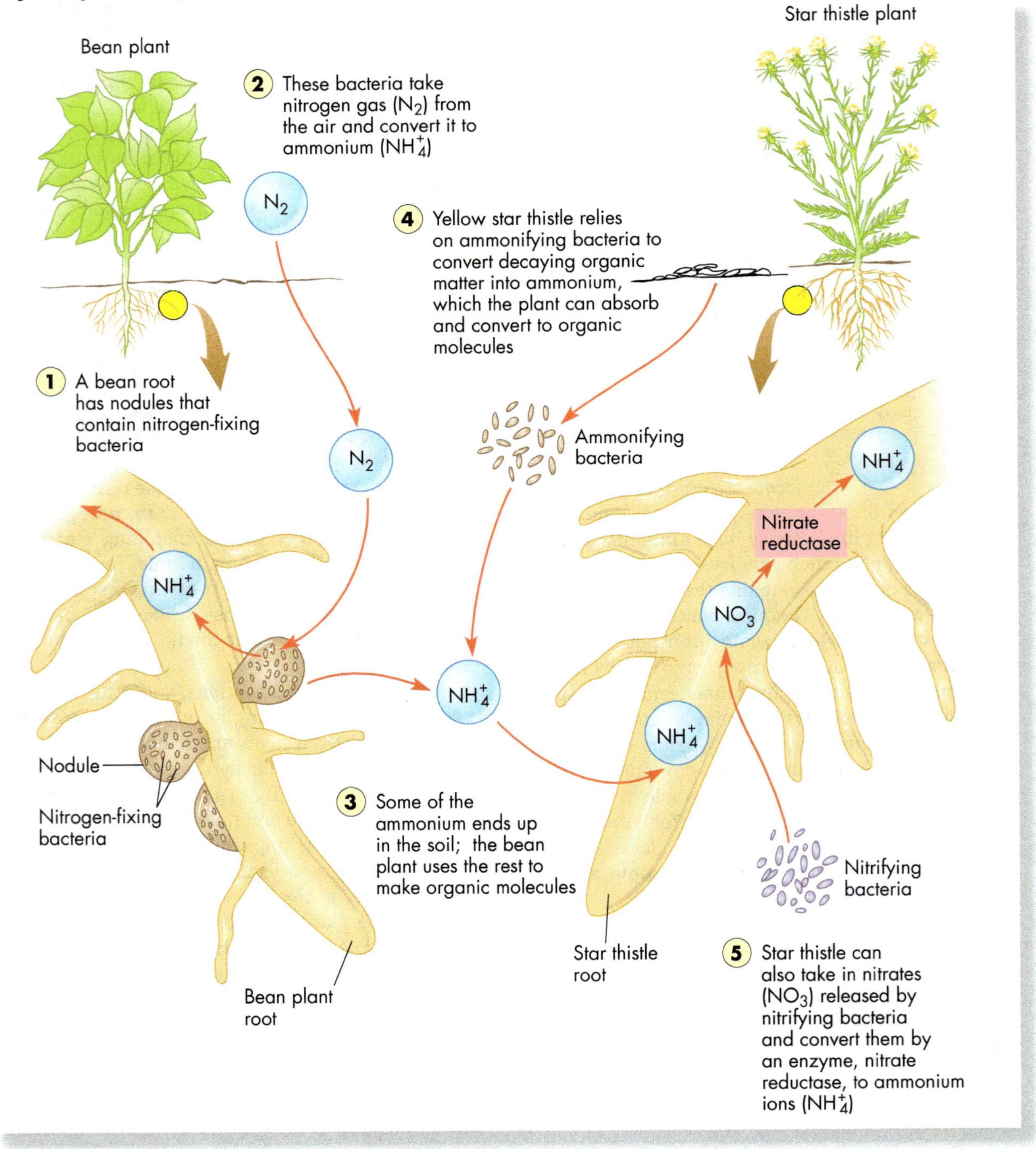

soil
the portion of Earth's surface consisting of disintegrated rock and decaying organic material

humus
the decomposing organic matter in soil

loam
the best agricultural soil, consisting of deep layers of soil with a high mineral and humus content and a mixture of particle sizes

secondary compound
a molecule that helps ensure plant survival by repelling, killing, or interfering with the normal activities of plant-eating organisms

glues adjacent cells together and helps young plants make strong cell walls. Cooks use commercial pectin in jams and jellies to make them jell.

Potassium regulates osmosis in plant cells such as guard cells and also helps activate enzymes, including some involved in protein synthesis. Potassium deficiency causes mottled or burnt-edged leaves. Magnesium is a macronutrient because magnesium atoms are components of chlorophyll molecules and cofactors of many kinds of enzymes. Phosphorus occurs in the backbone of DNA and RNA molecules, in ATP and other high-energy compounds, and in membrane phospholipids. Because phosphorus, like sulfur, is a mobile nutrient in the plant, it moves from older leaves to newly developing leaves and shoots. This explains why phosphorus deficiency is first evident in the plant's oldest leaves (for example, in phosphorus-deficient tomato seedlings, whose older leaves turn purple). Because sulfur is an important component of two amino acids, plants require it for building most proteins, as well as for manufacturing some fats and coenzymes.

Micronutrients

Plants require only small amounts of micronutrients for healthy growth and so deficiencies of the micronutrients are rare. Iron, for example, occurs in several energy-harvesting proteins within the mitochondria and is also involved in making chlorophyll. A tomato plant with an iron deficiency makes too little chlorophyll to mask the yellow pigments in leaves. A deficiency of copper (needed in chloroplasts and certain enzymes) can cause deformed stems, leaves, and fruits in many plant species. Chlorine deficiency stunts growing tomato roots and fruits and causes the entire plant to wilt. Because zinc plays a role in protein synthesis, zinc-deficient apple and peach trees become stunted and grow miniature leaves.

Soil: The Primary Source of Minerals

Plants get their carbon, oxygen, and hydrogen from air and water but obtain all the other macro- and micronutrients from the soil. For this reason, soil composition has a major influence on the kinds of plants and, indirectly, the kinds of animals that can live in a particular region of our planet.

Soil is a mixture of organic and inorganic material that starts with bedrock—relatively unbroken, unweathered rock. Over time, water, wind, heat, and cold disintegrate the bedrock and produce the inorganic parts of soil, from coarse sand and silt to small clay particles (Fig. 25.8). Bacteria, fungi, algae, lichens, and plants extract minerals from rocks, sand, silt, and clay and use them to synthesize organic materials. When the organisms die, the decomposing organic matter in the soil becomes humus.

Air spaces between soil particles are crucial to plant life. In a soil with a mixture of particle sizes, the spaces contain about half water and half air, with the water forming a continuous film over the soil particles. The size of soil particles, however, determines how much water the soil can hold: Soils made up primarily of coarse sand tend to hold water poorly, while soils made up mostly of fine clay particles hold water tenaciously. Plants such as cacti, which grow in rapidly draining sandy soils, absorb water quickly and store it, while plants that grow well in clay soils have roots resistant to dense, soggy soil deprived of air. The best agricultural soils are deep layers of loam, with a high mineral and humus content and a mixture of particle sizes.

LO4 How Plants Defend Themselves Against External Threats

Yellow star thistle plants have wicked spines surrounding every flower and spaced along the stems. The energy it takes to produce all these spines is apparently well spent: Deer, cows, and most other herbivores shun the nasty weeds in favor of plants that are easier to eat. Plants, after all, can't run away from hungry pests—but they aren't defenseless. Yellow star thistle spines are just one kind of protective barrier. Others include spiky leaf hairs and sticky sap, various toxic compounds, and walling off injured areas.

Chemical Protection

The star thistle spines are an excellent deterrent, but a starving animal may well persevere despite the pain in its mouth. The spines (or their equivalents on other plants) are therefore only a first line of defense. A second line of defense includes chemical warfare. Horses that eat star thistle develop a nervous disorder called "chewing disease" that is fatal once symptoms appear—drowsiness, unsteady gait, lack of coordination, and restlessness. This condition

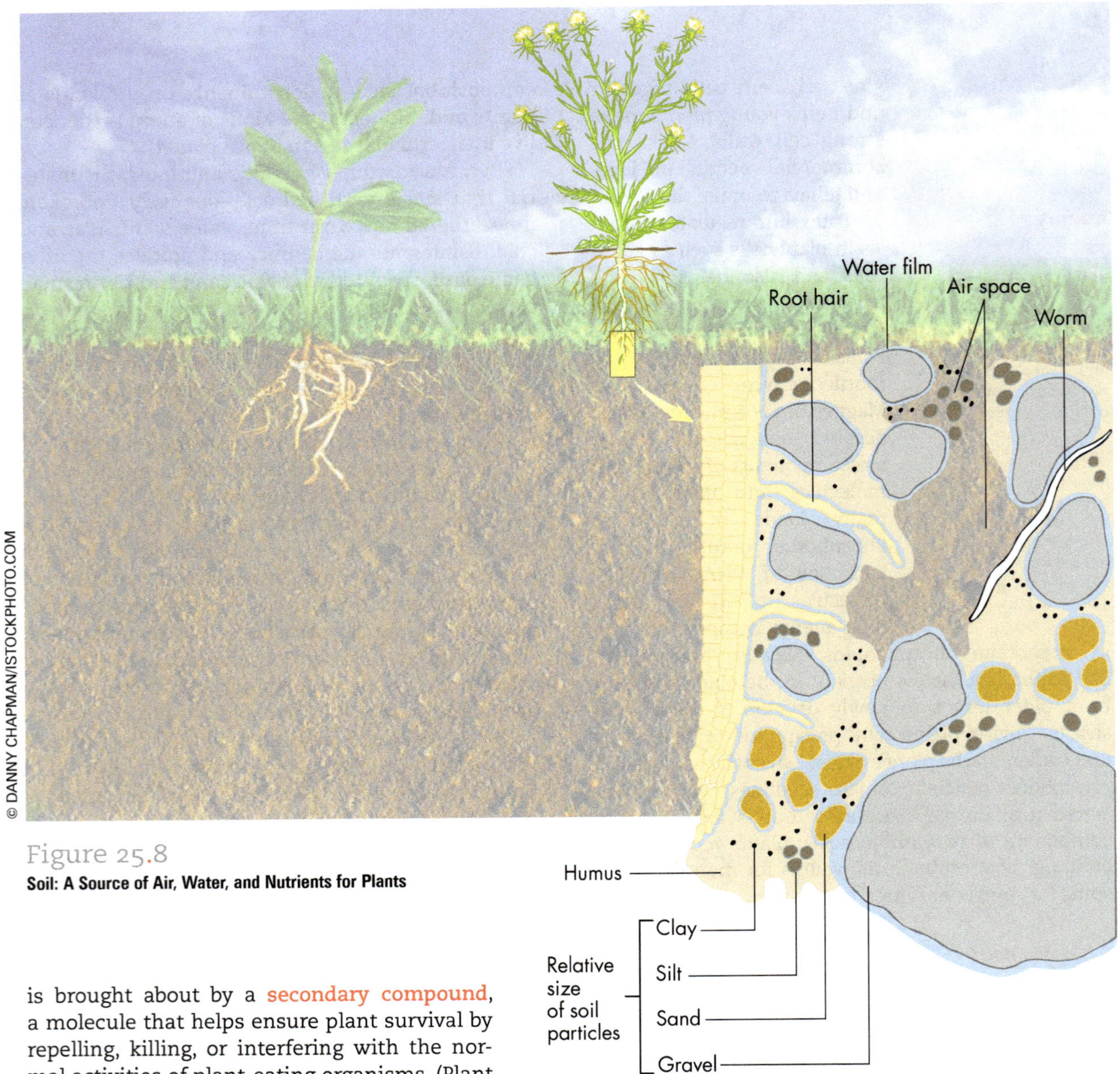

Figure 25.8
Soil: A Source of Air, Water, and Nutrients for Plants

is brought about by a secondary compound, a molecule that helps ensure plant survival by repelling, killing, or interfering with the normal activities of plant-eating organisms. (Plant primary compounds are biological molecules necessary for normal growth and regulation.) Weed experts don't yet know which secondary compound in yellow star thistle can kill a horse. But plant biologists have discovered that in some tree species, up to 50 percent of the plant matter is defensive secondary compounds! These are so effective that in a temperate forest in a typical year, predators (including insects) consume only about seven percent of the total leaf surface area.

Over 10,000 plant chemicals are toxic to animals, and these secondary compounds include such familiar substances as cyanide, camphor, cocaine, caffeine, nicotine, and tannins. Camphor and tannins can discourage insect pests from eating a plant's leaves or laying eggs on them. The nicotine in tobacco can poison various insect predators. Cyanide compounds in birdfoot trefoil, a member of the pea family, can kill snails that feed on its leaves. Some plants generate compounds that mimic insect hormones, disrupting insect metamorphosis. And recent evidence shows that compounds given off by microbes and injured plant cells can induce the plant to respond, making antibiotics that kill attacking fungi and bacteria.

In response to insect damage, many plants increase their production of secondary compounds, in some cases reaching ten percent of the total weight of fruits or vegetables. Since these compounds can often harm people as well as insects, it's best to stay away from diseased or damaged celery, apples, peanuts, and other produce. Some, such as safrol—found in the leaves of the sassafras plant, whose roots used to flavor root beer—can even cause cancer.

compartmentalization
the walling-off process of damaged areas of a plant that prevents invaders from gaining access to healthy tissues

herbicide
a chemical that kills plants

Many secondary compounds are harmful to animals, but a few have given rise to important human medicines. These include the pain reliever aspirin from willow bark, the heart medicine digitalis from foxglove, the main ingredient in oral contraceptives from yams, cancer treatments from periwinkle flowers and the Pacific yew tree, and the pain reliever morphine from poppies. Many scientists expect to find additional medicines in as-yet-unstudied plant species, and this is a primary reason they are fighting to preserve vast tracts of unexplored rain forests.

Walling Off Injured Areas

If protective chemicals are a second line of defense, plants have a third line, as well: a growth response that protects them from viruses, bacteria, fungi, animals, or simple physical injury, such as loss of a limb in a storm. A plant's rigid cell walls prevent new cells from migrating into a wounded area and healing the damage. Plants can, however, wall off the damaged area—a response that prevents invaders from gaining access to healthy tissues. This walling-off process is called compartmentalization. It involves the production of toxic secondary compounds in the invaded area and the plugging of nearby xylem and phloem tubes with thick saps or resins that prevent invaders from spreading to other parts of the plant (Fig. 25.9). Once the injury is walled off, the tree can survive and grow.

Figure 25.9
Compartmentalization in Plants: A Response to Wounding

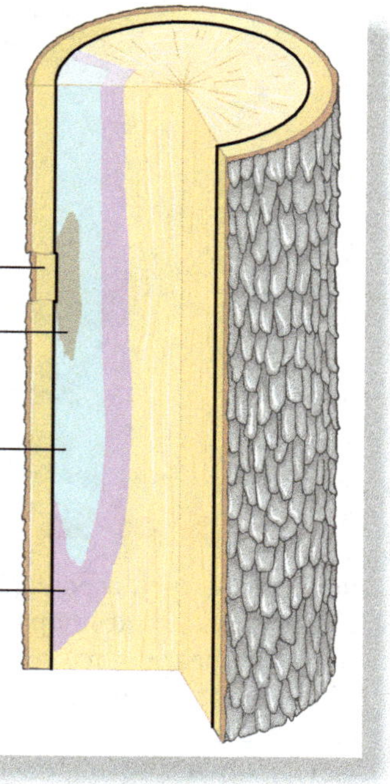

LO5 Weed Control and Super Weeds

One of Dr. Di Tomaso's jobs is advising farmers and ranchers about the various methods for controlling yellow star thistle and other problem weeds. These include tilling the soil and burning the weeds, but the main approach is applying herbicides, chemicals that kill plants. Most herbicides target biochemical processes carried out by plants but not animals. For example, the herbicide 2,4-D is an analog of the plant growth regulator auxin. The 2,4-D causes phloem to grow so quickly that it becomes plugged and blocks substance transport, leading to death. Another herbicide is glyphosate, sold by Monsanto under the brand name Roundup.

Glyphosate is a simple molecule based on glycine, the smallest amino acid (Fig. 25.10a). In the soil, it immediately binds tightly to clay particles and breaks down. When sprayed on a plant, however, glyphosate binds to and inhibits an enzyme that helps build the amino acids phenylalanine, tyrosine, and tryptophan (Fig. 25.10b). Do these amino acids sound familiar? Look back at Figure 19.2 to find the answer: Two of them are essential amino acids; they occur in all of your proteins but you lack the enzymes necessary to make them, and so you have to acquire them in your diet. Humans don't have the enzyme that is blocked by glyphosate, and that's why Roundup can kill plants but has virtually no effect on people or other animals.

When Dr. Di Tomaso sprays glyphosate on star thistle, he finds that the foliage quickly absorbs the chemical and it moves into plant tissues through the cell cytoplasm pathway (shown as the pale blue path in Fig. 25.3), and then enters the vascular system. Once in the phloem, glyphosate translocates by means of pressure flow, moving from photosynthesizing leaves to lateral meristems, areas of active growth. (This is why the label on glyphosate advises the user to apply it on a sunny morning: If the plant is not photosynthesizing rapidly, the pressure flow mechanism is not operating and the herbicide does not reach the tips of the roots and shoots before it quickly breaks down into harmless compounds in the soil.) Once in the meristems, glyphosate blocks the plant's ability to make the three amino acids it requires for survival. The plant immediately stops growing and, lacking new protein synthesis for growth and repair, it slowly begins to degrade. Soon the plant yellows and dies from dehydration.

Figure 25.10

How the Herbicide Glyphosate Works

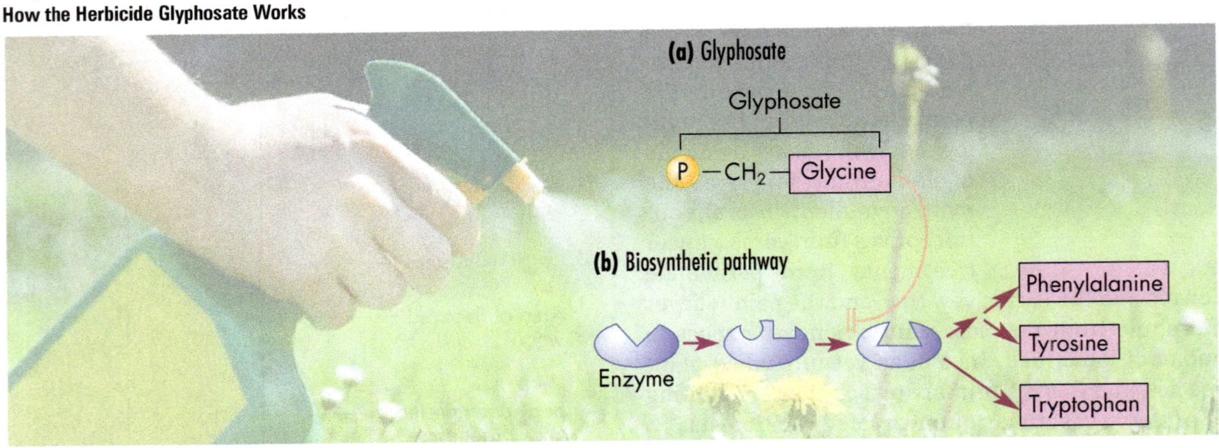

The Fear of Super Weeds

Roundup is such a powerful weapon against weeds that Monsanto developed "Roundup-ready" varieties of soybeans, corn, and other crops. The splicing of genes for glyphosate resistance into crop species creates plants that can survive a spraying of the herbicide, while weeds growing in or around the fields die off.

However, some plant researchers and most anti-biotechnology groups have expressed concern over the potential escape of Roundup-resistance genes into wild plant populations. Say, for example, that an artichoke grower planted "Roundup-ready" artichokes (which are essentially domesticated thistles) and through rare natural hybridization, the genes jumped into nearby populations of yellow star thistle weeds. Could the result be "super weeds," weeds now resistant to glyphosate and no longer controllable with the formerly powerful tool?

No, says Dr. Di Tomaso. In the first place, he explains, no company would ever invest enough research funds in a small, specialty crop such as artichokes to engineer it with Roundup-resistance genes. And secondly, he says, even if star thistle somehow did acquire Roundup resistance, "it's not going to become some even worse, horrifically spiky monster plant with deadly fruits and leaves like poison oak. The term 'super weed' is very misleading. "It really means you would just have to switch to some other herbicide."

If there is a true danger of herbicide resistance genes transferring from crops to weeds, he says, it would involve crops grown widely enough to be engineered with Roundup resistance and then also planted in areas that happen to have closely related weeds. For example, sorghum is a widely grown cereal grass and is closely related to the weed johnson grass, and the huge commercial crop canola is closely related to wild mustard. The chances are greater of a gene transfer between pairs like this, says Di Tomaso, so it's important for sorghum and canola oil farmers to be aware of the local weed populations and to avoid planting genetically modified crops if they are present.

"I think the scientific community needs to get together," he concludes, "to develop a set of guidelines for what are realistic fears and what is just misinformation."

Who knew?

While it's true that water moves in a one-way path through the xylem from roots to leaves, it is not true that this is the only path for water movement. Water can also move from leaves to roots through the phloem.

Super Weed Villain? Uma Thurman portrays a *Batman* villain, "Poison Ivy," who developed an immunity to all natural toxins and diseases, making her nearly unstoppable.

THE DYNAMIC PLANT

25 Chapter in Review

guttation
upward pressure from plant roots producing tiny droplets of water that rim leaves early in the morning

root pressure
force that is generated when fluid moves up xylem cells; root pressure can force water out of the leaves, where it accumulates in tiny droplets

capillary action
the tendency of water to move upward in a thin tube; studies show that capillary action occurs in a plant's narrow xylem vessels

transpiration-cohesion-tension
the mechanism scientists propose for water transport in plants, which explains how a column of water molecules is "pulled" up the stem

transpiration
the loss of water from plants by evaporation, mainly through the stomata on stems and leaves

cohesion
the tendency of like molecules to cling to each other

tension
the force upon the unbroken chain of water molecules in the xylem tubes of plants as they are pulled upward from above

translocation
in plants, the transport of solutes in phloem cells

pressure flow hypothesis
the best current hypothesis for phloem transport, which asserts that phloem transports substances from "sources," or areas of production, to "sinks," or areas of use, down a pressure gradient established via osmosis

macronutrient
(Gr., *makros* = large + L., *nutrire* = to nourish) an inorganic chemical element, such as nitrogen, potassium, calcium, phosphorus, magnesium, or sulfur, required in large amounts for plant growth

micronutrient
(Gr., *mikros* = small + L., *nutrire* = to nourish) a nutrient, such as the minerals iron, copper, zinc, chlorine, manganese, molybdenum, and boron, required in small amounts for plant growth

nitrogen fixation
the assimilation of atmospheric nitrogen by certain prokaryotes into biologically usable nitrogenous compounds

legume
member of the pea family, including peas, beans, alfalfa, clover, and lupine

nodule
a swelling on the roots of legumes and some other plants that houses nitrogen-fixing bacteria

soil
the portion of Earth's surface consisting of disintegrated rock and decaying organic material

humus
the decomposing organic matter in soil

Introduction Dr. Joseph Di Tomaso is a weed expert who works to help control yellow star thistle, a major infester of open land in California. No one can say precisely why a weed grows so vigorously, but most weeds grow rapidly; produce seeds abundantly; spread by stems, roots, and seeds; and can adjust to a wide range of habitats. Many, including star thistle, are also imported "aliens" with no natural predators. This plant, in addition, has a long taproot that allows it to pull water from deep soil layers during dry spells. Star thistle is a good case history for studying how plants transport water, minerals, and sugars; how they meet their nutritional needs; and how genetic engineering has led to fears of uncontrollable super weeds.

LO1 Water Balance in Plants

How does water reach the tops of tall plants? The transpiration-cohesion-tension mechanism can explain how plants transport water to heights over 20 inches. When a water molecule evaporates from a leaf via transpiration, the entire column of water molecules is pulled up the xylem tube and new water molecules are pulled into the roots. A hole in the xylem can allow air bubbles to break the chain. The rapid evaporation from leaves on a very hot, dry day can outstrip water absorption by roots, and the xylem walls can be sucked inward and/or the water chain can break, leading to plant wilting. The swelling or deflating of stomata, based on potassium ion levels and osmotic water gain or loss, controls the rate of transpiration from leaves.

How Water Enters Root Hairs and Passes into the Vascular Cylinder

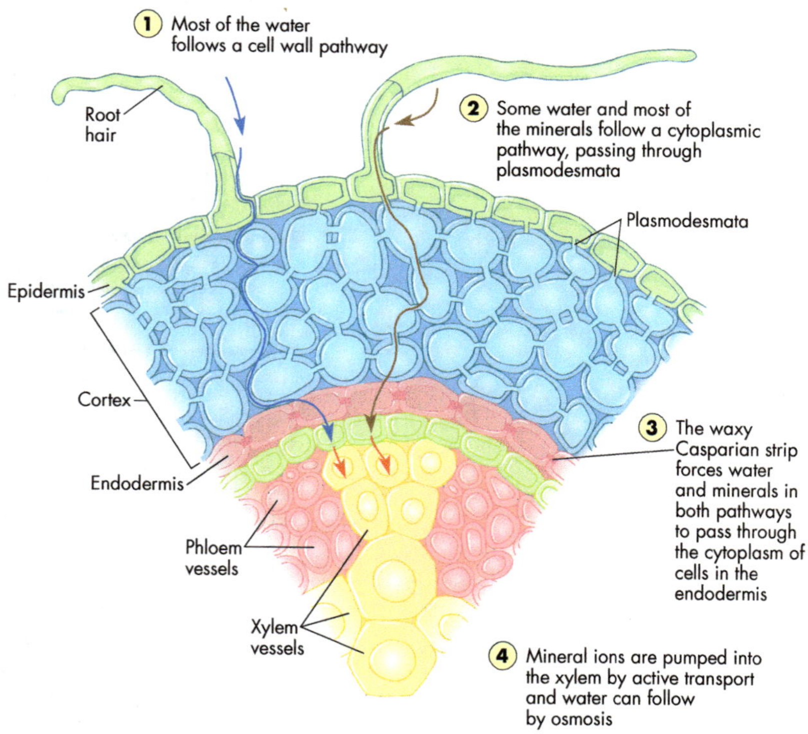

LO2 Movement of Nutrients Throughout the Plant

Plants need to move around sugars and other organic nutrients, depending on their production and use levels in different plant parts. The pressure flow hypothesis is the idea that phloem transports substances from areas of production to areas of use or storage. When storage cells in the root (or growing plant parts needing sugar for metabolic energy requirements) remove sugars from the phloem, the solute concentration drops, water flows out of the phloem into the xylem by osmosis, and it is carried upward again by transpiration.

loam
the best agricultural soil, consisting of deep layers of soil with a high mineral and humus content and a mixture of particle sizes

secondary compound
a molecule that helps ensure plant survival by repelling, killing, or interfering with the normal activities of plant-eating organisms

compartmentalization
the walling-off process of damaged areas of a plant that prevents invaders from gaining access to healthy tissues

herbicide
a chemical that kills plants

Some nutrients, such as calcium, are transported only in xylem. Others, such as sulfur and phosphorous, move in the phloem. Plants can have mineral deficiencies, and these show up in various distinctive ways in the leaves and stems.

LO3 Plant Nutrients In about 1600, Dutch researcher Jan Baptista van Helmont found that virtually all of a tree's weight comes from some source other than the soil. Modern scientists worked out that source: photosynthesis using carbon and oxygen from the air and hydrogen from water. These elements account for 96 percent of a plant's dry weight, but the plant can't live without the other 4 percent, the macronutrients and the micronutrients. Plants get most macro- and micronutrients from the soil.

Nutients (Percent of Dry Weight)*	Location/Function
Macronutrients	
Carbon (45.0)	In all organic compounds
Oxygen (45.0)	In most organic compounds, including all sugars and carbohydrates
Hydrogen (6.0)	In all organic compounds
Nitrogen (1.0–4.0)	In proteins, nucleic acids, chlorophyll, and coenzymes
Potassium (1.0)	Involved in activating enzymes, protein synthesis, and regulation of osmotic balance
Calcium (0.5)	In cell walls and starch-digesting enzymes; regulates cell membrane permeability
Magnesium (0.2)	In chlorophyll and many cofactors
Phosphorus (0.2)	In nucleic acids, some coenzymes, ATP, and some lipids
Sulfur (0.1)	In proteins, some lipids, and coenzyme A
Micronutrients	
Iron (0.01)	In electron transport molecules; involved in synthesis of chlorophyll
Chlorine (0.01)	Essential to photosynthesis
Manganese (0.005)	Essential to photosynthesis; activates many enzymes
Boron (0.002)	Unknown
Zinc (0.002)	In some enzymes; involved in protein synthesis
Copper (0.0006)	In chloroplasts and some enzymes
Molybdenum (0.0001)	Essential to nitrogen fixation and assimilation
Elements Essential to Only Some Plants	
Silicon (0.25–2.0)	In the cell walls of grasses and horsetails
Sodium (trace)	Essential to a few desert and salt-marsh species
Cobalt (trace)	Essential to nitrogen fixation

* Concentration of nutrients expressed as percentage of dry weight in a typical plant. Actual proportions vary greatly from species to species.

LO4 How Plants Defend Themselves Against External Threats Plants have physical defenses, such as the spines on a yellow star thistle, and chemical defenses, including secondary compounds such as cyanide, nicotine, caffeine, and tannins that poison or discourage animal predators. Plants can also wall off injured areas through compartmentalization.

LO5 Weed Control and Super Weeds Weeds can be pulled, burned, or tilled under, but the most common control method is spraying with herbicides, including glyphosate (Roundup), which blocks an enzyme that plants need to build three essential amino acids. Genetic engineers have spliced glyphosate-resistance genes into corn, soybeans, and other crops. This has raised major concerns over the escape of these genes into weed populations and the creation of uncontrollable super weeds.

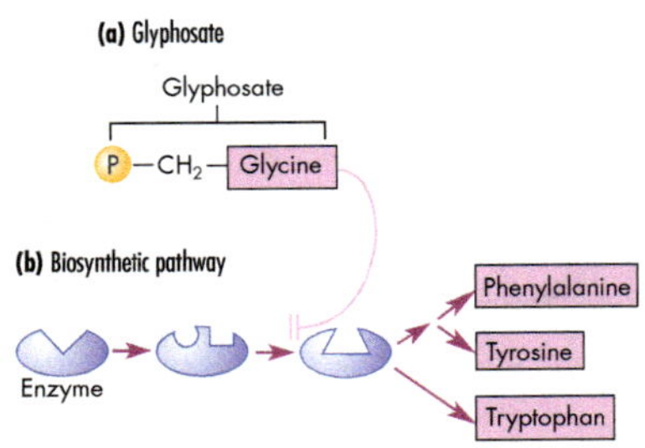